W9-BEL-538

Sociology

Sociology

Understanding a Diverse Society

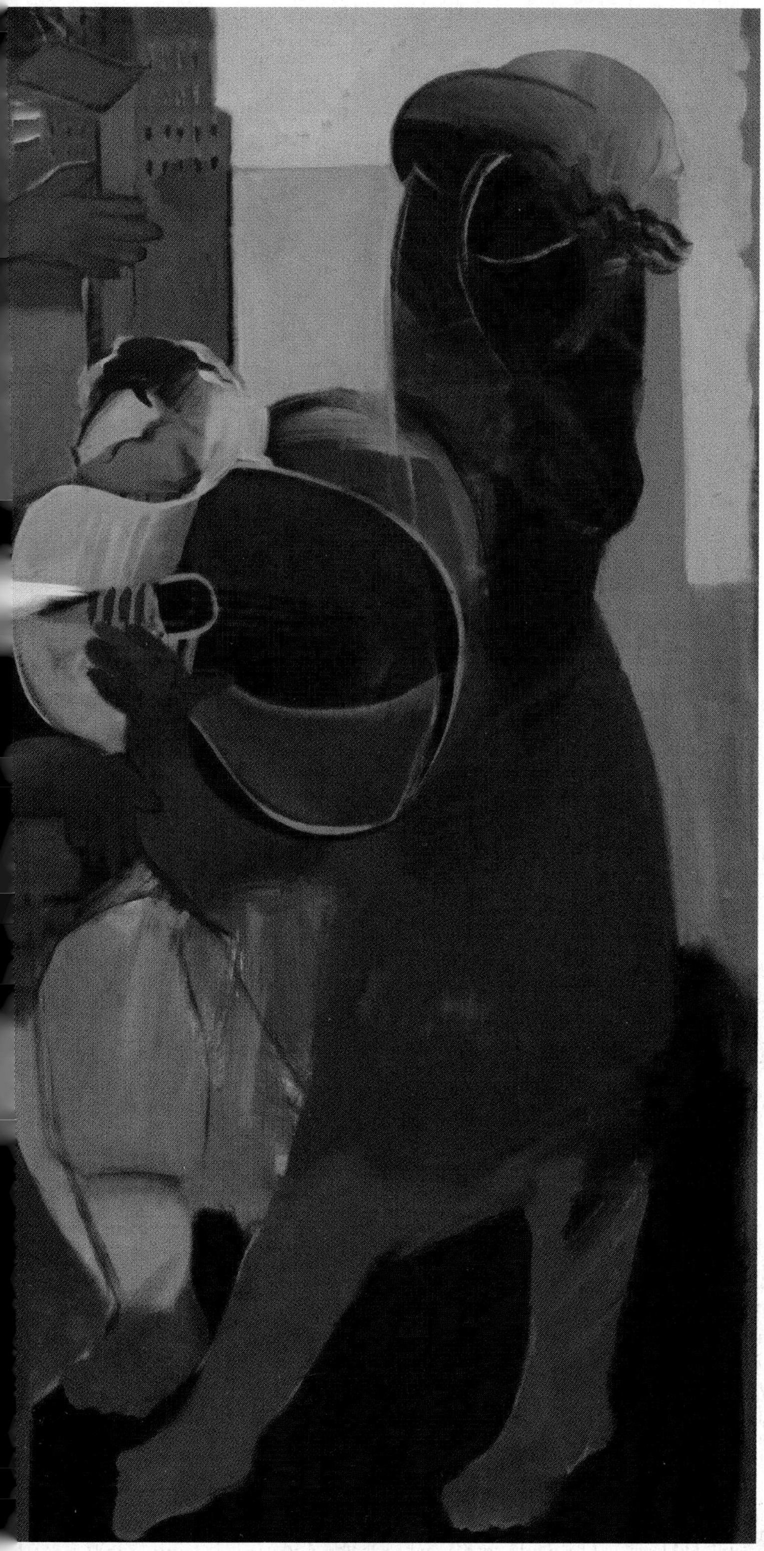

MARGARET L. ANDERSEN
University of Delaware

HOWARD F. TAYLOR
Princeton University

Australia • Canada • Denmark • Japan • Mexico • New Zealand
Philippines • Puerto Rico • Singapore • South Africa • Spain
United Kingdom • United States

Publisher: Eve Howard
Development Editor: Robert Jucha
Assistant Editor: Ari Levenfeld
Editorial Assistant: Bridget Schulte
Marketing Manager: Diane McOscar
Signing Representative: Ron Shelly
Marketing Assistant: Kelli Goslin
Project Editor: Jerilyn Emori
Print Buyer: Karen Hunt
Permissions Editor: Susan Walters
Production Editor: Dusty Friedman/The Book Company
Text Designer: Baugher Design Inc.
Photo Researcher: Stephen Forsling
Copy Editors: Robert Fiske, Anne Harris
Illustrators: Impact Publications, C.H. Wooley and Associates
Compositor: Thompson Type
Cover Designer: Baugher Design Inc.
Cover Image: Hessam Abrishami
Cover Printer: Phoenix Color Corp.
Printer/Binder: World Color Book Services/Versailles

Cover and chapter art: The cover and title-page painting *(The Entertainment)* and the details of paintings used in the interior are from the work of Hessam Abrishami, whose colorful art reflects the diversity of society. Hessam's passion for art began in the eighth grade. Later, as a teenager, he won two drawing contests, which encouraged him to pursue his passion and to obtain a master's degree in fine arts. His work is exhibited in Asia, Europe, and the United States.

For permission to use material from this text, contact us:
Web: www.thomsonrights.com
Fax: 1-800-730-2215
Phone: 1-800-730-2214

Printed in the United States of America
2 3 4 5 6 7 03 02 01 00

Library of Congress Cataloging-in-Publication Data
Andersen, Margaret L.
Sociology : understanding a diverse society / Margaret L. Andersen, Howard F. Taylor
p. cm.
Includes bibliographical references and index.
ISBN 0–534–56664–2
1. Sociology. 2. Pluralism (Social sciences) 3. Pluralism (Social sciences)--United States. 4. United States--Social conditions. I. Taylor, Howard Francis, 1939– . II. Title.
HM585.A53 2000
301--dc21 99-33633

Wadsworth/Thomson Learning
10 Davis Drive
Belmont, CA 94002-3098
USA
www.wadsworth.com

International Headquarters
Thomson Learning
290 Harbor Drive, 2nd Floor
Stamford, CT 06902-7477
USA

UK/Europe/Middle East
Thomson Learning
Berkshire House
168-173 High Holborn
London WC1V 7AA
United Kingdom

Asia
Thomson Learning
60 Albert Street #15-01
Albert Complex
Singapore 189969

Canada
Nelson/Thomson Learning
1120 Birchmount Road
Scarborough, Ontario M1K 5G4
Canada

Dedication

*We dedicate this book to Sarah and Jessica Hanerfield,
Patricia E. Taylor, Carla Taylor Brown, and Roger Brown
with hope for a future marked by both diversity and justice.*

MARGARET L. ANDERSEN

HOWARD F. TAYLOR

BRIEF CONTENTS

PART ONE
Introducing the Sociological Perspective

PART TWO
Individuals in Society

PART THREE
Social Inequalities

PART FOUR
Social Institutions

PART FIVE
Social Change

TABLE OF CONTENTS

PART ONE
Introducing the Sociological Perspective

PART TWO
Individuals in Society

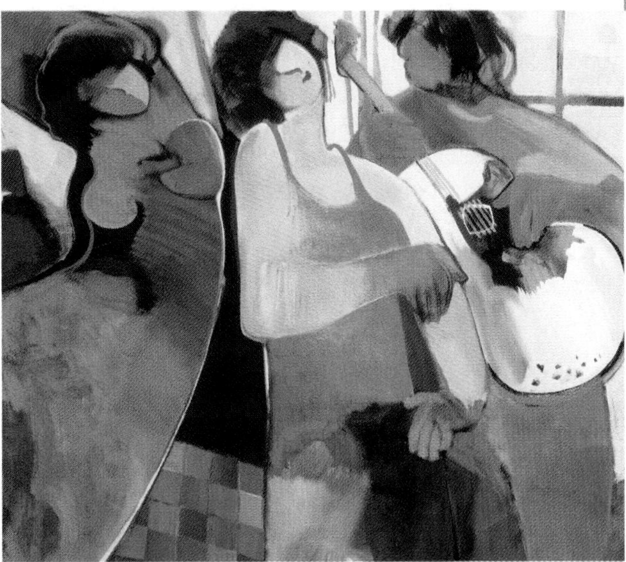

CHAPTER 11
RACE AND ETHNICITY 276

CHAPTER 12
GENDER 310

CHAPTER 13
AGE AND AGING 342

PART FOUR
Social Institutions

CHAPTER 14
FAMILIES 372

CHAPTER 15 EDUCATION 400

CHAPTER 16 RELIGION 424

CHAPTER 17 WORK AND THE ECONOMY 452

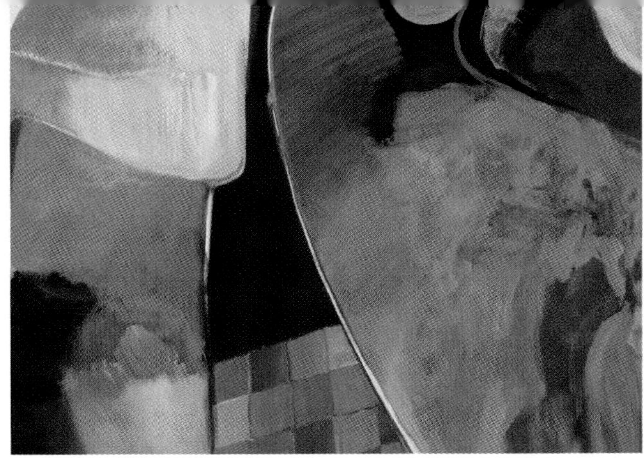

BOXES

Sociology in Practice

Doing Sociological Research

Understanding Diversity

Analyzing Social Issues

MAPS

MAPPING AMERICA'S DIVERSITY

VIEWING SOCIETY IN GLOBAL PERSPECTIVE

IN THIS BOOK, we want to teach students the basic concepts and theories of sociology and to do so with a full recognition of the experiences of diverse groups in society. With a focus on making current research and theory accessible to students, we have introduced sociology by drawing on the best contemporary and classical scholarship. Emphasizing recent research and balance in the presentation of different theoretical perspectives, we have written the book to show students the excitement and fascination that a sociological perspective brings to the study of society.

The book introduces students to the basic principles of sociology and does so in a way that provides a balanced approach to sociological work. Unlike other introductory books that provide an encyclopedic overview of the field, this book takes an integrated approach, linking basic sociological concepts to theory and research and providing students and professors with pedagogical features that are integrated with the themes in the text. Thus, critical thinking questions that challenge students to apply sociological concepts to their own observations of the social world are presented throughout each chapter (see the feature "Thinking Sociologically"), and basic principles of sociology are explored by helping students see the difference that a sociological perspective gives to common understandings about the world (see the feature "Debunking Society's Myths"). Similarly, throughout the text, students are shown how the sociological perspective can be applied to understanding contemporary social issues.

We present a balanced approach to sociological theory, with chapters exploring how different theoretical frameworks each inform different aspects of the fascinating subjects of sociological investigation. Students are helped in this regard by having tables in most chapters that compare different theoretical viewpoints, showing how each differently illuminates certain concepts and principles of the sociological imagination. We provide students with an authoritative, yet accessible, presentation of some of the most current, as well as classical, sociological ideas, keeping an eye on how this work is important to understanding an increasingly global, rapidly changing, and diverse society. Whereas other introductory books present sociology as an encyclopedic patchwork of different studies, concepts, and themes, we have provided a fresh approach that maintains a firm grounding in the central principles and theoretical frameworks of sociology but brings a fresh and updated approach—one that integrates an understanding of the diversity within society with the core of the sociological imagination.

A Focus on a Diverse Society

We see understanding diversity as central to sociology and, with that in mind, have made the study of diversity central to the book. Unlike other introductory books that have added diversity to a preexisting approach to sociology, we have taken a fresh new approach by being inclusive of diversity throughout our presentation of sociological research, theory, and substantive topics. Because we see diversity as part of the fabric of society, you will find diversity woven throughout the text, at the same time that it is included in many of the book's special features.

We see diversity as forming a rich tapestry of group experiences that result from all the different factors that shape society. These factors include race, ethnicity, class, and gender, as well as age, religion, sexual orientation, and region of residence, to name some, but we also see diversity as contributing to the rich texture of society through the diverse cultures and identities of different groups in society. We stress positive aspects of a diverse society as well as its problems. Although we do not treat members of disadvantaged groups solely as victims, we nonetheless clearly show how the structure of society has a different impact upon the life chances of various groups in society.

Features and Pedagogical Aids

The special features of this book flow from the book's basic mission: to provide students with a current and comprehensive understanding of the sociological perspective; to comprehend the basics of sociological research and theory; to be able to apply the sociological perspective in analyzing society; to appreciate and understand the increasing significance of diversity within the United States; and to comprehend the implications of living in an increasingly global and technological society. Thus, several features are built into this book. The features of the book are intended to help students develop critical thinking so that they can apply abstract concepts to observed experiences in their everyday life and so they can learn how to interpret different theoretical paradigms and approaches to sociological research questions.

Fostering Critical Thinking Skills

For example, the feature "Thinking Sociologically" (which appears in each chapter) takes a concept from the chapter and asks students to think about this concept in relationship to something they can easily observe in an exercise or class discussion. The feature "Debunking Society's Myths" takes certain common assumptions and shows students how the sociological perspective would inform such assumptions and beliefs. These features could also be developed as group projects.

Unparalleled Integration of Web-Based Resources

Instructors will find that several technology-based teaching enhancements are integrated throughout the book, making this book the best conceived in using the tools of the Internet for teaching and learning introductory sociology. Many of the graphs within the chapters are based on web sites, and each chapter includes a list (at chapter's end) of web-based resources that students and faculty can use to explore data and information pertinent to the chapter topic. We have deliberately selected sites that we know are stable and provide the latest information or resources on a given subject.

At the end of each chapter, the feature "Sociology and Social Policy" is designed to explore current public issues and show students how the sociological perspective can provide different viewpoints about such issues. This feature also integrates web-based instruction by giving instructors and students the option of using informative web sites to aid discussion of the question and to search on key terms within each question. Several relevant web sites are provided in the text for each issue that will provide students with a starting point when exploring these issues.

An additional tool for using the Internet in sociology research is InfoTrac College Edition. Throughout the text, at the bottom of the page, are search terms that may be used in conducting an InfoTrac College Edition search. InfoTrac College Edition is a powerful online library, providing access to the articles contained in more than 700 periodicals and journals.

An Extensive and Content-Rich Mapping Program

Also in our book, each chapter includes a map feature (usually two maps): One map feature is called "Mapping America's Diversity" and the other, "Viewing Society in Global Perspective." These maps have multiple instructional value, not the least of which is instructing students about world and national geography. But the maps have been designed primarily to visually portray the differentiation by country or by state on certain key social facts. For example, in Chapter 4, we show the dispersion of the population under five years of age, both nationally and worldwide. Students can use this visually presented information to ask questions about how the age distribution of the population might be related to such factors as immigration, poverty, or global stratification. In other cases, maps are drawn to show such things as the penetration of U.S.-based culture throughout the world, such as the map in Chapter 3, which shows how commonly fast-food franchises have infiltrated the entire world.

High-Interest Theme Boxes

We use four high-interest box themes that embellish our focus on diversity and sociological research throughout the text. One is "Understanding Diversity," which is used to further explore the approach to diversity taken throughout the book. In most cases, these boxes provide personal narratives or other information designed to teach students about the experiences of different groups in society. Since many of these boxes are also written as first-person narratives, they can invoke student empathy toward groups other than those to which they belong—something we think is critical to teaching about diversity. We hope to show students the connections between different races, classes, and other groups that they otherwise find difficult to grasp.

The box feature "Doing Sociological Research" is intended to show students the diversity of research questions that form the basis of sociological knowledge and, equally important, how the question one asks influences the method used to investigate the question. We see this as an important part of sociological research—that how one investigates a question is determined as much by the nature of the question as by allegiance to a particular methodological strategy. Some questions require a more qualitative approach; others, a more quantitative approach. In developing these boxes, we ask, "What is the central question a sociologist is asking?" "How did they explore this question using sociological research methods?" "What did they find?" and "What are the implications of this research?" We deliberately selected those that show the full and diverse range of sociological theories and research methods, as well as the diversity of sociologists themselves.

Our third box feature, "Sociology in Practice," is designed to show the application of the sociological perspective in different contexts. Thus, we show examples where sociologists have testified before Congress, advised presidents, changed an organization, or elucidated a cultural phenomenon, such as the "Dead Heads" (see Chapter 3). Finally, a fourth box theme, "Analyzing Social Issues," provides a more journalistic style of discussing contemporary social issues, which is intended to show students an underlying sociological imagination. We think students will find all the boxes readable, interesting, and a diversion that supplements the textual material.

In-Text Learning Aids

In addition to the features already described, an entire set of learning aids within each chapter promotes student mastery of the sociological concepts.

- Chapter Outlines. A concise chapter outline at the beginning of each chapter provides students with an overview of the major topics to be covered.
- Key Terms. Major terms and concepts are in boldface when first introduced in the chapter. A list of the key terms is found at the end of the chapter, making study more effective.
- Chapter Summary. A point-by-point chapter summary immediately follows each chapter to focus on the most significant points covered in that chapter.
- Suggested Readings. An annotated list of suggested readings is found at the end of each chapter as a source for further study.
- A glossary and complete bibliography for the entire text is found at the back of the book.

Organization of the Book

Sociology: Understanding a Diverse Society is organized into five major parts: "Introducing the Sociological Perspective," "Individuals in Society," "Social Inequality," "Social Institutions," and "Social Change."

Part I, "Introducing the Sociological Perspective," introduces students to the unique perspective of sociology, differentiating it from other ways of studying society, particularly distinguishing it from the individualistic framework students tend to assume. Within this section, Chapter 1 introduces students to the sociological perspective, using the framework provided by C. Wright Mills's the sociological imagination. This chapter also reviews the development of sociology as a discipline, with a focus on the classical frameworks of sociological theory, but also incorporating the perspectives and contributions of those sometimes excluded from the history of sociological thought, such as W. E. B. Du Bois and Jane Addams. Chapter 2 introduces students to the basics of sociological research. Balanced in its approach to both qualitative and quantitative research, this chapter helps students see the connection between questions asked and appropriate research strategies. More than other introductory texts, the chapter provides a detailed presentation of the methodology of different research approaches and helps students think about the ethical dimensions and practical applications of sociological research.

In Part II, "Individuals in Society,"students learn some of the core concepts of sociology, beginning with the study of culture (Chapter 3), moving on to the significance of socialization (Chapter 4), and then to understanding the complexity of social interaction and organization. Chapter 5, "Society and Social Interaction," differentiates micro- and macroanalysis in sociology, introducing students to the nuances in the study of social interaction and moving on to the complex structure of society and social institutions. In Chapter 6, we study the social structure of groups and organizations, using sociology to understand the complex processes of group formation, organizational dynamics, and the bureaucratization of society. Unique to this book is a separate chapter on sexuality (Chapter 7), placed in this section to help students see the connection between some of the most intimate features of individual experience and the social structure of society. The study of sexuality is a new and emerging field in sociology. Often linked only to the study of gender in other texts, we wanted the subject of sexuality to have a place of its own so that students would associate it not only with the study of gender (often implicitly associated only with women), but also with the social and historical organization of society. Finally, this section includes a chapter on deviance (Chapter 8)—looking at the unique way that sociologists frame their studies of deviance and centering the presentation on the difference in functionalist, conflict, and symbolic interactionist approaches to deviant behavior. Instructors will notice that we include the study of crime and criminal justice both in this chapter and in a later chapter on "Power, Politics, and Authority" (Chapter 18). In this chapter, we focus on the commission of crime and its relationship to social deviance; in the later chapter, we include the study of criminal justice institutions (namely, prisons, the police, courts, and the law) as part of the system of authority in society.

In Part III, "Social Inequalities," each chapter explores a particular dimension of stratification in society. Beginning with the significance of class, Chapter 9 provides an overview of basic concepts central to the study of social stratification, as well as current sociological perspectives on class inequality, poverty, and welfare. Chapter 10 follows with a particular emphasis on understanding the significance of global stratification—meaning the inequality that has developed among, as well as within, different nations throughout the world. Throughout this text, we see globalization as a process that is transforming many societies, including the United States; here we emphasize the global process of developing inequality and explore its consequences worldwide. Chapter 11, "Race and Ethnicity," is a comprehensive review of the significance of race and ethnicity in society. Although these concepts are integrated throughout the book because of our focus on diversity, they also require particular focus by looking at how race and ethnicity differentiate the experiences of diverse groups in society. Likewise, although the study of gender is integrated throughout this text, Chapter 12 focuses on gender as a central concept in sociology—one that is closely linked to systems of stratification in society. Finally in this section, we examine "Age and Aging" (Chapter 13) not only in the context of the significance of aging across the life course, but also as a system linked to the advantages and disadvantages of different age groups.

Part IV, "Social Institutions," includes six chapters, each on a basic institution within society. Beginning with Chapter 14 on "Families," these chapters explore the basic structure of social institutions and examine how different theoretical perspectives within sociology help us interpret different dimensions of people's experiences within social institutions. Chapter 15 on "Education" examines the institution of education, viewing it both as a functional part of society and as an institution connected to systems of inequality. Chapter 16 on "Religion" studies the significance of religious belief systems in society, as well as looking at the institutional structure of religious organizations—both in the United States and in a comparative framework. Chapter 17 on "Work and the Economy" begins with an analysis of the significance of different economic systems in the organization of society and proceeds through the study of contemporary patterns in the social organization of work. Chapter 18, "Power, Politics, and Authority," discusses several institutions associated with the state: government, the military, the police, courts, and the law. Framed by a discussion of power and authority, this chapter takes these basic sociological concepts as its core and explores their significance in these different, but overlapping, systems of authority. The last chapter in this section, Chapter 19, "Health Care," examines the social organization of health care and the sociological dimensions of health and sickness among diverse groups in society.

In Part V, "Social Change," we include three chapters, beginning with "Population, Urbanism, and Environment" (Chapter 20). In this chapter, we connect a presentation of the basics of demography to the phenomena of growing urbanism in society and increasing environmental problems, including problems of population growth and pollution. In Chapter 21, we examine "Collective Behavior and Social Movements" as significant sources of social change in society. Finally, in Chapter 22, "Social Change in Global Perspective," we examine the broad dynamics of social change, emphasizing the broad patterns of chance associated with modernization, globalization, and technological development. Although certainly we could have included additional chapters, we have tried to provide a comprehensive introduction to the basics of sociology, without losing the detail and richness that comes from in-depth study of the different subjects examined throughout the book.

e-Andersen and Taylor: The Electronic Version of *Sociology: Understanding a Diverse Society*

A student may purchase the electronic version of *Sociology: Understanding a Diverse Society,* called e-Andersen and Taylor, for their computer from **www.wizeup.com,** or call 212-324-1300. Students have the option to buy the entire text, or just the chapters they need. e-Andersen and Taylor is the entire content of *Sociology: Understanding a Diverse Society,* hot-linked Internet exercises, quizzes, and resources as well as great software tools to aid students in their note taking and study skills. Updates on content and software will also be available after purchase free of charge. e-Andersen and Taylor allows you to study more effectively and create a customized book with your electronic notes, successful web sites, and professor's lectures and priorities in one convenient place, the student's computer.

Supplements

Supplements for the Instructor

Instructor's Manual. Written by Jan Abu-Shakrah from Portland Community College, this supplement offers the instructor chapter-specific lecture outlines, lecture suggestions to facilitate in-class discussion, student activities, worksheets that the instructor can copy and use as handouts, suggestions for further readings, and Internet exercises.

Test Items. The Test Items, composed by Lee Frank, Community College of Allegheny County–South Campus, consists of one hundred multiple-choice questions and twenty true–false questions for each chapter, all with page references and rejoinders. It also includes ten short essay questions and five extended essay questions for each chapter.

World-Class Computerized Test Bank

Available in Both Macintosh and Windows Platforms. With computerized testing, you can quickly create and print a test containing any combination of questions from the test bank. You can also add your own questions, edit the provided questions, print "scrambled" versions of the same test by changing the sequence of the questions, as well as maintain a grade book. Also available online.

Introduction to Sociology Acetates. Acetates are available to help you prepare lecture presentations for your course.

PowerPoint. Available free to adopters, these book-specific slides can be downloaded from the book's web site in either Macintosh or Windows platforms.

Soclink 2000 CD ROM. Soclink 2000 is an easy-to-use interface that instructors can use to create customized lecture presentations for their students. Soclink includes a searchable database of thousands of pieces of art and media, including a photo gallery of all Wadsworth Introduction to Sociology, Marriage & Family, and Social Problems titles, unique CNN video clips, and Sociology PowerPoints. Soclink 2000 gives the instructor the ability to post his or her presentations on the Web and import information from lecture notes. Soclink 2000 is free to adopters.

Demonstrating Sociology: ShowCase Presentational Software. This is a software package for instructors that allows them to analyze data in front of a classroom. It is a powerful, yet easy-to-use statistical analysis package that enables professors to show students how sociologists ask and answer questions using sociological theory. A resource book accompanies it with detailed "scripts" for using ShowCase in class. It is available free only to adopters of any Wadsworth sociology textbook. This supplement is for Windows users with CD-ROM capability only.

Tips for Teaching Sociology. Prepared by Jerry Lewis of Kent State University, this booklet contains tips on course goals and syllabi, lecture preparation, exams, class exercises, research projects, and course evaluations. It is an invaluable tool for first-time instructors of the introductory course and for veteran instructors in search of new ideas.

Wadsworth Sociology Video Library

CNN Sociology Today Video Volumes I & II. CNN Sociology Today video series are hour-long videos consisting of forty-five minutes of footage from stories originally broadcast on CNN within the last several years. Each video is broken into segments ranging in length from two to seven minutes. In turn, each segment is preceded by an introduction setting the segment into a sociological framework. Immediately after each segment, there are several discussion questions of a sociological nature related to the video students just saw. These questions are perfect as the starting points for classroom discussion.

Customized Videos. In addition to the CNN video series, instructors may choose from two customized videos complied from films offered through Films for the Humanities. *Discovering Sociology* and *Sociology in Our Times* consist of short clips that focus on diversity, culture, and current social issues. Both tapes are excellent tools to spark in-class discussion or facilitate lectures.

Supplements for the Student

Study Guide. Written by Kathryn Dennick-Brecht of Robert Morris College, the Study Guide that accompanies the text includes chapter outlines matching that of main text, detailed sentence chapter outlines, lists of key terms from chapter and practice tests, twenty-five multiple-choice questions with rejoinders and page reference, true–false questions, and several essay questions.

Interactions Sociology CD-ROM. Interactions Sociology CD-ROM based exclusively on this text is an interactive tool for students that includes the following features:

- A tutorial for your students with self-quizzes
- Interest-generating multimedia presentations of chapters with audio, photos, and selected videos
- An electronic glossary of key terms and concepts
- Sociological charts and graphs
- A direct link to the Internet

Doing Sociology Software and Workbook, 3rd edition. This software–workbook package shows students what it takes to do real sociological research, using the same data and techniques used by professional researchers. The step-by-step approach in the workbook includes explanations of basic research concepts and methods, expanded exercises, and suggestions for independent research projects, effectively guiding students through the research process and offering them a real sense of what sociologists do. This supplement is for IBM-compatible computers only.

Practice Tests. These practice tests, written by Marlese Durr of Wright State University, contain twenty-five multiple-choice, ten true-false, and five fill-in questions to help the student master course material.

Surfing Sociology. Surfing Sociology is a full-color, tri-fold brochure containing sixty-seven URLs covering the following general areas of sociology: general resources, organizations, theory and methods, race and ethnicity, gender, marriage and family, culture, socialization, social stratification, deviance, education, religion, work, population, social change, and career resources. It also contains a full panel with information about InfoTrac College Edition.

Sociology Internet Companion. Designed by Gene House, a programmer as well as a sociology professor, the Sociology Internet Companion is a disk containing a general introduction to the Internet as well as an extensive "links" section that points students toward several sociology-related sites on the Internet. The introductory section is written primarily for students who have never used the Internet before: They will learn exactly what the Internet is, how it came into being, how to access it from home or school, what a browser is, what a search engine is, how to access e-mail, even how to start building their own home pages. Once they are ready to use the Internet, students can use the links section on the disk in conjunction with a browser and access numerous web sites that are conveniently organized by standard sociology topics.

Investigating Change in American Society Software and Workbook. This workbook–software package allows students to analyze and manipulate huge data sets drawn from the U.S. Census. Each chapter consists of exercises specifically tailored to the data sets. The Student Chip software included with the workbook is a user-friendly statistics program that comes with a tutorial on how to use it and how to analyze data. This supplement is available on both Macintosh and IBM platforms.

Web-Based Resources and Supplements

Virtual Society. Wadsworth's Sociology Resource Center is your companion Web site for *Sociology: Understanding a Diverse Society.* You should bookmark this Web site at the beginning of your Introduction to Sociology course in order to gain full benefit of its many features.

http://sociology.wadsworth.com

At Virtual Society you can find a career center, "surfing" lessons (tips on how to find information on the Web), links to great sociology web sites, and many other selections.

You can also access the Andersen and Taylor web site through Virtual Society. (Click on the Introductory Sociology tab, then click on the cover for Andersen and Taylor.) Features of the Andersen and Taylor website include:

- Hypercontents: Chapter-by-chapter resources available on the Internet.
- Chapter Quizzes: Interactive quizzes for each chapter in the text.
- Student Guide to InfoTrac College Edition: Critical thinking exercises and suggested readings on the InfoTrac College Edition online library, including extra InfoTrac College Edition search terms and articles highlighted specifically for this book.
- Join the Forum: Online bulletin board.
- Book-specific PowerPoint presentations available and easily downloadable in both Macintosh and Windows formats.
- Chapter-by-chapter links to other pertinent websites.

InfoTrac College Edition. If the instructor requests the addition of InfoTrac College Edition in adopting *Sociology: Understanding a Diverse Society* students will receive shrink-wrapped with the text, a special pass-code. The pass-code allows access to InfoTrac College Edition for three months during the time students are enrolled in this course. InfoTrac College Edition is an online source of both scholarly and popular articles from more than 900 periodicals. This is a fantastic resource for conducting research, writing papers, or stimulating discussions in the classroom.

Student Guide to InfoTrac College Edition http://sociology. wadsworth.com. This online supplement, prepared by Tim Pippert, Augsburg College, contains exercises and suggested readings on InfoTrac College Edition. It consists of critical thinking questions using InfoTrac College Edition (an online library containing more than 900 journals and periodicals) for each of the following standard topics in sociology: culture, socialization, deviance, social stratification, race and ethnicity, gender, aging, family, economy/work, education, politics/government, health/medicine, population, social change, and religion. It is accessible via the web sites of Wadsworth introductory sociology texts, as well as from the Virtual Society home page.

Let's Go Sociology: Travels on the Internet, 2nd edition. This book, by Joan Ferrante, Northern Kentucky University, expands on the Internet Home Library found in that book and includes a URL directory with the college sociology student in mind. It is a resource guide—an Internet yellow pages—with URLs that enable its users to do the background research needed to complete most assignments from their homes and residence halls.

Online Products

Web Tutor. Web Tutor is designed to help students learn key concepts in introductory to sociology online. Web Tutor uses multimedia and the Internet to present students with important concepts and to help them learn in a variety of different ways. Since different students learn in different ways, Web Tutor is ideal for laying the foundation for understanding sociology. Some of the features of Web Tutor include the following:

- A link to InfoTrac College Edition
- Flashcards with optional audio still images and video
- Internet exercises online
- Online quizzing
- Web links
- Tutorials
- Audio glossaries

ACKNOWLEDGMENTS

Working on a project as large as this requires the support and help of a large number of people, and we worry that in acknowledging some, we will inadvertently omit someone who answered a key question, provided a good reference, lent an ear, or helped in some other important way. There are many who over the seven years we worked to complete this book have provided advice, friendship, support, and technical assistance. We thank particularly Bob and Gloria Albright, Maxine Baca Zinn, Serina Beauparlant, Audrey Blackburn, Anne Bowler, John Cavanaugh, Patricia Hill Collins, Ken Haas, Arlene Hanerfeld, Valerie Hans, Phil Herbst, Elizabeth Higginbotham, Carol Hoffecker, Carla Howery, Lionel Maldonado, Angela March, Carole Marks, Buffy and Tony Miles, Joanne Nigg, Kim Logio-Rau, Kathleen Slevin, Catherine Simile, Randall Stokes, Kathleen Tierney, and Juan Villamarin. We also thank Alan McClare, the original editor for this book, for his support over the years.

We want to offer a special thanks to Grant M. Farr of Portland State University. Professor Farr not only served as an insightful reviewer of the manuscript he also acted as a consultant in the writing of Chapter 10, "Global Stratification." Thanks also to Tim Pippert, who helped select the InfoTrac College Edition search terms found throughout the book.

We give our very special thanks to Wei Chen for her brilliance in developing all the maps in the book, as well as providing the information on web sites, even after completing her studies at the University of Delaware. Thanks also to Cathleen Brooks, Heather Smith, and Taj Carson for their expert research assistance and to Norman Andersen for initially imagining and designing the conceptual art for the book.

Blanche Anderson, AnnaMarie Brown, Ann Draper, Cindy Gibson, Lisa Huber, Susan Phipps, and Cindy Sterling also helped with many of the details that producing this book required and we appreciate their support. We thank Jeffrey Quirico and Mel Schiavelli in the Provost's Office at the University of Delaware for providing funds for research assistance. Thanks, as well, to the staff of Morris Library; without their help, especially in developing online resources, it would have been much more difficult to complete this book.

We have been especially lucky to work with an extraordinary editorial team at Wadsworth. We particularly thank Eve Howard, our editor, and Bob Jucha, our development editor, for their enthusiasm, hard work, and good humor as we worked together on this project. Thanks as well to Jerilyn Emori, Stephen Forsling, Dusty Friedman, and Ari Levenfeld for expertly overseeing different aspects of the huge production effort.

Our special and heartfelt thanks go to our spouses, Richard Morris Rosenfeld and Patricia Epps Taylor, for having endured through the many years we worked on this book. Although we were frequently preoccupied with the book and the project often interfered with time we would have otherwise had with them, their love, fun, and friendship have sustained us.

Many reviewers over the years have provided commentary that has strengthened this book. We thank the following people:

Jan Abushakrah
Portland Community College

Susan Albee
University of Northern Iowa

Angelo Alonzo
The Ohio State University

Elena Bastida
University of Texas–Pan American

Rebecca Brooks
Kent State University

Valerie Brown
Cuyahoga Community College East

Russell Buenteo
University of South Florida

Jeffery Burr
State University of New York at Buffalo

Kathy Dennick-Brecht
Robert Morris College

Marlese Durr
Wright State University

John Ehle
Northern Virginia Community College

Jess Enns
University of Nebraska–Kerry

Kevin Everett
Radford University

Grant Farr
Portland State University

Michael Goslin
Tallahassee Community College

Richard Halpin
Jefferson Community College (New York)

C. Allen Haney
University of Houston

Dean Harper
University of Rochester

Gary Hodge
Collin County Community College

Matt Huffman
George Washington University

Wanda Kaluza
Camden Community College

Edward Kick
University of Utah

Keith Kirkpatrick
Victoria College (Texas)

Koorps Mahamoudi
Northern Arizona University

Susan Mann
University of New Orleans

Patrick Mcguire
University of Toledo

Beth Mintz
University of Vermont

Charles Norman
Indiana State University

Tracy Orr
University of Illinois–Champaign

Carol Ray
San Jose State University

David Redburn
Furman University

Michael Smith
St. Philip's College, Alamo Community College District

Larry Stern
Collin County Community College

Glenna Van Metre
Wichita State University

Mark Winton
University of Central Florida

We also wish to thank the following individuals who provided valuable insights into the teaching of introductory sociology and technology tools for teaching by participating in surveys and focus groups within the past year.

William Camp
Luzerne County Community College

Carole M. Carroll
Middle Tennessee State University

Robert Futrell
University of Kansas

Thomas Gieryn
Indiana University

Michael Macy
Cornell University

Patrick J.W. McGinty
University of Missouri, Columbia

Richard E. Miller
Navarro College

Carrie Uihleim Niles
Marshall University

Robert Perry
Johnson County Community College

John Reynolds
Florida State University

Luis Salinas
University of Houston

Ira M. Wasserman
Eastern Michigan University

Mary Lou Wylie
James Madison University

Maliha Zulfacar
California Polytechnic State University, San Luis Obispo

Sociology

CHAPTER 1

Developing a Sociological Perspective

IMAGINE that you had been switched with another infant at birth. How different would your life have been? What if your accidental family was very poor . . . or very rich? How might this have affected the schools you attended, the healthcare you received, the possibilities for your future career? What if you had been raised in a different religion—how would this have affected your beliefs, values, and attitudes? Taking a greater leap, what if you had been born another sex or a different race? What would you be like now?

We are talking about changing the basic facts of your life—your family, social class, education, religion, sex, and race. Each has major consequences for who you are and how you will fare in life. These factors play a major part in writing your life script. Social location (meaning, one's place in society) establishes the limits and possibilities of a life.

Consider this:

- Children raised in single-parent families are more likely than those from two-parent families to hold low-paying jobs in their adult lives (Bibliarz and Raftery, 1993).
- People with more education have better health than others (Ross and Wu, 1995).
- Men who work in jobs traditionally defined as "women's work" behave in ways that emphasize their masculinity; doing so brings rewards since they tend to be promoted faster than similarly qualified women in the same occupations (Williams, 1995).
- The physical transformations that female impersonators must make to deliver a convincing performance are more easily mastered than the behavioral changes they have to make (Tewksbury, 1994).
- An African American can expect to earn 10 percent less in personal income than a White American with the same education, occupation, age, and marital status (Thomas, 1993; Oliver et al., 1995).
- Members of the Catholic Church are less likely to commit suicide than Protestants, although this does not hold for Catholics who have been divorced (Burr et al., 1994).

These conclusions, drawn from current sociological research, describe some of the consequences of particular social locations in society. Although we may take our place in society for granted, our place has a profound effect on our chances in life. The power of sociology is that it teaches us to see how society influences our lives and the lives of others, and it helps us explain the consequences of different social arrangements.

Sociology is the study of human groups. As such, sociologists analyze the significance of diversity in society.

What Is Sociology?

Sociology is the study of human behavior in society. Sociologists are interested in the study of people and have learned a fundamental lesson: All human behavior occurs in a societal context. That context—the institutions and culture that surround us—shapes what people do and think. In this book, we will examine the dimensions of society and analyze the elements of social context that influence human behavior.

Sociology is a scientific way of thinking about society and its influence on human groups. Observation, reasoning, and logical analysis are the tools of the sociologist, coupled with knowledge of the large body of theoretical and analytical work that has been done by sociologists and others. Sociology is inspired by the fascination people have for the thoughts and actions of other people, but it goes far beyond casual observations about people and events. It builds upon observations that are *objective* and *accurate* to create analyses that are reliable and can be validated by others.

Every day, the media in its various forms (television, film, video, and print) bombard us with social commentary. Whether it is Oprah Winfrey, Ted Koppel, or Rush Limbaugh, media commentators and celebrities provide endless opinion about the various and sometimes bizarre forms of behavior in our society. Sociology is different. Sociologists may study the same subjects that the media examine—such as domestic violence, religious cults, race relations, or wealth and poverty, just to name a few—and many sociologists make appearances on the media, but sociologists use specific research techniques and well-tested theories to examine and explain social issues. Indeed, sociology can provide the tools for testing whether the things we hear and believe about society are actually true. Much of what one hears in the media and elsewhere about society, delivered as it may be with perfect earnestness and sincerity, is misstated, misguided, sometimes completely wrong. The lessons and practices of sociology can be applied to media pronouncements and other pieces of social lore to find objective understanding of social issues that are often clouded by people's opinions.

Whether or not you become a professional sociologist, the perspective of sociology is useful in everyday life since it helps you see the societal context of human behavior and, in so doing, reveals the underlying basis for many social issues, problems, and current events. In this chapter, we introduce students to the sociological perspective and to some of the essential concepts and theories that guide the discipline.

Sociology: A Unique Perspective

The subject matter of sociology is all around us. The routines of everyday life, the social problems we see, the rapid changes taking place in society—these are the topics of sociological study. Psychologists, anthropologists, political scientists, economists, and others also study social behavior and social change. Along with sociologists, these disciplines make up the *social sciences*. The difference between sociology and these other disciplines is not in the topics that each studies, but in the perspective each discipline brings to the subject at hand.

Psychology analyzes individual behavior. Sociologists share the interest of psychologists in individuals, but whereas the unit of analysis for psychology is the individual and groups, for sociologists, it is society—the whole configuration of group life. In a society that cherishes individualism and individual rights, there is a tendency to develop psychological explanations of all behavior. Thus, when people think about why people behave as they do, they tend to give psychological answers—such as that people behave differently because of personality differences or because their motivations are different from others. From a sociological point of view, psychological explanations are not wrong, just incomplete. Sociologists explain people's behavior as arising not only from motives and attitudes that are internal to the person, but also from the social context in which people live. One person's behavior might be attributed to personality—a psychological explanation—but when we realize that there are consistent patterns of thought and behavior across the whole society, it is necessary to seek a larger perspective. Sociological inquiry extends to the largest social unit of all, society itself.

Anthropology is the study of all human cultures. Sociology and anthropology both study people, but the fields are not the same. Anthropologists see culture as the basis for society; sociologists see culture as a piece of a complex configuration of other social systems that together compose society. Anthropologists also tend to study cultures other than their own, whereas sociologists are more likely to study a society of which they are a part. Anthropologists are also more likely to study faraway and remote cultures, although some anthropologists now also study various cultural dimensions of U.S. society and many sociologists do research in different countries. The primary difference in the two disciplines is not so much where sociologists and anthropologists do their work but the central emphasis that anthropologists place on culture.

Economics and political science are the other pillars of the social sciences. *Political science* is the study of politics, including political behavior, political philosophy, and the organization of government and political parties. *Economics* scrutinizes the production, distribution, and consumption of goods and services. Political scientists and economists are each interested in a specific set of social institutions, those that shape economic and political behavior. Sociologists are interested in all **social institutions,** which they define as established and organized systems of social behavior with a recognized purpose. In a nutshell, sociologists are interested in how social forces of all kinds—cultural, economic, political (to name a few)—affect human behavior.

Another discipline related to sociology is social work. Many students enter the field of sociology because they want to help people and think they might become social workers. Sociology and social work are closely allied, but they are not the same. *Social work* is an applied field that makes use of the lessons of the social sciences to serve people in need. Social work draws heavily from psychological thinking and from studies of families, but it typically addresses people's problems on an individual basis, using counseling as a major tool. Studying sociology is a good way to prepare for a career in social work since it delivers a perspective that will help a social worker better understand the plight of various individuals, but sociologists do not see individual solutions as adequate for addressing society's problems. Sociologists are likely to argue for societal-level changes to address the broad-based problems that we face.

To understand the differences among the social sciences, consider how practitioners in each of these fields might study the family. Psychologists would be interested in how individual personalities are formed within families; for example, they might look at abnormalities in psychological functioning that result from problematic family dynamics. Anthropologists would be curious about the diverse family structures that develop in different cultures. Political scientists might scrutinize how a policy decision would affect different families, or they might examine how voting behaviors and political opinions are passed on within families. Economists expend great effort interpreting family consumer patterns and assessing how changes in the economy affect employment, the mainstay of family economics. Social workers would be most concerned with delivering social services to families in need. Sociologists, on the other hand, would be interested in how families are shaped by society—their size, structure, opportunities, and so on—and how changes in society are affected by changes in family structure.

The boundaries between the social sciences are not rigid, and you should not overdraw the differences among them. Each is capable of making significant contributions to our knowledge of people, and some of the best work in each field draws on work in others. Many sociologists engage in interdisciplinary research, seeking insight by using the perspectives of multiple disciplines. Sociology is also linked to several new academic fields that study groups that have previously been ignored or misperceived by scholars in the traditional disciplines. African American studies, Chicano and Latino studies, Asian American studies, Native American studies, women's studies, lesbian and gay studies, and Jewish studies have drawn from sociological ideas and in return have enriched the research and theory that sociologists use to understand the lives of diverse people.

Sociology in Practice

Sociology is not merely fascinating, it is also useful. One of the most basic applications of sociology is its teaching mission—using sociological research and theory to educate people about the diverse groups that make up society. This is no small task. The media emit an endless stream of conventional and sometimes prejudicial explanations of social issues. Learning to think like a sociologist can teach one to challenge unexamined assumptions about society, values, and people. Sociology teachers want students to think about society from new vantage points. Open-mindedness is the first step to learning the sociological perspective.

Sociologists have used their sociological perspective for many purposes. When the publication of *The Bell Curve* created great public controversy over race and intelligence, Howard Taylor, an expert on race and intelligence testing (and co-author of this book), appeared on TV's *Nightline* in a debate with the author Charles Murray. In the mid-1990s, when some universities in Hong Kong first adopted policies against sexual harassment, Margaret Andersen (the other author of this book) used her knowledge about gender and violence against women, as well as her experience as a university administrator, to conduct workshops on sexual harassment for faculty and administrators in Hong Kong. Sociologists have testified before Congress on topics as diverse as teenage pregnancy, social welfare reform, AIDS funding, and street violence. Sociologists sometimes provide expert testimony in criminal or civil legal cases. They may consult with organizations trying to develop fair employment practices, or they may work with community action groups trying to solve local problems of housing, safety, or education. The practical applications of sociology are as diverse as the people who become sociologists. What sociologists hold in common is a commitment to rigorous study as the path to understanding society and making a difference in people's lives. You will see some of the different ways that sociologists have put their knowledge to use in the feature of this book called "Sociology in Practice." These different applications are the fruits of one of sociology's most potent tools, a habit of mind called the sociological imagination.

The Sociological Imagination

Think back to the vignette that opened this chapter in which you were asked to imagine yourself having grown up under completely different circumstances. Our goal in that passage was to make you feel the stirring of the **sociological imagination**—the ability to see the societal patterns that influence individual and group life. The beginnings of the sociological imagination can be as simple as the pleasures of watching people or wondering about how society influences people's lives. Indeed, many students begin their study of sociology because "they are interested in people." Sociologists convert this curiosity into the systematic study of how society influences different people's experience within it.

C. Wright Mills (1916–1962) was one of the first to write about the sociological imagination. In his classic book, *The Sociological Imagination* (1959), Mills wrote that the task of sociology was to understand the relationship between individuals and the society in which they lived. Sociology should be used, Mills argued, to reveal how the context of society shapes our lives; he described this as understanding the intersection between biography and history. Mills was not just emphasizing the importance of history. He thought that to understand the experience of a given person or group of people, one had to have knowledge of the social and historical context in which people lived.

To visualize the junction of biography and history, think about the stress that many people experience today in trying to manage the multiple demands of a family and a job. In recent years, changes in society have led to what

BOX SOCIOLOGY IN PRACTICE

1.1 Careers in Sociology

IT IS HARD to estimate how many sociologists there are in the United States since people often hold this degree and work with sociological skills but call themselves something different. There are more than 13,000 members of the American Sociological Association, the national professional organization. About 80 percent of these are employed as teachers in colleges and universities; 20 percent are employed in government positions, private organizations, and public agencies. Sociologists have an important role in education, in community and public services, and in the development of public policy.

Most students see sociologists in their roles as educators—an important part of the work that sociologists do. An education in sociology can help you think logically and analytically about society and its problems. Many students who take sociology courses do not become professional sociologists, but the skills it gives them prepare them for a variety of careers. Sociology teaches valuable analytical skills, such as quantitative reasoning, understanding those with different cultural and social orientations from oneself, and dissecting the different causes of a social problem. Sociology also helps you appreciate the experiences of diverse groups in society.

In addition to their work as teachers, sociologists have become more active in the formation of public policy. Their research and theory are often the basis for changes in public policy. Sociologists are frequently called upon as experts for the U.S. Congress, by the media, by state and federal agencies, and by private organizations to consult on a variety of subjects. For example, when the Family/Medical Leave Policy of 1993 was passed, sociologists gave briefings to congressional staff members on the sociological research on family–work connections; when the Family/Medical Leave Act passed in 1993, sociologists from the American Sociological Association also provided briefings for the press on the implications of this bill. Sociological research documenting the gender and race bias in a widely used test of job aptitudes (called the General Aptitude Test Battery, or GATB) resulted in the Department of Labor and Department of Health and Human Services stopping use of the test in every state until it was revised to ensure more fairness (Hartigan and Wigdor, 1989).

Other examples of the impact of sociological research on social policy are numerous. Sociologists use their sociological imaginations to affect change in a variety of areas. As an example, feminist scholars, Latino sociologists, African American sociologists, and Asian American sociologists have been leaders in the movement to make education more multicultural. Some sociologists hold positions of political office; sociologists also work in the offices of U.S. representatives and senators doing background research on the various issues being addressed in the political process. Several sociologists have been direct advisors to a U.S. president. Other sociologists have worked in their communities to deliver more effective social services. Some are employed in business organizations and social services where they use their sociological training to address issues like poverty, crime and delinquency, population studies, substance abuse, violence against women, family social services, immigration policy, and any number of other important issues.

These are but a few examples of the ways that sociology has been used to direct social change. Sociology is a rigorous intellectual subject for study, but in the tradition of its founders, sociology also has an orientation toward progressive social change.

some call *social speedup:* Those who work are working longer hours; women are more likely to be working outside the home; middle-class families need more than one earner to keep a certain standard of living; less advantaged families need more than one earner just to make ends meet. These multiple demands cause stress. People may respond by organizing their lives better, refusing additional demands, exercising to relieve stress, or turning to drugs and alcohol for relief. These personal solutions may offer momentary relief, but they do not solve the basic problem, which is that changes in society are placing additional strains on people. The origins of these problems lie beyond personal lives, and so do the solutions. Our attempts to solve these problems on a personal basis may even create new problems—drug and alcohol addiction, juvenile delinquency, or family breakup. The various parts of this complex picture are put together by exercising the sociological imagination, which permits us to see that these are problems arising from a social context, not individual characteristics. Sociologists are certainly concerned about individuals, but they direct their attention to the social and historical context that shapes the experiences of individuals and groups.

A fundamental concept for organizing the sociological imagination is the distinction Mills made between troubles and issues. **Troubles** are privately felt problems that spring from events or feelings in one individual's life. **Issues** affect large numbers of people and have their origins in the institutional arrangements and history of a society (Mills, 1959). This distinction is the crux of the difference between individual experience and **social structure,** defined as the organized pattern of social relationships and social institutions that together constitute society. Issues shape the context within which troubles arise. Sociologists employ the sociological imagination to understand how issues are shaped by social structures.

Mills used the example of unemployment to explain the meaning of troubles versus issues. When a person becomes unemployed, he or she has a personal trouble. In addition to financial problems, the person may feel a loss of identity, may become depressed, may lose touch with former work associates, or may have to uproot a family and move. Being out of work also causes hardships that can spread to other people, such as the worker's children. The problem of unemployment, however, is deeper than the experience of one person. Unemployment is rooted in the structure of society; this is what interests sociologists. What causes unemployment? Who is most likely to become unemployed at different times? How does unemployment affect an entire community (for instance, when a large plant shuts down) or an entire nationwide group (such as the corporate downsizing of older workers)? Sociologists know that unemployment causes personal troubles, but understanding unemployment is more than understanding one person's experience. It requires understanding the social structural conditions that influence people's lives.

The specific task of sociology, according to Mills, is to comprehend the whole of human society—its personal and public dimensions, historical and contemporary, and its influence on the lives of human beings. As Mills wrote,

Personal troubles felt by individuals who are experiencing problems; social issues arise when large numbers of people experience problems that are rooted in the social structure of society.

> The sociological imagination enables its possessor to understand the larger historical scene in terms of its meaning for the inner life and external career of a variety of individuals. . . . The first fruit of this imagination—and the first lesson of the social science that embodies it—is the idea that the individual can understand his experience and gauge himself only by locating himself within his period, that he can know his chances in life only by becoming aware of those of all individuals in his circumstances (Mills, 1959: 3–5).

Mills had an important insight: People often feel that the situation they are in is larger than their experience alone because they feel the influence of society in their lives. This is not always readily apparent to people. During the Great Depression, as an example, many were ruined by economic collapse. With such widespread distress, many began to understand that they had been the victims of a disaster far larger than their personal failure. Social forces had influenced their lives in the most drastic manner. Many turned their energy into activism aimed at correcting what was perceived to be an unfair economic order. Now, sociology has a special contribution to make—revealing the forces at work that are shaping the society of the twenty-first century.

Revealing Everyday Life

June Jordan, the noted African American poet and essayist, says that, if you understand the sociological perspective, you can never be bored (1981: 100). Jordan made her comment while reflecting on her experience as a student in introductory

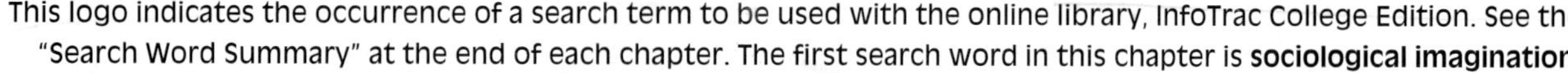

This logo indicates the occurrence of a search term to be used with the online library, InfoTrac College Edition. See the "Search Word Summary" at the end of each chapter. The first search word in this chapter is **sociological imagination.**

sociology. While she was in college, Jordan felt alienated from most of what she studied because she never saw herself, her people, her street, or her community reflected in her college studies. Sociology, however, was fascinating to her because it revealed patterns in everyday life that helped her understand the society she lived in.

Why would Jordan think that sociology relieves boredom? Think of a boring situation—for example, waiting in an airport terminal for an overdue flight. To the sociologist, even this situation can be brimming with observable phenomena. Sociologists might observe the rituals people engage in when they meet and greet each other. They might note how two men in turbans interact differently from two women in police uniforms. They might observe how workers behave in different jobs, perhaps inspiring thoughts about how work conditions influence workers' attitudes about work and about each other. The keen sociological observer might also detect differences in who is employed in what jobs and may perceive the influence of gender and race on employment patterns at the airport.

Arlie Hochschild, a contemporary sociologist, has observed how flight attendants interact with passengers and with each other. She concluded that flight attendants are trained to produce certain feelings in passengers, such as feeling as if you are a guest in someone's home, not hurtling through space at 400 miles per hour! Hochschild also noted that flight attendants are evaluated by supervisors according to how well they succeed in producing the desired emotional state. Based on her research, she wrote a fascinating book on how service industries expect employees to "manage" the emotions of others (Hochschild, 1983). The result, argues Hochschild, is the commercialization of human feeling. Perhaps Hochschild got the initial idea for her study while sitting in an airplane stuck on the tarmac, trying not to be bored!

Revelations about everyday life are part of the appeal of sociology, but sociology is much more than simply thinking about things seen around you. Sociology uses rigorous methods of research to examine its assumptions and conclusions. To see the difference between social musing and sociological research, consider how one would approach a common explanation of inequality in the United States. Many people believe that poor people do not get ahead because they do not want to work hard enough. A sociologist may believe this to be so, but would not assume it to be so. Instead, the sociologist would first determine what facts can be shown to be true. Do people who are less well off have a work ethic that is different from those who are better off? Probing this question, sociologists have found that attitudes toward work are a poor explanation of the likelihood that someone will be poor or on welfare (Santiago, 1995; Wagner, 1994). Forced to look to other factors to explain why there are poor people in society, sociologists have studied educational and job opportunities, and have asked whether the labor market and the educational system give some people better opportunities than others. They do. By seeking to determine a systematic explanation of why there is inequality, sociologists are led beyond commonsense explanations. They usually discover that the most compelling explanations of behavior are found in social factors, not just individual states of mind.

A big difference between sociology and common sense is that *sociology is an empirical discipline.* The **empirical** approach to knowledge requires that conclusions be based on careful and systematic observations—not on previous assumptions, but on what one observes. For empirical observations to be useful to other observers, they must be gathered and recorded rigorously. It is not enough to say, "Many people failed to vote this year." A contribution to sociology is made when we know what percentage of people voted and how many fewer voted in this election than in the last one. It is not enough to say, "People did not vote; they must have been disgusted." The sociologist questions voters, interviews nonvoters, calculates how many people have the same reason for not voting, determines how many people from diverse groups voted—all before presenting the findings. The findings will show the complexity of reasons why people vote or not.

Unlike those who just use common sense, sociologists are obliged to constantly reexamine their assumptions and conclusions. Only careful, unprejudiced observations add to the fund of sociological knowledge. Commonsense explanations may be widely shared and sometimes accurate, but until they are subjected to rigorous, objective analysis, they remain common sense—a mix of casual logic, folk wisdom, rumor, and prevailing prejudices. Although the specific methods that sociologists use to examine different problems vary (see Chapter 2), the empirical basis of sociology is what distinguishes it from mere opinion or other forms of social commentary.

Debunking in Sociology

The power of sociological thinking is that it helps us see everyday life in new ways. Sociologists question actions and ideas that are usually taken for granted. Peter Berger (1963) calls this process **debunking.** Debunking refers to looking behind the facades of everyday life—what Berger called the "unmasking tendency" of sociology (1963: 38). In other words, sociologists look at the behind-the-scenes patterns and processes that shape the behavior they observe in the social world.

Using the example of schooling, we can see how the sociological perspective debunks taken-for-granted assumptions about education. A commonsense perspective sees education as helping people learn and get ahead. A sociological perspective on education, however, would reveal something more. Sociologists have concluded that more than learning takes place in schools; other social processes are at work. Social cliques are formed where some students are "insiders" and others are excluded. Young school children acquire not just formal knowledge, but also the expectations of society and people's place within it. Race and class conflicts are often played out in schools. Relative to boys, girls

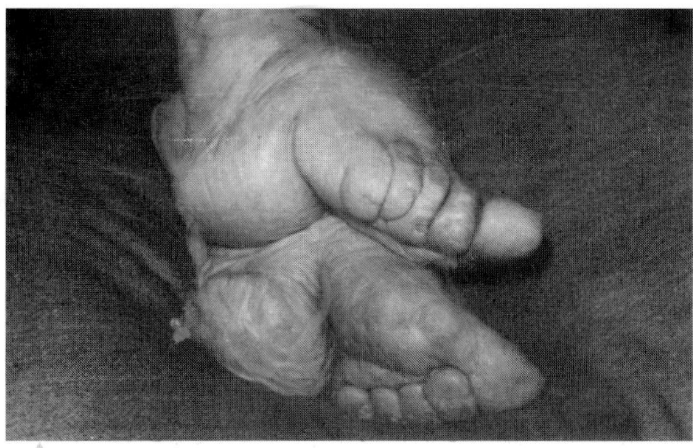

Cultural practices that seem bizarre to outsiders may be taken for granted or defined as appropriate by insiders. The ancient practice of footbinding in China may appall some, but so might contemporary practices of body piercing—popular and stylish among young people.

are often short-changed by the school system—receiving less attention and encouragement (American Association of University Women, 1992; Sadker and Sadker, 1994). Poor children seldom have the same resources in schools as middle-class or elite children, and they are often assumed to be incapable of doing schoolwork and are treated accordingly. The somber reality is that schools may actually stifle the opportunities of some children rather than launching all children toward success—in contrast to common belief.

Sociological inquiry requires taking nothing for granted; however, as people grow up in a society, they learn to regard the attitudes and behaviors that occur in their society as "natural," making it difficult to see these ready-made understandings. Taking a sociological perspective requires stepping outside this natural attitude. A **natural attitude** develops when members of a society accept social expectations and absorb social information at face value without questioning what they see (Schutz and Luckmann, 1973). To acquire a sociological perspective, one must shake off the natural attitude and refuse to take anything for granted. If you engage comfortably in a social situation, the social forces shaping the situation may become invisible to you. Thinking sociologically requires disembarking from one's "natural" way of thinking.

THINKING SOCIOLOGICALLY

Identify an activity that you do every day, something about which you have a *natural attitude*. If this activity were being observed by someone from a different society, how might he or she describe it?

This is sometimes easier to do when looking at a culture or society different from one's own. Consider how behaviors that are unquestioned in one society may seem positively bizarre to an outsider. For a thousand years in China, it was usual for the elite classes to bind the feet of young girls to keep the feet from growing bigger—a practice allegedly derived from one of the emperor's concubines. Bound feet were a sign of delicacy and vulnerability. A woman with large feet (defined as more than 4 inches long!) was thought to bring shame to her husband's household. The practice was supported by the belief that men were highly aroused by small feet even though men never actually saw the naked foot. If they had, they might have been repulsed since a woman's actual foot was U shaped and often rotten and covered with dead skin (Blake, 1994).

Outside the social, cultural, and historical context in which it was practiced, footbinding seems bizarre, even dangerous. Feminists have pointed out that Chinese women were crippled by this practice, making them unable to move about freely and making them more dependent on men (Chang, 1991). This is an example of outsiders debunking a practice that was taken for granted by those within the culture. Many behaviors that ordinarily go unquestioned can be debunked by an alert, imaginative, and careful observer.

Suppose that you were to use the debunking process to examine practices that typically go unquestioned in U.S. culture. Strange as the practice of Chinese footbinding may seem to you, how might someone from another culture view the implantation of silicon in women's breasts to enhance their size? Or piercing the tongue, nose, or ears and filling the hole with jewelry? Or wearing a completely useless cloth necktie snugly around one's neck? Or as Victoria's Secret does, mailing catalogs full of glossy pictures of women in lacy underwear to millions of consumers every month? These practices of contemporary U.S. culture are taken for granted by many, just as Chinese footbinding was. Until these cultural processes are debunked, seen as if for the first time, they might seem normal. One can step away from cultural familiarity to glimpse the social structures behind them, revealing these practices in their sociological context.

In debunking the taken-for-granted world, sociologists find their subject matter all around them. They carefully observe social behavior and reveal different levels of meaning that are ordinarily hidden from consciousness in everyday life. Simply put, sociology is appealing to those who find human beings fascinating.

Establishing Critical Distance

Racehorses are fitted with blinders that block their sideways vision so that they will not be distracted or frightened by the other animals thundering alongside them. Longtime social habits and automatic responses to familiar situations are like blinders on our sociological vision. Seeing past the cultural beliefs established in any society requires stepping back, detaching oneself from unquestioned engagement in everyday life and establishing *critical distance*. Detachment, or pulling off the blinders, is not always easy to achieve. This requires stepping back from the taken-for-granted assumptions provided in everyday life—a process of establishing some critical distance from what is expected of us. Sociology provides a way to do this.

Under what circumstances are people best able to develop this critical distance? A person has to be able to become detached from the situation at hand, to some ex-

B • O • X 1.2 DOING SOCIOLOGICAL RESEARCH

Debunking the Myths of Black Teenage Motherhood

News of teen mothers who deliver their babies in secret, perhaps then killing them or leaving them at someone's door, gives a compelling image of the seriousness of this problem. Most teen births occur without such fanfare although public concern about teenage motherhood is a key part of discussions about poverty, family values, and child welfare. The common view is that many of the social and economic problems of today's poor result from female-headed households, households where "babies are having babies" and where a cycle of poverty is generated because of the high rate of teen pregnancy.

What truth is there to this popular image? This is what sociologist Elaine Bell Kaplan wanted to know in doing her research on African American teen mothers. Kaplan knew that there was a stereotypical view of Black teen mothers: that they had grown up in fatherless households where their mothers had no moral values and no control over their children. The myth of Black teenage motherhood also depicts teen mothers as unable to control their sexuality, as having children to collect welfare checks, and as having families who condone their behavior. Looking beyond the public myth, Kaplan wanted to understand how these young women responded to a world where poverty, racism, and gender norms intertwine such that these young women face a huge array of economic and social problems.

Kaplan did extensive research in two communities in the San Francisco Bay Area—East Oakland and Richmond, both communities with a large African American population and typical of many inner-city, poor neighborhoods. Once thriving Black communities, East Oakland and Richmond are now characterized by high rates of unemployment, poverty, inadequate schools, crime and drug-related violence, and high numbers of single-parent households. Having grown up herself in Harlem, Kaplan knew that communities like those she studied have not always had these problems, nor have they condoned teen pregnancy. To do her research, she spent several months in the community, working as a volunteer in a community teen center that provided educational programs, daycare, and counseling to teen parents, and "hanging out" with a core group of teen mothers. She did extensive interviews with thirty-two teen mothers, supplementing them when she could with interviews with their mothers and, sometimes, the fathers of their children.

The richness of her study can be found only by reading the book and hearing the full experience of the women she writes about. She found that these teen mothers adopt strategies for survival that help them cope with their environment even though these same strategies do not help them overcome the magnitude of problems they face. Unlike what the popular stereotype suggests, she did not find that the Black community condones teen pregnancy; quite the contrary, the teens felt embarrassed and stigmatized by being pregnant and experienced tension and conflict with their mothers who saw their pregnancy as disrupting the hopes they had for their daughters' success. When the women had to go on welfare, they also felt embarrassed, often developing strategies to hide the fact that they depended on welfare to make ends meet. These conclusions are directly counter to the public image that such women do not value success and live in a culture that promotes welfare dependency.

Kaplan also interprets the teen mothers' lives in the context of adolescent development. Instead of simply stereotyping them as young and tough, she sees them as struggling to develop their own gender and sexual identity. Like other teens, they are highly vulnerable, searching for love and aspiring to create a meaningful and positive identity for themselves. But failed by the educational system and locked out of the job market, the young women's struggle to develop an identity is compounded by the disruptive social and economic conditions in which they live.

Her research is a fine example of how sociologists debunk some of the commonly shared myths that surround contemporary issues. Carefully placing her analysis in the context of the social structural changes that affect these young women's lives, Kaplan provides an excellent example of how sociological research can shed new light on some of our most pressing social problems.

SOURCE: Kaplan, Elaine Bell. 1996. *Not Our Kind of Girl: Unraveling the Myths of Black Teenage Motherhood.* Berkeley: University of California Press.

tent, to develop a sociological perspective. Some people live in circumstances that make this easier for them. *Marginal people* are those who share the dominant culture to some extent but are blocked from full participation within it because of their social status. One need not be a marginal person to be a sociologist, but marginality has often provided the critical distance necessary to inspire a thriving sociological imagination. For example, many of the leaders of the civil rights movement during the 1950s and 1960s studied sociology. Jesse Jackson, Martin Luther King, Jr., and Ella Baker (one of the major leaders of the student civil rights movement in the American South) were sociology majors who used their sociological skills to help students organize and understand the social forces that produced racism (Carson, 1981). They understood that problems like racial inequality have their origins in the structure of society, not just in the minds of individuals. Most likely, their experiences as African American women and men helped them acquire a sociological imagination and defy the time-honored beliefs in U.S. society that supported racial segregation. They benefited from the critical distance their lives had provided them as they analyzed how the structure of society produced racism.

The role of critical distance in developing a sociological imagination is well explained by the early sociologist Georg Simmel (1858–1918). Simmel was especially interested in the role of strangers in social groups. Strangers have a position both inside and outside social groups; they are part of a group without necessarily sharing the group's assumptions and points of view. Simmel wrote that strangers are "fixed within a particular spatial group . . . but whose position in this group is determined, essentially, by the fact that he has not belonged to it from the beginning" (1950: 402). Strangers are both close to and distant from the group and its beliefs. For that reason, the stranger can sometimes see the social structure of a group more readily than can those who are thoroughly imbued with the group's world view. The stranger is "not tied down . . . by habit, piety, and precedent" (Simmel, 1950: 405) and takes nothing for granted. Since strangers have not yet acquired the assumptions and expectations of the group, they can be more objective about the group's behavior.

Simmel's work suggests that a combination of nearness and distance illuminates the sociological perspective. One must have enough critical distance to remain aloof from the group's own definition of the situation, but be near enough to share in the group's understanding of its situation. This unique positioning of the mind—what Patricia Hill Collins (1990) in writing about African American women has called "the outsider within"—is the foundation of the sociological imagination.

Sociologists are, of course, not typically strangers to the society they study, nor do they have to be marginal to acquire the sociological imagination. One can acquire critical distance through simple willingness to question the forces that shape social behavior. Often, in fact, sociologists become interested in things because of their own biographical experiences. The biographies of sociologists are rich with examples of how their personal lives informed the questions they ask. Among sociologists, you will find former ministers and nuns now studying the sociology of religion, women who have encountered sexism in their lives now studying the significance of gender in women's and men's experiences, rock-and-roll fans who study the significance of popular music in contemporary culture, and sons and daughters of immigrants now analyzing race and ethnic relations. There are countless other examples where a person's biography has shaped his or her sociological curiosity (see the box, "Understanding Diversity: Becoming a Sociologist").

Sociologists' work may be informed by their personal lives, but personal considerations must not outweigh professional rigor. Doing sociology requires rising above personal feelings about a research interest while preparing oneself to see society for what it is. The term for this clarity of vision is *objectivity*. Sociology does not deny feelings and opinions, but its importance rests on the ability of sociologists to be objective and systematic in describing and analyzing society.

To cultivate a sociological imagination, the learner must master the tools of sociological research and theory. One must learn to observe accurately and record one's observations carefully. Subsequent analysis will benefit greatly from an awareness of the theoretical structures that sociologists have devised earlier to make sense of the mass of sociological information that has been collected. Sociology is not simply the development of informed opinion about social issues. It involves rigorous study, as we will see in the material to follow on sociological theory and research methods. As one learns to develop original and dependable findings, one also becomes more capable of assessing the credibility of other people's conclusions.

Sociology produces a critical attitude toward society because it probes the development and consequences of societal arrangements. Some sociologists are more critical of the existing society than others; some are more conservative, but there is a long tradition of using sociological knowledge for purposes of social change. In doing sociological work, one often discovers unsettling facts about society. People who have a strong attachment to the status quo are unlikely to acquire or enjoy the sociological perspective. They may even resist it, finding its perspectives and conclusions threatening to their sense of well-being, privilege, or comfort in the world (Berger, 1963). People who are bigoted and unwilling to question their beliefs are not likely to be good sociologists. The sociological perspective puts beliefs in context, including prejudiced beliefs. Sociologists do not accept things at face value, but instead apply the tools of sociological research and theory to determine how beliefs and behavior develop. Those who take an absolutist stance on social issues will generally find the sociological perspective threatening, even subversive. As the sociologist Peter Berger writes, "Sociological understanding is always potentially dangerous to the minds of . . . guardians of the public order, since it will always tend to relativize the claim to absolute rightness upon which such minds like to rest" (1963: 48).

Discovering Unsettling Facts

In studying sociology, it is critical to examine the most controversial topics and to do so with an open mind, even when you see the most disquieting facts. Acquiring a sociological perspective can be unsettling. Revealing troubling facts about society is the price of seeing society for what it is. Consider the following:

- Since 1970, the gap between Black and White median family income in the United States has remained unchanged. Many believe that the gap has narrowed. It has not (U.S. Bureau of the Census, 1997a).
- Women with a college degree earn less on average than men with only some college experience (U.S. Bureau of the Census, 1997b).
- The United States has the highest rate of imprisonment of all nations in the world (Mauer, 1997a).
- Despite the idea that Asian Americans are a "model minority," poverty among Asian American families is higher than that among White American families and has increased in recent years; among certain Asian American groups, namely, Laotians and Cambodians, the poverty rate is as high as two-thirds of all families (Lee, 1994).
- Infant mortality in the United States, especially among African Americans and Hispanics, exceeds that of many impoverished nations (Children's Defense Fund, 1997).
- Half of all marriages formed in the 1990s will last under 7.2 years (U.S. Bureau of the Census, 1997c).

These facts provide unsettling evidence of persistent problems in the United States, problems that are embedded in society, not just in individual behavior. Sociologists try to reveal the social factors that shape society and that structure the chances of success for different groups. Doing so often challenges the widely held view that assumes that all people have a fair chance to succeed if they only try hard enough. The sociological perspective reveals instead that there are currents underlying ordinary life that carry people along paths not of their own choosing. Some never get the chance to go to Harvard; others are likely never to go to jail. These divisions persist because of people's placement within society.

Sociologists do not study only the disquieting side of society. Doing sociology can also be fun. Sometimes, sociologists ask questions for no reason except that they are curious. There are many insightful and intriguing studies of unusual groups, including tattoo artists (Sanders, 1989), cyberspace users (Turkle, 1995), transvestites (Bullough, 1993), and jazz musicians (Faulkner, 1971). Other sociologists study questions that affect everyday life, such as how women's employment affects the balance of power in families (Komter, 1989; Pyke, 1994), how urban gangs behave (Sánchez-Jankowski, 1991), how businesses develop among

---

BOX 1.3

UNDERSTANDING DIVERSITY
Becoming a Sociologist

INDIVIDUAL biographies often have a great influence on the subjects sociologists choose to study. The authors of this book are no exception. Margaret Andersen, a White woman, now studies the sociology of race and women's studies; Howard Taylor, an African American man, studies race, social psychology, and especially, race and intelligence testing. Here, each of them writes about the influence of their early experiences on becoming a sociologist.

Margaret Andersen • As I was growing up in the 1950s and 1960s, my family moved from California to Georgia, then to Massachusetts, and then back to Georgia. Moving as we did from urban to small-town environments and in and out of regions of the country that were very different in their racial character, I probably could not help becoming fascinated by the sociology of race. Oakland, California, where I was born, was highly diverse; my neighborhood was mostly White and Asian American. When I moved to a small town in Georgia in the 1950s, I was ten years old, but I was shocked by the racial norms I encountered. I had always loved riding in the back of the bus—our major mode of transportation in Oakland—and could not understand why this was no longer allowed. Labeled by my peers an outsider, because I was not southern, I painfully learned what it meant to feel excluded just because of "where you are from."

When I moved again to suburban Boston in the 1960s, I was defined by Bostonians as a southerner and ridiculed because of it. Nicknamed "Dixie," I was teased for how I talked. Unlike in the south, where, despite strict racial segregation, Black people were part of White people's daily lives, Black people in Boston were even less visible. In my high school of 2500 or so students, Black students were rare. To me, the school seemed little different from the strictly segregated schools I had attended in Georgia. My family soon returned to Georgia, where I was an outsider again; when I later returned to Massachusetts for graduate school in the 1970s, I worried about how a "southerner" would be accepted in this "Yankee" environment. Because I had acquired a southern accent, I think many of my teachers thought I was not as smart as the students from other places.

These early lessons, which I may have been unaware of at the time, must have kindled my interest in the sociology of race relations. As I explored sociology, I wondered how the concepts and theories of race relations applied to women's lives. So much of what I had experienced growing up as a woman in this society was completely unexam-

recent immigrants to the United States (Portes and Rumbaut, 1996), and how racism has changed in recent years (Feagin and Vera, 1995). The subject matter of sociology is vast. Some work illuminates odd corners of society; other work addresses urgent problems of society that may affect the lives and futures of millions.

Optimism and Pessimism: Keeping One's Balance

The sociological perspective forces an observer to confront the inequities of current social structures. This can be depressing since it points to the complex and sometimes intractable causes of society's problems; however, sociology can also be empowering since it provides perspective on the problems society faces. By analyzing problems in their societal context, sociology can provide outlines for social change. This can be the basis both for meaningful work and for alleviating some of the difficult problems of society.

Although not all sociologists study social problems, sociology emerges from a reformist and humanitarian tradition, and most sociologists believe that social changes can transform our society for the better. More than most people, sociologists know how hard it is to change a society, yet they know that change is possible and that it begins with an understanding of the sociological basis for the problems we face. In this way, sociologists are optimists, but optimism must not crowd out realism.

Consider the question of racial progress. By many measures, race relations have improved over the past thirty years. The African American and Latino[1] middle classes have expanded; there is greater integration of diverse groups in work and social settings; African Americans, Asian Americans, and Latinos have a greater presence in professional and influential positions. Opinion polls also show that White people exhibit less prejudice against racial minorities. Still, the magnitude and meaning of these trends should not be misunderstood. Racial inequality has not disappeared. Despite some changes for the better, Asian American, African American, American Indian, and Latino families suffer from high rates of unemployment and poverty. Many sociologists argue that a new "underclass" has been created, consisting of those with extremely limited resources and little chance of economic or social success, including poor White Americans (Wilson, 1996; Wilson, 1987; Massey and

[1]Latino refers generically to all groups of Spanish descent in the Americas; thus, it includes Chicanos (those of Mexican descent born in the United States), Puerto Ricans, Cubans, and groups from Latin America and Central America in the United States. In this text, it is used interchangeably with Hispanic. Latina is the feminine form and is used when referring specifically to women; see the "Note on Language" in the preface to this book.

ined in what I studied in school. As the women's movement developed in the 1970s, I found sociology to be the framework that helped me understand the significance of gender and race in people's lives. To this day, I write and teach about race and gender, using sociology to help students understand their significance in society.

Howard Taylor • I grew up in Cleveland, Ohio, the son of African American professional parents. My mother, Murtis Taylor, was a social worker and the founder and then president of a social work agency. She is well known for her contributions to the city of Cleveland and was an early "superwoman," working days and nights, cooking, caring for her two sons, and being active in many professional and civic activities. I think this gave me an early appreciation for the roles of women and the place of gender in society though I surely would not have articulated it as such at the time.

My father was a businessman in an all-Black life insurance company. He was also a "closet scientist," always doing experiments and talking about scientific studies. He encouraged my brother and me to engage in science, so we were always experimenting with scientific studies in the basement of our house. In the summers, I worked for my mother in the social service agency where she worked—as a camp counselor and other jobs. Early on, I contemplated becoming a social worker, but I was also excited by science. As a young child, I acquired my father's love of science and my mother's interest in society. In college, the one field that would gratify both sides of me, science and social work, was sociology. I wanted to study human interaction, but I also wanted to be a scientist, so the appeal of sociology was clear.

At the same time, growing up African American meant that I faced the consequences of race everyday. It was always there, and like other young African American children, I spent much of my childhood confronting racism and prejudice. When I discovered sociology, in addition to bridging the scientific and humanistic parts of my interests, I found a field that provided a framework for studying race and ethnic relations. The merging of two ways of thinking, coupled with the analysis of race that sociology has long provided, made sociology so fascinating to me.

Today, my research on race and intelligence testing and Black leadership networks seems rooted in these early experiences. I do quantitative research in sociology and see sociology as a science that reveals the workings of race, class, and gender in society.

Denton, 1993). African Americans, Latinos, and Native Americans still lag behind White Americans in educational attainment and political influence. Racism, though perhaps more subtle in some settings, is still a feature of everyday life for even the most successful minorities. Although White Americans may express less direct prejudice, they also say they are tired of hearing about racism, perhaps because they are less sensitive to the nuances and consequences of racism than those who experience it themselves (Blauner, 1989; Feagin and Vera, 1995; Bobo and Kluegel, 1993).

How should we feel about the extent of racial progress—optimistic or pessimistic? Excessive optimism can discourage the kind of critical analysis that is essential in sociological thinking (Killian, 1971), but being overly pessimistic also has its consequences, primarily by discouraging people from trying to make positive changes. Sound sociology strives for balance—enough optimism not to lose the sense that change is possible and enough pessimism to be able to face the facts of difficult social issues. Acquiring a sociological imagination and recognizing the social structures that shape social behaviors can be disillusioning, but objectivity and realism can also inspire the insights and energetic action that can help solve society's problems.

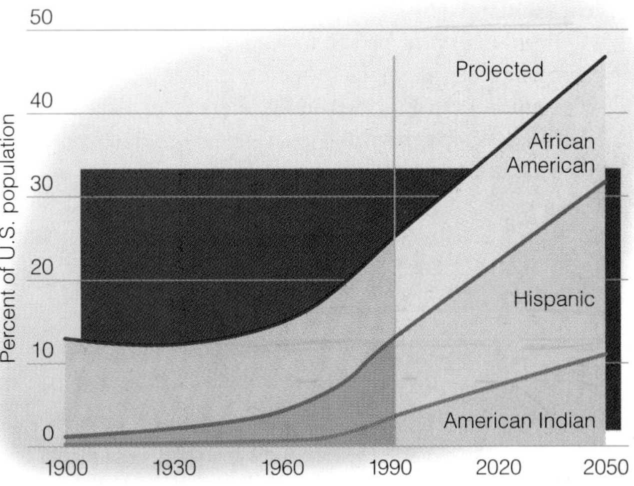

FIGURE 1.1 Share of Minorities in the U.S. Population

SOURCE: Jeffery S. Passel and Barry Edmonston, "Immigration and Race: Recent Trends in Immigration to the United States" (Washington, DC: The Urban Institute, 1992), table 3; and Bureau of the Census, *Current Population Reports P-25*, no. 1092 (Washington, DC: GPO, 1992), table 1.

The Significance of Diversity

The analysis of diversity is one of the central themes of sociology. Differences among groups, especially differences in how groups are treated, are significant in any society, but they are particularly compelling in a society as diverse as the United States. What is the experience of different groups in society? How are groups defined by others? How do societies dispense rewards and punishments to different groups? The answers to these questions generate a picture not just of diversity in society, but of society itself.

Perhaps the most basic lesson of sociology is that people are shaped by the social scene immediately around them. In a country like the United States, with so much cultural diversity, people will share some experiences, but not all. Experiences not held in common can include some of the most important influences on social development, such as language, religion, and the traditions of family and community. The richness of cultural diversity in this nation makes the analysis of how cultures thrive and interact timely and important. In part, diversity refers to the many different groups that together compose the United States; understanding diversity means recognizing this diversity and making it central to sociological analyses.

Diversity: A Source of Change

Today, the United States includes people from all nations and races. In 1900, one in eight Americans was not White, today, racial and ethnic minority groups compose one in four Americans. In California, 45 percent of the population is African American, American Indian, Asian American, and Hispanic; in Texas, 40 percent; in many southern states, African Americans alone make up one-quarter to one-third of the population. Overall, 26 percent of the population is African American, American Indian, Asian American, or Latino. These broad categories themselves are internally diverse, including, for example, those with long-term roots in the United States, as well as Cuban Americans, Salvadorans, Cape Verdeans, Filipinos, and many others. By the year 2010, 38 percent of the population under eighteen years old will be members of minority groups (U.S. Bureau of the Census, 1997c). Immigration is also higher than at any time since the early twentieth century and has become a critical social issue, though not for the first time (Figure 1.1).

Defining Diversity

In this book, *we use diversity to refer to the variety of group experiences that result from the social structure of society. Diversity is a broad concept that includes studying group differences in society's opportunities, the shaping of social institutions by different social factors, the formation of group and individual identity, and the process of social change. Diversity includes the study of different cultural orientations, although diversity is not exclusively about culture. Understanding diversity is not a simple matter despite the recent popularity of the concept, but understanding diversity is critical to understanding society because the population is increasingly diverse and because fundamental patterns of social change and social structure are increasingly patterned by diverse group experiences. Thus, understanding society requires recognizing and analyzing diversity—its sources, its effects, and its impact on the experiences and interrelationships of groups in society. Race, class, and gender are fundamental to understanding diversity since they have been so critical to shaping social institutions in the United States. They are not, however, the*

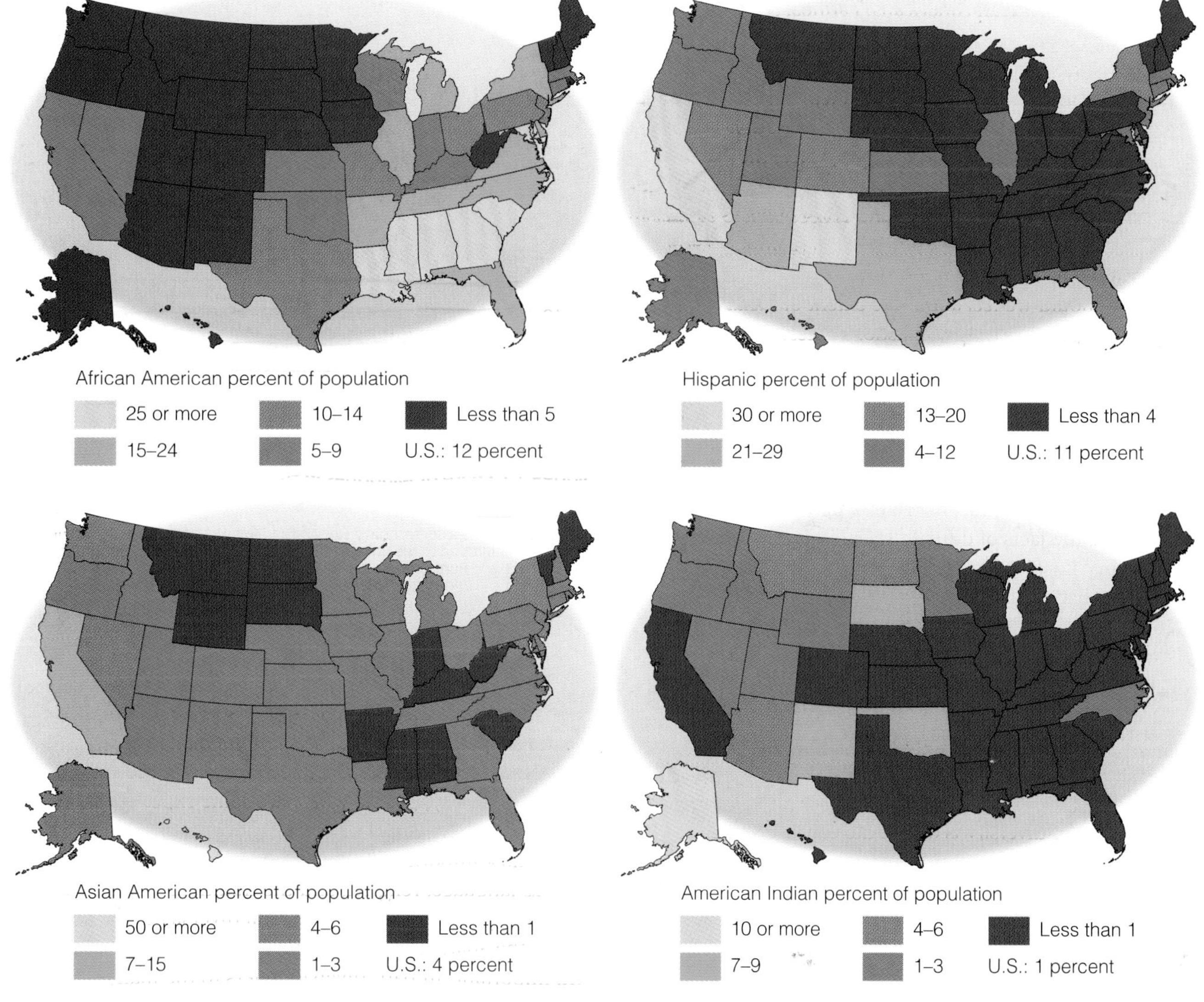

MAP 1.1 Mapping America's Diversity: Minorities in the United States

Only Hawaii has more than 50% Asian American/Pacific Islanders.
DATA: From the U.S. Bureau of the Census. 1997. *Statistical Abstracts of the United States, 1997.* Washington, D.C.

only sources of diversity. Age, nationality, sexual orientation, and region of residence, among other social factors, also differentiate the experience of diverse groups in the United States. And, as the world is increasingly interconnected through global communication and a global economy, the study of diversity must also encompass a global perspective—that is, an understanding of the international connections existing across national borders and the impact of such connections on life throughout the world.

Understanding diversity is necessary to analyze social institutions, a basic objective of sociological thought. Sociologists know that different groups have widely different experiences within the major institutions in this society. For example, if you are rich, White, well insured, and enter a hospital on the referral of your family doctor, you are likely to have a very different experience than if you enter the hospital poor, Latino, and uninsured. Educational experiences are likewise uneven for groups with different social characteristics. If your family has a long history of attending elite educational institutions, there is a good chance that you too will attend an elite school. This advantage is built into the system since elite schools like Princeton, Harvard, Stanford, and others typically give preference to children of alumni who are applying for admission. Even though the common perception is that Asian American students have an advantage in college admissions, a study of Princeton admissions found that children of alumni had an admission rate of 48 percent, Asian American students had a rate of 14 percent, and White students had a rate of 17 percent (Takagi, 1992). This and other studies show how privilege and disadvantage are structured into social institutions.

The study of diversity within society is also important because it lies at the heart of so many difficult social problems. Using sociological analysis to alleviate social problems is one of the compelling values of a sociological perspective. Racial tensions, inequalities between social classes, and

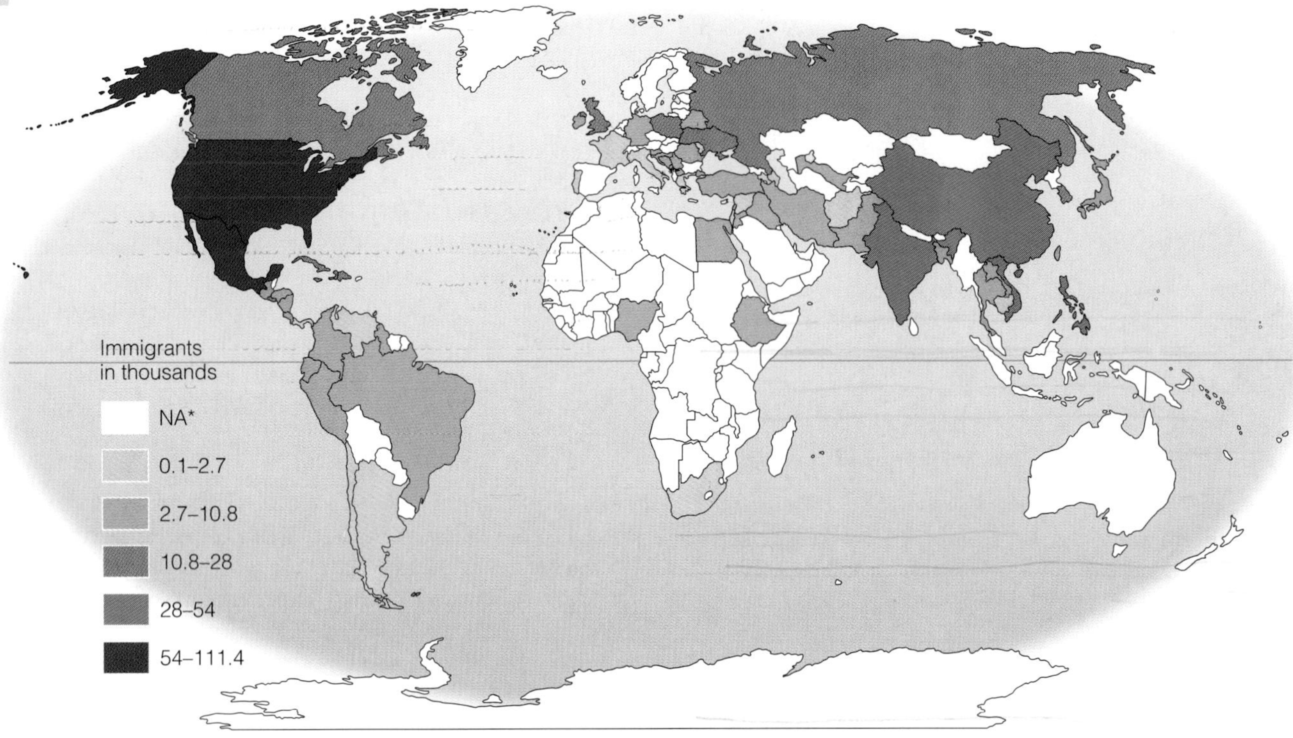

MAP 1.2 Viewing Society in Global Perspective: U.S. Immigrants by Country of Origin

*The total number of immigrants from these countries is 4,700.
DATA: From the U.S. Bureau of the Census. 1997. *Statistical Abstracts of the United States, 1997.* Washington, D.C.

differences in men's and women's experiences in society are at the root of many contemporary problems. Other forms of diversity, such as religious, generational, or regional differences, also produce social conflicts.

Diversity is not just about the study of problems. The complex weave of society is a source of great cultural richness and human achievement. Numerous sociological studies also show how people respond to oppression; these studies provide a portrait of how people mobilize even against extraordinary odds, such as in studies of the civil rights movement (Morris, 1984; Barnett, 1993) and the gay and lesbian rights movement (D'Emilio, 1983; Gamson, 1995a). Sociologists have also shown how, in less dramatic and visible ways, even the poorest people use human creativity and social activism to challenge their situation; this is, for example, reflected in work about the survival strategies of poor women and their children (Naples, 1992; Ladner, 1971/1995). In each case, sociologists have shown how groups use their social position to generate social change. We also include features that help you see how sociology can be practically used (see, especially, the "Sociology in Practice" boxes) and how it can be used to analyze current social issues.

THINKING SOCIOLOGICALLY

What are some of the sources of *diversity* on your campus? What questions might a sociologist pose about the significance of diversity in this environment?

class, race, and gender

The Significance of Class, Race, and Gender

We use a broad definition of diversity in this book. Some of the most important dimensions of diversity in the United States are class, race, and gender—concepts central to sociological thought. In almost all societies, mechanisms are in place that permit some groups to acquire more resources and power than other groups. A longstanding and important concept in sociology is that of **social class,** defined as the position groups hold relative to the economic, social, political, and cultural resources in society. Social class (often simply called class) is a complex concept, one we will examine later in this book. The analysis of class is one of the gifts that sociology has given to other fields of study. Social class is a significant source for different experiences—both in the United States and world wide. As we will see, analyzing social class and its consequences for different groups' chances in life is a major contribution of sociology.

Sociologists have also long studied race and ethnic relations, not only to understand specific group experiences but also to discover how racial prejudice and inequality are perpetuated in society. Sociologists have documented the connections between race and the organization of the labor market (Bonacich, 1972), segregation in cities (Massey and Denton, 1993), and the shaping of public attitudes (Bobo and Kleugel, 1993; Feagin and Vera, 1995), as well as the role of race and ethnicity in shaping people's identity and cultural practices (Nagel, 1994; Waters, 1990). Without the

Understanding diversity is important in a society comprising so many different groups, each with unique, but interconnected, experiences.

sociological research on this subject, we would know far less about the experiences of different groups in society. Furthermore, sociologists have been careful to note the different experiences *within* racial and ethnic groups that are the result of social class, along with other factors, such as gender, age, and sexual orientation. Understanding diversity thus requires attention to all the forms of racial and ethnic diversity that comprise different group experiences.

Gender is another source of diversity in society and one that has an impact on virtually every aspect of life. Defined as the social expectations associated with being a man or a woman, gender shapes attitudes and values (Beutel and Marini, 1995), influences the development of individual and group identity, guides occupational choices (Reskin and Padavic, 1994), and even affects the amount of leisure time we have (Hochschild, 1997; Schorr, 1991). A vast amount of research has been done recently on the subject of gender, some of which we will examine in Chapter 12. Here we note that gender is one of the significant sources for diverse experiences, and it is central, therefore, to understanding how society operates.

These three sociological variables—class, race, and gender—are some of the strongest influences on the outcomes of people's lives. As a result, the analysis of class, race, and gender—as factors that shape all dimensions of people's lives—is part of understanding diversity in the United States and other societies. Furthermore, class, race, and gender exert their influence not independently, but interactively: The configuration of all three together creates a unique recipe for social experience. Being a man, for example, is not the same for all men; class and race may affect some men's lives more than their gender, but gender is never insignificant in men's experiences. Class, race, and gender form overlapping categories of experience for all groups; thus, an Asian American woman's experience is shaped by her gender, race, and social class simultaneously. She is confronted with the expectations attached to being a woman *and* an Asian American *and* given the resources of those in her social class (whether that be middle class, working class, rich, or poor); moreover, her identity as an Asian American woman will be different from that of an Asian American man. For the individual and for society as a whole, class, race, and gender, as well as other social categories (such as age, sexual orientation, religion, having a disability, nationality), overlap. Experience in society is the blend of all these social influences. Each affects the others.

The inequalities created by class, race, and gender in this society, as well as other nations, have certainly produced victims of social injustice. Poverty, physical and emotional abuse, poor healthcare, and victimization by crime are just some of the consequences of inequality. Studying class, race, and gender, however, is not just the study of victimization. Although sociologists have to be realistic about the impact of class, race, and gender inequality, studying these factors is not just about disadvantage. Prejudice tends to make people associate race and poverty, but attention to the intersections of class and race, along with gender, reveals the differentiation of experience within different groups.

"Actually, Lou, I think it was more than just my being in the right place at the right time. I think it was my being the right race, the right religion, the right sex, the right socioeconomic group, having the right accent, the right clothes, going to the right schools . . ."

The social factors that influence an individual's social status, such as age, race, gender, and class, all interconnect, although a person may feel one or another as more salient at a given point in time.

In addition, a long record of achievement is found among all groups in the United States. Women, people of color, and those from the working class have contributed enormously to the making of this society, not only in the labor they have provided in building the nation, but also in the cultural traditions that have sometimes stemmed from the most oppressive conditions (such as the blues, stemming from the slave experience). These achievements contribute to our national vitality and help us see that studying diversity is not just about studying victims, but is also about studying the different ways that people respond to living in different situations. Class, race, and gender do not affect just those who are victimized. They affect the experiences of all groups in society, though in different ways. White men are shaped by their gender and race as much as are Latinos, Native Americans, African Americans, and White women. Understanding how these categories of experience overlap is critical to understanding patterns of social inequality in the United States.

Diversity in Global Perspective

The primary focus on diversity in this book is diversity within the United States. Nonetheless, no society can be understood apart from the global context that now influences the development of all societies. The social and economic system of any one society is increasingly intertwined with those of other nations around the world. Coupled with the increasing ease of travel and telecommunication, this means that a global perspective is necessary to understand change both in the United States and in other parts of the world.

For many, understanding diversity means seeing things from a global perspective. Indeed, many sociologists use a cross-cultural or comparative perspective to understand the developments they observe in any given society. In this book, we introduce a global perspective to understand those aspects of U.S. society that are most influenced by worldwide developments. The U.S. economy, as one obvious example, extends far beyond the nation's borders. Processes such as technological development, the distribution of natural and economic resources, unemployment, and poverty simply cannot be understood anymore without analyzing the global basis of contemporary society. As an example, immigration—a process dramatically affecting social systems in the United States—can be fully understood only by also observing international patterns of work, migration, and political and economic change. Or in another example, something as seemingly simple as what toys children play with can be fully understood only by analyzing the international economic system in which these toys are produced and distributed. Throughout this book, we will use a global perspective to understand some of the developments that are shaping contemporary life in the United States.

The Development of Sociology

Like the subjects it studies, sociology is itself a social product. It is an intellectual framework deeply influenced by the social and historical conditions in which it is produced. Sociology first emerged in western Europe during the eighteenth and nineteenth centuries. In this period, the political and economic systems of Europe were rapidly changing. Monarchy, the rule of society by kings and queens, was disappearing. After almost ten centuries, the end came for feudalism, in which common people were obligated to the aristocrats who owned all usable land. Religion as the system of authority and law was giving way to scientific authority. At the same time, capitalism grew. Contact between different societies increased, and worldwide economic markets developed. The conservative, traditional ways of the past were giving way to a new social order. The time was ripe for a new understanding.

The Influence of the Enlightenment

The **Enlightenment** in eighteenth- and nineteenth-century Europe had an enormous influence on the development of modern sociology. Also known as the Age of Reason, the Enlightenment was characterized by faith in the ability of human reason to solve society's problems. Intellectuals believed that there were natural laws and processes in society, to be discovered and used for the general good. This belief was linked to the development of modern science, which was gradually supplanting traditional and religious explanations for natural phenomena with theories confirmed by experiments.

A second strand of thought that emerged during the Enlightenment and that continues to influence sociology today is **positivism**—a system of thought in which accurate observation and description is considered the highest form of knowledge, as opposed to, say, religious dogma or poetic inspiration. Positivists believed that society could be studied using the techniques of the natural sciences. The modern scientific method, which informs sociological research, grew out of positivism.

Humanitarianism was another product of the Enlightenment. Based on the ideals espoused by the French Revolution, whose motto was "liberty, equality, and brotherhood," humanitarianism rests on the principle that human reason can successfully direct social change for the betterment of society. The thread of social reform that has always been a part of sociology can be traced to the humanitarian movement. Along with technological progress and giant increases in industrial capacity, the Industrial Revolution brought with it poverty and other urban problems that foreshadowed those we face today. Sociology developed as a way of understanding and alleviating these problems. Not all early sociologists had a specific agenda for change; still, the ideals of the Enlightenment—positivism, humanitarianism, and faith in human reason—continue to permeate sociological thought (Nisbet, 1970).

The Development of Sociology in Europe

The earliest sociologists promoted a vision of sociology grounded in careful observation. Auguste Comte (1798–1857), who coined the term *sociology,* believed that just as science had discovered the laws of nature, sociology could discover the laws of human social behavior. It was Comte who first elaborated the positivist basis of sociology. He thought that once we understood the laws of behavior, we would be able to solve society's problems. Comte declared that scientific knowledge developed in stages, astronomy being first, followed by physics, chemistry, biology, and then sociology. Each was progressively more complex, with sociology, according to Comte, being the most highly evolved since it involved the study of a complex and changing system—society. Comte thought that sociology would become the "Queen of the Sciences."

Alexis de Tocqueville (1805–1859), a French citizen, traveled to the United States as an observer beginning in 1831. The book that followed from his experience, *Democracy in America,* is an insightful analysis of U.S. society and its democratic culture and structure. De Tocqueville thought that democratic and egalitarian values in the United States not only influenced American social institutions for the better but also transformed personal relationships. Less admiringly, he felt that in the United States the tyranny of kings had been replaced by the "tyranny of the majority." He was referring to the ability of a majority to impose its will on everyone else in a democracy. De Tocqueville also felt that despite the emphasis on individualism in American culture, Americans had little independence of mind and that individualism made people self-centered and anxious about their social position (Collins and Makowsky, 1972).

Another early sociologist, one commonly omitted from the history of sociology, is Harriet Martineau (1802–1876). One of eight children of a successful British surgeon and deaf from her teen years, Martineau was well educated and established herself as a liberal writer in the British press before she embarked on a long tour of the United States in 1834. Like de Tocqueville, Martineau was fascinated by the newly emerging culture in America. Her book, *Society in America* (1837), is an analysis of the social customs that she observed. This important work was overlooked for many years, probably because the author was a woman. It is now recognized as a classic. Martineau also wrote the first sociological methods book, *How to Observe Manners and Morals* (1838), in which she discussed how to observe behavior when one is a participant in the situation being studied. It is a classic work on the research method called participant observation (see Chapter 2). Martineau was also the original English translator of Comte's book *Positive Philosophy.*

As one of the earliest observers of American culture, Harriet Martineau used the powers of social observation to record and analyze the social structure of American society. Long ignored for her contributions to sociology, she is now seen as one of the founders of early sociological thought.

Martineau was a strong feminist and abolitionist whose travels in the United States were the cause of considerable controversy. She declared that the subservience of women in the United States was comparable to slavery, and in her home country of England, she compared women to servants. Although she was encouraged by the emphasis on independence in American culture, she feared that slavery would tear the country apart. She endured many threats on her life instigated by her views as an abolitionist and her public comments on the status of women. Threats finally forced her to restrict her travels to the northern United States, where she continued to insist on the right of women to speak their minds. Her work is a powerful analysis of American society and what she observed as the subordination of women within it (Hoecker-Drysdale, 1992; Rossi, 1973).

Classical Sociological Theory

Of all the contributors to the development of sociology, the giants of the European tradition were Emile Durkheim, Karl Marx, and Max Weber. The works of these scholars form a classical tradition in sociology that continues to be important.

Emile Durkheim. During the early academic career of the Frenchman Emile Durkheim (1858–1917), France was in the throes of the notorious Dreyfus Affair. Alfred Dreyfus was a Jewish captain in the French army who was

Emile Durkheim established the significance of society as something larger than the sum of its parts. Social facts stem from society but have profound influence on the lives of people within society.

scapegoated by the Army and the government in an attempt to cover up a spy scandal. Amid much public controversy, Dreyfus was convicted of treason in 1894, subjected to the humiliating ritual of having his badges of rank stripped from his uniform, and sent to Devil's Island. With the ritual disranking, the commotion surrounding the affair quieted for a while. It exploded again when the military court rejected exonerating evidence. For years, France was bitterly split over the case, which rested on little more than clumsily forged documents and antisemitism (hatred of Jews).

Durkheim, himself Jewish, was fascinated by how the public degradation of Dreyfus in 1894 seemed to calm and unify a large segment of the divided French public. Durkheim later wrote that public rituals have a special purpose in society, creating *social solidarity,* referring to the bonds that link the members of a group. Some of Durkheim's most significant works explore the question of what forces hold society together and make it stable.

According to Durkheim, people in society are glued together by belief systems (Durkheim, 1912/1947). Religion is an example of a belief system (see Chapter 16). The rituals of religion and other institutions symbolize and reinforce the sense of belonging that insiders in a society or group feel. Public ceremonies create a bond between people in a social unit. Drawing on this insight, Durkheim made a creative leap that helps explain social deviance. Deviant behavior is defined as actions that others perceive to be in violation of the customary ways of doing things (Chapter 8). Durkheim argued that deviance has a purpose in society—that by labeling some people as deviant, the society confirms a sense of normalcy and belonging in others. People express social solidarity when they condemn deviant behavior. Identifying some behaviors as deviant gives definition to nondeviant or "normal" behaviors. Durkheim thought that deviance, like public rituals, sustains moral cohesion in society.

Durkheim viewed society as an entity larger than the sum of its parts. He described this as society *sui generis* (thing in itself), meaning that society is a subject to be studied separate from the sum of the individuals who compose it. Society is external to individuals, yet its existence is internalized in people's minds—that is, people come to believe what society expects them to believe. Durkheim conceived of society as an integrated whole—each part contributing to the overall stability of the system. His work is the basis for *functionalism,* an important theoretical perspective that we will return to later in this chapter.

One of Durkheim's contributions was his conceptualization of the *social.* Durkheim conceptualized **social facts** as those social patterns that are *external* to individuals. Things like customs and social values exist outside individuals, whereas psychological drives and motivation exist inside people. Social facts, therefore, are not to be explained by biology or psychology, but are the proper subject of sociology and are its reason for being. Durkheim held that social facts, though having their existence outside individuals, nonetheless pose constraints on individual behavior. "The individual is dominated by a moral reality greater than himself: namely, collective reality" (Durkheim, 1897/1951: 4). Durkheim's major contribution was the discovery of the social basis of human behavior. He proposed that a social system could be known through the discovery and analysis of social facts, and that adding to the fund of reliable social facts was the central task of the sociologist (Bellah, 1973; Coser, 1977; Durkheim, 1938/1950).

Karl Marx. It is hard to imagine another scholar who has had as much influence on intellectual history as Karl Marx (1818–1883). Along with his collaborator, Friedrich Engels, Marx not only changed intellectual history, but world history, too.

Karl Marx analyzed capitalism as an economic system with enormous implications for how society is organized, in particular how inequality between groups stems from the economic organization of society.

Marx's work was devoted to explaining how capitalism shaped society. Capitalism is an economic system based on the pursuit of profit and the sanctity of private property. Marx used a class analysis to explain capitalism, describing capitalism as a system of relationships among different classes. Under capitalism, the capitalist class owns *the means of production,* the system by which goods are produced and distributed. To say that the capitalist class owns the means of production does not just mean that they own property, but that they own the actual system by which wealth is accumulated—the factories and machinery where goods are produced, the ships, railroads, and airlines by which goods are distributed, the banks and financial institutions by which profits are managed, the communications systems by which ideas supporting capitalist values are disseminated. In Marx's view, profit, the goal of capitalist endeavors, is produced through the exploitation of the working class. Workers sell their labor in exchange for wages, while capitalists make certain that wages are worth less than the goods the workers produce. The difference in value is the profit of the capitalist. In the Marxist view, the worker–capitalist class system is inherently unfair since the entire system rests on workers getting less than they give.

Marx thought that the economic organization of society was the most important influence on what humans think and how they behave. He found that the beliefs of the common people tended to support the interests of the capitalist system, not the interests of the workers themselves. Why? Because the capitalist class controls not only the production of goods but also the production of ideas. They own the publishing companies, endow the universities where knowledge is produced, and control who is allowed to become important and influential.

We will return to Marx's theories in Chapter 9. For now, the important point is that Marx considered society to be shaped by economic forces. Laws, family structures, schools, and other institutions all develop, according to Marx, to suit economic needs under capitalism. Like other early sociologists, Marx took social structure as his subject rather than the actions of individuals. It was the *system* of capitalism that dictated people's behavior. Marx had an evolutionary concept of societal development and change, but different from Durkheim, he saw social change arising from tensions inherent in a capitalist system—the conflict between the capitalist and working classes.

Marx's ideas are often misperceived by U.S. students because communist revolutionaries throughout the world have claimed Marx as their guiding spirit. It would be naive to reject his ideas solely on political grounds. Much that Marx predicted has not occurred—for instance, he claimed that the "laws" of history made a worldwide revolution of workers inevitable, and this has not happened. Still, he left us an important body of sociological thought springing from his insight that society is systematic and structural, and that class is a fundamental dimension of society that shapes social behavior.

The early American sociologists were struck by the contrasts in wealth and poverty that characterized the early twentieth century and the patterns of urbanization and industrialization that were emerging at the time.

Max Weber used a multidimensional approach to analyzing society, interpreting the economic, cultural, and political organization of society as together shaping social institutions and social change.

Max Weber. Max Weber (1864–1920; pronounced "Vay-ber") was greatly influenced by Marx's work, but whereas Marx saw economics as the basic organizing element of society, Weber theorized that society had three basic dimensions: political, economic, and cultural. According to Weber, a complete sociological analysis must recognize the interplay between economic, political, and cultural institutions (Parsons, 1947). Weber is credited with developing a multidimensional analysis of society that goes beyond Marx's more one-dimensional focus on economics.

Weber also theorized extensively about the relationship of sociology to social and political values. He did not believe there could be a value-free sociology since values would always influence what sociologists considered worthy of study. Weber thought sociologists should acknowledge the influence of values so that ingrained beliefs might not interfere with objectivity. Weber professed that the task of sociologists is to teach students the uncomfortable truth about the world. Faculty should not use their positions to promote their political opinions, he felt; rather, they have a responsibility to examine all opinions, including unpopular ones, and use the tools of rigorous sociological inquiry to understand why people believe and behave as they do.

An important concept in Weber's sociology is **verstehen** (meaning, "understanding"; pronounced, "ver-shtay-en"). Verstehen refers to understanding social behavior from the point of view of those engaged in it. Weber believed that to understand social behavior one had to understand the meaning that a behavior held for social actors. He did not believe sociologists had to be born into a group to understand it (in other words, he didn't believe "it takes one to know one"), but he did think sociologists had to develop some *subjective* understanding of how other people experience their world. One of Weber's major contributions was the definition of **social action** as a behavior to which people give meaning (Gerth and Mills, 1946; Parsons, 1951b; Weber, 1962). In other words, social behavior is more than just action; people do things in a context and use their interpretive abilities to understand and give meaning to their action. Weber's work emphasized this interpretive dimension of social life.

As we will see in subsequent chapters, Weber's work is far-reaching. He was interested in the rise of bureaucracy in the modern world, the nature of power and authority, the operation of class systems, the foundation of the work ethic in capitalist society, and the role of religion in supporting other social institutions. Weber continues to have a profound impact on sociology.

The Development of American Sociology

American sociology was built on the earlier work of Europeans, but unique features of U.S. culture contribute to its distinctive flavor. Pragmatism (the belief in practicality) influenced sociologists in the United States to value social planning. Less theoretical and more practical than their European counterparts, American sociologists believed that if they exposed the causes of social problems, they could alleviate some of the consequences, which are measured in human suffering.

Sociology and Social Change. Early sociologists in both Europe and the United States conceived of society as an organism, a system of interrelated functions and parts that work together to create the whole. This perspective is called the **organic metaphor.** Sociologists saw society as constantly evolving, like an organism. The question many of the early sociologists asked was to what extent humans could shape the evolution of society.

Some were influenced in this question by the work of British scholar Charles Darwin (1809–1882), who had revolutionized biology when he identified the process by which evolution creates new species. **Social Darwinism** was the application of Darwinian thought to society. According to the Social Darwinists, the "survival of the fittest" is the driving force of social evolution. In Britain, Herbert Spencer (1820–1903) conceived of society as an organism that evolved from simple to complex in a process of adaptation to the environment. Spencer believed that society was best left alone to follow its natural evolutionary course. Because he had faith that evolution always took a benign course toward perfection, he advocated a *laissez-faire* (that is, hands-off) approach to social change. Social Darwinism was a conservative mode of thought; it assumed that the current arrangements in society were natural and inevitable. Although Social Darwinism departed from some of sociology's general emphasis on social reform, it had an important influence on the development of sociology, especially with its characterization of society as a system that was both stable and changing—consistent in many ways, yet constantly evolving.

American sociologists were much involved in this debate about human intervention in the course of social evolution. The American sociologist William Graham Sumner (1840–1910) was an ultra-conservative thinker who claimed that survival of the fittest justified the inequalities in society. The rich and powerful, he felt, were those who were most successful in the competition for scarce resources. He called millionaires the "bloom of a competitive society" (cited in Hofstadter, 1944: 58). Utterly convinced that society had an evolutionary path to follow, Sumner thought that tampering with the social system would bring certain disaster.

Unlike Sumner, most other early sociologists in the United States took a hands-on approach. Lester Frank Ward (1831–1914) thought sociology could be used to engineer social change. He called this *social telesis*—the idea that human intervention in the natural evolution of society would advance the interests of society. Ward and the reform-minded sociologists of the early twentieth century were optimistic about the ability of sociology to address society's problems.

When early American sociologists identified industrialization and urbanization as the cause of problems, they were not content with that knowledge for its own sake, but wanted to use it to create change. This activist tradition has left an important legacy for contemporary sociologists. **Applied sociology** is the use of sociological research and theory in solving real human problems.

The Chicago School. Nowhere was the emphasis on application more evident than at the University of Chicago, where a style of sociological thinking known as the *Chicago School* developed. Two tendencies characterized the pioneers of the Chicago School. First, they were interested in how society shaped the mind and identity of people. Second, they used social settings as human laboratories in which sociologists could do scientific studies that would address human needs.

Charles Horton Cooley (1864–1929), though at the University of Michigan most of his career, was one of the foremost theorists of the Chicago School. Always interested in the formation of individual identity, Cooley theorized that it developed through people's understanding of how they are perceived by others. This was further elaborated in the work of George Herbert Mead (1863–1931), who like Cooley was interested in specifying the processes that linked the individual to society. Cooley and Mead saw the individual and society not as separate entities but as interdependent. Individuals developed through the relationships they established with others. Society, the ever-changing web of social relationships, existed because it was imagined in the minds of individuals. We return to the work of Cooley and Mead in Chapter 4 when we discuss socialization, the process by which individuals are shaped to fit within society.

The focus on the individual espoused by the Chicago School is distinctively American, and reflects the importance of individualism in U.S. culture. The concern for the individual also stemmed from perceptions about modern society, including the belief that individuals were increasingly isolated and that community feeling was being lost. In other words, focusing on the individual was a way to approach society as a whole. In the same period at the University of Chicago, *social psychology* was being founded. Social psychology studies the social basis of psychological behavior and supplies a bridge between the study of the individual, which is psychology's realm, and the study of society.

Sociologists at the University of Chicago worked close to home using the city they lived in as a living laboratory. Race riots, immigrant ghettoes, poverty, and delinquency were all features of "modern" life in prewar Chicago, seeding the work of the Chicago theorists. The five-volume study of Polish immigrants in Chicago produced by W. I. Thomas (1863–1947) and his colleague Florian Znaniecki (pronounced, "Znan-yecki"; 1882–1958) remains a classic study of ethnicity, immigration, neighborhood segregation, and urban life (Thomas, 1918, 1919/1958). W. I. Thomas promoted the use of personal documents to reveal how people interpreted their experiences. There are legends of Thomas sifting through the garbage of Polish immigrants to retrieve personal letters, which he used to study their social relationships and to learn how they felt about their lives in the New World. Thomas also thought that the influence of society was so strong that people would behave according to what they thought was true—even if confronted with evidence to the contrary. Thomas's famous dictum (developed in collaboration with Dorothy Swaine Thomas, his wife) was, "If men define situations as real, they are real in their consequences" (Thomas and Thomas, 1928); this is called the *definition of the situation*.

Robert Park (1864–1944) was another founder of sociology from the University of Chicago. Originally a journalist who worked in several midwestern cities, Park was an experienced witness to urban problems. Park was interested in how other races interacted with Whites. At the turn of the century, he was a regular correspondent and advisor to the African American leader Booker T. Washington. He was also fascinated by the sociological design of cities, noting that cities were typically sets of concentric circles. The very rich and the very poor lived in the middle, ringed by slums and low-income neighborhoods. Progressing to the outer rings, the residents became increasingly affluent (Collins and Makowsky, 1972; Coser, 1977; Park and Burgess, 1921). Although this particular geometric shape to cities has changed with increased suburbanization and the concentration of poverty in inner cities, Park would still be intrigued by sociological research showing how boundaries are defined and maintained in urban neighborhoods. You might notice this yourself in cities. A single street crossing might delineate a Vietnamese neighborhood from an Italian one, a White affluent neighborhood from a barrio. The social design of cities continues to be a major subject of sociological research.

Jane Addams, the only sociologist to win the Nobel Peace Prize, used her sociological skills to try to improve people's lives. The settlement house movement provided social services to groups in need, while also providing a social laboratory in which to observe the sociological dimensions of problems like poverty.

Many of the early sociologists of the Chicago School were women whose work is only now being rediscovered. Jane Addams, Marion Talbot, Edith Abbott, and others produced excellent work at the University of Chicago although the sexist practices of that time excluded them from the careers available to men. Jane Addams (1860–1935), close friend of and collaborator with W. E. B. Du Bois, was one of the most renowned sociologists of her day. She was a leader in the settlement house movement, which provided community services and did systematic research designed to improve the lives of slum dwellers, immigrants, and other dispossessed groups. Hull House, one of the settlement homes where Addams worked, was well known in this movement. Educated at the University of Chicago, Addams wrote eleven books and hundreds of articles. Although the University of Chicago was progressive in admitting women when it opened in 1892, it barred hiring women as full professors. Addams, the only practicing sociologist ever to win a Nobel Peace Prize (in 1931), never had a regular teaching job. Instead, she used her skills as a research sociologist to develop community projects that assisted people in need (Deegan, 1988).

Other women, including Edith Abbott and Sophinista Breckenridge, worked closely with some of the men at the University of Chicago, completing numerous studies on education, housing, working women, urban children, and other subjects. Some women held appointments in the department; others were employed as administrators. George Herbert Mead was among the men who publicly acknowledged their admiration for the work of these women, yet women were frequently marginalized in the profession, and their contributions have been largely neglected (Deegan, 1990). Many women, excluded from the universities, made their contribution in the newborn profession of social work rather than as academic sociologists. Over time, theoretical sociology and social work became highly gender-segregated, with men in universities doing abstract, theoretical work that was better paid and women doing applied work for smaller financial rewards.

The Segregated Academy. Two prominent sociologists (one Black and one White) have written, "As long as sociology has been in existence in the United States, despite the segregation

An insightful observer of race and culture, W. E. B. Du Bois was one of the first sociologists to use community studies as the basis for sociological work. His work, long excluded from the "great works" of sociological theory, is now seen as a brilliant and lasting analysis of the significance of race in the United States.

of American intellectual life, there has been a powerful tradition of the study of the Black community by Blacks" (Blackwell and Janowitz, 1974: xxi). African American thinkers whose work has not been properly recognized are now being credited for their contribution to the history and evolution of the discipline. They made their contributions despite forbidding obstacles to their success, including exclusion from White universities where they would have been more likely to have the resources for teaching and research that facilitate professional success.

W. E. B. Du Bois (1868–1963; pronounced "due boys") was a prominent Black scholar, a co-founder in 1909 of the NAACP (National Association for the Advancement of Colored People), a prolific writer and one of America's best minds. He also studied for a time in German, hearing several lectures by Max Weber. His PhD from Harvard was the first awarded by Harvard to a Black person in any field. While he was an Assistant Professor of Sociology at the University of Pennsylvania (1896–1898), he was assigned to a research project investigating neighborhood conditions in the Black community in Philadelphia. Du Bois was required to live in the settlement he studied, a situation he found to be harsh and unsatisfactory. His book from this study, *The Philadelphia Negro,* in collaboration with Isabel Eaton, a White sociologist from Columbia University, is a classic, one of the first empirical community studies and one of Du Bois's many collaborations with women sociologists. Du Bois was censured for denying the inferiority of African Americans, harassed for his leftist pronouncements by the U.S. government (which rescinded his passport in 1959), and indicted as an alleged "foreign agent." Optimistic in his youth, he grew bitter over racial issues in America by the end of his life. In 1961, at the age of ninety-three, he emigrated to Ghana, where he died in 1963 (Lewis, 1993).

Du Bois's vision of sociology differed in some ways from that of the White European–American mainstream. He was deeply troubled by the racial divisiveness in society, writing in a classic essay published in 1901 that "the problem of the twentieth century is the problem of the color line" (Du Bois, 1901: 354). Like many of his women colleagues, he envisioned a community-based, activist profession committed to social justice (Deegan, 1988). At the same time, like Weber, he believed in the importance of a scientific approach to sociological questions. His own works are exemplars of careful, rigorous sociology. W. E. B. Du Bois thought that sociology should be used to study the pressing issues of the time. Although he did not think sociology was value-free, since he believed convictions always directed one's studies, he did think sociology should use the most rigorous tools of sociological research to ensure accuracy and fairness.

Oliver Cromwell Cox is another example of an African American theorist whose work has been largely neglected. Like Max Weber, he had strong comparative and historical interests. Cox analyzed the racial prejudice, discrimination, and segregation of American society from the perspective of an outsider. He was born in Trinidad and educated and lived in the United States. Cox was especially interested in the origins of capitalism and what we now call *world systems theory*—theoretical explanations that take a global approach to understanding the relationship of social systems to economic markets and political structures (Hunter and Abraham, 1987). Added to his minority status, his Marxist orientation led many sociologists to shun him despite the known influence of Marx on the development of sociological thought.

The history of sociology as a discipline shows that, despite its liberal leanings, sociology is not immune from the patterns of exclusion found in other areas of society. African American sociologists have long been present in the discipline of sociology though they have only rarely been recognized. When E. Franklin Frazier (1894–1962) became president of the American Sociological Society (now the American Sociological Association, or ASA) in 1948, it was the first time a Black person had been elected; this has happened only one time since, when William Julius Wilson was elected in 1990.

The sociological profession is more diverse than it has ever been. Initiatives like the Minority Fellowship Program of the American Sociological Association and MOST (Minority Opportunity Through School Transformation) provide research support and financial assistance to minorities pursuing their education in sociology. Professional caucuses like the Association for Black Sociologists, Sociologists for Women in Society, and the Section on Latino Sociology work to promote the inclusion of women and minorities in the discipline, to make the discipline more attentive to the experiences of these groups, and to advocate social policies that will transform society on behalf of these groups. Members of these caucuses—men, women, Whites, and people of color—are deeply committed to the creation of an inclusive profession, one that works for justice, the development of the field, and the betterment of society. In doing so, they carry on the work of sociologists like Du Bois, Addams, and others who were denied the same opportunities.

Theoretical Frameworks in Sociology

The founders of sociology have established theoretical traditions that ask basic questions about society and inform sociological research. For many students, the idea of theory seems dry since it connotes something that is only hypothetical and divorced from "real life." To the contrary, sociological theory is one of the tools that sociologists use to interpret real life. Sociologists use theory to organize their empirical observations, to produce logically related statements about observed behavior, and to relate observed social facts to the broad questions sociologists ask: How are individuals related to society? How is social order maintained? Why is there inequality in society? How does social change occur? Sociologists work within theoretical traditions to answer these questions. The major theoretical frameworks—there are several—make different assumptions about these questions and about society, and orient sociologists in specific ways toward new and perennial sociological questions.

Within sociology, there is not a single theoretical perspective. Different theoretical perspectives make different assumptions about the nature of society. In the realm of *macrosociology* are theories that strive to understand society as a whole. Durkheim, Marx, and Weber were macrosociological theorists. Theoretical frameworks that center on face-to-face social interaction are known as *microsociology*. Some of the work that has been derived from the Chicago School—research that studies individuals and group processes in society—is microsociological. Although sociologists draw from diverse theoretical perspectives to understand society, three broad traditions form the major theoretical perspectives that they use: functionalism, conflict theory, and symbolic interaction.

Functionalism

Functionalism has its origins in the work of Durkheim, who, you will recall, was especially interested in how social order is possible, or how society remains relatively stable. **Functionalism** interprets each part of society in terms of

BOX 1.4

Key Sociological Concepts

As you build your sociological imagination, you will need certain key concepts to begin understanding how sociologists view human behavior. *Social structure, social institutions, social change*, and *social interaction* are not the only sociological concepts, but they are fundamental to grasping the sociological imagination.

Social Structure

Earlier, we defined **social structure** as the organized pattern of social relationships and social institutions that together constitute society. This is a critical, though abstract, concept in sociology. Social structure is not a "thing," but it refers to the fact that social forces, not always visible to the human eye, guide and shape human behavior. It is as if there were a network of forces embracing people and moving them in one direction or another. Acknowledging that social structure exists does not mean that humans have no choice in how they behave, only that those choices are largely conditioned by one's location in society. We examine this concept further in Chapter 5.

Social Institutions

In this book, you will also learn about the significance of **social institutions**, defined as established and organized systems of social behavior with a particular and recognized purpose. The family, religion, marriage, government, and the economy are examples of major social institutions. More than individual behavior, social institutions are abstractions, but they guide human behavior nonetheless. Mills understood that social institutions confront individuals at birth as given systems of behavior and that they transcend individual experience. Understanding the character of social institutions is one of the products of the sociological imagination.

Social Change

As you can tell, sociologists are also interested in the process of **social change**. As much as sociologists see society as producing certain outcomes, they do not see society as fixed or immutable, nor do they see human beings as passive recipients of social expectations. In other words, sociologists do not take a determinist approach to human behavior—as if all is fixed and without any possibility for change, or even surprise. Society exists because people interact with one another in socially meaningful ways. Society is, after all, an abstraction, real only in that it reflects the total pattern of people's social interaction.

Social Interaction

Sociologists see **social interaction** as behavior between two or more people that is given meaning. Through social interaction, people react and change, depending on the actions and reactions of others. Society results from social interaction; thus, people are active agents in what society becomes. Since society changes as new forms of human behavior emerge, change is always in the works. As you read this book, you will see that these key concepts—social structure, social institutions, social change, and social interaction—are central to the sociological imagination.

how it contributes to the stability of the whole. As Durkheim suggested, functionalism conceptualizes society as more than the sum of its component parts. Each part is functional for society—that is, contributes to the stability of the whole. The different parts are primarily the institutions of society, each of which is organized to fill different needs, and each of which has particular consequences for the form and shape of society. The parts are each then dependent on one another. Each of these institutions has consequences for the form and shape of society.

The family, for example, serves multiple functions: at its most basic level, the family has a reproductive role; within the family, infants receive protection and sustenance; as they grow older, they are exposed to the patterns and expectations of their culture; across generations, the family supplies a broad unit of support and enriches individual experience with a sense of continuity with the past and future. All these aspects of family can be assessed in terms of how they contribute to the stability and prosperity of society. The same is true for other institutions. The economy supplies a framework for the production and distribution of goods and services. Religion propagates a set of beliefs, which has proved to be one of the greatest sources of stability within societies. Together, these institutions compose society.

The functionalist framework emphasizes the consensus and order that exist in society, focusing on social stability and shared public values. From a functionalist perspective, *disorganization* in the system leads to change since societal components must adjust to achieve stability. This is a key part of functionalist theory—that when one part of society is not working (or dysfunctional), it affects all the other parts and creates social problems. Change may be for better or worse; changes for the worse stem from instability in the social system, such as a breakdown in shared values or a social institution no longer meeting people's needs (Collins, 1994; Eitzen and Baca Zinn, 1998; Turner, 1974).

Functionalism was the dominant tradition in sociology for many years, and one of its major figures was Talcott Parsons (1902–1979). In Parsons's view, all parts of a social system are interrelated. Different parts of society have different functions, the basic ones of which are (1) adaptation to the environment, (2) goal attainment, (3) integrating members into harmonious units, and (4) maintaining basic cultural patterns.

Parsons labeled this scheme AGIL: A stands for adaptation, G stands for goal attainment, I stands for integration, and L stands for latent pattern maintenance (Parsons, 1951a). Each of these functions of social behavior creates a system of interrelated parts—the parts being these behavioral goals. To Parsons, society worked like a machine that would always move to social stability. If one part was malfunctioning, the other parts would quickly adjust to return the system to equilibrium (Parsons, 1968), as when a disaster strikes a community, but people organize to meet one another's needs.

Functionalism was further developed by Robert Merton (born in 1910). Merton elaborated an important point in functionalist theory: Social practices have consequences for society that are not always immediately apparent, nor are they necessarily the same as the stated purpose of a given practice. Earlier we saw that Durkheim found rituals to produce group solidarity, a function that may be quite remote from the apparent purpose of a ritual, such as the celebration of religious beliefs; thus, solidarity is one function of ritual (Merton, 1968). Merton suggested that human behavior had both manifest and latent functions. **Manifest functions** are the stated and open goals of social behavior. **Latent functions** are the unintended consequences of behavior (Merton, 1968). For example, reforming social welfare programs may have the manifest function of reducing federal budget expenditures, but the policy may also have the latent function of increasing crime (as people have to support themselves through illegitimate means) or increasing homelessness and street violence.

THINKING SOCIOLOGICALLY

What are the *manifest functions* of grades in college? What are the *latent functions*?

Functionalism emphasizes that society works together as a whole system. This insight is an important one although some aspects of functionalism have been severely criticized by other sociologists. C. Wright Mills, whose ideas were examined at the beginning of this chapter, was especially critical of Talcott Parsons. Mills, a political radical, was not alone in arguing that functionalism's emphasis on social stability is inherently conservative, and that functionalism understates the roles of power and conflict in society. Critics also dispute the explanation of inequality offered by functionalism. Functionalists contend that racial, gender, and class inequality persist because they make a functional contribution to society by confirming the status of more powerful groups and creating a system for the distribution of societal resources. Critics of functionalism do not see such benevolence in the presence of inequality in society, arguing that functionalists are too accepting of the status quo. Functionalists would counter this argument by saying that, regardless of the injustices that inequality produces, inequality serves a purpose in society: It provides an incentive system for people to work and promotes solidarity among groups linked by common social standing.

Conflict Theory

Conflict theory emphasizes the role of coercion and **power,** a person's or group's ability to exercise influence and control over others, in producing social order. Whereas functionalism emphasizes cohesion within society, conflict theory emphasizes strife and friction. Derived from the work of Karl Marx, conflict theory pictures society as fragmented into groups that compete for social and economic resources. Social order is maintained not by consensus, but by domination, with power in the hands of those with the greatest political, economic, and social resources. When consensus exists, according to conflict theorists, it is attributable to people's being united around common interests, often in opposition to other groups (Dahrendorf, 1959; Mills, 1956).

According to conflict theory, inequality exists because those in control of a disproportionate share of society's resources actively defend their advantages. The masses are not bound to society by their shared values, but by coercion at the hands of the powerful. In conflict theory, the emphasis is on social control, not consensus and conformity. Groups and individuals advance their own interests, struggling over control of societal resources. Those with the most resources exercise power over the others; inequality and power struggles are the result. Conflict theory gives great attention to class, race, and gender in society because these are seen as the grounds of the most pertinent and enduring struggles in society.

Whereas functionalists find some benefit to society in the unequal distribution of resources, conflict theorists see inequality as inherently unfair, persisting only because groups who are economically advantaged use their social position to their own betterment. Their dominance even extends to the point of shaping the beliefs of other members of the society, by controlling public information and having major influence over institutions like education and religion. From the conflict perspective, power struggles between conflicting groups are the source of social change. Typically, those with the greatest power are able to maintain their advantage at the expense of other groups.

Conflict theory has been criticized for neglecting the importance of shared values and public consensus in society while overemphasizing inequality and social control. Like functionalist theory, conflict theory finds the origins of social behavior in the structure of society, but it differs from functionalism in emphasizing the importance of power.

Symbolic Interaction

The third major framework of sociological theory is **symbolic interaction theory.** Rather than conceptualizing society in terms of abstract institutions and values, symbolic interactionists consider immediate social interaction to be the place where "society" exists. Because of the human capacity for reflection, people give meaning to their behavior, and this is how they interpret the meaning of different behaviors, events, or things that are significant for sociological study. Because of its emphasis on face-to-face contact, symbolic interaction theory is a form of microsociology, whereas functionalism and conflict theory are more macrosociological. Derived from the work of the Chicago School, symbolic interaction theory analyzes society by addressing the subjective meanings that people impose on objects, events, and behaviors. Subjective meanings are given primacy because, according to symbolic interactionists, people behave based on what they *believe,* not just on what is objectively true; thus, society is considered to be *socially constructed* through human interpretation (Berger and Luckmann, 1967; Blumer, 1969; Shibutani, 1961). Symbolic interactionists see meaning as constantly modified through social interaction. People interpret one another's behavior, and it is these interpretations that form the social bond.

To illustrate a symbolic interactionist interpretation of human behavior, let us consider a recent study of African American mothers and sickle cell anemia. Sickle cell anemia is the result of a defective gene whose gene product causes impaired circulation of the blood. The disease is painful, debilitating, and can result in early death. Many African American mothers who are told they carry the sickle cell trait (making their offspring susceptible to full-blown sickle cell anemia should they be fathered by someone also carrying the sickle cell trait) nevertheless decide to have children. The sociologist who studied this issue found that the mothers construct two important definitions of the situation. First, the women have acquired a mistrust of the medical system, so they choose to define the warnings they are given about sickle cell anemia as unreliable. Second, motherhood confers high status among African Americans, and many women are unwilling to forgo motherhood even if they carry the sickle cell gene. This is particularly true among low-income women who have few other routes to elevated status (Hill, 1994). In the end, the mothers' definition of the situation is more influential in determining their behavior than the objective presence of a potentially dangerous condition.

Symbolic interactionists see social order as being constantly negotiated and created through the interpretations people give to their behavior. To interpret society, symbolic interactionists seek not simply facts but "social constructions," the meanings attached to things, whether those are concrete symbols (like a certain way of dress or a tattoo) or nonverbal behaviors. To a symbolic interactionist, society is highly subjective; that is, it exists in the imaginations of people even though its effects are real. This is a point not to be taken lightly; think, for example, of the number of people who deny that the Holocaust ever happened despite convincing evidence to the contrary. The emphasis on subjectivity has been an important source of criticism of symbolic interaction theory. Functionalists and conflict theorists, on the other hand, consider society to have an objective reality.

Functionalism, conflict theory, and symbolic interaction theory are by no means the only theoretical frameworks in sociology. For some time, however, they have provided the most prominent general explanations of society. Each has a unique view of the social realm. None is a perfect explanation of society, yet each has something to contribute. Functionalism gives special weight to the order and cohesion that usually characterizes society. Conflict theory emphasizes the inequalities and power imbalances in society. Symbolic interactionism emphasizes the meanings that humans give to their behavior. Most sociologists working on current studies examine their data from a variety of theoretical approaches. Together, these frameworks provide a rich, comprehensive perspective on society, individuals within society, and social change (Table 1.1).

TABLE 1.1 *THREE SOCIOLOGICAL FRAMEWORKS*

	Functionalism	Conflict Theory	Symbolic Interaction
	View of Society		
	objective; stable; cohesive	objective; hierarchical; fragmented	subjective; imagined in the minds of people
	Basic Questions		
Relationship of individual to society	individuals occupy fixed social roles	individuals subordinated to society	individual and society are interdependent
View of inequality	inevitable; functional for society	result of struggle over scarce resources	inequality demonstrated through meaning of status symbols
Basis of social order	consensus among public on common values	power; coercion	collective meaning systems; society created through social interaction
Source of social change	social disorganization and adjustment to achieve equilibrium; change is gradual	struggle; competition	ever-changing web of interpersonal relationships and changing meaning systems
	Major Criticisms		
	a conservative view of society that underplays power differences among and between groups	understates the degree of cohesion and stability in society	has a weak analysis of inequality and tends to ignore material differences between groups in society; overstates the subjective basis of society

Diverse Theoretical Perspectives

Sociological theory is in transition. The traditions introduced here—functionalism, conflict theory, and symbolic interaction, inform sociological work and are discussed throughout this book. There are also other theoretical orientations that we have not covered.

Contemporary sociological theory has been greatly influenced by the development of *feminist theory*. Feminist theory analyzes the status of women in society with the purpose of using that knowledge to better women's lives. Attention to the women in society by feminist theorists has created vital new knowledge about women and has also transformed what we understand about men. Feminist scholars have shown that traditional theoretical perspectives have distorted or ignored women's experiences and have therefore given an inadequate picture of society. When Karl Marx investigated social class, for example, labor was analyzed exclusively in the context of wage earners, ignoring completely women's unpaid work in the home. Traditional Marxist thinkers have therefore been ill-equipped to analyze the social class of housewives. In another example, Weber argued that rationality would increasingly replace tradition as the basis of modern social relationships. Feminists have argued that this is a male-oriented prediction, made by Weber without taking into account the central role of emotion in the maintenance of social relationships—an area of social life traditionally tended by women. We will return to the status of feminist theory in Chapter 12. Feminist scholarship in sociology, by focusing on the experiences of women, provides new ways of seeing the world and contributes to a more complete view of society.

Other theoretical perspectives are also used by sociologists. *Exchange theory*, for example, argues that the behavior of individuals is determined by the rewards or punishments they receive in day-to-day interaction with others (Homans, 1961). Exchange theory has been especially useful in analyzing subjects like social networks. *Rational choice theory* is a relatively recent development used by political scientists and economists as well as sociologists. Rational choice theory posits that the choices human beings make are guided by reason. Society is seen as the sum of these individual decisions and actions (Smelser, 1992b).

Many contemporary theorists are increasingly influenced by **postmodernism**—a strand of thinking now influencing many disciplines. Postmodernism is based on the idea that society is not an objective thing; rather, it is found in the words and images—or *discourses*—that people use to represent behavior and ideas. Postmodernist theory challenges the idea that anything, including society, is objectively real. Postmodernists think that images and texts reveal the underlying ways that people think and act. Postmodernist studies typically involve detailed analyses of images, words, film, music, and other forms of popular culture. In a civilization like ours, saturated by the imagery of the mass media, postmodernist analysis of cultural images can pro-

postmodernism

vide deep insight into the society. Postmodernist thinkers see contemporary life as involving multiple experiences and interpretations; thus, they avoid categorizing and simplifying human experience into broad concepts like institutions or society (Rosenau, 1992).

Sociology studies many diverse topics, and sociologists have developed many theoretical orientations. Some sociologists rely primarily on one theoretical perspective and choose the topics they study accordingly. Others draw from multiple theoretical perspectives, guided by whichever one seems best able to explain a question at hand. Someone studying the sociology of families, for example, might draw from postmodernism to examine the significance of cultural imagery in producing popular definitions of family life. A sociologist who wanted to know how families are affected by economics might use conflict theory to study the stresses that work can produce in families. Feminist theory might reveal how an individual's experience of the family is shaped by gender. Relying on a single theory can produce an incomplete explanation, so sometimes several theoretical models may be used in research. Whatever theoretical framework one uses, theory should be evaluated in terms of its ability to explain observed social facts. The sociological imagination is not a single-minded way of looking at the world. It is the ability to observe social behavior and interpret that behavior in light of the societal influences that shape it. Theoretical tools that contribute to insight are valuable whether they are modern or classical.

CHAPTER SUMMARY

- *Sociology* is the study of human behavior in society, and the *sociological imagination* is the ability to see societal patterns that influence individual and group life. Sociology is an *empirical* discipline, relying on careful observations as the basis for its knowledge. The *debunking* tendency in sociology refers to the ability to look behind taken-for-granted things, looking instead to the origins of social behavior.
- One of the central insights of sociology is its analysis of social diversity and inequality. Understanding *diversity* is critical to sociology because it is necessary to analyze social institutions and because diversity shapes most of our social and cultural institutions.
- Sociology emerged in western Europe during the Enlightenment and was influenced by the values of *critical reason, humanitarianism,* and *positivism. Auguste Comte,* one of the earliest sociologists, emphasized sociology as a positivist discipline. *Alexis de Tocqueville* and *Harriet Martineau* developed early and insightful analyses of American culture.
- *Emile Durkheim* is credited with conceptualizing society as a social system and with identifying social facts as patterns of behavior that can be explained as external to individuals. *Karl Marx* showed how capitalism shaped the development of society; he theorized economic systems as basic to the structure of society. *Max Weber* thought that society had to be explained through cultural, political, and economic factors. Weber also wrote that sociologists should use *verstehen*—defined as the ability to see things from the point of view of others—to understand social behavior.
- Most American sociologists believed in using sociological research and theory to solve social problems. The *Chicago School* is characterized by its concern with the relationship of the individual to society and the use of society as a human laboratory. Numerous scholars are identified with this school, including *Charles Horton Cooley, George Herbert Mead, W. I. Thomas, Florian Znaniecki, Robert Park,* and *Jane Addams.*
- Many women and African American sociologists have made important contributions to the development of the discipline even though their work has gone largely unrecognized until very recently. *W. E. B. Du Bois* used his sociological training to develop extensive studies of race in American society.
- There are three major theoretical frameworks in sociology: *functionalism, conflict theory,* and *symbolic interaction theory.* Functionalism emphasizes the stability and integration in society. Conflict theory sees society as organized around the unequal distribution of resources, held together through power and coercion. Symbolic interaction theory emphasizes the role of individuals in giving meaning to social behavior, thereby creating society. Different theoretical perspectives are used by sociologists, depending on the questions they are trying to answer. All theories should be evaluated in terms of their ability to explain observed social facts.

KEY TERMS

applied sociology
conflict theory
debunking
empirical
Enlightenment
functionalism
humanitarianism
issues
latent functions
manifest functions
natural attitude
organic metaphor
positivism
postmodernism
power
social action
social class
Social Darwinism
social facts
social institution
social structure
sociological imagination
sociology
symbolic interaction theory
troubles
verstehen

THE INTERNET: A Tool for the Sociological Imagination

Technology is increasingly an important resource for sociological work and it is transforming how sociologists do their work. Using the Internet to locate the vast amounts of information now available is a good skill to learn as you develop a sociological imagination. Internet resources can help you find published research on the topics covered in this book. The Internet is also a vehicle for the delivery of vast amounts of information which sociologists find important in doing their work. On the Internet you can find federal data on income, education, and occupation, as well as surveys of public attitudes on many subjects and fact sheets from various organizations (such as on elder abuse, child welfare, or criminal justice statistics, just to give a few examples).

The Internet service which is the most widely used and the service for which the Internet exercises in this book are designed is the World Wide Web, usually referred to simply as the Web. The Web is a means of gathering information stored on computers throughout the world. Many institutions, groups and individuals maintain Web pages and post a vast variety of information on them. This book in the end-of-chapter exercises like the one immediately following makes use of the Web as a resource for gathering information to help answer important sociological questions.

Three particular features of this book introduce students to using the Internet in sociological studies. First, at the end of each chapter is a list, Resources on the Internet, that will lead you to the Internet addresses (URLs) for some of the major sites with information on that chapter's subject. A good place to begin browsing the Internet is from the Web site for this book.

Virtual Society: The Wadsworth Sociology Resource Center at
http://sociology.wadsworth.com

Use this as a starting point, but feel free to conduct your own research using search engines such as Yahoo, Hotbot or AltaVista to find the information you are looking for. It is also helpful to utilize the links available at some of the better constructed web sites.

Second, the feature Sociology and Social Policy: Internet Exercises is designed to help students think about how a sociological perspective helps you understand many contemporary debates about social issues. These exercises are constructed to make you think about the application of sociological research and theory to matters of public importance—matters that are widely debated in the media, the political realm, and in other social institutions. The "Sociology and Social Policy" feature poses a question based on certain key concepts highlighted in the text. Students can then search the Internet to find published research and other information on the subject at hand. This should then inform whatever position you take in debating these different issues.

The third Web-based Internet feature unique to this textbook is the InfoTrac College Edition: Search Word Summary. InfoTrac College Edition is an online university library with access to over 600 periodicals. To get started, point your Web browser to: **http:www.infotrac-college.com/wadsworth** then enter your account ID number when it is requested. You may also link to InfoTrac College Edition from the Virtual Society Home Page.

Resources on the Internet:

American Sociological Association Home Page
http://www.asanet.org

The American Sociological Association is a national organization of sociologists. Their home page provides information on annual meetings, resources available for sociological research, information on careers in sociology, special reports on current initiatives in social policy, and information on how students can get involved in this sociological organization, including the Minority Fellowship Program.

Society for the Study of Social Problems
http://funnelweb.utcc.utk.edu/~SSSP

This national professional group sponsors an annual meeting and publishes one of the major journals in sociology, *Social Problems.*

Regional Sociological Associations

There are several regional sociological associations, all of which welcome student participation:

Eastern Sociological Society:
http://wings.buffalo.edu/ess/

Midwest Sociological Society:
http://www.drake.edu/MSS/home.html

Pacific Sociological Society:
http://www.csus.edu/psa/

Southwestern Sociological Association:
http://www.sfasu.edu

Southern Sociological Association:
http://ww.MsState.edu/org/SSS/sss.html

Sociology and Social Policy: Internet Exercises

Sexual abuse is a good example of the distinction C. Wright Mills makes between personal troubles and public issues. Obviously, to the person experiencing abuse (and, perhaps, to the abuser), this is a personal trouble. Sociologists would add that sexual abuse is a public issue. It is more common than one would like to believe and is rooted in some of the social structures of society (the power of men over women, conflict within the family, the use of children as sexual objects, and so forth). Using the Infotrac resources that come with this book, or using the library, search the sociological research literature on sexual abuse, and identify some of the structural factors that make this a public issue. How does your investigation illustrate Mills's two concepts of personal troubles and public issues?

Internet Search Keywords:

sexual abuse
public issue
social structures
sexual abuse, men
sexual abuse, women
abuse in families
child sexual abuse

Web sites:

http://www.jimhopper.com/abstats
A sociological study on the lasting effects of child abuse, especially in men. This site also indicates a number of related links.

http://www.safersociety.org/index.htm
An index of new books with abstracts on sexual abuse of children and women.

http://www.childsexualabuse.org/
Web site of the National Foundation to Prevent Child Sexual Abuse.

http://www.cs.utk.edu/~bartley/sacc/childAbuse.html
A study on the symptoms, consequences and prevention of child sexual abuse.

http://www.ippf.org/newsinfo/ma98/women.htm
A report on violence against women by the International Planned Parenthood Federation.

InfoTrac College Edition: Search Word Summary

class, race, and gender	post modernism
diversity in society	sociological imagination

In order to learn more about these central topics in sociology, you can coduct an electronic search using InfoTrac College Edition. To aid in your search and to gain useful tips, see the Student Guide to InfoTrac College Edition on the Virtual Society web site:
http://sociology.wadsworth.com

INTERACTIONS—A SOCIOLOGY CD-ROM: CONCEPTS FOR THIS CHAPTER

Go to the Wadsworth Sociology CD-ROM for further study on the concepts in this chapter. The CD-ROM also includes quizzes and additional activities to expand your learning experience.

SUGGESTED READINGS

Becker, Howard. 1986. *Writing for Social Scientists.* Chicago: University of Chicago Press.

This engaging book is not only an excellent discussion of how to improve one's writing, it is also a good guide to sociological thinking. Premised on the idea that poor writing produces errors in thinking, Becker's book will help students at all levels improve their sociological perspective and their ability to communicate it.

Blackwell, James E., and Morris Janowitz. 1974. *Black Sociologists: Historical and Contemporary Perspectives.* Chicago: University of Chicago Press.

This collection provides an overview of the contributions of Black sociologists to their profession, and it is an important reminder of the work of those who have too often been overlooked in the history and analysis of sociological theory.

Deegan, Mary Jo. 1988. *Jane Addams and the Men of the Chicago School, 1892–1918.* New Brunswick, NJ: Transaction Books.

Deegan's discussion of the career of Jane Addams is an important addition to the history of sociology. She shows how women were excluded from the early years of sociology, but also points to the important contributions Addams and other women reformers made in the development of sociological thought.

Goetting, Ann, and Sarah Fenstermaker (eds.). 1995. *Individual Voices, Collective Visions: Fifty Years of Women in Sociology.* Philadelphia: Temple University Press.

This series of autobiographical essays by contemporary women sociologists of different ages reveals the connection between women's lives and their work as sociologists.

Johnson, Allan G. 1997. *The Forest and the Trees: Sociology as Life, Practice, Promise.* Philadelphia: Temple University Press.

This brief introduction to sociology provides an excellent overview of the basic concepts of the field and the application of sociology to interpreting contemporary social issues. The book also grounds its discussion in an awareness of the diversity among different groups in the United States.

Lewis, David Levering. 1993. *W. E. B. Du Bois—Biography of a Race.* New York: H. Holt.

This excellent biography of Du Bois introduces students to the life of a man who, though long overlooked, is one of the century's greatest intellectuals and sociologists.

Mills, C. Wright. 1959. *The Sociological Imagination.* New York: Oxford University Press.

This book is a classic statement of the meaning and significance of the sociological perspective.

CHAPTER 2

Doing Sociological Research

SOCIOLOGICAL research is the tool sociologists use to answer questions. There are a variety of ways that sociologists do research, all of which involve rigorous observation and careful analysis. The method that you use depends on the kind of question you are asking.

Suppose that you wanted to know if gang violence had increased in recent years. You might turn to official police records and measure the number of arrests made for gang-related crimes. If you wanted to take your question further and ask if gang violence was related to the unemployment rate for youth in inner cities or the rate at which young people drop out of school, you might collect additional data and then analyze the relationship between the crime data, unemployment rates, and school dropout rates in given cities. These questions and their answers would require both a sociological imagination and some research tools, including the ability to gather and interpret data, as well as some statistical and computer skills that you would likely need to analyze the relationship among the factors you have identified.

Suppose, on the other hand, that you were interested not so much in the extent of gang violence or its relationship to unemployment and education, but instead wanted to know more about the dynamics of gang behavior—who is in gangs, what encourages them to join, how gang members experience their association with gangs, and what gangs mean to the women and men who join them. This research question would require an approach different from the one described earlier—one that would introduce you to gang members (and perhaps others in the community where they exist). With this question, you would need to get more of an insider's view of gang members, as Mártin Sánchez Jankowski (1991) did when he studied gangs in several major U.S. cities. Being careful not to put yourself in danger, you might hang out with gangs for a time, interviewing members and observing their behavior.

These different approaches to studying gangs reflect some of the different research methods sociologists use to answer the questions they ask. Some questions require more direct involvement with those whom one is studying; other methods involve more quantitative skills. Either way, the chosen research method must be appropriate to the sociological question being asked.

Let us consider another example. Suppose you want to know if racial prejudice in the United States has increased or declined in recent years. You might use large national opinion polls that explore a variety of public attitudes, including racial prejudice. If these surveys included information on educational level, occupation, age, and gender, you could examine how these factors are related to the existence of prejudice. If, however, you wanted to know whether prejudice increases or declines when different groups interact, opinion polls would not be the best tool. Polls ask what people think, whereas this question would require sound evidence of how people's attitudes change depending on their relations with other groups. To examine prejudice between, say, Asian Americans and African Americans, you could design a study of those who have extensive contact with people in the other group and compare them with those whose contact with the other group is minimal or nonexistent. If you take care to ensure that the groups are in other ways similar, then the differences in prejudice you observe between the two study populations would be due, at least in part, to the extent of intergroup contact.

In this chapter, we will examine the research methods that sociologists use to answer the different kinds of questions they ask. As you will see, research is fundamental to the sociological enterprise—whether it is the analysis of surveys or interviews; direct observation of people in diverse settings; scrutiny of media documents, images, and historical records; or any of the other methods sociologists use. Research is an engaging and demanding process. It requires skill, careful observation, and the ability to think logically about the things that spark your sociological curiosity. Sociologists ask many different questions, and their techniques for answering questions are limited only by the horizon of the sociological imagination. All research methods follow certain general rules, which we will examine in the upcoming sections.

The Research Process

When sociologists do research, they engage in a process of discovery. Like scientists who do research in a laboratory, sociologists organize their research questions and procedures systematically. Their laboratory is the social world. Through research, they organize their observations and interpret them.

Sociology and the Scientific Method

Sociological research derives from the **scientific method,** originally defined and elaborated by the British philosopher Sir Francis Bacon (1561–1626). The scientific method involves several steps in a research process, including observation, hypothesis testing, analysis of data, and generalization. Since its beginnings, sociology has attempted to adhere to the scientific method. To the degree that it has succeeded, sociology is a science; yet, there is also an art to developing the sociological imagination. The process is difficult to capture as a simple series of steps or procedures.

The research process involves several operations that can be performed on the computer, such as entering data in numerical form and writing the research report.

For more than a hundred years, there has been lively debate about whether sociology is a "true" science. Many argue that it is indeed a science, albeit a "little science" (Mazur, 1968). Sociology aspires to be both scientific and humanistic, whereas there are purists who contend that it can be only one or the other. The crux of the scientific method is observation and the testing of theory. Science is **empirical,** as we noted in the first chapter, meaning it is based on careful and systematic observation, not just conjecture. It builds upon objective experiences and data rather than abstractions and subjective impressions. Sociological studies may be based on surveys, observations, and many other forms of analysis, but always they depend on an empirical underpinning. Although some sociological studies are highly quantitative and statistically sophisticated, others are qualitatively based—that is, based on more interpretive observations, not statistical analysis. Both quantitative and qualitative studies are empirical. Some sociological works are not empirical; purely theoretical work is the exception that comes to mind. Even purely theoretical work, however, is subject to empirical testing.

Sociological knowledge is not the same as philosophy or personal belief. Philosophy, theology, and personal experience can certainly deliver insights into human behavior, but at the heart of the scientific method is the notion that a theory must be testable. This requirement distinguishes science from purely humanistic pursuits like theology and literature. Imagination and fascination with humanity may drive the creation of theories in theology and literature, as much as in the social sciences; however, aesthetic theories are not directly testable by empirical study. Sociological theories must endure rigorous testing and observation.

One wellspring of sociological insight is **deductive reasoning,** wherein specific conclusions are derived from general principles. One might reason deductively that since Catholic doctrine forbids abortion, Catholics answering a survey would therefore be less likely than other religious groups to support abortion rights. **Inductive reasoning** reverses this logic, deriving general conclusions from specific observations. Observing that most of the demonstrators protesting abortion in front of a family planning clinic are evangelical Christians, one might perceive that, in general, religious identification appears to correlate with attitudes toward abortion, perhaps inspiring a research project (this subject is discussed in Chapter 16). Deductive and inductive reasoning are both based on the importance of observation and logical analysis. With deductive reasoning, one is more likely to set up a hypothesis prior to making one's observations; with inductive reasoning, one's observations lead to the formation of a theory.

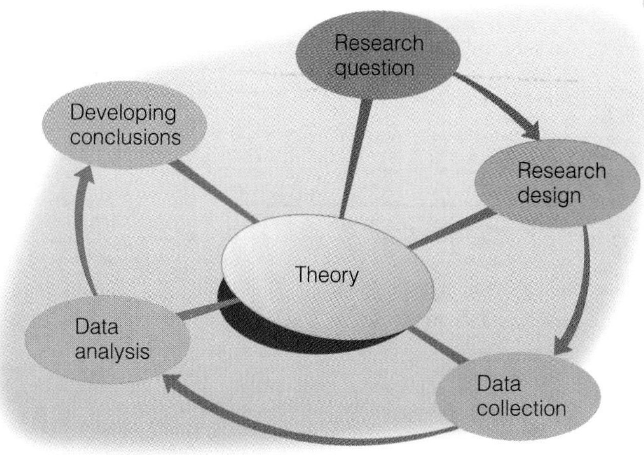

FIGURE 2.1 The Research Process

Developing a Research Question

Sociologists use different methods to explore different questions, but all sociological research is an organized practice that can be described in a series of steps (see Figure 2.1).

The first step in sociological research is to develop a research question. One source of questions is past research. For any number of reasons, the sociologist might disagree with a research finding and decide to carry out further research or develop a detailed criticism of the prior research. Someone else's work may suggest an idea, seem incomplete, or appear to be seriously in error. Criticism of past research has been an important inspiration for many new investigations, and the research of other workers can often be refined. The best research frequently inspires other sociologists to study a question further.

Developing a sociological research question typically involves reviewing the existing literature on the subject, such as past studies and research reports. A subject interesting enough to spur research has probably attracted the attention of other researchers, and their work can help define a research question or guide a researcher toward an original path. Digital technology has vastly simplified the task of reviewing the literature. Researchers who once had to burrow through paper indexes and card catalogs to find material relevant to their studies can now scan much larger swaths of literature in far less time, using online databases. The catalogs of most major libraries in the world are accessible via the Internet, as are specialized indexes, discussion groups, and other research tools developed to assist sociological researchers. Increasingly, the journals that report new sociological research are also available online and can often be electronically read in full text. Although electronic books are only now starting to become available online through companies such as Barnes and Noble, Versabook, and Rocket-ebook, most scanning through research today is still done at the library with paper books.

Sometimes a review of past research may cause you to wonder if the same result would be found again if the study were repeated, perhaps with a different group being examined. Research that is repeated exactly, but on a different group of people or in a different time or place, is called a **replication study.** As an example, suppose earlier research found that women managers have fewer opportunities for promotion than men. You might want to know if this still holds several years later. You would then replicate the original study, probably using a different group of women and men managers, but asking the same questions that were asked earlier. A replication study can tell you what changes have occurred since the original study was conducted, and

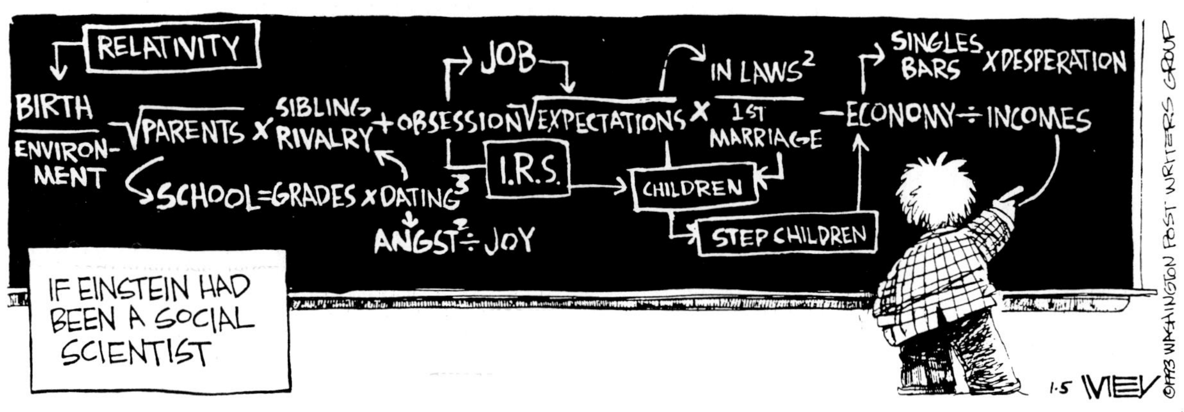

may also refine the results of the earlier work. Research findings should be reproducible; if research is sound, other researchers who repeat a study should get the same results unless, of course, change has occurred in the interim.

Sociological research questions can also come from casual observation of human behavior. Perhaps you have observed the seating patterns in your college dining hall and wondered why people sit with the same group day after day. Does the answer point to gender, age, race, or political views? When men and women brush past each other, men generally turn toward women, and women generally turn away from men (Basow, 1992; De Paulo, 1992). Is this example of body language a clue to the relative status of men and women? African Americans and Whites have different conversational styles; they interrupt each other differently, maintain eye contact differently, and respond to casual greetings differently (De Paulo, 1992). What effects do these differences have on Black–White interactions?

Social policy or practical goals are very often the source of research questions. As politicians have debated welfare reform in recent years, they have often indulged in careless and inaccurate pronouncements about welfare recipients. Sociologists have studied the welfare system to get firm answers to questions such as whether welfare recipients become dependent on welfare over a long period (no) or whether welfare encourages women to have more children (no). Most people who receive welfare tend to do so for two years or less, and researchers find no correlation between the level of welfare benefits and the number of children (Butler, 1996; Duncan and Hoffman, 1990; Harris, 1996; this research is discussed further in Chapter 9). Such research can provide the basis for making sound policy judgments, especially on matters otherwise informed only by strong political views.

Finally, sociological theory can be a source of research questions. Sociological research is often developed to test an existing theory. A theory can stand only as long as research continues to support it. This is what makes sociology a scientific endeavor. Some sociological theories have withstood years of challenges. Other theories are in vogue only a short time before the weight of additional research causes them to be rejected. Commonly, a new theory is substituted for the one rejected. The new theory then undergoes the same tests as the old one.

An example of the rise and fall of a theory is seen in the idea, once considered respectable, that Black Americans were intellectually inferior to White Americans and that the cause was genetic. Otto Klineberg (1935) was the first to show that, from the 1930s on, northern Blacks on average actually outperformed southern Whites on intelligence tests. This discovery meant that social environment was more important than an alleged Black–White, genetically based intelligence difference. Today, Blacks on average have slightly lower test scores overall than Whites. Sociologists consider this difference to be primarily because of differences in social, educational, and economic environments and inadequate methods for testing intelligence (Fischer et al., 1996; Kincheloe et al., 1996; Lewontin, 1996; Taylor, 1980, 1992, 1995). The research supporting this view is discussed in the chapters on race and ethnicity (Chapter 11) and education (Chapter 15). The point is that in the face of hard data, a theory had to be scrapped and a new theory substituted in its place. When a topic is the subject of public controversy—and the controversy about race and intelligence has flared hot and often in this century—it is essential that the truth be discovered by careful, rigorous research studies.

We saw in the previous chapter that sociologists have devised several theoretical frameworks to interpret what they observe. One's theoretical orientation can channel what questions are asked and how they are answered. For example, suppose you want to understand why some women are more likely to be poor than men. You may have concluded already that conflict theory offers the best approach to your question. You might then do a study that examines, among other things, differences in the work done by men and women, how much they earn, how much they spend on themselves, and how much they spend on their children. If you want an especially precise answer, you might separate the data for Latinos, Whites, Asians, Native Americans, and African Americans since poverty is known to vary among different racial groups.

Symbolic interaction theory is a more microsociological approach. If you wanted to know how people of different racial and class backgrounds interact with each other on the street, symbolic interaction would suggest that you observe individuals rather than whole populations. You might study eye contact, the distances people maintain between themselves, and how often they touch each other during conversation. You would then investigate the meanings people attach to these behaviors, perhaps by secretly arranging violations of usual behaviors. You could have an associate talk to people while standing too close, or staring too intently, or refusing eye contact completely. Observing the responses would give clues to what these behaviors mean within a culture.

Creating a Research Design

Once you have developed a research question, you must decide how to proceed with your investigation. This is the next step in doing sociological research. A **research design** is the overall logic and strategy underlying the research. Sociologists engaged in research may distribute questionnaires, interview people, or make direct observations in a social setting or laboratory. They might analyze cultural artifacts, such as magazines, newspapers, television shows, or other media. Some do research using historical records. Others base their work on the analysis of social policy. All these are forms of sociological observation. Research design consists of choosing the observational technique best suited to a particular research question.

Suppose that you wanted to study the career goals of student athletes. In reviewing past studies, you might have found an article that discussed how athletics is related to the development of men's identities (Schacht, 1996; Messner,

1996, 1989). Perhaps this sparked your curiosity about how athletics shapes women's identities. You may also have read an article in your student newspaper reporting that the rate of graduation for women college athletes is much higher than the rate for men athletes and wondered if women athletes are better students than men athletes. You decide to ask, specifically, whether women's coaches emphasize academic work more than men's coaches.

Your research design lays out a plan for investigating these questions. It includes details of your study. Which athletes will you focus on? How will you study them? To begin, you will need to get sound data on the graduation rates of the groups you are studying to verify that your assumption of better rates among women athletes is actually true. You then might need to develop a way to measure the academic encouragement students get from coaches. You might observe interactions between coaches and student athletes, recording what coaches say about class work. As you proceed, you would probably refine your research design, and even your research question. Perhaps, you think, the differences between men and women are not so great when the men and women play the same sports, or perhaps the differences depend on other factors, such as how the students were recruited and what kind of financial support they get. Do graduation rates vary for women and men depending on the kind of college they attend? If you want to answer this question, you have to build a comparison of college types into your research design and study different colleges. *The details of your research design flow from the specific questions you ask.*

Some research designs involve the testing of hypotheses. A **hypothesis** (pronounced "hy-POTH-i-sis") is a prediction, a tentative assumption that one intends to test. If you have a research design that calls for the investigation of a very specific hunch, you might formulate a hypothesis to test or, as is true in many sociological studies, a series of hypotheses ("hypothesis" is the singular form; "hypotheses" is the plural). Hypotheses are often formulated as if–then statements. For example:

> *hypothesis:* If a person's parents are racially prejudiced, that person will, on average, be more prejudiced than a person whose parents are free of prejudice.

This is a hypothesis, not a demonstrated fact or finding. Having phrased a hypothesis, the sociologist must then determine if it is true or false. If a hypothesis is shown to be correct, that tends to support (though not necessarily prove) the hypothesis and the reasoning behind it. To test the preceding example, one might take a large sample of people and determine their level of prejudice via interviews or some other mechanism. One would then determine the level of prejudice of their parents. According to the hypothesis, one would expect to find an association between bigoted children and bigoted parents and between nonbigoted children and nonbigoted parents. If the association is found, the hypothesis is supported. If no such association is found, then the hypothesis would be incorrect.

Not all sociological research follows the model of hypothesis testing. Some is more exploratory or open-ended, or based on more unstructured observations of human behavior, as we will see. One's research design must be appropriate to the question for which an answer is sought.

Any research design includes a plan for how **data** will be gathered. (*Note:* "data" is the plural form; one says, "data are used . . . ," not "data is used. . . .") Quantitative data are numerical and can be analyzed using statistical techniques. Qualitative data are not numerical although they can be logically interpreted and analyzed. Sociologists often try to convert their observations into a quantitative form, using techniques that are discussed later in the section on "Statistics in Sociology." Quantifying the responses to questionnaires or interviews, for example, helps categorize the data, makes it possible to analyze qualitative responses statistically, and removes an element of subjectivity from the interpretation of research results. Research data are the heart of sociological research; the research design specifies what kinds of data an investigator plans to use.

A **variable** is something that can have more than one value or score; a variable can be relatively straightforward, such as age or income, or a variable may be more abstract, like social class or degree of prejudice. In much sociological research, variables are analyzed to understand how they influence each other. With proper measurement techniques and a good research design, the relationships between different variables can be discerned.

Sociologists frequently design research to test the influence of one variable on another. Most of the time, they are trying to establish cause and effect. The **independent variable** is the one that the researcher wants to test as the presumed cause of something else. The **dependent variable** is the one upon which there is a presumed effect. Sociological variables are often sensitive to a variety of influences, and trying to control for the influence of factors other than the one being studied is a challenge to the creativity of the researcher. Outside factors can generally be managed by good research design. In fact, the brilliance of a research design may reside in how well outside influences are excluded and in how focused the data are on a single issue. At the same time, it is important to know the context in which the research is conducted because this can influence the results. Intervening influences (called intervening variables) can change the effect of an independent variable on a dependent variable. Figure 2.2 illustrates the relationship between independent, dependent, and intervening variables.

Research designs may be cross-sectional or longitudinal. In a *cross-sectional study,* something is studied at a single point in time. Using the example of student athletes and academic performance, a cross-sectional study might examine the grades of a wide selection of athletes in a given year, subdividing the group by how much they participated, what sports they played, where they played, and so on. In a *longitudinal study,* something is studied at two or more points in time. A longitudinal study of athletes might look at how many of those who start out at one time with aspirations to become athletic stars actually achieve greatness at a later time. Longitudinal research designs can measure change over a given period and can measure a particular change

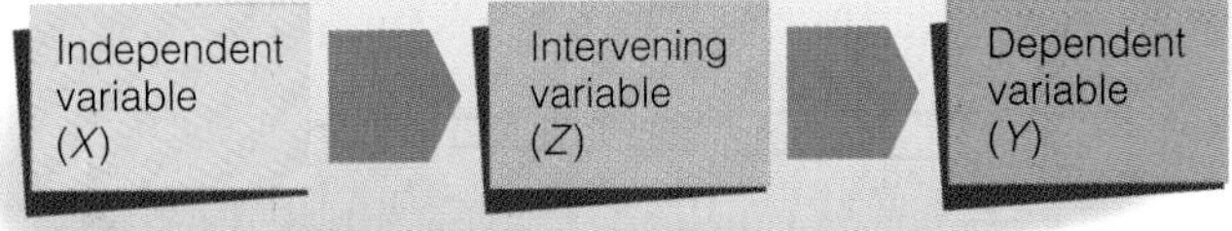

FIGURE 2.2 The Analysis of Variables

Sociological research seeks to find out whether some independent variable (X) affects an intervening variable (Z) which in turn affects a dependent variable (Y).

against other variables, such as whether academic performance by athletes is affected by won–loss record. Longitudinal research requires more time to complete than cross-sectional studies, since the researcher has to capture data (often the same data that inform a cross-sectional study) at different points in time. These studies may go on for very long periods, even decades, with data constantly being added and compared to earlier data.

One thing to be decided when designing a research project is how a social phenomenon should be measured. Units of measurement for social behavior are not always as obvious as the physical measurements scientists depend on. Some social phenomena might be easily quantified, such as the rate at which people vote or how many people in a group vote. Other social phenomena are not easily quantified. The degree of a person's prejudice, for example, cannot be measured in the same way as the temperature outside. Determining how one will measure social facts is an extremely important stage in research design.

Sociological research proceeds via the study of concepts. A **concept** is any abstract characteristic or attribute that can be potentially measured. Prejudice is a concept, as are some of the terms introduced in the first chapter, such as social class, power, and social institution. These are not things that can be seen directly, although they are key concepts in the field of sociology. When sociologists want to study concepts, they must develop measures for "seeing" them.

To enable the measurement of concepts, researchers devise **operational definitions,** meaning any definition that spells out how a concept is to be measured (Babbie, 1998). One might measure social class by defining class in terms of income, educational level, and occupation. Prejudice might be measured by designing a questionnaire in which respondents rate the qualities of different social groups—the answers would tell less about the groups than about the respondents. Operational definitions are typically based on a series of **indicators,** something that points to or reflects an abstract concept. Income is an indicator of the concept of social class; so is educational level. An indicator is a measure of something real that corresponds in some useful way to a concept. The use of indicators can be found in the map, "Viewing Society in Global Perspective: The Human Development Index." Here the United Nations has taken a number of indicators, including life expectancy, educational attainment, and the standard of living, and combined them into a single measure, known as an *index,* of an abstract concept—in this case, human development (United Nations, 1997). The index of human development provides a single measure of a more abstract concept (human development), and the resulting quantitative view allows us to see at a glance the differing levels of overall well-being for people around the world.

Great care must be taken when developing operational definitions of sociological concepts because they determine what indicators will be used, and it is these measurements that will be used to reach conclusions about the concept. The appropriateness of the operational definitions critically affects the accuracy of the conclusions.

THINKING SOCIOLOGICALLY

If you wanted to conduct research that would examine the relationship between student alcohol use and family background, what *indicators* would you use to measure these two variables: alcohol use and family background? How might you *design* your study?

The **validity** of a measurement is the degree to which it accurately measures or reflects a concept. One step in research design is deciding whether to develop new procedures or to borrow from research done before. Other researchers may have investigated the same topic and designed questionnaires or other research instruments that measure the concept one is working on. In this case, a researcher must make a decision about the validity of earlier research. Is it accurate? Is it up to date? In order to ensure the validity of their findings, researchers usually measure more than one indicator for a particular concept. If two or more chosen indicators of a concept give similar results, it is likely that the measurements are giving an accurate—that is, valid—depiction of the concept.

Sociologists also must be concerned with the **reliability** of their research results. A measurement is reliable if a repeat of the measurement gives the same result. If a person is given a test daily, and every day the test gives different results, then the reliability of the test is poor. One way to ensure that sociological measurements are reliable is to use measures that have proved to be sound in past studies. Another technique is to have a variety of people carry out the data gathering so as to make certain that the results are not skewed by the tester's appearance, personality, and so forth. Surveys related to sexual matters offer an example of the challenge of retrieving reliable data. Respondents are sometimes uncomfortable answering questions about private matters, and the answers they give may be skewed depending on whether the questions are presented in written form or in a personal interview, or whether the interviewer is a man or woman. The researcher must be sensitive to all factors that affect the reliability of a study.

Sometimes sociologists want to gather data that would almost certainly be unreliable if the subjects knew they were being studied. An example would be a professor who wants to measure student attentiveness by observing how many notes are taken during class. Students who know they are being scrutinized magically become more diligent. *Unobtrusive measures* are those designed so that the people being studied cannot react to the fact of being studied. A professor

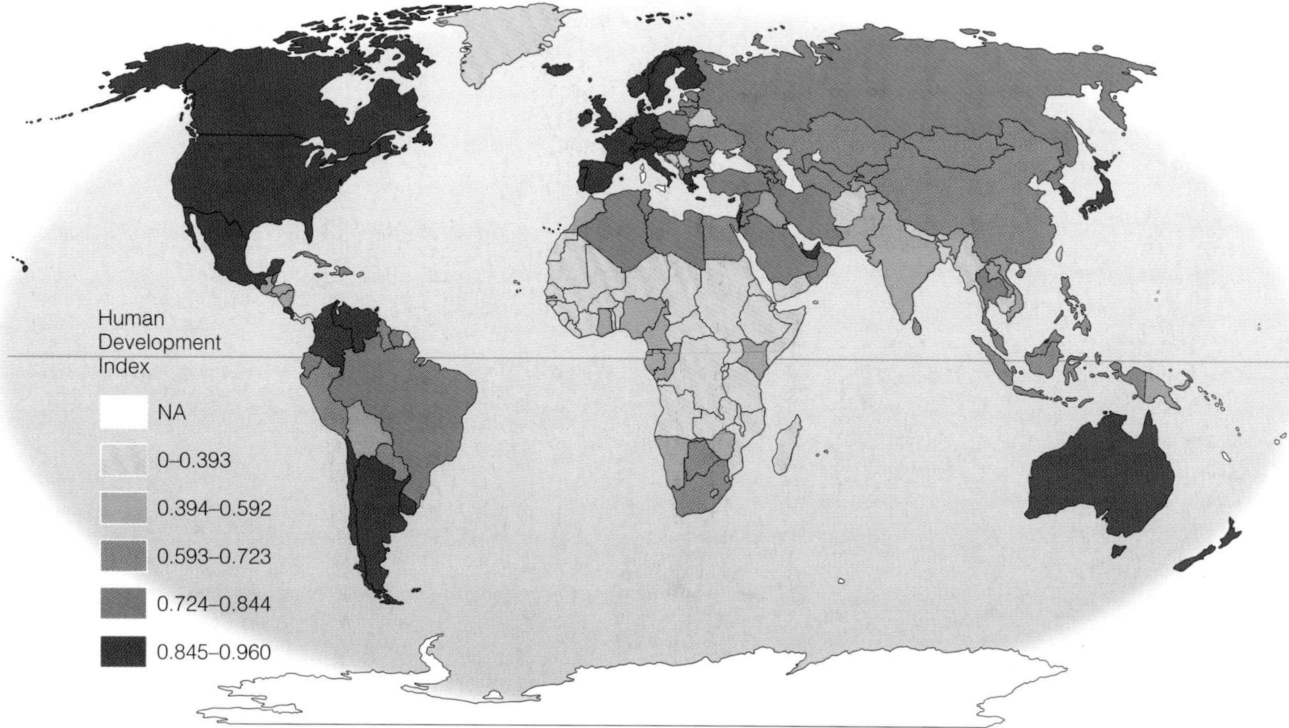

MAP 2.1 Viewing Society in Global Perspective: The Human Development Index

The Human Development Index is an indicator developed by the United Nations used to show the differing levels of well-being in nations around the world. The index is calculated using a number of measures, including life expectancy, educational attainment, and standard of living.

DATA: Human Development Index, UN online files at http://www.undp.org/undp/hdro/hdi1.htm

might measure notetaking unobtrusively by appointing someone in the back row to observe the students in front of them. Unobtrusive measures are desirable but difficult to arrange. The act of observation can be hard to hide, and one must be careful not to violate ethical standards and laws governing the subject's privacy or right of consent.

Consider another example: Suppose you wanted to measure the amount of alcohol consumption on the part of college professors. One way to do so would be to count the number of beer cans and liquor bottles in their garbage cans over a period of a month or two. Looking through someone's garbage might not be your idea of doing research, but sociologists have been known to go to great lengths to collect data! The garbage measure might not be totally valid, however, since some of the beverages could have been consumed by guests. You could correct for this to some extent by monitoring the number of visitors and devising some way to calibrate how much they might be expected to drink. Rummaging through garbage cans is an unobtrusive measure. It is also ethically unsound, since research subjects are entitled to give their *informed consent,* indicating their willingness to participate in a research study. Ethical issues are critical in research design and are discussed further at the end of this chapter.

Even excellent research designs are sometimes thwarted by the realities of field work. Officials guarding necessary data may refuse access or cooperation. Some steps in the research program may prove to be impossible. Surprising results may change initial assumptions, which is no reason to stop; in fact, it is the best reason to go on! Researchers make adjustments and change direction when new information arises. Designing research is a craft requiring a good sociological imagination, careful and logical thought, the ability to master details, the stamina and focus to execute a well-designed plan of action, and the willingness to be surprised.

Gathering Data

After research design comes data collection. During this stage, the researcher interviews people, observes behaviors, or collects facts that throw light on the research question. When sociologists gather original material, the product is known as *primary data.* Examples include the answers to questionnaires or notes made while observing group behavior. Often, sociologists rely on *secondary data,* that which has already been gathered and organized by some other party. A large collection of data is known as a data set or database and would include national opinion polls, census data, national crime statistics, or data from an earlier study made available by the original researcher. Secondary data may also come from official sources, such as university records, city or county records, national health statistics, or historical records. Historical archives are a source of secondary data that can give insight into social change or give historical perspective on contemporary issues.

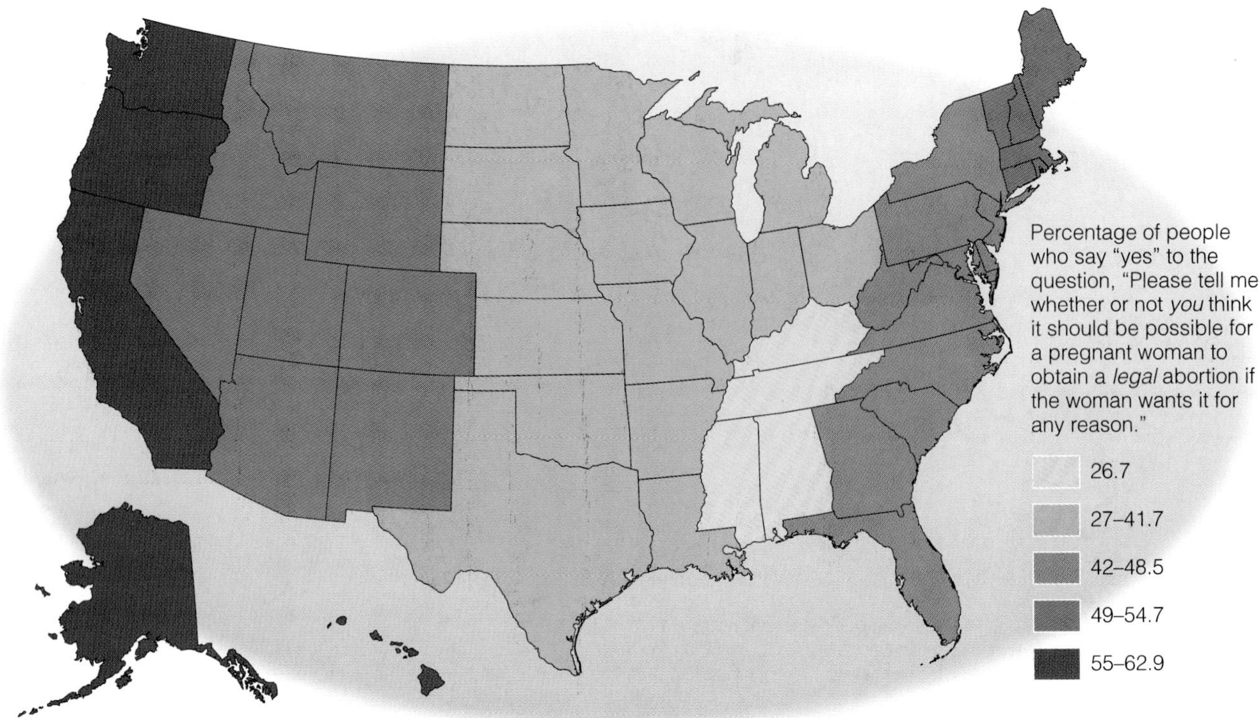

MAP 2.2 Mapping America's Diversity: Regional Attitudes Toward Abortion
Attitudes on various public opinions can vary by region. The General Social Survey shows that there are significant regional differences in attitudes toward legal abortion.

DATA: General Social Survey, 1994.

Since 1972, the National Opinion Research Center (NORC) at the University of Chicago has annually conducted the General Social Survey (GSS), an important source of secondary data frequently used by sociologists. A sample of about 1500 Americans is surveyed each year, producing a gold mine of data for secondary analysis. Questions are asked on a large range of topics. Basic demographic data are gathered, such as occupation, religion, and hours worked, as well as opinions on topics such as abortion, marijuana use, consensual homosexual sex, welfare spending, gun control, and capital punishment.

Analyzing these kinds of data, sociologists have been able to track how attitudes toward such topics have changed since 1972. For example, according to the General Social Survey, the percentage of Americans favoring abortion rights has increased steadily from a low of 33 percent in 1978 to 46 percent in 1994, although more recent data from Gallup (Roper Organization, 1996) suggest that the latest percent estimate may be somewhat less, depending on how the survey question is worded. Such data can be further analyzed to show variation by such things as race, gender, educational level, income, or regional origin, among other factors. The map "Mapping America's Diversity: Regional Attitudes Toward Abortion" shows the relationship between region and attitudes toward abortion; it shows, for example, that people in the Pacific region are considerably more likely to favor abortion than are people in the central South (National Opinion Research Center, 1996).

Another major source of raw material for sociologists has been census data generated by the U.S. Bureau of the Census. Census data yield detailed information on standards of living, income, educational attainment, language spoken in the home, birth and death rates, and many other specific items, broken down by race and ethnic group, gender, region, age, and many other variables. The census is a giant database from which a great deal of information about the United States can be extracted. Sociologists have used it extensively in their research.

An ongoing issue affecting the use of census data is the census undercount problem. Certain groups are more likely to be undercounted than others, such as transients and homeless people, individuals who fear the government (criminals, legal and illegal immigrants), those who dislike or distrust the government, and those who are unable to fill out a questionnaire because of illiteracy, mental illness, or other causes. It is known, for example, that members of minority groups are undercounted, particularly African Americans and Hispanics. One reason is that minorities tend to be less trusting of government bureaucracies such as the Census Bureau, with the result that their levels of compliance are lower than that of the general population. Sociologists need to estimate the extent of the undercount in census data and take it into account in their research.

Because many federal, state, and local government programs distribute money based in part on census data, the undercount problem has potentially serious repercussions.

Taking the census involves interviewing people in what is called a survey. Here, the interviewer from the Census Bureau is asking such survey questions of another person, called the respondent.

Cities and special-interest groups have been particularly active in addressing the inaccuracies in the U.S. census. If the census undercounts a city, then it will be shortchanged when public funds are distributed. In the early 1980s, the mayors of Detroit and New York City (cities with a high percentage of racial and ethnic minorities), actually sued the federal government over the census, claiming that their cities were particularly victimized by undercounting and seeking to raise the level of funds distributed based on census data. This court case, plus several similar ones, continued for some years in litigation until 1996, when the government's lawyers decided that estimates of undercounting could not be used and thus a city could not sue for more governmental funding on this basis (Greenhouse, 1996b).

Different kinds of data have different strengths and weaknesses. Public opinion polls, for example, are very useful for gauging changes in public attitudes. They use large samples, are repeated regularly, and are widely available. Poll data are limited, however, in that they give no context for people's responses—complex issues are often compressed into simple questions that generate abbreviated responses. There is also a tendency for people to respond to poll questions with the answers they think are most socially acceptable, a syndrome that subtracts from the validity of the poll. For example, poll data show a decline in group prejudice, as measured by the number of people who say they support equal opportunity. That decline does not necessarily translate into approval of social programs designed to contribute to equal opportunity, however. Two sociological researchers refined the question by asking more specifically whether people in the United States support equal opportunity and whether they support race-specific programs designed to improve the economic opportunities of African Americans and Latinos. Although the majority favor equal opportunity and oppose segregation, an overwhelming majority of White Americans oppose programs specifically designed to assist racial and ethnic minorities. Indeed, despite a steady decline in White prejudice, Black and White opinion is highly polarized on whether race-specific policies, such as affirmative action programs, should be supported (Bobo and Kluegel, 1993).

Analyzing the Data

After data have been collected, whether through direct observation or the use of existing data, they must be analyzed. **Data analysis** is the process by which sociologists organize collected data in order to discover the patterns and uniformities that the data reveal. The analysis may be statistical or qualitative. Survey data require statistical treatment, whereas a study based on an analysis of rap music lyrics would likely have a large qualitative component. When the data analysis is completed, conclusions and generalizations can be made.

If data have been gathered through a survey or interviews, they will likely have to be converted into numerical form. *Coding* is the process by which data are categorized and put into a form suitable for numerical analysis. Careful coding is critical since inaccuracies would pollute the data set. If interviews were done, tapes may have to be transcribed (transferred to print by word processing) and their information organized. A researcher may even have a coding scheme for quantifying or otherwise organizing the information from interviews.

Data analysis is labor intensive, but it is also an exciting phase of research. Here is where research discoveries are made. Sometimes while pursuing one question, a researcher will stumble across an unexpected finding, something sociologists refer to as *serendipity* (Merton, 1957). A serendipitous finding is something that emerges from one's study that was not anticipated, perhaps discovering an association between two variables that one was not looking for or discovering some pattern of behavior that was outside the scope of one's research design. Such findings can be minor sidelines to one's major conclusions or, in some cases, lead

These individuals are interviewing people by phone in a survey and at the same time entering the responses into the computer for analysis.

to major new discoveries. They are a part of the excitement of doing sociological research.

Computers have greatly facilitated the process of analyzing data. Computational operations that were once done by hand, perhaps over weeks, can now be performed by computer in seconds. Especially valuable is the ability to perform complex operations on large data sets, those containing many thousands or even millions of data points. Once, large data sets were stored on index cards. Data sorted by date might be duplicated and then laboriously resorted by region. Nowadays, computers can resort data by any number of variables instantly. Once data are entered in a spreadsheet or statistical program, they can be stored, retrieved, sorted, compared, and analyzed with ease. Perhaps most useful of all, they can be shared since digital data are easily reproduced, transported, and merged with other data. Government data are now instantly accessible on the Internet, and increasingly, researchers are creating public electronic archives for their own data and that of colleagues working in similar areas.

Reaching Conclusions and Reporting Results

When the data are gathered and analyzed, the background reading is done, and the files are organized, the researcher arrives at the final stage in research: developing conclusions, relating findings to sociological theory and past research, and reporting the findings in the professional literature.

BOX 2.1 SOCIOLOGY IN PRACTICE

What Causes Violence?

A BOMB in Olympic Park, drive-by shootings, nightly news of murders in the cities, children carrying guns to school: These all raise fears that the United States is an increasingly violent society. What causes the violence that we witness in society? The American Sociological Association (ASA), the national professional organization for sociologists, has addressed this question and identified violence as an important area for new sociological research. Using funds from the Spivack Program in Applied Social Research and Social Policy, the ASA sponsored a national workshop on the social causes of violence, asking these experts to develop new research questions that would shed new light on the causes of violence. With this new research, sociologists hope to identify social policies that would help reduce the violence in society.

What questions did the experts think were most important to address?

- How do we explain violence within and among different racial and ethnic groups? Too often statistical data on violence look only at two categories—White and Black. Other groups are vulnerable to violence, and we need to know more about them.
- How and why are poverty and inequality linked to violence? Sociologists already know much about this question, but are societies with more rigid social class boundaries—or those moving in that direction—most likely to experience increased violence?
- Why is violence prevalent among young people? Young people seem increasingly vulnerable to violence, yet this is an area that has not been studied much.
- What are the characteristics of different communities that foster or discourage violence? Sociologists know that social context (a community, a group, a society) is an important factor in shaping behavior, but little research has been done looking at communities, not just violent offenders, as an explanation of violence. Similarly, how do social institutions, especially families, schools, and workplaces, encourage or discourage violent behavior?

- What is the effect of the media on the promulgation of violence? This is a much debated question, especially since violence is increasingly used as a form of entertainment; there is somewhat more research evidence showing that violence in the media causes at least some violence, especially in children.
- How do different people, groups, and institutions respond to violence in effective or counterproductive ways? Answering this question might identify constructive ways to reduce violence.

These questions will help to craft future research for sociologists. If you were an expert on this ASA panel, what questions would you pursue? The answers will be important since they can show federal, state, and local groups where best to direct their resources if they want to reduce violence.

SOURCE: Levine, Felice J., and Katherine J. Rosich. 1996. *Social Causes of Violence: Crafting a Science Agenda.* Washington, DC: American Sociological Association.

An important question researchers will ask at this stage is whether their findings are generalizable. **Generalization** is the ability to draw conclusions from specific data and be able to apply them to a broader population. One asks: Do my results apply only to those people who were studied, or to the world beyond? Assuming that the results have wide application, one can then ask if the findings refine or refute existing theories. One can also ask if the research has direct application to social issues. Using the earlier example of prejudice and intergroup contact, if you found that increased contact between groups lessened group prejudice, what programs for reducing prejudice might your study suggest?

As you make conclusions from research, you typically relate your conclusions to the theoretical questions you asked from the start, and you are also likely to think about new questions that your research inspires. You might also think about the implications of your study, either for social policy or for further understanding of the issue you have addressed (see, for example, the box "Sociology in Practice: What Causes Violence?"). An important part of your research, however, is sharing your results with others.

Researchers report their results in several ways, among them discussing their research with others—either students in a classroom, colleagues at a regional or national conference, or perhaps in the media. Results are typically also reported in writing, perhaps via publication in a professional sociology journal, such as the *American Sociological Review,* the *American Journal of Sociology, Social Problems,* or *Gender & Society*. Professional journals generally come monthly, bimonthly, or quarterly. Research may also be published as a book, in which case it is called a *research monograph;* many sociologists also write books for more general audiences, sometimes even producing two books based on the same research—one that is more technical and methodologically detailed and another version that interprets the results in a more accessible style for a general readership.

Another alternative is to publish the research report for limited distribution, say, to interested colleagues; certain members of the U.S. Congress; or to some federal, state, or local agency. The research report must state clearly what results were obtained, what methodology was used, what past studies were considered, what theory or theories may have been tested, and what recommendations are made. The report should contain a section specifying why and in what contexts the research and conclusions are important. Finally, the report should evaluate the overall quality of the data and results, discuss any ethical issues raised by the research, and if appropriate, suggest what further research on the topic is necessary.

The Tools of Sociological Research

There are several tools sociologists use to gather data. Among the most important are survey research, participant observation, controlled experiments, content analysis, historical research, and evaluation research. These are not the only sociological research methods. Others are used less often, and innovative scholars can even invent entirely new ones, but the research tools we are about to examine are the ones that have consistently proved themselves most useful.

The Survey: Polls, Questionnaires, and Interviews

Whether in the form of a questionnaire, interview, or telephone poll, surveys are among the most commonly used tools of sociological research. Typically, a survey questionnaire will solicit data about the respondent, such as income, occupation, educational level, age, race, and gender, coupled to questions that throw light upon a particular research subject. In a *closed-ended questionnaire,* people must reply from a list of possible answers, like a multiple-choice test; *open-ended questionnaires* allow the respondents to elaborate on their answer.

Questionnaires are typically distributed to large groups of people. A fundamental requirement of successful questionnaires is an efficient mechanism for getting the questionnaires back. The *return rate* is the percentage of questionnaires returned of all those distributed. A low return rate introduces possible bias since the small number of people may not be representative of the whole group. The researcher must be clever about following up the original distribution and inducing recipients to complete and return the questionnaire.

Like questionnaires, interviews provide a structured way to ask people questions. They may be conducted face to face, by phone, even by electronic mail. Interview questions may be open-ended or closed-ended, though the open form is particularly accommodating if respondents wish to elaborate.

As a research tool, surveys offer characteristic advantages that are linked to their structured formats. The orderly presentation of questions and answers makes it possible to ask specific questions about a large number of topics and then to perform sophisticated analyses in search of patterns and relationships among variables. The disadvantages of surveys arise from their rigidity. Whereas data analysis may be greatly simplified by prescribed responses ("a,b,c,d, or none of the above"), the approved responses may not accurately capture the opinions of the respondent; furthermore, written surveys and even those conducted by phone fail to capture nuances in behavior and comportment that would be obvious and informative to a direct observer.

Finally, what people say and what they do are not always the same. Survey researchers must take pains to get answers that are truthful. Respondents may disguise their true opinion if it is not the most socially acceptable answer. During the Democratic presidential primaries in 1988, there was a period when Jesse Jackson performed substantially better in the exit polls than in the actual polls. (Exit polls are surveys of voters taken as they leave the voting place.) People voted one way and then minutes later claimed to have voted another. Jesse Jackson was closer to the presidency than any African American had ever been, and many White voters answering exit polls wanted to show that they were not racist and that they did not disapprove of an African American candidate, so they claimed to have voted for Jackson, although they had not. On issues where respondents may feel social pressure to answer one way or another, or to conceal their true opinion, it may be necessary to arrange the study so as to allow respondents to give their answers anonymously.

Participant Observation

A unique and interesting way for sociologists to collect data and study society is to actually become part of the group they are studying. This is the method of **participant observation.** Two roles are played at the same time: subjective *participant* and objective *observer*. Usually, the group is aware that the sociologist is studying them, but not always. The group might be a youth gang, religious sect, class of first-year medical students, or denizens of a corner bar. Participant observation is sometimes called field research, a term derived from anthropology. Like anthropologists, participant observers typically go to the places where research subjects are found and, to some degree, adopt their ways. "Entering the field," as it is called, can be among the most illuminating stages in the maturing of a sociologist.

Participant observation combines subjective knowledge gained through personal involvement and objective knowledge acquired by disciplined recording of what one has seen. The subjective component supplies a dimension of information that is completely lacking in survey data and similar techniques. The potency it adds to interpretations of human interaction has proved to be well worth the burden of managing the rich and sometimes subjective raw material.

Street Corner Society (1943), a classic work by sociologist William F. Whyte, documents one of the first participant observation studies ever done. Whyte studied the Cornerville Gang, a group of Italian men whose territory was a street corner in Boston in the late 1930s and early 1940s. Although not Italian, Whyte learned to speak the language, lived with an Italian family, and then infiltrated the gang by befriending the gang's leader, the pseudonymous "Doc." Doc was Whyte's *informant*, a person with whom the participant observer works closely to learn about the group. For the duration of the study, Doc was the only gang member who knew that Whyte was doing research on the group.

Most social scientists of the 1940s thought gangs were socially disorganized, deviant groups, but Whyte's study showed otherwise. He found that the Cornerville Gang, and by implication other urban street-corner gangs, was actually a highly organized minisociety, with its own social hierarchy, morals, and practices, and its own punishments for deviating from the norms or rules of the gang. The high degree of social organization in gangs has recently been shown again by Sánchez Jankowski's (1991) study of urban street gangs, mentioned at the beginning of this chapter.

Whyte made an interesting observation about the relationship between social rank and bowling scores that illustrates the depth of the social order. All gang members bowled together each week. Interestingly, the men's scores

The men in this bar, much like Jelly's Bar on Chicago's South Side as studied by sociologist Elijah Anderson, reveal status differences among themselves such as "regulars" versus "winos."

matched up fairly well with their social rank in the gang—from Doc, the leader, to the second-in-command ("Mike"), and on down to the lowest-ranking and least influential gang member, "Alec." If a low-ranking member got "out-of-line" and bowled "too well," he was sanctioned by "razzing" or ridicule.

Elijah Anderson, an African American sociologist who completed his doctoral work at the University of Chicago, did a participant observation field study of a local bar called Jelly's on Chicago's South Side (Anderson, 1976). During the study, Anderson hung out in the bar and on the corner with working and nonworking Black men. Unlike Whyte's subjects, the men at Jelly's knew they were being studied. Anderson socialized with them, drank with them, talked with them, and most of all, listened to them. As with any sort of participant observation field work, Anderson did not merely share their idle hours—after returning home from Jelly's, he wrote extensive field notes. Participant observation research is very time-consuming. Field work takes time, recording observations takes time, and both are at the expense of other professional and personal activities. The work may be as enjoyable as it is enlightening, but it takes great dedication to do successfully.

Like the earlier study by Whyte, Anderson's study of Jelly's Bar found a status hierarchy. The men themselves ranked their associates as either "regulars," "hoodlums," or "wineheads." The regulars prided themselves on having steady jobs and looked down on those who did not. They tended to be older men, mostly married or "shackin' up." Although a considerable amount of their leisure time was spent at Jelly's, they were likely to disclaim the importance of Jelly's in their social life. Hoodlums were less accepting of conventional values than the regulars. The hoodlums often intimidated the regulars and others, and their activities were known to involve stickups, burglaries, fencing stolen property, and running numbers. Many had jail records. They

participant observation

hung around Jelly's corner at all hours of the day and night. The wineheads were the lowest-status group, a kind of residual category within the larger group into which both regulars and hoodlums could, and often did, fall. In better days, wineheads had been successful hoodlums or regulars, but because of age, alcohol, or both, they had lost their social standing and were given no deference by other groups.

Debunking Society's Myths

Myth: People who are just hanging out together and relaxing don't care much about social differences between them.

Sociological research: Even casual groups, like the Black men who hung out at Jelly's Bar and the street-corner gang in William Foote Whyte's research, have organized social hierarchies. That is, they make distinctions within the group that give some higher status than others (Anderson, 1976; Whyte, 1943).

An interesting controversy in sociology is the insider–outsider debate, which raises the issue of whether the researcher should or should not be an actual member of the group, society, or culture that the researcher is studying.

The power of participant observation studies was shown dramatically in still another study by a Black woman sociologist, Judith Rollins, who studied Black domestic workers (Rollins, 1985). Rollins examined the relationships between Black domestic workers and their White employers by posing as a maid and hiring herself out to employers in the Boston, Massachusetts, area. She discovered that Black female domestics were treated by their employers as virtually invisible and as inferior. White employers, women as well as men, often openly discussed the domestics with other Whites as if the Black woman herself were not even present. One woman employer turned down the heat in the house when she left even though Rollins was expected to stay there. Through participant observation, Rollins was able to observe directly and intimately how social class status combines with race and gender to shape the world of the Black domestic worker.

There are a few built-in weaknesses to participant observation as a research technique. We already mentioned that it is very time-consuming. Mártin Sánchez Jankowski spent ten years doing field research to collect his data on White, Latino, and African American urban gangs. Once they leave the field, participant observers have to cull data from vast amounts of notes. Finally, participant observation studies usually focus on fairly small groups, and care must be taken not to generalize too widely from such a group. These limitations aside, participant observation has been the source of some of the most arresting and valuable studies in sociology.

Controlled Experiments

Controlled experiments are highly focused ways of collecting data and are especially useful for determining a pattern of cause and effect. To conduct a controlled experiment, two groups are created, an *experimental group,* which is exposed to the factor one is examining, and the *control group,* which is not. All other conditions must be equal for both groups. In a controlled experiment, external influences are either eliminated or *equalized,* that is, held constant. This is necessary in order to establish cause and effect.

Suppose you wanted to study whether violent television programming causes aggressive behavior in children. You could conduct a controlled experiment to investigate this question. The behavior of children would be the dependent variable; the independent variable is exposure to violent programming. To investigate your question, you would expose an experimental group of children (under controlled conditions) to a movie containing lots of violence (martial arts, for example, or gunfighting). The control group watches a movie that is free of violence. Aggressiveness in the children is measured twice: a *pretest* measurement made before the movies are shown, and a *posttest* measurement made afterward. You would take pre- and post-test measures on both the control and the experimental groups.

In one such series of studies (Berkowitz, 1974, 1984), an experimental group of children saw a seven-minute film clip showing boxers fighting (the clip was from the movie *The Champion,* starring Kirk Douglas). This movie showed an "accepted" form of violence, namely, boxing, not some deviant form such as chain-saw murder. A control group of children watched a neutral film clip. Afterward, aggressiveness was measured by seeing how willing each child was to deliver a painful electric shock to another child, and the amount, in volts, of the electric shock. Aggressiveness was also measured before seeing the movie, on both groups, to establish a baseline measure of aggressiveness. It was found that the children in the experimental group delivered more electric shocks and more voltage per shock than the control group children.

In another example of a controlled experiment, researchers wanted to know if viewing pornography makes men more violent toward women. The pornography industry has argued against a link between violence and pornography, but many feminist organizations argue the opposite. Who is right?

An insightful experiment was conducted to answer this question (Donnerstein and Berkowitz, 1981). Men were assigned randomly to three separate groups. Two were experimental groups, and one was a control group. Each was shown different footage.

- The control group saw a neutral film.
- The first experimental group saw an erotic film showing consensual sex between a man and a woman.
- The second experimental group saw a film that combined sex and violence, showing a man tying up a woman, tearing her clothes off, slapping her, and then raping her.

The researchers measured the effects of the movies on the men by testing the willingness of the men in each group to deliver an electric shock to a woman who posed as a subject for the experiment but who was secretly working with the researchers. A questionnaire given after the experiment was over confirmed that almost all men did not know that the woman was in alliance with the researchers.

There were striking results. The men who watched the neutral film or the purely erotic film had little willingness to shock the woman secretly collaborating in the experiment. The men who saw the sexually violent film were willing to deliver considerably more electric shocks and shocks of higher intensity. The conclusion was that sexually violent pornography causes men to behave with more aggression and hostility toward women; purely erotic film does not. Since all the men were randomly assigned across the three groups beforehand, you cannot conclude that the men in the group seeing sexual violence were more inclined beforehand to behave in a hostile way toward women.

Among its advantages, a controlled experiment can establish causation, and it can zero in on a single independent variable. On the down side, controlled experiments can be artificial. They are for the most part done in a contrived laboratory setting (unless it is a "field" experiment), and they tend to eliminate many real-life effects. Analysis of controlled experiments includes making judgments about how much the artificial setting has affected the results.

Content Analysis

Researchers can learn a vast amount about a society by analyzing cultural artifacts such as magazine stories and magazine ads, TV commercials, novels, biographies, soap operas, movies, popular music, including "rap" songs, and all other kinds of cultural communication. **Content analysis** is a way of measuring via such cultural artifacts what people write, say, see, and hear. The researcher studies not people but the communications they produce as a way of creating a picture of the society.

context analysis

Content analysis is frequently used to measure cultural change and to study different aspects of culture—manners, morals, slang, customs, and so on (Lamont, 1992). Sociologists also use content analysis as an indirect way to determine how social groups are perceived—they might examine, for example, how Asian Americans are depicted in television dramas or how women are depicted in advertisements.

Simonds (1988) applied content analysis to the "true confessions" in *True Story* magazine between 1920 (when the magazine began) and 1985. In this magazine, which is geared toward working-class White women, there is a prototypical story plot that involves a female protagonist married to a man who continually gets drunk and beats her. The woman enters an affair out of loneliness, becomes pregnant, and then has an abortion or miscarriage, all without her husband's knowledge.

Simonds tracked the change in frequency of this type of story from the inception of *True Story* in 1920. She also examined how the women in these stories dealt with blame and guilt. Simonds found seventy-five stories between 1920 and 1985 that fit within this genre. The number of such stories per decade was as follows:

Decade	Number of Stories
1920–1929	7
1930–1939	6
1940–1949	0
1950–1959	9
1960–1969	12
1970–1979	26
1980–1985	15 (This projects to 30 stories for the complete decade.)

We see that such stories were few in the 1920s and 1930s, decreasing all the way to zero in the 1940s, the years of World War II and its aftermath. Stories about taboo subjects such as pregnancy resulting from extramarital affairs became more frequent in the 1950s and 1960s, reaching high points in the 1970s and the first half of the 1980s. Among the researcher's interests was finding out whether the women protagonists blamed themselves for their predicaments or whether blame was attributed elsewhere. Simonds found that, in general, the women in these stories were likely to blame themselves. The researcher concluded that *True Story* reflected the powerlessness and desperation of working-class women. She also found that the stories allowed working-class readers to see their lives reflected in popular literature, whereas other mainstream women's magazines featured only higher-class women. Simonds concluded that articles like the ones in *True Story* gave working-class women a sense of expression and belonging. Similar analyses show that content analysis is a valuable tool for analyzing changes over time in portrayals of women's social roles (Reinharz, 1992).

Children's books have been the subject of many content analyses, in acknowledgment of their impact on the development of youngsters. Three sociologists compared the portrayal of Black Americans in children's books from the 1930s to the present (Pescosolido et al., 1997). The researchers focused on four things: (1) how frequently the images appeared, (2) how they had changed over time, (3) the portrayal of interaction between Black adults, and (4) whether multicultural portrayals were consistent across different kinds of children's books. The authors made four important findings. First, they found that there was declining representation of Black Americans from the 1930s through the 1950s with practically no representation from 1950 through 1964; beginning in 1964, there was increasing representation until the mid-1970s when the appearance of Black characters leveled off. Second, they found that the images of Black Americans do change significantly over time; interestingly, in the 1960s—a period of much racial protest—Black Americans were mostly portrayed in "safe," distant images. Third, they found few portrayals of Black adults in intimate, egalitarian, or interracial relationships. Finally, they found that award-winning books were more likely to include Black representation. By the way, a related study examining images of women in award-winning children's books found that Black-authored books are more likely than White-authored books to depict women as independent, nurturing, active, persistent, and attached to their communities (Clark et al., 1993). Together, these studies show how something as seemingly innocent as children's literature can produce distorted images of various groups in the population.

Content analysis has the advantage of being unobtrusive. The research has no effect on the subject being studied, since the cultural artifact has already been produced. Content analysis is limited in what it can study, however, since it is based only on mass communication—either visual, oral, or written. It cannot tell us what people really think about these images or whether they affect people's behavior. Other methods are required to answer those questions.

Comparative and Historical Research

Many sociologists are interested in how societies compare to one another, or how societies change and develop over long periods. For them, either comparative or historical research is the most appropriate research method. In comparative research, sociologists examine how features in two or more societies shape social behavior. For example, after comparing eight industrial countries, sociologists determined that poverty was lowest in the countries that had the most equal employment patterns between men and women and the lowest number of single parents (Casper et al., 1994). These kinds of *cross-cultural studies* (investigations that compare different societies) can help identify the features of society that produce specific sociological outcomes.

Historical research examines sociological themes over time. It is commonly done in historical archives, which come in many varieties. The official archives of government bodies are a deep and vast source of information, but compelling data may also be found in less obvious places—church records, town archives, the collected minutes and proceedings of societies, even private diaries. Oral histories have been especially illuminating, most dramatically in revealing the unknown histories of groups that have been ignored or misrepresented in other historical accounts. The sources of this sort of material are critical to its quality and applicability. For example, when developing an account of the spirituality of American Indians, one would be misguided to rely solely on the records left by Christian missionaries or U.S. Army officials. These records would give a useful picture of how Whites perceived Indian religion, but they would be a very poor source for discovering how Indians understood their own spirituality. The writings of a slave owner can deliver fascinating insights into slavery, but a slave owner's diary will certainly present a different picture of slavery as a social institution than will the written or oral histories of former slaves themselves.

Official records can also be misleading since they do not capture the experiences of all groups. Sociologists must be discriminating in how much credence they give to historical material, and they must be creative in discovering the material that brings light to the darkened past. Handled properly, comparative and historical research is rich with the ability to capture long-term social changes and is the perfect tool for sociologists who want to ground their studies in historical or comparative perspectives.

Evaluation Research

Many sociologists use their research to make social policy recommendations. On the opposite side of the coin, research is used to evaluate the effectiveness of social policy that has already been implemented. **Evaluation research** assesses the effect of policies and programs on people in society. If the research is intended to produce policy recommendations, then it is called policy research.

Suppose you want to know if an educational program is actually improving student performance. You could design a study that measured the academic performance of two groups of students, one that participates in the program and one that does not. If the academic performance of students in the program is better, and if the groups are alike in other ways (they are often matched to accomplish this), you would conclude that the program was effective.

Alternatively, you might want to discover which of several intervention programs is most effective in improving academic performance. You might use grades to compare the effectiveness of study skills, seminars, required study halls, or tutoring in math and science. In a study like this, academic performance is a dependent variable, and the intervention programs are independent variables.

A final example: Suppose you wanted to know whether reforms in child support laws actually increase well-being among children. In an important and timely study, sociologists Irwin Garfinkel and Sara McLanahan (1992) examined how forcing delinquent fathers—"deadbeat dads"—to pay

cross-cultural studies

child support affects the economic and emotional environment of children. The variables examined by Garfinkel and McLanahan included changes for the better in the financial status of the child's household, increased amount of contact between fathers and their children, and reduction in the amount of conflict between the parents. Along similar lines, a study of Princeton University women students who were themselves children of divorce found that daughters were more likely to have positive feelings toward their fathers if the father paid child support as scheduled (Cosse, 1995).

Evaluations of social policy can be controversial when there is disagreement about what indicators are appropriate measures of success. A policy designed to assist disadvantaged families may be judged a "success" by one group if it increases some measure of personal comfort or happiness in the families. Another group may consider well-being to be irrelevant, judging success solely by budgetary criteria—how much public money did the program save in the long run by creating healthier, better-adjusted citizens? Effective evaluation research anticipates and prepares to answer expected arguments during the research design stage.

Prediction, Sampling, and Statistical Analysis

Not all sociological research is based on scientific precision, but all sociologists learn to understand some basic scientific concepts and techniques. Central to scientific work is the ability to predict certain outcomes. Likewise, knowing how to draw a scientific sample and organize one's data into meaningful statements is critical to sociological studies. In this section, we review the basic concepts of prediction and probability in sociology, the importance of sampling, and some elementary uses of statistics in sociological research.

Prediction and Probability

The essence of science is prediction and explanation. If we understand the principles of a physical system, we can predict how it will behave. We can predict the outcome of chemical reactions and the time that the moon will rise, but can we ever predict what humans will think or say or do under specified conditions, or are human beings too unique and too mercurial to conform to scientific predictability? The answer to both questions is, roughly, yes. Human social, behavioral, and attitudinal characteristics can be measured, and within a margin of error can be predicted. This notion is a cornerstone of sociology; however, there are qualifications. Humans are not mechanical beings, and there is a special quality to the predictions we make about people.

Sociology analyzes, explains, and predicts human social behavior in terms of trends and probabilities. A **probability** is the likelihood that a specific behavior or event will occur. We cannot predict that a specific individual will die within the next five years, yet we can predict how *likely* it is that a person will die in the next five years knowing only his or her age, and we can predict how many people in a large population will die in the next five years to within a rather fine margin of error. Age is an excellent predictor of the chances, or odds, of dying: The older you are, the greater your chances of dying within the next five years. Actuarial tables commissioned by insurance companies use probability statistics and a wide range of data to tabulate life expectancies. Drawing on empirical data, probabilities have been calculated for the life expectancy of an African American, thirty-seven-year-old, tobacco-smoking, married, college-educated, self-employed male from a midwestern industrialized state (his life expectancy goes up significantly if he quits smoking). The probability that a person will live another five years is calculated by determining the number of people in the general population of a given age who actually die in that amount of time. If 85 percent of all people die within five years of the time they turn seventy, then we know there is an 85 percent probability that a seventy-year-old person will die within five years.

Many sociological conclusions are stated in terms of probability. Here are two examples:

- If you are a Black male, aged sixteen to twenty-nine, you are three times more likely to be arrested for a given offense than a White male of the same age from the exact same neighborhood and the same socioeconomic background, for the same offense and the same prior arrest record (Federal Bureau of Investigation, 1997; Chambliss, 1988).
- If you are a divorced woman with children, there is a 39 percent chance that you live in poverty (Kurz, 1995).

These statements reflect general patterns; however, it is not possible to say with certainty that a specific person will adhere to the pattern. This difference between pattern and prediction can be seen in research on violence. Data show that nearly all men who murder their wives have previously assaulted them (Browne and Williams, 1993); yet, sociologists cannot predict which of the men who assault their wives will go on to murder them. Sociologists can predict patterns of violence, but they cannot predict specific incidents of violence (Goleman, 1995).

Sampling

Sociologists typically study groups, but often the groups they want to study are so large or so dispersed that research on the whole group is impossible. To construct a picture of the entire group, they take data from a subset of the group and extrapolate to get a picture of the whole. A **sample** is any subset of a population. A **population** is a relatively large collection of people (or other units) that a researcher studies and about which generalizations are made. Suppose a sociologist wants to study the students at your school. All the students together constitute the population being studied. A survey could be done that reached every student, but conducting a detailed interview with every student would be highly impractical. If the sociologist wants the sort of information that can be gathered only during a personal interview, she would study only a portion, or sample, of all

sample

the students at your school. This is how major national surveys like the Gallup Poll work. You or your family may never have been contacted by Gallup pollsters, yet the picture of the public that they derive from their sample accurately reflects the opinions of the larger population.

How is it possible to draw accurate conclusions about a population by studying only part of it? The secret lies in making sure that the sample is *representative* of the population as a whole. The sample should have the same mix of people as the larger population, and in the same proportions. If the sample is representative, and if it is large enough to overcome statistical quirks, then the researcher can generalize what she finds from the sample to the entire population. For example, if she interviews a sample of one hundred students and finds that 10 percent of them are in favor of a tuition increase, and if the sample is representative of the population, then she can conclude that about 10 percent of *all* the students at your school are in favor of a tuition increase. Note that a sample of five or six students would probably result in generalizations of poor quality because the sample is not large enough to be representative.

Samples should have the same overall composition as the population from which the sample is drawn. That way, the sample is likely to be representative. If the population at your school is 60 percent female and 40 percent male, the survey sample should have the same female-to-male ratio. In a representative sample of exactly one hundred people, approximately sixty of those people would have to be female and approximately forty would have to be male. When a sample is *not* representative—if it contains, say, eighty females and twenty males—then it is said to be *biased.* A sampling bias is a deviation from an expected or proper value that results from a systematic error in sampling procedure. Even if the researcher got an even split between females and males—if she ended up with exactly fifty of each in the sample of one hundred—the sample would still not be representative. It might look "fair," but it would be biased in the direction of too many males. Of course, you won't always know the exact representation of different groups in the population you are studying, but in trying to draw representative samples, sociologists try to reflect the population they are studying in their sample and to be aware of any potential limitations in a given research design.

A biased (nonrepresentative) sample can lead to grossly inaccurate conclusions. A famous example is the Gallup Poll done in 1948 just before the November presidential election. After surveying U.S. voters, Gallup confidently predicted that Thomas E. Dewey, the Republican candidate and governor of New York, would win the election over Harry Truman, the Democratic candidate. Faith in the Gallup Poll was strong enough that one newspaper, the *Chicago Tribune,* went to press late on the night of the election with a banner headline wrongly proclaiming Dewey the winner! Despite the polling data, Harry Truman took office as the thirty-third president of the United States.

How could the Gallup researchers have made such an embarrassing error? The answer is that they inadvertently used a biased sample. In the three prior presidential elections, the Gallup Poll organization had used a *quota sample*—a set number of people (the quota) were polled in a selection of key counties across the nation—and had successfully predicted the winners of the elections. The same sample was used again in 1948, but in the 1930s and 1940s, people had begun to migrate in great numbers from rural farms and towns to the cities. The United States had become less rural and more urban, and also more ethnically diverse. Because the Gallup organization did not revise their quota sample to reflect the change in the U.S. population, rural people were *overrepresented* in the poll and urban people were *underrepresented.* Urban people proved more likely to vote Democratic than rural people, Gallup missed the changing trend, and the photo of a smiling president-elect Truman holding aloft the headline "Dewey Wins!" has become a holy icon reminding researchers of the consequences of botched data gathering.

A smiling Harry S. Truman, who was elected president of the United States in 1948, holds up the infamous morning newspaper headline, based on an erroneous Gallup Poll result, stating incorrectly that his opponent (Dewey) has won the election.

A more recent example of sampling bias comes from the widely discussed reports on women, sexuality, and love done by Shere Hite. Her first study, *The Hite Report: A Nationwide Survey of Female Sexuality* (1976) was based on the responses of 4500 women who answered a questionnaire originally distributed to 100,000 women. Hite's conclusion that women were largely dissatisfied with their sexual relationships with men made her quite a celebrity. Her books have been the subject of numerous talk shows, and she has made many media appearances, but her sampling simply does not stand up to scientific scrutiny—a point, by the way, with which she agrees. Her survey sample was highly self-selected, based only on women belonging to women's organizations, and her questions, according to other social scientists, were ambiguous and overly general. Although 4500 responses sounds like a large number, it represents only 4.5 percent of the 100,000 questionnaires sent out—a very low *return rate*. Thus, fully 95.5 percent of the 100,000 women did *not* respond to the survey. Such a low return rate alone is enough to bias the sample to a very great degree.

Hite's follow-up book, *The Hite Report: Women and Love* (1987), also based on the earlier survey, concludes that the overwhelming majority of women are emotionally harassed by men, unhappy with their love lives, and desirous of more verbal closeness with their male partners. She said, famously, that 70 percent of women married five years or longer have extramarital affairs, a figure generally scorned by more careful practitioners. Poor sampling has not stopped sales of her books; her first book alone earned the author $2.5 million (Wallis, 1987). Her research has hit a nerve, but that does not make her pronouncements reliable, at least not in a scientific way. They may trigger some insight or suggest grounds for a new study, but they are not generalizable.

How does a researcher ensure that a sample is representative? The best way is to make certain that the sample population is selected randomly. A scientific **random sample** gives everyone in the population an equal chance of being selected. The following example demonstrates the usual routine for creating a scientific random sample. Assume that your school has exactly 10,000 undergraduates enrolled, both part time and full time. You decide to do a survey, but you have time to interview only one hundred students. One hundred is the sample size (in professional notation, N = 100). To make it a scientific random sample, you would follow these steps:

1. Assign everyone a number from 1 to 10,000.
2. Pick one hundred numbers at random. You could do this by laboriously numbering 10,000 slips of paper and plucking 100 of them from a (large) hat, but most samples today are created using a computer and a random number generator.

Translating everyone into numbers and picking at random gives everyone an equal chance of being chosen for the sample. The sample will therefore be truly representative of the population. Other techniques for gathering random samples, such as tossing darts at a list of names, may introduce unexpected elements of nonrandomness that are eliminated by deploying the computer-generated number technique.

Researchers occasionally have need to deliberately *oversample* some respondents. In certain years, for example, the General Social Survey oversamples Black respondents to be sure that enough Black respondents are in the sample to allow for detailed analysis of Black public opinion. Blacks are 12 percent of the population, and in a sample of 1500, there would be about 180 Black respondents. This is an adequate sample size for some information, such as overall comparisons of Blacks and Whites, but it is not large enough to do detailed analyses that break down the Black population into subsets: rich, poor, old, young, and so on. So the researcher would probably want a larger sample of Black persons, namely, an oversample of Blacks. Likewise, if you were doing a survey of sexual attitudes on your campus using a sample of one hundred students, you might want to oversample lesbian and gay respondents. Gays and lesbians are likely to be 3 to 10 percent of the population (and not all of that percentage will voluntarily identify themselves). Without oversampling, the number of gay or lesbian students likely to appear in a random sample of one hundred are too few to allow meaningful analysis of gay and lesbian attitudes.

Quite often, striking and controversial research findings prove to be distorted by inadequate sampling. The so-called man-on-the-street survey, much favored by radio and TV news reports, is the *least* scientific type of sample, and the least representative. Generally, one cannot find a representative sample of people, all of whom happen to be in one particular place at one time. A sample of people in a lively pub, for example (a common venue for man-in-the-street interviews), is likely to underrepresent the old, the young, most parents of young children, the studious, the sober, and people who work evenings, as well as many other groups. It is likely to *over*represent people who go to lively pubs. TV commentators often erroneously say their interviewees were picked at "random," but it is not random in the scientific sense of the word.

Statistics in Sociology

Not all sociological research is quantitatively based, but quantitative research is an important part of sociology, and all sociologists must have at least basic quantitative skills to interpret the research in their field. Although a growing number of sociologists work at high levels of mathematical sophistication, all sociologists need to understand at least some fundamental statistical concepts.

The fundamental statistical tools of research are the percentage, rate, mean, and median. A **percentage** is the same as parts per hundred. To say that 22 percent of U.S. children are poor tells you that for every one hundred children randomly selected from the whole population, approximately twenty-two will be poor. A **rate** is the same as parts per some number, such as per 10,000 or 100,000. The homicide rate in 1996 was 7.4 killed per 100,000 (Federal Bureau of Investigation, 1997). That means that for every 100,000 in the population, approximately 7.4 were murdered. In a population of 200,000, the number of murders expected would be $2 \times 7.4 = 14.8$. A rate is meaningless without knowing the numerical base on which it is based—it is always the number per some larger number.

Debunking Society's Myths

Myth: Violence is at an all-time high.

Sociological research: Although rates of violence are high in the United States, at least twice before—1930 and 1980—the murder rate was at a comparable level. It is also true, however, that rates of violence in the United States are higher than in any other industrialized nation (Levine and Rosich, 1996).

A **mean** is the same as an average. Adding a list of fifteen numbers and dividing by 15 gives the mean. The **median** is often confused with the mean but is actually quite different. The median is the midpoint in a series of values arranged in numerical order. In a list of fifteen numbers arrayed in numerical order, the eighth number (the middle number) is the median. If there were sixteen numbers, the median would be the number half way between the eighth and ninth numbers. In some cases, the median is a better measure than the mean because the mean can be skewed by extremes at either end. Another often-used measure is the **mode,** which is simply the value that appears most frequently in a set of data.

Let's illustrate the difference between mean and median using national income distribution as an example. Suppose that you have a group of ten people. Two make $10,000 per year, seven make $40,000 per year, and one makes $1 million per year. If you calculate the mean, it comes to $130,000. The median, on the other hand, is $40,000—a figure that more accurately suggests the income profile of the group. That single millionaire dramatically distorts, or skews, the picture of the group's income. If we want information about how the group in general lives, we are wiser to use the median income figure as a rough guide. Note that here the mode is the same as the median: $40,000.

Sociologists frequently choose to relate different variables to each other. **Correlation** is a widely used technique for analyzing the patterns of association, or correlations, between pairs of variables like income and education. We might begin with a questionnaire that asked for annual earnings (Y) and level of education (X). Correlation analysis delivers two dimensions of information: It tells us the *direction* of the relationship between X and Y, and the *strength* of that relationship. The direction of a relationship is positive (that is, a positive correlation exists) if X is low when Y is low *and* if X is high when Y is high. But there is also a correlation if Y is low when X is high (or vice versa); this is a negative or inverse correlation, but it is a correlation, nonetheless. The strength of a correlation is simply how closely or tightly the variables are correlated or associated, regardless of the direction of correlation. For example, if all those persons who were low in education (X) were also low in earnings (Y), then the correlation is strong and positive, but if most (but not all) of those low in education were low in earnings, then the correlation would still be positive, but less strong.

When interpreting correlations, one must realize that a correlation does not necessarily imply cause and effect. A correlation is simply an association, one whose cause must be explained by other means than simple correlation analysis. A *spurious correlation* exists when there is no meaningful causal connection between apparently associated effects. A famous example is the finding that the number of women reporting physical assaults by their husbands goes up on Super Bowl Sunday. You might conclude, wrongly, that watching the Super Bowl causes husbands to be violent. In fact, this is a spurious correlation. The real correlation is between wife beating and alcohol consumption. Men watching the Super Bowl are prone to drink more than usual. It is the increased consumption of alcohol that correlates meaningfully with the rise in wife beating: The more men drink, on average, the more likely they are to become violent with their spouses.

TABLE 2.1 *A CROSS-TABULATION OF OPINION ON GUN CONTROL BY GENDER*

Question asked: "Do you feel that the laws covering the sale of firearms should be made more strict, less strict, or kept as they are now?"

	More strict	Less strict	Kept as they are	No opinion	Totals
Women	75%	4%	20%	—	100%
Men	59%	10%	31%	1%	100%

Another widely used method for analyzing sociological data is *cross-tabulation,* in which data are broken down into subsets for comparison. Table 2.1 reports on a Gallup Poll that asked people, "Do you feel that the laws covering the sale of firearms should be made more strict, less strict, or kept as they are now?" The answers are broken down by gender. This table cross-tabulates the answers, "more strict," "less strict," "kept as they are," and "no opinion," with the gender of the respondent, male or female. The table is thus a cross-tabulation of opinion on gun control by gender (Roper Organization, 1994).

Did men and women answer the question differently? The cross-tabulation shows that they did: A large majority of women (75 percent) thought the laws should be more strict, 4 percent said less strict, and 20 percent thought they should be kept as they are. In contrast, fewer men, though still a majority at 59 percent, said the laws should be more strict, only 10 percent said they should be less strict, and 31 percent thought they should be kept as they are. Thus, women are more likely than men to think that gun control is not as strict as it should be. Survey answers might differ not only by gender, but also by race, class, region, or an infinite variety of other variables. By cross-tabulating, the researcher is able to discover such relationships among the variables.

THINKING SOCIOLOGICALLY

Suppose that you asked a *random sample* of students on your campus if they approved or disapproved of premarital sex. What questions would you ask, and what would you expect to find? Now suppose you extended your sample to include a campus very different from your own (perhaps one with more commuting adult students or a religiously affiliated college). What *hypotheses* would you suggest if you were planning to do one cross-tabulation by age and one by religion?

The Use and Misuse of Statistics

Statistical information is notoriously easy to misinterpret, willfully or accidentally. Unfortunately, statistics put forward by interest groups are often of little real value because they are incomplete, out of context, or misinterpreted. Worse than misinterpretation are statistics that are actually falsified. "Fudging" is the term for falsifying statistics, and it includes any tampering with actual data. Fudging is not only a misuse of statistics, it is intellectual vandalism, and in some cases it is even a criminal offense.

Examples of some statistical mistakes include:

- *Citing a correlation as a cause.* A correlation reveals an association between things, nothing more. Correlations do not necessarily indicate that one thing causes the other. A warning often used by sociological researchers is: *Correlation is not proof of causation.*
- *Overgeneralizing.* Statistical findings are limited by the extent to which the sample group actually reflects or represents the population from which the sample was obtained. Generalizing beyond the population is a misuse of statistics. Studying only men, and then generalizing conclusions to both men and women, would be an example of overgeneralizing.
- *Building in bias.* In a famous advertising campaign, public taste tests were offered between two soft drinks. A wily journalist verified that in at least one site, the brand sold by the sponsor of the test was a few degrees colder than its competitor when it was given to testers, which helped its scores. Statistics derived from this test were bunk. In a similar but less loathsome vein, bias can be built into a questionnaire by little more than careless wording.
- *Faking data.* Perhaps one of the worse misuses of statistics is actually making up, or faking, data. A famous instance of this occurred in a study of separated identical twins (Burt, 1966). The researcher wished to show that despite their separation, the twins remained similar in certain traits such as measured intelligence, thus suggesting that their (identical) genes caused their striking similarity in intelligence. It was later shown that the data were fabricated (Kamin, 1974; Hearnshaw, 1979; Taylor, 1980, 1995; Mackintosh, 1995).
- *Using data selectively.* Sometimes a survey includes many questions, but the researcher reports on only a few of the answers. Doing so makes it quite easy to misstate the findings. Often, researchers will not report findings that show no association between variables, but these can be just as telling as associations that do exist. For example, researchers on gender differences typically report the differences they find between men and women, but seldom publish their findings when the results for men and women are identical. This tends to exaggerate the differences between women and men, and falsely confirms certain social stereotypes about gender differences.

One particularly common manifestation of the selective use of data is the suppression of findings that do not fit a general pattern; yet, sometimes it is the anomalous finding that is most interesting or revealing—such as the lone dissenter in a drug-ridden school who is not on drugs. Unlike legal experts who are expected to argue a particular point of view to make a case, scientists, including sociologists, attempt to describe a phenomenon and then understand it, not just convince an audience that they are right. Statistics are like snapshots used as evidence in court. Context is critical—when, where, and how was the snapshot taken, and by whom? What exactly does it show? Adding, deleting, or distorting the snapshot in any way is unacceptable. Statistics are numerical snapshots of reality, and great care must be taken that they are not presented in a way that gives a distorted picture.

Is Sociology Value-Free?

The topics dealt with by sociology are often controversial. People have strong opinions about social questions, and they may have deeply felt commitments. In some cases, the settings for sociological work are highly politicized. Imagine spending time in an urban precinct house to do research on police brutality or doing research on AIDS and sex education in a conservative public school system. Under these conditions, can sociology be scientifically objective? How do researchers balance their own political and moral commitments against the need to be objective and open-minded? Sociological knowledge has an intimate connection to political values and social views. Often, the very purpose of sociological research is to gather data as a step in creating social policy. Can sociology be value-free? Should it be value-free?

This is an important question without a simple answer. Most sociologists do not claim to be value-free, but they do try to produce objective research. It must be acknowledged that researchers make choices throughout their research that can influence their results. The problems sociologists choose to study, the people they decide to observe, the research design they select, and the outlets through which they choose to disseminate their results can all be influenced by the values of the researcher; thus, it is appropriate to ask: Do politics poison research?

Political Commitments and Sociological Research

Sometimes a political component is a valuable instigator to research. For many years, sociological research was *androcentric*—that is, it centered on the experiences of men. Without necessarily being aware of it, sociologists, almost all of whom were men, often studied only males in surveys or looked at their subjects through the perspectives and experiences of men. The result was that much research either excluded women altogether or did not represent their experience accurately. Feminist scholars have asked if the concepts and theories generated from men's experiences are entirely valid when applied to women's lives. Based on these stirrings, new questions have been asked, new data generated, and new theories and concepts developed that have transformed sociological thought (Andersen, 1997). In this case, political motivations actually improved the scientific grounding of sociology by removing an artificial constraint on its breadth. Likewise, Asian American, African American, Native American, and Latino studies, as well as gay and lesbian studies, have raised new questions for sociology and have generated insights that extend the frontier of sociological knowledge.

This does not mean that research should be driven by politics. The result would be work that reproduced researchers' opinions rather than testing their theories. The goal is not to confirm our political values, but to generate new knowledge. As Weber pointed out, this may require questioning the status quo. Sociology cannot be perfectly value-free, but researchers have an obligation to make their research as objective and value-free as possible. Sociologists should not deny the existence of their own values and biases; they should make them known. They should also ask many questions, be open-minded, and listen to the evidence.

Research is produced in specific social, historical, political, and economic contexts. The context can influence research findings. For example, can scientists at a major chemical company objectively assess the impact of chemical pollution on the environment? They can if they have the freedom and independence to pursue all possible research questions; however, if the company does not allow publication of results unless they are favorable to the company's interests, and if it suppresses incriminating research, then it is unlikely that the organization will produce objective research. To preserve objectivity, it is necessary that investigators be free to ask questions without political constraint.

Controversy would seem to be an unmistakable signal that something deserves more study, but it can be a heavy constraint on the freedom of a researcher. During the late 1980s, a study of adolescent attitudes about sexuality was aborted because officials in the Bush administration, fearing that asking young people about sex would make them more promiscuous, cut off the funding. As the box "Doing Sociological Research: Freedom in Research" shows, this incident seriously violated scientific standards for open inquiry. To produce objective research, one must have independence to reach scientific conclusions.

BOX 2.2 DOING SOCIOLOGICAL RESEARCH

Freedom in Research

SOCIOLOGICAL research is often the foundation for knowledge about current social issues. Without the ability to conduct open research, these issues can be misunderstood. Academic freedom is a concept that protects the rights of researchers to pursue knowledge, no matter how controversial. Without academic freedom, knowledge is compromised. In the early 1990s, sociologists became concerned when the federal government refused to fund a survey of teen sexual attitudes and behaviors. Although the proposed research had received favorable reviews from the social scientists who reviewed the research proposal, the government refused to fund the study, arguing that it would encourage sexual promiscuity. In response, the Executive Officer of the American Sociological Association, Dr. Felice Levine, wrote an editorial discussing the importance of open inquiry in research. Dr. Levine wrote:

> We are now confronted with a difficult situation where policymakers, in the face of experiencing discomfort about a message (in this instance, the realities about sexual patterns and practices), limit the production of sound knowledge and essentially beat up on the messenger. How sad indeed for the integrity of science, for the health and welfare of society, for our national commitment to procedural fairness, and for our very trust in public officials. . . . Whatever are our individual areas of research, teaching, policy, or practice, as sociologists and as concerned citizens, we are all adversely affected by acts and maneuvers aimed at imposing a political agenda on what is essentially a scientific activity. . . . What is at issue is not the merits of this or any other prevention program, but the fragility of support for research generally and social research in particular on topics (like sexual behavior) that create discomfort or challenge strongly held beliefs or assumptions. . . . As a learned society and discipline, we need to protect the free exploration of and flow of ideas and the advancement of knowledge about all topics of importance, whether or not these topics are harmonious with the values of certain interests or groups. . . . In the face of the political violence of 1968 surrounding the assassinations of Martin Luther King and Robert Kennedy, there was a compelling Herb Block cartoon of the Lincoln Memorial with President Lincoln holding his head in his hands in grief and dismay. As news spread about the cancellation of the ATS [American Teenage Study] study and the likely threat to all social research on sexual behavior, I found myself walking past the statue of Albert Einstein, who sits on the lawn of the National Academy of Sciences (just steps from the Lincoln Memorial). He too for just one moment seemed to be holding his head in disbelief.

SOURCE: Levine, Felice J. 1991. "The Open Window." *Footnotes* (December): p. 2, American Sociological Association. Used by permission.

Claiming to be value-free when you are not is probably more damaging to your research credibility than making your commitments clear. Sociologists should take care not to hide their assumptions and should try to make apparent their possible sources of bias, including identifying sponsors of research, should they exist (such as government agencies and corporate sponsors). Of course, we are not always conscious of the assumptions we make. Those who produced androcentric research studies, for example, were probably unaware that they were doing so. Only with the development of feminism did scholars question the assumptions they were making about women. In the spirit of objectivity, scholars should try to be aware of how their experiences, values, and commitments shape the research they do. Adopting fresh points of view is probably the best way to do this. Systems of privilege, for example, may be least visible to those who benefit from them; to understand how inequality affects all groups, we would want to see how the oppressed experience the world rather than studying society solely through the eyes of the dominant class. Open debate and the inclusion of new perspectives are essential to producing objective sociological knowledge.

The Insider–Outsider Debate

Assuming that you can overcome your personal values in favor of objectivity, another question emerges: Is it possible to understand a person or group whose experience is greatly different from your own (Merton, 1972)? Do you have to be an "insider" in order to understand the people you are studying? Does one have to be a woman to understand the experience of women? Does one have to be Asian American to understand what it is like to be Asian American in the United States?

No one will ever have another's experience. If only insiders could understand the experiences of a group, we would never be able to empathize with others or learn from their experiences. In fact, insiders can have a distorted understanding of their own experience, the sort of distortion that is overcome by stepping away from taken-for-granted assumptions.

Max Weber (1864–1920) argued that sociologists could understand others through *verstehen,* a form of understanding achievable when researchers imagine themselves in the place of their subject (Chapter 1). Verstehen, like empathy, is a way of sharing in the experiences of other people. Verstehen is an essential part of doing good research, just as empathy is an essential part of human relationships.

The insider–outsider debate has been especially seen in questions about studies of race. Can White sociologists do meaningful research on the experience of oppressed groups? Latinos, African Americans, Asian Americans, and Native Americans as research subjects may be suspicious of Whites, making them reserved in interviews, or they may fail to absorb Whites fully into their social interactions. Some have concluded that minority researchers are better positioned to generate new insights and hypotheses, especially ones that challenge the traditional frameworks on which scholarship about minorities has been based (Baca Zinn, 1979).

Other dilemmas are posed for minority scholars, however, as the Chicana sociologist Maxine Baca Zinn learned when she was doing field research about Chicano families. Being Chicana herself, she faced expectations from the Chicano community that White scholars might not have met—such as the belief that she should maintain her connection to the community after the study was over and that she should provide assistance, financial and other, to community members.

Regardless of one's identity, doing research requires trust between the researcher and the study population. Although we can never completely share the experiences of others, we can cultivate our ability to understand and interpret those experiences. Luckily so—otherwise we could not comprehend cultures other than our own, and social understanding, including sociological knowledge, would be impossible.

Research Ethics

Sociological research often raises ethical questions. In fact, ethical considerations of one sort or another exist with any type of research. In a survey, the person being questioned is often not told the purpose of the survey or who is funding the study. Is it ethical to conceal this type of information? In controlled experiments, deception is often employed: Many experiments depend on respondents giving natural, unconsidered responses to staged situations. Researchers often reveal the true purpose of an experiment after it is completed in a session called a *debriefing.* The deception is therefore temporary. Does that lessen the potential ethical violation?

Participant observation research, if done without the knowledge of the people being studied, is also a form of deception. Whyte's street-corner study is an example. A participant observer may choose to work covertly if it appears that the known presence of a researcher would affect the behavior of the group, tainting the observer's findings. None of the Cornerville Gang except Doc knew they were being studied. Years later, however, the gang members found out about the study, read the book, and became angry at Whyte for "fooling" them and trading on their relationship to get famous. The discontent of the Cornerville Gang confirms that an ethical issue exists. Do researchers have the right to hide their intentions? Did Whyte have the right to decide that the irritation of the Cornerville Gang (which was not unexpected) was less important than the needs of the study? Did gaining the collaboration of the gang leader, Doc, bring with it the right to study the rest of the group secretly?

When Judith Rollins did her research on Black women domestic workers, she posed as a maid and did not tell her employers that she was a PhD student writing her dissertation. Had she done so, her study would have been ruined since surely the subjects would have acted differently toward her. Was this unethical? Most sociologists would say no, as long as no harm came to her subjects. The definition of "harm," however, can be debated. Some of Judith Rollins's subjects might have been quite embarrassed if they had recognized themselves in her study, especially if they felt that others might recognize them, too. Does that count as mental harm? Most sociologists would say no, or that any harm caused was extremely minor. The professional code of ethics is clear that if a research subject is at risk of physical, mental, or legal harm, then the subject must be informed of the rights and responsibilities of both researcher and subject. Sociologists also take measures not to identify their subjects through the use of pseudonyms or without the use of names at all.

THINKING SOCIOLOGICALLY

Based on your knowledge of Judith Rollins's research, would you say it is *value-free*? Drawing on the concept of *verstehen* from Max Weber, would you say one has to be a Black person to understand Rollins's observations of Black domestic workers and the employers?

A very important ethical dilemma emerges when sociologists study topics that may be outside the law. What if legal authorities pursuing a criminal investigation demand the release of confidential data and the identities of research subjects? This has been put to the test in several cases. The most recent involved a sociology graduate student, Rick Scarce, who was doing research on animal rights activists. Some animal rights groups have resorted to violence, bombing laboratories and threatening scientists who use animals in their studies. One of Scarce's informants was accused of vandalizing laboratories. A federal grand jury demanded that Scarce testify about what the accused activist had revealed to him in interviews. Scarce refused and was jailed for more than six months for contempt of court (Monaghan, 1993).

The law is not yet clear on whether sociologists must reveal the identities of their subjects or hand over their data to investigating authorities. Journalists have long claimed a special legal right to protect their sources. Sociologists and other scholars are claiming the same rights to protect their research subjects. The courts have not decisively acknowledged the rights of either. The American Sociological Association has taken the position that guarantees of privacy and confidentiality to research subjects are essential to open scientific inquiry, and therefore must be protected under the law. At this point in the development of the law, judges are making decisions on a case-by-case basis whether scholars and journalists can be forced to testify against people from whom they have acquired information (American Sociological Association, 1993). Such cases show that questions about research ethics has two sides—protecting the rights of individuals, on one side, and guaranteeing the possibility of open scientific inquiry, on the other, particularly when the subject is controversial or sensitive.

CHAPTER SUMMARY

- Sociological research is the tool used by sociologists to answer questions. The research method one uses depends on the question asked.
- Sociological research is derived from the *scientific method,* meaning that it relies on empirical observation and, sometimes, the testing of *hypotheses.* The research process involves several steps: developing a research question, designing the research, collecting data, analyzing data, and developing conclusions. Different designs are appropriate to different questions, but sociologists have to be concerned about the validity, reliability, and generalization of their studies. *Validity* refers to whether something accurately measures the concept being studied. *Reliability* is whether the same results would be found by another researcher using the same measurement. *Generalization* means being able to make claims that extend beyond the specific observations of a given research project.
- The most common tools of sociological research are *surveys* and *interviews, participant observation, controlled experiments, content analysis, comparative* and *historical research,* and *evaluation research.* Each has its own strengths and weaknesses. Surveys, for example, tend to be more generalizable than participant observation, but they are unable to capture the subtle nuances in social behavior and its meaning that participant observation can.
- Sociologists use certain basic statistical concepts, including *percentages* and *rates.* The *mean* in statistics is the same as an average. The *median* represents the midpoint of an array of values. The *mode* is the most common value. *Correlation* and *cross-tabulation* are statistical procedures that allow sociologists to associate two or more different variables. In interpreting statistics, it is important to know what research methods were used and how the statistics were produced.
- Can sociology be value-free? Although no research in any field can always be value-free, sociological research nonetheless strives for *objectivity* while recognizing that the values of the research may have some influence on the work. Value considerations as well as political agendas can often result in corrected and refocused research with new insights.
- There are ethical considerations in doing sociological research, such as whether one should collect data without letting research subjects know they are being observed. Other ethical dilemmas include whether researchers have the right to hold their data in confidence and withhold the names of research subjects from officials—an issue society is wrestling with at this moment.

KEY TERMS

concept
content analysis
controlled experiment
correlation
data
data analysis
deductive reasoning
dependent variable
empirical
evaluation research
generalization
hypothesis
independent variable
indicator
inductive reasoning
mean
median
mode
operational definition
participant observation
percentage
population
probability
random sample
rate
reliability
replication study
research design
sample
scientific method
validity
variable

THE INTERNET: A Tool for the Sociological Imagination

Resources on the Internet:

Virtual Society: The Wadsworth Sociology Resource Center at:
http://sociology.wadsworth.com

Visit this site to find additional learning tools, including interactive quizzes, links to related web sites, and an easy link to InfoTrac College Edition.

ASA Code of Ethics
http://www.asanet.org/ecoderev.htm

The Code of Ethics of the American Sociological Association is a professional statement of the ethical conduct expected of sociologists, including ethical research practices.

Sociology and Social Policy: Internet Exercises

In 1996, the U.S. Bureau of the Census announced a plan to conduct the year 2000 *census* by using sampling rather than trying to count every member of the population. The plan was to count 90 percent of the population and use *sampling* techniques to determine the number remaining. This plan triggered a public debate about whether sampling was an adequate way to accurately describe the U.S. population, a debate ultimately resolved by a Supreme Court decision not allowing the use of sampling in the census. Proponents point out that only 63 percent of the U.S. population returned their forms in the prior 1990 census, and they argue that this is also a cost-saving measure. Opponents argued that such a plan would aggravate the problem of the census undercount. There are important implications of this decision since much depends on population data collected during the census, including the number of representatives one has in government and the apportionment of federal funds. Based on what you have learned in this chapter, do you think it would be a good idea or a bad idea to conduct the census by using a sample instead of trying to count every person?

Internet Search Keywords:

U.S. Census
polling
2000 census
U.S. Bureau of the Census

Web sites:

http://tiger.census.gov/
Mapping service powered by U.S. census data that offers a variety of illustrated maps.

http://www.census.gov/
Excellent web site on the U.S. Census Bureau.

InfoTrac College Edition: Search Word Summary

content analysis
cross-cultural studies
participant observation
sample

In order to learn more about these central topics in sociology, you can conduct an electronic search using InfoTrac College Edition. To aid in your search and to gain useful tips, see the Student Guide to InfoTrac College Edition on the Virtual Society web site:
http://sociology.wadsworth.com

INTERACTIONS—A SOCIOLOGY CD-ROM: CONCEPTS FOR THIS CHAPTER

Go to the Wadsworth Sociology CD-ROM for further study on the concepts in this chapter. The CD-ROM also includes quizzes and additional activities to expand your learning experience.

SUGGESTED READINGS

Babbie, Earl. 1998. *The Practice of Social Research,* 8th ed. Belmont, CA: Wadsworth.

One of the most widely used texts on research methods and techniques, this book is both comprehensive and readable.

Dunier, Mitchell. 1992. *Slim's Table: Race, Respectability, and Masculinity.* Chicago: University of Chicago Press.

Based on participant observation in an inner-city cafeteria, this book examines the friendships among a group of Black men, their ethical concerns, and their concepts of themselves. It debunks some of the media-based stereotypes of inner-city Black men and is also a model of participant observation research.

Ladner, Joyce A., ed. 1973. *The Death of White Sociology.* New York: Vintage Books.

This book offers a classic collection of articles by social and behavioral scientists on the biases against minorities that can be part of the scientific research process. Written in the early 1970s, these essays were prophetic and highlighted the importance of multiculturalism in sociological research.

Lurie, Alison. 1967. *Imaginary Friends.* New York: Avon.

A novel about two sociologists who do participant observation research in a bizarre religious cult, this fictional account is an interesting and entertaining examination of research ethics and the process of participant observation.

Nachmias, Chava. 1997. *Social Statistics for a Diverse Society.* Thousand Oaks, CA: Pine Forge Press.

By showing the relevance of statistical analysis for understanding issues that emerge in a diverse society, this book provides a good introduction to statistics as used in sociological research.

Reinharz, Shulamit. 1992. *Feminist Methods in Social Research.* New York: Oxford University Press.

Reinharz's book reveals how a variety of feminist approaches has resulted in new research questions as well as new findings. Includes a discussion of the past "invisibility" of women as research subjects.

CHAPTER 3
Culture

IN ONE contemporary society known for its advanced culture, many women would not consider going to work in the morning without encasing their legs in nylon sleeves. The nylon covers them from waist to toe and is believed to make their legs more attractive. If, as often happens, the delicate nylon filaments tear, the women are embarrassed. Many women put on their leg sleeves during a morning ritual that includes smearing oily pigments on their faces to achieve striking, unnatural hues, and dusting their faces with powder to eliminate the skin's natural shininess. The hair on the scalp is slathered with gels and liquids and then modeled into decorative forms. The hair may be augmented with knots; strips of cloth; and small items of hardware made of plastic, wood, or metal. Beads, metals, and glittering objects are often inserted into holes punched into the women's earlobes. Plastic may be glued onto the fingernails and then adorned with bright colors, paint, and sparkles. All this effort is expended in the belief that it makes women more attractive to others.

Most men in this culture begin their day by scraping their faces with sharp-bladed tools to remove all traces of facial hair. The scraping tools of the wealthiest men may be embellished with silver and brass, but most men use plain metal or plastic tools that are sold in local markets. After removing the hair from their faces, the men douse their skin with healing spirits that are usually scented with fragrances thought to be sexually provocative. Some men modify this ritual by arranging their facial hair in elaborate designs, even packing the hair above their lips with wax and modeling it into dramatic shapes. Men usually slather the hair on their scalps with much the same substances used by women, but they generally do not augment their hair designs with the ornamental objects commonly used by women. Individuals incur social penalties if they do not engage in these rituals. Men who appear in public without removing the day's growth of facial hair are considered unkempt. Women who do not wear nylon sleeves on their legs may be considered ugly or unfeminine, and they may jeopardize their chances of being appointed to the highest-paying and most prestigious jobs. Reflecting the importance of these cultural adornments, elaborate markets have developed to sell these products. People in this culture exchange millions of the tokens they are paid for work to pay for these bodily decorations.[1]

From outside the culture, these practices seem strange, yet few within the culture think the rituals are anything but perfectly ordinary. They sometimes wish they did not have to go through the elaborate morning ritual, and many forgo the elaborate grooming and costuming on weekends. Most of the time, people neither analyze these rituals nor spend time thinking about their meaning.

You have surely guessed that the practices described here are taken from U.S. culture. When viewed from the outside, cultural habits that seem perfectly normal take on a strange

[1]These illustrations are adapted from a classic article on the "Nacirema"—American, backwards—by Horace Miner, 1956.

aspect. The Tchikrin people—a remote culture of the central Brazilian rain forest—paint their bodies in elaborate designs. Painted bodies communicate meanings to others about the relationship of the person to his or her body, society, and the spiritual world. The designs and colors symbolize the balance the Tchikrin think exists between biological powers and the integration of people into the social group. The Tchikrin associate hair with sexual powers, and lovers get a special thrill from using their teeth to pluck an eyebrow or eyelash from their partner's face (Turner, 1969). To the Tchikrin people, these practices are no more exotic than the morning rituals we adhere to in the United States.

To study culture, to analyze it and measure its significance in society, we must separate ourselves from judgments such as "strange" or "normal." We must see a culture as it is seen by insiders, but we cannot be completely taken in by that view. One might say we should know the culture as insiders and understand it as outsiders.

Defining Culture

Culture is the complex system of meaning and behavior that defines the way of life for a given group or society. It includes beliefs, values, knowledge, art, morals, laws, customs, habits, language, and dress, among other things. Culture includes ways of thinking as well as patterns of behavior. Observing culture involves studying what people think, how they interact, and the objects they make and use. As stated by two sociologists, "Culture appears to be 'built into' all social relations, constituting the underlying assumptions and expectations on which social interaction depends" (Wuthnow and Witten, 1988: 50).

In any society, culture defines what is perceived as beautiful and ugly, right and wrong, good and bad. Culture helps hold society together. It gives people a sense of belonging, instructs them in how to behave, and tells them what to think in particular situations. Culture gives meaning to society.

Culture is both material and nonmaterial. **Material culture** consists of the objects created in a given society—its buildings, art, tools, toys, print and broadcast media, and other tangible objects. In the popular mind, material artifacts constitute culture since they can be collected in museums or archives and analyzed for what they represent. These objects are significant because of the meaning they are given. A temple, for example, is not merely a building, nor is it only a place of worship. Its form and presentation signify the religious meaning system of the faithful.

Nonmaterial culture is the norms, laws, customs, ideas, and beliefs of a group of people. Nonmaterial culture is less tangible than material culture, but it has a strong presence in social behavior. A good example of nonmaterial culture comes from reports following the earthquake in Kobe, Japan, in 1995. In the wake of this major disaster, many nations and organizations sent food, medical supplies, and other forms of disaster relief to Japan. The Japanese, in the midst of great need, refused to accept most of these donations. Why? The answer lies in the cultural beliefs of the Japanese people. In their culture, accepting help would have created a sense of obligation, thereby violating national pride and cultural beliefs in self-reliance. Refusing aid may have seemed incomprehensible to outsiders, but within Japanese culture the behavior made perfect sense (Nigg, 1995).

Characteristics of Culture

Across societies, certain features of culture are noted by sociologists. These different characteristics of culture are examined here.

1. Culture is shared. Culture would have no significance if people did not hold it in common. Culture is not idiosyncratic; rather, it is collectively experienced and collectively agreed upon. The shared nature of culture makes human society possible.

The shared basis of culture may be difficult to see in complex societies where groups have different traditions, perspectives, and ways of thinking and behaving. In the United States, for example, different racial and ethnic groups have unique histories, languages, and beliefs—that is, different cultures. Even within these groups, there are diverse cultural traditions. Latinos, for example, comprise many groups with distinct origins and cultures. Still, there are features of Latino culture, such as the Spanish language and some values and traditions, that are shared. The different groups constituting Latino culture also share a culture that is shaped by their common experiences as minorities in the United States. Similarly, African Americans have created a rich and distinct culture that is the result of their unique experience within the United States. What identifies African American culture are the practices and traditions that have evolved from both the U.S. experience and African traditions. Placed in another country, such as one of the African nations, African Americans would likely recognize elements of their culture, but they would also feel culturally distinct as Americans.

Within the United States, culture varies by age, region, gender, ethnicity, religion, class, and other social divisions. A person growing up in the South is likely to develop different tastes, modes of speech, and cultural interests than a person raised in the West. Despite these differences, there is a common cultural basis to life in the United States. Certain symbols, language patterns, belief systems, and ways of thinking are distinctively American and form a common culture even though great cultural diversity exists.

culture

Cultural traditions represent the different values and beliefs of diverse groups. Here, Hannukah, Christmas, Ramaddan, and Kwanza are celebrations of diverse cultures.

2. **Culture is learned.** Cultural beliefs and practices are usually so well learned that they seem perfectly natural, but they are learned nonetheless. How do people come to prefer some foods to others? How is musical taste acquired? Sometimes, culture is taught through direct instruction. A parent teaching a child how to use silverware is one example of direct instruction; children learning songs, myths, and traditions in school is another.

Much of culture is learned indirectly through observation and imitation. Think of how we learn what it means to be a man or a woman. Although we may never be explicitly taught the "proper" roles for men and women, we learn what is expected from observing others. These expectations change over time and vary with different contexts, as we will see when we explore gender more thoroughly in Chapter 12.

A person becomes a member of a culture through both formal and informal transmission of culture. Until the culture is learned, the person will feel like an outsider. The process of learning culture is referred to by sociologists as *socialization*, a subject we will return to in Chapter 4.

3. **Culture is taken for granted.** Because culture is learned, members of a given society seldom question the culture of which they are a part, unless for some reason they become outsiders or establish some critical distance from the usual cultural expectations. People engage unthinkingly in hundreds of specifically cultural practices every day; culture makes these practices seem "normal." If you suddenly stopped participating in your culture and questioned each belief and every behavior, you would soon find yourself feeling detached and perhaps a little disoriented; you would also become increasingly ineffective at functioning within your group. Little wonder that tourists stand out so much in a foreign culture. They rarely have much knowledge of the culture they are visiting and, even when they are well informed, typically approach the society from their own cultural orientation.

Think, for example, of how you might feel if you were a Native American student in a predominantly White classroom. Many, though not all, Native American people are raised to be quiet and not outspoken. If students in a classroom are expected to assert themselves and state what is on their minds, a Native American student may feel awkward, as will others for who these expectations are contrary to their cultural upbringing. If the professor is not aware of these cultural differences, he or she may penalize students who are quiet, resulting perhaps in a lower grade for the student from a different cultural background. Culture binds us together, but lack of communication across cultures can have negative consequences, as this example shows.

4. Culture is symbolic. The significance of culture lies in the meaning it holds for people. **Symbols** are things or behaviors to which people give meaning; the meaning is not inherent in a symbol but is bestowed by the meaning people give it. The U.S. flag, for example, is literally a piece of cloth. Its cultural significance derives not at all from the cloth of which it is made, but from its meaning as a symbol of freedom and democracy.

That something has symbolic meaning does not make it any less important or influential. In fact, symbols are powerful expressions of human life. You may remember the significance that the Union Jack (the British flag) had, flown at half-mast over Buckingham Palace, following Princess Diana's death. Tradition holds that the flag of the House of Windsor should fly over the palace only when the Queen is there. Never before had the Union Jack been flown over Buckingham Palace, except to symbolize the Queen's presence. Flying the flag at half-mast was a major symbolic expression of public grief.

The meaning of symbols depends on the cultural context in which they appear. Flying a flag at half-mast over Buckingham Palace would have not been so significant had it been a common tradition. Likewise, symbols mean different things in different contexts. The U.S. flag, for example, has been used by the Shiite rulers of Iran as a despised symbol of American imperialism. A flag tattooed on someone's arm has a meaning different from a flag insignia stitched on a military uniform. Similarly, a cross on a church altar has a meaning different from a cross burning in someone's front yard.

One of the interesting things about culture is the extent to which symbolic attachments guide human behavior. For example, people stand when the national anthem is sung, and they may be emotionally moved by displays of the cross or the Star of David. Under some conditions, people organize mass movements to protest what they see as the defamation of important symbols, such as the burning of a flag or the burning of a cross. The significance of the symbolic value of culture can hardly be overestimated. Learning a culture means not just engaging in particular behaviors but learning their symbolic meanings within the culture.

5. Culture varies across time and place. Culture develops as humans adapt to the physical and social environment around them. Since this environment varies from one society to another, and since human beings use their creative imagination to develop cultural solutions to the challenges they face, culture is not fixed from one place to another. In the United States, for example, there is a strong cultural belief in scientific solutions to human problems; consequently, many think that problems of food supply and environmental deterioration can be addressed by scientific breakthroughs, such as gene splicing to create high-yield tomatoes or bacteria that eat oil spills. In another cultural setting, other solutions may seem preferable; indeed, a religious culture might think that gene splicing trespasses on divine territory, and in many cultures science may be seen as creating more problems than it solves (Harding, 1998).

Because culture varies from one setting to another, the meaning systems that develop within a culture must be seen in their cultural context. **Cultural relativism** is the idea that something can be understood and judged only in relationship to the cultural context in which it appears. This does not make every cultural practice morally acceptable, but it suggests that without knowing the cultural context, it is impossible to understand why people behave as they do.

An understanding of cultural relativism gives insight into some controversies, such as the international debate about the practice of clitoridectomy. In a clitoridectomy (sometimes called female circumcision), all or part of a young woman's clitoris is removed, usually not by medical personnel, often in very unsanitary conditions, and without any pain killers. Sometimes, the lips of the vagina may also be sewn together. Human rights and feminist organizations have documented this practice in more than forty countries on the African continent and the southern part of the Arabian Peninsula and Persian Gulf (Slack, 1988). This practice is most frequent in cultures where women's virginity is highly prized and where marriage dowries depend on some accepted proof of virginity (Slack, 1988; Morgan and Steinem, 1980).

From the point of view of Western cultures, clitoridectomy is genital mutilation and an outrageous example of violence against women; for many feminists, it also represents the persistence of patriarchal (male-dominated) practices. Many feminist organizations have called for international intervention to eliminate the practice of clitoridectomy (Morgan and Steinem, 1980), but Western feminists have also debated whether their disgust at this practice should not be balanced by a reluctance to impose Western cultural values on other societies. Should cultures have the right of self-determination, or should cultural practices that maim or kill people be treated as violations of human rights? This controversy is unresolved. The point is to see that understanding a cultural practice like clitoridectomy requires knowing the cultural values on which it is based. This does not make the practice right, but it illuminates why it occurs, thereby making us better able to recommend effective change.

Not only does culture vary from place to place, it also varies over time. As people encounter new situations, the culture that emerges is a mix of the past and present. Second-generation immigrants to the United States provide a good example. Raised in the traditions of their culture of origin, children of immigrants typically grow up with both the traditional cultural expectations of their parents' homeland and the cultural expectations of a new society. Adapting to the new society can create conflict between generations, especially if the older generation is intent on passing along their cultural traditions. The children may be more influenced by their peers and may choose to dress, speak, and behave in ways that are characteristic of their new society but unacceptable to their parents.

To sum up: Culture is concrete in that we can observe the cultural objects and practices that define human experience. Culture is abstract in that it is a way of thinking, feeling, believing, and behaving. Culture links the past and the present because it is the knowledge that makes us part of human groups. Culture gives shape to human experience.

Humans and Animals: Is There a Difference?

It is cultural patterns that make humans so interesting. If culture is what makes us human, what does that say about the difference between humans and other animals? Some animals use tools, develop language, and form social structures—that is, patterns of behavior that are organized, relatively stable, and persistent over time. Is this evidence of culture? Like humans, animals adapt to their environment. Many species return to traditional nesting sites each year. Does this indicate that animals, like humans, are capable of learning? The distinction between humans and animals is not as clear as was once believed.

Tool-making was at one time considered the dividing line between human and animal societies, but animals have now been observed to make tools. Jane Goodall, famed observer of chimpanzees, lived for many years among the chimps in the Gombe Stream Chimpanzee Reserve in Tanzania, Africa, and practically became a part of their society. She learned to recognize each chimp and was able to simulate the expressions and gestures that they used to communicate (Goodall, 1990). She observed chimpanzees stripping the leaves from a stick, licking the end to make it sticky, and then using the tool they had made to catch termites from their nests. She argued that tool use among the chimpanzees was purposeful and learned behavior, not just instinctual. Apparently then, humans are not unique just because they make and use tools. Of course, the tools that humans make are unequaled in complexity: Think of the difference between using a twig to catch termites and making an automobile. The difference between humans and animals as toolmakers is not an absolute distinction but a matter of degree.

Animals have also developed systems of communication that they use for a variety of purposes, such as attracting mates and alerting others to danger. Some of these communication systems are quite elaborate, even involving rudimentary forms of language. Dolphins, for example, are now believed to have a surprisingly complex auditory language. The essential question is whether animals develop language-based cultures, as humans do.

The most dramatic examples of language capability among animals have come from studies of nonhuman primates. Gorillas and chimpanzees do not have the vocal apparatus necessary to make the sounds of human speech, but under the tutelage of Beatrice and Allen Gardiner (1969), a female chimpanzee named Washoe acquired a vocabulary of 160 standard signs in Ameslan, the American Sign Language system (McGrew, 1992; Wallman, 1992). By the time Washoe was two years old, she had independently put the words she knew into simple sentences.

Using a keyboard equipped with sound, Koko is able to produce English words. Koko is able to use complex sentences and to communicate her wants and thoughts to her human teacher.

Lucy, another chimp, learned more than 100 signs, which she reportedly used to joke, swear, tell lies, and teach others. One critic has argued that what were initially thought to be jokes or lies were simply errors made by Lucy (Terrace, 1980). A far more advanced case of language learning is that of Koko, a female gorilla who has learned more than 500 Ameslan signs and has used 1,000 signs at least once. Koko uses more complex sentences than Lucy and Washoe, saying such things to her teacher as "Hurry, go drink," while pointing to a vending machine, or "Want apple eat want" (Patterson, 1978: 454). Koko showed further social development when she adopted a small kitten that she named All Ball. She played affectionately with the cat and at one point signed to her teacher, "Koko love visit Ball" (Vessels, 1985: 110).

Despite evidence of what could be called social skills among animals, none have developed the level of language or tool-making that is characteristic of human culture. Even after several years of intense instruction for several hours every day, no primate or other animal has been able to acquire the level of linguistic skill that is routinely expected of a human three-year-old. Scientists generally conclude that animals lack the intelligence required to develop the elaborate symbol-based cultures common in human societies.

Studying animal groups reminds us of the interplay between biology and culture. Human biology sets limits and provides certain capacities for human life and the development of culture. Similarly, the environment in which humans live establishes the possibilities and limitations for human society. Nutrition, for instance, is greatly influenced by environment, thereby affecting human body height and weight. Not everyone can skate like Tara Lipinsky, drive a golf ball like Tiger Woods, leap like Michael Jordan, or play baseball like Cal Ripken, but with training and conditioning, people can alter their physical abilities. Biological limits exist, but cultural factors have an enormous influence on the development of human life.

The Elements of Culture

As we have seen, culture consists of both material objects and abstract thoughts and behavior. In sociological work, several elements of culture are of particular interest: language, norms, beliefs, and values.

Language

Language is a set of symbols and rules that, put together in a meaningful way, provides a complex communication system (Cole, 1988). The formation of culture among humans is made possible by language. Learning the language of a culture is essential to becoming part of a society, and it is one of the first things children learn. Indeed, until children acquire at least rudimentary command of language, they seem unable to acquire other social skills. Language is so important to human interaction that it is difficult to think of life without it; indeed, as one commentator on language has said, "Life is lived as a series of conversations" (Tannen, 1990: 13).

Think about the experience of becoming part of a social group. When you enter a new society or a different social group, you have to learn its language to become a full member of the group. This includes not only learning the spoken language of the overall society, but also any special terms of reference used by the group. New members of a street gang, for example, must learn new words and their meanings before being accepted as full-fledged gang members. Similarly, entering any profession requires understanding the "terms of the trade." Lawyers, for example, have their own vocabulary and their own way of constructing sentences. The intricate prose rendered by lawyers is called (not always kindly) "legalese"; anyone who is not adept in the language of the law, both in reading and in writing it, is likely to be at a disadvantage in any legal proceeding. Likewise, becoming a sociologist requires learning the words and concepts that sociologists use to communicate their ideas. Becoming a part of any social group—a friendship circle, fraternity or sorority, professional organization, or any other group—involves learning the language the group uses. Those who do not share the language of a group cannot participate fully in its culture.

Language systems are fluid and dynamic, and language evolves in response to social change. Think, for example, of how the introduction of computers and other electronic systems of communication has affected the English language. People now talk about needing "downtime" and providing "input." Only a few years ago, had you said you were going to "fax" someone or "click on" something, people would have thought you were speaking in some private code. Even now, those without exposure to advanced electronic technologies might not understand what you mean by some of these terms. Other cultural changes are also reflected in the ordinary language we use. As the monetary system of capitalism has evolved, many of its terms have entered daily jargon, such as "buying into" something, "banking on it," and getting to the "bottom line." These examples show how language evolves through social change.

Does Language Shape Culture? Language is clearly a big part of culture. Edward Sapir (writing in the 1920s) and his student, Benjamin Whorf (writing in the 1950s), thought that language was central in determining social thought. Their theory, the **Sapir–Whorf hypothesis**, was that language determines other aspects of culture since language provides the categories through which social reality is defined and constructed. In other words, Sapir and Whorf thought that language determines what people think and perceive because language forces people to perceive the world in certain terms (Sapir, 1921; Whorf, 1956). It is not that you think or perceive something first and then think of how to express it, but that language itself determines what you think or perceive.

If the Sapir–Whorf hypothesis is correct, then speakers of different languages have different perceptions of reality. Whorf compared the language of Hopi Indians with English to illustrate his theory. For example, the single Hopi word *masaytaka,* "flying things," is all-inclusive and can mean airplane, insect, or pilot. In English, three different words are used for these three things. A speaker of the Hopi language may be unaware of distinctions among the three objects that would be linguistically obvious in another language system.

Whorf also used the example of the construction of time to illustrate cultural differences in how language shapes perceptions of reality. He noted that the Hopis conceptualize time as a slowly turning cylinder, whereas English-speaking people conceive of time as running forward in one direction at a uniform pace. Linguistic constructions of time shape how the two different cultures think about it and therefore how they think about reality. In Hopi culture, events are located not in specific moments of time, but in "categories of being"—as if everything is in a state of becoming, not fixed in a particular time and place (Carroll, 1956). To the Hopi, a house "houses"; a stone "stones." In contrast, the way English sentences are constructed locates things in a definite time and place. For example, European languages place great importance on verb tense—things are located unambiguously in the past, present, or future.

Recent critics do not think that language is as deterministic as Sapir and Whorf proposed—language does not single-handedly dictate the perception of reality—but there is no disagreement that language has a strong influence on culture. Most scholars now see two-way causality between language and culture. Asking whether language determines culture or vice versa is like asking which came first, the chicken or the egg. Language and culture are inextricable. Each shapes the other, and to understand either, we must know something of both.

Consider again the example of time. Contemporary Americans think of the week as divided into two parts: weekdays and weekends. The words *weekday* and *weekend* reflect the way we think about time. When does a week end? Having language that defines the weekend encourages us to think about the weekend in specific ways. It is a time for rest, play, chores, and family. In this sense, language shapes the way we think about the passage of time—we look forward to the weekend, we prepare ourselves for the work week—but the language itself (the very concept of the weekend) stems from patterns in the culture—specifically, the work patterns of advanced capitalism. Language and culture shape each other.

Concepts of time in American society are tied to the work ethic. The rhythm of life in the United States is the rhythm of the workplace. "9 to 5" is the name for an entire lifestyle. The capitalist work ethic makes it morally offensive to merely "pass the time"; instead, time is to be managed. Concepts of time in preindustrial, agricultural societies follow a different rhythm. In agricultural societies, time and calendars are based on agricultural and seasonal patterns; the year proceeds according to this rhythm, not arbitrary units of time like weeks and months. In preindustrial societies, the line between work and private life is less clear than in industrialized societies, where clock time replaces seasonal time, and productivity is measured in the units of labor, that is, "man hours" and the "hourly wage" (Genovese, 1972).

Social Inequality in Language. The significance of language in culture is particularly apparent in how patterns of race, gender, and class inequality are reflected in language. The language of any culture reflects and reinforces attitudes that are characteristic of the culture. Language can also reproduce these inequalities through the stereotypes and assumptions that may be built into what people say (Moore, 1992). As shown in the box "Understanding Diversity: The Social Meaning of Language," our names for different groups of people reflect assumptions about those groups. Mexican American (hyphenated or not), Latino/Latina, Hispanic, and Chicano/Chicana all have different meanings, and may be interpreted differently by different groups. What someone is called can really matter since it imposes an identity on that person. This is why the names for various racial and ethnic groups have been so heavily debated. Even conventions like whether to capitalize the word *Black* in Black American can reflect assumptions about the status of different groups. You may have noticed that in this book, the authors capitalize Black whenever the label is used as a proper noun. Likewise, the word *White* is capitalized when used to refer to the specific group of White Americans. This convention is a relatively new one and still debated: Do you treat the color designator as an adjective (as if only color marks the group's experience), or do you recognize that these labels describe particular group experiences and, therefore, treat the name like a proper noun (as you would American Indian, Latino, and Jewish American)? The standard reference book for writing style, *The Chicago Manual of Style*, now includes capitalization of Black and White as an acceptable style, illustrating that even the formal rules of language can change and are affected by social and political debate.

Language reflects the social value placed on different groups, which is why it is so demeaning when derogatory terms are used to describe a group or group member. Throughout the period of Jim Crow segregation in the American South, Black men, regardless of their age, were routinely referred to as "boy" by Whites. Calling a grown man "boy" is an insult; it diminishes his status by defining him as childlike. Referring to a woman as a "girl" has the same effect. If girls become young women when they reach puberty, why are young women—even well into their twenties—routinely referred to as "girls"? Like calling a man, "boy," this diminishes women's status. Who are you likely to consider a serious intellectual or leader, a girl or a woman? Many women, and not just feminists, believe that being called a "girl" is belittling.

Debunking Society's Myths

***Myth*: Debates about what to call people of color are not terribly significant.**

***Sociological perspective*: Language carries with it great meaning that reflects the perceived social value of diverse groups.**

Note, however, that terms like "girl" and "boy" are pejorative only in the context of dominant and subordinate group relationships. Whereas African American men would be insulted if called "boy" by a White man, the term is not necessarily insulting when African American men refer to each other that way. Similarly, African American women often refer to each other as "girl" in informal conversation. The term *girl*, used between those of similar status, is not seen as derogatory. When used by someone in a position of dominance, as when a man calls his secretary a "girl," it is demeaning. Likewise, terms like "dyke," "fag," and "queer" are terms lesbians and gay men sometimes use without offense in referring to each other even though the same terms are offensive to lesbians and gays when used about them by others. By reclaiming these terms as positive within their own culture, lesbians and gays build cohesiveness and solidarity (Due, 1995). These examples show that power relationships between groups supply the social context for the connotations of language.

The perceived social status of groups can be found in the offensive terms sometimes used to describe different groups, particularly women and racial–ethnic groups. Labels can also stigmatize people, as in the phrase "lower class." Although meant to reflect the relative placement of a group in the hierarchy of power and economic resources, the phrase also carries the implication that the group is "lowlier" than the ones above it. For this reason, the phrase "working class" is often preferred although it, too, has problems since it suggests that other people do not work.

Racist dimensions in language can also be seen in the way certain historical events are described. Native American victories during the nineteenth century are typically described as "massacres"; comparable victories by White settlers are described in heroic terms. The statement that Columbus "discovered" America implies that Native American societies did not exist before Columbus "found" the Americas. Indians are frequently referred to as "savages," whereas White colonists are called "settlers." These are just a few examples of how racism permeates language (Moore, 1992). Likewise, the word *tribe* to describe nations of African or Native American people suggests primitive, backward societies. Even the term *American* is problematic; it is used synonymously with "United States" when the Americas actually include Canada, Mexico, Central America, and South America. To say that the culture of the United States is

BOX 3.1

UNDERSTANDING DIVERSITY

The Social Meaning of Language

LANGUAGE reflects the assumptions of a culture and develops in the context of the society's history and social institutions. Language is important in how we see our society and different groups within it. Language even carries with it political and social attitudes about different groups. This is especially evident in thinking about the language used to describe racial groups in the United States.

The concept of "race" is a cultural construct. That is, the meaning of race is found in the culture; it is not fixed categories of people. Many of us were taught in school that there are three races: Caucasian, Negroid, and Mongoloid, but this division of the world's population into three racial groups reflects the historical European conquest of Asia and Africa. In South Africa, in contrast, school children have been taught that there are four races: White, Colored, Indian, and Black. These group designations, which reflect the racial divisions in South African society, seem rather arbitrary. Their significance comes from the social relationships groups have with each other, not from physical, national, or cultural differences.

The naming of different racial groups also reflects the assumptions made about different groups. In the contemporary United States, the terms used to define different groups are cultural definitions that change over time, depending on the group's own definition of itself and the imposition of labels by the dominant culture. In the 1960s, "Black American" replaced the term *Negro* because the civil rights and Black Power movements inspired Black pride and the importance of self-naming (Smith, 1992). Earlier "Negro" and "colored" were used to define African Americans. These terms had replaced the earlier term *darky*, which had replaced the even earlier term *Black*. African American is now preferred by many who previously called themselves Black American. This is a conscious choice intended to recognize the origins of African American people in Africa.

A fundamental question is whether African Americans should be referred to by color at all. Skin pigmentation is as varied among African Americans as it is among White Americans. After all, is it people's color that is their significant defining feature? Currently, it is popular to refer to all so-called racial groups as "people of color." This phrase was derived from the phrase "women of color," created by feminist African American, Latina, Asian American, Native American, and women to emphasize their common experiences. Some people find the use of "color" in this label offensive since it harkens back to the phrase "colored people," a phrase generally seen as paternalistic and racist since it was a label used by dominant groups to refer to African Americans prior to the civil rights movement. When a group names itself, is it less offensive?

The point here is not that terms themselves designate a particular definition of a group, but that the implications of language (including names) emerge from specific historical and cultural contexts. Using the term *women of color* in the 1990s has a different and more positive meaning than the earlier term *colored women* because the people who chose the term are recognizing the ties they have in common.

The terms *Latino* and *Hispanic* are similarly constructed though political and social discourse. We use these terms interchangeably in this book, but recognize that they homogenize the experience of groups who have very different histories; thus, the general category renders the experiences of some groups invisible. "Hispanic" and "Latino" lump together Mexican Americans, island Puerto Ricans, U.S.-born Puerto Ricans, as well as people from Honduras, Panama, El Salvador, and other Central and South American countries. "Hispanic" and "Latino" point to the shared experience of those from Latin cultures, but like the terms *Native American* and *American Indian*, they tend to lose sight of the specific experiences of unique groups, such as the Sioux, Nanticoke, Cherokee, Yavapai, or Navajo.

In this book, our use of language is not perfect, nor is it always consistent.[1] We have tried to be sensitive to the feelings and ideas of different groups. To avoid using a long string of prefixes each time a given group is named, we sometimes had no choice but to use general group terms. We recognize that the language we use is fraught with cultural and political assumptions, and that what seems acceptable now may be offensive later. Perhaps the best way to solve this problem is for different groups to learn as much as they can about one another, becoming more aware of the meaning and nuances of naming and language and more conscious of the racial assumptions embedded in the language we use. Greater sensitivity to the language used in describing different group experiences is an important step in promoting better intergroup relationships.

[1] See also the Preface of this book.

"American" culture suggests that the United States encompasses the other Americas or that the other cultures of the Americas are not worth notice.

In a sense, language constructs our sense of truth. Think of the different meanings implied in describing urban "riots." Are they riots or insurrections? Whose point of view does each version take? Calling such an event a riot evokes an image of unruly behavior by ghetto residents; describing such an event as an insurrection implies political revolt by oppressed people. Depending on the language used, a different meaning is constructed.

Sexism in language works much the same way. Calling a person who heads a committee or an office a "chairman" suggests that only men can hold this position. Using "man" to represent all human beings likewise suggests that we can generalize to all people from men's experiences. These practices make women seem like exceptions to a rule. Despite protests by some that this is a trivial issue, using male pronouns does affect people's perception of reality. Researchers have found that men, more than women, tend to think that women are not included when terms like "man" and "mankind" are used to mean all people and when "he" is used as the all-purpose, second-person pronoun. Studies find that when college students look at job descriptions written in masculine pronouns, they are more likely to conclude that women are not qualified for the job than when they read more gender-neutral descriptions (Hyde, 1984; Switzer, 1990). Similarly, a term like "woman doctor" suggests that the gender of the doctor is something exceptional and noteworthy. The term *career girl* suggests that a career is something exceptional for a person of that gender; furthermore, note that no one speaks of a "career boy"! Students might be interested in observing how sexist language is used in descriptions of men and women in newspaper reports and articles. Women are frequently described in terms of their clothing and appearance, even when their appearance is totally irrelevant to the purpose of the article. An example is found in a magazine article about the first all-woman team to compete in the America's Cup sailing trials. The article described the women as a "pony-tailed band of yachting fraternity interlopers" (Edwards, 1995: 26)!

In sum, language can reproduce racist and sexist thinking. At the same time, changing the language that people use can, at least to some extent, alter social stereotypes and thereby change the way people think. The Sapir–Whorf hypothesis claims that language shapes how people define social reality. Changing what people are called can change what others think of them and what they think of themselves. This is why questions such as whether people should be called Latino, Mexican American, or Chicano are debated so strongly.

Norms

Another component of culture are social norms. **Norms** are the specific cultural expectations for how to behave in a given situation. Society without norms would be chaos; with norms in place, people know how to act, and social interactions are consistent, predictable, and learnable. There are norms governing every situation. Sometimes they are *implicit*; that is, they need not be spelled out for people to understand what they are. For example, when entering a line, there is an implicit norm that you should not barge in front of those ahead of you. Implicit norms may not be formal rules, but violation of these norms may nonetheless produce a harsh response. Implicit norms may be learned through specific instruction or by observation of the culture; they are part of a society's or group's customs. Norms are *explicit* when the rules governing behavior are written down or formally communicated. Typically, there are specific sanctions for violating explicit norms.

In the early years of sociology, William Graham Sumner (1906), whose work we described in Chapter 1, identified two types of norms: folkways and mores. **Folkways** are the general standards of behavior adhered to by a group. You might think of folkways as the ordinary customs of different group cultures. Men's wearing pants and not skirts is an example of a cultural folkway. Other examples are the ways that people greet each other, decorate their homes, and prepare their food. Folkways may be loosely defined and loosely adhered to, but they nevertheless structure group customs and implicitly govern much social behavior. Elijah Anderson's research on urban Black street culture (see the box, "Doing Sociological Research: Streetwise") gives an example of the cultural folkways that have developed in a particular urban neighborhood.

Mores (pronounced "more-ays") are strict norms that control moral and ethical behavior. Mores provide strict codes of behavior, such as the injunctions, legal and religious, against killing others and committing adultery. The mores are often upheld through rules or **laws**, the written set of guidelines that define right and wrong in society. Basically, laws are formalized mores. Violating mores can bring serious repercussions.

When any social norm is violated, sanctions are typically meted out against the violator. **Social sanctions** are mechanisms of social control that enforce norms. The seriousness of a sanction depends on how strictly the norm is held. Violations of folkways carry lighter sanctions than violations of mores. Dressing in an unusual way (that is, violating the cultural folkways of dress) may bring ridicule but is usually not seriously punished. In some cultures, however, the rules of dress are strictly held. Islamic law requires that women appearing in public must have their bodies cloaked and their faces veiled. The sanctions for violating mores can be severe—whipping, branding, banishment, even death.

Sanctions include both rewards and punishments. When children learn social norms, for example, correct behavior may elicit positive sanctions—the behavior is reinforced through praise, approval, or an explicit reward. Early on, for example, parents might praise children for learning to put on their own clothes; later, children might get an allowance if they keep their rooms clean. Bad behavior earns negative sanctions, such as getting spanked or grounded.

In society, negative sanctions may be mild or severe, ranging from subtle mechanisms of control, such as ridicule,

BOX 3.2 DOING SOCIOLOGICAL RESEARCH

Streetwise

SOCIOLOGISTS study culture in a variety of ways. Some study cultural artifacts themselves, analyzing images in the media, for example, or the use of particular objects in a given social setting. Others use a comparative view of culture, asking broad questions about the relationship of culture to other forms of social organization. Some study language in great detail, analyzing how language structures social interaction and reflects the social order in a given society. Even this comprehensive chapter cannot capture all the different ways sociologists study and interpret culture.

One example of a detailed study of the culture of a particular community comes from Elijah Anderson, an African American sociologist. Anderson was interested in the street culture of two bordering communities in the city of Philadelphia. One of them, which he named "The Village," is one of the most culturally and ethnically diverse areas of the city and includes middle-class and working-class residents. Neighboring this community is "Northton," an urban Black ghetto, fraught with all the problems of unemployment, crime, and street violence that characterize urban poor areas. (Sociologists typically give pseudonyms to the people and places they study to protect the privacy and anonymity of research subjects.)

A resident of The Village himself, Anderson had not initially planned to study these communities, but he became increasingly fascinated by the cultural norms that shaped each community and its relationship to its neighbors. He began his study informally, as he says, "simply experiencing everyday life" (1990: ix), but he soon began more formal *participant observation* research, taking detailed field notes and carefully interviewing residents. As a participant observer, he spent many hours on the street, talking with and listening to the people in both neighborhoods. Anderson observed people in the ordinary places where they gathered: street corners, laundromats, bars, parties, carryout stores. All told, his research extended over a period of fourteen years; he has, to date, published two books based on his study (*A Place on the Corner*, University of Chicago Press, 1978, and *Streetwise*, University of Chicago Press, 1990).

In a study so rich with detail, it is impossible to summarize a single finding, but Anderson began by asking, "How do people get along here? What is the culture of the local public spaces? How are social changes affecting residents of each community?" He was interested in learning about the cultural norms that shaped people's public behavior. Both of these communities have been affected by social structural changes in the economy. Rising unemployment, poverty, and the exodus of major companies from urban areas leave people in the urban ghetto, especially young people, without work opportunities. They adopt street life, according to Anderson, as a way of supporting themselves, often through illegal activities.

Anderson found that there is a particular etiquette for surviving on the street in these places, and he chronicles the profound "street wisdom" that people acquire. This is demonstrated through people's comportment and by the repertoire of knowledge people acquire in order to travel the streets safely. By watching how people carried themselves on the street, hearing what they said, and asking them questions about the "rules" of the street, Anderson was able to develop a rich analysis of the culture of these neighborhoods.

He situates this culture within the broad social structural changes affecting these two urban communities. The ghetto has been left economically depleted by the loss of manufacturing jobs in the city; those who could leave have gone; others remain either unemployed or in low-wage, dead-end jobs. The loss of a strong middle class in the ghetto has undermined traditional leadership patterns in the community and has created tensions between younger and older residents. Anderson observed that youths mock their elders for "not understanding how the world works"; likewise, the old do not understand the perceived absence of the work ethic, decency, a church life, and values among the young.

Anderson also found that two overlapping cultures—one a more gentrified neighborhood and the other an urban poor ghetto—co-existed peacefully. Still, residents in both communities were alert to strangers on the street and used specific etiquette to maintain distance between themselves and strangers. Middle-class people, however, find it increasingly difficult to live in this area. The White "yuppies" who have recently moved here do not share the progressive vision of living in a racially mixed neighborhood as was true of the White liberals who moved here in the 1970s.

Anderson concludes that there are multiple victims of the changing economy here: "the yuppie who is mugged and the kid who does it; the old head who loses the respect of the kid, who impregnates the teenage girl, who goes on welfare, which raises taxes, which drives out local companies, which causes unemployment, which causes homelessness, which causes crime, which depresses property values and drives out middle-class residents, which further isolates the poor and the criminal. Each of the groups is frustrated in some way in getting itself together to fulfill the promise of the local area, a promise that is frustrated by increasingly broad structural factors" (1990: 250–251).

Based on: Anderson, Elijah. 1990. *Streetwise: Race, Class, and Change in an Urban Community*. Chicago: University of Chicago Press.

to overt forms of punishment, such as imprisonment, physical coercion, and even death. Enforcement of the segregationist practices in the American South offers a historical example of how norms are enforced. Segregation was controlled through strict social norms that governed everyday forms of behavior. In one horrendous incident in 1955, Emmett Till, a fourteen-year-old Black youth from Chicago who was visiting relatives in Mississippi, was shot, beaten to death, mutilated, and then dumped in the Tallahatchie River after he whistled at a White woman and called her "baby" in a local store. Segregationist norms were so strong at the time that the courts acquitted the White men accused of his murder (Carson et al., 1987). The teen's violation of the social norms was more strictly punished than the act of murder.

One of the ways to study social norms is to observe what happens when they are violated. Once you become aware of how social situations are controlled by norms, you can see how easy it is to disrupt situations where adherence to the norms produces social order. **Ethnomethodology** is a technique for studying human interaction by deliberately disrupting social norms and observing how individuals respond; the idea is that the disruption of social norms helps one discover the normal social order (Garfinkel, 1967). In a famous series of ethnomethodological experiments, college students were asked to pretend they were boarders in their own homes for a period of fifteen minutes to one hour. They did not tell their families what they were doing. The students were instructed to be polite, circumspect, and impersonal; to use terms of formal address; and to speak only when spoken to. After the experiment, two of the participating students reported that their families treated the experiment as a joke; another's family thought the daughter was being extra nice because she wanted something. One family believed that the student was hiding some serious problem. In all the other cases, parents reacted with shock, bewilderment, and anger. Students were accused of being mean, nasty, impolite, and inconsiderate; the parents demanded explanations for their son's and daughter's behavior (Garfinkel, 1967). Through this experiment, the student researchers were able to see that even the informal norms governing behavior in one's home are carefully structured. By violating the norms of the household, the norms were revealed.

Ethnomethodological research teaches us that society proceeds on an "as if" basis. That is, society exists because people behave *as if* there were no other way to do so. Most of the time, people go along with what is expected of them. Culture is actually "enforced" through the social sanctions applied to those who violate social norms. Usually, specific sanctions are unnecessary because people have learned the normative expectations. When the norms are violated, their existence becomes apparent.

Beliefs

As important as social norms are the beliefs of people in society. **Beliefs** are shared ideas held collectively by people within a given culture. Shared beliefs are part of what binds people together in society. Beliefs are also the basis for many of the norms and values of a given culture. Examples in the United States are the belief in God and the belief in democracy. These are beliefs that are widely held and cherished.

Some beliefs are so strongly held that people find it difficult to cope with ideas or experiences that contradict them. Someone who devoutly believes in God may find atheism intolerable; those who believe in reincarnation may seem irrational to those who think life ends at death. Similarly, those who believe in magic may seem merely superstitious to those with a more scientific and rational view of the world.

Whatever beliefs people hold, they orient us to the world. They provide answers to otherwise imponderable questions about the meaning of life. Beliefs provide a meaning system around which culture is organized (see Chapter 16). Whether belief stems from religion, myth, folklore, or science, it shapes what people take to be possible and true. Although a given belief may be unprovable or even logically impossible, it nonetheless guides people through their lives.

Sociologists study belief in a variety of ways, depending on their theoretical orientation. Functionalists, for example, see beliefs as a functional component of society in that they integrate people into social groups. Conflict theorists, on the other hand, interpret beliefs as potentially competing world views; those with more power are able to impose their beliefs on others. Symbolic interaction theorists are interested in how beliefs are constructed and maintained through the social interaction people have with each other. Each of these theoretical orientations provides different insights into the significance of beliefs for human culture and society.

Values

Deeply intertwined with beliefs are the values of a culture. **Values** are the abstract standards in a society or group that define ideal principles. Values define what is desirable and morally correct; thus, values determine what is considered right and wrong, beautiful and ugly, good and bad. Although values are abstract, they provide a general outline for behavior. Freedom, for example, is a value held to be important in U.S. culture, as is equality. Values are ideals, forming the abstract standards for group behavior, but they are also ideals that may not be realized in every situation.

Values can be a basis for cultural cohesion, but they can also be the source of conflict. For example, the political conflict over abortion is a conflict over values. Pro-choice groups value reproductive freedom and women's right to control their bodies. Those opposed to abortion value the life of the fetus over the woman's right to choose whether or not to terminate a pregnancy. Abortion law has attempted to strike a balance among the values behind competing rights: the right to privacy, the right of the state to protect maternal health, and the right of the state to protect developing life. The abortion issue puts the values of different groups into direct conflict with each other. You can see this value conflict played out in the language the different

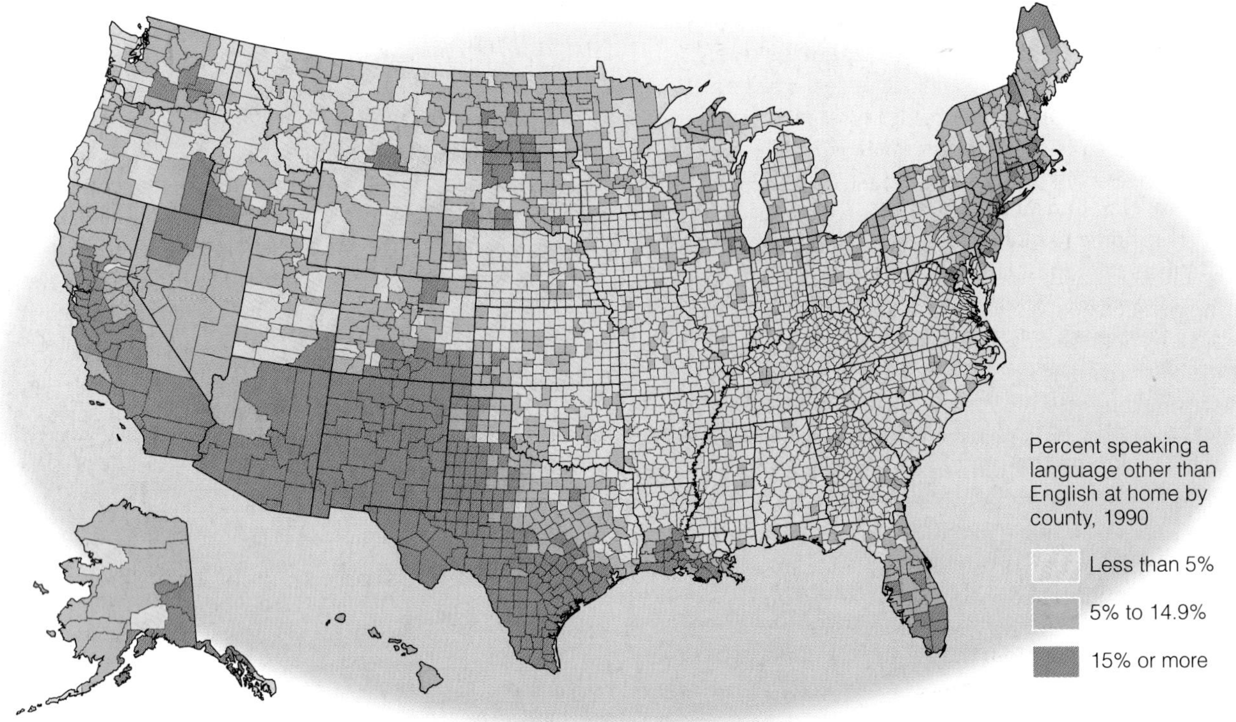

MAP 3.1 Mapping America's Diversity: Percent Speaking Language Other Than English at Home

SOURCE: Doyle, Rodger, 1994. *Atlas of Contemporary America: Portrait of a Nation.* New York: Facts on File, p. 77.

groups use to describe themselves. By adopting the label "pro-life," those opposed to abortion lay claim to moral high ground by suggesting that those who are pro-choice are somehow "anti-life." The political slogans on both sides of this conflict use language that subtly promotes their value position, exploiting the fact, pointed out earlier, that language has the power to define how groups are perceived by themselves and by others.

Values guide the behavior of people in society. Norms are the reflection of underlying values. For example, faithfulness within marriage is a value expressed in a variety of cultural norms, such as the prohibition against dancing too intimately with another's spouse. An example of the impact that values can have on people's behavior comes from an American Indian society known as the Kwakiutl (pronounced "kwok-ee-ud-ul"), a group from the coastal region of southern Alaska, Washington State, and British Columbia. The Kwakiutl developed a practice known as *potlatch*, in which wealthy chiefs would periodically pile up their possessions and give them away to their followers and rivals (Benedict, 1934; Harris, 1974). The object of potlatch was to give away or destroy more of one's goods than did one's rivals. The potlatch reflected Kwakiutl values of reciprocity, the full use of food and goods, and the social status of the wealthiest chiefs in Kwakiutl society. (By the way, chiefs did not lose their status by giving away their goods because the goods were eventually returned in the course of other potlatches. At times, they would even burn large piles of goods, knowing that others would soon replace their accumulated wealth through other potlatches.)

Contrast this practice with the patterns of consumption in the United States. Imagine the CEOs of major corporations regularly gathering up their wealth and giving it away to their workers and rival CEOs! In the contemporary United States, *conspicuous consumption* (consuming for the sake of displaying one's wealth) celebrates values opposite those of the potlatch: High-status people demonstrate their position by accumulating more material possessions than those around them (Veblen, 1899). Indeed, conspicuous consumption encourages consumerism throughout the society—people feel as though they have to "keep up with the Joneses." Values and norms together guide the behavior of people in society. It is necessary to understand the values and norms operating in a situation to understand why people behave as they do.

Cultural Diversity

It is rare for a society to be culturally uniform. As societies develop and become more complex, different cultural traditions appear. The more complex the society, the more likely its culture will be internally varied and diverse. The United

States, for example, hosts enormous cultural diversity stemming from religious, ethnic, and racial differences, as well as regional, age, gender, and class differences. Currently, more than 8 percent of people in the United States are foreign born. In a single year, immigrants from more than one hundred countries come to the United States (Portes and Rumbaut, 1996; U.S. Bureau of the Census, 1997: 53). New York City is home to more Irish than Dublin, Ireland, and more Italians than most cities in Italy. Recently, the number of Arab, Caribbean, Mexican, and Southeast Asian immigrants to the United States has risen. As Figure 3.1 shows, whereas earlier immigrants were predominantly from Europe, now Latin America and Asia are the greatest sources of new immigrants. One result is a large increase in the number of U.S. residents for whom English is the second language (see Table 3.1), as shown in "Mapping America's Diversity." Cultural diversity is clearly a characteristic of contemporary American society.

The richness of American culture stems from the many traditions that different groups have brought with them to this society, as well as from the cultural forms that have emerged through their experience within the United States. Jazz, for example, is one of the few musical forms that originated in the United States. Jazz is an *indigenous* art form, one originating in a particular region or culture; however, jazz also has roots in the musical traditions of the slave community and African cultures. Since the birth of jazz, cultural greats like Ella Fitzgerald, Count Basie, Duke Ellington, Billie Holiday, and numerous others have not only enriched the jazz tradition, but have also influenced other forms of music, including rock and roll.

Debunking Society's Myths

Myth: Americans share a single culture that is derived from western European traditions.

Sociological perspective: The United States is an increasingly diverse society, and many of its dominant cultural traditions have been derived from African Americans, Latinos, and other groups whose experience is often neglected.

Native American cultures have likewise enriched the culture of our society. One example is the influence of the Iroquois on our style of government. As early as the late sixteenth century, the Iroquois of northern New York and Canada had established a system of governance based on the principles of self-rule, federation, and representation. Local bodies contributed to a central governing body that protected them. Although the democratic form of government adopted by American colonists is based on the models of ancient Greece and Rome, the particular form of our federal government—its division into executive, judicial, and legislative branches—may have been influenced by the example of the Iroquois nation (Allen, 1986).

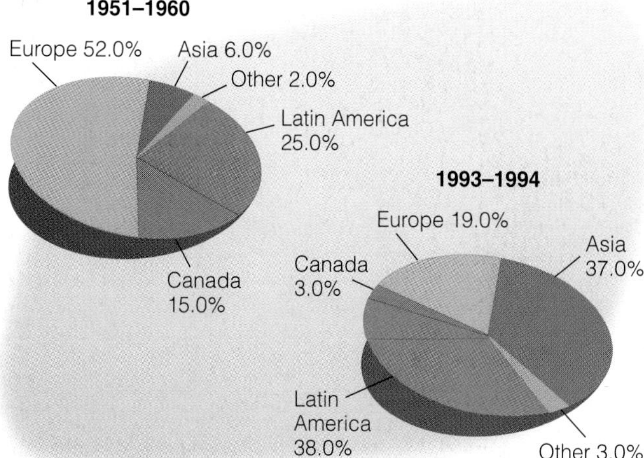

FIGURE 3.1 U.S. Immigrants by Region of Origin

SOURCE: U.S. Immigration and Naturalization Service. 1995. *1994 Statistical Yearbook of the Immigration and Naturalization Service.* Washington, DC: U.S. Government Printing Office.

TABLE 3.1 *ENGLISH SPOKEN HERE, BUT ALSO . . .*

Diversity within the United States is reflected in the growing numbers of those who speak a language other than English at home. Since 1980, that has increased from 23 million to almost 32 million (14 percent of all residents over 5 years old). This table shows the different languages spoken at home and the changes in those numbers since 1980.

Language used at home	Total speakers in 1980	Total speakers in 1990	Percent change
Spanish	11,549,000	17,339,000	50.1%
French	1,572,000	1,703,000	8.3%
German	1,607,000	1,547,000	−3.7%
Italian	1,633,000	1,309,000	−19.9%
Chinese	632,000	1,249,000	97.7%
Tagalog	453,000	843,000	86.6%
Polish	826,000	723,000	−12.4%
Korean	276,000	626,000	127.2%
Vietnamese	203,000	507,000	149.5%
Portuguese	361,000	430,000	13.09%
Japanese	342,000	428,000	25.0%
Greek	410,000	388,000	−5.4%
Arabic	227,000	355,000	57.4%
Hindu (Urdu and related)	130,000	331,000	155.1%
Russian	175,000	242,000	38.5%
Yiddish	320,000	213,000	−33.5%
Thai	89,000	206,000	131.6%
Persian	109,000	202,000	84.7%
French Creole	25,000	188,000	654.1%
Armenian	102,000	150,000	46.3%
Navajo	123,000	149,000	20.6%
Hungarian	180,000	148,000	−17.9%
Hebrew	99,000	144,000	45.5%
Dutch	146,000	143,000	−2.6%

SOURCE: U.S. Bureau of the Census. 1997. *Statistical Abstract of the United States.* Washington, D.C.

The potlatch ceremony led by Nimpkish chief Ray Cranmer during an opening celebration of the U'mista Cultural Centre in British Columbia, is a normative practice symbolizing the value of reciprocity among some Indian groups.

With such great variety, how can the United States be called one culture? The culture of the United States, including its language, arts, food customs, religious practices, and dress, can be seen as the sum of the diverse cultures that constitute this society. Altogether, these behaviors and beliefs constitute "American" (U.S.) culture, but cultural diversity is a major feature of this culture as a whole.

Dominant Culture

Two concepts from sociology help us understand the complexity of culture in a given society: the notions of the *dominant culture* and of *subcultures*. The **dominant culture** is the culture of the most powerful group in society. It is the cultural form that receives the most support from major institutions and that constitutes the major belief system. Although the dominant culture is not the only culture in a society, it is commonly believed to be "the" culture of a society despite the other cultures present. Social institutions in the society perpetuate the dominant culture and give it a degree of legitimacy that is not shared by other cultures. Quite often, the dominant culture is the standard by which other cultures in the society are judged.

A dominant culture need not be the culture of the majority of people; rather, it is simply the culture of the group in society that has enough power to define the cultural framework. As an example, think of a college or university that has a strong system of fraternities and sororities. On such a campus, the number of students belonging to fraternities and sororities is probably a numerical minority of the total student body, but the cultural system established by the Greeks may dominate campus life nonetheless. Cultural domination by a minority is strikingly illustrated in South Africa, where Black South Africans are 90 percent of the population, but the country's laws and dominant culture stem from White South African traditions—a group that is a small minority (10 percent) of the population.

In a society as complex as the United States, it is hard to isolate a single dominant culture although there is a widely acknowledged "American" culture that constitutes the dominant culture. Stemming from middle-class values, habits, and economic resources, this culture is strongly influenced by such instruments of culture as television, the fashion industry, and Anglo-European traditions, and includes such diverse elements as fast food, Christmas shopping, and professional sports.

The dominant culture has recently become the subject of much political debate, as the nation responds to increasing diversity within the population. Groups sensitive to the interests of minority cultures have criticized the dominant culture for being based so exclusively on White, male, European cultural traditions. Political conservatives have come to the defense of the status quo, labeling the entire dispute "the culture wars" (Hunter, 1991; Gitlin, 1995).

Political debate about the dominant culture is reflected in the movements in many states to establish English as the official language. Through public referenda and legislative action, at least seventeen states have adopted laws designating English as the official state language, and federal legislation has also been proposed. Proponents of this concept insist that English is already the de facto official language of America, and that mastery of English is essential for educational and economic success and should therefore be promoted as the common tongue. To opponents, this movement represents an anti-immigrant backlash and springs from growing hostility toward groups perceived as outsiders (Horton and Calderón, 1992; Adams and Brink, 1990; Crawford, 1992). Sociological research has found that those most likely to support English as the official language of the United States are those who support a "melting pot" ideal and who think there should be common identity as "American" (Frendreis and Tatalovich, 1997). We can expect this struggle to continue for some years as the number of foreign-born Americans continues to grow.

Subcultures

Subcultures are the cultures of groups whose values and norms of behavior are somewhat different from those of the dominant culture. Subcultures arise because groups live in different conditions or because they bring different cultural patterns to society. Subcultures typically share some elements of the dominant culture and co-exist within it though the subculture may be quite separated from the dominant culture. Some subcultures actually retreat from the domi-

subculture

nant culture, such as the Amish, some religious cults, and some communal groups. In these cases, the subculture is actually a separate community that lives as independently from the dominant culture as possible. Other subcultures may co-exist with the dominant society, and members of the subculture may participate in both the subculture and the dominant culture. "Deadheads," the followers of the musical group The Grateful Dead, are a colorful example. Deadheads have a unique culture that includes common language, a flamboyant manner of dress featuring tie-dyed designs, and quite often, drug use. Before the death of Jerry Garcia, lead guitarist in the band, many Deadheads dropped out of the dominant culture and followed the band on tour for months; others lived within the norms of the dominant culture, working as professionals and living in the suburbs, but donning their Deadhead gear when the band came to town. Upon Garcia's death in 1995, thousands mourned their loss with spontaneous and, later, planned gatherings to celebrate their culture (see "Sociology in Practice: Touring with the Grateful Dead" for an example of how one sociologist has studied the Dead.)

Sometimes subcultures are imposed, as when groups are excluded from participation in the dominant group. Prohibiting a group from expressing its own culture may drive the culture underground, resulting in a *culture of resistance*, one that is a specific challenge to the dominant culture. In this case, cultural exclusion produces cultural solidarity. The history of Asian Americans in California provides an example. Immigration of Chinese and Japanese laborers to the West Coast near the turn of the century generated racist fears about the "Yellow Peril" threatening Western civilization. In 1882, the United States passed the Chinese Exclusion Act. In 1905, White workers in California formed the Asiatic Exclusion League, proclaiming that "the preservation of the Caucasian race upon American soil . . . necessitates the adoption of all possible measures to prevent or minimize the immigration of Asiatics to America" (Takaki, 1989: 201). In response, the San Francisco school board, with the support of the state governor and the president of the United States, Theodore Roosevelt, directed "all Chinese, Japanese, and Korean children to attend the 'Oriental School'" (Takaki, 1989: 201). Forced into their own schools, excluded from mainstream trades, neighborhoods, and professional associations, Asian Americans developed their own neighborhoods, associations, and social clubs, fostering a high degree of solidarity among themselves; ironically, they were then accused by Whites of being "clannish."

A later example of a subculture prodded into existence as the result of exclusion from the dominant society is found in the history of rhythm and blues. The exclusion of Black musicians from the White-dominated music industry in the 1940s and 1950s led to the creation of what was known as "race music." Race music was produced on independent Black labels and sold primarily to Black audiences. Studios like Stax Records in Memphis produced a vast repertoire of music and supported a large number of Black artists. As race music gathered a following, White artists and producers copied it, recording and marketing it through White-owned recording companies and giving little or none of the huge profits to the original artists and songwriters. There are numerous examples of Black musicians who signed contracts for as little as $100, while giving away their rights to subsequent royalties. Classic Black musicians, such as Big Mama Thornton, Sam Cooke, and countless others whose music is still heard and reproduced on current labels seldom received more than a pittance for their work. In the late 1950s and early 1960s, Pat Boone, Elvis Presley, and many others made fortunes recording songs originally written and performed by Black artists. These rerecordings, called "covers," were developed for sale to White markets. Even though Black artists received neither money nor social recognition for their artistic contributions, race music thrived as a subcultural expression of Black experience. Race music gave birth to rhythm and blues, and it was R&B that spawned rock and roll.

For years, loyal followers of the band The Grateful Dead, known as Deadheads (a classic subculture), gathered where the band was playing, displaying their representative cultural symbols: tie-dyed clothing, beads, and other festive attire.

THINKING SOCIOLOGICALLY

Identify a group on your campus that you would call a *subculture*. What are the distinctive *norms* of this group? Based on your observations of this group, how would you describe its relationship to the *dominant culture* on campus?

Subcultures also develop when new groups enter a society. Such is the case with the numerous and diverse immigrant cultures characteristic of the United States, both now and in the past. Puerto Rican immigration to the U.S. mainland, for example, has generated distinct Puerto Rican subcultures within many urban areas. Although Puerto Ricans also partake in the dominant culture, their unique heritage is part of their subcultural experience. Parts of this culture are now entering the dominant culture. Salsa music, now heard on mainstream radio stations, was created in the late 1960s by Puerto Rican musicians who wanted to "create a cultural sound distinctively representative of their New York barrio [neighborhood] experience" (Padilla, 1989: 349). The themes in salsa reflect the experience of barrio people and mix the musical traditions of other Latin music, including rumba, mambo, and cha-cha. As with other subcultures, the boundaries between the dominant culture and the subculture are permeable, thus resulting in cultural change as new groups enter society.

BOX 3.3 SOCIOLOGY IN PRACTICE

Touring with the Grateful Dead

ONE OF the attractions of being a sociologist is the vast scope for creativity in applying the sociological imagination. Consider the work of sociologist Rebecca Adams, a trained survey researcher whose areas of study are gerontology (the study of aging) and friendship. A member of the faculty at the University of North Carolina at Greensboro, Adams once attended a Grateful Dead concert, by invitation of her husband's boss. At the time, Adams was not a "Deadhead" herself, but had heard the Dead as a college student. At the concert, she saw some of her students, many of whom were probably surprised to see her. After the run of shows, a group of students asked her to supervise their independent research project surveying Deadheads. This was the beginning of a new dimension to Adams's career.

She subsequently toured with the Dead and extensively analyzed the Deadhead subculture, becoming a Deadhead herself along the way. She has been interviewed by numerous rock music and mainstream publications, has appeared on the television program *Entertainment Tonight*, and has participated in the production of a syndicated PBS video on the Grateful Dead (for which she provides the narration). She has regularly published articles about Deadhead culture in fan magazines and is currently writing a book about Deadheads. As she puts it, she has been adopted by the Deadhead community and is considered by Deadheads to be a "tribal representative" to the world outside. She has received thousands of e-mail messages from Deadheads, many of whom have ended up helping her with her project in one way or another.

When Jerry Garcia died in 1995, Adams was consulted by senior members of the subculture to write about grieving. This drew on her expertise in the sociology of aging, where she knew the research on mourning; her sociological knowledge also helped her understand her own grief about Garcia's passing though it did not ease her sadness. Elders in the subculture also talked with her about the future of the subculture, wondering how it would be sustained in the absence of the group leader.

Adams notes that people in the Deadhead subculture come disproportionately from middle- and upper middle-class families; thus, they are reasonably well educated. Deadheads make good sociologists, Adams says, because they tend to take a critical perspective on society and are analytical about their own experience. Adams writes that Deadheads actually have the perspective of symbolic interaction though they would not call it by that name. Symbolic interactionists see that we all create the world we live in as we interact with each other through gestures, language, symbols, and other cultural habits. She sees this reflected in Jerry Garcia's comments about Deadheads. He says, "We didn't invent them, and we didn't invent our original audience. In a way, this whole process has kind of invented us" (cited in Adams, 1995: 17). Adams interprets this to mean that Garcia understood the sociological notion that people create multiple realities in their minds. Deadheads understand that the world is socially constructed, and they actively question it, challenge it, and thus change it.

Drawing on the work of sociologist Georg Simmel, Adams sees different pairs of social forms patterning Deadhead interaction and culture. These include individuality and unity, differentiation and integration, conflict and cooperation, egoism and experientialism, and freedom and order. She describes the research process she uses in studying Deadheads as not that different from the experience of attending a Grateful Dead show. She writes, "Just as Deadheads take mental journeys during improvisational jams and then returned to the reassuring structure of familiar songs, I took conceptual journeys and then return to the comfort of my research design" (1995: 18). Adams's work as a practicing sociologist is surely one of the most intriguing applications of the sociological imagination and an illustration of the diverse ways that sociologists do their work.

SOURCES: Interview with Rebecca Adams, December 21, 1995; Adams, Rebecca G. 1991. "Mourning for Jerry: We Haven't Left the Planet Yet" in *Garcia: A Grateful Celebration* (Dupree: Diamond News), pp. 32–37; and Adams, Rebecca G. 1995. "The Sociology of Jerry Garcia." *Garcia: Reflections* 1: 16–18.

In a highly multicultural society, it is not unusual to see contrasting cultural elements existing side by side.

Countercultures

Countercultures are subcultures created as a reaction against the values of the dominant culture. Members of the counterculture reject the dominant cultural values, often for political or moral reasons, and develop cultural practices that explicitly defy the norms and values of the dominant group. Nonconformity to the dominant culture is often the hallmark of a counterculture.

"Women's communities" are examples of countercultures that have been created to resist the dominant culture. Women's communities are based on woman-centered and feminist social networks, relationships, and cultural activities. Although they are typically not geographically separate from the mainstream society, specific places, such as feminists' bookstores, community activist groups, and women's shelters, promote a feeling of belonging that some women may find absent in the dominant culture. For those within women's communities, this solidarity is a relief from the male domination and harassment that they find in the dominant society (Taylor and Rupp, 1993). The counterculture of the Hippies in the 1960s is another example. Hippies deliberately developed styles of dress, manners of speech, and social practices (such as advocacy of the use of certain drugs, such as marijuana and hashish) that symbolized their rejection of the dominant culture and its values, especially the perceived emphasis on conformity and "straight" behavior.

Some countercultures directly challenge the dominant political system. The contemporary militia movement is an example. Not only have those in this movement withdrawn from society into isolated compounds that they have created in rural areas, but they have also armed themselves for what they perceive as a countercultural revolution against the authority of the U.S. government. Like other countercultures, militia groups have unique modes of dress, a commonly shared worldview, and a distinctive lifestyle.

Countercultures may also develop around cultural pursuits, such as music. Rap is a contemporary example. Originally developed in African American inner-city neighborhoods, rap describes and analyzes the economic and political conditions from which it springs although many outside this urban culture may hear it as simply dangerous, violent, and misogynist (woman hating). Although some rap is harsh, commentators have analyzed rap music as a form of countercultural expression that tells us much about the living experiences of young, urban African American men. Unable to express themselves freely within the dominant culture, young Black Americans see rap music as a means of self-expression. Ironically, it is also big business, with young Black consumers and, increasingly, White consumers spending an estimated $23 billion on it per year (Lusane, 1993; Gray, 1993; Ransby and Matthews, 1993; Kuwahara, 1992; Powell, 1991). It is music that provides a social commentary, just as folk music and antiwar songs in the 1960s articulated the values of a subculture resisting the dominant societal values. As happened with some of the countercultural elements of the hippie lifestyle, some parts of rap culture have been adopted by the dominant culture, like styles of dress, dance forms, and some talk; thus, countercultures can be a source of cultural change—a point we return to later in this chapter.

Ethnocentrism

Ethnocentrism is the habit of seeing things only from the point of view of one's own group. To judge one culture by the standards of another culture is ethnocentric. A White American who sees Latino culture as deficient because it is different from Anglo culture is engaging in ethnocentric thinking. An ethnocentric perspective prevents you from understanding the world as it is experienced by others, and it can also lead to narrow-minded conclusions about the worth of diverse cultures. Any group can be ethnocentric, including members of racial minority groups if they believe their way of living is superior to others. Any group that sees the world only from its own point of view is engaging in ethnocentrism.

A great example of ethnocentrism has recently been documented by a Japanese sociologist studying American perceptions of Japan. Professor Akuto conducted a survey in several major U.S. cities in which he asked Americans to name the most famous Japanese people. When he tallied his results, he discovered that the three most famous "Japanese" people to Americans were Yoko Ono (an American citizen who has not lived in Japan in decades), Bruce Lee (the deceased Chinese kung fu star from Hong Kong), and Godzilla—a fictional monster from the Marshall Islands (Akuto, 1994; cited in Reid, 1994). Considering the significance of Japan in the international economy and its

The Harlem Renaissance produced some of the most distinctive and accomplished artists, including some of the greatest jazz musicians ever. In this photo, taken by Art Kane in Harlem in 1958, we see the musicians gathered on 126th street.

For a list of the artist's names in this photo, see the last page of this book.

importance for contemporary American society, it is amazing to learn that the American people are so ethnocentric, at least collectively, that not a single Japanese citizen appeared among the top three in this list. This extraordinary example shows how ethnocentric thinking is related to cross-cultural ignorance. How many prominent Japanese people can you now name?

Ethnocentrism can build group solidarity, but it also discourages intercultural or intergroup understanding. A good example is seen in the problems of *nationalism*, the sense of identity that arises when a group exalts its own culture over all others and organizes politically and socially around this principle. Nationalist groups tend to be highly exclusionary, rejecting those who do not share their cultural experience and judging all other cultures to be inferior. Nationalist movements tend to use extreme ethnocentrism as the basis for nation-building. Taken to its limits, ethnocentrism can lead to overt political conflict, war, even *genocide* (the mass killing of people based on their membership in a particular group). The conflict between the Hutus and the Tutsis in Rwanda is an example of ethnocentrism taken to such an extreme. In 1994, at least 800,000 people were killed during 100 days of mayhem, when the ruling ethnic group, the Hutus, declared it the obligation of Hutus loyal to the government to slaughter all Tutsis and any Hutus who resisted government orders. The massacre that followed wiped out 75 percent of the Tutsis population. Mass rape and looting accompanied the massacre (Gourevitch, 1996). This genocidal war seems almost incomprehensible, like that in Kosovo, but it shows the potency of ethnocentric thinking, especially when such strongly held beliefs are accompanied by a power structure endorsing homicidal actions as "ethnic cleansing." More recently, in the Spring of 1999, Serbian forces exiled and killed many ethnic Albanians fleeing Kosovo, Yugoslavia, in a similar operation of "ethnic cleansing".

A concept related to ethnocentrism is **androcentrism**, which refers to modes of thinking that are centered only in men's experiences. Like ethnocentrism, androcentrism limits one's perspective to a single group, offering at best a limited and partial view of the world. In recent years, feminist scholars have criticized knowledge that has been based only on the experiences of White men; proponents of multicultural education, including those in African American and Latino studies, gay and lesbian studies, and women's studies, have argued that studying culture, history, and society only from the perspective of any one group distorts the diverse experiences of everyone else and may well misrepresent the favored group as well. Multicultural education is intended to remedy ethnocentric and androcentric thinking.

The Globalization of Culture

The infusion of Western culture throughout the world seems to be accelerating, as the commercialized culture of the United States is marketed on a worldwide basis. One can go to quite distant places in the world and see familiar elements of U.S. culture, whether it be McDonald's in Hong Kong, Coca Cola in South Africa, or Disney products in western Europe, as the map "Viewing Society in Global Perspective" shows. From films to fast food, the United States dominates international mass culture, largely through the influence of capitalist markets, as conflict theorists would argue. The diffusion of a single culture throughout the world is referred to as **global culture**. Despite the enormous diversity of cultures worldwide, fashions, foods, entertainment, and other cultural values are increasingly dominated by U.S. markets, thereby creating a more homogenous world culture. Global culture is increasingly marked by capitalist interests, and it squeezes out the more diverse folk cultures that have been common throughout the world (Rieff, 1993). One result is that tourists can step off the Star Ferry in Hong Kong and, instead of eating Chinese food, have a familiar Big Mac—just as they can now in most major cities of the world. Some sociologists have argued that the feeling of belonging to a community vanishes, as global cultural diffusion occurs (Albrow et al., 1994). Sociologists are increasingly turning their attention to how

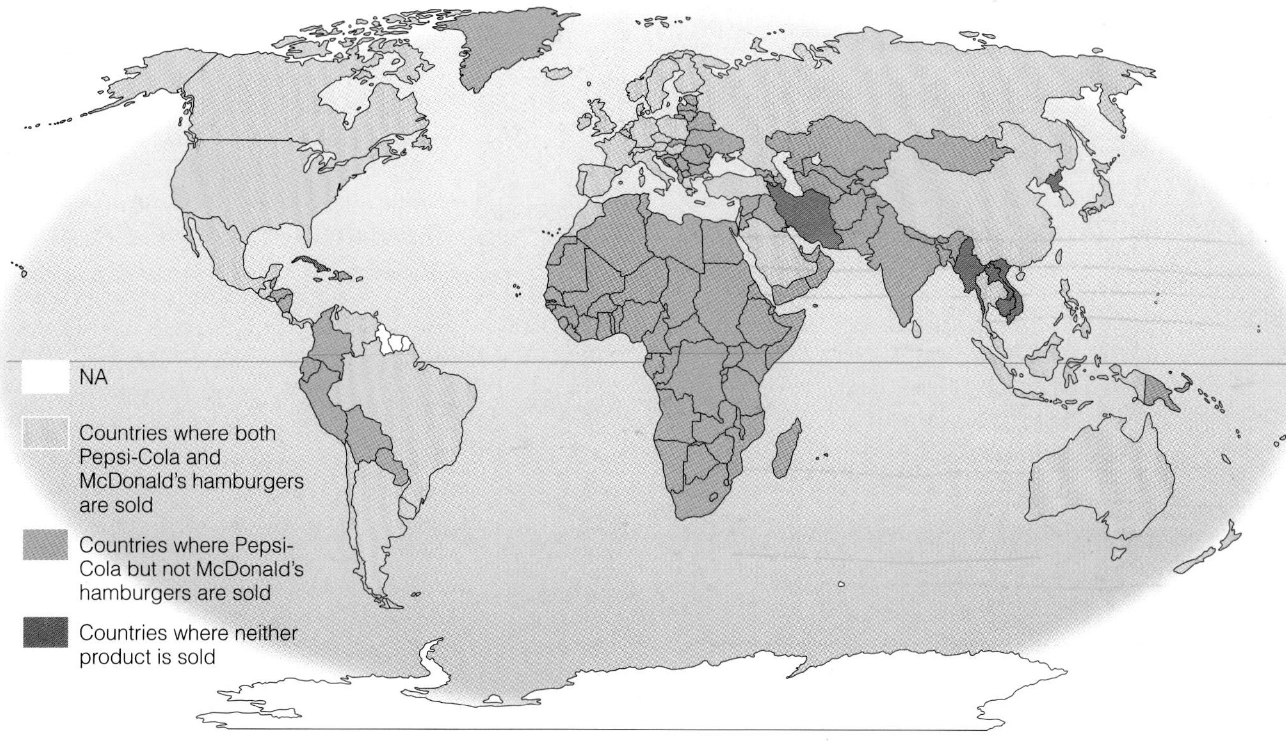

MAP 3.2 Viewing Society in Global Perspective: American Cultural Penetration

SOURCE: Doyle, Rdger, 1994. *Atlas of Contemporary America: Portrait of a Nation.* New York: Facts on File, p. 134.

the development of global culture is linked to other forms of social change (Griswold, 1994; Robertson, 1992).

Popular Culture

Some aspects of culture pervade the whole society, such as common language, general patterns of dress, and dominant value systems. **Popular culture** includes the beliefs, practices, and objects that are part of everyday traditions. This includes mass-produced culture, such as popular music and films, mass-marketed books and magazines, large-circulation newspapers, and other parts of the culture that are shared by the general populace (Mukerji and Schudson, 1986). Popular culture is distinct from *elite culture,* that which is shared by only a select few, but highly valued. Unlike elite culture (sometimes referred to as "high culture"), popular culture is mass-produced and mass-consumed. Popular culture has enormous significance in the formation of public attitudes and values, and plays a significant role in shaping the patterns of consumption in contemporary society.

The distinction between elite and popular culture is relatively new. The industrialization of modern life in late nineteenth-century America fragmented public life, creating greater cultural divisions between people and establishing a more distinct hierarchy within cultural institutions. Even such simple things as having hierarchical seating arrangements in theaters or the development of separate theaters for the "truly refined" reflect changes in the socioeconomic arrangements of society. During the early twentieth century, as a larger middle and upper middle class emerged, the distinction between elite and mass culture widened. Whereas in the early nineteenth century Shakespeare and other classic writers were read by a wide spectrum of the reading public, now they are considered more refined (Levine, 1984).

This trend has continued although there is some blurring of the distinction. Although elites may derive their culture from expensive theater shows, opera performances, or private libraries, millions of "ordinary" citizens get their primary cultural experience from book-of-the-month clubs, Harlequin romances, mass-produced magazines, and popular television shows. In common discourse, the very word *cultured* is associated with elite cultural taste.

Different groups partake of popular and elite culture in different ways, as Figure 3.2 shows (DiMaggio and Ostrower, 1990). These patterns lead sociologists to conclude that cultural taste and participation in the arts are socially structured. First, participation in the elite forms of culture is expensive and, therefore, shaped by the class status of different groups. Second, familiarity with different cultural forms stems from patterns of historical exclusion, as well as integration into networks that provide information about the arts. As Figure 3.2 shows, Black Americans are much

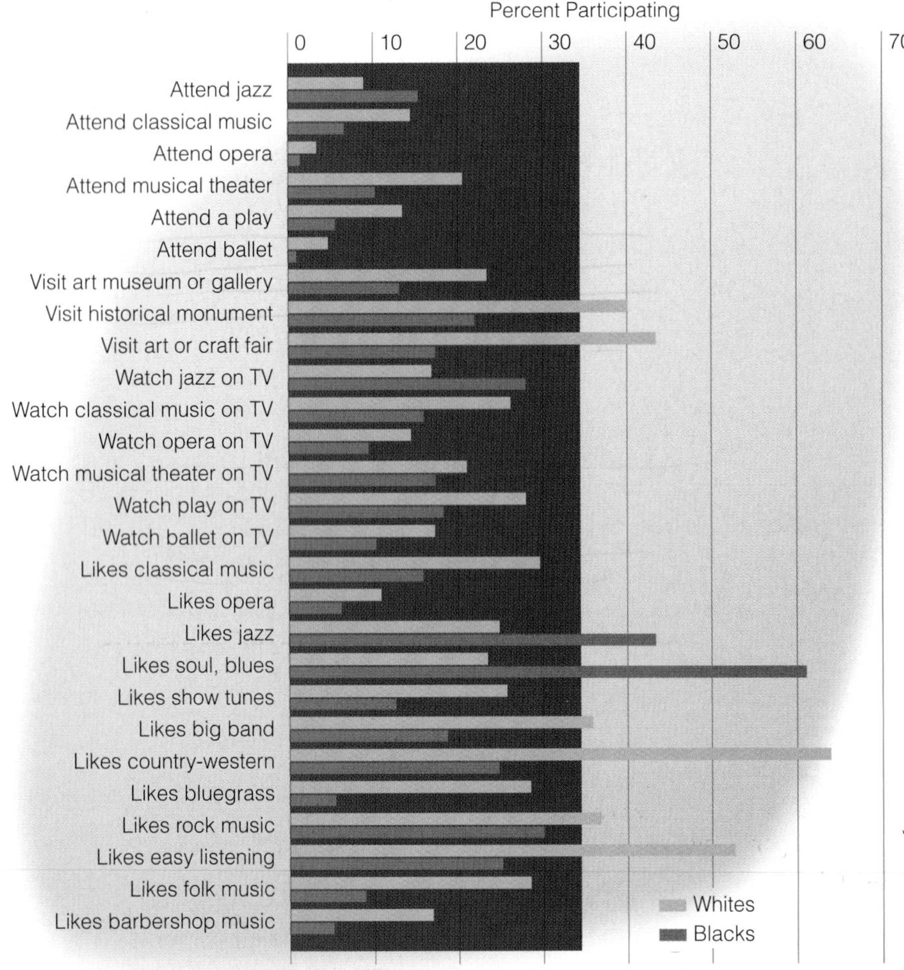

FIGURE 3.2 Participation in the Arts

Popular culture is defined by its mass appeal. The percentage of those who enjoy different forms of popular culture, however, differs significantly by race.

DATA: From DiMaggio, Paul and Francie Ostrower. 1990. "Participation in the Arts by Black and White Americans." *Social Forces* 68 (March): 759.

more likely than White Americans to attend jazz concerts and listen to soul, blues, rhythm-and-blues music, and other historically African American musical forms. Aside from racial differences, only a small percentage of Whites and Blacks participate in the more elite forms of culture; far more popular is mass culture, especially radio and television.

The Influence of the Mass Media

Popular culture is characterized by mass distribution. The term **mass media** refers to those channels of communication that are available to very wide segments of the population—the print, film, and electronic media (radio and television). The mass media have extraordinary power to shape public perceptions. In an era when the presentation of complex political and social issues are reduced to "sound bites" and "photo opportunities," the mass media increasingly have the ability to shape public information.

Television, for example, is a powerful force for transmitting cultural values. The majority of adults now get most of their news from TV (Lee and Solomon, 1990). The increasing popularity of radio talk shows has had a significant impact of the public's political values. In 1985, there were about 300 all-talk radio stations; now that number exceeds 1100. One in five adults listens regularly and one-third say it is their favorite source of political information (Sifry and Cooper, 1995). The conservative ideology of many of these shows, which play upon the theme of little people being dominated by big government and the presumed power of liberals, feminists, and minorities, has contributed to a strong shift to conservativism in U.S. politics.

George Gerbner, a communications analyst, suggests that television is the national religion of the United States. Television is the common basis for social interaction in a widely dispersed and diverse national community. The power of television to captivate the public was vividly demonstrated when Princess Diana of England died. Millions of people worldwide watched her funeral on television, and within hours, Elton John's tribute to Princess Diana, the song "Candle in the Wind," was broadcast on television and radios around the world, becoming so popular that it quickly became the highest-selling recording ever, generating more than $16 billion for charity.

Gerbner sees television as having more than just influence. He argues that television generally portrays the most homogeneous view of culture and is resistant to change (Gerbner, 1978). For most Americans, leisure time is dominated by television. Ninety-eight percent of all homes in the United States have at least one television—more than have telephone service (U.S. Bureau of the Census, 1997: 566). The average person watches television for thirty-one hours per week. Young people between the ages of five and eighteen will spend more time watching television than they will spend in school; for people in their midtwenties, television is the third most time-consuming activity of the day, second only to work and sleep (Davis, 1990). The power of television to reach the public was especially apparent during the O. J. Simpson trial. When the verdict was read, 107.7 million people, 57 percent of the population, watched the verdict on television. The nation virtually came to a halt. Airplane flights were held, stock trading on Wall Street dropped by two-thirds, and long-distance telephone calls in the minutes surrounding the verdict dropped almost 60 percent below normal (Kleinfield, 1995).

The mass media have enormous power to shape public opinion and behavior. For example, content analyses of news programs show that the largest percentage of time is spent reporting stories about crime. This can influence people's perception and fear of crime; indeed, sociologists have found that people's fear of crime is directly related to the time they spend watching television or listening to the radio. Furthermore, during recent years, when the crime rate has actually fallen, the amount of time spent reporting crime has actually increased (Angotti, 1997; Chiricos et al., 1997).

Although people tend to think of the news as authentic and true, news is actually manufactured in a complex social process. From a sociological perspective, it is not objective reality that determines what news is presented and how it is portrayed, but commercial interests, the values of news producers, and perceptions of what matters to the public. The media shape our definition of social problems by determining the range of opinion or information that is defined as legitimate and by deciding which experts will be called on to elaborate an issue (Gans, 1979; Gitlin, 1983).

Now an icon in popular culture, Ally McBeal, the principal character on the television show of the same name, is a symbol to many of the values and ideals of young professional women.

Racism and Sexism in the Media

Many sociologists have argued that the mass media promote narrow definitions of who people are and what they can be. What is considered beauty, for example, is not universal; ideals of beauty change as cultures change and depend upon what certain cultural institutions promote as beautiful. Aging is not beautiful, youth is; light skin is promoted as more beautiful than dark skin (regardless of one's race) although being "tan" is seen as more beautiful than being pale. In African American women's magazines, the models typified as most beautiful are generally those with clearly Anglo features—light skin, blue eyes, and straight or wavy hair. These depictions have fluctuated over time; in the early 1970s, for example, there was a more Afrocentric ideal of beauty—darker skin, "Afro" hairdos, and African clothing. Today, the images of African American women have returned to more Anglocized depictions of beauty. European facial features are also pervasive in the images of Asian and Latino women appearing in U.S. magazines. The media communicate that only certain forms of beauty are culturally valued; these ideals are not somehow "natural"; they are constructed by those who control cultural and economic institutions (Kassell, 1995; Wolf, 1991; Barthel, 1988; Chapkis, 1986).

Images of women and racial and ethnic minorities in the media are similarly limiting. Content analyses of television reveal that during prime time, men are a large majority of the characters shown. On soap operas, women are cast either as evil or good, but naive. One study of commercials on MTV finds that in music videos, women characters appear less frequently, have more beautiful bodies, are more physically attractive, wear more sexy and skimpy clothing, and are more often the object of another's gaze than their male counterparts (Signorielli et al., 1994).

Even though African Americans and Hispanics watch more television than Whites do, they are a small proportion of TV characters, generally confined to a narrow variety of character types, depicted in stereotypical ways. Latinos are often stereotyped as criminals or passionate lovers. African American men are most typically seen as athletes and sports commentators, criminals, or entertainers; African American women, in domestic or sexual roles or as sex objects. It is difficult to find a single show where Asians are the principal characters—usually they are depicted in silent roles as domestics or other behind-the-scenes characters. More often, they are in the background as domestic workers or extras. Native Americans make occasional appearances where they usually are depicted as mystics or warriors; typically, if presented at all, Native Americans are marginal characters stereotyped as silent, exotic, and mysterious. Jewish women are generally invisible on popular TV programming, except when they are ridiculed in stereotypical roles (Kray, 1993). Class stereotypes abound, as well, with working-class men typically portrayed as being ineffectual, even buffoonish (Butsch, 1992; Dines and Humez, 1995). On television, people of color are also far more likely to be found in comedies than dramas. African American women, especially, are most typically seen as characters in comedy shows. In dramas, African American and Latino men are typically the sidekicks, not the major characters.

THINKING SOCIOLOGICALLY

Watch a particular kind of television show (situation comedy, sports broadcast, children's cartoon, or news program, for example) and, using *content analysis*, make careful written notes on the depiction of different groups in this show. How often are women and men, boys and girls shown? How are they depicted? You could also observe the portrayal of Asian Americans, Native Americans, African Americans, or Latinos. What do your observations tell you about the cultural ideals that are communicated through *popular culture*?

Women and racial–ethnic minorities are also seriously underrepresented in and on network news, one of the most important outlets of information about society and culture. They are underrepresented as news commentators, as correspondents, as news directors, and as news subjects (Rhode, 1995; Gibbons, 1992; Ziegler and White, 1990). Although by the 1990s, women's representation among network news reporters had increased, they are only 15 percent of network news reporters and 20 percent of print journalists (though they are 68 percent of journalism school graduates). Men write two-thirds of the front-page stories in the news and provide 85 percent of television reporting; only 3 percent of newspaper executives are women. Women of color are even further underrepresented, providing only 2 percent of broadcast media stories. Men also provide 85 percent of quotes or references in the media, are 75 percent of those interviewed on TV, and are 90 percent of the most recently cited pundits—even on issues that involve women (Rhode, 1995). With limited coverage of issues important to women and racial–ethnic groups, the public can hardly be well informed about race and gender as social and political issues. This has led sociologists to conclude that the news media reflect the White male social order and primarily support the "public, business, and professional upper middle-class sectors of society" (Gans, 1979: 61). It is therefore not surprising that many minorities are dissatisfied with how they see themselves represented in the media (McAneny, 1994; see Figure 3.3).

Television is not the only form of popular culture that influences public consciousness about gender and race. Music, film, books, and other industries play a significant role in molding public consciousness. What images are produced by these cultural forms? Studies of rock videos are instructive. Even with the relatively recent introduction of Black Entertainment Television (BET) and other shows focusing on Black and Latino artists, more than three-fourths of rock videos feature White male singers or bands led by White men. Rock videos more often depict White men as the center of attention; women and African American or Hispanic men are rarely shown in the foreground. Typically, White women are depicted as trying to get the attention of men; African Americans are more likely than Whites to be seen singing and dancing.

Do these images matter? In an example of the influence of popular culture on people's beliefs, researchers designed experiments to assess the influence of listening to heavy-metal rock on men's attitudes toward women. They found that exposure to heavy-metal rock music increases men's stereotyping of women, including their acceptance of violence against women, regardless of the lyrical content. This same study also found that men reported more sexual arousal from hearing "easy-listening classical music" than they did from hearing heavy-metal music! (St. Lawrence and Joyner, 1991). More than mere cultural curiosities, the artifacts and images that cultural institutions promote have an enormous effect on people's behavior and values.

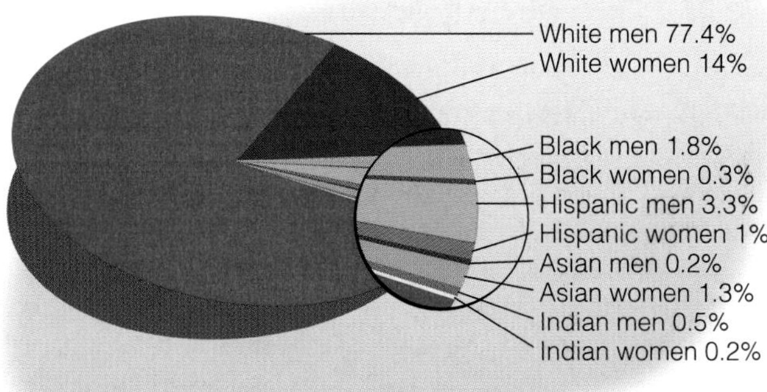

FIGURE 3.3 Race and Gender of Television News Directors

Despite their growing presence in the media, women and racial minorities are still a small proportion of those who produce and direct the news.

DATA: From Sheila Gibbons. 1992. *Media Report to Women* (Fall): 5.

Theoretical Perspectives on Culture

Sociologists study culture in a variety of ways, asking a variety of questions about the relationship of culture to other social institutions and the role of culture in modern life. One important question for sociologists studying the mass media is whether these images have any effect on those who see them. Do the media create popular values or reflect them? The **reflection hypothesis** contends that the mass media reflect the values of the general population (Tuchman, 1979). The media try to appeal to the most broad-based audience, so they aim for the middle ground in depicting images and ideas. Maximizing popular appeal is central to television program development; media organizations spend huge amounts on market research to uncover what people think and believe and what they will like. Characters are then created with whom people will identify. Interestingly, the images in the media with which we identify are distorted versions of reality. Real people seldom live like the characters on television although part of the appeal of these shows is how they build upon and then mythologize the actual experiences that people have.

The reflection hypothesis assumes that images and values portrayed in the media reflect the values existing in the public, but the reverse can also be true—that is, the ideals portrayed in the media also influence the values of those who see them. As an example, social scientists have studied the stereotyped images commonly found in children's programming. Among their findings, they have shown that the children who watch the most TV hold the most stereotypic gender attitudes (Signorielli, 1991). Although there is not a simple and direct relationship between the content of mass media images and what people think of themselves, clearly these mass-produced images have a significant impact on who we are and what we think.

Culture and Group Solidarity

Many sociologists have studied particular forms of culture and have provided detailed analyses of the content of cultural artifacts, such as images in certain television programs or genres of popular music. Other sociologists take a broader view by analyzing the relationship of culture to other forms of social organization. Beginning with some of the classical sociological theorists (see Chapter 1), sociologists have studied the relationship of culture to other social institutions. Max Weber looked at the impact of culture on the formation of social and economic institutions. In his classic analysis of the Protestant work ethic and capitalism, Weber argued that the Protestant faith rested on cultural beliefs that were highly compatible with the development of modern capitalism. By promoting a strong work ethic and a need to display material success as a sign of religious salvation, the Protestant ethic indirectly but effectively promoted the interests of an emerging capitalist economy. (We revisit this issue in Chapter 16.) In other words, culture influences other social institutions.

Many sociologists have also examined how culture integrates members into society and social groups. Functionalist theorists, for example, believe that norms and values create social bonds that attach people to society. Culture therefore provides coherence and stability in society. Robert Bellah and his colleagues have examined this idea in their book *Habits of the Heart* (1985), which became a national bestseller. Among other things, Bellah concluded that the culture of individualism in the United States has promoted a sense of separation from others. Bellah calls for new cultural forms, such as a commitment to community service, as a way of resolving the isolation and detachment that he thinks characterizes the modern period in the United States.

Classical theoretical analyses of culture have placed special emphasis on nonmaterial culture—the values, norms, and belief systems of society. Sociologists who use this perspective emphasize the integrative function of culture; that is, its ability to give people a sense of belonging in an otherwise complex social system (Smelser, 1992a). In the broadest sense, they see culture as a major integrative force in society, providing societies with a sense of collective identity and commonly shared world views.

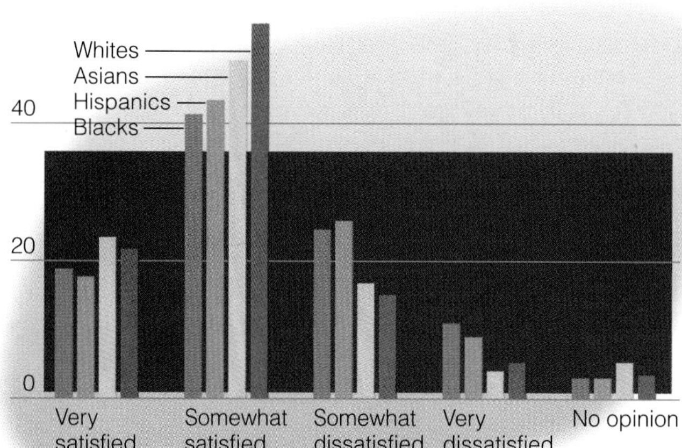

FIGURE 3.4 Satisfaction with the Media View of Minorities

The question asked: "Think for a moment about subjects in the news such as education, politics, business and society—that affect all Americans. How satisfied are you with the job the national television news does of including (respondent's ethnic group) in its regular news coverage of these stories?

DATA: From McAneny, Leslie . 1994. "Ethnic Minorities View of the Media's View of Them." *The Gallup Poll Monthly* (August), p. 37.

Culture, Power, and Social Conflict

Whereas the emphasis on shared values and group solidarity drives one sociological analysis of culture, conflicting values drives another. Conflict theorists (see Chapter 1) have analyzed culture as a source of power in society. From a conflict-oriented perspective, culture is dominated by economic interests. A few powerful groups are seen to be the major producers and distributors of culture, as Figure 3.5 shows. Disney Cap Cities not only controls Disneyland and Disney World, but many stores, the ABC television and radio networks, several motion picture groups, book publishers,

TABLE 3.2 *THEORETICAL PERSPECTIVES ON CULTURE*

According to:

Functionalism	Conflict Theory	Symbolic Interaction	New Cultural Studies
Culture . . .			
integrates people into groups	serves the interests of powerful groups	creates group identity from diverse cultural meaning systems	is ephemeral, unpredictable and constantly changing
provides coherences and stability in society	can be a source of political resistance	changes as people produce new cultural meaning systems	is a material manifestation of a consumer-oriented society
creates norms and values that integrate people in society	is increasingly connected by economic monopolies	is socially constructed through the activities of social groups	is best understood by analzying its artifacts—books, films, television images

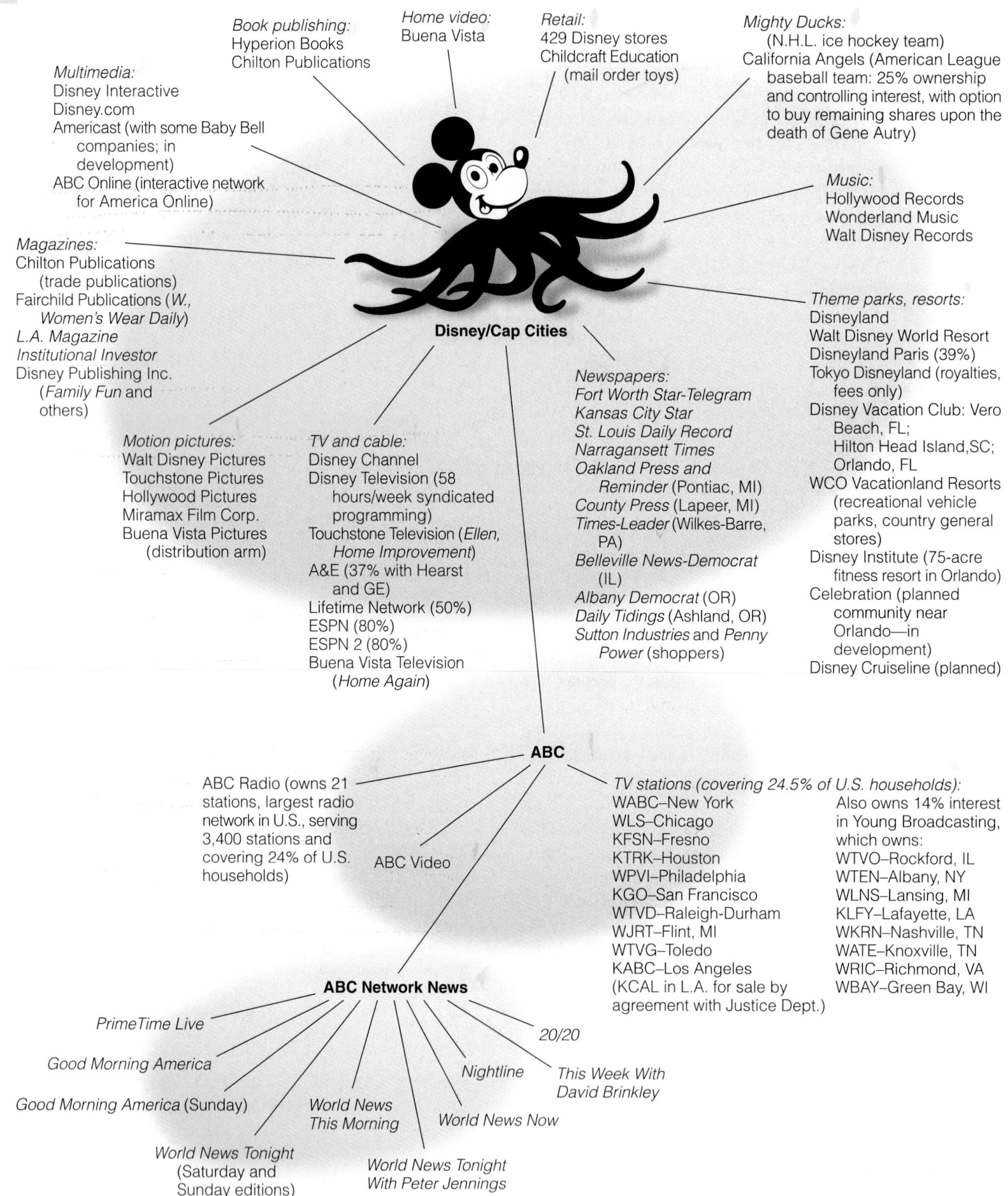

FIGURE 3.5 Who Owns Culture?

As an example of cultural hegemony, *this chart shows the reach of one major cultural industry—Disney/Cap Cities—via its ownership of other cultural producers.*

SOURCE: Miller, Mark Crispin. 1996. "Free the Media." *The Nation* 262 (June 3): 25–26.

newspapers, and other entertainment media. Similar octopuslike holdings characterize other commercial purveyors of popular culture (Miller, 1996), revealing that the production of popular culture is in just a few hands.

Conflict theorists see contemporary culture as produced within institutions that are based on inequality and capitalist principles. As a result, the cultural values and products that are produced and sold promote the economic and political interests of the few—those who own or benefit from these cultural industries. This is especially evident in the study of popular culture, which is marketed to the masses by entities with a vast economic stake in the dissemination of their products. Conflict theorists conclude that the cultural products most likely to be produced are those that are consistent with the values, needs, and interests of the most powerful groups in society. The evening news, for example, is typically sponsored by major financial institutions, oil companies, and automobile makers. Conflict theorists then ask how this commercial sponsorship influences the content of the news. If the news were sponsored by labor unions, would conflicts between management and workers always be defined as "labor troubles," or might newscasters refer instead to "capitalist troubles" (Lee and Solomon, 1990)?

Conflict theorists see culture as increasingly controlled by economic monopolies. Whether it is books, music, films, news, or other cultural forms, monopolies in the communications industry (where culture is increasingly located) have a strong interest in protecting the status quo. As media conglomerates swallow up smaller companies and drive out smaller, less efficient competitors, the control that economic monopolies have over the production and distribution of culture becomes enormous. Mega-communications companies then influence everything—from the movies and television shows you see to the books you read in school.

Sociologists refer to the concentration of cultural power as **cultural hegemony** (pronounced "heh-JEM-o-nee"), defined as the pervasive and excessive influence of one culture throughout society. Cultural hegemony creates a homogeneous mass culture. It is a means through which powerful groups gain the assent of those they rule. The concept of cultural hegemony implies that culture is highly politicized, even if it does not appear so. Through cultural hegemony, those who control cultural institutions can also control people's political awareness because they create cultural beliefs that make the rule of those in power seem inevitable and right. As a result, political resistance to the dominant culture is blunted (Gramsci, 1971).

Not only does the mass marketing of culture have a political effect, it has an interesting economic side as well. Those who produce cultural products in a capitalist society must constantly produce new needs in order to expand economic markets. The mass media report on changing fashions, thereby encouraging people to buy more clothes. The latest recordings receive extensive radio time, sending listeners to the CD store to be culturally current. Cornel West, a contemporary African American philosopher, argues that the development of a pervasive market culture like that of contemporary American society produces an "addiction to stimulation" (hooks and West, 1991: 95). The purveyors of culture thrive by constantly producing and feeding new needs in the public.

Conflict theorists point out, however, that culture can also be a source of political resistance. The development of new cultural beliefs and practices can challenge dominant group control over others. Reclamation of indigenous culture is one of the ways that groups such as Black Africans in Zimbabwe have overthrown colonial rulers. Within the United States, the American Indian Movement (AIM) used the resuscitation of traditional Indian culture as a central organizing tool in their campaign to reassert the independence of Native Americans.

A final point of focus for sociologists studying culture from a conflict perspective lies in the concept of **cultural capital**, which refers to the cultural resources that are socially designated as being worthy (such as knowledge of elite culture) and that give advantages to groups possessing such capital (Bourdieu, 1977; DiMaggio, 1982). This idea has been most developed by the French sociologist Pierre Bourdieu (1984), who sees the appropriation of culture as one of the ways that groups maintain their social status. Bourdieu argues that members of the dominant class have distinctive lifestyles that mark their status in society; their ability to display this cultural lifestyle signals their importance to others; that is, they possess cultural capital. From this point of view, culture has a role in reproducing inequality between groups. Those with cultural capital use it to improve their social and economic position in society; sociologists have found a significant relationship, for example, between cultural capital and grades in schools—that is, those from the more well-to-do classes (those with more cultural capital) are able to parlay their knowledge into higher grades, thereby reproducing their social position by being more competitive in school admissions and, eventually, in the labor market (DiMaggio, 1982). Likewise, Lamont's study (1992) of American and French upper middle-class men shows how cultural status is used to mark their moral worth; Lamont argues that the upper middle class uses their cultural capital to proclaim their social value to others.

THINKING SOCIOLOGICALLY

Sociologists argue that those with *cultural capital* have more educational advantage than others. As you have gone through school, have you seen evidence of this? What effect does it have on different groups?

New Cultural Studies

Especially productive when applied to the study of culture has been symbolic interaction theory—a perspective that analyzes behavior in terms of the meaning people give it. The concept of culture is central to this orientation in that symbolic interaction emphasizes the interpretive basis of social behavior, and culture provides the interpretive framework through which behavior is understood.

Symbolic interaction also emphasizes that culture, like all other forms of social behavior, is socially constructed.

That is, culture is produced through social relationships and in social groups, such as the media organizations through which culture is disseminated. People do not just passively submit to cultural norms; rather, they are active, skillful makers of culture. Culture is not one-dimensional; it contains diverse elements and provides people with a wide range of choices from which to select how they will behave (Swidler, 1986). Culture, in fact, represents the creative dimension of human life.

In recent years, a new interdisciplinary field known as *cultural studies* has emerged that builds on the insights of the symbolic interaction perspective in sociology. Sociologists who work in cultural studies are often critical of classical sociological approaches to culture, arguing that the classical approach has overemphasized nonmaterial culture—that is, ideas, beliefs, values, and norms. The new scholars of cultural studies find that material culture has increasing importance in modern society (Crane, 1994). This includes cultural forms that are recorded through print, film, artifacts, or the electronic media. Postmodernist theory has greatly influenced new cultural studies. Recall from Chapter 1 that postmodernism is based on the idea that society is not an objective thing; rather, it is found in the words and images that people use to represent behavior and ideas. Given this orientation, postmodernism calls significant attention to the artifacts produced in modern culture.

Classical theorists have tended to study the unifying features of culture; cultural studies researchers tend to see culture as more fragmented and unpredictable. To them, culture is a series of images—images that can be interpreted in multiple ways depending on the viewpoint of the observer. From the perspective of new cultural studies theorists, the ephemeral and rapidly changing quality of contemporary cultural forms is reflective of the highly technological and consumer-based culture on which the modern economy rests. Their fascination is in part with the illusions that such a dynamic and rapidly changing culture produces.

Cultural Change

In one sense, culture is a conservative force in society; it tends to be based on tradition and is passed on through generations, conserving and regenerating the values and beliefs of society. Culture is also increasingly based on institutions that have an economic interest in maintaining the status quo. People are also often resistant to cultural change since familiar ways and established patterns of doing things are hard to give up; nevertheless, cultures do change. Culture is dynamic, not static, and it develops as people respond to changes in their physical and social environment.

Culture Lag

Culture lag refers to the delay in cultural adjustments to changing social conditions (Ogburn, 1922). Some parts of culture may change more rapidly than others; thus, one aspect of culture may "lag" behind another. Rapid technological change is often attended by culture lag since some elements of the culture do not keep pace with technological innovation. In today's world, we have the technological ability to develop efficient, less polluting rapid transit, but changing people's transportation habits is difficult.

When culture changes rapidly, or someone is suddenly thrust into a new cultural situation, the result can be **culture shock**, the feeling of disorientation that can occur when one encounters a new or rapidly changed cultural situation. Imagine the experience of Soviet cosmonaut Sergei Krikalev, who in 1992, returned to earth after 313 days in space, during which time the Union of Soviet Socialist Republics had ceased to exist. On a less spectacular level, returning home after visiting a foreign country can make one's "native" cultural habits seem suddenly strange. Even moving from one cultural environment to another within one's own society can make a person feel out of place and alienated. The greater the difference between cultural settings, the greater the culture shock.

Sources of Cultural Change

There are several causes of cultural change, including (1) a change in the societal conditions, (2) cultural diffusion, (3) innovation, and (4) the imposition of cultural change by an outside agency. Let us examine each.

Cultures change in response to changed conditions in the society. Economic changes, population changes, and other social transformations all influence the development of culture. A change in the makeup of a society's population may be enough by itself to cause a cultural transformation. An example is the appearance of the generation known as Baby Boomers. In the decade and a half following World War II, there was a marked rise in the U.S. birthrate, and it was on this rising tide that the Baby Boomers arrived. The explanation for this population boom is generally considered to be people's desire to make families before or after separations caused by war (the Second World War, the Korean War, and to a lesser extent, the Vietnam War). Because of their size, this generation, now in middle age and aging, exerts a great influence throughout the culture. The Baby Boomers have drawn the concentrated attention of marketers and have driven the development of a variety of consumer industries. They are now older, more affluent, more influential, and more economically powerful than ever, and they are an increasingly powerful political majority. The Baby Boomers were expected to have fewer children than prior generations, but we are now seeing that many simply delayed childbearing. We are in the midst of a "Baby Boomlet," another birth spurt that will cause cultural change as the children of the Boomers, already christened "yuppie puppies," grow up.

Cultures change through cultural diffusion. **Cultural diffusion** is the transmission of cultural elements from one society or cultural group to another. In our world of instantaneous communication, cultural diffusion is swift and widespread. This is evident in the degree to which worldwide cultures have been westernized. Cultural diffusion also

cultural lag

occurs when subcultural influences enter the dominant group. Dominant cultures are regularly enriched by minority cultures. An example is the influence of Black and Latino music on other musical forms, and in expressions like "dissing," a term from Black urban street culture that is now part of the common parlance. Cultural diffusion is one of the things that drives cultural evolution, especially in a society such as ours that is lush with diversity.

Cultures change as the result of innovation, including technological developments. The discovery of new knowledge leads to cultural developments that would be unlikely to evolve in the absence of research and education. The modern kitchen, for example, reflects many such changes, including automatic dishwashers, microwaves, food processors, and other appliances. These innovations have fostered dramatic changes in household organization and cultural practices like cooking; in the popularity of microwaves, we can see how technological advances match changes in modern life. Harried schedules lead to quick meals. The employment of women outside the home and the increasing likelihood that a household will include at least two workers has meant that cultural habits of the past, like the family dinner prepared in the afternoon and shared in the evening, may be disappearing. Innovations in computing have had a giant effect on contemporary culture. The smallest laptop or handheld computer today weighs hardly more than a few ounces, and its capabilities rival that of computers that filled entire buildings only twenty years ago. Computing infiltrates every dimension of life. It is hard to overestimate the effect of innovation on contemporary cultural change. Technological innovation is so rapid and dynamic that one generation can barely maintain competence with the hardware of the next. Answering machines, cellular phones, and fax machines were glamorous devices just a few years ago. Today, they are as routine as electric can openers. In such a rapidly changing technological world, it is hard to imagine what will be common in just a few years.

Cultural change can be imposed. Change can occur when a powerful group takes over a society and imposes a new culture. The dominating group may arise internally, as in a political revolution, or it may appear from outside, perhaps as an invasion. When an external group takes over the society of a "native," or indigenous, group, as White settlers did with Native American societies, they typically impose their own culture while prohibiting the indigenous group from expressing its original cultural ways. Manipulating the culture of a group is a way of exerting social control. Many have argued that public education in the United States, which developed during a period of mass immigration, was designed to force White, northern European, middle-class values onto a diverse immigrant population that was perceived to be potentially unruly and politically disruptive; likewise, the schools run by the Bureau of Indian Affairs have been used to impose dominant group values on Native American children (Snipp, 1996).

Resistance to political oppression often takes the form of a cultural movement that asserts or revives the culture of an oppressed group; thus, cultural expression can be a form of political protest. Nationalist movements—those that identify a common culture as the basis for group solidarity—are an example. Black nationalism in the United States, or Afrocentrism, is the idea that African Americans should be united by their African heritage and that this should be the basis for cultural identity and political mobilization. This is not an entirely new phenomenon. Black nationalism was a strong theme in Black protest movements in the 1920s and surfaced again in the 1970s as the Black Power movement emphasized that "Black is beautiful," encouraging African Americans to celebrate their African heritage with Afro hairstyles, African dress, and African awareness. Cultural nationalism has also inspired a similar movement among Latinos—*La Raza Unida* (meaning, "the race, or the people, united"), which promoted solidarity among Chicanos. As people acquire new means of cultural expression, they can come to question the existing value systems and challenge dominant cultural forms. Cultural change can promote social change, just as social change can transform culture.

CHAPTER SUMMARY

- *Culture* is the complex and elaborate system of meaning and behavior that defines the way of life for a group or society. It is shared, learned, taken for granted, symbolic, and emergent and varies from one society to another.
- The major elements of culture are *language, norms, beliefs,* and *values.* Becoming part of a social group requires learning the special language associated with that group. Language shapes our perception of reality.
- *Norms* are rules of social behavior that guide every situation and may be formal or informal. When norms are violated, social sanctions are applied. *Beliefs* are strongly shared ideas about the nature of social reality. *Values* are the abstract concepts in a society that define the worth of different things and ideas.
- As societies develop and become more complex, cultural diversity can appear. The United States is highly diverse culturally, with many of its traditions influenced by immigrant cultures and the cultures of African Americans, Latinos, and Native Americans. The *dominant culture* is the culture of the most powerful group in society. *Subcultures* are groups whose values and cultural patterns depart significantly from the dominant culture.
- *Popular culture* is the beliefs, practices, and objects that are part of everyday traditions, including the *mass media.* The mass media have an increasing influence on cultural values and images, both reflecting and creating cultural values.
- Sociological theory provides different perspectives on the significance of culture. *Functionalist theory* emphasizes the influence of values, norms, and beliefs on the whole society. *Conflict theorists* see culture as influenced by economic interests and power relations in society. *Symbolic interactionists* emphasize that culture is socially constructed. This has influenced new cultural studies, which interprets culture as a series of images that can be analyzed from the viewpoint of different observers.
- As cultures change, *culture lag* can result, meaning that sometimes cultural adjustments are out of synchrony with each other. Persons who experience new cultural situations may experience *culture shock.* There are several sources of *cultural change,* including change in societal conditions, cultural diffusion, innovation, and the imposition of change by dominant groups.

KEY TERMS

androcentrism	language
beliefs	laws
counterculture	mass media
cultural capital	material culture
cultural diffusion	mores
cultural hegemony	nonmaterial culture
cultural relativism	norms
culture	popular culture
culture lag	reflection hypothesis
culture shock	Sapir–Whorf hypothesis
dominant culture	social sanctions
ethnocentrism	subculture
ethnomethodology	symbols
folkways	values
global culture	

THE INTERNET: A Tool for the Sociological Imagination

Resources on the Internet:

Virtual Society: The Wadsworth Sociology Resource Center at
http://sociology.wadsworth.com

Visit this site to find additional learning tools, including interactive quizzes, links to related web sites, and an easy link to InfoTrac College Edition.

Center for Media Education
http://www.cme.org/cme/

This non-profit organization is dedicated to improving the quality of the media, especially for children. The web site includes links to press releases on information about the media, as well as new work on the use of interactive technology by children.

National Coalition of Television Violence
http://www.nctvv.org/

This web site provides news and information about violence on television, including recent media stories about television violence.

Media Watch
http://www.mediawatch.com/

An organization dedicated to challenging racism and sexism in the media; the web site includes archives on the organization's work, as well as lists of the educational videos distributed by the organization.

Sociology and Social Policy: Internet Exercises

Culture may seem like a relatively neutral concept, but in recent years, it has become the basis for numerous social conflicts and policy debates. The movement to make English the official language of the United States and related debates about *bilingual education* (such as the 1998 referendum in California to ban bilingual education) signify the significance of culture in the creation of social policy. What are the arguments for and against bilingual education, and how is this policy debate linked to increasing diversity in the United States?

Internet Search Keywords:

bilingual education	diversity
cultural conflicts	English official language
California Referendum 1998	immigration
social policy	

Web sites:

http://www.nabe.org/
Web site for the National Association for Bilingual Education. Contains most recent initiatives on bilingual education.

http://www.ncbe.gwu.edu/
National Clearinghouse for Bilingual Education. This site includes all NCBE publications, plus databases, journal articles, and many links to related sites.

http://www.ceousa.org/
The Center for Equal Opportunity offers information on issues relating to racial preferences, immigration and assimilation, and multicultural education.

http://www.us-english.org/
The U.S. English web site includes information on legislation to make English the official language.

http://www.ins.usdoj.gov/
United States Immigration and Naturalization Service web site.

InfoTrac College Edition: Search Word Summary

culture	norm
culturalism	subculture

In order to learn more about these central topics in sociology, you can conduct an electronic search using InfoTrac College Edition. To aid in your search and to gain useful tips, see the Student Guide to InfoTrac College Edition on the Virtual Society web site:
http://sociology.wadsworth.com

INTERACTIONS—A SOCIOLOGY CD-ROM: CONCEPTS FOR THIS CHAPTER

Go to the Wadsworth Sociology CD-ROM for further study on the concepts in this chapter. The CD-ROM also includes quizzes and additional activities to expand your learning experience.

SUGGESTED READINGS

Asian Women United of California (ed.). 1989. *Making Waves: An Anthology of Writings By and About Asian American Women.* Berkeley: University of California Press.

The essays in this anthology provide a rich perspective on the cultural experiences of Asian American women. Much of the book is literary; some of it is social and political analysis. It is a good introduction to the experiences of different Asian American women in the context of race, class, and gender oppression.

Dyson, Michael Eric. 1993. *Reflecting Black: African-American Cultural Criticism.* Minneapolis: University of Minnesota Press.

This collection of essays provides an African American perspective on contemporary U.S. culture. The essays include telling analyses of topics as diverse as rap music, presidential campaigns, and contemporary Black leaders. Dyson is a major social philosopher whose work is an important part of African American cultural thought.

Gitlin, Todd. 1995. *The Twilight of Common Dreams: Why America is Wracked by Culture Wars.* New York: H. Holt.

A sociologist who studies the mass media examines the social conditions that have resulted in conflict over defining U.S. culture. Gitlin locates the conflicts over culture in the growing inequality within the United States.

Griswold, Wendy. 1994. *Cultures and Societies in a Changing World.* Thousand Oaks, CA: Pine Forge Press.

Intended for undergraduates, this book examines basic concepts in the sociology of culture, with a special eye to understanding the linkage of culture to the global economy.

Harris, Marvin. 1974. *Cows, Pigs, Wars, and Witches: The Riddles of Culture.* New York: Vintage.

Marvin examines several cross-cultural practices and beliefs with attention to explaining the rational basis for seemingly contradictory or irrational forms of behavior.

Kephart, William H. 1993. *Extraordinary Groups: An Examination of Unconventional Life Styles.* New York: St. Martin's.

This book offers an analysis of the several major U.S. subcultures, including the Amish, the Oneida community, and the Shakers. Each culture is examined as a whole social system, including the socialization of the young, religious beliefs, family systems, and normative expectations about dress and other customs.

Lamont, Michèle. 1992. *Money, Morals, and Manners: The Culture of the French and the American Upper-Middle Class.* Chicago: University of Chicago Press.

Based on interviews with professional and managerial people, Lamont's book examines how upper middle-class groups use culture to mark their social status. The comparative framework shows the significance of cultural context for understanding how groups define and present themselves.

Simonds, Wendy. 1992. *Women and Self-Help Culture: Reading Between the Lines.* New Brunswick, NJ: Rutgers University Press.

This book, based on a content analysis of popular self-help books, develops a sociological perspective on popular culture that shows the significance of self-help culture in forming women's identities and shows how a generalized culture influences individual lives.

Wolf, Naomi. 1991. *The Beauty Myth: How Images of Beauty Are Used Against Women.* William Morrow and Co., New York.

Written for a popular audience, but well informed by sociological perspective, Wolf's book analyzes "the beauty myth"—the idea that women must be beautiful. She shows how the beauty myth is used against women's advancement and how it injures women's self-esteem and physical well-being.

CHAPTER 4
Socialization

SOME people want to be alone. Hermits, high-country trappers, and recluses contrive to be by themselves. The desire for privacy, even extreme privacy, is perfectly routine, yet consider how exotic it would be to have *never* had contact with other people—not for one moment in your entire life. Assume that somehow you acquired enough food to persist, enough shelter to survive, but that you had no one to teach you to talk or hide or even stand. If all of your social training and social experience were subtracted, what would be left?

Feral children is the name given to a few rare individuals who seem to have been raised in the complete absence of human contact (*feral* means "like a wild beast"). There have been several accounts of abandoned infants raised by wolves and other animals in the last couple of hundred years. Often the tales are fantasies, or they are tainted by unscientific observations and enthusiasm for a wild tale. In some cases, the "wild child" was simply severely retarded. However it happens, people are fascinated by the notion of a person who is "purely natural," untouched by social forces.

One of the first carefully documented cases of a feral child was that of a twelve-year-old boy named Victor who in 1800 was captured in the Luxembourg Gardens of Paris. How the boy ended up in the forests was never discovered. When first seen, he was naked, on all fours, scavenging for acorns and roots. Known as the Wild Boy of Aveyron, Victor was treated as a spectacle and paraded before curious crowds as a specimen of what people thought then was "natural man." Many people hoped that after a few months of intensive education, the boy would be able to provide information about his past life (Lane, 1976). That was not to be.

Jean-Marc-Gaspard Itard, resident physician at the National Institute for Deaf Mutes, attempted to educate the boy. With heroic diligence, Itard taught Victor to recognize a few written words although Victor was never able to say much more than "milk" and "Oh God" (in French). Victor seemed to acquire some understanding of abstract concepts, such as gratitude, remorse, and the desire to please, but he never became more than a fraction of a fully developed person.

To some scientists who observed him, Victor seemed to be deaf and, in the term of the day, an idiot. Others have seen in Victor a man without his social dimension. To the twentieth-century onlooker Lucien Malson, Victor's case showed that there is no such thing as human nature: "[Man] has or rather is a history" (1972: 9). What would modern scientists and psychologists have learned if Victor had appeared before them? Perhaps a more recent event would tell us.

In November of 1970, case workers at the Los Angeles County welfare department opened a file on a nearly blind woman who wandered into the wrong department, leading her ancient mother and her apparently autistic daughter. The blind woman sought assistance for herself, but the caseworkers focused on the young girl. At first, they thought the girl was six years old. In fact, she was thirteen, although she weighed only 59 pounds and was 4 feet 6 inches tall. She was small and withered, unable to stand up straight, incontinent, and severely malnourished. Her eyes did not focus, she had two nearly complete sets of teeth, and there was a strange ring of callouses around her buttocks. She could not talk.

As the case unfolded, it was discovered that the girl, who became famous under the pseudonym Genie, had been kept in nearly total isolation for most of her life. The following description is from the first scientific report about Genie:

> In the house Genie was confined to a small bedroom, harnessed to an infant's potty seat. Genie's father sewed the harness himself; unclad except for the harness, Genie was left to sit on that chair. Unable to move anything except her fingers and hands, feet and toes, Genie was left to sit, tied-up, hour after hour, often into the night, day after day, month after month, year after year (Curtiss, 1977: 5).

At night, she was put in another restraining device—a handmade sleeping bag that held her arms stationary—and was placed in a crib. If she made a sound, her father beat her. She was given no toys and was allowed to play only with two old raincoats and her father's censored version of *TV Guide*. (He had deleted everything "suggestive," such as pictures of women in bathing suits.)

Genie's mother, timid and blind, was also victimized by her husband. Three weeks before entering the welfare office, she finally found the courage to escape from him. Her entry into the real world set off a storm of media attention. Shortly afterward, Genie's father committed suicide.

Genie was studied intensively by scientists interested in language acquisition and the psychological effects of extreme confinement. They hoped that her development would throw some light on the question of nature versus nurture—that is, are we the product of our genes or our social training? Genie's case did not settle the question. After intense language instruction and psychological treatment, Genie developed some verbal ability and, after a year, showed progress in her mental and physical development; yet, the years of isolation and severe abuse took their toll. In the 1990s, Genie was living in a home for retarded adults (Rymer, 1993).

Feral children force us to consider how much of our development as human beings is social. Human genes may confer skin and bone and brain, but our humanity is in our behavior and our understanding. It is by learning the values, norms, and roles of our culture that we acquire and develop our sense of self. Sociologists refer to this learning process as socialization.

The Socialization Process

Socialization is the process through which people learn the expectations of society. **Roles** are the expected behavior associated with a given status in society. We explore roles further in Chapter 6, but it is important to know here that roles are learned through the socialization process. Through socialization, people absorb their culture—customs, habits, laws, practices, and means of expression. Socialization is the basis for **identity**—how one defines oneself. Identity is both personal and social. To a great extent, it is bestowed by others since we come to see ourselves as others see us. Socialization also establishes **personality,** defined as the relatively consistent pattern of behavior, feelings, and beliefs in a given person.

The socialization experience differs for individuals depending on factors like race, gender, and class, as well as more subtle factors like attractiveness and personality. Women and men encounter different socialization patterns as they grow up since each gender brings with it different social expectations (see Chapter 11). Likewise, growing up Jewish, Asian, Latino, or African American involves different socialization experiences, as the box "Understanding Diversity" shows. In this example, a Korean American man reflects on the cultural habits he learned growing up in two cultures, Korean and American. His comments reveal the strain when socialization involves competing expectations, a strain that can be particularly acute when a person grows up within different, even if overlapping, cultures.

Through socialization, people internalize cultural expectations and in turn pass these expectations on to others. *Internalization* occurs when behaviors and assumptions are learned so thoroughly that people no longer question them, but simply accept them as correct. Through socialization, one internalizes the expectations of society. The lessons that are internalized can have a powerful influence on attitudes and behavior. For example, someone socialized to believe that homosexuality is morally repugnant is unlikely to be tolerant of gays and lesbians. If such a person, let's say a man, experiences erotic feelings about someone of the same sex, he is likely to have deep inner conflicts about his identity. Similarly, someone socialized to believe that racism is morally repugnant is likely to be more accepting of diverse groups. Of course, people can change the cultural expectations they learn. New experiences can undermine narrow cultural expectations. Attending college, to give one example, often has a liberalizing effect, supplanting old expectations with new ones generated by exposure to the diversity of college life.

BOX 4.1 UNDERSTANDING DIVERSITY

My Childhood: An Interview with Bong Hwan Kim

CHILDHOOD is a time when children learn their gender, as well as their racial and ethnic identity. This excerpt from an interview with Bong Hwan Kim, a Korean American man, is a reflection on growing up and learning both Korean and American culture.

I came to the United States in 1962, when I was three or four years old. My father had come before us to get a Ph.D. in chemistry. He had planned to return to Korea afterwards, but it was hard for him to support three children in Korea while he was studying in the United States, and he wasn't happy alone, so he brought the family over. He got a job at a photographic chemicals manufacturing company in New Jersey, where he still works after almost 30 years. He never made it past the "glass ceiling." I view him as a simple person who must have been overwhelmed by the flip side of the American dream they never tell you about. There was never a place for him in America except at home with the family or maybe at the Korean church. Both of my parents were perpetual outsiders, never quite comfortable with American life.

The Bergenfield, New Jersey, community where I grew up was a blue-collar town of about 40,000 people, mostly Irish and Italian Americans. I lived a schizophrenic existence. I had one life in the family, where I felt warmth, closeness, love, and protection, and another life outside—school, friends, television, the feeling that I was on my own. I accepted that my parents would not be able to help me much.

I can remember clearly my first childhood memory about difference. I had been in this country for maybe a year. It was the first day of kindergarten, and I was very excited about having lunch at school. All morning, I could think only of the lunch that was waiting for me in my desk. My mother had made *kimpahp* [rice balls rolled up in dried seaweed] and wrapped it all up in aluminum foil. I was eagerly looking forward to that special treat. I could hardly wait. When lunch bell rang, I happily took out my foilwrapped *kimpahp.* But all the other kids pointed and gawked. "What is *that*? How could you eat *that*?" they shrieked. I don't remember whether I ate my lunch or not, but I told my mother I would only bring tuna or peanut butter sandwiches for lunch after that.

I have always liked Korean food, but I had to like it secretly, at home. There are things you don't show to your non-Korean friends. At various times when I was growing up, I felt ashamed of the food in the refrigerator, but only when friends would come over and wonder what it was. They'd see a jar of garlic and say, "You don't *eat* that stuff, do you?" I would say, "I don't eat it, but my parents do; they do a lot of weird stuff like that."

My father spelled his name "Kim Hong Zoon." The kids at school made fun of his name. "Zoon?" they would laugh. They called me "Bong" because "Bong Hwan" was too hard to pronounce.

As a child you are sensitive; you don't want to be different. You want to be like the other kids. They made fun of my face. They called me "flat face." When I got older, they called me "Chink" or "Jap" or said "remember Pearl Harbor." In all cases, it made me feel terrible. I would get angry and get into fights. Even in high school, even the guys I hung around with on a regular basis, would say, "You're just a Chink" when they got angry. Later, they would say they didn't mean it, but that was not much consolation. When you are angry, your true perceptions and emotions come out. The rest is a façade.

They used to say, "We consider you to be just like us. You don't *seem* Korean." That would give rise to such mixed feelings in me. I wanted to believe that I was no different from my white classmates. It was painful to be reminded that I was different, which people did when they wanted to put me in my place, as if I should be grateful to them for allowing me to be their friend.

I wanted to be as American as possible—playing football, dating cheerleaders. I drank a lot and tried to be cool. I had convinced myself that I was "American," whatever that meant, all the while knowing underneath that I'd have to reconcile myself, to try to figure out where I would fit in a society that never sanctioned that identity as a public possibility. Part of growing up in America meant denying my cultural and ethnic identity, and part of that meant negating my parents. I still loved them, but I knew they could not help me outside the home. Once when I was small and had fought with a kid who called me a "Chink," I ran to my mother. She said, "Just tell them to *shut up.*" My parents would say that the people who did things like that were just "uneducated." "You have to study hard to become an educated person so that you will rise above all that," they would advise. I didn't really study hard. Maybe I knew somehow that studying hard alone does not take anyone "above all that." I have lost contact with everyone from my East Coast life except for a few Korean American friends from the Korean church our family attended. . . .

When I got to college, I experienced an identity crisis. Because I was no longer involved in sports and because I was away from home, I no longer felt special or worthwhile, and I became very depressed. I hated even getting up in the mornings. Finally, I dropped out of school.

I decided to go to Korea, hoping to find something to make me feel more whole. Being in Korea somehow gave me a sense of freedom I had never really felt in America. It also made me love my parents even more. I could imagine where they came from and what they experienced. I began to understand and appreciate their sacrifice and love and what parental support means. Visiting Korea didn't provide answers about the meaning of life, but it gave me a sense of comfort and belonging, the feeling that there was somewhere in this world that validated that part of me that I knew was real but few others outside my immediate family ever recognized.

SOURCE: Edited by Karin Aguilar-San Juan. Copyright © 1994. *The State of Asian America.* Boston: South End Press. Reprinted with permission of South End Press.

Through the socialization process, young children learn the values of their culture; these values shape our relationships with other people.

Examining the socialization process helps us see the degree to which our lives are *socially constructed,* meaning that the organization of society and the life outcomes of people within it are the result of social definitions and processes. For example, values and social attitudes are not inborn, but emerge through the interactions we have with others. This is what is meant by the expression that human beings are born *tabula rasa,* meaning a "blank slate." From a sociological perspective, it is clear that what a person becomes is more owing to his or her social experiences than his or her innate (inborn or natural) traits although innate traits are to some extent part of the picture. For example, a person may be born with a great capacity for knowledge, but if not provided with a good education, that person is unlikely to achieve his or her full potential and may not be recognized as intellectually gifted.

Socialization as Social Control

The sociologist Peter Berger pointed out that not only do people live in society, but society also lives in people (Berger, 1963). Socialization is, therefore, a mode of social control. Since socialized people conform to cultural expectations, socialization gives society a certain degree of predictability. Patterns are established that become the basis for social order. Although few people match the cultural ideal exactly, most of us fit comfortably within society's expectations.

To understand how socialization is a form of social control, imagine the individual in society as surrounded by a series of concentric circles (see Figure 4.1). Each circle is a layer of social controls, ranging from the most subtle, such as the expectations of others, to the most overt, such as physical coercion and violence. Usually, coercion and violence are not necessary to extract conformity because learned beliefs and the expectations of others are enough to keep people in line. These forces can be subtle since even when a person disagrees with others, he or she can feel pressure to conform and may experience stress and discomfort in choosing not to conform (Asch, 1955). People learn through a lifetime of experience that deviating from the expectations of others invites peer pressure, ridicule, and other social judgments that remind one of what is expected.

Conformity and Individuality

Saying that people conform to social expectations does not mean there is no individuality. We are all unique to some degree. Our uniqueness arises from different experiences, different patterns of socialization, the choices we make, and the imperfect ways that we learn our roles; furthermore, people resist some of the expectations society has of them. Sociologists warn against seeing human beings as totally passive creatures since people interact with their environment in creative ways (Wrong, 1961); yet, most people conform, although to differing degrees. Socialization is profoundly significant, but this does not mean that people are robots; instead, socialization emphasizes the adaptations people make as they learn to live in society.

Debunking Society's Myths

Myth: **We are all individuals, free to be whoever and whatever we want.**

Sociological perspective: **Although there is an emphasis on individualism in U.S. culture, what we become is largely influenced by the socialization process.**

Some people conform too much, for which they pay a price. Men have a lower life expectancy and higher rate of accidental death because of the stressful and risky behavior associated with the masculine role (Harrison et al., 1992; Kimmel and Levine, 1992). Social psychologists have also found that women with the most traditionally feminine identities tend to have a low opinion of themselves and a high rate of depression (Thornton and Leo, 1992; Tinsley et al., 1984). Although most of these studies examine primarily White women, a recent study has found the same to be true among Native American women. Native American women with more traditional gender role orientations are significantly more depressed, have lower self-esteem, experience greater role conflict, and are less satisfied with life than are Native American women with a more balanced gender role orientation (Napholz, 1995). Socialization into the traditional men's roles encourages aggression, independence, and competitiveness. Men and women who try to balance both masculine and feminine personality characteristics without excessive emphasis on one or the other enjoy greater mental health and more positive self-images than those who adopt only one gender type or the other (Baffi et al., 1991; Downey, 1984; Spence et al., 1975).

Socialization and Self-Esteem

Self-esteem is the value a person places on his or her identity. How much value one sees in oneself is affected by the socialization process and, in a deep way, by how one's group is seen by society. This is well illustrated by studies of African American children.

In a classic study in the 1930s, two African American psychologists, Kenneth Clark and Mamie P. Clark, studied children's preferences for Black or White dolls. They found that, given a choice, Black children preferred White dolls to Black dolls. At the time, this preference was taken to be a reflection of the negative self-images that Black children acquire growing up in a racially segregated society (Clark and Clark, 1939). The Clark study was a pivotal part of the legal argument provided to the Supreme Court in 1954 in the *Brown v. the Board of Education* case, in which the court ruled that segregation of public facilities was unconstitutional. (This was the first time that social science research was formally cited in a Supreme Court case to support a legal argument.)

A replication of the Clarks' experiment in the 1980s added interesting discoveries. Just as the Clarks had done in the earlier study, two psychologists presented African American and White preschoolers with dolls that were exactly alike except for skin color and hair texture (Powell-Hopson and Hopson, 1988). Using questions the Clarks developed, they asked the children to choose the doll that:

1. You want to be
2. You would like to play with
3. Is a nice doll
4. Looks bad
5. Is a nice color
6. You would take home if you could

In the replication study, initially 65 percent of the African American children and 74 percent of the White children chose the White doll to play with; 76 percent of the African American children and 82 percent of the White children chose the Black doll as the one that looked bad.

The replication then went one step further than the original experiment. The researchers intervened in the experiment by giving positive reinforcement to children who chose the Black doll. These children were allowed to sit in the front of the room with the researchers while the other children had to sit in the back. The researchers chose the Black dolls themselves, and they read a story to the children that depicted Black children in a very positive light. The researchers also had the children hold up the Black dolls and repeat positive adjectives like "pretty," "nice," "handsome," and "smart." Fifteen minutes later, the children put the dolls away and were asked the original questions. This time, 68 percent of the African American children chose the Black doll to play with, as did 67 percent of the White children, *completely reversing the trend* that existed before the children received positive reinforcement about Black dolls. The researchers concluded that positive reinforcement about being Black helps African American children develop a positive self-image.

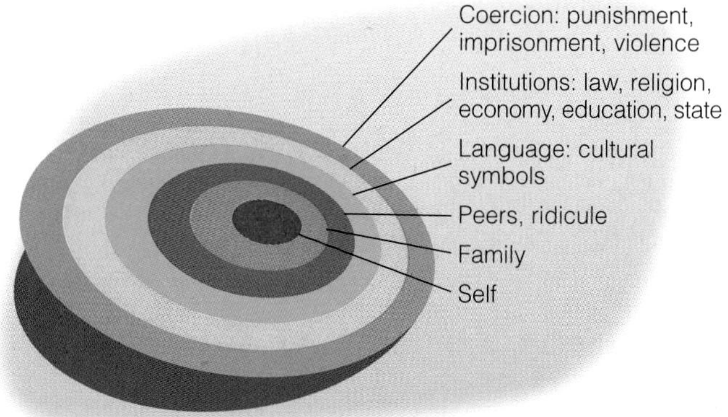

FIGURE 4.1 Socialization as Social Control

This study also documents an important sociological point: How you see yourself and the groups to which you belong depends to a great extent on how others in society define you. At the same time, however, people choose the groups against whom they judge themselves, with interesting consequences. One might think, for example, given their devaluation in the dominant culture, that African Americans would have lower self-esteem than White Americans; however, the reverse is true. That is, African Americans, overall, have high self-esteem compared to White Americans because they use other African Americans, not the dominant group, as their reference group—that is, those most likely to value being African American (Rosenberg, 1981).

At the same time, it is also true that African Americans have a relatively low sense of what is called *personal efficacy*—that is, the sense of being an effective person. How is this seeming contradiction explained? Sociologists argue that it is because African Americans are more likely to base their self-esteem on relationships within their own families, whereas their sense of personal efficacy is affected by the discrimination they experience in society (Hughes and Demo, 1989; Rosenberg, 1981). In other words, the societal context is an important factor in understanding the development of self-esteem.

The societal context can be a battleground of conflicting social expectations. For instance, a woman living in poverty may have internalized the expectations associated with traditional motherhood and homemaking; yet, if she does not work because she wants to stay at home to take care of her children, she may be condemned as a lazy "welfare queen" by the dominant culture. A wealthy woman who stays at home to take care of her children may be respected as an ideal mother. Even though both women may be trying to

Some social groups encourage conformity more than others. Here 40,000 couples from the Unification Church ("Moonies") are being married in Seoul in a mass wedding ceremony.

fulfill the same idealized role, they will not receive an equal payoff in positive reinforcement. Plainly, the expectations of society can be contradictory, often adding emotional stress for those already burdened by their lack of social and economic advantages.

The Consequences of Socialization

Socialization is a life-long process with consequences that affect how we behave toward others and what we think of ourselves. First, *socialization establishes self-concepts.* How we think of ourselves is the result of the socialization experiences we have over a lifetime. Our self-esteem and identity are established through the socialization we experience.

Second, *socialization creates the capacity for role-taking* or, put another way, for seeing ourselves as others see us. Socialization is fundamentally reflective; that is, it involves self-conscious human beings seeing and reacting to the expectations of others. The capacity for reflection and the development of identity are ongoing. As we encounter new situations in life, we are able to see what is expected and to adapt to the situation accordingly. Of course, not all people do so successfully. This can become the basis for social deviance (explored in Chapter 8) or for many common problems in social and psychological adjustment.

Third, *socialization creates the tendency for people to act in socially acceptable ways.* Through socialization, people learn the normative expectations attached to social situations and the expectations of society in general. As a result, socialization creates some predictability in human behavior and brings some order to what might otherwise be social chaos.

Finally, *socialization makes people bearers of culture.* Socialization is the process by which people learn and internalize the attitudes, beliefs, and behaviors of their culture. At the same time, socialization is a two-way process—that is, a person not only is the recipient of culture, but also is the creator of culture who passes cultural expectations on to others. The main product of socialization, then, is society itself. By molding individuals, socializing forces perpetuate the society into which individuals are born. Beginning with the newborn infant, socialization contributes to the formation of a self; the self learns roles and rules, and thus outfitted, the self becomes a bearer of culture, passing on all that it has acquired.

Theories of Socialization

Knowing that people become socialized does not explain how it happens. Different theoretical perspectives explain socialization, including psychoanalytic theory, object relations theory, social learning theory, and symbolic interaction theory. Each perspective carries a unique set of assumptions about socialization and its effect on the development of the self.

Psychoanalytic Theory

Psychoanalytic theory originates in the work of Sigmund Freud (1856–1939). Among Freud's greatest contributions was the idea that the unconscious mind shapes human behavior. Freud developed the technique of *psychoanalysis* to help discover the causes of psychological problems in the recesses of troubled patients' minds.

Psychoanalytic theory depicts the human psyche in three parts: the id, the superego, and the ego. The **id** consists of deep drives and impulses. Freud was particularly absorbed by the sexual component of the id, which he considered an especially forceful denizen of the unconscious mind. The **superego** is the dimension of the self that represents the standards of society. The superego incorporates or internalizes acquired values and norms—in short, society's collective expectations. According to Freud, an ordered society requires that people repress the wild impulses generated by the id; thus, the id is in permanent conflict with the superego. The superego represents what Freud saw as the inherent repressiveness of society. People cope with the

Freud, Sigmund

tension between social expectations (the superego) and their impulses (the id) by developing defense mechanisms, typically repression, avoidance, or denial (Freud, 1923/1960, 1930/1961, 1901/1965). Suppose someone has a great desire for the wrong person or for another person's property. This person might refuse to admit this (repression); or acknowledging the impulse, might avoid the opportunity for temptation (avoidance); or might indulge in misconduct, believing it was not misconduct (denial).

One way that people cope with the tension between the superego and the id is through development of the **ego**, the third component of the self in Freud's theory. The ego is the seat of reason and common sense. The ego plays a balancing act between the id and the superego, adapting the desires of the id to the social expectations of the superego. In one Freudian analysis, "A person with a strong ego is one who is able to accomplish the best, most realistic, compromise between the conflicting demands of the id and the superego. To give in to the id is to indulge in infantile and immature behavior. To be totally at the mercy of the superego is to be rigid and repressive" (Cuzzort and King, 1980: 27).

In psychoanalytic theory, the conflict between the id and the superego occurs in the subconscious mind; yet, it shapes human behavior. We get a glimpse of the unconscious mind in dreams and in occasional slips of the tongue—the famous "Freudian slip" that is believed to reveal an underlying state of mind. For example, someone might intend to say, "There were six people at the party," but instead say, "There were sex people at the party," revealing that either the memory of the party or the person he or she is speaking with is causing warm thoughts to lurk in the unconscious mind. Subconscious desires are forever present and vying for preeminence. Psychoanalytic theory locates the forces that shape the self in the unconscious mind (Freud, 1901/1965).

By understanding the self, we can understand the self in society, according to the psychoanalytic school. The id demands gratification for biological desires, the superego forces constant awareness of how society is perceiving one's actions, and the ego negotiates an uneasy peace between the two parties. As the stabilized three-part self interacts with others, the rules of socialization emerge, with the ego bridging the gap between the primal id and the socialized superego.

Freud's work has never been free from controversy, and among contemporary psychoanalysts, numerous schools of thought provide different interpretations of Freud's work and rival therapies for assisting patients. Some sociologists have criticized Freud's work for not being generalizable since he worked with only a small and nonrepresentative group of clients.

Freud has also been heavily criticized by feminist scholars who find his work to be based on sexist assumptions. Feminists have been especially critical of Freud's concept of "penis envy" in women. Freud argued that adult women become jealous and resentful of men because of their unconscious wish for a penis. Although some have argued that Freud was stating this metaphorically, using the penis to symbolize male power, many feminists have found a literal interpretation to be consistent with Freud's ideas that women were sexually and psychologically immature. Feminists have criticized Freud's theory for being too male centered (that is, androcentric), so much so that women are defined as psychologically maladjusted. On the other hand, some feminists have adopted a psychoanalytic perspective in their work, as we will see later. Though they reject the sexist dimensions of Freud's arguments, they use the concept of the unconscious to explain numerous dimensions of women's and men's experiences (Chodorow, 1994).

Psychoanalytic theory is an influential and popular way to think about human personality. We often speak of what motivates people, as if motives were internal, unconscious states of mind that direct human behavior. We think we can understand a person if we know what really makes him or her tick, as if the behavior we observe is not what actually constitutes the person's being. People also typically think that understanding someone's true self requires knowing about his or her childhood experience—a reflection of Freud's contention that personality is fixed at a relatively early age. Psychoanalysis has an enormous influence on how people think about human behavior.

To sum up, the psychoanalytic perspective sees human identity as relatively fixed at an early age, in a process greatly influenced by one's family. It sees the development of social identity as an unconscious process, developed from dynamic tensions between strong instinctual impulses and the social standards of society. Most important, psychoanalysis sees human behavior as directed and motivated by underlying psychic forces that are largely hidden from ordinary view.

Object Relations Theory

Psychoanalytic theory has been modified by a school of thought known as **object relations theory.** Placing less emphasis on biological drives, object relations theorists contend that the social relationships experienced by children determine the development of the adult personality. As in classical Freudian theory, the processes invoked by object relations theory are largely unconscious. Key concepts in object relations theory are *attachment* and *individuation,* the making and breaking of the bond with parents. Infants may be strongly attached to the parent who is their primary caregiver. As they grow older, they learn to separate themselves from their parents, or individuate, both physically and emotionally. They become a free-standing individual, but early attachments to the primary caregivers persist.

Within sociology, one of the most widely known applications of object relations theory is provided by Nancy Chodorow, a feminist sociologist and practicing psychoanalyst who has used object relations theory to explain how gender shapes personalities. Specifically, Chodorow asks, "Why do women learn to mother and men, generally speaking, do not?" Chodorow's answer connects the process of personality formation directly to the division of labor in the family. The modern family, Chodorow argues, has an *asymmetrical division of labor*—women "mother," and men do not. In addition, she argues, women's work as mothers is devalued by society. It is not defined as real work and does not receive the same social and economic support of work outside the family (Chodorow, 1978, 1994).

Chodorow agrees with the basic premises of psychoanalytic theory that personality is shaped by early, unconscious forces. She also agrees with object relations theory in that she sees attachment and individuation as important formative processes; however, she adds that children identify with their same-sex parent, which means that boys and girls individuate (separate themselves from their parents) differently. Boys identify with the father; the asymmetrical division of labor separates the father from the home and full-time nurturing; therefore, boys form personalities that are emotionally detached, independent, less oriented toward other people. Girls, on the other hand, identify with the mother; the mother's role is one of close attachment to the members of the household; therefore, girls develop personalities based on attachment and an orientation toward others.

Chodorow argues that men and women carry their patterns of separation and attachment into their adult lives. The unconscious processes associated with individuation and attachment create the gendered personalities that we think of as "masculine" and "feminine." Indeed, much research has shown that women are more likely to be oriented toward others and to seek and sustain affiliations; men are more likely to repress their attachments to others and to be more individualistic in their orientation toward the world (Chodorow, 1978).

One of the criticisms of Chodorow's work is that it is based on a single kind of family—the traditional, nuclear family. This question has caused scholars to question whether her theory would apply in other family contexts (Joseph, 1981; Lorber et al., 1981). This recently has been examined through a study of Chicano families. Chicano families are characterized by *familism*—large families, multigenerational households, high value placed on family unity, and high level of interaction between family and kin (Baca Zinn and Eitzen, 1997). Despite the differences between this family form and the nuclear family analyzed by Chodorow, her framework is useful in analyzing the experiences of Chicana mothers and their daughters. The identity of Chicana mothers revolves around family and home, and they are more likely to identify with their daughters than their sons. Chicanas do not practice exclusive mothering; rather, mothering figures may include grandmothers, aunts, or godmothers. Young Chicanas identify with their mothers, but they also see themselves as an extension of other women in the family system. In addition, the cultural representation of women within Chicano culture as sacred and self-sacrificing makes Chodorow's point about gender identification and attachment particularly salient for Chicanas (Segura and Pierce, 1993).

Although Chodorow's work was based on a particular kind of family, it appears that her theory has value in describing the reproduction of gender in other families. How would her theory explain personality formation in father-absent families? Chodorow would respond by saying that the absence of the father only exacerbates the tendency for boys to develop detached concepts of themselves. Chodorow's work has also been criticized for presuming that motherhood is universally devalued among all groups, which is not always true. African American families, for example, place a particularly high value on motherhood (Joseph, 1981; Collins, 1990). This does not mean we have to discard Chodorow's theory. It only suggests that one could predict, based on her theory, that the greater valuing of motherhood coupled with greater involvement of men in early child care will produce less gender-stereotyped personalities. In other words, her work has interesting practical implications, namely, that if men were to acquire more mothering skills and participate more in the daily tending of the family, the result would be a society where men and women have less gender-stereotyped personalities.

TABLE 4.1 *THEORIES OF SOCIALIZATION*

	Psychoanalytic Theory	Object Relations Theory	Social Learning Theory	Symbolic Interaction Theory
How each theory views:				
Individual learning process	unconscious mind shapes behavior	infants identify with the same-sex parent	people respond to social stimuli in their environment	children learn through taking the role of significant others
Formation of self	self (ego) emerges from tension between the id and the superego	self emerges through separating oneself from primary caretaker	identity is created through the interaction of mental and social worlds	identity emerges as the creative self interacts with the social expectations of others
Influence of society	societal expectations are represented by the superego	division of labor in the family shapes identity formation	young children learn the logical principles that shape the external world	expectations of others form the social context for learning social roles

Social Learning Theory

Whereas psychoanalytic theory places great importance on the unconscious processes of the human mind, **social learning theory** considers the formation of identity to be a learned response to social stimuli (Banders and Walters, 1963). Social learning theory emphasizes the societal context of socialization. Identity is regarded not as the product of the unconscious, but as the result of modeling oneself in response to the expectations of others. According to social learning theory, behaviors and attitudes develop in response to reinforcement and encouragement from those around them. Social learning theorists acknowledge the importance of early childhood experience, but they think that the identity people acquire is based more on the behaviors and attitudes of others than on the interior landscape of the individual.

Early models of social learning theory regarded learning rather simplistically in terms of stimulus and response. People were seen as passive creatures who merely responded to stimuli in their environment. This mechanistic view of social learning was transformed by the work of the Swiss psychologist Jean Piaget (1896–1980), who believed that learning was crucial to socialization, but that imagination also had a critical role. He argued that the human mind organizes experience into mental categories he called *schema,* which are modified and developed as social experiences accumulate. Schema might be compared to a person's understanding of the rules of a game. Humans do not simply respond to stimulus, but actively absorb experience and figure out what they are seeing to construct a picture of the world.

Piaget proposed that children go through distinct stages of cognitive development as they learn the basic rules of reasoning. They must master the skills at each level before they go on to the next (Piaget, 1926). In the initial *sensorimotor stage,* children experience the world only directly through their senses—touch, taste, sight, and sound. Next comes the *preoperational stage,* in which children begin to use language and other symbols. Children in the preoperational stage cannot think in abstract terms, but they do gain an appreciation of meanings that go beyond their immediate senses. They also begin to see things as others might see them. Third, the *concrete operational stage* is when children learn logical principles regarding the concrete world. This stage prepares for more abstract forms of reasoning. In the *formal operational stage,* children are able to think abstractly and imagine alternatives to the reality in which they live.

Piaget's work stresses the significance of conscious mental processes in social learning. He envisioned the human mind as highly creative and adaptive. Socialization in this framework is a process of developing new ways of thinking about and imagining the world. Most important, Piaget conceptualized humans as actively creating their mental and social worlds. The emphasis in social learning theory is on the influence of the environment in socializing people. The mind mediates the influence of the environment. Just as children learn appropriate behavior when they are rewarded or punished, social learning theory holds that behavior can be changed by altering the social environment (Bandura and Walters, 1963).

Social learning theory emphasizes how people model their behaviors and attitudes on that of others.

Symbolic Interaction Theory

Social learning theory is primarily represented in sociology by the theoretical perspective known as *symbolic interaction theory,* introduced in Chapter 1. According to symbolic interaction theory, human actions are based on the meanings we attribute to things; these meanings emerge through social interaction (Blumer, 1969). To explain further, people learn identities and values through the socialization process as they learn the social meanings that different behaviors imply. Learning to become a good student, then, means taking on the characteristics associated with that role. Because roles are socially defined, they are not real like objects or things, but are real because of the meanings people give them. As did Piaget, symbolic interactionists see the human capacity for reflection and interpretation as having an important role in the socialization process.

For symbolic interactionists, meaning is constantly reconstructed as people act within their social environments. The *self* is what we imagine we are; it is not an interior bundle of drives, instincts, and motives. Because of the importance attributed to reflection in symbolic interaction theory, symbolic interactionists use the term *self,* rather than the term *personality,* to refer to a person's identity. Symbolic interaction theory emphasizes that human beings make conscious and meaningful adaptations to their social environment; from a symbolic interactionist perspective, identity is not something that is unconscious and hidden from view, but is socially bestowed and socially sustained (Berger, 1963).

Symbolic interactionists think people become who they are through their interactions with others. Socialization from this perspective is a fluid and dynamic process, dependent on experiences with others. Although symbolic interactionists see

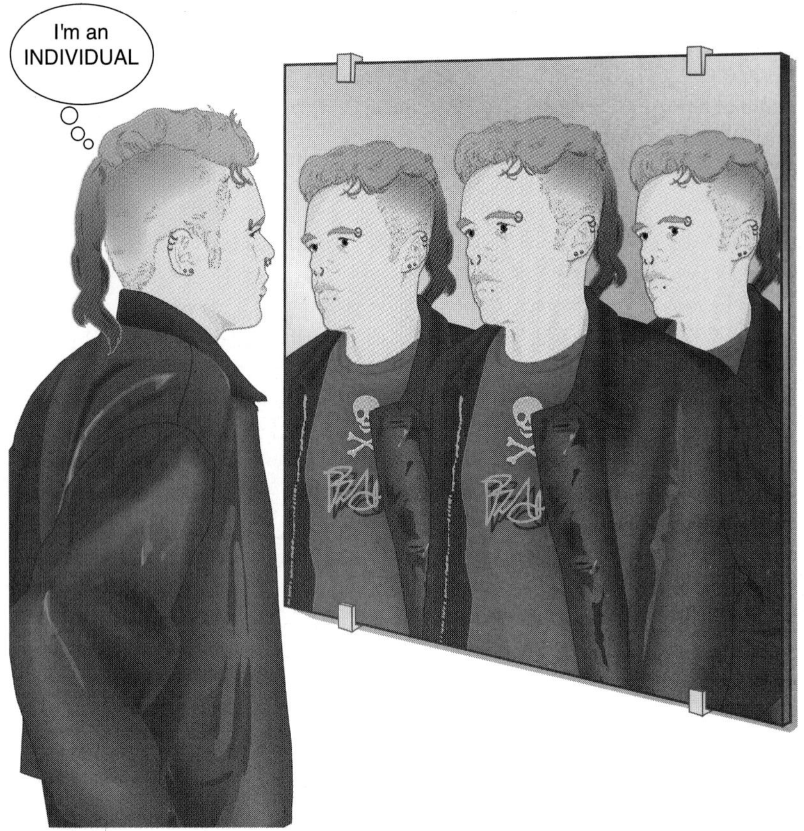

The Looking-glass Self
The looking-glass self refers to the process by which we see ourselves as others see us.

childhood as a very influential period in the human life cycle, they see the self as evolving over the life span. In other words, socialization is an ongoing process, not something fixed at an early age, as psychoanalysts contend.

Two theorists have greatly influenced the development of symbolic interactionist theory in sociology. Charles Horton Cooley (1864–1929) and George Herbert Mead (1863–1931) were both sociologists at the University of Chicago in the early 1900s (see Chapter 1). Cooley and Mead saw the self developing in response to the expectations and judgments of others in their social environment.

Charles Horton Cooley postulated the **looking-glass self** to explain how a person's conception of self arises through reflection about relationships to others (Cooley, 1902, 1909). The development of the looking-glass self emerges from (1) how we think we appear to others; (2) how we think others judge us; and (3) how the first two make us feel—proud, embarrassed, or other. The looking-glass self involves *perception* and *effect;* the perception of how others see us, and the effect of others' judgment on us.

How others see us is fundamental to the idea of the looking-glass self. In seeing ourselves as others do, we respond to the expectations others have of us. This means that the formation of the self is fundamentally a social process—one based in the interaction people have with each other, as well as the human capacity for self-reflection. One of the unique features of human life is the ability to see ourselves through others' eyes. People can imagine themselves in relationship to others and develop a definition of themselves, accordingly. From a symbolic interactionist perspective, the reflective process is key to the development of the self. If you grow up with others thinking you are smart and sharp-witted, chances are you will develop this definition of yourself. If others see you as dull-witted and withdrawn, chances are good that you will see yourself this way.

George Herbert Mead agreed with Cooley that children are socialized by responding to other's attitudes toward them. According to Mead, social roles are the basis of all social interaction. Roles are sets of expectations that govern a person's relationship with people and society (see Chapter 6). When people occupy roles, they are expected to fulfill the expectations associated with those roles. Roles can be thought of as the intersection between the individual and society. By occupying different roles, a person acquires the expectations associated with each role.

Taking the role of the other is the process of imagining oneself from the point of view of another. To Mead, role-taking is a source of self-awareness. As people take on new roles, their awareness of self changes. According to Mead, identity emerges from the roles one plays. He explained this process in detail by examining childhood socialization, which he saw as occurring in three stages: the imitation stage, the play stage, and the game stage (Mead, 1934). In each phase of development, the child becomes more proficient at taking the role of the other. In the first stage, the **imitation stage,** children merely copy the behavior of those around them. Role-taking in this phase is nonexistent since the child simply mimics the behavior of those in the surrounding environment without much understanding of the social meaning of the behavior. Although the child in the imitation stage has little understanding of the behavior being copied, he or she is learning to become a social being. For example, think of young children who simply mimic the behavior of people around them. In Mead's analysis, this behavior is one of the ways that children begin to learn the expectations of others.

In the second stage, the **play stage,** children begin to take on the roles of significant people in their environment, not just imitating but incorporating their relationship to the other. Instead of just mimicking others' behavior, the child has learned to understand more about context and the nuances of meaning that different behaviors represent. Especially meaningful is when children take on the role of **significant others,** those with whom they have a close affiliation. A child pretending to be his mother may talk to himself as the mother would. The child begins to develop self-awareness, seeing himself or herself as others do.

In the third stage of socialization, the **game stage,** the child becomes capable of taking on a multitude of roles at the same time. These roles are organized in a complex system that gives the child a more general or comprehensive view of the self. In this stage, the child begins to comprehend the system of social relationships in which he or she is located. The child not only sees himself or herself from the perspective of a significant other, but understands how people are related to each other and how others are related to him or her. This is the phase where children internalize (incorporate into the self) an abstract understanding of how society sees them.

Mead compared the lessons of the game stage to a baseball game. In baseball, all roles *together* make the game. The pitcher does not just throw the ball past the batter as if they were the only two people on the field; rather, each player has a specific role, and each role intersects with the others. The network of social roles and the division of labor in the baseball game is a social system, like the social systems children must learn as they develop a concept of themselves in society.

In the game stage, children learn more than just the roles of significant others in their environment. They also acquire a concept of the **generalized other**—the abstract composite of social roles and social expectations. In the generalized other, they have an example of community values and general social expectations that adds to their understanding of self; however, children do not all learn the same generalized other. Depending on one's social position (that is, race, class, gender, region, or religion), one learns a particular set of social and cultural expectations.

If the self is socially constructed through the expectations of others, how do people become individuals? Mead answered this by saying that the self has two dimensions: the "I" and the "me." The I is the unique part of individual personality, the active, creative, self-defining part. The me is the passive, conforming self, the part that reacts to others. In each person, there is a balance between the I and the me, similar to the tension Freud proposed between the id and the superego. Mead differed from Freud, however, in his judgment about when identity is formed. Freud felt that identity was fixed in childhood and henceforth driven by internal, not external, forces. In Mead's version, social identity is always in flux, constantly emerging (or "becoming") and dependent on social situations. Over time, identity is stabilized as one learns to respond consistently to common situations.

Expectations associated with given roles change as people redefine situations and as social and historical conditions change; thus, the social expectations learned through the socialization process are not permanently fixed. For example, as more women enter the paid labor force and as men take on additional responsibilities in the home, the expectations associated with motherhood and fatherhood are changing. Men now experience some of the role conflicts that women have faced in balancing work and family. As the roles of mother and father are redefined, children are learning new socialization patterns; however, traditional gender expectations maintain a remarkable grip. Despite many changes in family life and organization, young girls are still socialized for motherhood, and young boys are still socialized for greater independence and autonomy.

Agents of Socialization

Socialization agents are those who pass on social expectations. Everyone is a socializing agent since social expectations are communicated in countless ways and in every interaction people have, whether intentionally or not. When people are simply doing what they consider "normal," they are communicating social expectations to others. When you dress a particular way, you may not feel you are telling others they must dress that way, yet when everyone in the same environment dresses similarly, some expectation about appropriate dress is clearly being conveyed. People feel pressure to become what society expects of them even though the pressure may be subtle and unrecognized.

Socialization does not occur simply between individuals; it occurs in the context of social institutions. Recall from Chapter 1 that institutions are established patterns of social behavior that persist over time. Institutions are a level of society above individuals. Many social institutions shape the process of socialization, including, as we will see, the family, the media, peers, religion, sports, and schools.

The Family

For most people, the family is the first source of socialization; through families, children are introduced to the expectations of society. Children learn to see themselves through their parents' eyes; thus, how parents define and treat a child is crucial to the child's developing sense of self.

An interesting example of the subtlety in familial socialization comes from a study comparing how U.S. and Japanese mothers talk to their children (Fernald and Morikawa, 1993). Observers watched mothers from both cultures speak to their infant children, aged six, twelve, and nineteen months. Both Japanese and U.S. mothers simplified and repeated words for their children, but strong cultural differences were evident in what they said. U.S. mothers focused on naming objects for their babies: "Is that a *car*?" "Kiss the *doggy*." Japanese mothers were more likely to use their verbal interactions with children as an opportunity to practice social routines (such as "give me" and "thank you"). The behavior of the Japanese mothers implied that the name of the object was less important than the polite exchange. Japanese mothers were also more likely to use sounds to represent the objects, such as "oink-oink" for a pig, or "vroom-vroom" for a vehicle. U.S. mothers were more likely to use the actual names for objects. A U.S. mother might say, "See the car? Nice color!" whereas a Japanese mother might say, "Here! It's a vroom-vroom! I give it to you. Now you give this to me. Give me. Yes! Thank you." The researchers saw these interactions as reflecting the beliefs and practices of each culture. Japanese mothers used objects as part of a ritual of social exchange, emphasizing polite routines, whereas U.S. mothers focused on labeling things. In each case, the child receives a message about what is most significant in the culture.

What children learn in families is, of course, not uniform—even though families pass on the expectations of a given culture. Within a given culture, however, families may be highly diverse, as we will see in Chapter 14. Some families may emphasize educational achievement more than others; some may be more permissive, whereas others emphasize strict obedience and discipline. Even within families, children may experience different expectations, based on factors like gender or birth order (being first, second, or third born). Living in a family experiencing the strain of social problems (such as alcoholism, unemployment, domestic violence, or teen pregnancy) also affects how children are socialized; the specific effects of different family structures and processes are the basis for ongoing and extensive sociological research.

As important as the family is in socializing the young, it is not the only socialization agent. As children grow up, they encounter other socializing influences that join the family in molding the child, sometimes in ways that might even contradict family expectations. Parents who, for example, want to socialize their children in less gender-stereotyped ways might be frustrated by the influence of the media, which promotes highly gender-typed toys and activities to boys and girls. These multiple influences on the socialization process create a reflection of society in us.

The Media

Increasingly, the media are important agents of socialization. Counting only television, young children spend, on average, more hours with the media than they do in school. Add to that the print messages received in books, comics, newspapers, and even billboards, plus images from film, music, video games, and radio, and you begin to see the enormous influence the media have on the values we form, our images of society, our desires for ourselves, and our relationships with others. These images are powerful throughout our lifetimes (as we have seen in the chapter on culture), but many worry that their effect during childhood may be particularly deleterious.

Take the issue of violence. In 1995, two men ran into a New York City subway station, squirted a flammable liquid into the token booth, and lit it with a match. The explosion blew the cubicle to pieces, burning the token clerk over 80 percent of his body. He died two weeks later. The public quickly connected this to a scene from the movie *Money Train,* which had been released just the week before (McFadden, 1995). The only difference was that the film was fiction, and in the film the token clerk was not harmed by the violence. The high degree of violence in the media has led to the development of a rating system for televised programming.

No doubt, violence is extensive in the media. Analysts estimate that by age eighteen, the average child will have witnessed at last 18,000 simulated murders on television. The extent of violence during prime time is high enough—about sixteen violent acts, including two murders, in each evening's prime-time programming—but observers estimate that in children's programming (such as morning cartoons), violence is three times that on prime-time shows (Krasny, 1993; Signorielli, 1991).

What effect does televised violence have on children? Some research finds that children imitate the aggressive behavior they see on television and in films. News reports of children engaging in violent acts as they imitate the action of cartoon heroes would seem to support this conclusion. Numerous studies have found a relationship between watching televised violence and engaging in aggressive behavior (National Institute of Mental Health, 1982). Other studies say that claims of televised violence leading children to engage in aggressive, violent behavior are exaggerated. They argue that children understand that these images are only entertainment and that they are able to see it as fantasy (Thompson and Hickey, 1994; Hodge and Tripp, 1986). Most likely, children are not influenced by the images of televised and filmed violence alone, but also by the broader social context in which children live. Children do not watch television in a vacuum—they live in families where they learn different values and attitudes about violent behavior and they observe the society around them, not just the images they see in fictional representations. We live in a very violent society; images of violence in the media only reflect the violence in society (Figure 4.2). Although they may reproduce it, the media are not the sole determinant of violent behavior; nonetheless, the media are one of the many sources of socialization.

Every 8 seconds	a child drops out of school.
Every 10 seconds	a child is reported abused or neglected.
Every 15 seconds	a child is arrested.
Every 24 seconds	a child is born to an unmarried mother.
Every 34 seconds	a child is born into poverty.
Every 35 seconds	a child is born to a mother who did not graduate from high school.
Every 48 seconds	a child is born without health insurance.
Every minute	a child is born to a teen mother.
Every 2 minutes	a child is born at low birthweight.
Every 3 minutes	a child is born to a mother who received late or no prenatal care.
Every 4 minutes	a child is arrested for drug abuse.
Every 5 minutes	a child is arrested for a violent crime.
Every 5 minutes	a child is arrested for an alcohol-related offense.
Every 17 minutes	an infant dies.
Every 92 minutes	a child is killed by firearms.
Every 2 hours	a child is a homicide victim.
Every 4 hours	a child commits suicide.

FIGURE 4.2 Moments in America for Children

SOURCE: Children's Defense Fund. 1997. *The State of America's Children: Yearbook 1997.* Washington, DC: Children's Defense Fund, p. xi.

media violence

The media expose us to numerous images that shape our definitions of ourselves and the world around us. What we think of as beautiful, sexy, politically acceptable, and materially necessary is strongly influenced by the media. If every week, as you read a weekly newsmagazine, someone shows you the new car that will give you status and distinction, or if every weekend, as we watch televised sports, someone tells us that to really have fun we should drink the right beer, it is little wonder that we begin to think that our self-worth can be measured by the car we drive and that parties are seen as better when everyone is drunk. The values represented in the media, whether they are about violence, racist and sexist stereotypes, or any number of other social images, have a great effect on what we think and who we come to be.

Peers

Peers are those with whom you interact on equal terms. Friends, fellow students, and co-workers are examples of peer groups. Among peers, there are no formally defined superior and subordinate roles although status distinctions commonly arise in peer group interactions.

Peers are enormously important in the socialization process. For children, peer culture is an important source of identity. Through interaction with peers, children learn concepts of self, gain social skills, and form values and attitudes. The influence of peers persists through adolescence and adulthood; without peer approval, most people find it hard to feel socially accepted.

Particularly among the young, the approval and disapproval of peer groups are a significant source of the socialization process.

THINKING SOCIOLOGICALLY

Think about the first week that you were attending college. What expectations were communicated to you and by whom? Who were the most significant *socialization agents* during this period? Which expectations were communicated formally and which informally? If you were analyzing this experience sociologically, what would some of the most important concepts be to help you understand how one "becomes a college student"?

Peer relationships vary among different groups. Girls' peer groups, for example, tend to be closely knit and egalitarian; boys' peer groups tend to be more hierarchical, with evident status distinctions between members. Girls are more likely than boys to share problems, feelings, fears, and doubts. Beware of generalizations, however. In some all-girl groups, status distinctions form, and one study found that 40 percent of boys, a substantial fraction, have friendships with a high degree of intimacy. In general, however, boys are more prone to share activities than feelings (Corsaro and Eder, 1990; Youniss and Smollar, 1985).

As agents of socialization, peers are important sources of social approval, disapproval, and support. This is one of the reasons why groups without peers of similar status are often at a disadvantage in various settings, such as women in male-dominated professions or minority students on predominantly White campuses. Being a "token" or an "only," as it has come to be called, places unique stresses on those in settings where there are relatively few peers from whom to draw support. This is one reason why those who are minorities in a dominant group context often form same-sex or same-race groups for support, social activities, and the sharing of information about how to succeed in this environment.

Religion

Religion is another powerful agent of socialization, and religious instruction contributes greatly to the identities children construct for themselves. Children tend to develop the same religious beliefs as their parents; switching to a religious faith different from the one in which you were raised is rare (Hadaway and Marler, 1993). Even those who renounce the religion of their youth are deeply affected by the attitudes, self-images, and beliefs instilled by early religious training. Very often those who disavow religion return to their original faith at some point in their life, especially if they have strong ties to their family of origin and after they form families of their own (Wilson, 1994).

Religious socialization shapes the beliefs that people develop. An example comes from studies of people who believe in creationism. They have typically been taught to believe in creationism over a long period—that is, they have been specifically socialized to believe in the creationist view of the world's origin and to reject scientific explanations. Sociological research further finds that socialization into creationist beliefs is more likely to be effective among people who grow up in small-town environments where they are less exposed to other influences. Those who believe in creationism are also likely to have mothers who have filled

Through religious rituals, such as the traditional Jewish Seder at Passover, children learn the religious values of their families.

the traditional homemaker's role (Eckberg, 1992). This shows the influence that social context has on the religious socialization people experience.

Religious socialization influences a large number of beliefs that guide adults in how they organize their lives, including beliefs about moral development and behavior, the roles of men and women, and sexuality, to name a few. As examples, the most devout individuals tend to be the most traditional about behavior and about men's and women's roles (Thornton et al., 1983); those with strong religious beliefs also tend to be more conservative in their sexual attitudes (Beck et al., 1991; Cochran and Beeghley, 1991; Sheeran et al., 1993). A majority of those who say that religion is very important in their lives also believe strongly that premarital sex is wrong (Roper Organization, 1994), and those raised in less religious families are more likely to live with sexual partners outside of marriage when they become adults (Thornton et al., 1992).

Other effects of religious socialization can also be observed. Research on urban African American men and women shows that as socialization into religious faith increases, so does the level of educational attainment (Brown and Gary, 1991). It is unlikely that there is a direct cause-and-effect relationship (that is, that being religious causes a person to seek higher education), but it shows the influence that religious socialization can have on other values.

Sports

Sports are another setting for many lessons in socialization. Through sports, men and women learn concepts of self that stay with them in their later lives. Sports are also where many ideas about gender differences are formed and reinforced (Theberge, 1997, 1993). For men, success or failure as an athlete can be a major part of a man's identity; even for men who have not been athletes, knowing about and participating in sports is an important source of men's gender socialization. Men learn that being competitive in sports is considered a part of "manhood." Indeed, the attitude that "sports builds character" runs deep in the culture. Sports are supposed to pass on values like competitiveness, the work ethic, fair play, and a winning attitude. Along with the military and fraternities, sports are considered to be where one learns to be a man.

Debunking Society's Myths

Myth: **Sports are basically played just for the fun of it.**

Sociological perspective: **Although sports are a form of entertainment, playing sports is also a source for socialization into gender roles.**

Michael Messner's research on men and sports reveals the extent to which sports shape masculine identity. Messner interviewed thirty former athletes: Latino, Black, and White men from poor, working-class, and middle-class backgrounds. All of them spoke of the extraordinary influence of sports on them as they grew up. Not only are sports a major source of gender socialization, but working-class, African American, and Latino men often see sports as their only possibility for a good career even though the number of men who succeed in athletic careers is a minuscule percentage of those who hold such hopes.

Messner's research shows that, for most men, playing or watching sports is often the context for developing relationships with fathers, even when the father is absent or emotionally distant in other areas of life. Older brothers and other male relatives also socialize young men into sports. For many of the men in Messner's study, the athletic accomplishments of other family members created uncomfortable pressure to perform and compete although on the whole they recalled their early sporting years with positive emotions. It was through sports relationships with male peers, more than anyone else, however, that the men's identity was shaped. As boys, the men could form "safe" bonds with other men; still, through sports activity, men learned homophobic attitudes (that is, fear and hatred of homosexuals) and rarely developed intimate, emotional relationships with each other (Messner, 1992).

Sports historically have been less significant in the formation of women's identity though this is changing. Athletic prowess, highly esteemed in men, is not tied to cultural images of "womanliness"; quite the contrary, women who excel at sports are often stereotyped as lesbians and may be ridiculed for not being "womanly" enough. These stereotypes are a form of social control since they reinforce traditional gender roles for women (Blinde and Taub, 1992a; Blinde and Taub, 1992b). With athleticism and fitness on the rise as components of female beauty, however, the sociological impact of women's participation in sports will likely change. For example, when the U.S. women's hockey team won the Olympic gold in 1998, many speculated that this would have a strong impact on young girls' desire to be strong and athletic.

Current research finds that women in sports develop a strong sense of bodily competence—something typically denied to them by the prevailing, unattainable cultural images of women's bodies. Sports also give women a strong sense of self-confidence and encourage them to seek challenges, take risks, and set goals (Blinde et al., 1993). In addition, women report that sports are frequently the background for forming bonds with other women. Women competing together develop a sense of group power; women athletes also say that playing sports heightens their awareness of gender inequalities off the playing field though sports participation has not been shown to correlate with activism on women's issues (Blinde et al., 1994).

Research in the sociology of sports shows how activities as ordinary as shooting baskets on a city lot, playing on the soccer team for one's high school, or playing touch football on a Saturday afternoon can convey powerful cultural messages about our identity and our place in the world. Sports are a good example of the power of socialization in our everyday lives.

Sports are important agents of socialization; through sports girls and boys learn the values that different teams and coaches emphasize.

Schools

Once young people enter kindergarten (or, even earlier, day care), another process of socialization begins. At home, parents are the overwhelmingly dominant source of socialization cues. In school, teachers and other students are the source of expectations that encourage children to think and behave in particular ways. The expectations encountered in schools vary for different groups of students. These differences are shaped by a number of factors, including teachers' expectations for different groups and the resources that different parents can bring to bear on the educational process. The parents of children attending elite, private schools, for example, often have more influence on school policies and classroom activities than do parents in low-income communities. In any context, studying socialization in the schools is an excellent way to see the influence of gender, class, and race in shaping the socialization process.

For example, research finds that teachers respond differently to boys and girls in school. Boys receive more attention than girls from teachers. Even when teachers' respond negatively to boys who are misbehaving, they are calling more attention to the boys (Bailey and Campbell, 1991; Sadker and Sadker, 1994). Social class stereotypes also affect teachers' interactions with students. Teachers are likely to perceive working-class children and poor children as less bright and less motivated than middle-class children; teachers are also more likely to define working-class students as troublemakers (Bowditch, 1993). These negative appraisals are *self-fulfilling prophecies,* meaning that the expectations they create often become the basis for actual behavior; thus, they affect the odds of success for children. (More is said about this in Chapter 14.)

THINKING SOCIOLOGICALLY

Visit a local day-care center, preschool, or elementary school and observe children at play. Record the activities they are involved in, and note what both girls and boys are doing. Do you observe any differences between boys' and girls' play? What do your observations tell you about socialization patterns for boys and girls?

Boys also receive more attention in the curriculum than girls. The characters in texts are more frequently boys; the accomplishments of boys are more likely to be portrayed in classroom materials; and boys and men are more typically depicted as active players in history, society, and culture (Bailey and Campbell, 1991; Sadker and Sadker, 1994). This is called the *hidden curriculum* in the schools—the informal and often subtle messages about social roles that are conveyed through classroom interaction and classroom materials—roles that are clearly linked to gender, race, and class.

Debunking Society's Myths

Myth: Schools are primarily places where young people learn skills and other knowledge.

Sociological perspective: There is a hidden curriculum in schools where students learn expectations associated with race, class, and gender relations in society.influenced by the socialization process.

In schools, boys and girls are quite often segregated into different groups, with significant sociological consequences. Differences between boys and girls become exaggerated when they are separated into groups based on gender and when they are defined as distinct groups (Thorne, 1993). Seating children in separate gender groups or sorting them into play groups based on gender heightens gender differences and greatly increases the significance of gender in the children's interactions with each other. Equally important is that gender becomes less relevant in the interactions between boys and girls when they are grouped together in common working groups though gender does not disappear altogether as an influence. Barrie Thorne, who has observed gender interaction in schools, concludes from her observations that gender has a "fluid" character, and that gender relations between boys and girls can be improved through conscious changes that discourage gender separation. (See the box "Sociology in Practice" on the next page.)

While in school, young people acquire identities and learn patterns of behavior that are congruent with the needs of other social institutions. School is typically the place where children are first exposed to a hierarchical, bureaucratic environment. School not only teaches them the skills of reading, writing, and other subject areas, it is also where children are trained to respect authority, be punctual, and follow rules—thereby preparing them for their future lives as workers in organizations that value these traits. Schools emphasize conformity to societal needs although not everyone internalizes these lessons to the same degree (Bowles and Gintis, 1979; Lever, 1978). Research has found, for example, that working-class school children form subcultures in school that resist the dominant culture. Rejecting the mental labor associated with school values, working-class children are inadvertently learning the expectations associated with working-class jobs. As a result, the subculture of the school fits the needs of the class system (Willis, 1977).

Growing Up in a Diverse Society

Understanding the institutional context of socialization is important to understanding how socialization affects different groups in society. Socialization makes us members of our society. It instills in us the values of the culture and brings society into our self-definition, our perceptions of others, and our understanding of the world around us. Socialization is not, however, a uniform process, as the different examples developed in this chapter show. In a society as complex and diverse as the United States, no two people will have exactly the same experiences. We can find similarities between us, often across vast social and cultural differences, but variation in social contexts creates vastly different socialization experiences. Furthermore, current

BOX 4.2 SOCIOLOGY IN PRACTICE

Building Cooperative Social Relations among Children

BARRIE Thorne's research on children's interaction in school found that children establish boundaries between themselves that reflect the race and gender divisions in society. Based on her research findings and her observations of teachers' practices, she offers the following suggestions for improving cross-gender and cross-race relationships among children.

1. *When grouping students, use criteria other than gender or race.* Thorne notes that, when left to themselves, children will separate by gender and, often, race. Although teachers should not devalue same-gender and same-race relationships, she suggests that teachers should promote more intergroup interaction.
2. *Affirm and reinforce the values of cooperation among all children, regardless of social categories.* Promoting a sense that the class works together as a group creates more inclusive relationships between different groups.
3. *Whenever possible, organize students into small, heterogeneous, and cooperative work groups.* Small group instruction tends to promote more cooperative interaction among students. When groups focus on shared goals, they are also more likely to value each other.
4. *Facilitate children's access to all activities.* Thorne suggests that teachers need to make a point of teaching the same skills to everyone rather than letting children segregate into activities that promote one skill for one group and another skill for another group. Without this intervention, children tend to engage in gender-typed activities.
5. *Actively intervene to challenge the dynamics of stereotyping and power.* Even when groups are integrated, they may interact based on tensions and inequality. If teachers pretend that race makes no difference, these tensions will simply be reproduced. Instead, teachers should help students learn to interact with each other across these differences and to explore the nature and meaning of cultural and racial differences and the dynamics of racism.

ADAPTED FROM: Barrie Thorne. 1993. *Gender Play: Girls and Boys in School.* New Brunswick, NJ: Rutgers University Press.

BOX 4.3 DOING SOCIOLOGICAL RESEARCH

Learning About Race

MOST people think of race as a feature of identity that primarily shapes the experiences of people of color (Latinos, Native Americans, African Americans, and Asian Americans). Ruth Frankenberg wanted to think about race a little differently. She asked how race shapes the experience of White Americans, thereby turning questions about race on their head to see how a society structured by racial inequality shapes the experiences of those in more racially privileged groups. To put it another way, Frankenberg asked, "How is Whiteness constructed?"

To answer this question, she conducted interviews with thirty White women, many of whom identified with feminism and who came from a variety of social backgrounds. Her sample is not a random one, nor is it intended to represent a particular universe of White women. Because she wanted detailed life histories, the number of her interviews is relatively small; this allowed her to talk to each subject in greater depth, thereby hearing more nuance in what they reported. As a result, her interviews were long (three to eight hours apiece), done usually over two or more sessions. Her book draws heavily on the interviews and includes some of the dialogue between her and her research subjects.

One of the limitations of her study is its restriction to a sample of White women. Frankenberg did this deliberately. As a feminist, she saw feminism as a movement that worked on behalf of all women; she saw her White feminist friends as women who were well meaning and not consciously racist. She wanted to understand what feminist women of color meant when they accused White feminists of being racist. Her research questions were (1) How does racism shape White women's lives? (2) How do White women reproduce racism? and (3) How can White women resist racism?

Based on her interviews, Frankenberg identifies three different ways that the women think about race. *Essentialist racism* espouses the idea of race as a basic biological difference between people; few of the women in Frankenberg's study thought this way although Frankenberg points out how other modes of thinking about race grow from this earlier formulation.

Color-evasive/power-evasive thinking is represented by those who purport not to see race differences between people at all, as in thinking "we're all just people." This thinking, popularly thought of as being "color-blind," is problematic according to Frankenberg because it ignores the power differences that exist between groups as the result of racial domination. Frankenberg also writes that in treating race as if it did not exist, this thinking does not implicate White people in racism; in other words, thinking this way alleviates guilt that White people may feel about racism, but it presumes that racism arises from nowhere—not from the specific historical actions of White people. Those who evade speaking about race also tend to view people's status purely as a matter of individual achievement; they do not think in terms of social structural contexts.

Frankenberg identifies the third way that the women think about race as *race-cognizant* thinking. Here race is conceptualized as difference, but as historical, social, cultural, or political difference, not biological difference. Race-cognizant thinkers see race as making a difference in people's lives and as structuring U.S. society. These women are more likely to be political activists, but they also see themselves as differing significantly from the mainstream of society. They are more likely than the other women to grapple with questions about White complicity in the structure of racism in society.

One of the most fascinating parts of Frankenberg's research is her discussion of the women's views of "White" culture. Most of the women describe being White as being without culture. "Whiteness" is described as something that is amorphous and indescribable, something that can be seen only in contrast to other identities that are marked by race, ethnicity, region, and class; yet, White culture is the unspoken norm. In the women's way of thinking, Whiteness stands for sameness, whereas race stands for difference; as a result, "Whiteness often stood as an unmarked marker of others' differences" (Frankenberg, 1993: 198). Still, Whiteness is linked to domination in the women's narratives. Frankenberg concludes that Whiteness is most clearly identifiable to those whom it excludes. She concludes that Whiteness is a "relational" category—that is, although it is not uniform, it generates norms, ways of thinking, and daily practices, and it is based in specific historical contexts of the domination of others. The book Frankenberg published, *White Women, Race Matters: The Social Construction of Whiteness,*[1] is complex because of its theoretical grounding in the analysis of discourses (taken from postmodernist theory), but it provides a strong analysis of the contemporary politics of race.

[1]Ruth Frankenberg won the American Sociological Association's Jessie Bernard Award in 1995—an award given for the best scholarly work on the sociology of women and gender. This award is given annually and is named in honor of Jessie Bernard—one of the first feminist sociologists who over her long career contributed greatly to the study of gender, families, and women's lives. Jessie Bernard (1903-1996) was a founding mother of feminist scholarship in sociology.

transformations in the U.S. population are creating new multiracial and multicultural environments where young people grow up. Schools, as an example, are in many places being transformed by the large number of immigrant groups entering the school system. In such places, children come into contact with other children from a variety of different groups. This creates a new context where children form their social values and learn their social identities.

One task of the sociological imagination is to examine the influence of different contexts on socialization. Where you grow up; how your family is structured; what resources you have at your disposal; your racial–ethnic identity, gender, and nationality—all these shape the socialization experience. Growing up White, female, and middle class, for example, is quite different from growing up Latino, male, and working class. Even within a given group, there are great differences in socialization patterns. Growing up as an African American, middle-class woman will produce results quite different from growing up African American and poor. Socialization experiences for all groups are shaped by a number of factors that intermingle to form the context for socialization.

To probe deeper into the effect of social context on socialization, Clyde Franklin has studied the socialization experiences of young African American men. He finds that the messages they receive through socialization are often contradictory. Within the African American community, young men learn that self-sufficiency and independence are highly valued components of masculine identity, but they may find these values to be unattainable in society. The degree to which they can attain self-sufficiency varies by social class. African American men who grow up in a disadvantaged class encounter many impediments to attaining the ideals of self-sufficiency and independence. Young men may then turn to violence and aggression as ways of asserting their manhood when more legitimate means of doing so are inaccessible (Fordham, 1996; Franklin, 1994).

Other groups also experience the complexities of multiple expectations that are inevitable in a society as diverse as the United States. Try to imagine the experience of those who grow up with mixed racial identities in a society where racial identity is such an important marker of how others define you and how you define yourself (Root, 1992, 1996; Tizard, 1993). In this society, one is presumed to be Black, White, Latino, Asian, *or* Native American. These one-dimensional labels are inadequate for multiracial people; yet, the labels are confronted every day when multiracial people fill out a job application, respond to what people call them, or decide what student organizations to join. A person who is Latino and Black, Asian and White, or Native American and Black may experience contradictory expectations from the dominant society and the home community—compounded by the contradictory expectations that normally occur within either community by itself. In the words of the sociologist Michael Thornton (himself an individual of mixed racial identity), "Individuals are expected to locate themselves 'accurately' within established racial structures (such as current census categories), finding where society places them and reconciling this placement with what they want to be. . . . Society defines race as distinctive and homogenous; multiracial people experience it as multidimensional" (1995: 97–98).

The socialization process is clearly patterned by factors like class, race, gender, religion, regional background, sexual preference, age, and ethnicity, to name a few, but none of these characteristics singly can define the socialization experience. Certainly, some may be more salient at given times than others. A Latina may, for example, be particularly cognizant of her racial identity as she is growing up, but she does not experience it as separate from her gender identity. Race and gender together are integral to her identity, as they are for all people. She may be more conscious of one or the other at different points in her life, but she is never completely unaffected by either.

The point is that you cannot think in narrow terms when thinking like a sociologist. Those with sociological imaginations have to recognize the diversity in human experience, both within and across different cultures, and must understand how these diverse experiences shape the values and identities of the great variety of people who inhabit the world.

Socialization Across the Life Cycle

Socialization begins the moment a person is born. As soon as the sex of a child is known, parents, grandparents, brothers, and sisters greet the infant with different expectations, depending on whether it is a boy or a girl. Socialization does not come to an end as we reach adulthood; rather, it continues through our lifetime. As we enter new situations, and even as we interact in familiar ones, we learn new rules and undergo changes in identity.

Childhood

During childhood, socialization establishes one's initial identity and values. In this period, the family is an extremely influential source of socialization, but as we have seen, experiences in school, peer relationships, sports, religion, and the media also have a profound effect. Children acquire knowledge of their culture through countless subtle cues that provide them with an understanding of what it means to live in society.

Socializing cues begin as early as infancy, when parents and others begin to describe their children according to their perceptions of them. Frequently, these perceptions are derived from the cultural expectations parents have for children, and they typically vary depending on whether the child is a boy or a girl. In one classic study, parents were asked just twenty-four hours after the birth of their first child to describe the baby. Although the researchers controlled for the physical size of the infant to ensure that there were no objective differences between the boys and the girls, parents of girls described their babies as "softer" and smaller than did parents of boys. Boy infants were described as strong, alert, and well-coordinated (Rubin et al., 1974). Researchers have recently repeated this study with a new twist by comparing how both parents and children describe

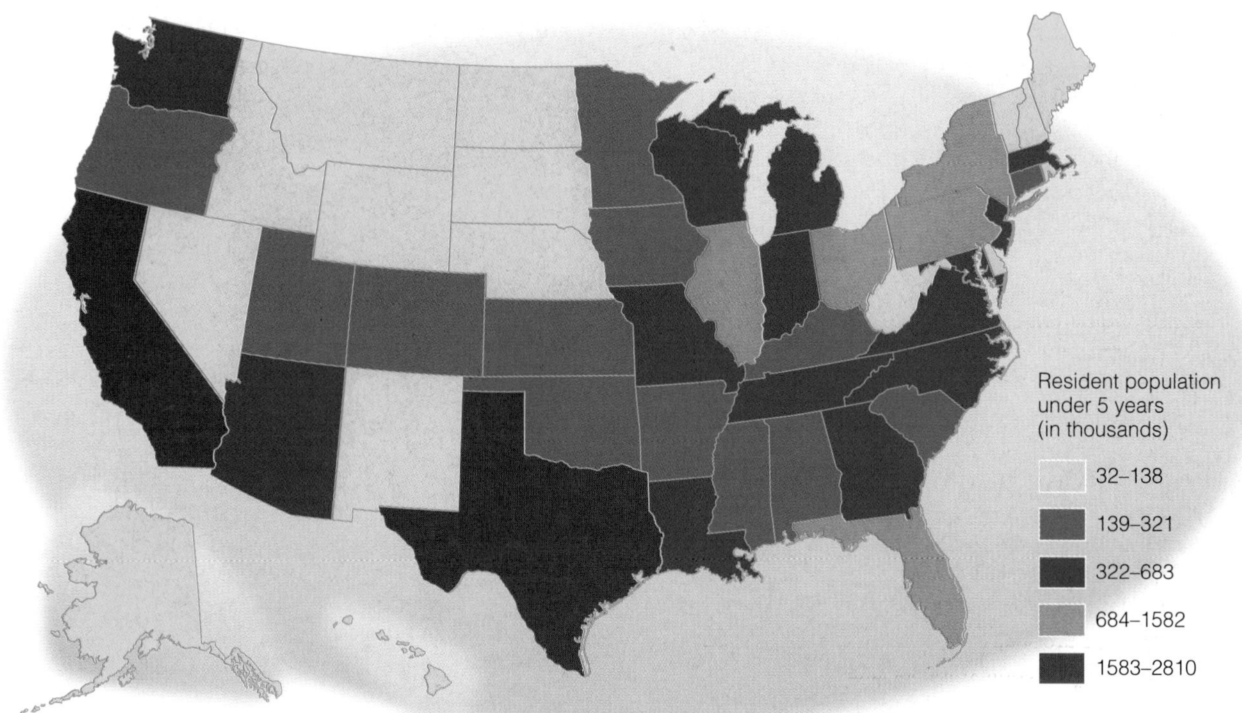

MAP 4.1 Mapping America's Diversity: Children under Five: Where They Live

The socialization of children occurs in the context of specific social influences, including such things as region of residence, national origin, race, social class, and immigrant status. As this map shows, children vary as a proportion of the population in different regions of the United States. Lying behind this distribution of the population are different patterns of immigration and migration, different economic and educational opportunities, and varying levels of social support services for children and their parents.

DATA: From the Bureau of the Census 1997. *Statistical Abstract of the United States, 1997.* Washington, DC: U.S. Government Printing Office, p. 33.

newborn infants; interestingly, they found that children describe infants in even more gender-stereotyped ways than the adults (Stern and Karraker, 1989).

The lessons of childhood socialization come in a myriad of ways, some more subtle than others. In an example of how gender influences childhood socialization, researchers observed mothers and fathers who were walking young children through public places. Although the parents may not have been aware of it, both mothers and fathers were more protective toward girl toddlers than toward boys. Parents were more likely to let boy toddlers walk alone, but held girls' hands, carried them, or kept them in strollers. The children were not the only ones learning gender roles; the researchers also observed that when the child was out of the stroller, the mother was far more likely to be pushing the empty stroller than was the father (Mitchell et al., 1992). In countless ways, sometimes subtle, sometimes overt, we learn society's expectations.

Much socialization in early childhood takes place through play and games. Games that encourage competition help instill the value of competitiveness throughout someone's life; likewise, play with other children and games that are challenging give children important intellectual, social, and interpersonal skills. Extensive research has been done on how children's play and games influence their identities as boys and girls. Generally, the research finds that boys' play tends to be rougher, more aggressive, and involve more specific rules; boys are also more likely to be involved in group play. Girls, on the other hand, engage in more conversational play (Moller et al., 1992). Sociologists have concluded that the games children play significantly influence their development into adults.

Another enormous influence on childhood socialization is what children observe of the adult world. Children are keen observers of their surroundings, and what they perceive will influence their self-concept and how they relate to others. This is vividly illustrated by research revealing how many adult child abusers were themselves victims of child abuse (Fattah, 1994). Children become socialized by observing the roles of those around them and internalizing the values, beliefs, and expectations of their culture.

Adolescence

Only recently has adolescence been thought of as a separate phase in the life cycle. Until the early twentieth century, children moved directly from childhood roles to adult roles. It was only when formal education was extended to all

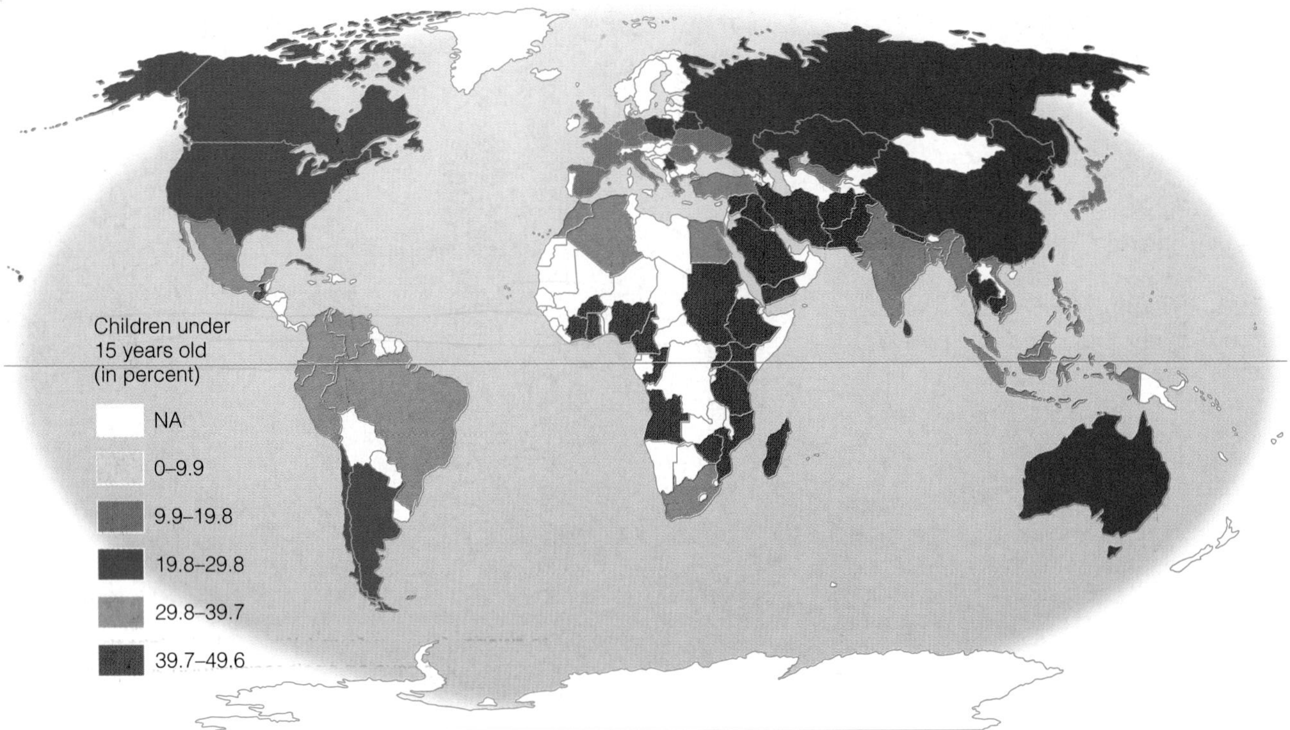

MAP 4.2 Viewing Society in Global Perspective: The World's Children

Throughout the world, the proportion of children as a percentage of the population of a given country tends to be higher in those countries that are most economically disadvantaged and most overpopulated.

DATA: From the Bureau of the Census 1997. *Statistical Abstract of the United States, 1997.* Washington, DC: U.S. Government Printing Office, p. 832.

classes that adolescence emerged as a particular phase in life—the phase when young people are regarded as no longer children, but not yet adults. There are no clear boundaries to adolescence although it is generally taken to last from junior high school until the time one takes on adult roles by getting a job, marrying, and so forth. Adolescence can include the period through high school and extend right up through college graduation.

Erik Erikson (1980), the noted psychologist, stated that the central task of adolescence is the formation of a consistent identity. Adolescents are trying to become independent of their families, but they have not yet moved into adult roles. Conflict and confusion can arise as the adolescent swings between childhood and adult maturity. Some argue that adolescence is a period of delayed maturity. Although society expects adolescents to behave like adults, they are denied many of the privileges associated with adult life: They cannot vote, drink alcohol, or marry without permission, and sexual activity is condemned. The tensions of adolescence have been blamed for numerous social problems, such as drug and alcohol abuse, youth violence, and the school dropout rate, to name a few.

The values that young people learn are a good barometer of social change across generations. In the 1970s and 1980s, high school seniors identified a good marriage and good family life as their major goals. Throughout the 1980s and 1990s, making money gained importance as a life goal of young people. In the same period, the number of young people who consider environmental issues to be important rose although only 24 percent say that being involved in programs to help clean up the environment is important to them. Fewer young people than before are interested in helping others in difficulty although interest in careers like public service and teaching has grown (Conger, 1988; Higher Education Research Institute, 1996; Shea, 1995).

Patterns of adolescent socialization vary significantly with race, gender, and social class. Youths from the upper and middle classes are more likely to base their friendships on shared activities and interests, often changing friendships as their activities change. Working-class youths are more likely to base their friendships on loyalty and stability; their friendships determine their activities, not the other way around. Class differences among adolescents can also be found in their subcultures. Middle-class and elite students tend to see the "ideal student" as someone involved in many different activities, and they form their subcultures accordingly. Working-class students, who have much less control over school resources and activities, are more likely to develop subcultures based on defiance against rules, authority, and academic work (Eckert, 1988; Lesko, 1988). Working-class youth subcultures, at least among young men, are also more likely to focus on fighting and exchanging insults (a highly developed skill, often done with great humor) since these are activities over which they have some control and that assert their masculine camaraderie. Studies of street cultures among young African American men, for example, find that the main concerns of street gangs are toughness, excitement, cleverness, and autonomy; moreover, skill in various verbal activities, such as ritual insulting ("playing the dozens," talking about "yo momma," and so on), storytelling, and telling jokes is a mark of high status in these groups (Anderson, 1990).

The adolescent subcultures of young women also show marked class differences. White working-class women are more likely than middle-class women to see themselves as nonconformists at school although their resistance to school authority is less visible than that of young men. Young working-class women may defy authority by withholding their participation in class, passing notes, reading magazines, and daydreaming. Within their peer subcultures, they often emphasize the open pursuit of pleasure, including sexual pleasure, and they are less concerned with being "nice" and "ladylike" than is typical of middle-class women's peer subcultures (Griffin, 1985; Wulff, 1988).

Differences in adolescent socialization are associated with significantly different life paths. There are forks in the road before an adolescent. Staying in school is one such fork. Of all young adults, Asian Americans are most likely to be in school only or to combine school and work. Hispanic, American Indian, and African American young people are most likely to be both out of school and out of work (Figure 4.3). Whether adolescents are in school, on the job, both, or neither can have striking effects on the values and self-concepts they develop. Socialization is dependent on social context, which has a deciding influence on where young people find themselves later in life. The different lessons they acquire in class, on the job, or on the street corner teach them how to think and act in society. As Figure 4.3 shows, there are significant differences in the context of young people's lives, including whether one works and attends school, only works, or only attends school.

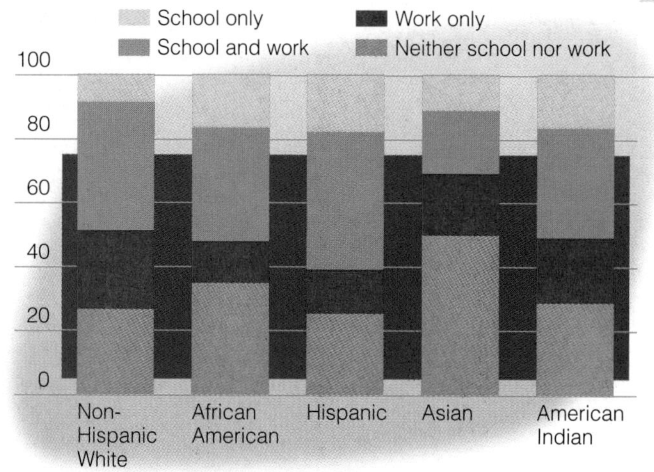

FIGURE 4.3 The Context of Childhood Socialization

Socialization does not occur in a vacuum. Young people's activities at work and at school provide a context in which their values and life chances are formed. The data shown here indicate the very different contexts in which young people of different races, aged sixteen to twenty-four, are found. What effect might these different experiences have on one's socialization as a young person?

DATA: From O'Hare, William P. 1992. "America's Minorities—The Demographics of Diversity." *Population Bulletin* 47 (December): Washington, DC: Population Reference Bureau, p. 31.

Adulthood and Old Age

Socialization does not end when one becomes an adult. Building on the identity formed in childhood and adolescence, **adult socialization** is the process of learning new roles and expectations in adult life. More so than at earlier stages in life, adult socialization involves learning behaviors and attitudes appropriate to specific situations and roles.

Youths entering college, to take an example from young adulthood, are newly independent and have new responsibilities. In college, one acquires not just an education but also a new identity. Those who enter college directly from high school may encounter conflicts with their family over their newfound status. Older students who work and attend college may experience difficulties (defined as role conflict; see Chapter 6) trying to meet dual responsibilities, especially if their family is not supportive. Research on returning adult women students finds that women who lack the support of family and friends have the greatest emotional difficulty adjusting to the multiple roles of student, mother, and wife (Suitor, 1988). Meeting the new demands may require the returning student to develop different expectations about how much she can accomplish, or different priorities about what she will attempt—in other words, a change in values and attitudes, reflecting a new stage in her socialization.

THINKING SOCIOLOGICALLY

Find a group of adults (young or old) who have just entered a new stage of life (getting a new or first job, getting married, becoming a grandparent, retiring, entering a nursing home, and so forth), and ask them to describe this new experience. Ask questions such as what others expect of them in this new role, how these expectations are communicated to them, what changes they see in their own behavior, and what expectations they have of their new situation. What do your observations tell you about *adult socialization*?

Adult life is peppered with events that may require the adult to adapt to new roles. Marriage, a new career, starting a family, entering the military, getting a divorce, or dealing with death in the family all transform an individual's previous social identity. Divorce is a good example of a forced transition from one identity to another. Marital status is a significant part of one's overall social role, and getting divorced will bring changes in how a person is treated by others. Social relationships a person had while married may change or disappear completely. It is little surprise that newly divorced people often feel disoriented and out of place. Divorce may also have an impact on others in the family, especially youngsters, who undergo their own ordeal of new socialization (Demo and Acock, 1988).

Another part of learning a new role is **anticipatory socialization,** the learning of expectations associated with

Every culture has important rites of passage that mark the transition from one phase in the life course to another. Here different cultural traditions distinguish the rites of passage associated with marriage—a traditional Nigerian wedding (upper left); a young American couple (upper right); a Shinto (Japanese) bride taking a marital pledge by drinking sake (lower left); and a newlywed Orthodox Christian couple in Eritrea (lower right).

a role one expects to enter in the future. One might rehearse the expectations associated with being a professor by working as a teaching assistant, taking a class in preparation for becoming a father, or attending a summer program to prepare for entering college. Anticipatory socialization allows a person to foresee the expectations associated with a new role and learn what is expected in that role in advance.

In the transition from an old role to a new one, individuals often vacillate between their old and new identities as they adjust to fresh settings and expectations. The coming-out process for gays and lesbians supplies an interesting example. *Coming out* is the process of identifying oneself as gay or lesbian. This can be either a public coming out or a private acknowledgment of the homoerotic feelings that one has. Coming out can take years. It generally means coming out to a few people at first, people from one's family or friends who are seen as likely to have the most positive reaction; thus, coming out can be selective. Coming out is rarely a single event but occurs in stages on the way to developing a new identity—one that is not only a new sexual identity, but also a new sense of self (Due, 1995).

Do people entering new roles become exactly what is expected of them? Sometimes they do. Women and minorities who acquire power and prestige are often accused of becoming like White men. White men have long had a near monopoly on positions of power, and their attitudes and behaviors usually define the culture of large organizations. Newcomers may be advised to conform to the same attitudes and behaviors in order to succeed. How thoroughly they adopt the new role depends on many factors, such as whom they identify with, how they define themselves, and what they receive as rewards and punishments within the organization. Always, in adulthood and earlier, there is variation in the extent to which people internalize new social expectations.

Passage through adulthood involves many transitions. In our society, one of the most difficult transitions is the passage to old age. We are taught to fear aging in this society, and many people spend a lot of time and money trying to keep themselves looking young. Unlike many societies, ours does not revere the elderly but instead devalues them, making the aging process even more difficult. As we will see in Chapter 13, growing old in a society with such a strong emphasis on youth means encountering social stereotypes about the old, adjusting to diminished social and financial resources, and sometimes living in the absence of social supports, even when facing some of life's most difficult transitions, such as declining health and the loss of loved ones. Still, many people experience old age as a time of great satisfaction and enjoy a sense of accomplishment connected to work, family, and friends. The degree of satisfaction during old age depends to a great extent on the social support networks established earlier in life. Strong social networks have a positive effect on the ability of people to cope with stressful life events; sociologists have found that older women report higher levels of emotional support from such networks than men; men tend to experience greater social isolation when they are old—an indication of how the gender expectations one learns earlier in life continue to have an influence over the life course (Gibson, 1996).

Rites of Passage

A **rite of passage** is a ceremony or ritual that marks the passage of an individual from one role to another. Rites of passage define and legitimize abrupt role changes that begin or end each stage of life. The ceremonies surrounding rites of passage are often dramatic and infused with awe and solemnity. Examples include graduation ceremonies; weddings; and religious affirmations, such as the Jewish ceremony of the bar mitzvah for boys or the bas mitzvah for girls, confirmation for Catholics, and adult baptism for many Christian denominations.

Formal promotions or entry into some new careers may also include rites of passage. Completing police academy training or being handed one's diploma are examples. Such rites usually include family and friends, who watch the ceremony with pride; people frequently keep mementos of these rites as markers of the transition through life's major stages. Bridal showers and baby showers have been analyzed as rites of passage. At a shower, the person who is being honored is about to assume a new role and identity—from young woman to wife or mother (Cheal, 1988). Rites of passage entail public announcement of the new status, for the benefit of both the individual and those with whom the newly anointed person will interact. In the absence of such rituals, the transformation of identity would not be formally recognized, perhaps leaving uncertainty in the rising youngster or the community about the individual's worthiness, preparedness, or community acceptance.

Sociologists have noted that contemporary society has no formalized rite of passage marking the transition from childhood to adulthood. As a consequence, the period of adolescence is attended by ambivalence and uncertainty. As adolescents hover between adult and child status, they may not have the clear sense of identity that a rite of passage can provide. Although there is no universal ceremony in our culture by which one is promoted from child to adult, some subcultures do mark the occasion. Among the wealthy, the debutante's coming-out celebration is a traditional introduction of a young woman to adult society. Latinos may celebrate the *quinceañera* (fifteenth birthday) of young girls. A tradition of the Catholic Church, this rite recognizes the girl's coming of age, while also keeping faith with an ethnic heritage. Dressed in white, she is introduced by her parents to the larger community. Formerly associated mostly with working-class families in the barrios, the *quinceañera* has also become popular among affluent Mexican Americans, who may match New York debutante society by spending as much as $30,000 to $50,000 on the event (McLane, 1995; Williams, 1990).

The quinceañera *is a traditional rite of passage among Latinas, marking the transition to womanhood at age fifteen.*

Resocialization

Most of the transitions people experience in their lifetimes involve continuity with the former self as it undergoes gradual redefinition. Sometimes, however, adults are forced to undergo a radical shift of identity. **Resocialization** is the process by which existing social roles are radically altered or replaced (Fein, 1988). Resocialization is especially likely when people enter institutional settings where the institution claims enormous control over the individual. Examples include the military, prisons, monastic orders, and some cults (see also Chapter 6 for a discussion of *total institutions*). When military recruits enter boot camp, they are stripped of personal belongings, their heads are shaved, and they are issued identical uniforms. Although military recruits do not discard their former identities, the changes brought about by becoming a soldier can be dramatic and are meant to make the military, not one's family, friends, or personal history, primary. The military represents an extreme form of resocialization, one in which individuals are expected to subordinate their identity to that of the group. In such organizations, individuals are interchangeable, and group consensus (meaning, in the military, unanimous, unquestioned

As part of their resocialization, the women and men who are new students at the Virginia Military Institute (VMI) are stripped of their former identities, with close cropped hair and identical clothing. Called "rats," they endure extreme physical excercise and the taunts of older students.

subordination to higher ranks) is an essential component of group cohesion and effectiveness. Military personnel are expected to act as soldiers, not as individuals. Understanding the importance of resocialization upon entry to the military helps us understand such practices as "the rat line" at VMI (Virginia Military Institute) where members of the senior class taunt and harass new recruits.

Resocialization often occurs when people enter hierarchical organizations that require them to respond to authority on principle, not out of individual loyalty. The resocialization process promotes group solidarity and generates a feeling of belonging. Participants in these settings are expected to honor the symbols and objectives of the organization; disloyalty is seen as a threat to the entire group. In a convent, for example, nuns are expected to subordinate their own identity to the calling they have taken on, a calling that requires obedience both to God and to an abbess.

Resocialization may involve degrading initiates physically and psychologically with the aim of breaking down or redefining their old identity. They may be given menial and humiliating tasks, and expected to act in a subservient manner. Social control in such a setting may be exerted by peer ridicule or actual punishment. Fraternities and sororities offer an interesting everyday example of this pattern of resocialization. When fraternities and sororities induct new members, the pledges are often given onerous, meaningless chores, and they may be forced to behave in comically degrading, and sometimes life-threatening, ways. Depending on the traditions of the group, resocialization practices range from mildly sadistic mischief to hazardous behavior, such as hazing. Intense resocialization rituals, whether in jailhouses, barracks, convents, or sorority and fraternity houses serve the same purpose: imposing some sort of ordeal to cement the seriousness and permanence of new roles and expectations.

The Process of Conversion

Resocialization also occurs during what people popularly think of as *conversion*. A conversion is a far-reaching transformation of identity, often related to religious or political beliefs. People usually think of conversion in the context of cults, such as the Heaven's Gate group, but it happens in other settings as well.

The Autobiography of Malcolm X provides an example of conversion spanning religion, politics, and social convictions. Malcolm X writes that in his youth he was a petty criminal. Believing in the dominant White culture's view of Black people, he defined himself in negative, self-destructive terms. In prison, he encountered the Nation of Islam, a political and religious movement that radically altered his sense of self and his views on race in America. At first, his new beliefs caused him to hate White people, whom he saw as the oppressors of Black America. Later, following his trip to Mecca, he refined his beliefs, analyzing racial oppression in the context of international capitalism. His analysis became a foundation of Black protest in America. His conversion is an example of how identity can be reconstructed through membership in social groups, especially groups like the Nation of Islam that set out explicitly to "rescue" and absorb individuals who may knowingly or unknowingly long for deeper political or spiritual purpose.

The Brainwashing Debate

More extreme examples of resocialization are seen in the phenomenon popularly called *brainwashing*. In the popular view of brainwashing, converts have their previous identities totally stripped; the transformation is seen as so complete that only deprogramming can restore the former self. Potential candidates of brainwashing include people who enter religious cults, prisoners of war, and hostages. Sociologists have examined brainwashing to illustrate the process of resocialization. As the result of their research, sociologists have cautioned against using the word *brainwashing* when referring to this form of conversion. The term implies that humans are mere puppets or passive victims whose free will can be taken away during these conversions (Robbins, 1988). In religious cults, however, converts do not necessarily drop their former identity.

Sociological research has found that the people most susceptible to cult influence are those who are the most suggestible, primarily young adults who are socially isolated, drifting, and having difficulty performing in other areas (such as in their jobs or in school). Such people may choose to affiliate with cults voluntarily. Despite the widespread belief that people have to be deprogrammed to be freed from the influence of cults, many people are able to leave on their own (Robbins, 1988). As we will see further in Chapter 16, so-called brainwashing is simply a manifestation of the social influence people experience through interaction with others. Even in cult settings, socialization is an interactive process, not just a transfer of group expectations to passive victims.

Prisoners of war, such as these Bosnian men, are typically subjected to the extremes of resocialization: deprived of personal objects and forced to live in extreme conditions, they become quite dependent on their captors.

Forcible confinement and physical torture can be instruments of extreme resocialization. Under conditions of severe captivity and deprivation, a captured person may come to identify with the captor; this is known as the *Stockholm Syndrome.* In such instances, the captured person has become dependent on the captor. Upon his or her release, the captive frequently needs debriefing. Prisoners of war and hostages may not lose free will altogether, but they do lose freedom of movement and association, which makes prisoners intensely dependent on their captors, and therefore vulnerable to the captor's influence. The Stockholm Syndrome can help explain why some battered women do not leave their abusers. Dependent on their abuser both financially and emotionally, battered women often develop identities that keep them attached to men who abuse them. In these cases, outsiders often think the women should leave instantly, whereas the women themselves may find leaving difficult, even in the most abusive situations.

In sum, resocialization involves establishing a radically new definition of oneself. The new identity may seem dramatically different from the former one, but the process by which it is established is much the same as the ordinary socialization process—a process that, as we have seen, is a critical part of society.

CHAPTER SUMMARY

- Cases of total social isolation demonstrate the importance of social interaction in shaping human characteristics. *Feral children* are those who have been raised in the absence of human contact. *Socialization* is the process by which human beings learn the social expectations of society.
- Socialization is powerful because it creates the expectations that are the basis for people's attitudes and behaviors; it also shows that society and the people within society are socially constructed. Through socialization, people conform to social expectations although people still express themselves as individuals.
- Several theoretical perspectives are used to understand socialization, including *psychoanalytic theory, object relations theory, social learning theory,* and *symbolic interaction theory*. Psychoanalytic theory sees the self as driven by unconscious drives and forces that interact with the expectations of society. Object relations theory explains the development of the self as the result of individuation and attachment in relationship to parenting figures. Social learning theory sees identity as a learned response to social stimuli. Symbolic interaction theory sees people as constructing the self as they interact with the environment and give meaning to their experience. Charles Horton Cooley described this process as the *looking-glass self.* Another sociologist, George Herbert Mead, described childhood socialization as occurring in three stages: imitation, play, and games.
- *Socialization agents* are those who pass on social expectations. They include the family, the media, peers, sports, religious institutions, and the schools, among others. The family is usually the first source of socialization. The media also influence people's values and behaviors. *Peer groups* are an important source of individual identity; without peer approval, most people find it hard to be socially accepted. Schools also pass on expectations that are also influenced by gender, race, and other social characteristics of people and groups.
- Socialization also influences how different groups are valued and value themselves. Studies of race and *self-esteem* show that more positive images in the dominant culture do enhance people's definitions of themselves. Socialization continues through the lifetime although childhood is an especially significant time for the formation of identity. Adolescence is also a period when peer cultures have an enormous influence on the formation of people's self-concepts. *Rites of passage* are ceremonies or rituals that symbolize the passage from one role to another. *Adult socialization* involves the learning of specific expectations associated with new roles.
- *Resocialization* is the process by which existing social roles are radically altered or replaced. It can take place in an organization that maintains strict social control and demands that the individual conform to the needs of the group or organization.

KEY TERMS

adult socialization
anticipatory socialization
ego
feral children
game stage
generalized other
id
identity
imitation stage
looking-glass self
object relations theory
peers
personality
play stage
psychoanalytic theory
resocialization
rite of passage
role
self-esteem
significant others
socialization
socialization agents
social learning theory
superego
taking the role of the other

THE INTERNET: A Tool for the Sociological Imagination

Resources on the Internet:

Virtual Society: The Wadsworth Sociology Resource Center at
http://sociology.wadsworth.com

Visit this site to find additional learning tools, including interactive quizzes, links to related web sites, and an easy link to InfoTrac College Edition.

Children's Defense Fund
http://childrensdefense.org

An organization dedicated to improving children's lives, especially poor and minority children.

National Black Child Development Institute
http://www.nbcdi.org

An organization that serves as an advocate for improving the lives of Black children and their families.

Sociology and Social Policy: Internet Exercises

Concerns about the influence of *media violence* on young people have led many to advocate more restrictions on the contents of the media, including films, television programs, and music. Some want more censorship; others want a stronger system of warning labels, such as the television chip (V chip) that allows parents to make certain programs and channels unavailable to their children. Others argue that violence in the media does not cause violence in society since the media only reflect violence that already exists. Based on the understanding of socialization you have acquired from this chapter, what do you think is the effect of violent images in the media on the socialization experiences of young children? What social policies would you suggest regarding the content of the media?

Internet Search Keywords:

media violence
censorship
entertainment violence

Web sites:

http://www.crin.org/media.htm
The web site for Child Rights Information Network includes important information on organizations, conventions, and publications on children and the media plus many links to related sites.

http://www.mediascope.org/clearing.htm
Mediascope's Clearinghouse. Contains the results of the National Television Violence Study 1994–1995 and an extensive collection of resources on social and public health issues in mass media.

http://www.cln.org/themes/media_violence.html
Community Learning Network's *Violence in the Media Theme Page* includes publications from a variety of sources on the topic.

http://www.ccp.ucla.edu/archive.htm
Three UCLA Television Violence Monitoring Reports from 1994 to 1997.

http://ericps.ed.uiuc.edu/npin/respar/texts/media.html
Web site for National Parent Information Network at University of Illinois. Contains a number of publications on the impact of media violence on children and parental control.

InfoTrac College Edition: Search Word Summary

socialization
media violence
Freud, Sigmund

In order to learn more about these central topics in sociology, you can conduct an electronic search using InfoTrac College Edition. To aid in your search and to gain useful tips, see the Student Guide to InfoTrac College Edition on the Virtual Society web site:

http://sociology.wadsworth.com

INTERACTIONS—A SOCIOLOGY CD-ROM: CONCEPTS FOR THIS CHAPTER

Go to the Wadsworth Sociology CD-ROM for further study on the concepts in this chapter. The CD-ROM also includes quizzes and additional activities to expand your learning experience.

SUGGESTED READINGS

Due, Linnea. 1995. *Joining the Tribe: Growing Up Gay & Lesbian in the '90s.* New York: Doubleday.

Written for a popular audience and based on many personal narratives, this book examines the sociological dimensions of growing up lesbian or gay.

Kotlowitz, Alex. 1991. *There Are No Children Here.* New York: Anchor Books.

Kotlowitz's account of two young Black brothers growing up in a Chicago housing project is a moving portrayal of the experience of growing up amid poverty, racism, and violence.

Lee, Joann Faung Jean. 1991. *Asian Americans: Oral Histories of First to Fourth Generation Americans from China, the Philippines, Japan, India, the Pacific Islands, Vietnam, and Cambodia.* New York: The Free Press.

This collection of writings by Asian Americans provides first-hand accounts of the different experiences of growing up in the United States.

Riley, Patricia. 1993. *Growing Up Native American: An Anthology.* New York: Morrow.

This is a collection of personal narratives about the experience of growing up Native American. The collection reveals both the influence of growing up in Indian culture and the diversity of cultures among American Indians.

Thompson, Becky, and Sangeeta Tyagi. 1996. *Names We Call Home: Autobiography on Racial Identity.* New York: Routledge.

A collection of contemporary autobiographical essays, this book explores the various dimensions of racial identity formation for diverse groups in the United States. Its sociological focus is interwoven with moving personal statements.

Thorne, Barrie. 1993. *Gender Play: Girls and Boys in School.* New Brunswick, NJ: Rutgers University Press.

Based on direct observation of children in schools, this book shows how the play of children shapes their gender identities. The book includes an analysis of how race, age, and ethnicity shape identity in children and has suggestions for teachers on how to develop improved interaction between boys and girls.

CHAPTER 5

Society and Social Interaction

THE WRITER and poet John Donne wrote more than three centuries ago that "No man is an island, entire of itself; every man is a piece of a continent, a part of the main." As humans, we are constantly involved with other humans. Even when we are alone, we think about other people. When we interact with others, we do so in patterned and often predictable ways.

A parent and teenager argue about what the teen should wear to a party. A police officer has a man spread-eagle against an automobile while searching him for drugs. A group of business-women confer over lunch about a recent sale. A physics professor lectures to students about the nature of the universe. A congregation listens to a sermon and then prays together. All these highly diverse actions have something in common: They are all regulated, to a greater or lesser degree, by the elements of society—groups, statuses, roles, and social institutions. These elements guide the formation of human society.

In some ways, society is just the sum of all the individual interactions people have with one another, but it includes more than that. Society takes on a life of its own—the sum of people's behavior, but somehow more than that. This is one of the basic ideas that guide sociological thinking. Society is patterned on humans and their interactions, but it is something that endures in itself and takes on shape and structure.

In this chapter, we examine the different pieces of society, beginning with the study of social interaction and proceeding from this close-up level of society to studying the larger forces that hold society together—social institutions and social structures. We will see that human interaction is guided by social forces, at the same time that humans shape the social institutions that characterize different societies. We proceed by discussing the different elements of society—groups, statuses, roles, and the social bonds that people form through social interaction. In the latter part of the chapter, we study the grandest levels of society—social institutions. As you proceed through the chapter, you will begin to see the complexity of diverse societies—how they are arranged and how they are held together.

What Is Society?

In Chapter 3, we studied culture as one of the forces that holds society together. Culture refers to the general way of life, to norms, customs, beliefs, and language. Human **society** is a system of social interaction that includes both culture and social organization. Within a society, members have a common culture even though there may also be great diversity within. In society, people think of themselves as distinct from other societies, maintain ties of interaction, and have a high degree of interdependence. The interaction that they have, whether based on harmony or conflict, is one of the elements of society. That is, social interaction is how human beings communicate with each other, and in so doing, they form a social bond.

Sociologists use the term **social interaction,** or social action, to mean behavior between two or more people that is meaningful. Social interaction involves more than simply acting; it involves communication—the conveyance of information to a person by any means. Social action may be simple (a word, wave, or threatening gesture) or complex (speaking, organizing a social movement, or forming a family). That social interaction is meaningful is what distinguishes it from instinctual behavior.

Social interaction is the foundation of society, but society becomes more than a collection of individual social actions. Emile Durkheim, the classical sociological theorist, described society as *sui generis*—a Latin phrase meaning "a thing in itself, of its own particular kind." To sociologists, seeing society "sui generis" means that society is more than just the sum of its parts. Durkheim saw society as an organism, something comprised of different parts that work together to create a unique whole. Just as a human body is not just a collection of organs but is alive as a whole organism, society exists as a whole entity rather than as a disorderly collection of individuals, groups, or institutions.

To imagine this, think of how a photographer views a landscape. The landscape is not just the sum of its individual parts—mountains, pastures, trees, or clouds—although each of these contributes to the whole. The power and beauty of the landscape is that all its parts *relate* to each other, some in harmony, some in contrast, to create a panoramic view. The photographer who tries to capture this view will likely use a wide angle lens—a method of seeing that captures the breadth and comprehensive scope of what the photographer sees. Similarly, sociologists try to picture sociology as a whole—seeing its individual parts, but also recognizing the relatedness of these parts and their vast complexity. Their wide angle lens is the sociological imagination.

Microanalysis and Macroanalysis

Like photographers, sociologists use different lenses to see the different parts of society. Some views are more microscopic—that is, they focus on the smallest, most immediately visible parts of social life, such as specific people interacting together. This is called *microanalysis.* Other views are more macroscopic—that is, they try to comprehend the whole of society, how it is organized and how it changes. This is called *macroanalysis.* Each view provides a distinct vision of society, and both reveal different dimensions to society.

Some sociologists study the patterns of social interactions that are relatively small, less complex, and less differentiated—the *microlevel* of society. Studying a small group, such as your friendship group or your family, is an example; a clique (or subgroup) that forms within your own friendship group is another example. Perhaps you are a member of one or more such cliques at this time, or perhaps you are a leader within your group. All these kinds of behavior would be of interest to someone studying the microlevel of society. The study of interpersonal attraction—what makes people attracted to someone and not attracted to someone else, a topic discussed later in this chapter—would also be studied at the microlevel.

Some sociologists are interested in the microlevels of society like group interaction; others are interested in the broadest views of society—how they are organized and how they change. Sociologists who study the *macrolevel* of society study the large patterns of social interactions that are vast, complex, and highly differentiated. Sociologists studying this level of society might look at a whole society or compare different societies. The broader framework of social problems in our society, such as poverty, the homeless, and urban crime, are all macrolevel problems. When the sociologist analyzes such problems as these, he or she is analyzing society from what is called a macrolevel perspective. In this chapter, we proceed by starting with the microlevel of social life, continuing through macrolevel sociology. The idea is to help you see the most immediate ways that social forces influence people's behaviors, proceeding through larger questions of how society is held together and organized.

Social organization is the term sociologists use to describe the order established in social groups at any level. Specifically, social organization is the order that brings regularity and predictability to human behavior. Social organization is present at every level of interaction, from the smallest groups to the whole society, but appears in social roles and social institutions.

Groups

At any given moment, each of us is a member of many groups simultaneously, and we are subject to their influence: family, friendship groups, athletic teams, work groups, office staffs, racial and ethnic groups, and so on. Groups impinge on every aspect of our lives and are a major determinant of our attitudes and opinions on everything from child care, politics, and the economy to our view on the death penalty.

To sociologists a **group** is a collection of individuals who:

1. Interact and communicate with each other
2. Share goals and norms
3. Possess a subjective awareness of themselves as "we," that is, as a distinct social unit

To be a group, the social unit in question must possess all three of these characteristics together. We will examine the nature and behavior of groups in greater detail in Chapter 6.

In sociological terms, not all social units are groups. *Social categories* are people who are lumped together based on one or more shared characteristics. Examples of social categories are teenagers (an age category), truck drivers (an occupational category), and millionaires (an economic category). Some social categories can also form a social stratum or even a class, as we shall see in Chapter 9; the homeless are one example. Ethnic and racial groups may be either categories or groups, depending upon the amount of "we" feeling. When "we" feeling is high, racial and ethnic categories are indeed groups.

All people nationwide who are watching a TV program at 8 o'clock Wednesday evening form another distinct social unit, an *audience*. They are not a group since they do not interact with one another, nor do they possess an awareness of themselves as "we." If many of the same viewers come together in a TV studio, where they interact and develop a "we" feeling, then they would constitute a group.

Finally, *formal organizations* are highly structured social groupings that form to pursue a set of goals. Bureaucracies, such as business corporations or municipal governments, as well as formal associations, such as the PTA, are examples of formal organizations. A deeper analysis of bureaucracies and formal organizations appears in Chapter 6.

Judging on the basis of dress and behavior, the woman in the foreground is of a lower status in this society's social structure than the woman (in red) in the background.

Statuses

Within groups, people occupy different statuses. **Status** is an established position in a social structure that carries with it a degree of prestige (that is, social value). A status is a rank in society. For example, the position "vice president of the United States" is a status, one that carries very high prestige. "High school teacher" is another status; it carries less prestige than vice president of the United States, but more prestige than, say, "cab driver." Statuses occur within institutions. "High school teacher" is a status within the education institution. Other statuses in the same institution are student, principal, and school superintendent.

Typically, an individual occupies many statuses simultaneously. The combination of statuses composes the individual's **status set**—the complete set of statuses occupied by a person at a given time (Merton, 1968). An individual may occupy different statuses in different institutions; simultaneously, a person may be a daughter (in the family institution), bank president (in the economic institution), voter (in the political institution), church member (in the religious institution), and treasurer of the PTA (in the education institution). Each of these statuses may be associated with a different level of prestige.

Sometimes the multiple statuses of an individual are in conflict with one another. **Status inconsistency** exists where the statuses occupied by a person bring with them significantly different amounts of prestige. For example, someone trained as a lawyer, but working as a cab driver, experiences status inconsistency. Some recent immigrants from Vietnam and Korea have experienced status inconsistency. Many refugees who had been in high-status occupations in their home country, such as teachers, doctors, and lawyers, could find work in the United States only as grocers, sales clerks, and jobs of similar status. A relatively large body of research in sociology has demonstrated that status inconsistency can lead to stress and depression (Taylor et al., 1997; Blalock, 1991; Min, 1990; Taylor, 1973a; Taylor and Hornung, 1979; Hornung, 1977; Jackson and Curtis, 1968).

Achieved statuses are those attained by virtue of independent effort. Most occupational statuses—police officer, pharmacist, or boat builder—are achieved statuses. In contrast, **ascribed statuses** are those occupied from the moment a person is born. Your biological sex is an ascribed status; yet, even ascribed statuses are not exempt from the process of social construction. Race is an ascribed status. For most individuals, race is fixed at birth although an individual with one light-skinned African American parent and one White parent may appear to everyone to be White and may go through life as a White person. This is called "passing" although this term is used less often now than several years ago. Ascribed status is therefore not always perfectly unambiguous, as in the case of individuals with *biracial* or *multiracial* status.

Gender, too, although typically thought of as fixed at birth, is a social construct. You can be born female or male (an ascribed status), but becoming a woman or a man is the result of social behaviors associated with your ascribed status. In this sense, gender is a social construction. People who cross-dress, have a sex change, or develop some of the

achieved and ascribed status

characteristics associated with the other sex are good examples of how gender is achieved. As we will see later, however, you do not have to see these exceptional behaviors to note how gender is achieved. People "do" gender in everyday life—that is, they put on appearances and behaviors that are associated with their presumed gender (West and Zimmerman, 1987).

The line between achieved and ascribed status may be hard to draw. Social class, for example, is determined by occupation, education, and annual income—all of which are achieved statuses—yet, job, education, and income are known to correlate strongly with the social class of one's parents; hence, one's social class status is at least partly—though not perfectly—determined at birth. Social class status is an achieved status that includes an inseparable component of ascribed status as well.

Although people occupy many statuses at one time, it is usually the case that one of these statuses is dominant, overriding all other features of the person's identity. This is called the person's **master status.** The master status may be imposed by others, or a person may define his or her own master status. A woman judge, for example, may carry the master status "woman" in the eyes of many; thus, she is seen not just as a judge, but as a woman judge. Being in a wheelchair is another example of a master status; people may see this, at least at first, as the most salient part of one's identity, ignoring other statuses that define one as a person. As a result, they may come to see someone in a wheelchair as helpless and unable to care for himself or herself. Because master statuses override other identities, they are often the basis for prejudice and stereotypes.

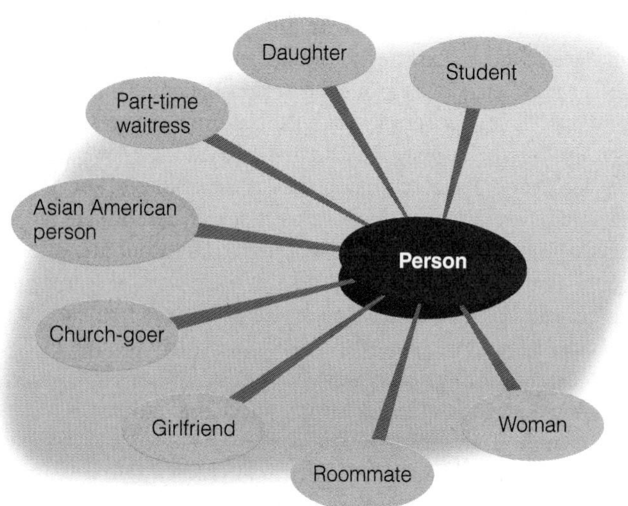

FIGURE 5.1 Examples of Roles in a College Student's Role Set

THINKING SOCIOLOGICALLY

Make a list of terms that describe who you are. Which of these are *ascribed statuses,* and which are *achieved statuses?* What do you think your *master status* is in the eyes of others? Does one's master status depend on who is defining you? What does this tell you about the significance of social judgments in determining who you are?

It is possible for a master status to completely supplant all other statuses in the person's status set. For example, when a person has AIDS, that person's health condition becomes the master status. AIDS becomes the defining criterion for the person's status in society. The person then becomes stigmatized by the master status. Not all master statuses are negative; positive-defining master statuses also exist, such as hero or saint.

Roles

A **role** is the expected behavior associated with a particular status; thus, a role is a collection of expectations that others have for a person occupying a particular status. Statuses are occupied; roles are acted or "played." The status of police officer carries with it numerous expectations; this is the "role" of police officer. Police officers are expected to uphold the law, pursue suspected criminals, assist victims of crimes, and so on. Usually, people behave in their role as others expect them to, but not always. When a police officer commits a crime, such as physically brutalizing someone just arrested, he or she has violated the role; role expectations have not been met.

Just as an individual may occupy several statuses at any one time, an individual will also typically occupy many roles. A person's *role set* includes all the roles occupied by the person at a given time. These roles may clash with each other, a situation called **role conflict,** wherein two or more roles are associated with contradictory expectations. Notice that in Figure 5.1, some of the roles diagrammed for this college student may conflict with each other. Can you speculate about which might and which might not?

In our society, one of the most common forms of role conflict arises from the dual responsibilities of job and family. The parent role demands extensive time and commitment; so does the job role. Time given to one of these roles is time taken away from the other. Although the norms pertaining to working women and men are rapidly changing, it is still true that women are more often expected to uphold traditional role expectations and are more likely to be held responsible for minding the family when job and family conflict. The sociologist Arlie Hochschild captured the predicament of today's women when she described the "second shift": A working mother spends time and energy all day on the job, only to come home to the "second shift" of family and home responsibilities—sometimes responsibilities delegated to the man of the house, who encounters less well-formed role expectations that he will take on those responsibilities, and who is therefore more likely to leave the jobs undone (Hochschild, 1998). Hochschild has further found that the demands of family work, coupled with the demands of jobs, have resulted in a serious time bind for both men and women. She has found that some companies have instituted "family-friendly" policies, designed to reduce the conflicts generated by the "second shift." Ironically, however, in her study, she found that few workers take advantage of such programs as more flexible hours, paid maternity leave, and job sharing—except for the on-site child care that actually allowed them to work more! Hochschild concludes that many women and men find their work more satisfying than their family life. At the same time, despite the progressive company policies, management still expanded its demands on workers. The consequence is a new blurring of role expectations where "work becomes home and home becomes work" (Hochschild, 1997) or, as one of Hochschild's reviewers says, "Now work is where the heart is" (Shapiro, 1997: 64).

Hochschild's studies point to the conflict between two social roles: family roles and work roles. Her new research is illustrative of a different sociological concept: **role strain,** a condition wherein a single role brings conflicting expectations. Different from role conflict, which involves tensions *between* two roles, role strain involves conflicts within a single role. In Hochschild's study, the work role has not only the expectations traditionally associated with work, but also the expectation that one "love" one's work and be as devoted to it as to one's family. The result is role strain—or the time bind that Hochschild analyzes.

The role of student often involves role strain. For example, students are expected to be independent thinkers, yet often they feel that they are required to simply repeat what a professor tells them. The tension between the two competing expectations is an example of role strain. This is contrasted with the role conflict some students encounter when they prepare to pursue their educations away from home. A common experience of first-generation college students, whether Asian, African American, White, Native American, or Latino, is that their families, though proud of their educational achievement and wishing for their success, have the expectation that the student will remain close to home, fulfilling the role of the dutiful child as it has been filled in the family in the past. If the student chooses a school far from home (often the best educational option), he or she is liable to experience resistance from the parents. The result is role conflict—a conflict between success and tradition, in a sense.

The origins of role conflict and role strain are social structural. As C. Wright Mills pointed out in his discussion of the sociological imagination, sociologists study the connection between individual lives and society. Role conflict and role strain are experienced at the individual level, but their origins are societal—they originate in the expectations embedded in specific social roles. Those experiencing role conflicts may feel they are encountering a personal problem. Understanding that personal problem requires a sociological perspective.

This excellent photograph shows three aspects of society and social interaction: First, men in a foraging society, hunting tools at the ready; second, socialization (the young boy is being trained in hunting techniques); and third, the role of gender in social organization (in this society, men are most likely to do the hunting).

Theories About Analyzing Social Interaction

Groups, statuses, and roles form a web of social interaction. The interaction people have with one another is a basic element of society. Sociologists have developed different ways of understanding social interaction. Functional theorists are primarily interested in social roles and how they are related to the needs of society. As we saw in Chapter 1, functionalist theorists find great consensus in society around basic values, and they see this consensus as part of what makes society cohesive. Other sociologists analyze social interaction using different theoretical frameworks. Here we detail five of them: the social construction of reality, ethnomethodology, impression management, social exchange, and the study of interaction in cyberspace. The first three of these come directly from the symbolic interaction perspective.

The Social Construction of Reality

What holds society together? This is a basic question for sociologists, one that, as we will see at the end of this chapter, has long guided sociological thinking. Sociologists note that society cannot hold together without something that is shared—a shared social reality. Some sociological theorists have argued convincingly that what is shared is for the most part socially constructed, namely, that there is little actual reality beyond that which is produced by the process of social interaction itself.

This is the principle of *the social construction of reality,* the idea that our perception of what is real is determined by the subjective meaning that we attribute to an experience, a principle that is central to symbolic interaction theory (Berger and Luckmann, 1967; Blumer, 1969; Jones and Davis, 1965; Taylor et al., 1997). In this way, the process of social interaction and the subjective meanings we give to things arising from it actually determine what we "see" as actual fact. Hence, there is no objective "reality" in itself. Things do not have their own intrinsic meaning; we subjectively impose meaning on things.

Children do this routinely. For example, upon seeing a marble roll off of a table, the child attributes causation (meaning) to the marble: The marble rolled off the table "because it wanted to." Such perceptions carry into adulthood: The man walking down the street who then accidentally walks smack into a telephone pole at first thought glares at the pole, as though the pole somehow caused the accident! He inadvertently attributes causation and meaning to an inanimate object—the telephone pole (Heider, 1958; Taylor et al., 1997).

There is considerable evidence that people do just that—they will force meaning on something when it serves them to see or perceive what they want to perceive—even if it seems to someone else to be contrary to actual fact. They then come to believe that what they perceived is indeed "fact." A classic and convincing study of this is Hastorf and Cantril's (1954) study of Princeton and Dartmouth students who watched a film of a game of basketball between the two schools. Both sets of students watched the same film. The students were instructed to watch carefully for rule infractions by each team. The results were that the Princeton students reported twice as many rule infractions involving the Dartmouth team than the Dartmouth students saw. The Dartmouth students saw about twice as many rule infractions by Princeton than the Princeton students saw! Remember that they all saw exactly the same movie. Each group of students reported exactly what they saw—actual violation of the rules. We see what we want to see, as a result of the social construction of reality.

Our perceptions of what reality is are determined by what sociologists call our *definition of the situation:* We observe the context in which we find ourselves and then adjust our attitudes and perceptions accordingly. The sociological theorist W. I. Thomas embodies this idea in his well-known statement, that *situations defined as real are real in their consequences* (Thomas, 1966/1931). The Princeton and Dartmouth students saw different "realities" depending on what college they were attending, and the consequences (the perceived rule infractions) were very real to them.

The definition of the situation is a principle that can affect even so "factual" an event as whether or not an emergency room patient is perceived to be dead by the doctors there. In his research in the emergency room of a hospital, Sudnow (1967) found that patients who arrived at the emergency room with no discernible heartbeat and not breathing were treated differently by the attending physician depending on the patient's age. A person in his or her early 20s or younger was not immediately pronounced "dead on arrival" (DOA). Instead, the physicians spent a lot of time listening for and testing for a heartbeat, stimulating the heart, examining the patient's eyes, giving oxygen, and administering other stimulations in attempts to revive the patient. If no lifelike responses were obtained, the patient was pronounced dead. Older patients, however, were on the average less likely to receive such extensive procedures. The older person was examined less thoroughly and often was pronounced dead on the spot with only a stethoscopic examination of the heart. In such instances, how the physicians defined the situation—how they socially constructed reality—was real indeed in consequence for the patient!

Understanding the social construction of reality helps one see many aspects of society in a new light. Race and gender are significant influences on social experience because people believe them to be so. Indeed, society is constructed based on certain assumptions about the significance of race and gender; these assumptions have guided the formation of social institutions, including what work people do, how families are organized, and how power is exercised. In society, some occupations are afforded more value than others; thus, people who hold these jobs may be perceived as more valued. The social construction of reality perspective shows how significant the formation of meaning is in guiding all aspects of human behavior.

Ethnomethodology

As already discussed in the chapter on culture, our interactions are guided by rules that we follow. These rules are the norms of social interaction. Society cannot hold together without norms, but what rules do we follow? How do we know what these rules or norms are? As we saw earlier, an approach in sociology called ethnomethodology is a clever technique for finding out. **Ethnomethodology** (Garfinkel, 1967), after "ethno" for "people" and "methodology" for mode of study, is a technique for studying human interaction by deliberately disrupting social norms and observing how individuals attempt to restore normalcy. The idea is that to study norms, one must first break those norms since the subsequent behavior of the people involved will reveal just what the norms were in the first place.

Ethnomethodology is based on the premise that human interaction takes place within a consensus and, further, that interaction is not possible without this consensus. According to Garfinkel, this consensus will be revealed by people's background expectancies, namely, the norms for behavior that they carry with them into situations of interaction (Zimmerman, 1992). It is presumed that these expectancies are to a great degree shared, and thus studying norms—by deliberately violating them—will reveal the norms that most people bring with them into interaction. The ethnomethodologist argues that you cannot simply walk up to someone and ask him or her what norms they have and use because most people will not be able to articulate what they are. We are not wholly conscious of what norms we use even though they are shared. Ethnomethodology is designed to "uncover" those norms.

Ethnomethodologists often use ingenious procedures for assessing norms by thinking up clever ways to interrupt "normal" interaction. For example, William Gamson, a sociology professor, had one of his students go into a grocery store where jelly beans, normally priced at that time at 49 cents per pound, were on sale for 35 cents. The student engaged the saleswoman in conversation about the various candies and then asked for a pound of jelly beans. The saleswoman then wrapped them and asked for 35 cents. The rest of the conversation went like this:

> Student: Oh, only 35 cents for all those nice jelly beans? There are so many of them. I think I will pay $1.00 for them.
>
> Saleswoman: Yes, there are a lot, and today they are on sale for only 35 cents.
>
> Student: I know they are on sale, but I want to pay $1.00 for them. I just love jelly beans, and they are worth a lot to me.
>
> Saleswoman: Well, uh, no, you see, they are selling for 35 cents today, and you wanted a pound, and they are 35 cents a pound.
>
> Student (voice rising): I am perfectly capable of seeing that they are on sale at 35 cents a pound. That has nothing to do with it. It is just that I personally feel that they are worth more, and I want to pay more for them.
>
> Saleswoman (becoming quite angry): What is the matter with you? Are you crazy or something? Everything in this store is priced more than what it is worth. Those jelly beans probably cost the store only a nickel. Now do you want them or should I put them back?

At this point, the student became quite embarrassed, paid the 35 cents, and hurriedly left (Gamson and Modigliani, 1974).

The point here is that the saleswoman approached the situation with a presumed consensus, a consensus that becomes revealed by its deliberate violation by the student. The puzzled saleswoman took measures to attempt to normalize the interaction, to even force it to be normal. By so doing, she revealed her expectatations, that is, her norms. The technique used by the student, an ethnomethodological technique, reveals what norms people unconsciously use in everyday interaction and conversation. That is the purpose of ethnomethodology.

Ethnomethodology, as we also saw in Chapter 3, notes that society may be more precarious in its stability than is generally believed. By disrupting taken-for-granted realities, ethnomethodologists reveal the social forces that are at work in shaping human behavior, but they also show how easily these forces can be challenged. By just stepping outside the expected norms of behavior, these forces become readily apparent. Most of the time, however, people do not challenge social norms; rather, they continue to engage in the expectations others have of them, either consciously or unconsciously. As a result, society is maintained.

Impression management is a technique we all use to manipulate others' perceptions of us. What impressions is this person "giving off"? As a consequence, what perceptions would you have about this person if the two of you were to meet? What impressions might this person be giving in terms of gender? Age? Social status? Creativity? Politics? Others?

Impression Management and Dramaturgy

Another way of analyzing social interaction is to study **impression management,** a term coined by symbolic interaction theorist Erving Goffman (1959). Impression management is a process by which people control how others will perceive them. For example, a student handing in a term paper late may wish to give the instructor the impression that it was not the student's fault, but was because of uncontrollable circumstances ("my computer hard drive crashed," "my dog ate the last hard copy," and so on). The impression that one wishes to "give off" (to use Goffman's phrase) is that "I am usually a very diligent person, but today—just today—I have been betrayed by circumstances." Impression management can be seen as a type of con game. A person willfully attempts to manipulate the other's impression of him or her. Goffman regarded everyday interaction as a series of attempts to con the other. In fact, trying in various ways to con the other is, according to Goffman, at the very center of much social interaction and social organization in society.

Perhaps this cynical view is not true of all social interaction, but we do present different "selves" to others in different settings. The settings are, in effect, different stages upon which we act as we relate to others. For this reason, Goffman's theory is sometimes called the *dramaturgy model* of social interaction, a way of analyzing interaction that assumes that the participants are actors on a stage in the drama of everyday social life. People present different faces (give off different impressions) on different stages (in different situations or different roles) with different others. To your mother, you present yourself as the dutiful, obedient daughter, which may not be how you present yourself to a friend. Perhaps you think acting like a diligent student makes you seem like a jerk, so you hide from your friends that you are really interested in a class or enjoy your homework. Analyzing impression management reveals that we try to con the other into perceiving us as we want to be perceived. The box "Doing Sociological Research" shows how impression management can be involved in many settings, including the everyday world of the hair salon.

A clever study by Albas and Albas (1988) demonstrates just how pervasive impression management is in social interaction. The Albases studied how students interacted with one another when the instructor returned graded papers during class. Some students got good grades ("aces"), others got poor grades ("bombers"), but both employed a variety of devices (cons) to maintain or give off a favorable impression. For example, the aces wanted to show off their grades, but they did not want to appear to be braggarts, so they casually or "accidentally" let others see their papers. In contrast, bombers hid or covered their papers to hide their poor grades, said they "didn't care" what they got, or simply lied about their grades.

Analysis of impression management shows how subject to the influence of others we all are. Although we may protest, claiming we are all "individuals," our individuality is shaped by the numerous social forces we encounter in society.

Social Exchange

Another way of analyzing social interaction is through the social exchange model. The *social exchange model* of social interaction holds that our interactions are determined by the

B • O • X 5.1 DOING SOCIOLOGICAL RESEARCH

"Doing Hair, Doing Class"

When you begin to study social interaction, you will see that you can study it in many places, including places you would not ordinarily think of as locations for sociological research. Debra Gimlin did this when she began to observe the interaction that takes place in hair salons—interaction that, at least in some salons, she noticed is often marked by differences in the social class status between clients and stylists.

Her research question was how do women attempt to cultivate the cultural ideals and beauty, and in particular, how is this achieved through the interaction between hair stylists and their clients. She did her research by spending more than 200 hours observing social interaction in a hair salon. She watched the interaction between clients and stylists, and conducted interviews with the owner, the staff, and twenty women customers. During the course of her fieldwork, she recorded her observations of the conversations and interaction in the salon, frequently asking questions of patrons and staff. In the salon she studied, the patrons were mostly middle and upper middle class, the stylists, working class; all the stylists were White, as were most of the clients.

"Beauty work," as Gimlin calls it, involves the stylist bridging the gap between those who seek beauty and those who define it; her (or his) role is to be the expert in beauty culture, bringing the latest fashion and technique to clients. Beauticians are also expected to engage in some "emotion work"—that is, they are supposed to nurture clients and be interested in their lives; they are often put in the position of having to sacrifice their professional expertise to meet clients' wishes.

According to Gimlin, since stylists are typically of lower class status than clients, this introduces an element into the relationship between stylists and clients that stylists negotiate carefully in their routine social interaction. Hairdressers emphasize their special knowledge of beauty and taste as a way of reducing the status differences between them and their clients. They also try to nullify the existing class hierarchy by conceiving an alternative hierarchy, not one based on education, income, or occupation (as the usual class hierarchy is), but on the ability to style hair competently. Thus, stylists describe clients as perhaps "having a ton of money," but unable to do their hair or know what looks best on them. Stylists also try to nullify status differences by appearing to create personal relationships with their clients even though they never see them outside the salon. Stylists become confidantes with clients, often telling them highly personal information about their lives. Gimlin concludes that beauty ideals are shaped in this society by an awareness of social location and cultural distinctions. As she says, "Beauty is . . . one tool women use as they make claims to particular social statuses" (1996: 525).

The next time you get your hair cut, you might observe the social interaction around you and ask how class, gender, and race shape interaction in the salon or barbershop that you use. Try to get someone in class to corroborate with you so that you can compare observations in different salon settings. Would you expect the same dynamic in a salon where men are the clients and the stylists; do Gimlin's findings hold in settings where the customers and stylists are not White or where they are all working class? In doing so, you will be studying how gender, race, and class shape social interaction in everyday life.

SOURCE: Gimlin, Debra. 1996. "Pamela's Place: Power and Negotiation in the Hair Salon." *Gender & Society* 10 (October): 505–526.

rewards or punishments that we receive from others (Homans, 1974; Blau, 1986; Thibaut and Kelly, 1959; Cook et al., 1988; Levine and Moreland, 1998). An interaction that elicits approval from another (a type of reward) is more likely to be repeated than an interaction that incites disapproval (a type of punishment). According to the exchange principle, one can predict whether a given interaction is likely to be repeated or continued by calculating the degree of reward or punishment inspired by the interaction.

Often it is necessary to assess the situation in terms of a difference between reward and punishment. If the reward for an interaction exceeds the punishment for it, then a potential for *social profit* exists and the interaction is likely to occur or continue. If the reward is less than the punishment, then the action will produce a social loss (negative profit) and will be less likely to occur or continue. Exchange theorists analyze human interaction in terms of concepts like reward, punishment, profit, and loss, in addition to the ratio of inputs to outputs—people want to get from an interaction at least as much as they put into it, resulting in an *equitable* interaction. Calculations of inputs versus outputs constitute a measure of social costs versus rewards. A reward is in effect an approval for conformity; a punishment is a sanction against deviance. Social exchange thus tends to encourage conformity and discourage deviance, and in this way acts as a force toward cohesion in everyday interaction in society.

Rewards can take many forms. They can include tangible gains like gifts, recognition, and money, or subtle everyday rewards like smiles, nods, and pats on the back. Similarly, punishments come in many varieties, from extremes like public humiliation, beating, banishment, or execution, to gestures as subtle as a raised eyebrow or a frown. For example, if you ask someone out for a date and the person says yes, you have gained a reward, and you are likely to repeat the interaction. You are likely to ask the person out again, or to ask someone else out. If you ask someone out, and he or she glares at you and says, "No way!" then you have elicited a punishment that will probably cause you to shy away from repeating this type of interaction with that person.

Exchange theory has been used to probe the perpetuation of racist and sexist attitudes. Research has shown that antiwoman attitudes and antiwoman stereotypes among men persist much longer if they are constantly rewarded by one's social group or are reinforced by other aspects of one's social milieu (Levine and Moreland, 1998; Cook et al., 1988). In contrast, when such attitudes and stereotypes are punished in social exchange, they are unlikely to persist. If a man exits high school with a few antiwoman attitudes and enters a college that has a strong culture of feminist sophistication, evidence of casual sexism in his interactions is likely to earn quick scorn among disapproving classmates.

Interaction in Cyberspace

When people interact and communicate with one another by means of personal computers—through what have come to be called "chat rooms," computer bulletin boards, virtual communities, e-mail, and other computer-to-computer interactions—they are engaging in *cyberspace interaction*. In language that has rapidly become popular only within the last few years, one creates a virtual reality as a result of such activity. Virtual reality is the computer control of human sense; what the individual perceives to be real is what is created by and on the computer. When two or more persons share a virtual reality experience via communication and interaction with each other, they are engaged in **cyberspace interaction** (Turkle, 1995; Cartwright, 1994).

Cyberspace interaction has characteristics that distinguish it from ordinary face-to-face interaction. First, certain kinds of nonverbal communication are eliminated. Thus, when interacting with someone via a chat room or other such computer channel, facial expressions such as smiles, laughter, and the raised eyebrow are eliminated. Also eliminated are hand motions, voice pitch, and other such nonverbal modes of communicating with others. Second, one is free to become a different self. One can literally create a new identity, in fact, several new identities: One can become a cowboy, a cowgirl, a dragon, and even a sex partner in what has come to be called cybersex. A person can even instantly "become" the other gender; men can interact as (virtual) women, and women as men. In creating this new virtual self, interaction becomes anonymous, and one's secrets are hidden from the other, including one's appearance, dress, and personal data such as social security number, city and state address, and other characteristics that would otherwise identify one's actual (as opposed to virtual) self.

In this respect, cyberspace interaction is the application of Goffman's principle of *impression management* and *dramaturgy*. The person can put forward a totally different and wholly created self, or identity. One can "give off," in Goffman's terms, any impression that one wishes to and, at the same time, know that one's true self is protected by anonymity. This gives the individual quite a large and free range of roles and identities from which to choose and that one can become. As predicted by symbolic interaction theory, of which Goffman's is one variety, the reality of the situation grows out of the interaction process itself. What one becomes during chat room interaction is a direct outgrowth of the chat room interaction; it is the interaction process that produces the reality. This is a central point of symbolic interaction theory: Interaction creates reality.

Whether cyberspace interaction is ultimately beneficial or harmful to the person has become the subject of much research (Turkle, 1995). One researcher has noted that people can develop extremely close and in-depth relationships with each other as a result of their interaction in cyberspace (Parks, 1996). Another has noted that individuals who have engaged in virtual interaction with each other sometimes establish some other form of contact, such as communication by telephone or by post office mail, and sometimes, they actually meet face to face. Such a subsequent face-to-face meeting often leads to an end to the relationship! In fact, a number of marriages have broken up as a result of cyberspace interaction or cybersex on the part of a spouse [Fernandes-y-Freitas, 1996]. This implies that should one be engaging in cyberspace interaction and communication with someone else, it may be better to keep it that way. Cyberspace interaction and the anonymity connected with it represent a new kind of relationship in society. Attempts to shift it back to one of the older forms of relationship, such as face-to-face interaction, may meet with failure (Fernandes-y-Freitas, 1996).

Cyberspace interaction has resulted in a new subculture in our society—in fact, a totally new social order, including new social structure as well as a culture, according to some (Turkle, 1995; Casalegno, 1996). There has been a transition from the use of large mainframe computers as machines to perform mathematical and statistical calculations (what Turkle calls the *culture of calculation*) to the use of more user-friendly computers (PCs) to perform calculations and provide experiences such as interaction in cyberspace (the *culture of simulation*).

This new *cyberculture* is a subculture, as sociologically defined (in Chapter 3). It has certain rules or norms, its own language, a set of beliefs, and practices or rituals—in short, all the elements of a culture. These rules are that one is encouraged to develop a new identity or set of new identities while engaging in cyberspace interaction; that certain negative forms of interaction are permitted and even encouraged, such as aggression, intolerance, and exclusion (Casalegno, 1996); and that one maintains a kind of "frontier mentality" in the same sense that early Americans did in their push West to establish the new frontier. A frontier is new and exciting, and cyberculture contains the belief that virtual interaction is indeed a new frontier in our society.

A Study in Diversity: Forms of Nonverbal Communication

As the interchange of everyday life between people, social interaction involves both verbal and nonverbal communication. *Verbal communication* consists of spoken and written language, and includes a conversation with the person next to you, an exchange of letters, or a telephone conversation between you and a friend overseas. When communication takes place, interaction usually does too although not always. A TV commercial is an example of verbal communication without interaction: The communication is one way, from the advertiser to you, with no communication from you to the advertiser—unless you write or call them, in which case an interaction occurs.

Nonverbal communication is conveyed by nonverbal means such as touch, tone of voice, and gestures. A punch in the nose is a nonverbal communication. So is a knowing glance. A surprisingly large portion of our everyday communication with others is nonverbal although we are generally only conscious of a small fraction of the nonverbal "conversations" in which we take part. Consider all the nonverbal signals that are exchanged in a casual chat: body position, head nods, eye contact, facial expressions, touching, and so on. These nonverbal performances are of special interest to sociologists exploring the state of the individual in society. The box "Understanding Diversity" gives an example of how subtle communication "on the street" can be.

Studies of nonverbal communication, like verbal communication, show it to be much influenced by social forces, including the relationships between diverse groups of people. The meanings of nonverbal communications depend heavily upon race, ethnicity, social class, and particularly, gender, as we shall see (Wood, 1994). In a society as diverse as the United States, understanding how diversity shapes communication is an essential part of understanding human behavior. Sociologists, psychologists, social psychologists, anthropologists, and linguists have classified nonverbal communication into the following four categories: touch, paralinguistic, kinesic, and use of personal space or proxemics (Gilbert et al., 1998; Taylor et al., 1997; Ekman, 1982; Argyle, 1975).

Debunking Society's Myths

Myth: What you say is more important than how you say it.

Sociological perspective: Nonverbal forms of communication, such as gazes, gestures, and tone of voice, can communicate meaning even when it is unspoken.

Touch

Touching, also called *tactile communication,* involves any conveyance of meaning through touch. It may involve negative communication (hitting, pushing) as well as positive (shaking hands, embracing, kissing). These actions are defined as positive or negative by the ethnic cultural context. An action that is positive in one culture can be negative in another. For example, shaking the right hand in greeting is a positive tactile act in the United States, but the same action in East India or certain Arab countries would be an insult. A kiss on the lips is a positive act in most cultures, yet if you were kissed on the lips by a stranger, you would probably consider it a negative act, perhaps even repulsive. The vocabulary of tactile communication changes with social and cultural context.

Patterns of tactile communication are strongly influenced by gender. Parents vary their touching behavior depending upon whether the child is a boy or a girl. Boys tend to be touched more roughly; girls, more tenderly and protectively. This pattern continues into adulthood, where women touch each other more often in everyday conversation than do men. Touching appears to have different meanings to women and men. Women are on the average more likely to touch and hug as an expression of emotional support, whereas men touch and hug more often to assert power, or to express sexual interest (Wood, 1994). Clearly, there are also instances where women also touch in order to express

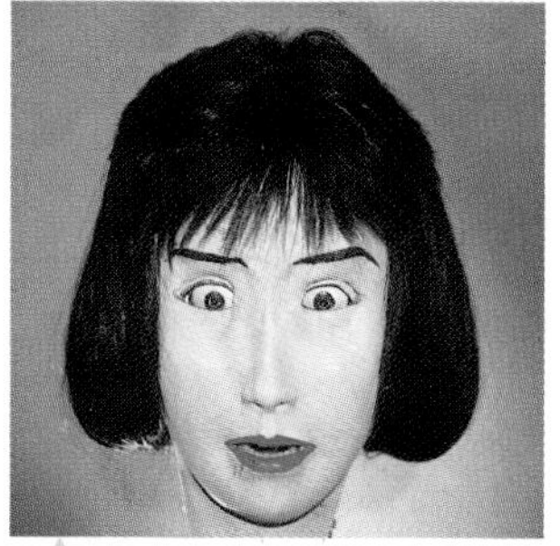
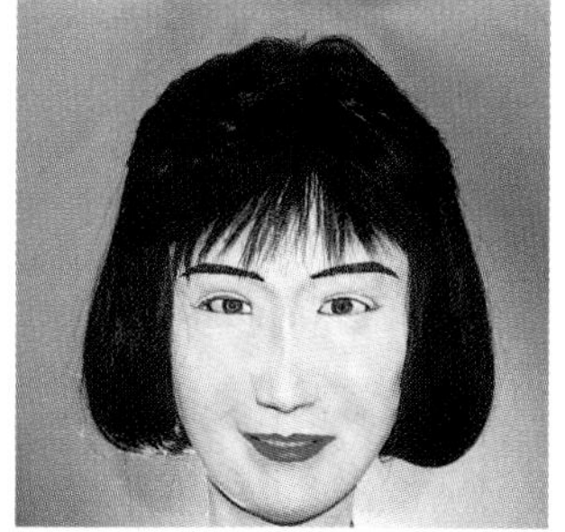
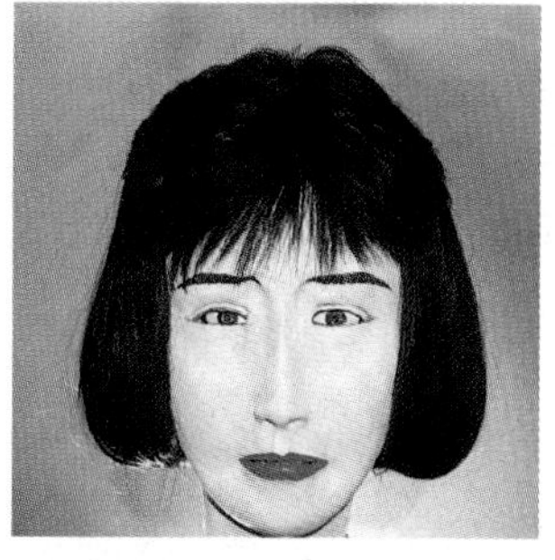
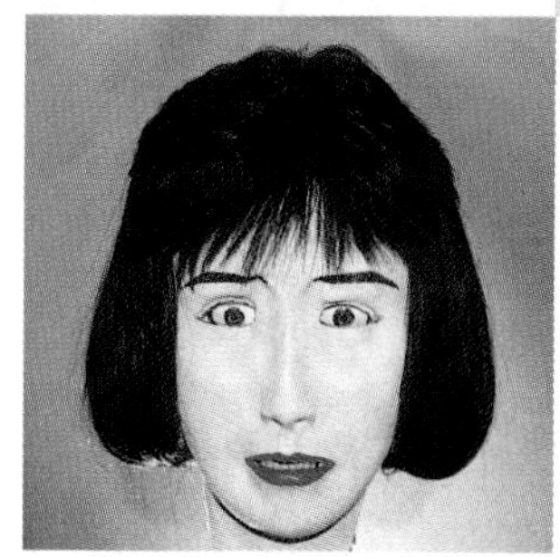

Facial expressions, a form of nonverbal communication, can be created mechanically by computer, as shown in these photographs. Can you guess what emotions are being conveyed by each of the four different faces?

sexual interest and also dominance, but research shows that in general, with some exceptions, for women touching is a supporting activity, and for men is a dominance-asserting activity (Wood, 1994; Tannen, 1990).

Professors, male or female, may pat a man or woman student on the back as a gesture of approval; students will rarely do this to a professor. This shows the effect of status differences. Male professors touch students more often than do female professors. This shows the additional effect of gender. Men often approach women from behind and let their hands rest upon the woman's shoulders; women are less likely to do the same to men. Male doctors touch female nurses more often than nurses touch doctors. Since doctors are disproportionately male and nurses disproportionately female in our society (and most others), this illustrates the combined influences of gender and status. Male bosses touch their female secretaries far more often than female secretaries touch their male bosses (Gilbert et al., 1998; Tannen, 1990). As more women are promoted to high-status positions, we may see changes in how often men and women of higher and lower rank touch each other in the workplace. In the context of a power relationship (that is, one where one person by

BOX 5.2

UNDERSTANDING DIVERSITY

"Passing Behavior"

SOCIOLOGIST Eli Anderson's book *Streetwise* (Anderson, 1990) notes how people use space to interact with strangers in the street. Even the deceptively simple decision to pass a stranger on the street involves a set of mental calculations. Is it day or night? Are there other people around? Is the stranger a child, a woman, a White man, a teenager, or a Black man? Each participant's actions must be matched to the actions and cues of the other. The following field note of Anderson's illustrates how well tuned strangers can be to each other and how capable of subtle gestural communication:

> It is about 11:00 on a cold December morning after a snow fall. Outside, the only sound is the scrape of an elderly white woman's snow shovel on the oil-soaked ice of her front walk. Her house is on a corner in the residential heart of the Village, at an intersection that stands deserted between morning and afternoon rush hours. A truck pulls up directly across from the old lady's house. Before long the silence is split by the buzz of two tree surgeons' gasoline-powered saws. She leans on her shovel, watches for a while, then turns and goes inside. A middle-aged white man in a beige overcoat approaches the site. His collar is turned up against the cold, his chin buried within, and he wears a Russian-style fur-trimmed hat. His hands are sunk in his coat pockets. In his hard-soled shoes he hurries along this east–west street approaching the intersection, slipping a bit, having to watch each step on the icy sidewalk. He crosses the north–south street and continues westward.
>
> A young black male, dressed in a way many Villagers call "street-ish" (white high-top sneakers with loose laces, tongues flopping out from under creased gaberdine slacks, which drag and soak up oily water; navy blue "air force" parka trimmed with matted fake fur, hood up, arms dangling at the sides) is walking up ahead on the same side of the street. He turns around briefly to check who is coming up behind him. The white man keeps his eye on the treacherous sidewalk, brow furrowed, displaying a look of concern and determination. The young black man moves with a certain aplomb, walking rather slowly.
>
> From the two men's different paces it is obvious to both that either the young black man must speed up, the older white man must slow down, or they must pass on the otherwise deserted sidewalk.
>
> The young black man slows up ever so slightly and shifts to the outside edge of the sidewalk. The white man takes the cue and drifts to the right while continuing his forward motion. Thus in five or six steps (and with no obvious lateral motion that might be construed as avoidance), he maximizes the lateral distance between himself and the man he must pass. What a minute ago appeared to be a single-file formation, with the white man ten steps behind, has suddenly become side-by-side, and yet neither participant ever appeared to step sideways at all.

SOURCE: Anderson, Eli. 1990. *Streetwise: Race, Class and Change in an Urban Community*. Chicago: University of Chicago Press, pp. 217–218.

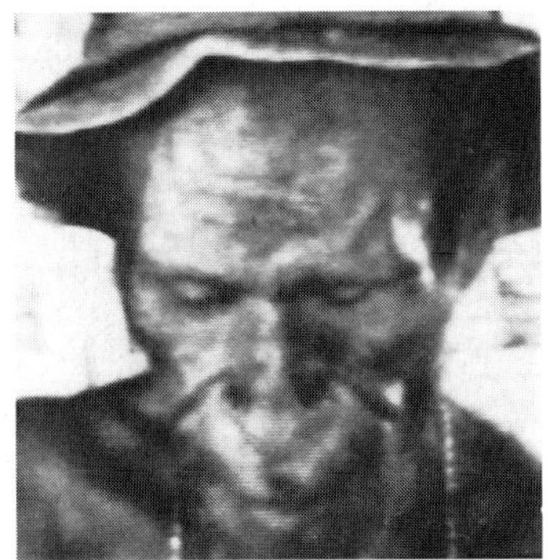

Researcher Paul Ekman (1982) has demonstrated that these facial expressions of New Guinea tribesmen are almost identical to those given by college students in the United States to the same four situations depicting anger, happiness, sadness, or disgust. This suggests that some emotions are conveyed by facial expressions in the same way across different cultures, rather than in different ways. Can you identify which expression goes with which emotion in these four photographs?

virtue of his or her status has power over another), these learned patterns can constitute sexual harassment, which is discussed further in Chapters 6 and 17.

Paralinguistic Communication

Paralinguistic communication is the component of communication that is conveyed by the pitch and loudness of the speaker's voice, its rhythm, emphasis, and frequency, and the frequency and length of hesitations. In other words, it is not what you say, but how you say it. A baby's cry communicates effectively to its parent: Its pitch, loudness, and frequency convey specific meanings to experienced parents, such as hunger, anger, or discomfort. The baby is communicating paralinguistically.

The exact meaning of paralanguage, like that of tactile communication, varies with the ethnic and cultural context. For some people under some circumstances, a pause may communicate emphasis; for others, it may indicate uncertainty. A high-pitched voice may mean a person is excited—or that the person is lying. During interactions between Japanese businessmen, long periods of silence often occur. Unlike U.S. citizens, who are experts in "small talk," and who try at all costs to avoid periods of silence in conversation, Japanese people do not need to talk all the time and regard periods of silence as desirable opportunities for collecting their thoughts (Fukuda, 1994). Unprepared U.S. businesspeople in their first meetings with Japanese executives often think, erroneously, that these silent interludes mean the Japanese are responding negatively to a presentation. More traveled Western executives learn to master the art of paralinguistic communication as practiced in Japan. Even though some find the Japanese mode of conversation highly uncomfortable, getting used to it is a key tool in negotiating successfully with the Japanese. The fate of a deal may depend on a glance, an exhalation, or a smile. Americans often consider paralinguistics to be a minor aspect of the conversation, with much greater attention paid to the verbal transaction; the Japanese, conversely, often consider the paralinguistics to be more important than the verbal communication as a source of information (Mizutami, 1990).

The formal grammar of some languages includes paralinguistic elements. In Chinese, variations in pitch have grammatical meaning. A sentence in Chinese can mean entirely different things if the speaker's voice is pitched high, medium, or low. The rapid pitch changes of spoken Chinese explain why the language has a "sing-song" quality to the Western ear. The English language does not include paralinguistic elements in its formal grammar, but conversational flourishes like a silent stare or an aggressive whisper may be as important as syntax in making the meaning of a communication understood.

People often reveal their true feelings and emotions by paralinguistic slips. Emotions tend to "leak out" even if a person tries to conceal them, hence the term *nonverbal leakage* (Ekman and Friesen, 1974; Ekman, 1982; Ekman, Friesen, and O'Sullivan, 1988; Taylor et al., 1997; Gilbert et al., 1998). People who are lying often betray themselves through paralinguistic expressions of anxiety, tension, and nervousness. Research has shown that when a person is lying, the pitch of his or her voice is higher than when the same person is telling the truth. The difference is small and hard to detect simply by listening, but electronic analysis of vocal pitch can reveal lying with considerable, though not perfect, accuracy (Ekman, Friesen, and Scherer, 1976; Krauss, Geller, and Olson, 1976; Ekman, Friesen, and O'Sullivan, 1988).

Kinesic Communication

Kinesic communication involves gestures, facial expressions, and body language. Waving hands, crossed arms, and extended legs all transmit meanings. Add hundreds of facial expressions, the infinite nuances of eye contact, and kinesic communication forms a crucial part of nonverbal communication. Facial expressions are gender related and reflect dominance patterns in society. For example, White middle-class women have been socialized to smile a lot, and to do so even if they are not happy (Halberstad and Saitta, 1987). Men are taught not to do this and to generally avoid facial expressions that convey emotion (Wood, 1994).

Meanings conveyed by kinesis are usually different in different cultures and ethnic subcultures. For example,

Mexicans and Mexican Americans may display the right hand held up, palm inward, all fingers extended, as an obscene gesture meaning "screw you many times over." This provocative gesture has no meaning at all in Anglo (White) society, except insofar as Mexican American traditions have been adopted by White people—a type of cultural diffusion.

Ignorance of the meanings that gestures have in a society can get you in trouble. Eye contact, a category of kinesic communication, is especially prone to misinterpretation. People who grow up in urban environments learn to avoid eye contact on the streets. Staring at someone for only two or three seconds can be interpreted as a hostile act, if done man to man (Anderson, 1990). If a woman maintains mutual eye contact with a strange man for more than merely two or three seconds, she may be assumed by the man to be sexually interested in him. On the other hand, in sustained conversation with nonstrangers, women maintain mutual eye contact for longer periods than do men (Gilbert et al., 1998; Wood, 1994).

White teachers in ghetto schools often find that when talking to Black, Hispanic, or Native American children standing directly in front of them, the child stares at the ceiling, the wall, or the floor. The White teacher often interprets this as lack of attention or interest; the children in these groups, however, may have been socialized to feel that sustaining eye contact during conversations with elders is a sign of disrespect. Similarly, averting the eyes during conversation is a sign of respect among the Japanese and many Japanese Americans, even among adults (Fukuda, 1994).

Conversely, some gestures retain the same meaning across different cultures, ethnic groups, and societies. The hand gestures for "stop," "come here," "go away," and "good-bye" are the same in the United States across racial, socioeconomic, and other group subcultures, and they are the same in other societies as well. The obscene gesture of an extended middle finger, palm inward ("flipping the finger" or "flipping the bird") is now virtually universal across all ethnic groups, races, social classes, and both genders both inside and outside the United States. The U.S. gestures for "OK," "shame on you," and "crazy" are also widespread both inside and outside U.S. society (Gilbert et al., 1998). Certain facial expressions are culturally universal, too: The facial expressions for anger, happiness, sadness, and even disgust appear to be recognized in all cultures, from the tribal peoples of New Guinea to the White middle class of the United States. These expressions appear to be used equally by both men and women in these cultures, which means that certain modes of kinesic communication—these kinds of facial expressions—are the same across genders for these different cultures (Gilbert et al., 1998; Ekman, 1982).

This homeless man carries his belongings with him, conspicuously displayed in order to establish his own particular proxemic bubble—that area around him representing his own private space, which goes where he goes.

Use of Personal Space

Meaning is conveyed by the amount of space between interacting individuals; this is referred to as **proxemic communication.** As with much nonverbal communication, the persons involved are generally not conscious of the proxemic messages they are sending. Generally, the more friendly a person feels toward another, the closer he or she will stand. In casual conversation, friends stand closer to each other than strangers. People who are sexually attracted to each other stand especially close, whether the sexual attraction is gay, lesbian, or heterosexual (Taylor et al., 1997).

According to anthropologist E. T. Hall (1966; Hall and Hall, 1987), we all carry around us a *proxemic bubble* that represents our personal three-dimensional space. When people we don't know enter our proxemic bubble, we feel threatened and may take evasive action. Friends stand close; enemies tend to avoid interaction and keep far apart. According to Hall's theory, we attempt to exclude from our private space those we do not know or do not like even though we may not be fully aware that we are doing so.

You may have noticed that in elevators, people tend to take up positions that maximize the average distances between them: Two people tend to stand in diagonally opposite corners; three people tend to stand in the shape of a triangle; four people tend to gravitate toward the four corners. A fifth person getting on the elevator will stand in the exact middle, with arms held rigidly at his or her sides. In

proxemic communication

an enclosed space like an elevator, everyone is obliged to stand inside everyone else's proxemic bubble. People in this situation maximize the interpersonal distance in an attempt to stay outside everyone else's bubble—and to keep everyone outside their own bubble.

Proxemic interaction varies strongly by cultural differences and also by gender. This can be clearly seen when people in different racial and ethnic groups interact. The proxemic bubbles of different groups have different sizes. Hispanic people tend to stand much closer to each other than White middle-class Americans; their proxemic bubble is on the average smaller. Similarly, African Americans also tend to stand close to each other while conversing. Interaction distance is quite large between White British males—their average interaction distances can be as much as several feet.

THINKING SOCIOLOGICALLY

Try the following little experiment yourself. Enter an empty elevator and go to a far (back) corner. Wait for another person to enter the elevator. If *proxemic communication theory* is correct, and if you do not already know the person, this person will tend to go to a diagonally opposite front corner of the elevator. (This would be an illustration of the *proxemic bubble* principle.) Next, instead of staying put in your corner, take a few small steps toward the other back corner. (You will have to get your nerve up a bit to do this!) What does the other person now do? If the theory of proxemics is correct, he or she will take some sort of evasive action, such as move toward the opposite front corner or toward the door. If the person does not move at all, then you should move even closer, and observe what the other person does. Repeat the experiment several times, with different people, with people whose gender is different from yours and whose race is different from yours, and see what happens.

Proxemic interactions also differ between men and women (Taylor et al., 1997; Tannen, 1990). Women of the same race and culture tend to stand closer to each other in casual conversation than do men of the same race and culture. The space between Black women is very close. Hispanic women also stand close. White men stand quite far apart. When a Middle Eastern man engages in conversation with a White middle-class U.S. man, the Middle Eastern man tends to move toward the White American, who tends to back up. You can observe the negotiations of proxemic space at cocktail parties or any other setting that involves casual interaction (Basow, 1992; Tannen, 1990; Sussman and Rosenfeld, 1982; Halberstad and Saitta, 1987). Look at how people array themselves in relation to each other and visualize the bubbles that surround them. As June Jordan told us in Chapter 1, with a sociological perspective, you can never be bored.

Interpersonal Attraction and the Formation of Pairs

Perhaps the most interesting interactions between humans are pairings, including friendships, romances, and sexual pairings. Pairings constitute one of the most fundamental processes in society. How do pairs form? Are pairings influenced by social structure, or are they as random as chance encounters? You will not be surprised to learn that formation of pairs has a strong social structural component—that is, it is patterned by social forces. The study of pairings is clearly central to our continuing interest in the question: How do people form the social bonds that hold society together?

Humans have a powerful desire to be with other human beings; in other words, they have a strong need for affiliation. We tend to spend about 75 percent of our time with other people when doing all sorts of activities—eating, watching TV, studying, doing hobbies, working, and so on. Overall, women reveal this affiliative tendency somewhat more than men (Basow, 1992). Persons who lack all human contact are very rare in the general population, and their isolation is usually rooted in psychotic or schizophrenic disorders. Extreme social isolation at an early age causes severe disruption of mental and emotional development.

The affiliation tendency has been likened to *imprinting*, a phenomenon seen in newborn or newly hatched animals, who attach themselves to the first living creature they encounter, even if it is of another species (Lorenz, 1966). Studies of ducks and squirrels show that once the young animal attaches itself to a human experimenter, the process is irreversible. The young animal prefers the company of the human to the company of its own species. A degree of imprinting may be discernible in human infant attachment, but researchers note that in humans the process is more complex, more changeable, and more influenced by social factors (Brown, 1986).

Somewhat similar to affiliation is interpersonal attraction, a nonspecific positive response toward another person. Attraction figures in ordinary day-to-day interaction and varies from mild attraction (such as thinking your grocer is a "nice person") all the way to deep feelings of love.

Debunking Society's Myths

***Myth:* Love is purely an emotional experience that you cannot predict or control.**

***Sociological perspective:* Whom you fall in love with can be predicted beyond chance by such factors as proximity, how often you see the person, how attractive you perceive him or her to be, and whether you are similar (not different) to him or her in social class, race-ethnicity, religion, age, and educational aspirations.**

According to one view, attractions fall on a single continuum ranging from hate to strong dislike to mild dislike to mild liking to strong liking to love. Another view is that attraction and love are two different continua, able to exist separately. By this view, you can actually like someone a whole lot, but not be in love. Conversely, it is held that you can feel passionate love for someone (with its associated strong sexual feelings and intense emotion), yet not really

Konrad Lorenz, the animal behaviorist, shows that adult ducks that have imprinted on him the moment they were hatched will follow him anywhere, as though he were their mother duck.

"like" the person. Have you ever been in love with someone you did not particularly like?

Can attraction be scientifically predicted? Can persons be identified with whom you are most likely to fall in love? The surprising answer to these questions is a loud, although somewhat qualified, yes. Most of us have been raised to believe that love is impossible to measure and certainly impossible to predict scientifically. We think of love, especially romantic love, as ephemeral, mysterious—a lightning bolt. Countless novels and stories support this view, but extensive research in sociology and social psychology suggests otherwise: In a probabilistic sense, love can be predicted beyond the level of pure chance. Let us take a look at some of these intriguing findings.

Proximity

A strong determinant of your attraction toward others is whether you live near them, work next to them, or have frequent contact with them. You are more likely to form friendships with people from your own city than with people from a thousand miles away. You are more likely to be attracted to someone on your floor, your residence hall, or your apartment building than to someone even two floors down or two streets over. Such is the effect of proximity in the formation of human friendships.

Segal (1974) demonstrated this effect in a study of recruits at a police academy. Seating for classes and seminars at the academy was alphabetical, with recruits sitting alongside the person next in the alphabet. Segal found that when the police officers were listed alphabetically at the end of their training, the list also effectively captured friendship pairings. The major determinant for choosing friends was relative position in the alphabet, which dictated who sat next to whom in class. The recruits chose as friends those sitting near them. Segal found that proximity had a stronger effect than all other factors, including race, socioeconomic background, age, religion, and nationality.

Mere Exposure Effect

Our attraction to another is greatly affected by how frequently we see him or her or even that person's picture. Have you ever noticed when watching a movie that the central character seems more attractive at the end of the movie than at the beginning? This is particularly true if you already find the person very attractive when the movie begins. Have you ever noticed that the fabulous-looking person sitting next to you in class looks better every day?

You may be experiencing *mere exposure effect*—the more you see someone, the more you like him or her. This works even if you only see the person in photographs. For example, in studies where people are repeatedly shown photographs of the same faces, the more often a person sees a particular face, the more he or she expresses liking for that person (Zajonc, 1968). There are two qualifications to the effect. First, "overexposure" can result when a photograph is seen too often. The viewer becomes "saturated" and ceases to like the pictured person more with each exposure. (Some celebrities—certainly not all—are careful not to allow themselves to become "overexposed" on talk shows, lest the public tire of them.) Second, the initial response of the viewer can determine how much liking will increase. If one starts out liking someone, seeing that person more will increase the liking for that person; however, if one starts out disliking the pictured person, the amount of dislike tends to remain about the same, regardless of how often one sees the person (Taylor et al., 1997).

Perceived Physical Attractiveness

We hear that beauty is only skin deep. Apparently, that is deep enough. To a surprisingly large degree, the attractions we feel toward people of either gender are based on our perception of their physical attractiveness. A vast amount of research over the years has consistently shown the importance of attractiveness in human interactions: Adults react more leniently to the bad behavior of an attractive child than to the same behavior of an unattractive child (Dion, 1972; Taylor et al., 1997). Teachers evaluate cute children of either gender as "smarter" than unattractive children with identical academic records (Taylor et al., 1997; Clifford and Walster, 1973). In studies of mock jury trials, attractive defendants, male or female, receive lighter sentences on average than unattractive defendants convicted of the same crime (Gilbert et al., 1998; Sigall and Ostrove, 1975).

Of course, standards of attractiveness vary between cultures and between subcultures within the same society. As we saw in Chapter 3, that which is highly attractive in one culture may be repulsive in another. In the United States, there is a maxim that you can never be too thin. This cultural belief has been cited as a major cause of eating disorders such as anorexia nervosa and bulimia, especially among White women (Taylor et al., 1997; Wolf, 1991), although less so among African American, Hispanic, and Native American women (Thompson, 1994). The maxim itself is a source of oppression for women in U.S. society; yet, it is clearly highly culturally relative. In other cultures, plumpness is sexy, as was the case in England and France in the Middle Ages. Among certain African Americans, chubbiness in women is considered attractive. Such women are called "healthy" and "phatt" (not "fat"), which means the same as "stacked," or well built. Similar cultural norms often apply in U.S. Hispanic populations. The skinny woman is considered ugly, not sexy.

Although standards of attractiveness vary from culture to culture, there is considerable agreement within a culture about who is attractive. For example, people ranking photographs of men and women of their own race tend to rank the attractiveness of the pictured individuals very similarly (Hatfield and Sprecher, 1986). This is equally true when men rank photos of women and when women rank photos of men.

Studies of dating patterns among college students show that the more attractive one is, the more likely one will be asked on a date, but there are two very important exceptions to this finding:

1. Those regarded as "exceptionally" attractive get asked out for dates slightly less often than those rated as somewhat less attractive. This is the so-called *movie star effect*. Although you may want to ask a beautiful acquaintance out for a date, he or she is so attractive that you assume either the person already has plenty of dates or the person wouldn't be interested in you. So you do not ask, avoiding the risk of rejection.
2. Here is the good news for those of us who are not movie stars: Physical attractiveness predicts only the early stages of a relationship. When one measures relationships that last a while, other factors come into play, principally religion, political attitudes, social class background, and race. Perceived physical attractiveness may predict who is attracted to whom initially, but other variables are better predictors of how long a relationship will last (Berscheid and Reis, 1998; Hill et al., 1976).

Broadly, we tend to be attracted to individuals whom we (and others of our group) regard as physically attractive. Of course, when thinking about our friends and those we love, we are aware that there are things far more important than physical attractiveness; nevertheless, platitudes about human relationships cannot dispute research findings showing that perceived physical attractiveness, measured reasonably well within a culture or subculture, is an important dimension of human interaction.

Similarity

"Opposites attract," you say? Not according to the research. We have all heard that people are attracted to their "opposite" in personality, social status, background, and other characteristics. Many of us grow up believing this to be true; yet, if there is one thing that the research tells us about interpersonal attraction, it is that, with few exceptions, we are attracted to those who are similar or even identical to us in socioeconomic status, race, ethnicity, religion, perceived personality traits, and general attitudes and opinions. "Dominant" people tend to be attracted to other dominant people, not to "submissive" people. Couples tend to have similar opinions about political issues of great importance to them, such as attitudes about abortion, crime, and urban violence. Overall, couples tend to exhibit strong cultural or subcultural similarity.

There are exceptions, of course. We often fall in love with the exotic—the culturally or socially different. Novels and movies return endlessly to the story of the young, White, preppy woman who falls in love with a Hell's Angel biker, but such is by far the exception and not the rule. When it comes to long-term relationships, including both friends and lovers, heterosexual, gay, and lesbian humans vastly prefer a great degree of similarity even though, if asked, they might not admit it.

In fact, the less similar a heterosexual relationship is with respect to race, social class, age, and educational aspirations (how far in school the person wants to go), the quicker the relationship is likely to break up (Berscheid and Reis, 1998; Stover and Hope, 1993; Hill et al., 1976). An especially interesting qualification to all the research on similarity and attraction presents itself, however, in the matter of interracial dating. Although people tend to date within their own race, nationality, or ethnicity, a large number of interracial couples today enjoy long-lasting relationships. Similarity research sheds light on these relationships as well. The research tends to show that for interracial couples, similarity in characteristics other than race (social class background, religion, age, and educational aspirations) tends to predict how long the relationship will last. In general, the more similar the couple is in characteristics other than race, the longer, on average, the interracial relationship will last. The less similar the couple, the shorter the relationship (Berscheid and Reis, 1998; Stover and Hope, 1993; Eagly et al., 1989).

Most romantic relationships, regrettably, come to an end. On campus, relationships tend to break up most often during gaps in the school calendar, such as Christmas recess and spring vacation. Summers are especially brutal on relationships formed during the academic year. Breakups are seldom mutual. Almost always, only one member of the pair wants to break off the relationship, whereas the other wants to keep it going. This sad truth means that the next time someone tells you that their breakup last week was "mutual," you know they are probably lying or deceiving themselves (Hill et al., 1976).

Social Institutions and Social Structure

So far, we have been studying the more microlevels of society. At the macrolevel of society, sociologists are interested in the role of social institutions in constituting society. Just as sociologists see social forces shaping behavior at the microlevel, so do they see social forces molding society at the macrolevel.

Social Institutions

Societies are identifiable by their cultural characteristics and the social institutions of which they are composed. A **social institution** is an established and organized system of social behavior with a recognized purpose. The term refers to the broad systems that organize specific functions in society. Social institutions are a weave of behaviors, norms, and values, as the box "Analyzing Social Issues" demonstrates, using baseball as an example of a social institution. As a whole, institutions are organized to meet various needs in society. For example, the family is an institution that provides for the care of the young and the transmission of culture. Religion is an institution that organizes the sacred beliefs of a society. Education is the institution through which people learn the skills needed to live in the society. Institutions channel human action in socially approved ways; consequently, people do not create new educational systems every time they want to learn something. Institutions persist through time.

The concept of social institutions is an abstract one. Unlike group behavior, institutions cannot be directly observed, but their impact and structure can be seen nonetheless. An institution is a patterning of social relationships that exist as distinct from individuals or specific groups. Social institutions have an existence all their own. Although abstract, they impinge on the behavior of people and groups, and they persist long after these people and groups are gone. Take the example of education. You and your classmates form a social group—one whose behavior can be directly observed. As a group, you are part of a broader educational institution—both a specific school and, at the societal level, the institution of education. When sociologists speak of institutions, they mean more than the specific school that you attend although a school will have an institutional structure and patterns of relationships that can be sociologically analyzed. At the broadest level of society, however, education (and higher education, in particular) is a social institution—one that includes all components of education in the society, including not just the shape of given institutions, but also the norms, values, and beliefs that guide education at the societal level.

The major institutions in society include the family, education, work and the economy; the political institution (or state), religion, and health care, as well as institutions such as the mass media, organized sports, and the military. These all represent institutions—they are all complex structures that exist for a rather explicit purpose. Together, institutions perform certain purposes that are necessary for society to proceed. Functionalist theorists have identified

BOX 5.3 • ANALYZING SOCIAL ISSUES

Baseball as a Social Institution

MANY PEOPLE speak of baseball as the "quintessential American institution." They are referring to its distinct place in American culture and its uniqueness as a feature of American life. There is another way of thinking about baseball as an institution as illustrated here by sociologist Robert Bellah and his colleagues:

> The very idea of institutions is often repugnant to Americans. But whatever their conscious attitude, Americans are deeply fascinated by the moral drama of institutions, at least when they understand them, or think they do, as in the case of sports.
>
> Consider baseball, the national pastime. Tens of millions of fans depend for the excitement of a season not only upon the practice of the sport but on the institution of the leagues, with their complex athletic, economic, and legal rules. The drama of the annual pennant races is what it is only because the skills of players and teams are supported and guided by the less visible structure of coaches, umpires, accountants, and contracts. Equally crucial is the moral infrastructure of collective honor, loyalty, and devotion to the sport. For many fans the drama of baseball is heightened at those moments when the larger institutional patterns come into view—especially in moments of crisis, as when a team is separated from a city long identified with it, or when scandal shocks the public's sense of the honor and propriety that ought to govern the sport.
>
> It is expected that star players are extraordinarily well paid; but when they allow their own image or their own self-indulgence to become more important than their contribution to the team, the enjoyment of the game changes to moral outrage, as it does when owners appear to be acting for private gain at the expense of the honest life of the sport. Why this indignation? At such moments the public acknowledges that moral norms are woven throughout baseball; shared indignation expresses the fans' tremendous moral identification with the sport as an institution. Clearly baseball in such moments is not being understood as a neutral device for individual satisfaction—if it were, who would care about scandals? Rather, baseball, with its purposes, codes, and standards, is a collective moral enterprise, an institution in the full sense, and many Americans care deeply about it. As an institution, baseball is more than the actual players and organizations who play the game during any given season. That is why we can see the sport as sometimes succeeding, sometimes failing, in becoming what baseball really ought to be.

SOURCE: Bellah, Robert N., Richard Madsen, William M. Sullivan, Ann Swidler, and Steven M. Tipton. 1991. *The Good Society*. New York: Alfred A. Knopf.

these purposes as follows (Aberle et al., 1950; Parsons, 1951a; Levy, 1949):

1. *The socialization of new members of the society.* As discussed in Chapter 3, the young in a society need to be taught its ways, its values, its norms—in short, its culture. It is the "function" of the family, as well as the education institution, to do so. Other institutions as well contribute to the socialization of the young, such as religious organizations, education, and increasingly, the mass media.

2. *The production and distribution of good and services.* The economy is generally given as the institution that performs this set of tasks. All societies, according to functional theory, must produce, distribute, and consume goods and services—food, clothing, shelter, recreation and leisure, and other goods and services. Political institutions also have a role in performing these tasks—by regulating commerce and economic policy.

3. *Replacement of the membership.* All societies must have a means of replacing its members who die, move or migrate away, or otherwise leave the society. The family and other heterosexual pairings provide a society with the means for replacement of the population. The political institution or government also performs this task, as by its policies on immigration and birth control, for example.

4. *The maintenance of stability and existence.* As already noted in the first chapter, one of the major assertions of functional theory is that certain institutions within a society assure the stability and continuance of the society. The government does this by both passing laws and then establishing law enforcement agencies, such as the police. Most societies have some sort of police force. Similarly, many societies, certainly our own, has a military force that exists to protect, stabilize, and continue the society. Other institutions also contribute to the stability of society through shared values and instituting social control.

5. *Providing the members with an ultimate sense of purpose.* Societies accomplish this task by having national anthems, patriotism, and the like, in addition to providing—through institutions such as the family, religion, and education—basic values and moral codes. When people in society hold a common sense of purpose, the society is highly cohesive, but even societies that are internally quite diverse are held together by certain commonly shared values and assumptions.

Functionalists see these societal needs as universal, although societies do not perform them in the same way or by means of the same institutions. This is what makes societies distinct.

In contrast to functional theory, conflict theory further notes that because conflict is inherent in social existence, the institutions of society do not provide for all its members equally. Some members are provided for better than others, thus demonstrating that institutions affect people with differential power, granting more power to some social groups than to others. Racial and ethnic minorities in a society possess considerably less power and less of society's benefits than does the dominant group. In most societies known to anthropologists and sociologists, women occupy lower social status on the average than do men. There are few exceptions to this. Thus, on average, women have less political power, wealth, and prestige than do men in society. Similarly, power is not equally distributed across social class strata; generally, the lower one's social class, the less one's political power, wealth, influence, and prestige. How institutions affect the person, as well as the benefits that accrue to the individual from the institution, depends upon the person's race-ethnicity, gender, and social class status, among other factors. In the United States, one can say that major institutions are shaped by race-ethnicity, class, and gender since these factors have such a strong role in the institutional fabric of this society.

Social Structure

At both the micro- and macrolevel, we have seen how social behavior is patterned. Sociologists use the term **social structure** to refer to the organized pattern of social relationships and social institutions that together compose society. The social structure of society is observable in the established patterns of social interaction and social institutions. Social structural analysis is a way of looking at society in which the sociologist analyzes the patterns in social life that reflect and produce social behavior. Social structures are not immediately visible to the untrained observer; nevertheless, they are present, and they affect all dimensions of human experience in society.

Social class distinctions are an example of a social structure. Class shapes the access that different groups have to the resources of society, and it shapes many of the interactions people have with each other. People may form cliques with those who share similar class standing, or they may identity with certain values that are associated with a given class, for example, middle-class values. Class then forms a social structure—one that shapes and guides human behavior at all levels, no matter how overtly visible or invisible this structure is to someone at a given time.

Like social institutions, the concept of social structure is an abstract one, yet, it is critical to sociological analysis. Social structures form invisible patterns—invisible at least until your sociological imagination begins to see the patterned behaviors that mark human interaction in society.

The philosopher Marilyn Frye aptly describes the concept of social structure in her writing. Using the metaphor of a birdcage, she writes that if you look closely at only one wire in a cage, you cannot see the other wires. You might then wonder why the bird within does not fly away. Only when you step back and see the whole cage instead of a single wire do you understand why the bird does not escape. Frye writes:

> It is perfectly obvious that the bird is surrounded by a network of systematically related barriers, no one of which would be the least hindrance to its flight, but all of which, by their relations to each other, are as confining as the solid walls of a dungeon. It is now possible to grasp one reason why oppression can be hard to see and recognize:

> one can study the elements of an oppressive structure with great care and some good will without seeing or being able to understand that one is looking at a cage and that there are people there who are caged, whose motion and mobility are restricted, whose lives are shaped and reduced (Frye, 1983: 4–5).

Frye's analysis focuses on oppressive social structures that confine and exploit people. Oppression and social structure are not the same thing although many people find existing social structures oppressive. Just as a birdcage is a network of wires, society is a network of social structures, both micro and macro.

THINKING SOCIOLOGICALLY

Using Marilyn Frye's analogy of the birdcage, think of a time when you believed your choices were constrained by *social structure*. When you applied to college, for example, could you go anywhere you wanted? What social structural conditions guided your ultimate selection of schools to attend?

What Holds Society Together?

What holds societies together? That is a central question in sociology, one that was first addressed by Emile Durkheim. He argued that people in society had a **collective consciousness,** defined as the body of beliefs that are common to a community or society and that give people a sense of belonging and a feeling of moral obligation to its demands and values. According to Durkheim, it is collective consciousness that gives groups social solidarity; it gives members of a group the feeling that they are part of one society.

Where does the collective consciousness come from? Durkheim argued that it stems from people's participation in common activities, such as work, family, education, and religion—in short, society's institutions. Goals, values, and beliefs emanate from the institutions of a society, and individuals aligned with common institutions develop a sense of common purpose, providing solidarity in society.

Mechanical and Organic Solidarity

According to Durkheim, there are two different kinds of social solidarity, or holding together: mechanical and organic, each the basis of a different form of societal solidarity. **Mechanical solidarity** arises when individuals play similar roles within the society; people feel bonded to the group in less complex societies because everyone in the group is so similar. Individuals in societies marked by mechanical solidarity share the same values and hold the same things sacred. This potent source of cohesiveness is weakened when a society becomes differentiated into more complex systems of work behavior. Contemporary examples of mechanical solidarity are rare because most of the societies of the world have been absorbed in the global trend to greater complexity and interrelatedness. Native American groups prior to European conquest were bound together by mechanical solidarity; indeed, many Native American groups are now trying to regain the mechanical solidarity on which their cultural heritage rests, but they are finding that the superimposition of White institutions on Native American life interferes with the adoption of traditional ways of thinking and being, and hence prevents mechanical solidarity from gaining its original strength.

In contrast to societies where individuals play the same roles, we find societies marked by **organic solidarity** (also called contractual solidarity), in which individuals play a great variety of different roles, and unity is based on role differentiation, not similarity. The United States and other industrial societies are examples. A society built on organic solidarity is cohesive because of its differentiation. Roles are no longer necessarily similar, but they are necessarily interlinked—the performance of multiple roles is necessary for the execution of society's complex and integrated functions.

Debunking Society's Myths

Myth: Society is held together because people share common values.

Sociological perspective: Some societies are held together by commonly shared values, such as in mechanical solidarity; other societies, marked by organic solidarity, are held together by the interrelated but different roles that are part of the division of labor.

Durkheim described this state as the **division of labor,** defined as the systematic interrelatedness of different tasks that develops in complex societies. The labor force within the contemporary U.S. economy, for example, is divided according to the kinds of work people do. Within any division of labor, tasks become distinct from one another, but they are still woven together into a whole.

The division of labor is a central concept in sociology because it represents how the different pieces of society fall together. The division of labor in most contemporary societies is often marked by gender, race, and class divisions. In other words, if you look at who does what in society, you will see that women and men tend to do different things in society; this is the gender division of labor. This is cross-cut by the racial division of labor, the pattern whereby those in different racial–ethnic groups tend to do different work in society. At the same time, the division of labor is also marked by class distinctions, with some groups providing work that is highly valued and rewarded, and others doing work that is devalued and poorly rewarded. As you will see throughout this book, gender, race, and class intersect and overlap in the division of labor. Thus, women of color tend to be clustered in certain forms of work that typically disadvantage them by virtue of their race, gender, *and* class. Likewise, a White, working-class man is situated in a unique position in the division of labor where, although he may benefit to some extent from being a White man in society, his class undercuts this, leaving him in a structurally disadvantaged status relative to others in the society.

The significance of the division of labor in analyzing solidarity in society is apparent throughout Durkheim's work. Societies marked by mechanical solidarity have a relatively simple division of labor; those marked by organic solidarity have a complex division of labor. Durkheim thought that the collective consciousness was strongest in societies marked by mechanical solidarity. Indeed, in such societies, the collective consciousness may be a greater feature of individual identity than individual consciousness. People's existence under mechanical solidarity is governed by social imperatives, and there is little room for deviation or diversity. Under organic solidarity, by contrast, the collective consciousness is eased. Values like individualism are more likely to flourish, and it becomes more difficult to speak of a uniform set of values that guide the entire social order. Tensions arise as diverse groups vie for their place in such a society, and finding "common ground" becomes more difficult. In the absence of the unifying forces associated with mechanical solidarity, the unity that emerges from the differentiation and division of labor becomes critical.

Durkheim's thinking about the origins of social cohesion can bring light to contemporary discussions over "family values." Some want to promote traditional family values as the moral standards of society. Is such a thing necessarily good, or even possible? The United States is an increasingly diverse society, and family life differs among different groups. It is unlikely that a single set of family values can be the basis for social solidarity; nevertheless, groups in this society are bound together through the division of labor, even when the bond of shared values is not strong. Although contemporary society is enormously diverse, it can still have the cohesion of a unified society because its people are interrelated through social institutions.

Gemeinschaft and Gesellschaft

Different societies are held together by different forms of solidarity. Some societies are characterized by what sociologists call **gemeinschaft,** a German word that means "community"; others are characterized as **gesellschaft,** which literally means "society" (Tönnies, 1887 [1963]). Each involves a type of solidarity or cohesiveness. Those societies that are gemeinschafts ("communities") are characterized by a sense of "we" feeling, a moderate division of labor, strong personal ties, strong family relationships, and a sense of personal loyalty. In gemeinschaft, the sense of solidarity between members of the society arises from personal ties; small, relatively simple social institutions; and a collective sense of loyalty to the whole society. People tend in a gemeinschaft society to be well integrated into the whole, and social cohesion comes from deeply shared values and beliefs (often, sacred values). In such a society, social control need not be imposed externally since control comes from the internal sense of belonging that members share.

In contrast, in societies marked by gesellschaft, there is an increasing importance placed on the secondary relationships people have—that is, those that are less intimate and more instrumental, such as work roles instead of family or community roles. Gesellschaft is characterized by less prominence of personal ties, a somewhat diminished role of the nuclear family, and a lessened sense of personal loyalty to the total society. This does not mean that the gesellschaft lacks solidarity and cohesion, for indeed it can be very cohesive, but the cohesion comes from an elaborated division of labor, greater flexibility in social roles, and the instrumental ties that people have to one another.

Social solidarity under gesellschaft is weaker than in the gemeinschaft society, however. Gesellschaft societies are marked by an elaborate division of labor; gesellschaft is more likely than gemeinschaft to be torn by class conflict since class distinctions are less prominent, though still present, in the gemeinschaft. Racial–ethnic conflict is also more likely within gesellschaft societies since the gemeinschaft tends to be ethnically and racially very homogeneous; often it is characterized by only one racial or ethnic group. This means that conflict between gemeinschaft societies, such as ethnically based wars, can be very high since both groups have a strong internal sense of group identity that may be intolerant of others.

In sum, complexity and differentiation are what make the gesellschaft cohesive, whereas similarity and unity cohere the gemeinschaft society. In a single society, such as the United States, you can conceptualize the whole society as gesellschaft, with some internal groups marked by gemeinschaft. Our national motto seems to embody this idea: E pluribus unum—unity within diversity.

Types of Societies: A Global View

In addition to comparing how different societies are bound together, sociologists are interested in how social organization evolves in different societies. Over time and across different cultures and continents, societies are distinguished by different forms of social organization—forms that evolve from the relationship of a given society to its environment and from the processes that the society develops to meet basic human needs. Such simple things as the size of a society can also shape its social organization, as do the different roles that men and women engage in as they produce goods, care for the social old and young, and pass on societal traditions. Societies also differ according to their resource base: whether they are predominantly agricultural or industrial, for example, and whether they are sparsely or densely populated.

Thousands of years ago, societies were small, sparsely populated, and technologically limited. In the competition for scarce resources, larger and more technologically advanced societies dominated smaller ones. Today, we have arrived at a society that is global, with highly evolved degrees of social differentiation and inequality, notably along class, gender, racial, and ethnic lines (Lenski et al., 1998). Sociologists distinguish six types of societies based on the complexity of their social structure, the amount of overall cultural accumulation, and the level of their technology. They are *foraging, pastoral, horticultural, agricultural, industrial,* and *postindustrial* (see Table 5.1). Examples of all these types of society can still be found on earth though all but the most isolated societies are moving toward the latter, industrial and postindustrial stages of societal development.

These different societies vary in the basis for their organization and the complexity of their division of labor. Some, like foraging societies, are subsistence economies, where men and women hunt and gather food, but accumulate very little. Others, such as pastoral societies and horticultural societies, develop a more elaborate division of labor as the social roles that are needed for raising livestock and farming become more numerous. With the development of agricultural societies, production becomes more large scale, and strong patterns of social differentiation, sometimes in the form of a caste system or slavery, develop.

One of the driving forces behind the development of these different societies is the development of technology. All societies utilize technology to assist with human needs; technology may be as simple as a rough-hewn shovel or as elaborate as computing technology. Either way, the technology both develops from the society's form and shapes the possibilities for human life. The most elaborated technologies occur in the forms of societies that are most common today: industrial and postindustrial society.

These Bedouins (a pastoral society) are engaged in an economic transaction in this desert market.

Industrial Societies

An *industrial society* is one that uses machines and other advanced technologies to produce and distribute goods and services. The so-called Industrial Revolution began only 200 years ago when the steam engine was invented in England, delivering previously unattainable amounts of mechanical power for the performance of work. Steam engines powered locomotives, factories, and dynamos, transforming societies as the Industrial Revolution spread. The growth of science led to advances in farming techniques such as crop rotation, harvesting, and ginning cotton, as well as industrial-scale projects like dams for hydroelectric power. Joining these advances were developments in medicine, new techniques to prolong and improve life, and the emergence of birth control to limit population growth.

TABLE 5.1 *TYPES OF SOCIETIES*

	Economic base	Social organization	Examples
Foraging societies	economic sustenance dependent on hunting and foraging	gender is basis for social organization, although division of labor is not rigid; little accumulation of wealth	Pygmies of Central Africa
Pastoral societies	nomadic societies, with substantial dependence on domesticated animals for economic production	complex social system with an elite upper class and greater gender role differentiation than in foraging societies	Bedouins of Africa and Middle East
Horticultural societies	society marked by relatively permanent settlement and production of domesticated crops	accumulation of wealth and elaboration of the division of labor, with different occupational roles (farmers, traders, craftspeople, etc.)	Aztecs of Mexico; Incan empire of Peru
Agricultural Societies	livelihood dependent on elaborate and large-scale patterns of agriculture and increased use of technology in agricultural production	caste system devlops that differentiates the elite and agricultural laborers; may include system of slavery	American South, pre-Civil War
Industrial societies	economic system based on the development of elaborate machinery and a factory system; economy based on cash and wages	highly differentiated labor force with a complex division of labor and large formal organizations	19th and 20th century United States and western Europe
Postindustrial societies	information-based societies in which technology plays a vital role in social organization	education increasingly important to the division of labor	contemporary United States, Japan, and others

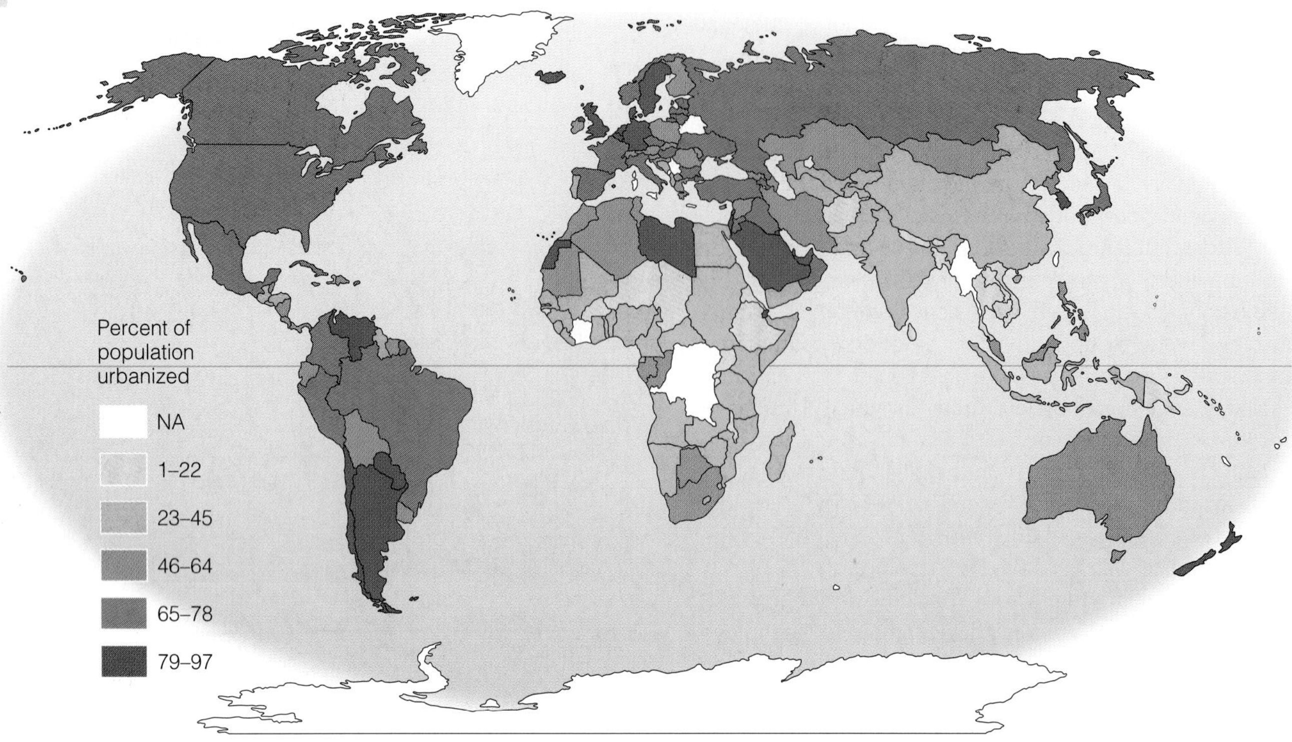

MAP 5.1 Viewing Society in Global Perspective: Global Urbanization

DATA: From the Population Division of the United States Secretariat, 1997. *World Urbanization Prospects: The 1996 Revision.* New York: United Nations (http://www.un.org/Depts/unsd/social/hum-set.html)

Unlike agricultural societies, industrial societies rely upon a highly differentiated labor force and the intensive use of capital and technology. Large formal organizations are common. The task of holding society together, falling more on institutions such as religion in preindustrial societies, now falls more upon the high division-of-labor institutions such as the economy and work, government and politics, and large bureaucracies.

Within industrial societies, the forms of gender inequality that we see in contemporary U.S. society tend to develop. With the advent of industrialization, societies move to a cash-based economy, with labor performed in factories and mills paid on a wage basis and household labor remaining unpaid. This introduced what is known as the *family-wage economy,* one in which families become dependent on wages to support themselves, but work within the family (housework, childcare, and other forms of household work) is unpaid and, therefore, increasingly devalued (Tilly and Scott, 1978). In addition, even though women (and young children) worked in factories and mills from the first inception of industrialization, the family-wage economy is based on the idea that men are the primary breadwinners; therefore, a system of inequality in men's and women's wages was introduced—an economic system that continues today to produce a wage gap between men and women.

Industrial societies tend to be highly productive economically with a large working class of industrial laborers. People become increasingly urbanized as they move from farm lands to urban centers or other areas where factories are located, as we see in the map "Viewing Society in Global Perspective: Global Urbanization." Immigration is common in industrial societies, particularly as industries are forming where there is a high demand for more, cheaply paid labor. The transition to industrialization also has a strong effect on social institutions. Families become less the center of production, instead becoming dependent on industrial systems for jobs and wages. With the movement of economic activity outside the family, other institutions begin to take on the responsibilities once performed within the family. Schools, for example, take on the role of educating the young; indeed, school calendars reflect the needs of the industrial society, training children to specialize in "vocational" tracks, if they are destined to remain in the working class, and teaching the values of punctuality, conformity to a time clock, and other routines that are necessary to the smooth functioning of an industrial order.

Industrialization has brought many benefits to U.S. society—a highly productive and efficient economic system, expansion of international markets, extraordinary availability of consumer products, and for many, a good working wage. Industrialization has, at the same time, however, also produced some of the most serious social problems that our nation faces: industrial pollution, an overdependence on consumer goods, wage inequality and job dislocation for millions, and problems of crime and crowding in urban areas; the portrait of population density depicted in the map "Mapping America's Diversity: Population Density in the United States" illustrates where

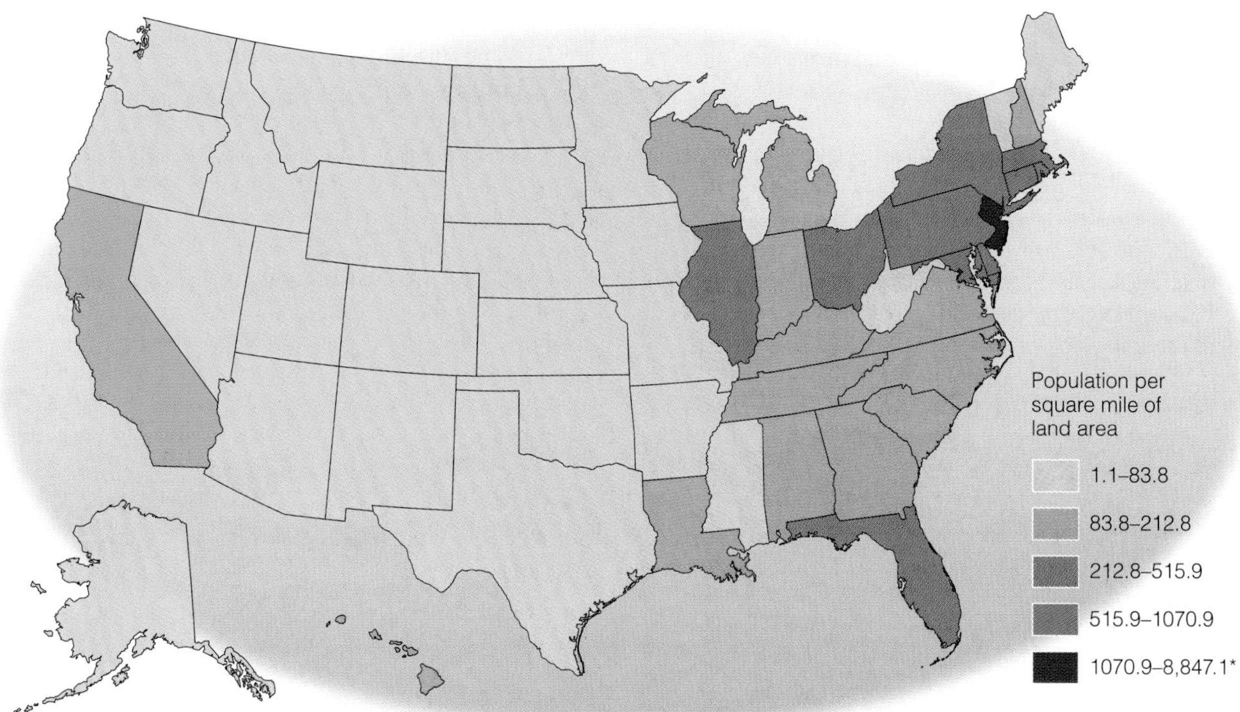

MAP 5.2 Mapping America's Diversity: Population Density in the United States

*District of Columbia is the only entity within this range with a population density of 8, 847. 1.

DATA: From the Bureau of the Census. 1997. *Statistical Abstracts of the United States, 1997.* Washington, DC: U.S. Government Printing Office, p. 29.

crowding is most extreme. Understanding the process of industrialization and its accompanying process of urbanization is a major avenue for sociological research and is explored further in Chapter 19.

Postindustrial Societies

In the contemporary era, a new type of society is emerging. Whereas most twentieth-century societies can be characterized in terms of their generation of material goods, *postindustrial society* is economically dependent on the production and distribution of services, information, and knowledge (Bell, 1973). Postindustrial societies are information-based societies in which technology plays a vital role in the social organization of society. The United States is fast becoming a postindustrial society, and Japan may be even further along. In the postindustrial society, many of the workers provide services such as administration, education, legal services, scientific research, and banking, or they engage in the development, management, and distribution of information, such as computer use and design. Central to the economy of the postindustrial society are highly advanced technologies such as computers, robotics, genetic engineering, and laser technology. Multinational corporations globally link the economies of postindustrial societies.

The transition to a postindustrial society has a strong influence on the character of social institutions. Educational institutions acquire paramount importance in the postindustrial society, and science takes an especially prominent place. For some, the transition to a postindustrial society means more discretionary income for leisure activities—tourism, entertainment, and relaxation industries (spas, massage centers, and exercise) become more promiment—at least for those in certain classes. For others, the transition to postindustrialism can mean permanent joblessness or the need to hold down more than one job to simply make ends meet.

An information-based technology is the primary characteristic of the postindustrial society, exemplified here by this worker, who is surrounded by the latest in computer hardware.

Workers without highly technical skills may be dispossessed in such a society, and millions may find themselves stuck in low-paid, unskilled work.

The United States is suspended between the industrial and postindustrial phases. Manufacturing jobs are still a major segment of the labor force though they are in decline since most workers are employed in the service sector of the economy (the sector involving the delivery of services and information, not the actual production of material goods). Postindustrial societies are also increasingly dependent on a global economy since goods are still produced, but they tend to be produced in economically dependent areas of the world for consumption in the wealthier nations. As a result, the world, not just individual societies, becomes characterized by poverty and inequality resulting from the social structure that postindustrialism produces.

As we continue to study the different elements of society—groups, social interaction, social institutions—you should keep these broad processes of societal development in mind. They form the stage on which social structures are produced. As our society becomes increasingly interrelated with those in other parts of the world, it is important to understand the major social forces that affect the character of contemporary society.

CHAPTER SUMMARY

- An individual is typically a member of several groups simultaneously. *Groups* are of varying types, as are social groupings (social categories, audiences, and formal organizations). *Status* is a hierarchical position in a structure; a *role* is the expected behavior associated with a particular status. Since the typical individual occupies many statuses as well as the many roles associated with him or her, these can often conflict, thus producing *status inconsistency* or *role strain,* either of which can produce stress and unpleasantness for the individual.
- *Social interaction* takes place in society within the context of social structure and social institutions. Social interaction is analyzed in several ways, including the *social construction of reality* (we impose meaning and reality on our interactions with others); *ethnomethodology* (deliberate interruption of interaction in order to observe how a return to "normal" interaction is accomplished); *impression management* (wherein a person "gives off" a particular impression in order to achieve certain goals), as in *cyberspace interaction;* and *social exchange* (wherein one engages in reward and punishment interactions in order to achieve one's goals).
- Interaction can be verbal or nonverbal. Forms of nonverbal communication are *touch, paralinguistic, kinesic,* and *use of personal space.* The study of nonverbal communication reveals much rich information about diverse gender, cultural, and racial–ethnic variations in human interaction.
- People form pairings on the basis of several things, among them proximity; repeated exposure to another; perceived physical attractiveness; and similarity in age, socioeconomic status, race, ethnicity, religion, perceived personality traits, general attitudes and opinions, and educational aspirations. Similarities on such things attract; opposites do not. Such variables also tend to predict how long an interracial relationship will last.
- According to theorist Emile Durkheim, society with all its complex social organization and culture, is held together, depending upon overall type, by one of two kinds of cohesion or solidarity: *mechanical solidarity* (based on individual similarity) and *organic (contractual) solidarity,* based on a *division of labor* among dissimilar individuals. Two other forms of social organization also contribute to the cohesion of a society: *gemeinschaft* (characterized by cohesion based on friendships and loyalties) and *gesellschaft* (characterized by cohesion based on complexity and differentiation).
- Societies across the globe vary in type, as determined mainly by the complexity of their social structures, their division of labor, and their technologies. From least to most complex, they are foraging, pastoral, horticultural, agricultural, industrial, and postindustrial societies.

KEY TERMS

achieved status
ascribed status
collective consciousness
cyberspace interaction
division of labor
ethnomethodology
gemeinschaft
gesellschaft
group
impression management
master status
mechanical solidarity
organic solidarity
proxemic communication
role
role conflict
role strain
social institution
social interaction
social organization
social structure
society
status
status inconsistency
status set

THE INTERNET: A Tool for the Sociological Imagination

Resources on the Internet:

Virtual Society: The Wadsworth Sociology Resource Center at
http://sociology.wadsworth.com

Visit this site to find additional learning tools, including interactive quizzes, links to related web sites, and an easy link to InfoTrac College Edition.

The Psychology of Cyberspace
http://www.rider.edu/users/suler/psycyber/psycyber.html

This site includes an on-line book about social interaction and cyberspace, with links to other sites exploring the implications of cyberspace for interpersonal interaction.

Center for Social Informatics

http://www-slis.lib.indiana.edu/CSI/

The Center for Social Informatics at Indiana University does research on the influence of technology and computerization on society; the web site includes information on publications, conferences and other resources.

Sociology and Social Policy: Internet Exercise

In their well-received book, *Habits of the Heart,* sociologists Robert Bellah and his colleagues argued that the individualistic orientation of people in the United States has created a society in which people find it difficult to sustain their commitments to others. The authors suggest that the nation needs to develop new traditions—traditions that would unite people through common community service activities. Identify one volunteer project in your community, and learn what you can about its purpose, its volunteers, and its effect. At the same time, search the literature on the focus of this project (such as *Habitat for Humanity* and low-income housing or AIDS and the AIDS Caregivers Support Network), and identify the sociological dimensions of this policy issue. Based on what you have learned, would you agree or disagree with Bellah and his colleagues that service-oriented groups promote a sense of community?

Internet Search Keywords

community service
volunteer
Habitat for Humanity
community projects
social policy
philanthropy

Web sites:

http://www.servenet.org/
Site includes daily service news, online resources, and volunteer opportunities.

http://www.libertynet.org/nol/natl.html
Online resource center sponsored by Philadelphia's Liberty Net to provide information and ideas about neighborhood revitalization as well as to create a national network of activists working on the problem.

http://www.habitat.org/
Habitat for Humanity International web site.

http://www.philanthopy.iupui.edu/
Indiana University Center on Philanthropy home page.

http://www.policy.com/issuewk/97/1222/index.html
This site explores trends in giving, tips for givers, and more.

http://www.pin.org/
Home page for the Public Involvement Network. This group collects and distributes how-to information about citizen participation and related activities.

InfoTrac College Edition: Search Word Summary

achieved and ascribed status
proxemic communication
social interaction

In order to learn more about these central topics in sociology, you can conduct an electronic search using InfoTrac College Edition. To aid in your search and to gain useful tips, see the Student Guide to InfoTrac College Edition on the Virtual Society web site:
http://sociology.wadsworth.com

INTERACTIONS—A SOCIOLOGY CD-ROM: CONCEPTS FOR THIS CHAPTER

Go to the Wadsworth Sociology CD-ROM for further study on the concepts in this chapter. The CD-ROM also includes quizzes and additional activities to expand your learning experience.

SUGGESTED READINGS

Bell, Daniel. 1973. *The Coming Crisis of Postindustrial Society: A Venture in Social Forecasting.* New York: Basic Books.

This classic work anticipated some of the problems of America's continuing transition from industrial to postindustrial society.

Gilbert, Daniel T., Susan T. Fiske, and Gardner Lindzey, eds. 1998. *The Handbook of Social Psychology,* 4th ed. New York: Oxford University Press and McGraw-Hill.

This is a comprehensive summary of research in some of the areas covered in this chapter, such as forms of nonverbal communication and interpersonal attraction.

Goffman, Erving. 1959. *The Presentation of Self in Everyday Life.* Garden City, NY: Doubleday.

Goffman's analysis is the classic analysis of impression management and how people negotiate their individual identity through social interaction.

Hendrick, S. S. and C. Hendrick. 1992. *Romantic Love.* Newbury Park, CA: Sage.

The Hendricks' provide a comprehensive and readable discussion of research on romantic love and the formation of dyads on the basis of love.

Lenski, Gerhard, Jean Lenski, and Patrick Nolan. 1995. *Human Society: An Introduction to Macrosociology,* 7th ed. New York: McGraw-Hill.

This book offers an expansive account of different types of societies in detail, with clear discussions of Durkheim and other theorists.

Tannen, Deborah. 1990. *You Just Don't Understand: Women and Men in Conversation.* New York: William Morrow.

Tannen's discussion of communication styles between women and men became a popular best-seller. Based on social science research, it shows how gender differences in communication lead to misunderstanding and conflict between men and women.

Turnbull, Colin. 1972. *The Mountain People.* New York: Simon and Schuster.

The *Ik* were an isolated people living as a separate society in Uganda. This is a powerful account of how the Ugandan government completely destroyed *Ik* society by moving them onto a government-controlled reservation. Comparisons to the historical treatment of Native Americans are made.

CHAPTER 6

Groups and Organizations

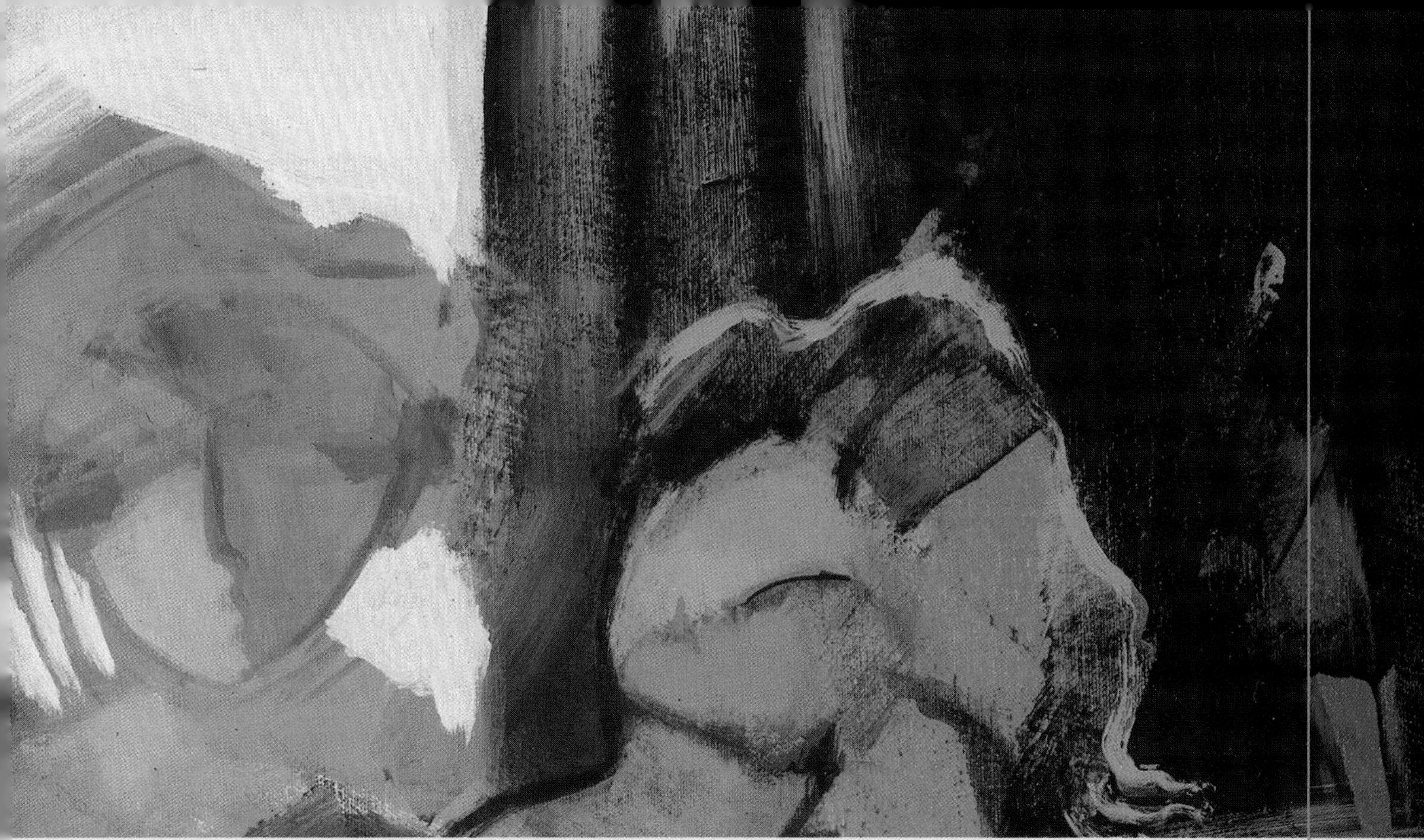

TWELVE citizens sit together in an elevated enclosure, like a choir loft, and silently watch a drama unfold before them, day after day. They are respectfully addressed by highly paid professionals: lawyers, judges, expert witnesses. Their job, ruling on the innocence or guilt of a defendant or the settlement of a legal claim, was once the prerogative of kings. Their decision may mean freedom or incarceration, fortune or penury, even life or death.

The jurors all hear the same presentations about the evidence. They are given identical instructions by the judge when the time arrives for a decision to be made, yet the jurors' chamber is frequently a battleground, the site of bitter disputes, tears, bullying, flights of reason and unreason. Juries have been the focus of much research, in part because they fill such a vital role in our society and in part because within this curiously artificial, yet intimate, group of random strangers can be found a wealth of interesting sociological phenomena. Jury verdicts and jury deliberations show the same inescapable influences of status, race, and gender that affect the rest of society. Juries are, in many ways, society in miniature (Hans and Vidmar, 1986).

A vast folklore exists among trial lawyers about jurors and jury performance. During jury selection, lawyers on both sides are entitled to eliminate a number of potential jurors with no explanation required. Many lawyers who have great faith in their ability to judge jurors consider jury selection to be the most important part of a trial: By choosing the jurors, they are choosing the verdict. Consider some of the folk wisdom clung to by trial lawyers: Farmers believe in strict responsibility, whereas waiters and bartenders are forgiving (Belli, 1954); bankers are hard on robbery, but easy on white-collar crime (Campbell, 1972); avoid wage earners and the clergy; select married women. A guideline for Dallas prosecutors advised against selecting "Yankees . . . unless they appear to have common sense" (Guinther, 1988: 54).

Folklore, appropriately analyzed, can provide hypotheses for scientific testing. High-powered legal teams now make room for a new breed of legal specialist—the trial consultant trained in sociological techniques who contributes nothing but juror analysis as part of the jury selection process. Such was the case in the well-publicized murder trial of football player O. J. Simpson. The jury found Simpson not guilty because, according to one writer, prosecutor Marcia Clark ignored social science consultants (Toobin, 1996). The level of analysis goes beyond simply identifying the likely bias of a given juror. Juries are *groups,* and groups behave differently from individuals. Understanding group behavior is critical to predicting the performance of a jury. For instance, it is possible to make an educated prediction about who in a jury will become the most influential (Hans and Martinez, 1994). Researchers have found that people with high status in society do the most talking in jury deliberations and are thought by other jurors to be the most helpful in reaching a verdict (Cohen and Zhou, 1991; Berger and Zelditch, 1985).

Factions form during jury deliberations, and if jury analysts expect a difficult decision, they can attempt to influence how fractionalized juries will resolve their disputes based on sociological and psychological data about small group decision making (Hastie et al., 1983). For instance, jurors are much less likely to defect from large factions than from small ones. The larger the faction, the less willing a juror will be to defy the weight of group opinion; furthermore, the amount that jurors talk changes as faction size changes. Dominant jurors in the majority dominate more as the minority faction gets smaller. Putting together all the data that have been gathered about juries and other small groups, trial lawyers can attempt to script the deliberations. Of course, jurors are not robots, jury deliberations are not perfectly predictable, and jury-sculpting lawyers are in great danger of trading old prejudices for new as they apply sociological results to actual people. Studies have nevertheless repeatedly shown that jury verdicts correlate not just with the evidence but also with jury composition, especially its racial–ethnic and gender composition (Hans and Martinez, 1994; Brown, 1986; Toobin, 1996; Fukurai, 1993; Nemeth, 1980).

What does this say about the state of justice in our legal system, when guilt or innocence depends not only on the facts but also on the jury? Like society as a whole, groups are subject to social influences, and understanding these influences can be essential to understanding a process like the operation of social justice. Analyzing the nature of groups is a step toward understanding how groups interact to form society. Sociologists therefore study groups of different kinds, from the smallest and most intimate to large groups known as organizations. This chapter introduces students to the study of groups and the social processes and dynamics that are characteristic of group behavior. Whether a relatively small group, like a jury, or a large organization, such as a major corporation, groups are influenced by sociological forces.

This jury of twelve persons is voting on something (it looks unanimous), perhaps on whether or not to acquit a defendant of a serious crime. The social pressures in a jury are extremely strong, making the lone "hold-out" person a very unlikely event.

Types of Groups

Each of us is a member of many groups simultaneously. We have relationships with family, friends, team members, and professional colleagues in groups. Within these groups are gradations: We are generally closer to our siblings than our cousins; we are intimate with some friends, merely sociable with others. If we count up all of our group associations, ranging from the powerful associations that define our daily lives to the thinnest connections about which we have any feeling (other pet lovers, fans of Madonna, other company employees), we will uncover connections to literally hundreds of groups.

What is a group? As always in sociology, we want to define the term with precision. A **group** is a collection of individuals who (recall from Chapter 5) interact with one another, share goals and norms, and have a subjective awareness as "we." To be considered a group, a social unit must have all three characteristics. The boundaries of the definition are necessarily hazy. Consider three superficially similar examples: The individuals in a line waiting to board a plane are unlikely to have a sense of themselves as a group; they are unlikely to think of themselves as "we the people of Flight 175." A line of prisoners chained together and waiting to board a bus to the penitentiary is likely to have a thicker sense of common feeling. A line of Olympic athletes boarding a bus that will take them to their morning's events is certain to have a sense of group feeling: "We are world-class athletes, going off together to perform."

As will be remembered from the previous chapter, certain gatherings are not groups in the strict sense, but may be *social categories* (for example, teenagers, truck drivers) or *audiences* (everyone watching a movie). The importance of defining a group is not so that we can perfectly diagnose whether a social unit is a group—an unnecessary endeavor—but to help us understand the behavior of people in society. As we inspect groups, we are able to identify characteristics

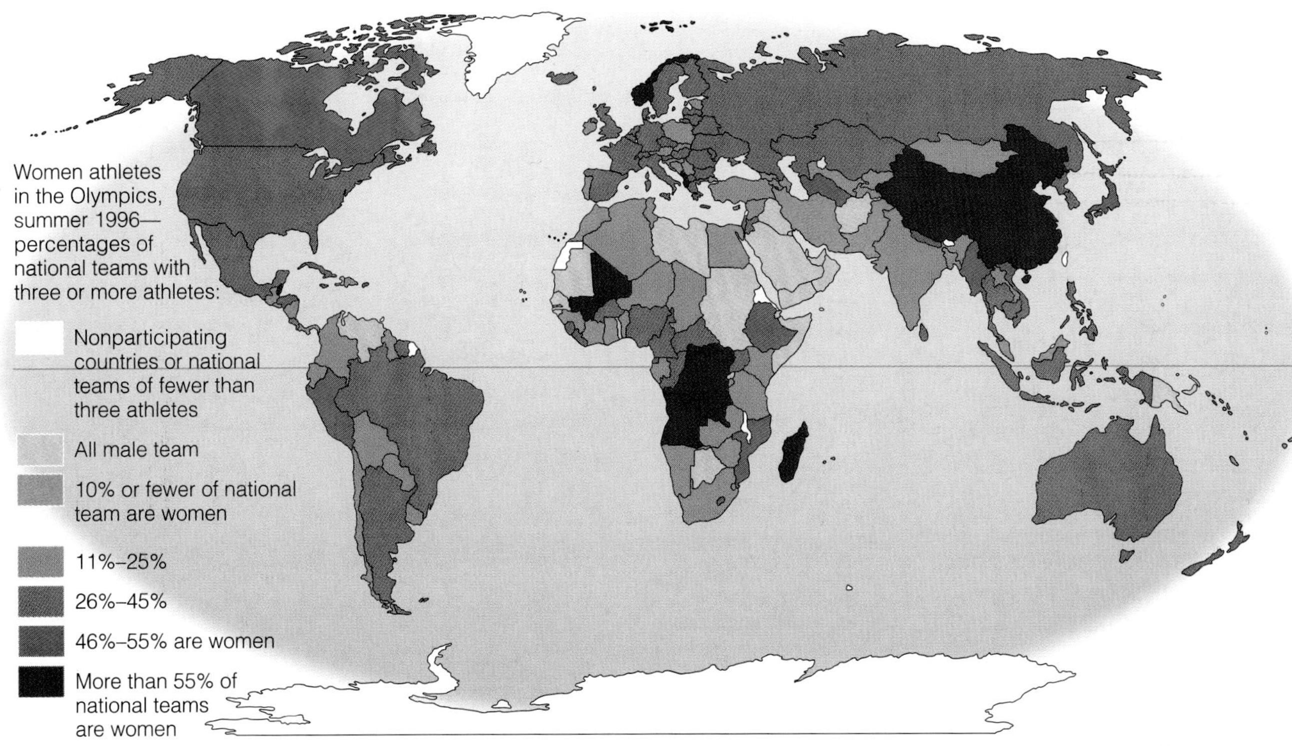

MAP 6.1 Viewing Society in Global Perspective: Going for the Gold

SOURCE: Seager, Joni. 1997. *The State of Women in the World Atlas*, 2e. The Penguin Group.

that reliably predict trends in the behavior of the group, and even the behavior of individuals in the group. We can begin to cultivate orderly thinking about groups by analyzing the range of variables that come into play. Is the group large or small? Is participation in the group intensely emotional or a pallid experience? Does the group convene frequently or rarely? Is the group formally or informally organized?

The study of groups has application at all levels of society, from the attraction between people who fall in love to the characteristics that make some corporations drastically outperform their competitors. The aggregation of individuals into groups has a transforming power, and sociologists understand the social forces that make these transformations possible.

Dyads and Triads: Group Size Effects

Even the smallest groups are of acute sociological interest. A **dyad** is a group consisting of exactly two people. A **triad** consists of three people. This seemingly minor distinction, first scrutinized by the German sociologist Georg Simmel (1858–1918), can have critical consequences for group behavior (Simmel, 1902). Simmel was interested in discovering the effects of size on groups, and he found that the mere difference between two and three spawned entirely different group dynamics (the behavior of a group over time).

Imagine two people standing in line for lunch. First one talks, then the other, then the first again. The interaction proceeds in this way for several minutes. Now a third person enters the interaction. The character of the interaction suddenly changes: At any given moment, two people are interacting more with each other than either is with the third. When the third person wins the attention of one of the other two, a new dyad is formed, supplanting the previous pairing. The group, a triad, then consists of a dyad (the pair that is interacting) plus an *isolate*.

Triadic segregation is what Simmel called the tendency for triads to segregate into a pair and an isolate. A triad tends to segregate into a *coalition* of the dyad against the isolate. The isolate then has the option of initiating a coalition with either of the members of the dyad. This choice is a type of social advantage, leading Simmel to coin the principle of *tertius gaudens*, a Latin term meaning "the third one gains."

Triadic interactions often end up as two against one (Caplow, 1968). You may have noticed this principle of coalition formation in your own conversations. Perhaps two friends want to go to a movie you do not want to see. You appeal to one of them to go instead to a minor league baseball game. She wavers and comes over to your point of view—you have formed a coalition of two against one. The friend who wants to go to the movies is now the isolate. He may recover lost social ground by trying to form a new coalition, such as by suggesting a new alternative (going bowling or going to a different movie). This flip-flop interaction may continue for some time, demonstrating another observation by Simmel: A triad is an *unstable* social grouping, whereas dyads are relatively stable. The minor distinction between dyads and triads—one person—has important consequences since it changes the character of the interaction within the group. Simmel is known as the discoverer of **group size**

One of the best examples of the primary group is that consisting of parent and child.

These are members of a large class of graduating seniors—a secondary group.

effects—the effects of group number on group behavior. Later in this chapter, we will encounter several examples of how dramatically group size can change behavior.

Primary and Secondary Groups

Charles Horton Cooley (1864–1929), a famous sociologist of the Chicago School of sociology, introduced the concept of the **primary group**—defined as a group consisting of intimate, face-to-face interaction and relatively long-lasting relationships. Cooley had in mind the family and early peer group. In his original formulation, "primary" was used in the sense of "first," the intimate group of the formative years (Cooley, 1909 [1967]). The insight that there was an important distinction between intimate groups and other groups proved extremely fruitful. Cooley's somewhat narrow concept of family and childhood peers has been elaborated over the years to include a variety of intimate relations as examples of primary groups.

Primary groups have a powerful influence on an individual's personality or self-identity. The effect of family on an individual can hardly be overstated. The weight of peer pressure on school children is particularly notorious. Street gangs are a primary group. The camaraderie formed among Marine Corps units in boot camp is another classic example of a primary group.

Primary groups are defined in part by their contrast with **secondary groups,** those that are larger in membership, less intimate, and less long-lasting. Secondary groups tend to be less significant in the emotional life of people. Secondary groups include all the students at a college or university, all the people in your neighborhood, and all the people in a bureaucracy or corporation. Secondary groups occasionally take on the characteristics of primary groups. The process can be accelerated in situations of high stress or crisis. For example, when a neighborhood meets with catastrophe, such as a flood, people who may know each other only fleetingly often come to depend on each other and in the process become more intimate. The group of neighbors becomes, for a time, a primary group. The ritualistic abuse of military boot camp—manufactured stress expertly applied—is calculated to accelerate the formation of primary groups among the overstressed soldiers, strengthening their bonds as a unit.

Primary and secondary groups serve different needs. Primary groups give people intimacy, companionship, emotional support. These are termed **expressive needs** (also called socioemotional needs). Family and friends share and amplify your good fortune, rescue you when you misbehave, and cheer you up when life looks grim. Primary groups are a major influence on social life and an important source of social control. They are also a dominant influence on your likes and dislikes, preferences in clothing, political views, religious attitudes, and other characteristics. Many studies have shown the overwhelming influence of family and friendship groups on religious and political affiliation, as illustrated in the "Doing Sociological Research" box "Sharing the Journey," in which sociologist Wuthnow (1994) analyzes the significance of primary group membership in contemporary society. Compare the apparel of your circle of friends and see if there is evidence of primary group influence in your choice of dress—commonly regarded as the banner of individuality, but much the result of group behavior.

Secondary groups serve **instrumental needs** (also called task-oriented needs). Athletic teams form to have fun and win games. Political groups form to raise funds and bend the will of the legislature. Corporations form to make profits, and employees join corporations to make a living. Needless to say, intimacies can develop in the act of fulfilling instrumental needs, and primary groups may also devote themselves to meeting instrumental needs. The true distinction between primary and secondary groups is less in why the groups form than in how strongly the participants feel about one another, and how dependent they are on the group for sustenance and identity. Both primary and secondary groups are indispensable elements of a life in society.

THINKING SOCIOLOGICALLY

Name a *primary group* and a *secondary group* to which you belong. Compare the two groups in terms of their size, degree of intimacy, and the nature of the interaction you have within them. Are these groups primarily *expressive, instrumental,* or both?

The distinction between primary and secondary groups is not a perfect one; some groups have elements of both. People in a work group, for example, typically considered by sociologists to be a secondary group, can develop personal and intimate ties. Likewise, primary groups can feel isolating and callous, as is sometimes the case in problematic families. For sociologists, primary and secondary groups are ideal types—that is, the dichotomy is one made for purposes of analyzing the group processes that are usually, even if not universally, found in groups of a certain kind.

Reference Groups

Primary and secondary groups are groups to which one belongs. **Reference groups** are those to which you may or may not belong, but that you use as a standard for evaluating your values, attitudes, and behaviors (Merton and Rossi, 1950). Reference groups are generalized versions of role models and are, in this respect, not "groups" in the sense that the individual interacts within (or in) them. Do you pattern your behavior on that of sports stars, musicians, military officers, or business executives? If so, those models are reference groups for you. A well-rounded person will

BOX 6.1 DOING SOCIOLOGICAL RESEARCH

Sharing the Journey

MODERN society is often characterized as remote, alienating, and without much feeling of community or group belongingness. This image of society has been carefully studied by sociologist Robert Wuthnow, who noticed that in the United States, people are increasingly looking to small groups as a place where they can find emotional and spiritual support and where they find meaning and commitment, despite the image of society as an increasingly impersonal force. Wuthnow began his research by noting that, even with the individualistic culture of U.S. society, small groups play a major role in this society; he saw the increasing tendency of people to join recovery groups, reading groups, spiritual groups, and a myriad of other support groups.

Wuthnow began his research by asking some specific questions, including, "What motivates people to join support groups?" "How do these groups function?" and "What do members like most and least about such groups?" His broadest question, however, was to wonder how the wider society is being influenced by the proliferation of small, support groups.

To answer these questions, a large research team of fifteen scholars designed a study that included both a quantitative and a qualitative dimension. They distributed a survey to a representative sample of more than 1000 people in the United States. Supplementing the survey were interviews with more than 100 support group members, group leaders, and clergy. The researchers chose twelve groups for extensive study; researchers spent six months to three years tracing the history of these groups, meeting with members and attending group sessions.

Based on this research, Wuthnow concludes that the small group movement is fundamentally altering U.S. society. Forty percent of all Americans belong to some kind of small group. As the result of people's participation in these groups, social values of community and spirituality are undergoing major transformation. People say they are seeking community when they join small groups—whether the group be a recovery group, a religious group, a civic association, or some other small group. People turn to these small groups for emotional support more than for physical or monetary support.

Wuthnow argues that large-scale participation in small groups has arisen in a social context in which the traditional support structures in U.S. society, such as the family, no longer provide the sense of belonging and social integration that they provided in the past. Geographic mobility, mass society, and the erosion of local ties all contribute to this trend. People still seek a sense of community, but they create it in groups that also allow them to maintain their individuality. In voluntary small groups, different from, for example, the family, you are free to leave if the group no longer meets your needs.

Wuthnow also concludes that these groups represent a quest for spirituality in a society when, for many, traditional religious values have declined. As a consequence, support groups are redefining what is sacred. They also replace explicit religious tenets imposed from the outside with internal norms that are implicit and devised by individual groups. At the same time, these groups reflect the pluralism and diversity that characterizes society. In the end, they buffer the trend toward disintegration and isolation that people often feel in mass societies.

SOURCE: Wuthnow, Robert. 1994. *Sharing the Journey: Support Groups and America's New Quest for Community.* New York: Free Press.

have a number of reference groups at one time and will emulate them with varying degrees of fidelity.

Imitation of reference groups can have both positive and negative effects; in the latter case, they represent *negative reference groups.* Members of a Little League baseball team may revere major league baseball players and thus attempt to imitate laudable behaviors like tenacity and sportsmanship, but young baseball fans are also liable to be exposed to tantrums and fights, tobacco spitting, and scandals of the sort that befall many, such as drug problems and domestic violence. This illustrates that the influence of a single reference group can be both positive and negative.

Research has shown that identification with reference groups can have a strong effect on self-evaluation and self-esteem. People are exultant when their favorite team wins a game. Wearing a team's logo emblazoned on a sweatshirt demonstrates pride in one's reference group and pride in one's self: My team is a winner, therefore I am a winner (Felson and Reed, 1986; Rosenberg and Simmons, 1971).

In the early days of desegregation, it was thought that all-Black schools contributed to a negative self-evaluation among Black students. Desegregation was expected to raise the self-esteem of Black children (Clark and Clark, 1947). In some cases, it did, but later research has also found that more important than desegregation was identification with a positive reference group. When racial or ethnic groups were consistently and methodically presented in a positive way, as in pluralistic and multicultural educational programs designed to increase pride in Black culture, the self-esteem of the children was greater than that of Black children in integrated programs that had no pluralistic component. The same has been found for Latino children enrolled in Latino cultural awareness programs. Plainly, the representation of racial and ethnic groups in a society can have a striking positive or negative effect on children who are acquiring their lifetime set of group affiliations (Steele, 1996, 1992; Steele and Aronson, 1995; Banks, 1976).

In-Groups and Out-Groups

When groups have a sense of themselves as "us," there will be a complementary sense of other groups as "them." The distinction is commonly characterized as *in-groups* versus *out-groups.* The concept was originally elaborated by the early sociological theorist W. I. Thomas (Thomas, 1931). College fraternities and sororities certainly exemplify "in" versus "out." So do families. The same can be true of the members of your high school class, your sports team, your racial group, your gender, and your social class. Members of the upper classes in the United States sometimes refer to one another as PLUs—"people like us" (Frazier, 1957; Taylor, 1992).

The emotional charge attached to in-groups and out-groups can be dramatic and has been the focus of much research, ranging from work on families and gangs to the sociology of racism and nationalism. The term *out-group* has a negative ring that is no accident. Generally, people are prone to regard out-group persons with suspicion, whereas in-group members are favored with liking and trust.

Attribution theory is the principle that *dispositional attributions* are made about others (what the other is "really like") under certain conditions, such as out-group membership. Thomas F. Pettigrew has summarized the research on attribution theory, showing that individuals commonly generate a significantly distorted perception of the motives and capabilities of other people's acts based on whether that person is an in-group or out-group member (Pettigrew, 1992). Pettigrew describes the misperception as *attribution error,* meaning errors made in attributing causes for people's behavior to their membership in a particular group, such as a racial group. Attribution error has several dimensions, all tending to favor the in-group over the out-group.

1. Observing improper behavior by an out-group member, onlookers are likely to attribute the deviance to the *disposition* of the wrongdoer. *Disposition* refers to the perceived "true nature" of the person, often considered to be genetically determined. (Example: A White person sees a Hispanic person carrying a knife, and attributes this behavior to his or her "inherent tendency" to be violent, or a Hispanic person perceives all White people as inherently racist.)
2. When the *same* behavior is exhibited by an in-group member, the perception is commonly that the act is due to the *situation* of the wrongdoer, not to the in-group member's disposition. (Example: A White person sees another White person carrying a knife and concludes that the weapon must be carried as protection in a dangerous area.)
3. If an out-group member is seen to perform in some laudable way, the behavior is often attributed to a variety of special circumstances, and the out-group member is seen as "exceptional." Members of the in-group may say the person just had good luck or that his or her motivation was unusually high; perhaps, they might say, the behavior was not so special after all.
4. An in-group member who performs in the same laudable way is given credit for a worthy disposition.

In a word, we perceive people in our in-group positively and those in out-groups negatively. What we disapprove of in an out-group individual we ascribe to a flaw in their group; what we approve of in the same person we ascribe to luck or some special circumstance. Rather than give up our prejudices about out-groups, we are more likely to invent explanations for nonstereotypic behavior that preserve our original bias.

Typical attribution errors include misperceptions between racial groups and also between men and women. If one is White, one may perceive other Whites in positive terms and Blacks, Latinos, and even Native Americans in relatively less positive or negative terms. If a White policeman shoots a Black or Latino, a White individual, given no additional information, is likely to assume that the victim instigated the shooting, whereas a Black individual is more likely to assume that the policeman fired unnecessarily, perhaps because he is dispositionally assumed to be a racist (Kluegel and Bobo, 1993; Bobo and Kluegel, 1991). A related phenomenon has been seen in men's perceptions of women co-workers. Meticulous behavior in a male is perceived positively and is seen by the male as "thorough"; in a female, the exact *same* behavior is perceived negatively and is considered "picky." Behavior applauded in a male as "aggressive" is condemned in a woman exhibiting the same behavior as "pushy" or "bitchy" (Wood, 1994).

Social Networks

As already noted, no individual is a member of only one group. Social life is far richer than that. Membership in several groups provides links between groups and groups overlap. A **social network** is a set of links between individuals or between other social units, such as bureaucratic organizations or entire nations. One could say that people belong to several networks, or one could acknowledge the interlocking nature of networks and say that there is a single, large, highly textured network constituted of all society's relations. The texture within this web is provided by identifiable units within the network—networks within networks. Your group of friends, for instance, or all the people on an electronic mail list to which you subscribe are social networks.

The network of people you know, rather than those merely linked to you in some impersonal way, is probably most important to you. A large mass of research indicates that people get jobs via their personal networks more often than through formal job listings, want ads, or placement agencies (Granovetter, 1974). This is especially true for high-paying, prestigious jobs. Getting a job is more often a matter of *who* you know than what you know. Who you know, and who they know in turn, is a social network that may have a marked effect on your life and career.

Networks form with all the spontaneity of other forms of human interaction (Granovetter, 1973; Mintz and Schwartz, 1985; Wasserman and Faust, 1994; Fischer, 1981; Knoke, 1992). One's family usually forms the first social network in a person's life. Later other networks evolve, such as social ties within one's neighborhood, professional contacts, and associations formed in fraternal, religious, occupational, and volunteer groups. Schooling, for instance, gives birth to many networks—our old school friends, our high school graduating class, our college class, our alumni group. Networks to which you are only weakly tied (you may know, say, only one person in some key network) provide you with access to that network; hence, the sociological principle that there is "strength in weak ties" (Granovetter, 1973).

Networks based on race, class, and gender form with particular readiness. This has been particularly true of job networks. The person who leads you to a job is likely to have a social background similar to yours. Recent research indicates that the "old boy network"—any network of White, well-positioned men—is less important than it used to be, although it is certainly not gone. The diminished importance of the old boy network is because of the increasing prominence of women and minorities in business organizations. Still, as we will see later in this chapter, women and minorities are considerably underrepresented in corporate life, especially in high-status jobs (Gerson, 1993; Collins-Lowry, 1997; Reskin and Padavic, 1994).

The Small World Problem

Networks can reach around the world, but how big is the world? How many of us have remarked, "My, it's a small world," upon discovering that someone we just met is a friend of a friend? Research into what has come to be known as the *small world problem* has shown that networks make the world a lot smaller than you might think.

Small world researchers Travers and Milgram wanted to test whether a document could be routed to a complete stranger more than 1000 miles away using only a chain of acquaintances (Travers and Milgram, 1969). If so, how many steps would be required? The researchers organized an experiment in which approximately 300 "senders" were all charged with getting a document to one "receiver," a complete stranger. The receiver was a Boston stockbroker. The senders consisted of three groups: 100 Nebraskans with stock holdings, 96 Nebraskans chosen at random, and 100 Bostonians chosen at random. Everyone in the study was given the receiver's name, address, occupation, alma mater, year of graduation, his wife's maiden name, and hometown. They were asked to send the document directly to the stockbroker if they knew him on a first-name basis. Otherwise, they were asked to send the folder to a friend, relative, or acquaintance known on a first-name basis who might be more likely than the sender to know the stockbroker.

How well do you think this informal mail system worked? How many intermediaries do you think it took, on average, for the document to get through? (Most people estimate from twenty to hundreds.) About one-third of the documents arrived at the target. This was quite good, considering that the senders did not know the target person. The average number of contacts was only 6.2! Not as efficient as the post office, but then the U.S. Postal Service is the biggest bureaucracy on earth, whereas the small world network is nothing more than an accident of human interaction. The Boston group of senders reached the stockbroker in fewer steps than the Nebraskans—4.4 steps versus 5.7—but the difference seems negligible considering that the Bostonians were within 25 miles of the stockbroker, whereas the Nebraskans were 1300 miles away.

It really is a small world. One inference of the study is that you are probably no more than five acquaintance links (or fewer) removed from *any* citizen of the United States—a lawyer in New York, a grocer in Los Angeles, or even the president of the United States. The world is even smaller than suggested in the movie and play entitled "Six Degrees of Separation," which was based in part on the Travers and Milgram study.

A recent study of Black national leaders by Taylor and associates shows that Black leaders form a very closely knit network, one considerably more dense than longer-established White leadership (Jackson et al., 1995, 1994; Taylor, 1992; White leadership networks have been examined by Mills, 1956; Domhoff, 1967; Kadushin, 1994; Moore, 1979; and Alba and Moore, 1982). Included in the study of Black leadership networks were members of Congress, mayors, business executives, military officers (generals and full colonels), religious leaders, civil rights leaders, media personalities, entertainment and sports figures, and others. The study found that when considering only personal acquaintances—not indirect links involving intermediaries—one-fifth of the entire national Black leadership network is included. This means that the Black leadership network is *forty times* more closely connected than White leadership networks. There are forty times more direct links in the Black network. The Black network had greater *density*. Add one intermediary, the friend of a friend, and almost *three-quarters* of the entire Black network is included. Reasons for this include the relatively smaller nationwide numbers of Black leaders, and also the sense of "family" that often exists among influential African Americans nationwide (Taylor, 1992). Any given Black leader can generally get in touch with three-quarters of all other Black leaders in the entire country, regardless of job or leadership type, either by knowing them personally (as a "friend") or via only one common acquaintance (as a "friend of a friend"). Virtually everyone in the elite network of national Black leaders can reach any other Black leader in no more than two steps. The world is as small as the networks you are in.

Social Influence

The groups in which we participate exert tremendous influence on us. We often fail to appreciate how powerful these influences are. For example, who decides what you should wear? Do you decide for yourself each morning, or is the decision already made for you by fashion designers, role models, and your peers? Consider how closely your hair length, hair styling, and choice of jewelry has been influenced by your peers. Did you invent your baggy pants, your cornrows, or your blue blazer? People who label themselves "nonconformist" often conform rigidly to the dress code of their in-group. This was true of the Beatniks in the 1950s, the hippies and Black militants of the early 1960s and 1970s, the punk rockers of the 1970s and 1980s, and the grunge and club-kid culture of the 1990s. Groups can be endlessly inventive, conjuring new styles in clothing, slang, courting habits; yet, once invented, the new styles are not merely disseminated, but also enforced.

The influences of our youth extend to adulthood. The choices of political party among adults (Republican, Democratic, or Independent) correlate strongly with the party of one's parents—the family primary group. Seven out of ten people vote with the party of their parents (Jennings and Niemi, 1974; Taylor et al., 1997); furthermore, most people share the religious affiliation of their parents.

We like to think we stand on our own two feet, immune to a phenomenon as superficial as group pressure. The conviction that one is impervious to social influence results in what social psychologist Philip Zimbardo calls the *Not-Me Syndrome:* When confronted with a description of group behavior that is disappointingly conforming and not individualistic, most individuals counter that some people may be like that, "but not me" (Zimbardo et al., 1977). But sociological experiments often reveal a dramatic gulf between what people think they will do and what they actually do.

The Bystander Intervention Problem

A branch of research termed *bystander intervention studies* examines the dynamics of when and how people come to the aid of someone in trouble. One motive for such studies is that the findings are so consistent, fascinating, and distressing. A large number of studies show that in a situation where a person on the street is in danger (getting mugged, having a heart attack, or the like), the more bystanders there are observing the episode, the *less* likely that any of them will call for the police or an ambulance (Connelly, 1989; Dovidio, 1984; Amato, 1990; Latané and Darley, 1970). Consider what this means. Practically everyone would proclaim their willingness to aid a stranger in dire need of help. The claims are sincere, yet the fund of goodwill seems to evaporate when it is a gathering rather than an individual that witnesses trouble.

There is debate about why crowds respond so poorly to someone in distress and why aid is delivered less often as crowd size increases. It may be that two bystanders each think the other will help; three bystanders are even more certain that one of the other bystanders will jump in, and so on. This is one example of the group size effect. The failure of bystanders to offer help correlates directly with an increase in group size. Pressure to conform may also contribute. As more people fail to act, the pressure to conform (that is, not help) increases. In either case, social forces cause individuals to act in a way that, under other circumstances, they would judge dishonorable, even humiliating.

THINKING SOCIOLOGICALLY

Drop some books in the street. Does anyone help you pick them up? Try it several times, noting how many people there are around. Do you notice a *bystander group size effect*? Are you less likely to get help when there are more people around?

The Asch Conformity Experiment

We see in the previous section that social influences are evidently strong enough to make us behave in ways that would cause us discomfort on later examination. Are they strong enough to make us disbelieve our own senses? In a classic piece of work known as the Asch Conformity Experiment, Solomon Asch showed that even simple objective facts cannot withstand the distorting pressure of group influence (Asch, 1951, 1955).

Examine the two figures in Figure 6.1. Which line on the right is equal in length to the line on the left? Line B, obviously. Could anyone fail to answer correctly?

In fact, Solomon Asch discovered that social pressure of a rather gentle sort was sufficient to cause an astonishing rise in the number of wrong answers. Asch lined up five students at a table and administered the test shown in the

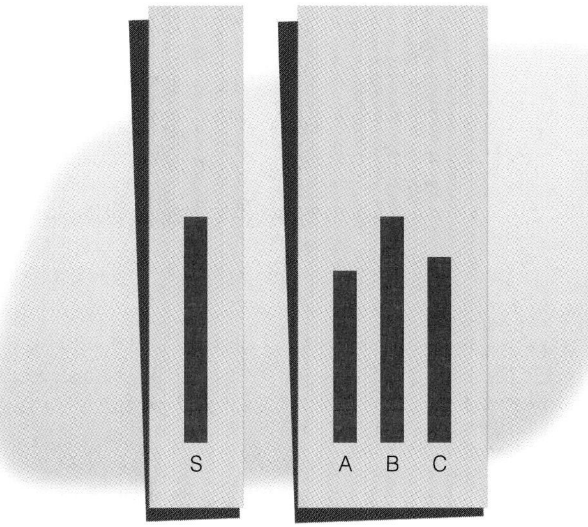

FIGURE 6.1 Lines from Asch Experiment

SOURCE: S. Asch, 1955

figure. Unknown to the fifth student, the first four were confederates—collaborators with the experimenter who only pretended to be participants. For several rounds, the confederates gave correct answers to Asch's tests. The fifth student also answered correctly, suspecting nothing. Then the first student gave a wrong answer. The second student gave the same wrong answer. Third, wrong. Fourth, wrong.

In Asch's experiment, fully *one-third* of all students in the fifth position gave the same wrong answer as the confederates at least half the time. Forty percent gave "some" wrong answers. Only one in four consistently gave correct answers in defiance of the invisible pressure to conform.

Line length is not a vague or ambiguous stimulus. It is clear and objective. Wrong answers from one-third of all subjects is a very high percentage. The subjects fidgeted and stammered while doing it, but they did it nonetheless. Those who did not yield to group pressure showed even more stress than those who yielded to the (apparent) opinion of the group.

Would you have gone along with the group? Perhaps, perhaps not. Sociological insight grows when we acknowledge the fact that a third of all participants—one in three—will yield to the group. The Asch experiment has been repeated many times over the years, with students and nonstudents, old and young, in groups of different sizes, and in different settings (Taylor et al., 1997). The results remain essentially the same. One-third to one-half of the subjects make a judgment contrary to objective fact, yet in conformity with the group.

There is also little gender difference in patterns of conformity. Women conform neither more nor less frequently than men although female subjects tend to conform to an otherwise all-male group more often than females to an all-female group; men conform to an all-female group only *slightly* less than to an all-male group. This reveals the effect of status on conformity. Individuals accorded higher status in society (in this case, males) tend to have more influence on small groups than those of lesser status (females). The same effect has been shown to some extent for other status mismatches involving race and class. **Status generalization** is the term used when the status hierarchy in a society has a measurable effect on behavior within a closed group (Berger and Zelditch, 1985).

It will probably be no surprise that group size has a systematic effect on conformity. Up to a point, the larger the group, the greater the pressure to conform. A test subject confronted with only one person giving wrong answers is unmoved by pressure to conform. The subject gives correct answers. The pressure to conform inexorably rises as the number of confederates increases, leveling off at about seven, after which the results remain steady, with a fixed percentage bending to the alleged opinion of the group.

The Milgram Obedience Studies

What are the limits of social pressure? In terms of moral and psychological issues, judging the length of a line is a small matter. What happens if an authority figure demands obedience—a type of conformity—even if the task is something the test subject finds reprehensible? A chilling answer emerged from the now-famous Milgram Obedience Studies, done from 1960 through 1974 by Stanley Milgram (Milgram, 1974).

In this study, a naive research subject entered a laboratorylike room and was told that an experiment on learning was to be conducted. The subject was to act as a "teacher," presenting a series of test questions to another person, the "learner." Whenever the learner gave a wrong answer, the teacher would administer an electric shock.

The test was relatively easy. The teacher read pairs of words to the learner, such as:

blue	box
nice	house
wild	duck

The teacher then tested the learner by reading a multiple-choice answer, such as:

blue: sky ink box lamp

The learner had to recall which term completed the pair of terms given originally, in this case, blue box.

If the learner answered incorrectly, the teacher was to press a switch on the shock machine, a formidable-looking device that emitted an ominous hum when activated (see Figure 6.2). For each successive wrong answer, the teacher was to increase the intensity of the shock by 15 volts.

The machine bore labels clearly visible to the teacher: Slight Shock, Moderate Shock, Strong Shock, Very Strong Shock, Intense Shock, Extreme Intensity Shock, Danger: Severe Shock, and lastly, XXX at 450 volts. As the voltage rose, the learner responded with squirming, groans, then screams.

The experiment was rigged. The learner was a confederate. No shocks were actually delivered. The true purpose of the experiment was to see if any "teacher" would go all the way to 450 volts. In Milgram's words:

> Observers of the experiment agree that its gripping quality is somewhat obscured in print. For the subject, the situation is not a game; conflict is intense and obvious. On one hand,

Milgram's Obedience Study

(a)

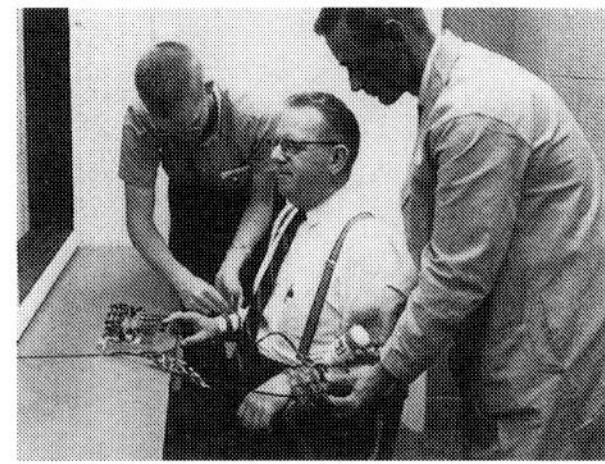
(b)

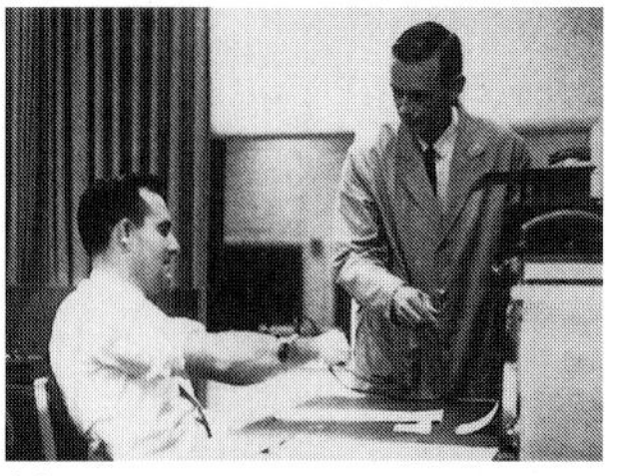
(c)

(d)

FIGURE 6.2 Milgram's Setup

These photographs show how intimidating—and authoritative—the Milgram experiment must have been. The first picture (a) shows the formidable-looking shock generator. The second (b) shows the role player, who pretends to be getting the electric shock, being hooked up. The third (c) and fourth (d) show an experimental subject (seated) and the experimenter (standing, in lab coat). The fourth picture (d) shows a subject terminating the experiment prematurely, that is, before giving the highest shock level (voltage). A large majority of subjects did not do this and actually went all the way to the maximum shock level.

SOURCE: Milgram, 1974, p. 25.

> the manifest suffering of the learner presses him to quit. On the other, the experimenter, a legitimate authority to whom the subject feels some commitment, enjoins him to continue. To extricate himself from the situation, the subject must make a clear break with authority (Milgram, 1974: 4).

If the subject tried to quit, the experimenter responded with a sequence of prods:

> "Please continue."
> "The experiment requires that you continue."
> "It is absolutely essential that you continue."
> "You have no other choice, you *must* go on."

In the first experiment, fully 65 percent of the volunteer subjects—two-thirds—went *all the way* to 450 volts on the shock machine.

Milgram was astonished. Before carrying out the experiment, he had asked a variety of psychologists, sociologists, psychiatrists, and philosophers to guess how many subjects would actually go all the way to 450 volts. The opinion of these consultants was that *one-tenth of 1 percent* (or one in a thousand) would actually do it.

What would you have done? Remember the Not-Me Syndrome. Think about the experimenter saying, "You have no other choice, you *must* go on." Most people claim they would refuse to continue as the voltage escalates. The power of this experiment derives in part from how starkly it highlights the difference between what people *think* they will do and what they *actually* do.

Milgram devised a series of additional experiments in which he varied the conditions to find out what would cause subjects *not* to go all the way to 450 volts. He moved the experiment from an impressive laboratory to a dingy basement to counteract some of the tendency for people to defer to a scientist conducting a scientific study. He had the learner complain of a heart condition. Still, well over half of the subjects delivered the maximum shock level. Speculating that women might be more humane than men (all prior experiments used only male subjects), Milgram did the experiment again using only women subjects. The results? Exactly the same. Class background made no difference. Racial and ethnic differences had no detectable effect on compliance rate.

At the time that the Milgram experiments were conceived, the world was watching the trial in Jerusalem of World War II Nazi Adolf Eichmann. Millions of Jews, Gypsies, homosexuals, and communists were murdered between 1939 and 1945 by the Nazi party, led by Adolf Hitler. As head of the Gestapo's Jewish section, Eichmann oversaw the deportation of Jews to concentration camps and the mass executions that followed. Eichmann disappeared after the war, was abducted in Argentina by Israeli agents in 1961, and was transported to Israel, where he was tried and ultimately hanged for crimes against humanity.

The world wanted to see what sort of monster could have committed the crimes of the Holocaust, but a jarring picture of Eichmann emerged. He was slight and mild mannered, not the raging ghoul that everyone expected. The psychiatrists who examined him found him to be sane. He insisted that although he had indeed been a chief administrator in an organization whose product was mass murder, he was guilty only of doing what he was told to do by his superiors. He did not hate Jews, he said. In fact, he had a Jewish half-cousin whom he hid and protected. He claimed, "I was just following orders."

How different was Adolf Eichmann from the rest of us? The political theorist Hannah Arendt dared to suggest in her book *Eichmann in Jerusalem* (1963) that evil on a giant scale is banal—that it is not the work of monsters, but an accident of civilization. Arendt argued that to find the villain, we need to look into ourselves. We may all have a little bit of Eichmann in us, in that part of us that dutifully does what we are told. The response to Arendt's thesis was heated. The controversy continues to this day.

Milgram described the dilemma revealed by his experiments as a conflict between conscience and authority. How often have you found yourself, for example, being in a group of friends when someone makes a comment or does something that you find objectionable? Do you challenge it or just shrug and go along, thinking it better to say nothing than to create conflict? Suppose, for example, a White teacher makes a racist statement to a White student; should the student say something or ignore it? Often other Whites ignore it, in effect implicitly colluding with the person who said it. In sociological terms, this and the Milgram experiments reveal the power of social influence at its most potent. Our social world is shaped by such influences. Rarely are we confronted with instances as ethically unnerving as the Milgram shock experiments, yet the consequences of social influence can affect our affairs at all levels, from family decision making to international affairs.

Debunking Society's Myths

Myth: People are just individuals who make up their own minds about how to behave.

Sociological perspective: The Asch and Milgram experiments show quite conclusively that people get profoundly influenced by group pressure, often causing them to make up their minds contrary to objective fact.

Groupthink

Wealth, power, and experience are apparently not enough to save us from social influences. **Groupthink,** as described by I. L. Janis, is the tendency for group members to reach a consensus opinion, even if that decision is downright stupid (Janis, 1982). Janis reasoned that because major government policies are often the result of group decisions, it would be fruitful to analyze the group dynamics that operate at the highest level of government—for instance, in the Office of the President of the United States. The president makes decisions based on group discussions with his advisors. The president is human and thus susceptible to group influence. To what extent have past presidents and their advisors been influenced by their mode of decision making—group consultation—rather than just the facts?

Janis investigated five ill-fated presidential decisions, all the products of group deliberation: the decision of the Naval High Command in 1942 not to prepare for attack by Japan; President Truman's decision to send troops to North Korea in 1951; President Kennedy's attempt to overthrow Cuba by launching an invasion at the Bay of Pigs in 1962; President Johnson's decision in 1967 to increase the number of U.S. troops in Vietnam; and finally, the fateful decision by President Richard M. Nixon's advisors in 1972 to rob Democratic party headquarters at the Watergate apartment complex, launching the famed "Watergate affair."

All were group decisions, and all were fiascoes. For example, the Bay of Pigs invasion was a major humiliation for the United States, a covert outing so ill conceived it is hard to imagine how it survived discussion by a group of foreign policy experts. Fifteen hundred Cuban exiles trained by the CIA to parachute into heavily armed Cuba landed in an impassibly dense 80-mile swamp far from their planned drop zone with inadequate weapons and incorrect maps. A sea landing was demolished by well-prepared, prewarned defenders. In retrospect, it is impossible to imagine how the plan, even without its tragic foul-ups, could have succeeded in toppling Cuban President Fidel Castro.

The men who advised President Kennedy to undertake the invasion were not stupid. In fact, many considered them the brightest policy team ever assembled—"the best and the brightest" as they sometimes had been called. Secretary of State Dean Rusk was a past president of the Rockefeller Foundation. Secretary of Defense Robert McNamara was a gifted statistician and past president of the Ford Motor Company. McGeorge Bundy, Special Assistant for National Security, had been Dean of Arts and Sciences at Harvard University. How could such a smart team perpetrate such a blunder?

Janis discovered a common pattern of misguided thinking in his investigations of presidential decisions. He surmised that outbreaks of groupthink had several things in common:

1. *An illusion of invulnerability.* "With such a brilliant team, and such a nation, how could *any* plan fail," thought those in the group.
2. *A falsely negative impression of those who are antagonists to the group's plans.* Fidel Castro was perceived to be clownish, and Cuban troops were supposed to be patsies. In truth, the defenders at the Bay of Pigs were actually Russian-trained troops, and Castro has remained in power to this day, several decades after the failed invasion.
3. *Discouragement of dissenting opinion.* As groupthink takes hold, dissent is equated with disloyalty. This can even discourage dissenters from voicing their objections.
4. *An illusion of unanimity.* In the aftermath, many victims of groupthink recall their reservations, but at the moment of decision, there is a prevailing sense that the entire group is in complete agreement.

Groupthink is not inevitable when a team gathers to make a decision, but it is common and appears in all sorts of groups, from student discussion groups to the highest councils of power (Flowers, 1977; McCauley, 1989; Aldag and Fuller, 1993). A striking illustration of the pressure toward consensus is seen in the behavior of juries. Four polling methods are used for most jury decisions: verbal go-around, a show of hands, verbal dissent only, and secret ballot. Of these, secret ballot is the most likely to result in deadlock, presumably because anonymity absolves jurors of their perceived responsibility to create a consensus even if they have doubts (Hawkins, 1960).

Debunking Society's Myths

Myth: A group of experts brought together in a small group will solve a problem according to their collective expertise.

Sociological perspective: Groupthink can lead even the most qualified people to make disastrous decisions since people in groups in the United States tend to reach consensus at all costs.

groupthink

Risky Shift

The term *groupthink* is commonly associated with group decision making whose consequences are not merely unexpected but disastrous. Another group phenomenon, **risky shift** (also called *polarization shift*), may help explain why the products of groupthink are frequently calamities. Have you ever found yourself in a group engaged in a high-risk activity that you would not do alone? When you created mischief as a child, were you not usually part of a group? If so, you were probably in the thrall of risky shift—the tendency for groups to weigh risk differently than individuals.

Mr. A, an electrical engineer, who is married and has one child, has been working for a large electronic corporation since graduating from college 5 years ago. He is assured of a lifetime job with a modest, though adequate, salary, and liberal pension benefits upon retirement. On the other hand, it is very unlikely that his salary will increase much before he retires. While attending a convention, Mr. A is offered a job with a small, newly founded company that has a highly uncertain future. The new job would pay more to start and would offer the possibility of a share in the ownership if the company survived the competition of the larger firms.

Imagine that you are advising Mr. A. Listed here are several probabilities or odds of the new company's proving financially sound.

Please check the lowest probability that you would consider acceptable to make it worthwhile for Mr. A to take the job.

— The chances are 1 in 10 that the company will prove financially sound.
— The chances are 3 in 10 that the company will prove financially sound.
— The chances are 5 in 10 that the company will prove financially sound.
— The chances are 7 in 10 that the company will prove financially sound.
— The chances are 9 in 10 that the company will prove financially sound.
— Place a check here if you think Mr. A should not take the new job no matter what the probabilities.

FIGURE 6.3 Risky Shift Questionnaire

Wallach, Michael A., Nathan Cogan and Darryl Berry. 1968. "Group Influence on Individual Risk Taking," Darwin Cartwright and Alvin Zander, eds. *Group Dynamics: Research and Theory*, 3rd ed., New York: Harper & Row. p. 432.

Risky shift was first observed by James Stoner (1961). Stoner gave study participants descriptions of a situation involving risk, such as the one in Figure 6.3, in which an engineer must choose between job security and potentially lucrative but risky advancement.

The participant is then asked to decide how much risk the engineer should take. Before performing his study, Stoner believed that individuals in a group would take *less* risk in a group than individuals alone, but he found that after his groups had engaged in open discussions, they favored greater risk than before discussion. His research stimulated hundreds of studies using males and females, different nationalities, different tasks, and other variables (Pruitt, 1971; Johnson et al., 1977; Blaskovitch, 1973; Hong, 1978; Taylor et al., 1997). The results are complex. Most but not all group discussion leads to greater risk-taking. In subcultures that value caution above daring, as in some work groups of Japanese and Chinese firms, group decisions are *less* risky after discussion than before. The shift can occur in either direction, driven by the influence of group discussion, but there is generally some kind of shift, in one direction or the other, rather than no shift at all.

What causes risky shifts? The most convincing explanation is that *deindividuation* occurs. Deindividuation is the sense that one's self has merged with a group. In terms of risk-taking, one feels that responsibility (and possibly blame) is borne not only by oneself but by the group. The greater the number of people in a group, the greater the tendency toward deindividuation. In other words, deindividuation is a group size effect. As groups get larger, trends in risk-taking are amplified.

Streaking, or running nude in a public place, is more common as a group activity than as a strictly individual one. This illustrates how the group can provide the persons in it with deindividuation, or merging of self with group. This allows the individual to feel less responsibility or blame for his or her actions.

THINKING SOCIOLOGICALLY

Think of a time when you engaged in some risky behavior. What group were you part of, and how did the group influence your behavior? How does this illustrate the concept of *risky shift*? Is there more risky shift with more people in the group?

Formal Organizations and Bureaucracies

A **formal organization** is a large secondary group, highly organized to accomplish a complex task or tasks and to achieve goals in an efficient manner. Many of us belong to various formal organizations: work organizations, schools, political parties, to name a few. Organizations are formed to accomplish particular tasks and are characterized by their relatively large size, compared, for example, to a small group like a family or a friendship circle. Often organizations consist of an array of other organizations. The federal government is a huge organization, comprising numerous other organizations, most of which are also vast. Each organization within the federal government is also designed to accomplish specific tasks, be it collecting your taxes, educating the nation's children, or regulating the nation's transportation system and national parks.

To sociologists, organizations are fascinating examples of how behavior becomes embedded in social structures. An organization is in some sense only a collection of people, but organizations develop cultures and routinized practices. The culture of an organization may be reflected in certain symbols, values, and rituals. Some organizations develop their own language and styles of dress. The norms can be subtle, such as men being expected to wear long sleeve shirts or women being expected to wear stockings, even on hot summer days. It does not take explicit rules to regulate this behavior; comments from co-workers or bosses may be enough to enforce such organizational norms. Some work organizations have recently instituted a practice called "casual day"—one day per week, usually Friday, when workers can dress less formally. Casual day has its own set of organizational norms—norms that retailers have capitalized upon by developing new lines of clothing that are marketed as office "casual wear"—and at quite a price to workers!

Within organizations, people typically conform to expected patterns of behavior. Secretaries do not challenge bosses, at least not openly; leaders encourage loyalty to the organizations, sometimes by expecting financial contributions to the organization, such as by expecting college alumni to donate money to the school. Newcomers in the organization are expected to learn the organization's rules. New student orientations at colleges and universities and management training seminars in work organizations are examples of this.

Organizations tend to be persistent although they are also responsive to the broader social environment where they are located (DiMaggio and Powell, 1991). Organizations are frequently under pressure to respond to changes in the society, incorporating new practices and beliefs into their structure as society itself changes. Business corporations, as an example, have had to respond to increasing global competition; they do so by expanding into new international markets, developing a globally focused workforce, and trimming costs by "downsizing" (that is, eliminating workers and various layers of management).

Organizations can be tools for innovation, depending on the organization's values and purpose. Rape crisis centers are examples of organizations that originally emerged from the women's movement because of the perceived need for services for rape victims. Rape crisis centers have, in many cases, changed how police departments and hospital emergency personnel respond to rape victims. By having advocated changes in rape law and services for rape victims, rape crisis centers have generated change in other organizations as well (Fried, 1994; Schmitt and Martin, 1995).

Types of Organizations

Sociologists Blau and Scott (1974) and Etzioni (1975) classify formal organizations into three categories, distinguished by their types of membership affiliation: normative, coercive, and utilitarian.

Normative Organizations. People join normative organizations to pursue goals that they consider personally worthwhile. They obtain personal satisfaction but no monetary reward for being in such an organization. In many instances, the person joins the normative organization for the social prestige that it offers him or her. Many are service and charitable organizations. Such organizations are often called **voluntary organizations**, and they include organizations such as the PTA, Kiwanis clubs, political parties, religious organizations, the NAACP, B'Nai Brith, LaRaza, and other similar voluntary organizations that are concerned with specific issues. Civic and charitable organizations (such as the League of Women Voters) and political organizations (such as the National Organization of Women, also known as NOW) reflect the fact that for decades women have been excluded from traditionally all-male voluntary organizations, such as the Kiwanis and Lions. Like other service and charitable organizations, these have been created to meet particular needs, ones that members see as not being served by other organizations.

Gender, class, race, and ethnicity all play a role in who joins what voluntary organizations. Social class is reflected in the fact that many people do not join certain organizations simply because they cannot afford to join; joining a professional organization, as one example, can cost hundreds of dollars each year. Those who feel disenfranchised, however, may join grassroots organizations—voluntary organizations that spring from specific local needs that people think are unmet. A tenants' organization may form to protest rent increases or lack of services, or a new political party may emerge from people's sense of alienation from existing party organizations. African Americans, Latinos, and Native Americans have formed many of their own voluntary organizations, in part because of their historical exclusion from traditional White voluntary organizations. African American fraternities and sororities are a classic example. They emerged as select organizations of African American college students. Like White fraternities and sororities, they have a strong service orientation, but unlike White groups, they were created in a context of racial inequality and have long taken racial justice as a central organizing theme (Giddings, 1988, as illustrated in the box "Understanding Diversity" on the Delta Sigma Theta sorority).

The NAACP (National Association for the Advancement of Colored People), founded in 1909 by W. E. B. Du Bois (recall him from Chapter 1), and the National Urban League, are two other large national organizations that have historically fought racial oppression on the legal and urban fronts, respectively. La Raza Unida, a Latino organization devoted to civic activities as well as combating racial–ethnic oppression, has a large membership with Latinas holding major offices. In fact, such voluntary organizations dedicated to the causes of people of color have in recent years had more women in leadership positions in the organizations than have many standard White organizations. Similarly, Native American voluntary organizations such as AIM (American Indian Movement) have recently boasted increasing numbers of women in leadership positions (Feagin and Feagin, 1993; Snipp, 1989).

Coercive Organizations. Coercive organizations are characterized by membership that is largely involuntary. Prisons are an example of organizations that people are coerced to "join" by virtue of their being punished for a crime. Similarly, mental hospitals are coercive organizations: People are placed in them, often involuntarily, for some form of psychiatric treatment. In this respect, prisons and mental hospitals are similar in many respects in their treatment of inmates or patients. They both have strong security measures such as guards, locked and barred windows, and high walls (Goffman, 1961; Rosenhan, 1973). Sexual harassment and sexual victimization is common in both prisons and mental hospitals.

The sociologist Erving Goffman has described coercive organizations as total institutions. A **total institution** is an organization cut off from the rest of society where individuals who reside there are subject to strict social control (Goffman, 1961). Total institutions include two populations: the "inmates" and the staff. Within total institutions, the staff exercises complete power over inmates (for example, nurses over mental patients, guards over prisoners). The staff administers all the affairs of everyday life, including basic human functions like eating and sleeping. Rigid routines are characteristic of total institutions, thus explaining the common complaint by those in hospitals that they cannot sleep since nurses routinely enter their rooms at night, whether or not the patient needs treatment.

Formal rules of behavior are also typical of total institutions: How you dress, when you eat, even when you use the bathroom are all regulated by those in charge. Goffman wrote that individuals placed in total institutions completely surrender their rights to privacy and are frequently subjected to what he called *mortification of the self*—actions taken by the staff to make the inmates surrender their former identity, such as by expecting or requiring them to dress in public, turn over personal possessions, or urinate in bedpans. Also called "degradation ceremonies," Goffman described such actions as intended to support the organizational structure, even if at the cost of human dignity.

Utilitarian Organizations. The third type of organization named is utilitarian. These are large organizations, either for profit or nonprofit, that are joined by individuals for specific purposes, such as monetary reward. Large business organizations that generate profits (in the case of for-profit organizations) and salaries and wages for their employees (as with either for-profit or nonprofit organizations) are these kinds of organizations. Examples of large for-profit organizations include General Motors, Microsoft, and Procter & Gamble. Examples of large nonprofit organizations that pay salaries to employees are colleges, universities, and the organization

BOX 6.2

UNDERSTANDING DIVERSITY

The Deltas: Black Sororities as Organizations

DELTA Sigma Theta sorority is one of the largest Black women's organizations in the world, with more than 125,000 members. Its membership has included such significant Black women leaders as Mary Church Terrell, Sadie T. M. Alexander, Patricia Roberts Harris, Barbara Jordan, Leontyne Price, and Dorothy Height. Like other Black women's sororities, the unique history of race and gender discrimination gives this organization an identity very different from that of White sororities. In her history of Delta Sigma Theta, whose members are referred to as "Deltas," Paula Giddings, a contemporary African American scholar, writes that organizations:

> . . . must be able to adapt to changing environments: in this case, the ever changing exigencies of race relations and the attitudes toward women. Consequently, Delta has had to alter its purposes and goals throughout the years—and must continue to do so—as well as its internal structure to accommodate them. This makes the Black sorority a particularly dynamic organization.
>
> At the same time, however, changes cannot occur too abruptly or without the consensus of an increasingly diverse constituency. For like other social movement organizations, its viability is dependent on the growth of its membership, which in turn, is largely determined by the number of its members who feel that the sorority's goals are in harmony with their own. This makes it complex as well as concentrated, it invites apathy—the kiss of death for a social movement organization.
>
> The challenge of the sorority, one made all the more difficult by the pathos of the Black women's experience in North America, is to maintain that sense of sisterhood while striving, organizationally, for a more general purpose: aiding the Black community as a whole through social, political, and economic means. As one can see from the rules that govern the social movement organization, this idea can be a difficult one to realize. But the effort to resolve the tension between the goals of organization, and those of the sisterhood, through strengthening social bonds within the context of social action has been an interesting and engaging experiment. It is one that can be seen as a model for Black organizational life, and which adds contour and dimension to the history of Black women in this country.

SOURCE: Giddings, Paula. 1994. *In Search of Sisterhood: Delta Sigma Theta and the Challenge of the Black Sorority Movement.* New York: William Morrow.

that manufactures the Scholastic Assessment Ability Test (SAT), Educational Testing Service (ETS). Joining a utilitarian organization is usually a matter of choice although people may join to make a living.

Some utilitarian organizations may also be coercive organizations. This is especially the case as various organizations have become increasingly privatized. Mental hospitals may be owned by a large for-profit chain; although now run by states or the federal government, some have suggested that prisons should become privatized, and some cities, such as Hartford, Connecticut, have turned their public schools over to private corporations to see if they can be better managed.

Bureaucracy

As formal organizations develop, many develop into a **bureaucracy**, a type of formal organization characterized by an authority hierarchy, a clear division of labor, explicit rules, and impersonality. Bureaucracies are notorious for their unwieldy size and complexity, as well as for their reputation of being remote and cumbersome organizations that are highly impersonal and machinelike in their operation. The federal government is a good example of a cumbersome bureaucracy that many see as overwhelmed by its sheer size. Numerous other formal organizations have developed into huge bureaucracies: IBM, Disney, the Environmental Protection Agency, universities, hospitals, state motor vehicle registration systems, some law firms, and many of the organizations in which most people work and many people have to conduct their affairs.

The early sociological theorist Max Weber (1925 [1947]) analyzed the classic characteristics of the bureaucracy. These characteristics represent what he called the *ideal type* bureaucracy—a model that is rarely seen in reality but that defines the typical characteristics of a social form. The characteristics of bureaucracies are:

1. *High degree of division of labor and specialization.* The notion of the specialist embodies this criterion. Bureaucracies ideally employ specialists in the various positions and occupations, and these specialists are responsible for a specific set of duties. Job titles and job descriptions define the nature of each such position. Sociologist Charles Perrow (1994, 1986) notes that many modern-day bureaucracies have hierarchical authority structures and an elaborate division of labor.

2. *Hierarchy of authority.* In bureaucracies, positions are arranged in a hierarchy so that each position is under the supervision of a higher position. Such hierarchies are often represented in an organizational chart, a diagram in the shape of a pyramid that shows relative rank of each position plus the lines of authority between each. These lines of authority are often called the "chain of command," and show not only who has authority over whom, but who is responsible to whom, and how many positions are responsible to a given position.

3. *Rules and regulations.* All the activities in a bureaucracy are governed by a set of detailed rules and procedures. These rules are designed, ideally, to cover almost every possible kind of situation and problem that might arise, including hiring, firing, salary scales, rules for sick pay and absences, and the everyday operation of the office.

4. *Impersonal relationships.* Social interaction in the bureaucracy is supposed to be guided by instrumental criteria, such as the organization's rules, rather than by social or emotional criteria, such as personal attractions or likes and dislikes. The ideal is that the application of rules objectively will minimize such matters as personal favoritism—such as giving someone a promotion simply because you like him or her a lot, or firing someone because you do not like him or her personally. Of course, as we will see, sociologists have pointed out that bureaucracy has another face—the social interaction that keeps the bureaucracy working and that often involves interpersonal friendships and social ties, typically among those who are taken for granted in these organizations, such as secretaries (Page, 1946).

5. *Career ladders.* Candidates for the various positions in the bureaucracy are supposed to be selected on the basis of specific criteria, such as education, experience, and standardized examinations. The idea is that advancement through the organization becomes a career for the individual. Theoretically, one advances as one achieves the skills associated with the bureaucracy although movement in an organization is influenced by many factors, including one's gender, race, level of education, class background, and age, to name some of the more salient.

6. *Efficiency.* Bureaucracies are designed to coordinate the activities of a large number of people in pursuit of organizational goals. Ideally, all activities have been designed to maximize this efficiency. The whole system is intended to keep social emotional relations and interactions at a minimum and instrumental interaction at a maximum.

These characteristics of bureaucracies have given them the reputation of being huge, remote organizations, more intent on perpetuating themselves than serving people's needs. There seems to be some tendency for organizations to become bureaucratized as they grow bigger. This reflects the impulse toward greater specialization in modern society and the move to greater efficiency.

Bureaucracy's Other Face

All the preceding characteristics of Weber's "ideal type" are general defining characteristics. Rarely do actual bureaucracies meet this exact description. This is because bureaucracy has, in addition to the ideal characteristics of structure, an *informal structure*—social interactions in bureaucratic settings that ignore, change, or otherwise bypass the formal structure and rules of the organization. Sociologist Charles Page (1946) coined the term *bureaucracy's other face* to describe this condition.

This other face is an informally evolved culture. It has evolved over time as a reaction to the formality and impersonality of the bureaucracy; thus, secretaries "bend the rules a bit" when asked to do something more quickly than usual for a boss they like and bend the rules in another direction

bureaucracy

for a boss they do not like by slowing down or otherwise sabotaging the boss's work. As a way around the cumbersome formal communication channels within the organization, the informal network or "grapevine" often works better, faster, and sometimes even more accurately. As with any culture, the informal culture in the bureaucracy has its own norms or rules. One is not supposed to "stab friends in the back," as by "ratting on" them to a boss or spreading a rumor about them that is intended to somehow hurt them, such as by getting them fired. Just as with any norms, there is deviation from the norms, and stabbing someone in the back is not uncommon.

Bureaucracy's other face can also be seen in the workplace subcultures that develop, even in the most massive bureaucracies. Some sociologists interpret the subcultures that develop within bureaucracies as people's attempts to humanize an otherwise impersonal organization. Keeping photographs of one's family and loved ones in the office, placing personal decorations on one's desk (if you are allowed), and organizing office parties are some of the ways that people resist the impersonal culture of bureaucracies. Of course, this informal culture can also become exclusionary, increasing the isolation that some workers feel at work. Gay and lesbian workers may feel left out when other workers gossip about people's dates or discuss family weddings; minority workers may be excluded from the casual conversations in the workplace that connect people to one another. In male-dominated organizations, women may be left out of the informal banter that signifies men's belonging in the organization.

Sexual harassment can also become an aspect of the informal culture of the workplace. Sexual harassment has occurred if someone in a higher status within the organization makes unwanted sexual overtures, verbal or nonverbal, toward someone in a relatively lower status position. Sexual harassment is certainly not part of the formal Weberian model. Sexual harassment used to be thought of as an unharmful and sometimes playful mode of interacting, but recent years have seen increasing formalization of the rules about just what constitutes sexual harassment and what does not; nonetheless, it is a major problem in our society (Langelou, 1993; for further discussion of sexual harassment, see Chapter 17).

The informal norms that develop within the modern-day bureaucracy often cause worker productivity to go up or down, depending on the norms and how they are informally enforced. The classic Hawthorne Studies, so named because they were carried out at the Western Electric telephone plant in Hawthorne, Illinois, in the 1930s (Roethlisberger and Dickson, 1939), discovered that small groups of workers developed their own ideas—their own norms—about how much work they should produce each day. The formal rules of the organization specified exactly what they were supposed to be producing, but the informal norms were developed to circumvent the formal norms. In the original study, male workers in what was called the "bank wiring observation room" were supposed to produce a certain number of completed wiring boards or "banks" per day. If someone produced too many such completed tasks in a day, he would make the rest of the workers "look bad" and thus run the risk of having the organization raise its expectations of how much work the group might be expected to produce. Because of this, anyone producing too much was informally labeled a "rate buster," and that person was punished by some act, such as punches on the shoulder (called "binging") or by group ridicule ("razzing"). By the same token, one could be accused of producing too little, in which case he was labeled a "chiseler" and punished in the same way by either binging or razzing.

Such norms, and the punishments or sanctions that go along with them, are part of the developed informal organizational culture. As with any culture, it is complex with highly developed systems of norms and sanctions. The informal culture, bureaucracy's other face, continues today in a manner not unlike that culture initially discovered in the early Hawthorne studies (Perrow, 1986).

Problems of Bureaucracies

The criteria just given for the classic organization or bureaucracy represent the ideal. Problems have, however, developed in contemporary times that grow out of the nature of the complex bureaucracy. Two of these problem areas have already been discussed, with respect to the groups that form in organizations: the occurrence of risky shift in work groups and the development of groupthink. Additional problems include a tendency to ritualism and the potential for alienation on the part of those within the organization.

Ritualism. Rigid adherence to rules can produce a slavish following of them, whether or not it accomplishes the purpose for which the rule was originally designed. The rules become ends in themselves rather than means to an end. Often the filling out of endless paper forms is an example of ritualism.

A now classic example of the consequences of *organizational ritualism* has come to haunt us: the explosion of the Space Shuttle *Challenger* on January 28, 1986. People in the United States became bound together at that single, historic moment. Many remember where they were and exactly what they were doing when they heard about the tragedy. The failure of the now-famous "O-ring" gaskets on the solid fuel booster rockets of the shuttle caused the catastrophic explosion. It was revealed later that the O-rings became brittle at below-freezing temperatures, as was the temperature at the launch pad the evening before the *Challenger* took off.

Why did the managers and engineers at NASA (National Aeronautics and Space Administration) allow the shuttle to take off given these prior conditions? The managers had all the information about the O-rings prior to the launch; furthermore, they had been warned against it by engineers. Diane Vaughan (1996), in a detailed analysis of the decision to launch involving extensive interviews with the managers and engineers who were directly involved, uncovered both risky shift and organizational ritualism within the organization. Vaughan documents the incremen-

The horror of the explosion of the Space Shuttle in 1986 is seen in the faces of the observers on the left. All seven astronauts died in the explosion. Sociological researchers such as Diane Vaughan (1996) attribute the disaster to an ill-formed launch decision based on group interaction phenomena such as risky shift, ritualism, groupthink, and the routinization of deviance.

tal descent into poor judgment supported by a culture of high-risk technology. The NASA insiders, confronted with signals of danger, proceeded as if nothing was wrong when they were repeatedly faced with the evidence that something was indeed *very* wrong. They in effect *normalized* their own behavior, an occurrence of risky shift, so that their actions became acceptable to them, representing nothing out of the ordinary. This is an example of organizational ritualism.

No safety rules were actually broken. No single individual was at fault. The story is not one of evil but of the ritualism of organizational life in one of the most powerful bureaucracies in the United States. It is a story of rigid group conformity within an organizational setting and of how deviant behavior is redefined, that is, socially constructed, to be perceived as normal. Organizational procedures in this case, or rituals, come to dominate such that the means toward goals become the goals themselves. Vaughan's analysis is a powerful warning about the hidden hazards of bureaucracy in a technological age.

Alienation. The stresses on rules and procedures within bureaucracies can result in a decrease in the overall cohesion of the organization, with the result that the individual becomes psychologically separated from the organization and its goals. This is a state of what is called alienation. Alienation results in increased turnover, tardiness, absenteeism, and overall dissatisfaction with the organization.

Alienation can be widespread in organizations where workers have little control over what they do, or where workers themselves are treated like machines—employed on an assembly line, doing the same repetitive action for an entire work shift. Alienation is not restricted to manual labor, however. In organizations where workers are isolated from others, where they are expected only to implement rules, or where they think they have little chance of advancement, alienation can be common (see Chapter 17). As we will see, some organizations have developed new patterns of work to try to minimize worker alienation and, therefore, enhance their productivity. These new forms of organization are attempts to overcome some of the problems associated with excessive bureaucratization.

The McDonaldization of Society

Sometimes the problems and peculiarities of bureaucracy can have effects on the total society. Such has been the case with what George Ritzer (1996) has called the *McDonaldization of society,* a term coined from the well-known fast-food chain. Ritzer noticed that the principles that characterize fast-food organizations are increasingly coming to dominate more and more aspects of U.S. society, indeed, of societies around the world. "McDonaldization" refers to the increasing and ubiquitous presence of the fast-food model in most of the organizations that shape daily life; work, travel, leisure, shopping, health care, education, and politics have all become subject to McDonaldization. Each of these industries is based on a principle of high and efficient productivity, based on a highly rational social organization, with

Evidence of the "McDonaldization of society" can be seen everywhere, perhaps including on your own campus. Shopping malls, food courts, sports stadiums, even cruise ships reflect this trend toward standardization. Ask yourself, how does this process affect the development of culture and social structure?

workers employed at low pay, but customers experiencing ease, convenience, and familiarity.

McDonaldization can be seen in the permeation of shopping malls and fast food into every aspect of life. Fast-food restaurants not only spring up around campus communities, but increasingly campus student centers look like fast-food courts in shopping malls. (The Marriott Corporation now provides food to more than 500 colleges and universities.) McDonald's and other fast foods are now found at highway rest stops, airports, subway stations, and even on some airline flights.

Ritzer argues that McDonald's has been such a successful model of business organization that other industries have adopted the same organizational characteristics, so much so that their nicknames associate them with the McDonald's chain: McPaper for *USA Today,* McChild for child-care chains like Kinder-Care, McDoctor for the drive-in clinics that deal quickly and efficiently with minor health and dental problems.

Ritzer identifies four dimensions of the McDonaldization process: efficiency, calculability, predictability, and control:

1. *Efficiency* means that things move from start to completion in a streamlined path. Steps in the production of a hamburger are regulated so that each hamburger is made exactly the same way—hardly characteristic of a home-cooked meal. Business can be even more efficient if the customer does the work once done by an employee; thus, automated teller machines mean that customers do the work once done by tellers (and without the personal contact). In fast-food restaurants, the claim that you can "have it your way" really means that you assemble your own sandwich or salad.

2. *Calculability* means there is an emphasis on the quantitative aspects of products sold—size, cost, and the time it takes to get the product. At McDonald's, branch managers must account for the number of cubic inches of ketchup used per day; likewise, ice cream scoopers in chain stores measure out predetermined and exact amounts of ice cream, unless machines do it for them. Workers are monitored for how long it takes them to complete a transaction; every bit of food and drink is closely monitored by computer, and everything has to be accounted for. Employee training manuals may even instruct workers to greet customers within three seconds of their entering the shop. Indeed, Ritzer argues, quantity comes to replace quality, and businesses distinguish themselves by the speed with which they can deliver products, not the quality of their goods.

3. *Predictability* is the assurance that products will be exactly the same, no matter when or where they are purchased. Eat an Egg McMuffin in New York, and it will taste just the same as an Egg McMuffin in Los Angeles or Paris! The setting where you eat it will also be much the same, and since workers too are predictable in this model, you may be met with a carefully scripted greeting—one that is seemingly personal, but uniformly followed—as if Dave Thomas, owner of Wendy's restaurant, was about to join you for your burger at any moment!

4. *Control* is the primary organizational principle that lies behind McDonaldization. People's behavior, both customers and workers alike, is reduced to a series of machine-like actions. Ultimately, efficient technologies replace much of the work that humans once did. People are also carefully monitored and watched in these organizations since uncertainty in human behavior will produce inefficiency and unpredictability. At one national credit card chain, managers routinely listen in on telephone calls being handled by service workers; in other settings, computers might monitor the speed with which workers handle a particular function. Sensors on drink machines can actually cut off the liquid flow to ensure that each drink is exactly the same size.

McDonaldization clearly brings many benefits. There is a greater availability of goods and services to a wide proportion of the population, instantaneous service and convenience to a public with less free time, predictability and familiarity in the goods bought and sold, and standardization of pricing and uniform quality of goods sold, to name a few. However, this increasingly rational system of goods and services also spawns irrationalities. Ritzer argues that, as we become more dependent on the familiar and taken for granted, there is the danger of dehumanization. People lose their creativity, and there is little concern with the quality of goods and services, thereby disrupting something fundamentally human—the capacity for error, surprise, and imagination. Even with increasing globalization and the opportunities it provides to expose ourselves to diverse ways of life, McDonaldization has come to characterize other societies, too. The tourist can travel to

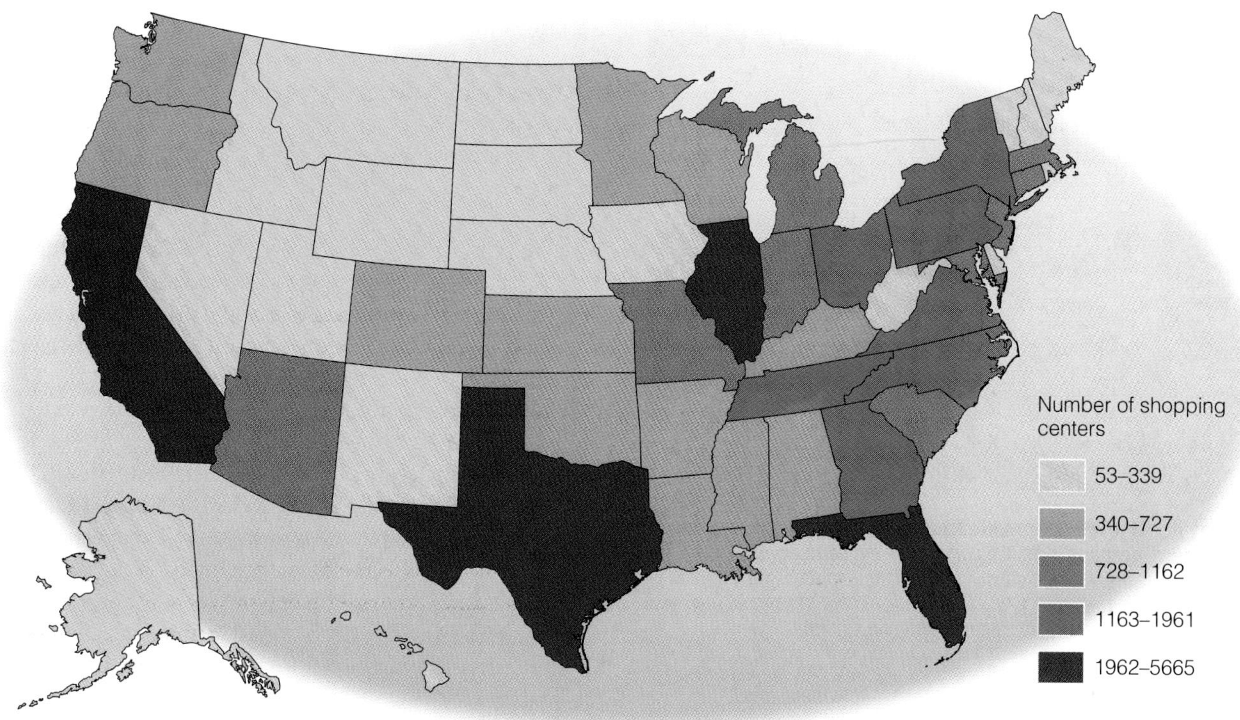

MAP 6.2 Mapping America's Diversity: The McDonaldization of Society

DATA: From the Bureau of the Census, 1997. *Statistical Abstract of the United States* Washington DC: Government Printing Office, p. 776.

the other side of the world and taste the familiar Chicken McNuggets or a Dunkin' Donut!

Ritzer builds his theory from the classical perspective of Max Weber who predicted that, as society progressed, it would become increasingly characterized by formal rationality. Weber meant that, as history proceeded, people's behavior would be increasingly guided by rational systems—rules, regulations, and formal structures—that no longer necessitated people being guided by abstract value systems from which they would decide how to act. Instead, people would follow efficient means to an end. Weber saw danger in this tendency, however. The "iron cage of rationality," as he called it, would trap people as they moved from one rational organization to another—from rational educational organizations to rational work organizations to rational recreational settings. Ultimately, Weber thought, these structures would deny people their creativity and humanity since the ultimate rational system is also one where human labor is replaced by machines.

Notice, as you go through your daily life, the extraordinary presence of the process Weber and Ritzer have observed. In what areas of life do you see the process of McDonaldization? How has it influenced the national culture? In a society with as much cultural diversity as the United States, what does it mean that we have developed a system for the production and distribution of goods that relies on such uniformity? These questions will help you see how McDonaldization has permeated U.S. society and will help you think about formal organizations as a sociologist would.

New Global Organizational Forms: The Japanese Model

The problem of bureaucratization and increasing rationality in social systems has encouraged some to look beyond Western society bureaucracy for new organizational forms that will potentially avoid some of the problems of dehumanization that sociologists have noted (Perrow, 1994, 1986). At one time, managers used something called *Theory X* (Argyris, 1990) to guide organizational forms. Theory X characterized U.S. bureaucracies and postulated that by nature people hate their jobs and want to avoid working and responsibility; this theory suggested that people do not care about the needs of the organization, only about their own personal needs. As a result, according to this perspective, people will work only when they are rigidly controlled and monitored, as with time clocks, and with a system of promotions and salaries. This organizational model lies behind the classical bureaucracy and, though not so crudely put, still dominates in many work organizations and management perspectives in the United States and parts of western Europe.

Other organizational perspectives, known as *Theory Y*, suggest that people are passive and irresponsible because of their experience with the organization. By nature, people have a desire to work, and also to be creative and to take

A Japanese work group, illustrating close and expressive interaction despite an outwardly formal appearance.

on responsibility. This theory predicts that an organization can operate efficiently and produce if the self-direction of the individual, called *self-actualization,* is permitted expression (Argyris, 1990). Management's task, under Theory Y, is to organize things so that people can accomplish their own personal goals while furthering the organization's goals as well.

In recent years, U.S. competition with Japan has spawned interest in Japanese styles of management. *Theory Z,* as it was first known (Ouchi, 1981), is an organizational form used by many Japanese corporations and adopted by some U.S. companies. This perspective values long-term employment, interpersonal trust, and above all, close personal relationships. In this respect, Theory Z advocates the direct opposite of what the classical Western bureaucracy and Theory X advocate. Theory Z places considerable emphasis upon primary group relationships; thus it has created a form of work known as "quality circles," in which small groups of workers meet periodically with managers to discuss ways the organization can work better.

By this style of management, managers and their companies are encouraged to take a long-term orientation toward their problems and products. Theory Z encourages personal friendships among co-workers. Managers are encouraged to move around to different organizational positions rather than become narrow specialists. Finally, they are encouraged to manage by "walking around," and develop closer relationships with other employees. In sum, Theory Z attempts to reduce the common problems of ritualism and alienation.

This style of management emerges from the particular characteristics of Japanese culture, one that emphasizes lifelong employment in the same company, company loyalty, and complete attachment to one's work. Whether it can be successfully transported to the United States and other Western countries, with their unique cultural traditions, is questionable. Sociological research finds some support for the idea that workers are less alienated and experience greater solidarity among themselves in firms that practice more participative management, compared to workers where all power rests with management. Participative management does not, however, bring workers the same levels of satisfaction as forms of work where workers make their own decisions based on their skills and long-term employment (Hodson, 1996).

Current pressures on U.S. business to enhance profitability through mergers, downsizing, and speedup in productivity, militate against the styles of management associated with Japanese culture. Whereas once workers would be hired into an organization with the expectation that they would be employed there for life, this is no longer true. "Lean and mean," as the popular business slogan proclaims, is highly incompatible with a form of organization that values human relationships and strong ties to other workers.

Diversity: Race, Gender, and Class in Organizations

The hierarchical structuring of positions within organizations results in the concentration of power and influence with a few individuals at the top. Since organizations tend to reflect patterns within the broader society, this hierarchy, like that of society, is marked by inequality in race, gender, and class relations. Although the concentration of power in organizations is incompatible with the principles of a democratic society (Perrow, 1986), discrimination against women and minorities still occurs, as was vividly revealed in 1996 when senior executives at Texaco were secretly taped during a company meeting using strong racial epithets. Texaco, like other major organizations, was scrutinized for having a leadership hierarchy that was discriminatory toward African Americans. The Department of Labor found widespread disparities at Texaco in the promotion rate for White and Black workers—a pattern that is repeated in most work organizations (Eichenwald, 1996; Collins, 1989; McGuire and Reskin, 1993). Only days after the racial epithets were revealed in the national press, Texaco settled for $115 million a major racial discrimination suit brought by Black employees.

Traditionally, within organizations, the most powerful positions are held by White men of upper social class status. Women and minorities, on average, occupy lower positions in the organization. Although very small numbers of minorities and women do get promoted, there is typically a "glass ceiling" effect, meaning that women and minorities may be promoted but only up to a point. The glass ceiling acts as a barrier to the promotion of women and minorities into higher ranks of management, as discussed further in Chapter 17. What are the barriers that prevent more inclusiveness in the higher ranks of organizations?

Sociological research finds that organizations are sensitive to the climate in which they operate. The more egalitarian the environment in which a firm operates, the more equitable its treatment of women and minorities will be. In particular, sociologists have found that those industries with the strongest ties to the national government tend to be more favorable in their treatment of women and minorities. Put a little differently, the greater the involvement of the federal government in a given industry, the more favorable jobs and earnings are for Black men and women and White women (Beggs, 1995). This suggests that organizations respond to external pressures and that a federal climate that emphasizes gender and race equality will encourage organizations to take a more proactive role in hiring and promoting women and minorities.

Other studies find that patterns of race and gender discrimination persist throughout organizations, even when formal barriers to advancement have been removed. Many minorities are now equal to Whites in education, particularly in organizational jobs that require advanced graduate degrees such as the Master of Business Administration (MBA). Still, White men in organizations are more likely to receive promotions than African American, Hispanic, and Native American workers with the *same* education (DeWitt, 1995; Zwerling and Silver, 1992). In such cases, lack of promotion of the minority person cannot then be attributed to a lack of education. The same thing often happens to both White and minority women in organizations: Women are less likely to receive promotions than a White male who is of the same education, and sometimes even of *less* education. Studies find that women actually change jobs more frequently within organizations than do men, but these tend to be lateral moves; whereas for men job changes are more likely to mark a jump from a lower to a higher level in the organization, thus constituting a promotion. Studies consistently find that women are held to higher promotion standards than men; for men, the longer they are in a position in an organization, the more likely they will be promoted, but the same is not true for women (Smith, 1999; White and Althauser, 1984).

Things work the same way with respect to people being discharged or fired. Studies (DeWitt, 1995; Zwerling and Silver, 1992) show quite clearly that Black federal employees (men and women) were more than *twice as likely* to be dismissed as were their White counterparts (see also Chapter 9). This disparity was regardless of education, and also regardless of occupational category, pay level, type of federal agency, age, performance rating, seniority, or attendance record. The main reasons cited in the studies for such disparities were lack of network contacts with the "old boy" network and racial bias within the organization. These studies were particularly well designed because they took into account quite a few factors besides race-ethnicity that are often given as reasons for not promoting minorities. The studies showed quite conclusively that with all these other factors taken into account, it is still race that is an important reason for the way people are treated in organizations today. The studies strongly suggest that racism still thrives in the bureaucracy.

Debunking Society's Myths

Myth: **Programs designed to enhance the number of women and minorities in organizational leadership are no longer needed since discriminatory barriers have been removed.**

Sociological perspective: **Research continues to find significant differences in the promotion rates for women and minorities in most organizational settings. Even with the removal of formal discriminatory barriers, organizational practices persist that block the mobility of these workers.**

A classic study by Rosabeth Moss Kanter (1977) shows how the structure of organizations leads to obstacles in the advancement of groups that are tokens in the organizational environment. Kanter demonstrated how the hierarchical structure of the bureaucracy negatively affects both minorities and women who are underrepresented in the organization. In such cases, they represent *token* minorities or women; they feel put "out front" and under the all-too-watchful eyes of their superiors. As a result, they often suffer severe stress (Jackson et al., 1994, 1995; Spangler et al., 1978; Kanter, 1977; Yoder, 1991). Such token women and minorities find it difficult to gain credibility with not only their superiors, who are for the most part White males, but with their co-workers as well, who often accuse them of getting the position simply because they are women, a minority, or both. It is a widespread phenomenon in universities and colleges that not only minorities, but women as well, are accused of being admitted simply because of their gender or their race, even in instances when this has clearly not been the case. This has stressful effects on the person and shows that tokenism can have very negative consequences.

Social class, in addition to race and gender, plays a part in determining people's place within formal organizations. Middle- and upper-class employees in organizations make higher salaries and wages, and are more likely to get promoted than are people of less social class status, even for individuals who are of the same race or ethnicity (Blau and Meyer, 1987). This even holds for persons coming from families of less social class status but who are themselves as well educated as their middle- and upper-class co-workers. This means that their lower salaries and lack of promotion cannot necessarily be attributed to lack of education on their part. In this respect, their treatment in the bureaucracy only perpetuates rather than lessens the negative effects of the social class system in the United States. Blau and Meyer

(1987) conclude that the social class stratification system produces major differences in the opportunities and life chances of individuals, and that the bureaucracy simply carries these differences forward. Class stereotypes also influence hiring practices in organizations. Personnel officers look for people with "certain demeanors," a code for those who convey middle-class or upper middle-class standards of dress, language, manners, and so on, which some people may simply be unable to afford.

Patterns of race, class, and gender inequality in organizations persist at the same time that many organizations have become more aware of the need for a more diverse workforce. Responding to the simple fact of more diversity within the working-age population, organizations have developed human relations experts who work within organizations to enhance sensitivity to diversity. Such "diversity training" has become commonplace in most large organizations, a reflection of the significance of diversity in today's society. As the box "Sociology in Practice" shows, however, it is not clear how much effect such programs have on actual organizational change. These good intentions, aside, however, most organizations remain highly patterned by race, gender, and class.

Functional, Conflict, and Symbolic Interaction: Theoretical Perspectives

All three major sociological perspectives—functional, conflict, and symbolic interaction—are exhibited in the analysis of formal organizations and bureaucracies. The functional perspective, based in this case on the early writing of Max Weber, argues that certain functions, called *eufunctions* (meaning "positive functions") characterize bureaucracies and contribute to the overall unity of the bureaucracy. The bureaucracy exists in order to accomplish these eufunctions, such as efficiency, control, impersonal relations, and a chance for the individual to develop a career within the

B • O • X • ANALYZING SOCIAL ISSUES

6.3

Managing Diversity

NUMEROUS organizations, recognizing the growing diversity of the U.S. population, have responded by creating diversity training workshops to improve intergroup relations within the workforce. The following excerpt discusses the significance of such programs:

> The alarm went off in the late 1980s, when demographers alerted corporations that by the year 2000 white males would make up only 45 percent of America's workforce. Companies concluded that they needed a way to manage the new "multiculturalism," so they turned to their human-resource departments and consultants and decided what was needed was . . . "diversity training." In practice, those buzzwords meant sending workers to lectures and workshops to learn to work with colleagues of diverse cultures and races.
>
> If diversity training has become ingrained in corporate America—75 percent of large companies use it—the Texaco episode shows that it remains a controversial way of changing the workplace. Done well, it can promote corporate harmony. But more often, critics say, diversity training is ineffectual in altering employee behavior; at its worst, it exacerbates the very stereotyping and divisiveness it is meant to alleviate. Dubious training techniques and unskilled "consultants" have proliferated; anyone can, and does, hang out a diversity-training shingle. And too many companies merely pay lip service to their programs, a sop to minority and female employees or as a perceived buffer to any discrimination lawsuits. If the Texaco case shows anything, says Roosevelt Thomas, one of the leading diversity consultants and originator of the now famous multicolored jelly-bean metaphor, it's that companies need to change to a "culture of inclusiveness," not just run training courses.
>
> Critics say the very heart of many programs—highlighting group differences to create awareness—ends up alienating and offending employees. Until 1993 the U.S. Department of Transportation sponsored training programs in which participants where addressed as "jerks and jerkettes," blacks and whites were encouraged to exchange epithets, and men were groped while running a gantlet of women. This was extreme, but such "blame and shame" sessions abound. They operate on the no-pain, no-gain theory that multicultural harmony will emerge only after a period of discomfort. "In the name of diversity these seminars have turned people against each other," says Harris Sussman, a Cambridge diversity consultant. They often reinforce rather than ameliorate stereotypes. If managers are instructed that Asians prefer indirect communication, says Stephen Paskoff, an Atlanta consultant, they may treat all Asians that way, "trivializing their individuality."
>
> If there's a lesson, it's that there is no substitute for management's commitment. Former Hoechst Celanese CEO Ernest Drew went from factory to factory to pitch diversity. Top Xerox execs meet regularly with minority representatives within the company. Some experts say executives should simply tell workers they expect total adherence to the law on discrimination and harassment—and any behavior short of that will be punished. It's also crucial that companies set up clear channels to hear discrimination complaints, fostering accountability among managers. Diversity is a fact of life whether corporations like it or not. Managing it, however, seems to require more than a few dubious metaphors.

SOURCE: Reibstein, Larry. 1996. "Managing Diversity." *Newsweek*, November 23, p. 50.

bureaucracy. As we have seen, however, bureaucracies develop the other face (informal interaction and culture, as opposed to formal or "bureaucratic" interaction and culture) as well as the problems of ritualism and alienation of the person from the organization. These latter problems are called dysfunctions (negative functions), which have the consequence of contributing to the disunity and lack of harmony in the bureaucracy.

The conflict perspective argues that the hierarchical or stratified nature of the bureaucracy in effect encourages rather than inhibits conflict among the individuals within it. These conflicts are between superior and subordinate, as well as between racial and ethnic groups, men and women, and people of different social class backgrounds. This actually hampers the smooth and efficient running of the bureaucracy.

The symbolic interaction perspective underlies two management theories, Theory Y and Theory Z. Symbolic interaction stresses the role of the self in any group, and especially how the self develops as a product of social interaction. Theory Y advocates increased interaction between superior and subordinate as a way of "actualizing" the self and, as a result, reducing the disconnection between individual and organization as well as other organizational problems and dysfunctions. Theory Z argues that increased interaction between superior and subordinate, based on the Japanese organization model of executives "walking around" and interacting more on a primary group basis, will reduce organizational dysfunctions. Theory Z also advocates participatory management. The focus of both Theory Y and Theory Z on changes in social interaction illustrates their connection to symbolic interaction theory.

CHAPTER SUMMARY

- Groups are a fact of human existence and permeate virtually every facet of our lives. Groups are of several types. *Primary groups* form the basic building blocks of social interaction in society. *Reference groups* play a major role in forming our attitudes and life goals, as do our relationships with *in-groups* and *out-groups*. *Social networks* partly determine things such as whom we know and the kinds of jobs we get. Networks are responsible for our perception that it is a "small world." Leadership networks, such as that among U.S. national African American leaders, are extremely closely connected, or very dense.
- The social influence groups exert upon us is tremendous, as seen by the Asch conformity experiments. The Milgram experiments demonstrated that the interpersonal influence of an authority figure can cause an individual to act against his or her deep convictions. Bystanders observing an emergency tend not to intervene to help as the number of visible bystanders around is greater, thus demonstrating a *group size effect*.
- *Groupthink* can be so pervasive that it adversely affects group decision making. *Risky shift* similarly often compels individuals to reach decisions that are at odds with their better judgment.
- *Formal organizations* are of several types, such as normative, coercive, or utilitarian. Groups occur within organizations, and thus individuals within organizations are subject to group effects, such as groupthink and risky shift. Weber typified *bureaucracies* as organizations with an efficient division of labor, an authority hierarchy, rules, impersonal relationships, and career ladders. Bureaucratic rigidities often result in organizational problems such as ritualism and alienation. The "McDonaldization of society" has resulted in greater efficiency, calculability, and control in many industries, probably at the expense of some individual creativity.
- Some new organizational forms have attempted to reduce worker alienation by adopting Japanese styles of management instead of traditional top-down management styles. Whether such practices can thrive in a different cultural context is to date uncertain.
- Formal organizations perpetuate inequality of race-ethnicity, gender, and social class. Blacks, Hispanics, and Native Americans are less likely to get promoted, and more likely to get fired, than Whites of comparable education and other qualifications. Women experience similar effects of inequality, especially negative effects of tokenism such as stress and lowered self-esteem. Finally, persons of less than middle-class origins make less money and are less likely to get promoted than a middle-class person of comparable education.
- Functional, conflict, and symbolic interaction theory highlight and clarify the analysis of organizations by specifying both organizational functions and dysfunctions (the functional perspective); by analyzing the consequences of hierarchical, gender, race, and social class conflict in organizations (the conflict perspective); and, finally, by studying the importance of social interaction and integration of the self into the organization (the symbolic interaction perspective).

KEY TERMS

attribution theory
bureaucracy
dyad
expressive needs
formal organization
group
group size effect
groupthink
instrumental needs
primary group
reference group
risky shift
secondary group
social network
status generalization
total institution
triad
triadic segregation
voluntary organization

THE INTERNET: A Tool for the Sociological Imagination

Resources on the Internet:

Virtual Society: The Wadsworth Sociology Resource Center at
http://sociology.wadsworth.com

Visit this site to find additional learning tools, including interactive quizzes, links to related web sites, and an easy link to InfoTrac College Edition.

The Kevin Bacon Game
http://www-usacs.rutgers.edu/~cainan/kbacon/

This Internet game is an illustration of the small world effect in social networks.

McDonaldization of Society
http://www.sociology.net/mcdonald/

This is the home page for Ritzer's book, and it provides additional information about and links to discussion and illustration of this concept.

Sociology and Social Policy: Internet Exercise

Many organizations have tried to foster a more positive organizational climate for all members by sponsoring diversity workshops and other training efforts designed to make people more aware of racial, ethnic, and gender differences that affect people's interaction in the workplace. Advocates of diversity training say that enhanced awareness and sensitivity toward others makes a more positive environment for all members of the organization. Critics say that, well meaning as such programs are, they do not deal with fundamental problems of equity in the workplace and have done little to advance women and racial–ethnic minorities into senior positions in such firms. Identify one such diversity training program in an organization and outline its goals and techniques. Do you think such a program is effective in creating more opportunities for women and minorities?

Internet Search Keywords:

minorities and women in the workplace
minorities in the workplace
diversity training

Web sites:

http://www.diversitydtg.com/
Home page for the Diversity Training Group. Offers information on workplace diversity and gender equity as well as providing consulting and training services.

http://www.equalopportunity.on.ca/
A Canadian web site featuring discussion and resources on workplace diversity and equal opportunity.

http://www.nmci.org/
National MultiCultural Institute home page provides articles, statistics, and conferences information on diversity workforce and cultural understanding as well as links to other databases.

http://www1.od.nih.gov/ohrm/oeo/wdi/wdi1.htm
The Workplace Diversity Initiative by National Institutes of Health.

http://cc.ysu.edu/diversity/mframe.htm
Web site for Partners for Workplace Diversity, a cooperative effort among organizations to develop responses to diversity issues in workplaces.

InfoTrac College Edition: Search Word Summary

bureaucracy
groupthink
Milgram's Obedience Study
social network

In order to learn more about these central topics in sociology, you can conduct an electronic search using InfoTrac College Edition. To aid in your search and to gain useful tips, see the Student Guide to InfoTrac College Edition on the Virtual Society web site:
http://sociology.wadsworth.com

INTERACTIONS—A SOCIOLOGY CD-ROM: CONCEPTS FOR THIS CHAPTER

Go to the Wadsworth Sociology CD-ROM for further study on the concepts in this chapter. The CD-ROM also includes quizzes and additional activities to expand your learning experience.

SUGGESTED READINGS

Aaronson, Elliot. 1992. *The Social Animal,* 6th ed. New York: Freeman.

This is a classic and readable discussion of human beings who interact, conform, get influenced, become aggressive, act out prejudices, like and love, and in many other ways socially interact in order to survive.

Collins-Lowry, Sharon M. 1996. *Black Corporate Executives.* Philadelphia: Temple University Press.

Based on extensive interviewing of successful Black executives, this research examines the experiences of some of the first Black managers to succeed in major corporations. Collins-Lowry's work shows the continuing significance of race in structuring experiences within corporate organizations.

Daniels, Arlene Kaplan. 1988. *Invisible Careers: Women Civic Leaders from the Volunteer World.* Chicago: University of Chicago Press.

Although women worldwide have played a large part in voluntary organizations, the largely ignored role of women in America's volunteer organizations is analyzed in this work, and attention is given to women's participation as a career.

Ferguson, Kathy E. 1984. *The Feminist Case Against Bureaucracy.* Philadelphia: Temple University Press.

The author argues convincingly that the traditional Weberian model of bureaucracy that embodies the notion of hierarchy reflects a paternalistic and patriarchal notion of social structure and organization.

Kanter, Rosabeth Moss. 1977. *Men and Women of the Corporation.* New York: Basic Books.

This classic study shows the significance of gender in structuring opportunities for women in corporations. Although more than twenty years old, the book's insights continue to be supported by subsequent research on gender dynamics in work organizations.

Ritzer, George. 1996. *The McDonaldization of Society.* Newbury Park, CA: Pine Forge Press.

This is an entertaining account of the fast-food approach to bureaucratic organization. The book highlights the influence of this model of bureaucracy upon many kinds of bureaucracies in American society, as upon the structure of the community health clinic.

Vaughan, Diane. 1966. *The Challenger Launch Decision: Risky Technology, Culture, and Deviance in NASA.* Chicago: The University of Chicago Press.

Vaughan's research explores how ordinary behavior and ritualism within organizations can result in disastrous consequences, such as the *Challenger* space shuttle explosion.

CHAPTER 7

Sexuality and Intimate Relationships

A VISITOR from another planet might conclude that people in the United States are obsessed with sex. A glimpse of MTV shows men and women gyrating in sexual movements. A stroll through a shopping mall reveals expensive shops selling delicate, skimpy women's lingerie. Popular magazines are filled with images of women in seductive poses trying to sell every product imaginable. Even bumper stickers brag about sexual accomplishments. People dream about sex, form relationships based on sex, fight about sex, and spend money to have sex.

There is ample evidence of sexual permissiveness in the United States. Young people are initiating sexual activity at earlier ages than before. The availability of birth control has separated sex from procreation, meaning that people can engage in sex just for the fun of it. Diversity in sexual partnerships is becoming more evident.

Some think that our society is overly promiscuous, that sexuality is rampant when it should be confined to the family. Sex education in schools has been challenged on the grounds that teaching young people about sex will incite them to be more promiscuous. Others say sex education is essential for the prevention of pregnancy and AIDS. In the early 1990s, the government canceled a national survey of sexual attitudes and behaviors among teenagers that had been proposed for funding by the National Institute of Child Health and Human Development, despite recommendations from scientists and federal funding agents that the study should proceed. The survey was to be the first large national study of adolescent sexual behavior in recent years, intended to provide information useful for developing public policies on AIDS prevention, teenage pregnancy, sexually transmitted disease, and a host of other social–sexual issues. The study was canceled on the grounds that asking teenagers about sex would cause them to be more sexually active. Sex is clearly a subject that polarizes political constituencies on various issues, such as abortion, teenage pregnancy, birth control, family policies, AIDS education, and gay and lesbian rights.

Others argue that our society is not sexually free enough. They point to repressive attitudes directed against people of diverse sexual orientations. Sexual repression is indeed common. Gay-bashing runs the gamut from joking and ridicule to extreme violence,

Sex symbols, such as Madonna in the middle of this photo, are common in a society where sexuality is also marketed as a commodity. One might ask if making sex an object of consumption is exploitative or if it has the effect of liberating people from repressive sexual values.

even murder, against those believed to be lesbian or gay. Prominent women in the public limelight who are not visibly attached to a man are often accused of being lesbians, regardless of their actual sexual identity—with the implication that this identity alone affects someone's qualifications for leadership. Sexual repression affects heterosexuals, as well. Men and women are still held to a double standard about sexuality, although the difference has lessened somewhat. Women who openly engage in multiple sexual relationships may be defined as loose and immoral, whereas men's having multiple partners is taken as a badge of manhood; thus, "scoring" is a positive achievement for men, a sign of success in a competition. The origin of the term in sports is probably no coincidence. In a sexually repressive culture, some people grow up associating sex with shame, and many find open discussions of sex embarrassing. Sex is rarely treated as a serious topic for public discussion, and people are often cautious about talking about it too openly. In our society, even people who study sex scientifically are sometimes suspected of prurient interests (Troiden, 1987).

Is our society repressive or permissive? The answer is both. Neither alone characterizes the intricate and sometimes contradictory character of contemporary sexual behavior and attitudes. Sexuality is a complex social phenomenon. This chapter examines the social significance of sexuality and how sexual experience is shaped by society and culture. Using the sociological imagination, we are able to see that sexual behavior and attitudes are shaped by the institutional structure of society. The expression of human sexuality, like other forms of social behavior, is socially structured.

Is Sex Natural?

Sex would seem to be utterly natural. Pleasure and, sometimes, the desire to reproduce are reasons people have sex, but sexual relationships develop within a social context, and the social context establishes what sexual relationships mean, how they are conducted, and what social supports are given (or denied) to particular kinds of sexual relationships. *Sexuality, although experienced as a bodily phenomenon, is socially defined and patterned.*

From a sociological point of view, little in human behavior is purely natural, as we have seen in previous chapters. Behavior that appears to be natural is usually that which is accepted by cultural customs and is sanctioned by social institutions. Sex is a physiological experience, but it is not physiology alone that makes sex pleasurable. Sexuality creates intimacy between people. People engage in sex not just because it feels good, but also because it is socially meaningful. Sex is an important part of our social identity. How we enjoy sex is shaped as much by cultural influences as by physical possibilities for arousal and pleasure. Void of a cultural context and the social meanings attributed to sexual behavior, people might not find certain behaviors sexy and others a turnoff, nor would they attribute the emotional commitments, psychological interpretations, spiritual meanings, and social significance to sexuality that it has in different human cultures.

We can see the social and cultural basis of sexuality in a number of ways:

1. *Human sexual attitudes and behavior vary in different cultural contexts.* Culture affects whether we define certain sexual behaviors as normal or deviant. For example, if the culture allowed men to wear dresses, would there be a category of people known as cross-dressers or transvestites? Likewise, you might ask why those who choose same-sex partners are judged to be sexually deviant. From a sociological point of view, it is not because such behavior is unnatural or inherently wrong, but because of the cultural assumptions made about same-sex partners.

If sex were purely natural behavior, sexual behavior would also be uniform from one society to another, but it is not. Sexual behaviors considered normal in one society may be seen as peculiar in another. In some societies mouth-to-mouth kissing is considered unhealthy and disgusting. For example, the common practice of "French kissing" in the United States is by no means universal; some societies would consider putting your tongue in another's mouth repulsive (Tiefer, 1978). The expression of sexual feeling is influenced by its cultural setting.

Consider the hijras (pronounced HIJ-ras) of India, a religious community of men who are born male but come to think of themselves as neither men nor women; the hijras' role is a third gender identity. Hijras dress as women and may marry men, although they typically live within a hijra communal subculture. Men are not born hijras, but they become so. As male adolescents, they have their penis and testicles cut off in an elaborate and prolonged cultural ritual. This rite of passage marks the transition to becoming a hijra. Hijras' perceive that sexual desire results in the loss of spiritual energy; their emasculation is seen as proof that they are beyond such desires. Whereas Western culture emphasizes dichotomous and separate identities for women and men, Hindu religion values the ambiguity of in-between sexual categories. Hinduism holds that all persons contain both male and female principles within themselves, and many Hindu gods are sexually ambiguous. Hijras are believed to represent the power of man and woman combined, although they are impotent themselves. At Indian weddings, they may ritually bless the newly married couple's fertility. They also perform at celebrations following the birth of a male child, a cherished event in Indian society (Nanda, 1998).

Hijras are not transvestites, hermaphrodites, or sex impersonators, as Westerners might think. Although at times

The hijras of India are a sexual minority group; this "man" is considered to be a "third gender"—neither man nor woman. Hijras provide a good illustration of the socially constructed basis of sexuality.

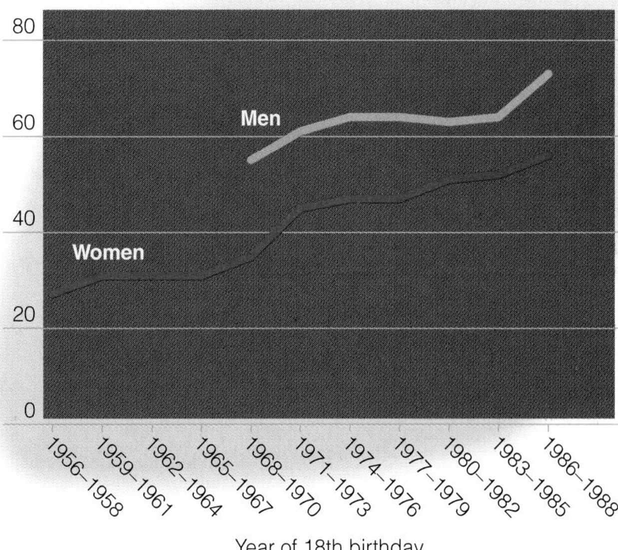

FIGURE 7.1 Sex in Teenage Years

SOURCE: Reproduced with the permission of the Allan Guttmacher Institute from The Alan Guttmacher Institute, *Sex and America's Teenagers.* New York: AGI, 1994, p. 22.

they are treated as social outcasts, they are also important figures who represent the values of Hindu gods. They can be understood only within the context of Hindu culture.

2. *Sexual attitudes and behavior change over time.* Fluctuations in sexual attitudes are easy to document. For example, public opinion polls show that young people are more permissive in their sexual values than were young people in the past, although tolerance of casual sex has decreased somewhat in recent years (Astin, 1992; Shea, 1996). At the same time, national polls show that more people think premarital sex in general is acceptable (Roper Organization, 1995a). Behavioral changes are harder to document since people in this culture generally consider sex to be a private matter. The evidence shows that young people are having sex at an earlier age (see Figure 7.1), that more people experience sex before marriage (Janus and Janus, 1993), and that people have more sex partners over their lifetimes than was likely true in the past (Laumann et al., 1994).

In long-term historical perspective, sexual behavior has changed dramatically. During the Victorian period in nineteenth-century England, dominant sexual attitudes are reputed to have been very repressive. It was in this period that Sylvester Graham invented the graham cracker, thinking that its blandness would reduce people's carnal desires and make them think pure thoughts. In the same period, Queen Victoria coined the phrases *light* and *dark* meat for chicken and other fowl since she thought mention of the words *breast* and *leg* would cause men to be overcome with lust and faint.

Although attitudes may change, a range of attitudes will always exist, with the sexually conservative on one end and the sexually liberal on the other. For example, at a conference of the Christian Women's National Concerns held in the 1980s, speakers emphasized the importance of women wearing good sturdy bras made of solid material since the sight of an erect nipple could induce a man to uncontrollable lust. Silky lace bras, they said, were inappropriate, especially in brisk weather when the cold makes the nipples stand out (Ehrenreich et al., 1987: 134). Contrast this with the widespread availability of "men's magazines" showing women in various nude and suggestive sexual poses, and you see the range of sexual attitudes in society today.

3. *Sexual identity is learned.* Like other forms of social identity, sexual identity is acquired through socialization and the ongoing relationships we have with other people. Information about sexuality is transmitted culturally and becomes the basis for what we know of ourselves and others. Where did you first learn about sex? For that matter, what did you learn? For some, parents are the source of information about sex and sexual behavior. For many, peers have the strongest influence on sexual attitudes. Learning about sex—however it occurs—is part of the socialization process since it is an important dimension of learning social norms and mores.

Sexual socialization takes place at a very early age, long before young people become sexually active. In their talk and play, school children learn sexual attitudes and behaviors. One study of fourth and fifth graders found that through classroom play children learn sexual "scripts" that are the basis for later adolescent and adult behaviors (Thorne and Luria, 1986). A **script** is a learned performance. Like an actor, one learns one's part by learning the script that dictates how one is expected to act. **Sexual scripts** teach us what is appropriate sexual behavior for a person of our gender (Schwartz, 1998; Laws and Schwartz, 1977). From an early age, children learn sexual scripts by playing roles; they play doctor as a way of exploring their bodies, or they pretend to be mommy and daddy so that they can hug and kiss. Role-playing teaches children social norms about sexuality, as well as norms about marriage and gender relationships. For instance, sexual talk among young boys socializes them into masculine roles. Sociologists who have studied peer cultures have concluded that sexual joking is common among young boys and creates a bond between them (Lyman, 1987; Fine, 1986). As we will see when we study gender more thoroughly in Chapter 12, the roles learned in youth have a significant impact on our relationships as men and women later in life. The sexual socialization that occurs early also profoundly influences our sexual attitudes and behavior throughout life.

THINKING SOCIOLOGICALLY

Observe a popular children's film (such as *Pocahantas*, *Aladdin*, or *Balto*) and describe the *sexual scripts* that are suggested by the film. What do these sexual scripts tell you about the *social construction of sexual identity*?

4. *Social institutions channel and direct human sexuality.* Social institutions, such as religion, education, or the family, define some forms of sexual expression as more legitimate than others. For example, being heterosexual is a more "privileged" status in society than being gay or lesbian; heterosexuals are given more institutional privileges than lesbians and gays. Examples of institutional privileges for heterosexual couples include the right to marry and, as a result, to have mutual employee health care benefits and the option to file joint tax returns. Although many employers are now extending benefits to domestic partners regardless of their sexual orientation, there are far more institutional supports for heterosexual relationships. Numerous laws, religious doctrines, and family and employment policies explicitly and implicitly promote heterosexual relationships; thus, what we come to see as "natural" or chosen, is actually promoted by the social structure of society.

Laws, for example, direct sexuality into socially legitimate forms by defining some forms of sexual behavior as lawful and others as criminal. Most states have laws prohibiting "crimes against nature," which can include oral and anal sex, even though surveys tell us that these practices are quite common (Laumann et al., 1994). Luckily for those who engage in oral and anal sex, these laws are rarely enforced, especially not against those in heterosexual relationships.

Sex is also influenced by the economic institutions of society. Sex sells. In the capitalist economy of the United States, sex is used to hawk everything from cars and personal care products to stocks and bonds; moreover, sex itself is bought and sold as some people, particularly poor women, are forced to sell sexual services for a living. Sometimes the economics of sex and the law intersect as quasi-legal sex trades are regulated in the form of red-light districts, selective enforcement, and even outright state licensing.

Although sexuality is generally considered a private matter, and many think state authority should have no jurisdiction over private life, governments routinely intervene in people's sexual lives—in the United States and elsewhere. Particularly among the poor, state-supported agencies may exert their influence on sexuality and reproduction. Social welfare workers, for example, may pressure poor women to use birth control or to be sterilized. Poor women and women who receive welfare have very high rates of sterilization. Sterilization rates are highest among African American, Puerto Rican, and Native American women (Horton, 1995); it is estimated that 24 percent of Native American women of childbearing age are sterilized. Although that fact in itself does not indicate government intervention in the reproductive choices women make, numerous studies have found that the sterilization of poor and minority women often occurs with inadequate or no consent (Nsiah-Jefferson, 1989; Cushman et al., 1988). State workers, including case workers and clinic doctors, may not communicate the consequences of sterilization clearly to patients; many women believe that sterilization can be reversed. Low-income women may also be coerced into sterilization by government funding schemes that pay for sterilization, but not other reproductive options. Physicians may also imply that the women have no alternative to sterilization if they want continued state support. All these practices compromise the degree of "choice" poor and minority women have.

Public policies also regulate sexual and reproductive behaviors. Prohibiting federal spending on abortion, for example, eliminates reproductive choices for women dependent on state or federal aid. Government decisions about which reproductive technologies to endorse influence the choices of birth control technology available to men and to women. Government funding, or lack thereof, for sex education, can influence how people understand sexual behavior. In these and other ways, the government intervenes in people's sexual and reproductive decision making. This fact challenges the idea that sexuality is a private matter and shows how social institutions direct sexual behavior.

To summarize, human sexual behavior is situated in a cultural and social context. The culture defines certain sexual behaviors as appropriate or inappropriate. Human sexuality also varies in expression across culture and changes over time. Like other forms of social identity, sexual identity is learned. Social institutions channel and direct human sexual activity.

The Sexual Revolution

With the coming of the twentieth century, sexual ideals and sexual norms began to change. The **sexual revolution** refers to the widespread changes in men's and women's roles and a greater public acceptance of sexuality as a normal part of social development. Although perhaps most obvious in the late twentieth century, this movement originates in social and historical changes in the late nineteenth and early twentieth

sexual revolution

century, particularly changes in how sexuality came to be understood and how women's and men's roles were emerging.

Debunking Society's Myths

Myth: **Over time in a society, sexual attitudes become more permissive.**

Sociological perspective: **All values and attitudes develop in specific social contexts; change is not always in a more permissive direction (D'Emilio and Freedman, 1988).**

Scientific Studies of Sex

The development of **sexology**—the scientific study of sex and sexuality— has brought profound changes to the understanding of sexuality. Early sexologists changed the cultural understanding of sex from something believed to be ordained by religious belief to something subject to social science research.

The Influence of Freud. There is no one more significant in the scientific study of sex than Sigmund Freud (1856–1939). His influence was not only felt in academic, scientific, and psychoanalytic quarters, but also in the general public. His message that the formation of sexual identity is a basic part of personality development and his analysis of the role of sexuality still underlie much of the public understanding of sex (see also Chapter 4). Although many of the specific ideas he developed about sex, especially about the sexuality of women, have been subsequently criticized, he opened the door for scholars and scientists to make sex the subject of serious study.

Freud presented a developmental model of sexuality, postulating that sexual expression originated in childhood and developed over the life cycle. During the transition to adulthood, in Freud's view, several stages of psychosexual development occurred. Freud thought that sexual energy, or *libido,* was the driving force behind all human endeavors, generating the tension and excited state that leads to creativity in artistic and intellectual expression (Freud, 1923 [1960]). Upon translation of his works from German in the 1920s, Freud was widely read by the public, and his writings became the common language of sexuality.

A myth propagated by Freud was that women were capable of two kinds of orgasm: vaginal and clitoral. Vaginal orgasm, according to Freud, was more mature; clitoral orgasm was immature. He advised that all women should transfer their center of orgasm to the vagina during phallic intercourse. Male sexual penetration, according to Freud's argument, was therefore required to make women's sexual response complete. Freud has since been soundly criticized for building male-oriented value judgments into his interpretations of female sexuality. Later research by Masters and Johnson (1966) showed decisively that there is no distinction between clitoral and vaginal orgasm; yet, for half a century, the emphasis on vaginal orgasm left many women sexually unsatisfied. During the period in which Freud's ideas were most popular, especially through the 1940s, 1950s, and early 1960s, psychotherapists reported that the single most common reason for women to seek clinical therapy was their fear that they were frigid (Chesler, 1972). Feminists who are critical of Freud conclude that his work buttressed male interests; his emphasis on male penetration made it seem that women required a penis to have a "mature" sexual experience even though women are quite capable of experiencing sexual pleasure without male penetration. Much of Freud's interpretation of women's sexuality was based on the assumption that men are more sexually and psychologically mature, which has been the cause of severe feminist criticism.

Although his work has been criticized on a number of fronts, Freud established the groundwork for our modern understanding of sexuality and social life. One major consequence of Freud's work and that of other sexologists was that science replaced religion as the authority on sexual morals.

Havelock Ellis. Havelock Ellis (1859–1939), one of the most influential of the early sexologists, saw sexual dysfunction as the result of psychological, not organic problems. He depicted some forms of sexuality as pathologically rooted. He described what he thought were pathological roots of sexuality, often tingeing his work with the sexist assumptions of the times. He thought, for example, that lesbian women had turned their emotions upside down by rejecting the proper passive place of female sexuality; he thought lesbians made the mistake of loving the public world, instead of men. He associated lesbianism with insanity, arguing that the professional women who were emerging during the 1920s and 1930s were particularly prone to this "disease." Ellis's interpretations, in which anything but heterosexual, monogamous sexuality was regarded as "sexual deviance," were extremely popular and became the prevailing wisdom for the first part of the twentieth century.

The Kinsey Reports. The first major national surveys of sexual behavior were the Kinsey reports, published in the 1940s and 1950s. Alfred Kinsey was a zoologist at Indiana University and head of the Institute for Sex Research. He was the first to give a detailed picture of the sexual activities of the U.S. public—or at least what people reported. Kinsey published two highly influential works: *Sexual Behavior in the Human Male* (1948) and *Sexual Behavior in the Human Female* (1952).

Kinsey's research was based on a national sample of 11,000 interviews, but his sample was not representative. The research subjects were all White, relatively well educated, and middle class. The interviewers and staff members were all White, heterosexual, Anglo-Saxon, Protestant men since Kinsey believed they represented the "yardstick of morality" (Irvine, 1990: 44). Kinsey also refused to hire interviewers with ethnic names, arguing that this would make rapport with the research subjects difficult. He claimed that he did not want to make any judgments about sexual practices, nor condemn any behavior, and that he wanted his research to be objective and value-free, but he would have

Heterosexual love is supported by a wide array of institutional practices and dominant cultural values.

been wise to have used a more diverse and representative sample. Kinsey also did not question the race and class stereotypes prevalent at the time, trusting that the scientific method alone would control for bias. His research also had the built-in limitation that it was based on self-reports about sexual habits. Information based on self-reports is notoriously unreliable, especially when the topic is a delicate one, but it was the best information Kinsey could get.

Even acknowledging the limitations of his work and the bias in his sample, the Kinsey reports were nevertheless the first comprehensive and nationally based studies of sexual practices. They are still influential in describing sexual behavior in the U.S. population. It was Kinsey who first reported (in the 1940s) that 33 percent of women and 71 percent of men engaged in premarital intercourse despite public belief to the contrary. Later surveys have shown that now these figures have grown to 70 percent of women and at least 80 percent of men having sex before marriage; some estimate that as many as 97 percent of men have sex before marriage (Hunt, 1974; Laumann et al., 1994; Lips, 1993). Kinsey also found that the sexual activities of younger husbands and wives was more varied than was typically believed. Oral sex, a variety of coital positions, and genital touching were common among the couples he interviewed. Kinsey's reports also documented that many women were more sexually adventurous than was expected. Prior to 1920, sex was often considered one of a woman's less enjoyable duties; many women remained fully clothed during sex. Kinsey found that women were enjoying sex and experimenting with different sexual practices, although he also found that, generally speaking, women desired less sex than men; men were also much more likely than women to desire extramarital affairs. Overall, Kinsey reported a weakening of sexual taboos and a more permissive sexual atmosphere.

Kinsey was the first to deliver evidence that a significant proportion of the population was homosexual. He reported that, based on his interview data, 37 percent of men had experienced homosexual contact resulting in orgasm at some point in their lives. As we will see, the extent of homosexual behavior is still much debated.

When the Kinsey studies were done, widespread societal changes were taking place in the sexual practices and beliefs of the U.S. public. Women were increasingly independent, moving into cities, remaining single longer, living on their own, and working in the public sector. These changes were more pronounced among White, middle-class women—those most likely to have been included in Kinsey's sample. For the White middle class, the moralistic sexual ideology of the Victorian period was giving way to a more secular view of sex. Marriage was increasingly defined as a partnership between a man and a woman, not simply a pairing in which the woman was expected to obey the man. Sexuality and marriage were increasingly seen as the province of experts. Marriage manuals instructed women and men on ways to achieve sexual fulfillment although, in retrospect, much of their advice now seems antiquated and silly. Kinsey's findings helped liberate people from the shame and guilt that many felt about their sexual practices. His reports questioned the absolute distinctions made between heterosexual and homosexual people; as a result, he laid the foundation for some of the sexual liberation movements that followed.

The Masters and Johnson Studies. Kinsey's work laid the groundwork for a second series of important studies of human sexuality—the Masters and Johnson research (1966). The work of William Masters and Virginia Johnson is still the primary basis for understanding physiological sexual response (heart rate, blood pressure, orgasm, and so on), and their research broke the taboos on studying sex that had existed before. Their work is still the most extensive and thorough study of human sexual response (in a physical sense) ever done.

Like Kinsey's research, the Masters and Johnson study is flawed by a nonrepresentative sample. The subjects they worked with included 510 married couples and 57 single people, but they chose only people who they thought were "respectable"—White, middle-class, well-educated men and women, the majority of them married couples. Like Kinsey, Masters and Johnson carried their own race and class biases into the design of their sample. They thought that well-educated people would be better able to visualize the sexual experiments and communicate fine details of sexual response. As a result, their sample was purposefully weighted toward people from higher socioeconomic backgrounds (Irvine, 1990).

Masters and Johnson actually began their studies using prostitutes as their research subjects since they believed that they would be the only women who would participate in such a project. This turned out to be a boon. In the early stages of the research, the prostitutes, who obviously had wide experience in sexual matters, made many recommendations for refining the research, including the suggestion that a woman be added to the research team. They were collaborators as much as research subjects, demonstrating methods for sexual stimulation and techniques for elevating or controlling sexual tensions. Many of the techniques they demonstrated have since been used in sex therapy programs, marital counseling, and sex manuals (Irvine, 1990).

Masters and Johnson conducted detailed observations of sexual activity, based on studies of human subjects who were paid for the time they spent in a university laboratory. Their research methods were technologically sophisticated, allowing them to observe the physiological and psychological responses to sexual stimulation. Their research subjects engaged in a number of explicit sexual activities in the lab, with sexual response observed and measured, including during orgasm (Masters and Johnson, 1966).

Masters and Johnson documented the physiological responses that men and women experience during sexual excitement, but perhaps the greatest significance of their work is that it dispelled a number of prevalent social myths about sexuality—most notably the myth that men and women experienced sex in radically different ways. Masters and Johnson defined sex as a natural bodily function, and they asserted that women as well as men had a right to sexual pleasure. Despite the enormously liberating impact of their research, however, Masters and Johnson were themselves sexual conservatives. They emphasized the importance of sex within marital units only—one of the reasons why their research sample was biased so strongly in favor of married couples (Irvine, 1990).

More contemporary studies of sexuality have not investigated the physiological character of sexuality so much as analyzed the social dimensions of sexual behavior. In a major national survey of sexual practices and attitudes (Laumann et al., 1994), as we will see when we examine contemporary sexual attitudes and behaviors, sociologists have examined the social context of sexuality, linking such things as attitudes toward sex, and sexual behavior itself, to the changes in the social climate.

Sex and Social Change

Many of the changes associated with the sexual revolution have been changes in women's behaviors. Fewer women today are virgins at marriage, and women are more likely than in the past to describe their sex lives as active and satisfying. Women are having more sex with more partners. Both men and women initiate sex at a younger age and before marriage. Essentially, the sexual revolution has narrowed the differences in the sexual experiences of men and women.

The sexual revolution has been strongly influenced by political movements in recent years, especially feminism and the gay rights movement. Feminists have provided an analysis that links sexuality to the status of women in society. The women's movement also interprets sexuality in the context of power relationships between men and women. **Sexual politics** refers to the link feminists argue exists between sexuality and power, and between sexuality and race, class, and gender oppression. Feminists have argued one cannot have an equal or satisfying sexual relationship if one is powerless in a relationship or is defined as the property of someone else. Sexual politics does not just refer to power within individual relationships, however. It also refers to how the sexual exploitation of women is linked to the distribution of power in society, as reflected in the high rates of violence against women, the cultural treatment of women as sex objects, and the exploitation of women's work. The frequency of sexual harassment in the workplace (further discussed in Chapter 17) is one indication of how women's sexuality is linked to the expression of men's power in the world of work; although men are, on occasion, sexually harassed at work, this occurs far less frequently than the sexual harassment of women (Charney and Russell, 1994).

The gay and lesbian liberation movement has also put sexual politics at the center of the public's attention by challenging gender role stereotyping and sexual oppression. The gay and lesbian movement has profoundly changed understandings of gay and lesbian sexuality. Gay and lesbian scholars have argued, and many now concur, that homosexuality is not the result of psychological deviance or personal maladjustment, but is one of several alternatives for happy and intimate social relationships. The work of feminists and gay and lesbian activists has had a profound impact on how sociologists and the public think about sex and human relationships. Some of this research is examined further in a later section of this chapter.

The sexual revolution has also been significantly influenced by the development of new contraceptive technologies. The widespread availability of contraceptives brings new possibilities for sexual freedom. Sex is no longer necessarily linked with reproduction, and new sexual norms associate sex (both within and outside family relationships) with intimacy, emotional ties, and physical pleasure (D'Emilio and Freedman, 1988). These sexual freedoms are not equally distributed among all groups, however. For women, sex is still more closely tied to reproduction than it is for men since women are still more likely to take the responsibility for birth control. One way to see this is to note how many different contraceptive devices are designed for women, whereas only two—the condom and vasectomy—are designed for men. Poor women are more likely to have to sell their sexuality for economic survival, as are young men and women who, most typically, are runaways from abusive home situations. Young women runaways are particularly likely to have left homes where they were being sexually abused (McCormack et al., 1986).

The sexual revolution has also produced more commercialized sex. Sex has been defined as a commodity. Definitions of sexuality in the culture are heavily influenced by the advertising industry, which narrowly defines what is considered "sexy." Thin women, White women, and rich women are all depicted as more sexually appealing in the mainstream media. Images defining "sexy" are also explicitly heterosexual. The commercialization of sex also uses women

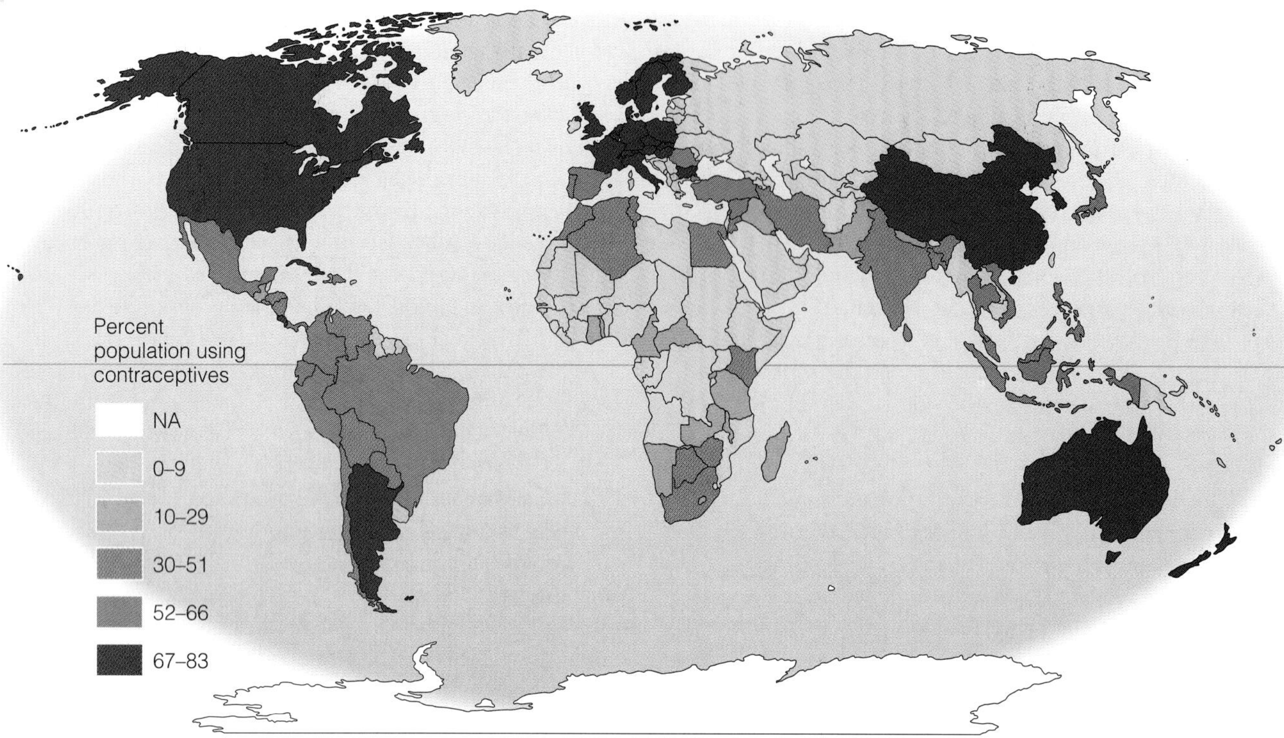

MAP 7.1 Viewing Society in Global Perspective: Contraceptive Use Worldwide

DATA: From the United Nations, *The State of the World's Children 1996.* New York: Oxford University Press, pp. 92–93.

and, increasingly, men in demeaning ways. Although the sexual revolution has in some ways removed sexuality from its traditional constraints, the inequalities of race, class, and gender still shape the social definitions of sexuality.

Historical changes in patterns of sexual relationships reveal that political movements intended to change sexual ideas and practices thrive during periods when other aspects of society are undergoing rapid change (D'Emilio and Freedman, 1988). The feminist movement and the gay and lesbian movements have emerged as gender roles are rapidly changing. Women have become a greater part of the labor force, they are marrying later, and they are achieving higher levels of educational attainment, all of which have transformed relations between women and men. The political mobilization of large numbers of lesbian women and gay men and the willingness of many to make their sexual identity public have also raised public awareness of the civil and personal rights of gays and lesbians. These changes make it possible for other changes in intimate relations to occur.

Contemporary Sexual Attitudes and Behavior

Public attitudes toward sex in the United States are considerably more progressive than they have been in the past. People are more sexually liberal, and there is greater tolerance for diverse sexual lifestyles and practices. Three-quarters of the U.S. public believe that greater openness about sex and the human body is a good thing. A huge majority of the public (87 percent) now approve of sex education in the schools—an increase from 65 percent in 1970; however, many remain decidedly conservative about sexual matters. Forty percent still think it is wrong to have sex before marriage although this is less than the 68 percent who thought so in 1969. Thirty-eight percent think that there should be laws forbidding the distribution of pornography—even though one-quarter of all people in the United States and 46 percent of men between 18 and 45 say they have enjoyed erotic magazines, X-rated movies, adult book stores, and nude stage shows (U.S. Bureau of Justice Statistics, 1996; Hugick and Leonard, 1991).

Changing Sexual Values

How liberal one is on sexual matters varies depending on the issue and change over time. In 1969, only 21 percent of the public approved of premarital sex; now 55 percent do (Roper Organization, 1995a). At the same time, the U.S. public has also become more accepting of gays and lesbians although the public is strongly divided on this issue. In 1977, 43 percent of the public thought that homosexual relations between consenting adults should be legal; by the late 1990s, that figure had increased to 48 percent. At the same time, only 27 percent think gay marriages should be recognized as valid although a very large majority—84 percent—think that gays should have equal rights in the workplace (Moore, 1996; Roper Organization, 1996).

Over time college students have also changed their opinions about sex. In 1974, 46 percent agreed strongly or somewhat that it was all right for people who really like each other to have sex even if they have known each other only for a very short time. By the 1990s, this had changed to 51 percent, then changed again to 43 percent by 1995 (Astin et al., 1987; Astin, 1992; Shea, 1996). Indeed, there are significant gender differences among young people on many topics regarding their sexual behavior. Figure 7.2 also gives some indication of the consequences of the gap in women's and men's descriptions of reasons for their first sexual experience. The gender gap in sexual attitudes may explain some of the personal conflicts that men and women experience about sex.

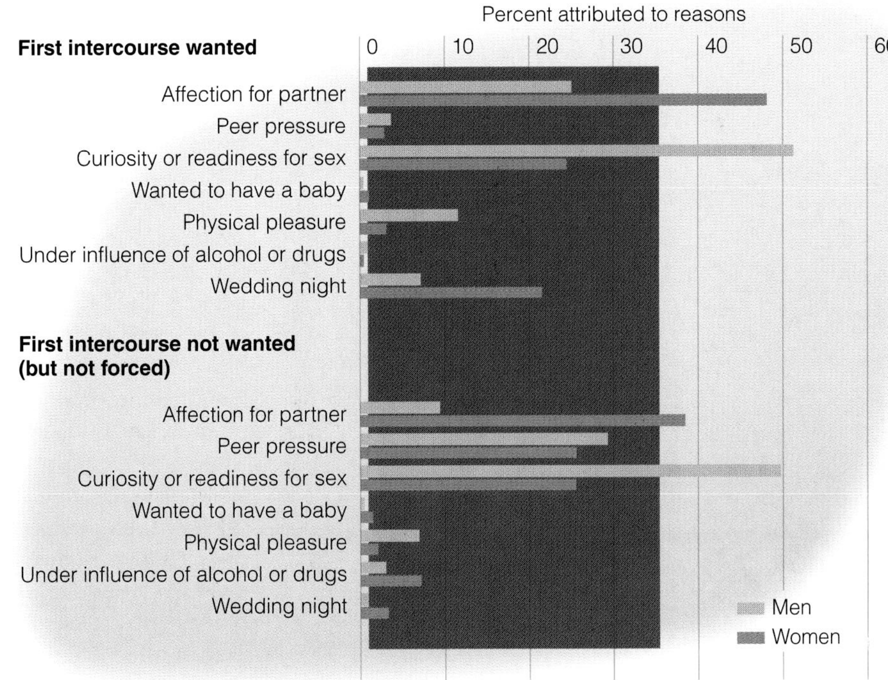

FIGURE 7.2 Reasons for Sexual Intercourse

DATA: From Michael, Robert T., John H. Gagnon, Edward O. Laumann, and Gina Kolata. 1994. *Sex in America: A Definitive Survey.* Boston: Little, Brown, and Company, p. 93.

Attitudes about sex vary significantly across an array of social background characteristics. On many sexual attitudes, there are significant differences in opinion between women and men, and there are some differences between racial groups. Women are more likely than men—39 versus 31 percent—to think that homosexual relations should be legal. White Americans are slightly more likely to think so than Black Americans and Latinos although the difference is small—36 percent for Whites versus 33 percent for Blacks and Latinos. Women are more likely than men to think that premarital sex is wrong (53 versus 39 percent), and older people are more likely than younger people to think it is wrong. Whites are also more likely than other groups to think premarital sex is wrong (Hugick and Leonard, 1991).

Sexual attitudes are also shaped by age and education. For example, 55 percent of those from ages sixteen to twenty-nine believe homosexual relations between consenting adults should be legal, compared to 49 percent of those age thirty to forty-nine and 41 percent of those age fifty to sixty-four and only 24 percent of those sixty-five and older. These differences likely reflect not only the influence of age per se, but also of historical influences on different generations. Better educated people are more likely to think homosexuality is an acceptable lifestyle and to recognize gay marriages as having the same rights as traditional marriages although there is little difference between college graduates and those with no college experience in support for extending civil rights laws to include homosexuals (Roper Organization, 1996).

Opinions about sexual matters also vary significantly by religious identification. Protestants are more likely than Catholics to think premarital sex is wrong; 70 percent of evangelicals, compared to 37 percent of nonevangelicals, think premarital sex is wrong. Perhaps surprisingly, Catholics are more likely than Protestants to support legalizing homosexual relations; some of this difference is attributable to the large number of evangelical Protestants included in the general category. Whether they are Catholic, Protestant, or Jewish, those who are the most religiously observant are the most conservative in their sexual attitudes, although among all Protestants and Catholics, including evangelicals, there has been a decrease in those opposing premarital sex (Petersen and Donnenwerth, 1997; Hugick, 1991).

Public opinion on matters like teen pregnancy, AIDS, child care, women's place in the workplace, and abortion rights tap underlying sexual value systems, often generating public conflicts that are indicative of these underlying differences. In general, sexual liberalism is associated with greater education, youth, urban lifestyle, and political liberalism on other social issues. In other words, the meaning that sex has for different groups shapes their support for or resistance to other social issues. The fact that sexual attitudes differ by factors like gender, age, education, class, religion, and other sociological variables shows that sexual meaning systems are shaped by the social and political systems within society.

Sexual Practices of the American Public

Actual sexual practices are hard to document. What we know about sexual behavior is typically drawn from surveys. Most of these surveys ask about sexual attitudes, not actual behavior. When researchers gather explicit data about sexual behaviors, they encounter the problem that what people say they do may differ significantly from what they really do.

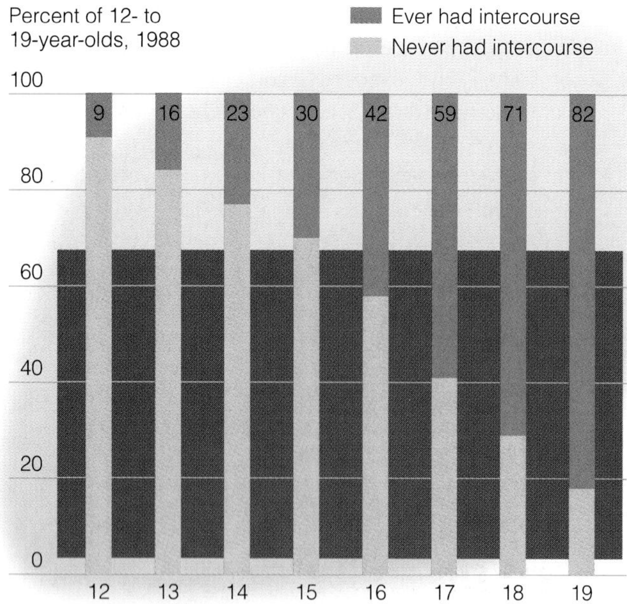

FIGURE 7.3 Sex Among Teenagers

SOURCE: Reproduced with the permission of the Alan Guttmacher Institute from The Alan Guttmacher Institute *Sex and America's Teenagers.* New York: AGI, 1994, p. 19.

The most recent survey of the sexual practices of the U.S. public was conducted by a team of sociologists in the early 1990s (Laumann et al., 1994; Michael et al., 1994). Led by Edward Laumann at the University of Chicago, these researchers surveyed a national random sample of nearly 3500 adults. Although there are some limitations to their study (for example, they do not analyze gay, lesbian, or bisexual behavior separately from heterosexual behavior), this is the latest comprehensive study of sexual behavior that utilizes careful social science research methods. This study and others tell us several things about the sexual practices of the American public.

1. *Young people are becoming sexually active earlier.* Sex is rare among very young teenagers, but it is common in the later teenage years (see Figure 7.3). Black Americans tend to begin sexual activity sooner than Whites although the difference is shrinking. By age fifteen, half of young Black men say they have had sex; White and Hispanic men do not report this level of sexual activity until they are seventeen years old. By age twenty, 80 percent of men have had sex, and 76 percent of women have (Alan Guttmacher Institute, 1994; Harlap et al., 1991).

2. *The proportion of young people who are sexually active has increased,* particularly among young women. In the 1940s, when Kinsey's data were first collected, 33 percent of women and 71 percent of men reported that they had engaged in premarital sexual intercourse by age twenty-five. By the 1990s, 97 percent of men said they had had premarital sexual intercourse compared to 73 percent of women. The proportion of sexually active teenagers has risen even in the short period since the mid-1980s. Most of this increase is because of an increase in the number of non-Hispanic, White teenagers who are sexually active. The largest increase in sexual activity is among White teenagers with higher family incomes; Blacks, Hispanics, and those from low-income families have not significantly increased their sexual activity in the last ten years (Alan Guttmacher Institute, 1994; Forrest and Singh, 1990).

3. *Having only one sex partner in one's lifetime is rare.* The common perception that people have more sex partners than they did in the past seems to be true, according to the Laumann study. Laumann's study asked different age groups how many sexual partners they had in their lifetime. Those over fifty typically reported having five or more sex partners over their lifetime; those thirty to fifty years old reported the same number, but Laumann and his associates argued that the similarity in numbers is only because the younger group had less time to "accumulate" partners (Laumann et al., 1994).

Other studies have found that men report more sexual partners than women although some of this difference is probably attributable to overreporting by men and underreporting by women. There are no reported differences in the number of sexual partners people have by race, education, or region of residence, but the average number of partners for people in metropolitan areas is slightly higher than for those living in other areas. About 22 percent of adults report having no sexual partners in the past year. Abstinence from sex is somewhat more prevalent among women than men (Smith, 1991).

4. *A significant number of people have extramarital affairs.* Most of the public strongly believes in sexual fidelity, about three-quarters say that it is always wrong for a married person to have sex with someone other than the person's spouse; yet, extramarital affairs are fairly common. Laumann's study found that about 25 percent of married men and 15 percent of married women have extramarital affairs, fairly consistent with other studies (Laumann et al., 1994; Reinisch, 1990), although there is likely to be significant misreporting of these data given the secrecy surrounding such behavior. The likelihood of infidelity seems to increase with age—and, presumably, the number of years married (Laumann et al., 1994; Smith, 1991).

5. *A significant number of people are lesbian or gay.* Popular estimates are that about 10 percent of the population is gay; however, measuring the extent of homosexuality is difficult, and there is debate on the accuracy of this estimate. Some use the estimate of 10 percent of the population although that figure is hard to verify. National surveys now report, as did Kinsey in the 1940s, that a substantial number of people have some homosexual experience in their lifetime. Kinsey reported that 37 percent of men had homosexual contact at some time in their lives; current surveys report that 22 percent of men and 17 percent of women do. Having a homosexual experience, however, is different from adopting a gay or lesbian identity. In his study, Laumann distinguished between those who had engaged in any same-sex behavior, felt desire for someone of the same sex, and identified as a homosexual. He found that 2.8 percent of men and 1.4 percent of women identified themselves as homosexual, but 9 percent of men and 4 percent of women

reported having had a sexual experience with someone of the same sex; likewise, more men and women reported having experienced same-sex desire than the number reporting an identity as gay. This helps explain why Laumann reports a smaller percentage of gay people in the population than has been reported in other surveys, such as the Janus survey that reports 9 percent of men and 5 percent of women as gay (Janus and Janus, 1993).

Whatever the actual extent of gay and lesbian experience, it is more common than many think. No doubt, the social stigma associated with being gay or lesbian also causes underreporting. Sociologists have concluded that gays and lesbians are a substantial group in the population—one that is typically misunderstood, discriminated against, and incorrectly portrayed in much of the popular and traditional social science literature.

6. *For those who are sexually active, sex is relatively frequent.* On average, almost two-thirds of Americans report having sex a few times per month or two to three times per week (Laumann et al., 1994). In some studies, men report having sex more often—a statistical oddity if one presumes that most sex is between men and women. This difference is explained by differences in men's and women's reporting behavior: Men tend to overreport both their number of sex partners and how often they have sex (Smith, 1991). Lesbians have sex far less frequently than other couples and are less likely to have sex outside their relationship than are either gay men or heterosexual men and women (Blumstein and Schwartz, 1983). People in their twenties report having sex most often, confirming popular stereotypes. Contrary, however, to stereotypical images about single life, it is married and cohabiting people in their twenties who have sex most often (Laumann et al., 1994). Comparing groups across racial, religious, or educational lines, there are no significant differences in the frequency of sex. Age and marital status are the most significant predictors of the frequency of sex (Laumann et al., 1994).

7. Despite people's active sexual lives and progressive sexual attitudes, *people are not very well informed about sex.* A recent national survey of knowledge about sex from the Kinsey Institute found that the majority of those polled could not answer certain basic questions about sex. The survey included questions about whether a woman could get pregnant during her menstrual cycle (the answer is yes), whether women preferred men with larger-than-average penises (the answer is no), and other knowledge people had about the public's sexual habits and practices.

THINKING SOCIOLOGICALLY

Using some of the *survey data* that has been discussed in this chapter, select a set of questions that tap contemporary *sexual attitudes* among young people. Ask these questions to a *sample* of men and women on your campus. What do your results reveal about the influence of gender on attitudes about sex?

The study found that many people believe things about sexuality that are simply not true. As one example, when asked the average length of a man's erect penis, almost one-third were wrong. Men were more likely than women to know the correct answer (5 to 7 inches), and young people did better than those over sixty years of age. More men said the average length was 8 to 12 inches, whereas more women said less than 4 inches. The majority (70 percent) also thought that menopause caused women to be less interested in sex, which is not true. In this survey, men were generally more knowledgeable than women, and people with higher levels of education or higher incomes or people between thirty and forty-four years of age were more informed about sexual practices than others—namely, both those younger than thirty and those older than forty-four. Interestingly, people from the Midwest were most knowledgeable about sex; those in the South and Northeast were least knowledgeable (Reinisch, 1990).

Sex: Diversity, Social Organization, and the Global Context

Patterns of sexuality reflect the social organization of society. In a society structured along lines of race, class, and gender inequality, one can expect that sexual relations will reflect such patterns, as we will see shortly. At the same time, although it may seem unusual to think of sexuality in a global context, such an international perspective raises new questions for the study of sexuality from a sociological perspective.

The Influence of Race, Class, and Gender

Some of the most interesting sociological studies of sexuality examine the connections between sexuality and race, class, and gender relations. Each of these important sociological factors influences how sexuality is experienced. All three are linked to various social stereotypes about sexuality, as we will see. At the same time, gender, race, and class relations in society form a context in which sexual relationships are formed.

Gender, examined in detail in Chapter 12, establishes numerous patterns of behavior between men and women. An important sociological point is that *sexual behavior follows gendered patterns that stem from definitions of masculinity and femininity in the culture.* As Janet Lee's research shows, discussed in the box "Doing Sociological Research: Menstruation and Sexuality: Adolescent Socialization," young girls learn early how they and their bodies are defined in the culture. Often, they learn to be secretive and shamed by natural bodily processes, as her interviews with young women reveal.

Cultural definitions of what is sexually appropriate differ significantly for women and men. The "double standard" is the idea that men are expected to have a stronger sex drive than women. Although this attitude is weakening somewhat, men are still stereotyped as sexually overactive, whereas women are depicted as having relatively less sexual feeling of their own and as needing men for sexual satisfaction. Women who openly violate this cultural double standard by being openly sexual are then cast in a negative light as "loose," as if the appropriate role for women is the opposite of loose, say, "secured" or "caged." The double standard forces women into polarized roles: as "good" girls or "bad" girls. The belief that women who are raped must have asked for it rests on such images of women as "temptresses," as if men would not overpower women if women did not act like temptresses. Contrary to popular belief, men do not have a stronger sex drive than women. Men are, however, socialized more often to see sex in terms of performance and achievement, whereas women are more likely socialized to associate sex with intimacy and affection. One consequence is that, unlike men, women are more likely to have their first sexual experience with someone with whom they are romantically involved (Lips, 1993).

For men, sexuality is also frequently associated with power. Being unable to have an erection is referred to as impotence, meaning "without power." Gender expectations also emphasize passivity for women and assertiveness for men in sexual encounters; men traditionally have been expected to initiate sexual activity. These norms are changing but still shape the sexual behavior of many.

In addition to gender, *sexual politics are integrally tied to race and class relations in society*. You can see this in the sexual stereotypes that are associated with racism and class inequality. Sexual stereotypes of different racial–ethnic groups illustrate how sexuality is linked to systems of oppression. Latino women are stereotyped as either "hot" or "virgins"; African American men are stereotyped as overly virile; Asian American women are stereotyped as compliant and submissive, but passionate.

The nation vividly witnessed the persistence and convergence of racial and sexual stereotypes during the confrontation between Anita Hill and Clarence Thomas during Thomas's Supreme Court confirmation hearings. Clarence Thomas defended himself by saying he was being subjected to a "high-tech lynching," referring to the fact that Black men have been vulnerable to false accusations of sexual assault; Anita Hill was subjected to other gender and race stereotypes, namely, the persistent image of Black women as promiscuous and that women, in general, are false accusers (Freedman, 1992). Class relations also produce sexual stereotypes of women and men. Working-class and poor men may be stereotyped as "dangerous," whereas working-class women are disproportionately likely to be labeled "sluts."

BOX 7.1 DOING SOCIOLOGICAL RESEARCH

Menstruation and Sexuality: Adolescent Socialization

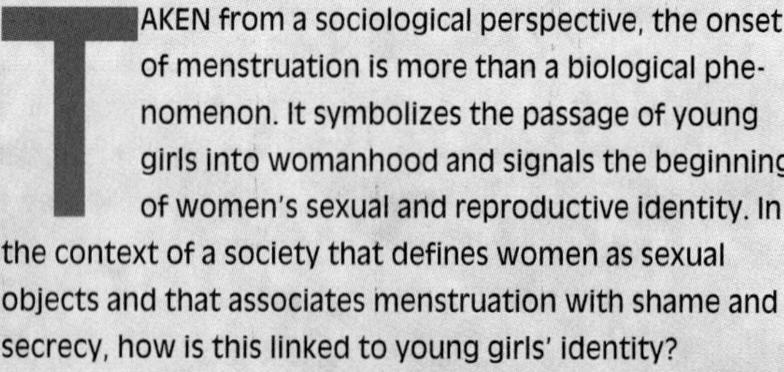

TAKEN from a sociological perspective, the onset of menstruation is more than a biological phenomenon. It symbolizes the passage of young girls into womanhood and signals the beginning of women's sexual and reproductive identity. In the context of a society that defines women as sexual objects and that associates menstruation with shame and secrecy, how is this linked to young girls' identity?

This is what sociologist Janet Lee asked in her study of adolescent girls. Lee's perspective was phenomenological—that is, she wanted to know how menstruation was defined and understood by the women. Her sample was small—forty-four women, most of them White ($N = 30$), but also including three Jewish women; two African Americans; one biracial woman; three Asian Americans; two Mexican Americans; and three women each from Nepal, Malaysia, and Iran. Her sample is not random or representative, so her study is only exploratory. The interviews are also retrospective accounts of menstruation since she interviewed the women as adults. You might ask yourself how her findings would differ had she interviewed young girls at the time they were experiencing menstruation for the first time.

In the interviews, the women talked about how ambivalent menstruation made them feel about their bodies. As young girls, they felt they had to hide this emerging evidence of their sexuality and bodily presence. Their words repeated images of contamination, coupled with a sense of alienation from their bodies—as if menstruation was something happening to them, like an invasion of the self. Menstruation also symbolized their changing relationship to boys. The women reported that their prior camaraderie with boys disappeared and that they could no longer "be one of the boys." Boys also teased them about menstruation, making them self-conscious and embarrassed. When the girls began to menstruate, a double standard emerged for boys' and girls' sexuality. The girls were told by their mothers now to be ladylike and demure even though boys might still be "sowing their wild oats." At the same time, however, menstruation produced stronger bonds between women, something Lee interprets as a form of resistance to the powerlessness women have experienced in a society where their bodies are objectified and sexualized.

Lee's research shows the significance of cultural expectations in defining otherwise physiological events, and it also indicates the influence that definitions of gender relations have on our experiences of our bodies and our sexual identities. The experiences she describes also show us the sexual socialization that takes place during adolescence and the social meanings on which adult sexual identities rest.

SOURCE: Lee, Janet. 1994. "Menarche and the (Hetero)Sexualization of the Female Body." *Gender & Society* 8 (September): 343–362.

The linkage of sexuality to race, class, and gender relations is also demonstrated by the fact that class, race, and gender hierarchies have historically been justified based on claims that people of color and women are sexually promiscuous and uncontrollable. During slavery, for example, the sexual abuse of African American women was one way that slaveowners expressed their ownership of African American people. Access to slave women's sexuality was seen as a right of the slaveowner. Under slavery, racist and sexist images of Black men and Black women were developed to justify the system of slavery. Black men were stereotyped as highly sexed, lustful beasts whose sexuality had to be controlled by the "superior" Whites. Black women were also depicted as sexual animals who were openly available to White men. These beliefs were some of the arguments used to explain why African people should be enslaved. Lynching was another mechanism of social control, used to maintain White men's power over a system of extreme racial exploitation. A Black man falsely accused of having had sex with a White woman could be murdered without penalty to his killers; for many years, including well into the twentieth century and after the end of slavery, lynching was a method of terrorism by which the Black community was told to keep in their place (Genovese, 1972; Jordan, 1968).

Similarly, sexual abuse was part of the white conquest of American Indians. Historical accounts show that the rape of Indian women by white conquerors was common (D'Emilio and Freedman, 1992), as is the rape of women following other wars and military conquests (Brownmiller, 1975), such as the widespread rape of Bosnian women that has taken place in the 1990s. In the United States, images of American Indian women using their sexual wiles to negotiate treaties between Whites and American Indians are part of the cultural mythology (Tuan, 1984).

These historical patterns continue to influence contemporary images. Advertisements commonly depict women of color in bondage, as animals, or as an exotic part of nature. These condescending images strongly imply that women of color need to be tamed or can be treated as pets (Tuan, 1984; Collins, 1990). Contemporary sexual stereotypes still define African American men as sexually threatening; similar stereotypes depict Latin men as "hot lovers" who are similarly sexually uncontrollable. Asian American women are stereotyped as passive and subservient, which is a way of keeping them in their place, since to define them as active and independent would be to liberate them from their racial–sexual status.

Finally, the linkage of sexuality to systems of inequality is also seen in the fact that *poor women and women of color are the groups most vulnerable to sexual exploitation.* It is these women who may have to resort to selling their sexuality because of their lack of economic resources. Becoming a prostitute, or otherwise working in the sex industry (as a topless dancer, striptease artist, pornographic actress, or other sex-based occupation) is often the "court of last resort" for women with limited options to support themselves. Women who sell sex are also condemned for their behavior, but less so for their male clients—further illustration of how gender stereotypes mix with race and class exploitation. Why, for example, are women, not usually men, arrested for prostitution? A sociological perspective on sexuality helps one see the social conditions of inequality that shape race, class, and gender relations in our society.

A Global Perspective on Sexuality

Once you understand the connection between sexuality and systems of inequality within the United States, you can use a similar perspective to develop a more global perspective on sexuality. Returning to a point made early in this chapter, if you thought of sexuality as only a "natural" thing, without a broader sociological perspective, you would probably never stop to think about global dimensions of sexuality. As you develop a sociological perspective, many new, fascinating questions about sexuality in an international context surface.

An obvious question is how sexuality is expressed in different cultural contexts. Cross-cultural studies of sexuality show that different sexual norms, like other social norms, develop differently within cultural meaning systems. Take sexual jealousy. Perhaps you think that seeing your sexual partner becoming sexually involved with another would "naturally" evoke jealousy—no matter where it happened. Researchers have found this not to be true. In a study comparing patterns of sexual jealousy in seven different nations (Hungary, Ireland, Mexico, the Netherlands, the United States, Russia, and the former Yugoslavia), researchers found significant cross-national differences in the degree of jealousy when women and men saw their partners kissing, flirting, or being sexually involved with another person (Buunk and Hupka, 1987). Likewise, tolerance for gay and lesbian relationships varies significantly in different societies around the world. Such cross-cultural studies can make you more sensitive to the varying cultural norms and expectations that apply to sexuality in different contexts. Different cultures simply view sexuality differently. In Islamic culture, for example, women and men are viewed as equally sexual although women's sexuality is seen as potentially disruptive and needing regulation (Mernissi, 1987).

Thinking about the global dimensions of sexuality also reveals intricate connections between sexuality and the international economy. As the world has become more globally connected, an international sex trade has flourished—one that is linked to economic development, world poverty, tourism, and the subordinate status of women in many nations. The *international sex trade,* sometimes also referred to as the "traffic in women" (Rubin, 1975) refers to the use of women, worldwide, as sex workers in an institutional context where sex itself is a commodity, bought and sold in an international marketplace where women are sex workers and where they are used to promote tourism, to cater to business and military men, and to support a huge industry of nightclubs, massage parlors, and teahouses (Enloe, 1989).

In Thailand, for example, men as tourists outnumber women by a ratio of three to one. The international sex trade in Thailand involves plane loads of businessmen who come to Thailand as tourists, sometimes explicitly to buy sexual companionship. Hostess clubs in Tokyo similarly cater to corporate men; one fascinating study of such clubs, where the researcher worked as a hostess in a Tokyo club, shows how the men's behavior in these settings is linked to the expression of their gender identity among other men (Allison, 1994). Sociologists see the international sex trade as part of the global economy, contributing to the economic development of many nations and supported by the economic dominance of certain other nations. The international sex trade has been strongly implicated in such problems as the spread of AIDS worldwide, as well as the exploitation of women in parts of the world where they have limited economic opportunities (Leuchtag, 1995; Lamas, 1996; Wawer et al., 1996; Ford and Koetsawang, 1991).

Sexuality and Sociological Theory

Putting sex in a theoretical context may seem abstract and dull (and probably not your idea of a good time!), but thinking about sexuality in an analytical framework reaches into the heart of many current social controversies. Is sex something that should be restricted to the family? Should prostitution be legal? Is being gay or lesbian a matter of choice or biological destiny? Any one of these or other questions about sexuality could be the subject of sociological study. Sociological theory puts an analytical framework around the study of sexuality, examining its connection to social institutions and current social issues. How do the major sociological theories frame an understanding of sexuality?

Sex: Functional or Conflict-Ridden?

Two of the major sociological frameworks—functionalist theory and conflict theory—take divergent paths in interpreting the social basis of human sexuality. Functionalist theory with its emphasis on the interrelatedness of different parts of society tends to depict sexuality in terms of how it contributes to the stability of social institutions. Functionalist theorists might, for example, discuss sexual norms that restrict sex to within marriage as how society encourages the formation of families. Similarly, functionalist theorists would interpret beliefs that give legitimacy to heterosexual behavior, but not homosexual behavior, as maintaining a particular form of social organization, one where gender roles are easily differentiated and where nuclear families are defined as the dominant social norm. From this point of view, regulating sexual behavior is functional for society because it prevents the instability and conflict that more liberal sexual attitudes supposedly generate.

The functionalist perspective is useful in understanding many contemporary public issues. Some have called for a return to "family values," meaning a return to more traditional nuclear families and less tolerance for the diverse family forms that have emerged in recent years. In so doing, these people are arguing for a uniformity in values that they see as necessary for social order.

Conflict theorists, on the other hand, see sexuality as part of the power relations and economic inequality in society. Power is the ability of one person or group to influence the behavior of another. Instead of arguing for a return to traditional and commonly shared values, conflict theorists argue that sexual relations are linked to other forms of subordination, namely, race, class, and gender inequality, that associate sex with power. According to conflict theorists, sexual violence, such as rape or sexual harassment, is the result of power imbalances, specifically between women and men. At the same time, because conflict theorists see economic inequality as a major basis for social conflict, they tie the study of sexuality to economic institutions. As we saw earlier in discussing the international sex trade, conflict theorists would link sexual exploitation of this sort to poverty, the status of women in society, and the economics of international development and tourism (Enloe, 1989). Conflict theorists do not see all sexual relations as oppressive; indeed, sexuality is an expression of great social intimacy. In connecting sexuality and inequality, conflict theorists are developing a structural analysis of sexuality, not condemning sexual intimacy.

Because both functionalism and conflict theory are macrosociological theories—that is, they take a broad view of society, seeing sexuality in terms of the overall social organization of society—they do not tell us much about the social construction of sexual identities. This is where the third major sociological framework, symbolic interaction, is important.

The Social Construction of Sexual Identity

Sociologists tend to see sexual identity, whatever its form, as created within specific cultural, historical, and social contexts. Drawing from symbolic interaction theory, the **social construction perspective** interprets sexual identity as learned, not inborn. To symbolic interactionists, culture and society, and less biology, shape sexual experiences since patterns of social approval and social taboos make some forms of sexuality permissible and others not (Connell, 1992; Lorber, 1994).

Biology and Sexual Identity. Is there a biological basis to sexual identity? Sociologists would say perhaps, but even if there is this influence, which many doubt, sociologists think that social experiences are far more significant in shaping any form of sexual identity (Connell, 1992; Lorber, 1994). Sociologists are also quite critical of reports about a so-called gay gene or biological cause for homosexual behavior. A study released by scientist Simon LeVay in the 1990s, which received much public attention, is a case in point.

LeVay claims to have found anatomical differences in the brains of homosexual and heterosexual men (LeVay, 1991, 1993). LeVay had sliced open the brains of forty-one corpses and found that a region of the hypothalamus was two to three times larger in the brains of heterosexual men than it was in the brains of homosexual men (LeVay, 1991).

This was taken as evidence for a physiological basis for homosexuality.

On closer examination, however, many scholars raised serious questions about the validity of these conclusions. The study involved only forty-one subjects, nineteen of whom were reported to be homosexual men who had died of AIDS; sixteen other men (six of whom were heterosexuals who had died of AIDS and ten of unknown sexual orientation who had died of other causes); and six women not identified by sexual orientation (one of whom had died of AIDS). One criticism is that the sample size was too small to make the sweeping conclusions that were reported. Studies with small samples can be informative, especially as exploratory studies, but you cannot make sweeping generalizations from studies with very small samples, as LeVay has done.

A second serious criticism was about how the sexual identities of the subjects were determined; this was not disclosed by the scientist. The study compared the corpses of men who were presumed to be homosexual and who had died of AIDS with the corpses of men presumed to be heterosexual (some of whom had died of AIDS). Both groups were compared to the six women. Critics of the study pointed out that the HIV virus is known to attack brain cells so that possibly HIV, not a biological disposition to homosexuality, had attacked the hypothalamus in the men identified as homosexuals.

Third, critics noted that the researcher found several exceptions to his findings—a serious problem especially with so few subjects to begin with (Rist, 1992). Nonetheless, few media stories reported the many criticisms that scientists made of this research. The public was left with the impression that there was strong scientific evidence to support the claim that homosexuality is biologically caused, when that evidence, on closer inspection, was potentially quite flawed.

A true scientific test of the hypothesis that homosexuality is biologically caused would require a strict standard of evidence. One could study identical twins who were the offspring of gay parents to see if, controlling for the environments in which they are raised, both turned out to be gay. For example, a 1993 study claiming to have found a genetic basis for homosexuality among men was based on a study of self-acknowledged homosexual men and their relatives. The authors acknowledge, however, that there are no studies of biological relatives raised apart (Hamer et al., 1993). Despite the authors' claim that they had identified a strong link between sexual orientation and genetics, the fact that the research subjects grew up in similar environments is a serious limitation of their research.

When LeVay's study was published in *Science* magazine, a major scientific journal widely read by the public, every major newspaper carried a story about the article, as did many radio and television broadcasts. News reports, however, rarely report the scientific methodology of research studies, leaving little basis for careful assessment of popular research conclusions. To sociologists, one of the interesting things about media headlines regarding the biological basis for homosexuality is their reception. Public beliefs that there is a biological cause to homosexuality are typically stronger and more entrenched than the current scientific evidence supports (see Figure 7.4). In fact, there is far more support in the sociological research literature for a social basis for the development of homosexuality, but these studies are rarely, if ever, reported in the media (Gagnon, 1995; Gagnon, 1973; Caulfield, 1985; Money, 1995, 1988). One cannot help speculating that if the evidence for the social basis of homosexuality received such loud acclaim, the public might be less inclined to think that sexual identity has a biological origin.

TABLE 7.1 *SEXUALITY: NATURAL OR SOCIAL CONSTRUCTION?*

	Biological Determinism	Social Construction
Sexual identity	fixed at birth	emerges through social experience
Sexual relationships	natural expressions of biological predisposition	provide the basis for forming sexual identities
Social institutions	based on "natural" sexual/social order	support dominant group sexual identities
Social change	social order relatively fixed with individual variation	change comes as people mobilize social movements to establish rights for minority groups

Debunking Society's Myths

Myth: Men and women are naturally attracted to each other.

Sociological perspective: Although there may be some biological ("natural") basis for sexual orientation, sexual identity is learned through social interaction in the context of social norms, values, and institutions.

On balance, despite widely cited periodic reports about a biological basis for homosexuality, there is not yet compelling evidence for this conclusion. Even were a biological basis for sexual identity to be found, sociological research suggests that social influences interact with biological influences, such that the environment plays a very influential role in the creation of the various dimensions of our identity. Although sociologists would not necessarily rule out a biological influence on sexual identity, they are far more interested in how sexual identity is constructed through one's social experiences.

The Social Basis of Sexuality. From a social construction perspective, one's sexual identity is not innate (that is, inborn); rather, it develops through social experiences. This is also supported by intriguing research on those born with mixed sex characteristics and those who undergo sex change operations. John Money, a nationally recognized medical scientist, has conducted extensive research on the biological and cultural bases of sexual identity (see also Chapter 12). Money concludes that humans are born "psychosexually neutral" at birth (Money, 1988, 1995). In other words, their capabilities for sexual expression are open-ended and multiple. It is our social and cultural experience that shapes and directs our sexual identities, and sexual identity is constructed over the life course, not just determined at birth.

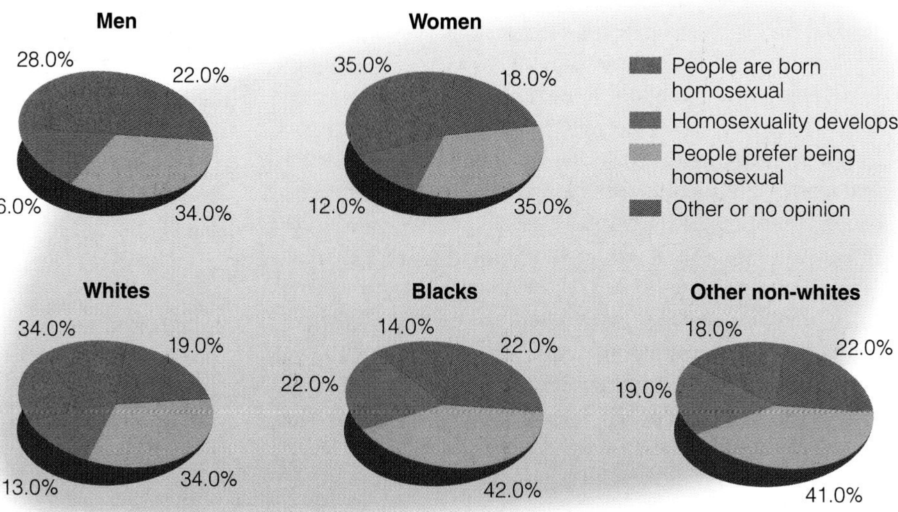

FIGURE 7.4 Attitudes on the Causes of Homosexuality

DATA: From Moore, David W. 1993. "Public Polarized on Gay Issues." *The Gallup Poll Monthly* (April). Storrs, CT: Roper Organization, pp. 30–34.

From a social constructionist perspective, different sexual identities are possible, each of them learned through the process of socialization. *Heterosexuals* are those who are sexually attracted to members of the other sex; *homosexuals* are those attracted to the same sex; and *bisexuals* are those who are attracted to members of both sexes. **Sexual orientation** is how individuals experience sexual arousal and pleasure. Debate about the terms *sexual orientation* and *sexual preference* is further evidence of the significance of social meaning in talking about sexual identities. Although some use these two terms interchangeably, gays and lesbians have argued that the term *sexual preference* makes it seem as though sexual orientation is a mere choice, whereas the term *sexual orientation* implies something more deeply rooted in their identity development. As we will see, whether sexual orientation is a matter of biology or culture, or both, is much debated—within and outside the gay community.

The social construction perspective considers no form of sexual identity more natural than any other; instead, this perspective explains all forms of sexual identity as created through the self-definitions that people develop through relationships with others. From the social constructionist point of view, sexual identity develops from exposure to different cultural expectations and, in the case of heterosexual identity, adherence to dominant cultural expectations. Those who develop heterosexual identities have experiences and receive social supports that reinforce heterosexual identity. Adrienne Rich, a feminist poet and essayist, has called this **compulsory heterosexuality,** the idea that heterosexual identity is not a choice; rather, she says, institutions define heterosexuality as the only legitimate form of sexual identity and enforce it through social norms and sanctions, including peer pressure, socialization, law, economic policies, and at the extreme, violence (Rich, 1980).

THINKING SOCIOLOGICALLY

Keep a diary for one week and write down as many examples of homophobia and heterosexism as you observe in routine social behavior during the week. What do your observations tell you about how *compulsory heterosexuality* is enforced?

Few ever question their heterosexual identity as anything other than "normal," but when you consider the extent of institutional supports for developing a heterosexual identity, it is not surprising that it is taken for granted by most. In other societies, however, where cultural supports for multiple forms of sexual identity are stronger, distinctions between heterosexual and homosexual behavior may not be so rigid. For example, the *Xaniths* of Oman, a strictly gender-segregated Islamic society, are men who are homosexual prostitutes. Different from male prostitutes in Western culture, they have a specific and valued social function: preserving the sexual purity of women. Women in this society are traditionally prohibited from mingling with men in public or having sex prior to marriage; they are veiled and robed in public. For legal purposes, women are considered lifelong minors. Xaniths, on the other hand, are biological males, who mingle freely with women and men, but maintain the legal status of men. Xaniths provide sexual outlets for men who are not married even though the men who have sex with Xaniths are not considered homosexuals. Xaniths themselves are seen as "reverting to manhood" if they later marry a woman (and they often do), so long as they provide evidence of having taken her virginity. In this culture, no man is seen as culminating his manhood until he marries and "deflowers" his bride (Lorber, 1994).

This example shows how fluid the construction of sexual identity can be. Within the United States, the social construction of sexual identity is also revealed by studies of the "coming out" process. **Coming out**—the process of defining oneself as gay or lesbian—is a series of events and redefinitions in which a person comes to see herself or himself as having a gay identity. In coming out, a person consciously labels that identity either to oneself or others (or both). Coming out is usually not the result of a single homosexual experience. If it were, there would be far more self-identified homosexuals than there are since researchers find that a substantial portion of both men and women have some form of homosexual experience at some time in their lifetime. A person may even change between gay, straight, and bisexual identities (Rust, 1993). This indicates, as the social construction perspective suggests, that identity is created, not fixed, over the course of a life.

Current research shows that the development of sexual identity is not necessarily a linear or unidirectional process, with persons moving through a defined sequence of steps or phases. Although most people learn stable sexual identities, over the course of one's life, sexual identity evolves. Heterosexual identity is thus strongly encouraged by the dominant culture, and young people growing up shape their sexual identity in this context. Sometimes, one's sexual identity may change. For example, a person who has always thought of himself or herself as a heterosexual may decide at a later time that he or she is gay or lesbian. In more unusual cases, people may actually undergo a sex change operation, changing their sexual identity in the process. In the case of bisexuals, a person might adopt a dual sexual identity.

Although sexual identity is socially constructed, individuals generally experience their sexual identities as stable, perceiving changes in their identity as part of the process of discovering their sexuality and often reinterpreting earlier life events as indications of some preexisting identity. From a sociological point of view, changing one's sexual identity is not a sign of immaturity, as if someone has missed a so-called normal phase of development. Change is a normal outcome of the process of identity formation (Rust, 1993). Unlike animal behavior, human sexual behavior is "uniquely characterized by its overwhelmingly symbolic, culturally constructed, non-procreative plasticity" (Caulfield, 1985: 344). The "plasticity" of human sexuality means it is not fixed, but, rather, emerges through social experience.

People do not necessarily move predictably through certain sequences in developing a sexual identity. Although they may experience certain "milestones" in their identity development, some people move back and forth over time between lesbian or gay identities and heterosexual identities. Some experience periods of ambivalence about their identity and may switch back and forth between lesbian, heterosexual, and bisexual identity over a period (Rust, 1993). Some people may engage in lesbian or gay behavior, but not adopt a formal definition of oneself as lesbian or gay. Certainly many gays and lesbians never adopt a public definition of themselves as gay or lesbian, instead remaining "closeted" for long periods, if not entire lifetimes.

This shows how sensitive sexual identity is to its societal context. Changing social contexts (including dominant group attitudes, laws, and systems of social control), relationships with others, and even changes in the language used to describe different sexual identities all affect people's self-definition. As an example, political movements can encourage people to adopt a sexual identity that they previously had no context to understand. Further, studies of gay and lesbian academics who are publicly "out" have shown that there are negative consequences for them, in terms of job opportunities, promotions, and inclusion in social networks (Taylor and Raeburn, 1995). Such a context can discourage those with lesbian or gay identities from publicly acknowledging this.

When Matthew Shepard, a young gay college student, was brutally beaten and murdered by two young men in Laramie, Wyoming, many organized candlelight vigils to protest this hate crime and other homophobic acts of violence.

Understanding Gay and Lesbian Experience

Sociological understanding of sexual identity has developed largely through new studies of lesbian and gay experience. Long thought of only in terms of social deviance (see Chapter 8), gays and lesbians have been much stereotyped in traditional social science, but the feminist and gay liberation movements have discouraged this approach, arguing that gay and lesbian experience is merely one alternative in the broad spectrum of human sexuality. Now there is a growing research literature examining different aspects of gay life (Nardi and Schneider, 1997; Seidman, 1994; Stein and Plummer, 1994). *Queer theory* is a new perspective that interprets various dimensions of sexuality as thoroughly social and constructed through institutional practices.

The institutional context for sexuality within the United States, as well as other societies, is one where homophobia permeates the culture. **Homophobia** is the fear and hatred of homosexuality. It is manifested in prejudiced attitudes toward gays and lesbians, as well as in overt hostility and violence directed against people suspected of being gay. Homophobia is a learned attitude, as are other forms of negative social judgments about particular groups. Homophobia is deeply embedded in people's definitions of themselves as men and women. Boys are often raised to be "manly" by repressing so-called feminine characteristics in themselves. Being called a "fag" or a "sissy" is one of the peer sanctions that socializes a child to conform to particular gender roles. Similarly, verbal attacks on lesbians are a mechanism of

homophobian

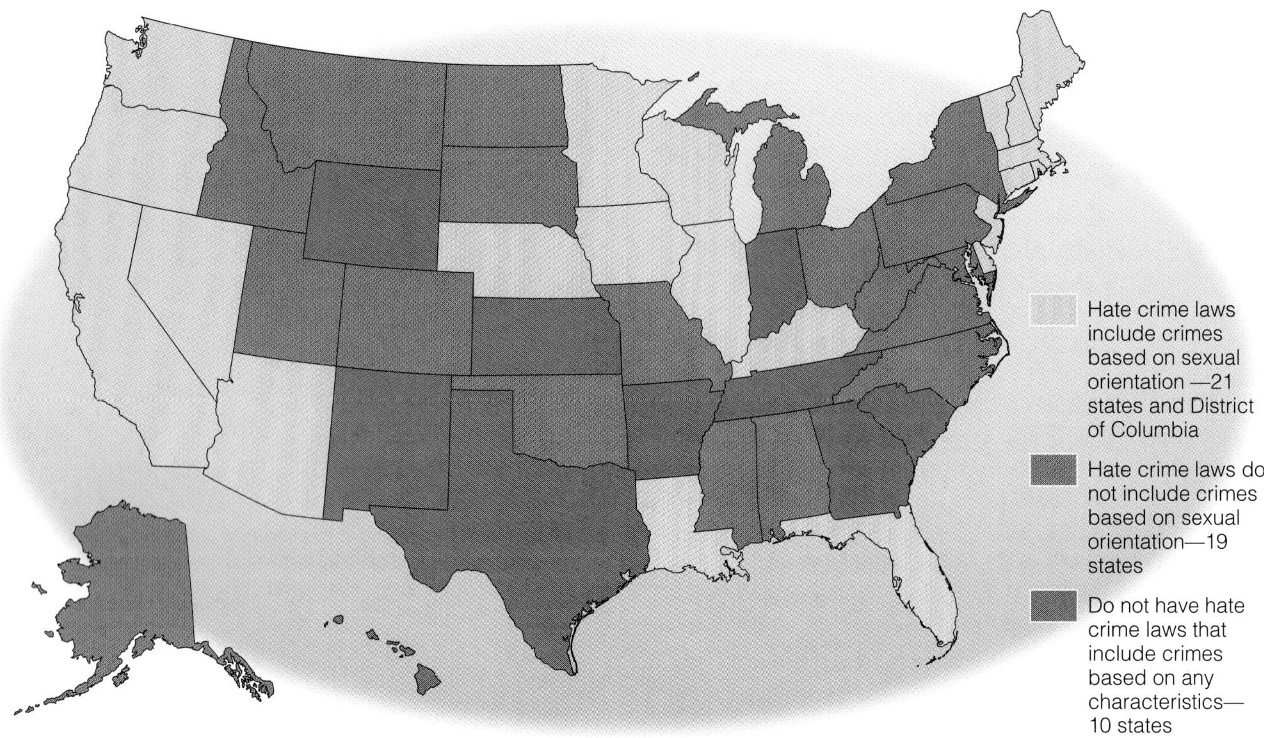

MAP 7.2 Mapping America's Diversity: Hate Crime Legislation

SOURCE: National Gay and Lesbian Task Force. July, 1998. Washington, D.C.

social control since ridicule can be interpreted as encouraging social conformity.

Homophobia produces many misunderstandings about gay people. One such misunderstanding is the idea that gays have a desire to seduce straight people. This belief has been expressed in the debate about allowing gays and lesbians into the military. Fears that living in close quarters or having to shower near a gay person would lead to seduction reflect the belief that gays and lesbians cannot control their sexuality. This belief persists even though no evidence exists that homosexual men are more likely to commit sex crimes than heterosexual men; homosexual men convicted of child molestation are, however, almost seven times more likely to be imprisoned than are heterosexual men who are convicted as child molesters and have the same past criminal record (Walsh, 1994).

Fears that gay and lesbian parents will have negative effects on their children are also unsubstantiated by research, and most children of gay parents grow up with a heterosexual identity (Lewin, 1984). The ability of parents to form good relationships with their children is far more significant in children's social development than is their parents' sexual orientation (Joseph, 1984).

Other myths about gay people are that they are mostly White men with large discretionary incomes who work primarily in artistic areas and personal service jobs (like hairdressing). This stereotype prevents people from recognizing that there are gays and lesbians in all racial–ethnic groups, some of whom are working class or poor, and who are employed in a wide range of occupations (Gluckman and Reed, 1997). Some lesbians and gays are also old people even though the stereotype defines them as mostly young or middle aged (Smith, 1983). These different misunderstandings reveal just a few of the many unfounded myths about gays and lesbians. Support for these attitudes comes from homophobic attitudes, not from empirical truths.

Heterosexism refers to the institutionalization of heterosexuality as the only socially legitimate sexual orientation. Heterosexism is rooted in the belief that heterosexual behavior is the only natural form of sexual expression and that homosexuality is a perversion of "normal" sexual identity. Heterosexism is reinforced through institutional mechanisms that project the idea that only heterosexuality is normal. Institutions also structure the unequal distribution of privileges to people presumed to be heterosexual. Businesses and communities, for example, rarely recognize the legal rights of those in homosexual relationships, although this is changing. Heterosexism is an institutional phenomenon although people's beliefs can reflect heterosexist assumptions. Thus a person may be quite accepting of gay and lesbian people (that is, not be homophobic), but still benefit from heterosexual privileges or engage in heterosexist practices that have the unintentional effect of excluding lesbians and gays, such as by talking with co-workers about their dating activities on the assumption that everyone is interested in a heterosexual partner.

The heterosexist framework of social institutions means that gay and lesbian relationships lack the institutional sup-

heterosexism

port of heterosexual relationships. In the absence of such socially sanctioned institutions, lesbians and gays have invented their own rituals and communities to support these relationships. Whereas heterosexual relations get many social supports that encourage their formation and continuation, the durability of gay and lesbian relationships depends more on the quality of the relationship than on institutional support mechanisms. Those who remain in the closet deny themselves this support system, unless they have organized their lives in such a way that they maintain two separate lives, one gay and one outwardly straight (Meyer, 1990).

The absence of institutionalized roles for lesbians and gays affects the roles they adopt within relationships. Despite popular stereotypes, gay partners typically do not assume roles as the dominant or subordinate sexual partner; they are more likely to adopt roles as equals. Gay and lesbian couples are also more likely than heterosexual couples to both be employed, another source of greater equality within the relationship. Researchers have also found that the quality of relationships among gay men is positively correlated with the social support the couple receives from others (Smith and Brown, 1997; Metz et al., 1994; Harry, 1979).

Lesbians and gays are a minority group in our society, one that has been denied equal rights in the society. In sociology, the term **minority group** has a specific meaning. Minority groups are not necessarily numerical minorities; minority groups are those with similar characteristics (or at least perceived similar characteristics) that are treated with prejudice and discrimination (see also Chapter 11). Minority groups also encounter significant prejudice directed against them. Gays and lesbians are minority groups in society in that they have been denied equal rights and are singled out for negative treatment by dominant groups. As a minority group, gays and lesbians have organized to advocate for their rights and to be recognized as socially legitimate citizens. Some organizations and municipalities have enacted civil rights protections on behalf of gays and lesbians, typically prohibiting discrimination in hiring against gays. Gays and lesbians won a major civil rights victory in 1996 when the Supreme Court ruled (*Romer v. Evans*) that they cannot be denied equal protection of the law. This ruling overturned the ordinances that many states and municipalities had passed that specifically denied civil rights protections to lesbians and gays (Greenhouse, 1996a). At the same time, however, the U.S. Congress passed a law in 1996 (the "Defense of Marriage Act") that allows states not to recognize same-sex marriages that other states might legally define as legitimate marriages.

Sex and Social Issues

In studying sexuality, sociologists tap into some of the highly contested social issues of the time. Reproductive technology, abortion, teen pregnancy, and pornography are all subjects of public debate and controversy, and are important in the

The gay rights movement has brought the issue of sexual orientation "out of the closet," making lesbians and gays more visible in society and enabling sociologists to ask new questions about sexuality and its relationship to social behavior.

formation of contemporary social policy. Debates about these issues hinge in part on attitudes about sexuality and are also shaped by race, class, and gender relations.

Although most people think of sex in terms of interpersonal relationships, sexual norms are deeply intertwined with various social problems in society. Social issues like reproductive technology, pornography, and teen pregnancy are good examples of personal troubles that have their origins in the structure of society—recall the distinction made by C. Wright Mills between personal troubles and social issues. These different issues show us how the social structure of society contributes to some of our nation's social problems.

Birth Control

The availability of birth control is now less debated than it was in the not too distant past, but this important reproductive technology has been strongly related to the position of women in society (Gordon, 1977). Reproduction has been controlled by men who have made most of the decisions, legal and scientific, about whether birth control will be available or not. To this day, it is mostly men who define the laws and make the scientific decisions about what types of birth control will be available. The social position of women also means that women are more likely seen as responsible for reproduction since that is a presumed part of their traditional role. At the same time, changes in birth control technology have also made it possible for women to change their roles in society since breaking the link between sex and reproduction has freed women from some traditional constraints.

The right to birth control is a recently won freedom. It was not until 1965 that the Supreme Court, in *Griswold v. Connecticut,* defined the use of birth control as a right, not a crime. This ruling applied originally only to married people;

unmarried people were not extended the same right until 1972, in the Supreme Court decision, *Eisenstadt v. Baird.* Today, birth control is routinely available over the counter, but there is heated debate at the highest levels of public policy about whether access to birth control should be curtailed for the young and information about it restricted—even as sociological data show that U.S. youths are experimenting with sex at younger ages and therefore increasing the risk of teenage pregnancy and AIDS. Some argue that sexual activity at this age must be controlled and that increasing access to birth control and knowledge about how to use it would encourage more sexual activity among the young.

Class and race relations have also had a role in shaping birth control policy. The birth control movement in the United States began in the midnineteenth century. During this period, increased urbanization and industrialization removed the necessity for large families, especially in the middle class, since fewer laborers were needed to support the family. The ideology of the birth control movement reflected these conditions. Early feminist activists like Emma Goldman and Margaret Sanger also saw birth control as a way of freeing women from unwanted pregnancies and allowing them to work outside the home if they chose, rather than being obliged to work in the home in support of a large family. As the birthrate fell among White upper-class and middle-class families in this period, these classes feared that immigrants, the poor, and racial minorities would soon outnumber them. The dominant class feared "race suicide"; they imagined that immigrants and the poor would overpopulate the society, leaving the White middle and upper class as a numerical minority. The eugenics movement of the early twentieth century grew from this social climate.

Eugenics sought to apply scientific principles of genetic selection to "improve" the offspring of the human race. It was explicitly racist and class based, calling, among other things, for the compulsory sterilization of those who eugenicists thought were unfit. Margaret Sanger herself made eugenics part of her birth control campaign although she wanted poor women to have fewer children so that their children would not be exploited as workers. Eugenicist arguments, though they make scientific claims, are typically based on popular social stereotypes more than scientific fact. They appeal to a public that fears the social problems that emerge from racism, class inequality, and other injustices, but, rather than attributing these problems (such as crime and poor educational achievement) to the structure of society, eugenics puts the blame on the genetic composition of certain groups—groups that are among the least powerful in society.

Contemporary arguments that differences in educational attainment can be attributed to genetic differences raise the specter of a new eugenics movement—one perhaps more damaging than that of the early twentieth century since modern technologies now make genetic engineering possible when earlier it could only be imagined. Socially identifiable groups can now be screened for "genetic disorders." Coupled with new forms of racism that have evolved from contemporary conservatives, old eugenicist arguments have been reinvigorated, linking race, gender, and ethnicity to an assortment of phenomena, including propensity to crime, mental illness, homosexuality, alcoholism, and poor educational achievement (Duster, 1990). Echoes of eugenicism can be heard in the calls for sperm banks reserved for "smart" people where eager parents could create "designer children" free of defects and stigmatized traits.

New Reproductive Technologies

Developments in the technology of reproduction have ushered in new sexual freedoms, but they have also raised questions for social policy. Practices like surrogate mothering; in vitro fertilization; and new biotechnologies of gene splicing, cloning, and genetic engineering mean that reproduction is no longer inextricably linked to biological parents (Rifkin, 1998). A child may be conceived through means other than sexual relations between one man and one woman; one woman may carry the child of another; offspring may be planned through genetic engineering. A sheep can be cloned (genetically duplicated); are humans next? With such developments, those who could not otherwise have children (such as infertile couples, single women, or lesbian couples) are now able to do so, thereby raising new questions. To whom are such new technologies available? Which groups are most likely to sell reproductive services, which groups to buy? What are the social implications of such changes?

There are not simple answers to such questions, but sociologists would point first to the class, race, and gender dimensions of these issues. Poor women, for example, are far more likely than middle-class or elite women to sell their eggs or offer their bodies as biological incubators. Those groups that can afford new, extremely expensive methods of reproduction may do so at the expense of women whose economic need places them in the position of selling themselves for economic support. Sociologists also point to the potential for such technologies to produce a new eugenics movement. Sophisticated prenatal screening makes it possible to identify fetuses with presumed defects. Might the society then try to weed out those perceived as undesirables—the disabled, certain racial groups, certain sexes? If boys and girls are differently valued, one sex may be more frequently aborted, a practice that is frequent in India and China—two of the most populous nations on earth. State policy in China, for example, because of population pressures, encourages families to have only one child. Since girls are less valued than boys, the aborting and selling of girls is common; indeed, it has created a market for the adoption of Chinese baby girls in the United States.

Breakthroughs in reproductive technology raise especially difficult questions for makers of social policy. There are no traditions to guide us on questions that would have been considered outlandish and remote just a few years ago. Our thinking about sexuality and reproduction will have to evolve. Although the concept of reproductive choice is im-

eugenics

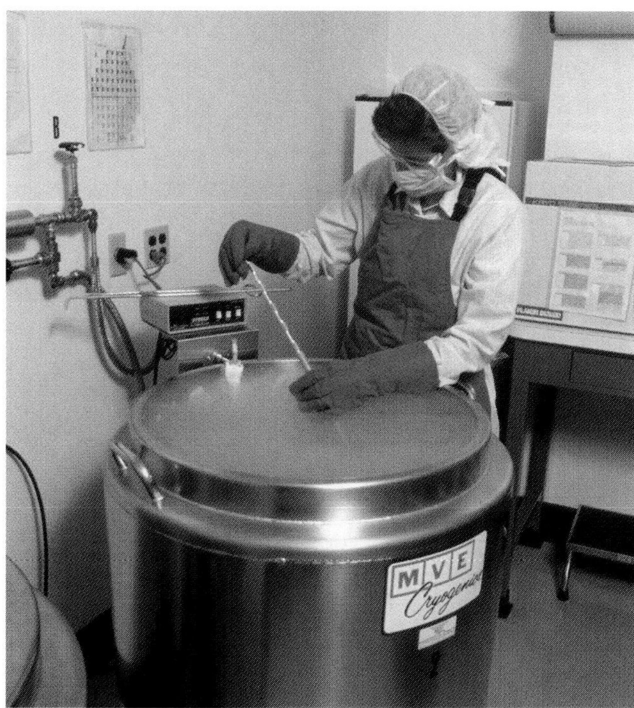

New reproductive technologies, such as this sperm bank, raise new ethical questions about sexuality, reproduction, race, class, and gender. Does the technological capability to engineer human life mean that another eugenics movement could arise?

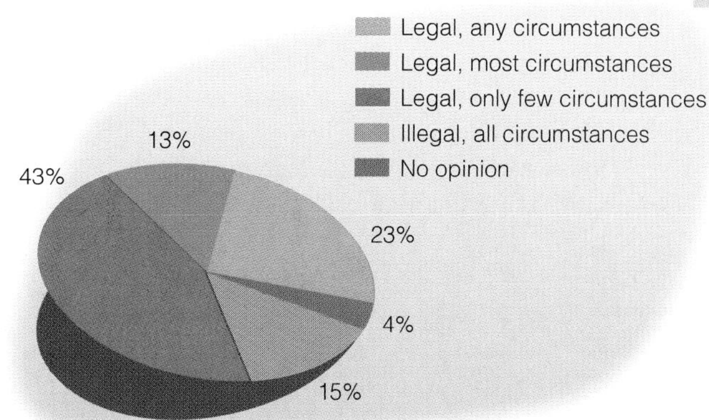

FIGURE 7.5 Attitudes Toward Abortion

DATA: From The Gallup Poll. 1996. (September): p. 110. Roper Organization, Storrs, CT.

portant to most people, choice is conditioned by the constraints of race, class, and gender inequalities in society. Like other social phenomena, sexuality and reproduction are shaped by their social context.

Abortion

Abortion is one of the most seriously contended political issues in recent years. The majority of the U.S. public clearly support abortion rights under at least some circumstances. Now, the majority of the public think abortion should be legal—at least in some circumstances; 15 percent think it should be illegal (see Figure 7.5). Public support for abortion rights has remained relatively constant over the last twenty years.

The right to abortion was first established in constitutional law by the *Roe v. Wade* decision in 1973. In *Roe v. Wade,* the Supreme Court ruled that at different points during a pregnancy, separate but legitimate rights collided: the right to privacy, the right of the state to protect maternal health, and the right of the state to protect developing life. To resolve this conflict of rights, the Supreme Court ruled that pregnancy occurred in trimesters. In the first, women's right to privacy without interference from the state prevailed; in the second, the state's right to protect maternal health took precedence; in the third, the state's right to protect developing life prevailed. In the second trimester, the government cannot deny the right to abortion, but it can insist on reasonable standards of medical procedure. In the third, abortion may be performed only to save the life or health of the mother. More recently, the Court has allowed states to impose restrictions on abortion but it has not, to date, overturned the legal framework of *Roe v. Wade.*

Data on abortion show that it occurs across social groups although certain patterns do emerge. The abortion rate has declined somewhat since 1980 (from a rate of 35.9 per 1000 women to 31.1). Young women are the most likely to get abortions although fewer teenagers are having abortions than in years prior. Black women have the highest rate of abortion (53.4 per 1000 women), compared to White women (20.4; National Center for Health Statistics, 1998). Perhaps surprisingly, Roman Catholic women have the same rate of abortion as other women. Women who described themselves as born-again or evangelical Christians are half as likely to have abortions as other women although they still accounted for one in six women who have abortions. The majority of women who have abortions say they do so because they cannot afford to have a baby or because having a child will interfere with their work or schooling. Studies also find that six in ten of those having abortions say they were using contraceptives when they became pregnant although they were likely not using them correctly (Henshaw and Silverman, 1988; Henshaw and Kost, 1996; U.S. Bureau of the Census, 1997).

The abortion issue provides a good illustration of how sexuality has entered the political realm. Sociological research on abortion rights activists and antiabortion activists demonstrates that each group holds very different views about sexuality and the roles of women. Antiabortion activists tend to believe that giving women control over their fertility breaks up the stable relationships in traditional families. They tend to view sex as something that is sacred, and they are disturbed by changes that make sex less restrictive. Antiabortion activists also tend to believe that making contraception available to teens encourages promiscuity. Abortion rights activists, on the other hand, see women's

The debate over abortion rights has brought issues of sexual behavior into the realm of social policy. Research on pro-choice and right-to-life activists finds that, in general, the two groups differ significantly in educational attainment, social class, and family structure.

control over reproduction as essential for their independence; they encourage women to take control of their own lives. Abortion rights activists also tend to see sex, not just as a sacred act, but as an experience that develops intimacy and communication between people who love each other. They support the freedom of women to enter and leave sexual relationships, whereas most antiabortion activists (certainly not all) are more likely to advocate women remaining within traditional relationships with their husbands. The abortion debate can thus be interpreted not only as a struggle over abortion but also as a battle over different sexual values and a referendum on the nature of men's and women's relationships (Luker, 1984).

Pornography

There is little social consensus about the acceptability and effects of pornography. Part of this debate is about defining what is obscene. The legal definition of obscenity is one that changes over time and in different political contexts. The U.S. Supreme Court offered the first set of comprehensive guidelines about obscenity in 1957. In this decision (*United States v. Roth*), material was defined as obscene if, by applying contemporary community standards, the court concluded that the dominant theme of the material taken as a whole seemed designed to appeal to prurient interests in the average person. Two subsequent decisions added to this definition of obscenity. First, in 1962, the Supreme Court added the criterion that the material must be patently offensive (*Manual Enterprises v. Day*). Two years later, in *Jackobellis v. Ohio,* the Court added a third criterion that obscenity is material utterly without redeeming social value. This made it difficult to prove obscenity since what is considered redeeming is a highly subjective judgment. In 1973, it became easier to legally prove obscenity when the Court ruled in *Miller v. Ohio* that material is obscene when it lacks substantial literary, political, artistic, or scientific value. This decision also determined that community standards could refer to the state or local community, not a single national standard. Many think that since the 1973 ruling, sexually explicit material has increased dramatically because often the communities in which outlets for pornography appear are already economically vulnerable communities. Although residents may organize to oppose the presence of pornography, the fact that pornography appears in their neighborhoods and not well-to-do communities tells you about the power of different social groups to control their lives.

Public agitation over pornography has divided into those who think it is solidly protected by the First Amendment, those who want it strictly controlled, those who think it should be totally banned for moral reasons, and those who think it must be banned because it harms women. Even within the feminist movement, there has been much debate on this issue. Many feminists hold firmly to the principle of freedom of expression guaranteed by the First Amendment. Others argue that pornography must be treated differently from other forms of expression because it is actually a form of violence against women—against the women who appear in pornographic materials and against

Sexually explicit images are easily available on the Internet, raising new policy questions about balancing the values of First Amendment freedoms of free speech and access to information with questions about censorship.

pornography

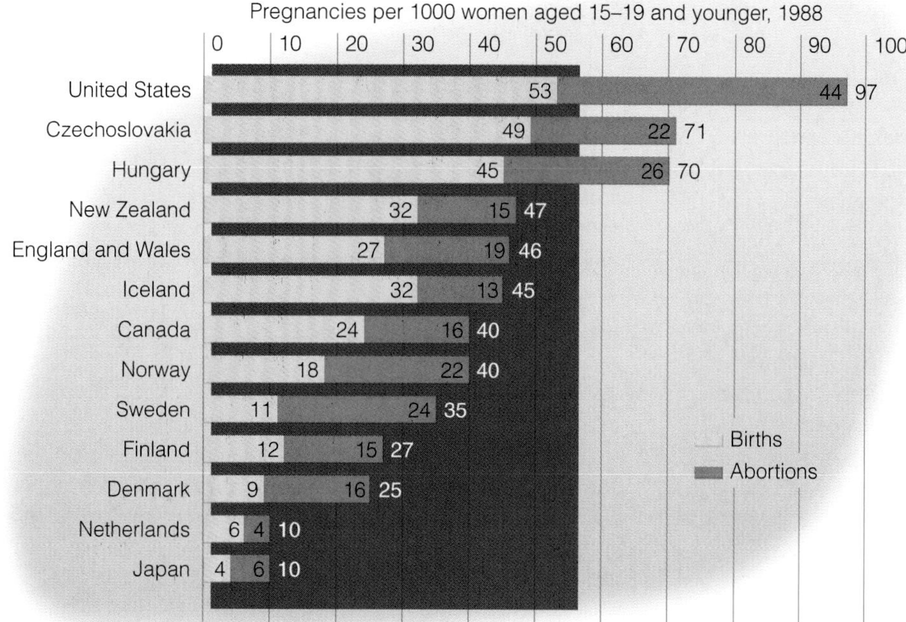

FIGURE 7.6 Teen Pregnancy: An International Perspective

SOURCE: The Alan Guttmacher Institute. 1994. *Sex and America's Teenagers.* New York: The Alan Guttmacher Institute, p. 76.

all women because of the effect pornography has on the overall social climate. Some feminists also argue that outlawing pornography leaves only women who are already economically dependent on this industry in a more vulnerable position. Without pornography as an industry where they can earn money, they may be forced into other illegal enterprises. This debate within the feminist movement is a fair reflection of the extent to which the public is divided on the issue of pornography.

Two-thirds of the U.S. public say they are concerned about the proliferation of pornography; 48 percent of women and 25 percent of men say there should be laws forbidding the distribution of pornography, and almost half (48 percent) think pornography leads people to commit rape (U.S. Bureau of Justice Statistics, 1996; Thompson et al., 1990). Are they right? Recall the discussion from Chapter 2 that reported on studies showing that exposure to violent pornography does have an effect on sexual attitudes. After exposure to violent pornography, men are more likely to see victims of rape as responsible for their assault, less likely to regard them as injured, and more likely to accept the idea that women enjoy rape. These effects are found, however, only after exposure to pornography that is both sexually explicit and violent (Donnerstein et al., 1987). There is little evidence to support the assertion that exposure to pornography in general increases sexual promiscuity or sexual deviance.

Despite public concerns about pornography, a strong majority (two-thirds) think pornography should be protected by the constitutional guarantees of free speech and a free press. At the same time, people believe that pornography dehumanizes women. Women tend to be more negative about pornography than men, perhaps because pornography is generally geared toward men. Men are more likely than women to see pornography as a form of sex education and as a means for releasing sexual tension. Some people say pornography can improve people's sex lives (Thompson et al., 1990). The best predictors of people's attitudes about pornography and its censorship, beyond gender per se, are other attitudes. Those holding the strongest gender role stereotypes, sexual conservatives, and religious traditionalists are those most opposed to the distribution of pornography (Woodrum, 1992; Fisher et al., 1994). The controversy about pornography is not likely to go away since it taps so many different sexual values among the public.

Teenage Pregnancy

Each year a half million teenage girls have babies in the United States. This country has the highest rate of teen pregnancy in the world even though levels of teen sexual activity around the world are roughly comparable (see Figure 7.6). Teens accounted for almost 13 percent of all births (11 percent of White births, 23 percent for African Americans, 23 percent for American Indians, 17 percent for Hispanics, and almost 6 percent for Asian Americans and Pacific Islanders). Since 1970, the rate of teen pregnancy for White teens has increased; for African Americans, it declined over the same period. Marriage rates for pregnant teens have also dropped, meaning that most babies born to teens will be raised by single mothers—a departure from the past when teen mothers more often got married (U.S. Bureau of the Census, 1997; Alan Guttmacher Institute, 1994). What concerns people about teen parents is that they earn half as much in their lifetime as women who wait to have children until they are twenty. Teen mothers are twice as likely to be poor as other mothers (Children's Defense Fund, 1997; Edelman, 1987; Alan Guttmacher Institute, 1994). In short, teen parents are among the most vulnerable of all social groups.

Teenage pregnancy correlates strongly with poverty, lower educational attainment, joblessness, and health problems although it is important to point out that teen pregnancy does not necessarily cause poverty since most teen mothers were poor prior to their becoming pregnant (Luker, 1996). Teen mothers face numerous risks, however. They have a greater incidence of problem pregnancies and are more likely to deliver low-birth-weight babies, a condition associated with a myriad of other health problems. Teen parents face chronic unemployment. A teen parent has only a 50 percent probability of graduating from high school. Many continue to live with their parents although this is more likely among Black teens than among Whites. Given the higher rates of poverty among African Americans, the costs of adolescent pregnancy then fall most heavily on those families least able to help, as we saw in the "Doing Sociological Research" box in Chapter 1 (Kaplan, 1996).

Although teen mothers feel less pressure to marry than they did in the past, if they raise their children alone, they suffer the economic consequences of raising children in female-headed households—the poorest of all income groups. Teen mothers report that they do not marry because they do not think the fathers are ready for marriage. Sometimes their families also counsel them against marrying precipitously. These young women are often doubtful about men's ability to support them; they want men to be committed to them and their child, but they do not expect their hopes to be fulfilled (Farber, 1990). As a result, most feel resigned to their plight (Ladner, 1986). Both Black and White teen parents see marriage as an impediment to their career ambitions although Black teen mothers are more likely than Whites to continue attending school (Farber, 1990; Trent and Harlan, 1990).

Why do so many teens become pregnant given the widespread availability of birth control? One-third of teenagers use no contraceptive the first time they have sexual intercourse, but typically delay using contraceptives until several months after they become sexually active. In fact, researchers have found that most teens delay the use of birth control until after they first suspect they are pregnant. About one-

Debunking Society's Myths

Myth: Teens have babies so that they can collect bigger welfare checks.

Sociological perspective: Research has found that, looking across the different states, there is no correlation between welfare benefits and the number of children; actually, having higher welfare benefits hastens the exit from poverty (Butler, 1996; see also Chapter 9).

B • O • X 7.2 **SOCIOLOGY IN PRACTICE**

Report to Congress: Pregnancy Outside Marriage

IN 1995, A TEAM of sociologists produced a report for Congress on childbearing outside marriage. Congress had mandated the report as part of a provision requiring the secretary of Health and Human Services to submit to Congress an analysis of the causes, consequences, and possible preventive measures for childbearing out of marriage. The report challenged a number of previously held assumptions about pregnancy outside marriage, including:

- Economic factors and attitudes about marriage, sex, and childbearing are the causes of increased out-of-wedlock births, not welfare benefits.
- Most women bearing children out of marriage are not teenagers although out-of-wedlock birth is disproportionately high for teenagers.
- The majority of out-of-wedlock births are to White women.
- Poorly educated and less affluent men and women are less likely to marry, but not less likely to have children.
- Unmarried couples experiencing a pregnancy today are much less likely to marry than such couples twenty-five to thirty years ago.
- The population at risk for nonmarital pregnancy is much larger today than in the past.
- Unmarried women who are sexually active are less likely than married women to use contraceptives.
- Abortions among unmarried pregnant women decreased substantially between 1980 and 1991.
- Evidence linking welfare benefits to increases in nonmarital births is inconsistent; when a link is found, it is small.

Included on the team were sociologists from a host of institutions: Daniel T. Lichter, Pennsylvania State University; Linda M. Burton, Pennsylvania State University; Martha R. Burt, The Urban Institute; Arlan D. Thornton, University of Michigan; Brent C. Miller, Brigham Young University; and Sara S. McLanahan, Princeton University. This effort shows the potential impact that sociological research can have on public policy. Sound public policies are based on sound factual research. Particularly when the subject is one surrounded by many social myths, sociological research can provide the evidence that allows policy makers to make reasoned judgments, not prejudicial ones, about social issues. Students who are interested in reading the full report can order it from the U.S. Government Printing Office.

SOURCE: American Sociological Association. 1995. "New Report Explodes Myths on Nonmarital Pregnancy." *Footnotes* 23 (November): 1 ff.

quarter continue not to use contraceptives when they are sexually active, but half of teen pregnancies occur within six months of the time that the teen first engages in sexual intercourse. In recent years, the percentage of teens using birth control has increased (especially condom use), although the pill is still the most widely used method (44 percent), followed by the condom (38 percent of birth control users), and next, injectable birth control (10 percent) (Alan Guttmacher Institute, 1998; Homes for the Homeless, 1996).

Sociologists have argued that the effective use of birth control requires a person to identify himself or herself as sexually active (Luker, 1975). Teen sex, however, tends to be episodic; teens who have sex on a couple of special occasions may not identify themselves as sexually active and may not feel obliged to take responsibility for birth control. Despite the large number of teens initiating sex at an earlier age, social pressure continues to discourage them from defining themselves openly, or even privately, as sexually active.

At the same time, teen pregnancy is integrally linked to the gender expectations of men and women in society. Some teen men consciously avoid birth control, thinking it takes away from their manhood. Teen women often romanticize motherhood, thinking that becoming a mother will give them social value they do not otherwise have; for teens in disadvantaged groups, motherhood confers a legitimate social identity on those otherwise devalued by society (Horowitz, 1995). Although their hopes about motherhood are not realistic, they are an indicator of how pessimistic the teenagers feel about their lives, which are often marked by poverty, a lack of education, and few good job possibilities (Ladner, 1986). That young women romanticize motherhood is not surprising in a culture where motherhood is defined as a cultural ideal for women, but the ideal can seldom be realized when society gives mothers little institutional or economic support.

Teen parents face unique challenges, particulary in continuing their own education, supporting their children, and finding adequate child care.

Review of these different social issues shows how sociological study of sexuality can help us understand some of our difficult social problems. As the box "Sociology in Practice: Report to Congress: Pregnancy Outside Marriage" shows, sociologists have actively supported research on sex and related topics (like teen pregnancy and AIDS awareness) as a way of grappling with these difficult social issues.

CHAPTER SUMMARY

- Sexual relationships develop within a social and cultural context. Sexuality is learned through socialization and channeled and directed by social institutions, and reflects the race, class, and gender relations in society.
- *Sexology* is the scientific study of sex. Freud was one of the most influential early writers on sex. He showed the social dimension of sexual identity. Many of his ideas, especially those involving women's sexuality, have subsequently been criticized. Studies like those of Alfred Kinsey and William Masters and Virginia Johnson have added much to our understanding of human sexual response.
- Contemporary sexual attitudes vary considerably by social factors like age, gender, race, and religion. Sexual behavior has also changed in recent years. Young people initiate sex sooner. Although the public believes in marital fidelity, many have extramarital affairs. The number of gays and lesbians in the population, though debated, is significant. Despite the *sexual revolution,* Americans are not well informed about sex. Cultural expectations about sexuality are also deeply influenced by gender expectations in the society.
- Sociologists see sexual identity, whatever its form, as constructed through socialization experiences. They have moved away from studying gay and lesbian experience as deviant behavior. There is still debate about the underlying origins of homosexuality although evidence for a biological basis of homosexuality is not as strong as the evidence for social causes. *Homophobia* is the fear and hatred of homosexuals. *Heterosexism* is the institutionalization of heterosexuality as the only socially legitimate *sexual orientation.*
- Reproduction is also shaped by social relations. Social issues surrounding birth control, abortion, and new reproductive technologies are situated within the sexual, class, and racial politics of society. Finally, sexuality is related to some of the most difficult social problems —including reproductive technologies, pornography, and teen pregnancy.

KEY TERMS

coming out
compulsory heterosexuality
eugenics
heterosexism
homophobia
minority group
script
sexology
sexual orientation
sexual politics
sexual scripts
sexual revolution
social construction perspective

THE INTERNET: A Tool for the Sociological Imagination

Resources on the Internet:

Virtual Society: The Wadsworth Sociology Resource Center at
http://sociology.wadsworth.com

Visit this site to find additional learning tools, including interactive quizzes, links to related web sites, and an easy link to InfoTrac College Edition.

Alan Guttmacher Institute
http://www.agi-usa.org/

An organization that produces numerous reports on sexual activity and contraceptive practices

The Kinsey Institute
http://www.indiana.edu/~kinsey/

An institute supporting research on human sexuality

Sociology and Social Policy: Internet Exercises

Imagine that you have been asked to chair a commission that will make recommendations to the Congress about reducing *teen pregnancy*. Your report should include a statement of the causes and consequences of teen pregnancy, as well as recommendations for social policy on teen pregnancy. Because your commission involves political appointments, different points of view will emerge in your discussion, but you should be prepared to defend the policies you recommend with sound research. What policies would you recommend?

Internet Search Keywords:

teen pregnancy
birth control
social policy
sex education
reproductive health
sexuality

Web sites:

http://www.teenpregnancy.org/
Web site for the National Campaign to Prevent Teen Pregnancy.

http://www.advocatesforyouth.org/
This site works to help teenagers make informed decisions about sexuality.

http://www.agi-use.org/
The home page of the Alan Guttmacher Institute. AGI seeks to help individuals with reproductive health choices.

http://www.noappp.org/
The National Organization on Adolescent Pregnancy, Parenting and Prevention provides a forum for addressing adolescent pregnancy.

http://www.rci.rutgers.edu/~sxetc/
This is a teen-produced web site that answers questions from teens about sex.

http://babynet.ddwi.com/tlc/pregnancy/teenfact.html
A web site that discusses health risks of teenage pregnancy.

InfoTrac College Edition: Search Word Summary

eugenics	pornography
heterosexism	sexual revolution
homophobia	

In order to learn more about these central topics in sociology, you can conduct an electronic search using InfoTrac College Edition. To aid in your search and to gain useful tips, see the Student Guide to InfoTrac College Edition on the Virtual Society web site:
http://sociology.wadsworth.com

INTERACTIONS—A SOCIOLOGY CD-ROM: CONCEPTS FOR THIS CHAPTER

Go to the Wadsworth Sociology CD-ROM for further study on the concepts in this chapter. The CD-ROM also includes quizzes and additional activities to expand your learning experience.

SUGGESTED READINGS

D'Emilio, John, and Estelle B. Freedman. 1988. *Intimate Matters: A History of Sexuality in America.* New York: Harper and Row.

This is a comprehensive examination of the historical context of sexual behavior and attitudes. The book shows how sexuality is often the basis for political and social conflicts, and it examines the relationship of race, class, and gender inequality to the history of sexual practices and ideologies.

Duster, Troy. 1990. *Backdoor to Eugenics.* New York: Routledge.

Duster explores new genetic technologies, looking at the social dimensions of these new techniques. He warns against the misuse of technologies if they are linked to specific race and class groups in the society. The book also provides an excellent historical overview of the earlier eugenics movement and its connection to race and class inequality in the United States.

Irvine, Janice M. 1990. *Disorders of Desire: Sex and Gender in Modern American Sexology.* Philadelphia: Temple University Press.

This discussion of the social context of sex research offers a sociological account of changing attitudes about sexuality and its social-historical context.

Luker, Kristin. 1996. *Dubious Conceptions: The Politics of Teenage Pregnancy.* Cambridge: Harvard University Press.

Luker analyzes social policy about teenage pregnancy, with a keen eye to how gender, class, and race shape public beliefs and social policy on teen pregnancy. Her book also provides current and comprehensive discussion of the causes of teenage pregnancy, some of which are counter to dominant impressions.

Michael, Robert T., John H. Gagnon, Edward O. Laumann, and Gina Kolata. 1994. *Sex in America: A Definitive Survey.* Boston: Little, Brown and Company.

Written for a wide audience, this book presents a sociological overview of sexual behavior in the 1990s. Based on a recent survey of sexual practices, it is the most current available research on sexual habits of the American public.

Nardi, Peter M., and Beth Schneider. 1997. *Social Perspectives on Lesbian and Gay Studies.* New York: Routledge.

This collection of articles provides a sociological perspective on new research on gay and lesbian issues. It includes classic readings with the work of young scholars who are redefining questions about sexual identity, community, and social change.

Schwartz, Pepper, and Virginia Rutter. 1998. *The Gender of Sexuality.* Thousand Oaks, CA: Pine Forge Press.

This review of sociological perspectives on sexuality links the study of gender roles to sexuality.

CHAPTER 8

Deviance and Social Control

IN THE EARLY 1970s, an airplane carrying forty members of an amateur rugby team crashed in the Andes Mountains in South America. The twenty-seven survivors, teenagers and some of their relatives and friends from Uruguay's elite class, were marooned at 12,000 feet in freezing weather and deep snow. There was no food except for a small amount of chocolate and some wine. A few days after the crash, the group heard on a small transistor radio that the search for them had been called off.

Scattered in the snow were the frozen bodies of dead passengers. Preserved by the freezing weather, these bodies became, after a time, sources of food. At first, the survivors were repulsed by the idea of eating human flesh, but as the days wore on, they agonized over the decision about whether to eat the dead crash victims, eventually concluding that they had to eat if they were to live.

In the beginning, only a few ate the human meat, but soon the others began to eat, too. One married couple held out the longest, but when they began to think about wanting another child if they found their way home, they too ate. The group experimented with preparations as they tried different parts of the body. They developed elaborate rules about how, what, and whom they would eat. Some could not bring themselves to cut the meat from the human body, but would slice it once someone else had cut off large chunks. They all refused to eat certain parts—the lungs, skin, head, and genitals; no one was expected to eat an immediate friend or relative. The survivors also developed other uses for the bodies, including making warm, insulated "socks" from human skin (Read, 1974; Henslin, 1993).

Eventually, the group sent out an expedition of three survivors to find help. The group was rescued, and the world learned of their ordeal. Their cannibalism (the eating of other human beings) was generally accepted as something they had to do to survive. To many, it even seemed like a triumph over extraordinary hardship; at the very least, although people might have been repulsed by the story, the survivors' behavior was understood as a necessary adaptation to their life-threatening circumstances. The survivors also maintained a sense of themselves as good people even though what they did profoundly violated ordinary standards of socially acceptable behavior in most cultures in the world.

Was the behavior of the Andes crash survivors socially deviant? Were the people made crazy by their experience, or was this a normal response to extreme circumstances? Does the behavior of the survivors seem understandable given the extremity of the circumstances?

Compare the Andes crash to another case of human cannibalism. In 1991, in Milwaukee, Wisconsin, Jeffrey Dahmer pleaded guilty to charges of murder after he killed at least fifteen men in his home, then chopped them up, cooked them, and ate some of their body parts. Dahmer had lured the men, eight of them African American, two White, and one a fourteen-year-old Laotian boy, to his apartment where he murdered and dismembered them. For those he considered most handsome, he boiled the flesh from their heads so that he could save and admire their skulls. Dahmer was seen as a total social deviant, someone who violated every principle of human decency. Even hardened criminals were disgusted by

Dahmer. He was killed in prison by another inmate in 1994. Why was his behavior considered so deviant when that of the Andes survivors was not?

The answer can be found by looking at the situation in which these behaviors occurred. The same behavior—eating other human beings—is considered reprehensible in one context and acceptable in another. For the Andes survivors, eating human flesh was essential for survival; Dahmer, on the other hand, murdered his victims. From a sociological perspective, the deviance of cannibalism resides not just in the act itself but also in the social context in which it occurs.

For many, Dahmer was simply a sociopath. Psychological explanations are not without merit, but a sociological perspective reveals more. In Dahmer's case, racism and homophobia contributed to Dahmer's atrocity. Most of his victims were African American; he was White, and people who knew him described him as a racist (Celis, 1991). Even more telling is the reaction of the police to a tragic incident that occurred before Dahmer's crimes were discovered. The fourteen-year-old Laotian victim actually escaped from Dahmer and was seen wandering around naked, bleeding, crying, and looking dazed. When a Black woman in the neighborhood notified the police, they returned the boy to Dahmer's custody, interpreting the episode as a homosexual lovers' quarrel. Dahmer then murdered the boy. Many people saw this as police disregard for homosexuals and police dismissal of the complaints of Black city residents; homophobia and racism in this case cost the boy his life (Miller, 1991). At the time, Dahmer was on probation for an earlier sexual assault against a teenage boy. His probation officer never visited Dahmer in his apartment because she had a heavy caseload, and Dahmer lived in a high-crime, low-income neighborhood. These sociological factors allowed Dahmer's deviance to continue unnoticed for some time.

Strange, unconventional, or nonconformist behavior is often understandable in its sociological context. Consider suicide. Are people who commit suicide crazy, or might their behavior be explained? Are there conditions under which suicide is acceptable? Should someone who commits suicide in the face of a terminal illness be judged differently from a despondent person who jumps from a window? Is mental illness itself solely a psychological problem, or do social factors contribute? These are the kinds of questions probed by sociologists who study deviance.

Defining Deviance

Sociologists define **deviance** as *behavior that is recognized as violating expected rules and norms.* Deviance is more than simple nonconformity; it is behavior that departs significantly from social expectations. There is subtlety in the sociological perspective on deviance that distinguishes it from commonsense understandings of the same behavior. First, the sociological definition of deviance stresses social context, not individual behavior. Sociologists see deviance in terms of group processes, definitions, and judgments, not just as unusual individual acts. Second, the sociological definition of deviance recognizes that not all behaviors are judged similarly by all groups. What is deviant to one group may be normative to another; understanding what society sees as deviant also requires understanding the context that determines who has the power to judge some behaviors as deviant and others not. Third, the sociological definition of deviance recognizes that established rules and norms are socially created, not just morally decreed or individually imposed. Sociologists emphasize that deviance lies not just in behavior itself, but in the social responses of groups to behavior by others.

Sociological Perspectives on Deviance

Once you have a sociological perspective on deviance, you are likely to see things a little differently when you observe someone behaving in an unusual way. Sociologists sometimes use their understanding of deviance to explain otherwise ordinary events—such as cigarette smoking (Markle and Troyer, 1979), coffee drinking (Troyer and Markle, 1984), or teen smoking and alcohol use (Logio-Rau, 1998). Most commonly, sociologists examine behavior like mental illness, suicide, delinquent behavior, and substance abuse as examples of deviance, but sociological studies of deviance have also been used to explain how people respond to social stigmas (like having a disability) or how people become deviant, such as in prostitution, drug dealing, or crime. Many sociological studies of deviance examine criminal behavior, some of which is explored in this chapter; it is also discussed again later in Chapter 18. Deviant behavior varies in its severity, as well as in how ordinary or unusual the behavior might be.

Sociologists distinguish between different types of deviance: formal and informal. *Formal deviance* is behavior that breaks laws or official rules. Crime is an example. There are formal sanctions against formal deviance, such as imprisonment and fines. *Informal deviance* is behavior that violates customary norms (Schur, 1984). Although such deviance may not be specified in law, it is judged to be deviant by those who uphold the society's norms. A good example is the body piercing that is popular among young people. Although there are no laws against such a practice, it violates common norms about dress and appearance, and is judged by many to be socially deviant even though it is quite fashionable for others.

Note that the sociological definition requires that deviance be recognized as such. The study of deviance can be divided into the study of why people violate laws or norms and the study of how society reacts, including the labeling process by which deviance comes to be recognized. *Labeling theory* is discussed in detail later, but it is important to point out here that the meaning of deviance is not just in the breaking of norms or rules; it is also in how people react to those behaviors. This dimension of deviance—the societal reaction to deviant behavior—suggests that social groups actually create deviance "by making the rules whose infraction constitutes deviance, and by applying those rules to particular people and labeling them as outsiders" (Becker, 1963: 9).

The Context of Deviance. Some situations are more conducive than others to the formation of deviant behavior. Sociologists have underscored that even the most unconventional behavior can be understood if we know the context in which it occurs. Behavior that is deviant in one circumstance may be normal in another, or behaviors may be ruled deviant only when performed by certain people. For example, people who break gender stereotypes may be judged as deviant even though their behavior is considered normal for the other sex. Heterosexual men and women who kiss in public are the image of romance; lesbians and gay men who dare even to hold hands in public are seen as flaunting their sexual orientation. White students who associate only with people of their own race are not remarked upon; African American, Latino, or Asian students who choose to form organizations or social groups with people of their own group are accused of separatism and may be blamed for the perpetuation of racism.

The definition of deviance can also vary over time. Date rape, for example, was not considered social deviance until recently. Because it was not considered deviant, it was not even named. Women were presumed to mean "yes" when they said "no," and men were expected to "seduce" women through aggressive sexual behavior. Even now, women who are raped by someone they know may not think of it as rape (Johnson et al., 1992). If they do, they may find that prosecuting the offender is difficult because others do not think of it as rape. Studies have found, for example, that students think that date rape is justified if the victim was wearing provocative clothing (Johnson, 1995; Cassidy and Hurrell, 1995). Those who hold more traditional attitudes about gender roles are also more likely to excuse men's aggression in date rape and to define the situation as something other than rape (Scritchfield, 1989). These examples show that the definition of deviance derives not only from what one does, but also from who does it, when, and where.

The sociologist Emile Durkheim argued that one reason acts of deviance are publicly punished is that the social order is threatened by deviance; judging those behaviors as deviant and punishing them confirms general social standards. Therein lies the value of widely publicized trials, public executions, or the historical practice of displaying a wrongdoer in the stocks or pillory. The punishment affirms the collective beliefs of the society, reinforces social order, and inhibits future deviant behavior, especially as defined by those with the power to judge others. One could argue in this vein that the widely publicized trials of Timothy McVeigh and Terry Nichols, convicted of the Oklahoma City bombing, had the effect of solidifying Americans' belief in patriotic behavior.

Violent acts of deviance, such as the bombing of the U.S. embassy in Kenya in 1998, can also reaffirm public beliefs, such as beliefs about national loyalty, nonviolence, or justice.

Durkheim argued that societies actually *need* deviance to know what presumably normal behavior is. In this sense, Durkheim considered deviance "functional" for society (Durkheim, 1951 [1897]; Erikson, 1966). Consider Megan's law—the law requiring that convicted sex offenders be identified to community residents where offenders seek to live after their release from prison. On the face of it, this protects the safety of residents and their children, but sociologists would argue that it also affirms the standards for presumably normal sexual behaviors even though some in the community may privately engage in other forms of sexual deviance. One of Durkheim's important insights was the idea that deviance produces social solidarity; if societies did not have deviance, they would have no way to integrate and socialize people into social norms.

Another example of Durkheim's point comes from a widely publicized stoning that took place in Afghanistan in 1996. A man and a woman who had been caught as adulterers were stoned to death, based on an interpretation of Islamic law by contemporary extremists. Thousands of spectators enthusiastically observed as the woman was put in a sandpit with only her chest and head above ground; the man was blindfolded as he faced the Muslim cleric and others who joined in the stoning. After the couple's death, those who witnessed the ritual stoning commented, "It was a good thing—the only way to end this kind of sinning," and "No, I didn't feel sorry for them at all; I was happy to see Sharia [the Muslim code] being implemented." Such stonings are not uncommon in Muslim Afghanistan, and often only a suspicion of adultery, rather than "hard" evidence, is necessary. How could such cruel deaths be seen so enthusiastically? Durkheim would answer that the stoning, by publicly condemning deviant behavior, is how the community reaffirms its values and promotes social solidarity.

This contemporary stoning is a classic example of deviance defining so-called normal behavior.

The Influence of Social Movements. The perception of deviance may also be influenced by *social movements*—networks of groups that organize to support or resist changes in society (see Chapter 21). With a change in the social climate, formerly acceptable behaviors may be newly defined as deviant. Smoking, for instance, was once considered glamorous, sexy, and "cool." Now, smokers are scorned as polluters and misfits. Despite strong lobbying by the tobacco industry, regulations against smoking have proliferated. The change in public attitudes toward smoking results as much from social and political movements as it does from the known health risks (Markle and Troyer, 1979). Likewise, the emergence of groups like MADD (Mothers Against Drunk Driving) and SADD (Students Against Drunk Driving) has changed the perception of drunk driving. Before these movements gathered strength, few thought that drunk driving was a good idea, but neither did most people regard driving after too many drinks as a criminal or antisocial act. In just a few years, the legal penalties and social stigma for drunk driving have shot up. A new and positive social role has also been produced: the designated driver.

Social movements can also organize to remove the deviant label from certain behaviors. The gay and lesbian movement is a good example. Whereas gay and lesbian behavior traditionally has been defined as deviant, this movement has encouraged people to see gay and lesbian relationships as legitimate. Organizing to recognize gay marriage or to extend employment benefits to domestic partners, people in this movement have not only advocated new rights for lesbian women and gay men, but have challenged the public label of gays and lesbians as deviant.

Moral entrepreneurs are people who organize a social movement to reform how a behavior is perceived and dealt with by society. They create new categories of deviance by imbuing some behaviors with moral value (Becker, 1963). An example is the movement in the 1920s that led to the establishment of the juvenile justice system. Moral reformers, subsequently called the "child savers," objected to treating juvenile offenders the same as adults. They identified juvenile delinquency as a new category of deviance and advocated the segregation of juvenile and adult offenders, arguing that juvenile delinquents needed reform, not punishment. They campaigned for the removal of juvenile offenders to environments where they could be taught to be sober, industrious, and ambitious, while receiving training for technical and lesser-skilled jobs (Platt, 1977).

The influence of these moral entrepreneurs remains today. Whereas adult criminals encounter the full weight of state disapproval, juveniles are seen as needing state protection. Recently, however, a new interpretation of juvenile delinquency has begun to evolve. Modern moral entrepreneurs, political conservatives who are "tough on crime," have organized to make the courts treat some juvenile criminals like adults, subject to the adult criminal justice system, instead of applying separate and more lenient standards to

Moral entrepreneurs bring acts of deviance to public attention often through symbolic appeals to public morality, such as in this memorial tribute to Polly Klaas, an 11-year old girl who was kidnapped from a slumber party and then murdered. To many this represented the threat of growing violence in society, even in small, seemingly secure towns, such as that where Polly Klaas lived.

juveniles. The point is that deviant categories can be produced by groups who mobilize around specific issues and change how deviant behavior is defined, who is defined as deviant, and how society deals with deviance.

Most people find deviant behavior inherently fascinating. Perhaps because it violates social conventions or because it sometimes involves unusual behavior, deviance captures the public imagination. Commonly, however, the public understands deviance as the result of individualistic or personality factors. Many people see deviants as crazy, threatening, "sick," or in some other ways inferior, but sociologists see deviance as influenced by society—by the same social processes and institutions that shape all social behavior.

Deviance as Social Behavior. Many people routinely engage in such deviant acts, never thinking of themselves as deviant or having others label them such. Consider the common practice of employing private housekeepers. Perhaps you will recall when President Clinton twice nominated women for U.S. attorney general (Kimba Wood and Zoë Baird), but each had to withdraw because of controversy over their employment of undocumented workers. When this happened, it called attention to the fact that many middle-class people were lawbreakers employing undocumented domestic workers or failing to pay Social Security taxes on the money they pay to regular babysitters. These people do not consider themselves deviant, even when they openly acknowledge their deeds. Moreover, even in the face of such public controversy and recognition of the illegality of this behavior, few have changed their behavior. This tells you that certain widespread behaviors, even illegal ones, are often not considered deviant and are seldom punished. In addition, sometimes the *majority* of a population may engage in deviance, even extreme deviance, such as the Nazi Holocaust in Germany or the human massacres in Rwanda in the 1990s.

People commonly think that deviance is irrational behavior. Deviance, however, may be a positive and rational adaptation to a situation. Think of the Andes survivors. Was their action irrational, or was it an inventive and rational response to a dreadful situation? To use another example, are gangs the result of the irrational behavior of maladjusted youth, or are they predictable responses to a social context? A sociological study of Puerto Rican girls' gangs in the United States sheds light on this question. Given the class, race, and gender inequality faced by many Puerto Ricans in the United States, some young Puerto Rican girls live in a relatively confined social environment. They have little opportunity for educational or occupational advancement, and their community expects them to be "good girls" and to remain close to their families. Joining a gang is one way to reject the roles in which the girls are confined (Campbell, 1987). Are these young women irrational or just doing the best they can to adapt to their situation? Sociologists interpret their behavior as an understandable adaptation to conditions of poverty, racism, and sexism.

Selling drugs provides another illustration of deviance as a rational response to some situations. Many think of drug dealers as degenerate and corrupt, threatening the social fabric and value system of U.S. society. This may well be true. However much harm drugs produce, it is possible to view drug dealing as rational economic behavior in a society where individual entrepreneurship, the pursuit of profit, and material affluence are prized social attributes. Dealing drugs, although defined as deviant, can be seen as just another way of achieving the American dream.

Although deviance is a violation of social norms, it is not always behavior of which others disapprove. In some subcultures or situations, deviant behavior is encouraged and praised. Have you ever been egged on by friends to do something that you thought was deviant, or done something you knew was wrong, but that was encouraged by your peers? Take cheating on an exam, for example. Most students know that cheating is wrong, yet many cheat, even openly justifying their behavior by claiming that "everybody does it," so why not? Many would also argue that the reason so many college students drink excessively is that the student subculture encourages them to do so—even though students know it is harmful. Similarly, the juvenile delinquent regarded by school authorities as defiant and obnoxious is rewarded and praised by peers for the very behaviors that school authorities loathe. Much deviant behavior occurs, or escalates, because of the social support it receives from the group in which it develops. Gang rape, for example, the most brutal form of rape, occurs in part because men are encouraged by other men during a group assault (Brownmiller, 1975).

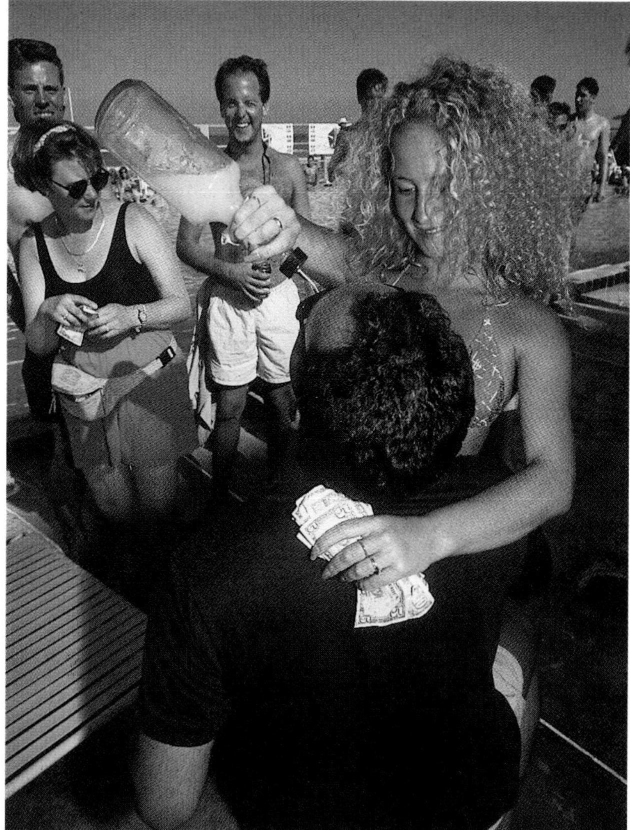

Deviance can be encouraged in certain peer subcultures, such as the widespread phenomenon of binge drinking among college students.

Some behavior patterns defined as deviant are also surprisingly similar to so-called normal behavior. Is a heroin addict who buys drugs with whatever money he or she can find so different from a business executive who spends a large proportion of his discretionary income on alcohol? Each may establish a daily pattern that facilitates drug use; each may select friends based on shared interests in drinking or taking drugs; and each may become so physically, emotionally, and socially dependent on their "fix" that life seems unimaginable without it. Which of the two is more likely to be considered deviant?

Deviance is not just weird, pathological, irrational, offputting, or unconventional behavior. Sociologists who study deviance understand it solely in the context of social relationships and society. They define deviance in terms of existing social norms and the social judgments people make about one another. Sometimes behavior that is deviant in one context may be perfectly normal in another (for example, men wearing earrings or women wearing boxer shorts). Indeed, sometimes deviant behavior can be indicative of changes that are taking place in the cultural folkways. Thus, whereas only a few years ago having pink or green hair may have been extraordinarily deviant (and, to some, it still is!), among some groups it is highly fashionable.

deviant behavior

Biological Explanations of Deviance

Some of the earliest attempts to explain deviance, particularly criminal behavior, centered on biological explanations. During the early part of the twentieth century—a time when many new immigrant groups were coming into the United States—biological explanations of crime and deviance thrived. These explanations were often (though not always) linked to racist and sexist explanations of group differences, and tend to be popular during periods of widespread change in race and gender relations. Even now, when much positive attention is focused on racial and cultural diversity in the population and when gender relations have undergone widespread transformation, there is a resurgence in biological explanations of crime and deviant behavior.

One of the earliest to espouse a biological basis for deviance was Cesare Lombroso (1835–1909). Lombroso believed that some people were born deviants and that deviant and criminal tendencies were associated with certain physical traits, such as small cranial capacity, thinning hair, receding forehead, and large ears. Lombroso believed that crime and deviance represented the survival of "primitive traits" in the modern world; he argued that women, Black Americans, and American Indians were less evolved than White men and more susceptible to primitive urges. He also thought that women had less varied mental capacity than men and were more passive and sedentary, leaving them open to negative influences. In his words, "Even the female criminal is monotonous and uniform compared with her male companion, just as in general woman is inferior to man" (Lombroso, 1920: 122). Ridiculous as these arguments seem now, they reinforced the extremely racist and sexist beliefs of the 1920s, and the arguments were widely believed. (Interestingly, although Lombroso's specific conclusions are thoroughly discredited, sociologists nevertheless credit him with introducing the use of scientific measurement to the study of crime and deviance.)

Biological explanations of deviance are still with us, although not in the same crude form offered by Lombroso. Two social scientists, James Wilson and Richard Herrnstein (1985), have argued in support of biological explanations for human behavior, particularly criminal and deviant behavior. They assert that crime is most often committed by people with low IQs and psychological impairments that prevent them from deferring gratification or resisting the impulse to commit crimes. Wilson and Herrnstein also argue that there may be a biological predisposition to crime. They think that men, for example, are more likely to become criminals than women because they have a greater, hormonally driven tendency toward aggression. Wilson and Herrnstein acknowledge the interaction between biology, psychology, and social conditions, saying that social and cultural arrangements have some influence on sex differences in crime, but they maintain that social scientists have underestimated biological influences on criminal and deviant behavior.

A key part of the Wilson–Herrnstein thesis is that IQ is determined significantly by genetics and that criminality is determined significantly by IQ. Critics point out that the connection they rely on between genetics and IQ is, at best, tenuous (Hauser et al., 1995; Gould, 1994; Kamin, 1986). As we will see in Chapter 15, IQ tests are significantly biased by race and gender. Even separated-twin studies, often cited as "proof" of a strong biological basis for IQ and other measures of aptitude, are biased by not accounting for many of the common environmental circumstances in which twins typically grow up (Taylor, 1980; Hauser et al., 1995). Critics of biological explanations of crime and deviance also note that biological arguments are typically used only to explain the problems of poor people and minority groups, but seldom are they used to examine the crimes of the middle class, including white-collar and organized crime. For example, no one argues that the crimes of stockbrokers and lawyers are caused by low IQ. Using biological arguments only to explain the behavior of the most disadvantaged implies that such groups are somehow "biologically inferior"—an assumption laden with class and race prejudices.

Biological explanations of deviance reflect a strongly held popular assumption that there is something fundamentally different between those who are deviant and those who are not. They also offer an easy explanation for complex social problems. Although there are certainly some biological differences between groups in society, attributing complex social phenomena primarily to biological causes oversimplifies and distorts the sociological processes at work, especially the elaborate social patterns that frame the development and labeling of deviance.

Debunking Society's Myths

Myth: **Deviance is bad for society because it disrupts normal life.**

Sociological perspective: **Deviance stabilizes society; by defining some forms of behavior as deviant, people are affirming the social norms of groups.**

Psychological Explanations of Deviance

Sociology also goes beyond explanations of deviance that root it in the individual personality. The various branches of psychological theory have generated several explanations for deviance. Psychoanalytic theory (see Chapter 4) explains deviance as the product of deep-rooted personality structures originating in a person's conflicted relationship with his or her parents. Violence, for example, is interpreted from a psychoanalytic perspective as the acting out of hostilities toward a parent. A sociopath is seen as acting out urges rooted in early childhood experience that have made the person dysfunctional in later life. Deviance, from this point of view, is explained as unconscious, perhaps self-destructive, behavior springing from the internal drives of individuals. The frustration–aggression hypothesis also sees deviant behavior as occurring when a person is frustrated by not achieving desired goals by legitimate means.

Psychological explanations of deviance emphasize individual factors as the underlying cause of deviant behavior. Sociologists have been critical of this perspective not because it is wrong but because it is incomplete. By locating causes of deviance within individuals, psychological explanations tend to overlook the context in which deviance is produced. Individual motivation simply does not explain the social patterns that sociologists observe in studying deviance. Why is deviance more common in some groups than others? Why are some more likely to be labeled deviant than others, even if they engage in the same behavior? How is deviance related to patterns of inequality in society? To answer these kinds of questions requires a sociological explanation. Sociologists do not ignore individual psychology but integrate it into an explanation of deviance that focuses on the social conditions surrounding deviant behavior.

People commonly interpret acts of deviance, particularly those that are especially harmful, as the behavior of people who are "sick" or sociopathic. This reaction is what sociologists call the **medicalization of deviance,** which refers to explanations of deviant behavior that interpret deviance as the result of individual pathology or sickness. The medical slant may be expressed as a metaphor, such as when deviant behavior is attributed to a "sick" state of mind (Conrad and Schneider, 1992) and where the solution is to "cure" deviance through individual treatments such as intensive therapy for sex offenders.

THINKING SOCIOLOGICALLY

Ask some of your friends to explain why rape occurs. What evidence of the *medicalization of deviance* is there in your friends' answers?

Medicalizing deviance also emphasizes the physical or genetic roots of deviant behavior. Alcoholism provides an example. There is some evidence that there may be a genetic basis to alcoholism. Certainly alcoholism has medical consequences and can be partially understood in medical terms, but viewing alcoholism solely from a medical perspective ignores the social causes that influence the development and persistence of this behavior. Medical treatment alone does not solve the problem; the social relationships, social conditions, and social habits of alcoholics must be altered, or the behavior is likely to recur.

Attention deficit disorder (ADD) is another example. People diagnosed with ADD are unable to concentrate for long periods and may have a learning disability. Although ADD seems to have physiological causes, there may also be environmental causes for this behavior—namely, the overstimulation, stress, and frenzy of modern life. Treating it only as a medical problem and not considering the social factors associated with this condition may lead to inadequate solutions to the problem, at both the individual and societal level.

Sociologists criticize the medicalization of deviance for ignoring the effects of social structures on the development of deviant behavior. From a sociological perspective, deviance originates in society, not in individuals. Changing the incidence of deviant behavior requires changes in society, in addition to changes in individuals. Deviance, to most sociologists, is not a pathological state, but an adaptation to the social structures in which people live. Factors like family background, social class, racial inequality, and the social structure of gender relations in society produce deviance, and these factors must be considered to explain deviance.

Sociological Theories of Deviance

Sociologists have drawn upon several major theoretical traditions to explain deviant behavior, including functionalism, conflict theory, and symbolic interaction theory.

Functionalist Theories of Deviance

Recall that *functionalism* is a theoretical perspective that interprets all parts of society, including those that may seem dysfunctional, as contributing to the stability of the whole. At first glance, deviance seems dysfunctional for society. Functionalist theorists argue otherwise. They contend that deviance is functional because it creates social cohesion. Branding certain behaviors as deviant provides contrast with behaviors that are considered normal, giving people a heightened sense of social order. According to functionalist theorists, norms are meaningless unless there is deviance from the norms; thus, deviance is necessary to clarify what society's norms are. Group coherence then comes from sharing a common definition of legitimate, as well as deviant, behavior. The collective identity of the group is affirmed when those who are defined as deviant are ridiculed or condemned by group members (Erikson, 1966).

To give an example, think about how gay men and lesbian women are defined by many people as deviant. Although lesbians and gay men have rejected this label, labeling homosexuality deviant is one way of affirming the presumed normality of heterosexual behavior. Labeling

TABLE 8.1 *SOCIOLOGICAL THEORIES OF DEVIANCE*

Functionalism	Symbolic Interaction	Conflict Theory
deviance creates social cohesion	deviance is a learned behavior, reinforced through group membership	dominant classes control the definition of and sanctions attached to deviance
deviance results from structural strains in society	deviance results from the process of social labeling, regardless of the actual commission of deviance	deviance results from inequality in society, including that of class, race, and gender
deviance occurs when people's attachment to social bonds is diminished	those with the power to assign deviant labels produce deviance	elite deviance goes largely unrecognized and unpunished

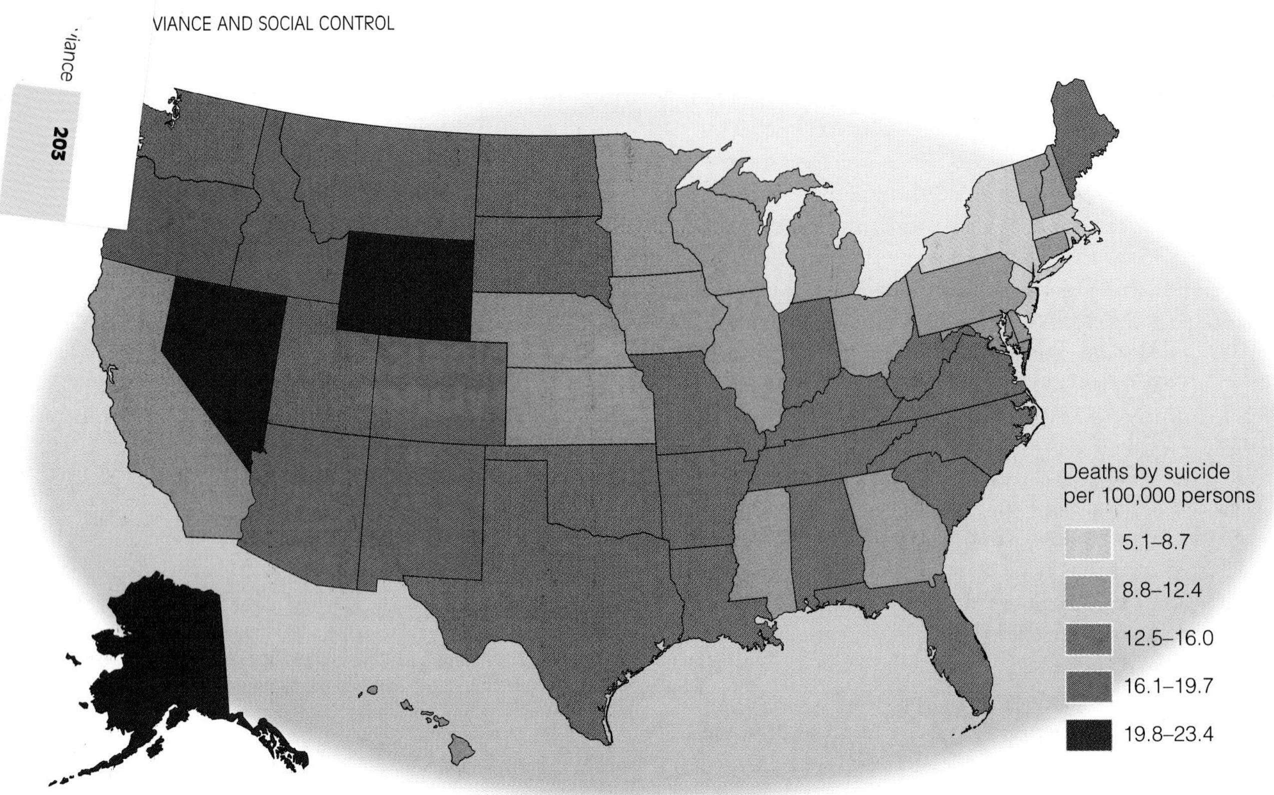

MAP 8.1 Mapping America's Diversity: Suicide Rates by State

DATA: From the U.S. National Center for Health Statistics, "Monthly Vital Statistics Report," 1994, p. 99.

someone else an outsider is, in other words, a way of affirming one's "insider" identity.

Durkheim: The Study of Suicide. Functionalist perspectives on deviance stem originally from the work of Emile Durkheim. Recall that one of Durkheim's central concerns was how society maintains its coherence (or social order). Durkheim saw deviance as functional for society because it produces solidarity among society's members. He developed his analysis of deviance in large part through his analysis of suicide. Through this work, he made a number of important sociological points. First, he criticized the usual psychological interpretations of why people commit suicide, turning instead to sociological explanations with data to back it up. Second, he emphasized the role of social structure in producing deviance. Third, he pointed to the importance of people's social attachments to society in understanding deviance. Finally, he elaborated the functionalist view that deviance provides the basis for social cohesion. Let us examine these points in the context of his studies of suicide.

Most people think of suicide as a symptom of psychological distress. Durkheim was the first to argue that the causes of suicide were to be found in social factors, not individual matters. Observing that the rate of suicide in a society varied with time and place, Durkheim looked for causes linked to time and place rather than emotional stress. Durkheim argued that suicide rates are affected by the different social contexts in which they emerge. He looked at the degree to which people feel integrated into the structure of society and their social surroundings as social factors producing suicide.

Debunking Society's Myths

Myth: People who commit suicide are crazy or sick.

Sociological perspective: Some people who commit suicide may suffer from psychological problems, but sociologists explain suicide in a social context rather than looking just at factors internal to the individual deviant.

Durkheim analyzed three types of suicide: what he called anomic suicide, altruistic suicide, and egoistic suicide. **Anomie,** as defined by Durkheim, is the condition existing when social regulations in a society break down; the controlling influences of society are no longer effective, and people exist in a state of relative normlessness. **Anomic suicide** occurs when the disintegrating forces in the society make individuals feel lost or alone. Teenage suicide is often cited as an example of anomic suicide. Studies of college campuses, for example, trace the cause of campus suicides to feelings of loneliness and a sense of hopelessness (Westefeld et al., 1990).

Another example is provided by the Navajo Indians—a group with one of the lowest suicide rates in the United States. This is most likely because Navajo cultural values strongly integrate members into the group. These values are

Durkheim and suicide

a defense against feelings of social disintegration. Even though Navajos experience a high rate of poverty, which in any society is often attended by feelings of isolation and desperation, strong cultural attachments militate against the taking of one's own life. The low incidence of suicide among Navajos is unusual among Americans Indians since other American Indian groups have very high suicide rates (Young, 1990); according to Durkheim's perspective, this would be because their group ties are not as strong as those among the Navajo. Traditional cultural attachments are likely to have been disrupted by structural patterns like urbanization and joblessness. Durkheim's analysis also helps explain a new trend: a rising rate of suicide among African American youth (Durant and Louden, 1986). Disadvantaged by the larger society, these youth experience anomie, leading to higher rates of suicide.

Altruistic suicide, the second type analyzed by Durkheim, occurs when there is excessive regulation of individuals by social forces. An example is someone who commits suicide for the sake of a higher cause, as did Buddhist monks who set themselves on fire to protest the Vietnam war. Another example is the practice of *suttee,* the practice of burning or burying a woman alive with her deceased husband—a practice stemming from the Hindu belief that a woman is responsible if her husband predeceases her. Orthodox Traditional Hinduism portrayed women as "sexually unreliable" and incapable of leading chaste lives without a husband to control them; burning women whose husbands predecease them is a classic illustration of excessive regulation by social forces (Stein, 1978). Altruistic suicide results when individuals are excessively dominated by the expectations of their social group. People who commit altruistic suicide subordinate themselves to collective expectations even when the result is death. As with anomic suicide, altruistic suicide is rooted in social structural conditions, not just in the psychology of individual people.

Egoistic suicide occurs when people feel totally detached from society. This helps explain the high rate of suicide among the elderly in the United States. Those between seventy-five and eighty-four years of age have one of the highest rates of suicide, presumably because the elderly lose many of their functional ties to society (McIntosh, 1992; Lester and Yang, 1992; Meehan et al., 1991). Ordinarily, people are integrated into society by work roles, ties to family and community, and other social bonds. When these bonds are weakened, the likelihood of egoistic suicide increases. Many elderly people have lost these ties, making them most susceptible to egoistic suicide. At a finer level of statistical detail, older men are more likely to commit suicide than older women, which would be explained by Durkheim as the result of men's greater social isolation. Older women are more likely to be integrated into family and community relationships. For older men, retirement, unemployment, living alone, separation, divorce, and widowhood are risk factors for suicide (Canetto, 1992).

Egoistic suicide also explains some of the differences in rates of suicide between married and unmarried people.

Strong ties among the Navajo produce social integration, resulting in the fact that the Navajo have one of the lowest suicide rates of any group in the United States.

Marriage integrates people into primary groups—the marital couple, the family of in-laws, perhaps close friends of the spouse. People attached to others through marriage have a strong group identification, which Durkheim would point to as a reason why they are less likely to commit suicide. Egoistic suicide occurs in the absence of these group bonds. Similarly, men are more likely to commit suicide than women, although it is important to note that women are more likely to *attempt* suicide but not succeed. The concept of egoistic suicide also contributes to an explanation of the general pattern of higher suicide rates among men. Men are less likely to be embedded in the social relationships of care and responsibility than women, leaving them more prone to commit egoistic suicide. Durkheim would also be interested in how suicide patterns among men and women are related to their gender roles. Men tend to use more lethal methods than females, probably because their socialization includes a greater acclimation to violence and danger.

Durkheim's major point is that suicide is a social, not just an individual, phenomenon. Recall from Chapter 1 that Durkheim sees sociology as discovering the social forces that influence human behavior. As individualistic as suicide might seem, Durkheim discovered the influence of social structure even here.

Durkheim's discussion of anomie, developed in the context of suicide, has implications beyond suicide per se. This term refers not to an individual state of mind, but to social conditions—a state in which social regulations have broken down. Anomie is reflected in how individuals feel, but its origins are in society. When behavior is no longer regulated by common norms and values, individuals are left without moral guidance (Durkheim, 1897/1951; Coser, 1977). Norms governing individual behavior never disappear entirely, nor is there total social breakdown, but when there is a weakening of social regulations or when people or

Some acts of deviance may be functional in the sense that they support certain social norms, such as in how prostitution supports cultural norms associating sex with commercialism and defining women as sexual objects.

groups are caught between two sets of irreconcilable norms, anomie results (Nisbet, 1970). For example, a young student may feel caught between the strong expectations of a devoutly religious family that he return home for religious holidays and the equally strong expectations of his peers that he be part of the crowd and enjoy spring break on a sunny beach. If both sets of norms have equally strong influence on the person, the person will experience anomie and may feel adrift or without roots.

Sociologists have linked the condition of anomie to a variety of social problems, particularly deviant behavior. One example, juvenile delinquency, can be explained as the response of young people caught between conflicting expectations. Adolescents are, by virtue of their age, suspended between the norms of childhood and adulthood. The anxieties of adolescence, often aggravated by racial and class inequality, can produce delinquency. The causes of these problems can therefore be traced not just to individual states of mind but to social conditions of loosened or conflicting social norms. This has been confirmed by studies that have linked crime and delinquency to the sense of cohesion within communities. Urban communities with a greater sense of trust, cohesion, and common values, even though they may be marked by poverty, have lower rates of violence than comparably poor communities without a sense of cohesion (Sampson et al., 1997).

Merton: Structural Strain Theory. The functionalist perspective on deviance has been further elaborated by the sociologist Robert Merton (1910–). Merton's **structural strain theory** traces the origins of deviance to the tensions caused by the gap between cultural goals and the means people have to achieve those goals. Merton noted that societies are characterized by both *culture* and *social structure.* Culture establishes goals for people in society; social structure provides, or fails to provide, the means for people to achieve those goals. In a well-integrated society, according to Merton, people use accepted means to achieve the goals society establishes. In other words, the goals and means of the society are in balance. When the means are out of balance with the goals, deviance is likely to occur.

To explain further, a collective goal in U.S. society is the achievement of economic success; the legitimate means to do so are education and jobs, but not all groups have equal access to those means. The result is structural strain that produces deviance. According to Merton, lower-class individuals are most likely to experience these strains since they internalize the same goals and values of the rest of society but have blocked opportunities for success (Merton, 1938). Structural strain theory therefore helps explain the high correlation that exists between unemployment and crime. Rather than blaming poor and unemployed criminals for not having the values of the middle class, structural strain theory explains their deviant behavior as the *product* of those values, which are internalized but not achievable. Illegitimate means become the only option for success.

To illustrate his theory, consider the case of prostitution. From the perspective of structural strain theory, the prostitute has accepted the cultural values of the dominant society—she (or, in some cases, he) wishes to be economically successful and accepts the commercialization of sexuality that characterizes the dominant culture. If located in a situation with few other options for economic support, she turns to prostitution because of these structural strains, not because of holding values different from the dominant culture.

Merton elaborates his theory to categorize how different individuals adapt to social systems, depending on whether or not they accept the society's goals and means. Those who accept the goals *and* the means are social conformists. Those who accept the goals but not the means are deviant; these *innovators* develop creative, though illegitimate, means to achieve the goals set by society. Cheating on an exam would be an example. Those who accept the means but not the goals are *ritualists;* they engage in socially legitimate behavior but have no sense of purpose or direction. The bureaucrat who sticks to the rules at all costs is an example. Merton would identify this type as deviant because the person has lost sight of the cultural goals and has no overall purpose, other than conformity to the rules. Those who accept neither the goals nor the means of the society are *retreatists,* or social pariahs; they retreat fully from the social system. Examples include skid-row alcoholics and hard-core drug addicts. Finally, there are those who reject both the goals and the means of society, but substitute other goals and means; these are the *politically rebellious.*

Social Control Theory. Taking functionalist theory in another direction, Travis Hirschi has developed social control theory to explain why deviance occurs. **Social control theory** posits that deviance occurs when a person's (or group's) attachment to social bonds is weakened (Hirschi, 1969;

Gottfredson and Hirschi, 1990). According to this view, most of the time people internalize social norms because of their attachments to others. People care what others think of them and, therefore, conform to social expectations because they accept what people expect. You can see here that social control theory, like the functionalist framework from which it stems, assumes the importance of the socialization process in producing conformity to social rules. When that conformity is broken, deviance occurs.

Social control theory assumes that there is a common value system within society and that breaking allegiance to that value system is the source of social deviance. The focus in this theory is then on such questions as how are deviants attached to common value systems, and what are the situations that break people's commitment to these values. Social control theory suggests that most people probably feel some impulse toward deviance at some times, but that the attachment to social norms prevents them from actually participating in deviant behavior. When conditions arise that break those attachments, deviance occurs. This explains why sociologists find that juveniles whose parents exercise little control over violent behavior and who learn violence from aggressive peers are most likely to engage in violent crimes (Heimer, 1997); perhaps this helps you understand how two teenaged boys, alienated from the dominant peer culture in their school, could murder twelve high school students and a teacher, plant bombs throughout their school, and kill themselves in Littleton, Colorado.

School violence, although difficult for people to comprehend, can be understood as a form of deviance. One factor leading to the shootings at Columbine High School was the formation of a deviant peer group, the Trench Coat Mafia, two of whose members shot and killed twelve students and a teacher before also killing themselves.

Social control theory could also be used to explain why women are generally less likely to commit crime than men. Traditionally, women are socialized to be more sensitive to the needs of others; this may lead to more conforming behavior. Of course, men conform, too, but their conformity may take a different course: They may be socialized to take risks or to be more aggressive. Social control theory relates these patterns to the likelihood that one might engage in crime. Those who are most attached to conventional social norms, from the point of view of social control theory, are those least likely to become deviant.

Functionalism: Strengths and Weaknesses. Functionalism emphasizes that social structure, not just individual motivation, produces deviance. Functionalists argue that social conditions exert pressure on individuals to behave in conforming or nonconforming ways. Types of deviance are linked to one's place in the social structure; thus, a poor person blocked from economic opportunities may use armed robbery to achieve economic goals, whereas a stockbroker may use insider trading to achieve the same end. Functionalists acknowledge that people choose whether or not to behave in a deviant manner, but believe that they make their choice from among socially structured options. The emphasis in functionalist theory is on social structure, not individual action.

Functionalists also point out that what appears to be dysfunctional behavior may actually be functional for the society. An example is the fact that most people consider prostitution to be dysfunctional behavior; from the point of view of an individual, that is true—it demeans the women who engage in it, puts them at physical risk, and subjects them to sexual exploitation. From the view of functionalist theory, however, prostitution is functional for the society in that prostitution supports and maintains a social system that links women's gender roles with sexuality, associates sex with commercial activity, and defines women as passive sexual objects and men as sexual aggressors. In other words, what appears to be deviant may actually serve other purposes for society.

Critics of the functionalist perspective argue that it does not explain how norms of deviance are established to begin with. Despite its analysis of the ramifications of deviant behavior for society as a whole, functionalism does little to explain why some behaviors are defined as normative and others as illegitimate. Questions such as who determines social norms and upon whom such judgments are most likely to be imposed are seldom asked by those using a functionalist perspective. Functionalists see deviance as having stabilizing consequences in society, but they tend to overlook the injustices that labeling someone deviant can produce. Others would say that the functionalist perspective too easily assumes that deviance has a positive role in society; thus, functionalists rarely consider the differential effects that the administration of justice has for various social groups. The tendency in functionalist theory to assume that the system works for the good of the whole too easily ignores the inequities in society and how these inequities are reflected in patterns of deviance. These issues are left for sociologists who work from the perspectives of conflict theory and symbolic interaction.

Conflict Theories of Deviance

Recall that *conflict theory* emphasizes the unequal distribution of power and resources in society. It links the study of deviance to social inequality. Based on the work of Karl Marx (1818–1883; see Chapter 1), conflict theory sees a dominant class as controlling the resources of society and using its power to create the institutional rules and belief systems that support its power. Like functionalist theory, conflict theory is a macrostructural approach; that is, both theories look at the structure of society as a whole in developing explanations of deviant behavior.

Conflict theory posits that the economic organization of capitalist societies produces deviance and crime. Because some groups of people have access to fewer resources in capitalist society, they are forced into crime to sustain themselves. Conflict theorists explain the high rate of crime among the poorest groups, especially economic crimes like theft, robbery, prostitution, and drug selling, as a result of the economic status of these groups. Rather than emphasizing values and conformity as a source of deviance, as do functional analyses, conflict theorists see crime in terms of power relationships and economic inequality (Grant and Martínez, 1997).

Conflict theorists also point out that the upper class can better hide crimes they commit because affluent groups have the resources to mask their deviance and crime. As a result, a working-class man who beats his wife is more likely to be arrested and prosecuted than an upper-class man who engages in the same behavior. In addition, those with greater resources can afford to buy their way out of trouble by paying bail; hiring talented, expensive attorneys; or even resorting to bribes.

Punishing deviance is a form of social control in society, such as the public hanging of women alleged to be witches in Salem, Massachusetts, in colonial America.

THINKING SOCIOLOGICALLY

Are there any conditions under which you can imagine committing a *crime*? What would lead you to do it? What would prevent you from doing it? What do your answers reveal about the different sociological perspectives that are used to explain deviance—*functionalism, labeling theory,* and *conflict theory*?

Conflict theorists expand our view of crime and deviance, revealing, for example, the significance of corporate crime—crime that is committed within the legitimate context of doing business. Appropriating profit based on the exploitation of the poor and working class is built into the fabric of capitalist society. **Elite deviance** is the term used to refer to the wrongdoing of wealthy and powerful individuals and organizations (Simon, 1995). Also known as corporate crime, elite deviance includes such things as illegal deals made for the purpose of corporate profit, illegal campaign contributions, corporate scandals that endanger or deceive the public but profit the corporation, and government actions that abuse public trust, like secret arms deals or lying about covert activity (Calavita et al., 1997). Sociologists see elite deviance as stemming from the institutional connections among the power elite (see Chapter 1), namely, the intersection of powerful systems of business, government, and military interests.

According to conflict theory, the ruling groups in society develop numerous mechanisms to protect their interests. Conflict theorists argue that law, for example, is created by elites to protect the interests of the dominant class; thus, law, supposedly neutral and fair in its form and implementation, works in the interest of the most well-to-do (Spitzer, 1975). Another way that conflict theorists see dominant groups as using their power is through the excessive regulation of populations that are a potential threat to affluent interests. The current support among government leaders for new prison construction is an example of this. Periodically sweeping the homeless off city streets, especially when there is a major political event or other elite event occurring, is another example. Conflict theory has also produced analyses of institutions that purportedly "treat" deviants (prisons, mental hospitals, detention homes, for example), but routinely fail those they are intended to help.

Conflict theory emphasizes the significance of social control in managing deviance and crime. **Social control** is the process by which groups are brought into conformity with dominant social expectations. Controlling social deviance is one way that dominant groups control the behavior of others.

A dramatic historical example of social control is the treatment of "witches" during the Middle Ages in Europe and during the early colonial period in America (Ben-Yehuda, 1986). In the Middle Ages, the Catholic Church was the preeminent social institution in Europe. Its laws and practices reigned supreme in most European societies.

Any behavior that contradicted the power of the church was subject to strict social control. Women who deviated from the religious practices of the church were frequently labeled witches and severely persecuted. Often, these women were healers and midwives whose views were at odds with the authority of the exclusively patriarchal hierarchy of the church—then the ruling institution. The church defined witchcraft as stemming from women's carnal urges. People believed that witches, acting as the agents of Satan, castrated men and used their organs in satanic rituals. Witches were also believed to steal semen from sleeping men and to cast spells upon them to make their sexual organs disappear. Witch-hunts, a term still used today to refer to the aggressive pursuit of those who dissent from prevailing political and social norms, were a means of confirming church authority.

Many of the women accused of witchcraft were single or widowed—women who were not publicly attached to men. In the colonial period of the United States, most women persecuted as witches were unattached to particular families and communities—women whose very independence caused suspicion. From this angle, the persecution of witches represents the social control of men over women, as well as the social control of the church over more secular worldviews and practices (Erikson, 1966).

We do not have to look to past centuries to see how social control operates. One implication of conflict theory, especially when linked with labeling theory, is that the power to define deviance confers an important degree of social control. *Social control agents* are those who regulate and administer the response to deviance, such as the police and mental health workers. Members of powerless groups may be defined as deviant for even the slightest infraction against social norms, whereas others may be free to behave in deviant ways without consequence. Oppressed groups may actually engage in more deviant behavior, but it is also true that they have a greater likelihood of being labeled deviant and incarcerated or institutionalized, whether or not they have actually committed an offense. This is evidence of the power wielded by social control agents.

When powerful groups hold stereotypes about other groups, the less powerful people are frequently assigned deviant labels. As a consequence, the least powerful groups in society are subject most often to social control. You can see this in the patterns of arrest data. Poor people are more likely to be considered criminals and are therefore more likely to be arrested, convicted, and imprisoned than middle- and upper-class people. The same is true of Latinos, Native Americans, and African Americans. Sociologists point out that this does not mean that these groups are somehow more criminally prone; rather, they take it as evidence of the differential treatment of these groups by the criminal justice system.

Conflict theorists argue that those with the least power are most likely to be labeled criminal by more powerful authorities, which, as we will see shortly, links conflict theory to another theory—labeling theory. Although it is the poor and members of racial minority groups who are most likely to appear in the arrest data, conflict theorists would argue that this is because of the greater vulnerability of these groups to police power and injustice in the application of justice.

Conflict Theory: Strengths and Weaknesses. The strength of conflict theory is its insight into the significance of power relationships in the definition, identification, and handling of deviance. Conflict theory links the commission, perception, and treatment of crime to inequality in society. It offers a powerful analysis of how the injustices of society produce crime and result in different systems of justice for disadvantaged and privileged groups. It is not without its weaknesses, however. Critics point out that laws protect most people, not just the affluent, as conflict theorists argue. In addition, although conflict theory offers a powerful analysis of the origins of crime, it is less effective in explaining other forms of deviance. For example, how would conflict theorists explain the routine deviance of middle-class adolescents? They might point out that much middle-class deviance is driven by consumer marketing (note the profits made from the accoutrements of deviance—rings in pierced eyebrows, alternative music, or punk dress), but economic interests alone cannot explain all the deviance observed in society. As Durkheim argued, deviance is functional for the whole of society, not just those with a major stake in the economic system.

Symbolic Interaction Theories of Deviance

Whereas functionalist and conflict theories are macrosociological theories, certain microsociological theories of deviance look directly at the interactions people have with one another as the origin of social deviance. *Symbolic interaction theory* holds that people behave as they do because of the meanings people attribute to situations (see Chapter 1). This perspective emphasizes the meaning systems surrounding deviance, as well as how people respond to those meanings. Symbolic interaction emphasizes that deviance originates in the interaction between different groups and is defined by society's reaction to certain behaviors. Symbolic interactionist theories of deviance originate in the perspective of the Chicago School of sociology, examined earlier in Chapter 1.

W. I. Thomas and the Chicago School. W. I. Thomas (1863–1947), one of the early sociologists from the University of Chicago, was among the first to develop a sociological perspective on social deviance. Thomas explained deviance as a *normal response to the social conditions in which people find themselves.* He called this *situational analysis,* meaning that people's actions and the subjective meanings attributed to these actions, including deviant behavior, must be understood in social, not individualized, frameworks. Much influenced by his women students in the Chicago School (Deegan, 1990), Thomas was one of the first to argue that delinquency was caused by the social disorganization brought on by slum life and urban industrialism; he saw it as a problem of social conditions, not individual character.

Differential Association Theory. Thomas's work laid the foundation for a classic theory of deviance: differential association theory. **Differential association theory** interprets deviance, including criminal behavior, as behavior one learns through interaction with others (Sutherland, 1940; Sutherland and Cressey, 1978). Edwin Sutherland argued that becoming a criminal or a juvenile delinquent is a matter of learning criminal ways within the primary groups to which one belongs. To Sutherland, people become criminals when they are more strongly socialized to break the law than to obey it. Differential association theory emphasizes the interaction people have with their peers and others in their environment. Those who "differentially associate" with delinquents, deviants, or criminals learn to value deviance. The greater the frequency, duration, and intensity of their immersion in deviant environments, the more likely that they will become deviant.

Consider the career path of con artists and hustlers. Like any skilled workers, they have to learn the "tricks of the trade"; they do so from mentors and teachers. Hustlers seldom work alone. A new recruit becomes part of a network of other hustlers, from whom the recruit learns the norms of this deviant culture (Prus and Sharper, 1991). Sociologists have also noted that crime tends to run in families, but rather than seizing on a genetic explanation for crime, they explain that youths in such circumstances are socialized to deviant behavior (Miller, 1986).

Differential association theory offers a compelling explanation for how deviance is culturally transmitted—that is, people pass on deviant expectations through the social groups in which they interact. This explains how deviance may be passed on through generations, or may be learned in particular families or peer groups.

Critics of differential association theory have argued that this perspective tends to blame deviance on the values of particular groups. Differential association has been used, for instance, to explain the higher rate of crime among the poor and working class, arguing that this is because they do not share the values of the middle class. Such an explanation, critics say, is class biased, both because it overlooks the deviance that occurs in the middle-class culture and among elites and because it overstates the degree to which disadvantaged groups actually share the values of the middle class, but cannot necessarily achieve them through legitimate means (a point, you will remember, made by Merton's structural strain theory). Still, differential association theory offers a good explanation of why deviant activity may be more common in some groups than others, and it emphasizes the significant role that peers play in encouraging deviant behavior.

Labeling Theory. One of the most significant theories about deviance, one that best illustrates the symbolic interaction perspective, is labeling theory. **Labeling theory** interprets the responses of others as the most significant factor in understanding how deviant behavior is both created and sustained (Becker, 1963). Labeling theory stems from the work of W. I. Thomas, who wrote, "If men define situations as real, they are real in their consequences" (Thomas, 1928: 572). In this case, being labeled deviant has serious social consequences. Once applied, the deviant label is difficult to shed. This is well illustrated in the encounter described by Michael Dyson, a prominent African American philosopher, when bank officials presumed he was a criminal when he tried to get a cash advance on his credit card.

Linked with conflict theory, labeling theory shows how those with the power to label an act or a person deviant and to impose sanctions—such as police, court officials, school authorities, experts, teachers, and official agents of social institutions—wield great power in determining societal understandings of deviance. When they apply a label, it sticks; furthermore, since deviants are handled through bureaucratic organizations, bureaucratic workers "process" people according to rules and procedures, seldom questioning the basis for those rules or being willing or able to challenge them; thus, bureaucrats are unlikely to linger over the issue of whether someone labeled deviant actually deserves that label although they use their judgments and discretion in deciding whether to apply the label or not. This leaves tremendous room for all kinds of social influence and prejudice to enter the decision of whether someone is considered deviant or not (Cicourel, 1968; Kitsuse and Cicourel, 1963; Margolin, 1992).

Beyond this, once the label is applied, it is difficult for the deviant to recover a nondeviant identity. To give an example, once a social worker or psychiatrist labels a client mentally ill, that person will be treated as mentally ill, regardless of his or her actual mental state. Avowals by the accused that he or she is mentally sound are typically taken as evidence of the illness! Likewise, anger and frustration about the label are taken as further support for the diagnosis. Once labeled, a person may have great difficulty changing his or her classification; the label itself has consequences. To illustrate this, you might think of a time when someone falsely accused you of something and how he or she treated you, regardless of your innocence. A person need not have actually engaged in deviant behavior to be labeled deviant; yet, once applied, the label has social consequences. Labeling theory helps explain why convicts released from prison have such high rates of *recidivism* (return to criminal activities). Convicted criminals are formally and publicly labeled wrongdoers. They are treated with suspicion ever afterward. This is why ex-convicts have great difficulty finding legitimate employment: The label *criminal* or *ex-con* defines their future options.

Labeling theory points to a distinction often made by sociologists between primary, secondary, and tertiary deviance. **Primary deviance** is the actual violation of a norm or law. **Secondary deviance** is the behavior that results from being labeled deviant, regardless of whether the person has previously engaged in deviance. A student labeled a troublemaker, for example, might accept this identity and move from being merely mischievous to engaging in escalating delinquent acts. In this case, the person accepts the deviant label and acts in accordance with that role. **Tertiary deviance** occurs when the deviant fully accepts the deviant role, but rejects the stigma associated with it, as when lesbians and gays proudly display their identity (Lemert, 1972; Kitsuse, 1980). These distinctions recognize that simply breaking a rule in itself might not be considered deviant, as we saw in the example of the middle class employing domestic workers without paying Social Security taxes. When others choose to ignore the violation of norms or laws and consider the behavior ordinary, the behavior is seldom recognized or treated as deviant.

In addition, a person may become deviant simply as the result of accepting a label placed on him or her, even if he or she did not engage in the deviant behavior initially. Even an innocent person falsely imprisoned may learn in prison how to think, act, and feel like a criminal, based on immersion in a criminal environment. The same infection with deviant attitudes occurs when relatively harmless teenage boys are labeled juvenile delinquents. Whatever their original standing, people may become deviant merely by virtue of being labeled deviant by society.

THINKING SOCIOLOGICALLY

Perform an experiment by doing something deviant for a period. Make a record of how others respond to you, and then ask yourself how *labeling theory* is important to the study of *deviance*. Then take your experiment a step further and ask yourself how people's reactions to you might have differed had you been of another race or gender. You might want to structure this question into your experiment by teaming up with a classmate of another race or gender. You could then compare responses to the same behavior by both of you. *A note of caution:* Do not do anything illegal or dangerous; even the most seemingly harmless acts of deviance can generate strong (and sometimes hostile) reactions, so be careful in planning your experiment!

Deviant Identity. Another contribution of labeling theory is the understanding that deviance refers not just to something one does, but to something one becomes. **Deviant identity** is the definition a person has of himself or herself as a deviant. The formation of a deviant identity, like other identities, involves a process of social transformation in which a new self-image and new public definition of a person emerges. Most often, deviant identities emerge over time (Lemert, 1972). A drug addict, for example, may not think of herself as a "junkie" until she realizes she no longer has any nonusing friends. In this example, the development of a deviant identity is gradual, but deviant identities can also develop suddenly. A person who becomes disabled as the result of an accident may be given a deviant label; no longer does that person conform to society's definition of "normal." Although the person has done no wrong and has actually had no choice about his or her condition, society applies a stigma to disability; people respond differently to the disabled person. Someone who enters this status, even if it happens suddenly, has to adjust to a new social identity.

Deviant Careers. In the ordinary context of work, a *career* refers to the sequence of movements a person makes through different positions in an occupational system (Becker, 1963); a **deviant career** refers to the sequence of movements people make through a particular subculture of deviance. Deviant careers can be studied sociologically like any other career. Within deviant careers, people are socialized into new "occupational" roles and encouraged, both materially and psychologically, to engage in deviant behavior. The concept of a deviant career emphasizes that there is a progression through deviance: Deviants are recruited, are given or denied rewards, and are promoted or demoted. For example, hospitalized mental patients are often rewarded with comfort and attention for "acting sick," but punished when they act normally—for instance, if they rebel against the boredom and constraints of institutionalization. Acting the role fosters their career as a "mentally ill" person (Scheff, 1966). As with legitimate careers, deviant careers involve an evolution in the person's identity, values, and commitment over time. Deviants, like other careerists, may have to demonstrate their commitment to the career to their superiors, perhaps by passing certain tests of their mettle, such as when a gang expects new members to commit a crime, perhaps even shoot someone.

Deviant careers are also sustained through people's reactions to particular behaviors. This explains why being caught and labeled deviant may actually reinforce, rather than deter, one's commitment to a deviant career. For example, a first arrest on weapons charges may be seen as a rite of passage that brings increased social status among peers. Whereas those outside the deviant community may think that arrest is a deterrent to crime, it may actually encourage a person to continue along a deviant path. Punishments administered by the authorities may even become badges of honor within a deviant community. Similarly, labeling a teenager "bad" for behavior that others think is immoral may actually encourage the behavior to continue since the juvenile may take this as a sign of success as a deviant.

Like anyone else, deviants may experience career mobility—that is, they may move up or down in rank within the deviant community. Male prostitution, for example, is a career organized around a hierarchy of illicit sexual services. Men or boys are recruited into prostitution at different ranks: as street hustlers, bar hustlers, or escorts. Some may become specialists (in sadomasochism, cross-dressing, or catering to pederasts—those who molest children sexually—for example). Newcomers often acquire mentors who train them in the deviant lifestyle. Someone who "learns the ropes" is more likely to continue in a deviant career. As one street hustler reports of his mentor:

> He showed me everything. He said that if you see something different you want, or you want to do this or you want to do that, you want to learn about the hustlers, I'll show you. And he showed me bookstores, showed me the streets, the corners, the hustlers, the johns. He just showed me everything. (Luckenbill, 1986: 288)

Deviant Communities. The preceding discussion indicates an important sociological point: Deviant behavior is not just the behavior of maladjusted individuals; it often takes place within a group context and involves group response. Some groups are actually organized around particular forms of social deviance; these are called **deviant communities.** Like subcultures and countercultures, deviant communities maintain their own values, norms, and rewards for deviant

Some deviance develops in deviant communities, such as the Skinheads shown here marching in a Ku Klux Klan rally protesting the Martin Luther King, Jr. holiday. Such right-wing extremist groups have become more common in recent years.

behavior. Joining a deviant community closes one off from conventional society and tends to solidify deviant careers since the deviant individual receives rewards and status from the in-group. Disapproval from the out-group may only enhance one's status within. Deviant communities also create a worldview that solidifies the deviant identity of their members. They may develop symbolic systems, such as emblems, forms of dress, publications, and other symbols that promote their identity as a deviant group. Gangs wear their "colors"; prostitutes have their own vocabulary of tricks and johns; skinheads have their insignia and music. All are examples of deviant communities. Ironically, subcultural norms and values reinforce the deviant label both inside and outside the deviant group, thereby reinforcing the deviant behavior.

Some deviant communities are organized specifically to provide support to those in presumed deviant categories. Groups like Alcoholics Anonymous, Weight Watchers, and various "12-step" programs help those identified as deviant overcome their deviant behavior. These groups, which can be quite effective, accomplish their mission by encouraging members to accept their deviant identity as the first step to recovery.

The Problem with Official Statistics. Because labeling theorists see deviance as produced by those with the power to assign labels, they question the value of official statistics as indicators of the true extent of deviance. Reported rates of deviant behavior are themselves the product of socially determined behavior, specifically the behavior of identifying what is deviant. Official rates of deviance are produced by people in the social system who define, classify, and record certain behaviors as deviant and others, not. Labeling theorists are more likely to ask how behavior becomes labeled deviant than they are to ask what motivates people to become deviant (Kitsuse and Cicourel, 1963).

Labeling theorists do not think that official rates of deviance reflect the actual incidence of deviance; they are far more likely to see the official rates as reflecting social judgments than the actual commission of crimes or deviant acts. Take the example of suicide. Officially reported suicide rates are based on records typically produced in a coroner's office where someone determines and records the cause of death. Suicide carries a stigma, and staff members in a coroner's office possess stereotypes about who is likely to commit suicide that influence how a death is recorded. As a result, a determination of suicide is less likely for upper-class people. Mentally ill people, especially from the working class, are more likely to be recorded as suicides than those not labeled mentally ill. A terminally ill middle-class person who takes his or her own life may not be recorded as a suicide in deference to the family. Other factors, such as religious affiliation and nationality, as well as unofficial interference by interested parties, may also influence whether a death is recorded as a suicide. As one sociologist has concluded, "The more socially integrated an individual is, the more he and his significant others will try to avoid having his death categorized as a suicide" (Douglas, 1967: 209).

AIDS provides another example. When AIDS first emerged, it was highly stigmatized because of its perceived association with gay men. Obituaries of AIDS victims seldom noted that the death was because of AIDS; more typically, obituaries reported only that the person died following a long illness. Many other examples of distortions that appear in official statistics have been revealed by labeling theorists. Official rape rates are underestimates of the actual extent of rape, not only because of victims' reluctance to report, but also because rapes that end in death are classified as murder; certain instances of rape are less likely to be "counted" as rape by police, such as if the victim is a prostitute, was drunk at the time of the assault, or had a prior relationship with the assailant. Moreover, rapes resulting in death are classified as homicides and therefore do not appear in the official statistics on rape. There is also enough evidence of discriminatory treatment by the police to know that official statistics showing higher arrest rates for African American men than other groups tell us as much about police behavior as about the behavior of African American men. Given these problems, any official statistics must be interpreted with caution.

Labeling Theory: Strengths and Weaknesses. The strength of labeling theory is its recognition that the judgments people make about presumably deviant behavior have powerful social effects. Labeling theory does not, however, explain why deviance occurs in the first place. Labeling theory may illuminate the consequences of a young man's violent behavior, but it does not explain the actual origins of the behavior. Bluntly, labeling theory does not explain why some people become deviant and others do not. Although it gives focus to the behaviors and beliefs of officials in the enforcement system, labeling theory does not explain why those officials define some behaviors as deviant or criminal, but not others. This shortcoming in the analysis of deviance has been carefully scrutinized by conflict theorists who place their analysis of deviance within the power relationships of race, class, and gender.

Forms of Deviance

Although there are many forms of deviance, the sociology of deviant behavior has focused heavily on subjects like mental illness, social stigmas, and crime. As we review each, you will also see how the different sociological theories about deviance contribute to understanding each subject. In addition, you will see how the social structural context of race, class, and gender relationships shape these different forms of deviance. Race, class, and gender are not just individual attributes; they are patterns of relationships that are supported by social institutions and social ideologies; consequently, they are an important part of the social context in which different forms of deviance emerge and from which people make judgments about who is deviant and who is not. Beyond this, the social institutions that handle deviance (courts, prisons, and mental institutions, for example) are also structured along lines of race, class, and gender inequality; this is examined further in Part 3 of this book where major social institutions are studied in more detail.

Mental Illness

Sociological explanations of mental illness look to the social systems in which mental illness is defined, identified, and treated even though it is typical for many to think of mental illness only in psychological terms. This has several implications for understanding mental illness. Functionalist theory suggests, for example, that by recognizing mental illness, society also upholds normative values about more conforming behavior. Symbolic interactionists tell us that mentally ill people are not necessarily "sick," but are the victims of societal reactions to their behavior. Some go so far as to say there is no such thing as mental illness, only people's reactions to unusual behavior; from this point of view, people learn faulty self-images and then are cast into the role of patient when they are treated by therapists. Once someone becomes a "patient," he or she plays the role, as expected by those who reinforce it (Szasz, 1974).

Labeling theory, combined with conflict theory, suggests that those people with the fewest resources are most likely to be labeled mentally ill. This is substantiated by data on mental illness. Women, racial minorities, and the poor all suffer higher rates of reported mental illness and more serious disorders than groups of higher social and economic status.

Sociologists give two explanations for the correlation between social status and mental illness. On the one hand, the stresses of being in a low-income group, being a racial minority, or being a woman in a sexist society contribute to higher rates of mental illness; the harsher social environment is a threat to mental health. On the other hand, the same behavior that is labeled mentally ill for some groups may be tolerated in others. For example, behavior considered crazy in a homeless woman (who is likely to be seen as "deranged") may be seen as merely eccentric or charming when exhibited by an elite person. To illustrate this, you might ask yourself what would have happened to an African American man who exhibited the same violent and threatening behavior that many people reported of John du Pont—the wealthy heir of the du Pont fortune who shot and killed Olympic champion wrestler Dave Schultz on his estate in 1996. For years, people observed his eccentric and crazy behavior, but did nothing about it (Longman, 1996).

In recent years, the relationship that sociologists have found between social class and mental illness has informed the study of mental illness among the homeless (Tessler and Dennis, 1992). By some estimates, 10–15 percent of the homeless are mentally ill; others say as much as a third (Snow, 1992; Wright, 1988). The difference is partially a result of differences in how mental illness is measured. Even taking the highest estimates, however, mental illness among the homeless is not as pervasive as is popularly believed. Some homeless people are indeed on the streets as the result of the deinstitutionalization movement in the 1980s, in which mental health facilities turned out many individuals who have proved unable to care for themselves (Jencks, 1994). Most homeless people, however, are on the streets because of deteriorating economic conditions; the major causes of homelessness are structural factors like unemployment, housing shortages, poverty, and insufficient health care (Elliott and Krivo, 1991; Snow et al., 1986; Snow, 1992). Still, the experience of homelessness is so stressful that it can produce mental illness. As one person wrote, "homelessness itself is an insane condition" (Wright, 1988: 185).

Patterns of mental illness also reflect gender relations in society. Women have higher rates of mental illness than men although men and women differ in the kinds of mental illnesses they experience (Horton, 1995). Most sociologists conclude that the higher rates of mental illness for women stem from women's role in society. Women with the most stereotypical female identities (that is, passive, dependent, and subservient) are most likely to suffer from dissatisfaction, anxiety, and low self-esteem—all conditions that may produce mental illness (Gove et al., 1984; Thornton and Leo, 1992). Learned personality traits developed by women are joined by other sources of stress that can lead to mental illness (Walters, 1993). Poverty, unhappy marriages, physical and sexual abuse, and the stress of rearing children all contribute to higher rates of mental illness for women. Employed women, generally speaking, have better mental health than women who work only in their homes; studies of Mexican

American women (Chicanas) have found that employed women whose husbands help most in the home have better mental health than those with little support (Saenz et al., 1989). Women's learned gender roles also make them more likely than men to seek help when they are distressed; thus, they are more likely to be labeled mentally ill and to appear in mental illness statistics since they are more likely to have sought treatment (Schur, 1984). Labeling theory also suggests that women are more likely to be labeled mentally ill than men because this label is one way that dominant groups can exert social control over others. The frequency with which physicians label women's complaints to be "psychologically grounded" is evidence of this trend.

There is a disproportionate amount of mental illness among racial minority groups in society, pointing again to a correlation between mental health and group status in society (Aponte, 1994; Chin, 1993). The patterns of mental illness are not the same for all minority groups, however. Mexican Americans, for example, have relatively low rates of mental illness; Puerto Ricans have higher rates of mental illness than non-Hispanic Whites. Some sociologists claim that the higher rates of mental illness among African Americans and other racial groups are the result of the stresses of living in a racially conflicted society, as well as the tendency for White psychiatrists to overdiagnose mental illness among racial minorities, when the real problems are not psychological but social structural (Loring and Powell, 1988). This problem is exacerbated by the very low number of minority psychiatrists and psychologists.

The concentration of African Americans, Latinos, and Native Americans among the poor also contributes to high rates of mental illness. In such cases, poverty, not race per se, contributes to the stress and depression. Sociological studies have demonstrated, for example, that poverty leads to persistent anxiety and depression among children, regardless of the child's race (McLeod and Shanahan, 1993). Patterns of inequality in society affect the treatment of mental illness in different populations. Lower-income people are less able to afford expensive psychiatric care; consequently, they may delay treatment, and their illness may persist and become aggravated over time. Well-to-do people who can afford private care may be far more likely to recover from mental illness than someone admitted through the emergency room of a county hospital. Like other forms of deviance, mental illness—its incidence, expression, and treatment—reflects other conditions in society.

Debunking Society's Myths

Myth: Mental illness is an abnormality best studied by psychologists.

Sociological perspective: Mental illness follows patterns associated with race, class, and gender relations in society and is subject to a significant labeling effect; those who study and treat mental illness benefit from a sociological perspective.

Social Stigmas

A **stigma** is an attribute that is socially devalued and discredited. Some stigmas result in people being labeled deviant. The experiences of people who are disabled, disfigured, or in some other way stigmatized are studied in much the same way as other forms of social deviance. Like other deviants, people with stigmas are stereotyped and defined only in terms of their presumed deviance.

Think, for example, of how disabled people are treated in society. Their disability can become a **master status** (Chapter 5), a characteristic of a person that overrides all other features of the person's identity (Goffman, 1963). Physical disability can become a master status when other people see the disability as the defining feature of the person; a person with a disability becomes "that blind woman" or "that disabled guy." Persons with a particular stigma are often seen to be all alike. This may explain why stigmatized individuals of high visibility are often expected to represent the whole group. When Magic Johnson was found to be HIV-positive, he was quickly accepted as a symbolic leader on behalf of research and education on AIDS. Master statuses can also be oppressive—an ever-present possibility when someone is seen as representing his or her entire group in a world of prejudices.

People who become suddenly disabled often have the alarming experience of their new master status rapidly erasing their former identity. They may be treated and seen differently by people they knew before. A master status may also prevent people from seeing other parts of a person. A person with a disability may be assumed to have no meaningful sex life, even if the disability is unrelated to sexual ability or desire. Sociologists have argued that the negative judgments made about people with stigmas tend to confirm the "usualness" of others (Goffman, 1963b: 3). For example, when welfare recipients are stigmatized as lazy and undeserving of social support, others are indirectly promoted as industrious and deserving.

Stigmatized individuals are measured against a presumed norm and may be labeled, stereotyped, and discriminated against. In Erving Goffman's words, people with stigmas are perceived as having a *spoiled identity*. Seen by others as deficient or inferior, they are caught in a role imposed by the stigma. They may respond by bonding with others, perhaps even strangers, whom they see as sharing their trait, an acknowledgement of "kinship" or affiliation that can be as subtle as an understanding look, a greeting that makes a connection between two people, or a favor extended to a stranger whom the person sees as sharing the presumed stigma. Such public exchanges are common between various groups that share certain forms of disadvantage, such as people with disabilities, lesbians and gays, or members of other minority groups.

Stigmatized individuals may sometimes attempt to hide their stigma by "passing"—trying to appear like members of the nondeviant majority. Some light-skinned African Americans were able to pass as Whites during the height of segregation—sometimes for their entire lives. Similarly, gay people may stay in the closet rather than openly reveal themselves as gay. People may also try to manage stigmas by behaving so as to hide the stigmatized behavior from others. In an interesting application of this idea, Nancy Blum has studied the family caretakers of Alzheimer's victims. Caretakers may inadvertently collude with the victim by ignoring the signs of the disease, going along with the deception that nothing is wrong, or concealing the Alzheimer's victim from public view. By doing so, the caretaker participates in the management of the stigma, gaining a "courtesy stigma"—the stigma that someone bears by virtue of association with a stigmatized person (Goffman, 1963b; N. S. Blum, 1991).

Maintaining an assumed identity can be psychologically costly for these individuals, especially since they may be isolating themselves from communities where their deviant status is not stigmatized and where they can get help, such as support groups like 12-step programs. Stigmatized people may become social outcasts in both communities—the society at-large and the help community. If they deny their "deviant" status, those in the helping communities may see them as indirectly upholding social norms that define stigmas as deviant.

Although violent crime is widely depicted in the media, such as the arrest and trial of the "Unabomber"—Ted (Theodore J.) Kaczynski, violent crime has actually declined in the United States in recent years.

Crime and Criminal Justice

The concept of deviance in sociology is a broad one, encompassing many forms of behavior—both legal and illegal, ordinary and unusual. **Crime** is one form of deviance, specifically, behavior that violates specific criminal laws. **Criminology** is the study of crime from a social scientific perspective. Although this broad and comprehensive field of research can only be touched upon briefly here, it provides a vast window on the phenomenon of crime in society. The criminal justice system is discussed further in Chapter 18.

All the theoretical perspectives on deviance that we examined earlier contribute to our understanding of crime. Symbolic interaction helps us understand how people learn to become criminals, or come to be accused of criminality even when they may be innocent. Conflict theory suggests that disadvantaged groups are more likely to become criminal; the criminal behavior is traced to social conditions rather than the intrinsic tendencies of individuals. Functionalists contribute a unique understanding of crime, explaining that crime draws people together in a common sense of community. By singling out criminals as socially deviant and a threat to society, the group defines itself and its nondeviant behavior as good. Durkheim would see the nightly reporting of crime on television as a demonstration of this sociological function of crime.

Types of Crime

The stereotypical crimes one sees reported in the media distort our view of what constitutes crime. Crime comes in many forms, including the violent crimes that are more often reported, as well as hidden crimes that go largely unnoticed.

Personal and Property Crimes. Each year the Federal Bureau of Investigation (FBI) measures the extent of crime in the United States. Although these measurements are subject to the same biases in official statistics mentioned earlier, they are the major source of information on patterns of crime and arrest. In these *Uniform Crime Reports, personal crimes* are violent or nonviolent crimes directed against people. Included in this category are murder, aggravated assault, forcible rape, and robbery. As we see in Figure 8.1, aggravated assault is the most frequently reported personal crime. The rate of violent crime has actually been declining somewhat in recent years. Hate crimes, referring to assaults and other malicious acts directed against gay men, lesbian women, people with disabilities, and racial groups, are also personal crimes. This is a form of crime that has been increasing in recent years (Levin and McDevitt, 1993; Martin, 1995; Waxman, 1991). *Property crimes* involve theft of property, without threat of bodily harm. These include burglary (breaking and entering), larceny (the unlawful taking of property, but without unlawful entry), auto theft, and arson. Property crimes are the most frequent criminal infractions.

A third category of crime, so-called *victimless crimes,* violate laws but are not listed in the FBI's serious crime index. These include illicit activities, such as gambling, illegal drug use, and prostitution, in which there is no complainant. Enforcement of these crimes is typically not as rigorous as the enforcement of crimes against persons or property, although periodic crackdowns occur, such as the current trend toward mandatory sentencing for drug violations.

White-Collar Crime. Sociologists use the term *white-collar crime* to refer to criminal activities by persons of high social status who commit their crimes in the context of their oc-

white-collar crime

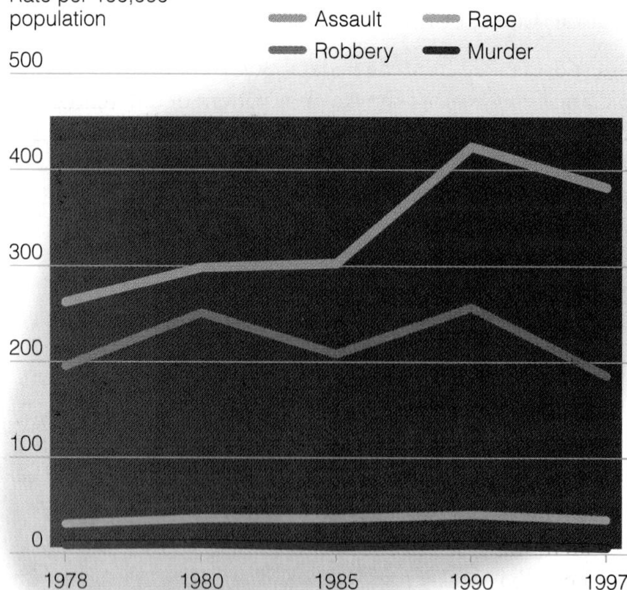

FIGURE 8.1 Violent Crime in the United States

SOURCE: Federal Bureau of Investigation. 1997. *Uniform Crime Reports*. Washington, DC: U.S. Department of Justice, p. 66.

cupation (Sutherland and Cressey, 1978). White-collar crime includes such activities as embezzlement (stealing funds from one's employer) and involvement in illegal stock manipulations.

White-collar crime seldom generates great concern in the public mind—far less than the concern directed at street crime. In terms of total dollars, however, white-collar crime is even more consequential for society. Scandals involving prominent white-collar criminals come to the public eye occasionally, like the falls of Ivan Boesky and Michael Milken, but white-collar crime is generally the least investigated and least prosecuted form of criminal activity. Some would argue, for example, that tobacco executives are guilty of crimes, given the known causal relationship between smoking and lung disease. From a sociological point of view, one of the interesting questions that stems from studies of deviant behavior is how those who engage in deviance "normalize" their behavior. Most tobacco executives, for example, likely believe they are doing nothing wrong; they are likely to believe they are pursuing good business practices even though that business may have serious consequences for public health and safety. From the perspective of conflict theory, sociologists argue that class bias lies at the heart of what is perceived as criminal behavior.

Organized Crime. *Organized crime* is crime committed by organized groups, typically involving the provision of illegal goods and services to others (Block, 1991). Organized crime is typically stereotyped as the "Mafia," but it can refer to any group that exercises control over large illegal enterprises, such as the drug trade, illegal gambling, prostitution, weapons smuggling, or money laundering. Organized crime syndicates are often based on racial or ethnic membership, with different groups dominating and replacing each other in different criminal "industries." A key concept in sociological studies of organized crime is that these industries are organized along the same lines as legitimate businesses. There are likely to be senior partners who control the profits of the business, workers who manage and provide the labor for the business, and clients who buy the services that organized crime provides. In-depth studies of this underworld are difficult, owing to its secretive nature, although some sociologists have penetrated underworld networks and provided fascinating accounts of how these crime worlds are organized (Block and Scarpitti, 1993).

Corporate and Governmental Crime. Even corporations and governments may engage in deviance—behavior that can be very costly to society. Sociologists estimate that the costs of corporate crime may be as high as $200 billion every year, dwarfing the take from street crime (roughly $15 billion), which most people imagine to be the bulk of criminal activity. Tax cheaters in business alone probably skim $50 billion a year from the IRS, three times the value of street crime. Taken as a whole, the cost of corporate crime is almost 6000 times the amount taken in bank robberies in a given year and 11 times the total amount for all theft in a year (Reiman, 1998).

Organizational deviance is wrongdoing that occurs within the context of a formal organization and is sanctioned by the norms and operating principles of the organization. Organizational deviance can occur within any kind of organization—corporate, educational, governmental, or religious. It exists once deviant behavior becomes institutionalized in the routine procedures of an organization. Individuals within the organization may participate in the deviant behavior with little awareness that their behavior is illegitimate. In fact, their actions are likely to be defined as in the best interests of the organization—business as usual. New members who enter the organization learn to comply with the organizational expectations or leave.

In the 1980s, Beech Nut baby foods proudly claimed that their "nutritionists prepare fresh-tasting vegetables, meats, dinner, fruits, cereals, and juices without artificial flavoring." Not only was there no artificial flavoring in the apple juice, there were no apples. Beech Nut was selling sugar water colored brown to resemble apple juice (Ermann and Lundman, 1992). No one ordered plant operators to make fake juice; rather, the owners insisted that the plant make a stronger return on its investment; corners were cut; the pursuit of higher profits resulted in organizational deviance. Most of the people in the production line probably never knew what was happening.

Sociological studies of organizational deviance show that this form of deviance is embedded in the ongoing and routine activities of organizations (Punch, 1996). Instead of conceptualizing organizational deviance as the behavior of bad individuals, sociologists see it as the result of people in organizations following rules and making decisions in more ordinary ways. As we saw in Chapter 6, people at NASA defined the flaws in the space shuttle *Challenger* technology as within the range of acceptable risk, just as Beech Nut executives defined continued sales of the sugar water as good business practice. Conformity to rules, not deviance, led to the *Challenger* disaster and to the sale of fake apple juice.

Studies of corporate crime and organizational deviance show that the perpetrators often go unpunished despite the harm their actions may cause. The abstract nature of organizations and the difficulty of fixing responsibility for collective actions make it hard to determine where guilt lies and what punishment is appropriate. These gray areas are exploited strategically by perpetrators defending themselves; they are likely to argue, often successfully, that their actions were not deviant, that is, not in violation of any clearly and widely understood norms. Organizations may be fined or required to pay punitive damages, but often these expenses are built into the organization's budget. It is rare that an act of corporate or governmental deviance is answered by the punishment of an individual in the same publicly visible ways that befall those who commit individual acts of deviance.

Race, Class, Gender, and Crime

Arrest data taken from the Federal Bureau of Investigation's *Uniform Crime Reports* show a clear pattern of differential arrest along lines of race, gender, and class. To sociologists the central question posed by such data is whether this reflects actual differences in the extent of crime among different groups or whether this reflects differential treatment by the criminal justice system. The answer is both.

Certain groups are more likely to commit crime than others, since crime is distinctively linked to patterns of inequality in society. Unemployment, for example, is one correlate of crime, as is poverty. In a direct test of the association between inequality and crime, sociologist Ramiro Martinez (1996) has explored the connection between rates of Latino violence and the degree of inequality in 111 U.S. cities. As explained in the box, "Doing Sociological Research," his research shows a clear link between the likelihood of lethal violence and the socioeconomic conditions for Latinos in these different cities.

BOX 8.1 DOING SOCIOLOGICAL RESEARCH

Latino Violence

ALTHOUGH homicide is reported as the third largest cause of death among Mexican and Cuban Americans, there have been few studies directly examining the causes of such killings. This has recently been remedied by sociologist Ramiro Martinez in a comprehensive study of Latino homicides.

Martinez notes that sociologists typically offer two explanations of the connection found between economic conditions and homicide. One is that high rates of poverty contribute to homicide since these difficult life conditions escalate into violence. The other explanation is that economic inequality between groups leads to frustration and alienation, thereby producing frustration and violence. Martinez began his research by asking how social and economic conditions affect the likelihood of homicide among Latinos. He wanted to know whether poverty per se or income inequality was a better predictor of homicide.

To do his research, he examined crime data from 111 U.S. cities with more than 5000 Latino residents and at least one Latino reported homicide. If poverty were the primary cause of Latino violence, Martinez would expect to find that Latino homicide rates would vary with shifts in the poverty level from city to city. To study the effect of economic inequality, he used two measures: (1) the distribution of family income, comparing Latinos and Anglos, and (2) income inequality *among* Latinos. Anglo-Latino income inequality is the ratio of Anglo to Latino median family income. As this gap increases, you would expect the number of killings to increase. Income inequality among Latinos is indicated by the Gini coefficient—a measure that shows how much Latino median family income is dispersed in each city.

Using these and other measures, Martinez found that Latinos in the cities studied had a substantially higher rate of poverty than Anglos; also, less than half the Latino population had graduated from high school. Contrary to initial expectations, however, a higher poverty rate is *not* associated with a higher homicide rate, but as the income gap within the Latino community increases, the rate of Latino killings increases as well. One of Martinez's major conclusions is that income inequality relative to other Latinos, not Anglos, has an important influence on homicides in the Latino community. He also found that educational attainment is the strongest predictor of homicide—that is, increased education (defined as graduation from high school) has the greatest effect on decreased killings.

Since this is one of the first studies of Latino homicide, Martinez argues that we need more race-specific research studies (those examining the unique experiences of diverse groups) since the pattern found here differs from what one would have predicted from previous studies that did not focus specifically on Latinos. Although Martinez found that economic inequality, not poverty per se, is responsible for high homicide rates among Latinos, he concludes that the changes in the class structure—with increasing class differentiation among Latinos, as well as other groups—leave the class-disadvantaged in inner cities vulnerable to high rates of violence and death.

SOURCE: Martinez, Ramiro, Jr. 1996. "Latinos and Lethal Violence: The Impact of Poverty and Inequality." *Social Problems* 43 (May): 131–146.

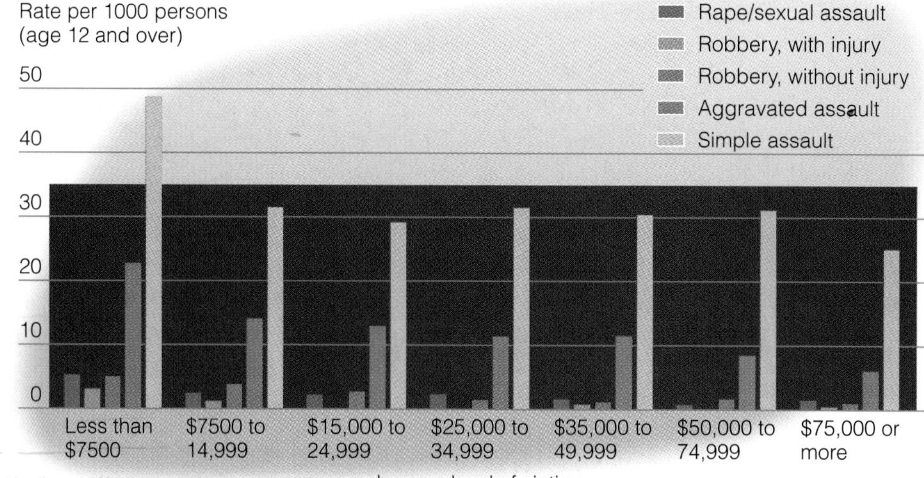

FIGURE 8.2 Victimization by Crime: A Class Phenomenon

SOURCE: U.S. Bureau of Justice Statistics. 1997. *Criminal Victimization 1996.* Washington, DC: U.S. Department of Justice, p. 5.

Sociologists use the analysis of socio-economic conditions to explain the commission of crime, but they also make the important point that prosecution by the criminal justice system is significantly related to patterns of race, gender, and class inequality. We see this in the bias of official arrest statistics, in treatment by the police, in patterns of sentencing, and in studies of imprisonment.

Race, Class, and Crime. One of the most important areas of sociological research on crime is the relationship between social class, crime, and race. Arrest statistics show a strong correlation between social class and crime, the poor being more likely than others to be arrested for crimes. Does this mean that the poor commit more crimes? To some extent, yes. Sociologists have demonstrated a strong relationship between unemployment, poverty, and crime (Hagan, 1993; Britt, 1994; Smith et al., 1992). The reason is simple: Those who are economically deprived often see no alternative to crime.

It is also the case that law enforcement is concentrated in lower-income and minority areas. People who are better off are further removed from police scrutiny and better able to hide their crimes. Although white-collar crime costs society more than street crime, it often goes undetected; when and if white-collar criminals are prosecuted and convicted, they typically receive light sentences—often a fine or community service instead of imprisonment; middle- and upper-income people may be perceived as less in need of imprisonment since they likely have a job and high-status people to testify for their good character. When the poor are accused of a crime, they are more likely to be prosecuted, convicted, and sentenced to prison. White-collar crime is simply perceived as less threatening than crimes by the poor. Class also predicts who will be most likely victimized by crime, with those at the lowest ends of the socioeconomic scale far more likely to be victims of violent crime (see Figure 8.2).

Paralleling the correlation between class and crime shown by arrest data is a strong relationship between race and crime. Minorities constitute 15 percent of the population of the United States, but are a little more than one in three of those people arrested for property crimes and almost one-half of those arrested for violent crimes. African Americans are more than twice as likely to be arrested for crime than are Whites. The rate of arrest among Hispanics is lower than that of African Americans, but it is still twice as high as that of Whites (U.S. Department of Justice, 1997). Native Americans and Asian Americans are exceptions, with both groups having relatively low rates of arrest for crime.

These data may seem to reinforce racial stereotypes, but sociologists have learned not to take these statistics at face value. Instead, they consider how poor and rich communities are policed and the social origins of crime to explain the differences in criminal behavior among groups. What do they find? Police have wide latitude in deciding when to enforce laws and make arrests. Their discretion is greatest when dealing with minor offenses, such as disorderly conduct. Sociological research has shown that police discretion is strongly influenced by class and race judgments, just as labeling theory would predict. The police are more likely to arrest persons they perceive as troublemakers, and they are more likely to make arrests when the complainant is White. They are less likely to arrest middle-class, White, and prominent citizens (Smith et al., 1985; Flowers, 1988). In addition, minority communities are policed much more intensively, which leads to more frequent arrests of those who live there.

Bearing in mind all the factors we have seen in this chapter that affect the official rates of arrest and conviction—the bias of official statistics, the influence of powerful individuals, discrimination in patterns of arrest, differential policing—there remains evidence that the actual commission of crime varies by race. Why? Again, sociologists find a compelling explanation in social structural conditions. Racial minority groups are far more likely than Whites to be poor, unemployed, and living in single-parent families. These social facts are all predictors of a higher rate of crime.

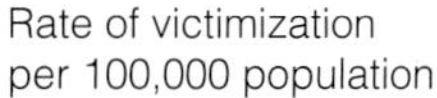

Some argue that the high rate of crime in minority communities is more a matter of social class than race (Wilson, 1987); indeed, crime rates are higher in low-income neighborhoods, whatever the racial composition. What's more, there is a higher percentage of young people in the African American and Latino population than in the population at large; since young people are more likely to commit crimes, this demographic fact contributes to the disparity in patterns of crime by race. Note, too, as Figure 8.3 shows, that African Americans are generally more likely to be victimized by crime.

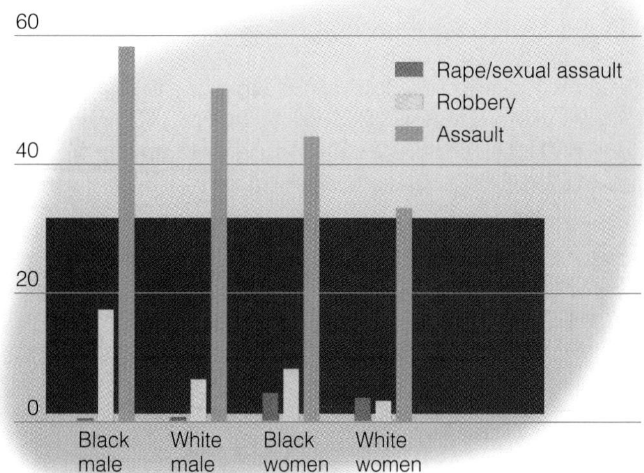

FIGURE 8.3 Crime Victimization (by race and gender)

SOURCE: U.S. Bureau of Justice Statistics. 1994. *Sourcebook of Criminal Justice Statistics.* 1994. Washington, DC: U.S. Department of Justice, p. 233.

Gender and Crime. Until recently, most sociological research on crime and deviance focused on men. Women's deviance was seen as uninteresting or unimportant or studied only with very stereotyped vision (such as only seeing women as helpers in crime). The development of feminist scholarship within sociology has brought new analyses of women, deviance, and crime.

Why do women commit fewer crimes than men? Although the number of women arrested for crime has increased slightly in recent years, the numbers are still small relative to men, except for a few crimes, such as fraud, embezzlement, and prostitution. Some argue that women's lower crime participation reflects their socialization into less risk-taking roles; others say that women commit crimes that are extensions of their gender roles—this would explain why the largest number of arrests of women are for crimes like shoplifting, credit card fraud, and passing bad checks.

Women's participation in crime has been increasing in recent years. Sociologists relate this to several factors. First, changes brought about by the women's movement have made women more likely to be employed in jobs that present opportunities for crimes, such as property theft, embezzlement, and fraud. Violent crime by women has also increased notably since the early 1980s, possibly because the images that women have of themselves are changing, making new behaviors possible. Most significant, crime by women is related to their continuing disadvantaged status in society. Just as crime is linked to socioeconomic status for men, so it is for women (Miller, 1986). Despite the achievements of the feminist movement, women remain in disadvantaged low-wage positions in the labor market. At the same time, changes in the social structure of families mean that more women are economically responsible for their children, without the economic support of men. Women often have to turn to illegitimate means for support, a trend that could be exacerbated by reductions in welfare support.

Women are also less likely to be victimized by crime than are men although victimization by crime among women varies significantly by race and age. Black women are much more likely than White women to be victims of violent crime; young Black women are especially vulnerable. Divorced, separated, and single women are more likely than married women to be crime victims. Regardless of their actual rates of victimization, women are more fearful of crime than men. Minority women and widowed, separated, and divorced women are the most fearful. Women's fear of crime increases with age even though the likelihood of victimization decreases with age (Gordon and Riger, 1989). Women's fear of crime, sociologist Esther Madriz argues, results from an ideology that depicts women as needing protection and identifies public spaces as reserved for men; women's fear of crime is then a system of social control that keeps women from enjoying the rights of full citizenship (Madriz, 1997).

TABLE 8.2 ***ARRESTS BY RACE**** *(percent of all arrests, 1997)*

Selected Crimes	White	Black	American Indian or Alaskan Native	Asian American or Pacific Islander
Total arrests	67.1	30.4	1.3	1.2
Murder	41.9	56.4	.7	1.0
Forcible rape	58.2	39.7	1.1	1.0
Robbery	41.2	57.1	.6	1.2
Aggravated assault	61.2	36.6	1.0	1.1
Burglary	68.0	29.6	1.0	1.3
Larceny-theft	64.7	32.4	1.2	1.7
Motor vehicle theft	58.1	39.0	1.1	1.8
Arson	73.2	24.9	.8	1.1
Forgery	68.0	32.2	0.6	1.2
Embezzlement	63.2	34.8	.5	1.5
Vandalism	73.0	24.7	1.3	1.0
Prostitution	57.9	40.4	.5	1.2
Sex offenses (not rape or prostitution)	74.0	23.6	1.2	1.2
Driving under the influence	86.3	10.9	1.5	1.3
Suspicion	64.0	33.9	1.5	0.6

*Hispanics appear in any of these categories.

SOURCE: Federal Bureau of Investigation. 1997. *Uniform Crime Reports for the United States 1997.* Washington, DC: U.S. Department of Justice, p. 240.

date rape

For all women, victimization by rape is probably the greatest fear. Although rape is the most underreported crime, until recently, it has been one of the fastest growing—something criminologists explain as the result of a greater willingness to report than in the past *and* an actual increase in the extent of rape (Federal Bureau of Investigation, 1997). More than 100,000 rapes are reported to the police annually although officials estimate that this is probably only about one in four of all rapes committed. Many women are reluctant to report rape because they fear the consequences of having the criminal justice system question them. Rape victims are least likely to report the assault when the assailant is someone known to them, even though a large number of rapes are committed by someone known by the victim. A disturbingly frequent form of rape is "date rape"—rape committed by an acquaintance or someone the victim has just met (see "Sociology in Practice: Stopping Date Rape"). Date rape is especially problematic on college campuses where surveys indicate that perhaps as many as 20 percent of college women have been sexually assaulted (Rivera and Regoli, 1987; Dolphin, 1988).

Sociological research on rape shows that it is clearly linked to gender relations in society. Rape is an act of aggression against women, and as feminists have argued, it stems from learned gender roles that teach men to be sexually aggressive. This is reflected in sociological research on convicted rapists who think they have done nothing wrong and who believe, despite having overpowered their victims, that the women asked for it (Scully, 1990). Feminists have argued that the causes of rape lie in women's status in society—that women are treated as sexual objects for men's pleasure. The relationship between women's status and rape is also reflected in data revealing who is most likely to become a rape victim. African American women, Latinas, and poor women have the highest likelihood of being raped, as do women who are single, divorced, or separated. Young women are also more likely to be rape victims than older women (U.S. Bureau of Justice Statistics, 1997). Sociologists interpret these patterns to mean that the most powerless women are also most subject to this form of violence.

As we will see further in Chapter 18 on the state, sociological studies consistently find patterns of differential treatment by the institutions that respond to deviance and crime in society. Whether it is in the police station, the courts, or in prison, race, class, and gender are highly influential in the administration of justice in this society. Those in the most disadvantaged groups are more likely to be defined and identified as deviant, regardless of their behavior, and, having encountered these systems of authority, are more likely to be detained and arrested, found guilty, and punished.

BOX SOCIOLOGY IN PRACTICE

8.2 Stopping Date Rape

DATE rape, also known as acquaintance rape, has been widely acknowledged as a social problem. Sociological research has exposed a number of myths about rape, including the myth that most rapes are committed by strangers. The research done by sociologists and other social scientists on the subject of rape has been used by organizations that provide information to college students about how they can avoid date rape. The Association of American Colleges (AAC), for example, has produced a pamphlet for students intended to educate them about the occurrence of date rape and to give them strategies to avoid it. Their advice has also been informed by sociological research about gender and power. Understand that victims do not "cause" date rape, but the AAC suggests that there are things you can do to keep yourself out of precarious situations. Many of their recommendations stem from understanding the dynamics of gender relations. Here are some of their suggestions for women:

- Be alert to unintended messages you may be sending; others may think you mean something you do not.
- Do not do anything you do not want to do just to be popular or pleasant or to avoid a scene.
- Be aware of situations in which you do not feel comfortable or where you may be at risk—for example, large parties where men greatly outnumber women.
- Avoid putting yourself in a vulnerable position; have your own transportation or money to get home, if needed.
- Examine your attitudes about money and power. If a date pays for you, does that influence your ability to say "No"?
- Avoid falling for lines like, "You would if you loved me." Someone who loves you will respect your feelings and choices; giving in, even if you do not want to, makes it difficult to argue that a rape was not consensual.
- If things start to get out of hand, protest loudly, leave, and go for help.
- Be aware that alcohol and drugs are often involved in acquaintance rape. They compromise your ability to make responsible decisions. If you choose to drink, do not rely on others to take care of you.
- Understand that it is never OK for someone to force himself on you, even if you are drunk or have had sex with the person before. If you do not consent, it is rape.

SOURCE: Adapted from Hughes, Jean O'Gorman, and Bernice R. Sandler. 1987. "'Friends' Raping Friends." Washington, DC: Project on the Education and Status of Women, American Association of Colleges.

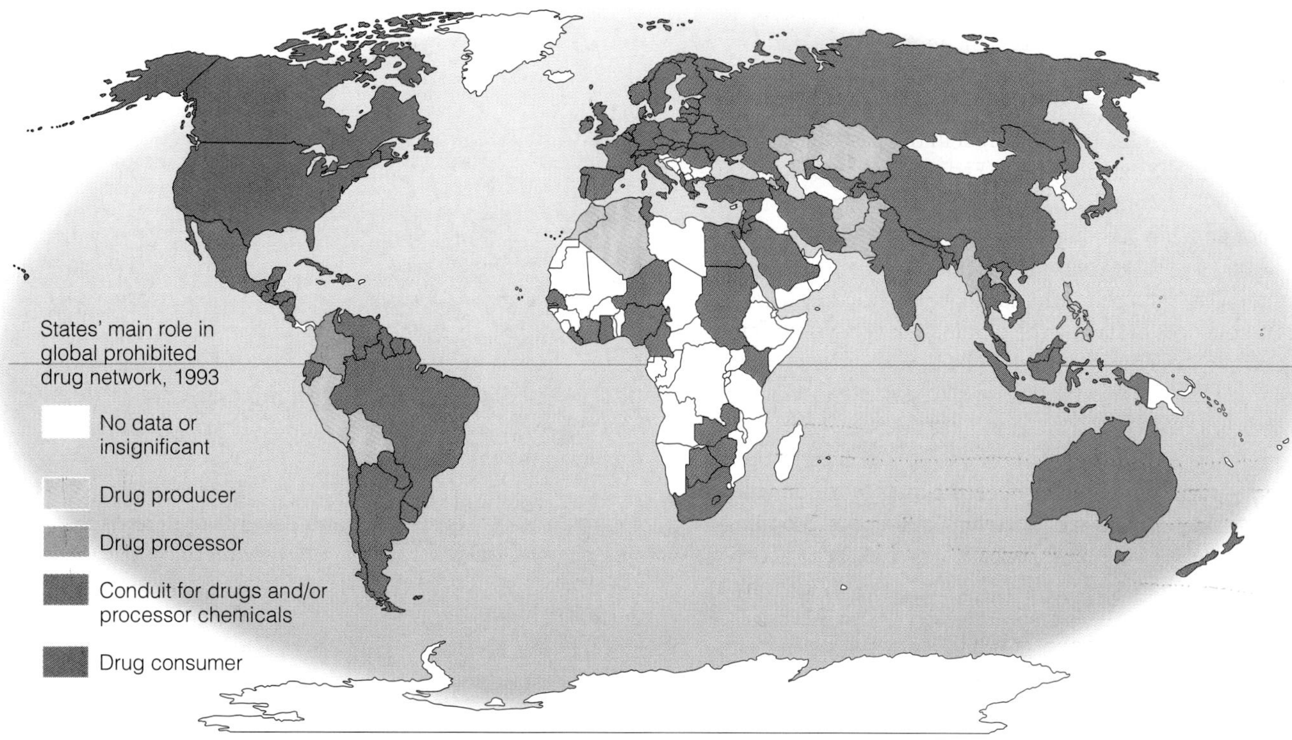

MAP 8.2 Viewing Society in Global Perspective: The Global Fix

SOURCE: "The Golden Fix (map)," from *State of the World Atlas, New Edition* by Michael Kidron and Ronald Segal. Copyright © 1995 by Michael Kidron and Ronald Segal, text. Copyright © 1995 by Myriad Editions Limited, maps and graphics. Used by permission of Viking Penguin, a division of Penguin Putnam Inc.

Deviance in Global Perspective

Increasingly, crime and deviance are crossing national borders so that understanding crime and deviance now requires a global perspective. Terrorism is a case in point. Motivated by political conflicts, often involving ethnic and religious conflict, terrorism has been the cause of some the world's most violent incidents in recent years. Bombings of buildings, airplanes, urban trains and buses, and other targets have become almost commonplace. They involve expressions of extremist political beliefs that stem from the many international conflicts of our current world history. Without understanding the political and economic relations that are the origins for such violence, such acts seem like the crazed behavior of violent individuals. Although sociologists in no way excuse such acts, they look to the social structural conflicts from which terrorism emerges as the cause of such criminal and deviant behavior.

Globalization also means that crime networks can more easily flourish across national borders. The same technological developments that have eased communication for legitimate business activities also enable illegitimate activities to thrive. Money acquired through illegal activity in one country can easily be transferred to another; likewise, transportation systems critical to the international exchange of illegal goods—whether drugs, weapons, or sexual services—link what were once distant and inaccessible places. Even gang behavior can now take place across national boundaries, as street gangs in one city may have wide-reaching connections into other criminal networks, sometimes on an international basis that links them to arms trading, drug dealing, gambling, or extortion (International Office of Justice Statistics, 1997). At the same time, poverty generated by the polarization between the haves and have-nots in the world means that, for some, criminal activity is the only way to survive; thus, crime is high in those nations marked by high rates of poverty and the social dislocation that promotes criminal activities.

We can see the effect of globalization on crime and deviance by considering the drug problem. The map "The Golden Fix" shows how many nations are involved, one way or another, in the international traffic in drugs. Some nations, like the United States, Australia, and parts of western Europe are vast markets for the consumption of illegal drugs. Other nations, such as Colombia, are known as major drug producers. Still others, like China, Brazil, and Mexico, have their role in this international drug economy as conduits for drug traffic and production. Thus, the person on the streets of New York or Amsterdam who is labeled deviant by virtue of being a "crackhead" is part of an international network of drug production and sales. Going even further, the profits on such drugs may end up in other places—the Cayman Islands, Nigeria, or Switzerland, where the profits from the drug trade are laundered in foreign bank accounts. Similar international trafficking in deviant activities is found in the sex trade—the international system whereby women's sexual services are bought and sold in an international marketplace mingling sex, money, and tourism.

CHAPTER SUMMARY

- *Deviance* is behavior that is recognized as violating expected rules and norms and that should be understood in the social context in which it occurs.
- Biological and psychological explanations of deviance place the cause of deviance within the individual. Early sociological explanations of deviance from the Chicago School identified social conditions as influencing deviant behavior. *Functionalist theory* sees deviance as functional for the society because it affirms what is acceptable by defining what is not. *Symbolic interaction theory* explains deviance as the result of meanings people give to various behaviors. *Differential association theory* interprets deviance as learned through social interaction with other deviants. *Labeling theory* argues that societal reactions to behavior produce deviance with some groups having more power than others to assign deviant labels to people. *Conflict theory* explains deviance in the context of unequal power relationships and inequality in society. Conflict theorists also see powerful groups in society as creating laws and other regulatory mechanisms for protecting dominant group interests.
- Studies of mental illness, social *stigmas,* and crime reveal some of the sociological factors that produce deviance, relating these phenomena to societal conditions. A stigma is an attribute that is socially devalued and discredited; those with stigmas are often treated like social deviants.
- *Criminology* is the study of crime from a social scientific perspective. Sociological studies of crime analyze the various conditions, including race, class, and gender inequality, that produce crime and shape how different groups are treated by criminal justice agents. Not only do conditions of inequality influence groups to participate in crime and deviance (though in different ways), but inequality also influences people's perceptions of deviance and how crime and deviance are treated within social institutions.

KEY TERMS

altruistic suicide
anomic suicide
anomie
crime
criminology
deviance
deviant career
deviant community
deviant identity
differential association theory
egoistic suicide
elite deviance
labeling theory
master status
medicalization of deviance
organizational deviance
primary deviance
secondary deviance
social control
social control theory
stigma
structural strain theory
tertiary deviance

THE INTERNET: A Tool for the Sociological Imagination

Resources on the Internet:

Virtual Society: The Wadsworth Sociology Resource Center
http://sociology.wadsworth.com

Visit this site to find additional learning tools, including interactive quizzes, links to related websites, and an essay link to Infotrac College Edition.

Federal Bureau of Investigation
http://www.fbi.gov/

A federal agency that publishes the Uniform Crime Reports annually, detailing information on crime and arrests

Bureau of Justice Statistics
http://www.ojp.usdoj.gov/bjs/

A division in the federal government that compiles information on subjects pertinent to the study of crime and deviance

Drug Enforcement Administration
http://www.usdoj.gov/dea/index.htm

A branch in the U.S. government that offers access to publications and policy statements about drug use and abuse

Sociology and Social Policy: Internet Exercises

On many college campuses, many students are binge drinkers—defined as drinking an excessive amount on a regular basis. To sociologists, binge drinking is deviant behavior even though it may be common on a given campus. In your sociological judgment, what causes student binge drinking? What are the policies on drinking on your campus, and how effective are they in reducing problems associated with binge drinking?

Internet Search Keywords:

binge drinking
social drinking
fraternity/sorority parties
alcoholism

Web sites:

http://www.gov.ab.ca/aadac/addictions/abc/binge_drinking.htm
Information sheet by the Alberta Alcohol and Drug Abuses Commission on definition, consequences, and prevention of binge drinking. (ABC's of binge drinking)

http://www.edc.org/hec/pubs/binge.htm
A national study on campus binge drinking by the Higher Education Center for Alcohol and Other Drug Prevention.

http://www.health.org
The award-winning web site—Prevline—by the National Clearinghouse for Alcohol and Drug Information

http://www.ohiohealth.com/fsalco.htm
OhioHealth fact sheet on definition of social drinking and how it can lead to alcoholism

http://www.niaaa.nih.gov
Web site for the National Institute on Alcohol Abuse and Alcoholism. It contains a variety of publications, databases, press releases, and other resources on alcohol abuse.

InfoTrac College Edition: Search Word Summary

date rape	Durkheim and suicide
deviant behavior	white-collar crime

In order to learn more about these central topics in sociology, you can conduct an electronic search using InfoTrac College Edition. To aid in your search and to gain useful tips, see the Student Guide to InfoTrac College Edition on the Virtual Society web site:
http://sociology.wadsworth.com

INTERACTIONS—A SOCIOLOGY CD-ROM: CONCEPTS FOR THIS CHAPTER

Go to the Wadsworth Sociology CD-ROM for further study on the concepts in this chapter. The CD-ROM also includes quizzes and additional activities to expand your learning experience.

SUGGESTED READINGS

Adler, Patricia A., and Peter Adler. 1997. *Constructions of Deviance: Social Power, Context, and Interaction,* 2nd ed. Belmont, CA: Wadsworth Publishing Co.

This anthology explores various dimensions of deviant behavior from a labeling theory perspective. Ranging from studies of card sharks and missing children to punks, the book covers a wide range of intriguing sociological analyses of deviant behavior.

Becker, Howard S. 1963. *Outsiders: Studies in the Sociology of Deviance.* New York: Free Press.

Becker gives a straightforward and important analysis of labeling theory to explain deviance. Using the examples of marijuana users and jazz musicians, among others, to explain labeling theory, he also develops the concepts of deviant careers and moral entrepreneurs. Highly readable and engaging, this book is a classic in the sociology of deviance.

Fries, Kenny. 1997. *Staring Back: An Anthology of Writers with Disabilities.* New York: NAL/Dutton.

Contributors to this anthology challenge the stigma associated with various disabilities and provide personal narratives about the experience of being disabled.

Goffman, Erving. 1963. *Stigma: Notes on the Management of Spoiled Identity.* Englewood Cliffs, NJ: Prentice Hall.

Goffman's book is a classic study. Using a symbolic interactionist perspective, he studies social responses to stigmas of various sorts and discusses social responses to stigmas and their effect on the individuals who have stigmas of one sort or another, including physical deformities and devalued social statuses.

Sánchez Jankowski, Mártin. 1991. *Islands in the Street: Gangs and American Urban Society.* Berkeley: University of California Press.

Based on ethnographies of gangs in several urban areas, this book provides a detailed sociological analysis of gangs and their relationship to the world around them. The book is inclusive of gangs of different racial and ethnic backgrounds, and provides a comprehensive analysis of gang behavior.

Scully, Diana. 1990. *Understanding Sexual Violence: A Study of Convicted Rapists.* Cambridge, MA: Unwin Hyman.

Based on interviews with prisoners convicted for rape, this book examines the societal factors that encourage sexual violence. Scully examines the views of women held by men convicted of rape and also studies what these men think about avoiding rape.

Simon, David R. 1995. *Elite Deviance,* 5th ed. Boston: Allyn and Bacon.

By examining the connections between corporate, government, and military institutions, the author explores the causes and consequences of elite deviance, defined as the wrongdoing of wealthy and powerful individuals and organizations.

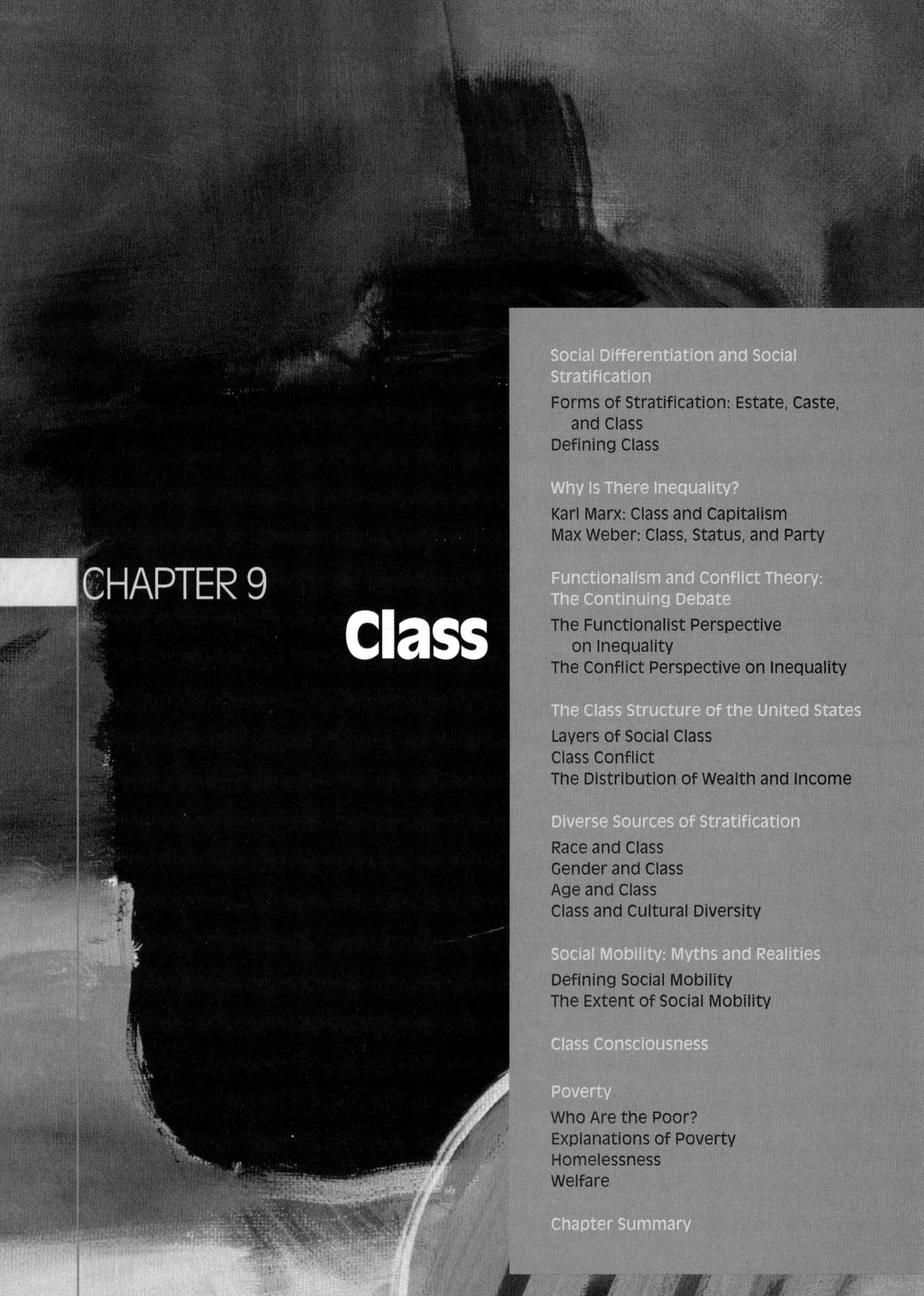

CHAPTER 9

Class

ONE AFTERNOON in a major U.S. city, two women go shopping. They are friends—wealthy, suburban women who shop for leisure. They meet in a gourmet restaurant and eat imported foods while discussing their children's private schools. After lunch, they spend the afternoon in exquisite stores—some of them large, elegant department stores; others, intimate boutiques where the staff knows them by name. When one of the women stops to use the bathroom in one store, she enters a beautifully furnished room with an upholstered chair, a marble sink with brass faucets, fresh flowers on a wooden pedestal, shining mirrors, an ample supply of hand towels, and jars of lotion and soaps. The toilet is in a private stall with solid doors; in the stall there is soft toilet paper and another small vase of flowers.

The same day, in a different part of town, another woman goes shopping. She lives on a marginal income earned as a stitcher in a textiles factory. Her daughter badly needs a new pair of shoes since she has grown out of last year's pair. The woman goes to a nearby discount store where she hopes to find a pair of shoes for under $15, but she dreads the experience. She knows her daughter would like other new things—a bathing suit for the summer, a pair of jeans, and a blouse, but this summer the daughter will have to wear hand-me-downs because medical bills over the winter have depleted the little money left after food and rent. For the mother, shopping is not a recreation but a bitter chore because it reminds her of the things she is unable to get for her daughter.

While this woman is shopping, she, too, stops to use the public bathroom. She enters a vast space with sinks and mirrors lined up on one side of the room and several stalls on the other. The tile floor is gritty and gray. The locks on the stall doors are missing or broken. Some of the overhead lights are burned out, so the room has dark shadows. In the stall, the toilet paper is coarse. When the woman washes her hands, she discovers there is no soap in the metal dispensers. The mirror before her is cracked. She exits quickly, feeling as though she is being watched.

Two scenarios, one society. The difference is the mark of a society built upon class inequality. The signs are all around you. Think about the clothing you wear. Are some labels worth more than others? Do others in your group see the same marks of distinction and status in clothing labels? Do some people you know never seem to wear the "right" labels? Whether it is clothing, bathrooms, schools, homes, or access to healthcare, the effect of class inequality is enormous, giving privileges and resources to some and leaving others barely able to get by.

Observing and analyzing class inequality is fundamental to sociological study. Sociologists want to know what features of society cause different groups to have different opportunities. How does the organization of society lead to an unequal allocation of society's resources, with people's share dependent on class position? Many basic features of life besides wealth are shaped by class, such as where you live, the education you receive, your likely age at first marriage, and the probable cause of your death. Other inequalities intersect with class, such as racial inequality, gender inequality, and age inequality. Why? These are the kinds of questions sociologists pursue in studying social inequality.

As Table 9.1 shows, great inequality divides society; nevertheless, most people think that in the United States equal opportunity exists for all. The tendency is to think that individuals are to blame for their own failures. Instead of understanding inequality as the result of social structural causes, most people think success and failure can be attributed to individual effort. Sociologists have found that those of higher class status are most likely to hold individualist explanations for people's lack of success. White Americans are also more likely to believe this, whereas African Americans and Latinos are more likely to attribute poverty and failure both to structural causes *and* to individual effort (Hunt, 1996; Kluegel and Smith, 1986). As a result, many people accuse welfare recipients of being cheaters or of not wanting to work—a sentiment that has driven recent changes in national welfare policy. At the same time, those who are rich are admired for their supposed initiative, drive, and motivation. Although sociologists respect distinguished individual achievements, they have found that by far the greatest cause for the disparities in material success is the organization of society.

TABLE 9.1 *INEQUALITY IN THE UNITED STATES*

- In 1997, nearly one-fifth (19.9%) of all American children lived in poverty, including 37 percent of all Hispanic and African American children.
- Eighteen percent of the U.S. population has no health insurance; the average cost of a day's stay in a hospital is $539—the equivalent of almost two weeks' pay for most African American and Hispanic men and/or most working woman.
- One night during a Fourth of July celebration, Lee Iacocca entertained guests aboard a $10 million yacht docked in New York Harbor. The boat has a teak deck that lifts to hold the owner's Rolls Royce; on the same night, 90,000 people slept on the streets of New York City.
- Black family income is 62 percent of White family income; Hispanic family income is 66 percent of White family income. The income gap between Whites and other groups has remained largely unchanged since 1970.
- Among women heading their own households, 35.1 percent live below the poverty line.
- In 1997, women working full time and year-round earned 74 percent of what men earned.
- One percent of the U.S. population controls 39 percent of total household wealth; looking only at financial wealth, the top 1 percent control 48 percent of the total.

SOURCES: U.S. Bureau of the Census, Current Population Reports, pp. 60–200. 1998a. *Money Income in the United States: 1997 (With Separate Data on Valuation of Noncash Benefits).* Washington, DC: U.S. Department of Commerce; Dalaker, Joseph, and Mary Naifeh. 1998. *Poverty in the United States: 1997.* Current Population Reports, pp. 60–201. Washington, DC: U.S. Department of Commerce; Wolff, Edward N. 1996. *Top Heavy: A Study of the Increasing Inequality of Wealth in America.* New York: Twentieth Century Fund Press; Dowd, Maureen. 1986. "Yachts and Names Crowd the Harbor." *The New York Times* (July 3): B3.

Social Differentiation and Social Stratification

All social groups and societies exhibit social differentiation. **Status,** as we have seen earlier, refers to a socially defined position in a group or society. **Social differentiation** is the process by which different statuses in any group, organization, or society develop. Think of a sports organization. The players, the owners, the managers, the fans, the cheerleaders, and the sponsors all have a different status within the organization. Together, they constitute a whole social system, one that is marked by social differentiation. The fans have a different status than the cheerleaders, the owners have a different status than the managers, and the sponsors have a different status than the players. Altogether, the organization is marked by the social differentiation between these different statuses in the organization.

Status differences can become organized into a hierarchical social system. **Social stratification** is a relatively fixed, hierarchical arrangement in society by which groups have different access to resources, power, and perceived social worth. Social stratification is a system of structured social inequality. To use the sports example, owners control the resources of the teams; players, although they may earn high salaries, do not control the team resources and have less power than the owners and managers. Sponsors (including individuals and corporations) provide the resources on which this system of stratification rests; fans are merely observers who pay to watch the teams play. Sports organizations include a system of stratification because the groups that constitute the organization are arranged in a hierarchy where some have more resources and power than others.

All societies seem to have a system of social stratification, although they vary in the degree to which stratification exists. Socialist societies are supposed to have minimal stratification; however, purely socialist societies are rare. Before the collapse of the Soviet Union in 1991, the Soviet socialist system claimed universal equality, but in reality was a highly stratified system of privileges and conveniences accessible only to a few. Some societies stratify only along a single dimension, such as age, keeping the stratification system relatively simple. Most contemporary societies are more complex, with many factors interacting to create social strata. In the United States, social stratification is strongly influenced by class, which is in turn influenced by such matters as one's occupation, income, and education, along with race, gender, and other influences like age, region of residence, ethnicity, and national origin.

Forms of Stratification: Estate, Caste, and Class

Stratification systems can be broadly categorized into three types: estate systems, caste systems, and class systems. An **estate system** of stratification is one in which the ownership of property and the exercise of power is monopolized by an elite who have total control over societal resources. Historically, such societies were feudal systems where classes were differentiated into three basic groups—the nobles, the priesthood, and the commoners. Commoners included peasants (usually the largest class group), small merchants, artisans, domestic workers, and traders. The nobles controlled the land and the resources used to cultivate the land, as well as controlling all the resources resulting from peasant labor.

Estate systems of stratification are most common in agricultural societies. Although such societies have been largely supplanted by industrialization, there are still examples of societies that have a small but powerful land-holding class ruling over a population that works mainly in agricultural production. Unlike the feudal societies of the European Middle Ages, however, contemporary estate systems of stratification display the influence of international capitalism. The "noble class" comprises not knights who conquered lands in war, but international capitalists or local elites who control the labor of a vast and impoverished group of people, such as in some South American societies where landholding elites maintain a dictatorship over peasants who labor in agricultural fields.

In a **caste system,** one's place in the stratification system is an *ascribed status* (see Chapter 5), meaning it is a quality given to an individual by circumstances of birth. The hierarchy of classes is rigid in caste systems and is often preserved through formal law and cultural practices that prevent free association and movement between classes.

The system of *apartheid* in South Africa was a stark example of a caste system. Under apartheid, the travel, employment, associations, and place of residence of Black Africans were severely restricted. Segregation was enforced using a pass system in which Blacks in White areas were obliged to account for themselves to White authorities. Interracial marriage was illegal, and Black Africans were prohibited from voting. In 1992, a Whites-only referendum overwhelmingly endorsed reform of the apartheid system, followed by the election of Nelson Mandela as president of the first postapartheid government. Economic power and most other social advantages still remain largely in the hands of the White minority, however.

Another historical caste system was Jim Crow segregation in the American South. "Jim Crow" refers to the laws and social practices that governed the segregated relations between Blacks and Whites in the American South from the end of slavery until the enforcement of contemporary civil rights protections. Before the civil rights movement reaped new legislation, Black Americans in the South were strictly segregated from Whites. Blacks and Whites attended separate schools, and Southern states prohibited interracial marriage. Black Americans were denied the right to vote, either legally or through contrived barriers like absurdly difficult "literacy tests" or expensive "poll taxes." Cultural mores enforced Jim Crow segregation in a variety of ways: Black people were obliged to sit at the back of buses; Whites referred to Blacks only by first name, but Blacks had to address Whites as "Mr.," "Miss," or "Mrs." These demeaning social practices, sometimes referred to as the "etiquette of race relations," defined the group status of Black Americans as inferior and secondary to White Americans.

A final example is the traditional caste system of India. Membership in one of the Indian caste groups was determined by birth. Each group was restricted to certain occupations, and the lowest group, "the untouchables," was prohibited from associating with the others. The Brahmans were the highest caste, nobles who were considered spiritually and socially superior to all other groups. (The term has been transmitted to American culture, where it is used to refer to the old elite families of New England—the "Boston Brahmans.") The Indian caste system is no longer mandated, but its effects remain. People retain attitudes about the alleged superiority and inferiority of different groups long after formalized distinctions are eradicated, which is one reason social inequalities and prejudices are so tenacious.

In class systems, stratification exists, but one's placement in the class system can change according to one's personal achievements; that is, class is to some degree

Jim Crow segregation was a caste system in which Blacks and Whites in the American South were kept separate in all public facilities.

The class status of different groups can be readily seen by looking at the context in different people's everyday lives.

dependent on *achieved status,* defined as status that is earned by the acquisition of resources and power, regardless of one's origins. Class systems are more open than caste systems, since position does not depend strictly on birth, and classes are less rigidly defined than castes, since the divisions are blurred by socially mobile individuals moving between one class and the next.

Despite the potential for movement from one class to another, in class systems like that found in the United States class placement is highly dependent on one's social background. Although ascription (the designation of ascribed status according to birth) is not the basis for social stratification in the United States, the class a person is born into has major consequences for that person's life. Patterns of inheritance; access to exclusive educational resources; the financial, political, and social influence of one's family; and similar factors all shape one's likelihood of achievement. Although there is no formal obstacle to movement through the class system, individual achievement is very much shaped by an individual's class of origin.

Defining Class

In common parlance, *class* refers to style or sophistication: "She's got a lot of class!" In sociological use, **class** refers to standing within the stratification system: **Social class** (or class) is the social structural position groups hold relative to the economic, social, political, and cultural resources of society. Class determines the different access groups and individuals have to these resources and puts groups in different positions of privilege and disadvantage. Members of a given class have similar opportunities and tend to share a common way of life. Class also includes a cultural component, since class shapes such things as language, dress, mannerisms, taste, and other preferences. Class is not just an attribute of individuals; it is a feature of society. The social structural arrangements of society determine class standing.

THINKING SOCIOLOGICALLY

Take a shopping trip to different stores and observe the appearance of stores serving different economic groups. What kinds of bathrooms are there in stores catering to middle-class clients? The rich? The working class? The poor? Which ones allow the most privacy or provide the nicest amenities? What fixtures are in the display areas? Are they simply utilitarian with minimal ornamentation, or are they opulent displays of consumption? Take detailed notes of your observations and write an analysis of what this tells you about *social class* in the United States.

The social theorist Max Weber described the consequences of stratification in terms of **life chances,** meaning the opportunities that people have in common by virtue of belonging to a particular class. To Weber, life chances include the opportunity for possessing goods, having an income, and having access to particular jobs. In general, life chances refer to the living conditions and life experiences a person or group has by virtue of belonging to a particular class.

To sociologists, life chances are the result of social structural arrangements, but the concept of life chances indicates how important these arrangements are for individuals. Put simply, class matters. Class standing determines how well one is served by social institutions. Healthcare, though problematic for many, is particularly inadequate for poor and working-class people, who are less likely to enjoy private care or corporate assistance. Class also influences access to quality education, which is critically important, since an elite education provides credentials and social networks that pay off over a lifetime. Class is also strongly related to political and social attitudes; those from higher income brackets are more likely to be Republican than Democrat. Even friendship is strongly influenced by class since friendships arise more frequently within class groups than across them (Wright and Cho-Donmoon, 1992). One's life chances are shaped by one's place in the class system.

Debunking Society's Myths

Myth: Unlike other societies where there are distinct social classes, most Americans are middle class and share similar values and life chances.

Sociological perspective: Although middle-class values form the dominant culture in the United States, this assertion overlooks the extent to which class differentiates the life chances of different groups.

Class is a structural phenomenon; it cannot be directly observed. Since sociologists cannot isolate and measure social class directly, they use other indicators to serve as measures of class. Remember from Chapter 2 that an *indicator* is something that points to a concept. By assessing indicators, we are able to examine concepts that are too abstract to be measured directly. A prominent indicator of class is income; other common indicators are education, occupation, and place of residence. These indicators do not define class by themselves, but they are often accurate measures of the class standing of a person or group. We will see that these indicators tend to be linked, at least to some extent. A good income, for example, makes it possible to afford a house in a prestigious neighborhood and an exclusive education for one's children. In the sociological study of class, indicators like income and education have had enormous value in revealing the outlines and influences of the class system.

As we will see, sociologists have different ways of conceptualizing class, depending on the frameworks (or model) they use for understanding the class system. These frameworks are derived from long-standing theoretical traditions in sociology, traditions that have stemmed from asking why inequality exists.

Why Is There Inequality?

Stratification occurs in all societies. Why? This question originates in classical sociology in the works of Karl Marx and Max Weber, theorists whose work continues to inform the analysis of class inequality today.

Karl Marx: Class and Capitalism

Karl Marx (1818–1883), whose work was also described in Chapter 1, provided a complex and profound analysis of the class system under capitalism—an analysis that, although more than one hundred years old, continues to inform sociological analyses and has been the basis for major world change. Marx defined classes in terms of their relationship to the **means of production,** defined as the system by which goods are produced and distributed. In Marx's analysis, two primary classes exist under capitalism: the *capitalist class*—those who own the means of production—and the *working class*—those who sell their labor for wages. There are further divisions within these two classes: the *petty bourgeoisie*—small business owners, managers (those whom you might think of as middle class) who identify with the interests of the capitalist class, but who do not own the means of production, and the *lumpenproletariat*—those who have become unnecessary as workers and are then discarded. (Today, these would be the underclass, the homeless, and the permanently poor.) Marx thought that with the development of capitalism, the capitalists and working class would become increasingly antagonistic (something he referred to as *class struggle*); as class conflicts became more intense, the two classes would become more polarized, with the petty bourgeoisie becoming deprived of their property and dropping into the working class. This analysis is still reflected in contemporary questions about whether the classes are becoming more polarized with the rich getting richer and everyone else worse off—a question examined further later in this chapter.

Much of Marx's analysis boils down to the consequences of a system based on the pursuit of profit. If goods were exchanged at the cost of producing them, no profit would be produced. Capitalist owners want to sell commodities for more than their actual value—more than the cost of producing them, including materials and labor. Because workers insert value into the system and capitalists extract value, Marx saw capitalist profit as the exploitation of labor. Marx believed that as profits became increasingly concentrated in the hands of a few capitalists, the working class would become increasingly dissatisfied. The basically exploitative character of capitalism, according to Marx, would ultimately lead to its destruction as workers organized to overthrow the rule of the capitalist class. Class conflict between workers and capitalists, he argued, was inescapable, with revolution being the inevitable result.

At the time Marx was writing, the middle class was small and consisted mostly of small business owners and managers. Marx saw the middle class as dependent on the capitalist class, but exploited by it, since the middle class did not own the means of production. He saw them as identifying with the interests of the capitalist class because of the similarity in their economic interests and their dependence on the capitalist system. Marx believed that the middle class failed to work in its own best interests because it falsely believed that it benefited from capitalist arrangements. Marx thought that in the long run the middle class would pay for their misplaced faith when profits became increasingly concentrated in the hands of a few, and more and more of the middle class dropped into the working class. Because he did not foresee the emergence of the large and highly differentiated middle class we have today, not every part of Marx's theory has proven true. Still, his analysis provides a powerful portrayal of the forces of capitalism and the tendency for wealth to belong to a few, while the majority work only to make ends meet. He has also influenced the lives of billions of people under self-proclaimed Marxist systems that were created in an attempt, however unrealized, to overcome the pitfalls of capitalist society.

Max Weber: Class, Status, and Party

Max Weber (1864–1920) agreed with Marx that classes were formed around economic interests, and he agreed that material forces (that is, economic forces) have a powerful effect on people's lives. He disagreed with Marx that economic forces are the primary dimension of stratification. Weber saw three dimensions to stratification: class (the economic dimension), status (or prestige, the cultural and social dimension), and party (or power, the political dimension). Weber is thus responsible for a multidimensional view of social stratification because he analyzed the connections between economic, cultural, and political systems. Weber pointed out that, although the economic, social, and political dimensions of stratification are usually related, they are not always consistent. A person could be high on one or two dimensions, but low on another. A major drug dealer is an example: high wealth (economic dimension) and power (political dimension), but low prestige (social dimension)—at least in the eyes of the mainstream society, even if not in other circles.

Weber defined *class* as the economic dimension of stratification—how much access to the material goods of society a group or individual has, as measured by income, property, and other financial assets. A family with an income of $100,000 per year clearly has more access to the resources of a society than a family living on an income of $25,000 per year. Weber understood that a class has common economic interests and that economic well-being was the basis for one's life chances. In addition, though, he thought that people were also stratified based on their status and power differences.

Status, to Weber, is the prestige dimension of stratification—the social judgment or recognition given to a person or group. Weber understood that class distinctions are linked to status distinctions—that is, those with the most economic resources tend to have the highest status in society, but not always. In a local community, for example, those with the most status may be those who have lived there the longest, even if newcomers arrive with more money. Physicians are a group who typically earn high incomes and enjoy high social prestige; in this case, status is related to economic standing, but status may be independent of income. Ministers and priests are accorded high social status, but do not typically earn high incomes; and as previously discussed, drug dealers may earn a lot of money but are not accorded high status by the dominant culture.

Finally, *party* (or what we would now call power) is the political dimension of stratification. It is the capacity to influence groups and individuals even in the face of opposition. Power is also reflected in the ability of a person or group to negotiate their way through social institutions. An unemployed Latino man wrongly accused of a crime, for instance, does not have much power to negotiate his way through the criminal justice system. By comparison, business executives accused of corporate crime can afford expensive lawyers and, thus, frequently go unpunished or, if they are guilty, serve relatively light sentences in comparatively pleasant facilities. Again, Weber saw power as linked to economic standing, but he did not think that economic standing was always the determining cause of people's power. To the contrary, he wrote that economic standing could be the result of having a strong power base. Although having power is typically related to also having high economic standing and high social status, this is not always the case. An arrogant boss may have money and power, but be held in low esteem by those who work for him. The point is that stratification does not rest solely on economics; political power and social judgments are important components of stratification systems. Weber refined the sociological analyses of stratification to account for the subtleties that can be observed when you look beyond the sheer economic dimension to stratification.

Marx and Weber explain different features of stratification. Both understood the importance of the economic basis of stratification, and they knew the significance of class for determining the course of one's life. Marx saw people as acting primarily out of economic interests, but Weber saw people's position in the stratification system as the result of economic, social, and political forces. Together, Marx and Weber provide compelling theoretical grounds for understanding the contemporary class structure.

Functionalism and Conflict Theory: The Continuing Debate

Marx and Weber analyzed what they observed in all societies, each of them developing a unique way to understand the development of persistence of inequality. Like more contemporary sociologists, they were trying to understand why differences existed in the resources and power that different groups in society hold. The question of why there is inequality persists and has been further developed by sociologists working from these classical traditions. Two of the major frameworks in sociological theory—functionalist and conflict theory—take quite different approaches to understanding inequality. (See Table 9.2).

The Functionalist Perspective on Inequality

Functionalist theory views society as a system of institutions that are organized to meet society's needs (see Chapter 1). The functionalist perspective emphasizes that the parts of society are in basic harmony with each other; society is characterized by cohesion, consensus, cooperation, stability, and persistence (Parsons, 1951a; Merton, 1957; Eitzen and Baca Zinn, 1998: 48). Different parts of the social system complement one another and are held together through social consensus and cooperation. To explain stratification, functionalists propose that the roles filled by the upper classes—such as governance, economic innovation, investment, and management—are essential for a cohesive and smoothly running society, and hence are rewarded in proportion to their contribution to the social order (Davis and Moore, 1945).

According to the functionalist perspective, social inequality serves an important purpose in society: It motivates people to fill the different positions in society that are needed for the survival of the whole. Those who articulate this perspective think that some positions in society are more functionally important than others. The most important positions require the greatest degree of talent or training. Rewards attached to those positions (such as higher income and prestige) are a way to ensure that people will make the sacrifices needed to acquire the training for functionally important positions (Davis and Moore, 1945). Social mobility thus comes to those who acquire the necessary talents and tools for success (such as education and job training). In other words, functionalist theorists see inequality as an element of a reward system that motivates people to succeed.

The functionalist perspective is well illustrated by Herbert Gans's (1991 [1971]) analysis of the functions of poverty. No one typically thinks of poverty as a "good" thing, least of all, Gans, but Gans delineates how poverty works to sustain an overall social system—one of the basic points in functional theory: that all things in society work toward the stability of the whole. Poverty, no matter how unwanted, has, according to Gans, both economic and social functions in society. Economically, poverty ensures that society's "dirty work" will be done, by relegating menial jobs to the poor. In addition, poverty works to the economic benefit of the affluent, by providing a source of cheap domestic labor and by providing the clientele for various illegal activities (drugs, cheap alcohol, prostitution) that benefit wealthy illegal entrepreneurs. Social mobility for some may also be financed at the expense of the poor, particularly notable in the case of slumlords and other merchants. An additional economic function is that the poor buy or use goods that others do not want, such as secondhand clothes, old automobiles, and surplus food; without the poor, society would need a different way of disposing of such surplus goods. Having a large number of poor also creates jobs—for those in social services and charitable foundations—as well as creating a reserve labor force.

Gans also identifies a number of social functions that poverty provides. They provide a category of social deviants—people whose existence can be used to uphold the legitimacy of mainstream social norms. By labeling the poor indolent or unworthy, others uphold a definition of themselves as socially valuable. Thus, the poor provide the basis for status comparisons. The wealthy may, for example, congratulate themselves for their charity, even though their largesse may benefit themselves (in the form of tax deductions and social recognition) more than those to whom their philanthropy is directed. Gans's presentation of the functions of poverty makes a strong sociological point: Although poverty is in one sense a problem for society, it is at the same time functional for society because it contributes to the overall stability of society. Such would be the perspective of functionalist theorists who, though they might object to the presence of poverty in society, would nonetheless see its unintended effects.

The Conflict Perspective on Inequality

Like functionalist theory, conflict theory sees society as a social system, but different from functionalism, conflict theory emphasizes that the system is held together through

TABLE 9.2 *FUNCTIONAL AND CONFLICT THEORIES OF STRATIFICATION*

Interprets:	**Functionalism**	**Conflict Theory**
Inequality	Inequality serves an important purpose in society by motivating people to fill the different positions in society that are needed for the survival of the whole.	Inequality results from a system of domination and subordination where those with the most resources exploit and control others.
Class structure	Differentiation is essential for a cohesive society.	Different groups struggle over societal resources and compete for social advantage.
Reward system	Rewards are attached to certain positions (such as higher income and prestige) as a way to ensure that people will make the sacrifices needed to acquire the training for functionally important positions in society.	The more stratified a society, the less likely that society will benefit from the talents of all its citizens, since inequality prevents the talents of those at the bottom from being discovered and used.
Classes	Some positions in society are more functionally important than others and are rewarded because they require the greatest degree of talent and training.	Classes exist in conflict with each other as they vie for power and economic, social, and political resources.
Life chances	Those who work hardest and succeed have greater life chances.	The most vital jobs in society—those that sustain life and the quality of life—are usually the least rewarded.
Elites	The most talented are rewarded in proportion to their contribution to the social order.	The most powerful reproduce their advantage by distributing resources and controlling the dominant value system.
Class consciousness/ ideology	Beliefs about success and failure confirm status on those who succeed.	Elites shape societal beliefs to make their unequal privilege appear to be legitimate and fair.
Social mobility	Upward mobility is possible for those who acquire the necessary talents and tools for success (such as education and job training).	There is blocked mobility in the system since the working class and poor are denied the same opportunities as others.
Poverty	Poverty serves economic and social functions in society.	Poverty is inevitable because of the exploitation built into the system.
Social policy	Since the system is basically fair, social policies should only reward merit.	Social policies should support disadvantaged groups by redirecting society's resources for a more equitable distribution of income and wealth.

conflict and coercion. From a conflict-based perspective, society comprises competing interest groups, some of whom have more power than others. Different groups struggle over societal resources and compete for social advantage. Conflict theorists argue that those who control society's resources also hold power over others. The powerful are also likely to act in ways that reproduce their advantage and they try to shape societal beliefs in ways that make their unequal privilege appear to be legitimate and fair. In sum, conflict theory emphasizes the friction in society rather than the coherence, and conflict theorists see society as dominated by elites.

From the perspective of conflict theory, derived largely from the work of Karl Marx, social stratification is based on class conflict and blocked opportunity. Conflict theorists see stratification as a system of domination and subordination in which those with the most resources exploit and control others. They also see the different classes as in conflict with each other, with the unequal distribution of rewards reflecting the class interests of the powerful, not the survival needs of the whole society (Eitzen and Baca Zinn, 1998). According to the conflict perspective, inequality provides elites with the power to distribute resources, make and enforce laws, and control value systems; elites use these powers in ways that reproduce inequality. Others in the class structure, especially the working class and the poor, experience blocked mobility.

Conflict theorists argue that the consequences of inequality are negative. From a conflict point of view, the more stratified a society, the less likely that society will benefit from the talents of its citizens, since inequality limits the life chances of those at the bottom, preventing their talents from being discovered and used. To the waste of talent is added the restriction of human creativity and productivity.

Implicit in the argument of each perspective is criticism of the other perspective. Functionalism assumes that the most highly rewarded jobs are the most important for society, whereas conflict theorists argue that some of the most vital jobs in society—those that sustain life and the quality of life, such as farmers, mothers, trash collectors, and a wide range of other laborers—are usually the least rewarded. Conflict theorists also criticize functionalist theory for assuming that the most talented get the greatest rewards. They point out that systems of stratification tend to devalue the contributions of those left at the bottom and underutilize the diverse talents of all people (Tumin, 1953). On the other hand, functionalist theorists contend that the conflict view of how economic interests shape social organization is too simplistic. Conflict theorists respond by arguing that functionalists hold too conservative a view of society and overstate the degree of consensus and stability that exists.

The debate between functionalist and conflict theorists raises fundamental questions about how people view inequality. Is it inevitable? Will there always be poor people? How is inequality maintained? Do people basically accept it? This debate is not just academic. The assumptions made from each perspective frame public policy debates. Whether the topic is taxation, poverty, or homelessness, if people believe that anyone can get ahead by ability alone, they will tend to see the system of inequality as fair and accept the idea that there should be a differential reward system. Those who tend toward the conflict view of the stratification system are more likely to advocate programs that emphasize public responsibility for the well-being of all groups, and to support programs and policies that result in more of the income and wealth of society going toward the needy.

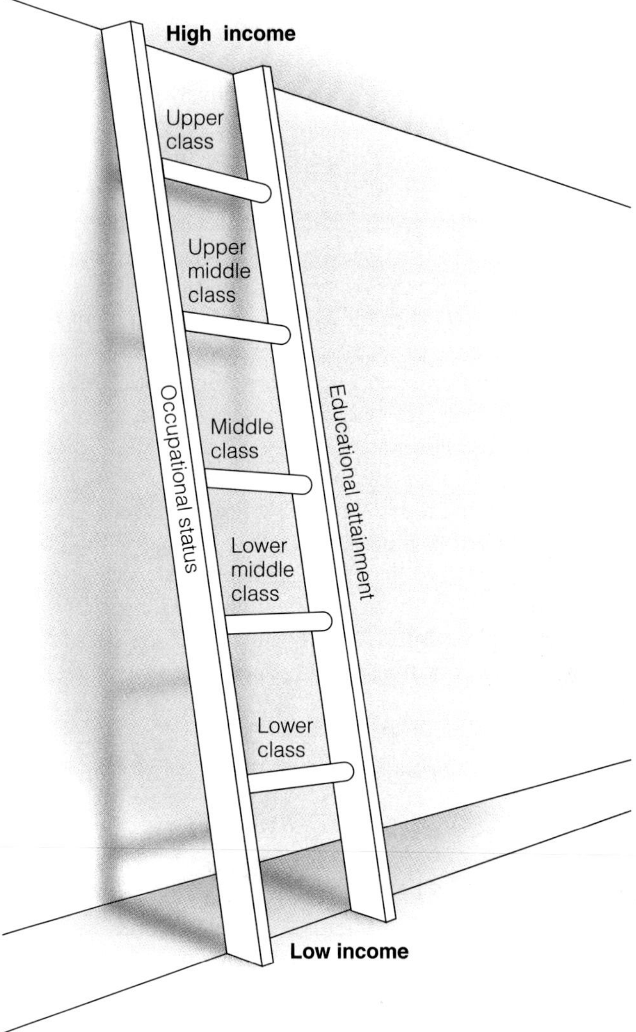

FIGURE 9.1 The Layered Model of Stratification

The Class Structure of the United States

The class structure of the United States is elaborate, arising from the interactions of old wealth, new wealth, intensive immigration, a culture of entrepreneurship and individualism, and in recent times, vigorous globalization (see Figure 9.1). One can conceptualize the class system as a series of layers, with different class groups arrayed up and down the rungs of the class ladder, each rung corresponding to a different level in the class system. Conceptualized this way, social class is the common position groups hold in a status hierarchy (Wright, 1979; Lucal, 1994); class is indicated by factors like levels of income, occupational standing, and educational attainment. People are relatively high or low on the ladder depending on the resources they have, whether those are education, income, occupation, or any of the other factors known to influence people's placement (or ranking) in the stratification system. Indeed, an abundance of sociological research has stemmed from the concept of **status attainment,** the process by which people end up in a given position in the stratification system. Status attainment research describes how such factors as class origins, educational level, and occupation produce class location, and it describes the extent to which people are able to move throughout the class system, as we will see in the section on social mobility later in this chapter.

This layered model of class suggests that in the United States, stratification is hierarchical, but somewhat fluid. Although there are different gradients in the stratification system, they are not fixed as they might be in a society where one's placement is sheerly a matter of the class into which one is born. In a relatively open class system such as the United States, people's achievements do matter, although the extent to which people rise rapidly and dramatically through the stratification system is less than the popular imagination claims. Some people do begin from modest origins and amass great wealth and influence (celebrities like Bill Gates, Oprah Winfrey, and millionaire sports heroes come to mind), but these are the exceptions, not the rule. Some people move down in the class system, but as we will see, most people remain relatively close to their class of origin. When people rise or fall in the class system, the distance they travel is usually relatively short, as we will see further in the section on social mobility.

The image of stratification as a layered system, with different gradients of social standing, emphasizes that one's **socioeconomic status (SES)** is derived from certain factors. Income, occupational prestige, and education are the three measures of socioeconomic status that have been found to be most significant in determining people's placement in the stratification system.

Income is the amount of money a person receives in a given period; socioeconomic status can be measured in part by income brackets. The **median income** for a society is the midpoint of all household incomes. In other words, half of all households earn more than the median income; half earn less. In the layered model of class, those bunched around the median income level are considered middle class. Sociologists may argue about which income brackets constitute middle-class standing, but they agree that income is a significant indicator of class, although not the only one. In 1997, median household income in the United States was $37,005, although the only households reaching this point were Black and White married-couple families and White families with male heads and no wife present (see Figure 9.2; U.S. Bureau of the Census, 1998).

Occupational prestige is a second important indicator of socioeconomic status. **Prestige** is the value assigned to people and groups by others. **Occupational prestige** refers to the subjective evaluation people give to jobs. To determine occupational prestige, sociological researchers typically ask nationwide samples of adults to rank the general standing of a series of jobs. These subjective ratings provide information about how people perceive the worth of different occupations. People tend to rank Supreme Court justice as one of the most prestigious occupations, followed by occupations such as physician, professor, judge, lawyer, and scientist. In the middle ranges are occupations like electrician, newspaper columnist, insurance agent, and police officer. The occupations typically considered to have the lowest prestige are farm laborer, maid or servant, garbage collector, janitor, and shoe-shiner (Davis and Smith, 1984). These rankings do not reflect the worth of people within these positions, but are indicative of the judgments people make about the worth of these jobs and their value to society.

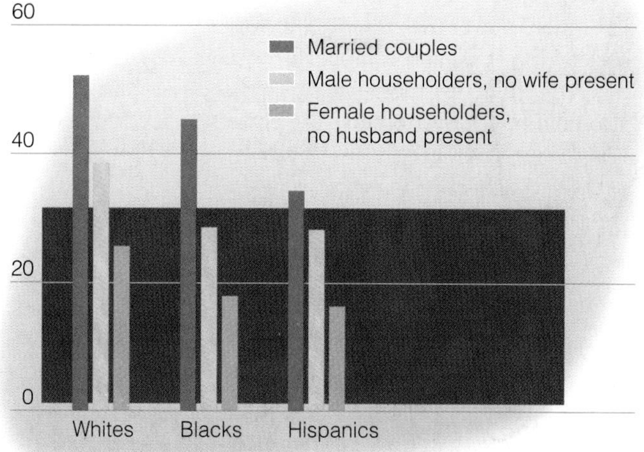

FIGURE 9.2 Median Income by Race and Household Status

SOURCE: U.S. Bureau of the Census. 1998a. *Money Income in the United States: 1997.* Washington, DC: U.S. Government Printing Office, pp. 2–4.

The final major indicator of socioeconomic status is **educational attainment,** typically measured as the total years of formal education. The more years of education attained, the more likely a person will have higher class status. The prestige attached to occupations is strongly tied to the amount of education the job requires—the more education people think is needed for a given occupation, the more occupational prestige people attribute to that job (Blau and Duncan, 1967; MacKinnon and Langford, 1994).

Layers of Social Class

Taken together, income, occupation, and education are good measures of people's social standing. How then do sociologists understand the array of classes in the United States? Using a layered model of stratification, most sociologists describe the class system in the United States as divided into several classes: upper, upper middle, middle,

The lifestyles of elites, such as at this private club, are beyond the reach of the vast majority of Americans, even though this lifestyle forms a cultural ideal.

lower middle, and lower class. Each class is defined by characteristics such as income, occupational prestige, and educational attainment. The different groups are arrayed along a continuum with those with the most money, education, and prestige at the top and those with the least at the bottom.

In the United States, the *upper class* owns the major share of corporate and personal wealth; it includes those who have held wealth for generations as well as those who have recently become rich. Only a very small proportion of people actually constitute the upper class, but they control vast amounts of wealth and power in the United States. Those in this class are *elites* who, in Marxist terms, own the means of production. They exercise enormous control throughout society. Most of their wealth is inherited.

Despite social myths to the contrary, the best predictor of future wealth is the family into which you are born. Each year, the business magazine *Forbes* publishes a list of the "Forbes Four Hundred"—the four hundred wealthiest families and individuals in the country. To be on the list published in 1997, your net worth had to be at least $475 million. Bill Gates, the richest person on the list, has an estimated worth of more than $39.8 billion, double that of the previous year. Of all the wealth represented on the Forbes 400 list, more than half is inherited. Those on the list who could be called "self-made" were not typically of modest origins; most inherited significant assets (*Forbes,* 1997; Sklar and Collins, 1997). Because elites control such gigantic amounts of wealth, they wield immense power in society.

THINKING SOCIOLOGICALLY

Suppose that you wanted to reduce inequality in the United States. Since you know that the transmission of *wealth* is one of the bases of *stratification,* would you be willing to eliminate the right to inherit property to achieve greater equality in America?

Those in the upper class with newly acquired wealth are known as the *nouveau riche.* Although they may have vast amounts of money, they are often not accepted into "old rich" circles. As we mentioned earlier, wealth is not the sole defining characteristic of the upper class. Social connections and family prestige can be as important as money at the pinnacle of the class structure.

The *upper middle class* includes those with high incomes and high social prestige. They tend to be well-educated professionals or business executives. Their earnings can be quite high indeed—successful business executives can earn millions of dollars a year. It is difficult to estimate exactly how many people fall into this group because of the difficulty of drawing lines between the upper, upper middle, and middle class. Indeed, the upper middle class is often thought of as "middle class" because their lifestyle sets the standard to which many aspire, but this lifestyle is actually unattainable by most. A large home full of top-quality furniture and modern appliances, two or three luxurious cars, vacations every year (perhaps a vacation home), high-quality college education for one's children, and a fashionable wardrobe are simply beyond the means of a majority of people in the United States.

Probably the largest group in the class system is the *middle class,* including those who fall within a given range above and below the median income figure. Some have suggested that those who earn between 75 and 125 percent of the median income level are middle class. In 1997, this would have included all households earning between $27,753 and $46,256 in that year. Many think this definition is too narrow since households earning three or four times as much as the high median income are typically also seen as middle class. Different income brackets may be used to draw the line around the middle class, but the concept is the same—the middle class is the group clustered around the median income point. The bulk of the middle class are white-collar workers although the category also contains many skilled craftspeople, individuals who are self-employed, and other types of workers.

One of the reasons the middle class is so difficult to define in the United States is that being "middle class" is more than just economic position. By far the majority of Americans identify as middle class even though there is wide variation in their style of life and the resources they have at their disposal. But the idea that America is an open-class system leads many to think that the majority have a middle class lifestyle since, in general, people tend not to want to recognize class distinctions in the United States. Thus, the middle class becomes the ubiquitous norm even though many who would call themselves middle class have a tenuous hold on this class position. As we will see in a section on race and class, recent years have also seen the growth of the African American middle class; other racial–ethnic groups, including Asian Americans and Latinos, also have a significant middle class—marked by jobs in higher status careers and good educational attainment.

In the hierarchy of social class, the *lower middle class* includes workers in the skilled trades and low-income bureaucratic workers. Examples are blue-collar workers (those in skilled trades who do manual labor) and many service workers, such as secretaries, hairdressers, waitresses, police, and firefighters. Medium to low income, education, and occupational prestige define the lower middle class relative to the class groups above it. The term *lower* in this class designation refers to the relative position of the group in the stratification system, but it has a pejorative sound to many people, especially to people who are members of this class. They are unlikely to refer to themselves as lower middle class, tending to prefer working class or middle class.

The *lower class* is composed primarily of the displaced and poor. People in this class have little formal education and are often unemployed or working in minimum wage jobs. People of color and women make up a disproportionate part of this class. The poor include the working poor—those who work for wages, but who do not earn enough to work their way out of poverty. The working poor constitute a large portion of those who are poor. Forty percent of the poor work; 10 percent work year-round and full time—a proportion that has generally increased over time (see Figure 9.3).

Recently, the concept of the "underclass" has been added to the lower class. The **underclass** includes those who have been left behind by contemporary economic developments and who are likely to be unemployed and therefore dependent on public assistance or crime for economic support. The underclass includes those with little or no opportunity for movement out of the worst poverty. Rejected from the economic system, those in the underclass may become dependent on public assistance or illegal activities. According to sociologist William Julius Wilson, structural transformations in the economy have left large groups of people, especially urban minorities, in highly vulnerable positions. Without work and unable to sustain themselves in an economy that has discarded them, such groups have formed a growing underclass—a development that has exacerbated the problems of urban poverty and created new challenges for social policy makers if we are to reverse this trend (Wilson, 1996; Wilson, 1987).

Class Conflict

The model of class described here defines classes in terms of a status hierarchy. Sociologists have also analyzed class according to the perspective of conflict theory, derived from the early work of Karl Marx. This perspective defines classes in terms of their structural relationship to other classes and their relationship to the economic system. Derived from conflict theory in sociology, this analysis emphasizes power relations in society, interpreting inequality as resulting from the unequal distribution of power and resources in society (see Chapter 1). Instead of seeing class simply as a continuum, as suggested earlier, sociologists who work from a conflict perspective see the classes as facing off against each other, with elites exploiting and dominating others. The key idea in this model is that class is not simply a matter of what individuals possess in terms of income and prestige; instead, class is defined by the relationship of the classes to the

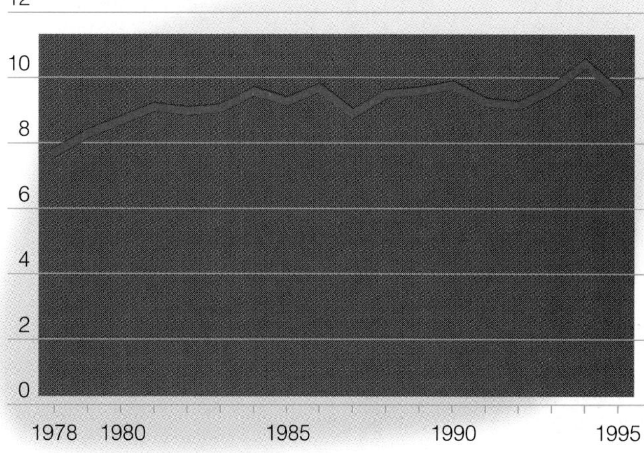

FIGURE 9.3 The Working Poor

SOURCE: U.S. Bureau of the Census. 1997. *Historical Poverty Tables—Persons.* World Wide Web site: http://www.census.gov/hhes/poverty/histpov/histpov18.html

larger system of economic production (Vanneman and Cannon, 1987; Wright, 1985).

From a conflict perspective, the position of the middle class in society is unique. The middle class, or the *professional-managerial class* includes managers, supervisors, and professionals. Members of this group have substantial control over other people, primarily through their authority to direct the work of others, impose and enforce regulations in the workplace, and determine dominant social values. Although, as Marx argued, the middle class is itself controlled by the ruling class, members of this class tend to identify with the interests of the elite. The professional-managerial class, though, is caught in a contradictory position between elites and the working class. Like elites, those in this class have some control over others, but like the working class, they have minimal control over the economic system (Wright, 1979). As capitalism progresses, according to conflict theory, more and more of those in the middle class drop into the working class, as they are pushed out of managerial jobs into working-class jobs or as professional jobs become organized more along the lines of traditional working-class employment.

Has this happened? Not to the extent Marx predicted whereby he thought that ultimately there would be only two classes—the capitalist and the proletariat—but, to some extent, this is occurring. Classes have become more polarized, with the well-off accumulating even more resources and the middle class seeing their median income (measured in constant dollars) falling (Wolff, 1996). Many who have lost their jobs because of corporate downsizing or mergers return to the labor force in positions that are less well paid and often without benefits. At the same time, corporate mergers concentrate more and more corporate wealth in the hands of a few. Perhaps the class revolution that Marx predicted has not occurred, but the dynamics of capitalism that he analyzed are unfolding before us.

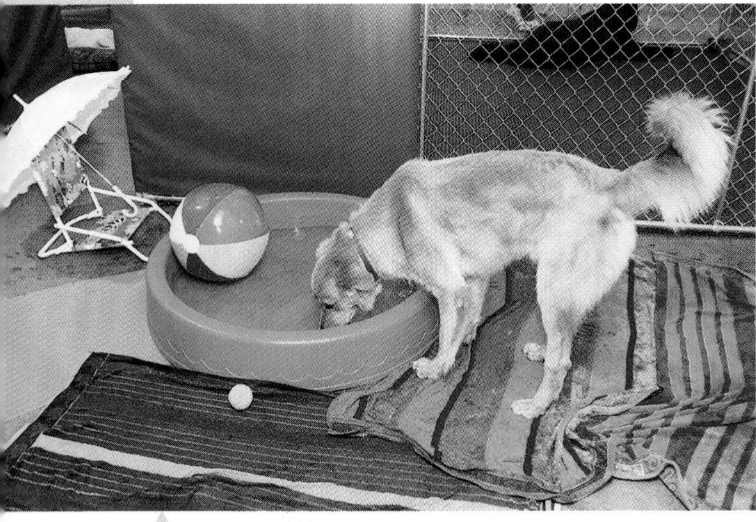

Sociologists have shown that the nouveau riche *can be ostentatious in their display of wealth. In this pet care center in San Francisco, even the family pet can engage in conspicuous consumption.*

Recall that Marx defined the working class as those who sell their labor for wages. This definition, taken from Marx, would include blue-collar workers and many white-collar workers. Members of the working class have little control over their own work lives; instead, they have to take orders from others. This concept of the working class departs from traditional "blue-collar" definitions of working-class jobs since it would include those like secretaries, salespeople, and nurses—any group that works under the rules imposed by managers or elites. Although those in the middle class may exercise some autonomy at work, typically the working class has little power to challenge decisions of those who supervise them, except insofar as they can organize collectively, as in strikes or other collective work actions.

Conflict theorists also tend to see the poor as under assault by society. The conflict inherent in a structured system of inequality leaves the poor especially vulnerable. Poor through no fault of their own, the poor are still blamed for their own poverty, especially as defined through the belief systems established about failure that are propagated by elites and the middle class. The trap of growing inequality that the poor find themselves within is well illustrated by Jeffrey Reiman's (1998) book on crime and inequality, aptly titled *The Rich Get Richer and the Poor Get Prison.* This view of poverty is not new. A political cartoon appearing during Lyndon Johnson's presidency in the 1960s, when federal antipoverty programs were labeled the War on Poverty, also reveals the class conflict built into society: The cartoon depicted a poor woman saying, "I hear they have declared a war on poverty. Does that mean they are going to shoot me?"

Whether one uses a layered perspective of class or a class conflict perspective, you can see that the class structure in the United States is a hierarchical one. People's class position gives them different access to jobs, income, education, power, and social status—all of which bestow further opportunities on some and deprive others of success. Although people do sometimes move from one class to another (though this is not the norm), the class structure is a system with boundaries and judgments built into it. Class conflict is built into the system. Thus, the middle and working classes shoulder much of the tax burden for social programs, producing resentment by these groups toward the poor. At the same time, corporate taxes have declined while tax loopholes for the rich have increased—an indication of the privilege that is perpetuated by the class system. Whatever features of the class system different sociologists study, they see class stratification as a dynamic process—one involving the interplay of access to resources, group judgments about other groups, and the exercise of power.

The Distribution of Wealth and Income

One thing that becomes clear with careful study of the U.S. class structure is that there is enormous class inequality in this society, and it is growing (see Figure 9.4). The elites control an enormous share of the wealth in society and exercise tremendous control over others. The gap between the rich and the poor is also increasing, while much of the middle class finds its class standing slipping. Consider this: Since the early 1970s, median income (in constant dollars) has actually declined. Since the 1970s, the top 5 percent of the population has also received a larger share of the total income (22 percent of all income), whereas the bottom fifth has received less (3.6 percent in 1997; U.S. Census Bureau, 1998). High compensation for CEOs explains some of this concentration at the high end. In one recent ten-year period, for example, salaries for CEOs increased by 212 percent, compared to an increase of only 53 percent for factory workers (Byrne, 1996; Byrne, 1991). The income gains were further augmented by tax reforms that have provided financial breaks to the highest earners. These trends verify the popular adage that the rich are getting richer and the poor, poorer. As the classes become more polarized, it is possible that there will be a weakening of the myth that the United States is basically a middle-class society.

When discussing the distribution of resources in society, sociologists make a distinction between wealth and income. **Wealth** is the monetary value of everything one actually owns. It is calculated by adding all financial assets (stocks, bonds, property, insurance, value of investments, and so on) and subtracting debts, which gives a dollar amount that is one's *net worth.* **Income** is the amount of money brought into a household from various sources (wages, investment income, dividends, and so on) during a given period. Unlike income, wealth is cumulative—that is, its value tends to increase through investment, and it can be passed on to the next generation, giving those who inherit wealth a considerable advantage in accumulating more resources.

Reliable data on the distribution of wealth are difficult to assemble. The government systematically monitors income data, but not wealth. The figures in the Forbes 400 list of the wealthiest Americans are only estimates, based on publicly disclosed corporate data, shrewd guesswork by market watchers and finance writers, and previous estimates updated by factoring in the overall performance of the economy. Measures of national income distribution are based on annually reported U.S. census data drawn from a sample of the population. The census reports the median income level of different groups, such as women, Hispanics, and Whites, and for all groups combined.

Where is all the wealth? The wealthiest 1 percent own 37 percent of all net worth. Indeed, the top 1 percent of the population possesses more wealth than the bottom 90 percent. That 1 percent controls 20 to 25 percent of all the nation's wealth; the top 10 percent controls 86 percent of the nation's wealth. This trend has led economists to conclude that "inequality is at its highest since the great leveling of wages and wealth during the New Deal and World War II" (Nasar, 1992: A17). There was a sharp increase in the concentration of wealth during the 1980s and 1990s—a trend that continues today; moreover, the concentration of wealth is higher in the United States than in any of the other industrialized nations (Ericksen, 1988; Wolff, 1996).

In contrast to the vast amount of wealth and income controlled by elites, a very large proportion of Americans (one-third) have no financial assets at all once debt is subtracted; many, in fact, have negative net worth, since their debt exceeds their assets (Wolff, 1996). The American dream of owning a home, a new car, taking annual vacations, and sending one's children to good schools—not to mention saving for a comfortable retirement—is increasingly unattainable by the majority. When you see the amount of income and wealth controlled by a small segment of the population, a sobering picture of class inequality emerges, as we can see in Figure 9.4.

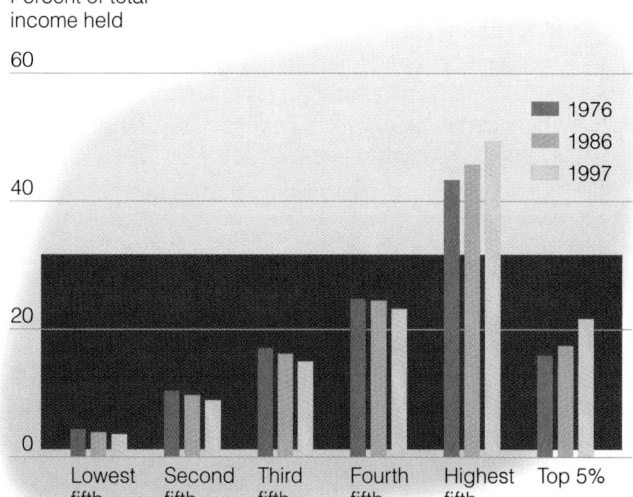

FIGURE 9.4 Share of Income Held by the U.S. Population

SOURCE: U.S. Bureau of the Census. 1998a. *Money Income in the United States: 1997.* Washington, DC: U.S. Bureau of Commerce, B6.

Popular legends, however, extol the possibility of anyone making it rich in the United States. The well-to-do are admired not just for their style of life, but also for their supposed drive and diligence. The admiration for those who rise to the top makes it seem like anyone who is clever enough and works hard can become fabulously rich. Despite the prominence of rags-to-riches stories in American legend, most wealth in this society is inherited. A few individuals—whether shrewd, swashbuckling, or lucky—make their way into the elite class by virtue of their own success, but this is rare. The upper class is also overwhelmingly White and Protestant. These elites travel in exclusive social networks that tend to be open only to those in the upper class. They tend to intermarry, their children are likely to go to expensive schools, and they spend their leisure time in exclusive resorts.

There is a myth in this country that most of the very rich are women, and that many of them are widows or descendants of wealthy families. Actually, only one-fifth of the nation's wealthiest people are women, and wealthy women are even more likely than wealthy men to have inherited their money (Baltzell, 1964; Barton, 1985; Ostrander, 1984). As the box "Doing Sociological Research: Black Wealth/White Wealth" shows, race also influences the pattern of wealth distribution in the United States; for every dollar of wealth held by White Americans, Black Americans have only 26 cents (Oliver and Shapiro, 1995).

Money and status alone do not tell the whole story of the significance of elites in the United States. Elites also exercise tremendous power. For example, elites own half of all personally held corporate stock in the United States, and they wield enormous influence over the political process by funding lobbyists, exerting their social and personal influence on other elites, and contributing heavily to political campaigns. The grip of the upper class on political power is also witnessed by the large number of multimillionaires in Congress. Just to be among the top 50 wealthiest members of Congress required a personal fortune of at least $2.4 million in 1996; those without great wealth are at a huge disadvantage in financing political campaigns (Broder, 1996).

Despite beliefs to the contrary, class divisions in the United States are real, apparently permanent, and becoming more marked. The elite are becoming better off, whereas the positions of the middle class, working class, and poor seem to be worsening. Many middle-class people feel that their way of life is slipping away. For example, home ownership is an increasingly remote possibility for many people, particularly those who do not have the advantage of large investments and family inheritance. Those who have studied patterns in the distribution of wealth and income have concluded that the concentration of wealth by the 1990s was more extreme in this nation than at any time since 1929 (Wolff, 1996). As just one example, John D. Rockefeller is typically heralded as one of the wealthiest men in U.S. history. But comparing Rockefeller to Bill Gates, in the value of today's dollars, Gates has already surpassed Rockefeller's riches (Myerson, 1998).

Many factors have contributed to the declining fortunes of the lower and middle classes in the United States, including the profound effects of national and global economic change. Economic restructuring is concentrating wealth increasingly in the hands of a few, while plant shutdowns and relocations have led to unemployment for many blue-collar workers, and reductions in state and federal spending have eliminated many government jobs—jobs that have traditionally been the route to middle-class standing for many workers. Economic downsizing, even by stalwart

BOX 9.1 DOING SOCIOLOGICAL RESEARCH

Black Wealth/White Wealth

MOST STUDIES of racial inequality focus on income inequality, but two sociologists have broadened our understanding of inequality by analyzing differences in the wealth of Black and White Americans. Melvin Oliver and Thomas Shapiro argue that income alone is insufficient as an indicator of people's economic well-being, since wealth (one's accumulated financial assets minus debt) is what gives people the ability to pass economic well-being on to the next generation.[1] Oliver and Shapiro begin their research by asking, "How has wealth been distributed? What changes have occurred in recent years? What are the implications of these changes for racial inequality?"

They were an ideal pair to do this research. Melvin Oliver is an African American sociologist, raised in a racially segregated community and in the first generation of his family to be college educated. Thomas Shapiro is a White American who grew up in an affluent White neighborhood and observed how this group benefits from the political structure, tax writeoffs, business development, and good housing. They base their research on a detailed statistical analysis of two national household surveys: the Federal Reserve Systems Survey of Consumer Finances and the Bureau of the Census Survey of Income and Program Participation. They supplement these quantitative data with in-depth interviews with Black and White middle-class families in two cities, Boston and Los Angeles. This dual research approach allows them to analyze statistical patterns in the relationships between household assets and various demographic variables (education, age, race, and income, for example), while also bringing in examples of people's experiences of trying to succeed.

Their study concludes that differences in income levels between Black and White Americans are not enough to explain the vast differences in wealth between these two groups. They find, for example, that Black Americans have one-half the net worth of their White counterparts. The long-term life prospects of Black households are substantially poorer than the prospects of Whites in similar income brackets.

Middle-class Blacks, Oliver and Shapiro find, earn 70 cents for every dollar earned by Whites, but Blacks have only 15 cents for every dollar of wealth held by the White middle class. Furthermore, even when controlling for the level of education, Whites at the top of the economic hierarchy have four times the assets of similarly situated Blacks.

How do they explain these disparities? They point to state policies that have impaired Black Americans from being able to accumulate wealth. Discriminatory housing policies, bank lending policies, tax codes, and other policies have disadvantaged Black Americans over time, making it impossible for them to accumulate the assets that give people a firm hold on economic well-being. Some of these discriminatory policies have ended, but many continue or their effects persist since accumulating wealth occurs over time. Black Americans were also restricted from developing businesses that would tap lucrative White markets. In the end, Oliver and Shapiro argue, "the cumulative effects of the past have cemented Blacks to the bottom of society's economic hierarchy" (1995: 5)—something they refer to as the *sedimentation of racial inequality*.

Oliver and Shapiro also find that patterns of social mobility and the ability to pass on wealth vary significantly between races. In the upper middle class, the White population has high occupational inheritance (that is, children move into the same occupational category as their parents), but Blacks from such favorable social origins are less able to pass this on to their children since they do not have the same level of wealth to optimize their children's life chances. Blacks are also more likely to be downwardly mobile; twice as many Blacks as Whites from white-collar, upper middle-class backgrounds fall into blue-collar positions. Blacks are less able to transmit assets from previous generations to the next generation, which is one of the main supports of White achievement. Reflecting on the American dream as described by Horatio Alger, who wrote of good-hearted people striking it rich, Oliver and Shapiro conclude, "In telling the mobility story, it is important to know the skin color of Horatio Alger's heroes" (1995: 169).

These scholars suggest the need for policies that promote the development of assets for all groups at the bottom of the social structure. In addition, they call for massive redistribution of wealth to reverse the historical advantage that some groups have had and now use to their cumulative benefit. Their rich and detailed analysis provides solid evidence of the continuing influence of both race and class in perpetuating inequality.

SOURCE: Oliver, Melvin L., and Thomas M. Shapiro. 1995. *Black Wealth/White Wealth: A New Perspective on Racial Inequality.* New York: Routledge.

[1] Oliver and Shapiro won the prestigious award for Distinguished Publication of the Year, given by the American Sociological Association in 1997.

corporations like IBM, has left many out of work or sent middle-class employees into jobs with lower pay, less prestige, and perhaps no employee benefits, such as they had come to expect given their levels of education and experience. At the same time, economic policies have been very favorable for the rich, amplifying the degree of class polarization.

Diverse Sources of Stratification

Class is only one basis for stratification in the United States. Factors like age, ethnicity, and national origin have a tremendous influence on stratification. Race and gender are two of the primary influences in the stratification system in the United States. In fact, analyzing class without also analyzing race and gender can be misleading. Race, class, and gender, as we are seeing throughout this book, are overlapping systems of stratification that people experience simultaneously. A working-class Latina, for example, does not experience herself as working class at one moment, Hispanic at another moment, and a woman, the next. At any given point in time, her position in society is the result of her race, class, *and* gender status.

To explain in another way: Class position is manifested differently, depending on one's race and gender, just as gender is experienced differently depending on one's race and class, and race is experienced differently, depending on one's gender and class. Depending on one's circumstances, at the moment, race, class, or gender may seem particularly salient at a given moment in a person's life. For example, a Black middle-class man stopped and interrogated by police when driving through a predominantly White middle-class neighborhood may at that moment feel his racial status as his single most outstanding characteristic, but at all times, his race, class, and gender influence his life chances. As social categories, race, class, and gender shape all people's experience in this society—not just those who are disadvantaged (Andersen and Hill Collins, 1998).

Class also significantly differentiates group experience within given racial and gender groups. Latinos, for example, are broadly defined as those who trace their origins to regions originally colonized by Spain. The ancestors of this group include both White Spanish colonists and the natives who were enslaved on Spanish plantations. Today, some Latinos identify as White, others as Black, and others by their specific national and cultural origins. The very different histories of those categorized as Latino are matched by significant differences in class. Some may have been schooled in the most affluent settings; others may be virtually unschooled. Those of upper-class standing may have had little experience with prejudice or discrimination; others may have been highly segregated into barrios and treated with extraordinary prejudice. Latinos who live near each other geographically in the United States and who are the same age and share similar ancestry, may have substantially different experiences based on their class standing (Massey, 1993). Neither class, race, nor gender, taken alone, can be considered an adequate indicator of different group experiences, as shown in the box "Understanding Diversity: Latino Class Experience."

Workers with minimal education and skills may need to hustle for minimum wage jobs, as is the case for these day workers in Austin, Texas.

Race and Class

The notable African American sociologist E. Franklin Frazier was one of the first sociologists to write about the relationship between race and class. Although the presence of a significant Black middle class is often seen as a relatively recent phenomenon, Frazier argued that the Black middle class goes all the way back to the small numbers of free Blacks in the eighteenth and nineteenth century (Frazier, 1957). In the twentieth century, the Black middle class expanded to include those who were able to obtain an education and become established enough in industry, business, or a profession that they could live comfortably, and in a few cases, affluently. These included teachers, doctors, nurses, and businesspeople. The Black middle class also historically included postal workers and railroad porters. Although Whites may not have considered these workers middle class, they had relatively high prestige within the Black community. The Black middle class was a class of its own—not comparable to the White middle class, but distinct within the Black community. Wages for Black middle-class and professional workers never matched those of Whites in the same jobs; furthermore, despite their status within Black communities, members of the Black middle class have been excluded from White schools, clubs, and social settings. Many sociologists conclude that the class structure among African Americans has existed alongside the White class structure, separate and different.

In recent years, both the African American and Latino middle class have expanded, primarily as the result of increased access to education and middle-class occupations for people of color (Collins, 1983; Durant and Louden,

1986; Landry, 1987). The persistence of racial discrimination and the recent arrival of these groups in the middle class means, however, that their hold on middle-class status is more tenuous than that of many middle-class Whites. For example, many in the Black and Latino middle class work in public sector jobs—positions that depend on continuing government support. During periods of economic recession or political conservatism, when there is considerable pressure to reduce federal spending, the jobs that are eliminated are likely to cause a significant thinning in the ranks of the Black and Latino middle class (Collins-Lowry, 1997; Collins, 1983; Silver, 1995). Although middle-class Blacks and Latinos may have economic privileges that others in these groups do not have, their class standing does not make them immune to the negative effects of race. Asian Americans have a significant middle class, but they have also been stereotyped as the most successful minority group because of their presumed educational achievement, hard work, and thrift. This stereotype is referred to as the *myth of the model minority* and includes the idea that a minority group must adopt alleged dominant group values to succeed. This myth about Asian Americans obscures the significant obstacles to success that Asian Americans encounter (Woo, 1998), and it ignores the hard work and educational achievements of other racial and ethnic groups. The idea that Asian Americans are the "model minority" also obscures the high rates of poverty among many Asian American groups (Lee, 1995).

Gender and Class

The effects of gender stratification further complicate the analysis of class. In the past, women were thought to derive their class position from their husband or father, but sociologists now challenge this assumption. In fact, men are more likely than women to exhibit a link between their own occupational status and that of their father (Roos, 1985). Measured by their own income and occupation, the vast majority of women would likely be considered working class. The median income for women, even among those employed full time, is far below the national median income level. (In 1997, when median income for men working year-round and full time was $35,248, the median income of women working year-round, full-time was $26,029; U.S. Bureau of the Census, 1998.) The vast majority of women work in low-prestige and low-wage occupations, even though women and men have comparable levels of educational attainment. Women's class status is abundantly clear in the aftermath of a divorce, when, typically, women's income drops significantly, while men's increases (Peterson,

BOX 9.2

UNDERSTANDING DIVERSITY

Latino Class Experience

LATINOS in the United States are a diverse population of many different groups, each with many different national origins, histories, and cultural backgrounds. This diversity is represented by the fact that Latinos do not agree among themselves on what they should be called; some prefer "Latino," others "Hispanic," and some prefer to be identified by their cultural origins, as in Chicanos, Cuban Americans, or Puerto Riqueños. Some think of themselves as White. Generational differences further add to the diversity among Latinos.

Sociologist Douglas Massey reminds us of this diversity in thinking about the diverse class experiences of Latinos. Massey writes that as a result of different histories, Latinos live in different socioeconomic circumstances. He states:

> They may be fifth-generation Americans or new immigrants just stepping off the jetway. Depending on when and how they got to the United States, they may also know a long history of discrimination and repression or they may see the United States as a land of opportunity where origins do not matter. They may be affluent and well educated or poor and unschooled; they may have no personal experience of prejudice or discrimination, or they may harbor stinging resentment at being called a "spic" or being passed over for promotion because of their accent. (1193: 7–8)

Massey also notes that levels of residential segregation and poverty vary across different Latino groups. He notes that as socioeconomic levels rise and as immigrants reach second, third, and later generations within the United States, the degree of segregation for Latinos progressively falls. Puerto Ricans have higher levels of segregation than other Latino groups, as well as some of the highest rates of poverty. The route of Latinos into this country also contributes to their class position. Some Latinos are indigenous to the United States—that is, their families owned land in the American Southwest that was colonized by White settlers; others are recent immigrants who enter the economy as low-wage workers with little opportunity for upward mobility.

Diversity among Latinos has many implications for understanding the experience of this population. Some are caught in the economic underclass, others are middle class; a few are among the nation's elites. Massey reminds us that factors such as class, historical origins, residential segregation, race, and migration patterns must be carefully analyzed to understand Latino experiences.

SOURCE: Massey, Douglas S. 1993. "Latino Poverty Research: An Agenda for the 1990s." *Social Science Research Council Newsletter* 47 (March): 7–11.

1996a, 1996b; Weitzman, 1996; Weitzman, 1985).

The measurement of class status for women is complicated by the question of how women's status is derived. Many women who have little or no income of their own consider themselves middle class by virtue of their husband's class status. How does one determine the class status of a White woman earning $25,000 per year as a secretary, but married to a business executive earning $85,000? What is her class status if she is divorced? Many stratification researchers have simply excluded women from research samples, arguing that the unit of stratification is the family, not the individual. With more women in the labor force and living independently from men, this assumption is problematic. It can be difficult to identify the "head" of a household when more married women are engaged in paid work and there are so many different forms of families in contemporary society. Even treating the family as the unit of stratification does not solve the problem completely since resources may not be equally shared within the family (Baxter, 1994; Morris, 1989).

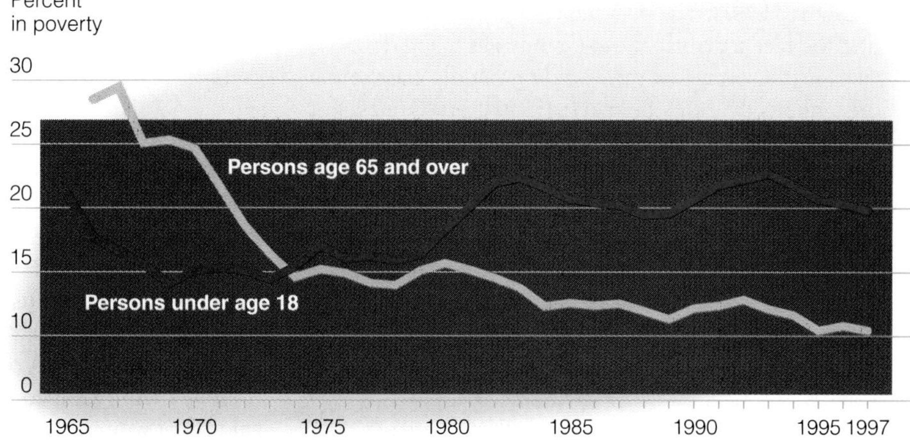

FIGURE 9.5 Poverty Among the Old and the Young

SOURCE: Dalaker, Joseph, and Mary Naifeh. *Poverty in the United States: 1997.* Current Population Reports, P60-201. Washington, DC: U.S. Department of Commerce, p. 2.

Class differences between women are also important since women are not a monolithic group. The problems faced by professional and managerial women are not the same as those experienced by women in low-wage, low-prestige work. Although both groups are influenced by their gender status, how each group experiences their gender depends on class, as well as race and position. It is important to see that race, class, and gender shape *everyone's* experience, not just those who have been victimized by inequality. The lives of White elite men are affected by the system of race, class, and gender relations just as much as the lives of poor Native American women, although obviously the effects will be quite different. The stratification system may simply be more invisible and seem more fair to those who benefit most from current arrangements.

Age and Class

Unlike race and gender, age changes over the course of one's life. Still, age is a significant source of stratification in the United States with different age groups experiencing quite different locations in the stratification system. We examine this further in Chapter 13, but note here that just being born in a particular generation can have a significant influence on one's life chances. The fears of current young, middle-class people that they will be unable to achieve the lifestyles of their parents show the effect that being in a particular generation can have on one's life chances. The effect of one's age on life chances is also not independent of the effect of race or gender. Being a young, African American boy, for example, puts you far more at risk of poverty than being an older, White American man.

It is now the case that the age group most likely to be poor are children, 20 percent of whom live in poverty in the United States. This represents a change from the recent past when the aged were the most likely to be poor (see Figure 9.5). Although many elderly people are now poor (10.5 percent of those 65 and over), far fewer in this age category are poor than was the case not many years ago—in 1970, 25 percent of those over 65 were poor (U.S. Bureau of the Census, 1980; Dalaker and Naifeh, 1998). This shift reflects the greater affluence of the older segments of the population—a trend that is likely to continue as the current large cohort of middle-aged, middle-class Baby Boomers grow older.

Class and Cultural Diversity

The study of class pays great attention to people's access to economic resources, but class has a cultural dimension as well. A particular class group is distinguishable not just by its access to money, prestige, and power, but also by the culture it develops and lives within. Cultural behaviors and values are never uniform within a class, but some general cultural patterns can be observed. As one sociologist has written:

> As a result of the class you are born into and raised in, class is your understanding of the world and where you fit in; it's composed of ideas, behavior, attitudes, values, and language; class is how you think, feel, act, look, dress, talk, move, walk; class is what stores you shop at, restaurants you eat in; class is . . . who our friends are, where we live and work, even what kind of car we drive, if we own one, and what kind of healthcare we receive, if any (Langston, 1998: 127–128).

The unequal distribution of economic resources is a quantitative index of class. The different cultures and ways of life found among different classes represent a more subjective aspect of class, stretching from friendships and recreations to how different classes communicate. Middle-class people tend to base their social life on friendships built through work and business contacts, whereas the working class is more likely to base its social life on the family (Gilbert and Kahl, 1987). Differences in leisure habits are in part because of the difference in resources available to class groups. One can see that economic wherewithal is likely to show itself in many cultural preferences. For example, liking good wine is not only an acquired taste; it requires money.

Dress is quite distinct by social class, and it is a marker that people use to make class distinctions. Classes also differ in their language. For example, the language of "the street" has its own vocabulary, idioms, syntax, intonation, and grammar—much of which would be unintelligible to class elites. Seen on its own terms, street language is efficient and meaningful, its vocabulary rich, but class-based judgments about language and other forms of culture are often based on false assumptions that stem from stereotypes that the dominant group holds about subordinate groups. There is a stereotype, for example, that anyone who speaks with a southern accent is ignorant or racist. This belief stems in part from the subordinated social and economic position of the South throughout much of the nation's history.

The cultural dimensions of social class are especially obvious if you think about the experience of entering a class setting different from your own background: You are likely to feel out of place. A middle-class person walking through a working-class community may find the environment uncomfortable and unpleasant. Unfortunately, the politics of class often lead those in the upper and middle classes to make disdainful judgments about the lifestyle of the working class and the poor. At the same time, working-class people who enter middle-class settings often find them inhospitable and alienating and may feel they have to cloak or eliminate their working-class habits to be accepted. One such experience is reported by a woman who was raised in the working class and is now a college faculty member. She describes her first experiences in college like this: "It was lonely and isolating in college. I lost my friends from high school—they were at the community college, vo-tech school, working, or married. I lasted a year and a half in this foreign environment before I quit college, married a factory worker, had a baby and resumed living in a community I knew" (Langston, 1998: 128).

The dominant culture supports White middle-class lifestyles and values more than other class values. As a result, succeeding in the middle-class world usually means abandoning (at least publicly) working-class mannerisms and habits. Barbara Ehrenreich writes that the lifestyles, habits, tastes, and attitudes of the middle class are so ubiquitous that it is difficult to see this group as a distinct class. As she puts it, "Its ideas and assumptions are everywhere, and not least in our own minds. Even those of us who come from very different social settings often find it hard to distinguish middle-class views from what we think we *ought* to think" (1989: 5). The pervasive influence of White middle-class culture also makes it difficult to fully accept those from different cultural backgrounds, even when they have achieved the economic standing of the middle class. The United States calls itself the "melting pot," but its dominant culture is distinctively White and middle class. The dominant culture projects the notion that people in the upper classes have more ability, making the entire class system seem legitimate even as it robs those in the working and lower classes of dignity. This is what the sociologists Richard Sennett and Jonathan Cobbs mean when they write of "the hidden injuries of class" (Sennett and Cobbs, 1973).

Social Mobility: Myths and Realities

There is a general belief in the United States that anyone can, by his or her own labor, move relatively freely through the class system. Is class mobility in the United States a fact or a myth? In this section, we will seek an answer to that question.

Defining Social Mobility

Social mobility is a person's movement over time from one class to another. Social mobility can be up or down, although the American dream emphasizes upward movement. Mobility can be either *intergenerational,* occurring between generations, as when a daughter rises above the class of her mother or father; or *intragenerational,* occurring within a generation, as when a person's class status changes as the result of business success (or disaster).

Debunking Society's Myths

Myth: The United States is a land of opportunity where anyone who works hard enough can get ahead.

Sociological perspective: Social mobility occurs in the United States, but less often than the myth asserts and over shorter distances from one class to another; most people remain in their class of origin.

As discussed earlier in this chapter, societies differ in the extent to which social mobility is permitted. Some societies are based on *closed class systems,* in which movement from one class to another is virtually impossible. In a caste system, for example, mobility is strictly limited by the circumstances of one's birth. At the other extreme are *open class systems,* in which placement in the class system is based on individual achievement, not ascription. In open class systems, there are relatively loose class boundaries, high rates of class mobility, and weak perceptions of class difference.

The class system in the United States is popularly characterized as an open class system where individual achievement,

social mobility

not birth, is the basis for class placement. Many in the United States revere so-called self-made men. H. Ross Perot, for example, the Texan who has drawn upon his personal fortune of $2.2 billion to run for president of the United States, is frequently described as someone who started from the poorest origins. In the Perot legend, cleverness and determination enabled Perot to amass one of the world's greatest fortunes. Such tales of rags to riches typically overlook the considerable advantages the future magnates have at the beginning. Despite his own claims to the contrary, Ross Perot actually came from a comfortably well-off family (Wright, 1992).

Success stories of social mobility do occur, but experiences of mobility over great distances are rare and given far greater prominence in the public imagination than the more usual story: Most people remain in the same class as their parents, and many actually drop. Interestingly, the cultural idols of social mobility are typically men. Women are less frequently depicted as having the entrepreneurial personality or motivational drive alleged to be a requirement for achieving the American dream. Social mobility in the United States is not impossible. Indeed, many have immigrated to this nation with the knowledge that their life chances are better here than in their country of origins. How much social mobility actually exists in the United States?

The Extent of Social Mobility

Sociological research has found that social mobility is much more limited than the American dream suggests. What mobility exists is greatly influenced for all groups by education. African Americans, in particular, are strongly committed to social mobility through education, and increases in educational attainment for African Americans account for a considerable portion of the gains they have made (Smith, 1989). In addition, the class status of social contacts (friends, relatives, neighbors) influences class placement for all groups (Hodge and Trieman, 1968; Jackman, 1983; Vanneman and Cannon, 1987); however, education and social contacts are themselves heavily influenced by a person's class of origin.

The social status of an individual is more likely to be influenced by factors that affect the whole society than by individual characteristics; most people who find themselves moving between classes can ascribe the cause to changes in the occupational system, economic cycles, and demographic factors, such as the number of college graduates in the labor force (Hout, 1988; Erikson, 1985; Vanneman and Cannon, 1987). Most of the time, people remain in the class where they started; the privileges and disadvantages of class reproduce themselves. What mobility exists is typically short in distance (Blau and Duncan, 1967; Ganzeboom, et al., 1991). In sum, social mobility is much more limited than the American dream of mobility suggests.

Upward Mobility. Those who are upwardly mobile are often expected to distance themselves from their origins. This may mean creating some distance from their community of origin, and can result in many conflicts—with family, with friends, and even within themselves. First-generation college students, for example, often find themselves torn between leaving home to go to school and remaining close to their family and community. Likewise, women who are raised to have greater attachments to family and community may feel pressure from these groups not to move away. White working-class women, for example, are likely to have been socialized to marry, not pursue careers. Even though parents may have encouraged working-class women to become educated and seek a career, when working-class women become successful, their families can be ambivalent about their success if success is seen as taking the women away from their family and community origins.

The experience of upward mobility varies significantly by race and gender. African American families, for example, are more likely than White families to prepare their daughters for careers and are less likely to encourage marriage as a primary goal. At the same time, African American working-class parents are likely to steer daughters toward traditionally female occupations. This underscores an important sociological point about mobility: It is not just an individual process but a collective effort that involves kin and sometimes even community (Higginbotham and Weber, 1992).

Those who have traveled a great distance on the social ladder tend to see their experience through the lens of their class of origin. A study of highly accomplished professional Jewish men who had grown up in working-class homes found that the men's occupational choices, self-images, and the way they understood their careers were deeply tied to their class and ethnic background (Kerp, 1986). This shows that even if one's class position changes, class origins continue to shape one's social experience.

Downward Mobility. The attention people in the United States give to upward mobility has tended to obscure the experience of downward mobility—movement down in the class system. Overall, the rate of upward mobility exceeds the rate downward, but the margin is much smaller than it was in the 1960s and 1970s (Hout, 1988), and downward mobility is becoming more common (Newman, 1988; Newman, 1993). As income distribution is becoming more skewed toward the top, many in the middle class are experiencing mobility downward. For the first time in American history, the middle class is experiencing a decline—levels of real income (that is, income that controls for the value of the dollar) are falling and the cost of living is increasing (especially the cost of housing). Although more wives are working, the additional income this brings to a family has not offset the decline in the value of wages when controlling for the value of the dollar (Gillman, 1988).

Doing better than one's parents has long been a hallmark of the American dream—a goal that many people achieved in this century as the economy of the United States grew. Today, many young people worry that they will be unable even to match the lifestyle of their parents. Those over age forty-five still consider surpassing one's parents very important (Roper Organization, 1987). Young people give less emphasis to this goal, but have not abandoned the American dream entirely.

Class Consciousness

The United States has typically been characterized as not a class-conscious society. **Class consciousness** is the perception that a class structure exists and the feeling of shared identification with others in one's class, others with whom one perceives common life chances (Centers, 1949). Notice that there are two dimensions to the definition of class consciousness: the idea that a class structure exists and one's class identification.

There has been a long-standing argument that Americans are not very class conscious because of the strong cultural belief in the possibility of upward mobility. Images of opulence also saturate popular culture, making it seem that material comfort and pleasure is openly available to anyone. Many working-class and middle-class Americans focus on getting ahead individually and have little concern for organizing around class interests. The alleged openness of the class system and the strong cultural belief that anyone can succeed militates against the development of strong class consciousness. Still, research finds that people in the United States do recognize class divisions and believe that classes are organized around opposing interests (Vannemann and Cannon, 1987). The expression of class consciousness in the United States is influenced by the myth that this is a classless society (DeMott, 1990). In fact, the myth of classlessness is one way that class inequality is supported, the presumption being that if everyone has the same chances of success, then the inequality that exists must be fair and just.

At some points in U.S. history, class consciousness has been higher than at other times. A significant labor movement in the 1920s and 1930s reflected a higher degree of class consciousness than currently exists. Certain historical developments have mitigated the formation of stronger class consciousness in the United States, undercutting any feelings of discontent that might otherwise arise. These include the formation of a relatively large middle class and the high standard of living. Racial and ethnic divisions within the working class also make strong alliances within this group less stable. And, finally, the presence of a strong welfare state, since the New Deal of the 1930s, lessens the growth of protest against perceived inequities (Piven and Cloward, 1971); it is too soon to tell if the recent trends toward abandoning this model of governmental responsibility for the poor will result in a movement toward greater consciousness of class inequities.

Class inequality in any society is usually buttressed by ideas that support (or actively promote) inequality. Beliefs that people are biologically, culturally, or socially different can be used to justify the higher position of some groups. If people believe these ideas, the ideas provide legitimacy for the system. Karl Marx used the term **false consciousness** to describe the class consciousness of subordinate classes who had internalized the view of the dominant class. Marx argued that the ruling class controls subordinate classes by infiltrating their consciousness with belief systems that are consistent with the interests of the ruling class. If people accept these ideas, which justify inequality, they need not be overtly coerced into accepting the roles designated for them by the ruling class.

Acceptance of the idea that anyone can get ahead by hard work varies by class, as you might expect. Of those with incomes over $50,000, 58 percent think the basis of wealth is strong individual effort; 46 percent of those earning under $20,000 believe this. In addition, of those who earn over $50,000 a year, 46 percent believe that lack of effort is the cause of poverty; among those earning incomes less than $20,000, 53 percent think poverty is caused by circumstances beyond the control of individuals (McMurray, 1990). Class status also operates directly on support for social policies, with high-status persons much less likely to support egalitarian policies (Heaton, 1987).

How much do people identify with their class location? This varies among different classes. There are no direct studies of class consciousness among elites since sociologists rarely include the top stratum of the social hierarchy in their studies (not so much the fault of sociologists as reflective of the ability of the elite to isolate themselves from public scrutiny). The upper class is class conscious, however, in the sense that they are a cohesive group (more so than other class groups) who are well aware of one another and are protective of their common interests (Domhoff, 1998, 1970). These are the people who hold institutional power in America. Although they may differ on some specific political and social issues, research finds that elites in recent years have become less supportive of government intervention in society's problems and more affirming of the role of the market economy in shaping society (Dye, 1990). Outside the elite, the working class is more class conscious than the middle class; that is, working-class people are more likely to perceive that their lives are controlled by others in the classes above them. In fact, sociologists have found that the single most important determinant of where one sees oneself in the class system is whether one does mental or manual labor (Vanneman and Cannon, 1987).

Poverty

Compared to most other nations in the world, the United States is affluent. Despite its relatively high average standard of living, however, poverty afflicts millions of people in the United States. Aside from imposing a grim quality of life on the poor, poverty is also the basis for many of society's problems. Poor healthcare, inadequate nutrition, failures in the education system, and crime are all related to poverty.

The federal government has established an official definition of poverty that is used to determine eligibility for government assistance and to measure the extent of poverty in the United States. The **poverty line** is the amount of money needed to support the basic needs of a household, as determined by government; below this line, one is considered officially poor. To determine the poverty line, the Social Security Administration takes a low-cost food budget (based

class consciousness

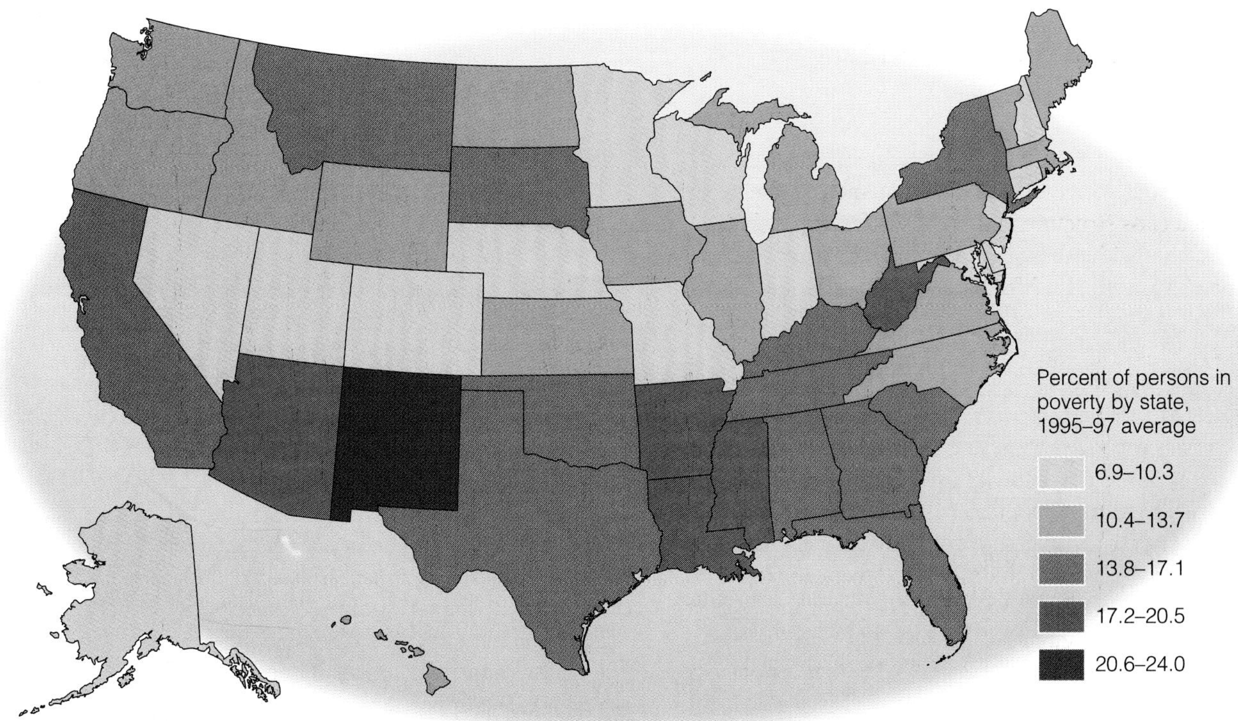

MAP 9.1 Mapping America's Diversity: Poverty in the United States

DATA: From U.S. Department of the Census. 1997. *Statistical Abstracts of the United States 1997.* Washington, DC: U.S. Department of Statistics, p. 477.

on dietary information provided by the Department of Agriculture) and multiplies by three, assuming that a family spends approximately one-third of its budget on food. The resulting figure is the official poverty line, adjusted slightly each year for increases in the cost of living. In 1997, the official poverty line for a family of four was $16,400.

Although a cutoff point is necessary to administer antipoverty programs, this definition of poverty can be misleading. A person or family earning $1 above the cutoff point would not be officially categorized as poor.

THINKING SOCIOLOGICALLY

Using the current federal *poverty line* ($16,555) as a guide, develop a monthly budget that does not exceed this income level and that accounts for all of your family's needs. For purposes of this exercise, assume that you head a family of four since that is the assumption on which the poverty level is based. Base your budget on the actual costs of such things in your locale (rent, food, transportation, utilities, clothing, and so forth). Don't forget to account for taxes (state, federal, and local), healthcare expenses, your children's education, car repairs, and so on. What does this exercise teach you about those who live below the poverty line?

Who Are the Poor?

In 1997, there were more than 35 million poor people in the United States, representing 13.3 percent of the population. Since the 1950s, poverty has declined in the United States, although since about 1978, it has been on the increase again, with a slight dip in recent years. Although the majority of the poor are White, disproportionately high rates of poverty are also found among Asian Americans, Native Americans, Black Americans, and Hispanics. Twenty-six percent of African Americans, 27 percent of Hispanics, 15 percent of Asians and Pacific Islanders, and 11 percent of Whites were poor (Dalaker and Naifeh, 1998; Hooper and Bennett, 1998), 1996 being the first year that Hispanic poverty exceeded that of African Americans. Among Hispanics, there are further differences among groups. Puerto Ricans—the Hispanic group with the lowest median income—have been most likely to suffer increased poverty, probably because of their concentration in the poorest segments of the labor market and their high unemployment rates (Tienda, 1989). Asian American poverty has also increased substantially since the 1980s, particularly among the most recent immigrant groups, including Laotians, Cambodians, Vietnamese, Chinese, and Korean immigrants; Filipino and Asian Indian families had lower rates of poverty (Lee, 1995). The federal government does not include Native Americans in its annual population surveys, but periodic reports indicate a very high rate of poverty among Native Americans—30.9 percent (U.S. Bureau of the Census, 1993c).

The vast majority of the poor have always been women and children, but their proportion of the poor has been increasing in recent years. The **feminization of poverty** refers to the increasing proportion of the poor who are women and children. This trend results from several factors, including the dramatic growth of female-headed households, a decline in the proportion of the poor who are elderly (not matched by a decline in the poverty of women and children),

feminization of poverty

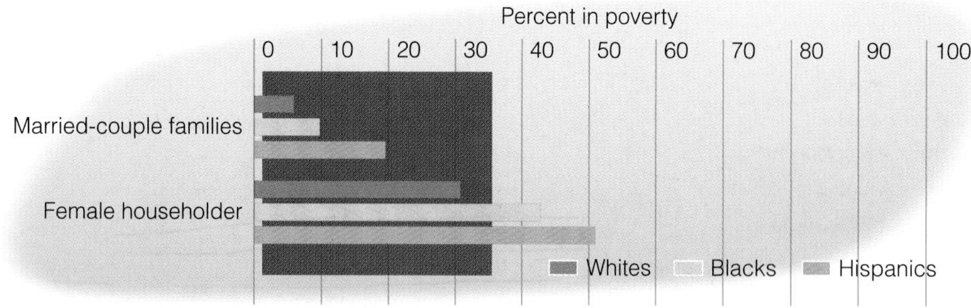

FIGURE 9.6 Poverty Status by Family Type and Race

SOURCE: Dalaker, Joseph, and Mary Naifeh. 1998. *Poverty in the United States: 1997.* Washington, DC: U.S. Department of Commerce, pp. 3–5.

and continuing wage inequality between women and men. The large number of poor women is associated with a commensurate large number of poor children. By 1997, 20 percent of all children (those under age eighteen) in the United States were poor, including 16 percent of White children, 37 percent of Black and Hispanic children, and 20 percent of Asian American children (U.S. Bureau of the Census, 1998; Dalaker and Naifeh, 1998).

More than half of all poor families are headed by women (see Figure 9.6). In recent years, wages for young workers have declined; since most unmarried mothers are quite young, one result is that the likelihood of a poor child being dependent on state subsidies has risen. Although this trend is offset by decreases in family size and rising rates of education among the young, the number of children who are persistently poor remains alarmingly high (Duncan and Rodgers, 1991).

Because of the divorce rate and generally little child support provided by men, women are increasingly likely to be without the contributing income of a spouse, and for longer periods of their lives. Women are more likely than men to live with children and to be financially responsible for them. However, women without children also suffer a high poverty rate, compounded in recent times by the fact that women now live longer than before and are less likely to be married than in previous periods.

Budget cutbacks in federal support programs and in federal employment have also contributed to the feminization of poverty. Currently, half of the poor receive no food stamps; 80 percent get no housing assistance (U.S. Bureau of the Census, 1996). Direct cuts in federal programs have also reduced the support available to poor women and their children. Government budget reductions also disproportionately affect women because women are more reliant than men on public sector jobs; women of color are particularly hard hit by such reductions (McLanahan et al., 1989). During periods of government reductions, minorities have been laid off at a rate 50 percent higher than nonminorities. This disparity holds regardless of level, education, federal agency, age, geographic location, performance rating, seniority, or attendance record (Eisenstein, 1984; Silver, 1995).

One of the marked changes in poverty is the location of poverty in metropolitan areas, particularly center cities (Wilson and Aponte, 1985). Poverty is still plentiful in rural areas, but by the mid-1980s, inner cities had become the home of most of the poor. Within cities, poverty rates are highest in the most racially segregated neighborhoods. For example, the income gap between Latinos and Anglos is higher in those metropolitan areas with the highest levels of residential segregation (Santiago and Wilder, 1991).

The poor, however, are not a monolithic group. They are racially diverse, including Whites, Blacks, Hispanics, Asian Americans, and Native Americans. They are diverse in age, including not just children and young mothers, but men and women of all ages, and especially a substantial number of the elderly, many of whom live alone. They are also geographically diverse, to be found in areas east and west, south and north, urban and rural.

Explanations of Poverty

Most agree that poverty is a serious social problem. There is far less agreement on what to do about it. Public debate about poverty hinges on disagreements about its underlying causes. Two points of view prevail: Some blame the poor for their own condition and some look to social structural causes to explain poverty. The first view, popular with the public and many policy makers, is that poverty is caused by the cultural habits of the poor. According to this point of view, behaviors like crime, family breakdown, lack of ambition, and educational failure generate and sustain poverty, a syndrome to be treated by forcing the poor to fend for themselves. Sociologists take a more structural view, seeing poverty as rooted in the structure of society, not in the morals and behaviors of individuals.

Blaming the Victim: The Culture of Poverty. Blaming the poor for being poor appeals to the myth that success requires only individual motivation and ability. Many in the United States adhere to this view and hence have a harsh opinion of the poor. This attitude is reflected in U.S. public policy concerning poverty, which is rather ungenerous compared to other industrialized nations. Although cheating on welfare is probably far less common than cheating on income taxes, welfare recipients are commonly portrayed as lazy and cheating the system. Those who blame the poor for their own plight typically argue that poverty is the result of early childbearing, drug and alcohol abuse, refusal to enter the labor market, and crime (See, 1991). Such thinking puts the blame for poverty on individual choices, not on societal problems. In other words, it blames the victim, not the society, for social problems (Ryan, 1971).

Likewise, the **culture of poverty** argument contends that poverty is its own cause. In this light, poverty is seen as a way of life that is transferred, like other cultures, from generation to generation. According to the culture of

culture of poverty

poverty argument, the major causes of poverty are welfare dependency, the absence of work values, and the irresponsibility of the poor. The argument originally stemmed from the work of anthropologist Oscar Lewis, who saw the behaviors of the poor as an adaptation to their marginal position in capitalist societies stratified by class. Although Lewis did not see cultural values as the underlying cause of poverty, he did argue that a culture of poverty evolved among the poor as they adapted to their feelings of despair and hopelessness (Lewis, 1966). The culture of poverty argument has now been adapted by conservative policy makers to argue that the actual causes of poverty are found in the breakdown of major institutions, including the family, schools, and churches.

What truth is there to the culture of poverty argument? To probe this question, first we might ask: Is poverty actually transmitted across generations? According to the culture of poverty argument, it is. Researchers, however, find only mixed support for this assumption. Children of poor parents have a 16 to 28 percent probability of being poor adults (Rodgers, 1995), but many of those who are poor remain poor for only one or two years. Their poverty occurs as the result of a household crisis, such as divorce, illness, unemployment, or parental death. Thus, only 22 percent of those poor in one year remain poor the next, although children and the elderly are less likely to exit poverty. Defining "chronic poverty" as remaining poor for two consecutive years, the U.S. Bureau of the Census reports that slightly under 5 percent of the poor are chronically poor. The public stereotype that welfare dependency is passed through generations is thus not well supported by data. More typical is cycling in and out of poverty. For example, 27 percent of those who have been out of poverty for one year will become poor again. Hispanics and African Americans have longer periods of poverty than White Americans; however, the majority of the poor are White (Dalaker and Naifeh, 1998; Eller, 1996; Shea, 1995; Stevens, 1994; Duncan et al., 1988).

A second question is: Do the poor want to work? The persistent public stereotype that they do not is central to the culture of poverty thesis. As stated by Lawrence Mead, a conservative political scientist, in an essay following the Los Angeles riots in 1992, "If poor adults behaved rationally, they would seldom be poor for long" (Mead, 1992: A19). This attitude presumes that poverty is the fault of the poor and that, if they would only change their values and adopt the American work ethic, then poverty would go away. What is the evidence for these claims?

Debunking Society's Myths

Myth: **If poor people would only get jobs, they could get out of poverty.**

Sociological perspective: **Forty percent of the poor work; the number of working poor has increased in recent years among those working year-round and full time.**

As the economy has moved away from manufacturing and toward a more service-oriented labor market, many in the blue-collar working class have experienced downward mobility.

Most of the able-bodied poor *do* work. In 20 percent of poor households, illness or disability prevents work, but of all poor persons, 41 percent work (U.S. Bureau of the Census, 1997b). The federal government itself has concluded that the proportion of workers who constitute the "working poor" has increased since 1980 to about 18 percent of the workforce. This means that 18 percent of the full-time workforce earn wages at or below the poverty line. If a single mother of three works forty hours per week, fifty-two weeks per year, at minimum wage, her annual income will still fall far below the poverty line.

Current policies that force those on welfare to work also tend to overlook how difficult it is for many to retain the jobs they are able to get. Poor women go off welfare to take jobs much more often than is usually acknowledged, but the dead-end jobs they get rarely provide a long-term solution to their problems. Most women soon leave these jobs and return to welfare, a pattern known as "welfare cycling." Studies of welfare cycling find that the women cannot support themselves and their families on the wages they earn from the poor jobs they are able to get; in many cases, paid employment also causes them (and, hence, their children) to lose the health insurance they receive as part of their welfare benefits. A typical pattern is for women to move back and forth between paid employment and welfare, often over a long period (Harris, 1996; Pavetti, 1993).

These facts indicate that public stereotypes about the poor not wanting to work are simply untrue. This leads inevitably to a political issue of the day. If the culture of poverty argument is incorrect, and yet is used to guide public policies, then the likely result will be policies that do not do what they are supposed to do. Perhaps even more important, public policy will fail to address the actual societal causes of poverty (Albelda and Tilly, 1996; Catanzarite and Ortiz, 1996).

Despite public stereotypes associating poverty with urban areas, much of the nation's poverty is rural and the majority of poor people are White.

Structural Causes of Poverty. From a sociological point of view, the underlying causes of poverty lie in the economic and social transformations taking place in the United States. Careful scholars do not attribute poverty to a single cause. There are many causes. Two of the most important are: *the restructuring of the economy,* which has resulted in diminished earning power and increased unemployment, and *the status of women in the family and the labor market,* which has contributed to the overrepresentation of women among the poor. Add to these underlying conditions the federal policies in recent years that have diminished social support for the poor in the form of welfare, public housing, and job training. Given these reductions in federal support, it is little wonder that poverty is so widespread.

The restructuring of the economy has caused the disappearance of manufacturing jobs, traditionally an avenue of job security and social mobility for many workers, especially African American and Latino workers (Baca Zinn and Eitzen, 1998). The working class has been especially vulnerable to these changes (Marks, 1991). Economic decline in those sectors of the economy where men have historically received good pay and good benefits has meant that fewer men are able to be the sole support for their families. Most families now need two incomes to achieve a middle-class way of life. The new jobs that are being created fall primarily in occupations that offer low wages and few benefits; they also tend to be filled by women, especially women of color, leaving women poor and men out of work (Andreasse, 1997; Smith, 1984). Such jobs offer little chance to get out of poverty. New jobs are also typically located in neighborhoods far away from the poor, creating a mismatch between the employment opportunities and the residential base of the poor.

The increase in the number of working poor people is the result of declining wage rates caused by transformations taking place within the economy. These changes fall particularly hard on young people, women, and African Americans and Latinos, who are the groups most likely to be among the working poor. For example, many Latino families are among the working poor; researchers find that they are strongly committed to work but are paid less than their Anglo counterparts (Santiago and Wilder, 1991). Approximately 15 percent of women working year-round and 13 percent of men fall below the poverty line. Among Hispanic men and women, approximately 25 percent working year-round have incomes below the poverty level. Among African Americans, 21 percent of men working year-round fall below the poverty level, and 31 percent of women, counting only those working year-round and full time (Dalaker and Naifeh, 1998).

The high rate of poverty among women is strongly related to women's status in the family and the labor market. For White women, divorce is one cause of poverty, although this does not hold for minority women since they are more likely than White women to be poor even within marriage (Catanzarite and Ortiz, 1996). Women's family status also means that women's child-care responsibilities make working outside the home on marginal incomes difficult. Women who work usually earn low wages, and affordable child care is unavailable to single mothers who want to work (Marks, 1991; Pearce and McAdoo, 1981). Wages in the United States are based on the **family wage system,** a wage structure based on the assumption that men are the breadwinners for families; therefore, men are typically paid higher wages than women. The family wage system ignores a fundamental change taking place in the economy—men are no longer the mainstay of family income. More women are now dependent on their own earnings, as are their children and other dependents. Whereas unemployment has always been considered a major cause of poverty among men, for women, wage discrimination has a major role. The median income for all women ($13,703 in 1997) is well below the poverty line (U.S. Bureau of the Census, 1998a). Given these data, it is little wonder that so many women are poor.

The paucity of affordable child care, coupled with the fact that it is usually women who are responsible for children, contributes greatly to women's poverty. Even though child-care workers receive low wages, the cost of child care consumes a huge proportion of the income most women earn. Many women with children cannot manage to work outside the home because it leaves them with no one to watch their children. In sum, women's traditional responsibility for child care joins with the status of women in the labor market as an additional structural cause of the high poverty rate.

The misery of the poor is not the only negative result of poverty caused by the structure of society. The persistence of poverty also increase tensions between different classes and racial groups. William Julius Wilson, one of the most noted analysts of poverty and racial inequality, has written, "The ultimate basis for current racial tension is the deleterious effect of basic structural changes in the modern American economy on Black and White lower-income groups, changes that include uneven economic growth, increasing technology and automation, industry relocation, and labor market segmentation" (1978: 154). This encapsulation demonstrates the power of sociological thinking by convincingly placing the causes of both poverty and racism in their societal context.

Receiving public assistance is not only an economic struggle, but means contending with the insults of bureaucratic institutions, such as the welfare system.

Homelessness

It is difficult to estimate the number of homeless people. Depending on how one defines and measures homelessness, the estimates vary widely. If you count the number of homeless on any given night, there may be about 760,000 homeless, but measuring those experiencing homelessness during one year, the estimates jump to 1.2 to 2 million people. If you further count those who have ever experienced homelessness, the estimates increase to between 5 and 10 million people. Although accurately counting homeless people is probably not possible, there is little doubt that homelessness substantially increased, probably more than doubled, over the past decade (National Coalition for the Homeless, 1998; Burt and Cohen, 1989; Link et al., 1994).

Whatever the actual number, there has certainly been an increase in the number of homeless in recent years. Some are homeless because of the movement in the 1970s to get mental patients out of institutional settings. The majority of such people returned to their homes, but it is estimated that between 20 and 30 percent of the homeless are emotionally disturbed individuals who would otherwise have been institutionalized; however, the great majority of the homeless are on the streets because of unemployment and eviction. Reduction in federal support for low-income housing also leaves many of those who previously would have fallen into this safety net literally out on the streets.

Families are the largest segment of the homeless population. More than one-third of the homeless are women; 58 percent of the homeless are under the age of thirty; 8 percent are preschool children. The homeless population also includes battered women, the elderly poor, single men and women, disabled people, and AIDS victims (Snow and Anderson, 1993).

The diversity of the homeless population makes it impossible to alleviate this problem with a single solution. Public policy responses to homelessness need to consider the different needs of the various homeless populations. Because homelessness, like other forms of poverty, has social structural causes, simple solutions that target the homeless as a deviant group are unlikely to have much positive result.

Welfare

A central issue in the public discussion of poverty is welfare, or public assistance to poor individuals and families. The federal program most people think of as welfare is titled Aid to Families with Dependent Children, or AFDC. This program was created in 1935 as part of the Social Security Act, based on the government's recognition that some people are victimized by economic circumstances beyond their control and are deserving of public assistance. The establishment of AFDC also acknowledged government responsibility for the well-being of families. Sixty years later in 1996, AFDC was eliminated, following months of national debate about welfare reform. With President Clinton's approval, the cash assistance that AFDC has provided to poor children was eliminated and replaced by block grants to states to administer their own welfare and work programs.

The new law also stipulates a lifetime limit of five years for people to receive welfare and requires all welfare recipients to find work within two years—a policy known as "workfare." Those who have not found work within two months of receiving welfare can be expected to perform community service jobs for free. In addition, the new welfare law denies payments to unmarried teen parents under eighteen years of age unless they stay in school and live with an adult. It also requires unmarried mothers to identify the fathers of their children or risk losing 25 percent of their benefits. This latter provision is an attempt to prosecute so-called deadbeat dads—those who do not provide financial support for their children (Pear, 1996). The new welfare law also denies a wide range of social support programs to immigrants—both legal and illegal.

National discussion of welfare reform has been clouded by myths about welfare that are directly challenged by sociological research. At the heart of welfare reform is the idea that poor people do not want to work and that providing public assistance creates dependence that discourages them from seeking jobs. Does workfare work? Sociologists find that, despite their looking for work, it is difficult for single mothers on welfare to find employment and that the jobs they do find do not provide livable wages; employed mothers may actually be worse off in terms of meeting basic needs. The low-wage work that welfare recipients find simply does not lift them out of poverty (McCrate and Smith, 1998). Throughout the 1980s and 1990s, political leaders argued that the antipoverty programs of the 1960s and 1970s contributed to the rates of poverty by creating dependency among the poor, although the poverty rate actually

BOX 9.3 SOCIOLOGY IN PRACTICE

Welfare: Myths and Realities

CURRENT debates about welfare reform are often based on fundamental misperceptions about the welfare system and those who receive it. Many initiatives to reform welfare are based on assumptions that sociological research has demonstrated to be wrong. Sociological research contributes to the welfare debate by informing the public about the facts about welfare, dispelling the myths that cloud a deeply opinionated and political discussion. Because of their knowledge about welfare issues, sociologists have served as consultants to the Congress, the White House, and to numerous community groups and organizations that are concerned about support for those in need. Here we assess some of the common myths about welfare in light of sociological research.

Myth: All welfare recipients are lazy.
Reality:

- Four out of ten welfare recipients work at paid jobs, either simultaneously combining work and welfare or cycling between work and welfare.
- Welfare recipients earn a substantial percentage of their total income; among those who combine work and welfare, earnings provide 60 percent of income.
- Attitudes toward welfare and low-wage work are not associated with the likelihood of welfare dependency.

Myth: Most of the poor are African Americans.
Reality:

- Although African Americans and Latinos have disproportionately high rates of poverty, 69 percent of the poor are White.
- Race does not predict whether a welfare recipient is likely to combine work and welfare.

Myth: Mothers on welfare have more children to increase the size of their welfare checks.
Reality:

- There is no causal relationship between the size of welfare benefits and the number of births by welfare recipients.
- Having more children reduces the likelihood that welfare recipients will combine work and welfare since welfare mothers are less likely to be employed as the number of children increases.
- Higher welfare benefits actually hasten the exit from poverty, when other sources of family income are also available.

Myth: Once someone goes on welfare, he or she never goes off; it becomes a way of life.
Reality:

- People who are persistently poor are about half the total number of poor people.
- Most people on welfare remain on welfare for a relatively short period, usually following a household crisis (death, unemployment, disability, and the like).
- Welfare cycling is the most common pattern; that is, people go on welfare for a relatively short time, exit, but return when jobs cannot sustain them.

Myth: People use their welfare checks to buy things they don't really need.
Reality:

- The average welfare payment in the United States was $378 per month for a family of three—an amount that would produce annual income that is only a small fraction of the poverty line were it the only source of income.
- Adding "in-kind" income (food stamps, rent subsidies, and so on) to the total income of poor families does not raise family income enough to lessen substantially the number of families considered poor.

Myth: The existence of federal welfare programs just encourages people to stay on welfare.
Reality:

- Persistent poverty, especially among children, has increased in recent years, a period when federal support programs were being cut back.
- Poverty dropped during the period when federal support to assist poor people was highest.

SOURCES: Butler, Amy C. 1996. "The Effect of Welfare Benefit Levels on Poverty Among Single-Parent Families." *Social Problems* 43 (February): 94–115; Duncan, Greg, J., and Saul D. Hoffman. 1990. "Welfare Benefits, Economic Opportunities, and Out-of-Wedlock Births Among Black Teenage Girls." *Demography* 27 (November): 519–535; Duncan, Greg J., and Willard Rodgers. 1991. "Has Children's Poverty Become More Persistent?" *American Sociological Review* 56 (August): 535–550; Harris, Kathleen Mullan. 1996. "Life After Welfare: Women, Work, and Repeat Dependency." *American Sociological Review* 61 (June): 407–426; Institute for Women's Policy Research. 1992. *Combining Work and Welfare.* Washington, DC: Institute for Women's Policy Research; Santiago, Anna M. 1995. "Intergenerational and Program-Induced Effects of Welfare Dependency: Evidence from the National Longitudinal Survey of Youth." *Journal of Family and Economic Issues* 16 (Fall): 281–306; U.S. Bureau of the Census, Current Population Reports, P60-198. 1997b. *Poverty in the United States: 1996.* Washington, DC: U.S. Department of Commerce; Wilson, William Julius, and Robert Aponte. 1991. "Urban Poverty." *Annual Review of Sociology* 11 (1985): 231–258.

declined during the period of greatest federal subsidy. The myth persists that the poor remain on welfare because they prefer it as a way of life (see the box "Sociology in Practice").

Critics of current social policy also worry that flooding the labor market with more low-income workers will only raise the level of poverty by taking jobs from those already employed. Sociological research suggests that current welfare reforms are not usually informed by the research on poverty and welfare. Higher levels of welfare benefits can actually hasten the exit from poverty since they give recipients the support they need to get job training and care for children while they seek work (Butler, 1996).

The public debate about welfare rages on—almost always without input from the subjects of the debate, the welfare recipients themselves. What do they think? Those who receive welfare think that it has negative consequences for them, but they say they have no other viable means of support. They have applied for welfare because they could not find work, were pregnant, or had small children. Most were forced to leave their last job because of layoffs or firings, or because the work was only temporary; few left their jobs voluntarily. Welfare recipients also say that the welfare system makes it hard to become self-supporting because the wages one earns while on welfare are deducted from an already minimal subsistence level. Furthermore, there is not enough affordable daycare for mothers to leave home and get jobs. The majority of welfare recipients think it has a bad effect on family life, not because welfare encourages family disorganization, but because they are ashamed to be on welfare and feel that it sets a bad example for their children. They are clear that the biggest problem they face regarding family well-being is not welfare, but lack of money. Contrary to the popular image of the conniving "welfare queen," the opinions of welfare recipients show that they have respect for family values and that they want to get off welfare, but they think welfare is their only option (Edin, 1991; Popkin, 1990).

Public suspicions have stigmatized welfare recipients in ways that other beneficiaries of government subsidies have not experienced. For example, Social Security supports virtually all old people, yet Social Security recipients are not stereotyped as dependent on federal aid, unable to maintain stable family relationships, or insufficiently self-motivated. The magnitude of federal spending on welfare programs, measured in real dollars and as a proportion of federal spending, is also a pittance compared to the spending on other, more popular subsidy programs such as Social Security—programs not identified with the poor. Sociologists conclude that the so-called welfare trap is not a matter of learned dependency, but a pattern of behavior forced on welfare recipients by the requirements of sheer economic survival. As a result, recipients have to supplement their income from other sources, such as assistance from family, friends, boyfriends, and absent fathers, and work (Edin, 1991; Edin and Lein, 1997).

CHAPTER SUMMARY

- *Status* refers to a socially defined position in a group or society. *Social stratification* is a relatively fixed hierarchical arrangement in society by which groups have different access to resources, power, and perceived social worth. All societies have systems of stratification, although they vary in how they are composed and how complex they are. Estate systems are those where power and property are held by a single elite class; in caste systems, placement in the stratification is by birth, whereas in class systems, placement is determined by achievement.
- Karl Marx saw class as primarily stemming from economic forces; Max Weber had a multidimensional view of stratification, involving economic, social, and political dimensions. Two theoretical perspectives are used in sociology to explain inequality—*functionalism* and *conflict theory*. Functionalists argue that social inequality motivates people to fill the different positions in society that are needed for the survival of the whole, claiming that the positions most important for society require the greatest degree of talent or training and are, thus, most rewarded. Conflict theorists see social stratification as based on class conflict and blocked opportunity, criticizing functionalist theory for assuming that the most talented are those who get the greatest rewards and pointing out that those at the bottom of the stratification system are least rewarded because they are subordinated by dominant groups. These perspectives are critical to understanding contemporary debates about social policy.
- *Social class* is the social structural position groups hold relative to the economic, social, political, and cultural resources of society. Social class can be imagined as a hierarchy, where income, occupation, and education are indicators of class, but classes are also organized around common interests and exist in conflict with one another. Sociological research on the class system shows that the United States is a highly stratified society, with an unequal distribution of wealth and income.
- Despite the presence of class inequality, the American dream is that people can move up in the class system based on their ability. This myth assumes that the United States has an open class system. Open class systems are those in which movement from one class to another is possible. *Social mobility* is the movement between class positions. It is less common than is believed. Education gives some boost to social mobility, but social mobility is limited.
- *Class consciousness* is the awareness that a class structure exists and the identification with others in one's class position. Perception of

the class system depends on one's class position. Race, class, and gender operate together to form stratification in the United States. Class groups are also distinguishable by the cultural behaviors and values they develop. Middle-class culture dominates in contemporary American society even though many are unable to attain this style of life.

- *Poverty* is extensive in the contemporary United States. Many factors result in an underestimation of the actual extent of poverty, but poverty is on the increase after having declined through the 1960s and 1970s. The majority of the poor are women and their children, but there are diverse groups among the poor. A myth about the poor is that they do not want to work. Such ideas blame the poor for their own situation, reflecting a belief in the culture of poverty thesis.
- The *culture of poverty* thesis is that poverty is the result of the cultural habits of the poor that are transmitted from generation to generation. Poverty is caused by social structural conditions, including unemployment, gender inequality in the workplace, and the absence of support for child care for working parents. In recent years, homelessness has increased. Public debate about poverty focuses on the welfare system. Welfare recipients are stigmatized in ways that other beneficiaries of government support are not.

KEY TERMS

caste system	occupational prestige
class	poverty line
class consciousness	prestige
culture of poverty	social class
educational attainment	social differentiation
estate system	social mobility
false consciousness	social stratification
family wage system	socioeconomic status
feminization of poverty	status
income	status attainment
life chances	underclass
means of production	wealth
median income	

THE INTERNET: A Tool for the Sociological Imagination

Resources on the Internet:

Virtual Society: The Wadsworth Resource Center at
http://sociology.wadsworth.com

Visit this site to find additional learning tools, including interactive quizzes, links related to web sites, and an easy link to *InfoTrac College Edition.*

U.S. Bureau of the Census
http://www.census.gov/

The U.S. Bureau of the Census publishes numerous reports that provide the best national data on income, poverty, and other measures that depict the class structure of the United States. In addition, the much-used reference book, *Statistical Abstracts of the United States,* is available online at this site.

National Coalition for the Homeless
http://nch.ari.net/numbers.html

This site provides information on the extent of homelessness, as well as bibliographies and other information pertinent to the study of homelessness.

Sociology and Social Policy: Internet Exercises

Much recent debate about welfare has focused on the desirability of requiring those receiving aid to work. How effective are *workfare* programs and do they reduce the need for *welfare?* As you examine this question, you will find different points of view, ranging from those promoting workfare to those who say such programs only take jobs from other low-income people, thereby increasing problems of inequality. Identify the different sides in this debate and see what empirical studies exist to examine the effectiveness of work requirements for welfare recipients.

Internet Search Keywords:

welfare	welfare to work
welfare reform	poverty
employment	economic inequality

Web sites:

http://wtw.doleta.gov/
Key information and current news about welfare-to-work programs.

http://www.welfareinfo.org/
The Welfare Information Network is a clearinghouse for information, policy analysis, and technical assistance on welfare reform.

http://www.acf.dhhs.gov/news/welfare/
This site by the Dept. of Health and Human Services provides links to fact sheets, resources, and information on many welfare-related issues.

http://www.apwa.org/
The American Public Welfare Association provides information on what is happening in the states regarding welfare.

http://povertycenter.cwru.edu/
This web site addresses problems of urban poverty and how social and economic changes affect low-income communities.

http://www.ssc.wisc.edu/irp/
Web site for the Institute for Research on Poverty at the University of Wisconsin.

http://www.nab.com/reentering/welfare.html
The National Alliance of Business web site offers information on private sector efforts in the welfare-to-work arena.

InfoTrac College Edition: Search Word Summary

class consciousness	feminization of poverty
culture of poverty	social mobility

In order to learn more about these central topics in sociology, you can conduct an electronic search using InfoTrac College Edition. To aid in your search and to gain useful tips, see the Student Guide to InfoTrac College Edition on the Virtual Society web site:
http://sociology.wadsworth.com

INTERACTIONS—A SOCIOLOGY CD-ROM: CONCEPTS FOR THIS CHAPTER

Go to the Wadsworth Sociology CD-ROM for further study on the concepts in this chapter. The CD-ROM also includes quizzes and additional activities to expand your learning experience.

SUGGESTED READINGS

Andersen, Margaret L., and Patricia Hill Collins. 1998. *Race, Class, and Gender: An Anthology,* 3rd ed. Belmont, CA: Wadsworth Publishing Company.

This anthology explores the intersections of race, class, and gender as systems of stratification and how they affect group experience in a number of social institutions. A widely used anthology, the book includes personal narratives, as well as analytical accounts, of race, class, and gender in the experiences of different groups.

Dujon, Diane, and Ann Withorn (eds.). 1996. *For Crying Out Loud: Women's Poverty in the United States.* Boston: South End Press.

The essays in this collection explore different dimensions of women's poverty, including the effect of recent welfare reforms on the class position of women and their children.

MacLeod, Jay. 1995. *Ain't No Makin' It: Aspirations and Attainment in a Low-Income Neighborhood.* Boulder, CO: Westview Press.

This ethnographic study of poor youth in a public housing project vividly demonstrates the impact of structural inequality on the lives of young men and women. Originally conducting his study in the 1980s, the author returned to the same project in the 1990s, showing in the new edition how contemporary changes in the class system have affected young people.

Moore, Joan, and Raquel Pinderhughes. 1993. *In the Barrios: Latinos and the Underclass Debate.*

This collection of articles examines specific U.S. cities and explains the development of the Latino underclass within them as the result of the city's role in the current economy. Moore and Pinderhughes also provide an excellent introduction that includes a discussion of Latino poverty and a brief history of Latino groups in the United States.

Oliver, Melvin L., and Thomas M. Shapiro. 1995. *Black Wealth/White Wealth: A New Perspective on Racial Inequality.* New York: Routledge.

By focusing their analysis on wealth, not just income, Oliver and Shapiro provide a compelling analysis of the source of continuing inequity in the class standing of African Americans and White Americans.

Quadagno, Jill. 1994. *The Color of Welfare: How Racism Undermined the War on Poverty.* New York: Oxford University Press.

Quadagno's book is a discussion of the effect of white backlash on the antipoverty programs that were established in the 1960s. Focusing on social policy, her work analyzes the influence of racism on movements for welfare reform.

Vanneman, Reeve, and Lynn Weber Cannon. 1987. *The American Perception of Class.* Philadelphia: Temple University Press.

This book, an analysis of class consciousness in the United States, argues that the perception of class depends on class position.

CHAPTER 10

Global Stratification

AS MOHAMMED Safari awakes in the single room in which his family is sleeping, he remembers that today is a special day. Today his country, Iran, plays the United States in the World Cup. But Mohammed's life is far different from the scene in Paris, or even other parts of Iran. Mohammed lives in the slums of Tehran, Iran's capital city, where his family has moved, having left their rural village five years ago. The soccer match is not the only thing on Mohammed's mind this day. Life is hard in the poor parts of the city. Tehran is crowded with many people who have left the countryside to move into the city hoping to find work, but there are few jobs. His father works periodically as an unskilled laborer in the construction that is going on in the city, while others of the family, including Mohammed and his brothers, sell small items such as cigarettes and chewing gum on the streets to support the family. Life is indeed difficult, but in some ways Mohammed and his family are lucky. At least they have a room in which to live.

Today, everybody will be following the soccer match. Since Mohammed's family does not have a television, or any hopes of getting one, he will try to find a shop with a television where he and many people will crowd around the window to watch. The family cannot usually afford more than bread and tea to eat. Sometimes they are able to have goat cheese and, at important celebrations, rice with lamb. They were poor in the village as well, but there they had a small garden and access to vegetables, eggs, and meat. In the city, these items are expensive, and Mohammed's family cannot afford them.

In this part of Tehran, most of the school-aged children like Mohammed do not go to school. This is, in part, because they must work to support the family, but also because there are no schools in the poor parts of the city. In addition, there are very few municipal facilities, no sewers, no clean water, few sanitation services, and very little housing. The Safaris are lucky. They have a cousin who had moved to the city before them, who allows them the use of a room in his family's compound. Family ties are very important in Iran, and people are expected to help their kinfolk.

There are thousands of people like the Safaris in the slums of Tehran and millions around the world in the poor sections of almost all the world's major cities. In some cases, like the Safaris in Iran, they are able to keep their families together and to eke out a living, but in other cases, these people live at extreme levels of poverty and desperation that are beyond the imagination of most in the United States. Mohammed and his family are at the bottom of the Iranian stratification system with little prospect for improvement. Although the average income in Iran, as estimated by using the per capita gross national product (GNP) is approximately $1000 per year, most of that wealth goes to a small elite. If others are lucky, they will make enough for the basic necessities in life, perhaps $20 a month.

How did Mohammed and his family become poor? They are not lazy or unskilled, rather their position is in part the result of the changes in Iran, including a misguided land reform effort in which modern agriculture has pushed peasant farmers off their land or into areas where the soil is poor and unproductive. Their situation is also the result of rapid urbanization that has overwhelmed the capacity of the government to deal with the enormous problems created by too many people in a limited area.

The Safaris' position at the bottom of Iranian society is also a result of world economic patterns. Iran is affected by world economic trends and forces. Iran is part of a global stratification system in which its position substantially is determined by its relationship to other countries in the world. Iran has great oil reserves, but oil is bought and sold on the world market, largely by European countries and Japan. These more powerful countries are able to control the price of oil and thus control Iran's economic fortunes. The rich industrialized nations import cheap oil from Iran and use Iran as a marketplace for Western goods. For this to work to the advantage of the developed countries, Iran must be kept from becoming too economically advanced, lest it develop industries that will compete with the West.

Iran belongs to a global stratification system in which the levels are made up of countries, not individuals. Iran is a middle-level country in this global system; many countries of the world are considerably poorer. As a result of its position in the world system, Iran, and Mohammed and his family, suffer.

The purpose of this chapter is to examine stratification from a global perspective. Chapter 9 has shown us how to understand the dimensions of stratification and what the stratification system looks like in the United States. Inequality within the United States must be understood in relationship to worldwide changes that are taking place. People often refer to the United States as the most powerful nation in the world. That status derives in part from the position of the United States at the top of the global stratification system. The position of the United States is the result of a global system of interdependent nations. The relative affluence of the United States means that American consumers have access to goods produced around the world. A simple glance at your clothing labels will show you this—"made in Indonesia," "made in Korea," "made in China" all show the linkage of the U.S. economy to systems of production around the world. The popular brand Nike, as just one example, has not a single factory in the United States, although its founder and CEO is one of the wealthiest people in America—worth $5.4 billion. Nike products are manufactured mostly in China and Indonesia, nations where human rights abuses of workers are commonly reported. In Vietnam, where one of Nike's manufacturing plants is located, young women provide much of the labor, working sixty-five hours per week for a total of $10 per week (Sanders and Kaptur, 1997). This chapter will look at the dynamics of global stratification. In this way, we will come to understand not only the class structure of the United States, but also the position of the United States and other countries in the world stratification system.

Global Stratification

As we look at the world today, we notice that there are not only rich and poor people, but also that there are rich and poor countries. There are countries that are well-off, countries that are doing so-so, and a growing number of countries that are poor and getting poorer. There is, in other words, a **global system of stratification** in which the units are countries, much like there is a system of stratification within countries in which the units are individuals or families. Just as we can talk about the upper-class or lower-class individuals within a country, we can also talk of the equivalent of upper-class or lower-class countries in this world system. One of the manifestations of global stratification is the great inequality in life chances that differentiates nations around the world. Recall that in Chapter 2 we examined the United Nations human development index. This index and other compilations of international well-being show the great inequities that stratification brings (see Map 10.1). Simple measures of well-being, including life expectancy, infant mortality, and access to health services reveal the consequences of a global system of inequality. And, the gap between the rich and poor is sometimes greater in those nations where the average person is least well off. No longer can any one of these nations be understood independently of the global system of stratification of which it is a part. Let us consider some of the dimensions and characteristics of this global system of stratification.

Rich and Poor

The global stratification system has several dimensions. As with stratification systems within countries, one of the dimensions of stratification between countries is wealth. There

Many common products marketed in the United States are produced in a global economy, sometimes by child laborers.

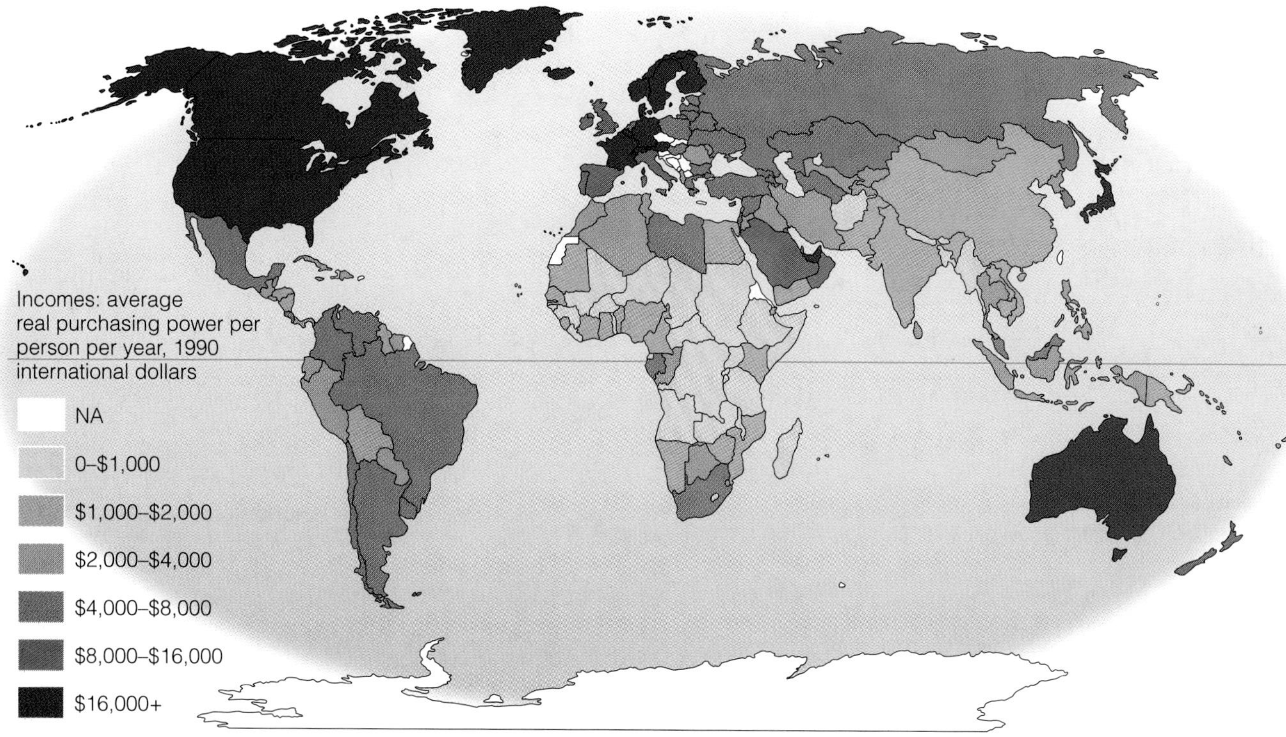

MAP 10.1 Viewing Society in Global Perspective: Rich and Poor

SOURCE: *State of the World Atlas, New Edition* by Michael Kidron and Ronald Segal. Copyright © 1995 by Michael Kidron and Ronald Segal, text. Copyright © 1995 by Myriad Editions Limited, maps and graphics. Used by permission of Viking Penguin, a division of Penguin Putnam Inc.

are enormous differences in the wealth of the countries at the top of the global stratification system and the wealth of the countries at the bottom. Though there are different ways to measure the wealth of nations, one of the most common is to use the **per capita gross national product** (per capita GNP). This measures the total volume of goods and services produced by a country each year, divided by the size of the population. This is sometimes used as a measure of average income, but it does not truly reflect what individuals or families receive in wages or pay. It is simply each person's annual share of their country's production.

Per capita GNP is reliable only in countries that are based on a cash economy. It does not measure informal exchanges or bartering in which resources are exchanged without money changing hands. These noncash transactions occur in all countries and do not get into the GNP calculation, but they are more common in developing countries. As a result, measures of wealth based on the GNP, or other statistics that count cash transactions, are less reliable among the poorer countries and may underestimate the wealth of the countries at the lower end of the economic scale.

The per capita GNP of the United States, which is one of the wealthier nations in the world though not the wealthiest on a per capita basis, was $28,020 in 1998. On the other hand, the per capita GNP in Mozambique, one of the poorer though not the poorest country in the world, was $80. The average person in the United States is 350 times wealthier than an average citizen of Mozambique using average GNP as the measure of wealth. This inequality between the rich and poor nations is increasing. Between 1950 and 1990, the per capita GNP of the world's wealthiest countries grew at 2.2 percent annually, whereas per capita GNP of the poorest countries did not grow at all (Durning, 1998).

Which are the wealthiest nations? Table 10.1 shows data from the World Bank listing the ten richest countries in the world measured by the annual per capita GNP in 1998. Luxembourg, with an annual per capita GNP of $45,360 in

per capita gross national product

The gap between the rich and poor worldwide can be staggering. At the same time that many struggle for mere survival, others enjoy the pleasantries of a gentrified lifestyle.

Global stratification means not only that there are enormous differences in the relative well-being of different countries throughout the world, but that within many nations large numbers of people live in slums, as in this shanty town of Haiti, one of the poorest countries in the world.

1998 US dollars, is the richest country in the world on a per capita basis. The United States ranks ninth. Of course, Luxembourg has a tiny population, whereas the United States has a much larger population. In terms of total wealth, the United States is the richest country in the world. It is easy to see that most of these wealthy countries are in Western Europe, with the addition of the United States, Japan, and Singapore. They are all industrialized countries, mostly urban, and with the exception of the United States and Japan, they have relatively small populations. Among the nations of the world, these countries represent the equivalent of the upper class on the dimension of wealth per person.

On the other hand, let us look at the ten poorest countries in the world, as shown in Table 10.2, again using 1998 per capita GNP as the measure of wealth. Several countries are even poorer than these countries, but they are not often listed in official data because they are too poor to report reliable statistics. In the Sudan, for instance, which is not listed in Table 10.2, experts estimate that the per capita GNP may be as low as $40! Note that the world's poorest countries are mostly in eastern or central Africa. These countries have become industrialized, are largely rural, and still depend heavily on subsistence agriculture. They have large populations and high fertility rates. Based on the dimension of wealth, these countries rank at the bottom of the world stratification system.

To make a more direct comparison between the nations of the world, Figure 10.1 shows the countries of the world in three categories: richest, middle income, and poorest. The average income of the richest countries—that is, the countries in western Europe, New Zealand, Australia, Japan, and North America (excluding Mexico), forty-one in all—is $22,304. In the poorest countries of the world—which are mostly in Africa and Asia, seventy-four in all—the average income per person is $420. The people in the forty-one richest countries, on the average, make fifty-three times more than the people in the eighty-one poorest countries.

Clearly, many countries in the world are very poor, whereas other countries are rich. Five countries in the world, all in eastern Africa, have an average annual income below $100. This does not mean that all the people in the rich countries are rich, or that all the people in the poor countries are poor. But on the average, people in the poor countries are much worse off than people in the rich countries. In many of these poor countries, the life of an average citizen is desperate, and starvation is a growing problem. More tragic, these are the countries with the largest populations. In a world with a population of nearly six billion, more than three billion—more than half the world's population—live in the poorest forty-five countries. We will look more closely at the nature and causes of world poverty later in this chapter.

TABLE 10.1 *TEN WEALTHIEST COUNTRIES BASED ON PER CAPITA GNP IN 1998 US DOLLARS*

Ten Richest Countries	Per Capita GNP
Luxembourg	$ 45,360
Switzerland	44,350
Japan	40,940
Norway	34,510
Denmark	32,100
Singapore	30,550
Germany	28,870
Austria	28,110
United States	28,020
Iceland	26,580

SOURCE: World Bank. 1998. *World Bank Atlas.* New York: The World Bank, pp.16–17. Used by permission.

TABLE 10.2 *THE TEN POOREST COUNTRIES BY PER CAPITA GNP IN 1998 US DOLLARS*

Ten Poorest Countries	Per Capita GNP
Niger	$ 200
Sierra Leone	200
Rwanda	190
Malawi	180
Burundi	170
Tanzania	170
Chad	160
Democratic Republic of the Congo	130
Ethiopia	100
Republic of Congo	80

SOURCE: World Bank. 1998. *World Bank Atlas.* New York: The World Bank, pp. 16–17. Used by permission.

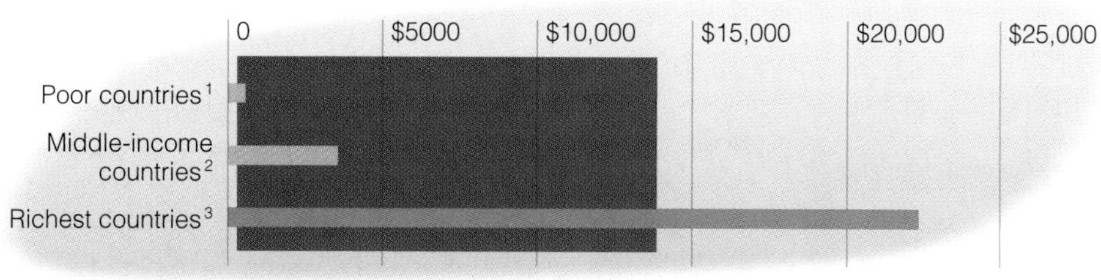

FIGURE 10.1 Average Income in Poor, Middle, and Rich Countries

[1]Countries with per capita GDP less than $999.
[2]Countries with per capita GDP between $1000 and $9999.
[3]Countries with per capita gross domestic product (GDP) over $10,000.

SOURCE: World Bank. 1998. *World Bank Atlas.* New York: The World Bank. Used by permission.

First, Second, and Third Worlds

Another dimension of world stratification combines levels of development with political orientation. This system is based on the logic of the politics of the Cold War and thus is changing. During the Cold War, the countries of the world were often divided into three groups: first-world countries, second-world countries, and third-world countries. **First-world countries** consisted of the industrialized capitalist countries of the world, including the United States, New Zealand, Australia, Japan, and the countries of western Europe. They were industrialized, and more important, all had a market-based economy and a democratically elected government. Although these countries are largely the same as the rich countries listed earlier, in this scheme, the important dimensions are their democratic forms of government and their market-based economies.

The **second-world countries** in this scheme were the socialist countries, which included the former Soviet Union, China, eastern Europe, Cuba, North Korea, and a few countries in other parts of the world. These countries had a communist-based government and a state-managed economy. Their governments were not democratically elected. Although less developed than the first-world countries, the second-world countries provided their citizens with such services as free education, health care, and low-cost housing, consistent with the principles of socialism.

The **third-world countries** in this categorization were the remaining countries of the world. The third world was usually pictured as those countries that were poor, underdeveloped, largely rural, and with high levels of poverty. The third-world economies were based on agriculture, often subsistence farming where people grew crops for their own consumption. The governments of the third-world countries were mostly autocratic dictatorships, with some exceptions, and though these countries were generally poor, wealth was concentrated in the hands of a tiny elite.

This system was based on the logic of the Cold War and is an imperfect system of categorizing nations, given world change. It is clear that there are more than three types of countries in the world. For instance, the oil-rich countries of the Middle East were not part of the first or second world, according to this scheme, but they also did not belong in the same category as the poor countries of Africa and Asia since they had considerably more wealth. In addition, some of the newly emerging states in Asia, such as Singapore and Korea, reached the income level of the first world, but have remained in the third-world category.

The collapse of the Soviet Union and the change in the governments of eastern Europe has led to the disappearance of almost all the so-called second-world countries. Although some countries still have a communist-based government, many of them, including China, are moving toward a market economic system. As a result, although these categories are still widely used, this system of categorizing the countries of the world is not perfect.

The Core and Periphery

Another dimension of stratification in the world system is **power.** By this we mean the ability of a country to exercise control over other countries or groups of countries. There are several kinds of power that countries can exercise over other countries, including military and political power. Here we are mostly concerned with economic power, that is, the power to control and influence the world economic system to one's advantage. This method of identifying the levels of countries in the world stratification system is based on the economic interrelationship between the countries of the world, and assumes that there is an integrated world economic system. More will be said about this world economic system later in the chapter.

Using power as a dimension, the countries of the world can be divided into three levels based on their position in the world economic system. At the top of the system are the **core countries.** The core countries are largely the same as the developed countries identified as the first world, but this dimension emphasizes their power in the world economic system. They are the countries that control and profit the most from the world system, and thus they are the "core" of the world system. These include the powerful nations of Europe, the United States, Australia, and Japan.

Surrounding the core countries, both structurally and geographically, are the **semiperipheral countries.** The semiperipheral countries are semi-industrialized and, to some degree, represent a kind of middle class. Their middle position is marked by their role in the world economic system. They play a middleman role, extracting profits from the poor countries and passing those profits on to the core countries. The semiperipheral countries include such countries as Spain, Turkey, and Mexico.

At the bottom of the world stratification system, in this model, are the **peripheral countries.** These are the poor, largely agricultural, countries of the world. Even though they are poor, they often have important resources that are exploited by the core countries. This exploitation, in turn, keeps them from developing and perpetuates their poverty. They have little power or influence in the world system.

THINKING SOCIOLOGICALLY

Observe the evening news for one full week, noting the countries that are featured in different news stories. What images are shown for each country and what events covered? What does this tell you about global stratification?

This categorizing system ends up looking much like the first-, second-, and third-world system. The difference is that it emphasized the power of each country in the world economic system as opposed to its political system or level of wealth. Clearly, the rich countries are also the powerful countries, and the poor countries are also the countries with little power. More will be said about the power dimension of world stratification when this chapter exams the major theories of global stratification.

Race and Global Inequality

In addition to class inequality, there is a racial component to world inequality, just as there is within many nations. This racial inequality in the world system can be seen in several ways. The rich core countries, those that dominate the world system, are largely European, with the addition of the United States and Japan. The populations of these countries, except for Japan, are mostly White. The poor countries of the world are mostly in Africa, Asia, or South America, where the populations are largely people of color. To put it differently, the countries with high standards of living, high levels of education, and the other positive characteristics have populations that are predominantly White. On the other hand, the countries that suffer a poor standard of living, have low levels of education, high death rates, and in general, are at the bottom of the world stratification system are predominantly people of color. On average, the differences in life chances and lifestyle between the countries of this world with White populations and the countries of the world with Black populations is enormous.

The most dramatic statistic that shows this enormous racial inequity is related to malnutrition and hunger. It is estimated that more than one billion people in the world suffer from malnutrition and hunger. The vast majority of these people are people of color—that is, those not of European descent (Uvin, 1998).

How did this racial inequality come about? On the surface, global capitalism is not explicitly racist, as were earlier forms of industrial capitalism. Yet, in fact, it is the rapid expansion of the global capital system that has led to the increase in racial inequality between nations. In the new capitalist system, a new world division of labor has emerged that is not tied to place, but can seek cheap labor anywhere. Cheap labor is usually found in non-Western countries. The exploitation of cheap labor has created a poor and dependent workforce that is mostly people of color, while the profits accrue to the wealthy owners, who are mostly White. This has resulted in a racially divided world. In addition, even within the industrialized nations, it is the importation of low-paid indigenous laborers that is used to drive labor prices down. As a result, racial stereotypes are created that see the labor of racial–ethnic groups as less valuable, perpetuating racist attitudes.

Some have argued further that the multinational corporation's exploitation of the poor peripheral nations has forced an exodus of unskilled workers from the impoverished nations to the rich nations, creating a flood of third-world refugees into the industrialized nations, thereby increasing racial tensions, fostering violence, and destroying worker solidarity (Sivanandan, 1995: 1).

Another way to look at racism around the world is offered in the work of Anthony Marx (1997). Comparing the United States, South Africa, and Brazil, Marx shows that how racial groups fare in a country depends on how race is defined, particularly in the early periods of national development. Racial categories are not just based on biology (see Chapter 11), but are constructed by national elites according to each country's history. As a result, countries define racial categories differently, and these definitions can have lasting effects on the ability of various racial groups to organize and fight for equality.

In the cases of South Africa, the United States, and Brazil, each country developed different sets of racial categories. Although all three of these countries have many people of mixed descent, the postcolonial elites chose to define race differently. In South Africa, the particular history of Dutch and English colonialism led to strongly drawn racial categories that defined people as "White," "colored," or "African." However, in the United States, given its history of slavery, the "one drop" rule was used, which defined anyone with any African heritage as Black, ruling out any category of mixed race.

Brazil presents yet a different case. The Brazilian elite chose to declare Brazil a racial democracy at the early stages of national development; racial differences were thought not to matter. Yet, rather than create an egalitarian society free of racism, this resulted in "Afro-Brazilians" remaining at the bottom of Brazilian society, while "Euro-Brazilians" remain at the top. Color continues to matter, and the ideal of a racial democracy accomplishes the opposite of what was intended. By not creating racial categories, the Brazilian elite robbed Afro-Brazilians of an identity and therefore an opportunity for collective action and group solidarity, as has been the case for African Americans in the United States and Black Africans in South Africa.

Ironically, Marx is suggesting that the creation of arbitrary racial labels in the long run may lead to more racial equality by providing an identity around which political mobilization can take place. An implication of this research for the United States would be that any attempt to weaken racial categories might weaken the political mobilization of people of color. For instance, in the debate over the racial categories to be used in the U.S. census, many in the Black civil rights establishment have argued strongly for not adding new multicultural or mixed-race categories since it would potentially weaken Black solidarity.

Consequences of Global Stratification

No matter how we divide the countries of the world, there are nations that are wealthy and powerful, and nations that are poor and powerless. This section looks at the consequences of this world stratification system, both for the countries as a whole and for the individuals in them.

Table 10.3 shows some of the basic indicators regarding the characteristics of the countries of the world divided by levels of wealth as measured by per capita GNP. The data for this table come from a variety of sources and were collected into a single data file by the United Nations Children's Fund (UNICEF, 1998). The richest countries are those with a per capita GNP over $10,000, the middle-income countries are those between $1000 and $10,000 per capita GNP, and the poorest countries are those below $1000. This table shows that there are considerable differences on a number of variables between the countries at each level of wealth.

Population

Consider the demographic differences, that is, differences in the size and characteristics of the population, among the rich, poor, and middle-income nations. The poorest countries comprise 3.6 billion people, over half the world's population! To put this into perspective, more than 62 percent of the people in the world live in countries where the average income is less that $1000 per year. In addition, these countries have the highest birthrates and the highest death rates. The total fertility rate, for instance, which is how many live births a woman will have over her lifetime at current fertility rates, shows that in the poorest countries women on the average have almost five children over their lifetimes. Because of this higher fertility rate, the populations of poor countries are growing faster than the populations of wealthy countries. This also means that in the poorer countries there is a high proportion of young children. The data also show that the populations in the poorest countries are mostly rural; only 36 percent of the people live in urban areas. In sum, the demographic characteristics of the low-income nations show that they have large, rapidly growing populations that are mostly rural.

The richest countries, on the other hand, have a total population of only approximately 825 million. That is, the richest countries of the world have only 14 percent of the world's population. In addition, the populations of the richest countries are not growing nearly as fast as the populations of the poorest countries. In the richest countries, women have about two children over their lifetime, and the populations of these countries are growing by only 1.2 percent. In fact, not reflected in the data in Table 10.3 is that many of the richest countries, including most of the countries of Europe, are experiencing population declines. With

TABLE 10.3 *CHARACTERISTICS OF NATIONS BY LEVELS OF PER CAPITA GNP IN 1998 DOLLARS*

Indicators of Basic Living Conditions	High-Income Nations (per capita GNP over $10,000)	Middle-Income Nations (per capita GNP between $1000 and $10,000)	Poorest Nations (per capita GNP less than $1000)
Health:			
Under five mortality rate (per 1000 births)	8	35	123
Life expectancy at birth (in years)	77	70	57
Percent of infants with low birthrate	6	9	15
Percent of people with access to safe water	97	83	57
Percent of children vaccinated for measles	85	85	68
Percent of people with adequate sanitation	77	78	44
Gender:			
Percent of females literate	88	81	51
Women's literacy as a percent of men's	96	91	70
Life expectancy for women as a percent of men's	108	107	106
Maternal death rate per 100,000 live births	16	160	768
Education:			
Percentage of adults literate	90	87	62
Primary school enrollment ratio	91	89	70
Basic Demographic Data:			
Crude death rate	8	8	13
Crude birthrate	14	23	36
Total fertility rate	2.0	3.0	4.9
Percent of population in urban area	80	59	36
Growth rate	1.2	1.54	2.4
Total population in thousands	825,157	1,325,986	3,587,052

SOURCE: Data based on *The State of the World's Children*, 1998, UNICEF. New York: 1998.

Although overpopulation is not restricted to the poorest nations, the highest birthrates and largest populations are in the poorest countries. India, although not one of the poorest nations, has the second highest population in the world with 843 people per square mile, compared with 76 per square mile in the United States.

a low fertility rate, the rich countries have proportionately fewer children, but they also have proportionately more elderly, which can also be a burden on societal resources. Table 10.3 also shows that the richest countries are largely urban, with 80 percent of their populations in urban areas.

Rapid population growth as a result of high fertility rates can make a large difference in the quality of life of the country. Countries with high birthrates are faced with the challenge of having too many children and not enough adults to provide for the younger generation. Public services, such as schools and hospitals, are strained in high birthrate countries, especially since these countries are poor to start out with. On the other hand, very low birthrates, as many rich countries are now experiencing, can also lead to problems. In countries with low birthrates, there are often not enough young people to meet labor force needs, and workers must be imported from other countries. The growing elderly population also puts a strain on retirement programs and retirement funds.

Although it is clear that the data show that poor countries have large populations and high birthrates and rich countries have smaller populations and low birthrates, does this mean that the large population results in the low level of wealth of the country or that high fertility rates keep countries poor? Scholars are divided on the relationship between the rate of population growth and economic development (Cassen, 1994; Demeny, 1991). Some theorize that rapid population growth and high birthrates lead to economic stagnation and that too many people keep a country from developing, thus keeping the country in poverty (Ehrlich, 1990). However, other researchers point out that some countries with very large populations have become developed (Coale, 1986). After all, the United States has the third largest population in the world at 265 million people (*World Bank Atlas*, 1998), and yet the United States is one of the richest and most developed nations in the world. India and China, the two nations in the world with the largest populations, are also showing significant economic development. Scholars now believe that even though in some situations large population and high birthrates can impede economic development, in general, fertility levels are affected by levels of industrialization, not the other way around. That is, as countries develop, their fertility levels decrease and their population growth levels off (Hirschman, 1994; Watkins, 1987).

Health

As Table 10.3 shows, there are also significant differences in the basic health standards of countries, depending on where they are in the global stratification system. The high-income countries have lower childhood death rates, higher life expectancies, and fewer children born underweight. In addition, most of the people in the high-income countries, but not all, have clean water and access to adequate sanitation. In the nations with per capita GNP over $10,000, for instance, fewer than 8 children die in the first 5 years of life for every 1000 children born and only about 6 percent of babies are born at low birthrates. People born today in these wealthy countries can expect to live about seventy-seven years, and women outlive men by several years. Except for some isolated areas or very poor areas of the rich countries, almost all people have access to clean water and acceptable sewer systems.

In the poor countries with annual per capita GNPs below $1000, the situation is completely different. Many children die within the first five years of life, people live considerably shorter lives, and fewer people have access to clean water and adequate sanitation. In the low-income nations, 123 out of 1000 children die in the first 5 years of life. Almost 15 percent of babies are born underweight, and the life expectancy of people born at this time is only fifty-seven years. Less than 50 percent of the people in the poor nations have adequate sanitation, and only 57 percent have access to clean water.

In the low-income countries, the problems of sanitation, clean water, childhood death rates, and life expectancies are all closely related. In many of the poor countries, drinking water is contaminated from poor or nonexistent sewage treatment. This contaminated water is then used to drink, to clean eating utensils, and to make baby formula. For adults, the water-born illnesses such as cholera and dysentery cause sickness, sometimes severe, but seldom result in death. However, children under five, and especially those under the age of one, are highly susceptible to the illnesses carried in contaminated water. A common cause of childhood death in countries with low incomes is dehydration caused by the diarrhea contracted by drinking contaminated water.

Education

In the high-income nations of the world, education is almost universal, and the vast majority of people have attended school at least at some level. Table 10.3 shows that 90 percent of the people in the high-income countries are listed as literate. Literacy and school enrollment are now taken for granted in the high-income nations. High-income countries require that most of their citizens be educated and literate. People in these high-income countries without a good education stand little chance of success.

In the middle- and lower-income nations, the picture is quite different. Elementary school enrollment, virtually universal in wealthy nations, is less common in the middle-income nations and even less common in the poorest nations. The UNICEF data summarized in Table 10.3 show that only 70 percent of elementary-age children in the low-income nations are in school. Sixty-two percent of adults in the low-income countries are listed as literate. This figure would be much lower except that China, a poor country with a large population, has a high rate of literacy and a high proportion of children in school. In fact, 18 percent of the nations of the world have literacy rates below 50 percent. However, education is improving greatly in the world, even in the poor areas. Only 6 percent of the nations report a school enrollment rate below 50 percent.

How do people survive who are not literate or educated? In much of the world, education takes place outside formal schooling. Just because many people in the poorer countries never go to school, this does not mean that they are ignorant or that they are uneducated. Most of the education in the world, in fact, takes place in family settings, in religious congregations, or in other settings where elders teach the next generation the skills and knowledge they need to survive. This type of informal education often includes basic literacy and math skills that people in these poorer countries need for their daily lives.

The disadvantage of this informal and traditional education is that though it prepares people for their traditional lives, it often does not give them the skills and knowledge needed to operate in the modern world. Therefore, as these poor countries and the people within them are confronted by the developing world, they often do not have the skills and knowledge to survive in these new environments. As a result, their low educational level limits their ability to adjust to changing world situations, particularly in an increasingly technological world, and it perpetuates their underdeveloped status.

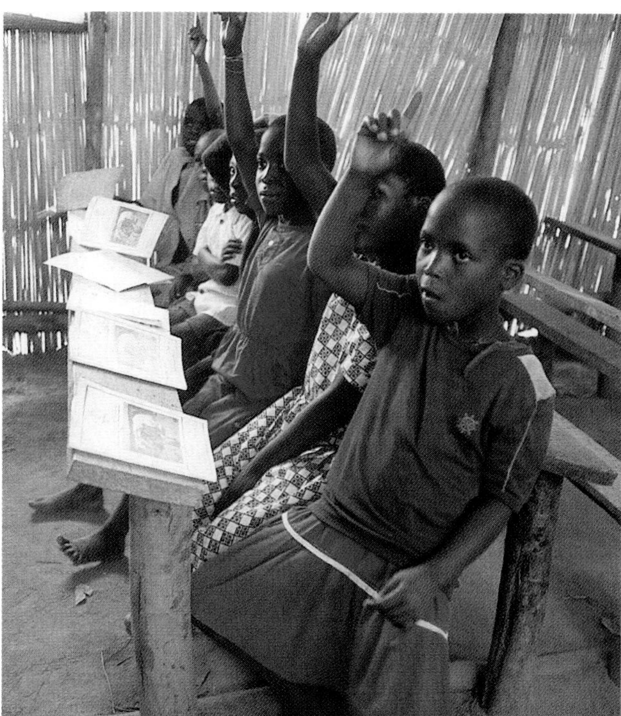

Educational attainment in the third world is much lower than in higher-income nations. Here, Rwandan school children show their eagerness to learn, even though without the facilities and resources more common in wealthier countries.

Gender

The position of a country in the world stratification system has a differential impact on the men and women in the country. As with inequality within a country, poverty is usually felt more by women than by men. Table 10.3 shows several variables regarding the situation of women in countries at the three levels of income. In the richest nations, 88 percent of the women are literate, women live 8 percent longer than men, and very few women die in childbirth. However, no serious sociologist would argue that there is gender equality in the industrialized countries. We know it is not so, but compared to women in the middle- and low-income countries, women in the wealthier countries have better health and education.

In the poorer countries, women suffer poorer health and less education. In the poorest nations, 768 women die in childbirth each year for every 100,000 births. Childbirth is often fraught with potential risks, but in the high-income countries, births are more often attended by trained midwives or other medical personnel, and therefore maternal mortality rates are lower. In the poorer countries of the world, such services are not available and many women die while giving birth. One thing constant in most countries, however, is that women outlive men in all three levels of countries although the longevity gap is less in the least developed nations.

In general, the developed nations have better sanitation, better health, more education, a lower birthrate, and smaller populations. The least developed countries of the world have poor health and sanitation, a much higher death rate, fewer people in school, and a large and growing population. The differences are profound and large, and indicate a markedly unequal world.

Theories of Global Stratification

Consider the following questions: How did this world inequality occur?

Why is it that so many of the countries in the world have fallen into poverty while other countries have developed and become wealthy? How do we understand the geographic distribution of wealth and poverty? Sociological explanations of world stratification generally fall into three camps. One sociological approach emphasizes the market forces within and between countries and argues that each country rises and falls according to its own efforts, resources, and energies. This *market-oriented approach* encompasses several specific theories (Appelbaum and Chambliss, 1997). It is both an explanatory approach, describing and explaining why things are as they are, and a normative approach, suggesting that this is the way things should be. The market-oriented approach is also the most common theoretical orientation of world development organizations that seek to help poor countries become more developed.

Modernization Theory

One early version of the market-oriented approach, sometimes called **modernization theory**, views the economic development of a country as a worldwide process involving nearly all countries that have been touched by technological change. According to the theory, a country becomes more "modernized" by increased technological development, and this technological development is dependent on other countries as well. The role of technological and economic development was first proposed by W. W. Rostow, who was an economic advisor to President Kennedy in the 1960s (Rostow, 1961). Modernization theory was developed to explain why some countries (such as those in Latin America) had achieved economic development and why some had not; it was also developed to guide the United States in helping unsuccessful countries succeed.

BOX 10.1 DOING SOCIOLOGICAL RESEARCH

Afghan Refugees

ONE OF the major results of the world stratification system is the increase in refugees. For a number of reasons, both political and economic, large numbers of people are fleeing their home country into surrounding countries, creating enormous problems for the host countries and the world relief agencies that house and feed them. These large refugee migrations also create political problems for the countries that house these refugees. Refugees are now a major problem in all continents.

One of the largest refugee migrations in the last two decades was the movement of Afghans into the surrounding countries of Pakistan and Iran. During the 1980s, more than three million Afghans left their homes in Afghanistan to move into Pakistan, and another two million moved into Iran. In Pakistan, the Afghan refugees were placed in large "camps," and the government of Pakistan, along with the United Nations High Commissioner for Refugees, established a large bureaucracy to deal with the issues of housing and feeding of the refugees, as well as with the political and security issues their presence brought.

Yet, despite the massive aid effort, little was known about the refugees, particularly, why they left Afghanistan, and if and when they might return. In addition, it was not known how they would accommodate to camp life. To answer these questions, a sociologist familiar with the Afghans was hired to conduct a survey among the Afghan refugees in Pakistan. However, standard survey techniques could not be used. For one thing, the Afghans were largely unable to read or write. In addition, the Afghans, though constituting one nation, were actually made up of several ethnolinguistic groups, each with its own language and political agenda. Finally, the refugee situation was a part of a guerrilla war that was fought out of the refugee camps so that security issues for the interviewers was problematic.

To conduct the research, a team of Afghans, many of whom had been professors at Kabul University before fleeing, was assembled from the refugee populations. This team conducted interviews with a large number of Afghan refugees. This research found a number of interesting facts about the Afghan refugees. For one, since the Afghans lived in extended family kin or tribal groups in Afghanistan, the decision to flee Afghanistan was made by the collective, not by individual families. By implication, this also meant that the decision to return would be made in the same way.

This research also found that the refugees who were able to survive were those who were embedded in kin, tribal, or other communal groups. The refugees who had lost their relatives or who had become separated from their tribe were those unable to survive in the camp situation. Unfortunately, this was particularly true for women and children. Many men were killed in the Afghan guerrilla war. As a result, many of the refugees were widows and children. In the chaos of the refugee camps, and in the highly patriarchal Afghan society, these women and children had a difficult existence without a male head of household.

SOURCE: Farr, Grant. 1993. "Refugee Aid and Development in Pakistan: Afghan Aid After Ten Years" in *Refugee Aid and Development: Theory and Practice*, Robert F. Gorman (ed.). Westport, CT: Greenwood Press.

Modernization theory sees economic development as a process by which traditional societies become more complex and differentiated, and it posits the steps by which this happens. For economic development to occur, modernization theory predicts, countries must change their traditional attitudes, values, and institutions. Specifically, economic achievement is thought to derive from attitudes and values that emphasize hard work, saving, efficiency, and enterprise. These values are said by the theory to be found in modern (developed) countries, but lacking in traditional societies. Traditional societies are perceived as fatalistic, not valuing hard work or thrift, and not controlling their fertility. Fatalism is particularly a hindrance to economic development, according to modernization theory, because people who are fatalistic accept their condition in life and do not work to change it. If poor people see their plight as the result of fate, they will be less inclined to try to improve their condition. Mohammed's Iran remains underdeveloped, modernization theory suggests, as a result of Iran's traditional customs and culture in which individual achievement is discouraged and where tradition and kin relations dominate.

Modernization theory, an outgrowth of functionalist theory, derives much of its thinking from the work of Max Weber. In *The Protestant Ethic and the Spirit of Capitalism* (1974 [1904]), Weber saw the economic development that occurred in Europe during the Industrial Revolution as a result of the values and attitudes of Protestantism. The Industrial Revolution took place in England and northern Europe, Weber argued, because the people of this area were hard-working Protestants who valued achievement and believed that God helped those who helped themselves.

Modernization theory is similar to the culture of poverty theory, which sees people as poor because they have poor work habits, poor time management, are not willing to defer gratification, and do not save or take advantage of educational opportunities (see Chapter 9). Modernization theory sees the people in poor countries as not saving enough and the governments of the poor countries as not investing in infrastructure development, such as building roads and schools. Countries are poor, in other words, because they have poor attitudes and poor institutions.

Modernization theory also presents a plan for development. Rostow originally saw development taking place in several stages, which he likened to an airplane taking off. In the first stage, the traditional stage, countries are poor, the people fatalistic and lacking a work ethic, and they do not save or invest. The airplane is still on the ground. But as the country begins to develop, as people begin to save, and value hard work and efficiency, the plane starts to move faster and faster down the runway. Eventually, the country takes off, and after several more stages reaches cruising altitude at which time it is fully modern, industrialized, and highly differentiated, and the people become high mass consumers, enjoying the fruits of their labor. At this time, the airplane, to continue the analogy, is on automatic pilot.

Modernization theory has been criticized on several fronts. Although it seems to explain the situation of some countries, it does not account for the development, or lack of development, in others. Critics point out that it blames countries for being poor when in fact there may be other causes outside their control. Specifically, although modernization theory considers the question of relationships between countries, it tends to ignore specific external factors that may be the main cause of whether or not a country develops or remains poor. The other countries of the world exert powerful pressures on countries that affect their economic development. The economic development of a country may be the result of its exploitation by other more powerful countries, for instance. Modernization theory does not sufficiently take into account the interplay and relationships between countries that can play an important role in a country's economic or social condition.

Another criticism of modernization theory is that it posits that governments should not be involved in making economic decisions or policies that restrict free trade or business activities. Developing countries, modernization theory says, are better off if they let the natural forces of competition determine which companies and individuals become successful and which do not. However, the experience of countries that have made rapid economic gains in the last few decades seems to show that government intervention may spur economic development rather than hindering it. In many of the newly developed countries, particularly those in East Asia, government policies and actions have played an important role in economic development by working directly with the private sector to enact specific export strategies and by restricting imports to allow indigenous industries to develop.

Debunking Society's Myths

Myth: **Overpopulation in third-world countries is destroying the environment and provoking massive immigration.**

Sociological perspective: **Expansion of export agriculture is the root cause of much of the environmental destruction and influences migration in the third world. Once the growth of export agriculture forces the poor to farm fragile lands, environmental problems are the result.**

Yet, modernization theory's focus on values and attitudes that favor hard work and savings partially explains why some countries have become successful. Japan is an example of a country that made huge strides in economic development in part because of a national work ethic (McCord and McCord, 1986). But the work ethic alone does not explain Japan's success. In sum, modernization theory may partially explain the value context in which some countries become successful and others do not, but it does not substitute for explanations that also look at the economic and political context of national development.

modernization theory

Dependency Theory

Although market-oriented theories may explain why some countries are successful, they do not explain why some countries remain in poverty or why some countries have not developed. It is necessary to look at issues outside the individual countries and to examine the ties and connections between countries. Drawing on the fact that many of the poorest nations are former colonies of European powers, another theory of world stratification focuses on the processes and results of European colonization and imperialism. This theory, called **dependency theory,** focuses on explaining the persistence of poverty in the world. It holds that the poverty of the low-income countries is a direct result of their political and economic dependence on the wealthy countries. Specifically, dependency theory argues that the poverty of many of the countries in the world is a result of their exploitation at the hands of the powerful countries. This theory is derived from the work of Karl Marx, who saw that a capitalist world economy would create an exploited class of dependent countries, just as capitalism within countries had created an exploited class of workers. However, it was V. I. Lenin (1939), writing several decades after Marx, who saw that the development of capitalism had led to the European colonization of the world.

A major modern proponent of this theory is Andre Gunder Frank (1969) who focused on poverty in Latin America. Although he never articulated a coherent theory, Gunder Frank called his approach the "development of underdevelopment." The so-called development schemes of the rich countries (read the United States), he said, resulted only in the continuation of underdevelopment and poverty in the poor nations. Not only are the economies of these poor countries being controlled by powerful countries, but the powerful countries also control and manipulate the social and political systems of the underdeveloped countries, Gunder Frank pointed out. The people of the poor countries have a clear choice, he said, between "underdevelopment or revolution."

Dependency theory begins by examining the historical development of this system of inequality. As the European countries began to industrialize in the 1600s, they needed raw materials for their factories and places to sell their product. To do this, the European nations colonized much of the world, including most of Africa, Asia, and the Americas. Colonization worked best for the industrial countries when the colonies were kept undeveloped so that they would not compete with the home country. For instance, India was a British colony for several hundred years. Britain bought cheap cotton from India, made it into cloth in British mills, and then sold it back to India, making large profits. Although India was able to make cotton into cloth at a much cheaper cost than the British, and very fine cloth at that, the British nonetheless did not allow India to develop its cotton industry. As long as India was dependent on Britain, Britain became wealthy and India remained poor.

During the period of colonization, dependency was created by the direct political and military control of the poor countries by powerful developed countries. Most colonial powers were European countries, but in fact other countries, particularly Japan and China, had colonies as well. However, the era of colonization came to an end soon after the Second World War, largely because it became too expensive to maintain large armies and administrative staffs in other lands. As a result, according to dependency theory, the powerful countries turned to other ways to control the poor countries and keep them dependent. Sometimes the powerful countries still intervene directly in the affairs of the dependent nations by sending troops or, more often, by imposing economic or political restrictions and sanctions. But other methods, largely economic, have been developed to control the dependent poor countries, such as price controls, tariffs, and especially, the control of credit.

The rich industrialized nations, according to dependency theory, are able to set prices for raw material produced by the poor countries at very low levels so that the poor countries are unable to accumulate enough profit to industrialize. As a result, the poor dependent countries must borrow from the rich countries. However, debt creates only more dependence. Many poor countries are so deeply indebted to the major industrial countries that they must follow the economic edicts of the rich countries that loaned them the money, thus increasing their dependency. This form of international control has sometimes been called **neocolonialism,** a form of control of the poor countries by the rich countries, but without direct political or military involvement.

Multinational corporations also play a role in keeping the dependent nations poor, dependency theory suggests. Although their executives and stockholders are from the industrialized countries, multinational corporations recognize no national boundaries and pursue business where they can best make a profit. Multinationals buy resources where they can get them the cheapest, manufacture their products where production and labor costs are the lowest, and sell their products where they can make the largest profits. From one point of view, multinational corporations are only following good business practices. Unfortunately, this strategy usually works to the detriment of the poor countries and to the advantage of the rich countries. Cheap resources and low-wage labor are found in poor countries, yet the profits end up in the core countries. Many fault companies such as Nike for perpetuating global inequality by taking advantage of cheap overseas labor to make large profits for U.S. stockholders. Another way to look at it is that Nike is in fact doing what it should be doing in a market system: trying to make a profit. Nonetheless, dependency theory views the multinationals as another way that poverty is maintained in the poor parts of the world.

Dependency theory focuses primarily on the connections between specific countries or groups of countries. It faults the powerful countries for causing poverty in the poor countries by exploiting them and keeping the poor countries dependent, either through colonization, neocolonialism, or the work of multinationals. The theory works well in explaining much of the poverty that was created in low-income countries in Africa and Latin America in the 1960s and 1970s. Yet the theory does not explain development in other parts of the world (Berger, 1986). One criticism of dependency theory is that many poor countries were never colonies, for example, Ethiopia. Some former colonies have also done well. Two of the greatest postwar success stories of economic development are Singapore and Hong Kong. Both of these countries were British colonies—Hong Kong until 1997—and clearly dependent on Britain, yet they have had successful economic development precisely because of their dependence on Britain. Other former colonies are also improving economically, including India.

Another criticism of dependency theory is that it is not clear that the involvement of multinational corporations always impoverishes nations or increases their dependency. Some nations in which multinationals have put factories to exploit cheap labor have eventually moved up the development ladder by educating their workforce and thus capturing more of the profits. In fact, some have argued that multinationals do as much if not more economic damage to the industrialized nations by pulling out jobs and sending profits overseas.

In sum, dependency theory corrects some of the shortcomings of modernization theory by looking at the interplay between countries. However, it is not always the case that dependence leads to exploitation and economic backwardness.

The international division of labor means that workers in semiperipheral nations typically provide labor for products that produce profits for core countries. The maquiladoras *are a series of assembly plants along the Mexican–U.S. border where many economic regulations are relaxed to encourage manufacturing and trade.* Maquiladoras *rely heavily on the labor of Mexican women with profits going mostly to the United States.*

World Systems Theory

Modernization theory examines the factors internal to an individual country, and dependency theory looks to the relationship between countries or groups of countries. Another approach to global stratification is called **world systems theory.** Like dependency theory, this theory begins with the premise that no nation in the world can be seen in isolation. Each country, no matter how remote, is tied in many ways to the other countries in the world. However, unlike dependency theory, world systems theory argues that there is a world economic system that must be understood as a single unit, not in terms of individual countries or groups of countries. This theoretical approach derives to some degree from the work of the dependency theorists and is most closely associated with the work of Immanual Wallerstein in *The Modern World System* (1974) and *The Modern World System II* (1990). According to this theory, the level of economic development is explained by understanding each country's place and role in the world economic system.

THINKING SOCIOLOGICALLY

Look at the labels in your clothes and note where your clothing was made. What does this tell you about the relationship of core, semiperipheral, and peripheral countries within world systems theory? What further information would reveal the connections among the country where you live and the countries where your clothing is made and distributed?

This world system has been developing since the sixteenth century. The countries of the world are tied together in many ways, but of primary importance are the economic connections in the world markets of goods, capital, and labor. All countries sell their products and services on the world market, and buy products and services from other countries. However, this is not a market of equal partners. Because of historical and strategic imbalances in this economic system, some countries are able to use their advantage to create and maintain wealth, whereas other countries that are at a disadvantage remain poor. This process has led to a global system of stratification, in which the units are not people, but countries.

World systems theory sees the world divided into three groups of interrelated nations, although Wallerstein originally talked of four groups of countries. We introduced these categories earlier in the chapter. At the center of the world economic system are the rich, powerful industrialized capitalist countries that control the system, called the core countries. Around these nations are the semiperipheral countries. These countries occupy an intermediate position in the world system and include such countries as South Korea, Mexico, and Turkey. The semiperipheral countries, which are at a middle level of income and partly industrialized, extract profits from the peripheral countries and pass the profits on to the core countries. Finally, there are the peripheral countries. These countries are poor, not industrialized, largely agricultural, and manipulated by the core countries that extract resources and profits from them. They are mostly in Africa, Asia, and Latin America.

This world economic system has resulted in a modern world in which some countries have obtained great wealth, and other counties have remained poor. The core countries, that is, the western European countries, the United States, and Japan, make themselves wealthy by exploiting the resources of the peripheral countries at low prices, world systems theory suggests. The core countries control and limit the economic development in the peripheral countries so as to keep the peripheral countries from developing and thus competing with the core on the world market and so that the core countries can continue to purchase raw materials at a low price. As a result, peripheral countries suffer low wages, inefficiency, a large dependence on agriculture, poverty, and high levels of inequality, and they depend on the export of raw materials to the core countries.

World systems theory focuses on the economic interconnection of the countries of the world and how these interconnections result in the exploitation of the peripheral countries. Though it was originally developed to explain the historical evolution of the world system, modern scholars now focus on the international division of labor and its consequences. The **new international division of labor** approach is an attempt to overcome some of the shortcomings in world systems theory by focusing on the specific mechanism by which differential profits are attached to the production of goods and services in the world market. The new international division of labor approach notes that products are now produced globally. A tennis shoe made by Nike is designed in the United States; uses synthetic rubber made from petroleum from Saudi Arabia; is sewn in Indonesia; is transported on a ship registered in Singapore, which is run by a Korean management firm using Filipino sailors; and is finally marketed in Japan and the United States. At each of these stages, profits are taken, but at a very different rate. Because of the intense competition in the world economic system, this approach suggests there is a drive to find the countries where wages are lowest, and therefore to perpetuate and even deepen the cycle of poverty in the peripheral countries.

THINKING SOCIOLOGICALLY

What are the major industries in your community? Find out in what parts of the world they do business, including where their product is produced. How does the international division of labor affect jobs in your region?

World systems theorists call this process a **commodity chain,** the network of production and labor processes by which a product becomes a finished commodity (Gerefi and Korzeniewicz, 1994). By following a commodity through its production cycle and seeing where the profits go at each link of the chain, one can identify which country is getting rich and which country is being exploited.

There are a number of criticisms of world systems theory. Certainly, it is useful to see the world as an interconnected set of economic ties between countries and to understand that these ties often result in the exploitation of poor countries. However, it is not at all clear that the system always works to the advantage of the core countries and to the detriment of the peripheral countries. For one, countries that were once at the center of this world system no longer occupy such a lofty position. England, for example, was once the most powerful nation in the world system, as were Holland and Italy before that. Yet these countries, though still part of the core, are no longer at the top of the global economy. World systems theory does not in itself account for changes in the position of countries in the world system.

In addition, critics also point out that the world economic system does not always work to the detriment of the peripheral countries and to the benefit of the core countries. Commodity chain theory finds that peripheral countries often benefit by housing low-wage factories and that the core countries are sometimes hurt when jobs move overseas. Low-wage sweatshops are found in all nations, not just the peripheral countries. In addition, some of the core countries, for instance, Germany, do not move factories to low-wage countries, but improve profitability by increasing efficiency. Nonetheless, world systems theory has provided a powerful tool for understanding global inequality.

World Poverty

Relative, Absolute, and Capability Poverty

One of the sad facts of global inequality is the growing presence and persistence of poverty in many parts of the world. There is poverty in the United States, but very few people in the United States live in the extreme levels of deprivation found in some of the poor countries of the world. In the United States, the poverty level is determined by the yearly income for a family of four that is considered necessary to maintain a suitable standard of living. As mentioned in Chapter 9, the official poverty line in 1997 (for a family of four) was $16,400. By this definition, 35.5 million Americans, or about 13.3 percent, were living in poverty in 1997 (Dalaker and Naifeh, 1998). This definition of poverty in the United States identifies **relative poverty.** The households in poverty in the United States are poor compared with other Americans, but when one looks at other parts of the world, an income of $16,400 would make a family very well-off.

On the other hand, **absolute poverty** is the situation in which human beings do not have enough money for basic survival. On the world level, there are several ways to define this kind of poverty. The United Nations (UN) defines world poverty in three ways. *World poverty* is defined as the situation in which individuals live on less than $365 a year. This means that people at this level of poverty live on approximately $1 a day. By this definition, 1.3 billion people, approximately one person in five, live in poverty (*World Bank Atlas,* 1998).

The UN also identifies a more severe level of poverty, which it calls *extreme poverty.* Extreme poverty is defined as the situation in which people live on less than $275 a year, that is, on less than 75 cents a day. There are 600 million people who live at or below this extreme poverty level. Many of these people are in very dire straits, and many are starving and dying.

However, money does not tell the whole story since many people in the poor countries do not always deal in cash. In many countries, people survive by raising crops

absolute poverty

Although geographically close, Tijuana, Mexico, and Rodeo Drive in Beverly Hills show that the contrasts in wealth that result from global stratification can be stunning.

for personal consumption and by bartering or trading services for food or shelter. These activities do not show up in the calculation of poverty levels that use amounts of money as the measure. As a result, the United Nations Development Program also defines what it calls **capability poverty.** By this it means the situation in which people lack the three essential human requirements to survive: to be well nourished, to be able to reproduce, and to be educated. The United Nations calculates this by looking at three statistics: the number of children under five who are underweight, the proportion of births that are attended by trained health personnel, and the proportion of women who are illiterate. By this definition of poverty, the number of people in the world who are in poverty climbs to 1.6 billion, or about 37 percent of the people in this world (UNDP, 1996).

Who Are the World's Poor?

Using the UN's definition of world poverty, of the 1.3 billion people in poverty, almost half, 515 million people, live in South Asia, mostly in the countries of India, Pakistan, and Bangladesh. In fact, 80 percent of the world's poor live in 12 countries, and 62 percent live in India and China (*World Bank Atlas,* 1998). There are 446 million people in poverty in East Asia and the Pacific. In Africa, 219 million people live in poverty, most below the Sahara, and 110 million people in Latin America are in poverty.

There are some differences in the nature of poverty in different areas of the globe. In Asia, the people in poverty live in rural areas of very high population density. Although there are many poor people in Asia, the situation in many of the Asian countries is improving. In sub-Saharan Africa, the poor live in marginal areas where poor soil, erosion, and continuous warfare have created extremely harsh conditions. Even though fewer people are in poverty in Africa, the situation, especially in East Africa, has grown worse in the last decade and is now very severe. Many people there are now starving.

Is world poverty increasing or declining? At the world level, poverty has definitely increased. In sub-Saharan Africa, the number of people in poverty rose from 180 million in 1987 to 219 million in 1993; in South Asia, the number of poor rose from 480 million to 515 million; and in Latin America and the Caribbean, the number rose from 91 million to 110 million. However, the absolute numbers do not tell the whole story. Viewed as a percentage of the total population, in some areas of the world, especially East Asia and parts of Latin America, the proportion of people in poverty has dramatically declined. Two countries, China and India, have reduced the percentage of people in poverty even though they still have more poor people than any other nations. Poverty in Africa, particularly East Africa, on the other hand, has increased dramatically in the last decade. The reasons for these changes will be discussed in the section on the causes of world poverty.

Women and Children in Poverty

There is no country in the world in which women are treated as well as men. As with poverty in the United States, women bear a larger share of the burden of world poverty. Some have called this **double deprivation**—in many of the poor countries women suffer because of their gender and because they disproportionately carry the burden of poverty. (This is similar to the *double jeopardy* effect of class and gender in the United States, to be discussed in Chapter 11.) In many of the very poor areas of the world, the poverty falls particularly hard on the women. For instance, in situations of extreme poverty, women have the burden of taking on much of the manual labor since in many cases the men have left to find work or food. The United Nations Commission on the Status of Women (1996) estimates that women constitute almost 60 percent of the world's population, perform two-thirds of all working hours, receive only one-tenth of the world's income, and own less than 1 percent of the world's wealth (United Nations Commission on the Status of Women, 1996).

In poor countries, women also suffer greater health risks. Although women outlive men in most countries, the difference in life expectancy is less in the countries in poverty. In fact, in some of the poorest countries, such as Bangladesh, men outlive women. Women's health suffers because of several factors. For one, fertility rates are higher in poor countries, as we saw in Table 10.3. As a result, women in poverty experience more pregnancies and childbirth. Poor women, therefore, spend a greater part of their lives pregnant, nursing, and raising small children than do women in the wealthier countries. These things take a toll on women's health and increase the risks of disease and death. Giving birth is a time of high risk to women, and women in poor countries with poor nutrition, poor maternal care, and the lack of trained birth attendants are at higher risk of dying during and after the birthing event.

Women also suffer in some of the countries in poverty because of traditions and cultural norms. Many of the poor countries are patriarchal, meaning that men have control over the household. As a result, in some situations of poverty, the women eat after the men, and boys are fed before girls. In poverty areas, women have been found to have higher rates of malnutrition than men, which leads to anemia and other diseases (Doyal, 1990). In conditions of extreme poverty, baby boys are fed before baby girls because boys have higher status than girls. As a result, female infants have a lower rate of survival than male infants.

Children are also hit particularly hard in the countries suffering from poverty. Scholars disagree on the causal relationship between fertility rates and poverty. Still, the highest fertility rates are found in the areas with the greatest poverty. It may seem strange that couples would have even more children in situations where many children are dying of starvation. Nonetheless, families in poverty, even extreme poverty, often do have many children. As a result, the burden of poverty falls particularly heavily on the children.

Children in poverty do not have the luxury of a childhood or an education. Schools are usually nonexistent in poor areas of the world, and families are so poor that they cannot afford for their children to attend schools in the first place. Children from a very early age are required to help the family survive by working or performing domestic tasks such as fetching water. In extreme situations, children at a young age work as beggars, young boys and girls are sold to work in sweatshops, and young girls are sold into prostitution by their families. This may seem unusually cruel and harsh by Western standards, but we cannot imagine the horror of starvation and the desperation that many families in the world must feel that would force them to take such measures to survive. Children in the poverty areas of the world therefore grow up fast. By the time they are five or six, or in some cases even younger, they must be working to support the family. By their early teens, they are usually on their own and must go out in the world to work.

It has become a vicious cycle. In poor countries, families feel they must have more children for their survival, yet having more children perpetuates the poverty. The United Nations estimates that there are 250 million children between the ages of five and fourteen in the paid labor force throughout the world (United Nations, 1996). Most of the children, 150 million, are in Asia, and 80 million are in Africa (United Nations, 1996). Many of these children work long hours in difficult conditions and enjoy few freedoms, making products (soccer balls, clothing, and toys, for example) for those who are much better off.

Another problem in the very poor areas of the world is homeless children. In many situations, families are so poor that they can no longer care for their children, and the children must go out on their own, even at young ages. Many of these homeless children end up in the streets of the major cities of Asia and Latin America. In Latin America, it is estimated that there are 13 million street children, some as young as six years old. Alone, they survive through a combination of begging, selling, prostitution, drugs, and stealing. They sleep in alleys or in makeshift shelters. Their lives are harsh, brutal, and short. They receive no formal education and have few prospects for a decent life.

Poverty and Hunger

How can you live on less than $1 a day? The answer is that you cannot, or at least you cannot live very well. Malnutrition and hunger are growing problems since many of the people in poverty cannot find or afford food. It is estimated that between 750 million and one billion people in the world are so poor that they are unable to obtain enough food to meet their nutritional needs (The Hunger Project, 1998). Each day, 30,000 people die as a consequence of chronic, persistent hunger (The Hunger Project, 1998).

In the report *The State of World Hunger* (1998), prepared by the World Hunger Program at Brown University, three situations are identified that result in people not having enough to eat. *Food shortage* occurs when there is not enough food available to feed a designated area. This area can be the world, a continent, a country, or a region of a country. *Food poverty* refers to the situation in which households cannot afford to purchase enough food to adequately feed the members of the household. *Food deprivation* refers to the inadequate food consumption of individuals to maintain a healthy life. These three concepts are interrelated. When there is a food shortage, households cannot find or purchase enough food, which leads to food deprivation among the household members. In some cases, food poverty occurs in households in regions where there are no food shortages, and in some situations, individuals can experience food deprivation when there is not food poverty.

Debunking Society's Myths

Myth: **There are too many people in the world and simply not enough food to go around.**

Sociological perspective: **Growing more food will not end hunger. If there were more just systems of distributing the world's food, hunger could be reduced.**

Why are people starving in the world? Is it that there is not enough food to feed all the people in the world? In fact,

TABLE 10.4 *THE PROPORTION OF CHRONICALLY UNDERFED PEOPLE*

Percentage Chronically Underfed	Sub-Saharan Africa	Near East North Africa	Middle America	South America	South Asia	East Asia	China	Total
1970	35	23	24	17	34	35	46	36
1975	37	17	20	15	34	32	40	33
1980	36	10	15	12	30	22	22	26
1990	37	5	14	13	24	17	16	20

SOURCE: Uvin, Peter. 1998. *The State of World Hunger.* World Hunger Program: Brown University. Website: http://www.worldhunger.org.

there is plenty of food grown in the world. The world's production of wheat, rice, corn, and other grains is sufficient to adequately feed all the people in the world. Much of the grain grown in the United States is stored and not used because we grow too much. The problem is that the surplus food in the world does not get to the truly needy. The people who are starving lack the exigencies for obtaining adequate food, such as arable land or a job that would pay a living wage. In many cases, these people, in the past, grew food crops and were able to feed themselves, but today so much of the best land has been taken over by agro-businesses that grow cash crops, such as tobacco or cotton, that subsistence farmers have been forced onto marginal lands on the flanks of the desert where conditions are difficult and crops often do not grow.

The location of hunger around the world is changing. Table 10.4 shows the change in the proportion of people who are chronically underfed in seven regions of the world

B • O • X 10.2

UNDERSTANDING DIVERSITY

Malnutrition in Burundi Camps

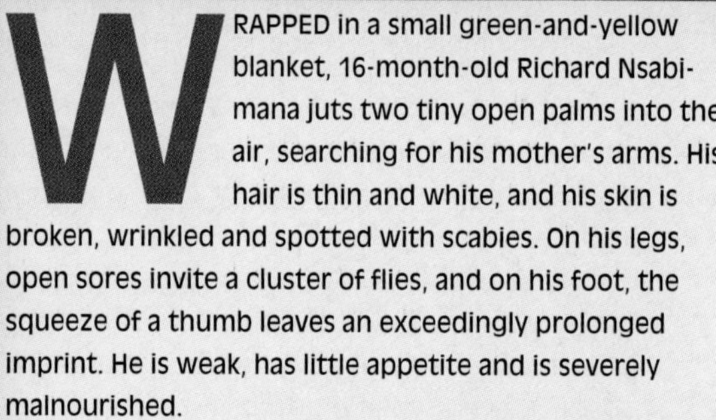

WRAPPED in a small green-and-yellow blanket, 16-month-old Richard Nsabimana juts two tiny open palms into the air, searching for his mother's arms. His hair is thin and white, and his skin is broken, wrinkled and spotted with scabies. On his legs, open sores invite a cluster of flies, and on his foot, the squeeze of a thumb leaves an exceedingly prolonged imprint. He is weak, has little appetite and is severely malnourished.

It is a condition by no means uncommon in Burundi, where the ghost of malnutrition has long since established itself in hills and homes throughout the tiny Central African nation's impoverished but verdant countryside, easily striking children whose access to food, health, sanitation and proper care is limited.

However, since the outbreak of civil war in 1993, levels of malnutrition in Burundi have only increased, climbing ever higher as the overall health situation deteriorates. Conditions are the worst by far in makeshift camps for the displaced. About 600,000 people—as many as 25 percent of whom are children—have been displaced inside Burundi during the past five years. In some of the worst-hit areas, particularly regroupment camps in Karuzi, rates of malnutrition for children have reached 18.7 percent, and rates of severe malnutrition 4.9 percent.

Malnutrition's causes are many and difficult to determine. "I don't know why he's sick," says Richard's mother, 24-year-old Madelene, cradling him in her arms. Richard has kwashiorkor, and its causes are rooted in a combination of factors: part diet, part health, part education, part access to food, part care, and all amplified by the civil war.

Before the war, Richard's family cultivated a small plot of land in the province of Karuzi, in central Burundi. The yield was relatively abundant, and they were able to sell the surplus of sweet potatoes, manioc and bananas in local markets. "Before the crisis," says Madelene, "we had just enough to make a good living." When civil war finally reached their fields late last year and military authorities forced them into a regroupment camp ostensibly for their own protection—all that changed. Their access to their fields, and therefore to their food, plummeted.

A month after entering the camp, Richard fell ill with diarrhea and intestinal worms. His parents are now sick as well: his father with malaria; his mother, like him, with scabies. The nearest health center is 3 kilometres away, but many times, Madelene says, they use traditional healers instead.

Their diet consists primarily of beans, porridge and supplementary food from an NGO. But, she says, "it is not enough." Like many families in Karuzi, they have been unable to cultivate their fields, partly due to sickness, partly to restrictions placed on their movement by authorities. And when Madelene is able to go back to her fields 2 km away, she gathers whatever food she can for her family. But, she says, "whatever is there is there by the grace of God."

The atmosphere in this camp, like in many others, has aggravated and compromised the nutritional status of the children of Burundi. Richard Nsabimana, unfortunately, is not alone.

SOURCE: *The State of the World's Children* 1998, Features, UNICEF Web site: (http://www.unicef.org/sowc98).

and for the world as a whole. This table shows that 36 percent of the people in the world were chronically underfed in 1970, but by 1990, only 20 percent were chronically underfed. This is a big improvement. By region, most areas of the world show a marked decrease in hunger. In China, for instance, 46 percent of the population were chronically underfed in 1970. Just twenty years later, only 16 percent of China's population were underfed. The only region of the globe where hunger has increased is sub-Saharan Africa.

Causes of World Poverty

What causes world poverty, and why is it that so many people are so desperately poor and starving? More to the point, why is poverty decreasing in some areas but increasing in others? We do know what does not cause poverty. Poverty is not necessarily caused by too rapid population growth although high fertility rates and poverty are related. In fact, many of the world's most populous countries, India and China, for instance, have large segments of their population that are poor, but these countries have begun to reduce poverty levels, even with very big populations. Poverty is not caused because people are lazy or uninterested in working. People in extreme poverty work tremendously hard just to survive, and they would work hard at a job if they had one. It is not that they are lazy; it is that there are no jobs for them.

Poverty is a result of a mix of a number of tragic causes. For one, the areas where poverty is increasing have a history of unstable governments or, in some cases, virtually no effective government at all to coordinate national development or plans that might alleviate extreme poverty and starvation. World relief agencies are reluctant to work in, or send food to, such countries as the Sudan, Afghanistan, or Somalia, where the national governments can guarantee neither the safety of relief workers nor the delivery of food and aid to where it should go. Because of lawlessness in some of these countries, food convoys are hijacked and roads are blocked by bandits or warlords. Those who suffer in these situations are the desperately poor.

In many of the countries with high proportions of poverty, the economies have collapsed and the governments have borrowed heavily to remain afloat. As a condition of these international loans, lenders, including the World Bank and the International Monetary Fund, have demanded harsh economic restructuring to increase capital markets and industrial efficiency. These economic reforms may make good sense for some and may lead these countries out of economic ruin over time, but in the short run, these imposed reforms have placed the poor in a precarious position since the reforms also called for drastically reduced government spending on human services.

However, poverty is not just a result of incompetent or bankrupted governments. As we discussed in the section on the theories of global stratification, poverty is also caused by changes in the world economic system. Although poverty has been a long-term problem and has many causes, the recent increase in poverty and starvation in Africa can be attributed in part to the changes in world markets beginning in the 1980s that favored Asia economically, but put sub-Saharan Africa and Latin America at a marked disadvantage. During the 1980s, the world entered a general recession that resulted in a sharp drop in commodity prices. For some of the Asian countries that had begun to industrialize, such as India, China, Indonesia, South Korea, Malaysia, and Thailand, this price drop was actually beneficial since it reduced the cost of raw materials. But for the commodity-producing nations in Africa and Latin America, the results of this economic downturn were disastrous. As prices declined, wages decreased, people lost jobs, and governments had less to spend on social services. In Latin America, the poor flooded to the cities, whereas in Africa, they did the opposite, fleeing to the countryside hoping to be able to grow subsistence crops. Governments had to borrow; many governments even collapsed or found themselves in such great debt that they were unable to help their own people. This worldwide compression in economic activity, in short, set off a collapse in national economies that put many people into marginal conditions, thus creating massive amounts of poverty and starvation.

Debunking Society's Myths

Myth: **The key to development for the third world lies in a renewed emphasis on exports and free trade.**

Sociological perspective: **Free trade policies and export agriculture have increased poverty throughout the third world.**

In sum, poverty has many causes. It is now a major global problem that not only affects the 1.3 billion or more people who are living in poverty, it also affects all people on the earth in one way or another. In some areas, poverty rates are declining as some countries begin to improve their economic situation; however, in other areas of the world, poverty is increasing, and countries are sinking into financial, political, and social chaos.

The Future of Global Stratification

As we have seen, many countries in the world are well-off, but many countries in the world are poor, some very poor. This global stratification system has been created over several centuries, but the conditions of extreme poverty and even starvation that we see in the world system at this time are relatively new. We must ask is the world getting better or worse, and what will happen in the future?

There is some good news. In some areas of the world, particularly East Asia, but also in Latin America, many countries have shown rapid growth and have emerged as developed countries. These countries are sometimes called the **newly industrializing countries** (NICs), and they include Korea, Malaysia, Thailand, Taiwan, and Singapore. In these countries, individuals have saved and invested, and the governments have invested in social and economic development. Since some of the NICs have large populations, their success demonstrates that economic development can occur in heavily populated countries. China has embarked on an aggressive policy of industrial growth, and India is also improving economically.

Yet for all the success stories, there are also nations that are not making it. These include nations on all continents. In many cases, governments have collapsed or are functioning only at minimal levels, the economy is bankrupted, the standard of living has plummeted, and people are starving. More disturbing is the situation in many areas of the world where the increase in ethnic hatred has led to mass genocide and forced millions of refugees from their homes. These situations have increased poverty and hunger. And these countries, some of which used to be relatively well-off, have experienced downward mobility in the global stratification system.

Of concern also are the fifteen newly created nations that emerged from the former Soviet Union. Some of these countries are economically stable and are developing robust economies. However, some of these newly created states face enormous economic and social problems that could threaten their very survival and impoverish millions of people.

Another factor is the continued growth of capitalism and of capital markets around the world. With the collapse of the Soviet Union and with China moving rapidly in the direction of capitalism, there is now virtually no opposition to the unfettered development of a world capitalistic system. The opening of new markets, the increasing global trade, the growth of multinational corporations, and the development of world financial markets will bring prosperity and wealth to many nations and to many individuals. The growing world market economy may allow some of the emerging countries that were once poor to move into the ranks of the rich nations and share in the newly created wealth, but whether this wealth will filter down to the people at the lower levels of society is another question.

The new push to further develop the world as one large capital market will also leave some countries behind, and therefore poverty and hunger will continue in many parts of the world. Market economies create opportunities to become wealthy, both for individuals and for nations. For those who can take advantage of these opportunities, the future looks bright, but many nations and individuals do not have this opportunity. Their conditions are so desperate that they do not have a chance to participate in the world market, except at a great disadvantage. These countries suffer from years of neglect; many owe debts to the world financial institutions; and they have no money for schools, roads, hospitals, or other social services. Before they can begin to think about industrial development, many of these countries must deal with the immediate problems of starvation, high death rates and birthrates, urban overcrowding, and ethnic conflict.

Capital markets also create poverty. As a result, the continued growth of the world capital market system will mean that some countries will prosper and some countries will not. In the future, the global stratification system will change. Some countries will become wealthy, but the outlook for the end of poverty and starvation is not good.

CHAPTER SUMMARY

- *Global stratification* is a system of inequality of the distribution of resources and opportunities between countries. A particular country's position is determined by its relationship to other countries in the world. The countries in the global stratification system can be categorized according to their per capital gross national product, or wealth. The world's countries can also be categorized as *first-*, *second-*, or *third-world* countries, which describes their political affiliation and their level of development. The global stratification system can also be described according to the economic power countries have.
- The poorest countries have more than half the world's population and have high birthrates, high mortality rates, poor health and sanitation, low rates of literacy and school attendance, and are largely rural. The richest countries constitute only 14 percent of the world's population and have low birthrates, low mortality rates, better health and sanitation, high literacy rates, high school attendance, and are largely urban. Although women in the wealthy countries are not completely equal to men, they suffer less inequality than do women in the poor countries.
- *Modernization theory* interprets the economic development of a country in terms of the internal attitudes and values. Modernization theory ignores that the development of a country may be due to its economic relationships with other more powerful countries. *Dependency theory* draws on the fact that many of the poorest nations are former colonies of European powers where colonial powers keep colonies poor and do not allow their industries to develop, thus creating dependency. *World systems theory* argues that no nation can be seen in isolation and that there is a world economic system that must be understood as a single unit. The economic core, the industrialized countries of Europe, North America, and Japan, exploit the periphery, which is made up of the poor countries of the world.
- *Relative poverty* means being poor in comparison to others. *Absolute poverty* describes the situation where people do not have enough to survive. *Capability poverty* refers to the situation in which people lack three essential human requirements to survive: nourishment, ability to reproduce, and an education. Poverty particularly affects women and children. Children in the very poor countries are forced to work at very early ages and do not have the opportunity for schooling. Street children are a growing problem in many cities of the world, especially in Latin America. Starvation is also a consequence of the global stratification system. Although the number of people experiencing malnutrition has decreased, especially in Asia, there are still many people in the world who do not have enough to eat.
- The future of global stratification is varied and depends on the country's position within the world economic system. Some countries, particularly those in East Asia—commonly referred to as *newly industrializing countries*—have shown rapid growth and emerged as developed countries. Many nations, though, are not making it. Governments collapse, countries suffer economic bankruptcy, the standard of living plummets, and people starve.

KEY TERMS

absolute poverty
capability poverty
commodity chain
core countries
dependency theory
double deprivation
first-world countries
global system of stratification
modernization theory
multinational corporations
neocolonialism
new international division of labor
newly industrializing countries
per capita gross national product
peripheral countries
power
relative poverty
second-world countries
semiperipheral countries
third-world countries
world systems theory

THE INTERNET: A Tool for the Sociological Imagination

Resources on the Internet:

Virtual Society: The Wadsworth Sociology Resource Center at
http://sociology.wadsworth.com

Visit this site to find additional learning tools, including interactive quizzes, links to related web sites, and an easy link to InfoTrac College Edition.

The United Nations
http://www.un.org/

The United Nations home page includes a wealth of information about social, economic, and political conditions around the world. Their publications on economic and social development are a good source of data about global stratification.

The United Nations Children's Fund
http://www.unicef.org/

The United Nations Children's Fund (UNICEF) site has information about the well-being of children around the world, including their annual report "The State of the World's Children."

The World Bank
http://www.worldbank.org/

The World Bank site includes publications and other information about the global economy and how different nations are faring.

Sociology and Social Policy: Internet Exercises

The persistence of poverty around the globe is a dilemma. On the one hand, plenty of food is grown each year to adequately feed all the people of the world and have food left over. On the other hand, 750 million people in the world suffer from a shortage of food, and approximately 30,000 people die each day from the consequences of *hunger* and *malnutrition* (World Hunger Project, 1998). Clearly, the food is not getting to the people who need it. Imagine that you are the head of a world agency that deals with food distribution. How would you go about making sure that those who are starving get adequate food? Should you encourage people to produce or grow their own food, or should food be imported into the poorer regions from the countries with food surpluses?

Internet Search Keywords:

hunger and malnutrition
World Hunger Project
food shortage
economic human rights
food distribution

Web sites:

http://www.brown.edu/Departments/World_Hunger_Program/
This web site is made available by the World Hunger Program to provide primary information regarding the causes of, and solutions to, hunger as well as links to other sites related to hunger.

http://www.cgiar.org/ifpri/
The web site by the International Food Policy Research Institute (IFPRI) contains rich resources on devising appropriate food policies and improving nutrition for the third-world populations.

http:www.foodfirst.org/
The Institute for Food and Development Policy ("Food First") introduces its publications and ongoing projects worldwide.

http://www.ifad.org/www.html
A list of web sites is provided by the International Fund for Agricultural Development dealing with hunger and poverty issues.

http://www.fian.org/
FoodFirst Information and Action Network advocates the right to feed oneself as a fundamental human right.

InfoTrac College Edition: Search Word Summary

Absolute Poverty
Per Capita Gross National Product
Modernization Theory

In order to learn more about these central topics in sociology, you can conduct an electronic search using InfoTrac College Edition. To aid in your search and to gain useful tips, see the Student Guide to InfoTrac College Edition on the Virtual Society web site:
http://sociology.wadsworth.com

INTERACTIONS—A SOCIOLOGY CD-ROM CONCEPTS FOR THIS CHAPTER

Go to the Wadsworth Sociology CD-ROM for further study on the concepts in this chapter. The CD-ROM also includes quizzes and additional activities to expand your learning experience.

SUGGESTED READINGS

Carnoy, Martin, Manuel Castells, Stephen S. Cohen, and Fernando Henrique Cardoso (eds.). 1993. *The New Global Economy in the Information Age: Reflections on Our Changing World.* University Park, PA: The Pennsylvania University Press.

This anthology examines various dimensions of the impact of the computer revolution on global stratification.

Marx, Anthony. 1997. *Making Race & Nation: A Comparison of the United States, South Africa, and Brazil.* New York: Cambridge University Press.

A comparative analysis of these three societies that analyzes racial and class stratification. The different histories of slavery, apartheid, and segregation in these nations create unique systems of race relations, but Marx argues that their similarities also illuminate the potential futures of race and class relations in all three societies.

McMichael, Philip. 1996. *Development and Social Change.* Thousand Oaks, CA: Pine Forge Press.

This book examines the development of the global development project, explaining the rise of the global marketplace and its role in the stratification of countries. A number of case studies are presented.

Schneider, Linda, and Arnold Silverman. 1997. *Global Sociology: Introducing Five Contemporary Societies.* New York: McGraw-Hill.

This book presents five case studies of societies at different levels in the global stratification system: Egypt, Mexico, Germany, Japan, and the !Kung bushmen. The history, culture, social structure, and social institutions are analyzed in each of the five societies so that comparisons can be drawn.

Shannon, Thomas Richard. 1989. *An Introduction to the World-System Perspective.* Boulder, CO: Westview Press.

This book is an analysis of the development of world systems theory from its beginning in the works of Marx and Lenin, including a section on the strengths and weaknesses of world systems theory and its application to the modern world.

Worsley, Peter. 1984. *The Three Worlds: Culture and World Development.* London: Weidenfeld and Nicolson Publisher.

This is one of the best accounts about the development of the global stratification system. Worsley shows how inequality in the global system developed during the agricultural revolution.

CHAPTER 11

Race and Ethnicity

YOU MIGHT expect a society based on the values of freedom, equality, and "brotherhood" not to be deeply afflicted by racial conflict, but think of the following:

- In Philadelphia, Black high school students beat a Cambodian student while onlookers cheered.
- In Detroit, Vincent Chin, a Chinese American, was clubbed with a baseball bat by a White autoworker and his son. Chin died within minutes from multiple skull fractures.
- A twenty-nine-year-old White female stockbroker was attacked while jogging in New York City's Central Park and was repeatedly raped and stabbed by a group of Black and Hispanic youths who had vowed earlier that evening to attack a White woman.
- In the spring of 1996, a pickup trick carrying Mexican immigrants was chased by California police and run off the road, and at least two Mexican occupants were severely beaten by police officers.
- Rodney King, a Black man, was pulled over in 1992 by White policemen. Seven of the officers viciously beat him with nightsticks while another shocked him with a Taser weapon. Captured on videotape, the officers were brought to trial. After several of the policemen were acquitted, a riot broke out. In the mayhem, two young Black men pulled Reginald Denny, a White male truck driver, from the cab of his truck and beat him mercilessly, breaking two bricks against his skull.
- In the 1990s, reminiscent of the terrorism that marked an earlier period of segregation in the South, there was a rash of church bombings and arson directed against Black churches, mostly rural, and mostly in the South. At the same time, periodic desecration of Jewish temples and cemeteries periodically flares up in diverse U.S. communities.

These ugly incidents, all true, have one thing in common—race. Along with gender and social class, race has fundamental importance in human social interaction, and it is an integral part of social institutions. Of course, the races do not always interact as enemies. Reginald Denny, the White truck driver beaten in the Los Angeles riots, was rescued by African American men—a fact that went largely unreported in the media. Nor is interracial tension always obvious. It can be as subtle as the White person who simply does not initiate interactions with African Americans and Latinos, or the elderly White man who almost imperceptibly leans backward at a cocktail party as a Japanese American man approaches him.

In everyday human interaction, as African American philosopher Cornel West has eloquently noted, race still matters, and matters a lot (West, 1993). What is race, and what is ethnicity? Why does society treat racial and ethnic groups differently, and why is there social inequality between these groups? How are these divisions and inequalities able to persist so stubbornly, and how extensive are they? These questions fascinate sociologists who do research on the causes and consequences of racial and ethnic stratification in our society. In this chapter, we examine these studies, with an eye to understanding the significance of racial and ethnic stratification in society. Just as class stratification differentiates people in society, according to the class privileges and disadvantages that they experience, so do **racial and ethnic stratification**—defined as the inequality between racial–ethnic groups in society—rest on privileges and disadvantages that accrue to groups based on their membership in different racial and ethnic groups.

Race and Ethnicity

Within sociology, the terms *ethnic, race, minority,* and *dominant group* have very specific meanings, different from the meanings these terms have in common usage. These concepts are important to developing a sociological perspective on racial and ethnic stratification.

Ethnicity

An **ethnic group** is a social category of people who share a common culture, for example, a common language or dialect; a common religion; and common norms, practices, customs, and history. Ethnic groups have a consciousness of their common cultural bond. Italian Americans, Japanese Americans, Arab Americans, Polish Americans, Greek Americans, Mexican Americans, and Irish Americans are examples of ethnic groups in the United States, but ethnic groups are also found in other societies, such as Serbians and Croatians in Bosnia-Herzegovina, the tribally-based HuTus and Tutsis in Rwanda, or ethnic Albanians in Kosovo.

An ethnic group does not simply exist because of the common national or cultural origins of a group, however. Ethnic groups develop because of their unique historical and social experiences. These experiences become the basis for the group's *ethnic identity*, meaning the definition the group has of itself as sharing a common cultural bond. Italian Americans, for example, did not necessarily think of themselves as a group with common interests and experiences prior to immigration to the United States. Originating from different villages, cities, and regions of Italy, Italian immigrants identified themselves by their family background and community of origin, but the process of immigration and the experiences Italian Americans faced as a group in the United States created a new identity for the group as a whole. They became Italian Americans, united by common political interests; similar cultural practices (including such things as religious observations, food, and language); residential isolation; connections to social, religious, and community-based organizations; and prejudice and discrimination as well. It is these social ties that created Italian Americans as an ethnic group, not simply their national origin. Indeed, many who now identify as Italian Americans have never been to Italy, and Italian Americans are of highly diverse social, regional, occupational, and educational backgrounds. Their commonly felt social and cultural bonds, arising from historical experiences in the United States, is what makes them an ethnic group (Alba, 1990; Waters, 1990).

Activities such as this Puerto Rican Day Parade in New York City reflect pride in one's group culture and result in greater cohesiveness of the group.

This band, playing for St. Patrick's day, an Irish holiday, contains people of varied racial–ethnic backgrounds.

The social and cultural basis of ethnicity is also seen in the fact that ethnic groups can develop more or less intense ethnic identification at different points in time. Ethnic identification may grow stronger when groups face prejudice or hostility from other groups. Perceived or real threats from other groups may unite an ethnic group around common political and economic interests. Ethnic unity can develop voluntarily, or may be involuntarily imposed when ethnic groups are excluded by more powerful groups from certain residential areas, occupations, or social clubs. These exclusionary practices strengthen ethnic identity.

Many have thought that ethnicity would decline in importance over time in the United States. Ethnicity no longer rigidly restricts one's choice of marriage partners, nor does it strictly dictate where one lives, how one votes, where one works, or who one's friends are; thus, in some ways ethnicity matters less than it did before but only somewhat

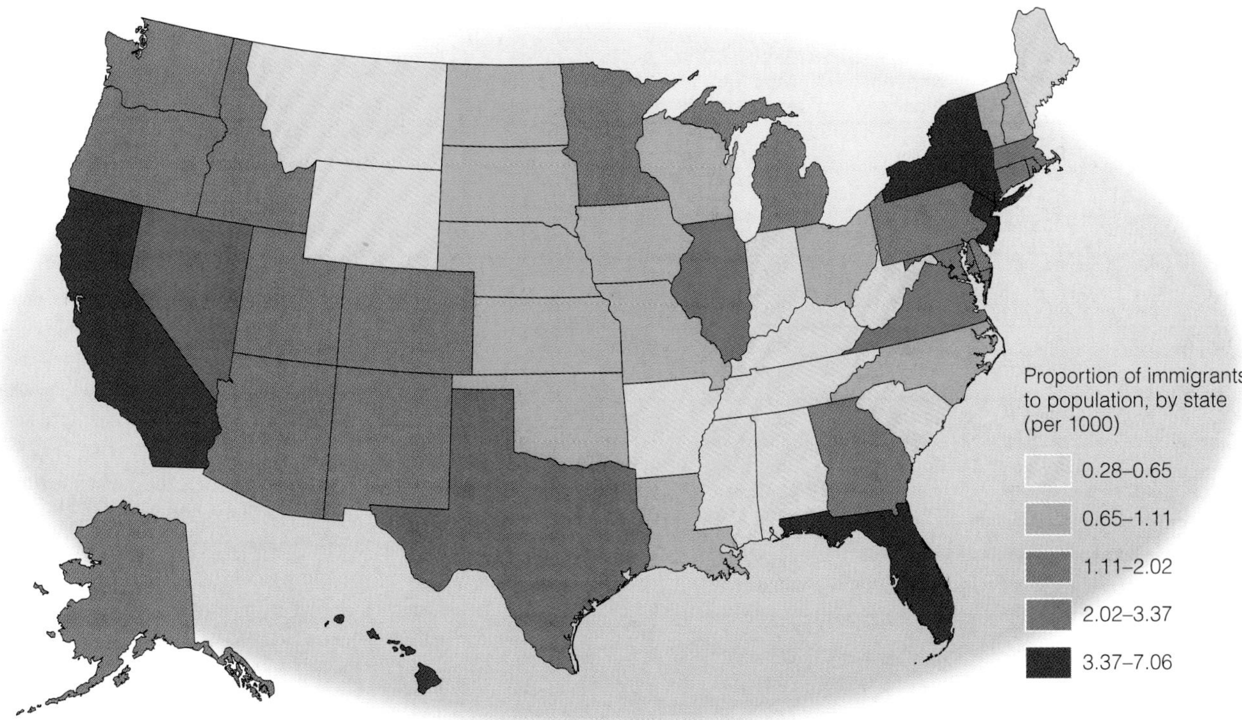

MAP 11.1 Mapping America's Diversity: The New Immigration

DATA: From U.S. Bureau of the Census. *Statistical Abstracts of United States, 1997.* Washington, DC: U.S. Government Printing Office.

less. Nonetheless, ethnic identity is still very important because of the strong basis for group identity it provides for many people. In a society as vast and diverse as the United States, ethnicity gives people a sense of a community, while making people feel unique, thereby preserving *both* a community and individual identity (Waters, 1990).

Race

Like ethnicity, race is a socially constructed category. Society assigns people to races not by logic or fact, but by opinion and social experience. In other words, how race is used to define groups is a social process. Although the meaning of race begins from perceived biological differences between groups, on closer examination, the assumption that racial differences are purely biological is problematic.

First, the categories used to presumably divide groups into races are not fixed, and they vary from society to society. Within the United States, laws defining who is Black have historically varied from state to state. North Carolina and Tennessee law historically defined anyone as Black who had even one great-grandparent who was Black (being "one-eighth" Black). In other southern states, having any Black ancestry at all defined one as a Black person—the so-called one drop (that is, of "black blood") rule. This is even more complex when we consider the meaning of race in other countries. In Brazil, a light-skinned Black person could well be considered White, especially if the person is of high

Tiger Woods, considered among the greatest golfers ever, lines up a putt. He is of Asian American and African American parentage. What race is he?

TABLE 11.1 *COMPARISON OF U.S. CENSUS CLASSIFICATIONS, 1890–1990*

Census Date	White	African American	Native American	Asian American	Other Categories
1890	White	Black Mulatto Quadroon Octoroon	Indian	Chinese Japanese	
1900	White	Black	Indian	Chinese Japanese	
1910	White	Black Mulatto	Indian	Chinese Japanese	Other
1920	White	Black Mulatto	Indian	Chinese Japanese	Other
1930	White	Negro	Indian	Chinese Japanese Filipino Hindu Korean	Mexican Other
1940	White	Negro	Indian	Chinese Japanese Filipino Hindu Korean	Other
1950	White	Negro	American Indian	Chinese Japanese Filipino	Hawaiian Other
1960	White	Negro	American Indian Aleut Eskimo	Chinese Japanese Filipino	Hawaiian Other
1970	White	Negro or Black	Indian (American)	Chinese Japanese Filipino Korean	Hawaiian Other
1980	White	Black or Negro	Indian (American) Eskimo Aleut	Chinese Japanese Filipino Korean Vietnamese	Hawaiian Guamanian Samoan Other
1990	White	Black or Negro	Indian (American) Eskimo Aleut	Chinese Japanese Filipino Korean Asian Indian Vietnamese	Hawaiian Guamanian Samoan Asian or Pacific Islander Other

SOURCE: Lee, Sharon. Routledge, 1993. Used by permission. "Racial Classification in the U.S. Census: 1890–1990." *Ethnic and Racial Studies* 16 (1): 75–94.

socioeconomic status, showing that "race" in Brazil is influenced by one's class. In fact, in Brazil, people are considered Black only if they are of African descent and have no discernible White ancestry. The vast majority of U.S. Blacks would not be considered Black in Brazil (Blalock, 1982b; Sowell, 1983). Mixed race people also defy the biological categories that are typically used to define race; is someone who is the child of an Asian mother and an African American father Asian or Black? The current proposal to introduce the category "multiracial" into the U.S. census reflects this issue (Wright, 1994; Waters, 1990). As Table 11.1 clearly shows, the census has dramatically changed its racial and ethnic classification since 1890, reflecting the fact that society's thinking about racial and ethnic categorization has not remained constant through time (Lee, 1993; Mathews, 1996).

Second, the biological characteristics that have been used to define different racial groups vary both within and between groups. Many Asians, for example, are actually lighter skinned than are many Europeans and White Americans, and regardless of their skin color, have been defined in racial terms as "yellow." Some light-skinned African Americans are actually lighter in skin color than some White Americans. Developing racial categories overlooks the fact that human groups defined as races are biologically much more alike than they are different.

Third, the biological differences that are presumed to define different racial groups seem somewhat arbitrary. Why, for example, do we differentiate people based on skin color, and not some other characteristic, like height or hair color or eye color? You might ask yourself how a society based on the presumed racial inferiority of red-haired people would compare to racial inequality in the United States. The likelihood is that, if a powerful group defined another group as inferior because of some biological characteristics and used their power to create social institutions that treated this group unfairly, a system of so-called racial inequality would result. In fact, there are very few biological differences between racial groups. Most of the variability in most biological characteristics is within, not between, racial groups (Lewontin, 1996).

Fourth, different groups use different criteria to define racial groups. To American Indians, being an American Indian depends upon proving Indian ancestry, but this varies considerably from group to group. Among some American Indians, one must be able to demonstrate 75 percent Indian ancestry to be recognized as American Indian; for other American Indians, demonstrating 50 percent Indian ancestry is sufficient. It also matters who defines racial group membership. The government makes Indian tribes prove themselves as tribes through a complex set of federal regulations (called the "federal acknowledgement process"); very few are actually given this official status. Official recognition by the government matters, though, since only those groups officially defined as Indian tribes qualify for health, housing, and educational assistance from the Bureau of Indian Affairs or are allowed to manage the natural resources on Indian lands and maintain their own system of governance (Snipp, 1989; Brown, 1993; Locklear, 1999).

In sum, a **race** is a group treated as distinct in society on the basis of certain characteristics, some biological, that have been assigned social importance in society. A race, because of presumed biologically or culturally inferior characteristics (as defined by powerful groups in the society), is singled out for differential and unfair treatment. It is not the biological characteristics per se that define racial groups, but how groups have been treated historically and socially. This is what it means to say that race is socially constructed.

Debunking Society's Myths

Myth: Racial differences are fixed, biological categories.

Sociological perspective: Race is a social concept, one in which certain physical or cultural characteristics take on social meanings that become the basis for racism and discrimination.

This definition of race emphasizes that what is important about race is not only physical or cultural differences per se, but how definitions of race are created and maintained by the most powerful group (or groups) in society and what these presumed group differences mean in the context of social and historical experience. As a result, who is defined as a race is often as much a political question as it is a biological one. Most White Americans probably think of Blacks as a race, but do not consciously think of themselves as having a racial identity or experience. Similarly, although they probably did not think of themselves as a race, Irish Americans in the early twentieth century were defined by more powerful White groups as a "race" that was inferior to White people. At that time, Irish people were not considered by many to even be "White" (Ignatiev, 1995)!

Thinking about the experience of Jewish people provides a good example of what it means to say that race is a socially constructed category. Jews are more accurately called an ethnic group because of their common religious and cultural heritage, but in Nazi Germany, Hitler defined Jews as a race. They were presumed to be biologically inferior to the group Hitler labeled the Aryans—white-skinned, blonde, tall, blue-eyed people. On the basis of this definition, which was supported through Nazi law, taught in Nazi schools, and enforced by the Nazi military, Jewish people were grossly mistreated: segregated, persecuted, and systematically killed.

The social construction of race has been elaborated in a new perspective in sociology known as racial formation theory (Omi and Winant, 1994). **Racial formation** is the process by which a group comes to be defined as a race; this definition is supported through official social institutions such as the law and the schools. This concept emphasizes the importance of social institutions in producing and maintaining the meaning of race; it also connects the process of racial formation to the exploitation of so-called racial groups. A good example comes from African American history. During slavery, African Americans were defined as three-fifths of a person, for purposes of deciding how slaves would be counted for state representation in the new federal government. The process of defining slaves in this way served the purposes of White Americans, not slaves themselves, and it linked the definition of slaves as a race to the political and economic needs of the most powerful group in society (Higginbotham, 1978).

The process of racial formation also explains how groups like Asian Americans, Native Americans, and Latinos have been defined as "races," despite the different experiences and nationalities of the groups composing these three categories. Race, like ethnicity, lumps groups together that may have very different historical and cultural backgrounds, but once so labeled, the group is treated as a single entity. This has recently been the case in the United States with Mid-Easterners: Lebanese, Iranians, Iraqis, Jordanians, and many others are lumped together into one group.

To sociologists, the important questions about the meaning of race are: Who defines groups as races and on what basis? What are the consequences of some groups being labeled racially different from others? In sum, race is a social and historical construct, one that is typically used by powerful groups in the society as the basis for political and economic exploitation.

Minority and Dominant Groups

Minorities are racial or ethnic groups, but not all racial or ethnic groups are minorities. Irish Americans, for instance, are not now minorities, although they once were in the early part of this century. A **minority group** is any distinct group in society that shares common group characteristics and is forced to occupy low status in society because of prejudice and discrimination. A group may be a minority on the basis of ethnicity, race, sexual preference, age, or class status, for example. A minority group is not necessarily a numerical minority, but is a group that holds low status relative to other groups in society—regardless of the size of the group. In South Africa, Blacks outnumber Whites ten to one, but until Nelson Mandela's election as president and the dramatic change of government in 1994, Blacks were an officially oppressed and politically excluded social minority under the infamous *apartheid* system of government. The group that assigns a racial or ethnic group to subordinate status in society is called the *dominant group* or *social majority*. In general, a racial or ethnic minority group has the following characteristics (Simpson and Yinger, 1985):

1. It possesses characteristics that are popularly regarded as different from those of the dominant group (such as race, ethnicity, sexual preference, age, or religion).
2. It suffers prejudice and discrimination by the dominant group.
3. Membership in the group is frequently ascribed rather than achieved, although either form of status can be the basis for being identified as a minority.
4. Members of a minority group feel a strong sense of group solidarity. There is a "consciousness of kind" or "we feeling." This bond grows from common cultural heritage and the shared experience of being the recipient of prejudice and discrimination.

5. Minority group marriages are typically, although not always, among members of the same group. The phenomenon of in-group marriage is called ethnic or racial *homogamy*.

Debunking Society's Myths

***Myth*: Minority groups are those with the least numerical representation in society.**

***Sociological perspective*: A minority group is any group, regardless of size, that is singled out in society for unfair treatment and generally occupies lower status in the society.**

Racial Stereotypes

Racial and ethnic inequality is peculiarly resistant to change. Racial and ethnic inequality in society produces racial stereotypes, and these stereotypes become the lens through which members of different groups perceive each other. We shall examine how stereotypes, the result of political and economic inequality, are used to maintain domination of one group over another, and how racial and ethnic stereotypes work together with gender and social class stereotypes to force persistent minority status onto some groups.

Stereotypes and Salience

In everyday social interaction, people tend to categorize other people. According to the most recent research (Taylor et al., 1997), the most common bases for such categorization are race, gender, and age. We *immediately* identify a stranger as Black, Asian, Hispanic, White, and so on; as a man or woman; and as a child, adult, or elderly person. After these categorizations come others based on social attributes, such as social class or attractiveness, followed by categories such as jock, nerd, obnoxious person, and so on.

Quick and ready categorizations help us process the huge amounts of information we receive about people with whom we come into contact. We quickly assign people to a few categories, saving ourselves the task of evaluating and remembering every discernible detail about a person. We learn that we are supposed to treat each person as a unique individual, but research over the years clearly shows that we tend not to. Instead, we routinely categorize people.

A **stereotype** is an oversimplified set of beliefs about members of a social group or social stratum that is used to categorize individuals of that group. It is based on the tendency of humans to categorize based on a small range of perceived characteristics. Stereotypes are presumed, usually incorrectly, to describe the "typical" member of some social group.

Stereotypes based on race or ethnicity are called *racial–ethnic stereotypes*. Examples are common: Asian Americans have been stereotyped as overly ambitious, sneaky, and clannish; at the turn of the century, Asians, particularly the Chinese, were stereotyped as dirty, ignorant, and untrustworthy. African Americans often bear the stereotype of being "typically" loud, lazy, naturally musical, and so on. Hispanics are stereotyped as lazy, oversexed, and for men, "macho." Such stereotypes, presumed to describe the "typical" member of a group, are of course factually inaccurate for the vast majority of the members of the group.

White ethnic groups have also been stereotyped: Jews are perceived as materialistic, clever, and unethical. The Polish are stereotyped as bumbling, unintelligent, and sports-minded. Italians are stereotyped as emotional, argumentative, and prone to crime. The Irish are stereotyped as prone to politics, drinking, and quarreling. No group in U.S. history has escaped the process of categorization and stereotyping.

Stereotypes change over time, sometimes quite dramatically. Table 11.2 shows how a number of racial–ethnic stereotypes changed between 1933 and 1982. In the early 1930s, Jews were stereotyped as shrewd, mercenary, grasping, and industrious. Note the presence of positive (for example, industrious) as well as negative (for example, grasping) stereotypes. By 1982, the stereotyping of Jews as shrewd, mercenary, and grasping had become less frequent, whereas the label "ambitious" became more common. Blacks were stereotyped in the early 1930s as superstitious, lazy, happy-go-lucky, and ignorant; these stereotypes became less frequent by 1982, but were still used (Essed, 1991; Dovidio and Gaertner, 1986).

The categorization of people into groups and the subsequent application of stereotypes is based on the **salience principle**, which states that we categorize people on the basis of what appears initially prominent and obvious—that is, salient—about them. Skin color is a salient characteristic; it is one of the first things that we notice about someone. Because skin color is so obvious, it becomes a basis for stereotyping. Gender and age are also salient characteristics of an individual, and thus serve as notable bases for group stereotyping.

THINKING SOCIOLOGICALLY

Observe several people on the street. What are the first things you notice about them (that is, what is *salient*?) Make a short list of these things. Do these lead you to *stereotype* these people? On what are your stereotypes based?

The choice of salient characteristics is culturally determined. In the United States, skin color, as well as hair texture, nose form and size, and lip form and size, have become salient characteristics. We use these features to categorize people in our minds on the basis of race. By contrast, in Brazil, race is somewhat less salient than in the United States. There is a considerably greater variety of skin color in Brazil, as well as an elaborate set of linguistic terms to describe different skin gradations, and categorization and stereotyping by race, although it exists, is less common. It is more common in Brazil to stereotype by education or in-

TABLE 11.2 *STEREOTYPES OF BLACKS AND JEWS IN FOUR GENERATIONS OF COLLEGE STUDENTS*

Stereotypes of Jews (% checking trait)

Trait	1933	1951	1967	1982
Shrewd	79%	47%	30%	—
Grasping	34	17	17	—
Sly	20	14	7	—
Mercenary	49	28	15	—
Materialistic	—	—	46	—
Industrious	48	29	33	—
Ambitious	21	28	46	—
Aggressive	12	—	23	—
Persistent	13	—	9	—
Talkative	13	—	3	—
Intelligent	29	37	37	—
Practical	—	—	19	—
Loyal to family ties	15	19	19	—
Very religious	12	—	7	—

Stereotypes of Blacks (% checking trait)

Trait	1933	1951	1967	1982
Superstitious	84%	41%	13%	6%
Very religious	24	17	8	23
Ignorant	38	24	11	10
Stupid	22	10	4	1
Naive	14	—	4	4
Lazy	75	31	26	13
Physically dirty	17	—	3	0
Slovenly	13	—	5	2
Unreliable	12	—	6	2
Happy-go-lucky	38	17	27	15
Pleasure-loving	—	19	26	20
Musical	26	33	47	29
Ostentatious	26	11	25	5
Sensitive	—	—	17	13
Gregarious	—	—	17	4
Talkative	—	—	14	5
Imitative	—	—	13	9

SOURCE: Dovidio, J. F. and S. L. Gaertner, eds., 1986. *Prejudice, Discrimination, and Racism.* New York: Academic Press.

come (Omi and Winant, 1994). In other cultures, religion may be far more salient than skin color. In Lebanon, whether one is Muslim or Christian is far more important than skin color; religion in Lebanon is salient and takes considerable priority over race.

What happens when one meets someone who does not fit the presumed stereotype? When faced with such a situation, the prejudiced or racially discriminating person will "explain away" the inconsistency. He or she will say that the person is a "special case" or is somehow "different"; hence, the person simply ignores or explains away anything that tends to contradict the held stereotype. As a consequence, the African American child who contradicts a prevailing stereotype and performs well in mathematics is perceived as a "special case"; the recently arrived Korean who speaks English without an Asian accent is perceived as somehow odd or perplexing; the Hispanic woman who is president of the company is perceived as "merely lucky."

The Interplay Between Race, Gender, and Class Stereotypes

Alongside racial and ethnic stereotypes, gender and social class are among the most prominent features by which people are categorized. In our society, there is a complex interplay between racial–ethnic, gender, and class stereotypes.

Among *gender stereotypes*, those based on a person's gender, the stereotypes about women are more likely to be negative than those about men. The "typical" woman has been traditionally stereotyped as subservient, flighty, overly emotional, overly talkative, prone to hysteria, inept at math and science, and so on. Many of these are *cultural stereotypes*; they are conveyed and supported by the cultural media—music, TV, magazines, art, and literature. Men too are painted in crude strokes, although usually not as negatively as women. Men in the media are stereotyped as macho, insensitive, pigheaded, and in situation comedies, as inept. Generally, men are depicted as wanting to have sex with as many women as possible in the shortest time available.

Social class stereotypes are based on assumptions about social position. Upper-class people are stereotyped (by middle- and lower-class people) as snooty, aloof, condescending, and phony. Some of the stereotypes held about the middle class (by both the upper class and the lower class) are that they are overly ambitious, striving, and obsessed with keeping up with the Joneses. Finally, stereotypes about lower-class people abound: They are perceived, by the upper and middle classes, as dirty, lazy, unmotivated, violent, and so on.

The principle of **stereotype interchangeability** holds that stereotypes, especially negative ones, are often interchangeable from one gender to the other, from one social class to another, and from one racial or ethnic group to another. Stereotype interchangeability is sometimes revealed in humor; ethnic jokes often interchange different groups as the butt of the humor, stereotyping them as dumb and inept. One traditional negative stereotype of African Americans is that they are inherently lazy. This stereotype has also been applied in recent history to Hispanics, Polish, Irish, and other groups. It has even been applied generally to all people perceived as lower class. In fact, "laziness" is often used to explain *why* someone is lower class or poor. Middle-class people are more likely to attribute the low status of a lower-class person to something *internal*, such as lack of willpower or laziness (Krasnodemski, 1996). Lower-class people are more likely to attribute their status to discrimination or poor opportunities, that is, to an *external* societal factor (Taylor et al., 1997; Kluegel and Bobo, 1993; Bobo and Kluegel, 1991).

Similarly, research shows that White Americans by a two-to-one margin blame the urban Black poor for their own predicament (they are "lazy"), whereas the urban Black poor by a similar margin blame discrimination and racism. The more racially prejudiced the White person, the more likely he or she will blame the lower status of Blacks on their own laziness and lack of willpower. Less prejudiced Whites are more likely to blame external causes such as the social structure, racism, and discrimination (Taylor et al., 1997; Bobo, 1989).

The same kinds of stereotypes have historically been applied to women. Many of the stereotypes applied to women in literature and the media, such as that they are childlike, overly emotional, unreasoning, and so on, have also been applied to African Americans, lower-class people, the poor, and earlier in this century, Chinese Americans. A common theme is apparent: Whatever group occupies the lowest social status in society at a given time (whether racial–ethnic minorities, females, or lower-class people) is negatively stereotyped, and often the same negative stereotypes are used between and among these groups. Finally, the stereotype is then used as an "explanation" for the observed behavior of a member of the stereotyped group, serving, incorrectly, as a justification for their lower status in society.

There are numerous examples. Blacks are seen to be disproportionately unemployed because they are inherently lazy. Hispanics are seen as disproportionately unemployed because *they* are inherently lazy. American Indians are seen as suffering from unemployment and other ills, such as a high suicide rate and alcoholism, because they lack motivation, and because *they* are inherently lazy. Italians are perceived as not performing well in elementary school because *they* are lazy. Other logical explanations for the condition of these groups (such as discrimination) are overlooked.

Prejudice, Discrimination, and Racism

Many people use the terms *prejudice, discrimination*, and *racism* loosely, as if they were all the same thing. Typically, in common parlance, people also think of these mostly in individualistic terms, as if the major problems of race were the result of individual people's bad will or biased ideas. Sociologists use more refined concepts to understand race and ethnic relations, distinguishing carefully between prejudice, discrimination, and racism.

Prejudice

Prejudice is the evaluation of a social group, and individuals within that group, based on conceptions about the social group that are held despite facts that contradict it and that involve both prejudgment and misjudgment (Allport, 1954; Pettigrew, 1971; Jones, 1997). Prejudices are usually defined as negative predispositions or evaluations, rarely positive. Thinking ill of people only because they are members of group X is prejudice. The negative evaluation arises solely because the person is seen as a member of group X, without regard to countervailing traits or characteristics the person may have. A negative prejudice against someone not in one's own social group is often accompanied by a positive prejudice in favor of someone who *is* in one's own group; thus, the prejudiced person will have negative attitudes about a member of an *out-group* (any group other than one's own), and positive attitudes about someone simply because he or she is in one's *in-group* (any group one considers one's own).

Most people disavow racial or ethnic prejudice, yet the vast majority of us carry around some prejudices, whether about racial–ethnic groups, men and women, old and young, upper class and lower class, or straight and gay. Prejudice is what social scientists call a variable—it varies from person to person. Five decades of research have shown definitively that people who are more prejudiced are also more likely to stereotype others by race or ethnicity, and often by gender, than those who are less prejudiced (Adorno et al., 1950; Jones, 1997; Taylor et al., 1997); nevertheless, everyone possesses some prejudices.

Prejudice based on race or ethnicity is called *racial–ethnic prejudice*. In-groups and out-groups in this case are defined along racial or ethnic lines. If you are a Latino and dislike an Anglo only because he or she is White, then this constitutes prejudice: It is a negative judgment ("prejudgment") based on race and ethnicity and very little else. In this example, Latino is the in-group and White is the out-group. If the Latino individual attempts to justify these feelings by arguing that "all Whites have the same bad character," then the Latino is using a stereotype as justification for the prejudice. Similarly, *gender prejudice* exists in our society. There are negative prejudices based on gender, and in-group and out-group are defined by gender. Finally, there is *class prejudice*, the negative evaluation of people solely on the basis of their social class, but little reference to anything else about them. Note that prejudice can be held by any group against another. Thus, African Americans can be prejudiced against Whites or against Hispanics and Asian Americans. Such prejudice does exist, although sociological research has found that recent concerns that there is a backlash of prejudice by African Americans against Hispanics and Asian Americans is no greater than prejudice by White Americans toward these groups (Cummings and Lambert, 1997).

Prejudice is also revealed in the phenomenon of **ethnocentrism**, which we examined in Chapter 3 on culture. Ethnocentrism is the belief that one's group is superior to all other groups. The ethnocentric person feels that his or her own group is moral, just, and right, and that an out-group—and thus any member of the out-group—is immoral, unjust, wrong, distrustful, or criminal. The ethnocentric individual uses his or her own in-group as the standard against which all other groups are compared. Generally, the greater the difference between groups, the more harshly the out-group will be judged by an ethnocentric individual and the more prejudiced that person will be against members of the out-group.

The Effects of Prejudice. Our judgments about other people can hardly escape being influenced by our feelings (prejudices) about their group as a whole. Several studies show this quite dramatically (Sweeney and Haney, 1992; Ugwuegbu, 1979). In one study, White students were asked to serve as jurors in a mock rape trial. Each read a hypothetical, but very realistic, account of the rape of a nineteen-year-old White woman by a twenty-one-year-old man. Four different accounts (experimental conditions) were used. In one account, the alleged rapist was a Black man, and the evidence against him was strong (both the female victim and an eyewitness identified the rapist). In another, the alleged rapist was Black and the evidence against him was weak (neither the female victim nor the eyewitness could identify the rapist). In two other accounts, the evidence followed a similar pattern, but the rapist was a White man. The results were interesting. On a scale of blame, the Black man was always blamed more than the White man, *regardless* of whether the evidence was weak or strong. The strength of the evidence made a slight difference: Both alleged rapists were perceived as more culpable when the evidence was strong as opposed to weak. The point is that race has an effect on the amount of blame attributed to the alleged rapist—*in addition to* the effect of the strength of the evidence. Race by itself has a clear and separate effect.

The results were similar no matter what the race of the juror, although a Black juror was somewhat less likely than

B • O • X 11.1

UNDERSTANDING DIVERSITY

Arab Americans: Confronting Prejudice

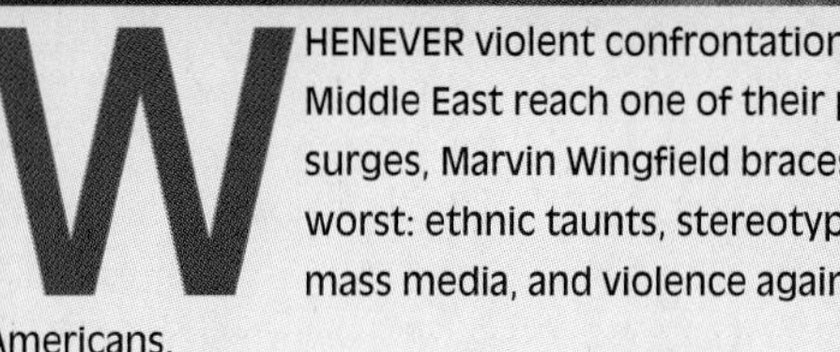

WHENEVER violent confrontations in the Middle East reach one of their periodic surges, Marvin Wingfield braces for the worst: ethnic taunts, stereotypes in the mass media, and violence against Arab Americans.

"The pattern seems to be that whenever there is a crisis in the Mideast, the incidence of hate crimes against Arab Americans increases," says Wingfield, coordinator of conflict resolution for the American-Arab Anti-Discrimination Committee (ADC) in Washington, D.C. "It becomes even more pronounced when the United States is involved directly."

The backlash, Wingfield notes, is simply one manifestation of centuries-old misperceptions in the West about the nature of Arab culture. Colonial arrogance, he says, fostered stereotypes of Arabs as camel-riding hedonists and devious traders. The more modern stereotype of the "Muslim terrorist" led initially to false assumptions about the Oklahoma City bombing.

Another offshoot of this arrogance, Wingfield points out, is the tendency of many white Americans to view other broad segments of the population as homogeneous. References to "the black community," for example, ignore the diversity among African Americans and their interests. Similarly, "the Arab World" is vast and varied. Most Arabs speak Arabic, and most are Islamic, but the Chaldeans prove that neither attribute is a prerequisite for "Arab-ness."

Part of the ADC's mission is to help combat stereotypes and reduce anti-Arab hate crimes through educational and political programs. A large part of this effort involves encouraging Arab Americans to play higher profile roles in their communities.

The ADC has prepared guidelines for community activists to follow in trying to increase the presence and awareness of Arab culture in school districts. Called "Working With School Systems," it describes various scenarios for involvement, ranging from volunteering at the classroom level to lobbying for district-wide curriculum development. One notable success occurred in Portland, Ore., where Arab American activists persuaded the local newspaper, *The Oregonian,* to include a special Mideast page in its annual "Newspapers in Education" issue.

"Our program is multifaceted," Wingfield says. "It's not a curriculum that can be taken into the schools. Instead, we have many different people doing many different things. We encourage people to take actions on their own horizons."

In metro Detroit, several programs through public schools and nonprofit organizations work to bridge differences between Arab Americans and the broader community. An organization called Arab-Jewish Friends runs an annual contest in which Arab and Jewish high school students co-write essays. It helps the students break down their own stereotypes by working with students from cultures with whom they have historically been at odds.

"We have them write about issues that are important to Arabs and Jews," says Jeannie Weiner, one of the contest's founders. "It's good for them to put these thoughts down in writing. And it's good for us because we can use the essays as tools to show what the kids are thinking."

Another project, directed by Wayne State University's Center for Peace and Conflict Studies, brought together students from high schools in suburban Dearborn, where metro Detroit's Arab-Muslim population is concentrated, to negotiate a contract for behavior among Arab and non-Arab students. They quickly found common ground in their dissatisfaction with the district's multicultural programs and petitioned school officials for classes in such topics as world religions.

Little came of the demands, but the students learned that they have more in common than they thought, says Mickey Petera, assistant director of the Peace and Conflict Studies Department. While she supports programs that break down barriers among older students and adults, Petera says she is particularly encouraged by Greenfield Union's efforts to reach students at earlier ages.

"We need to start at the elementary and preschool levels," she says, "and raise a generation of students that can get along."

SOURCE: *Teaching Tolerance* (Spring 1997): 49. This article is reprinted by permission of *Teaching Tolerance,* a publication of the Southern Poverty Law Center.

a White juror to attribute blame to the Black alleged rapist. Strikingly similar results have been obtained, when possible, in actual rape trials in real courts by comparing trials with defendants of a different race and trials with differing evidence strength (Loewen, 1982; Sweeney and Haney, 1992).

Racial and ethnic prejudice also has a marked effect on people's political views, and on how they vote. This is shown in studies that incorporate a separate measure of the amount of the individual's prejudice and relate this to voting preferences. For example, when Jesse Jackson, the Black minister and political activist, ran for the Democratic presidential nomination in 1984, he was opposed most by Whites who were highly prejudiced, whereas less prejudiced Whites were more likely to favor him (Sears et al., 1988). The difference was attributable to prejudice itself, not the objective advantages and disadvantages of the candidate.

Whether one is for or against affirmative action and similar programs, one's prejudice predicts fairly accurately how one will respond to a particular program, regardless of relevant, objective facts about the situation. For example, the more prejudiced a White person is, the more likely he or she will oppose busing to achieve racial balance in schools (Jessor, 1988; Kluegel and Bobo, 1993). Of course, prejudice is not the only reason one might oppose busing, but prejudice is part of the picture.

The same holds with regard to affirmative action policies on behalf of Asian Americans. The more strongly an individual is prejudiced against Asian Americans, the more that person will object to any governmental program that benefits Asian Americans, such as monetary reparations to the Japanese American families who were unjustly interned in concentration camps at the outset of U.S. entry into World War II. The same is true for other kinds of educational and political programs geared to benefit Asian Americans (Takaki, 1989; Takagi, 1992).

Prejudice and Socialization. Where does racial–ethnic prejudice come from? How do moderately or highly prejudiced people end up that way? People are not born with stereotypes and prejudices. Research shows that they are

BOX 11.2 DOING SOCIOLOGICAL RESEARCH

Children's Understandings of Race

IN A RACIALLY stratified society, people learn concepts about race that shape their interaction with others. Sociologists Debra Van Ausdale and Joe Feagin wanted to know how children understand racial and ethnic concepts, and how this influences their interactions with other children. These researchers noted that much of the social science literature studied older children or only looked at how race shaped children's self-concepts. Prior to this study, most knowledge about children's understandings of race came from experimental studies where children are observed in a laboratory or from psychological tests and interviews with children. Van Ausdale and Feagin wanted to study children in a natural setting—that is, one not contrived for purposes of research, so they observed children in school, during which time they systematically observed children's interactions with one another.

They observed three-, four-, and five-year olds in an urban preschool. Twenty-four of the children they observed were White; nineteen, Asian; four, Black; three, biracial; three, Middle Eastern; two, Latino; and three classified as "other." The children's racial–ethnic designations were provided by parents. The senior researcher (Van Ausdale) was in the classroom all day for five days a week over a period of eleven months. Van Ausdale says she observed one to three episodes involving significant racial or ethnic matters each day.

From these observations, the researchers conclude that young children use racial and ethnic concepts to exclude other children from play. Sometimes language is the ethnic marker children use; other times, it is skin color. The children also showed an awareness of negative racial epithets, even though they were attending a school that prided itself on limiting children's exposure to prejudice and discrimination and using a multicultural curriculum to teach students to value racial and ethnic diversity. At times, the children also used racial–ethnic understandings to include others—teaching other students about racial–ethnic identities. Race and ethnicity were also the basis for children's concepts of themselves and others. As an example, one four-year-old White child insisted that her classmate was Indian because she wore her long, dark hair in a braid. When the classmate explained that she was not Indian, the young girl remarked that maybe her mother was Indian.

In this and numerous other incidents reported in the research, the children showed how significant racial–ethnic concepts were in their interaction with others. Race and ethnic differences, as the researchers claim, are "powerful identifiers of self and other" (1996: 790). Despite the importance of race in the children's interactions with others, Van Ausdale and Feagin also noted a strong tendency for the adults they observed to deny that race and ethnicity were significant to the childen. One implication of their research is that, rather than denying the significance of race and ethnicity in social interaction, adults should become more attuned to how race and ethnicity shape social interaction—among children, as well as among older people—if they want to encourage more positive intergroup relations.

SOURCE: Van Ausdale, Debra, and Joe R. Feagin. 1996. "The Use of Racial and Ethnic Concepts by Very Young Children." *American Sociological Review* 61 (October): 779–793.

learned and internalized through the socialization process, including both *primary socialization* (family, peers, and teachers) and *secondary socialization* (such as the media). Children imitate the attitudes of their parents, peers, and teachers. If the parent complains about "Japs taking away jobs " from Americans, then the child grows up thinking negatively about the Japanese, including Japanese Americans. Attitudes about race are formed early in childhood, at about age three or four (Allport, 1954; Van Ausdale and Feagin, 1996). There is a *very* close correlation between the racial and ethnic attitudes of parents and those of their children (Ashmore and DelBoca, 1976). Later in life, peers will reinforce these negative feelings, for example, by jeering at Japanese American children on their way to school in the morning.

A major vehicle for the communication of racial–ethnic attitudes to both young and old is the media, especially television, magazines, newspapers, and books. For many decades, African Americans, Hispanics, Native Americans, and Asians were rarely presented in the media, and then only in negatively stereotyped roles. The Chinese were shown in movies, magazines, and early television in the 1950s as bucktoothed buffoons who ran shirt laundries. Japanese Americans were depicted as sneaky and untrustworthy. Hispanics were shown as either ruthless banditos or playful, happy-go-lucky "Pedros" who took long siestas. American Indians were presented as either villains or subservient characters like the Lone Ranger's famed sidekick, Tonto. Finally, there is the drearily familiar image of the Black person as subservient, lazy, clowning, and bug-eyed, a stereotypical image that persisted from the late nineteenth century all the way through the 1950s and early 1960s (Thibodeau, 1989).

The late 1960s, which included the Black Power movement and greatly increased civil rights activity in the country, saw some improvements in media stereotypes, which continued into the 1970s and 1980s. During this time, the media became somewhat less culpable in fostering racial and ethnic prejudice. Shows such as the Huxtable family on *The Cosby Show*, featuring an African American doctor (played by Bill Cosby), his lawyer wife (played by Phylicia Rashaad), their well-educated children, and their positive interactions with Whites and Asians, may have contributed to blunting anti-Black prejudice. The Huxtables were a well-to-do family headed by Black professionals, and thus could serve as positive role models for Black youth, but they also underscore what African Americans and other minorities in society have *not* attained. As one critic wryly noted from observing other Black sitcoms, Black people seem to be doing better on TV than in real life (Gates, 1992)!

Despite the Huxtables, careful content analyses have shown that improvements in the portrayal of minorities on television have been minimal. For example, positive cross-race interactions have always been infrequent, even into the early 1990s. During this period, positive interactions between Blacks and Whites (those in which Blacks and Whites interact as professional equals or friends) have been a mere 5 percent or less of total interactions, including dramas, comedies, and other types of television shows (Wei et al., 1990).

TV's The Cosby Show *attempted to show life in a professional African American family.*

Discrimination

Discrimination is overt negative and unequal treatment of the members of some social group or stratum solely because of their membership in that group or stratum. Prejudice is an attitude; discrimination is overt behavior. *Racial–ethnic discrimination* is unequal treatment of a person on the basis of race or ethnicity. Housing discrimination has been a particular burden on minorities. Many studies have been able to reproduce the situation that when two persons identical in nearly all respects (age, education, gender, social class, and other characteristics) both present themselves as potential tenants for the same housing, and if one is White and the other is a minority, the minority person will often be refused housing by a White landlord when the otherwise identical White applicant will not. A minority landlord refusing housing to a White person while granting it to a minority person of similar social characteristics is also discriminating, but "reverse discrimination" of this sort is far less frequent and far less of a problem in society (Feagin and Feagin, 1993; Feagin and Vera, 1995).

By similar reasoning, unfair and negative treatment of women compared to men of the same social characteristics is *gender discrimination*. Denying a job to a woman but giving it to a man of the same education, preparedness, test scores, and so on, is a common example of gender discrimination. In fact, as will be seen in Chapter 12, women in the United States are frequently refused a job that is then given to a man with *less* education or *lower* qualifications than the woman. If one is both a woman *and* a minority, one is often confronted by the *double jeopardy effect*, a form of double discrimination. In such cases, the discriminatory effects of both race and gender combine, and the overall effect is thus magnified.

Prejudice is an attitude; discrimination is behavior, and the two do not always go together. High prejudice does not always accompany high discrimination. A convincing example of high prejudice but low discrimination is seen in an early, now classic study by LaPierre (1934).

SOURCE: Used by permission of Mark Giaimo and the *Naples Daily News.*

A White male professor of sociology, LaPierre went on a tour of the United States with a young male Chinese student and his Chinese wife. At the time, prejudice against the Chinese, and against Asians in general, was very strong. They stopped at 66 hotels and motels and at 184 restaurants. Despite the widespread anti-Asian prejudice in the country, all but one of the hotels and motels gave the Chinese couple a room, and they were never refused service in any of the restaurants. And on many occasions, they did not show up at the same time as did LaPierre himself.

Some weeks later, LaPierre sent letters to each of the establishments that had served the Chinese couple, asking whether they would accept a Chinese person as a guest. Of the 128 establishments that replied, fully 92 percent said they would not do so. The Chinese couple received nearly perfect service at these establishments in person, but almost all the very same establishments said they would refuse service to anyone who was Chinese. The near absence of discrimination (behavior) was nonetheless accompanied by clear expressions of prejudice (attitude).

The discrimination affecting the nation's minorities takes a number of forms, for example, income discrimination and discrimination in housing. Discrimination in employment and promotion (discussed in Chapter 17) and discrimination in education (see Chapter 15) are two other forms of discrimination. Income discrimination is seen in the persistent differences in median income that appear when comparing different racial groups.

Although the median income of Black and Hispanic families has increased somewhat since 1950, the income gap between these groups and Whites has remained virtually unchanged since 1967, as can be seen from Figure 11.1. Furthermore, per capita income has grown at a faster rate for Whites. Yet even these median income figures tell only part of the story. Poverty among Blacks decreased from 1960 to 1970, but the poverty level has been somewhat steady since. In the late 1990s, for the first time, the rate of poverty among Hispanics exceeded that of Black Americans. The current poverty rate (the percent below the poverty level) is highest for Native Americans (35 percent), next highest for Hispanics (27 percent), then African Americans (26.5 percent), Asian Americans (14 percent), and least for White Americans (11 percent; Dalaker and Naifeh, 1998; Lee, 1995); in all these racial groups, children have the highest rate of poverty (see Figure 11.2 and Figure 11.3).

Housing discrimination is illegal under U.S. law. Title VIII of the Civil Rights Act of 1964 made it a federal policy to provide fair housing throughout the nation. This legislation bars public and private discrimination in the sale or rent of housing; yet in practice, this rule is often violated. Real estate agents or landlords may use such tactics as showing houses and apartments selectively to prospective buyers or renters; restrictive covenants may eliminate whole groups of prospective residents; and the policies of banks and mortgage companies may discourage lending to developers of minority housing. Banks and mortgage companies often withhold mortgages from minorities based on "red lining," an illegal practice in which an entire minority neighborhood is designated "no-loan." Racial segregation may also be fostered by *gerrymandering*, the calculated redrawing of election districts, school districts, and similar political boundaries in order to maintain racial segregation. As a result, **residential segregation**, the spatial separation of racial and ethnic groups into different residential areas, continues to be a reality in this country.

Racism

Prejudice is an attitude; discrimination, behavior. Racism includes both attitudes and behaviors. **Racism** is the perception and treatment of a racial or ethnic group, or member of that group, as intellectually, socially, and culturally inferior to one's own group. A negative attitude taken toward someone solely because he or she belongs to a racial or

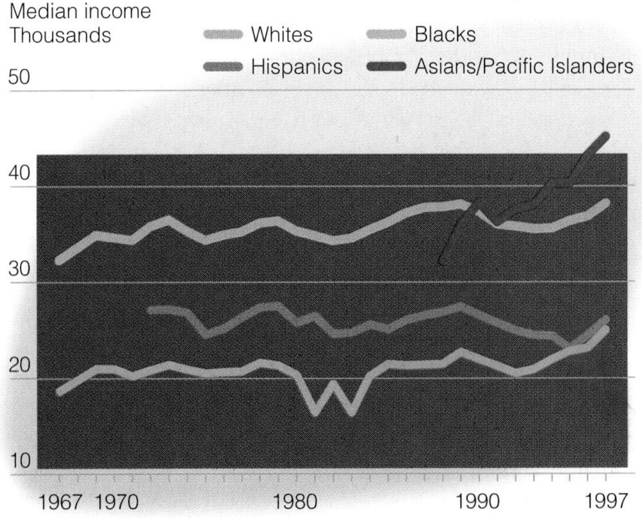

FIGURE 11.1 The Income Gap

SOURCE: U.S. Bureau of the Census. 1997. *Money Income in the United States: 1996.* Washington, DC: U.S. Department of Commerce, pp. 83–84; U.S. Bureau of the Census. 1996. *Median Income of Households, by Race and Hispanic Origin: 1967 to 1996.* Web site: http://www.census.gov/ftp/pub/hhes/income.

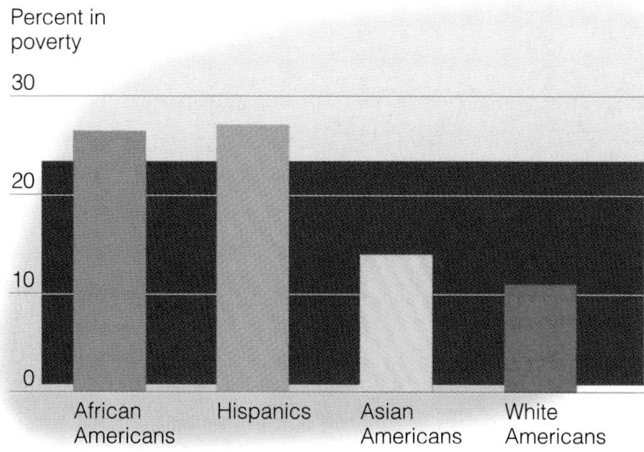

FIGURE 11.2 Poverty Among Racial Groups

SOURCE: Dalaker, Joseph, and Mary Naifeh. 1998. *Poverty in the United States: 1997.* Washington, DC: U.S. Department of Commerce, p. vii.

ethnic group is a prejudice, as we have already discussed. An attitude or prejudice is what you think and feel; a behavior, on the other hand, is what you do. Racism involves negative attitudes that are sometimes linked with negative behavior, though not always.

The castration and lynching of Black men and the lynching of Black women in the southern United States from the 1800s well into the 1950s is a clear and obvious case of racism. The killing through negligence of Chinese railway workers in the latter nineteenth century in this country is clear and obvious racism. The beating and maiming of foreigners by youthful skinheads, White power advocates, and neo-Nazis constitutes clear and obvious racism.

Dominative and Aversive Racism. Obvious, overt racism is called **dominative racism** (Jones, 1997), also sometimes called "old-fashioned racism." This form of racism exists in both individual and institutional forms, although it has declined in our society since the early 1960s (Schuman et al., 1985). Racism can also be subtle, covert, and nonobvious. This is called **aversive racism**. A White who avoids interaction with Latinos and African Americans is exhibiting aversive racism. Avoiding interaction is behavior. The action (avoidance) reflects disdain and aversion on the basis of race, hence the name *aversive*. This form of racism has remained at the same level over the last thirty years, with perhaps a slight increase (Kovel, 1970; Dovidio and Gaertner, 1986; Katz et al., 1986).

Examples of aversive racism have emerged from insightful studies of the nonverbal behavior of Whites as they interact with Blacks. Pettigrew (1985) has found that some Whites tend to sit farther away from Blacks than from other Whites when engaged in conversation. He also found that when talking to Blacks, some Whites use a friendly tone of voice less often and make eye contact less frequently, and they terminate interaction sooner than when talking with other Whites. Although subtle and nonverbal, such behaviors are nevertheless racist, since the behaviors involve treating Blacks and Whites differently, with Blacks treated more negatively. Similarly, Whites often reveal aversive racism in what they say to Blacks, Asians, or Hispanics and in what they say about them in private conversation—such as by mischievously imitating the speech of Blacks, Hispanics, or Asians (Jones, 1997; Feagin and Vera, 1995, Essed, 1991).

Aversive racism has its roots in how Whites tend to view racial–ethnic minorities, which is tinged with ambivalence. In a classic, much-quoted study of race from the early 1940s, *The American Dilemma*, Swedish economist and social scientist Gunnar Myrdal (Myrdal, 1944; cf. Jaynes and Williams, 1989) noted that Whites held contradictory beliefs about America. They subscribe to the beliefs of freedom and equality for all, yet they also harbor racial and ethnic prejudice and engage in racist behaviors. Most Whites today feel at least some degree of sympathy for the plight of American minorities, recognizing the role of discrimination and the effect of treating minorities as second-class citizens; yet, they also perceive minorities, especially Blacks, Hispanics, and Native Americans, and to a lesser extent Asians, as contributing significantly to their *own* difficulties. Many White Americans feel that they must bear a disproportionate amount of the burdens of society, such as greater work productivity, the costs of welfare, and the costs of other social services, because of negative qualities they attribute to minorities, such as laziness, lack of ambition, and criminal behavior. These negatives in minorities are perceived as inherent to their personalities and their subcultures. This leaves many Whites with simultaneous feelings of friendliness and disdain, goodwill and aversion. Aversive racism is one form of reaction to this ambivalence (Jones, 1977; Essed, 1991; Katz et al., 1986).

Institutional Racism. Racism is reflected not only in individual overt behavior, but also in society's institutions. **Institutional racism** is negative treatment and oppression of one racial or ethnic group by society's existing institutions based on the presumed inferiority of the oppressed group. Key to understanding institutional racism is seeing that dominant groups have the economic and political *power* to subjugate the minority group, even if they do not have the explicit intent of being prejudiced or discriminating against others. Power, or lack thereof, accrues to groups because of their position in social institutions, not just because of individual attitudes or behavior. The power that resides in society's institutions can be seen in such patterns as persistent economic inequality between racial groups (reflected in high unemployment among minorities, lower wages, and different patterns of job placement). Institutional racism is also seen in the criminal justice system.

African American and Hispanic people are arrested by police considerably more often than Whites and Asians. In fact, an African American or Hispanic wrongdoer is more likely to be arrested than is a White man who commits the *exact same crime* even though the White man shares the same age, socioeconomic environment, and even prior arrest record as the Black or Hispanic (Chambliss, 1988; Chambliss and Taylor, 1989, 1993). The statistical distortion does

institutional racism

Minorities are somewhat more likely to be arrested than Whites for the same offense. Does this reflect **institutional racism** *rather than any* **individual prejudice** *of the arresting police officer?*

not end with the arrest. Once in custody, an African American or Hispanic is more likely to be detained without bail, more likely to go to trial, and more likely to receive a stiffer sentence at the end of the trial than is a White man of similar social background and criminal record.

Institutional racism is also seen in educational institutions, such as in schools assigning Blacks and Latinos to the lower tracks than Whites with the *same* ability test scores, as will be seen in Chapter 15. In these instances, racism is a characteristic of the institutions, and not necessarily of the individuals within these institutions. This is why institutional racism can exist even without prejudice being the cause.

Consider this. Even if every White person in the country lost all of his or her prejudices and, even if he or she stopped engaging in individual acts of discrimination, institutional racism would still persist for some time. Over the years, it has become so much a part of U.S. institutions (hence, the term *institutional* racism) that discrimination can occur even when no single person is causing it. Existing at the level of social structure rather than at the level of individual attitude *or* behavior, it is "external" to the individual personality, and is thus a *social fact* of the sort about which the sociological theorist Durkheim wrote (Chapter 1).

If an employer prefers to hire people who graduated from the employer's university, and if that university has long denied entrance to minorities, then minorities will have little chance of being hired by that employer. If a medical school prefers to recruit children of its wealthy alumni, and when virtually all the alumni are White, then few minorities are likely to be admitted. As long as police departments continue to use their present height requirements, many otherwise qualified Japanese Americans—who are on average shorter than Whites—will be unable to become police officers. All these are examples of institutional racism. The "blame" for discrimination does not rest with the person but with the *institution*.

Debunking Society's Myths

***Myth*: The primary cause of racial inequality in the United States is the persistence of prejudice.**

***Sociological perspective*: Prejudice is one dimension of racial problems in the United States, but institutional racism can flourish even while prejudice is on the decline.**

The arena of everyday interactions reveals many subtle instances of institutional racism. In a thorough study of many automobile dealerships in the Chicago metropolitan area, it was found that salespeople who bargain with customers over the price of a new car make significantly lower final offers to White men than to women or Blacks (Ayers, 1990). Three White male researchers divided themselves among ninety car dealers, and then negotiated the best price for a new car. At roughly the same time (within a few days of one another), each of the dealerships was visited by one of three other testers—a White woman, a Black man, and a Black women—who bargained for the same car. Care was taken that each of the persons posing as buyers used the *same negotiation tactics* with the car salespersons, so the differences in final offer could not be attributed to different negotiating styles used by women versus men, or Blacks versus Whites. The pattern persisted regardless of the race and gender of the salesperson; only the race and gender of the potential buyer affected the final offer given by the salesperson.

The sticker price of the car they all bargained for was $13,465. White men received final offers that averaged $11,362. The White woman was given an average final price of $11,504. The Black man ended up with an average final offer of $11,783. The person who ended up with the highest (worst) final offer from the salesperson was the Black woman researcher, who was offered an average final deal of a whopping $12,237, nearly $1000 more than the final offer tendered to the White man. The investigators attribute these findings to subtle, institutional racism, as well as institutional sexism, operating together in the marketplace. Even though the auto salespersons in each instance may have been relatively free of prejudice, institutional racism in a subtle form nonetheless operated, resulting in the highest (worst) final offer to a Black woman.

Theories of Prejudice and Racism

Why do prejudice, as well as discrimination and the various forms of racism, exist? Two categories of theories have been advanced to seek to explain why. The first category consists of social psychological theories about prejudice. The second category consists of sociological theories of the types of racism, including institutional racism, as well as prejudice and discrimination.

Social Psychological Theories of Prejudice

Two social psychological theories of prejudice are the scapegoat theory and the theory of the authoritarian personality. The **scapegoat theory** argues that, historically, members of the dominant group in the United States have harbored various frustrations in their desire to achieve social and economic success (Feagin and Feagin, 1993). As a result of this frustration, they vent their anger in the form of aggression, and this aggression is directed toward some substitute that takes the place of the original perception of the frustration. Members of minority groups become these substitutes, that is, scapegoats. The psychological principle that aggression often follows frustration (from the frustration–aggression hypothesis of Dollard et al., 1939) is central to the scapegoat principle. For example, a White person who perceived that he was denied a job because "too many" Mexican American immigrants were being permitted to enter the country would be using Mexican Americans as a scapegoat if he became hostile (thus prejudiced) toward a specific Mexican American person, even if that person did not have the job in question and had nothing at all to do with the White person not getting the job.

The second theory argues that individuals who possess an authoritarian personality are more likely to be prejudiced against minorities than are nonauthoritarian individuals. The **authoritarian personality** (after Adorno et al., 1950, who coined the term) is characterized by a tendency to rigidly categorize other people, as well as tendencies to submit to authority, rigidly conform, be very intolerant of ambiguity, and be inclined to superstition. The authoritarian person is more likely to stereotype or rigidly categorize another, and thus readily place members of minority groups into convenient and oversimplified categories, or stereotypes. As a result, the authoritarian person who is a White man is likely to categorize (stereotype) all women as emotional or illogical, or "all Mexicans" as inherently violent. Minorities and women themselves may also reveal such tendencies toward stereotyping if they possess authoritarian personalities. Finally, there is some research that links high authoritarianism with high religious orthodoxy and extreme varieties of political conservatism (Bobo and Kluegel, 1991; Altemeyer, 1988).

Sociological Theories of Prejudice and Racism

Current sociological theory focuses more on explaining the existence of racism and its various forms, particularly institutional racism, although speculation about the existence of prejudice is also a component of these sociological formulations. The three sociological theoretical perspectives considered throughout this text have bearing on the study of racism, discrimination, and prejudice: functionalist theory, symbolic interaction theory, and conflict theory.

Functionalist Theory. Functionalist theory argues that in order for race and ethnic relations to be functional to society and thus contribute to the harmonious conduct and stability of that society, racial and ethnic minorities, and women as well, must assimilate into that society. **Assimilation** is a process by which a minority becomes socially, economically, and culturally absorbed within the dominant society. A first step in the functionalist assimilation process is to adopt as much as possible of the culture of the dominant society, particularly its language, its mannerisms, and its goals for success.

The rapidity of assimilation of a minority group into a society will depend among other things on the unique history of that minority group (different minority groups have different and diverse histories) as well as the desire of the group's members to assimilate. The functionalists argue that some groups (such as Asians) have assimilated more rapidly into U.S. society than others (such as Hispanics), both because of different histories as well as different desires for assimilation. The assimilation perspective is discussed later in more detail under "Patterns of Racial and Ethnic Relations."

Symbolic Interaction Theory. Symbolic interaction theory addresses several issues: first, the role of social interaction in reducing racial and ethnic hostility, and second, how race and ethnicity are socially constructed.

Symbolic interactionism asks the question: What happens when two people of different racial or ethnic origins come into contact with each other, and how can such interracial or interethnic contact reduce hostility and conflict? **Contact theory** (which originated with Allport, 1954; Cook, 1988) argues that interaction between Whites and minorities will reduce prejudice on the part of both groups if three conditions are met:

1. The contact must be between individuals of equal status; the parties must interact on equal ground. A Hispanic cleaning woman and the wealthier White woman who employs her may interact, but their interaction will not reduce prejudice. Instead, it is likely to perpetuate stereotypes and prejudices on the part of both.
2. The contact between equals must be sustained; short-term contact will not decrease prejudice. Bringing Whites together with Hispanics or Blacks or Native Americans for short weekly meetings will not erase prejudice. Daily contact on the job, between individuals of equal job status, will.
3. Social norms favoring equality must be agreed upon by the participants. Having African Americans and White skinheads interact on a talk show will probably not decrease prejudice; it might increase it.

Symbolic interaction theory is also the basis for understanding race and ethnicity as socially constructed categories. People are assigned to racial and ethnic groups, not just because of physical or cultural characteristics, but because of what these characteristics have come to mean in a racially stratified society. Neither ethnicity nor race is a fixed category; their meaning shifts in different social and historical contexts. White ethnic groups, for example, who simply thought of themselves as "White" have experienced a resurgence of ethnic consciousness and now proudly claim themselves as Irish American, Italian American, or Polish American. The social construction of race and ethnicity also creates new group identities, such as with Asian Americans who would otherwise think of themselves as Chinese American or Korean American; however, the discrimination and prejudice that these groups have faced in the United States has created a whole new racial–ethnic category—Asian American—one that lumps together their quite diverse histories.

Conflict Theory. The basic premise of conflict theory is that class-based conflict is an inherent and fundamental part of social interaction. To the extent that racial and ethnic conflict is tied to class conflict, conflict theorists argue that class inequality must be reduced to lessen racial and ethnic conflict in society.

The current "class versus race" controversy in sociology (reviewed in more detail later in this chapter under "Patterns of Racial and Ethnic Relations") concerns the question of whether class (namely, economic differences between races) or race (*caste* differences between races) is more important in explaining inequality and its consequences or whether they are of equal importance. Those focusing primarily on class conflict (such as sociologist William Julius Wilson, 1978, 1987, 1996) argue that class and changes in the economic structure are at times more important than race per se in shaping the life chances of different groups, although they see both as important. Wilson argues that being disadvantaged in the United States is more a matter of class, although he sees that this is clearly linked to race. Those focusing primarily on the role of race (for example, Willie, 1979; Feagin, 1991; Feagin and Feagin, 1993) argue the opposite: that race has been and is relatively more important than class—though class is still important—in explaining and accounting for inequality and conflict in society, and that directly addressing the question of race forthrightly is the only way to solve the country's racial problems.

A recent variety of the conflict perspective is **gendered racism** theory. This theory refers to the interactive or *combined* effects of racism and sexism in the oppression of women of color. An implication of this theory, argued by Essed (1991), is that the study of the position of women in society must be done *separately* within each racial or ethnic group, since the observed differences between women and men are not the same within the different racial or ethnic groups, and the experiences of women have not been the same, or even necessarily similar, in different racial or ethnic groups. The theory asserts not only that the effects of gender and race are intertwined and inseparable, but that, in addition, neither is separable from the effects of class. Class

BOX 11.3 SOCIOLOGY IN PRACTICE

Race, the Underclass, and Public Policy

SOCIOLOGISTS use their knowledge in a variety of ways to influence national social policy. One of the most influential to do so is William Julius Wilson, an African American sociologist, currently at the Harvard University School of Government. Wilson's work is so influential that in 1996 he was listed among *Time* magazine's twenty-five most influential Americans.

Wilson's work about the sociology of the underclass—which he defines to include African Americans, Asian Americans, Latinos, Whites, and others who are concentrated in urban poverty—has earned him a place as an advisor to President Clinton, as well as to the political campaigns of Jesse Jackson and Mario Cuomo. His work came to national prominence with the publication of *The Truly Disadvantaged: The Inner City, the Underclass, and Public Policy* (1987). Wilson argues that deep social and economic transformations have resulted in the formation of an urban underclass, permanently locked in poverty and joblessness. Differing from both liberals and conservatives, Wilson argues that the cause of this persistent poverty is neither overt racial discrimination, as liberals sometimes claim, nor the cultural deficiencies of racial groups, as conservatives claim. Rather, Wilson compellingly demonstrates that changes in the global economy have resulted in the concentration of a disenfranchised underclass, permanently jobless and located primarily in inner cities. Wilson argues that the flight of the more prosperous Black middle class from the inner city as traditional heavy industry declines, contributes to the concentration of poverty in the urban core.

In his policy-based work, Wilson argues that addressing the needs of the underclass cannot be based on race-specific policies. Arguing with contemporary conservatives, Wilson says he refuses to be intimidated by the rhetoric of conservatives. Instead, he says, "It's quite clear to me that we're going to have to revise discussion of the need for WPA-style jobs. Only these more structurally based programs, open to all in need, are likely to garner political support among the majority and to address the deep-seated problems that changes in the global economy have wrought" (*Time*, June 17, 1996: 57). With the publication of his new book, *When Work Disappears*, Wilson continues to be one of the most influential people in the nation. He has demonstrated how rigorous sociological thinking can make a difference in the national agenda for social policy.

SOURCES: Wilson, William Julius. 1996. *When Work Disappears: The World of the New Urban Poor.* Chicago: University of Chicago Press; Wilson, William Julius. 1987. *The Truly Disadvantaged: The Inner City, the Underclass, and Public Policy.* Chicago: University of Chicago Press; and "America's Most Influential People." *Time* (June 17, 1996): 56–57.

TABLE 11.3 *COMPARING SOCIOLOGICAL THEORIES OF RACE AND ETHNICITY*

	Functionalism	Conflict Theory	Symbolic Interaction
The racial order:	has social stability when diverse racial and ethnic groups are assimilated into society	is intricately intertwined with class stratification	is based on socially constructed meaning systems that assign groups of people to diverse racial and ethnic categories
Minority groups:	are assimilated into dominant culture as they adopt cultural practices and beliefs	have life chances that result from the opportunities formed by the intersection of class, race, and gender	form identity as the result of socio-historical change
Social change:	is a slow and gradual process as groups adapt to the social system	is the result of organized social movements and other forms of resistance to oppression	is dependent on the different forms of interaction that characterize intergroup relations (such as contact or competition)

conflict is seen as part and parcel of gender and race differences in this society, according to gendered racism theory (Collins, 1990; Andersen and Hill Collins, 1998).

Diverse Groups, Diverse Histories

The different racial and ethnic groups in the United States have arrived at their current social condition through histories that are similar in some ways, yet quite different in others. Their histories are related because of a common experience of White supremacy, economic exploitation, and political disenfranchisement. Native Americans were here first, yet they have been the victims of territorial expansion, armed conquest, genocide, and the destruction of their cultures. They were deprived of their land, forced to live on segregated reservations, or systematically exterminated. African Americans were forcibly abducted from West Africa—transported to this hemisphere under filthy inhuman conditions locked in the holds of slave ships and sold into slavery. Hispanics from both Mexico and Puerto Rico were imported to provide cheap labor, as were indigenous Hispanics. The Chinese were imported to labor on the Western frontier and were forced to endure a variety of hardships ranging from hunger and slavelike conditions to murder. Arriving Japanese met with high prejudice and discrimination, and even some who lived in the United States for several generations suffered the violation of having their property confiscated and their families forcibly moved to concentration camps after the United States entered into World War II against Germany and Japan.

Even White ethnic groups, many of whom arrived as indentured servants during the colonial era or arrived later expecting religious, political, and social freedom, found instead prejudice and discrimination. Such has been the story of the Irish, who fled domination by the British in Ireland and, in the 1840s, fled from the devastating effects of a famine in Ireland brought on partly by the potato blight. So severe were conditions in Ireland that half of Ireland's population (eight million in the early 1840s) disappeared—half of these dying from starvation and half, emigrating (Diner, 1996). Italians, Poles, and Jews have been excluded to varying degrees from full participation in U.S. society, have been denied rights and privileges, and have failed to receive full protection under the law. Assimilation of White ethnic groups from southern, central, and eastern Europe was slower than for groups from western and northern Europe (Nagel, 1996; Healey, 1995; Lieberson, 1980).

An historical perspective on each group follows, which will aid in understanding how prejudice, discrimination, and racism has operated throughout the history of U.S. society.

Native Americans

Native Americans have been largely ignored in the telling of U.S. history. The exact size of the indigenous population in North America at the time of the European's arrival in 1492 has been estimated at anywhere from one million to ten million people. Native Americans were here tens of thousands of years before they were "discovered" by Europeans. Discovery quickly turned to conquest, and in the course of

Cultural diffusion is evident in this Native American school kitchen, as seen by both the decorations and the food preparations.

the next three centuries, the Europeans systematically drove the Native Americans from their lands, destroying their ways of life and crushing various tribal cultures. Native Americans were subjected to the onslaught of European diseases. Lacking immunity to these diseases, Native Americans suffered a population decline considered by some to have been the steepest and most drastic of any people in the history of the entire world. Native American traditions have survived in many isolated places, but what is left is only a ghostly echo of the original 500 nations of North America (Nagel, 1996; Thornton, 1987; Snipp, 1989).

At the time of the first European contacts in the 1640s, there was great linguistic, religious, governmental, and economic heterogeneity among Native American tribes. Most historical accounts have underestimated the degree of cultural and social variety; however, between the arrival of Columbus in 1492 and the establishment of the first thirteen colonies in the early 1600s, the ravages of disease and the encroachment of Europeans caused a considerable degree of social disorganization. Sketchy accounts of Indian cultures by early colonists, fur traders, missionaries, and explorers underestimated the great social heterogeneity among the various Indian tribal groups, and also underestimated the devastating effect of the European arrival on Indian society.

Many Europeans were well aware of the rapid decline of the Indian population, and may even have counted on their disappearance in the years to come. Some contemporary historical accounts suggest that Whites may have entered into some treaties with Indians fully expecting that the terms would be moot since the Indians would soon be extinct (Thornton, 1987; Snipp, 1989).

By the year 1800, the number of Native Americans had been reduced to a mere 600,000, and by that time wars of extermination against the Indians were being conducted in earnest. Fifty years later, the population had fallen by another half. Indians were killed defending their land, or they died of hunger and disease taking refuge in inhospitable country. Four thousand Cherokee died in 1834 on a forced march from their homeland in Georgia to reservations in Arkansas and Oklahoma, a trip memorialized as the Trail of Tears. The Sioux were forced off their lands by the discovery of gold and the new push of European immigration. Their reservation was established in 1889, quite recently in history, and they were designated as wards, subjecting them to capricious and humiliating governmental policies. The following year, the U.S. Army, mistaking Sioux ceremonial dances for war dances, moved in to arrest the leaders. A standoff exploded into violence, during which federal troops killed 200 Sioux men, women, and children at the famous Wounded Knee massacre.

Today, about 55 percent of all Native Americans live on or near a reservation (lands set aside for their exclusive use); the rest are in or near urban areas (U.S. Bureau of the Census, 1997; Snipp, 1989). The reservation system has served the Indians poorly. A great many Native Americans now live in conditions of abject poverty, deprivation, and alcoholism, and they suffer massive unemployment (more than 50 percent among males). They are at the lowest rung of the socioeconomic ladder with the highest poverty rate. The first are now last, a painful irony of U.S. history.

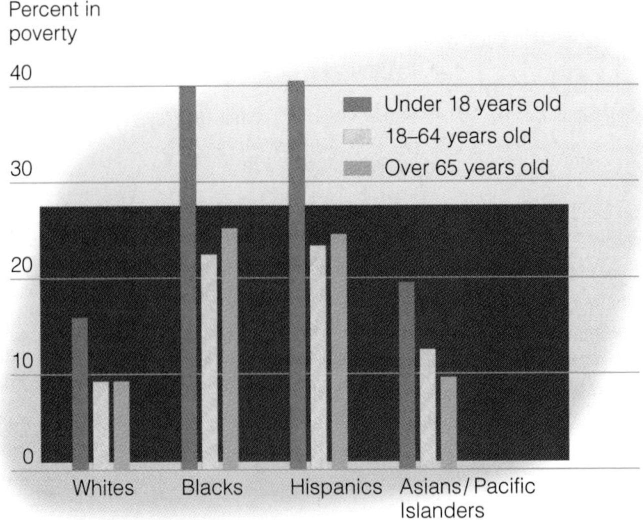

FIGURE 11.3 Race and Poverty (by age)

SOURCE: Dalaker, Joseph, and Mary Naifeh. 1998. *Poverty in the United States: 1997.* Washington, DC: U.S. Department of Commerce, pp. 3-5. Hooper, Linda M., and Claudette E. Bennett. 1998. *The Asian and Pacific Islander Population in the United States: March 1997 (Update)*: Washington, DC: U.S. Department of Commerce.

Many Native Americans have left the reservations to live in urban and suburban areas. Some have sought political power through public office and also by political activism such as the Red Power movement, AIM (American Indian Movement), or the NCAI (National Congress of American Indians). Entrepreneurship among Native Americans has increased somewhat in recent years, as the founding and management of successful casinos in Connecticut, New York, and other states has shown (Nagel, 1996).

African Americans

The development of slavery in the Americas is related to the development of world markets for sugar and tobacco. Slaves were imported from Africa to provide the labor for sugar and tobacco production and to enhance the profits of the slaveowners. It is estimated that somewhere between 20 million and 100 million Africans were transported under appalling conditions to the Americas. Thirty-eight percent went to Brazil; 50 percent to the Caribbean; 6 percent to Dutch, Danish, and Swedish colonies; and only 6 percent to the United States (Genovese, 1972; Jordan, 1969).

Slavery evolved as a *caste system* in which one caste, the slave owners, profited from the labor of another caste, the slaves. Central to the operation of slavery was the principle that human beings could be *chattel* (property). As an economic institution, slavery was based on the belief that Whites were superior to other races, coupled with a belief in a patriarchal social order. The social distinctions maintained between Whites and Blacks were castelike, with rigid categorization and prohibitions, rather than merely classlike, which suggests more pliant social demarcations. Vestiges of this caste system remain in many parts of the United States to this day.

The slave system also involved the domination of men over women. In this combination of patriarchy and White supremacy, White males presided over the property of White women as well as their human "property" of Black men and women. This in turn led to gender stratification among the slaves themselves, which reflected the White slaveowner's assumptions about the relative roles of men and women. Black women performed domestic labor for their masters and their own families. White men further exerted their authority in demanding sexual relations with Black women (Blassingame, 1973; Raboteau, 1978; White, 1985; Davis, 1981). The dominant attitude of Whites toward Blacks was paternalistic; they saw slaves as childlike and incapable of caring for themselves. More recent stereotypes of African Americans as childlike are traceable directly to the system of slavery.

Until recently, our knowledge of slavery has been distorted by its dependence on records and observations made by White male slaveowners. Recent research has been based on the accounts, records, and narratives of the slaves themselves. There is a widespread belief that slaves passively acquiesced to slavery. Scholarship shows this to be false. Instead, the slaves struggled to preserve both their culture and their sense of humanity and to resist, often by open conflict, the dehumanizing effects of a system that defined human beings as mere property. Slaves revolted against the conditions of enslavement in a variety of ways, from passive means such as work slowdowns and feigned illness to more aggressive means such as destruction of property, escapes, and outright armed rebellion. All the way to the early 1970s, it has been generally believed that active rebellions in the slave community were rare. In fact, they were frequent, numbering in the several hundreds (Rose, 1970; Blassingame, 1973). The slave rebellion on board the ship *Amistad*, en route from West Africa, which has only recently been given detailed scholarly coverage (Myers, 1998), occurred even before the ship landed. A movie of the same name was released in 1997.

After slavery was presumably ended by the Civil War (1861–1865) and the Emancipation Proclamation (1863), Black Americans continued to be exploited for their labor. In the South, the system of sharecropping emerged, an exploitative system in which the Black family tilled the fields for White landowners in exchange for a share of the crop. With the onset of World War I and the intensified industrialization of society came the Great Migration of Blacks from the South to the urban North. This massive movement, from 1900 through the 1920s and beyond, had a significant effect on the status of Blacks in society since there was now a greater potential for collective action (Marks, 1989).

In the early part of the century, the formation of Black ghettos had a dual effect, victimizing Black Americans with grim urban conditions while also encouraging the development of Black resources, including volunteer organizations, social movements, political action groups, and artistic and cultural achievements. During this period of the 1920s, Harlem in New York City became an important intellectual and artistic oasis for Black America. The Harlem Renaissance gave the nation great literary figures, such as Langston Hughes, Jesse Fausett, Alain Locke, Arna Bontemps, Zora Neale Hurston, Wallace Thurman, and Nella Larson (Bontemps, 1972; Rampersad, 1986, 1988). At the same time, many of America's greatest musicians, entertainers, and artists came to the fore, such as musicians Duke Ellington, Billie Holiday, Cab Calloway, and Louis Armstrong, and painter Hale Woodruff. The end of the 1920s and the stock market crash of 1929 brought everyone down a peg or two,

Novelist and essayist Dorothy West was the oldest living member of the Harlem Renaissance until her death at age 91 in 1998.

Langston Hughes (second row, third from left) and Dorothy West (bottom row, far right) and associates aboard the Europa, bound for Russia, June 1932.

Whites as well as Blacks, although in the words of Harlem Renaissance writer Langston Hughes, Black Americans at the time "had but a few pegs to fall" (Hughes, 1967).

Latinos

Latino Americans include Chicanos and Chicanas (Mexican Americans), Puerto Ricans, Cubans, and other recent Latin American immigrants to the United States. It also includes Latin Americans who have lived for generations in the United States and are thus not immigrants, but very early settlers. The population of Latino Americans has grown considerably over the past few decades, with Mexican Americans increasing the most. The terms *Hispanic* and *Latino* or *Latina* mask great structural and cultural diversity among the various Hispanic groups. The use of such inclusive terms also tends to ignore important differences in their respective entries into U.S. society: Mexican Americans through military conquest (1846–1848); Puerto Ricans through war with Spain (1898); and Cubans as political refugees fleeing since 1959 from a political regime, that of Fidel Castro, opposed by the U.S. government (Estrada et al., 1981; Bean and Tienda, 1987).

Mexican Americans. Before the Anglo (White) conquest, Mexican colonists had formed settlements and missions throughout the West and Southwest. In 1834, the U.S. government ordered the dismantling of these missions, bringing them under tight governmental control and creating a period known as the Golden Age of the Ranchos. Land then became concentrated into the hands of a few wealthy Mexican ranchers who had been given large land grants by the Mexican government. This economy created a class system within the Chicano community, consisting of the elite ranchers, mission farmers, and government administrators at the top; mestizos (small farmers and ranchers) as the middle class; a third class of skilled workers; and a bottom class of manual laborers, who were mostly Indians (Mirandé, 1985; Maldonado, 1997).

The Anglo conquest of Mexican lands was difficult in areas where the ranch owners were firmly established, but with the Mexican-American War of 1846–1848, Chicanos lost claims to huge land areas that ultimately became Texas, New Mexico, and parts of Colorado, Arizona, Nevada, Utah, and California. It has been argued that the Mexican-American War was the United States' first war of imperialism. The next twenty years witnessed steady growth in the influence and importance of U.S. business interests in Mexico.

White cattle ranchers and sheep ranchers enclosed giant tracts of land, thus cutting off many small ranchers, both Mexican and Anglo. It was at this time that Mexicans, as well as early settlers of Mexican descent, became defined as an inferior "race" that did not deserve social, educational, or political equality. This is an example of the racial formation process (Omi and Winant, 1994). Anglos believed that Mexicans were lazy, corrupt, and cowardly, launching stereotypes that would further oppress Mexicans and support a belief system used to justify the lower status of Mexicans, and to justify Anglo control of the land, which Mexicans were presumed to be incapable of managing (Moore, 1976). U.S. business relations with Mexico from the 1800s to the present have been a prime determinant of the experiences of Mexicans in the United States, including jobs and wages, immigration, residential segregation, and education (Bean and Tienda, 1987; Estrada et al., 1981; Maldonado, 1997).

Immigration policies of the 1920s disproportionately affected Mexicans. Whereas immigration from northern and western European countries was encouraged, immigration from eastern and southern Europe was discouraged, as was immigration from Mexico. Such peoples were considered too culturally dissimilar from "American stock" to be capable of assimilation.

During the twentieth century, advances in agricultural technology changed the organization of labor in the Southwest and West. Irrigation created year-round production of crops. Consequently, there was a new need for cheap labor to work in the fields. Migrant workers from Mexico came to be exploited as a cheap source of labor. Migrant work was characterized by low earnings, poor housing conditions, poor health, and extensive use of child labor. The extensive use of Mexican migratory laborers as field workers, domestic servants, and other kinds of poorly paid work continues, particularly in the Southwest (Amott and Matthaei, 1996).

Puerto Ricans. The island of Puerto Rico was ceded to the United States by Spain in 1899. In 1917, the Jones Act extended U.S. citizenship to Puerto Ricans, although it was not until 1948 that Puerto Ricans were allowed to elect their own governor. In 1952, the United States established the Commonwealth of Puerto Rico, with its own constitution. Following World War II, the first elected governor launched a program known as Operation Bootstrap that was designed to attract large U.S. corporations to the island of Puerto Rico, using tax breaks and other concessions. This program contributed to rapid overall growth in the Puerto Rican economy, although unemployment remained high and wages remained low. Seeking opportunity, unemployed farm workers began migrating to the United States. These migrants were interested in seasonal work, and thus a pattern of temporary migration characterized the early Puerto Rican entrance into the United States (Amott and Matthaei, 1996; Rodriquez, 1989).

Unemployment in Puerto Rico became so severe that the U.S. government even went so far as to attempt to reduce the population by some form of population control. Pharmaceutical companies experimented with Puerto Rican women in developing contraceptive pills, and the U.S. government actually encouraged the sterilization of Puerto Rican women. One source notes that by 1974, more than 37 percent of the women of reproductive age in Puerto Rico had been sterilized (Shapiro, 1985). More than one-third of these women have since indicated that they regret being sterilized since they were not made aware at the time that the procedure was irreversible.

Cubans. Cuban migration to the United States is recent in comparison to the other Hispanic groups. The largest migration has occurred since the revolution led by Castro in 1959; between then and 1980, more than 800,000 Cubans—one-tenth of the entire island population—migrated to the United States. The U.S. government defined this as a political exodus, facilitating the early entrance and acceptance of these migrants. Many of the early migrants had been middle- and upper-class professionals and landowners under the prior dictatorship of Fulencio Batista, but had lost their land during the Castro revolution. While in exile in the United States, some worked to overthrow Castro, often with the support of the federal government; yet, many other Cuban immigrants were of modest means, and like other immigrant groups came seeking freedom from political and social persecution and escape from poverty.

The most recent wave of Cuban immigration came in 1980, when the Cuban government, still under Castro, opened the port of Mariel to anyone who wanted to leave Cuba. In the five months following this pronouncement, 125,000 Cubans came to the United States—more than the combined total for the preceding eight years. The arrival of the Mariels has produced debate and tension, particularly in Florida, a major center of Cuban migration. The Cuban government had previously labeled the Mariels "undesirable"; some had previously been incarcerated in Cuba. They were actually not much different from prior refugees, but because they were so labeled, and also because they were forced to live in primitive camps for long periods after their arrival, they have been unable to achieve much social and economic mobility, in contrast to the fair degree of success enjoyed by the earlier migrants, who were on average more educated and much more settled (Amott and Matthei, 1996; Pedraza, 1996a; Portes and Rumbaut, 1996).

This Cuban restaurant in Calle Ocho reflects racial–ethnic residential segregation.

Asian Americans

Like Hispanic Americans, Asian Americans are from many countries and diverse cultural backgrounds, and cannot be lumped together under the single cultural rubric "Asians," as many are inclined to do. Asian Americans include migrants from China, Japan, the Philippines, Korea, and Vietnam, as well as more recent entrants from Cambodia and Laos.

Chinese. Attracted by the U.S. demand for labor, Chinese Americans began migrating to the United States during the mid-nineteenth century. In the early stages of migration, the Chinese were tolerated since they provided cheap labor. They were initially seen as good, quiet citizens, but racial stereotypes turned hostile when the Chinese came to be seen as competing with White California gold miners for jobs.

During the period 1865–1868, thousands of Chinese laborers worked for the Central Pacific Railroad. They were relegated to the most difficult and dangerous work, worked longer hours than the White laborers, and for a long time were paid considerably less than the White workers. It was Chinese laborers who performed the dangerous job of planting the stick of dynamite into rock, lighting the short fuse, and racing away. There were many accidents; the expression "You don't have a Chinaman's chance" originates in this period.

The Chinese were virtually expelled from railroad work near the turn of the century, and settled in rural areas throughout the western states. As a consequence, anti-Chinese sentiment and prejudice ran high in the West. This ethnic antagonism was largely the result of competition between the White and Chinese laborers for scarce jobs. In 1882, the federal government passed the Chinese Exclusion Act, which banned further immigration of unskilled Chinese laborers. Like African Americans, the Chinese, and Chinese Americans, were legally excluded from intermarriage (Takaki, 1989). The passage of this openly racist act, which was preceded by extensive violence toward the Chinese, drove the Chinese populations from the rural areas into the urban areas of the West. It was during this period that several Chinatowns were established by those who had been forcibly uprooted, and who found strength and comfort within enclaves of Chinese people and culture (Nee, 1973).

Japanese. Japanese immigration to the United States took place mainly between 1890 and 1924, after which passage of the Japanese Immigration Act forbade further immigration. Most of these first-generation immigrants (called Issei) were employed in agriculture or in small Japanese businesses. Many Issei were from farming families and wished to acquire their own land, but in 1913, the Alien Land Law of California stipulated that Japanese aliens could lease land for only three years, and that lands already owned or leased could not be bequeathed to heirs. The second generation of Japanese Americans (born in the United States of Japanese-born parents, or Nisei) became better educated than their parents, lost their Japanese accents, and in general became more "Americanized," that is, culturally assimilated. The third generation (Sansei) became even better educated and assimilated, yet still met with prejudice and discrimination, particularly where they were present in highest concentration, as on the West Coast from Washington to southern California (Glenn, 1986).

The Japanese suffered the maximal indignity of having their loyalty questioned when the federal government, thinking they would side with Japan after the Japanese attack on Pearl Harbor in December of 1941, herded them into concentration camps. By executive order of President Franklin D. Roosevelt, much of the West Coast Japanese American population, many of them loyal second- and third-generation Americans, had their assets frozen and their real estate confiscated by the government. A media campaign immediately followed labeling Japanese Americans "traitors" and "enemy aliens." Virtually all Japanese Americans in the United States had been removed from their homes by August of 1942, some to stay in relocation camps until as late as 1946. Relocation destroyed numerous Japanese families and ruined them financially (Glenn, 1986; Kitano, 1976; Takaki, 1989).

In 1986, the U.S. Supreme Court allowed Japanese Americans the right to file suit for monetary reparations. In 1987, legislation was passed awarding $20,000 to each person who had been relocated, and also offering an official apology from the U.S. government. One is motivated to contemplate how far this paltry sum and late apology could go in righting what many have argued was the "greatest mistake" the United States made during World War II.

Filipinos. The Philippine Islands in the Pacific Ocean fell under U.S. rule in 1899 as a result of the Spanish-American War, and for a while Filipinos could enter the United States freely. By 1934, the islands became a commonwealth of the United States, and immigration quotas were imposed upon Filipinos. More than 200,000 Filipinos immigrated to the United States during the recent period between 1966 and 1980, settling in major urban centers on the West and East Coast. More than two-thirds of those arriving were professional workers; their high average levels of education and skill have eased their assimilation. By 1985, there were more than one million Filipinos in the United States. Within the next thirty years, demographers project that they will become the largest group of Asian Americans in the United States, including Chinese Americans and Japanese Americans (Winnick, 1990).

Koreans. Many Koreans entered the United States in the late 1960s after amendments to the immigration laws in 1965 raised the ceilings on immigration from the Eastern hemisphere. The largest concentration of Koreans is in Los Angeles. As much as half of the adult Korean and Korean American population is college educated, an exceptionally high proportion. Many of the emigrants were successful professionals in Korea; upon arrival in the United States, they have been forced to take on menial jobs, thus experiencing the rigors of downward social mobility. This is especially true of those who migrated to the East Coast; however, nearly one in eight Koreans in the United States today owns a business; many own small greengrocer businesses. Many of these in turn are in predominantly African American communities, a source of ongoing conflict between African American residents and Korean grocers. This has fanned negative feeling and prejudice on both sides—among Koreans against African Americans, and among African Americans against Koreans (Chen, 1991).

Vietnamese. Among the more recent groups of Asians to enter the United States have been the South Vietnamese, who began arriving following the fall of South Vietnam to the communist North Vietnamese in 1975. These immigrants, many of them refugees fleeing for their lives, numbered about 650,000 in the United States in 1975. About a third settled in California. Many faced prejudice and hostility, resulting in part from the same perception that has dogged many immigrant groups before them: that they were competitors for scarce jobs. A second wave of Vietnamese immigrants arrived after China attacked Vietnam in 1978. As many as 725,000 arrived in America, only to face discrimination in a variety of locations. Tensions became especially heated when the Vietnamese became a substantial competitive presence in the fishing and shrimping industries in the Gulf of Mexico on the Texas shore, but many communities have welcomed them, and at last count, 95 percent of all Vietnamese heads of households were employed full time (Winnick, 1990; Kim, 1993).

Middle Easterners

Since the mid-1970s, immigrants from the Middle East have arrived in the United States. They have arrived from countries such as Syria, Lebanon, Egypt, and Iran. Contrary to popular belief, they speak no single language and follow no singular religion, and are thus ethnically quite diverse; some are Catholic, some are Coptic Christian, and many are Muslim. Many are from working-class backgrounds, but many were professionals—teachers, engineers, scientists, and other professionals, in their homeland. Some of these have been able to secure employment in their original occupations, but many have undergone the downward mobility—often experienced by present-day immigrant professionals—necessitated by taking lower-status jobs. Like immigrant populations before them, they are forming their own ethnic enclaves in cities and suburbs of this country as they purse the often elusive American dream (Ansari, 1991).

White Ethnic Groups

The story of White ethnic groups in the United States begins in the colonial era. White Anglo-Saxon Protestants (WASPs), who were originally immigrants from England, and also to some extent Scotland and Wales, settled in the "New World" (what is now North America) and were the first ethnic group to come into contact on a large scale with those already here, namely, Native American Indians.

WASPs came to dominate the newly emerging society earlier than any other White ethnic group. It was the WASPs who regarded the later immigrants from Germany and France as "foreigners" with "odd" languages, accents, and customs. For example, Benjamin Franklin thought that the Germans were too clannish and too insistent on keeping their language. He had strong opinions about other groups as well, such as the Dutch, and his opinions no doubt reflected (and influenced) the mood of the times. For example, he sought to bar the Germans, the Dutch, and other nationalities from interaction with the "old stock" Americans, who were largely English. He instituted programs to get the "foreign" groups to assimilate more quickly and become committed to American values, culture, and the political system. The tension between the old stock and the foreigners continued through the Civil War era around 1860, when the national origins of U.S. immigrants began to change (Handlin, 1951). Of all racial and ethnic groups in the United States during that time and since, only WASPs do not think of themselves as a nationality. The WASPs came to think of themselves as the "original" Americans despite the prior presence of Native American Indians, who were in turn described by WASPs as "savages."

The original WASP immigrants were skilled workers imbued with the Protestant Ethic—the desire to work hard and achieve much in life, particularly in material wealth. Max Weber (1905) argued that this work ethic has its roots in early Protestantism in England and western Europe, whereby the ability to work long hours was thought of as a religious sign that one had been chosen for later transport to heaven after death.

As immigrants from northern and western Europe and, eastern and southern Europe began to arrive, particularly during the mid- to late nineteenth century, WASPs began to direct prejudice and discrimination against many of these newer groups. Although the 1880s saw the Statue of Liberty first welcome those from foreign shores, ironically, at the same time WASPs founded the Daughters of the American Revolution (the DAR, founded in 1890), the Society of Mayflower Descendants (1894), and the Social Register (1887)—a formal listing of people and families of old-line inherited wealth and upper-class status in traditional White America (Baltzell, 1964, 1979).

The dominance of the WASPs in U.S. society has declined somewhat since 1960, when John Fitzgerald Kennedy, a Catholic, became the first non-WASP ever to be elected president. Since the late 1960s, there has been increased racial and ethnic awareness for other groups, especially African, Hispanic, Asian, and Native Americans. As a group, women, long discriminated against by the male WASP establishment, also began to assert social and political power, eroding the power of WASPs in the United States. Still, much of that dominance remains, as is evident in the popular use of the terms *race* and *ethnicity* to describe virtually everyone but them.

There were two waves of migration of White ethnic groups in the mid- and late nineteenth century. The first stretched from about 1850 through 1880, and included northern and western Europeans: English, Irish, Germans, French, and Scandinavians. The second wave of immigration was from 1890 to 1914 and included eastern and southern European populations: Italians, Greeks, Polish, Russians, and other eastern Europeans. The immigration of Jews to the United States extended for well over a century, but the majority of Jewish immigrants came to the United States during the period from 1880 to 1920.

Jewish immigrants are questioned at Ellis Island.

The Irish arrived in large numbers in the mid-nineteenth century as a consequence of food shortages and massive starvation in Ireland. During the latter half of the nineteenth century and the early twentieth century, the Irish in the United States were abused, attacked, and viciously stereotyped. It is instructive to remember that the Irish, particularly on the East Coast and especially in Boston, underwent a period of ethnic oppression of very great magnitude. A frequently seen sign posted in Boston saloons of the day proclaimed "no dogs or Irish allowed." German immigrants were similarly often stereotyped, as were the French and the Scandinavians. It is easy to forget that virtually all immigrant groups come through times of oppression and prejudice, although these periods were considerably longer for some groups than for others. As a rule, where the population density of an ethnic group was greatest, so too was the amount of prejudice, negative stereotyping, and discrimination to which that group was subjected.

More than 40 percent of the world's Jewish population lives in the United States, making it the largest community of Jews in the world. Most of the Jews in the United States arrived between 1880 and World War I, originating from the eastern European countries of Russia, Poland, Lithuania, Hungary, and Rumania. Jews from Germany arrived in two phases, the first just prior to the arrival of those from eastern Europe, and the second as a result of Hitler's ascension to power in the late 1930s in Germany. Because many German Jews were professionals who also spoke English, they assimilated more rapidly than those from the eastern European countries. Jews from both parts of Europe underwent lengthy periods of anti-Jewish prejudice, antisemitism, and discrimination, particularly on Manhattan's Lower East Side. Significant antisemitism still exists in the United States (Simpson and Yinger, 1985; Essed, 1991).

In 1924, the National Origins Quota Act was passed, one of the most discriminatory legal actions ever taken by the United States in the field of immigration. By this act, the first real establishment of *ethnic quotas* in the United States, immigrants were permitted to enter the United States only in proportion to their numbers already in the United States. Thus, ethnics who were already here in relatively high proportions (English, Germans, French, Scandinavians, and others, mostly western and northern Europeans) were allowed to immigrate in greater numbers to the country than were those from southern and eastern Europe, such as Italians, Poles, Greeks, and other eastern Europeans. Hence, the act discriminated against southern and eastern Europeans in favor of western and northern Europeans. It has been noted that the European groups who were discriminated against by the National Origins Quota Act tended to be those with darker skins.

Immigrants during this period were subject to literacy tests and even IQ tests given in English (Kamin, 1974). The act barred from immigration anyone who was classified as a convict, lunatic, "idiot," or "imbecile." On New York City's Ellis Island, non-English-speaking immigrants, many of them Jews, were given the 1916 version of the Stanford-Binet IQ test in English. Obviously, non-English-speaking persons taking this test were unlikely to score high. On the basis of this grossly biased test, governmental psychologist H. H. Goddard classified fully 83 percent of the Jews, 80 percent of the Hungarians, and 79 percent of the Italians as "feeble-minded"! It did not dawn on Goddard or the U.S. government that the IQ test, in English, probably did not measure something called "intelligence," as intended, but instead simply measured the immigrant's mastery of the English language (Kamin, 1974; Gould, 1981; Taylor, 1992, 1980).

Patterns of Racial and Ethnic Relations

Intergroup contact between racial and ethnic groups in the United States has evolved along a variety of identifiable lines involving specific actions on the part of both the dominant White group and the ethnic or racial minority. The character of the contacts has been both negative and positive, obvious and subtle, tragic and helpful. The features and forms of racial and ethnic relations that have received the most attention are minority culture, assimilation, pluralism, colonialism, segregation, and the interaction of class and race.

The Role of Minority Culture

Minority groups have certainly been oppressed in our society, but it would be erroneous to conclude that we can account for the condition of minorities in society simply by analyzing the victim status of Blacks, Latinos, Native Americans, or Asians. Assessing these groups as victims may inspire sympathy for their plight, but it also tends to perpetuate negative views of minorities by defining them as pitiable and pathological. By tending to ignore the strong cultural elements and the occupational, educational, economic, and cultural achievements of these groups, the **victimization perspective** takes attention away from the positive actions groups have taken in response to oppression, and how oppression has contributed to the development of group and cultural strengths.

The victimization perspective is evident in the long-standing claim that racial–ethnic groups think less of themselves because they have internalized the views of dominant groups. This idea was supported by early research (which we examined in Chapter 4) by psychologists Kenneth and Mamie Clark (Clark and Clark, 1947), who set out to determine if Black children in segregated schools had negative self-esteem. They based their findings on the now-famous technique of seeing whether Black children prefer White dolls over Black dolls when asked to express a preference. They claimed that the Black children showed a distinct preference for White dolls over the Black dolls, and from this evidence they concluded that Black children show self-rejection, which was taken to indicate negative self-images.

This study was cited in the famous 1954 *Brown v. Board of Education* Supreme Court decision, wherein the court ordered that schools must be desegregated "with all deliberate speed." The Court argued that segregated schools were "inherently unequal" and thus overturned an earlier Supreme Court decision, *Plessy v. Ferguson* (1896, 163 U.S. 537), which had established the infamous "separate but equal" doctrine for permitting segregated schools.

Although segregated schools may indeed be harmful to Black children, more recent research shows that negative self-esteem is not one of the harms (Steele and Aronson, 1995; Banks, 1976; Rosenberg and Simmons, 1971). In fact, contrary to popular belief, Black and Hispanic children in either segregated or integrated schools can actually have higher average self-esteem than Whites. The reason is that Blacks and Hispanics tend to base their self-evaluations on their own reference groups, that is, their own cultures, and not Whites and White society, as the original Clark and Clark research had erroneously assumed.

Research also shows that having high occupational and educational aspirations tend to lower self-esteem for Whites, whereas it raises self-esteem for Blacks (Harris and Stokes, 1978). Research also shows that Black and White girls have lower self-esteem than boys of their own race, but the self-esteem of Black girls is higher than that of White girls (Sadker and Sadker, 1994; Simmons et al., 1978). Whites similarly tend to see Hispanics as victims, but it has been demonstrated that when Latinos evaluate themselves with respect to Latino culture and Latino achievement, self-esteem is higher than that of comparable groups of Whites (Mirandé, 1985).

No doubt, some members of racial–ethnic groups do internalize negative definitions of themselves (Steele, 1992, 1996; Steele and Aronson, 1995), just as members of the dominant group sometimes develop low self-esteem, but focusing solely on the process of victimization leads you to ignore the many positive contributions that diverse groups have made to U.S. society, history, and culture.

THINKING SOCIOLOGICALLY

Write down your racial–ethnic background and list one thing that people from this background have positively contributed to U.S. society or culture. Also list one experience (current or historical) in which people from your group have been victimized by society. Discuss how these two things illustrate the fact that racial–ethnic groups both have been *victimized* and have made *positive contributions* to this society. Share your comments with others; what does this reveal to you about the connections between different groups of people and their experience as racial–ethnic groups in the United States?

An ESL (English as a second language) class teaches young Vietnamese students. How would this affect their assimilation *in an English-speaking country?*

Assimilation and Pluralism

A common question raised about African Americans, Hispanics, and Native Americans is "White immigrants made it, why can't they?" The question reflects the belief that with enough hard work and loyalty to the dominant White culture of the country, any minority can make it. This is the often-heard argument that African Americans, Hispanics, and Native Americans need only pull themselves up "by their own bootstraps."

The assimilation perspective dominated sociological thinking a generation ago and is still prominent in U.S. thought (Rumbaut, 1996a; Glazer, 1970). The assimilationist believes that to overcome adversity and oppression, the minority person need only imitate the dominant White culture as much as possible. In this sense, minorities must assimilate "into" White culture and White society. An early theory of Robert Park at the University of Chicago supported this view, holding that there were four stages in the attainment of complete equality for a minority: *accommodation* (the dominant culture allows the minority immigrant into its society); *acculturation* (the minority begins to learn the language and other elements of the dominant culture, including manners, customs, music, and so on); *assimilation* (the minority enters the economy and becomes upwardly mobile through hard work, education, and jobs); and finally, *amalgamation* (the minority group intermarries with the dominant group and produces offspring of both cultures, ethnicities, or races). Many Asian American groups have followed this pattern, and have thus been called the "model minority," but this label ignores the fact that Asians are still subject to prejudice, discrimination, and racism (Takaki, 1989; Lee, 1996; Woo, 1998).

There are four problems with the assimilation model. First, it fails to consider the time that it takes certain groups to assimilate. Those from rural backgrounds (Native Americans, Hispanics, African Americans, White Appalachians, some White ethnic immigrants) typically take much longer to assimilate than those from urban backgrounds. Second, the history of Black and White arrivals was very different, with lasting consequences. Whites came voluntarily; Blacks arrived in chains. Whites sought relatives in the new world; Blacks were sold and separated from relatives. For these and other reasons, the histories of African American and White experience as newcomers can hardly be compared, and their assimilation is hardly likely to follow the same course. Research over the last twenty-five years has thus corrected many other myths about African American history, such as the myth that all African culture was destroyed by slavery, that there is no such thing as Black culture, or that Blacks migrated voluntarily to the United States. Third, although they did indeed face prejudice and discrimination when they arrived in America, many White ethnic groups entered at a time when the economy was growing rapidly and their labor was in high demand; they were thus able to attain education and job skills. In contrast, by the time Blacks migrated to northern industrial areas from the rural South, Whites had already established firm control over labor and used this control to exclude Blacks from better-paying jobs and higher education. Fourth, assimilation is more difficult for people of color since skin color is an especially *salient* characteristic, ascribed and relatively unchangeable, as we have noted earlier in this chapter. White ethnics can change their names, but ethnics of color cannot easily change their skin color.

The assimilation model raises the question of whether it is possible for a society to maintain **cultural pluralism**, defined as different groups in society maintaining their distinctive cultures while also co-existing peacefully with the dominant group.

Some groups have explicitly practiced cultural pluralism; witness the Little Italies in U.S. cities. The Amish people of Lancaster County in Pennsylvania and also in north central Ohio, who travel by horse and buggy, use no electricity, and run their own schools, banks, and stores, constitute a good example of a relatively complete degree of cultural pluralism.

Cultural pluralism is the opposite of assimilation. Those groups that have managed to have cultural pluralism are

White ethnic groups, many of whom have also assimilated into different aspects of society. Italians in Little Italies, for example, maintain some residential segregation, hold on to cultural traditions, and in some cases, speak their own language, but in other ways they are completely assimilated in American life. Groups like the Amish are the exception. Black Muslims also maintain some degree of pluralism, practicing their own cultural habits and trying to maintain some degree of economic self-sufficiency. Some racial–ethnic groups have advocated the formation of separate societies within the larger society, and at least in terms of cultural practices, some have been successful in doing so, but whether pluralism would work as a model for the whole society, particularly given the broad range of diverse racial–ethnic groups within the United States, is questionable, even if people were to think it was desirable.

Colonialism

In the late 1960s and early 1970s, political activists and scholars advocated a power-based approach to the analysis of race and ethnic relations, replacing concerns with assimilation and pluralism. In his **domestic colonialism** model, sociologist Robert Blauner (1969, 1972) described the condition of U.S. Blacks as an internal colony. He argued that the situation and status of Blacks in the United States is like that of other colonized minorities around the world, except that colonization took the form of forced transport to the country colonizing them. Blauner identified four elements to domestic colonialism. First, there is forced and involuntary entry. Second, the affairs of the colonized group (the minority group) are administered and determined by the colonizers (the dominant group). Blauner argued that in this respect, ghettos were actually devices to achieve social control over minorities; ghettoes were thus colonies in themselves. Third, racism and its accompanying stereotypes are used to explain and justify the colonizer's domination over the minority group. Finally, the colonizers do not allow the minority to express its culture and values. The group history of the minority is rewritten from the colonists' point of view. The minority group, on the other hand, is likely to react by developing cultural nationalism—a form of pluralism—and by asserting its own culture, goals, ways of life, music, and identity. Cultural nationalism may lead to revolution. Blauner, who devised his theory in 1968 and 1969, used his analysis to explain the Black urban riots and revolutionary upheavals of the late 1960s and early 1970s.

Blauner's theory is applicable not only to Blacks but to other minorities as well, including Chicanos, Puerto Ricans, and to some degree other recent immigrant ethnic and racial groups such as Koreans and Filipinos. It may also apply to Native Americans, who are migrating in increasing numbers from rural to urban environments. The colonialism model places race and ethnic relations squarely into an analysis of group conflict and power relations, reflecting the attention to power—who has it and who does not—that characterizes the more radical perspectives that emerged from the Black power movement.

segregation

Segregation and the Urban Underclass

Segregation refers to the spatial and social separation of racial and ethnic groups. Minorities, who are often believed by the dominant group to be inferior, are compelled to live separately under inferior conditions and are given inferior educations, jobs, and protections under the law. Although *de*segregation has been mandated by law (thus eliminating *de jure segregation*, or legal segregation), *de facto segregation*—segregation "in fact"—still exists, particularly in housing and education.

THINKING SOCIOLOGICALLY

Using your community or school as an example, what evidence do you see of *racial segregation*? How might a sociologist explain what you have seen?

Segregation has contributed to the creation of an **urban underclass**, a grouping of people, largely minority and poor, who live at the absolute bottom of the socioeconomic ladder in urban areas (Massey and Denton, 1993). Though the proportion of middle-class Blacks has increased in the last two decades, so has the proportion of Blacks, Hispanics, and Native Americans who are in poverty. In 1997, 11 percent of Whites, 27 percent of Hispanics, 26.5 percent of Blacks, 14 percent of Asians, and fully 35 percent of Native Americans were below the official poverty level (see Figure 11.2). In a seminal study, Wilson (1987) attributes the causes of the urban underclass to economic and social structural deficits in society, while rejecting the "culture of poverty" explanation, which attributes the condition of minorities to their own presumed cultural deficiencies (Lewis, 1960, 1966; Moynihan, 1965). The problems of the inner city, such as joblessness, crime, teen pregnancy, welfare dependency, and AIDS, are seen to arise from social class inequalities, that is, inequalities in the structure of society, and that these inequalities have dire behavior consequences at the individual level—in the form of drug abuse, violence, and lack of education (Wilson, 1987; Sampson, 1987). Wilson argues that the civil rights agendas need to be enlarged, and that the major problem of the underclass, joblessness, needs to be addressed by fundamental changes in the economic institution.

Recently, Wilson (1996) noted a most dismaying finding: that most adults in many inner-city neighborhoods are not working in any typical week. In such environments, Wilson argues, people have little chance to gain the educational and social skills that would make them attractive to employers. Wilson advocates government-financed jobs and universal healthcare as important solutions to such conditions.

The Relative Importance of Class and Race

Which is more important in determining one's chances for overall success: social class or race-ethnicity? This is the "class versus caste" controversy. The question has been hotly debated in sociology. One view, put forward by Wilson in his controversial book *The Declining Significance of Race* (1978), contends that class has become more impor-

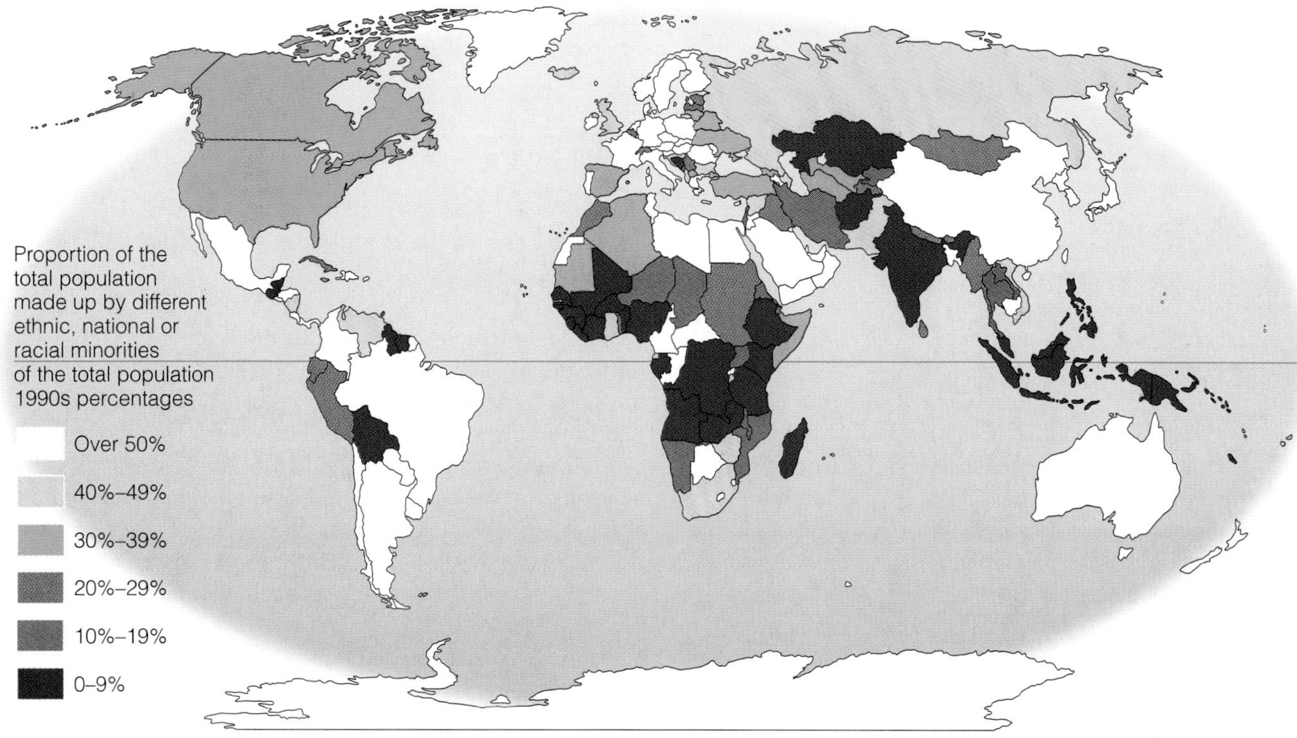

MAP 11.2 Viewing Society in Global Perspective: Ethnic Conflict Around the World

SOURCE: *The State of War and Peace Atlas* by Dan Smith. Copyright © 1997 by Dan Smith, text. Maps copyright © 1997 by Myriad Editions Ltd. Used by permission of Penguin Putnam Inc.

tant than race in determining Black access to privilege and power. In support of his argument, he cites the simultaneous expansion of the Black middle class and the Black urban underclass. Wilson does not argue that race is unimportant, but that the importance of social class is increasing as the importance of race is declining, even though race still remains important. In a more recent work, Wilson (1996) argues that both class and race combine to oppress not only many urban Blacks, but Whites and Hispanics, as well.

Among the critics of Wilson's argument, Willie (1979), Farley and Allen (1987), and Feagin (1991) argue that race is still critically important, which can be seen when one compares Blacks of a given social class to Whites of the same class. Blacks at every level of education earn less than Whites at the same education level. Blacks in blue-collar jobs earn only 83 percent as much as Whites in the same jobs with the same educations and class backgrounds. Blacks are less likely to graduate from college than Whites of comparable family socioeconomic background. Willie concludes that 90 percent of the income gap between Blacks and Whites is the result of race, that is because of low pay for Blacks with comparable levels of education and experience (Willie, 1979: 59; Jaynes and Williams, 1989).

Finally, as Blalock (1989, 1982b) has noted, in criticizing both the race versus class approaches, there are actually three contributing influences to be measured: the effects of class, the effects of race, and the effects of the statistical interaction of race and class, that is, the intersecting effects of race and class acting together. A statistical interaction effect is a *double jeopardy* effect of the type discussed in previous chapters. For example, a minority man may earn a lower salary than a White man because of his race; his salary may be lower still because he is of lower-class status; and finally, race and class acting in combination may produce even lower income than one would expect on the basis of race or class alone. This is what is meant by the statistical interaction effect.

Finally, the effects of race and class can interact with the effect of gender, as noted earlier, yielding a *triple* jeopardy effect. Being Black (or Latino or Native American), female, and underclass as well, subjects one to the maximal interaction effects of this triple jeopardy.

Attaining Racial and Ethnic Equality: The Challenge

The variety of racial and ethnic groups in the United States brings much to the society in cultural richness and diverse patterns of group and interpersonal interaction. But the inequities among these diverse groups also pose challenges for the nation in addressing questions of social justice. Those from other nations often see the United States as a racially divided nation, even though the problems of racial and ethnic conflict are not unique to Americans. Throughout the world, conflicts stemming from racial and ethnic differences are frequently the basis for economic inequities, cultural conflicts, and even war. Although the specific forms of racial and ethnic relations vary from society to society, many societies face the challenge of wondering how racial and ethnic inequality can be lessened.

Racial–Ethnic Conflict: A World Problem

Across the globe, many racial and ethnic groups, most often those of lower status, have sought self-determination—the right to choose their own leaders, politics, and lifestyle. This has led to severe conflicts between these minority groups and the dominant group and often between minority groups themselves.

When the former Yugoslavia was dissolved in 1991, people in Croatia, including those who were Muslims, desired self-determination and asserted their independence. Those who were from Serbia, the dominant group, resisted these efforts and declared war against both Croatians and Muslims. In what has become a complicated series of conflicts, the Croatians battled not only against the Serbians, but also against the Muslims, who were a Croatian minority group. Shortly after these conflicts began, the Serbians engaged in what has become notorious as the "ethnic cleansing" of Croatian and Muslim people—the systematic and genocidal oppression and elimination of an ethnic or religious group by means of murder, rape, harassment, forcible removal, and the confiscation of property. Many Croatians and Muslims have been summarily executed in their homes as their property was taken over. Finally, in early 1999, in a continued ethnic cleansing, Serbians forcibly ousted ethnic Albanians from Kosovo province in Serbia, causing them to seek refuge in neighboring Macedonia and Albania. It is likely that many Kosovar Albanians were murdered along the way.

In the African country of Rwanada, severe conflict broke out in 1994 between the Tutsi tribespersons, the minority, and the Hutu tribespersons, the dominant group. Ethnic cleansing in the form of open genocide took place as the Hutu majority engaged in a systematic and brutal program to exterminate the Tutsi population. Thousands of men, women, and children were shot and hacked to death by machetes. Entire families were wiped out. More than a million Tutsi refugees fled to neighboring countries such as Uganda and Zaire.

As television and other mass media continue to connect entire countries, as multinational corporations expand globally, and as the Internet widens, the economic and cultural interdependence among countries will continue to increase. Such is the nature of globalization. As a result, racial and ethnic conflicts in one country will affect the likelihood and nature of conflicts in other countries. In an earlier time, the absence of television and other mass communications media tended to isolate ethnic conflicts from discovery in other countries, and there was less international interdependence of such conflicts. As a consequence, racial and ethnic strife in the United States affects the nature of conflict and revolution in other countries, in the same way that "ethnic cleansing" in other countries affects the nature and form of ethnic and race relations, and people's perceptions and understandings of race and ethnicity, in the United States. Movements for racial justice in the United States have also become part of a worldwide framework for addressing racial and ethnic conflict in other parts of the world.

Civil Rights

The history of racial and ethnic relations in the United States shows several strategies to achieve greater equality. Political mobilization, legal reform, and social policy have been the basis for much social change in race relations, but there are continuing questions about how best to achieve a greater degree of racial justice in this society. The major force behind most progressive social change in race relations was the civil rights movement. Marked by the strong moral and political commitment and courage of participants, the civil rights movement is probably the single most important source for change in race relations in the twentieth century. The civil rights movement was based on the passive resistance philosophy of Dr. Martin Luther King, Jr., learned from the philosophy of *satyagraha* ("soul firmness and force") of Indian leader Mahatma (meaning, "leader") Ghandi. This philosophy encouraged resistance to segregation through nonviolent techniques such as sit-ins, marches, and appealing to human conscience in calls for brotherhood, justice, and equality.

The major civil rights movement in the United States intensified shortly after the 1954 *Brown v. Board of Education* decision when, in 1955, seamstress and NAACP secretary Rosa Parks, an African American, in Montgomery, Alabama, by prior arrangement with the NAACP, bravely refused to relinquish her seat in the "White only" section on a segregated bus when asked to do so by the White bus driver. At the time, the majority of Montgomery's bus riders were African American, and the action of Rosa Parks initiated the now famous Montgomery Bus Boycott, led by the then-young Reverend Martin Luther King, Jr. The boycott was successful in desegregating the buses, got more African American bus drivers hired, and catapulted Martin Luther King to the forefront of the civil rights movement. Impetus was given to the civil rights movement and the boycott by the brutal murder in 1954 of Emmett Till, a Black teenager from Chicago who was killed in Mississippi for whistling at a White woman (see Chapter 3).

The civil rights movement produced many episodes of both tragedy and heroism. In a landmark 1957 decision, President Dwight D. Eisenhower called out the national guard, after initial delay, to assist the entrance of nine Black students into Little Rock Central High in Little Rock, Arkansas. Sit-ins followed in which White and Black students perched at lunch counters until the Black students were served; the "freedom rides" forged on despite the murders of Viola Liuzzo, a White Detroit housewife, Andrew Goodman and Michael Schwerner, two White students, and James Chaney, a Black student. The murders of Goodman, Schwerner, and Chaney have been memorialized in the movies.

The civil rights movement culminated in the passage of the Civil Rights Bill in 1964, a federal law prohibiting discrimination on the basis of race, color, national origin, religion, or sex. This bill has laid the legal framework for antidiscrimination policies, as well as for further acts such as the Voting Rights Act of 1965 and the Fair Housing Act of 1968, all passed during the presidential administration of Lyndon Johnson. These laws all rest on the assumption that all people should be treated equally, with no regard for the racial, ethnic, and religious differences that may exist among them.

Black Power, Red Power: Radical Social Change

As the civil rights movement developed throughout the late 1950s and 1960s, a more radical philosophy of change also developed, as more militant leaders grew increasingly disenchanted with the limits of the civil rights agenda, which was perceived as moving too slowly. The militant Black power movement, taking its name from the book *Black Power* (published in 1967 by political activist Stokely Carmichael, later Kwame Touré, and Columbia University political science professor Charles V. Hamilton) had a more radical critique of race relations in the United States and saw inequality as stemming not just from moral failures, but from the institutional power that Whites had over Black Americans (Carmichael and Hamilton, 1967).

Malcolm X, before breaking with the Black Muslims (the Black Nation of Islam in America) and his religious mentor, Elijah Muhammad, and prior to his assassination in 1965, advocated a form of pluralism demanding separate business establishments, banks, churches, and schools for Black Americans. He echoed an earlier movement of the 1920s led by Marcus Garvey's back-to-Africa movement, the Universal Negro Improvement Association (UNIA).

The Black power movement of the late 1960s rejected assimilation and demanded instead self-determination and self-regulation of Black communities. Militant groups such as the Black Panther Party advocated fighting oppression with armed revolution. The U.S. government acted quickly, imprisoning members of the Black Panther Party and members of similar militant revolutionary groups and, in some cases, killing them outright (Brown, 1992).

The Black power movement also influenced the development of other groups who were influenced by the analysis of institutional racism that the Black power movement developed, as well as by the assertion of strong group identity that this movement encouraged. Groups like *La Raza Unida*, a Chicano organization, encouraged "brown power," promoting solidarity and the use of Chicano power to achieve racial justice. Likewise, the American Indian Movement (AIM) used some of the same strategies and tactics that the Black power movement had encouraged, as have Puerto Rican, Asian American, and other racial protest groups. Elements of Black power strategy were also borrowed by the developing women's movement, and Black feminism developed upon the realization that women, including women of color, shared in the oppressed status fostered by institutions that promoted racism (Collins, 1990). Overall, the Black power movement dramatically altered the nature of political struggle and race and ethnic relations in the United States.

Many of the leaders of the Black power movement were eliminated, either through assassination or imprisonment or other means, reducing its influence. Its significance, however, remains. Minister Louis Farrakhan, a protege of Malcolm X in the early 1960s, broke from the Black Muslim's Nation of Islam organization to found his own Black Muslim organization. The Black power movement also pushed those in the civil rights movement to a more radical analysis of race relations in the United States so that before his death even Dr. Martin Luther King, Jr., the major spokesperson for civil rights, articulated the need for radical economic change to address continuing racial inequality in the United States (Branch, 1988). The Black power movement and the movements it spawned changed the nation's consciousness about race and also forced academic scholars to develop a deeper understanding of how racism works in society and how fundamental it is to U.S. institutions. As a result, there were major revisions in sociological theories about race relations, some of which have been reviewed in this chapter.

The civil rights movement, which began in the mid-1950s and continued through the 1980s, resulted in significant social, economic, and political gains for Blacks, although gains in the 1960s have been somewhat offset by losses in the late 1970s and 1980s, giving an overall trend of "three steps forward yet two back" for advances of Blacks in U.S. society (Jaynes and Williams, 1989; Farley and Allen, 1987; Farley, 1984). The Black power movement of the late 1960s and early 1970s also resulted in certain educational gains even despite White opposition (Jaynes and Williams, 1989; Brown, 1992).

Race-Specific Versus Color-Blind Programs for Change

A continuing question from the dialogue between a civil rights strategy and more radical strategies for change is the debate between race-specific versus color-blind programs for change. *Color-blind policies* are those advocating that all groups be treated alike, with no barriers to opportunity posed by race, gender, or other group differences; equal opportunity is the key concept in color-blind policies. The maxim "equal pay for equal work" is a color- and gender-blind policy, meant to minimize the pay gap in the workplace that is now heavily influenced by race and gender.

Race-specific policies are those that recognize the unique status of racial groups because of the long history of discrimination and the continuing influence of institutional racism. Those advocating such policies argue that color-blind strategies will not work because Whites and other racial–ethnic groups do not start from the same position. Even given equal opportunities, continuing disadvantage produces unequal results. The tension between these two strategies for change is a major source for many of the political debates surrounding race relations now.

Affirmative action, a heavily contested program for change, is a race-specific policy for reducing job and educational inequality that has also had some success. Affirmative action means two things. First, affirmative action means

affirmative action

Demonstrators march in favor of affirmative action in education.

recruiting minorities from a wide base in order to ensure consideration of groups that have been traditionally overlooked, but at the same time not using rigid quotas based on race or ethnicity. Second, affirmative action means using admissions slots (in education) or set-aside contracts or jobs (in job hiring) to assure minority representation. The principle objection, heard from both sides of the racial line, is that either interpretation of affirmative action programs is, in effect, use of a quota.

Legal opinion on affirmative action is inconclusive. The Supreme Court decided in 1978 that race can be used as a criterion for admission to undergraduate or graduate and professional schools, or for job recruitment, as long as race is combined with other criteria, and as long as rigid racial quotas are not used. But in 1996, the University of California Board of Regents decided to eliminate race, but not social class, as a basis for admitting students to its campuses. Also in 1996, in *Hopwood v. Texas*, a panel of judges of the Fifth Circuit Court of Appeals in Texas ruled as unconstitutional the affirmative action admissions policy of the University of Texas, Austin, law school. To date, a higher panel of judges has upheld that decision (Higginbotham, 1998). Other legal decisions continue to challenge the continuation of affirmative action and related strategies. Educational scholarships specifically designated for minority students, for example, have been challenged in recent court decisions, although many argue that eliminating such programs would represent a backward move in reducing racial inequality in the United States.

Some policy analysts take a different approach altogether, saying that neither color-blind nor race-specific policies per se will reduce racial inequality and that what is needed is fundamental economic change. We saw in the work of William Julius Wilson (1996), one of the most influential sociologists in the nation, the argument that only massive, government-financed programs to create jobs for the unemployed, whatever their race, will eliminate the persistent problems of economic and racial inequality. Nonetheless, recent data have shown that Blacks admitted to selective colleges and universities under affirmative action programs reveal high rates of social and economic success after graduation. For example, the percent of graduating Blacks who went on to graduate school and law school was *higher* than the percent of whites from the same schools who did so (Bower and Bok, 1998).

No doubt, these diverse approaches to addressing racial injustice will continue to inform national debates on policies like affirmative action, equal employment, and race-based aid. Political mobilization of different groups will also continue to be an important mode for improving race and ethnic relations. Mass political actions such as the 1996 Million Man March in Washington, D.C., serve to pinpoint interest and effort toward the reduction of racial and ethnic inequalities in U.S. society.

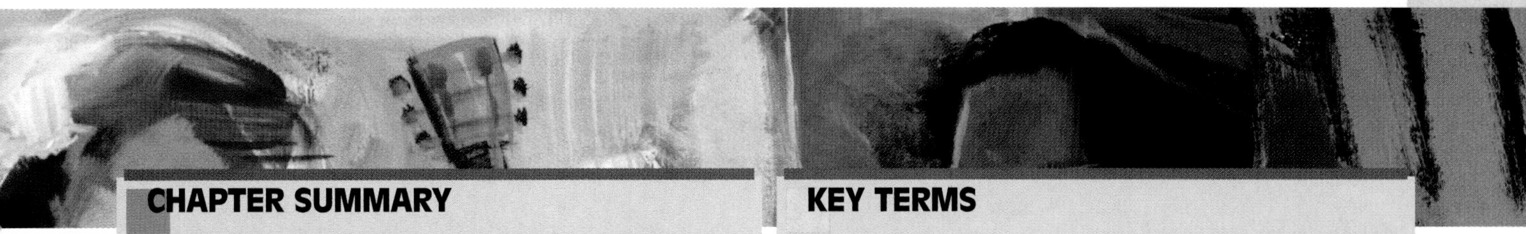

CHAPTER SUMMARY

- In virtually every walk of life, race matters. A *race* is a social construction based loosely on physical criteria, whereas an *ethnic group* is a culturally distinct group. A group is *minority* or *dominant* not on the basis of their numbers in a society, but on the basis of which group occupies lower average social status.
- *Stereotyping,* and *stereotype interchangeability,* reinforce racial and ethnic prejudices and thus cause them to persist in the maintenance of inequality in society. Racial and gender stereotypes are similar in dynamics in society, and both racial and gender stereotypes receive ongoing support in the media. Stereotypes serve to justify and make legitimate the oppression of groups based on race, ethnicity, and gender. Stereotypes such as "lazy" support attributions made to the minority that cast blame on the minority in question, thus in effect, removing some of the blame from the social structure.
- *Prejudice* is an attitude involving usually negative prejudgment on the basis of race or ethnicity. *Discrimination* is overt behavior involving unequal treatment. *Racism* involves both attitude and behavior. Racism can take either obvious (*dominative*) or subtle (*aversive*) forms. *Institutional racism* is unequal treatment, carrying with it notions of cultural inferiority of a minority, which has become ingrained into the economic, political, and educational institutions of society.
- Different theoretical positions have been developed in sociology to explain prejudice and racism and to explain different aspects of race and ethnic relations, including the social construction of race and ethnicity, intergroup conflict, and *gendered racism*.
- Historical experiences show that different groups have unique histories, although they are bound together by some similarities in the prejudice and discrimination they have experienced. Although each group's experience is unique, they are commonly related by a history of prejudice and discrimination.
- Sociologists analyze different forms of racial and ethnic relations, including the role of minority culture (and its role in the formation of self-esteem), *assimilation, domestic colonialism, segregation* and the *urban underclass,* the relative importance of social class versus race (caste), and the question of class–race interaction.
- Racial and ethnic conflict worldwide is frequently the source of war, genocide, and ongoing struggles for social justice. The civil rights movement is the basis for social change based on an equal rights philosophy. More radical activists have developed an analysis of *institutional racism* and the power relationships on which racism rests. Some programs for change rely on *color-blind strategies,* whereas others (such as *affirmative action* programs) propose *race-specific remedies* to address racial inequality. Some argue that only deep-rooted economic change will alleviate persistent racial stratification.

KEY TERMS

affirmative action
assimilation
authoritarian personality
aversive racism
contact theory
cultural pluralism
discrimination
domestic colonialism
dominative racism
ethnic group
ethnocentrism
gendered racism
institutional racism
minority group
prejudice
race
racial and ethnic stratification
racial formation
racism
residential segregation
salience principle
scapegoat theory
segregation
stereotype
stereotype interchangeablility
urban underclass
victimization perspective

THE INTERNET: A Tool for the Sociological Imagination

Resources on the Internet:

Virtual Society: The Wadsworth Sociology Resource Center
http://sociology.wadsworth.com

Visit this site to find additional learning tools, including interactive quizzes, links to related web sites, and an easy link to *InfoTrac College Edition.*

Racial Legacies and Learning
http://www.pbs.org/adultlearning/als/race/4.0/index.htm

This site, sponsored by the Public Broadcasting Service (PBS) and the American Association of Colleges and Universities (AAC&U), is directed at fostering dialogue between college campuses and communities in an effort to improve race relations. It includes important resources on cultivating diversity and numerous educational projects designed to improve race relations.

Center for Research on Women, University of Memphis
http://www.people.memphis.edu/~intprogs/crow/

The research center is dedicated to conducting, disseminating, and promoting research on women of color and southern women in the United States. The web site includes news and research about education, work, poverty, health, and other issues affecting women of color and southern women.

National Multicultural Institute

http://www.nmci.org/

This site includes resource materials, conference announcements, and employment possibilities in the area of multicultural work.

Sociology and Social Policy: Internet Exercises

Various methods have been suggested for reducing racial inequality in the United States. Some of these remedies are based on *color-blind* approaches, such as civil rights laws that prohibit employment practices that exclude people because of race. Other suggested remedies are *race specific,* like *affirmative action* programs that target racial groups for inclusion in employment or college admissions. Proponents of color-blind strategies base their policies on the argument that all groups should be treated alike, regardless of their historical, social, and cultural differences. Critics of this approach say that it cannot address the problems created by past discrimination and the unique experiences of different racial–ethnic groups. What are the arguments for and against each of these positions with regard to solving the problem of inequality among racial groups?

Internet Search Keywords:

racism
race relations
affirmative action

Web sites:

http://www.siop.org/
A comprehensive study on affirmative action by SIOP (Society for Industrial and Organizational Psychology)'s Committee on Ethnic and Minority Affairs.

http://www.essential.org/ussa/foundati/rr.html
The web site by the students of Color Strategy and Policy Department (SOCSPD) provides fact sheets on the status of student and faculty of color in higher education; affirmative action and minority scholarship information (available soon).

http://www.affirmativeaction.org/
Homepage of the American Association for Affirmative Action.

http://www.udayton.edu/~race/
The University of Dayton School of Law publishes statuses, cases, excerpts of law review articles, and annotated bibliographies related to racism and race in American law.

http://www.aadap.org/
The Americans against Discrimination and Preferences provides a list of the most recent articles concerning racial discrimination and their abstracts.

InfoTrac College Edition: Search Word Summary

affirmative action
institutional racism
racial formation
segregation

In order to learn more about these central topics in sociology, you can conduct an electronic search using InfoTrac College Edition. To aid in your search and to gain useful tips, see the Student Guide to InfoTrac College Edition on the Virtual Society web site:
http://sociology.wadsworth.com

INTERACTIONS—A SOCIOLOGY CD-ROM: CONCEPTS FOR THIS CHAPTER

Go to the Wadsworth Sociology CD-ROM for further study on the concepts in this chapter. The CD-ROM also includes quizzes and additional activities to expand your learning experience.

SUGGESTED READINGS

Brown, Elaine. 1992. *A Taste of Power: A Black Woman's Story*. New York: Pantheon Books.

This is the story of an African American woman who attained the leadership of the Black Panther party. In this engaging work, she discusses her liaison with Huey Newton, who led the party for a number of years, and who was in exile in Cuba while Brown led the party in his stead. Gender relations within the Black Panther Party are discussed in detail.

Daniels, Jessie. 1997. *White Lies: Race, Class, Gender and Sexuality in White Supremacist Discourse*. New York: Routledge.

By analyzing documents of the Ku Klux Klan and other contemporary white supremacist organizations, Daniels shows how racism, sexism, class, and sexuality connect in the ideology of White supremacy.

Massey, Douglas S., and Nancy A. Denton. 1993. *American Apartheid: Segregation and the Making of the Urban Underclass*. Cambridge: Harvard University Press.

This is an influential analysis of data on the urban underclass, including analysis of many ethnic groups and gender differences. Massey and Denton clearly show the multiple dimensions of institutionalized inequality in U.S. society.

Omi, Michael, and Howard Winant. 1994. *Racial Formation in the United States*, 2nd ed. New York: Routledge.

This important work advances racial formation theory, which emphasizes the social constructionist and historical processes by which a group comes to be defined as a "race," and deemphasizes the role of physical or biological traits. The book covers Latinos, American Indians, Asians, African Americans, and other such "racially formed" groups.

Pedraza, Silvia, and Rubén Rumbaut, eds. 1996. *Origins and Destinies: Immigration, Race, and Ethnicity in America*. Belmont, CA: Wadsworth.

This anthology explores diverse experiences of immigration in the United States. By examining different waves of immigration, from early White settlers to contemporary immigration, the book shows the significance of immigration in the U.S. institutions.

Portes, Alejandro, and Rubén G. Rumbaut. 1996. *Immigrant America*, 2nd ed. Berkeley, CA: University of California Press.

This is a very detailed account of diverse immigrant groups in the United States, with rich data demonstrating clearly both the similarities and the differences among immigrant groups.

Romero, Mary, Pierrette Hondagneu-Sotelo, and Vilma Ortiz. 1997. *Challenging Fronteras: Structuring Latina and Latino Lives in the United States*. New York: Routledge.

By reviewing the latest research on Latinas and Latinos, this book shows that Latinos are not a unidimensional group. The collection provides both historical background and contemporary analyses to explore the many facets of life for Latinos in the United States.

Root, Maria P.P. 1996. *The Multiracial Experience: Racial Borders as the New Frontier*. Thousand Oaks, CA: Sage Publications.

This anthology explores various issues confronting multiracial groups and shows how this growing category of people challenges existing ideas in the sociology of race relations.

Snipp, Matthew. 1989. *American Indians: The First of This Land*. New York: Russell Sage Foundation.

This book offers a definitive discussion of how American Indian ancestry is defined and how definitions vary by tribal unit. The role of the U.S. government's Bureau of Indian Affairs (BIA) in this process is discussed.

Waters, Mary C. 1990. *Ethnic Options: Choosing Identities in America*. Berkeley: University of California Press.

This book discusses how people will indicate their race or ethnicity for the U.S. census, and the advantages and disadvantages of the multiracial category.

Yoon, In-Jin. 1997. *On My Own: Korean Businesses and Race Relations in America*. Chicago: University of Chicago Press.

Based on extensive fieldwork among Korean immigrants in Chicago and Los Angeles, this book explores the social problems facing Korean immigrants and examines the tensions between them and African Americans in urban areas.

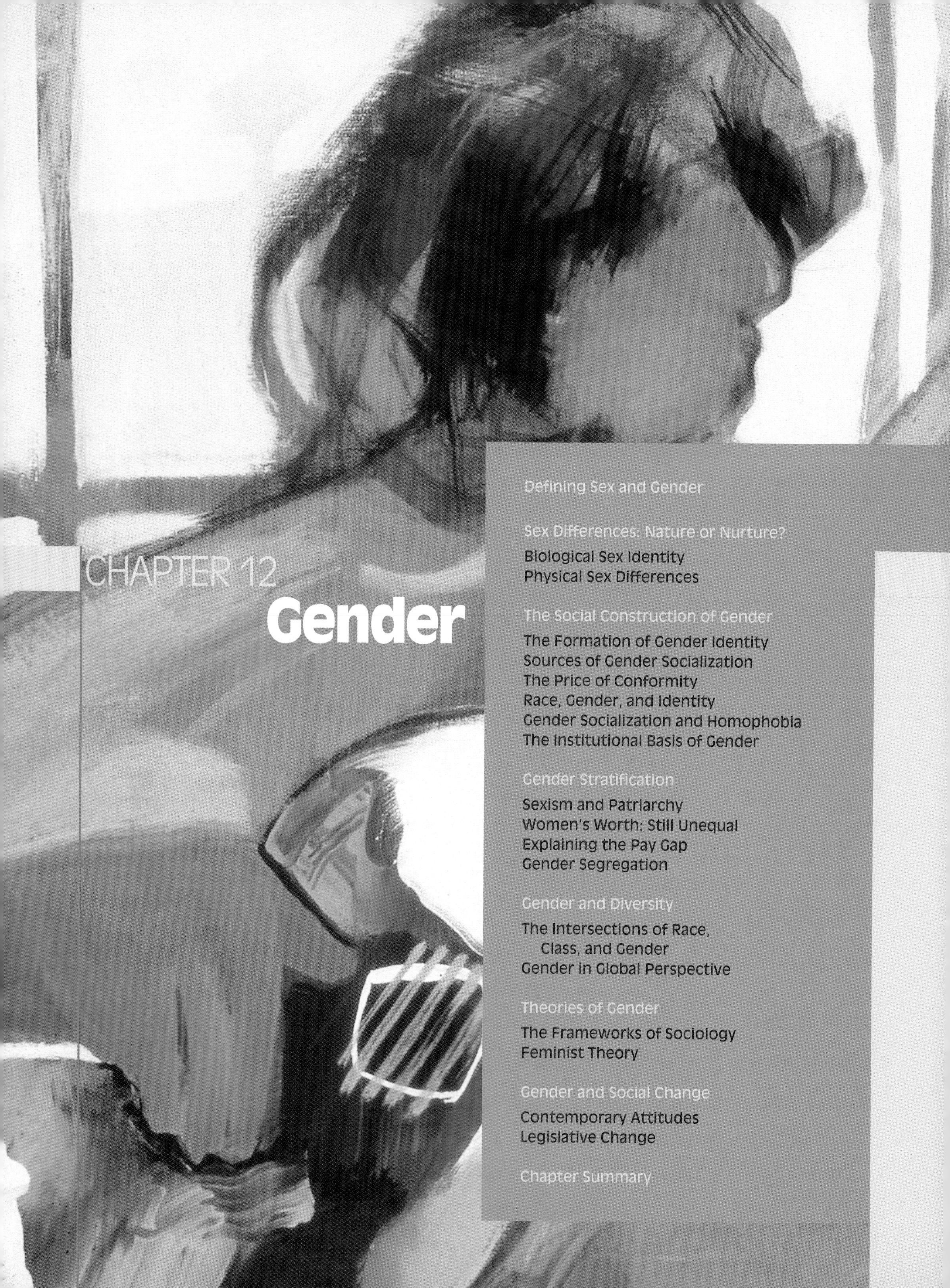

CHAPTER 12 Gender

IMAGINE suddenly becoming a member of the other sex. What would you have to change? First, you would probably change your appearance—clothing, hairstyle, and any adornments you wear. You would likely change your name since a name often identifies a person's gender. You would also have to change some of your interpersonal behavior. Contrary to popular belief, men talk more than women and are louder, more assertive, and more controlling. Women are more likely to laugh, express hesitance, and be polite (Tannen, 1990; Aries, 1987; Kemper, 1984). Gender differences also appear in nonverbal communication. Women use less personal space, touch less in impersonal settings (but are touched more), and smile more, even when they are not necessarily happy. Generally, women are also more attuned to nonverbal cues than men (Hall, 1984; Henley, 1977; Mayo and Henley, 1981; Basow, 1992), so if you became a woman, you might have to become more attentive. Finally, you might have to change many of your attitudes since men and women differ significantly on many, if not most, social and political issues (see Figure 12.1).

If you are a woman and became a man, perhaps the change would be worth it. You would probably see your income go up (especially if you became a White man). You would have more power in virtually every social setting. You would be far more likely to head a major corporation, run your own business, or be elected to a political office—again, assuming that you are White. Would it be worth it? As a man, you would be far more likely to die a violent death, and would probably not live as long. For young African American men (age 25–34) homicide is the leading cause of death; Hispanic and Native American men are far more likely to die by homicide than are women of any race. Men are also ten times more likely to be imprisoned than are women (U.S. Department of Health and Human Services, 1996; Mauer, 1997; Federal Bureau of Investigation, 1996).

If you are a man who becomes a woman, your income would likely drop significantly. More than thirty years after passage of the Equal Pay Act in 1963, men still earn 26 percent more than women, counting only those working year-round and full time (U.S. Bureau of the Census, 1998a). You would likely become resentful of a number of things since poll data indicate that women are more resentful than men about things like the amount of money available to live on, the amount of help they get from their mates around the house, how childcare is shared, and how they look. Women also report being more fearful on the streets than men. On the other hand, women are more satisfied than men with their roles as parents and with their friendships outside marriage; women are also more tolerant than men on a range of social and political attitudes (Roper Organization, 1995a, 1990).

For both women and men, there are benefits, costs, and consequences stemming from the social definitions associated with gender. As you imagined this experiment, you may have had difficulty trying to picture the essential change in your *biological* identity—but is this the most significant part of being a man or woman? As we will see in this chapter, nature determines whether you are male or female, but it is society that gives significance to this distinction. Sociologists see *gender* as a social concept; who we become as men and women is largely shaped by cultural and social expectations.

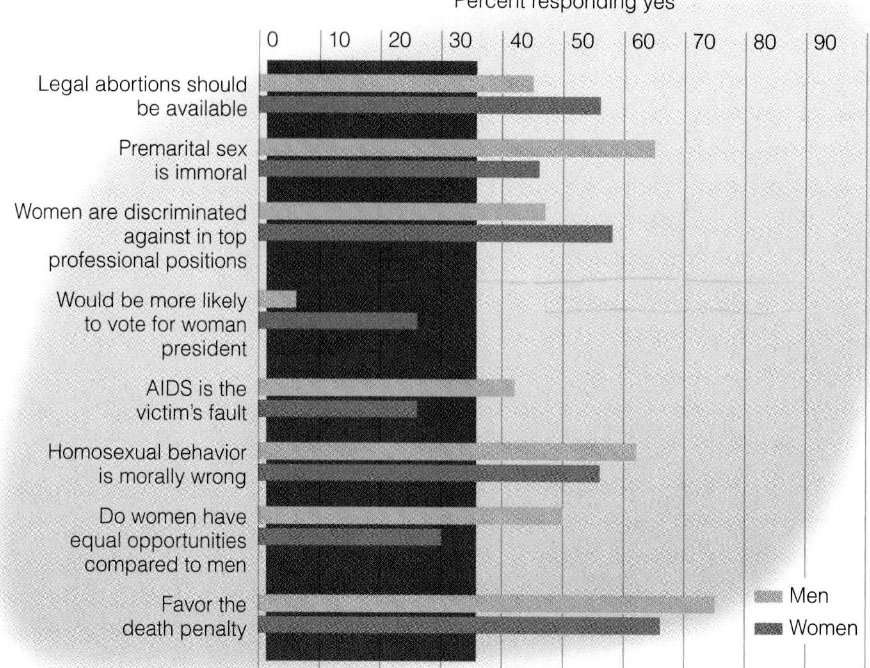

FIGURE 12.1 The Gender Gap in Attitudes

DATA: From the Roper Organization. 1995. *The 1995 Virginia Slims Opinion Poll.* Storrs, CT: Roper Organization; Roper Organization. 1990. *The 1995 Virginia Slims Opinion Poll.* Storrs, CT: Roper Organization; Gallup, George, Jr., and Frank Newport. 1991. "Large Majority Continue to Back AIDS Testing." *The Gallup Poll Monthly* (May): 28.

Defining Sex and Gender

Sociologists use the terms *sex* and *gender* to distinguish biological sex identity from learned gender roles. **Sex** refers to biological identity, male or female. For sociologists, the more significant concept is **gender**—the socially learned expectations and behaviors associated with members of each sex. This distinction emphasizes that behavior associated with gender is culturally learned. One is born male or female, but becoming a man or a woman is the result of social and cultural expectations that pattern men's and women's behavior.

From the moment of birth, gender expectations influence how boys and girls are treated. In fact, now that we can identify the sex of a child in the womb, gender expectations may begin before birth. Parents and grandparents may select pink clothes and dolls for baby girls, sports clothing and brighter colors for boys. Indeed, they have little choice to do otherwise since the clothing styles available on the markets are highly gender-stereotyped. Parents and others continue to treat children in stereotypical ways throughout their childhood. Girls may be expected to cuddle and be sweet, whereas boys are handled more roughly and given greater independence. Researchers have even found that young women respond more quickly to crying infants when they think the infant is a girl, indicating that girls tend to be more protected by others than boys (Stern and Karraker, 1989).

Parents, however, are not the only influence on children's development. Children actually have more gender-stereotyped perceptions of infants than their parents, and gender stereotyping increases as children grow into teenagers; thus, even as they develop their own gender identity, children influence the identities of their peers and the younger children around them (Vogel et al., 1991; Wasserman and Lewis, 1985).

The cultural basis of gender is especially apparent when we look at other cultures. Across different cultures, the gender roles associated with masculinity and femininity vary considerably. In Western industrialized societies, people tend to think of masculinity and femininity in dichotomous terms, with men and women decisively different, even defined as opposites. The view from other cultures challenges this assumption. The Navajo Indians, for example, offer interesting examples of alternative gender roles. Historically, the *berdaches* in Navajo society were anatomically normal men who were defined as a third gender considered to fall between male and female. Berdaches married other men, who were not considered berdaches. Those they married were defined as ordinary men; moreover, neither the berdaches nor the men they married were considered homosexuals, as they would be considered in many contemporary Western cultures (Nanda, 1990; Lorber, 1994).

We do not need to look at striking departures from familiar gender categories to see how differently gender can be constructed in different cultures. In the People's Republic of China, we can see a contrast of gender roles within a single society. The Chinese Marriage Law of 1950 formally defines marriage as a relationship between equal companions who share responsibility for childcare and the family; this is in sharp contrast to age-old roles of Chinese tradition (Evans, 1995). One would not expect to see marriage roles change overnight in China or anywhere else, but different cultural constructions of gender can co-exist in a single culture for many years. In contemporary China, as in many other societies, traditional roles persist side by side with modern roles.

Evidence of changing gender roles for women and men can be seen in many places. Only a few years ago, it would have been unimaginable for most people to think of women playing ice hockey, much less winning the Olympics! Likewise, many men have become more involved in child care, thereby changing their gender roles, as well.

There can also be substantial differences in the construction of gender across social classes or subcultures within a single culture. Within the United States, as we will see, there is considerable variation in the experiences of gender among different racial and ethnic groups (Baca Zinn et al., 1997; Baca Zinn and Dill, 1996; Demos and Segal, 1994; Chow et al., 1996). Looking at gender across cultures quickly reveals the social and cultural dimensions of something often popularly defined as biologically fixed (Brettell and Sargent, 1993).

Sex Differences: Nature or Nurture?

Despite the known power of social expectations, the belief persists that differences between men and women are biologically determined. Biology is, however, only one component in the difference between men and women. The important question in sociology is not whether biology or culture is more important in forming men and women, but how biology and culture interact to produce a person's gender identity. What is the interplay of nature and nurture?

Biological determinism refers to explanations that attribute complex social phenomena to physical characteristics. The argument that men are more aggressive because of hormonal differences (in particular, the presence of testosterone) is a biologically determinist argument. Although people popularly believe that testosterone causes aggressive behavior in men, studies find only a modest correlation between aggressive behavior and testosterone levels. Furthermore, changes in testosterone level do not predict changes in men's aggression (such as by "chemical castration," the administration of drugs that eliminate the production or circulation of testosterone). What's more, there are minimal differences in the levels of sex hormones between girls and boys in early childhood, yet researchers find considerable differences in aggression exhibited by boys and girls from an early age (Fausto-Sterling, 1992).

Biological Sex Identity

A person's sex identity is established at the moment of conception when the father's sperm provides either an X or a Y chromosome to the egg at fertilization. The mother contributes an X chromosome to the embryo. The combination of two X chromosomes makes a female. An X and a Y make a male. Under normal conditions, chemical events directed by the genes on the sex-linked chromosomes lead to the formation of male or female genitalia. **Fetal sexual differentiation** is the prenatal process by which biological differences between the sexes are established. Sometimes this process is compromised and biological sex identity is unclear, providing fascinating cases for social scientific study.

Hermaphroditism is a condition produced when irregularities in chromosome formation or fetal differentiation produce persons with mixed biological sex characteristics. In the most common form of hermaphroditism, the infant has either ovaries or testes, but the genitals are visually ambiguous or mixed. An example would be a child born with female chromosomes but an enlarged clitoris, making the child appear to be male. In other cases, a child may be a chromosomal male, but with an incomplete penis and no urinary canal. In other cases of hermaphroditism, physicians typically advise sex reassignment, including reconstruction of the genitals and hormonal treatment.

Case studies of such treatments reveal the extraordinary influence of social factors in shaping the person's identity. Parents of children who are given a sex reassignment are therefore advised not only to allow genital reconstruction, but also to give the child a new name, a different hairstyle, and new clothes—all intended to provide the child with the social signals judged appropriate to the new gender identity. One physician who has worked on such cases gives the directive to parents that they "need to go home and do their job as child rearers with it very clear whether it's a boy or a girl" (Kessler, 1990: 9). Indeed, in cases where adults choose to undergo a sex change operation, they are usually required to live for one year as if they were the sex they are becoming, so as to ensure that they can handle the social expectations associated with being the other gender.

Sometimes sex reassignments occur not from biological abnormalities but from accidental mutilation. In one well-known case involving one of a pair of twins, a biologically normal male, age seven months, had his penis burned off during a routine circumcision by a mismanaged electric current. The boy was recreated as a girl, including surgical and hormonal treatment. As the child grew up, "she" sometimes imitated "male" activities—trying to urinate standing up and mimicking the father's shaving, but sometimes the child was quite feminine, too. The child, however, had difficulty adjusting to her identity and at age fourteen chose to live as a man, undergoing extensive surgery to partially restore his penis and having a mastectomy. At age twenty-five, he married a woman and adopted her children.

Some have concluded that his case reveals that one's sense of sex identity is innate, but there is another side to this argument. When the child was growing up as a girl, her peers teased her mercilessly and refused to play with her; she also spent much time having her genitals scrutinized by doctors. By the time the child was a teenager, she was miserable, contemplated suicide, and was then told of what had happened earlier in life (Diamond and Sigmundson, 1997). Although some conclude that this proves the biological basis of sex identity, it also reveals the strong influence of social factors (such as peer ridicule) in managing one's gender identity. Thus, others conclude that the case is evidence for the strong interplay of culture and biology.

Consider the example of *transgendered* people. Transgendered people are those who deviate from the binary (that is, either/or) system of gender, including transsexuals, cross-dressers, and others; they are those who do not fit within the normative expectations of gender. Research on transgendered individuals shows that they experience enormous pressure to fit within the normative expectations. When they were young, for example, they would hide their cross-dressing. Those who change their sex as adults report enormous pressure particularly during their transition period, since others expect them to be one sex or the other. Most find that, whatever their biological sex, they are forced—from fear of rejection and the desire for self-preservation—to manage an identity that would fall into one category or another (Gagné and Tewksbury, 1998). Consider the case of Billy Tipton, a jazz musician who lived his entire life with everyone, including his adopted children, thinking he was a man. Not until the day he died at age seventy-four (in 1989) was it revealed that "he" was a female (Smith, 1998; Middlebrook, 1998). One cannot help marveling at the effort it would have taken to maintain this social identity and to

Jazz musician Billy Tipton, from the biography Suits Me *by Diane Wood Middlebrook. Tipton lived his life with most people thinking he was a man, when he was, in fact, female.*

imagine how his life would have been different, including what his career would have been, had it been known that "he" was a woman. Indeed, Tipton began his identity as a man in the 1930s—a time when an unwritten code in the jazz world kept women from being hired.

From a sociological perspective, biology alone does not determine gender identity. People must adjust to the expectations of others and the social understanding of what it means to be a man or a woman. A person may remain genetically one sex, socially the other—or perhaps something in between. In other words, there is not a fixed relationship between biological and social outcomes (Money, 1985). If you only see men and women as biologically "natural" states, you miss some of the fascinating ways that gender is formed in society.

Physical Sex Differences

Physical differences between the sexes do, of course, exist from birth. In addition to differences in anatomy, at birth boys tend to be slightly longer and weigh more than girls. As adults, men tend to have a lower resting heart rate, higher blood pressure, higher muscle mass and muscle density, and more efficient recovery from muscular activity (Stockard and Johnson, 1992). These physical differences contribute to the tendency for men to be physically stronger than women; however, the public is becoming accustomed to displays of women's athleticism and expects great performances from both men and women in world-class events like the Olympics. Women can achieve high degrees of muscle mass and muscle density through bodybuilding and can actually win over men in activities that require endurance and speed (Lips, 1993). As examples, women hold the record in speed for swimming the English Channel, and when Susan Butcher won the Iditarod dogsled race in Alaska, she dramatically proved that women can win one of the most grueling contests in the world.

Arguments based on biological determinism assume that differences between women and men are "natural" and, presumably, resistant to change. Like biological explanations of race differences, biological explanations of inequality between women and men tend to flourish during periods of rapid social change. In such times, biological explanations protect the *status quo* (existing social arrangements) by making it appear that the status of women or people of other races is "natural" and therefore should remain as it is. If social differences between women and men were biologically determined, we would also not find any variation in gender relations across cultures, but the extensive differences are well documented. We could not be who we are without biology. We would not be who we are without society and culture.

The Social Construction of Gender

As we saw in Chapter 4, socialization is the process by which social expectations are taught and learned. Through **gender socialization,** men and women learn the expectations associated with their sex. The rules of gender extend to all aspects of society and daily life. Gender socialization affects the self-concepts of women and men, their social and political attitudes, their perceptions about other people, and their feelings about relationships with others. Although not everyone is perfectly socialized to conform to gender expectations, socialization is a powerful force directing the behavior of men and women in gender-typical ways. For example, men who believe that the role of women is to act as wife and mother are not likely to share housework; women who believe they are incomplete unless they are attached to men are not likely to be very independent.

Even people who set out to challenge traditional expectations often find themselves yielding to the powerful influence of socialization. Women who consciously reject traditional women's roles may still find themselves inclined to act as hostess or secretary in a group setting. Similarly, men may decide to accept equal responsibility for housework, yet simply fail to notice when the refrigerator is empty or the child needs a bath—household needs they have been trained to let someone else notice (DeVault, 1991). These expectations are so pervasive that it is also difficult to change them on an individual basis. If you doubt this, try buying clothing or toys for a young child without purchasing something that is gender-typed, or talk to parents who have tried to raise their children without conforming to gender stereotypes and see what they report about the influence of such things as children's peers and the media.

THINKING SOCIOLOGICALLY

Visit a local toy store and try to purchase a toy for a young child that is not gender typed. What could you buy? What could you not buy? What does your experiment teach you about *gender role socialization*? (If you take a child with you, note what toys he or she wants and does not want; what does this tell you about how effective gender socialization is?)

The Formation of Gender Identity

One result of gender socialization is the formation of **gender identity,** defined as one's definition of oneself as a woman or man. Gender identity is basic to our self-concept and shapes our expectations for ourselves, our abilities and interests, and how we interact with others.

Two traits commonly associated with gender identity are *competition* and *dominance.* Generally, men develop a more competitive orientation than women, particularly when competition means winning over an opponent. Social psychological experiments show that women are generally more concerned with the interpersonal aspects of competitive situations, such as whether they are being socially acknowledged and whether the group approves of their behavior. Women also become less competitive when the person with whom they are competing is a man (Basow, 1992; Bunker et al., 1984; Gill, 1988).

Studies of dominance patterns also show significant differences by gender, especially in same-sex groups. In all-male groups, for example, boys use more commands and threats than in mixed-sex groups. Boys are less likely than girls to comply with others, and they are more physically aggressive. Girls in all-girl groups tend to give polite directives, express agreement with one another, and take turns speaking. Among adolescents and adults, women are more facilitating in interaction with others; that is, they encourage others to be included, whereas men tend to behave in ways that inhibit others (Maccoby, 1990; Stockard and Johnson, 1992).

Some cautions are in order when interpreting social psychological studies of gender. First, the findings of researchers are highly dependent on how they define the behavior being observed. Most studies, for example, find that boys are more "aggressive" than girls, with aggression defined rather narrowly to mean physical aggressiveness. Research conclusively finds that men and boys are more physically aggressive; however, some studies conclude that women and girls are more verbally aggressive (Stockard and Johnson, 1992). Second, the conventions for reporting research tend to amplify the appearance of gender stereotypes

gender socialization

since researchers tend to publish their results only when gender differences are found. As a result, research reports may overemphasize gender differences while denying attention to the many similarities between women and men. Third, no one completely conforms to the expectations passed on through socialization. Our widely different experiences and the creative ways we respond to social expectations are part of our uniqueness as individuals.

Sources of Gender Socialization

As with other forms of socialization, there are different *agents of gender socialization:* family, play, schooling, religious training, the mass media, to name a few. Gender socialization is reinforced whenever gender-linked behaviors receive approval or disapproval from these multiple influences.

Parents and Gender Socialization. Parents are one of the most important sources of gender socialization. Parents may discourage children from playing with toys that are identified with the other sex, especially when boys play with toys meant for girls. Parents tend to give boys and girls different chores around the house, with boys more likely to be given maintenance chores (like painting or mowing the lawn) and girls given more domestic chores (cooking, laundry, dishwashing). Of course, families improvise widely on these behaviors. For example, researchers have found that gender-linked assignment of chores is less marked in families where mothers are employed outside the home and in families from higher socioeconomic levels (Serbin et al., 1990). Fathers and grandfathers are also more likely than mothers and grandmothers to expect gender-appropriate behavior from their children (Antill, 1987; Basow, 1992; Lamb et al., 1986).

BOX 12.1 DOING SOCIOLOGICAL RESEARCH

Men, Gender, and Physical Disability

SOCIAL norms about masculine gender identity place much emphasis on the body. Bodies demonstrate the masculine characteristics of strength, toughness, and ability, with different bodies carrying different values in society, which in turn differentiates the status and prestige of different men. As Thomas Gerschick and Adam Miller argue in their research on men and physical disabilities, men with physical disabilities have to cope with the social pressures of masculine roles that call for strength and independence and those that associate physical disabilities with weakness and independence. Gerschick and Miller began their research by wanting to know how men with physical disabilities define their gender identity in the context of contradictory social expectations—that is, one set of expectations about masculinity as strong, aggressive, self-reliant, and independent, and another set of expectations that equate physical disability with weakness, passivity, dependence, and pity.

They based their research on in-depth interviews with a very small sample of men ($N = 10$), eight of them White, two of them African American. This is clearly not a representative sample, but it provides some initial insight into how men who are disabled construct their gender identity. Gerschick and Miller identified three dominant frameworks that the men use to form their gender identity: reformulation, reliance, and rejection.

Those who use a reformulation approach do not contest dominant standards for masculinity but recognize that they cannot meet these ideals. They try to distance themselves from the dominant definition. They find it difficult to do so, however. For example, although the men in their sample had the economic means to pay for personal assistance, many disabled people have to rely on public welfare for this support. This makes it difficult to contend with the definition of masculinity as independence.

Those who use a reliance framework are more concerned with meeting traditional definitions of masculinity. They are more conflicted about their masculine identity and feel conflicted about their ability to be perceived as masculine by others. They worry more about how they are perceived by others since they have internalized expectations that their masculinity is reflected in physical strength, athleticism, independence, and sexual prowess. They feel constantly reminded by others that they are unable to live up to this ideal.

The third framework Herschick and Miller identify is rejection of the dominant conception of masculinity and development of new standards of manhood. These men are more conscious of the fact that societal conceptions of masculinity, not themselves, are problematic. Of all those interviewed, they are the best able to construct alternative identities, emphasizing themselves as people who have independent standards of self-worth.

These different patterns of adaptation are ideal types—that is, the men use all these modes of adaptation but tend to emphasize one more than the others. Gerschick and Miller's research illuminates how dominant conceptions of masculinity subordinate some groups of men and shape people's ability to cope with situations that they may encounter. They also suggest that the disability rights movement has helped people with disabilities be more self-reliant in constructing gender identities. You might ask yourself how you think your gender identity is linked to your physical status.

SOURCE: Gerschick, Thomas J., and Adam Stephen Miller. 1994. "Gender Identities at the Crossroads of Masculinity and Physical Disability." *Masculinities* 2 (Spring): 34–55.

Gender socialization patterns in families also vary within different racial–ethnic groups and in different generations. Latinos, as an example, have generally been thought to be more traditional in their gender roles although this varies by generation and by the experiences of family members in the labor force. Gender expectations for Mexican and Puerto Rican women are more traditional in the older generations; the orientation of Mexican Puerto Rican women toward traditional roles is presumably one reason they are less likely to participate in the labor force. Younger Mexican and Puerto Rican women are less traditional in their attitudes and are more likely to be employed (Ortiz and Cooney, 1984).

Particular family experiences also influence patterns of gender socialization. For example, Mexican married couples who migrate to the United States tend to adopt more egalitarian family roles. This is not simply the result of living in a different culture, as many would believe, but is an adaptation to migration itself. Families are often separated during the early phases of migration; men may live in bachelor communities for a time, where they often learn to cook and clean for themselves. Once the family is reunited, the men do not necessarily discard these newly learned behaviors (Hondagneu-Sotelo, 1992, 1994). Other immigrant groups have similar experiences. In China, women have historically had low rates of labor force participation, but when they migrated to the United States, like other immigrant groups, women's work was necessary to support families. Once in the United States, even after the family's economic situation stabilized, Chinese American women tended to remain in the labor force. In other words, cultural norms about the desirability of women's working change as the result of actual experience (Geschwender, 1992).

Childhood play is a significant source of gender identity. The increasing participation of young girls and women in team sports is likely to have a significant impact on women's gender identity.

Childhood Play and Games. From the time they become aware of their surroundings, children are socialized to adopt behaviors and attitudes judged appropriate for members of their sex. Socialization comes not only from parents and other family members, but also from peers. Through play children learn patterns of social interaction, cognitive and physical development, analytical skills, and the values and attitudes of their culture. In one classic study, Janet Lever (1978) observed schoolyard play among fifth-grade children, including having children keep diaries describing their daily activities. She found that boys tend to play in larger groups, where the differentiation of the players' parts tends to be more complex. Girls are more cooperative in their play; boys, more competitive. Girls' play more often involves ritualistic activity (like chanting and jumping rope), whereas boys' play more often involves elaborate rules. When playing games with rules, girls are also more likely to ignore, bend, or remake the rules, whereas boys adhere more rigidly to established principles of play. Lever concluded that gender-typed forms of play do more to prepare boys for the hierarchical world they encounter in their adult lives, since the competencies they learn in childhood play are those they need for success in large, competitive, rule-oriented organizations, such as corporations. The greater likelihood that boys will play on team sports and in large groups also teaches boys how to compete and cooperate when they later work in group settings. The competencies that women learn as young girls, such as cooperation and flexibility, are not as valued in male-dominated groups and organizations although it could certainly be argued that they are no less important.

Subsequent research shows conclusions similar to Lever's, indicating that girls play more cooperatively when they are in same-sex groups than in play with boys (Neppl and Murray, 1997). Boys exert power over girls when they play together, and it is typically boys who establish the conditions of the play activities (Voss, 1997). A variety of other patterns are consistently found in children's play and games. Boys are encouraged to play outside more; girls, inside. Boys' toys frequently promote the development of militaristic values and tend to encourage aggression, violence, and the stereotyping of enemies—values rarely embedded in girls' toys. Children's literature, though somewhat improved in recent years, is still likely to show girls as less adventurous, in need of rescue, and in fewer occupations than men. Even children's rooms create a gendered world for boys and girls to grow up in. Systematic observation finds, for example, that girls' rooms are far more likely to be pink and are likely to contain more dolls, fictional characters, and children's furniture; boys' rooms, on the other hand, tend to feature bright colors and to have more sports equipment and vehicles (Pomerleau et al., 1990; Purcell and Stewart, 1990).

Schools and Gender Socialization. Schools are particularly strong influences on gender socialization because of the amount of time children spend in them. As we saw in Chapter 4, teachers often have different expectations for boys and

Gender socialization is influenced by peers, as well as by parents, the media, schools, and religious institutions.

girls. One study of elementary and middle school children found that boys called out answers eight times more often than girls. When they did so, teachers typically listened to the comment, but when girls called out answers, teachers usually corrected them and told them to raise their hands if they wanted to speak (Sadker and Sadker, 1994). Boys in school get more attention, even if it is sometimes negative attention. When teachers of either sex respond more to boys, both positively and negatively, they heighten boys' sense of importance (American Association of University Women, 1992).

Beginning in preschool and elementary school, classroom experience influences gender identity by teaching boys and girls different skills. Children entering school are usually already well socialized into gender roles, but the differences in their abilities are small in the early years. At the preschool level, girls tend to exceed boys in verbal skills, and boys slightly exceed girls in mechanical ability. By the time boys and girls leave school, however, there are great differences in their skills, interests, and abilities. Extensive research on this subject has generally concluded that teacher expectations, classroom interaction, the content of the curriculum, and the representation of men and women as teachers and school leaders all communicate to students that there are different expectations for women and men (American Association of University Women, 1992; Sadker and Sadker, 1994; Fiol-Matta and Chamberlain, 1994; Wetzel et al., 1993).

Religion and Gender Socialization. Religion is an often overlooked but significant source of gender socialization. The major Judeo-Christian religions in the United States place strong emphasis on gender differences, with explicit affirmation of the authority of men over women. In the New Testament of the Bible, for example, the apostle Paul famously urges "Wives, be subject to your husbands . . . for the husband is the head of the wife as Christ is the head of the church" (Ephesians, 4:22–23). In Orthodox Judaism, men offer a prayer blessing God for not having created them a woman or a slave. The patriarchal language of most Western religions and the exclusion of women from positions of religious leadership in some faiths also signify the lesser status of women in religious institutions.

Religious doctrines have a strong effect on the formation of gender identity, particularly among the most devout believers, who hold the most rigid attitudes toward traditional gender roles. Moreover, those in different religious denominations vary in their gender role attitudes, with evangelicals the most conservative and Jewish people and those unaffiliated with particular denominations, the most liberal. A study of college students furthermore showed that those who are most devout, especially among Whites, make the most gender-stereotyped occupational choices and are more likely to disapprove of married women having careers (Gay et al., 1996; Jelen, 1990; Rhodes, 1983). Other research finds that the more frequently people attend church, the less liberal they are in their attitudes toward women. The influence of religion on gender attitudes cannot be considered in isolation, however, since it is mediated by other factors. For example, it is among the less educated that religion tends to have the most influence on traditional gender attitudes (Barrish and Welch, 1980).

The Media and Gender Socialization. The media in their various forms (television, film, magazines, music, and so on) communicate strong, some would even say cartoonish, gender stereotypes. Blonde women are stereotyped as more beautiful and fun, but dumb, and even though they are only one-quarter of the White population, on TV, women are five times more likely than men to have blonde hair. Magazines contribute to the stereotypes; blondes are more than one-third of those shown in major women's magazines and more than 50 percent of those depicted in *Playboy*. There has been little change in the special place of blondes since the 1950s, except in *Playboy* where their number has actually increased substantially since then (Rich and Cash, 1993).

Despite some changes in recent years, television, the most pervasive communication medium, continues to depict highly stereotyped roles for women and men. Men on television heavily outnumber women: They are 65.4 percent of all prime-time television characters, whereas women are 34.6 percent. Men are not only more visible on television, they are more formidable. Women are more likely to appear in situation comedies; men outnumber women the most in action–adventure shows. Indeed, men are heavily stereotyped in strong, independent roles. The suggestion is that the world of women does not stretch much beyond domestic comedy, whereas men must have action and adventure (Merlo and Smith, 1994a; Davis, 1990).

Television also delivers unrealistic portrayals of women and men in terms of age and appearance. Nearly 54 percent of all women characters on television are between the ages of eighteen and thirty-four; in the actual population, only 28 percent of women are in this age group. The tendency to show young women on television is even more accentuated in action–adventure shows, where the percentage of women between eighteen and thirty-four is 75.8 percent! Women on television are four times more likely than men to be shown provocatively dressed—in nightwear, underwear, swimsuits, and tight clothing (Hall and Crum, 1994; Davis, 1990).

Even on news broadcasts, minority men and women are less likely to appear during prime news hours; instead, they most often appear in weekend and late night slots. Women of color as broadcasters are routinely given the "soft" human interest stories; White men almost always report the stories that people are supposed to take more seriously. Although many networks have added men and women of color as anchors, giving the appearance that equity has been achieved, the data show that women of color are only 3 percent of broadcast news executives and file only 2 percent of the reported stories (Rhode, 1995). On commercials, men are routinely depicted as more expert than women: Men are 90 percent of those providing voice-overs in commercials, equal to their representation as televised sports commentators (Rhode, 1994).

Social scientists debate the extent to which people actually believe what they see on television, but research with children shows that they identify with the television characters they see. Children report that they want to be like television characters when they grow up. Boys tend to identify with characters based on their physical strength and activity level; girls relate to perceptions of physical attractiveness (Reeves and Miller, 1978; Signorielli, 1989; Signorielli et al., 1994). Even with adults, researchers find that there is a link between seeing sexist images and such attitudes as lower acceptance of feminism, more traditional views of women, and attitudes supporting sexual aggression (MacKay and Covell, 1997; Garst and Bodenhausen, 1997).

Advertisements are another important outlet for the communication of gender images to the public—one that is especially noted for the communication of idealized, sexist, and racist images of women and men. Women in advertisements are routinely shown in poses that would shock people if the character were male. Consider how often women are displayed in ads dropping their pants, skirts, or bathrobe, or squirming on beds. How often are men shown in such poses? Men are sometimes displayed as sex objects in advertising, but not nearly as often as women. The demeanor of women in advertising—on the ground, in the background, or looking dreamily into space—makes them appear subordinate and available to men. One of the few places where women appear more frequently than men is in advertisements for drugs and where mothers are shown at home caring for sick children (Craig, 1992; Merlo and Smith, 1994b).

Other parts of popular culture are also a source of stereotypes contributing to gender socialization. Greeting cards, record jackets, books, songs, films, and comic strips all communicate images representing the presumed cultural ideals of womanhood and manhood. These popular products have an enormous effect on our ideas and self-concepts. To take one illustration, think about the impact of romance novels. Harlequin, only one of many romance publishers, is estimated to have more than 14 million loyal readers with sales in excess of 188 million books per year. In one major bookstore chain, Harlequin novels account for 30 percent of all paperback sales. These romances are marketed to appeal to working women, who read them for relaxation and escape. Interestingly, the plots reflect the powerlessness and depersonalization many women find at work, while fueling women's fantasies of escaping from these restraints. Heroines are portrayed as active and intelligent, struggling to win the recognition and love of their bosses (Rabine, 1985). Mass-marketed products like this shape our understanding of the possibilities open to ourselves and to others and reflect the values of the dominant culture. When women's independence is treated as a kind of fantasy, or when a love affair with one's boss is shown as the best avenue to recognition, popular culture is providing ideological support for the subordination of women in society.

Despite cultural changes in gender roles, young girls learn stereotypical ideals for feminine beauty early in life.

The Price of Conformity

A high degree of conformity to stereotypical gender expectations takes its toll on both men and women. The higher rate of early death among men from accidents and violence can be attributed to the stress and injury associated with the cultural definition of masculinity, which includes physical daring and risk-taking. The strong undercurrent of violence in today's culture of masculinity encourages men to engage in behaviors that put them at risk in a variety of ways (Kimmel and Levine, 1992). Sociologists are also finding that adhering to gender expectations of thinness for women and strength for men is related to a host of negative health behaviors, including eating disorders, smoking, and for men, steroid abuse. Furthermore, it is increasingly being shown that such problems are linked to a history of sexual abuse, a phenomenon that sociologists relate to the sexual oppression of women (Logio-Rau, 1998; Thompson, 1994).

Men also pay the price of overconformity if they too thoroughly internalize gender expectations that they must be independent, self-reliant, and not emotionally expressive. Men's gender socialization discourages intimacy among them, thereby affecting the quality of men's friendships, although in recent years, men are more likely to say they admire men who show a more sensitive side, and they say they have become better able to express their feelings (Roper Organization, 1995b). Men who have a greater balance of so-called masculine and feminine traits also tend to be more mentally and physically healthy than those who are more traditional in their orientation and self-concept (Thornton and Leo, 1992).

Conforming too much to gender expectations also has negative consequences for women. Women with the most stereotypical identities—those who are more passive, acquiescent, and dependent—experience the highest rates of depression and other forms of mental illness and poor health (Baffi et al., 1991; Tinsley et al., 1984). Women ranking most "feminine" on personality tests also tend to be more dissatisfied with their lives; they are more psychologically anxious and have lower self-esteem than other women (Thornton and Leo, 1992). On the other hand, women who balance their lives with multiple roles report more gratification, status security, and enrichment in their lives than do more traditionally oriented women (Gerson, 1985) even though holding multiple roles is stressful. Conformity to traditional gender roles denies women access to power, influence, achievement, and independence in the public world, and denies men the more nurturing, emotional, and other-oriented worlds that women have traditionally inhabited.

Luckily, some gender expectations relax with age. Older women describe themselves as more assertive, competent, and effective than do younger women (Sprock and Yoder, 1997). Men's attitudes toward women also tend to become more liberal as they age (Baker and Terpstra, 1986); both men and women report greater satisfaction with their lives as they get older. Generally, as people collect life experiences, they gain confidence in themselves and are more likely to challenge dominant cultural norms.

Race, Gender, and Identity

Because the experiences of race and gender socialization affect each other, men and women from different racial groups have differing expectations regarding gender roles; however, the differences are quite often not the ones people expect based on the gender stereotypes they have absorbed about other groups. For example, when asked to rate desirable characteristics in men and women, White men and women are more likely than Hispanic men and women to select different traits for men and women; this runs counter to the idea that Hispanics hold highly polarized views of manhood and womanhood. Comparing Hispanics, Whites, and African Americans, it is African Americans who are most likely to find value in both sexes displaying various traits such as being assertive, athletic, self-reliant, gentle, and eager to soothe hurt feelings (Harris, 1994). African American men and woman also show significant support for feminism and egalitarian views of men's and women's roles, although African American women are somewhat more liberal than African American men (Hunter and Sellers, 1998). There is little research on gender expectations among Asian American men and women, but the few studies available indicate that Asian American women are more likely than Asian American men to value egalitarian roles for men and women (Chia et al., 1994).

Gender expectations emerge from the specific experiences of different groups. African American women, like White women, are encouraged to become nurturing and other oriented, but they are also socialized to become self-sufficient, to aspire to an education, to want an occupation, and to regard work as an expected part of women's role. They are also expected to be more independent than White women (Ladner, 1995). This is specific to their experience as African American women.

African American women are also more likely than White women to reject the gender stereotyping of the dominant culture. These attitudes among African American women are probably related to the examples of their mothers, who were more likely than White women to have been employed and supporting themselves; the image they presented to their daughters and the lessons they passed on would have encouraged self-sufficiency (Wharton and Thorne, 1997; Collins, 1987; Dugger, 1988). The same trends are becoming true for White women as more have entered the labor force.

Among men, too, gender identity is affected by race. Latino men, for example, bear the stereotype of *machismo*—exaggerated masculinity. Although machismo is associated with sexist behavior by men, within Latino culture, it is also associated with honor, dignity, and respect (Mirandé, 1979). Maxine Baca Zinn argues that machismo is typically misinterpreted within the dominant culture. Although "macho" male behaviors do exist among Latinos, they are not the only way that Latinos interact with women. Researchers find, for example, that Latino families are rather egalitarian, with decision making frequently shared by men and women. To the extent that machismo exists, it is not just a cultural holdover from Latin societies, but can be how men defy their racial oppression (Baca Zinn, 1995). The definition of manhood among Latinos is more multidimensional than cultural stereotypes suggest.

African American men also define manhood in ways far more complex than simple stereotypes suggest. Self-determination, responsibility, and accountability to family and/or community are the attributes African American men most commonly associate with manhood; power over others is rated one of the least important traits (Hunter and Davis, 1992). African American men are also more likely than White men to emphasize the importance of their "breadwinning role," whereas White men are more likely to value the role of being nurturers as adults (Harris et al., 1994). African American men, especially those in the middle class, are also more liberal about gender roles than are White men (Cazenave, 1983). Men's roles in society, like women's, are conditioned by the social context of their experience. Gender identity is merged with racial identity for all people. For those in racial minority groups, this means that concepts of womanhood and manhood are shaped by the patterns of domination and exclusion that race and gender produce.

Gender Socialization and Homophobia

As we saw in Chapter 7, *homophobia* is the fear and hatred of homosexuals. Homophobia plays an important role in gender socialization because it encourages stricter conformity to traditional expectations, especially for men and young boys. Slurs directed against gays encourage boys to act more masculine as a way of affirming for their peers that they are not gay. As a consequence, homophobia also discourages so-called feminine traits in men, such as caring, nurturing, empathy, emotion, and gentleness. As one pair of researchers says, "The fear of being labeled homosexual serves to keep men within the confines of what the culture defines as sex-appropriate behavior, and it interferes with the development of intimacy between men" (Morin and Garfinkle, 1978: 41). Men who endorse the most traditional male roles also tend to be the most homophobic (Thompson et al., 1985).

The relationship between homophobia and gender socialization illustrates how socialization contributes to social control. Once people internalize societal expectations, they do not challenge or question the status quo. This is what is meant by the *social construction of gender:* What appears to be normal or customary is only that which people have been taught is normal. Gender has great significance in society, but the specific forms it takes are learned. Gender is therefore fluid, and since gender expectations are learned, they can be redefined and learned in new ways. There is little inherent in the social definition of women and men as gendered persons that could not be reconsidered and changed.

Although only a small percentage of women work in jobs traditionally defined as for men, a growing number of women are entering such nontraditional fields. Women in such jobs often quickly discover that these are gendered institutions.

The Institutional Basis of Gender

The process of gender socialization tells us a lot about how gender identities are formed, but gender is not just a matter of identity: Gender is embedded in social institutions. This means that institutions are patterned by gender, resulting in different experiences and opportunities for men and women. Sociologists analyze gender not just as interpersonal expectations, but as characteristic of institutions. This is what is meant by the term **gendered institution.** This concept means that entire institutions are patterned by gender. Gendered institutions are the total pattern of gender relations—stereotypical expectations; interpersonal relationships; and the different placement of men and women in social, economic, and political hierarchies of institutions. Schools, for example, are not just places where children learn gender roles, but the school itself is a gendered institution because it is founded on specific gender patterns. Seeing institutions as gendered reveals that gender is not just an attribute of individuals, but is "present in the processes, practices, images and ideologies, and distributions of power in the various sectors of social life" (Acker, 1992: 567).

As an example of the concept of gendered institution, think of what it is like to work as a woman in an organization dominated by men. Women in this situation report that there are subtle ways that men's importance in the organization is communicated, while women are made to feel like outsiders. Important career connections may be made in the context of men's informal interactions with each other—both inside and outside the workplace. Women may be treated as tokens or may think that company policies are ineffective in helping them cope with the particular demands in their lives. These institutional patterns of gender affect men, too, particularly if they try to establish more balance between their personal and work lives. To say then that work institutions are gendered institutions means that, taken together, there is a cumulative and systematic effect of gender throughout the institution.

The concept of gendered institutions also shows the limitation of thinking about gender only in terms of social roles. Gender roles, as we have seen, are the learned patterns of behavior associated with being a man or a woman. Significant as these roles are, the concept of gendered institutions goes further in explaining how gender patterns persist—even when people themselves try to change their roles. Gender is not just a learned role; it is also part of social structure, just as class and race are social structural dimensions of society. Notice that people do not think about the class system or racial inequality in terms of "class roles" or "race roles." It is obvious that race relations and class relations are far more than matters of interpersonal interaction. Race, class, and gender inequalities are experienced within interpersonal relationships, but they are also more. Just as it would seem strange to think that race relations in the United States are controlled by "race role socialization," it is also wrong to think that gender relations are the result of gender socialization alone. Most people understand that race relations are a matter of systems of privilege and inequality; likewise, gender is a system of privilege and inequality in which women are systematically disadvantaged relative to men. Gender, like race and class, involves institutionalized power relations between women and men, and involves unequal access to social and economic resources (Lopata and Thorne, 1978; Andersen, 1997).

To put the concept simply, socialization and roles cannot explain everything. Socialization affects how women and men choose among options, but the institutional basis of gender determines which options they will choose among. Gender shapes access to economic and political resources, as well as more personal dimensions like self-definition, relationships with others, and perceived worth. Studying gender as an institutional, or social structural, phenomenon also makes it more clear how gender intersects with systems of race and class relations. Latinas and Native American and African American women, for example, are oppressed by both race and gender and perhaps by class. White men on the whole are accorded more power, prestige, and economic resources than are women, but not all men share these advantages equally. As a group, Latino men are disadvantaged relative to White men and relative to some White women (see Figure 12.2). Understanding these complex relationships between gender, race, and class, and fully understanding the dynamics of gender in society thus require knowing about the system of *gender stratification.*

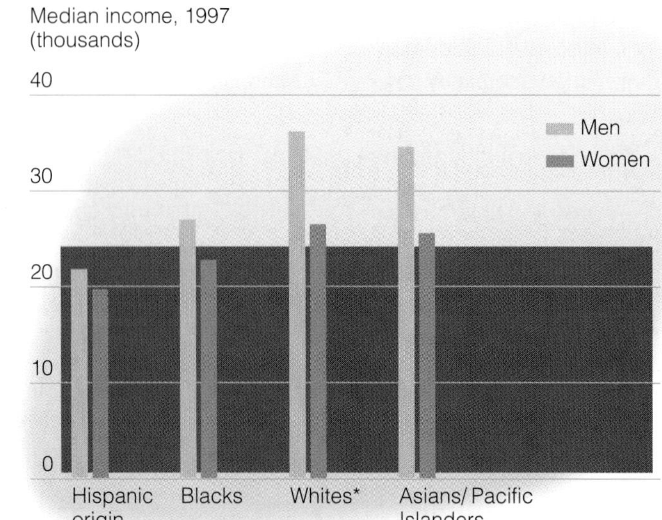

FIGURE 12.2 Median Income by Race and Gender

All data include only year-round, full-time workers.

*Includes some persons who identify as Hispanic.

DATA: From the U.S. Bureau of the Census. 1998a. *Money Income in the United States: 1997.* Washington, DC: U.S. Department of Commerce; Hooper, Linda, and Claudette E. Bennett. 1998. *The Asian and Pacific Islander Population in the United States: March 1997 (Update).* Current Population Reports, Series P20-512. Washington, DC: U.S. Department of Commerce.

Gender Stratification

Gender stratification refers to the hierarchical distribution of social and economic resources according to gender. Most societies have some form of gender stratification although the specific form varies from country to country. Comparative research finds that women are more nearly equal in societies characterized by the following (Chafetz, 1984):

- Women's work is central to the economy
- Women have access to education
- Ideological or religious support for gender inequality is not strong
- Men make direct contributions to household responsibilities, such as housework and childcare
- Work is not highly segregated by sex
- Women have access to formal power and authority in public decision making

In Sweden, where there is a relatively high degree of gender equality, the participation of both men and women in the workforce and the household (including childcare and housework) is promoted by government policies. Women also have a strong role in the political system. In a different example, women in Saudi Arabia have few rights. In Saudi Arabia, women are prohibited from working with men, and single women cannot go out alone without a chaperone. As late as 1990, the Saudi government did not permit women to drive cars. Even in countries without such drastic gender-based inequalities, numerous inequities between women and men exist, and in Sweden, women's earnings are still lower than men's and women tend to work in different occupations from men.

Gender stratification is, as the preceding list suggests, multidimensional. In some societies, women may be free in some areas of life, but not in others. In Japan, for example, women tend to be well educated and have high labor force participation. Within the family, however, Japanese women have fairly rigid gender roles and little reproductive freedom; condoms are the most common form of birth control in Japan and only in exceptional cases can Japanese women get access to birth-control pills. At the same time, however, the rate of violence against women in Japan in the form of rape, prostitution, and pornography is quite low relative to other nations (O'Kelly and Carney, 1986) even though women are widely employed as "sex workers"—in hostess clubs, bars, and sex joints (Allison, 1994). Patterns of gender inequality are most reflected in the wage differentials between women and men around the world, as Figure 12.3 shows.

Sexism and Patriarchy

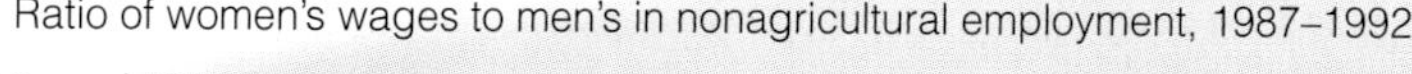

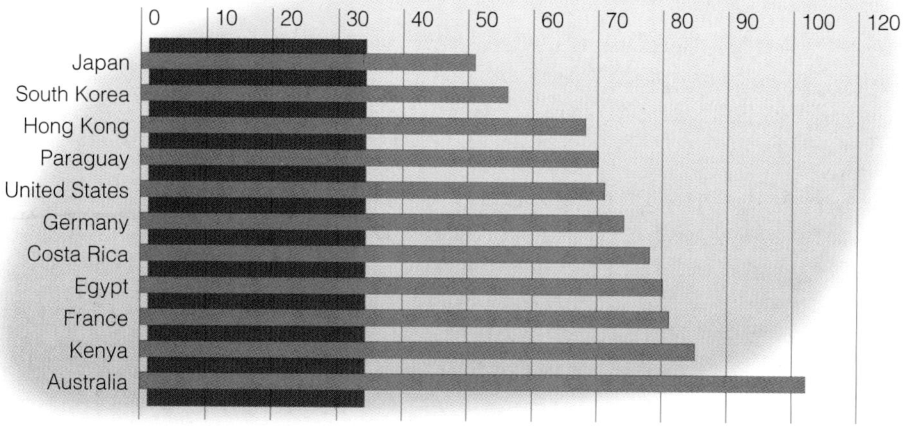

FIGURE 12.3 The Wage Gap: An International Perspective

DATA: From the Population Reference Bureau. 1995. *World's Women.* Washington, DC: Population Reference Bureau, p. 1.

Gender stratification tends to be supported by beliefs that treat gender inequality as "natural." An **ideology** is a belief system that tries to explain and justify the status quo. **Sexism** is an ideology, but it is also a set of institutionalized practices and beliefs through which women are controlled because of the significance given to differences between the sexes. Like racism, sexism distorts reality, making behaviors seem natural when they are rooted in entrenched systems of power and privilege. The idea that men should be paid more than women because they are the primary breadwinners reflects sexist ideology, but when this concept becomes embedded in the wage structure, people no longer have to believe explicitly in the original idea for the consequences of sexism to be propagated.

Sexism and racism tend to go hand in hand. Both generate social myths that have no basis in fact, but justify the continuing advantage of dominant groups over subordinates. A case in point is the belief that women of color are taking jobs away from men and are being hired more often and promoted more rapidly than others. First, this misrepresents the facts. Women rarely take jobs away from men since most women work in gender- and race-segregated jobs. Second, the idea that African American, Hispanic, and Asian women are advancing more rapidly than men because of affirmative action policies is simply untrue. The truth is that women, especially women of color, are burdened by obstacles to job mobility that are not present for men, especially White men; they have historically been less likely to get promotions, raises, and senior appointments, and that continues to be the case (Reskin and Padavic, 1994; McGuire and Reskin, 1993). This ideological myth, usually brought up in the context of programs designed to assist women and people of color, distorts the truth by making White men seem to be the victims of race and gender privilege. Sexism and racism are ideologies that support and defend the status quo.

Debunking Society's Myths

Myth: Because of affirmative action, Black women are taking a lot of jobs away from White men.

Sociological perspective: Sociological research finds no evidence of this claim. Quite the contrary, women of color work in gender- and race-segregated jobs and only rarely in occupations where they compete with White men in the labor market.

Aside from being a set of ideological beliefs, sexism is also part of the institutional structure of society. **Patriarchy** refers to a society or group in which men have power over women. **Matriarchy** is a society or group in which women have power over men. Among American Indians, the Cherokees had a Women's Council through which social and political decisions were made. It was women who made tribal decisions, including whether to wage war, bear arms, allow certain marriages, and other public policy matters (Allen, 1986). Anthropologists debate the extent to which there have been many matriarchal societies, but they are known to exist (Ledgerwood, 1995).

Patriarchy, on the other hand, is common throughout the world. In patriarchal societies, husbands have authority over wives in the private sphere of the family, and public institutions are also structured around male power. Men hold all or most of the positions of public power in patriarchal societies, whether as chief, president, CEO, prime minister, or other leadership positions. Forms of patriarchy vary from society to society. In some, it is rigidly upheld in both the public and private spheres. In these societies, women may be formally excluded from voting, holding public office, or working outside the home. In societies like the contemporary United States, patriarchy may be somewhat diminished in the private sphere (at least in some households), but the public sphere continues to be based on patriarchal relations.

In sum, gender stratification is an institutionalized system that rests on specific beliefs systems that support the inequality of men and women. Although theoretically one could have a society stratified by gender where women hold power over men, that is not how gender stratification has evolved. In the next sections, we examine different manifestations of gender stratification in the United States, especially as it involves women's economic position relative to men.

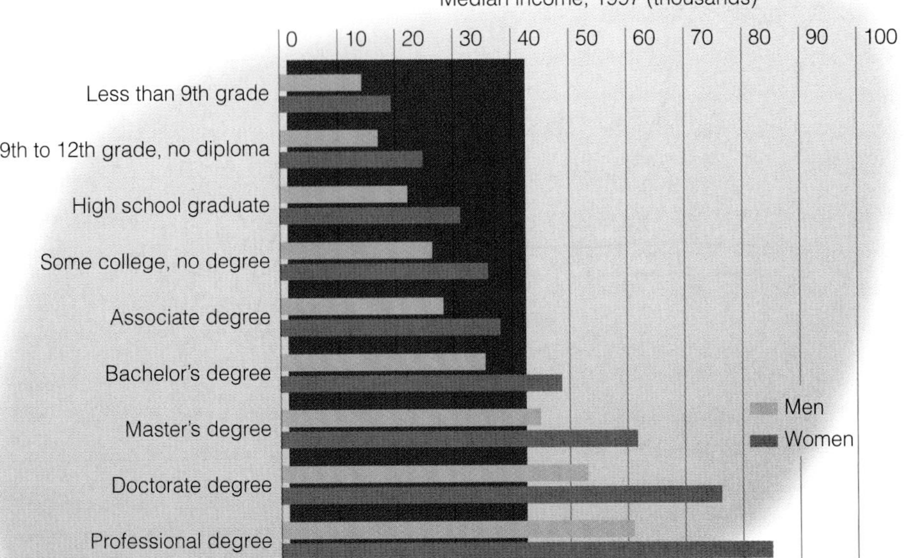

FIGURE 12.4 Education, Gender, and Income

DATA: From the U.S. Bureau of the Census. 1998a. *Money Income in the United States: 1997.* Washington, DC: U.S. Department of Commerce, pp. 31–33.

Women's Worth: Still Unequal

Gender stratification is especially obvious in the persistent earnings gap between women and men. Although the gap has closed somewhat since the 1960s, when women earned 59 percent of what men earned, women who work year-round and full time still earn, on average, only 74 percent of what men earn. Women with college degrees earn the equivalent of men who have only high school diplomas (see Figure 12.4). In 1997, the median income for women working full time and year-round was $26,029; for men, it was $35,248. Among all women (which includes part-time workers), the median income in 1997 was only $13,703; for men, it was $25,212 (U.S. Bureau of the Census, 1998a).

The income gap between women and men persists despite the increased participation of women in the labor force. The **labor force participation rate** is the percentage of those in a given category who are employed either part time or full time. By 1997, 60 percent of all women were in the paid labor force, compared with 75 percent of men. Since 1960, married women with children have nearly tripled their labor force participation. Two-thirds of mothers are now in the labor force, including more than half of mothers with infants. Current projections indicate that women's labor force participation will continue to rise. Men's labor force participation is expected to decline slightly (U.S. Department of Labor, 1998).

The labor force participation rate among women has changed most dramatically among White women in recent years, since African American women, some Latinas, and Asian American women have historically been more likely to work for pay than White women. Asian American women are the most likely to be employed, Puerto Rican and Cuban women the least. In general, however, the labor force participation rates of White women and women of color have converged in recent years.

Changes in family patterns in contemporary society mean that more women are the sole supporters of their dependents. The percentage of employed women reporting satisfaction with their job has declined since 1990 although the majority say they get satisfaction from their work. However, they do not work only for personal fulfillment. Only 8 percent of employed women say they work to have something interesting to do; an overwhelming 87 percent work to support themselves or their family or to bring in extra money (Roper Organization, 1995b).

Explaining the Pay Gap

Laws prohibiting gender discrimination have now been in place for more than thirty years. The Equal Pay Act of 1963 was the first federal law to require that men and women should receive equal pay for equal work, an idea that is supported by the majority of Americans. Wage discrimination is rarely overt. Employers do not tell women, "You'll be working in the same job as a man, but because you are a woman, I will pay you 70 percent of what he earns." Most employers do not even explicitly set out to pay women less than men, but despite good intentions and legislation on the books, differences in men's and women's earnings persist. Why? Research reveals three strong explanations for this continuing difference: human capital theory, dual labor market theory, and overt discrimination.

Human Capital Theory. Gender differences in wages are explained by human capital theory as the result of differences in the individual characteristics that workers bring to jobs. **Human capital theory** assumes that the economic system is fair and competitive, and that wage differences reflect differences in the resources ("human capital") that individuals bring to the job. Factors like age, prior experience, number of hours worked, marital status, and educa-

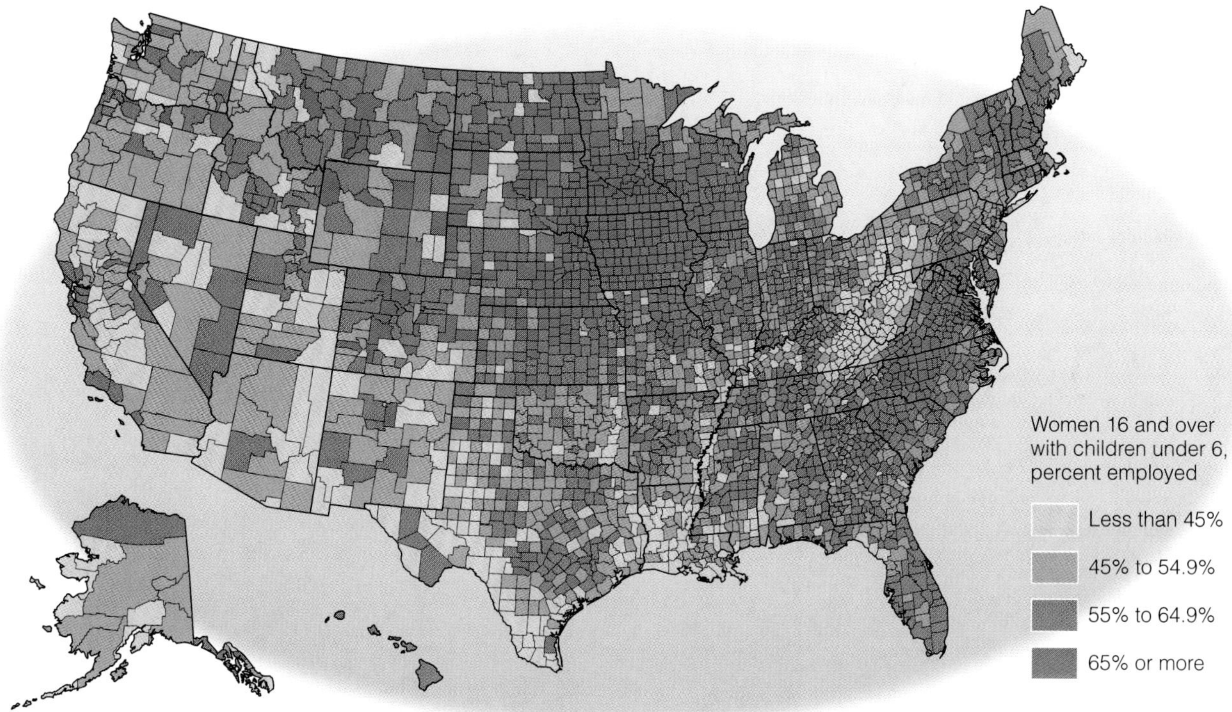

MAP 12.1 Mapping America's Diversity: Working Mothers

SOURCE: Doyle, Rodger, 1994. *Atlas of Contemporary America: Portrait of a Nation.* New York: Facts on File, p. 18.

tion are *human capital variables.* Human capital theory says that the extent to which human beings vary in these characteristics will influence their worth in the labor market. For example, higher job turnover rates or work records interrupted by child rearing and family responsibilities could negatively influence the earning power of women.

There is much evidence in support of the human capital explanation for the difference in men's and women's earnings since education, age, and experience do influence earnings; however, when we compare men and women at the same level of education, prior experience, and number of hours worked per week, women still earn less than men (Dill et al., 1987). Intermittent employment is also not as significant in explaining wage differences as human capital theory would lead one to expect (England et al., 1988). Although human capital theory explains some of the difference between men's and women's earnings, it does not explain it all. Sociologists have looked to other factors to complete the explanation of wage inequality.

The Dual Labor Market. A second explanation of gender differences in earnings is **dual labor market theory,** which contends that women and men earn different amounts because they tend to work in different segments of the labor market. The dual labor market reflects the devaluation of women's work since there generally are low wages in jobs where women are most concentrated. Although it is hard to untangle cause and effect in the relationship between the devaluation of women's work and low wages in certain jobs, once such an earnings structure is established, it is difficult to change. As a result, although equal pay for equal work may hold in principle, it applies to relatively few people since most men and women are not engaged in "equal work."

According to dual labor market theory, the labor market is organized in two different sectors: the *primary market* and the *secondary market.* In the primary labor market, jobs are relatively stable, wages are good, opportunities for advancement exist, fringe benefits are likely, and workers are afforded due process. Working for a major corporation in a management job is an example. Jobs in the primary labor market tend to be in large organizations where there is general stability, steady profits, and a rational system of management. Although in recent years even jobs in the primary labor market have been vulnerable to downsizing and corporate mergers, relative to other jobs, these have more advantages. The secondary labor market, on the other hand, is characterized by high job turnover, low wages, short or nonexistent promotion ladders, few benefits, poor working conditions, arbitrary work rules, and capricious supervision. Many of the jobs students take, such as in waiting tables, selling fast food, or cooking and serving fast food, fall into this category, but for students, unlike those permanently stuck in this labor market, these jobs are usually short term.

Women and racial–ethnic minorities are far more likely to be employed in the secondary labor market than in the primary labor market. Even within the primary labor market, there are two tiers. The first consists of high-status

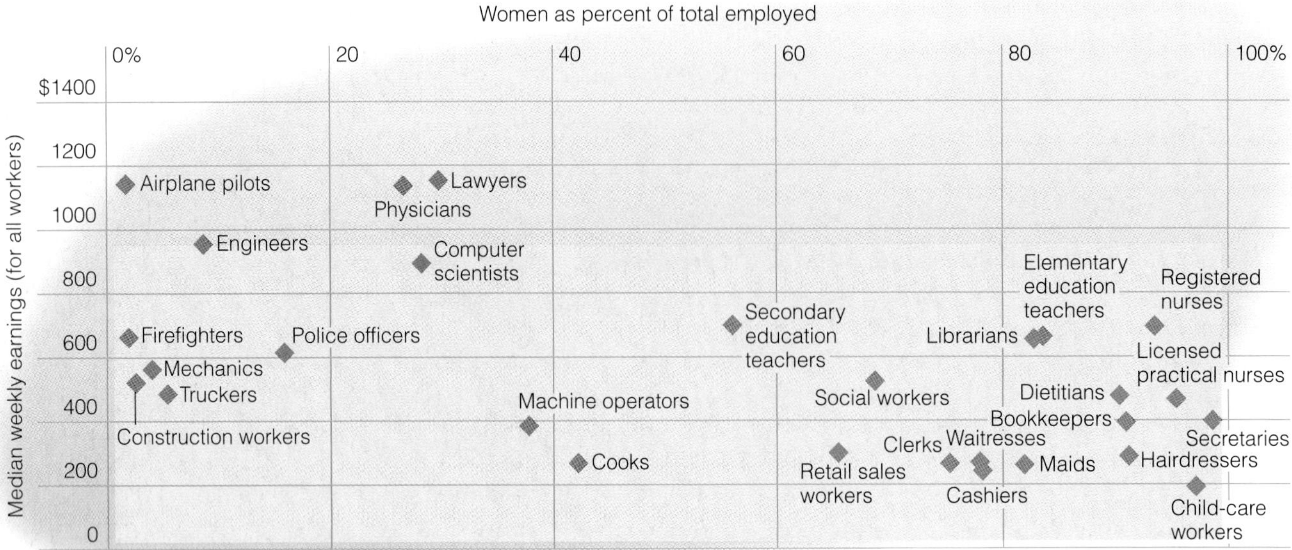

FIGURE 12.5 Earnings in Selected Occupations

This scattergram relates the percent of women in selected occupations to the average weekly earnings in these occupations. The scattergram plots the percent of women in an occupation (on the X axis, or bottom line of the figure), and the average weekly earnings in that occupation for all workers (on the Y axis, or the left side of the figure). Each dot in the scattergram thus represents a particular occupation. Those occupations to the right side are those with the highest percentage of women employed. For example, 97.1 percent of child-care workers are women; average weekly earnings in this occupation (in 1997) were $198 per week. On the left side of the scattergram, note that women are 1.4 percent of airplane pilots, who have an average weekly earning of $1138.

After studying the figure carefully, what conclusions would you draw about gender segregation and the wage gap?

DATA: From the U.S. Department of Labor. 1997. *Employment and Earnings.* Washington, DC: U.S. Department of Labor, pp. 171–176, 206–210.

professional and managerial jobs with potential for upward mobility, room for creativity and initiative, and more autonomy. The second tier comprises working-class jobs, including clerical work and skilled and semiskilled blue-collar work. Women and minorities in the primary labor market tend to be in the second tier. Although these jobs are secure compared to jobs in the secondary labor market, they are more vulnerable and do not have as much mobility, pay, prestige, or autonomy as jobs in the first tier of the primary labor market (Eitzen and Baca Zinn, 1998).

There is, in addition, an *informal sector* of the market where there is even greater wage inequality, no benefits, and little, if any, oversight of employment practices. Individuals may hire such workers, as private service workers or under-the-table workers who perform a service for a fee (painting, babysitting, car repairs, and any number of services). Businesses and corporations also employ such workers and can reap huge profits by not paying benefits and not providing compensation when workers are sick, injured, or disabled. Although there are no formal data on the informal sector, because much of it tends to be in an underground economy, it is likely that women and minorities form a large segment of this market activity, although White men are also disadvantaged by the instability and lack of protection in this sector of work.

According to dual labor market theory, wage inequality is a function of the structure of the labor market, not the individual characteristics of workers, as suggested by human capital theory. Because of the dual labor market, men and women tend to work in different occupations, and when working in the same occupation, in different jobs. This is referred to as **occupational segregation,** a pattern in which different groups of workers are separated into different occupations. Occupational segregation can be by gender, class, race, and other factors, as we discuss further in Chapter 17 on work. Wages tend to be linked to occupational segregation—there is a direct association between the number of women in given occupational categories and the wages paid in those jobs; in other words, the greater the proportion of women in a given occupation, the lower the pay (Terrell, 1992). Figure 12.5, a scattergram, illustrates this trend. At one extreme, occupations that are close to 100 percent female (private household child-care workers and dental hygienists) pay only half of what is paid in jobs that are at least 50 percent male; indeed, studies find that workers in jobs requiring nurturing social skills have the lowest pay, even when their education and experience are comparable to workers in other jobs (Kilbourne et al., 1994). Occupational segregation is exacerbated by race since the worst paid jobs are those with the largest numbers of African American,

Latino, and Native American women. Not surprisingly, the jobs where White men are most prevalent are the best paid (U.S. Department of Labor, 1998; Dill et al., 1987). The social structural analysis provided by dual labor market theory suggests that factors like women's exclusion from job networks, the size of given industries, and other factors in the environment of work organizations are significant in explaining the wage gap between women and men (Beggs, 1995; Kreft and DeLeeuw, 1994; Bartlett and Miller, 1988).

Overt Discrimination. A third explanation of the gender wage gap is overt discrimination. **Discrimination** refers to practices that single out some groups for different and unequal treatment. Despite the progress of recent years, overt discrimination continues to afflict women in the workplace. It is argued that men, especially White men, by virtue of being the dominant group in society, have an incentive to preserve their advantages in the labor market. They do so by establishing rules that distribute rewards unequally. Women pose a threat to traditional White male privileges, and men may organize to preserve their power and advantage (Reskin, 1988).

The discrimination explanation of the gender wage gap argues that dominant groups will use their position of power to perpetuate their advantage (Lieberson, 1980). There is some evidence that this occurs. Historically, White men used labor unions to exclude women and racial minorities from well-paying, unionized jobs, usually in the blue-collar trades. A more contemporary example is seen in the efforts of conservative groups, usually led by men, to dilute legislation that has been developed to assist women and racial–ethnic minorities. These efforts can be seen as an attempt to preserve group power.

Another example of overt discrimination is the harassment that women experience at work, including sexual harassment and other means of intimidation (see Chapter 17). Sociologists see such behaviors as ways for men to protect their advantages in the labor force. It is little wonder that women who enter traditionally male-dominated professions suffer the most sexual harassment; the reverse seldom occurs for men employed in jobs historically filled by women. Although men can be victims of sexual harassment, this is rare (see Chapter 17). Sexual harassment is a mechanism for preserving men's advantage in the labor force—a mechanism that also buttresses the belief that women are sexual objects for the pleasure of men.

Each of these explanations—human capital theory, dual labor market theory, and overt discrimination—contributes to an understanding of the continuing differences in pay between women and men. Wage inequality by gender is clearly the result of multiple factors that together operate to systematically disadvantage women in the workplace.

Gender Segregation

Gender segregation refers to the distribution of men and women in different jobs in the labor force; it is a specific form of occupational segregation. Despite several decades of legislation prohibiting discrimination against women in the workplace, most women and men still work in gender-segregated occupations (Wootton, 1997). That is, the majority of women work in occupations where most of the other workers are women, and the majority of men work mostly with men. Women also tend to be concentrated in a smaller range of occupations than men. To this day, more than half of all employed women work as clerical workers, sales clerks, or in service occupations such as food service workers, maids, health service workers, hairdressers, and child-care workers. Men are dispersed over a much broader array of occupations. Consider the fact that almost half of all women work either as teachers, retail sales workers, clerical workers, financial records processors, service workers, or machine operators. Women are 76 percent of teachers (not counting college and university teachers), two-thirds of retail sales workers, 92 percent of secretaries, 90 percent of financial records processors, and 95 percent of private service workers—stark evidence of the persistence of gender segregation in the labor force (U.S. Department of Labor, 1998).

Sociologists use a measure called the **index of dissimilarity** to measure the extent of occupational segregation. This measure indicates the number of workers who would have to change jobs in order to have the same occupational distribution as the comparison group. By current estimates, at least 53 percent of men (or women) would have to change occupations to achieve occupational balance by gender—a decline of almost 5 points from the percent who would have had to do so to achieve equality in the mid-1980s; this indicates that there is somewhat less gender segregation in the labor force, but the difference is small (Wootton, 1997). Most women, in fact, continue to work in occupations where two-thirds or more of the other workers are women (U.S. Department of Labor, 1998). This is especially true for women of color because gender segregation at work is further aggravated by race. Women of color, for instance, tend to be employed in occupations where they are segregated not only from men but also from White women, such as in domestic work in hotels. The occupations with the highest concentrations of women of color are among the worst paid of all jobs (U.S. Department of Labor, 1998).

The Devaluation of "Women's Work." Across the labor force, women tend to be located in those jobs that are the most devalued, causing some to wonder if the very fact that the jobs are held by women leads to devaluation of the job. Why, for example, is pediatrics considered a less prestigious specialty than cardiology? Why are elementary school teachers (83 percent of whom are women) paid less than airplane mechanics (99 percent of whom are men)? The association of elementary school teaching with children and its identification as "women's work" lowers its prestige and economic value. Indeed, if measured by the wages attached to an occupation, child care is one of the least prestigious jobs in the nation—paying on average only $202 per week in 1997, which would come out to $10,504 per year if you worked every week of the year (U.S. Department of Labor, 1998)—an annual income that is far below the federal poverty line!

Jobs that have historically been defined as "women's work" are some of the most devalued in terms of income and prestige, despite their importance for such things as nurturing children.

Only a small proportion of women work in occupations traditionally thought to be men's jobs (such as the skilled trades). The representation of women in skilled blue-collar jobs has increased fourfold since 1940, from 2 to 9 percent in 1997, still less than one in ten (U.S. Department of Labor, 1998). Likewise, very few men work in occupations historically considered to be women's work, such as nursing, elementary school teaching, and clerical work. Interestingly, men who work in occupations historically thought of as "women's work" tend to be more upwardly mobile within these occupations than are women who enter fields traditionally reserved for men (Williams, 1992; Zimmer, 1988).

Gender segregation in the labor market is so prevalent that most jobs can easily be categorized by whether they are considered "men's work" or "women's work." Occupational segregation reinforces the belief that there are significant differences between the sexes. Think of the characteristics of a soldier. Do you imagine someone who is compassionate, gentle, and demure? Similarly, imagine a secretary. Is this someone who is aggressive, independent, and stalwart? The association of each with a particular gender makes the occupation itself a "gendered occupation."

Gender Segregation and Gender Identity. For all women, perceptions of gender-appropriate behavior influence the likelihood of success at work. Even something as simple as wearing makeup has been linked to women's success in professional jobs (Dellinger and Williams, 1997). Particularly when men or women cross the boundaries established by occupational segregation, they are often considered to be gender deviants. They may be stereotyped as homosexual and may have their "true gender identity" questioned. Men who are nurses may be stereotyped as effeminate or gay; women Marines may be stereotyped as "butch." Social practices like these serve to reassert traditional gender identities, perhaps seeming to soften the challenge to traditionally male-dominated institutions that women's entry challenges (Williams, 1989).

As a result, many men and women in nontraditional occupations feel pressure to assert gender-appropriate behavior. Men in jobs historically defined as women's work may feel impelled to emphasize their masculinity, or if they are gay, they may feel even more pressure to keep their sexual identity in the closet. Such social disguises can make them seem unfriendly and distant, characteristics that can have a negative effect on their professional evaluations. Heterosexual women in male-dominated jobs may feel obliged to squash suspicions that they are lesbians or excessively "mannish," whereas lesbian women may be especially wary about having their sexual identity revealed. Studies have found that lesbian women are more likely to be open about their sexual identity at work when they work predominantly with women and have women as bosses (Schneider, 1984).

Internal Gender Segregation. Gender segregation also occurs within occupations, meaning that women not only usually work in different jobs from men, but even when they work within the same occupation, they are segregated into particular fields or job types. Medical doctors are a good example. Women physicians tend to be concentrated in the least prestigious specialties within medicine—pediatrics, obstetrics/gynecology, and general medicine—and are less concentrated in the more prestigious specialties, such as neurosurgery, cardiology, and oncology, where men predominate (American Medical Association, 1997). Even in less prestigious jobs, such as sales, women tend to do noncommissioned sales or sell products that are of less value than those men sell; among waiters and waitresses, women often work in lower-priced restaurants where they are likely to be tipped less than men (Hall, 1993).

Explanations of Gender Segregation. Why is gender segregation still so prevalent, and why do obstacles to mobility at work persist? One explanation is that women and men are socialized differently, and to some degree, they choose to go into different fields. There is some indication that women who are exposed to "masculine" tasks in childhood are more likely as adults to enter jobs that utilize these skills (Padavic, 1991). Conversely, many women shy away from traditionally "male" jobs because they believe that others will disapprove. Gender socialization certainly contributes to why men and women choose the occupations they do, but preference alone does not explain the gender segregation of women and men at work since, when given the opportunity, women will move into jobs traditionally defined as men's work. Although women in such jobs experience sexual harassment, paternalism, and other discouragement from male co-workers and supervisors, and although these

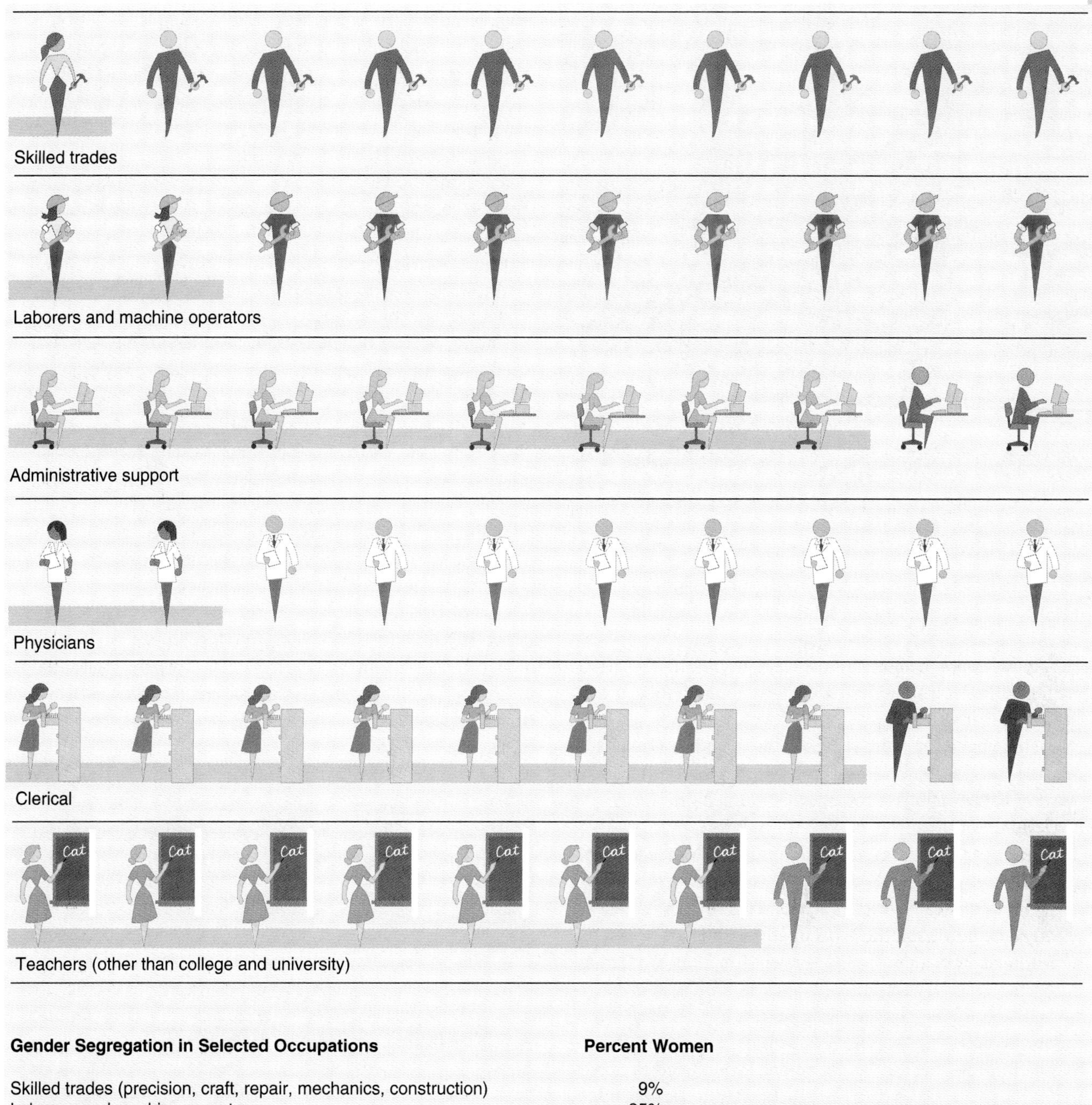

Gender Segregation in Selected Occupations	Percent Women
Skilled trades (precision, craft, repair, mechanics, construction)	9%
Laborers and machine operators	25%
Administrative support	88%
Physicians	26%
Teachers (other than college and university)	76%
Health assessment and treatment (RNs, dietitians, physicians' assistants)	87%
Retail sales	66%
Food service	57%
Personal service (hairdressers, child care, teachers' assistants)	87%

The U.S. Department of Labor regularly publishes data showing the occupational distribution of women and men in the labor force. These data reveal strong patterns of gender segregation, as illustrated by the selected data above. For example, women are 80 percent of clerical workers and 76 percent of elementary and secondary teachers, whereas men are 26 percent of physicians and 9 percent of workers in the skilled trades. You can find other information on gender segregation in the Department of Labor's publication, Employment and Earnings.

SOURCE: AUTH © 1995. The Philadelphia Inquirer. Reprinted with permission of Universal Press Syndicate. All rights reserved.

behaviors affect women's enjoyment of their work, they do not prevent women from being interested in these jobs (Reskin and Padavic, 1988).

A second explanation is that structural obstacles discourage women from entering male-dominated jobs and from advancing once they are employed. A well-known obstacle to even the most successful women has been dubbed the **glass ceiling.** This phrase evokes the subtle yet decisive barrier to advancement that women find in the workplace; they can see the top, they may even get close, but there is an invisible barrier beyond which they find that advancement stops. In 1995, the Glass Ceiling Commission, appointed by Senate majority leader Robert Dole, issued a report concluding that, despite three decades of policies meant to address inequality in the labor market, women and minorities are still substantially blocked from senior management positions (Glass Ceiling Commission, 1995)—a point often made by sociologists who have analyzed the treatment of women as tokens in male-dominated organizations (Kanter, 1977; Yoder, 1991). Interestingly, one study of men and women who are corporate presidents and CEOs has found that women who have risen to the top attribute their success to social networks, in addition to their abilities, whereas men attribute their success only to their own individual effort (Davies-Netzley, 1998).

The glass ceiling exists at every level where men and women work together but are unequally likely to rise to the next level of authority. Most women, in fact, can hardly hope to contend for higher positions, but instead hold jobs that are low in status, poorly paid, and offer little opportunity for advancement. Before women can encounter the glass ceiling, in the words of one researcher, they must pull away from the sticky floor (Berheide, 1992). Others have added that "glass walls" also prevent women from advancing, meaning that they tend to work in parts of work organizations where advancement is less likely. Secretaries, for example, may not be able to move to other professional jobs in an organization. And consider your average grocery store: Women are most likely to be found in the deli or bakery, whereas men are more likely found in produce and groceries, sections that traditionally have provided a better path to management (Reskin and Padavic, 1994).

Another structural barrier to the advancement of women, one located outside the workplace, is women's family responsibilities. These can prevent women from taking some desirable jobs, such as skilled trades that may require working a night shift. This issue never arises, however, for the large number of women, especially working-class and minority women, who have no choice but to sacrifice time with their family by taking whatever jobs they can find to support their households. In fact, among women who take on traditionally male jobs, economic need is their chief motivating factor (Padavic, 1991).

Despite the difficulties imposed, women continue to move into new areas of work and are advancing to some degree. Change is slow, however. As the participation of women in the labor force increases, they continue to hold primary responsibility for meeting the needs of home and family. This additional burden on women has been called "the second shift" and it can be the source of considerable stress (Hochschild, 1989). As we will see when we discuss the women's movement (briefly later in this chapter and also in Chapter 21), the changes that have occurred are the result of women and men organizing to advocate change in the workplace and to promote the development of policies designed to reduce gender inequality.

THINKING SOCIOLOGICALLY

Identify three women working in a field where men are the majority. Ask them about what influences the possibilities for their being promoted within the organization. Do they think a *glass ceiling* exists? Why or why not?

Gender and Diversity

Gender inequality does not exist in a vacuum. As the material discussed here shows, gender inequality overlaps with race and class inequality. At the same time, the experiences of women in the United States are increasingly affected by transformations occurring on a global level. Understanding the diversity of experiences that fall within the broad concept of "gender" is critical to a complete sociological view.

The Intersections of Race, Class, and Gender

The tendency to think of gender as referring only to White women has been one of the criticisms of the women's movement consistently articulated by women of color. Sociologists have to be careful not to generalize from White women's lives even though they have tended to be the norm for many research studies. Feminist scholarship, however, reminds us that working-class women of all races and

glass ceiling

women of color have experiences that are unique to their class, race, and gender position, just as middle-class White women's experience is predicated on their class, race, and gender status. Across class and race, women have many things in common because of the influence of gender in their lives, but their experiences also vary, depending on many other factors, including age, sexual orientation, religion, and any number of other social facts.

Understanding diversity among women (and men) means thinking about how gender shapes social experiences, but at the same time, how it intersects with other social systems. As we have seen throughout this book, race, class, and gender together influence all aspects of people's lives. Although at a given moment in a particular man's or woman's life, either gender, race, or class may feel more salient than the other factors, together each configures the experiences people have. This is what it means to say that race, class, and gender are different, but interrelated, dimensions of social structure (Andersen and Hill Collins, 1998). Each is manifested differently, depending on a group's location in the nexus of gender, race, and class relations. Thus, although we can say, for example, that the income of employed women is less than that of employed men, we have to be careful with such a generalization since the median income for White women exceeds the median income for Hispanic and African American men. Moreover, whereas people often cite the fact that women's income is 74 percent that of men's, among Hispanics and African Americans, women's income more nearly approximates that of men's in their same racial group (comparing those who work year-round, full time); thus, Black women's income is 85 percent of African American men's income, and Hispanic women's income is 91 percent of Hispanic men's income. Comparing women of color to white men, Black women earn 63 percent of White men's earnings; Hispanic women, 54 percent (U.S. Bureau of the Census, 1998a).

None of this is to say that gender is less significant, only that one has to be careful to understand the unique manifestation of experience that comes from different social locations. Likewise, age, sexual orientation, religious affiliation, along with other factors influence gender inequality. Consider, as an example, how Judaism influences the likelihood of men's supporting more feminist values, as Michael Kimmel points out in the box "Analyzing Social Issues." Developing an inclusive perspective means trying to understand the multiplicity of experience, at the same time that one uses the sociological imagination to comprehend the significance of these diverse social factors.

BOX 12.2 • ANALYZING SOCIAL ISSUES

Judaism, Masculinity, and Feminism

Michael Kimmel is a feminist sociologist whose scholarship has made major contributions to the study of gender and men's lives, in particular. In an early essay, Kimmel reflected on how his Jewish identity has, in Kimmel's words, provided the foundation for his participation in the struggle against sexism. Kimmel is not alone in linking Judaism and feminism since many studies have shown that Judaism is linked to more liberal social attitudes, including attitudes about gender issues.

Kimmel argues that the experience of Jewish people has three elements that support feminism despite the sexism in orthodox Jewish tenets. First, Jewish people have been outsiders, as Kimmel says, the symbolic "other," marginalized and oppressed by virtue of their ethnic status. As feminists have argued, being on the outside often helps one see oppression more clearly and to form alliances with other oppressed groups.

Second, Kimmel argues that Jewish men have been stereotyped as weak, emotional, and fragile—a stereotype that he says stems in part from the Jewish emphasis on literacy. As Kimmel writes:

> In my family, at least, to be learned, literate, a rabbi, was the highest aspiration one could possibly have. In a culture characterized by love of learning, literacy may be a mark of dignity. But currently in the United States literacy is a cultural liability. Americans contrast egghead intellectuals, divorced from the real world, with men of action—instinctual, passionate, fierce, and masculine. . . . The Jewish emphasis on literacy has branded us, in the eyes of the world, less than "real" men.

Such stereotypes can be the basis for consciousness-raising, helping Jewish men understand and appreciate the value of so-called feminine characteristics and also being aware of how sexism operates in their own oppression.

Finally, Kimmel argues that the historical experience of Jewish people gives them moral and ethical imperatives toward social justice. Seeing women beaten by husbands and lovers prompts Kimmel to the same moral imperative Jews associate with the Holocaust—"Never Again." He concludes:

> I see my Judaism as reminding me everyday of that moral responsibility, the *special* ethical imperative that my life, as a Jew, gives to me. Our history indicates how we have been excluded from power, but also, as men, we have been privileged by another power. Our Judaism impels us to stand against any power that is illegitimately constituted because we know only too well the consequences of that power. Our ethical vision demands equality and justice, and its achievement is our historical mission. (p. 80)

SOURCE: Kimmel, Michael S. 1987. "Judaism, Masculinity, and Feminism." *Changing Men* (Summer/Fall): 77–80.

Gender in Global Perspective

Increasingly, the economic condition of women and men in the United States is also linked to the fortunes of people in other parts of the world. The growth of a global economy and the availability of a cheaper industrial workforce outside the United States have meant that U.S. workers have become part of an international division of labor. U.S.-based multinational corporations looking around the world for less expensive labor often turn to the third world, and often they find that the cheapest laborers are women or children. The global division of labor is thus acquiring a gendered component, with women workers, usually from the poorest countries, providing a cheap supply of labor for the manufacturing of products that are distributed in the richer industrial nations (see also Chapters 10 and 17).

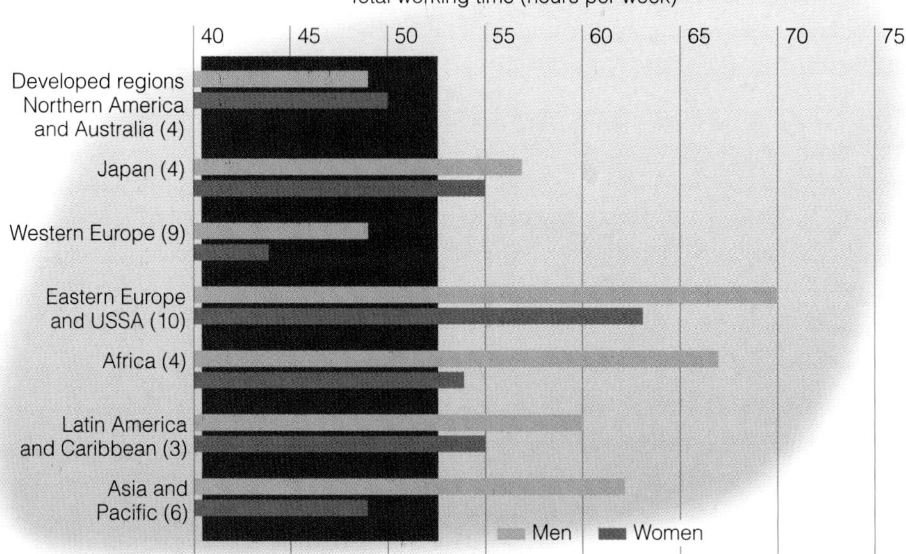

FIGURE 12.6 Women's Work Around the World?

Women in most regions spend as much or more time working than men when unpaid housework is taken into account.

Numbers in parentheses refer to the number of studies in each region.

SOURCE: United Nations. 1991. *The World's Women, 1970–1990: Trends and Statistics.* New York: United Nations.

Worldwide, women work as much or more than men. Worldwide, they receive 30 to 40 percent less pay, and own only 1 percent of all property. It is difficult to find a single place in the world where the workplace is not segregated by gender. Worldwide, women also do most of the work associated with home, child, and elder care. When household labor and child care are included in the measurement of productive labor, it is clear that women do more than half of the world's work (see Figure 12.6); yet, women's work in the home is usually unpaid and often considered to be economically unimportant. It is therefore a surprise to learn that the value of women's unpaid housework in the developed regions has been estimated to be as much as 30 percent of the gross national product (United Nations, 1995).

Gender segregation in the labor market intersects with racial segregation, with jobs employing primarily women of color being among the least paid in society.

Despite these worldwide trends, women's situations differ significantly from nation to nation. China is unusual in that there is far greater sharing of household responsibilities than is true in most other nations. In China, both women and men work long hours in paid employment; a typical work week is six days long or forty-eight hours. Eighty-two percent of women and 83 percent of men are in the paid labor force, and women are encouraged to stay in the labor force when they have children. There are extensive child-care facilities in China and a fifty-six-day paid maternity leave. Many work organizations have extended this paid leave to six months although women can lose seniority rights when they are on maternity leave (something that is illegal in the United States).

In contrast, Japan has marked inequality in the domestic sphere. Women are far more likely to leave the labor force upon marriage or following childbirth and Japanese women's identities are more defined by their roles at home, although, for many, this is changing. Still, compared to China, Japanese women more closely resemble the pattern in Britain and, to some extent, the United States, although they are less involved in paid employment than in either

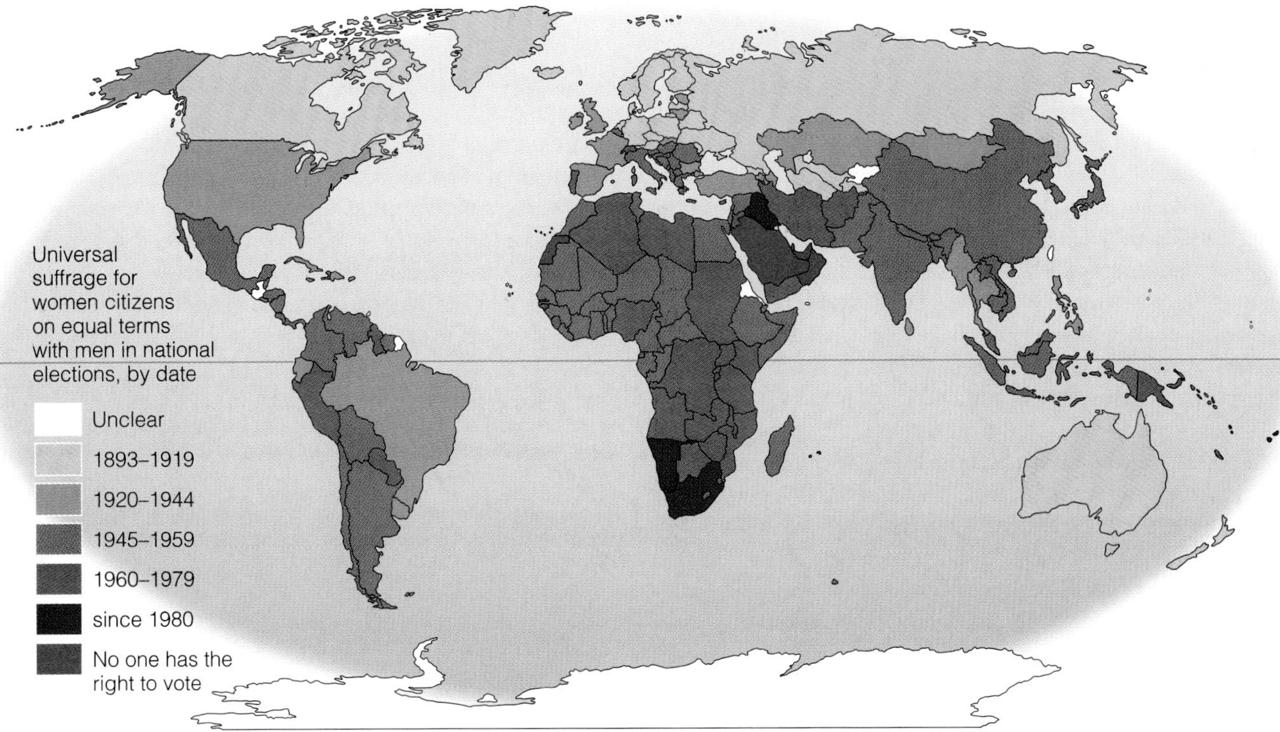

MAP 12.2 Viewing Society in Global Perspective: Women's Suffrage

In Kuwait, women are denied the vote. In Bhutan, most women and some men are not eligible to vote.

SOURCE: From Seager, Joni, 1997. *The State of Women in the World Atlas.* London: © Copyright Myriad Editions Limited, p. 88..

Britain or the United States. Ironically, comparing China, Japan, and Britain, researchers have found that Chinese women are the most discontent with what they perceive as gender injustice, whereas Japanese and British women express greater satisfaction with the more limited employment. This may seem surprising, given the greater gender equality of Chinese women with Chinese men, but sociologists explain it as the result of the gap Chinese women see between official ideologies of gender equality and their observations of continuing inequalities in promotions and other benefits of work (Xuewen et al., 1992).

Work is not the only measure by which the status of women throughout the world is inferior to that of men. Women are rarely in positions of top leadership, and in many sections of the world, the education of women lags behind that of men. Worldwide, illiteracy among men has fallen faster than it has for women; two-thirds of illiterate adults in the world are women (United Nations, 1995; Population Reference Bureau, 1995). A 1994 report from the U.S. State Department also found what it called "widespread abuse" of the world's women—including all forms of discrimination, high rates of violence and sexual abuse, forced sterilization, genital mutilation, and lack of legal protection (Greenhouse, 1994). International women's groups have responded by organizing feminist movements advocating greater security and safety for all women.

Theories of Gender

Why is there gender inequality? The answer to this question is important, not only because it makes us think about the experiences of women and men, but also because it guides attempts to address the persistence of gender injustice. The major theoretical frameworks in sociology provide some answers, but feminist scholars have also found that traditional perspectives in the discipline are inadequate to address the new issues that have emerged from feminist research.

The Frameworks of Sociology

The major frameworks of sociological theory—functionalism, conflict theory, and symbolic interaction—provide some answers to the question of why gender inequality exists, although, as we will see in the next section, feminist scholars have developed new, additional theories to address women's lives directly. Feminists, for example, have been criticized for interpreting gender as a fixed role in society. Functionalist theory purported that men fill instrumental roles in society, whereas women fill expressive roles, and presumed that this arrangement worked to the benefit of society (see Chapter 1). Feminists objected to such a characterization, arguing that this presumed that sexist arrangements were functional for society. Feminists view limiting women's role to expressive functions and men's to instrumental functions as dysfunctional, both for men and women. Although few contemporary functionalist theorists would make such traditionalist arguments, functionalism does emphasize people's socialization into prescribed roles as the major impetus behind gender inequality. Thus, conditions like wage inequality, a functionalist might argue, are the result of choices women make that may result in their inequality but, nonetheless, involve functional adaptation to the competing demands of family and work roles.

Conflict theorists, on the other hand, see women as disadvantaged by power inequities between women and men that are built into the social structure. This includes economic inequity, as well as women's disadvantage in political and social systems. Conflict theorists, for example, see wage inequality as produced from the power that men have historically had to devalue women's work and to benefit as a group from the services that women's labor provides. At the same time, conflict theorists have been much more attuned to the interactions of race, class, and gender inequality since they fundamentally see all forms of inequality as stemming from the differential access to resources that dominant groups in society have.

Functionalism and conflict theory tend to be macrosociological theories—that is, they focus on the broad institutional structure of society to explain gender relations. Symbolic interaction is more microsociological in that it tends to focus on direct social interaction as the context for understanding gender. From a symbolic interaction perspective, feminist scholars have developed what is known as "**doing gender**"—a theoretical perspective on gender that interprets gender as something that is accomplished through the ongoing social interactions people have with one another (West and Zimmerman, 1987; West and Fenstermaker, 1995). Seen from this framework, people "produce" gender through the interaction they have with one another and through the interpretations they have of certain actions and appearances. In other words, gender is not something that is an attribute of different people, as functionalists suggest; rather, it is constantly made up and reproduced through social interaction. When you "act like a man" or "act like a woman," you are confirming gender and reproducing the existing social order.

By implication, from this point of view, gender is relatively easy to change since all it would take for change to occur is for people to behave differently. This is one reason the theory has been criticized by those with a more macrosociological point of view, since they say that it underplays the significance of social structure and the economic and political basis for women's inequality. Critics of the "doing gender" perspective also say that it ignores the power differences and economic differences that exist based on gender, along with race and class, and they conclude that, although this perspective tells us much about how people reproduce gendered behaviors, it does not explain the structural basis of women's oppression (Collins et al., 1995).

The latest theoretical perspective on gender is gendered institutions theory, which we have already discussed. This perspective sees organizations as gendered in such a way that gendered expectations become built into social institutions, without people necessarily recognizing the specifically gendered outcomes that result. In this framework, sociologists do not study individual attitudes or roles, but dissect the structural patterns that construct gender in society.

THINKING SOCIOLOGICALLY

Try an experiment based on the example of changing genders that opens this chapter. For a period of twenty-four hours, note everything you would have to do to change your appearance and behaviors to become the other *gender*. Whenever possible, act in accordance with your new identity and record how others respond to you. What does your experiment tell you about *"doing gender"* and how gender is supported through social interaction?

These sociological frameworks have each provided direction in developing an understanding of the significance of gender in society. Feminist scholars, however, do not see functionalism, conflict theory, or symbolic interaction as adequate in addressing the complexities of women's lives. Although feminist sociologists have been especially influenced by the perspective of conflict theory and, in some cases, by symbolic interaction, they have developed other theoretical frameworks to suggest ways that we can more directly and comprehensively understand women's and, thereby, men's experiences.

Feminist Theory

Feminism has many meanings, but at the very least, it refers to beliefs and actions that support justice, fairness, and equity for all women, regardless of their race, age, class, sexual orientation, or other characteristics. *Feminist theory* refers to analyses that seek to understand the position of women in society for the purposes of bringing about liberating social changes. This link between theory and action is critical in feminist theory since feminists believe that theory is important not just for its own sake, but for purposes of generating change on behalf of all women. Four major frameworks have developed in feminist theory: *liberal feminism, socialist feminism, radical feminism,* and the *multiracial feminism.*

Liberal feminism emerged from a long tradition that began among British liberals in the nineteenth century. Harriet Taylor Mill and John Stuart Mill are the major early thinkers associated with this perspective. Liberal feminism argues that inequality for women originates in traditions of the past that pose barriers to women's advancement. It emphasizes individual rights and equal opportunity as the basis for social justice and social reform. The framework of liberal feminism has been used to support many of the legal changes required to bring about greater equality of women in the United States. Liberal feminists contend that gender socialization contributes to women's inequality because it is through learned customs that inequality is perpetuated. In the interests of social change, liberal feminism advocates the removal of barriers to women's advancement and the development of policies that promote equal rights for women.

Socialist feminism is a more radical perspective that discovers the origins of women's oppression in the system of capitalism. Because women constitute a cheap supply of labor, they are exploited by capitalism in much the way that the working class is exploited. In the view of some socialist feminists, such as Heidi Hartmann, capitalism interacts with patriarchy to make women less powerful both as women and as laborers. Socialist feminists are critical of liberal feminism for not addressing the fundamental inequalities built into capitalist-patriarchal systems. To these feminists, equality for women will come only when the economic and political system is changed.

Radical feminism interprets patriarchy as the primary cause of women's oppression. To radical feminists, the origins of women's oppression lies in men's control over women's bodies; thus, they see violence against women, in the form of rape, sexual harassment, wife-beating, and sexual abuse, as mechanisms that men use to assert their power in society. Radical feminists think that change cannot come about through the existing system since that system is controlled and dominated by men. Catharine MacKinnon, a contemporary feminist legal scholar, is a radical feminist who has argued that all major institutions, such as the state, the economy, and the family, are systems of men's power. Parting with liberal feminists who see state reform as holding the promise to free women (through legislative action and political participation), MacKinnon would argue that since "the state is male," it cannot be the source for liberating changes for women.

Most recently, **multiracial feminism** has developed new avenues of theory for guiding the study of race, class, and gender (Andersen and Collins, 1998; Baca Zinn and Dill, 1996; Chow et al., 1996). Although multiracial feminism is not a single theoretical perspective, it evolves from those who have pointed out that earlier forms of feminist thinking excluded women of color from analysis, making it impossible for feminists to deliver theories that informed us about the experiences of all women. The new scholarship on race, class, and gender studies the influence of all three—not only in the lives of women and men of color, but in the lives of White women and men as well. The sociologist Patricia Hill Collins, for example, has developed the concept of Black feminist theory as a framework for critical thinking about the lives of African American women (Collins, 1998, 1990).

Although multiracial feminism examines the interactive influence of gender, race, and class, this perspective posits race as a major factor influencing the different ways that women's and men's lives are constructed. Central to this perspective is that there is not a universal experience associated with being a woman; rather, different privileges and disadvantages accrue to women (and to men) as the result of their location in a racially stratified and class-based society. Much of the new scholarship emerging from multiracial feminism examines the lives of women of color and shows how race, class, and gender *together* shape the experiences of all women (Baca Zinn and Dill, 1996).

Feminist theory has developed in the context of the feminist movement; it is not theory for theory's sake, but is meant to be the basis for programs for social change. Each perspective provides unique ways of looking at the experiences of women and men in society. These theoretical orientations have been the bedrock upon which feminists have built their programs of social and political change.

TABLE 12.1 *FEMINIST THEORY: COMPARING PERSPECTIVES*

	Liberal Feminism	Socialist Feminism	Radical Feminism	Multiracial Feminism	Gendered Institutions	"Doing Gender"
Gender identity	learned through traditional patterns of gender role socialization	gender division of labor reflects the needs of a capitalist workforce	women's identification with men gives men power over women	women and men of color form an oppositional consciousness as a reaction against oppression	gender is learned in institutional settings that are structured along gender lines	gender is an accomplished activity created through social interaction
Gender inequality	inequality is the result of formal barriers to equal opportunity	gender inequality stems from class relations	patriarchy is the basis for women's powerlessness	race, class, and gender intersect to form a matrix of domination	organizations reproduce inequity by making gendered activity to be "business as usual"	people reproduce race and class inequality through assumptions they make about different groups
Social change	change accomplished through legal reform and attitudinal change	transformation of the gender division of labor accompanies change in the class division of labor	liberation comes as women organize on their own behalf	women of color become agents of feminist change through alliances with other groups	comes through policies that redesign basic structure of institutions	people can alter social relations through changed forms of social interaction

multiracial feminism

Gender and Social Change

Few lives have not been touched by the transformations that have occurred in the wake of the feminist movement. The women's movement has changed how women's issues are perceived in the public consciousness (for more on the women's movement, see Chapter 21). It has created access to work opportunities, generated laws that protect women's rights, and spawned organizations that lobby for public policies on behalf of women. Many young women and men now take for granted freedoms that their generation is the first to enjoy—including access to birth control, equal opportunity legislation, and laws protecting against sexual harassment, as well as an increased presence of women on corporate boards, increased athletic opportunities for women, more presence in political life, and greater access to child care, to name a few. These impressive changes occurred in a relatively short period. How have they affected public consciousness about gender?

Contemporary Attitudes

The public's heightened consciousness about sexism and its inappropriateness in a presumably fair and democratic society is probably one of the most significant results of the feminist movement. In recent years, public attitudes toward gender roles have changed noticeably, especially with regard to conceptions of the ideal lifestyle. Only a small minority of people disapprove of women being employed while they have young children (16 percent of women and 20 percent of men), and both women and men say it is not fair for men to be the sole decision maker in the household. Half of all women and men surveyed in 1995 stated that the ideal lifestyle was to be in a marriage in which husband and wife shared responsibilities, including work, housekeeping, and child care (Roper Organization, 1995b). Although the majority of women now want to combine work and families, they believe they will be discriminated against in the labor force if they do. More than two-thirds of all women feel that men have better job opportunities; 50 percent of men believe this (Gallup, 1993).

One of the most significant changes is the increased stress that women feel when they combine the demands of family and a job. Although most women and men think that society is less critical of women who work while raising children, women and men see both pros and cons to these situations. Both believe that children of employed mothers learn to be more responsible at an early age and that this provides them with positive role models, but there is an increase in the stress that women report as the result of combining family and work. Still, there has been a general decline in the numbers of people who think that having mothers work has a negative effect on children (Roper Organization, 1995b, 1990). Their belief is supported by sociological research, which finds that mother's employment does not in itself affect children's well-being. Having adequate financial resources is a far more important influence on children's development than whether a mother is working outside the home; moreover, employed mothers with high rates of job satisfaction have a positive effect on their children. Typically ignored in this public debate is the effect of fathers' work on children's well-being; interestingly, sociologists have found that fathers' working long hours is correlated with young children having behavior problems. Sociologists conclude from this research that the quality of parents' care for children is far more important than employment per se (Geschwender and Parcel, 1995; Parcel and Menaghan, 1994; Menaghan, 1993; Barnett and Marshall, 1992; Kalmijn, 1994).

Debunking Society's Myths

Myth: Having women work outside the home is harmful to the development of young children.

Sociological perspective: Research finds no evidence of harm to children because of mothers' employment; indeed, the additional financial resources can enhance children's well-being, both in terms of emotional development and educational opportunities. Stronger associations are found between behavior problems in young children and fathers working long hours outside the home.

People's beliefs about appropriate gender roles have evolved as women's and men's lives have changed. Less than half of men (47 percent) now believe that it is best for men to hold the provider role, compared to 69 percent who thought so in 1970. Men's support for women's roles in the family and at work, however, varies across different groups. Not surprisingly, women employed full time are most supportive of nontraditional gender roles; homemakers hold the most traditional views (Cassidy and Warren, 1997). Younger men and single men are more egalitarian than older, married men. Hispanic men are the least likely to support the idea of women's earning money. Some of the differences in the attitudes of African American, Hispanic, and White men are the result of differences in their experiences; minority men are more likely to have had employed mothers, and they have different educational and employment backgrounds from White men. Sociologists have concluded that these attitudinal differences reflect the economic necessity that minority men attribute to women's working (Wilkie, 1993; Blee and Tickamyer, 1995).

Debunking Society's Myths

Myth: The men most likely to support equality for women are White, middle-class men with a good education.

Sociological perspective: Although it is true that younger men tend to be more egalitarian than older men, African American men are the most likely to support women's equal rights and the right of women to work outside the home. On most measures of feminist beliefs, African American men tend to be more liberal than White men.

In spite of these changes, many men still believe in the traditional breadwinner role for men, even though only 15 percent of families are actually supported solely by a man (Wilkie, 1991). Old attitudes do not die easily; moreover, changed attitudes do not necessarily mean changed behavior. For example, although younger men and women are the most liberal on gender issues, in research simulations, college students still pick gender-specific toys for boys and girls (Fisher-Thompson, 1990). At the same time, surveys of contemporary women show that women are more critical of men than they were in the past. Gender attitudes change as society changes; thus, it is likely that further adjustments in the attitudes of men and women are on the way since it seems unlikely that the changes in men's and women's roles will reverse in the future. Attitudes, however, are only part of the problem of persistent gender inequality. There is still significant resistance to change in institutionalized gender practices. Social change requires more than changing individual attitudes. Supporting more egalitarian gender relations also means changing social institutions.

Following the Million Man March in 1996, which advocated men's taking responsibility for their families and communities, feminists debated whether this was a positive sign or just the reassertion of men's traditional roles at women's expense.

ED: "men's"? *comp*

Legislative Change

There is much legislation in place that prohibits, in theory, overt discrimination against women. In addition to the Equal Pay Act of 1963, the Civil Rights Bill of 1964, adopted as the result of political pressure from the civil rights movement, banned discrimination in voting and public accommodations and required fair employment practices. Specifically, Title VII of the Civil Rights Act of 1964 forbids discrimination in employment on the basis of race, color, national origin, religion, or sex. It is almost accidental that prohibitions against sex discrimination are part of this bill. The word *sex* was added as a last-ditch effort by conservative senators who thought the idea of including women was so ludicrous that it would defeat passage of the bill. They argued that women's place was in the home and that adoption of the bill would upset "natural" differences between the sexes. As one senator put it, "What would become of traditional family relationships. . . .You know the biological differences between the sexes. . . . The list of foreseeable consequences . . . is unlimited" (Senator Emanual Celler, Congressional Record, February 8, 1964, p. 2578). Some of the supporters of the bill also appealed to White racism to promote the bill's adoption, arguing that it would be wrong to elevate the rights of Black women over those of White women. As one senator put it, "Unless this amendment is adopted, the white women of this country would be drastically discriminated against in favor of a Negro woman" (Congressional Record, p. 2583; Dietch, 1993).

The passage of the Civil Rights Bill, and Title VII in particular, opened up new opportunities to women in employment and education. This was further supported by Title IX, adopted as part of the Educational Amendments of 1972, which forbids gender discrimination in any educational institution receiving federal funds. Title IX prohibits colleges and universities from receiving federal funds if they discriminate against women in any program, including athletics. Adoption of this bill radically altered the opportunities available to women students and laid the foundation for many of the co-educational programs that are now an ordinary part of college life. This law has been particularly effective in opening up athletics to women. In 1997, the U.S. Supreme Court further strengthened Title IX by refusing to rule on a case that supported the principles embedded in Title IX.

Passage of antidiscrimination policies does not, however, guarantee their universal implementation. Has equality been achieved? In college sports, men still outnumber women athletes by more than two to one, and there is still more scholarship support for male athletes than for women. Title IX allows institutions to spend more money on male athletes if they outnumber women athletes, but it also stipulates that the number of male and female athletes should be roughly proportional to their representation in the student body. Studies of student athletes show that, though there has been improvement in support for women's athletics since the implementation of Title IX, there is a long way to go toward equity in women's sports (Lederman, 1992).

Furthermore, some worry that in implementing Title IX, the scholarships that have been important in supporting the education of African American men, many of whom are supported through athletic scholarships, will be eroded in favor of sports that support White women. Since many colleges are responding to Title IX by reducing support for men's athletics, this is a serious concern and suggests how delicate the balance is in developing programs to support both race and gender equity (Greenlee, 1996).

In the workplace, a strong legal framework for gender equity is in place, yet equity has not been achieved. Since most women work in different jobs from men, principles of equal pay for equal work do not address all the inequities women experience in the labor market. **Comparable worth** is the principle of paying women and men equivalent wages for jobs involving similar levels of skill. This policy recognizes that men and women tend to work at different jobs. Comparable worth goes beyond the concept of equal pay for equal work by creating job evaluation systems that assess the degree of similarity between different kinds of jobs. Comparable worth schemes have been introduced only in a few places, but the states of Iowa and Minnesota did so for state employees, as did the city of San Jose, some private corporations, and a few other places. Where comparable worth plans have been implemented, women's wages have improved (Blum, 1991; Michel et al., 1989; Steinberg, 1992; Jacobs and Steinberg, 1990).

Many victories in the fight for gender equity are also now at risk. *Affirmative action* is a method for opening opportunities to women and minorities that specifically redresses past discrimination by taking positive measures to recruit and hire previously disadvantaged groups. Affirmative action has been especially effective in opening new opportunities for women although its effectiveness has been undermined by a national climate that is increasingly critical of affirmative action. The national discussion of affirmative action, fueled in part by conservative resistance to these policies, is a good illustration of what many would call the limitations of liberal philosophy. Because affirmative action promotes gender- and race-specific actions to remedy the effects of past discrimination, its opponents have argued that it constitutes "reverse discrimination." Those who want social policies that are gender and race blind—a classic liberal position, now articulated by conservatives, as well—find it difficult to support policies other than those that are gender and race blind. The problem is that gender and race inequities continue even though these inequities may not always result from conscious discrimination. Proponents of affirmative action argue that, as long as the structural conditions of gender and race inequality exist, there is still a need for race- and gender-conscious actions designed to address persistent injustices.

Some feminists argue that relying on government-sponsored programs to create meaningful change for women is simply not adequate. Radical feminists who see the government dominated by men think women should not rely on a male-dominated system to protect their interests. It is not enough, some feel, simply to change the laws; it is necessary to change the composition of the government and for women to act collectively on their own behalf.

One solution to the problem of gender inequality is to have more women in positions of public power. Is increasing the representation of women in existing institutions enough? Without reforming the sexism in the institutions, change will be limited and may generate benefits only for groups who are already privileged. Feminists advocate restructuring social institutions to meet the needs of all groups, not just those who already have enough power and privilege to make social institutions work for them. The successes of the women's movement demonstrate that change is possible, but change comes only when people are vigilant about their needs.

comparable worth

CHAPTER SUMMARY

- Sociologists use *sex identity* to refer to biological identity and *gender* to refer to the socially learned expectations associated with members of each sex. *Biological determinism* refers to explanations that attribute complex social phenomena to physical or natural characteristics. Studies of hermaphrodites (those of biologically mixed sex) show that biology alone does not produce gender differences. Biological and social systems are interrelated.
- *Gender socialization* is the process by which gender expectations are learned. One result of socialization is the formation of gender identity. Gender identity also develops alongside racial identity. Homophobia plays a role in gender socialization because it encourages strict conformity to gender expectations.
- *Gendered institutions* are those where the entire institution is patterned by gender. Sociologists analyze gender both as a learned attribute and as an institutional structure. *Gender stratification* refers to the hierarchical distribution of social and economic resources according to gender. Most societies have some form of gender stratification although they differ in the degree and kind.
- Gender stratification in the United States is obvious in the difference in men's and women's wages. There are several explanations for pay inequality between men and women, including human capital theory, dual labor market theory, and overt discrimination. *Human capital theory* explains wage differences as the result of individual differences between workers. *Dual labor market theory* refers to the tendency for the labor market to be organized in two sectors: the primary and secondary market. Jobs in the primary market are higher paying and more prestigious, and provide greater opportunity for advancement, than do those in the secondary market. Women and racial minorities tend to be concentrated in the secondary labor market. Continuing overt *discrimination* against women is another way that men protect their privilege in the labor market.
- *Gender segregation* refers to the unequal distribution of men and women in different job categories. Women are segregated both within and across occupations. Numerous structural barriers exist that discourage women's advancement at work; these barriers are popularly referred to as the *glass ceiling*. Gender inequality is a worldwide phenomenon.
- Different sociological theories have emerged from functionalism, conflict theory, and symbolic interaction to explain the position of women in society. *Liberal feminism, socialist feminism, radical feminism,* and *multiracial feminism* also contribute to our understanding of the status of women.
- Public attitudes about gender relations have changed dramatically in recent years. Women and men are now more egalitarian in their attitudes although women still perceive high degrees of discrimination in the labor force. A legal framework is in place to protect against discrimination, but legal reform is not enough to create gender equity.

KEY TERMS

biological determinism
comparable worth
discrimination
doing gender
dual labor market theory
feminism
fetal sexual differentiation
gender
gendered institution
gender identity
gender segregation
gender socialization
gender stratification
glass ceiling
hermaphroditism
human capital theory
ideology
index of dissimilarity
labor force participation rate
liberal feminism
matriarchy
multiracial feminism
occupational segregation
patriarchy
radical feminism
sex
sexism
socialist feminism

THE INTERNET: A Tool for the Sociological Imagination

Resources on the Internet:

Virtual Society: The Wadsworth Sociology Resource Center
`http://sociology.wadsworth.com`

Visit this site to find additional learning tools, including interactive quizzes, links to related web sites, and an easy link to *Infotrac College Edition.*.

Institute for Women's Policy Research
`http://www.iwpr.org/`

This is a Washington-based research organization, working to establish public policies for the benefit of women. The web site includes numerous publications from the Institute, which provide data and research on women's employment, poverty and welfare, work and family, and other matters.

U.S. Department of Labor: Women's Bureau
`http://gatekeeper.dol.gov/dol/wb`

The Women's Bureau is an important source of government data about women's employment. The site includes reference to numerous publications and fact sheets.

Menweb
`http://www.vix.com/menmag/`

This site includes interviews, references to articles, and citations to numerous books pertinent to men and feminism.

University of Maryland Women's Studies Program
http://www.inform.umd.edu/EdRes/Colleges/ARHU/Depts/WomensStudies/

This site, from one of the largest women's studies programs in the nation, provides extensive information on women, including reviews of films and books pertinent to studying women, information about women's studies and feminist organizations, and many other sites of interest to women's studies students.

Sociology and Social Policy: Internet Exercises

For many years, those who have supported *single-sex education* have noted that girls and women are more successful in single-sex schools. Supporters point out that in such settings women are not discouraged from pursuing subjects such as math and science, and they are more likely to emerge as leaders in the campus setting. In recent years, promoting single-sex education has resurfaced in a different context: the promotion of all-male schools for young African American boys. Developing all-male schools for African American boys, according to proponents, enhances the self-esteem of those otherwise ignored in schools and, as a result, improves their learning and opportunities for success. What are the arguments for and against single-sex education? Does the particular school context (whether women's colleges, military academies, or all-male schools for African American boys) change the argument? What do you conclude about the viability of single-sex education and its impact on educational success?

Internet Search Keywords:

single-sex education
gender issues
educational reform
African American boys
women's colleges

Web sites:

http://www.ed.gov/offices/OERI/PLLI/ipp.html
U.S. Department of Education web site, offers articles on women's colleges and historically Black colleges and universities.

http://www.education-world.com/
This site discusses current education topics.

http://www.rmcres.com/improve/index.html
A web page dedicated to educational improvement.

http://k12.school.net
This page provides educational information and links to schools.

http://education.indiana.edu/cas/tt/tthmpg.html
Home page of Teacher Talk. A site for teachers to discuss issues such as single-sex education.

InfoTrac College Edition: Search Word Summary

comparable worth	glass ceiling
gender socialization	multiracial feminism

In order to learn more about these central topics in sociology, you can conduct an electronic search using InfoTrac College Edition. To aid in your search and to gain useful tips, see the Student Guide to InfoTrac College Edition on the Virtual Society web site:
http://sociology.wadsworth.com

INTERACTIONS—A SOCIOLOGY CD-ROM: CONCEPTS FOR THIS CHAPTER

Go to the Wadsworth Sociology CD-ROM for further study on the concepts in this chapter. The CD-ROM also includes quizzes and additional activities to expand your learning experience.

SUGGESTED READINGS

Andersen, Margaret. 2000. *Thinking About Women: Sociological Perspectives on Sex and Gender,* 5th ed. Boston: Allyn and Bacon.

This is a comprehensive review of feminist studies about women. Widely read by undergraduates, it provides an understanding of the sociological analysis of women's lives and a basic introduction to feminist thinking.

Baca Zinn, Maxine, Pierrette Hondagneu-Sotelo, and Michael Messner. 1997. *Through the Prism of Difference.* Needham Heights, MA: Allyn and Bacon.

This anthology includes articles that explore race, class, and sexual diversity among women and men. Together, the essays challenge false generalizations about women and men as monolithic groups while revealing the structural forces that shape gender relations in the United States.

Cherry, Lynn, and Elizabeth Reba Weise, eds. 1996. *Wired Women: Gender and New Realities in Cyberspace.* Seattle: Seal Press.

This collection of essays explores numerous topics about how women are marginalized and often harassed in the high-tech environment of cyberspace. Covering topics as diverse as hackers, virtual sex, and women as cyberspace fans, the book links the study of gender to newly emerging technologies.

Collins, Patricia Hill. 1998. *Fighting Words.* Minneapolis: University of Minnesota Press.

Collins analyzes the experiences and perspectives on African American women by developing a theory of Black feminism. In this context, she also examines contemporary discussions of Afrocentrism and Black women's lives, the contributions of postmodernist theory, and a global perspective on Black feminism.

de la Torre, Adela, and Beatríz M. Pesquera. 1993. *Building with Our Hands: New Directions in Chicana Studies.* Berkeley: University of California Press.

The articles in this anthology reflect new feminist scholarship on Chicanas. The collection includes historical and contemporary pieces on images of Chicana women, their roles in work, families, and education.

Fausto-Sterling, Anne. 1992. *Myths of Gender,* 2nd ed. New York: Basic Books.

Fausto-Sterling is a biologist who uses her expertise to explore various biological myths about men's and women's lives. Presented in an engaging and accessible way, she provides a clear understanding of biological facts about such areas as sex differences between men and women, menopause, hormones, and health.

Kimmel, Michael S., and Michael A. Messner. 1998. *Men's Lives,* 4th ed. Needham Heights, MA: Allyn and Bacon.

This is an excellent collection of articles on men's experiences. It is notable for its inclusion of diverse groups, including racial minority groups and gay men. The subjects covered include families, work, sexuality, friendships, health, and sports, among others.

CHAPTER 13

Age and Aging

IT IS Thanksgiving Day and a family gathers for dinner. Represented at the table are five generations, from great grandmother to the newest grandchild. Same family, different generations—each facing experiences particular to his or her age group. Great grandma lives alone, but worries about how long she will be able to do so and how she will afford extended long-term care, should she need it. Grandmother and grandfather have just returned from an elderhostel where they combined their love of travel and lifelong learning; they are enjoying their retirement, but wonder how long their health will allow them to savor their older years. The mother is busily preparing last-minute parts of the meal; she is exhausted from having cooked every evening this week after coming home from her job. The father is watching football with his daughter who is home from college; during commercial breaks, they talk about her concerns about finding a good job when she graduates. He wonders if she will decide to live at home again once she graduates since rents in the city are so high. The family's teenaged son is in a surly mood; he'd rather be hanging out at the local mall with his friends. Another son and daughter-in-law are tending their own children, including a newborn. Their children represent the family's hopes for the future, although the parents wonder about the world these young children will face as they grow older. Reports of violence in the schools, overcrowded educational facilities, and the increasing cost of raising children make them wonder what the future holds for their children.

In a different family, or a different time, each generation would face distinct challenges, but their experiences would nonetheless be shaped by their age, like those in the family introduced here. Groucho Marx (1959: 4) once said, "Age is not a particularly interesting subject. Anyone can get old. All you have to do is live long enough." Sociologists, much as they might like Groucho, disagree. Yes, people inevitably grow older, but sociologists find age and its relationship to society to be fascinating. Sociologists use their sociological skills to understand the social aspects of aging.

Society differentiates people on the basis of age; thus, different age groups experience different life situations—situations that are further shaped by people's race, class, and gender. **Age stratification** refers to the hierarchical ranking of different age groups in society. Studying the experiences of differently situated generations reveals that there are unique social expectations and different life chances for different age groups in society. Simply being part of a given generation can shape much of your life experience, and being of a certain age also influences the opportunities available to you.

Understanding the sociology of age reveals that age matters, not just to individuals, but also to the structure of society. Sociologists understand that the age composition of a society makes a difference in the social issues that society faces and in how well social institutions serve different generations of people. Currently, for example, the United States has a population with an increasing proportion of older people—a phenomenon referred to as the "*graying of America.*" More older people are in the population than at any previous time in history—both in actual number and as a proportion of the total population. Since 1950, the number of those over age fifty has doubled and could reach eighty million by the year 2050. The aging of the so-called Baby Boom generation—those born in the late 1940s and 1950s—means that in the near future, this aging group will become the single largest age group in the country (see Figure 13.1).

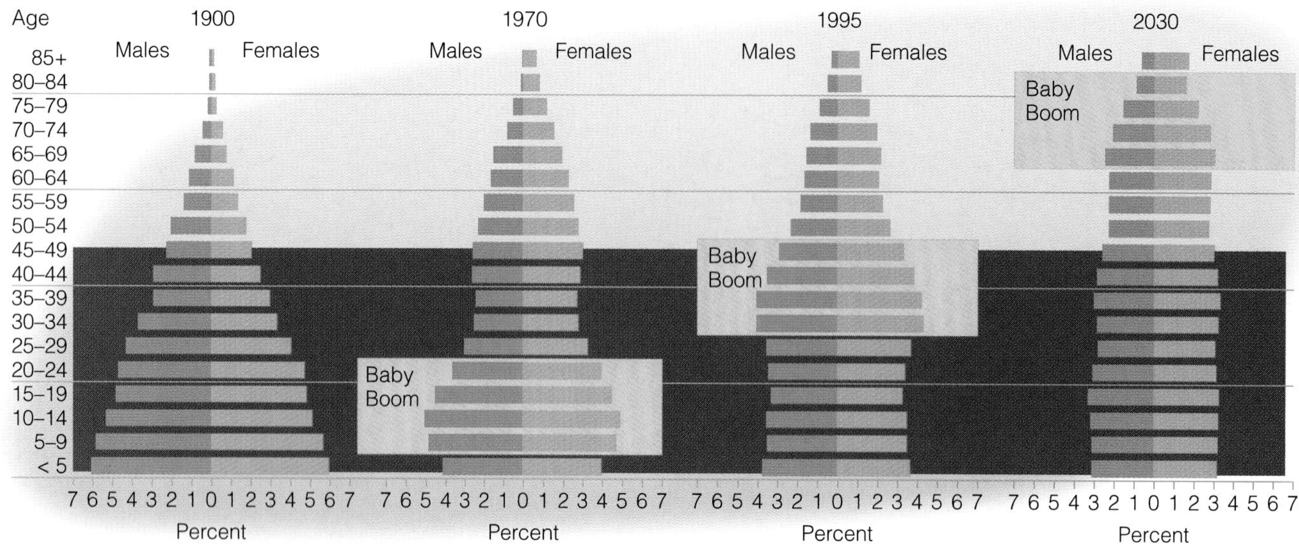

FIGURE 13.1 An Aging Society

SOURCE: U.S. Bureau of the Census. 1975. *Historical Statistics of the United States: Colonial Times to 1970* (Washington, DC: U.S. Government Printing Office); U.S. Bureau of the Census. 1993. *Current Population Reports,* Series P-25-1104. (Washington, DC: U.S. Government Printing Office).

To see the magnitude of this change recognize that in 1900 only 4 percent of the population was over age sixty-five; by the end of the century, it will be 13 percent, and by the year 2030, 20 percent. The number of those in the category of the "oldest old"—those over eighty-five years of age—is also likely to increase from 1 percent of the population in 1970 to 5 percent by the middle of the coming century. In fact, in only twenty-five years, the proportion of the U.S. population who are over sixty-five will be identical to that of Florida now (Krugman, 1996; Treas, 1995; U.S. Bureau of the Census, 1997a).

This grandmother, in her 70s, is teaching origami, the Japanese art of paper-folding, to her granddaughter. In Japanese culture, the elderly are held in high esteem and are regarded as possessing great knowledge.

At the same time, poverty among U.S. children is increasing; whereas not long ago, the elderly were the group most likely to be poor, now children are the most impoverished (see Figure 9.5 on page 241). These changes in the population structure will pose new dilemmas for providing social services to different population segments. How, for example, will the younger generation be able to care for their elders?

Debunking Society's Myths

Myth: **Old people are the group in society most likely to be poor.**

Sociological perspective: **Although there is significant poverty among the elderly, the highest rates of poverty now, unlike in the past, are among children (U.S. Bureau of the Census, 1997b).**

Under these changing social conditions, the analyses that sociologists provide of age and its relationship to society become even more critical. Some of the questions that sociologists ask about age and society include: How are people's experiences shaped by social expectations about age? How, for example, does society shape the social meaning of aging differently for women and for men? What stereotypes have developed about aging, and how are they communicated through popular culture? What opportunities and obstacles do you face in society, depending on your age generation?

Sociologists recognize that society shapes the different phases of life that people experience as they grow up and grow older. Sociologists also study the inequalities associated with different age groups and, within age groups, how

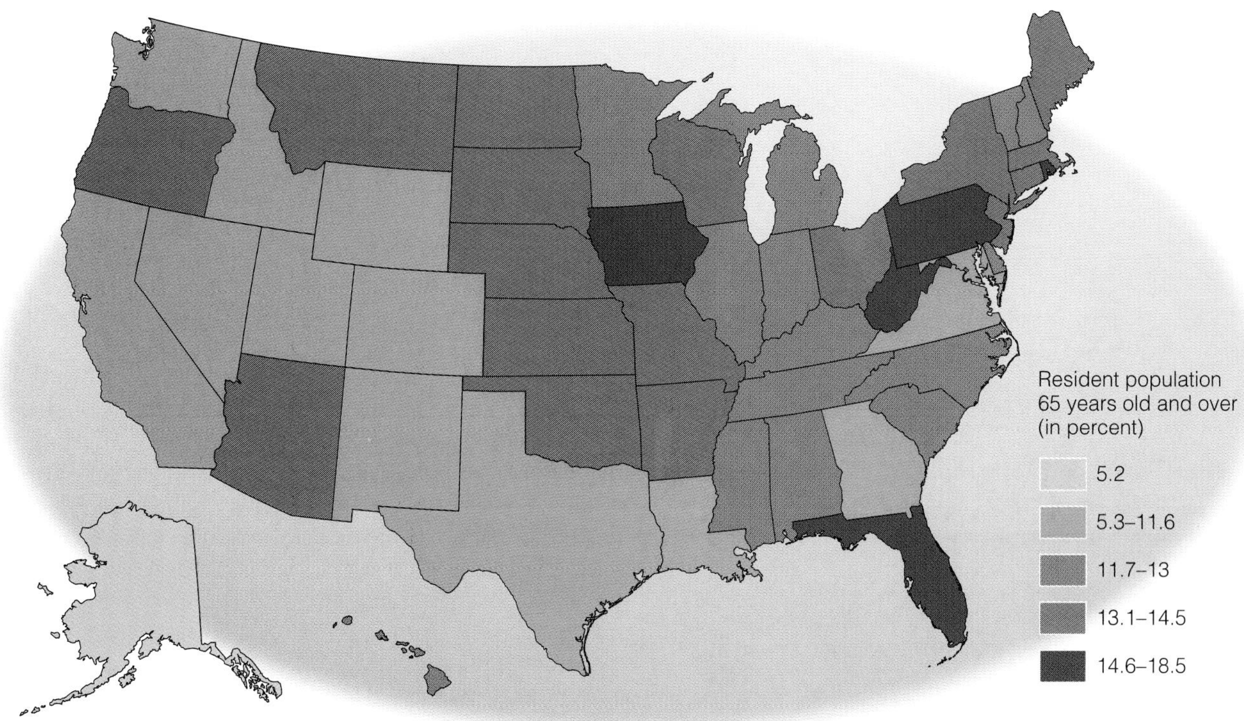

MAP 13.1 Mapping America's Diversity: Where the Aged Live

DATA: From the U.S. Bureau of the Census. 1997. *Statistical Abstract of the United States, 1997.* Washington, DC: U.S. Government Printing Office, p. 33.

class, race, and gender differentiate people's experiences of aging. Growing old in poverty versus growing old in affluence are markedly different experiences. In this chapter, we examine these different dimensions of the aging experience and consider how society meets the needs of different age groups. To begin this discussion, we first examine the social dimensions of aging.

The Social Significance of Aging

The process of aging involves numerous factors—physical, psychological, and social. Sociologists are most interested in the social dimensions of aging, and they also see social factors as important in comprehending the physical aspects of growing up and growing older. As you examine the physical process of aging, you will see that social factors have a great influence even on the process as "natural" as growing old.

The Physical Process of Aging

It is easy to think that aging is just a natural fact. Despite desperate attempts to hide gray hair, eliminate wrinkles, and reduce middle-aged bulge, aging is an inevitable process. The skin creases and sags, the hair thins, metabolism slows, and one's bones become more "hollow" by losing bone mass.

Growing old includes the wrinkling of skin, hair loss, and graying; a decrease in muscular and tissue elasticity; shrinkage of the disks in the spinal column (causing some elderly people to lose inches of height); susceptibility to heart disease; and some loss of kidney function. Changes in the senses also result from aging. Your vision is likely to become impaired, and your ability to hear lessens. Some older people lose their sense of touch, taste, and smell. As you grow older, you may also find your cognitive responses slowing down, though not necessarily.

Aging affects people differently since some people appear to age much more rapidly than others. Some sixty-year-olds look only forty, and some forty-year-olds look as haggard as a careworn sixty-year-old. Such differences are the result of biological and social factors, such as eating habits, stress, smoking habits, pollution in the physical environment, and many other such environmental factors. Lack of physical exercise can speed up the aging process. People who lead inactive lives tend to look, act, and feel older than people of the same age who are physically more active. Lifestyle choices do not supply a "fountain of youth," yet with an appropriate diet, exercise, and a positive attitude, some of the *physical* aspects of the aging process can be slowed down.

Older persons have slower psychomotor responses and reflexes. In this loss of physical capability, however, some benefits may appear. For example, older persons type at slower speeds than younger people, but they make fewer errors per 100 words typed (Salthouse, 1984). The older people get, the less accurate their short-term memory becomes, but they can remember events in the distant past with

At 75 years of age, former senator and astronaut John Glenn demonstrated that age was no necessary limitation to participation in space travel.

great accuracy. Social stereotypes of old people as "senile" have assumed that old people inevitably become forgetful and dumb, but this is not true. Aging can be accompanied by memory loss, both short term and long term, as well as some loss of reasoning ability, but this is not inevitable. Although some measures of intelligence may decline slightly with age, such as speed in solving math problems and speed in overall problem solving, other dimensions of intelligence may *increase* with age. Some artistic abilities, such as painting, have been shown to develop and flourish in later life, while the most prominent gift of age may be the sheer amount of accumulated knowledge and overall "wisdom."

The aging process also has psychological effects, which are linked to physical changes. For example, the hearing loss that accompanies advancing age can have psychological consequences. Many persons over the age of sixty-five suffer enough hearing loss that they think those around them are speaking in muffled tones. People who believe they are being whispered about are likely to find their social situation highly distressing and are liable to succumb to paranoia and depression. Whereas it was once believed that depression was a physical symptom of aging, many now believe that psychological and social causes are the basis for such depression since depression is not an inevitable consequence of aging. Experiencing stressful events and lacking adequate social support networks are most commonly related to depression among the elderly, just as they are related to depression among other age groups (Zisook et al., 1994).

Social stereotypes about the aged and their presumed "senility" have sometimes led people to overlook the physiological basis of some problems associated with aging. **Dementia** is the term used to describe a variety of diseases that involve some permanent damage to the brain. Dementia usually involves an impaired awareness of one's self and surroundings, mental disorientation, memory loss, and tendencies toward delusions and hallucinations. Some forms of dementia may be short term and are treatable. Others are irreversible and degenerative, the most common of which is Alzheimer's disease.

Alzheimer's disease is a degenerative form of dementia that involves neurological changes in the brain. New medical research is revealing more understanding of Alzheimer's disease. Once thought to be rare, Alzheimer's disease is now known to occur among approximately 10 percent of the population over aged sixty-five (Kail and Cavanaugh, 1996; Cavanaugh, 1997), and the number of diagnosed cases is increasing, primarily because of increased understanding and ability to diagnose this disease (Alzheimer's Association, 1996). Ten percent of all U.S. families are affected by Alzheimer's disease; 41 percent of those reporting having family members with Alzheimer's are themselves between the ages of eighteen and thirty-four. As the president of the Alzheimer's Association has said, "This is no longer an 'old age' problem. It's affecting Americans in the prime of their lives, whether they have the disease or take care of someone who does" (Alzheimer's Association, 1996: 1).

Alzheimer's disease is not necessarily age related although it is most common among older people. The major symptoms of Alzheimer's include gradual changes in a person's mental functioning, specifically a decline in memory (especially short-term memory), learning, attention, and judgment. Someone with Alzheimer's disease may be disoriented in time, have difficulty communicating with others, engage in inappropriate social behavior, and experience changes in personality. These symptoms are typically mild in the beginning, but typically get progressively worse.

Research now shows a possible genetic link to Alzheimer's disease (Sherrington et al., 1995), but regardless of its cause, Alzheimer's disease creates numerous social strains. Alzheimer's patients are difficult to care for, putting great stress on families and other caregivers (Barber and Paisley, 1995; Pearlin et al., 1994; Lyman, 1993). People with Alzheimer's disease typically lose memory of their most recent years, while retaining memories of their earlier years; this means they may not remember their spouses or children—those who are most likely to be caring for them and who therefore may suffer the emotional pain of having been forgotten by the person they love (Reid, 1994).

Social Factors in the Aging Process

From a sociological perspective, the social dimensions of aging are just as important, if not more so, in determining how particular people and groups experience the aging process. Take the changes in reproductive functioning as an example. Just as puberty brings on changes in the reproductive system of young people, midlife also brings major changes in the reproductive system. Most people think of midlife as a time when women's reproductive systems change, but men's reproductive systems change, too—there is just not a particular event like menopause that socially defines these changes. Nonetheless, for men at midlife, there is a decline in the quantity of sperm produced, although most remain fertile; some men may even produce children when they are quite old. As men age, prostate cancer also becomes more of a threat, and there is some reduction in men's testosterone levels after their midtwenties (Cavanaugh, 1997).

For women, *menopause* is the time when ovulation stops and women can no longer bear children. Simply stated, menopause is the biological cessation of the menstrual cycle, but menopause, like menstruation, is imbued with cultural and social meaning. Menopausal women are culturally depicted as cranky, unable to control their emotions, and highly prone to irritability and depression, *despite the evidence* that menopause is not related to serious depression among women. To the contrary, some studies find that the majority of women feel happy about the loss of childbearing ability; studies also find that a majority of menopausal women do not experience the stereotypical hot flashes and other physical symptoms generally thought to be associated with menopause (Fausto-Sterling, 1992). Although social stereotypes of menopause are fixed in many people's minds, the symptoms of menopause depicted in social stereotypes are less widespread than generally believed. How a woman experiences menopause depends not just on the physiological changes occurring in her body, but on how she adapts to the social attitudes that culturally define menopause.

Cultural understandings of events like menopause also change over time. Whereas for earlier generations menopause was something one experienced silently, as the Baby Boomer generation has approached menopause, it has become a widely discussed and recognized experience. New markets have developed to dispense advice to women about menopause, and women themselves have developed new support networks through which they share their feelings and experiences of menopause. For many of these women, menopause has been redefined from being a time of sadness and depression to being cause for celebration, humor, and self-affirmation. This changed attitude is well depicted in the slogan that was popularized in the 1990s and sometimes seen on bumper stickers: "Those are not hot flashes—they're power surges!" Attitudes change, and as a result, people's experience of the aging process can change as well.

Social influences on the physiology of aging are also evident in comparing the life expectancy of different groups. **Life expectancy** is the probable number of average years a particular group is likely to live, given aggregate statistical patterns; it is based on the age at which half the people born in a particular year die. Life expectancy is clearly shaped by several social factors, including gender, race, and social class. Women live longer on average than men, but the life expectancies of minorities of both genders are shorter than those of Whites. Although 75 percent of all White males will reach sixty-five, only 58 percent of all Black males will do so. Eighty-six percent of all White women will survive to age sixty-five or older (more than White men), whereas only 75 percent of Black women will do so (the same percentage as White men). Adding in the effects of social class further differentiates life expectancy patterns. The life expectancies of men and women who are poor are shorter than those who are middle and upper class. As a result, only about 60 percent of poor White males will reach the age of sixty-five—about the same as for Black males of all social classes. About 75 percent of poor White women will reach the age of sixty-five—about the same percent as all Black women and all White men. Only 65 percent of poor Black women, and only 48 percent of poor Black men—less than half—will reach age sixty-five (U.S. Bureau of the Census, 1997a).

These facts remind us that aging may be an inevitable process, but how it is experienced, indeed, how long one is even likely to live, is a social fact. Age, like class, race, and gender, is a structural feature of society. The age structure of society shapes people's opportunities, is the basis for social expectations, and plays an important part in the development of social change.

Age Stereotypes

Important as the physiological changes that accompany aging are, they pale beside the social and cultural aspects of aging. Although the physiology of aging proceeds according to biological processes, what it means to grow older is a social phenomenon. The definitions applied to different groups make the experience of growing older highly dependent on one's social circumstances. Much of the meaning of growing old in this society is embedded in *social stereotypes.* Recall that a stereotype is an oversimplified categorization of beliefs about the characteristics of members of a group (see Chapter 11).

Age stereotypes are preconceived judgments about what different age groups are like. Although stereotyping of the elderly in our society is more pervasive than stereotyping of youth, both groups are burdened by negative preconceptions. Studies routinely find that adults perceive teenagers to be irresponsible, addicted to loud music, lazy, sloppy, and so on—similar to stereotypes often directed against racial minorities. Just as derogatory language is sometimes used to describe members of minority groups, so too are pejorative terms (such as "curmudgeon," "old biddy," and "old fart") used to describe the elderly. Studies have also uncovered the common stereotyping of the elderly as senile, forgetful, set in their ways, meddlesome, conservative, inactive, unproductive, lonely, mentally dim, and uninterested in sex (Levin and Levin, 1980; Palmore, 1977, 1986). These stereotypes are largely myths, but they are widely believed (see Table 13.2 on page 360).

Stereotypes of aging can be both positive and negative. On the positive side, older people are viewed as all-knowing, happy, loving, and understanding; on the negative side, old people are viewed as senile, dependent, forgetful, complaining, and inflexible (Palmore, 1986). Negative stereotypes of adolescents are that they are lazy, slovenly, and aimless; positive stereotypes of children are that they are adorable, innocent, and charming.

Age stereotypes also differ for different groups. Older women are stereotyped as having lost their sexual appeal, quite contrary to the stereotype of older men as handsome or "dashing" and desirable. Gender is, in fact, one of the most significant factors in age stereotypes. Women may even be viewed as becoming "old" sooner than men since people describe women as "old" a decade sooner than they describe men as old (Stoller and Gibson, 1997: 77). Old age also brings a greater loss of social status for women; older women are seen as "over the hill." As Shevy Healey describes in the box "Analyzing Social Issues: Growing to be an Old Woman," these social definitions of gender and aging denigrate older women and associate beauty for women only with being young. For those who accept these social stereotypes, depression, low self-esteem, and perhaps even physical harm can result.

Age stereotypes are reinforced through popular culture. Advertisements depict women as needing creams and lo-

B • O • X 13.1 • ANALYZING SOCIAL ISSUES

Growing to Be an Old Woman

UNDER the guise of making themselves beautiful, women have endured torture and self-mutilation, cramped their bodies physically, maimed themselves mentally, all in order to please and serve men better, as men defined the serving, because only in that service could women survive.

Footbinding, for example, reminds us how false and relatively ephemeral those external standards of beauty and sexuality are. We can feel horror at what centuries of Chinese women had to endure. Do we feel the same horror at the process which determines that gray hair, wrinkled skin, fleshier bodies are not beautiful, and therefore ought to be disguised, pounded, starved to meet an equally unrealistic (and bizarre) standard set by the patriarchy? Women spend their lives accepting the premise that to be beautiful one must be young, and only beauty saves one from being discarded. The desperation with which women work to remove signs of aging attests only to the value they place upon themselves as desirable and worthwhile, in being primarily an object pleasing to men. Women's survival, both physical and psychological, has been linked to their ability to please men, and the standards set are reinforced over and over by all the power that the patriarchy commands. The final irony is that all of us, feminists included, have incorporated into our psyches the self-loathing that comes from not meeting the arbitrary standard.

As we alter and modify our bodies in the hopes of feeling good, we frequently achieve instead an awful estrangement from our own bodies. In the attempt to meet that arbitrary external standard, we lose touch with our own internal body messages, thus alienating ourselves further from our own sources of strength and power.

What does this have to do with aging and ageism? Having spent our lives estranged from our own bodies in the effort to meet that outer patriarchal standard of beauty, it is small wonder that the prospect of growing old is frightening to women of all ages. We have all been trained to be ageist. By denying our aging we hope to escape the penalties placed upon growing old. But in so doing we disarm ourselves in the struggle to overcome the oppression of ageism . . .

It is difficult to hold on to one's own sense of self, to one's own dignity when all around you there is no affirmation of you. At best there may be a patronizing acknowledgement; at worst, you simply do not exist.

The oppressed old woman is required to be cheerful. But if you're smiling all the time, you acquiesce to being invisible and docile, participating in your own "erasure." If you're not cheerful then you are accused of being bitter, mean, crabby, complaining! A real Catch-22.

Old people are shunted off to their own ghettos. Frequently they will say they like it better. But who would not when, to be with younger people is so often to be invisible, to be treated as irrelevant and peripheral, and sometimes even as disgusting.

We have systematically denigrated old women, kept them out of the mainstream of productive life, judged them primarily in terms of failing capacities and functions, and then found them pitiful. We have put old women in nursing "homes" with absolutely no intellectual stimulation, isolated from human warmth and nurturing contact, and then condemned them for their senility. We have impoverished, disrespected, and disregarded old women, and then dismissed them as inconsequential and uninteresting. We have made old women invisible so that we do not have to confront our patriarchal myths about what makes life valuable or dying painful.

Having done that, we then attribute to the process of aging *per se* all the evils we see and fear about growing old. It is not aging that is awful, nor whatever physical problems may accompany aging. What is awful is how society treats old women and their problems. To the degree that we accept and allow such treatment we buy the ageist assumptions that permit this treatment.

SOURCE: Healy, Sherry. 1986. "Growing to Be an Old Woman" in *Women and Aging*, edited by Jo Alexander et al., pp. 38–62, © Calyx Books.

tions to hide "the tell-tale signs of aging." Men are admonished to cover the patches of gray that appear in their hair, or to use other products to prevent baldness. Entire industries are constructed on the fear of aging that popular culture promotes. Face-lifts, tummy tucks, and vitamin advertisements all claim to "reverse the process of aging" even though the aging process is a fact of life.

THINKING SOCIOLOGICALLY

Visit a local place that sells greeting cards and systematically note the images on birthday cards for different age groups. What *stereotypes* are portrayed for aging women and aging men? Explain how this promotes *ageism* in society. How is ageism related to gender stereotypes?

The mass media, especially television and many popular magazines, play a strong role in perpetuating age stereotypes. Just as the media condition our views on gender and race, they present us with a distorted view of both youth and the elderly. Young people are portrayed as carefree; the elderly, as unhappy or evil. Sometimes the elderly are presented as childish, a common and harmful representation called *infantilization of the elderly*. In advertisements, for example, older people are frequently shown throwing temper tantrums or engaging in childlike forms of play.

Even though people over age sixty-five are 20 percent of the population, they are less than 5 percent of characters shown on television (Gerbner, 1993). Popular impressions are that, as the number of elderly is increasing, so is their representation on television. Research, however, finds this to be untrue. In fact, the percentage of characters on prime-time TV who are presumably over age sixty-five actually *decreased* between 1975 and 1990, as did the percentage of characters judged to be between fifty and sixty-four years of age. More women characters over fifty appeared than in the past, but they are still a small proportion of the whole (Ferraro, 1992; Robinson and Skill, 1995).

Stereotypes tend to be very persistent, but they are not fixed and can change. Two studies demonstrate this point well. In 1965, a team of researchers asked a middle-class, middle-aged sample to define the age range of "good-looking women." Ninety-two percent of the men and 82 percent of the women defined women as good looking between the ages of twenty and thirty-five. Twenty years later, when this study was repeated, there had been a significant change in both men's and women's attitudes about looks and aging, with only 33 percent of the men and 24 percent of the women saying that the cutoff age for being a good-looking woman was thirty-four (Neugarten et al., 1965; Passuth et al., 1984). Many popular movies and rock groups have stars who are over fifty, including now Robert Redford, Tina Turner, Mick Jaggar, Goldie Hawn, Paul McCartney, and others.

Even the popular icon of sexiness, the *Playboy* centerfold, has aged somewhat, though only slightly. From 1954 to 1980, the average age of a *Playboy* centerfold was 20.3 years old, inching up to 22.3 years of age between 1981 and 1989—still a very young ideal, but a change nonetheless. In recent years, *Playboy* has also shown a greater age range of women, even having Joan Collins appear seminude at the age of fifty (Harris et al., 1992). One can argue that showing older women nude is hardly a transformation of social stereotypes, but it does indicate that, even in the most unlikely places, stereotypes of age evolve as society changes.

Famous singer Tina Turner, born Anna Mae Bullock on November 26, 1939, in Nut Bush, Tennessee, clearly demonstrates that chronological age need not limit one's physical appearance. She is almost sixty years old in this picture.

Becoming aware of age stereotypes reveals a basic sociological idea: *Perceptions of aging are socially constructed.* Simone de Beauvoir, famous French essayist and feminist, expressed this well when she wrote, "Since it is the Other within us who is old, it is natural that the revelation of our age should come to us from outside—from others. We do not accept it willingly" (de Beauvoir, 1972: 284). de Beauvoir's point is a sociological one: Other people's definitions of aging affect how we think about aging, but people also resist the negative images others impose. In an illustration of this point, some have studied how people subjectively define their age, compared to their actual age. The results are fascinating. Twenty percent of people in their twenties report a subjective age (that is, how old they feel) older than

Culture is often transmitted from grandparent to grandchild, as demonstrated by this Iroquois Native American grandfather.

their chronological age, but most people in their sixties, seventies, and eighties report feeling younger than their chronological age. In addition, in this study, no one in their twenties said they look younger than their actual age, whereas most older people said they look younger than they are (Goldsmith and Heims, 1992). Perceiving one's "social" age as different from one's chronological age is one way to adapt positively to negative age stereotypes.

Age Norms

The social significance of aging is also apparent in the expectations people hold for different age groups. **Age norms** are explicit and implicit rules that spell out the expectations society has for the different age strata. Age norms define what you should or should not do according to your age. Age norms define some behaviors as appropriate for one age category but not for another; thus, the norms change as you age. Consider this: We have all been admonished, at one time or another, to "act your age." As one ages, one is expected to act more "mature," more "grown up." Age groupings are treated differently in society, and different expectations, or norms, go along with each. Young people are not supposed to be sexually active and are supposed to stay in school. Age norms that apply to the elderly include the expectation that they will retire from their job and be less active and publicly visible than those supposedly in their "prime." Many age norms become laws. You are not allowed to drive, drink, or vote until you reach a certain age; children are not permitted to skip school, and they are not supposed to work. Until recently, people had to retire at age sixty-five, although the Age Discrimination in Employment Act has eliminated the mandatory retirement age for many occupations (Lawlor, 1986; Bessey et al., 1991).

age norms

Age norms are not fixed. Like other norms, they change as society changes and people adjust to new social conditions. For example, becoming old in this culture has traditionally meant not working and being seen as not contributing to society. These norms are now changing as people recognize that older people can be productive members of society. Of course, the norms have never accurately described the experience of all people. Many working-class "retirees" continue to work because they need the economic resources to survive (Calasanti, 1992). A study of Mexican Americans has found, for example, that retirement for them often does not come as the culmination of a life of working. Because Mexican Americans have histories of work marked by unemployment, part-time work, and jobs with few retirement benefits, they are less likely to perceive themselves as retired even when they are no longer employed (Zsembik and Singer, 1990). This study shows that people's work experiences shape their perceptions of retirement. Groups whose employment histories are more erratic, less stable, and unsupported by retirement benefits are less likely to perceive themselves as retired even when they no longer work. Women, for example, who no longer work in the paid labor force do not perceive themselves as retired since their work as homemakers typically continues (Hatch, 1992).

Contemporary social changes have also disrupted the traditional norms that have distinguished different age groups. Older people now return to school and complete degrees; for many, marriage and childbearing comes much later in the life cycle than would have been true years ago. These changes have mixed up the various age norms that apply to a given generation. Ask a middle-aged person about the people in her social network and you may well find that one friend in her forties is having her first baby; another friend, the same age, may be sending children to college; still another is getting married for the first time; another has been recently divorced or is grieving over the death of a life partner.

This example illustrates how age norms change over time. New social conditions produce changes in the norms. For example, the increased presence of women in the labor market, especially among the White middle class, has meant a dramatic transformation of the norms associated with aging. Women no longer follow a prescribed life course in which marriage and children come first and work, second. Lifestyles are simply more diverse than ever before with some women reentering the labor market in middle age, others returning to school after their children are grown, others delaying child rearing until after their careers are well established, and an increasing, though small, number of people foregoing childbearing altogether (Boyd, 1989).

Age norms also differ within groups. Many racial–ethnic communities hold the elderly in a revered place, valuing their knowledge and perspective and giving them high social status as a result. In Native American cultures, as an example, the elderly have been accorded much respect, as illustrated in the box "Understanding Diversity: Grandparenting Among Native Americans" where the role of

Native American grandparents as cultural conservators is described. Similarly, other minority cultures emphasize great respect and care for elder parents although they do not always have the extended family networks that are sometimes assumed (Goodman, 1990; Stoller and Gibson, 1997).

In many other societies, the elderly also hold very high status. They are esteemed, and their judgment on important matters is sought and respected. In Samoa, as an example, old age is considered to be the best time of life, and the elderly are much revered. In many African nations, social status increases with age, and the elderly are regularly consulted in acknowledgment of their superior wisdom. Even in some highly industrialized societies, such as Japan, there is a long-standing tradition of respect for the elderly. This is further reinforced by the Japanese stratification system, which demands that servants, students, and children respect those in roles regarded as superior, such as masters, teachers, or parents. In Asian societies, the Confucian principle of filial duty emphasizes respect for the aged; to call an old man "lao" (as in a man named "Xu" becoming "Xu Lao") is a great compliment; Chinese children are also required to repay their parents with gratitude for bringing them up (Palmore and Maeda, 1985; Cohen and Eames, 1985; Kristof, 1996).

Age and Social Structure

All societies, including our own, practice **age differentiation,** the division of labor or roles in a society on the basis of age. Differentiation of roles by age is a feature of all societies, although the specific roles given to different age groups vary from society to society. For example, the Korongo of East Africa assign men and women of a similar age to live together in a single dwelling. In other societies, reaching a certain age may be the basis for assigning people to certain roles, such as young Zulu men of central and southern Africa who enter a military unit called an *impi,* where they receive training as warriors when they reach a certain age (Cohen and Eames, 1985).

In the United States, age differentiation can be seen in the different rights and privileges people have by virtue of their age. Youth in the United States in general enjoy only a subset of the rights of adults. The rights of youth to drink, drive, own property, work, and get married are all abridged. Sometimes the rights and restrictions associated with youth are implausibly contradictory or arbitrary. At age eighteen, for example, you cannot buy a beer in most states, but you can be sent overseas to fight and die for your country. In most cities, you cannot be a police officer or firefighter until

BOX 13.2

UNDERSTANDING DIVERSITY

Grandparenting Among Native Americans

SOCIAL roles develop in the context of cultural traditions and social institutions that vary among different groups. Being a grandparent is an example of a social role where there are different expectations among different racial and ethnic groups. The dominant culture defines grandparents as being indulgent, playful, and fun-seeking, but also hands-off. This role differs, however, among various groups in the society, and for some it is more important than for others. African American men, for example, tend to see grandfathering as a more central role in their identity than do most White men (Kivett, 1991). Communities where there tend to be multigenerational households may give grandparents a primary role in child rearing, whereas in other groups, grandparents are supposed to take an assisting, but hands-off approach to child care. In many cultures, grandparents are also defined as the dispensers of wisdom.

Joan Weibel-Orlando has studied grandparenting among Native American groups and found that, within these cultures, there are diverse roles for grandparents, all of which reflect the high esteem in which old people are held in Native American societies. Some of the styles she identifies are cultural conservator, ceremonial instructor, and custodian.

As cultural conservators, grandparents pass on the traditions of the group, providing cultural continuity and identity for young children as Native Americans. Storytelling can be an important way that this role is enacted since stories pass on the cultural beliefs of the group. Grandparents also teach young Native Americans a wide array of ceremonial activities—sun dances, rodeos, powwows, and memorial feasts. Through this instruction, children learn the values of the group. As custodians of young children, Native American grandparents also provide essential household labor; the gender division of labor most typically assigns this role to women.

Weibel-Orlando argues that some of these characteristics would be expected among other minority groups. The conservator role, for example, is one that provides cultural continuity for groups that fear a loss of their unique group identity in a society organized around different cultural values. For Native Americans, urbanization poses a particular threat to the maintenance of group traditions; therefore, the role of cultural conservator is particularly important. Similarly, Native American grandmothers' roles as primary caretakers is a common way that households provide childcare, particularly when there are limited other resources.

SOURCE: Based on Weibel-Orlando, Joan. 1990. "Grandparenting Styles: Native American Perspectives." In *The Cultural Context of Aging: Worldwide Perspectives.* New York: Bergin and Garvey, pp. 109–125.

you have reached the age of twenty-one. You cannot take a seat in the House of Representatives until you are twenty-five; a senator must be thirty; the president of the United States must be at least thirty-five. Perhaps these rules about age and political service are even more revealing of our society's views of age when you realize that they were made at a time when the average life expectancy for men was not much more than about forty years. Virtually all society's institutions—business, politics, church, education, and recreation—have age barriers that must be surpassed in order to enjoy the full benefits of the institution.

In addition to differentiating roles on the basis of age, societies also produce age hierarchies—systems in which some age groups have more power and better life chances than others. We have seen throughout this book that the United States is stratified into hierarchical groupings on the basis of class, race, gender, earnings, occupational rank, educational attainment, and other variables. Age, too, is the basis for stratification. *Age stratification*—the hierarchical ranking of age groups—exists because processes in society ensure that people of different ages differ in their access to society's rewards, power, and privileges. Indeed, in the United States, age is a major source of inequality.

Age is an *ascribed status*—that is, age is determined by when you were born. Recall that biological sex and, most often, race, are also ascribed statuses—that is, they are established at birth. Different from other ascribed statuses, which remain relatively constant over the duration of a person's life, age changes steadily throughout your life. Still, you remain part of a particular generation—something sociologists call an **age cohort**—an aggregate group of people born during the same period (Stoller and Gibson, 1997: 8). Cohorts are not homogenous. Within a given cohort, there will be considerable diversity on many dimensions: sexual orientation, gender, race, class, nation, ethnicity. How these cohorts are arrayed in a given society at a given time shapes the character of society and the social issues within it.

People in the same age cohort share the same historical experiences—wars, technological developments, economic fluctuations, although they might do so in different ways depending on other life factors. Living through the Great Depression, for example, shaped an entire generation's attitudes and behaviors, as did growing up in the 1960s, and as will being a member of the contemporary youth generation. Recall from Chapter 1 that C. Wright Mills saw the task of the sociological imagination as analyzing the relationship between biography and history. Understanding the experiences of different age cohorts is one way you can do this. People who live through the same historic period experience a similar impact of that period in their personal lives. The troubles and triumphs they experience and the societal issues they face are rooted in the commonality established by their age cohort. This does not mean there is no variation within age cohorts. As noted, other factors (like race, nationality, or gender) shape the experiences people have in society. Still, the shared historical experiences of age cohorts result in discernible generational patterns—in social attitudes and similarity of life chances (Stoller and Gibson, 1997).

Sociologists note that there is also a continuing interplay between age and social change. In society, successive groups of people grow up, grow old, and die, each group being replaced by the next age group coming up behind them. As people of different ages pass through social institutions, society itself changes. Each age group faces a unique slice of historical time, thereby facing their own challenges and social changes. To understand this, think of some of the significant facts of life that derive simply from being born at a certain time. In 1900, someone twenty years old would not likely look ahead to retirement since people seldom lived to what we would now consider retirement age. Many twenty-year-old women in 1900 could not expect to live beyond their childbearing years; indeed, as recently as 1970, the death rate from pregnancy was three times what it is today, although today it is still four times higher among Blacks than Whites (U.S. Bureau of the Census, 1997a). Now, however, a twenty-year-old can expect to spend almost one-quarter of his or her life in retirement—a significant social change (Riley, 1987).

Even without looking at so long a time span, you can see how different generations have to grapple with and respond to different social contexts. Someone born just after World War II would, upon graduation from high school or college, enter a labor market where there was widespread availability of jobs and, for many, expanding opportunities. Now, young people face a labor market where entry-level jobs in secure corporate environments are rare and where many are trapped in low-level jobs with little opportunity for advancement. Many young people worry, as a result, about whether they will be able to achieve even the same degree of economic status as their parents—the first time this has happened in U.S. history. Understanding how society shapes the experiences of different generations is what sociologists mean by saying that age is a structural feature of society.

Current generations—whether young, middle aged, or old—will be profoundly influenced by the graying of America. The proportion of old people in the population is increasing dramatically. In addition to this fact, consider the following:

- Racial minorities and ethnic groups are an increasing proportion of the older population (U.S. Bureau of the Census, 1997a).
- The proportion of the population classified as the oldest old (those over eighty-five) will also continue to grow (Treas, 1995).
- Women will continue to outnumber men in old age, especially among the oldest old (Treas, 1995).
- Since the educational status of the elderly is increasing rapidly, the historical gap in educational attainment between the old and the young will likely disappear by the middle of the twenty-first century (Uhlenberg, 1992).

These changes in the age structure of society will be reflected in social issues both now and in the future. How well we adapt to the graying of America may well be partially a result of how carefully we can analyze the potential effects of these population changes.

THINKING SOCIOLOGICALLY

Take any one of the social facts indicated in the preceding list and describe changes that you think might occur in the *age norms* as the result of this fact.

A Society Grows Old

Never before have so many people in the United States lived so long. This fact, in itself, has a number of implications for how society is organized and the issues to be faced in years to come. Currently structured into society are certain assumptions that guide the expectations and obligations that exist between generations of people. Sociologists speak of these assumptions as the *contract between generations*. Imagine this as a contract between your generation, your parents' generation, and your grandparents' generation. Not a formal contract, but a set of social norms and traditions, the contract between generations is the expectation that the first generation (say, your grandparents' generation) will grow up and raise the second generation (your parents' generation) who, in turn, produce a third generation (your generation). The expectation has been that each generation cares for the next, and the second or third generation will care for the first when they become old. Thus, the expectation has been that parents care for children who in turn care for their children; the children and grandchildren then care for their parents or grandparents once they are too old to care for themselves.

Currently, however, the shrinking size of families means that the proportion of elderly people is growing faster than the number of younger potential caretakers; moreover, as life expectancy has increased and people live longer, the traditional contract is upset. Women, who shoulder the work of elder care, can expect to spend more years as the child of an elderly parent than as the mother of children under eighteen (Watkins, 1987). Indeed, young people now can expect to spend more years caring for an elderly parent than raising their own children. As a result, family members can now expect to spend more time in intergenerational family roles than ever before.

Changes in family structure (which we will discuss in detail in Chapter 14) further alter the traditional patterns of intergenerational care. Childless couples cannot count on a younger family member caring for them in their older years. Those in single-parent families, particularly women, have the extra burden of caring not only for their own children, but perhaps also for an elderly parent. As people have delayed childbearing, they may even find that they are caring for their own children at the same time that they are caring for an elderly parent—certainly a source of stress that current social institutions are poorly designed to handle. Men and women in their middle ages may have to find ways, often with few institutional supports, to care for older, and perhaps ill, parents. Given the geographic mobility that has characterized modern life, they may have to do so over quite

Social activism is not limited to the young, as shown by these senior citizen demonstrators.

long distances. What we may be experiencing is a classic case of *culture lag* where the norms of care and support have not changed as rapidly as the composition of the population.

THINKING SOCIOLOGICALLY

Imagine yourself twenty years from now. How do you think your life will be affected by the *graying of America*?

At the same time, the growth of the welfare state in the twentieth century means that different generations now vie for the different entitlement programs that provide social and economic assistance. Public resources directed to the elderly have grown in recent years, both as the result of trying to combat poverty and as the result of effective lobbying by older people and the groups that represent them. This has led to improvements in the economic status and health care for the elderly, aided by the fact that the generation now growing older is generally better educated and financially more secure than previous generations of old people. As these trends have developed, the proportion of public funds directed to the elderly has increased. At the same time, resources to other groups, such as children and the poor, have declined. The result is a public perception that there is inequity in support for different generations, each with its own needs and interests (Bengtson and Achenbaum, 1993).

Currently, Social Security is one of the older and most successful national social policies. Established in 1935 under the Social Security Act, the Old Age, Survivors' and Disability Insurance commission created the Social Security system. Together with Medicare (established in 1966 as a national health-care system for those over age sixty-five), Social Security expenditures are 56 percent of the federal budget spent on social welfare (U.S. Bureau of the Census, 1997a). Can these expenses continue? The statistics are striking: In 1945, the Social Security system had thirty-five working people paying into the fund for every recipient drawing upon it. By the late 1970s, the ratio of wage earners to recipients was down to 3.2 to 1. Although this is a dramatic drop in the ratio of earners to recipients, recent estimates note that sufficient funds are available to pay those turning sixty-five in the late 1990s and soon after the year 2000, but without significant changes in Social Security policy, most agree that there will not be enough workers to support the number of retirees by about the year 2020 (Kingson and Quadagno, 1995).

For the most part, this looming social issue has been posed as a matter of fairness between generations. Social Security is frequently an issue in political campaigns, pitting young against old. Most people assume that more support for other groups could come only at the expense of the elderly. This debate is referred to as the question of **generational equity,** namely, the question of whether one age group or generation is unfairly taxed in order to support the needs and interests of another generation. There are those who argue that, as the proportion of individuals who are over age sixty-five increases, a disproportionate share of the burden of supporting them will fall upon those who are younger. Consequently, so this argument goes, those who are younger are not being treated equally, but are being treated unfairly. This debate has its origins in the aging of the population, budgetary crises in the federal government, growing health care costs, increased poverty among children, and as some sociologists note, a declining faith in social institutions on the part of the public (Kingson and Williamson, 1993).

This issue was first raised by an organization founded in 1984 called Americans for Generational Equity (the initials for which are, cleverly, AGE). This organization challenges the notion that the elderly are entitled to Social Security benefits. They argue that, since Social Security taxes people who are currently working in order to pay a monthly sum to those who are retired and do not work, the program is unfair to those who are working. The ever-increasing proportion of elderly retired persons will become increasingly dependent on an ever-decreasing proportion of working people. They also argue that, in many cases, those who are retired are wealthier than many of those who continue to work, and yet, the working are taxed to support the nonworking, who are not taxed to support their own lifestyle. They argue finally (as have some in Congress) that, at current rates of taxation and interest, the Social Security system will be bankrupt by the year 2030, so a new program must be put in place soon. The kind of program they propose would reduce, but not eliminate, the amount of benefits to the elderly and would involve a direct monetary contribution by the elderly into the system, in proportion to their own wealth and retirement income (Quadagno, 1989). Proposals such as this one are being debated in the U.S. Congress.

Sociologists who have assessed the generational equity debate have argued that this need not be a divisive issue. Studies of other nations find there is no necessary connection between high spending for older adults and lower spending for children and the poor (Pampel, 1994; Pampel and Adams, 1992). Other industrialized nations where governments have strong family allowance systems (government-sponsored programs to assist families across the years) do not target one group at the expense of another (Adams and Dominick, 1995). Instead, sociologists suggest that the problems associated with the graying of America could be addressed by having a changed public agenda—one that provides universal access to basic needs (income, housing, and health), a change that would require tax increases and some reduction in benefits (Kingson and Quadagno, 1995; Kingson and Williamson, 1991, 1993). To date, such a solution has not gained political appeal; instead, the nation's response may be to exacerbate stereotypes of the aged as "greedy geezers" and to blame the victim, rather than try to solve the problem. This debate about generational equity also has the risk of potentially increasing age prejudice in society, while addressing neither the problems of the elderly, children, nor the poor.

Growing Up/Growing Old: Aging and the Life Course

The phases in the aging process are familiar at least in name to all of us: childhood, youth and adolescence, adulthood, and old age. Together these strata make up the life span. Sociologists use a **life course perspective** to interpret the life span. The life course perspective connects people's personal attributes, the roles they occupy, the life events they experience, and their sociohistorical context, emphasizing that personal biography, sociocultural factors, and historical time are interrelated (Stoller and Gibson, 1997). Although psychologists and others study the life span, unique to sociology is the connection that sociologists make between individuals and the social contexts in which they live.

As we noted in Chapter 4 on socialization, the transitions to different phases in the life span are often marked by elaborate cultural rituals. These rituals that celebrate or memorialize events in the individual's life are called *rites of passage*. Birth, puberty, marriage, and death are each heralded by rites of passage—a christening or baptism, confirmation, the bar mitzvah and bas mitzvah for Jewish boys and girls at puberty. Among Mexican Americans, the *quinceañera* ceremony similarly marks puberty as the transition to adulthood (see Chapter 4).

The distinction between different phases in the life course is not, however, a rigid one. Without specific social markers to announce the passage from one phase of the life span to another, the distinctions between different phases blur. When, for example, does one become an adult? In most states, you are adult enough at age eighteen to enter the armed services and to vote, but not adult enough to drink—the legal drinking age in most states is twenty-one. Social change means that young people remain in school longer, marry and bear children later, and in some cases,

life course perspective

remain dependent upon a parent or parents for a longer time (see Table 13.1 on page 357). Similarly, "middle age," according to the traditional definition, used to be perceived as beginning when one's children grew up and left home. Now some middle-aged people may be just beginning college; others may already be grandparents. Indeed, given the high rate of teenage pregnancy, some become parents before they would traditionally have been defined as adults. Teenage pregnancy also means that some parents may become grandparents as early as their late twenties or early thirties—a time when other adults may still be in school. Grandparents themselves may still be in school—something that, not long ago, would have been extremely rare.

The traditional markers of adulthood (education, marriage, work) no longer as easily mark the change from one phase of life to another. Definitions of age are social products, and there are no rigid demarcations between age categories, merely approximations. Even old age can be seen as divided into two strata—the "young old" (around sixty-five to eighty-five) and the "oldest old" (eighty-five and older). Although the terminology is imprecise, sociologists have extensive knowledge of the general distinctions in the life span: childhood, youth and adolescence, adulthood and middle age, and finally, old age.

Childhood

The United States has typically been defined as a child-centered society. Many news commentators, historians, educators, and sociologists have noted that the United States is "youth oriented." The high valuation on youth in U.S. culture is especially evident in the mass media. The young are depicted in television commercials more often, and in a more positive light, than any other age group. Today's youth are generally shown as energetic, smart, and attractive. In the United States, more than in most other societies, to be young is revered—aged, reviled; youth is regarded as the prime of life, whereas middle age and old age are regarded as the long decline at the end of life.

Popular images of childhood also depict it as a period of play, fantasy, and freedom from responsibility. It was not always that way. Thinking of childhood as a separate stage of life dates back to the sixteenth and seventeenth centuries in Europe, when changes in the economy began to emphasize the family as a separate unit in society (Aries, 1962). At that time, society was well aware of the economic and occupational value of children. In the nineteenth century in the United States, children were used as additional laborers in the family and were given a variety of responsibilities. This is still true today among many immigrant groups, where children work to help support the family. It is also somewhat more pronounced today among the poor, whatever their race. In general, the lower the socioeconomic status of a family, the more likely the family will use its children as a source of labor.

The exploitation of child labor was so pervasive in the nineteenth century that people reacted against it. By the middle of the nineteenth century, at least among the middle classes, the idea that children were of limited economic value began to take shape. Instead of relying on their children to take care of them in later life, parents began to rely more on life insurance, pension plans, and other financial arrangements. All this resulted in the *sentimentalization* of children in the United States, meaning that children were seen as precious, but not generally defined as economically useful. Now, instead of seeing children as "functional" to the family, adults report that among the most important advantages of having children are the satisfaction of adults' desires for love and affection, and the pleasures of being part of a family (Hoffman and Norris, 1979). This tender attitude toward children no doubt buffers the realization that raising children is an expensive proposition. In return, parents demand from children only that they show respect, return love, and smile a lot. Parents who use their children to make money, such as parents of successful child actors and entertainers or those who enter young children in baby beauty contests, are viewed with suspicion. As one researcher put it, we live in a culture that defines children as economically useless but emotionally "priceless" (Zelizer, 1985).

Still, the image of childhood as a carefree time is not matched by reality for many children. In many respects, we live in a child-centered society; yet, the United States is becoming a more dangerous place for children. Violence takes a heavy toll on urban African American and Latino children. More than one-fourth of those living in shelters for the homeless are children (National Coalition for the Homeless, 1996). A huge number of children live in poverty, with their likelihood of doing so strongly linked to the race or ethnicity of the child. As Figure 13.3 shows, the proportion of Black and Hispanic children under the age of eighteen who are poor (40 percent for both groups) greatly exceeds the proportion of White and Asian children under eighteen who are poor—16 and 20 percent, respectively (U.S. Bureau of the Census, 1996).

Another astounding finding is that the United States ranks first among seventeen separate countries in child and youth poverty. A full 26 percent of those under the age of eighteen of all races and ethnicities in the United States live in poverty. Australia and Canada are next, followed by Ireland. Figure 13.4 shows the proportion of children in each country who are in poverty both before and after getting some form of government assistance to lift them out of poverty. The United States ranks highest in total child and youth poverty (26 percent) as well as those remaining in poverty after government assistance (22 percent). Switzerland, Denmark, Sweden, and Finland have the lowest child and youth poverty rates (3 percent for each country after government assistance).

Increasingly, society is also turning its attention to children as victims of sexual abuse and drug addiction. Estimates suggest that perhaps as many as 16 percent of girls under age eighteen are sexually abused (Russell, 1986). Drug addiction among children is also an increasing problem. One group of researchers have developed what they call an "index

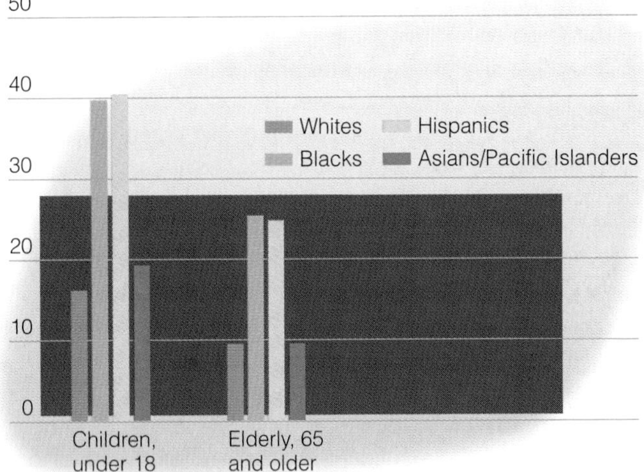

FIGURE 13.3 Age, Race, and Poverty

DATA: From Dalaker, Joseph and Mary Naifeh. 1998. U.S. Bureau of the Census, 1997. *Poverty in the United States: 1997.* Washington, DC: U.S. Department of Commerce, p. vii.

of social health"—a composite measure of various indicators of social problems. Looking specifically at the problems that affect youth—poverty, child abuse, infant mortality, suicide, drug abuse, and dropping out of school—this measure indicates that the social health of U.S. children and youth has seriously deteriorated in recent years (Miringoff, 1996).

Youth and Adolescence

Fast cars, loud music, arguing with parents, facing up to sexuality, dating, worrying about what to do when schooling is finished, and wondering "Who am I?" These are some of the concerns of adolescence. Adolescence is a relatively new category in the life cycle. Until the twentieth century, children moved directly into adult roles; there was no such thing as an in-between adolescent period. Adolescence came to be regarded as a separate stage of life as the period of formal education became longer, and as more people entered and completed high school. This delayed entry into adult roles, marking the period of adolescence as a particular stage in the life cycle.

Although the boundaries of adolescence are imprecisely defined, most regard the lower boundary as the transition from elementary school (sixth grade or its equivalent), or about age twelve, to junior high and high school. What has occurred over time in one institution (education) has structured how an age category is defined in another institution (the family). The upper boundary is usually at entry into some adult role, such as full-time employment, college, marriage, or parenthood, at around ages nineteen or twenty. The term *teenager* is often used to cover the period of adolescence. The notion of the preteen, which has come into usage in the last three to four decades, encompasses the ages of about nine through twelve years.

As we saw in Chapter 4, establishing an identity is a central concern in the adolescent period. Youths experiment during this period as they attempt to fit different social roles into a coherent whole. At the same time, adolescents attempt to establish personal autonomy and independence from authority figures such as parents and teachers. Paradoxically, during this period, society attempts to bring the behavior and attitudes of adolescents into line with adult standards while denying adolescents the privileges of adulthood. For example, adolescents may leave school in their midteens and set out on their own, but they will have an extremely difficult time finding steady, gainful employment, mainly on the basis of age. It is an oft-told tale: A young person is denied employment because she does not have "enough experience." She responds, quite reasonably, "But how can I get any experience if I can't get a job?" One result is very high rates of unemployment for the nation's teens who are looking for work—a particularly acute problem for teens from racial minority groups. At present, the national unemployment rate for sixteen- to nineteen-year-olds is 16 percent overall, 32.4 percent for Black youth, 21.6 percent for Hispanic youth, and 13.6 percent for White youth (U.S. Department of Labor, 1998).

Young people typically try to mark their unique identity through the establishment of *youth subcultures*—relatively distinct habits, customs, norms, and language that define youth in contrast to other generations. Youths participate in youth subcultures to different degrees, and youth subculture is crosscut by other cultures as well, such as African American, Hispanic, and Asian culture (Brake, 1985). Youth subcultures are also encouraged by the consumer markets that promote them.

Some youth subcultures reflect the alienation that young people feel from their families and schools. In one study of a youth subculture, researchers interviewed and observed young people who hung out regularly in a shopping mall. Known as "mall rats," these young people exhibited much alienation from both their family and their school, and used the mall as a neutral space where they could create their own mutually supportive community—one that researchers noted was significant to the young people although fragile in its stability and endurance (Lewis, 1989). Others, too, have noted that malls are increasingly places where young people congregate, mostly as an escape from their everyday life (Jacobs, 1984).

Many, certainly not all, young people partake of a variety of "escapist" entertainment, ranging from such mainstream diversions as rock, rap, and videos to the computer underground, which is populated mainly by late adolescents. Such pastimes constitute one element of youth subculture. Another element is style. Analysts of youth subculture define "style" in terms of its components, which include image, demeanor, and special vocabulary. Image is the impression delivered by hairstyle, jewelry, and dress. Demeanor is communicated by facial expression (pouting, nonchalant), gait

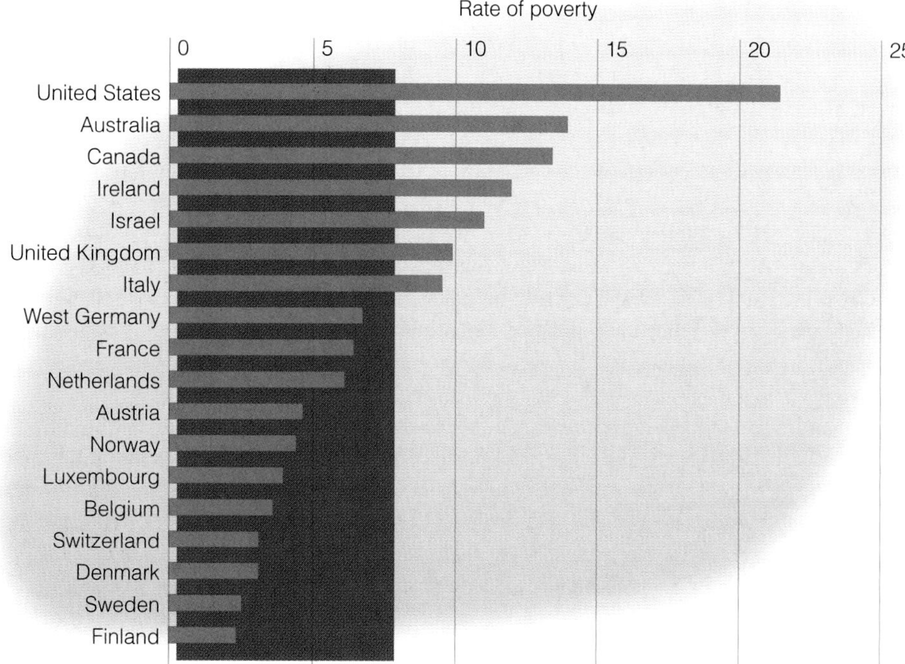

FIGURE 13.4 Child Poverty: A Global Perspective

SOURCE: Rainwater, Lee, 1995, *Doing Poorly: The Real Income of American Children in a Comparative Perspective.* Syracuse, NY: Maxwell School of Citizenship and Public Affairs.

(such as the shuffling, "cool" walk, traceable to African American and Latino youth in the 1940s and 1950s and spread, or culturally diffused, to White youth in the 1960s through the 1990s), and posture (the timeless adolescent slouch while sitting, popular even back in the 1950s).

Within youth subcultures, special vocabularies and manners of speaking, known as *argot,* also define youth autonomy and independence from adults. That which is now described as "hot" (an attractive man or woman, for example) might have been described as "cool" ten or fifteen years ago. This tends to confuse adults, as it is supposed to. Similarly, that which is perceived as good (an especially appealing music video, for example) is described as "bad." Older people who try to imitate the youth subculture will be seen by young people as silly; likewise, older people typically view youth subcultures as outlandish or whimsical. Both groups are implicitly acknowledging the significance that age has in defining our social identity.

Adulthood

The role of adult carries with it more responsibility and more rights and privileges than any other stage in the life cycle. It is also a period of change for people. Moving away from one's parents' home to college, the army, or an apartment or house of one's own are big changes. Currently, the high cost of living has raised the financial threshold on becoming independent, and many young adults are staying at the home of their parents for longer periods, even after entering adult roles.

Many people experience the transition into adulthood by finishing school, getting a job, marrying, and starting a family. These four activities, in order, used to be the norm. Increasingly, however, social conditions make it difficult for many people to progress to adulthood following this traditional path. Many people do not finish high school, some never marry or raise a family. Many spend much of their time unemployed, and recent changes in both the sequence of these steps and the length of time between them have also changed the nature of transition to adulthood. For one thing, teenage pregnancy outside marriage has increased enormously in the past three decades.

Becoming an adult is also taking longer than before. The percent of those who are in school while between the ages of twenty and twenty-one has increased since 1980 from 31 percent to about 45 percent today. Further, more persons in the young adult age range (eighteen to twenty-four years old) live with their parents (about 48 percent in 1980 versus 58

TABLE 13.1 *SLOWING THE TRANSITION TO ADULTHOOD*

	1980	1990
Percent aged 20–21 in school	31.0%	44.9%
Percent aged 18–24 living with parents	48%	58%
Median age at first marriage		
Women	21.8 yr	24yr
Men	23.6 yr	25.9 yr
Birthrate, women aged 15–19 (births per 1000 women)	53.0	45.2
Percent aged 16–19 in labor force		
Women	52.9%	51.3%
Men	60.5%	53.2%

DATA: From the U.S. Bureau of the Census. 1997. *Statistical Abstracts of the United States, 1997.* Washington, DC: U.S. Government Printing Office, pp. 58, 81, 106, 157, 397.

percent now). Women are marrying later and having children later. The median age at first marriage for women has increased from 21.8 years old to almost 24, and the median age at first child has risen from 24.3 years old to 26.3 (U.S. Bureau of the Census, 1997a). The average age at first child is rising despite the simultaneous trend of more teenage women having children, especially in urban areas. Finally, about half of today's married couples have lived together for a time before getting married (Sweet and Bumpass, 1992). In an earlier era, they would have married before setting up housekeeping; today, they are marrying some time after, contributing to the increase in the average age of marriage for both men and women. In sum, all such indicators show that the transition to adulthood is occurring at later ages.

As adulthood unfolds, the traditional norms of our society suggest that both men and women, but particularly men, should have achieved most of their life goals by the time they reach middle age. As popularly conceived, the *midlife crisis* is a time of trauma during which people, men in particular, become fixated on what they have failed to achieve in their work and family roles, or on the things they never attempted. The midlife crisis is reputed to spring from the feeling that something is missing. The media often typify such a man as the conservative accountant or executive who suddenly—to the consternation of his wife and children—buys a red Corvette, a black leather jacket, and several gold neck chains, in an attempt to attain a faster, less conservative lifestyle. Some men leave their middle-aged wives for younger women at this time. For women, the midlife crisis is popularly defined as troubling because it marks the end of the woman's role as mother. Children (if any) leave home, and the onset of menopause signals the end of the fertile years.

Despite all the talk about the midlife crisis for women and men, the truth is that the transitions experienced at midlife are very often not as painful as the popular image suggests. It is true that during this period many rethink their role in society and ask themselves what they have accomplished. They often dislike the physical signs of aging taking place in their bodies and their appearance. The bulk of research does not find adults having midlife crises—at least not any more so than in other periods of life (Kail and Cavanaugh, 1996). Midlife is actually experienced as happy and positive by most. Middle-aged, middle-class professional men in the United States enjoy high levels of power, prestige, and relative wealth. Although some people may use midlife as a time for reflection, most report high levels of satisfaction with their accomplishments (Hess and Markson, 1990). The events that cause stress in midlife are those that are unexpected (such as divorce, loss of a job, death of a partner), not the normative midlife events (menopause, children's leaving home, retirement, and so forth). Some men do fear retirement and the presumed loss of identity that goes along with one's occupation, especially those whose self-conception is fixed to their job. The same is true for women. If one *is* one's job, whether brick mason, judge, or professor, then retirement may be seen as the loss of one's self (Lewis, 1981; Hess and Markson, 1990).

Debunking Society's Myths

Myth: **As people reach middle age, they will all experience a midlife crisis.**

Sociological perspective: **For most people, midlife is no more or less traumatic than other age periods. Unexpected events (divorce, unemployment, death of a loved one) are more likely to create stress than normative life events, such as children leaving home or becoming a grandparent (Kail and Cavanaugh, 1996: 427).**

Women, like men, describe midlife as a time of satisfaction and freedom. Feelings of freedom from child care are widely reported, especially by women who are employed and pursuing careers. The so-called empty-nest syndrome is a phase of womanhood in which mothers supposedly miss their grown-up children and are plagued by loneliness. Actually, most women who have raised children describe this period as a happy time, contrary to reports of a decade ago popularizing the malady. Women who have devoted themselves exclusively to mothering roles are the most likely to experience some depression when the children leave home, whereas women with other roles, such as a career, tend to get greater enjoyment from the satisfactions of middle age.

For many women, middle age is a period of taking on new roles, such as a new job (or first job) or returning to school. Given changes in the workforce participation of women, many younger women will have held jobs their entire adult life up to middle age, but for an older generation of women who did not work outside the home, middle age can be a time when they, ironically, set out to develop a work role just as younger women do in early adulthood today. Like other phases in the life cycle, the resources one has available—both economic and social—have more to do with how one experiences midlife than some inherent crisis in growing older.

Retirement

Along with becoming a grandparent, one of the most significant markers of approaching old age is retirement from work. Many look forward to retirement as a time for increased freedom. How one experiences retirement, indeed whether one can retire at all, is the result of many social factors. Some approach retirement with some degree of apprehension and fear since it may symbolize bringing one's life's work to an end. For those whose careers have been the primary basis for their identities for many years, retirement can be a difficult period of adjustment. No longer does the person have a predetermined place to be for eight or more hours a day. The social interaction one had with friends and colleagues at work is absent. The amount of free time increases greatly. Once retired, many people experience stress in adjusting to their new lifestyles, new expectations, and new patterns of interaction within long-established marriages. For example, some wives report that their husband's increased participation in household activities is not a help but a nuisance that feels like an invasion of their domain; this changed division of labor in the household produces strains (Ward, 1992). For both men and women, the burdens of retirement are eased by maintaining social contacts derived from work and elsewhere (Palmore, 1977, 1986). Taking on a job when retired creates larger social networks, which is linked to better health (Wingrove and Slevin, 1991; Mor-Barak et al., 1992).

Many adults retire in their midsixties although certainly not all. By 1996, only 17 percent of all men over sixty-five, and 7 percent of all women over sixty-five were still in the labor force (U.S. Department of Labor, 1997). The overall number of retirees in the United States is growing, producing serious strains on the Social Security system, as we have seen.

The difficulties of retirement have generally been treated as men's problem, but the expanded role of women in the job market means that proportionately more women will retire in the coming years. Women are less likely to hold jobs that include retirement benefits although this is also true for men in less advantaged class positions. For both groups, lifelong experiences in the labor market affect one's experience at retirement. Women are approaching the same work patterns as men, but are not converging either in work experience or retirement. Women retire sooner following their spouses' retirement if they were also employed during child rearing; women who were not employed during child rearing retire more slowly than men (Henretta et al., 1993).

The experience of retirement varies for different racial, class, and gender groups. Research on minority experiences finds that their disadvantage in the labor force is reproduced in retirement; lower levels of job training and benefits mean that minorities have lower levels of postretirement income and are more reliant on public retirement funds, such as Social Security, than is true for Whites (Gibson and Burns, 1991). Women of all racial and ethnic groups have fewer resources in retirement than men (Hatch, 1990).

One of the difficulties retired women face is that most pension systems have been designed with men in mind. Having been designed at an earlier time, Social Security defines men as wage earners and women as dependents (Hill and Tigges, 1995; Myles, 1989). Like racial minority groups, women's disadvantage in the labor force is reproduced in retirement since retirement benefits are typically linked to income. Women also tend to receive less in Social Security, which is based on lifetime earnings. In the calculation of Social Security benefits, women who have left the labor force to care for children are disadvantaged. The formula for calculating average lifetime earnings allows workers to drop the five lowest years of earnings from the total, but if women (or men, for that matter) have taken off more than five years for caregiving, they receive lower benefits.

Retirement is often the beginning of a new life with new activities.

Old Age

Aging is not an entirely negative process (Gibson, 1996), but there is no doubt that old age is a difficult period, made worse by the inadequacy of social institutions to care for the aged. Negative images surround those who grow old, especially older women, but as Maggie Kuhn, organizer of the Gray Panthers says, "Old age is not a disease—it is strength and survivorship, triumph over all kinds of vicissitudes and disappointments, trials and illnesses" (1979). Many find being old to be a source of strength, reminding us that just because someone is no longer involved in the labor force does not have to mean he or she is unproductive. Jessie Bernard, as one example, was a sociologist widely known for her contributions to the sociology of women and the study of families. She died in 1996 at the age of ninety-three; some of her most extraordinary years were after her formal retirement. In one decade, the 1970s, when she was in her seventies, she published nine books, as well as holding a sit-in at a hotel bar that refused to serve women, including the women sociologists attending the annual meetings of the American Sociological Association.

Although the elderly undeniably face many problems, old age is for many a happy period. Research tends to show that being old looks a lot better to the old than to the young—every age begins to look better as you get closer to it. Some studies have found no differences among the young and the elderly on measures of satisfaction, morale, and general happiness, and there are even some studies showing that the elderly have a more positive self-concept than younger people. Gove et al. (1989) conducted a nationwide study that showed that older people, when compared to younger people, reported greater feelings of self-confidence, strength, intelligence, contentment, and organization.

Some critics of Gove's study argued that its results were because of what is called a *cohort effect*—that this particular age group, or cohort, has always felt positive relative to today's youth; in their younger days, they felt more favorably about themselves than today's youngsters at the same age; thus, the finding is a comparison of the qualities of these two different groups, but age is not necessarily relevant to the comparison or the reason for the differences. A difference between the two groups more strictly dependent on age would represent a true *age effect*. Gove argued that the results reflect some of both, with the age effect more important than the cohort effect.

Other researchers have also found that positive self-concept does indeed increase with increasing age (Costa, 1987). This means that the difference between old and young in self-concept (for those in the same cohort group) owes more to age than it does to the cohort effect. If the cohort effect were the main cause of the difference, then the same age group studied from youth to old age would have started high in positive self-concept and simply stayed that way.

Many of the perceived problems of old age have turned out to be myths (see Table 13.2). One of the biggest myths about the elderly is that they lose their interest in sex and their capacity for sexual relations. In fact, the *majority* of men and women remain sexually active well into their seventies, many even into their eighties. With advancing age, sexual partners become less available, but to the extent that partners are available, sexual interest and activity continue well into old age. The level of sexual activity early in life also plays a role in determining sexual activity in old age. Those who were most sexually active when younger continue to be so when they are older. Sexual activity actually increases the overall health and happiness of the elderly (Palmore, 1986).

Another myth about the elderly is that they are "set in their ways" and unable to adapt to the flow of change in the world around them. Studies show that the majority of the aged adjust quite well to changes, of which they face many of the most dramatic sort: children leaving home, the massive disruption of longtime habits caused by retirement, and the mutation of one's social circle as friends become infirm and even die. Aging itself is a process of change that the elderly must reckon with more than other populations, in part because the pace of aging increases as elders get close to the ends of their lives and their health finally begins to fail. It might even be said that adapting to change is a full-time undertaking for the old. Many believe that all old people become senile although this occurs to only about 10 percent of the elderly. Perhaps the most widely held belief is that the reasoning ability of the elderly becomes severely dulled—a belief contradicted by the evidence (Palmore, 1977, 1986).

No doubt one of the more difficult adjustments one makes in old age is becoming a widow or widower by losing a life partner—a situation more often faced by women since women, on average, live longer than men. Five times more women are widowed than men. Men who are widowed are more likely to remarry and to do so more quickly. Widowed

TABLE 13.2 *THE AGED: MYTHS AND REALITIES*

Myth	Reality
1. Most old people have no interest or capacity for sex.	Most old people retain sexual interest and capacity well into old age.
2. Most old people are very set in their ways and do not change.	The majority of the aged adjust to changes such as their children's leaving home, their own illness, and their own impending death.
3. Most old people feel miserable.	Some studies find no differences between the aged and younger persons in happiness, morale, and overall satisfaction with life; other studies actually find that the aged score higher on these.
4. Old people are usually senile.	Only a small minority of the elderly can be considered senile; only 10 percent suffer even a mild loss of memory.
5. Older workers are not as productive as younger ones.	On most measures of productivity, older workers are as productive as younger ones, despite some decline in perception and reaction speed.
6. Most old people live in poverty.	Although poverty among the elderly is a problem, compared to the general population, the aged are less likely to be poor and tend to have more sources of income.
7. Most old people are lonely.	Only slightly more aged indicate that they are lonely than do those in the general population; the vast majority of the aged indicate that they are not lonely.
8. Most old people end up in nursing homes and other institutions.	Less than 5 percent of all elderly are in a nursing home or other institution at any particular time.

SOURCE: Palmore, Erdman. 1977. "Facts on Aging." *The Gerontologist* 17: 315–320. Republished with permission of the Gerontological Society of America, 1030 15th Street, NW, Suite 250, Washington, DC 20005. *The Facts of Aging* (Figure), Palmore, Erdman, *The Gerontologist*, 1977. Vol. 17. Reproduced by permission of the publisher via Copyright Clearance Center, Inc.

men tend to know more people and have more acquaintances and work-related contacts than women; hence, women widows tend to suffer more from loneliness. As widowers, men participate in more organizations than women, and are more likely to own a car (which translates into a greater ability to get out and around). Finally, even during widowhood, men, on average, have higher incomes, reflecting the income differences between men and women across the entire age spectrum (Atchley, 1997). As the population of widows begins to include more women who have held their own jobs and have lived more independently, these patterns will likely change, however.

Although men have some advantages in widowhood, they also experience some disadvantages. Elderly widowed men are seven times more likely to die in a given year than married men in the same age bracket. Widowed men are three times more likely to die in a car accident than comparably aged married men, four times more likely to commit suicide, six times more likely to die of a heart attack, and ten times more likely to die from a stroke (Ward, 1990; Kucherov, 1981).

Having social support through extensive friendship and familial networks helps alleviate the stress experienced during widowhood, just as having friends lessens the negative impact of the problems of aging in general. In a study of Puerto Rican families in Boston, Sánchez-Ayéndez (1995) found that strong cultural norms encourage Puerto Rican children to support both parents and grandparents and parents to support older children. The families she studied strongly relied on neighbors, with reciprocity also stressed. As a result, the elderly in Puerto Rican communities, particularly widows, have a wider social support network than is often the case with White widows. This research confirms that importance of social support networks in alleviating the stress of aging.

Elder Care. Elder care in the United States is provided in two major ways: institutions for the elderly and private care in the home. Most of the care of older people in the United States is provided informally by families, in particular, mostly by women (Aronson, 1992). Family members provide 80 to 90 percent of long-term care for the elderly. Often, this is taken for granted work. Some estimate that for every $120 spent in publicly funded long-term care, families provide $287 of unpaid services (Glazer, 1990; Meyer, 1994).

Elder care tends to be confined to private sphere of households and reflects a gender division of labor that assigns women the responsibility for nurturing others. By contrast, sons and sons-in-law do not face the same expectations to be caregivers as do women. Despite social changes, men are still seen as having a primary commitment to work, and they are also perceived as being less able to anticipate and respond to elders' needs. Women also believe they are better at elder care than their husbands and brothers, but with the rapid increase in the older population that lies ahead, these social norms may have to change.

Many of the elderly live alone rather than with their families. The highest probability of living alone in old age is

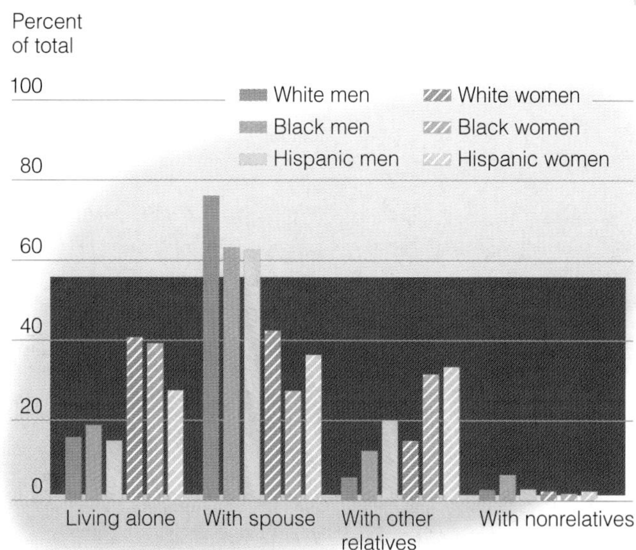

FIGURE 13.5 Living Arrangements of the Elderly

DATA: From the U.S. Bureau of the Census. 1994. *Marital Status and Living Arrangements: March 1994*. Series P20-484. Washington, DC: U.S. Government Printing Office, p. A11.

found among Black, White, and Hispanic women (see Figure 13.5; U.S. Bureau of the Census, 1995). At the same time, as noted by Sánchez-Ayéndez (1995), cultural norms among Puerto Rican families containing elderly women are likely to encourage use of informal support networks consisting of extended family as well as close friends to care for the elderly men as well as women. Race and ethnicity are powerful determinants of household structure, even controlling for age, sex, marital status and income. Blacks and Hispanics are twice as likely to live with other relatives, not including their spouses, as are Whites (U.S. Bureau of the Census, 1995).

In the past, it was relatively more common among Blacks than Whites for members of the extended family to live together, including grandparents and great grandparents. Now the Black elderly are about as likely to be living alone as the White elderly, suggesting that because of increased urbanization and some economic advances of Blacks over the last twenty-five years, Black families' treatment of the elderly has come to approximate White families' treatment of their elderly (Mutchler and Frisbie, 1987).

Many elderly continue to live in the homes where they once raised families. Most of these homes are free from mortgage debt, but many, having been built before 1940, are now in need of improvement, repairs, and remodeling requiring more money than the owners can afford to pay. Most elderly are greatly reluctant to be forced from their own homes by economic conditions or other circumstances. Whether voluntary or not, a move from one's home is physically and psychologically disruptive.

A relatively small percentage of the elderly are placed in long-term institutional care. Women are more likely to enter nursing homes than men, largely because of differences in life expectancy between the two groups and higher rates of

Many of the country's poor elderly must undergo minimal housing conditions as well as loneliness, as with this woman in a Washington, D.C. nursing home.

chronic illness among women. African Americans and Hispanics are also less likely to enter nursing homes than Whites, but they are three times as likely to rely on Medicaid. Approximately 1.7 million Americans live in the 15,000 nursing homes in the United States (U.S. Bureau of the Census, 1997a). Although this is only a small portion of all elderly (about one-tenth), many senior citizens worry it will happen to them. The decision to enter a nursing home is almost never made by the elderly person, but by a health-care professional or some other party who assumes control over the care of the older person.

Debunking Society's Myths

Myth: Nursing homes house most of the nations' elderly people.

Sociological perspective: Family members, most often women, provide almost all the care for elderly people in the United States; only about 5 to 10 percent of the elderly live in nursing homes.

The cost of nursing homes varies greatly, which tends to perpetuate social class differences. People of higher socioeconomic status tend to take up residence in higher-quality, higher-cost facilities or in modern retirement villages with restaurants, meals-on-wheels, and extensive social programs. The vast majority of nursing home residents are women, many of whom are very elderly (age eighty or more), ill, and frail. Elderly without living children or kin are the most likely to be admitted to a nursing home (Hess and Markson, 1990). Giving up one's residence, separating from one's family, and losing one's independence, coupled to the widespread belief that entering a nursing home is the final step before death, all give the elderly a bleak picture of such institutions.

The high cost of nursing homes also forces many, even from the middle class, into poverty. A congressional survey showed that half of all couples with one spouse in a nursing home become impoverished within six months of institutionalization. As many as 70 percent of single elderly patients reached poverty after only thirteen weeks in a nursing home. The average annual cost of a nursing home is $35,000. Since the median income for all elderly is under $20,000, it is easy to see that many of those who enter nursing homes also enter poverty (U.S. Bureau of the Census, 1997b).

The elderly and their families pay, on average, 44 percent of the total cost of nursing home care. **Medicaid** (a governmental assistance program for the poor) pays about 40 percent, and the remaining 16 percent is paid by either **Medicare** (a governmental assistance program for the elderly), other state and federal funds, or private health insurance (which pays less than 1 percent of the total cost). When long-term care is needed, Medicare covers *only* 2 percent of the total cost. The risk of poverty for those receiving long-term care in nursing homes is exceptionally high.

With the expansion of Medicaid and Medicare programs, the number of privately owned nursing homes has recently increased in comparison to those administered by charitable organizations (such as churches, the Salvation Army, and others). Federal subsidies contribute to the profitability of nursing homes and make them attractive to investors seeking tax shelters and long-term profits. This creates a problem pervasive throughout the health care industry, as corporate interests and the drive for profitability supplant affordable, humane care as the main interest of the institution. In the medical industry, this situation has caused attention to be focused on costs running out of control; in elder care, the focus has been on corruption. There have been many nursing home scandals in which owners embezzled funds from patients, and practiced medical fraud, Medicare fraud, and tax fraud. The more expensive institutions are generally better run and more closely scrutinized by the kin of the residents. It is the elderly poor in the second-tier institutions who are disproportionately victimized by scams and low-quality care.

The effects of the burden of care are apparent in the stress that women report from this role. In a study of Hispanic caregivers of Alzheimer's family members, sociologists have found that women's sense of burden and depression as caregivers is related to stronger adherence to cultural norms of filial support (Cox and Monk, 1993). Racial differences in patterns of elder care are also evident. Whites are more likely to institutionalize family members with Alzheimer's than are African American families; African Americans are also less likely to use formal support systems for elder care (Dungee-Anderson and Beckett, 1992); Cuban American Alzheimer's patients are also more likely to remain living in their daughters' homes; as a result, Cuban American daughters in such situations tend to be more depressed than their

White, non-Hispanic counterparts (Mintzer, 1992). Caring for a terminally ill person is also known to be detrimental to both the physical and psychological health of family members. The most likely to be so affected group are caregiving spouses and partners. These negative effects, however, can be mitigated by social factors like the quality of the relationship and the social support the caregiver receives (Kriegsman et al., 1994).

Elder Abuse. Physical and mental abuse of the elderly has only recently surfaced as a notable social problem. The National Center on Elder Abuse estimates that there are between 820,000 and 1,860,000 abused elders in the United States, but this organization acknowledges the difficulty of gauging the true extent of the problem. Elder abuse tends to be hidden in the privacy of families or behind institutional doors, and victims are reluctant to talk about their situations, so estimates are only approximations. What is known is that reports of elder abuse have increased since the mid-1980s. Whether elder abuse has actually increased over time or is simply reported more frequently is open to speculation (National Center on Elder Abuse, 1996; Tatara, 1996).

Further complicating the question of how widespread elder abuse has become is disagreement over how abuse should be defined. Elder abuse includes physical abuse, sexual abuse, emotional abuse, financial exploitation, neglect, abandonment, and self-neglect. Data on reported cases of elder abuse show that neglect is the most common form of abuse; half of all reported cases involve neglect. Physical abuse is the next most frequent form of abuse, accounting for 16 percent of all reported cases. Financial exploitation accounts for 12 percent of substantiated reports of elder abuse. The median age of elder abuse victims is 76.5 years; almost two-thirds are women. Sixty-five percent of victims in reported cases are White; 21 percent are Black. About 10 percent are Hispanic, with Asian Americans and Native American each making up less than 1 percent of known cases. Elderly individuals with prominent mental or physical disabilities are actually more likely to be victimized than those who are less disabled (National Center on Elder Abuse, 1996; Pillemer, 1988; Tatara, 1996).

Why are the elderly abused? One explanation is that having to care for the elderly is a major source of stress for the caregiver. Often the caregiver is a daughter or son who tries to hold a job as well as care for the elderly person. Research on elder abuse shows the following: The abusers are more likely to be women, middle aged, and (sadly) the daughter of the victim, in other words, the person most likely to be caring for the older person. Sons, however, are the ones most likely to be engaged in direct physical abuse, accounting for almost half of the known physical abusers. Sometimes the physical abuser is a husband where the abuse is a continuation of abusive behavior earlier in the marriage. In many cases of elder abuse, the *victim* is perceived by the abuser to be the major source of the stress. The same factors that affect family life in any generation contribute to the problem of elder abuse: Having few institutional support systems for elderly people and those who care for them results in social strains that produce frustration, violence, and neglect (Carson, 1995; Pillemer, 1988; Sengstock, 1991; Griffin and Williams, 1992).

This photograph by James Van Der Zee, a noted African American photographer, depicts the opulent context of some funerals.

Death and Dying

There is probably nothing sadder than watching a loved one die. At such a time, perhaps the last thing you would think to do is to analyze death sociologically. Still, if you were to engage your sociological imagination, you would find that the behaviors and events surrounding death have a clear sociological character. When it is known someone is going to die, people's behavior changes. As sociologists put it, a person can die a "social death" before actual biological death. The person dying starts to be treated as if he or she were "not there." People may talk about the person in the past tense, and the dying person may be perceived as a nonperson because of his or her physical, emotional, and communicative withdrawal. These tendencies are exacerbated by the dying person's placement in a hospital or nursing home (Glaser and Strauss, 1965; Sudnow, 1967).

Sociologists have also noted that people engage in *anticipatory grief,* beginning to grieve before someone actually dies. This is a form of resocialization in which the surviving person adjusts to life roles absent the other. Similarly, the rites of passage that follow death, such as funerals, provide a social transition for survivors. Funerals and other death rituals also reflect and reinforce the cultural values of social groups. By doing so, they can be seen as attempts to restore the break in social relationships that a death signifies (Kearl, 1989).

A sociological perspective on death reveals how death is socially structured. Patterns of stratification that reveal themselves in life are also apparent in death. Certain groups are more likely to die a violent death than others, namely, African American men, who are seven times more likely to die from homicide as White men. Infant death is also twice as likely to occur among racial minority groups than among White Americans (U.S. Bureau of the Census, 1997a). New medical procedures also produce new forms of inequality about who will live and die. With a shortage of organs available for heart, kidney, and lung transplants, who receives the transplant reflects social values about whose lives are considered most precious.

When a person dies, he or she dies within social institutions that are organized to handle death. Four-fifths of those who die on a given day die in hospitals. The social organization of death is especially apparent in the funeral home industry—an $8.5 billion business. Within this industry, funeral directors "manage" the death experience for others; no longer referred to as "undertakers," these people are seen as professionals—those who are skilled in the administration of death. As the box "Doing Sociological Research: Death's Work: A Sociology of Funeral Homes" shows, the bereaved are instructed by the funeral director on the ritual to come, including what choices to make, how to enter and exit the event, and what accoutrements to choose. Despite the appearance of condolences and sympathy, however, the funeral director is also a salesperson—someone who is part of an industry with a large market and considerable profit—the average cost of a funeral being well in excess of $7000 (Norrgard, 1992; Kearl, 1989).

Death was not always handled this way. Prior to the twentieth century, death was likely taken care of at home, and it was largely the work of women. Until the emergence

BOX 13.3 DOING SOCIOLOGICAL RESEARCH

Death's Work: A Sociology of Funeral Homes

DEVELOPING a sociological perspective reveals new ways of looking at seemingly ordinary events. The following analysis of funerals describes how sociologists, working from the perspective of functionalism, would analyze the manifest and latent functions of a funeral.

Most apparent of all the social shock absorbers of death is the funeral. The ritual disposal of the dead and social reintegration of the affected living are two of the few cultural universals known. But the cross-cultural variations in funerary observances are incredible. One can depart New Orleans–style, complete with marching brass bands and humorous graveside eulogies, or one can be the focal point of political protests, as when thousands march in the funeral processions of victims of political oppression.

Perhaps weddings are overrated as social events. It is at funerals that you meet the widest spectrum of people, where you see how many lives can be touched by a single individual. The funeral is the finished picture of a person, providing a ritual occasion when one reflects on the successes and shortcomings of a concluded biography. It also marks the endeavors of a generation: only the generation of the deceased can provide the frame of reference needed to grasp the principles to which one's biography was dedicated.

Funerals have the clearly apparent, manifest (Mandelbaum 1959) functions of disposing of the body, aiding the bereaved and giving them reorientation to the world of the living, and publicly acknowledging and commemorating the dead while reaffirming the viability of the group.

Although for some cultures and religious groups funerary ritual is explicitly directed for the dead, it is always a rite of passage for the principal survivors, a mechanism for restoring the rent in the social fabric caused by death.

There are less obvious, "latent" functions served by funerals. Symbolically dramatized in funerary ritual are reaffirmations of the extended kinship system. The restrictions and obligations of survivors (such as their dress, demeanor, food taboos, and social intercourse) serve to identify and demonstrate family cohesion. Also dramatized are the economic and reciprocal social obligations that extend from the family to the community and from the community to the broader society. In other words, the social bonds of the living are acted out, remembered, and reinforced in the minds of community members (Mandelbaum 1959). Not surprisingly, such functions have political significance. When, in 1982, 16,500 aborted fetuses were found in a container at the home of a Los Angeles man who ran a medical laboratory, three years of heated debate over their disposal followed. Antiabortionists sought permission to hold funeral services for the fetuses, claiming they were humans whose social membership had to be ritually reaffirmed, while a prochoice group, represented by the American Civil Liberties Union, argued that the remains were unwanted biological tissue and should be cremated without ceremony (*New York Times* 1985a).

Also reaffirmed in funerals are the social roles of the living. Claiming that the deceased was a winner in his or her roles reaffirms the system itself by having produced the opportunities for such a person to even create meaning. Why do we not speak ill of the dead? They are the roles they play, and to speak ill of them is to speak ill of ourselves. In the homosexual community, as the list of AIDS victims grows there is the sense that the traditional rituals are insufficient. When another member of New York City's People with AIDS Coalition dies, white helium baloons are released from St. Peter's Episcopal Church in Greenwich Village. Ashes of one partner are sometimes mixed with those of the other who predeceased him and then are dispersed in a place meaningful for the couple. And increasingly, the rainbow flag, the symbol of the annual Gay Pride Parade, is displayed on coffins (Dullea 1987).

SOURCE: Kearl, Michael C. 1989. *Endings: A Sociology of Death and Dying.* New York: Oxford University Press.

of undertaking as a profession (in the midnineteenth century), women were responsible for the care of the dead, particularly preparing the body for burial. Cultural beliefs about women's emotionality and caring supported this role. Women's care for the dead was also justified on religious grounds—society saw women as more devout than men. Since "laying out the body" was considered a sacred act, it logically followed that it should be women's duty. Some women in communities were seen as particularly skilled in this regard. Like midwives, they would be called upon to assist families in laying out the body. Known as "shrouding women," they washed and dressed the body, posed the body in the coffin, and constructed a "cooling board" on which the body laid, giving the impression of a restful sleep. When undertaking emerged as a profession in the late nineteenth century, women were used to adorn advertisements for embalming fluids and caskets. Draped over caskets in flowing clothes, women were supposed to lend an image of tranquility and beauty to death (Rundblad, 1995).

The **hospice movement**—where hospice workers, some of whom are volunteers, provide care for dying people and their families—developed as an alternative to hospital-based, technologically controlled death. The hospice movement has been a reaction to the impersonal forms of death that occur in more institutional settings. Rather than letting professional experts control death, hospice workers see control as more appropriately lying with the dying person and his or her family and friends. Ironically, as this movement has developed, many hospice workers have come to be defined as the experts in handling death, and the hospice movement has itself become a large and well-organized industry.

Age, Diversity, and Inequality

Throughout this chapter, we have seen how the experience of aging differs for various groups in society. Aging in itself can result in many problems—physical, psychological, social, and economic—but the effects of aging are compounded by the additional effects of other factors, most notably, class, race, and gender. Class, race, and gender shape the experience of aging in both positive and negative ways. Just as class inequality marks the experience of different groups of youth, so it does in old age. Racial differences in patterns of care, life expectancy, and economic resources make aging different for diverse groups in the United States. As we have seen, gender is also a significant factor in the aging experience. Not only are women reviled for aging, but they are more likely to be caregivers for both the old and the young.

Understanding diversity in the aging experience is not just a matter of studying disadvantage, although with attention to the sociology of race, class, and gender, you cannot help seeing the inequalities in life and death that mark the aging experience. As we have also seen, however, we also learn from different groups how to value the aged, with many racial communities historically placing more value on the elderly than has been true in the dominant culture. As the population ages, it will be useful to consider alternative patterns of care that have characterized different group practices since there will clearly be a need for new institutional and community practices to care for the large number of old people.

Age Groups as Minorities

Understanding diversity in aging is fundamental to sociological research; furthermore, concepts from the sociology of race can be used to analyze age stratification. The term *minority group* is one concept that is useful in interpreting age stratification. Recall that sociologists define a minority group as a group with relatively less status and power and fewer social and economic resources than more dominant groups (see Chapter 11). Age minority groups include both the old and the young—those groups with relatively the least power and the poorest life chances because of their age.

Seeing oneself as a minority group can also be the basis for mobilizing for group rights—a movement that is very apparent among the nation's older population. Groups like the American Association of Retired Persons (AARP), the Gray Panthers—a more radical group that has advocated, among other things, intergenerational living arrangements as a means for supporting old people—the Older Women's League (OWL), and other organizations have worked on behalf of old people to challenge age stereotypes and define the national agenda for old people. Like civil rights organizations, such groups are important advocates of social programs for the elderly.

Despite certain parallels, the similarities between the aged and racial and ethnic minorities should not be overstated; the aged differ from racial and ethnic minorities in important respects. For one thing, a person is in a racial or ethnic group for life, but aged for only part of his or her life; furthermore, as a group, the elderly have, on average, more political power than other minorities. A high percentage of the elderly vote—more than any other group—and the interests of the elderly are represented by powerful lobbying organizations with large and growing memberships, such as the AARP, which has a gigantic constituency of thirty-three million members. Unlike other minorities, the aged also receive far more subsidized support in the form of medical care for the elderly; insurance programs; senior discounts at supermarkets, restaurants, movies, hotels, and auto rentals; and a variety of other benefits that cumulatively are of great value to old people.

Age Prejudice and Discrimination

Age prejudice refers to a negative attitude about an age group that is generalized to all people in that group. Age prejudice is manifested in the stereotypes that we have seen of different age groups. You have probably heard elderly people taunted as "old geezers." As with any minority

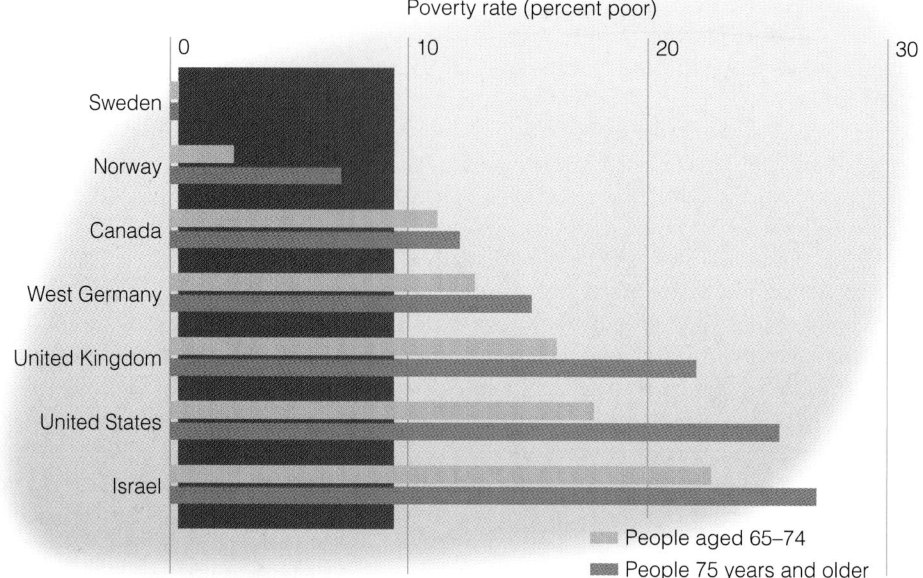

FIGURE 13.6 Poverty Among Older Persons: Selected Industrial Nations

SOURCE: Hedstrom, Peter, and Stein Ringen. 1990. "Age and Income in Contemporary Society." Pp. 77–104. In *Poverty, Inequality and Income Distribution in Comparative Perspective: The Luxembourg Income Study*, Timothy M. Smeeding, Michael O'Higgins, and Lee Rainwater (eds.). Washington, DC: The Urban Institute, p. 94.

group, a range of epithets has evolved to describe the elderly: geezer, curmudgeon, codger, fuddy-duddy, old biddy, old fart. The lobbying efforts of the elderly have also generated new stereotypes of the elderly as politically selfish and grasping, stereotypes that are surprisingly acceptable in social circles that would usually be quick to condemn such prejudice.

Prejudice against the elderly is so prominent in our culture that the myths given in Table 13.2 are seen as applicable to them as an accurate description of the elderly as a group. Physicians, nurses, and other health-care personnel tend to talk "baby talk" to the elderly. Doing so defines the elderly as childlike and incompetent, thus verifying their need for care. Just as for other groups who experience prejudice directed against them, bearing the labels that prejudice produces relegates people to a perceived lower status in society.

In Chapter 11, we defined racial–ethnic discrimination as behavior that singles people out for different treatment. **Age discrimination** is the different and unequal treatment of people based solely on their age. Some forms of age discrimination are illegal. The Age Discrimination Employment Act, first passed in 1967, but amended several times since, protects people from age discrimination in employment. No longer can an employer hire or fire someone based solely on age, nor can employers segregate or classify workers based on age. Age discrimination cases have become one of the most frequently filed cases through the Equal Employment Opportunity Commission (EEOC)—the federal agency set up to monitor violations of civil rights in employment.

Ageism is a term sociologists use to describe the institutionalized practice of age prejudice and discrimination. More than a single attitude or an explicit act of discrimination, ageism is structured into the institutional fabric of society. Like racism and sexism, ageism encompasses prejudice and discrimination, but it is also manifested in the structure of institutions. As such, it does not have to be intentional or overt to affect how age groups are treated. We have seen the consequences of ageism throughout this chapter: stereotypes of old and young that define their social worthiness, marked inequalities in the social and economic resources available to different age groups, institutional practices that manage how age groups are treated in society. Ageism in society means that, regardless of laws that prohibit explicit discrimination and regardless of individuals' positive efforts on behalf of old and young alike, people's age is a significant predictor of their life chances. Old people are stratified into some segments of society; resources are distributed in society in ways that advantage some age groups and disadvantage others; cultural belief systems devalue the elderly; society's systems of care are inadequate to meet people's needs as they grow old—these are the manifestations of ageism, which is a persistent and institutionalized feature of society.

Quadruple Jeopardy

For some groups, ageism is compounded by other factors. **Quadruple jeopardy** is a phrase referring to the simultaneous effects of being old, minority, female, and poor. The effects of age, race, gender, and social class intersect for persons who are old, minority, female, and poor. The status of old is lower than that of adult; the status of minority is lower than that of White; the status of women is lower than that of men; and the status of poor is lower than that of not poor. Anyone who is all four is placed in great jeopardy.

The effects of quadruple jeopardy can be seen in several areas. Income differences between groups reveal the risks of this multiple status. Overall, the economic condition of the elderly has improved over the last thirty years, with the

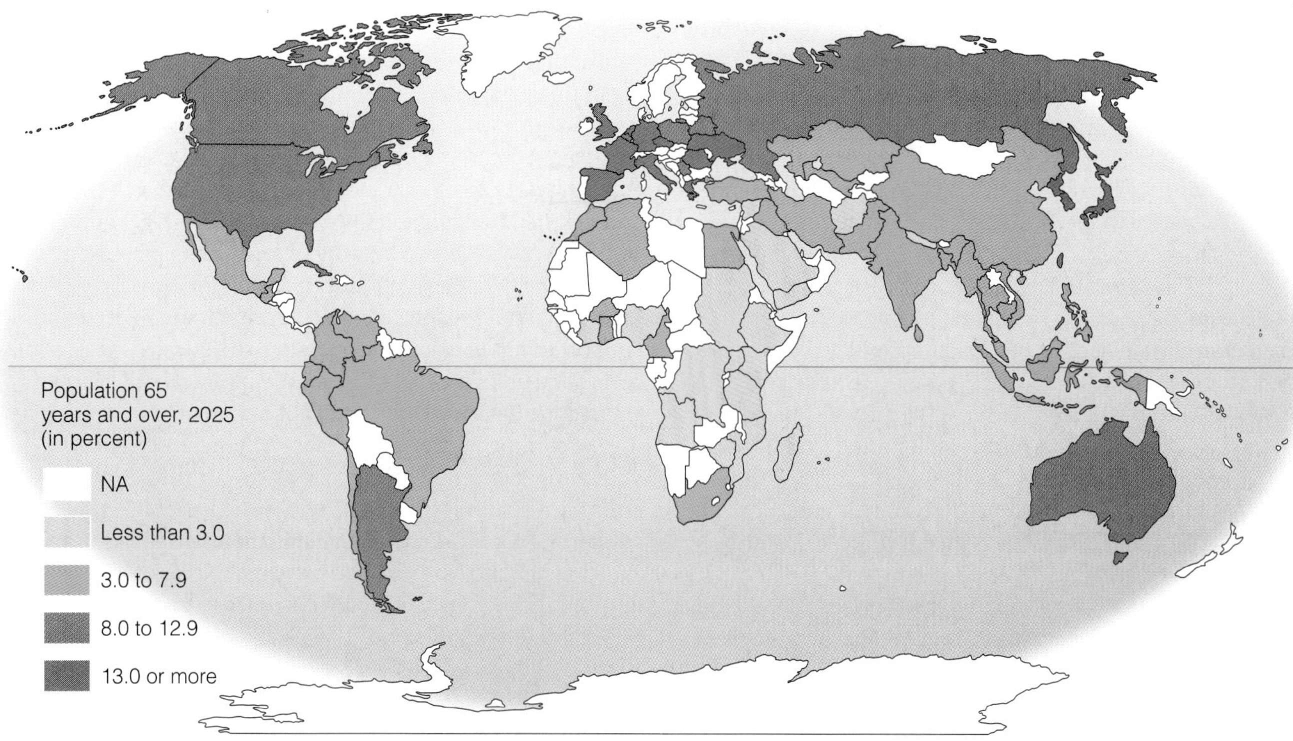

MAP 13.2 Viewing Society in Global Perspective: An Aging World

DATA: From Torrey, Barbara Boyle, et al., 1987: *An Aging World.* Washington, DC: U.S. Government Printing Office, pp. 2–3.

percentage of old people who are poor declining from 35 percent in 1959 to less than 11 percent in 1997. In combination with race and gender, however, age brings a particular risk of poverty. Poverty among the aged is more pronounced for African Americans, Latinos, and Native Americans than for Whites, and it is more pronounced for women than men. In 1997, 26.5 percent of all elderly Black people and 27.1 percent of all elderly Hispanics were poor, as contrasted with 11 percent of Whites and 10 percent of Asian/Pacific Islanders. For American Indians, more than *half* the elderly were below the poverty level. Fifteen percent of women over the age of sixty-five are poor compared to 11.6 percent of men (Dalaker and Naifehi, 1998). Furthermore, compared to other industrialized nations, the United States also has a poor record of alleviating poverty for its older residents (see Figure 13.6).

Health provides another example of the effects of quadruple jeopardy. Black and Hispanic women tend to have higher rates of health problems such as hypertension, diabetes, heart disease, and cancer, compared to White men and White women. Being poor further increases these rates. For unknown reasons, cancer of the esophagus is exceptionally high among middle-aged and older Black men—90 cases per 100,000 at age 60, whereas it is only 20 cases per 100,000 for Black women, White women, and White men of the same age (Horton, 1995; National Cancer Institute, 1990; Reynolds et al., 1992; Neighbors and Jackson, 1996). For Native Americans, especially those who are elderly, access to doctors, hospitals, and other health-care facilities is less than for any other social or ethnic group. In fact, travel times to health-care providers is substantially longer for Native Americans than for any other group (Cunningham and Cornelius, 1995; Inouye, 1993). These differences are perpetuated well into old age, and are perpetuated even further for those who are poor.

These data indicate the risks that are associated with age, race, gender, and class in the United States, and they illustrate what sociologists mean by saying that the aged, especially those who are also minority, poor, and female, are in quadruple jeopardy.

Age Seen Globally

As discussed earlier (in Chapter 5), societies vary in the extent of their industrialization, from societies that are nomadic or agrarian (such as the Bedouins of Saharan Africa or many settlements of Siberia) to those that are heavily industrial (such as the United States, England, France, and Germany) to the postindustrial (Japan). The treatment of the very old in society varies to some degree depending on the extent of industrialization of the society.

The elderly, on average, have greater social status in nomadic and agrarian societies, although there are exceptions. In such societies, those who have accumulated wealth during their lifetime (such as through raising crops and animals) gain high prestige and also economic and political power that continues into old age, which may be only forty or fifty years of age. When this is a man, he may be regarded as a patriarch and accorded considerable deference in the group, tribe, or town (Cohen and Eames, 1982; Stoller and Gibson, 1997).

In industrialized societies, advances in medicine and technology tend to result in longer life expectancy; as a consequence, a greater proportion of the population is elderly. Increased technology also has the effect of placing those who are relatively young into the workforce, with their education reflecting new and developing technologies. As a result, those who are older and who receive their education and training under an earlier technology may be less trained and thus less able to carry out more technologically advanced jobs and tasks. An effect of this process is a surplus of older individuals who are regarded as less economically useful in society (an effect predicted by functionalist theory). This in turn cases the elderly in such a society to be accorded lower social status and prestige. This process then results in the elderly having less social status in industrialized societies than in nomadic or agrarian societies.

There is one clear exception to the principle that the elderly have less social status in societies that are more industrialized, and that is the case of Japan. Japan is not only very highly industrialized, but is a postindustrial society, one that stresses the production and development of services (such as computer engineering and technology) more than goods (such as steel). In contrast to the industrialization principle, the elderly in Japan are accorded very high esteem and social status. Older men, in particular, are more likely than their age counterparts in the United States to remain in their jobs and to stay on as high executives and corporate heads and leaders well into old age. In fact, the elderly have been elevated to high status for many decades, even as Japan has moved from being an agrarian society to a highly industrialized and, now, postindustrial society. One explanation may be that the strong cultural tradition in Japan of high prestige for the elderly may have been powerful enough to overcome the effects of industrialization on lowering the status of the elderly. Nonetheless, as some researchers have noted, there is an increasing tendency for the elderly in Japan to give up positions of power to younger people, and in this respect, Japan may be undergoing social changes that render it somewhat more like the United States and other industrial or postindustrial societies (Treas, 1995; Palmore and Maeda, 1985).

Explaining Age Stratification

Why does society stratify people on the basis of age? Once again we find that the three main theoretical perspectives of sociological analysis—functionalism, conflict theory, and symbolic interactionism—offer different explanations. Functionalist sociologists ask whether the grouping of individuals contributes in some way to the common good of society. From this perspective, adulthood is functional to society because those who are adult and middle aged are seen as the group contributing most fully to society; the elderly are not. Functionalists argue that older people are seen as less useful and are therefore granted lower status in society. Youth are in-between. The constraints and expectations placed on youth—they are prohibited from engaging in a variety of "adult" activities, expected to go to school, not expected to support themselves—are seen to free them from the cares of adulthood and give them time and opportunity to learn an occupation and prepare to contribute to society.

According to the functionalist argument, the elderly voluntarily withdraw from society by retiring and lessening their participation in social activities such as church, civic affairs, and family. **Disengagement theory,** derived from functionalist theory, predicts that as people age, they gradually withdraw from participation in society and are simultaneously relieved of responsibilities. This withdrawal is functional to society because it provides for an orderly transition from one generation to the next; the elderly move aside so that the young can step in. The young presumably infuse the roles they take over from the elderly with youthful energy and stamina, while supporting themselves and providing for their families. As the elderly become less useful to society, they are rewarded less. According to the functionalist argument, the diminished usefulness of the elderly justifies their depressed earning power and their relative neglect in social support networks.

Conflict theory assesses the differences between age groups not in terms of what they contribute but in terms of what they want; the focus is on the competition over scarce resources that exists between age groups. Among the most important scarce resources are jobs. Unlike functionalist theory, conflict theory offers an explanation of why both youth and the elderly are assigned lower status in society. Conflict theory argues that barring youth and the elderly from the labor market is a way of eliminating these groups from competition, improving the prospects for workers who are adult and middle aged. Once removed from competition, both the young and the old have very little power, and like other minorities, they are denied access to the resources they need to change their situation. Conflict theory also helps explain the competition between groups that has marked concerns about generational equity, in particular the questions raised earlier about how the nation will take care of its growing elderly population (see Table 13.3).

A third approach to age stratification, symbolic interactionism, analyzes the different meanings attributed to social entities, in this case, the different age groups. Interactionists ask what symbolic meanings become attached to the different age groups and to what extent these meanings explain how society ranks them. As we saw in discussing age stereotypes, definitions of aging are socially constructed. More-

TABLE 13.3 *SOCIOLOGICAL THEORIES OF AGING*

	Functionalism	Conflict Theory	Symbolic Interaction
Age differentiation	contributes to the common good of society since each group has varying levels of utility in society	results from the different economic status and power of age cohorts	occurs in most societies, but the social value placed on different age groups varies across diverse cultures
Age groups	are valued according to their usefulness in society	compete for resources in society, resulting in generational inequities	are stereotyped according to the perceived value of different groups
Age stratification	results from the functional value of different age cohorts	intertwines with inequalities of class, race, and gender	promotes ageism, which is institutionalized prejudice and discrimination against old people

over, in some communities, the elderly may be perceived as having higher status than is true for the elderly in other groups. Symbolic interactionism therefore considers the role of social meanings in understanding the sociology of age. These meanings are cultural products. Age clearly takes on significant social meaning—meanings that vary from society to society for a given age group and that vary within a society for different age groups.

CHAPTER SUMMARY

- All societies practice some form of *age differentiation,* and the United States is no exception. *Age stratification* is the hierarchical ranking of different age groups in society. The age structure of the United States is rapidly changing, with a greater proportion of the population consisting of old people.
- The social aspects of aging are just as important as the physical aspects of aging. Certain physical processes are shaped by the social and cultural judgments made about aging in society. *Life expectancy,* as an example, is influenced by a number of social factors, notably race, gender, and social class.
- *Age stereotypes* pervade this society and burden the aged, in particular, with negative preconceptions about being old. Age stereotypes are compounded by the gender stereotypes that define the aging experience for men and women differently. *Age norms* spell out explicit and implicit expectations for different age groups although these norms change as social conditions change. Age is an ascribed status—established at birth. Different *age cohorts* in society share similar historical experiences. Each cohort's experience is also shaped by the age structure of society at a given time.
- As the nation's population grows older, the contract between generations that has established expectations for intergenerational care will likely change. Those who are young now can expect to spend a greater portion of their lives not only raising their own children, but also caring for their elderly parents.
- Sociologists use a *life course perspective* to relate people's life experience to their sociohistorical context. The different phases in the life span include childhood, youth and adolescence, adulthood, and old age. Within each generation, unique life events shape the sociological experience of these age groups.
- Old age brings with it both the good and the bad. The bad side is that the aged are treated as a social minority and tend to be shunted aside and ignored. The good side is that old age can be, and for many is, a period of great happiness and satisfaction, of renewed interest in education and perhaps a second career, and of increased self-evaluation.
- Both the youthful and the aged are treated as minorities in society, especially in industrialized societies. In many nomadic and agrarian societies, the elderly enjoy social status that is quite high. *Age discrimination* is the differential treatment of persons on the basis of age alone. *Age prejudice* refers to the tendency to have negative attitudes toward someone on the basis of age. *Ageism* is the institutionalized dimension of age prejudice and discrimination. It need not be conscious or intentional, but ageism shapes the experiences of diverse groups in society.
- *Quadruple jeopardy* is the combined effect of age, race, gender, and class. Those who are most disadvantaged in society—in income, education, health, and other areas—are those who are old, female, of a racial or ethnic minority, and economically disadvantaged.
- In attempting to answer the broad question of why society discriminates against both its youth and its elderly, three sociological theories lend some suggestions: Because the elderly are less useful, or "functional," to society *(functionalist theory);* because the elimination of both groups from competition in society frees up those who are adult and middle aged *(conflict theory);* and because both the youth and the elderly are infantilized via cultural symbols, such as language and popular culture *(symbolic interaction theory).*

KEY TERMS

age cohort
age differentiation
age discrimination
ageism
age norms
age prejudice
age stereotype
age stratification
Alzheimer's disease
dementia
disengagement theory
generational equity
hospice movement
life course perspective
life expectancy
Medicaid
Medicare
quadruple jeopardy

THE INTERNET: A Tool for the Sociological Imagination

Resources on the Internet:

Virtual Society: The Wadsworth Sociology Resource Center
http://sociology.wadsworth.com

Visit this site to find additional learning tools, including interactive quizzes, links to related web sites, and an easy link to *InfoTrac College Edition.*

World Health Organization Aging and Health Home Page
http://www.who.int/hpr/ahe/

This web site offers research information on the global dimensions of aging and health.

National Institute of Aging
http://www.nih.gov/nia

The home page of the U.S. government agency, provides information on the latest research on aging, links to additional sites, and information on funding for research on aging.

Section on Aging and the Life Course of the American Sociological Association
http://www.asanet.org/Sections/aging.htm

This research section is affiliated with the American Sociological Association that provides access to information on section activities, including news about the latest sociological research on aging and the life course.

Sociology and Social Policy: Internet Exercises

As the proportion of elderly people in the United States grows, questions about how to support this growing segment of the population are likely to increase in significance. Can the nation, for example, continue supporting *Social Security* at the same time that it needs support for other groups—for example, children, the most impoverished group in society? What are the major policy questions currently at stake in public discussions of Social Security?

Internet Search Keywords:

Social Security
senior citizens/elderly
child poverty

Web sites:

http:/www.ssa.gov/
Official web site of the Social Security Administration

http://www.ncscinc.org/
National Council of Senior Citizens provides its publications on seniority and Social Security plus related news and press releases.

http://cpmcnet.columbia.edu/news/childpov/newi0069.html
A report on increasing child poverty.

http://www.ssas.com
The web site of the Social Security Advisory Service faciliates access to important literatures and organizations.

InfoTrac College Edition: Search Word Summary

age norms
life expectancy
life course perspective

In order to learn more about these central topics in sociology, you can conduct an electronic search using InfoTrac College Edition. To aid in your search and to gain useful tips, see the Student Guide to InfoTrac College Edition on the Virtual Society web site: http://sociology.wadsworth.com

INTERACTIONS—A SOCIOLOGY CD-ROM: CONCEPTS FOR THIS CHAPTER

Go to the Wadsworth Sociology CD-ROM for further study on the concepts in this chapter. The CD-ROM also includes quizzes and additional activities to expand your learning experience.

SUGGESTED READINGS

Allen, Jessie, and Alan Pifer, eds. 1993. *Women on the Front Lines: Meeting the Challenge of an Aging America.* Washington, DC: The Urban Institute Press.

This book considers the social, economic, and political context of aging for women. The editors pay close attention to class, race, and ethnicity in the aging process. Also included is an analysis of policies and attitudes that affect older women.

Atchley, Robert C. 1997. *Social Forces and Aging,* 8th ed. Belmont, CA: Wadsworth.

This book, a summary of the sociology of aging, is one of the best accounts in the field of social gerontology.

Cox, Harold. 1993. *Later Life: The Realities of Aging,* 3rd ed. Englewood Cliffs, NJ: Prentice-Hall.

This textbook clearly reviews the entire range of the aging process, with attention to diverse groups, including American Indians, Japanese Americans, Blacks, and Hispanics.

Diamond, Timothy. 1992. *Making Grey Gold: Narratives of Nursing Home Care.* Chicago: University of Chicago Press.

Participant observer Diamond examines profit, care, and the exploitation of women of color in a nursing home.

Hess, Beth B., and Elizabeth W. Markson, eds. 1990. *Growing Old in America: New Perspectives on Old Age.* New Brunswick, NJ: Transaction Books.

This is a readable and representative selection of sociological research on the effects of aging.

Lyman, Karen A. 1993. *Day In, Day Out with Alzheimer's: Stress in Caregiving Relationships.* Philadelphia: Temple University Press.

Based on participant observation at several day-care centers for people with Alzheimer's, Lyman studies the stress associated with caregiving for Alzheimer's disease. She concludes that the organization of Alzheimer's day-care work, that is, how the system is designed, is actually more stressful for caregivers than the care itself.

Stoller, Eleanor Palo, and Rose Campbell Gibson. 1997. *Worlds of Difference: Inequality in the Aging Experience.* Thousand Oaks, CA: Pine Forge Press.

This anthology explores diverse patterns of aging, particularly given the influence of race, class, and gender on the aging process. Many of the articles are written as personal narratives; combined with a sociological perspective, they provide students with a compassionate view of diverse norms and social structures that shape the experience of growing old.

Sudnow, David. 1967. *Passing On: The Social Organization of Dying.* Englewood Cliffs, NJ: Prentice-Hall

A classic study in the sociology of death, this book examines, from a symbolic interaction perspective, how death is routinized in hospitals. The analysis reveals social perceptions of death and dying, as well as institutional practices that reveal the sociology of death.

Zelizer, Viviana A. 1985. *Pricing the Priceless Child: The Changing Social Value of Children.* New York: Basic Books.

Zelizer offers an insightful historical account of society's changing definitions and conceptions of childhood.

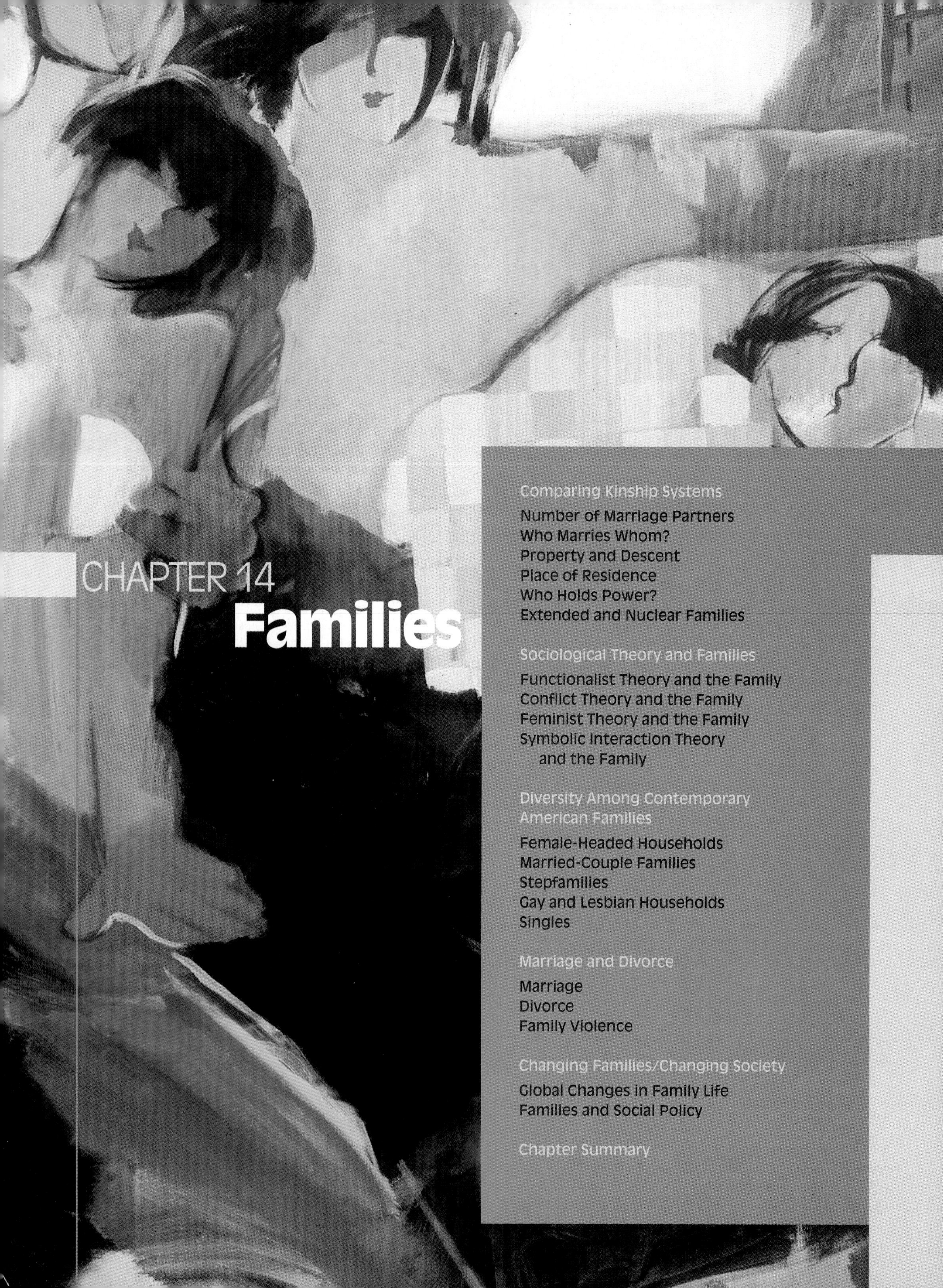

CHAPTER 14
Families

SUPPOSE you were to ask a large group of people in the United States to describe their families. What would they say? Many would describe divorced families—families that have been reconstituted following the breakup of a marriage. Some would describe *single-parent families,* most likely headed by a woman. Some would describe *stepfamilies* with new siblings and a new parent stemming from remarriage. Others would describe *gay or lesbian households* in which a single gay parent or a couple of the same sex is raising children. Still others would describe the so-called *traditional family* with two parents living as husband and wife in the same residence as their biological children. Also included would be *adoptive families* and families with foster children. The variety of descriptions would reflect the enormous diversity in families today. Indeed, families have become so diverse that it is no longer possible to speak of "the family" as if it were a single thing.

The traditional *family ideal*—a father employed as the major breadwinner and a mother at home raising children—has long been established within the dominant culture as the family to which we should all aspire. The family ideal has been communicated through a variety of sources, including the media, religious doctrine, and socialization within families. Few families now conform to this ideal, and the number that ever did is probably less than generally imagined (Coontz, 1992). As we can see in Figure 14.1, a large proportion of children do not live in traditional family arrangements. Families that do conform to this family type face new challenges, not the least of which may be the difficulty of living on one income or managing family affairs when both parents are employed. Many families also feel that they are under siege by other changes in the society. These are changes that are dramatically altering all family experiences, but that make the traditional family increasingly unattainable even for those who desire this lifestyle.

Families have an enormous influence on our personal lives and the relationships we form. The emotional climate within families ranges from joyous, loving, and intimate to traumatic, conflict-ridden, and sorrowful. The emotional grounding we acquire in a family is carried with us throughout our lives. Many live in loving families, where the family provides stability in relationships and nurturing care for its members. But the popular image of families as loving units that provide refuge from the impersonal, hectic world outside is not shared by all. Family affairs are believed to be private, but as an institution, the family is very much part of the public agenda. Many believe that "family breakdown" is the cause for society's greatest problems. Public policies shape family life both directly and indirectly, intentionally and unintentionally, and family life itself shapes the dynamics of behavior in other institutions. Family life is now being openly negotiated in political arenas, corporate boardrooms, and courtrooms, as well as in the bedrooms, kitchens, and "family" rooms of individual households.

Many view the changes taking place in family life as positive. Women have new options and often greater independence. Fathers are discovering that there can be great pleasure in domestic and child-care

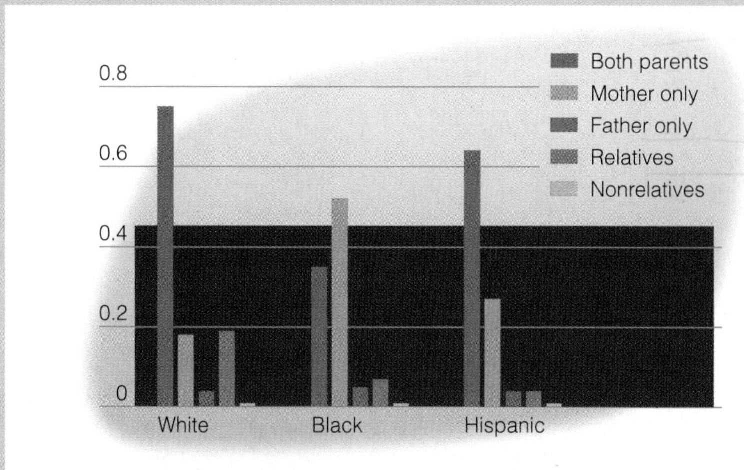

FIGURE 14.1 Living Arrangements of Children

DATA: From Lugaila, Terry A. 1998. *Marital Status and Living Arrangements: March 1997 (Update)*. Current Population Reports, Series P20-506. Washington, DC: U.S. Department of Commerce.

responsibilities. Change, however, also brings difficulties: balancing the demands of family and employment, coping with the interpersonal conflicts caused by changing expectations, and striving to make ends meet in families without sufficient financial resources. These changes bring new questions to the sociological study of families.

Sociologists address the family as a *social institution*. They are interested in how changes in the society affect the formation and character of families and how families, in turn, affect people's experiences in the society. People commonly believe that families are shaped by the personalities of family members. Sociologists recognize that individual personalities affect relationships with people both in and outside the family, and they know that personality is heavily influenced by family experience. Sociologists who study families, however, tend to focus on different issues, such as how families are affected by the economic, political, social, and cultural institutions of the society.

As a social institution, the family is an established social system that emerges, changes, and persists over time. Even with social change, the family, like other institutions, is somewhat stable. This does not mean that families are all the same; the economy is an institution, but that does not mean everyone is employed in the same kind of work organization. Studying the family as an institution simply recognizes that families are organized in socially patterned ways. Institutions are "there"; we do not make them up each time a new person is born or reinvent them each day upon waking, although people do make adaptive changes that result in institutions constantly evolving. Institutions shape and direct our action, and make it appear that the only options available to us are those deemed acceptable by society. Perhaps you find it hard to imagine living in a family where the husband is expected to have both a wife and a concubine or where children do not live with their parents. This is because the institutional structure of the family in this society does not support such family practices. Institutions shape both the form of families and the expectations that we have for family life.

Like other institutions, families are also shaped by their relationship to systems of inequality in society. Race, class, gender, and age stratification affect how society values certain families, and they influence the resources available to different families. In addition, these systems of stratification influence the power that individual members have within families. Children, for example, have fewer rights than adults; the very old are often relatively powerless within families. Heterosexual privileges also shape the resources available to certain families since our social institutions presume that families will be heterosexual. Those living in gay and lesbian families then have to invent new practices that are not typically supported by the institutional fabric of society.

The family is also intertwined with other social institutions in the society, such as the law, religion, and education. Changes in one institution affect the others, including families. For example, economic changes can greatly affect families through changes in patterns of employment: The employment status of a family member affects not only the financial condition of the household, but also the social dynamics within the family. Similarly, differences in religious values affect family behaviors, including reproduction, marriage, divorce, and sexual behavior.

The effect of other social institutions on the family even extends to things we might consider private. For example, the family is closely regulated by the government. Why else would one need a license to marry? Married, heterosexual couples can file joint tax returns, but gay and lesbian couples who own property together and share expenses and income cannot—an example of the government dictating what relationships can be considered a family. The poorest families are those most likely to experience the intrusion of government in their lives since for them state intervention in family life can be an everyday fact of life. Welfare mothers are now forced to work to receive state support; in the 1960s, the infamous "bedroom brigades" referred to state policies that prohibited women who were receiving welfare from having men stay in their homes overnight.

Given the diversity among families, how do we define the family? The family has traditionally been defined as a social unit of those related through marriage, birth, or adoption who reside together in officially sanctioned relationships and who engage in economic cooperation, socially approved sexual relations, and reproduction and child rearing (Gough, 1984). Changes in contemporary family life require some flexibility in this definition since not all families fit these conditions. Following divorce, for example, the family does not typically share a common residence. Some working parents leave their children in the care of others as they pursue seasonal or regional employment, sometimes far from their families (including across national borders), a pattern that has been most common among immigrant groups, racial–ethnic families, and now, transnational families (discussed later).

Female-headed families are more common among some groups than others. Twenty-three percent of Hispanic families are headed by women.

Defining the family has even proved difficult for the U.S. Census Bureau, the government organization officially responsible for counting and classifying families. Before 1980, the Census Bureau used a single definition of the family, routinely classifying the man as the head of household if he and his wife were living together. Since then, the Census Bureau has recognized that social changes have resulted in greater sharing of household responsibilities, making the term *head* increasingly inappropriate. The Census Bureau defines a **household** as all persons occupying a housing unit who may or may not be related, and a family as "a group of two persons or more (one of whom is the householder) related by birth, marriage, or adoption and residing together" (U.S. Bureau of the Census, 1993: B2). As social scientists have become more aware of the diversity in family life, they have refined their analyses to better show the different realities of family situations. Now, we can define the **family** broadly to refer to a primary group of people—usually related by ancestry, marriage, or adoption—who form a cooperative economic unit to care for any offspring (and each other) and who are committed to maintaining the group over time (adapted from Lamanna and Riedmann, 1997: 9).

Comparing Kinship Systems

Families are part of what are more broadly considered to be kinship systems. A **kinship system** is the pattern of relationships that define people's family relationships to one another. Kinship systems vary enormously across cultures and at different times; however, most societies recognize marriage in some form as a durable relationship between men and women, although not necessarily one that lasts an entire lifetime. In some societies, marriage is seen as a union of individuals; in others, marriage is seen as creating alliances between groups. The form families take in different societies varies enormously.

In some nations, such as Niger, Afghanistan, Nepal, and Bangladesh, more than half of all women are married before age nineteen, compared to less than 8 percent in the United States (Seager, 1997; U.S. Bureau of the Census, 1997b). In many societies, such as India, China, and others, many still follow traditional practices of arranged marriage, some even involving a broker whose job it is to conduct the financial transactions and arrange the marriage ceremonies (Croll, 1995). In many other societies, maintaining multiple marriage partners may be the norm, as we will see later in the discussion of polygamy and monogamy. At the heart of all these diverse family patterns are the social norms and structures associated with kinship systems.

Kinship systems can generally be categorized by the following features: *how many marriage partners are permitted at one time, who is permitted to marry whom, how descent is determined and how property is passed on, where the family resides,* and *how power is distributed* (O'Kelly and Carney, 1986).

Although it is rare, some cultures practice polygyny—the practice of a man marrying more than one woman.

Number of Marriage Partners

Polygamy is the practice of men or women having multiple marriage partners. Polygamy usually involves one man having more than one wife—technically referred to as *polygyny*. (People commonly use the term *polygamy* mistakenly when they are actually referring to *polygyny*.) *Polyandry* is the practice of a woman having more than one husband—an extremely rare custom.

Within the United States, polygynous marriage was once practiced among the Mormons. Their religious doctrine allowed men to have more than one wife. In 1890, polygyny was prohibited by the Mormon Church; it had been outlawed by the U.S. Congress earlier. The theological basis of polygyny is the Mormon belief that whole families should live as a unit under priestly authority in the afterlife. Mormon theology defines exultation (a state of extreme religious joy) as dependent on the number of children a man fathers and, for women, the number they bear; under polygyny, men were assured more children than they could have with only one wife. In reality, very few Mormons ever practiced polygyny; plural marriage had to be approved by the church and was mostly confined to the church leadership. It is practiced now only by a few Mormon fundamentalists (about 2 percent of the state population) who do so secretly and without official church sanction (Bachman and Esplin, 1992; Driggs, 1990; Brooke, 1998).

Although polygyny is most commonly associated with Mormons, other groups have practiced this form of marriage. Old Testament patriarchs had plural wives. Polygyny has also been practiced in some Muslim societies, such as Morocco, Tunisia, Egypt, and Kuwait, to name a few, although mostly among elites. Among Muslims, polygyny stems from the belief that men need more than one partner to satisfy their sexual tension. Women's sexuality, on the other hand, was traditionally defined as insatiable and uncontrollable; attaching a woman to just one man is a way to control women and restrict their presumed sexual capacity. Polygyny was also historically linked to high prestige among Muslim men since men with the most wives (and presumably the greatest sexual drive) held the highest social status (Mernissi, 1987).

Polygyny also has an economic function. It provides wealthy men who can afford multiple wives with a source of cheap labor: their wives. As a result, polygyny has been most common in agrarian societies where there is a need for a large and inexpensive labor force—societies where the birthrate is also high. Although polygyny may sound oppressive to women, it also had some positive features for women. Strong communities of women often emerged from such arrangements; having several women in a single household meant they also shared in household labor, although, since those with multiple wives were also wealthy, household servants would provide much of this work.

Monogamy is the marriage practice of a sexually exclusive marriage with one spouse at a time. It is the most common form of marriage in the United States and other Western industrialized nations. In the United States, monogamy is not only a cultural ideal, but is prescribed through law and promoted through religious teachings. Lifelong monogamy is not always realized, however, as evidenced by the high rate of divorce and extramarital affairs. Many sociologists characterize modern marriage as *serial monogamy* in which individuals may, over a lifetime, have more than one marriage, but maintain only one spouse at a time. Across different cultures, monogamy is not necessarily the norm, although it is the most common form of marriage in industrial societies. When you factor in the divorce rate and the common practice of extramarital affairs, however, sociologists estimate that strict monogamy is the norm in only about 20 percent of the world's cultures (Lamanna and Riedmann, 1997).

Who Marries Whom?

In addition to determining how many marital partners one can have, kinship systems determine whom one can marry. In contemporary U.S. society, people are expected to marry outside their own kin group; in other societies, people may be expected to marry within the kinship network. Among some Muslim people, the preferred form of marriage is between a man and his father's brother's daughter (a first cousin), thereby keeping family property within narrow bounds (Howard, 1989). Such a practice, though not generally prohibited by law, is considered deviant in the United States, even though it occurs occasionally.

Exogamy is the practice of selecting mates from outside one's group; the group may be based on religion, territory, racial identity, and so forth. **Endogamy** is the practice of selecting mates from within one's group. In the United States, neither exogamy nor endogamy is mandated by law, although some religious doctrines condemn marriage outside the faith. Even if certain forms of marriage are not explicitly outlawed, society establishes normative expectations about who is an appropriate marriage partner. The *incest*

endogamy

taboo, generally considered to be universal, for example, is a cultural norm forbidding sexual relations and marriage between certain kin. Research on incest shows, however, that despite this cultural taboo, incest can and does occur with alarming frequency in this society and others (Russell, 1986; Margolis, 1996; Adkins and Merchant, 1996). Many cultures have a tradition of *arranged marriages,* in which parents (or other elders) make rationally calculated choices about the appropriate marriage partner for their children. Arranged marriages work like business arrangements, perhaps even including a broker, in which a sum of money or property brought to the marriage may be part of the calculation about an appropriate marriage partner.

In general, people in the United States marry people with very similar social characteristics. For example, there is a clear tendency for people to select mates with similar social characteristics to their own—a pattern sociologists refer to as **homogamy.** Whether it is class, race, religion, or educational background, people tend to choose partners from a background similar to their own (Kalmijn, 1991). Although no laws in the United States prohibit marriage between people of different social rank, such marriages are atypical.

Most marriages are also between members of the same racial group. Sociologists have often used public attitudes about intermarriage as indicative of the public's support for or against integration since marriage is the ultimate loss of social distance between the so-called races. Attitudes toward intermarriage have changed dramatically. In 1972, 73 percent of White Americans disapproved of intermarriage, compared to 24 percent of Black Americans; in 1997, 33 percent of Whites and 17 percent of Blacks disapproved (Schumann et al., 1997). Still, interracial marriage is infrequent, although increasing. Interracial couples now constitute only about 2½ percent of married couples; this is, however, a fourfold increase since 1970 (U.S. Bureau of the Census, 1997b). Some groups are more likely to intermarry than others. African Americans and White Americans are the least likely to marry outside their group. Native Americans are the most likely to intermarry, with approximately one-third of Native Americans marrying outside their group (McLemore, 1994).

The growth of interracial marriage is also contributing to a more multiracial and multicultural society. The presence of mixed-race people also brings new issues into sociological focus. Most obviously, it highlights the inadequacy of seeing race as a fixed category. Mixed-race people must negotiate a complex route to identity formation, but their experience provides us with new perspectives on the meaning of race and culture in society (Root, 1996).

Patterns of intermarriage also vary by age and other social characteristics. Among African Americans, for example, younger men and women are more likely to marry someone of a different race than are older African Americans. Among Hispanics, Mexican Americans are least likely to marry outside their own group; Puerto Ricans, Cubans, and other Hispanics who are not Mexican are the most likely to marry outside their own group (Sung, 1990). Mexican Americans whose parents are natives of the United States are more likely to intermarry than are Mexican Americans of mixed parentage (that is, both parents Mexican American, but one born in the United States and the other in Mexico). The Mexican Americans most likely to marry other Mexican Americans are those who were themselves born in Mexico (Moore and Pachon, 1985; McLemore, 1994).

Despite laws against miscegenation (interracial relationships), the practice has been more common than generally agreed. The Westerinen family, pictured here, are the recent descendants of Thomas Jefferson and his African American slave, Sally Hemmings.

All these patterns indicate that choosing a mate is distinctively shaped by sociological factors. Although interracial marriages are not common, a tremendous amount of energy has historically been put into preventing them. **Miscegenation** refers to the mixing of the races through marriage. Antimiscegenation laws have prohibited marriage between various groups, including between Whites and African Americans and between Whites and Chinese, Japanese, Filipinos, Hawaiians, Hindus, and Native Americans. California enacted a law in 1880 prohibiting marriage between a White person and any "negro, mulatto, or Mongolian" (Takaki, 1989: 102). Laws against intermarriage were particularly rigid when the union involved Black American and Asian American men marrying White women; laws were less rigidly enforced when women of color married White men, revealing an intertwining of gender and race hierarchies in the regulation of interracial marriage (Pascoe, 1991).

Antimiscegenation laws are significant not only for how they regulated marriage, but also for the importance they have had in establishing definitions of racial groups. As we saw in Chapter 11, the concept of race is socially constructed. To prohibit marriage across racial groups, courts first had to decide who belonged to what race. Definitions of race varied from state to state. In North Carolina and Tennessee, for example, state laws prohibited the marriage of Whites to persons whose ancestry was one-eighth "Negro." Texas and Georgia prohibited marriage between Whites and a person with any "Negro" ancestry. A California law in 1880 prohibited marriage between Whites and Chinese. Later, when the goal was prohibiting marriages between Whites and Japanese

Americans, anyone considered "Mongolian" was not allowed to marry a White person. In 1934, when a debate arose about whether "Mongolians" included Filipinos, the courts cited an earlier precedent that defined Caucasian as excluding the Chinese, Japanese, Hindus, American Indians, and Filipinos; marriage between these groups and Whites was outlawed. Soon thereafter another California law forbidding marriage between Whites and "Malays" (Malaysians) was written to include Filipinos explicitly in this ban (Pascoe, 1991; Takaki, 1989). These laws were not declared unconstitutional until 1967—quite recently. The enactment of such restrictions provides examples of how the state regulates marriage according to the social norms of the time.

Property and Descent

Kinship systems also shape the distribution of property in society, most notably by prescribing how lines of descent are determined. In **patrilineal kinship systems,** family lineage (or ancestry) is traced through the family of the father. (The prefix *patri* means "of the father.") Offspring in patrilineal systems are typically given the name of the father. **Matrilineal kinship systems,** on the other hand, are those in which ancestry is traced through the mother. Among Native American groups, family ancestry is often traced through maternal descent (Allen, 1986). Among Jewish people, one is considered Jewish when born to a Jewish mother; others may convert, but orthodox Judaism limits Jewish identity to matrilineal descent. This practice evolved under early Talmudic law as the result of widespread persecution and military conquest of Jews. The prevalence of rape of Jewish women following military conquest made paternity difficult to establish; thus, a child's religious identity was established on the basis of the mother's faith.

Many kinship systems involve a mix of both systems. In **bilateral kinship systems,** descent is traced both through the father and the mother. Bilateral kinship is the practice in the United States, although there is a patrilineal bias in that children tend to take the name of the father; descent, however, is traced through both the mother and the father. The practice of children's taking the father's name is also changing, as more women are keeping their names. Children's names may be hyphenated, taken from the mother, taken from the father, or in rare cases, simply made up. Note, however, that even when the mother and father have different names, children most typically are given the name of the father.

Place of Residence

Residential patterns are also shaped by kinship systems. In the United States, newly married couples are expected to establish independent households if they can afford to do so. In many other societies, however, and among some groups within the United States, newly married couples often take up residence in the household of one spouse's parents. In **patrilocal kinship systems,** after marriage, a woman is separated from her own kinship group and resides with the husband or his kinship group. In **matrilocal kinship systems,** a woman continues to live with her family of origin; the husband resides with the wife and her family, although he does not give up membership in his own group. **Neolocal residence** is the practice of the new couple establishing their own residence. In most matrilocal societies, the husband retains his importance in the group of his birth and may exercise authority over his sisters and their children.

Who Holds Power?

Finally, marriage systems vary according to who holds power in the marriage. A **patriarchy** is a society or group where men have power over women; conversely, in a **matriarchy,** women have power over men. Patriarchal societies are far more common than matriarchal societies; indeed, as we saw in Chapter 3, scholars debate whether purely matriarchal societies have ever existed. Still, it would be wrong to conclude that patriarchy is somehow inevitable or universal since there are examples of societies where women have more power than they do in strongly patriarchal societies. **Egalitarian societies** are those where men and women share power equally. In egalitarian societies, women and men are equally valued by all societal members, have equal access to resources, and share decision making. Although women may have different roles than men in such societies, both men and women are perceived as contributing to the common good and are judged to have equal social worth (Leacock, 1978). Among Eskimos, as an example, women traditionally sewed fishing nets and men fished (as did some women), but the activity of both was culturally defined as critical to economic survival (Hensel, 1996).

In many Native American societies, for example, gender roles were far more egalitarian than in the societies of White colonizers. In some Native American groups, women held political power in the sense that they had a significant voice in decisions affecting the group as a whole. In some cases, there were councils of women with an equal role in decision making to councils of men; each group was seen as having different, but complementary, responsibilities. The colonists disrupted these forms of governance, finding that what they called "petticoat government" violated the norms of Christian patriarchy (Allen, 1986: 32). Within contemporary U.S. society, African American families, particularly those in the middle class, tend to be more egalitarian than White families, and they are more likely to promote the active involvement of fathers in childcare and household work; this may reflect the fact that African American families encourage the ideal of an equitable society in their children; White middle-class families are more likely to encourage family members to become individually successful (Willie, 1985).

Extended and Nuclear Families

Another important distinction is whether family systems are extended or nuclear. These concepts refer to the whole system of family relationships and whether the family resides in extended or relatively small household units.

patriarchy

Extended Families. **Extended families** are those in which a large group of related kin in addition to parents and children live together in the same household. Extended families are often common among groups that must share their labor and economic resources to survive. For example, extended families are common among the urban poor since they develop a cooperative system of social and economic support. "Kin" in such a context may refer to those who are not related by blood or marriage, but who are intimately involved in the family support system and are considered part of the family (Stack, 1974).

As an example, among African Americans, there are those whom sociologists have termed *othermothers.* These are "women who assist bloodmothers by sharing mothering responsibilities" (Collins, 1990: 119). Having othermothers is an adaptation to the demands of motherhood and work that characterize African American women's experience since they have always been likely to be employed and have families. This dual responsibility in families and in paid labor has meant that African American women have created alternative means of providing family care for children—a situation that the majority of all women now face. An othermother may be a grandmother, sister, aunt, cousin, or a member of the local community, but she is someone who provides extensive child care and receives recognition and support from the community around her. This is a unique form of an extended family that reflects the specific experience of African American women.

Family clans (large family groups that share common descent) in poor rural areas provide another example of extended kinship systems. Extended kinship networks place people in a system of mutual obligation and support, also creating protection against external threats. Extended family systems are common among Caribbean, Latin American, and African societies. Studies of African societies show, for example, that the extended family is a pervasive and durable family form that enables members to adapt to changing circumstances, including migration, upward mobility, and economic hardship (Shimkin et al., 1978).

The system of *compadrazgo* among Chicanos is another example of an extended kinship system. In this system, the family is enlarged by the inclusion of godparents, to whom the family feels a connection that is the equivalent of kinship. The result is an extended system of connections between "fictive kin" (those who are not related by birth but are considered part of the family) and actual kin that deeply affects family relationships among Chicanos (Baca Zinn and Eitzen, 1999). There are cultural traditions among Mexican Americans that foster strong ties between families and godparents, but culture alone does not explain the prevalence of extended family networks among Mexican Americans (Baca Zinn and Eitzen, 1999). Many of these families have migrated, and they rely on kin support for financial assistance, help finding housing and employment, and assistance in child care and household work (Angel and Tienda, 1982; Baca Zinn and Eitzen, 1999). Of course, not all Mexican American families have experienced migration. Middle-class Mexican Americans, like middle-class African Americans, are more likely to live in *nuclear families.*

Interestingly, extended families are also found at the very top of the socioeconomic scale. For example, family "compounds," such as the Kennedy estate in Hyannisport, Massachusetts, serve as community centers for extended kin groups (Baca Zinn and Eitzen, 1999). Among the elite, extended family systems preserve inherited wealth, whereas among the poor, extended family systems contribute to economic survival. In sum, extended families provide a means of adaptation to economic conditions that require great cooperation within families.

Nuclear Families. The **nuclear family** is one where a married couple resides together with their children. Like extended families, nuclear families develop in response to economic and social conditions. The origin of the nuclear family in Western society is tied to industrialization. Before industrialization, families were the basic economic unit of society, and large household units produced and distributed goods. Production took place primarily in the home, and all family members were seen as economically vital. There was no sharp distinction between economic and domestic life because household and production were one, whether in small communities or large plantation and feudal systems where slaves and peasants provided most of the labor. Women's role in the preindustrial family, though still marked by patriarchal relations, was publicly visible and economically valued. Women performed and supervised much of the household work, engaged in agricultural labor, and produced cloth and food. The work of women, men, and children was also highly interdependent. Although the tasks each performed might differ, together they were a unit of economic production.

With industrialization, paid labor was performed mostly away from the home—in factories and public marketplaces. The transition to wages for labor created an economy based on cash rather than domestic production. Families became dependent on the wages that workers brought home. Although single women were among the first to be employed in the factories, the shift to wage labor was accompanied by a patriarchal assumption that men should earn the "family wage" (that is, be the breadwinner); thus, men who worked as paid laborers were paid more than women (offering a great savings for industries employing women), and women became more economically dependent on men. At the same time, men's status was enhanced by having a wife who could afford to stay at home—a privilege seldom accorded to working-class or poor families. The family wage system has persisted and is reflected in the unequal wages of men and women today, and women's economic dependence on men within nuclear family units.

Another result of industrialization was the separation of the family and the workplace. The shift to factory production moved workers out of the household. This led to the development of dual roles for women as paid laborers and unpaid housewives; moreover, the invisibility of women's labor in the home led to a diminishment of their perceived status.

extended families

Racial–ethnic families have developed in the context of disruptions posed by the experiences of slavery, migration, and urban poverty. These experiences have created unique social conditions that affect the way that families are formed, their ability to stay together, the resources they have, and the problems they face. Chinese American laborers were explicitly forbidden to form families by state laws designed to regulate the flow of labor. Only a small number of merchant families were exempt from the law; thus, the development of Chinese American families was channeled not only by racial exclusion, but also by class differences.

Under slavery, African American families faced a constant threat of disruption. Marriages among slaves were not officially recognized in law since slaves were not considered to be citizens. Slaves formed families, nonetheless, although the children of slave parents were the legal property of the slaveholder. Slaveowners actually found it economically advantageous to allow the formation of slave families since the reproduction of new slaves added to their labor force. Slaveowners maintained strict control over slave life and, when it served their needs, would sell or separate family members. After slavery was abolished, African Americans had to reunite families by searching for those who had been relocated; they also faced new challenges in trying to hold families together under new conditions of rural poverty and urban unemployment. African American men could typically find only unskilled or seasonal employment; African American women were most likely to be employed as domestic workers, a more year-round occupation. As a result, African American women were often the steady providers for their families, resulting in the strong role of women in African American families (Gutman, 1976).

Debunking Society's Myths

Myth: **The tendency for there to be more female-headed households among African Americans is a holdover from slavery.**

Sociological perspective: **Most slave families formed family units even though they were unrecognized in law. The growth of female-headed households is the result of urban unemployment—a twentieth-century phenomenon (Gutman, 1976).**

Labor patterns have also affected the family experiences of other racial–ethnic groups. Families of racial–ethnic groups have been uprooted and even separated as workers travel to wherever laborers are needed. Many Mexicans who had settled in the Southwest, for instance, were displaced when Whites seized Mexican lands during the westward expansion of the United States. The loss of their land disrupted the family and kinship system they had developed, and they became more dependent on the dominant White society for economic survival. In the rapidly industrializing United States, many Mexican Americans were able to find work in the mines opening in the new territories or building the railroads spreading from the East toward the Pacific. Employers apparently thought that they had better control over laborers if their families were not there to distract them, so families were typically prohibited from being with the worker. One result was the development of prostitution camps, which followed workers from place to place. Some wives also followed their husbands to the railroad and mining camps (Dill, 1988).

When they are poor, families often find it necessary for the entire family to work to meet the economic needs of the household. Some groups have a family pattern in which there are often three earners spanning different generations within the family (children, parents, and grandparents, for example), such as many Cuban American families, one of the most economically successful immigrant groups. Migration to a new land and exposure to new customs and needs also disrupts traditional family values. Among Korean immigrants, for example, the majority of Korean wives are employed as full-time workers—a change from traditional Korean values, which expect a wife to be totally devoted to her family. Of course, not all changes in customs and values are bad; in this example, Korean women gained new opportunities and the possibility for expanded social roles. Immigrant women who enter the labor force often change their expectations about their gender role, but they also usually find that they must continue to perform traditional household tasks as well. Studies of Korean wives who are employed indicate that, after a while, they begin to feel an acute sense of injustice; their experience is probably shared by women in other immigrant families (Kim and Hurh, 1988: 162).

The experience of Chinese Americans is another example of how social policies shape the experience of family life. The Chinese Exclusion Act of 1882 prohibited the wives and children of resident Chinese laborers from entering the country. The result was an extraordinary sex imbalance in the Chinese American population—26.8 males for every female in 1890. The Chinese Exclusion Act was buttressed by the Immigration Act of 1924, which prohibited the entrance of any Chinese women to the United States, making family formation impossible; as a result, Chinatowns in the early twentieth century were primarily bachelor societies (Glenn, 1986; Takaki, 1989). These historical examples show how the ability to form and sustain nuclear families is directly linked to the economic, political, and racial organization of society.

THINKING SOCIOLOGICALLY

Interview three people from different family backgrounds (for example, a White, Black and Latino student or people from a stepfamily, a female-headed, and a male-headed household). Ask each about how his or her family is organized. Based on your interviews, how would you describe the *kinship system* of each family? What does each reveal about the *social structure* of families and how they adapt to social change?

Sociological Theory and Families

The complexity of family patterns makes it impossible to understand families from any singular perspective. Is the family a source of stability or change in society? Are families organized around harmonious interests, or are they sources of conflict and differential power? How do new family forms emerge, and how do people negotiate the changes that affect families? These and other questions guide sociological theories of the family.

Sociologists who study the family have used four primary perspectives in their analyses. These are *functionalism, conflict theory, feminist theory,* and *symbolic interaction* (see Table 14.1).

Functionalist Theory and the Family

According to functionalist theory, all social institutions are organized around the needs of society. Functionalism also emphasizes that institutions are based on the consensual values that members of the society share. Functionalist theorists interpret the family as filling particular societal needs, including socializing the young, regulating sexual activity and procreation, providing physical care for family members, and giving psychological support and emotional security to individuals. According to functionalism, families exist to meet these needs. Marriage, in the functionalist framework, is conceptualized as a mutually beneficial exchange wherein women receive protection, economic support, and status in return for emotional and sexual support, household maintenance, and the production of offspring (Glenn, 1987). At the same time, in traditional marriages, men get the services that women provide—housework, nurturing, food service, and sexual partnership. Functionalists also see families as providing care for children, who are taught the values that society and the family purport. In addition, functionalists see the family as regulating reproductive activity, including cultural sanctions about sexuality. (For more discussion of reproduction, see also Chapter 7.)

The nuclear family is the most common form of family structure in the United States.

According to functionalist theory, when societies experience disruption and change, institutions like the family become disorganized, weakening the social consensus around

TABLE 14.1 *THEORETICAL PERSPECTIVES ON FAMILIES*

	Functionalism	Conflict Theory	Feminist Theory	Symbolic Interaction
Families	meet the needs of society to socialize children and reproduce new members	reinforce and support power relations in society	are gendered institutions that reflect the gender hierarchies in society	emerge as people interact to meet basic needs and develop meaningful relationships
	teach people the norms and values of society	inculcate values consistent with the needs of dominant institutions	are a primary agent of gender socialization	are where people learn social identities through their interactions with others
	are organized around a harmony of interests	are sites for conflict and diverse interests of different family members	involve a power imbalance between men and women	are places where people negotiate their roles and relationships with each other
	experience social disorganization ("breakdown") when society undergoes rapid social changes	change as the economic organizations of society changes	evolve in new forms as the society becomes more or less egalitarian	change as people develop new understandings of family life

From the perspective of functionalist theory, the family fulfills many basic needs in society, including emotional needs and the socialization of children.

which they have formed. Currently, some interpret the family as "breaking down" under societal strains. Functionalist theory suggests that this breakdown is the result of the disorganizing forces that rapid social change has fostered.

Functionalists also note that, over time, other institutions have begun to take on some of the functions originally performed solely by the family. For example, as children now attend school earlier in life and stay in school for longer periods of the day, schools (and other caregivers) have taken on some of the functions of physical care and socialization originally reserved for the family. The family's share of these functions has been dwindling while other institutions have taken on more of the original functions of the family. Functionalists would say that the diminishment of the family's functions produces further social disorganization since the family no longer carefully integrates its members into society. To functionalists, the family is shaped by the template of society, and such things as the high rate of divorce and the rising numbers of female-headed and single-parent households are the result of social disorganization.

Conflict Theory and the Family

Conflict theory makes different assumptions about the family as an institution, interpreting the family as a system of power relations that reinforces and reflects the inequalities in society at large. Conflict theorists are especially interested in how families are affected by class, race, and gender inequality. This perspective sees families as the units through which the privileges, as well as the disadvantages of race, class, and gender, are acquired (Scarpitti et al., 1997). Conflict theorists view families as essential to maintaining inequality in society since they are the vehicles through which property and social status are acquired (Eitzen and Baca Zinn, 1999).

The conflict perspective also emphasizes that families in American society are shaped by capitalism. The family is vital to capitalism because it produces the workers that capitalism needs. Accordingly, within families, personalities are shaped to the needs of a capitalist system; thus, families socialize children to become obedient, subordinate to authority, and good consumers. Those who learn these traits become the kinds of workers and consumers that capitalism wants. Families also serve capitalism in other ways—for example, giving a child an allowance is how children learn capitalist habits involving money.

Whereas functionalist theory conceptualizes the family as an integrative institution (meaning it has the function of maintaining social stability), conflict theorists depict the family as an institution subject to the same conflicts and tensions that characterize the rest of society. Families are not isolated from the problems facing society as a whole. The struggles brought on by racism, class inequality, sexism, homophobia, and other social conflicts are played out within family life.

Feminist Theory and the Family

Feminist theory has contributed new ways of conceptualizing the family by focusing sociological analyses on women's experiences in the family and by making gender a central concept in analyzing the family as a social institution. Feminist theories of the family emerged initially as a criticism of functionalist theory. Feminist scholars argued that functionalist theory assumed that the gender division of labor in the household is functional for society. Feminists have also been critical of functional theory for assuming an inevitable division of labor within the family. In particular, they have criticized the work of the noted theorist Talcott Parsons for his analysis of men's and women's role in the family. Based on functional theory, Parsons wrote that men played *instrumental roles* in the family, meaning roles associated with being the economic provider; women, on the other hand, played *expressive roles,* those associated with affection, nurturing, and emotional support. Feminist critics argued that this conceptualization of the family was based on stereotypes about men's and women's roles and on proscriptions about what family life was supposed to be, not empirically accurate analysis of the sociology of the family.

Influenced by the assumptions of conflict theory, feminist scholars do not see the family as serving the needs of all members equally. Quite the contrary, feminists have noted how the family is one of the primary institutions producing the gender relations found in society. As one feminist scholar of the family has written, the hierarchies of gender that appear in society are "created, reproduced, and maintained on a day-to-day basis through interaction among members of a household. In a real sense, then, the debate about women's place in the family is actually a debate about women's place in society" (Glenn, 1987: 348). Feminist theory conceptualizes the family as a system of power relations and social conflict (Thorne, 1993). In this sense, it emerges from conflict theory, but adds the idea that the family is a gendered institution (see Chapter 12).

Symbolic Interaction Theory and the Family

Sociologists have also used *symbolic interaction* theory to understand families. Remember that symbolic interaction emphasizes that meanings people give to their behavior and that of others is the basis of social interaction. Those who study families from this perspective tend to take a more microscopic view of families. To develop a picture of the family, a symbolic interactionist might ask how different people define and understand their family experience. Symbolic interactionists also study how people negotiate family relationships, such as deciding who does what housework, how they will arrange child care, and how they will balance the demands of work and family life.

To illustrate, when people get married, they form a new relationship that has a specific meaning within society. The newlyweds acquire a new identity to which they have to adjust. Some changes may seem very abrupt—a change of name certainly requires adjustment, as does being called a husband or wife. Some changes are more subtle—how one is treated by others and the privileges couples enjoy (such as being a recognized legal unit). Symbolic interactionists see the married relationship as socially constructed; that is, it evolves through the definitions that others in society give it, as well as through the evolving definition of self that married partners make for themselves.

The symbolic interactionist perspective emphasizes the construction of meaning within families. Roles within families are not fixed, but rather evolve as participants define and redefine their behavior toward each other. Symbolic interaction is especially helpful in understanding changes in the family because it supplies a basis for analyzing new meaning systems and the evolution of new family forms over time. Each of the theoretical perspectives used to analyze families illuminates different features of family experiences.

Diversity Among Contemporary American Families

Today, the family is one of the most rapidly changing of all of society's institutions. Many demographic changes (that is, changes in the makeup of the population) are contributing to changes in family structure and family experience. Only about 10 percent of all families fit the so-called *family ideal:* the family as a nuclear unit with a father as head and two children present (U. S. Bureau of the Census, 1993). This family type was long seen as the most natural arrangement—heterosexuality was assumed, and a relatively fixed gender division of labor was taken as a given, with women's roles focused primarily on the home and men's on the public world of work (Andersen, 1991).

Diversity in families has increased since the 1950s, but when the family of today is compared to the family of the past, the comparison is often to a family that never existed (Coontz, 1992); nevertheless, nostalgic ideals can make a positive contribution since they influence how contemporary families think about family life, and since idealized portrayals of the past create symbols and values that people try to live up to. One indication of this was a survey conducted by *Rolling Stone* magazine, asking a national sample of adults to list the television shows that projected values they most wanted their children to learn. People listed "Father Knows Best" and "Leave it to Beaver," along with some more contemporary "family" shows, including "The Cosby Show" and "Family Ties" (Skolnick, 1991: 50).

THINKING SOCIOLOGICALLY

Identify two popular "family" shows on television. Devise a systematic way to conduct a content analysis of the family ideal portrayed by these shows. What do your observations reveal about how the *family ideal* is communicated through the popular media?

Careful historical study of families shows how specific family forms emerge as adaptations to new societal conditions. This idea is central to sociological analyses of the family: that *families are systems of social relationships that emerge in response to social conditions and that, in turn, shape the future direction of society.* There is no static or natural form for the family, no idealized state. The potency of the family ideal is revealed by the number of people who think that some family forms are more "natural" than others. This, in turn, leads people to think that any variation from this universal standard is deviant, and therefore deficient. These convictions are the reason that public debates about family values and family policies are so contentious. The legitimacy of diverse styles of life is at stake. As one sociologist has written, "the term [family] has substantial normative and ideological influence. The 'war over the family' . . . is a direct reflection of the tension between the economic (and racial) stratification of families and the ideological influence of a universal notion of the proper or 'normal,' family" (Hertz, 1986: 8).

Among other changes in the family, families today are also smaller than in the past. There are fewer births, and they are more closely spaced. Coupled with longer life expectancy, this means that childbearing and child rearing occupies a smaller fraction of the adult life of parents. Death, once the major cause of early family disruption, has been replaced by divorce. In earlier periods, death (often from childbirth) was more likely to claim the mother than the father of small children, and men in the past would have been more likely to raise children on their own after the death of a spouse. That trend is reversed, and it is now women who are more likely to be widowed with children (Rossi and Rossi, 1990).

Demographic and structural changes have resulted in great diversity in family forms. The composition of families varies substantially by race, social class, and age (see Figure 14.2). Family income also varies over the life span and among different types of families. Overall, married-couple families

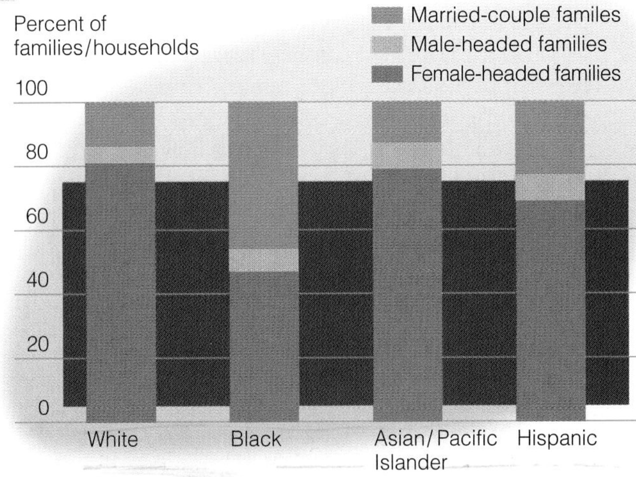

FIGURE 14.2 Diversity in U.S. Families

In the U.S. census, families are defined as two or more people living together and related by birth, marriage, or adoption; this does not include residential units where people are living together but not related.

DATA: From Casper, Lynne U., and Ken Bryson. 1998. *Household and Family Characteristics: March 1998 (Update).* Current Population Reports, Series P20-515. Washington, DC: U.S. Department of Commerce. Web site: http://www.census.gov/prod/3/98pubs/p20-515.pdf; Hooper, Linda M., and Bennett, Claudette E. 1998. *The Asian and Pacific Islander Population in the United States: March 1997 (Update),* Table 5. Washington, DC: U.S. Bureau of the Census, Current Population Reports. Web site: http://www.census.gov/population/socdemo/race/api97.

are three-quarters of household types, but this varies significantly by race. Single-parent households (typically headed by women), postchildbearing couples, gay and lesbian couples, and those without children are increasingly common.

Female-Headed Households

Perhaps one of the greatest changes in family life has been the increase in the number of families headed by women. One-fifth of all children live with one parent, almost three times as many as in 1960; half of all children can expect to do so at some point in their lives. Most single-parent households (88 percent) are headed by women. The odds of living in a single-parent household are even greater for African American and Latino children, although the number of households headed by women has risen across all racial groups (Lugaila, 1998; Bumpass and Sweet, 1989).

The two primary causes for the growing number of women heading their own households are the *high rate of pregnancy among unmarried teens* and the *high divorce rate,* with death of a spouse also contributing. Although the rate of pregnancy among both White and Black teenagers, married and unmarried, is lower today than it was in 1960, the proportion of teen births that occur outside marriage has increased tremendously. Unlike in the past, teen women who become pregnant are now less likely to marry so that the number of never married mothers is higher than in the past (U.S. Bureau of the Census, 1997b).

Teen mothers, regardless of race, are among the most economically and educationally disadvantaged groups in this society. They find it difficult to get jobs, their schooling is interrupted (if not discontinued), and they are very unlikely to receive child support (Kaplan, 1996). Teen fathers, also, are less likely to complete school and have lower earnings (Nock, 1998). The likelihood of poverty is great for teen parents, many of whom come from families that are already poor. (For more discussion on teen pregnancy, see Chapter 7.)

Debunking Society's Myths

Myth: **If teen women who get pregnant would marry the father of the child, a more stable family would be formed.**

Sociological perspective: **Since the social conditions that produce unstable families (poverty, unemployment, and low wages, to name a few) are even more likely among teens, marriage for young people is even more unstable than other marriages (Kaplan, 1996).**

Divorce, the second reason for the increase in the number of female-headed households, also contributes to the growing rate of poverty among women, as we saw in Chapter 9. Most women see a substantial decline in their income in the year following divorce (Hoffman and Duncan, 1988; Peterson, 1996a, 1996b; Weitzman, 1985). Women who remarry are often able to recover financially, but those who do not remarry experience a substantial economic setback (Stirling, 1989). Following divorce, most men pay very little support, either to their wives or to their children—the average outlay for child support is only $2,252 per year. Less than one-third of all divorced women receive any property settlement, with the rate even lower among African American and Hispanic women, who must rebuild their lives on incomes that are among the lowest for adults in the United States (U.S. Bureau of the Census, 1997b).

Another cause for the large number of female-headed households among African Americans and Latinos is the economic status of minority men. High unemployment among young African American men makes marriage less likely since, according to some sociologists, women are less likely to marry men who not economically stable. Further diminishing the pool of potential African American husbands is their relatively low life expectancy and their high rate of incarceration (Thornberry et al., 1997).

The rise in female-headed households has caused considerable alarm. Many people see this trend as representing a breakdown of the family and a weakening of social values. An alternative view, however, is that the rise of female-headed households reflects the growing independence of women, some of whom are making decisions to raise children on their own. Not all female-headed households are women who remain single, however; many are divorced and widowed women, whose circumstances may be quite different from those of a younger, never married woman.

Some claim that female-headed households are linked to problems like delinquency, the school dropout rate, poor self-image, and other social problems. Sometimes the cause of these troubles is attributed explicitly to the absence of men in the family. Sociologists, however, have not found the absence of men per se to be the basis for such problems; rather, it is the presence of economic pressure faced by female-headed households, compared to male-headed families, that puts female-headed households under great strain—with the threat of poverty being by far the greatest problem they face (see Figure 14.3). Among households headed by women with children, one-third live below the poverty line—31 percent among whites, 43 percent among Blacks, and 51 percent among Hispanics (Dalaker and Naifeh, 1998). It is not the makeup of households headed by women that is a problem, but the fact that they are so likely to be poor (Brewer, 1988).

Although the majority of single-parent families are headed by women, families headed by a single father are also increasing. Fifteen percent of children living with only one parent were living with their father, compared to only 8 percent in 1980 (Lugaila, 1998). Most of these are families where the father has gained custody of the children following a divorce or where the mother has died. Male-headed households have a lower poverty rate than female-headed households and are less likely to experience severe economic problems (see Figure 14.3).

Single fathers also tend to get outside help (typically from women), and to use daughters as mother substitutes, assigning them to do child care, housework, and manage household affairs (Grief, 1985). Mothering, however, is not exclusively a skill restricted to women. Single fathers feel that they are competent as parents and demonstrate more mothering behaviors than do married fathers. Compared to married fathers, single fathers spend more time with their children, report more sharing of feelings between father and child, are more likely to stay home with children when they are sick, and take a more active interest in their children's out-of-home activities (Risman, 1986; Risman, 1987).

Most sociological research on single parents has focused on female-headed households because there are so many of them. New research on unwed fathers, however, adds to our understanding of the difficulties these men face. Despite social stereotypes, many, if not most, unwed fathers try to take some responsibility for their children, but the likelihood of economic disadvantage, especially among young unwed fathers, is high. Most young unwed fathers are generally less well educated than other groups, have limited employment prospects, and are more likely to engage in crime (Lerman and Ooms, 1993). These conditions, particularly acute among African American men, make it difficult for these fathers to support their children. The sociological attention being given to the role of single and married fathers in caring for children and supporting families is increasing and is certain to have importance for social policy over the coming years.

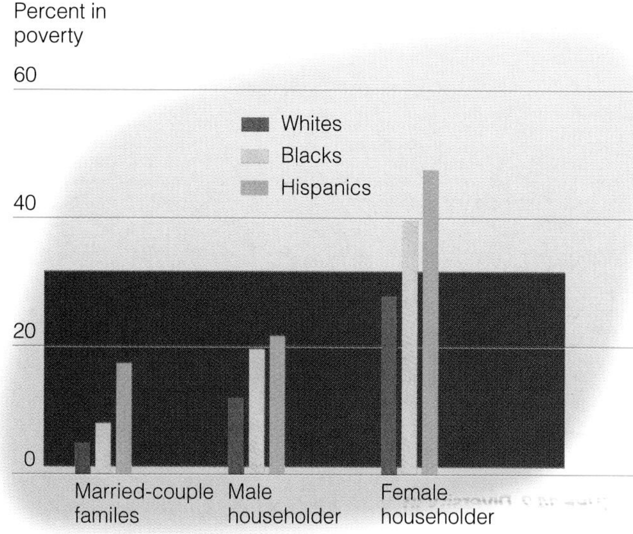

FIGURE 14.3 Poverty Status by Family Type and Race

DATA: From Dalaker, Joseph, and Mary Naifeh. 1998. *Poverty in the United States: 1997.* Current Population Reports, Series P60-201. Washington, DC: U.S. Department of Commerce, Table 3-C.

Married-Couple Families

Among married-couple families, one of the greatest changes in recent years has been the increased participation of women in the paid labor force. Indeed, as we saw in Chapter 9, families sustain a median income level by having both husband and wife in the paid labor force. Among most African American and many Latino and Asian American families, however, having both husband and wife employed is not a new trend since women of color have historically engaged in paid employment in addition to filling household roles. White working-class couples have also been more likely to have both partners employed. In fact, the family ideal of father employed, mother at home has been a mark of White, middle-class privilege. The striking differences in the historical employment experiences of middle-class, working-class, and minority women indicates that caution must be exercised when generalizing about the effect of women's increased labor force participation.

Even though increased labor force participation is most dramatic among White families, most families are experiencing substantial speedup in the amount of work that both men and women do. Economists estimate that after adding up the additional hours people work (in the form of more hours in the work week, more weeks worked per year, and hours worked in a second job), the average employed person works a month more per year than he or she did in 1970. Meanwhile, despite technological advances, the amount of time full-time housewives devote to their work has been unchanged over fifty years (Schorr, 1991).

Women especially have tended to experience the social speedup that exists as they end up working a "double day" of paid employment and unpaid work in the home. Arlie Hochschild has studied the effect of speedup on families. By interviewing men and women in dual-career families, Hochschild found that women who try to become "supermoms" report feeling numb and out of touch with their feelings. She found that women develop a variety of strategies for coping with the demands of work and family. Some

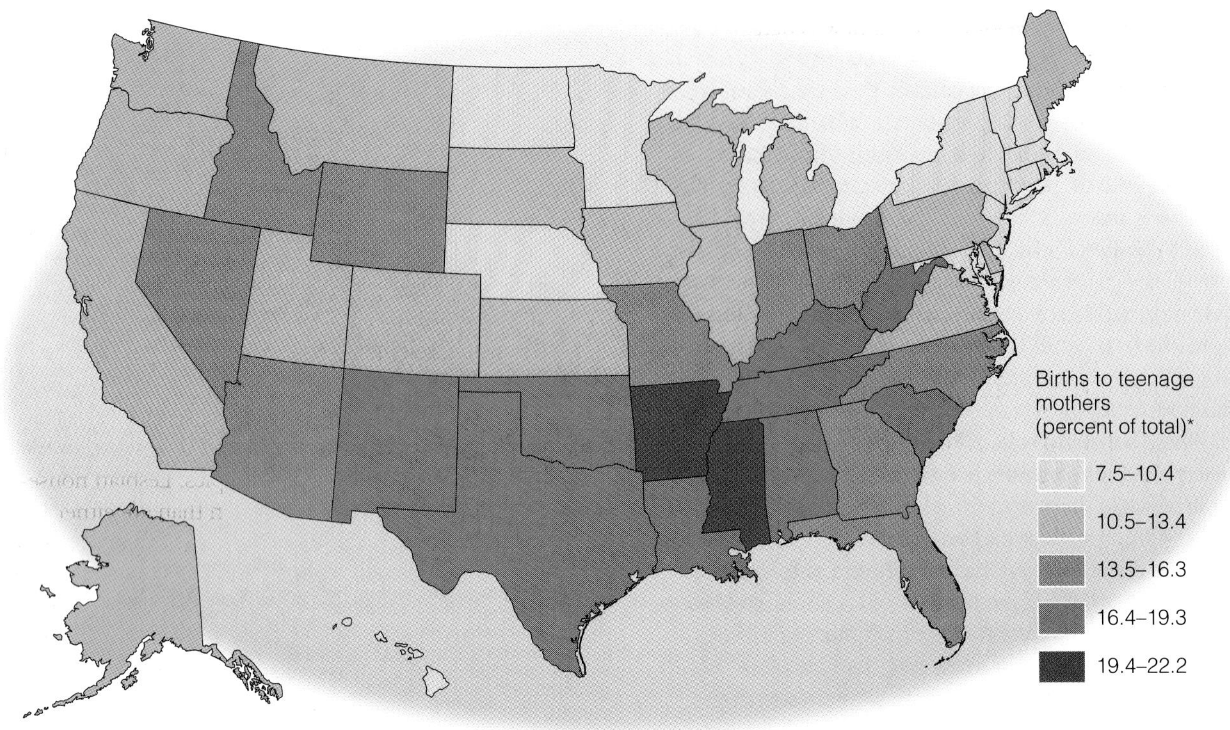

MAP 14.1 Mapping America's Diversity: Births to Teenage Mothers

*Represents registered births. Excludes births to nonresidents of the United States. Based on 100 percent of births in all states.

DATA: From the Bureau of the Census. 1997. *Statistical Abstract of the United States, 1996.* Government Printing Office: Washington DC, p. 79.

feign helplessness over certain tasks as a way to get men to do them. Others avoid conflict by taking on the work themselves. Some cut back on their expectations about what is necessary in the home or reduce their working hours. Often women turn to women friends and family members for help or, if they have the economic ability, pay other women to work for them. Some try to be supermoms. In each case, women typically pay the greater price in dealing with this added household stress (Hochschild, 1989). One of the ironies, which Hochschild discovered in her subsequent research, is that life in the family has for many come to seem more like work, whereas work is the place where many find personal fulfillment, recognition, and emotional support (see also Chapter 17 on work).

Women's labor force participation has created other changes in family life. One example is the number of married couples who have *commuter marriages,* an arrangement that typically arises when work requires one partner in a dual-earner couple to reside in a different city much of the time. Among the middle class, commuter marriage is associated with dual-career couples, separated by job transfers or jobs too distant for a daily commute. Among the working class, commuter marriages may involve periods of separation when one or both partners have to seek work in a place perhaps distant from the family's residence.

The common image of a commuter marriage is one consisting of a prosperous professional couple, each holding important jobs, flashing credit cards, and using airplanes like taxis; however, working-class and poor couples have done their share of long-distance commuting, ranging from agricultural workers who travel with the seasons to wherever labor is required to skilled laborers who have to leave their families to find jobs. Although their commute might be less glamorous than that of professional spouses, briefcases in hand, crossing paths in airports, they are commuting nonetheless. When all types of commuter marriages are included, this form of marriage is more prevalent than is typically imagined.

Studies of commuting relationships have found that women tend to be more satisfied with this arrangement than men, probably because it reflects the independence they have found in their own careers; however, commuting relationships are stressful. The cost of maintaining two households is a source of financial stress, augmented by the expense and difficulty of keeping the relationship together across long distances and long periods of separation. Sociologists have found that commuter marriages are most likely to last when the couples have been married longer, when there are no children (or they are grown), and when one (or both) partner has an already established career. Commuting relationships are also more likely to last if it is known that the arrangement is only temporary, even if not short term (Gerstl and Gross, 1987).

The movement of one or more members of a family in search of employment also strongly affects family relations. Although in the popular mind, migration occurs only among farm workers, migration more broadly includes any movement where people are seeking work in new geographic areas. This can include professional workers, people in the skilled trades, service workers, and domestic and farm laborers. Migration can have a strong impact on family life. Migration among Mexican immigrant families in the United States, for example, encourages more egalitarian relations within the family. This is the result not only of living in a new culture but also of changes in the family roles of men and women during migration. Faced with the usual routines of domestic life (cooking, child care, cleaning, and so on), but without each other to depend on, men and women engage in new behaviors. Women learn to act more assertively and autonomously; men learn to cook and clean up after themselves. These behavioral changes are accompanied by changes in the expectations that men and women have. When the family is reunited, both find that they have moved in a more egalitarian direction (Hondagneu-Sotelo, 1992, 1994).

This pattern of change has been found among many groups that migrate. Chinese women who immigrated to the United States to join their husbands, for example, initially had low rates of labor force participation since they held a traditional belief that "a woman's place was in the home." The difficulties of getting established after migration, however, produced a new attitude that married women have an obligation to work to help support their families (Geschwender, 1992).

Stepfamilies

Stepfamilies are becoming more common in the United States, matching the rise in divorce and remarriage. About 40 percent of marriages involve stepchildren. These families face a difficult period as they deal with the troubles that arise when two family systems blend, or when new people are introduced into an existing family system.

Parents and children discover that they must learn new roles when they become part of a stepfamily. Children accustomed to being the oldest child in the family, or the youngest, may find that their status in the family group is suddenly transformed. New living arrangements may require children to share rooms, toys, and time with people they perceive as strangers. The parenting roles of mothers and fathers suddenly expand to include more children, each with his or her own complex needs. Jealousy, competition, and demands for time and attention can make the relationships within stepfamilies tense. The problems are compounded by the absence of norms and institutional support systems for stepfamilies. Even the norms of language are vague in describing stepfamilies. When a woman's former husband remarries, how does the woman refer to the husband's new wife—someone who on visitation weekends may take care of the woman's children more than the ex-husband does (Coontz, 1992)? There are no norms to follow, and people have to adapt by creating new language and new relationships. Many develop strong relationships within this new kinship system; others find the adjustment extremely difficult, resulting in a high probability of divorce among remarried divorced couples with children (Baca Zinn and Eitzen, 1998).

Gay and Lesbian Households

The traditional heterosexual definition of the family has been challenged by the increased visibility of gay and lesbian lifestyles and a greater acceptance of gay and lesbian identity. Many gay and lesbian couples form marriages and living arrangements that have neither the official blessings nor the rewards of social institutions, but these arrangements still create long-term, primary relationships. Like other families, gay and lesbian couples share living arrangements and household expenses, make decisions as partners, and in many cases, raise children.

Researchers have found that gay and lesbian couples tend to be more flexible and less gender-stereotyped in their household roles than heterosexual couples. Lesbian households, in particular, are more egalitarian than are either heterosexual or gay male couples. Unlike heterosexual couples, lesbian couples seem to be unaffected by the amount of money each partner earns; money has little effect on the balance of power in the relationship or on the couple's feelings about each other. Among gay men, the highest earner usually has more power in the relationship. Some researchers have concluded from these results that, regardless of their sexual orientation, men have been socialized to believe that money equals power. Lesbians, on the other hand, tend to be critical of the power relationships between men and women in marriage and actively construct different household forms (Jacobs, 1997; Tanner, 1978; Taylor, 1980).

At the same time, lesbian families are shaped by some of the same structural features of society that shape the experience of heterosexual families. In an interesting examination of these social forces, sociologist Maureen Sullivan studied lesbian co-parents—those who are together raising children. She found that these couples tend to have more gender equality within their households, so long as both partners are employed. Even when the couple has made a conscious commitment to freeing themselves from traditional gender roles, however, patterns of employment shape the inequality that they experience. In cases of lesbian co-parents where one partner is the primary breadwinner and the other the primary caregiver for children, the partner staying at home becomes economically vulnerable, less able to negotiate her needs, and more devalued as a person and daily contributor—just like the traditional pattern where women stay home and men are the primary breadwinners! This study shows us that social structural arrangements, not just the values of domestic partners, shape the ability of individual people to influence family decisions, be seen as an equal partner, and have his or her needs satisfied (Sullivan, 1996).

Gay and lesbian couples also face unique domestic problems since the dominant culture often denies them benefits and privileges that heterosexual couples receive, such as shared health-care plans and other benefits that go along with a legally recognized marriage. An increasing number of businesses now allow same-sex partners to be enrolled in employee benefit plans, and some municipalities give legal recognition to gay marriage, but these remain the exception.

Gay and lesbian couples are also frequently the targets of hostility, as are their children when custody disputes arise over whether a gay parent has the right to custody. Although the law does not formally deny custody to lesbian mothers or gay fathers, researchers have found that judges tend to take sexual orientation into account in awarding custody even when there is not evidence of it creating adverse effects on the children (Duran-Aydintug and Causey, 1996; Crawford and Solliday, 1996; Zicklin, 1995).

Gay and lesbian marriages are producing new social forms and new social debates. Should gay marriage be recognized in law? One-third of all women think gay marriages should be recognized as having the same rights as traditional marriages; one-fifth of men think so (Roper Organization, 1996). Gay marriage is also far more acceptable in the eyes of younger people (those aged eighteen to twenty-nine) than to older groups, raising the question of whether over time social support for gay marriages will increase or whether, as these people age, they will shift their values. In the meantime, in most states and municipalities, gays and lesbians who form strong and lasting relationships do so without formal institutional support and, as a result, have been innovative in producing new support systems and new social definitions for their family arrangements. Sociologists have also revised their thinking about how families should be studied. Whereas family forms such as gay and lesbian relationships have been studied in the past as instances of social deviance, they are now studied as aspects of social diversity and are no longer measured against standards inaccurately assumed to be the norm.

Singles

Single people, including those never married and those widowed and divorced, today constitute 42 percent of the population—an increase from earlier years (the comparable figure in 1970 was 29 percent). Some of the increase is accounted for by the rising number of divorced people, but there has also been an increase in the number of those never married—24 percent of the population today compared to 16 percent in 1970 (Lugaila, 1998). The rise in the number of never married people is partially a statistical consequence of the fact that men and women are marrying at a later age. The typical age at first marriage for women is 24 years, for men 26 years, compared to 20.8 years for women and 23.2 years for men in 1970 (U.S. Bureau of the Census, 1997b). Longer life expectancy, higher educational attainment rates, and cohabitation all contribute to this longer time prior to first marriage.

As sexual attitudes have changed in the direction of greater permissiveness, many people find the same sexual and emotional gratification in single life as they would in marriage. Being single is also no longer the stigma it once was, especially for women. Single women were once labeled "old maids" and "spinsters"; now they have the image of being carefree, sexually active, unencumbered by family obligations, and free-thinking. Changes in social attitudes about single life seem to have had some effect on the happiness of singles as well. For many years, sociologists have documented a tendency for married people to report greater happiness than singles; in recent years, that trend has reversed. In particular, the rates of reported happiness among never married men have increased, and the rates of reported happiness among married women have declined (Glenn and Weaver, 1988). Among African Americans, however, married men and married women report higher life satisfaction than singles (Creighton-Zollar and Honnold, 1991). For all people, having strong social support networks is an important predictor of reported well-being (Acock and Hurlbert, 1993).

Among singles, Black women are likely to remain single longer than Black men or White women. Although Black women see marriage as a desirable goal, it is not regarded as an end in itself. Black women expect to work to support themselves and do not assume that someone else will be financially responsible for them. Many White working-class women are also raised to prepare themselves for work, but somewhat different from Black women, they see their work as contributing to family income, not as the potentially sole source of economic support (Higginbotham and Weber, 1992).

Cohabitation (living together) has become quite common among single people. More than three times as many couples live together without being married in the 1990s than was true in the 1970s. Sociologists estimate that a majority of people will have this lifestyle at some point in their lives; the likelihood is highest among urban residents. Research on cohabiting couples has shown that although these relationships are nontraditional in some ways, they do not differ much from traditional marriages. The roles people take within the household (including the division of household chores, the gender expectations couples have, and the degree of equality within the relationship) are similar to patterns among married heterosexual couples. Sixty-five percent of women in cohabiting relationships expect to marry their partners; only 38 percent of men have the same hope. After a time, the women adapt to the men's expectations since, although the majority of women in cohabiting relationships want marriage, only 35 percent think it is likely. At the same time, cohabiting women who have been married before are less eager to marry than cohabiting women who have never been married (Bumpass and Sweet, 1989; Blumstein and Schwartz, 1983; Spanier, 1983; Stafford et al., 1977).

Finally, a rising number of single people are remaining in their parents' homes for longer periods. Known as the "boomerang generation," these are young people who return home in their twenties when they would normally be expected to live independently. The increased cost of living means that many young people find themselves unable to pay their own way, even after marrying or getting an education. They economize by joining the household of their parents. Many from the rising number of pregnant unmarried women also stay in their parents' homes, forming extended households, often with multiple generations living in a single residence.

cohabitation

Marriage and Divorce

Although there is extraordinary diversity of family forms in the United States, the majority of people will still marry at some point in their lives. Indeed, the United States has the highest rate of marriage of any Western industrialized nation, as well as a high divorce rate. Even with a high rate of divorce, a majority of people place a high value on family life. People's opinions about their preferred family lifestyle have changed considerably in recent years, however. Half of all women and nearly half of all men (47 percent) now say that the most satisfying lifestyle would be to live in a marriage where the husband and wife share responsibilities for work, housekeeping, and childcare, but the public is not unified in this view. Thirty-seven percent of women and 36 percent of men still want a traditional marriage with the husband as the sole provider and the woman caring for the home and children. The number who wanted this lifestyle in the 1970s was half of both women and men (see Figure 14.4; Roper Organization, 1995b).

Marriage

The picture of marriage as a consensual unit based on intimacy, economic cooperation, and mutual goals is widely shared, although marital relationships also involve a complex set of social dynamics, including cooperation and conflict, different patterns of resource allocation, and a division of labor. Sociologists have to be careful not to romanticize marriage to the point that they miss these other significant social patterns within marriage.

Sociologists have, for example, long studied the power dynamics within marriage and have found, among other things, that the amount of money a person earns establishes that person's relative power within the marriage (such as the ability to influence decisions, the degree of autonomy and independence held by each partner, and who controls the expectations about family life). Wives who contribute more to the family's financial standing than their husbands have greater autonomy within the marriage than wives who contribute less financially (Blood and Wolfe, 1960; Ferree, 1984; Komter, 1989). A couple's expectations about the marriage can, however, mitigate this probability. If men view women's employment as a threat to their own role, then employment does not give wives more power within the marriage; on the contrary, a woman married to a man who views her employment negatively will be even less powerful within the marriage. Married couples who disagree about the wife's right to work tend to have the least happy marriages. Power within marriage is negotiated based on the attitudes and beliefs that the partners have about marriage and the resources that each partner brings to the household (Pyke, 1994). Contrary to popular belief, most marital conflicts are about finances and housework, not sexual jealousy (Dobash and Dobash, 1979; Link et al., 1990; Blair and Johnson, 1992).

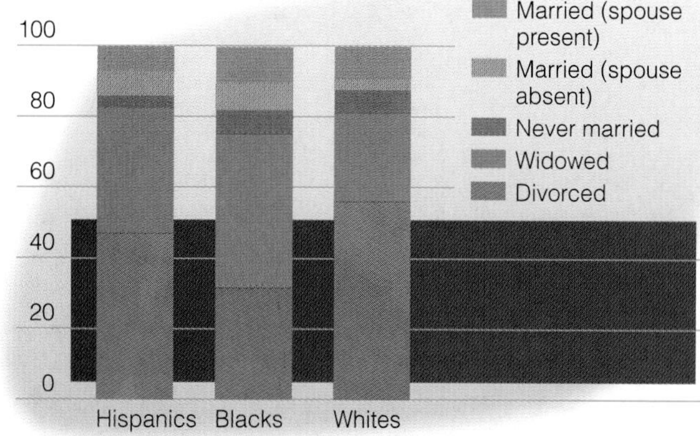

FIGURE 14.4 Marital Status of the U.S. Population

DATA: From Lugaila, Terry A. 1998. *Marital Status and Living Arrangements: March 1997 (Update)*. Washington, DC: U.S. Department of Commerce. Web site: http://www.census.gov/population/www/socdemo/hh-fam.html

Values within the institution of marriage are certainly changing. Most couples now agree that child care should be shared, although only 38 percent of couples think that housework should be shared. In general, men have more traditional expectations for marriage than women, although the degree of traditionalism varies for different aspects of the marital relationship. Men say they want to earn more money than their wives and maintain control over money matters in the household. Sociologists have found though that men and women are reluctant to give up traditional roles entirely. Further, husbands are, in fact, less traditional than wives expect they will be, and wives are more traditional than their partners expect. Spouses' misperceptions about each other's expectations are the source of some of the conflicts that occur between married couples (Hiller and Philliber, 1986).

Within marriage, the values of partners, as well as the roles they play, influence their experience of marriage. Research finds that, even with the ideal marriage defined as one based on consensus, harmony, and sharing, men and women have different experiences within marriage (Bernard, 1972; Thorne, 1993). Moreover, even with changes in marital roles, the amount of work people do in families varies significantly for women and men. Women do far more work in the home and have less leisure time (Baca Zinn and Eitzen, 1998). It is not unusual for employed mothers to be doing two jobs—the so-called second shift of doing the work at home after working all day in paid employment (Hochschild, 1989). Employed mothers now work an average of seventy-six to eighty-nine hours per week, counting paid labor, housework, and child care (Schorr, 1991). These estimates were based on studies of White married couples only; the inclusion of Black, Latino, and Asian couples would raise the number of hours since women in these families are more likely to work outside the home and hold more than one job. Women who do not work for wages spend about seventy hours per week in housework, about thirty of which are devoted exclusively to child care.

Research finds that, despite popular stereotypes, there is no deleterious effect on children who grow up in gay households.

Researchers also find that the arrival of a first child significantly increases the gender division of labor in households, with women increasing the housework they do and lessening their employment; there is far less effect for men (Sanchez and Thomson, 1997).

Men are also putting in longer hours, although primarily at work. Thirty percent of men with children under fourteen report working fifty or more hours per week at paid labor (Schorr, 1991). Are men more involved in housework? Yes and no. Men report that they do more housework, but they are devoting only slightly more of their time to housework than in the past. Estimates of the amount of housework men do vary, but studies generally find a large gap between the number of hours women give to housework and child care and the hours men give. Among couples where both partners are employed, only 28 percent actually share the housework equally. Fathers do more when there is a child in the house under two years of age; for the most part, what increases there have been in fathers' contributions to household work have been in the amount of child care they provide, not the housework they do. The end result is that men have about eleven more hours of leisure per week on average than women do (Hochschild, 1989; Press and Townsley, 1998).

Despite a widespread belief that young professional couples are the most egalitarian, studies find that there is little difference in the amount of housework that men do across social class (Wright et al., 1992). One offshoot of this pattern is that within marriage, women are more likely than men to want change, notably in the areas of domestic labor, children, sexual behavior, leisure activities, and finances (Komter, 1989). Almost two-thirds (61 percent) of women say the amount of work they have to get done during the day is a cause of stress, and half say that they feel resentment, at least sometimes, about how little their mate helps around the house and about their lack of free time (Roper Organization, 1995b). The idea of marriage as an equal partnership between two like-thinking individuals with unified interests is unsupported by these data.

People's expectations about family life are shaped by the situations in which they find themselves. In Chicano families, for example, women who are employed outside the home expect their husbands and children to help with the housework. The division of household labor changes when the wife is employed, but if her work is seasonal, she and her husband will likely return to the traditional division of labor in the home during the off-season. In all seasons, with assistance or without, the women view themselves as primarily responsible for the house (Zavella, 1987).

Divorce

The United States leads the world not only in the number of people who marry, but also in the number of people who divorce. More than sixteen million people have divorced but not remarried in the population today; more women are in this group than men since women are less likely to remarry following a divorce. Between 1960 and the 1990s, the rate of divorce more than doubled, although it has declined recently since its all-time high in 1980 (see Figure 14.5). There are now more than a million divorces per year; sociologists estimate that for every two marriages made today, one will end in divorce (Arendell, 1992; U.S. Bureau of the Census, 1997b).

The likelihood of divorce is not equally distributed across all groups. Divorce is more likely for couples who marry young (in their teens or early twenties). Second marriages are more likely to end in divorce than first marriages. The elevated rate of breakups among second marriages is the result of several factors. Those remarrying may have different attitudes about commitment, they may bring unresolved personal conflicts into the marriage, and some sociologists find that the strain produced by combining existing households is a contributing factor (Furstenberg, 1990).

Divorce also varies across race and class. For most groups, the divorce rate is higher among low-income couples, a pattern that reflects the strains that financial problems put on marriages. Divorce is somewhat higher among African Americans than among Whites, again partially because of financial problems—African Americans make up a disproportionately large part of the low-income group (Cherlin, 1981; Dalaker and Naifeh, 1998). The rate of divorce among Latinos is slightly lower than it is among Whites and Blacks; among Latinos, the divorce rate is highest among Puerto Ricans and lowest among Cubans. Sociologists have attributed this difference to the effect of higher educational attainment and greater economic security

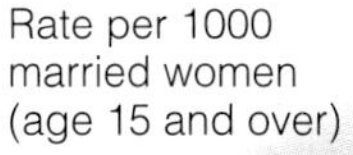

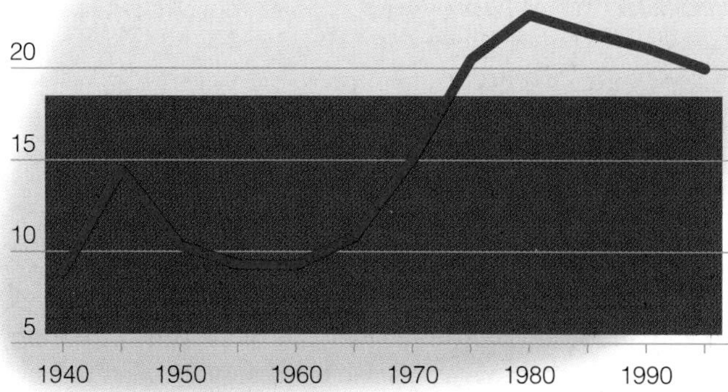

FIGURE 14.5 Divorce Rates, 1960–1995

DATA: From the Centers for Disease Control and Prevention. 1995. *Monthly Vital Statistics Report*, Vol. 43, March 22, p. 9; U.S. Bureau of the Census. 1997. *Statistical Abstracts of the United States, 1997.* Washington, DC: U.S. Department of Commerce, p. 105.

Despite the high divorce rate, marriage is a cultural ideal, and most people marry at some point in their lives.

among Cubans, in other words, pointing to the effect of social class on the likelihood of divorce.

In general, the divorce rate is highest at the extremes of the educational attainment continuum; that is, the divorce rates are highest for those without high school diplomas and those with a college education. There are exceptions to this pattern, however. Among Puerto Ricans, greater educational attainment seems to promote greater marital stability. Cuban and Mexican American experience, on the other hand, supports this pattern, since the less education these couples have, the more likely their marriage will end in divorce (Vega, 1991; Frisbie, 1986). Since educational level and class status are so intertwined, sociologists interpret these patterns as indicative of the different effects of class at each end of the class spectrum.

The U.S. public is well aware of the fragility of marriage. Still, the desire to marry remains high. Even following divorce, a majority want to remarry, although men are more likely to want to do so than women. Three-fourths of divorced men do remarry, compared to three-fifths of women, and men remarry faster. The likelihood that a woman will remarry following divorce depends on the number of children she has, with the odds of remarriage dropping with more children.

A number of factors contribute to the current high rate of divorce in the United States. Demographic changes (shifts in the composition of the population) are part of the explanation. The continuing rise in life expectancy, for example, has an effect on the length of marriages. In earlier eras, people died younger, and thus the average length of marriages was shorter. Some of the historical increase in divorce rates stems from the fact that, as people live longer, their marriages go on longer. In the later years, some marriages that would have ended with the death of a spouse may be expected to be dissolved by divorce. Still, this is not a complete explanation for divorce statistics since the rise in divorce rates exceeds the decline in death rates (Coontz, 1992).

Cultural factors also contribute to divorce, as functionalist theorists would predict. In the United States, individualism is highly valued. Within marriage, this translates into a high value placed on voluntary participation in marriage, based on individual satisfaction. Americans tend to see marriage in terms of its personal benefits. This cultural orientation toward individualism may predispose people to terminate a marriage in which they are personally unhappy. In another culture, indeed years ago in this society, marriage, no matter how difficult, may have been seen as inviolate and an unbreakable bond, whether one was happy or not.

A third contributing reason for the high rate of divorce is the change in women's roles. Women are now less financially dependent on husbands than they once were, even though they still earn less. As a result, the interdependence that bound women and men together as a marital unit is no longer as strong. Although most married women would be less well off without access to their husband's income, they could probably still support themselves. For men, the low likelihood that they will be required to pay both alimony and child support means that they have little to fear financially from breaking up their marriage. All these factors combine to make it easier for people to leave marriages that they find unsatisfactory.

Although divorce is emotionally painful and financially risky, it can nevertheless be a positive option. Research by Demi Kurz, examined in the box "Doing Sociological

Research: For Richer for Poorer: Mothers Confront Divorce," shows that, despite the economic struggles women experience following divorce, most are glad that their marriages ended. To people in unhappy marriages, divorce may be the only solution to a bad situation or at least the path to a more congenial lifestyle. In addition, the attitude that couples should stay together for the sake of the children is now giving way to a belief, supported by research, that a marriage with protracted conflict is more detrimental to children than divorce (Cherlin et al., 1991).

BOX 14.1 DOING SOCIOLOGICAL RESEARCH

For Richer for Poorer: Mothers Confront Divorce

IN RECENT years, many people have attributed the rising rate of divorce to the breakdown of family values. Presumably as family values weaken, people do not care so much about holding marriages together. Demi Kurz examines this assumption in her book *For Richer for Poorer: Mothers Confront Divorce,* asking how women experience divorce.

She based her study on a random sample of 129 women, 61 percent White, 35 percent African American, and 3 percent Latinas; all the women had at least one child. Their average age is thirty-five, and they were interviewed approximately two years after their divorce. The women fall into three class groups that Kurz identifies as middle class, working class, and poor, based on their education, occupation, income, and housing. Kurz suggests that popular images of divorce, especially the idea that individualism leads to divorce, stem from people's thinking only about the middle class. Bringing a more class-inclusive perspective to bear, Kurz exposes numerous aspects of divorce that are not commonly understood.

She finds four major factors that led to these women's marriages ending. One group reportedly ended the marriage because of personal dissatisfaction, but more specifically, they cited problems in the marriage resulting from the husband's gender role: his controlling behavior, his failure to be an adequate provider, or his emotional distance. Another group left the marriage because of violence; 70 percent of the women Kurz interviewed experienced violence from their husbands at least once. Of these, 37 percent said that violence took place more than three times, and 4 percent said that violence occurred during the separation only. The others experienced violence one, two, or three times. Nineteen percent of the women said they left their marriages because of the violence.

Another group of the women left their marriages because of "hard living," that is, their husband's heavy drinking, drug abuse, or extended absence from the family. Another significant group (19 percent) said their marriages ended because their husband was involved with another woman. Kurz concludes that the major causes of divorce are not the breakdown in family values, but dissatisfaction stemming from gender roles in the marriage, violence, "hard living," and infidelity.

Despite the emotional trauma of divorce, most of the women Kurz studied were glad their marriages had ended. Two-thirds expressed positive feelings about being divorced, 12 percent were ambivalent, and 25 percent still had negative feelings. Those with the most negative feelings said this was because of the difficulty the divorce created in their lives and the opportunities they had lost. For all the women, by far the greatest difficulty resulting from the divorce was surviving as a single parent and being solely responsible for their financial and emotional well-being. The women coped by getting substantial help from their families, taking on second jobs, and sometimes, accepting social welfare. The middle-class women were the most likely to experience a decline in their economic standing following divorce.

Most of the mothers reported unsatisfactory visitation arrangements with children after the divorce; 19 percent of the women thought the fathers remained caring and involved with their children, but the remainder thought the father was either underinvolved or overinvolved (having a negative influence) on their children. Those who wanted less visitation cited the conflict that occurred during visitation and indicated that violence, kidnapping, and problems with their husband's drug or alcohol use threatened both them and their children. Some thought the husband used visitation to harass them.

Kurz argues that those who attribute divorce simply to a breakdown in family values ignore the more complex factors involved, including important questions about gender, namely, who benefits from and who is harmed by marriage and divorce. Among the women Kurz studied, many saw their divorce as necessary to raising their children in a safe environment. The women also thought they were less harmed by the divorce than by the marriage.

Kurz analyzes divorce as part of a larger system of gender, class, and race relations. For women, divorce occurs in a context with significant violence against women, little social support (social and economic) for single parents and children, lack of equity for women in the labor market, and economic hardship that falls especially on divorced women and their children. Kurz concludes by suggesting several changes in social policy to alleviate the problem women face as the result of divorce. Among others, she reports the need to develop laws and practices that reduce violence in the marriage and during separations. In addition, policies to improve women's economic well-being will solve their biggest problem following divorce: supporting themselves and their children.

SOURCE: Kurz, Demi. 1995. *For Richer for Poorer: Mothers Confront Divorce.* New York: Routledge.

Studies of the effect of divorce on children have shown that a number of factors influence children's adjustment to divorce. Where conflict between the parents is greatest (before, during, and after the divorce), the children are most likely to have ongoing difficulties as the result of the breakup. Few children feel relieved or pleased by divorce; feelings of sadness, fear, loss, and anger are much more common, accompanied by desires for reconciliation and feelings of conflicting loyalties—feelings that can persist into adulthood (Cherlin et al., 1998). Most children, however, adjust reasonably well after a year or so; children's adjustment is greatly influenced by factors that precede the divorce per se. Marital discord is just as harmful, if not more so, for children (Furstenberg, 1998; Amato and Booth, 1997); in fact, the single most important factor influencing children's poor adjustment is marital violence and prolonged discord (Stewart et al., 1997; Arendell, 1998).

The behavior of the parents during divorce also influences how well the children adjust; the emotional strain on children is significantly reduced if the couple remain amicable and include the children in the discussion of a pending divorce (Kelly and Wallerstein, 1976; Tschann et al., 1989; Tschann et al., 1990). If both parents remain active in the upbringing of the children, the evidence shows that children do not suffer from divorce; especially important is the ability of the mother to be an effective parent after a divorce, and this, in turn, can be affected by the resources she has available and her ongoing relationship with the father (Buchanan et al., 1996; Simons et al., 1996; Furstenberg and Nord, 1985). Unfortunately, most children have little contact with the parent who does not reside with them (typically the father) following a divorce. Even when there is joint custody, it is the mother who most often retains primary responsibility for child care.

In the aftermath of divorce, many fathers become distant from their children. Sociologists have argued that the tradition of defining men in terms of their role as breadwinners minimizes the attachment they feel for their children; if the family is then disrupted, they may feel that their primary responsibility, as financial provider, is lessened, leaving them with a diminished sense of obligation to their children. A man may also distance himself as a way of minimizing or avoiding conflicts with his former wife. Salvaging a sense of masculinity may also play a role. Many fathers report feelings of emasculation when they are displaced from their home since acting as the head of a household is part of the definition of masculine success in this society. By their absence following a divorce, men may reassert another aspect of their masculine role—independence (Arendell, 1992).

THINKING SOCIOLOGICALLY

Identify a small sample of students in your school who have experienced *divorce* in their families. Design a short interview to test the research conclusions reported in this chapter on the effects of divorce on children. Do your results support the conclusions in this earlier research? What additional sociological insights does your research reveal?

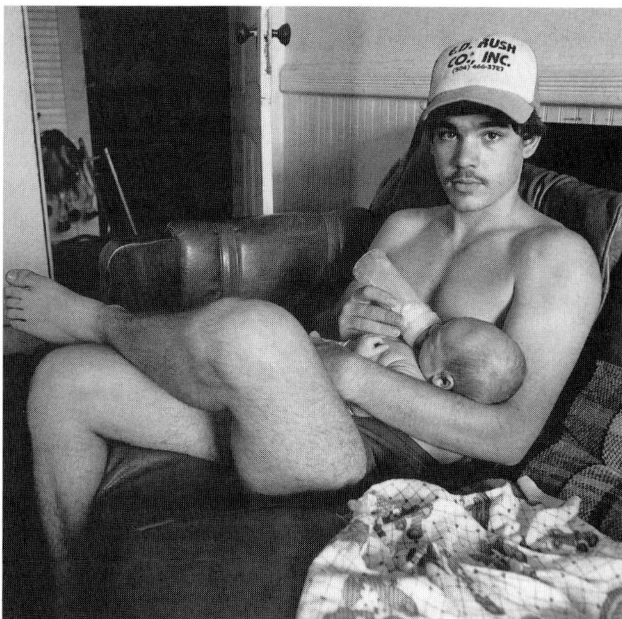

Changing family values mean that some men are more engaged in child care than in the past; the lives of single fathers have been understudied, but new studies find that single fathers may be more involved in the lives of their children than is perceived to be true.

One consequence of men's absence is the limited child support that families receive from fathers after divorce. One public opinion poll indicated that one-quarter of people in the United States say they would abandon their entire family for $10 million (Patterson and Kim, 1991: 66), but as one historian of the family has recently put it, "A large number of people are evidently willing to do the same thing for free, judging from the astonishing statistics on how few noncustodial fathers spend time with their children after divorce" (Coontz, 1992). Only three-quarters of the children eligible for child support following divorce actually receive any, most getting only a partial amount. The average child support payment for one year was $3193 for children of White women, $2200 for children of Hispanic women, and $2101 for children of Black women. White women and better-educated women are the most likely to receive whatever child support is awarded, because they have the resources to pursue the issue in court, whereas many women do not have these resources (U.S. Bureau of the Census, 1996c).

Debunking Society's Myths

Myth: Unwed fathers have little commitment to the care of their children.

Sociological perspective: Unwed fathers see their children more often than do divorced fathers who have remarried (Luker, 1996; McLanahan and Sandefur, 1994).

Violence against women in the family is the result, sociologists argue, of the differential power of men in the family and women's economic and emotional dependence upon them.

Family Violence

Generally speaking, the family is depicted as a private sphere where members are nurtured and protected away from the influences of the outside world. Although this is the experience of many, families can also be locales for violence, disruption, and conflict. Family violence is a phenomenon that, hidden for many years, has now been the subject of much sociological research.

Estimates of the extent of domestic violence are hard to come by and notoriously unreliable since the majority of cases of domestic violence go unreported. The American Medical Association has estimated that one in three women will be physically assaulted by their husbands at some time in their married lives; 24 percent of women are at some point hurt or injured by abuse (American Medical Association, 1998; Miller, 1994); a national survey has found that 30 of 1000 women reported severe wife beating (Gelles and Strauss, 1988). Studies also show that at least 10 percent of married women are raped by their husbands (Finkelhor and Yllo, 1985; Mahoney and Williams, 1998). In some instances, wives assault husbands, but this is far less frequent than husbands' assaulting wives (Finkelhor and Yllo, 1985; Brush, 1990; Gelles and Straus, 1988). Recently, partner violence in gay and lesbian relationships has also come to light, with researchers concluding that violence is equally prevalent among gays and lesbians as among heterosexual couples and involves similar issues about power and control. In gay and lesbian violence, however, silence around the issue may be even more pervasive given the marginalized status of gays and lesbians (West, 1998; Renzetti, 1992).

One of the most common questions asked about domestic violence is why victims stay with their abuser. The answers are complex and stem from sociological, psychological, and economic problems. Victims tend to believe that the batterer will change, but they also find they have few options; they may perceive that leaving will be more dangerous since violence can escalate when the abuser thinks he (or she) has lost control. Many women are also unable to support their children and their living expenses without a husband's income; mandatory arrest laws in cases of domestic violence exacerbate this problem since they may, despite their intentions, discourage women from reporting violence for fear their batterer will lose his job (Miller, 1997). Despite the belief that battered women do not leave their abuser, however, the majority do leave and seek ways to prevent further victimization (Gelles, 1993).

Sociological analyses of violence in the family have led to the conclusion that women's relative powerlessness in the family is at the root of high rates of violence against women. Two perspectives have been developed to explain violence in the family. One, the *family violence approach,* emphasizes that violence occurs in families because the society condones violence. According to this perspective, violence is endemic in society, and the general acceptance of violence is reflected in family relationships (Straus et al., 1980). A second perspective, *the feminist approach,* places inequality between men and women at the center of analyses of violence in the family, arguing that since most violence in the family is directed against women, the imbalance of power between men and women in the family is the source of most domestic violence. The feminist perspective also emphasizes the degree to which many women are trapped in violent relationships since they are relatively powerless within the society and may not have the resources to leave their marriage (Kurz, 1989; McCloskey, 1996).

Violence within families also strikes many children who are the victims of child abuse. Not all forms of child abuse are alike since some would consider repeated spanking to be abusive; others think of this as legitimate behavior. Child abuse, however, is behavior that puts children at risk and may include physical violence and neglect. As with battering, the exact incidence of child abuse is difficult to know, but in 1997, the number of reported cases of abuse was more than three million (National Committee to Prevent Child Abuse, 1998). Whereas men are the most likely perpetrators of violence in other forms of domestic abuse, with child abuse, women are the perpetrators as frequently as men.

Research on child abuse finds a number of factors associated with abuse, including chronic alcohol use by a parent, unemployment, and isolation of the family. Sociologists point to the absence of social supports—in the form of social services, community assistance, and cultural norms about the primacy of motherhood—as related to child abuse since most abusers are those with weak community ties and little contact with friends and relatives (Gelles and Strauss, 1988; Eitzen and Baca Zinn, 1998).

Incest is a particular form of child abuse, involving sexual relations between persons who are closely related. A history of incest has been related to a variety of other problems, including drug and alcohol abuse, runaways, delinquency, and various psychological problems, including the potential for violent partnerships in adult life. Studies find that fathers and uncles are the most frequent incestuous abusers and that incest is most likely in families where mothers are, for one reason or another, debilitated. In such families, daughters often take on the mothering role, being taught to comply with men's demands to hold the family together. Thus, scholars have linked women's powerlessness within families to the dynamics surrounding incest (Herman, 1981; Andersen, 1997).

Changing Families/ Changing Society

Like other social institutions, the family is in a constant state of change, particularly as new social conditions arise and as people in families adapt to the changed conditions of their lives. Some of these are changes that affect only a given family—the individual changes that come from the birth of a new child, the loss of a partner, divorce, migration, and other life events. These are the kinds of changes that C. Wright Mills referred to as "troubles" (see Chapter 1), although not all of them constitute troubling events. Some may even be happy events; the point is that they are changes that happen at the individual level, as individual people accommodate to the presence of a new child, adjust to a breakup with a long-term partner, or grieve the loss of a spouse. These microsociological events are interesting to sociologists, particularly when sociologists see common patterns in people's reactions and adjustments to such situations.

At the macrosociological level, other social changes affect families on a broad scale, and as Mills would have pointed out, many of the microsociological things that people experience in families have their origins in the broader changes affecting society as a whole. Migration, as we will see, is a good example. Certain economic, social, and political conditions may encourage people to move, and how they do so, where they go, what kind of work they seek, and who can go all affect family life. Strains in the society that affect the rate of marriage and divorce; women's changing roles in the workplace; and social policies supporting (or not) child care, battered women, or pregnant teens all affect the sociological situation of families—worldwide.

Global Changes in Family Life

Changes in the institutional structure of families are also being affected by the process of globalization. The increasingly global basis of the economy means that people often work long distances from other family members—a phenomenon that happens at all ends of the social class spectrum, although the experience of such global mobility varies significantly by social class. A corporate executive may accumulate thousands—even millions—of first-class flight miles, criss-crossing the globe to conduct business. A regional sales manager may spend most nights away from a family, likely staying in modestly priced motels, eating in fast-food franchises along the way. Truckers may sleep in the cabs of their tractor-trailers, after logging extraordinary numbers of hours of driving in a given week. Laborers may move from one state to the next, following the pattern of the harvest, living in camps away from families, being paid by the amount they pick.

These patterns of work and migration have created a new family form—the *transnational family,* defined as families where one parent (or both) live and work in one country while their children remain in their countries of origin. A good example is found in Hong Kong, where most domestic labor is performed by Filipina women, who work on multiple-year contracts, managed by the government, typically on a live-in basis. They leave their children in the Philippines, usually cared for by another relative, and send money home since the meager wages they earn in Hong Kong (the equivalent of about $325 per month) far exceed the average income of workers in the Philippines—on average, the equivalent of $1500 per year (Constable, 1997). This pattern is so common that the average Filipino migrant worker supports five people at home; one in five Filipinos directly depends on migrant workers' earnings.

One need not go to other nations to see such transnational patterns in family life. In the United States, Caribbean women and African American women have had a long history of having to leave their children with others as they sought employment in different regions of the country. Many Latinas are experiencing similar pulls on their need to work and their roles as mothers. Central American and Mexican women may come to the United States, working as domestics or other service jobs, while their children stay behind. Mothers may return to see them whenever they can, or alternatively, children may move for part of the year, but spend part of the year with other relatives in a different location. Sociologists studying transnational families find that mothers in such situations have to develop new concepts of their roles as mothers since their situation means giving up the idea that biological mothers should raise their own children. Many have expanded their definition of motherhood to include breadwinning, traditionally defined as the role of fathers. Transnational women also create a new sense of home, one not limited to the traditional understanding of "home" as a single place where mothers, fathers, and their children reside (Hondagneu-Sotelo and Avila, 1997; Das Gupta, 1997; Alicea, 1997).

Apart from transnational families, the high rate of geographic mobility that has emerged in the United States in the late twentieth century means that many U.S. families are geographically separated from their families of origin. Grandparents, sisters and brothers, aunts and uncles rarely live in the same community—a pattern that has profoundly affected systems of care since the extended family is dispersed too widely to provide child care and other forms of family support. Coupled with the presence of women in the labor force, there is, therefore, a great need for state-sponsored child care. In the United States, as the Baby Boom cohort grows older, there will also be an increased need for services for an aging population (see Chapter 13).

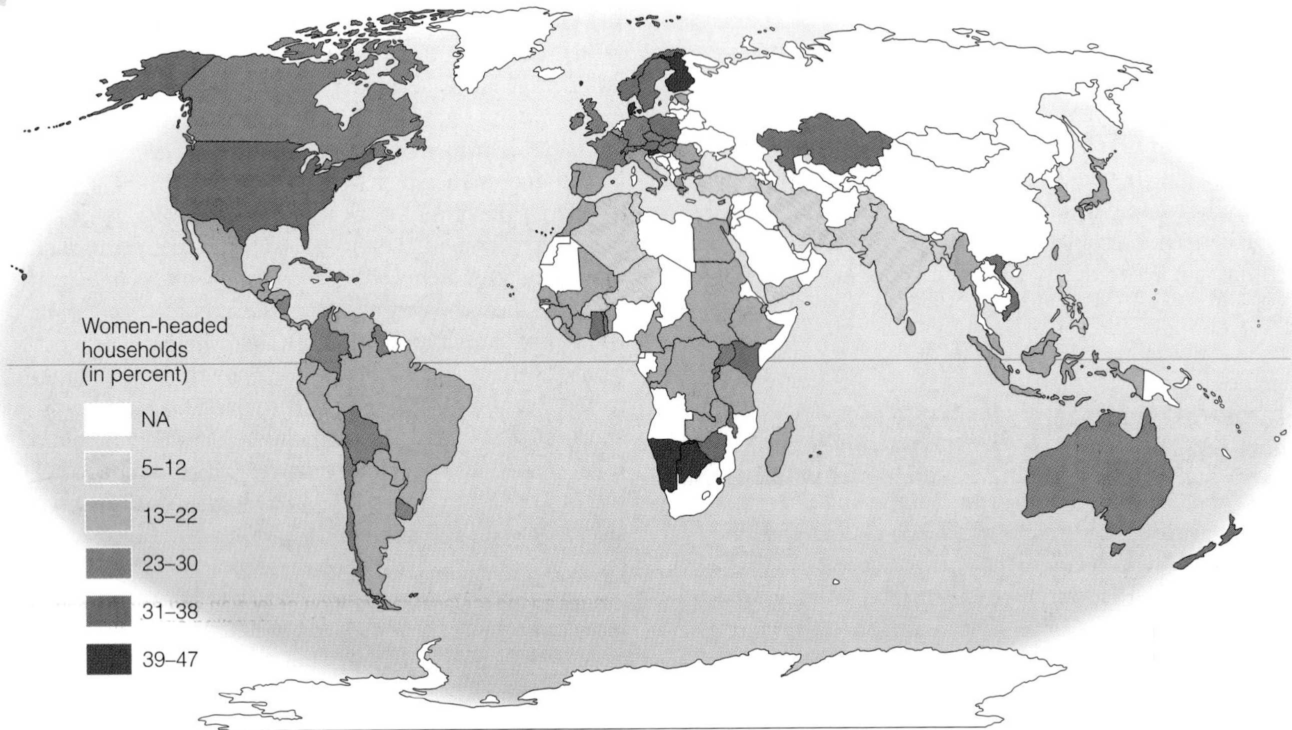

MAP 14.2 Viewing Society in Global Perspective: Women-Headed Households Around the World

DATA: From *Women of Our World*, 1998. Washington, D.C.: Population Reference Bureau.

To sociologists, global developments that produce new work patterns, wars that change the mix of women and men in a given region, and gender relations that change with new patterns of societal development are all fascinating processes that affect many dimensions of life, not the least of which is changes in the family as a social institution.

Families and Social Policy

Currently, family social policies are the subject of intense national debate. Should gay marriages be recognized by the state? What should society do to discourage teen pregnancy? What families should be given tax credits for the education of their children? Even welfare policy is embedded in assumptions about family life—that women are better off with men in the household and that women on welfare should work, not just stay home to care for their children. These and other heated debates are on the frontlines of national social policy and they require a sociological perspective to be fully understood.

Some claim the family is breaking down. Is this true? Many think it is, citing the diverse forms that families now take as evidence of the collapse of the family as an institution. What does family breakdown mean? First, those who assume there is family breakdown typically mean that the old ideal of the nuclear family with the man employed and the wife at home is no longer the dominant family experience. This assumes that there is one agreed-upon definition of the family—an assumption that is challenged by the research covered in this chapter. Beyond this, however, some also think that it is harder to hold families together, whatever their form. The idea of family breakdown assumes, however, that the family is no longer functioning. Most sociologists would say that this is misleading since the term *breakdown* confuses change and difference. The fact is that most people still marry, have children, and form cooperative household arrangements; nonetheless, the family is clearly changing, and new forms of the family are emerging.

Despite the changes that have taken place in the institution of the family, the ideal persists in public discussions about family values and is a potent element in national policy discussions about the family (Stacey, 1996). Many blame the family for the social problems our society faces. Drugs, low educational achievement, crime, and violence are often attributed to a crisis in "family values," as if rectifying these attitudes is all it will take to solve our nation's difficulties. The family is the only social institution that typically takes the blame for all of society's problems. Is it reasonable to expect families to solve social problems? Many sociologists would say that this is just another form of "blaming the victim" (Ryan, 1971). Families are afflicted by most of the structural problems that are generated by racism, poverty, gender inequality, and class inequality. Expecting families to solve the problems that are the basis for their own difficulties is like asking a poor man to save us from the national debt.

Among industrialized nations, the United States provides the fewest federally supported maternity and child-care policies (see Table 14.2). Balancing the multiple demands of work and family is one of the biggest challenges for most families. With more parents employed, having time from one's paid job to care for newborn or newly adopted children, tend to sick children, or care for elderly parents or other family members is increasingly difficult for parents. The Family and Medical Leave Act (FMLA), adopted by Congress in 1993, is meant to help address these conflicts. It requires employers to grant employees a total of twelve weeks in unpaid leave for any of the following reasons:

- To care for a child during the first twelve months after childbirth
- To care for an adopted child or foster child within the first twelve months of the placement
- To care for a spouse, child, or parent who has a serious health condition

TABLE 14.2 *MATERNITY LEAVE BENEFITS: A COMPARATIVE PERSPECTIVE ON NATIONAL POLICIES*

Country	Number of Weeks Maternity Leave	Percentage of Wages Covered	Provider of Coverage
Australia	6 (before birth); 6 (after)	100%	Applies only to federal workers and some states
Bulgaria	120 days (more days after first child)	100%	Social insurance
Canada	17	57% for 15 weeks	Unemployment insurance
Denmark	4 (before birth); 14 (after)	Cash benefits equal to average earnings	Government
France	6 (before birth); 10 (after)	84%	Social security
Germany	6 (before birth); 8 (after)	100%	Social security/ employer
Ireland	14	70%	Social security
Israel	12	75%	Sickness insurance
Italy	2 months (before birth); 3 months (after)	80%	Social insurance
Japan	14	60%	Health insurance/ social security
Poland	16 (18 for subsequent children)	100%	Social Insurance
Russian Federation	140 days	100%	Social security
Spain	16	75%	Social security
United Kingdom	12	90%	National insurance
United States	12	Unpaid	

SOURCE: United Nations. 1995. *The World's Women 1995: Trends and Statistics.* New York: United Nations, p. 137.

In recent years, there has been increasing mobilization by groups asking for legal recognition of lesbian and gay marriages; many companies have responded by extending employee benefits to domestic partners.

The Family and Medical Leave Act is the first law to recognize the need of families to care for children and other dependents. It is gender neutral, meaning that this right is available to men or women. The policy requires that employers continue their contribution to the employee's benefits during the leave; the employee must also be reinstated to the same or similar position upon return. In addition, the law states that it is not intended to subvert more generous policies, only to establish minimum standards for employers. This law is a major step forward in promoting some balance between work and family demands.

A number of conditions, however, limit the effectiveness of the FMLA, not the least of which is that the leave is unpaid, making it an impossibility for many employed parents. The law covers only employers who have fifty or more employees within a seventy-five mile radius. Employees have to have been employed by the granting employer at least one year or 1250 hours to be eligible. Although it provides twelve weeks of leave, employers may require employees to use other forms of leave first, including sick leave and accrued vacation time. Extensive documentation is also required to document the illness of someone for whom the employee needs to provide care. Finally, it is not yet clear by legal test of this law if it will apply to gay and lesbian workers (Gowan and Zimmermann, 1996; Auerbach, 1992; Rosenthal, 1993).

Family leave policies that give parents time off to care for their children or sick relatives are helpful but of little use to people who cannot afford time off work without pay. There is a need for national social policies to address not only family issues, but the problems that are the basis for difficulties in family life, such as unemployment, women's low wages, and programs to reduce violence. Because families are so diverse, different families need different social supports. Some policies will benefit some groups more than others—one reason that policymakers need to be sensitive to the diversity of family experiences.

Other policies are also needed. Social programs that promote contraceptive use, sex education, and family planning will help reduce the high rate of teenage pregnancy and the problems it causes. Policies also need to recognize the gender-specific character of people's experiences within families. Women still do the majority of family work. Policies can contribute to change in such matters in several ways. For example, legislation can create incentives for employers to allow men to take greater responsibility for family life—the institution of flexible work hours, for example, can permit more flexible parenting. Social policies cannot solve all the problems that families face, but they can go a long way toward creating the conditions under which diverse family units can thrive.

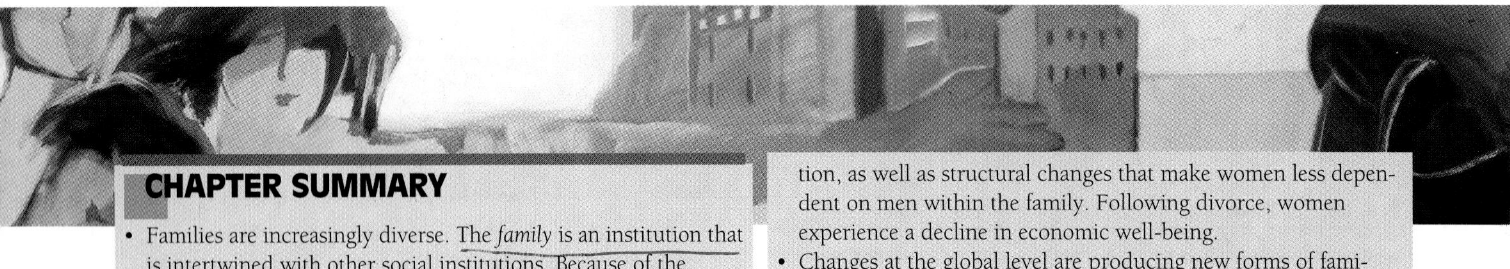

CHAPTER SUMMARY

- Families are increasingly diverse. The *family* is an institution that is intertwined with other social institutions. Because of the diversity of families, it is difficult to define the family as a single thing. The Census Bureau distinguishes families from households for purposes of enumerating the family characteristics of the U.S. population.
- All societies are organized around a *kinship system*. Different kinship systems are defined by how many marriage partners are allowed, who can marry whom, how descent is determined, family residence, and power relations within the family. *Extended family* systems develop when there is a need for extensive economic and social cooperation. The *nuclear family* arose as the result of Western industrialization that separated production from the home. In the United States, most people marry within their class and race background, although the number of interracial marriages is increasing.
- Sociologists have developed four theoretical perspectives to understand families: *functionalism, conflict theory, feminist theory,* and *symbolic interaction.* Functionalism emphasizes that families have the function of integrating members into society's needs. Conflict theorists see the family as a power relationship, related to other systems of power and inequality. Feminist theory emphasizes the family as a gendered institution and is critical of perspectives that take women's place in the family for granted. Symbolic interaction tends to take a more microscopic look at families, emphasizing such things as how different family members experience and define their family experience.
- One of the greatest changes in families has been the increase in female-headed households, which are more likely than others to live in poverty. The increase in women's labor force participation has also affected experiences within families, resulting in dual roles for women. *Stepfamilies* face unique problems stemming from the blending of different households. *Gay and lesbian households* are also more common and challenge traditional heterosexual definitions of the family. Single people make up an increasing portion of the population, due in part to the later age when people marry.
- The United States has the highest *marriage* and the highest *divorce* rates of any industrialized nation. The majority of people now prefer a lifestyle in which husband and wife work and share household responsibilities, although many retain more traditional values. The high divorce rate is explained as the result of a cultural orientation toward individualism and personal gratification, as well as structural changes that make women less dependent on men within the family. Following divorce, women experience a decline in economic well-being.
- Changes at the global level are producing new forms of families—*transnational families*—where at least one parent lives and works in a different nation than the children. Patterns of migration, war, and economic development have a profound effect on the social structure of families. Changing patterns of family relationships create new social policy needs.
- The family is often blamed for many of the social problems the nation experiences. *Social policies* designed to assist families should recognize the diversity of family forms and needs and the interdependence of the family with other social conditions and social institutions.

KEY TERMS

bilateral kinship system
cohabitation
egalitarian societies
endogamy
exogamy
extended families
family
homogamy
household
kinship system
matriarchy
matrilineal kinship system
matrilocal kinship system
miscegenation
monogamy
neolocal residence
nuclear families
patriarchy
patrilineal kinship system
patrilocal kinship system
polygamy

THE INTERNET: A Tool for the Sociological Imagination

Resources on the Internet:

Virtual Society: The Wadsworth Sociology Resource Center
http://sociology.wadsworth.com

Visit this site to find additional learning tools, including interactive quizzes, links related to web sites, and an easy link to *Infotrac College Edition.*

National Council on Family Relations
http://www.ncfr.org

This is a professional organization of researchers, teachers, and practitioners that sponsors research on families.

Family Violence Prevention Fund
http:www/igc.apc.org/fund/

A nonprofit organization focusing on policy and education about domestic violence. The web site includes research information, personal testimonies, and national contacts for other resources.

The National Committee to Prevent Child Abuse
http://www.childabuse.org/

This national organization surveys current trends in child abuse and reporting, and organizes chapters in each state. The web site includes annual reports and prevention information.

Sociology and Social Policy: Internet Exercises

Many major employers have extended employment benefits to gay and lesbian couples, recognizing that these unions are one of the diverse forms of family life in the contemporary United States. As a matter of social policy, such programs extend the same support to gay and lesbian domestic partners (with and without children) that married heterosexual couples receive. What reasons support the implementation of *domestic partner benefits,* and what are the consequences for businesses of doing this or not? As you develop your answer, think about how the broader societal context influences your analysis of such policies.

Internet Search Keywords:

domestic partner benefits
social policy
employment benefits
gay and lesbian families
employment
discrimination
workplace

Web sites:

http://www.cwru.edu/orgs/glba/GLBA.html
The web page for the Gay Lesbian Bisexual Alliance. This site provides a foundation for the issues facing gay and lesbian couples.

http://www.hrc.org/issues/workplac/dp
A web page produced by the Human Rights campaign discussing domestic partner benefits.

http://www.aclu.org/
The home page for the American Civil Liberties Union.

http://www.ngltf.org/
The National Gay and Lesbian task force offers current information on issues affecting domestic partnership.

InfoTrac College Edition: Search Word Summary

cohabitation
endogamy
extended families
patriarchy

In order to learn more about these central topics in sociology, you can conduct an electronic search using InfoTrac College Edition. To aid in your search and to gain useful tips, see the Student Guide to InfoTrac College Edition on the Virtual Society web site:
http://sociology.wadsworth.com

INTERACTIONS—A SOCIOLOGY CD-ROM: CONCEPTS FOR THIS CHAPTER

Go to the Wadsworth Sociology CD-ROM for further study on the concepts in this chapter. The CD-ROM also includes quizzes and additional activities to expand your learning experience.

SUGGESTED READINGS

Baca Zinn, Maxine, and Stanley Eitzen. 1998. *Diversity in American Families,* 5th ed. New York: Harper Collins.

This is the most comprehensive sociological analysis of families with a focus on the different family experiences across race, class, and gender.

Hochschild, Arlie. 1997. *The Time Bind: When Work Becomes Home and Home Becomes Work.* New York: Metropolitan Books.

Hochschild's study of a company with "family-friendly" policies explores the increasing stresses that time demands at work, for both women and men, put on family life. As a result, family life is more hectic, not the haven that the family ideal purports.

Hondagneu-Sotelo, Pierrette. 1994. *Gendered Transitions: Mexican Experiences of Immigration.* Berkeley: University of California Press.

Based on field research with Mexican immigrants, Hondagneu-Sotelo examines the effect of immigration on family experiences and, particularly, gender relations within the family.

Lerman, Robert I., and Theodora J. Ooms. 1993. *Young Unwed Fathers: Changing Roles and Emerging Policies.* Philadelphia: Temple University Press.

With an eye toward social policy, this anthology reviews current research on young unwed fathers and examines the sociological conditions underlying their experiences.

Stacey, Judith. 1996. *In the Name of the Family.* Boston: Beacon Press.

Stacey examines the family values debate, with a view to understanding the problems that contemporary families face. Using an analysis of gender, she looks at the changes in families brought about by increases in divorce rates, employed mothers, and economic pressures in different social classes.

Thorne, Barrie. 1992. *Rethinking the Family: Some Feminist Questions,* 2nd ed. Boston: Northeastern University Press.

This anthology of contemporary feminist thinking about families is a classic because of its theoretical perspective, articulated by Thorne in the introduction, and its sensitivity to diverse family forms, including different racial–ethnic families, gay families, and families in different social classes.

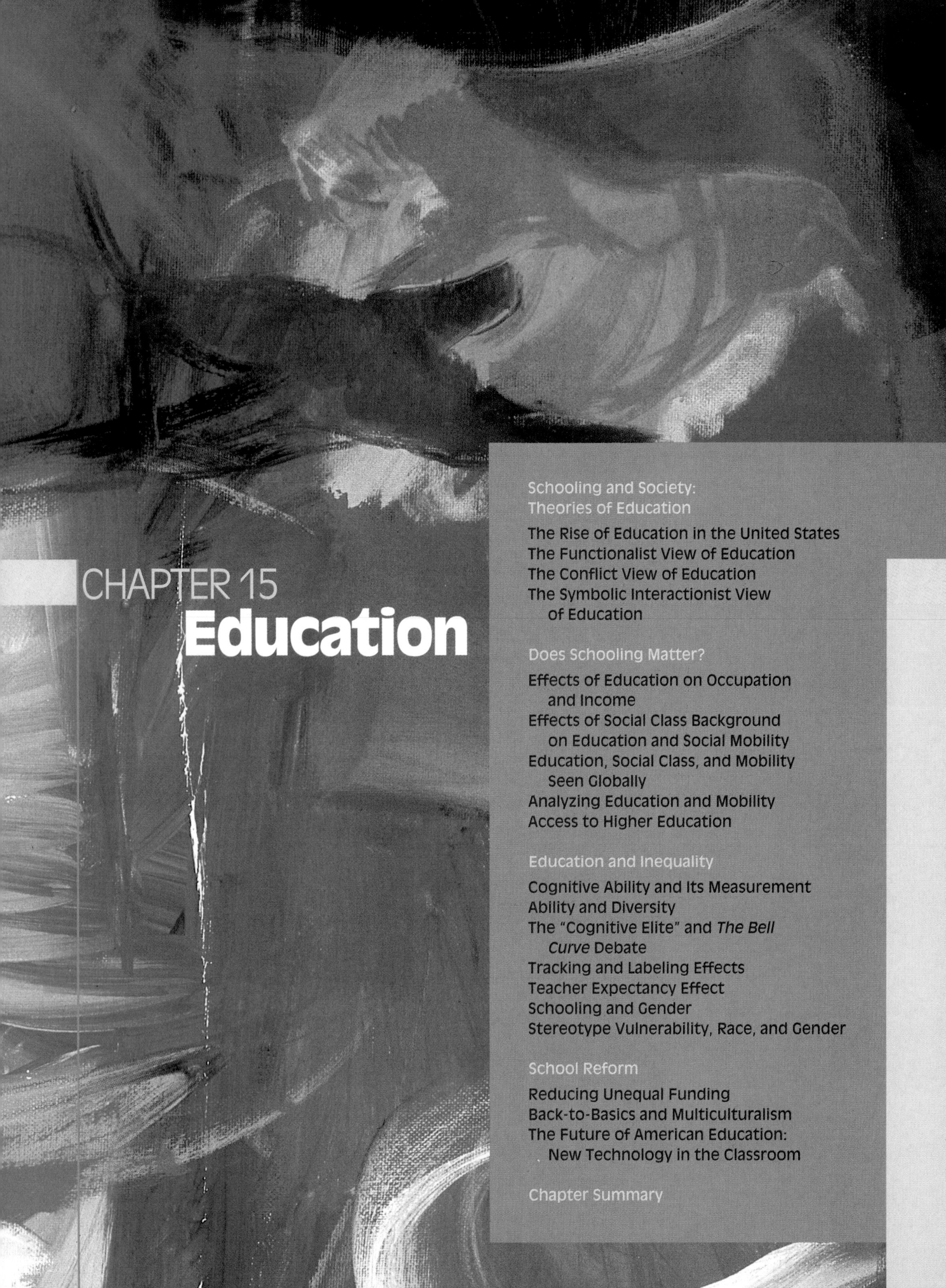

CHAPTER 15
Education

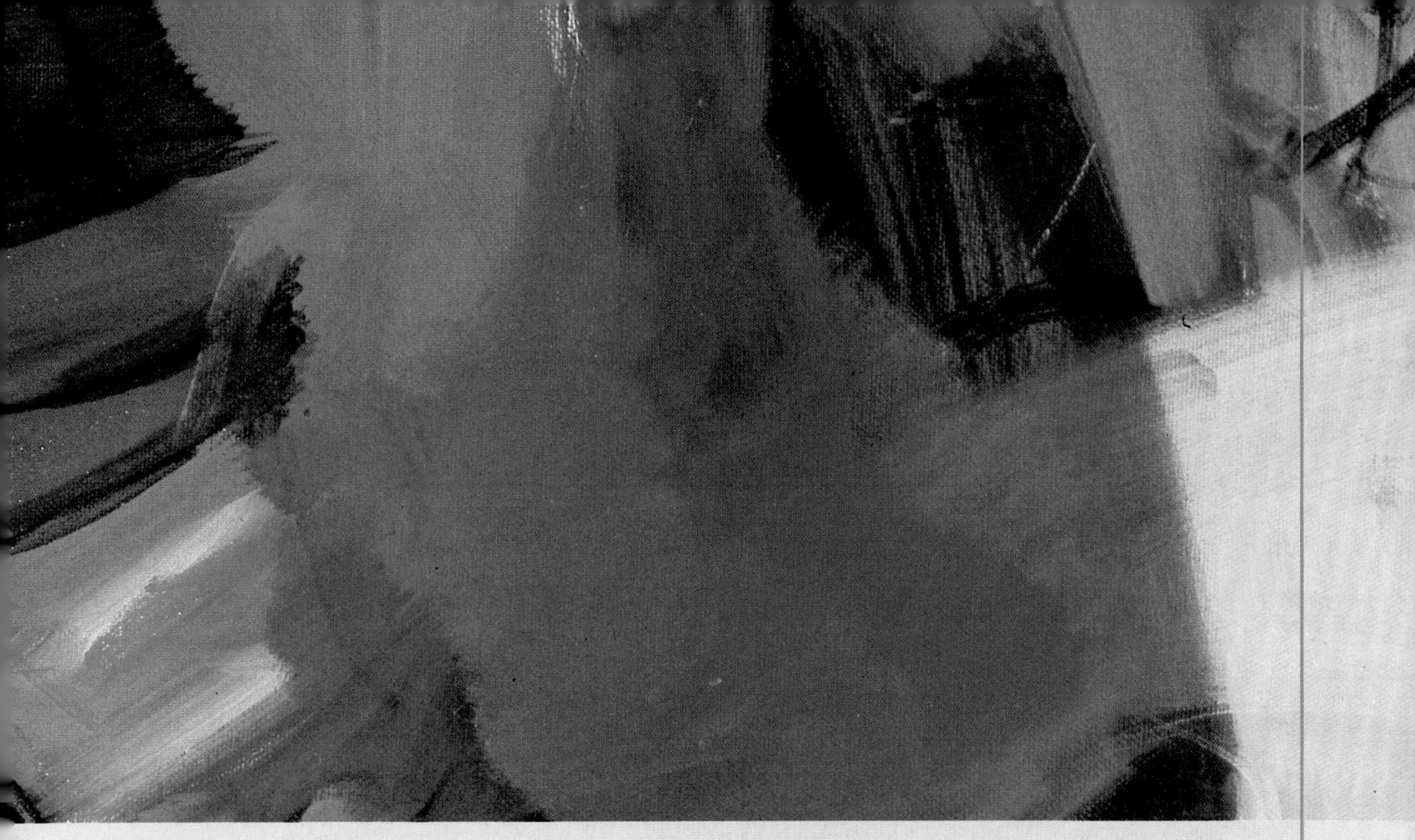

OLLIE Taylor (a pseudonym) is a bright eleven-year-old African American. He is growing up in a two-parent family in Boston. His father works fifty hours a week, but they are still poor. Although he receives considerable encouragement, love, and respect from his family, Ollie Taylor is certain that everything he attempts will fail. He is convinced that he is, in his own words, worthless. A psychologist, Dr. Tom Cottle of Boston, who interviewed Ollie Taylor over a three-year period, states conclusively that Ollie's feelings about himself can be traced directly to his school and not to his family or his peers. In Ollie Taylor's case, the feeling of worthlessness was significantly created by the system of **tracking**, the grouping, or stratifying, of students within the educational system.

In Ollie's own words:

> The only thing that matters in my life is school, and there they think I am dumb and always will be. I'm starting to think they are right. . . . Every word those teachers tell me, even the ones I like most, I can hear in their voice that what they're really saying is, "All right you dumb kids. I'll make it as easy as I can, and if you don't get it then, then you'll never get it." Upper tracks? Man, when do you think I see those kids? I never see them. . . . If I ever walked into one of their rooms, they'd throw me out before the teacher ever came in. They'd say I'd only be holding them back from their learning (Cottle, quoted in Persell, 1990: 82).

Stories like Ollie Taylor's are common. His case highlights one of the problems facing the American system of education: Should children be grouped (tracked) according to their ability, allowing them to learn at their most comfortable pace, or should they be grouped simply by age, perhaps retarding the progress of fast learners and keeping a pace too fast for the slow learners?

There are both positive and negative consequences to early tracking of students, a system discussed in detail later in this chapter. Ollie Taylor's case exemplifies the negative consequences, yet he may also benefit from the positive aspects of getting an education in the United States, even with tracking in place: As an African American, he is six times more likely to graduate from high school than he would have been in 1940. With a high school diploma, he will be more likely to get a good job than if he were to drop out of school before graduation. He will be less likely to join the ranks of the unemployed if he completes his education, and he will be more likely—though certainly not guaranteed—to earn a sufficient wage if he does so.

His chances of dropping out prior to graduation are, unfortunately, fairly high. They are considerably higher than for a White man of his age, but significantly lower than for a Hispanic young man. If he graduates, he will stand a better chance of avoiding the grim path of many urban Black male dropouts, and many White youth—a path leading to unemployment and possibly drug abuse, poor health, incarceration, and death at an early age.

The education institution in the United States is capable of sending a person down two entirely different paths. One leads to the perpetuation of racial, ethnic, socioeconomic, and gender inequalities in society; the other lessens these inequalities as the individual goes on to a good job, a decent income, and relative happiness. The system of education in this country spawns both outcomes, making or breaking the life prospects of millions of students like Ollie Taylor.

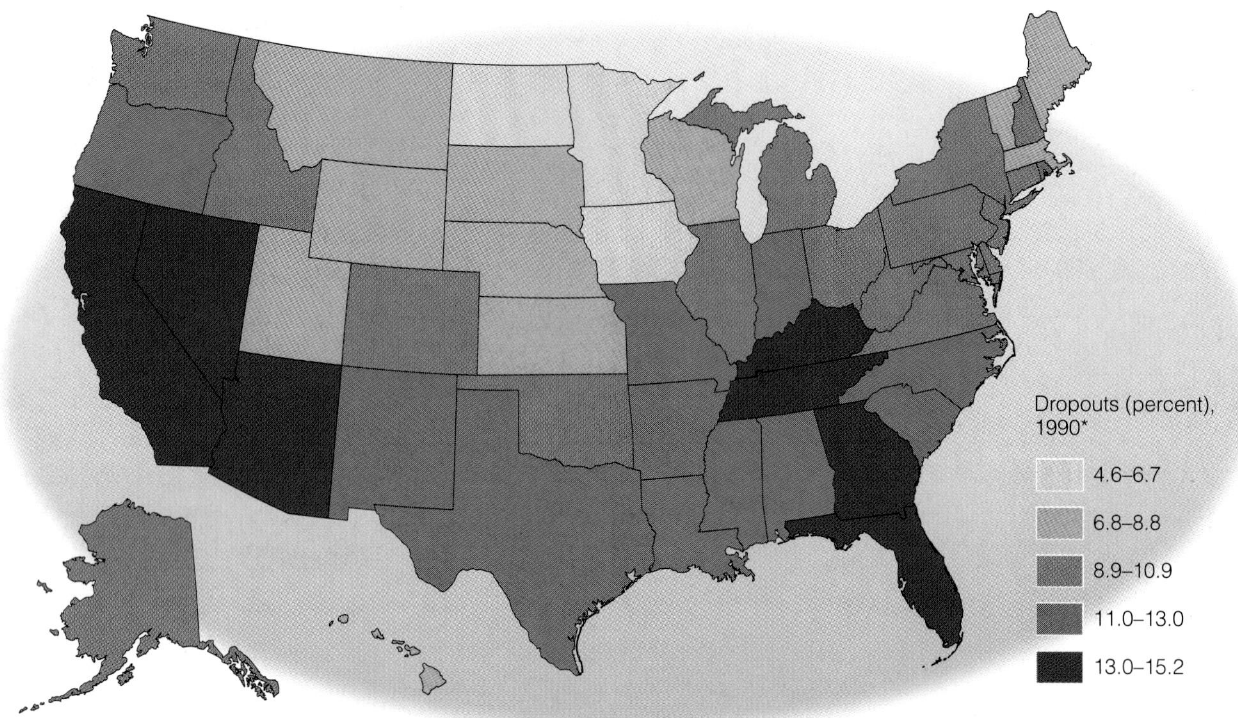

MAP 15.1 Mapping America's Diversity: School Dropout Rates

*For persons sixteen to nineteen years old. A dropout is a person who is not in school regularly and who has not completed the twelfth grade or received a general equivalency degree.

DATA: From the U.S. Bureau of the Census. 1997. *Statistical Abstracts of the United States 1997.* Washington, DC: U.S. Government Printing Office, p. 161.

Schooling and Society: Theories of Education

Education in a society is concerned with the systematic transmission of the society's knowledge. This includes teaching formal knowledge such as the "three R's," reading, writing, and arithmetic, as well as the conveyance of morals, values, and ethics. Education prepares the young for entry into society and is thus a form of socialization. Sociologists refer to the more formal, institutionalized aspects of education as **schooling.**

Debunking Society's Myths

Myth: To get ahead in society, all you need is an education.

Sociological perspective: Education is necessary, but not sufficient, for getting ahead in society. Success depends significantly on one's class origins; the formal education of one's parent or parents; and one's race, ethnicity, and gender.

Most of us have spent a significant portion of our lives within formal educational institutions of one sort or another. Nowadays, twelve years of schooling is usual, and some spend much more than that. Years in school include preschool, elementary, secondary, college, and postgraduate. Three out of ten Americans are either enrolled in some kind of education institution or employed in the field of education as teachers, administrators, secretaries, or janitors. More money is spent on education in the United States than on any other activity or institution except health care. In a single year, the United States spends more than $458 billion on all levels of education, including public, private, and parochial—about 8 percent of the gross national product (U.S. Bureau of the Census, 1997a).

The Rise of Education in the United States

Compulsory education is a relatively new idea. During the nineteenth century, many states did not yet have laws requiring education for everyone. Most jobs in the middle of the nineteenth century demanded no education or literacy whatsoever. Education was considered a luxury, available only to children of the upper classes (Cookson and Persell, 1985). Education was prohibited by law for slaves even after the Emancipation Proclamation passed in 1863.

By 1900, compulsory education was established by law in all states except for a few in the South, where Black Americans were still largely denied formal education of any kind (Higginbotham, 1978). State laws in the South and West have also in the past prohibited education for Hispanics, American Indians, and Chinese immigrants. State laws requiring attendance were generally enforced for White

one-room schoolhouse

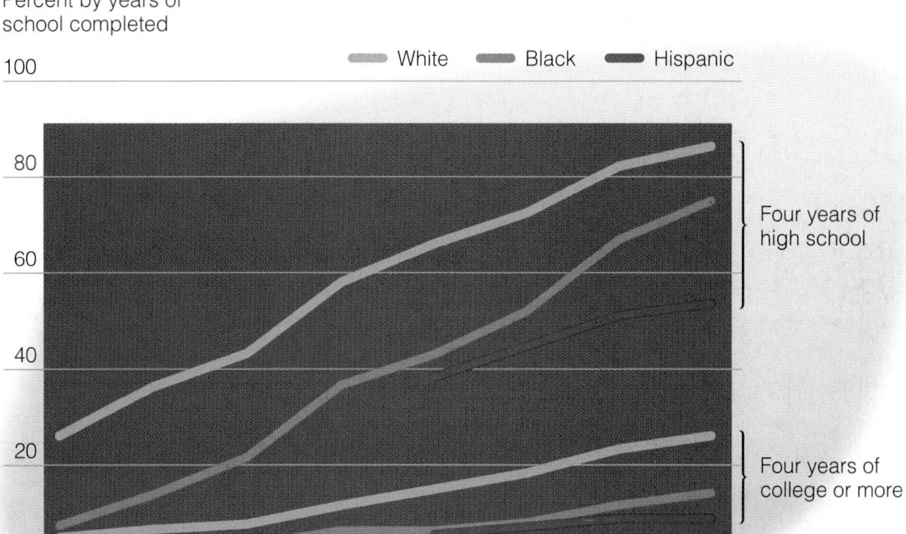

FIGURE 15.1 The Expansion of Education

DATA: From the National Center for Educational Statistics. 1997. *Digest of Educational Statistics.* Washington, DC: U.S. Department of Education, p. 17.

Americans at least through eighth grade. Education all the way through high school lagged considerably. In 1910, *less than 10 percent* of White eighteen-year-olds in the United States had graduated from high school. High school attendance rose steadily in the following decades, but by the 1930s, the number of eighteen-year-olds in this country who had attended high school was still less than half. It was not until as recently as 1950 that the number of young adults with a diploma rose above 50 percent. Today, almost 90 percent of those under thirty-five in the United States have at least a high school diploma (Parelius and Parelius, 1987; U.S. Bureau of the Census, 1997a).

Attendance in both high school and college has expanded dramatically since World War II, as shown in Figure 15.1. One factor contributing to the rise in college attendance was the G.I. Bill, which enabled well over twenty million veterans of World War II, the Korean War, and the Vietnam War to attend college. Federal loan programs and the expansion of community colleges have further contributed to the increase in college attendance and graduation. Within the past few years, however, federal and state aid has decreased, causing the growth of community colleges to slow considerably.

Figure 15.1 reveals another dramatic trend—the high school and college graduation rates have not been equal across racial groups. The high school graduation rate for Whites has increased steadily from 26 percent in 1940 to more than 80 percent in the late 1990s. For African Americans, the increase has been equally dramatic, but only parallel to the increase for Whites, from less than 9 percent in 1940 to about 75 percent in 1996. Although these trends are certainly encouraging, high school graduations for African Americans nevertheless consistently lag behind high school graduations for Whites (del Pinal and Singer, 1997; U.S. Bureau of the Census, 1997a).

As shown in Figure 15.1, the Hispanic high school graduation rate has fallen way behind both the African American and the White rates since 1970. Whereas 75 percent of African Americans aged twenty-five or older graduated from high school in 1990, only 53 percent of the Hispanics of this age did so. In general, the rate at which Hispanics have dropped out of high school has been consistently greater than that for Blacks since 1970. Before that time, the high school dropout rates for both groups were about the same (del Pinal and Singer, 1997).

Ethnic and cultural diversity in the public high school classroom are increasingly common.

College graduations have shown similar trends. The percentage of Whites who received a college degree (or higher) rose from 5 percent in 1940 to 26 percent in 1996. For Blacks and Hispanics, those completing four years or more of college increased from only 2 percent in 1940 to about 14 percent for Blacks and 9 percent for Hispanics in 1996. As with high school graduations, college completion for Hispanics has lagged behind that of Blacks since 1970. In general, the rate at which Hispanics have dropped out of college prior to graduation is greater than that of Blacks, and the Black student dropout rate from college is higher than that of Whites, particularly since 1980 (del Pinal and Singer, 1997).

The Functionalist View of Education

All known societies have an educational institution of some sort. In the United States, as in other industrialized societies, the education institution is large and highly formalized. In some other societies, such as pastoral societies, it may consist only of parents teaching their children how to till land and gather food. Under these circumstances, the family is both the education institution and the kinship institution.

Why does an education institution exist in the first place? What does it do for society? Functionalist theory in sociology answers these questions by arguing that education accomplishes certain consequences, or "functions," for a society. Among these functions are socialization, occupational training, and social control.

Socialization is brought about as the cultural heritage is passed on from one generation to the next. This heritage includes a lot more than only "book knowledge." It also includes moral values, ethics, politics, religious beliefs, habits, and norms—in short, culture. Schools strive to teach a variety of skills and knowledge, from history, literature, and mathematics to handcrafts and social skills, while also inculcating values such as school loyalty and punctuality. According to functionalist explanations, the importance of this kind of socialization to society explains why an education institution began and grew in society.

Occupational training is another function of education, especially in an industrialized society such as the United States. Less complex societies, such as the United States prior to the nineteenth century, passed on jobs and training from father to son, or more rarely, from father or mother to daughter. A significant number of occupations and professions today are still passed on from parent to offspring, particularly among the upper classes (such as a father passing on a law practice to his son) but also among certain highly skilled occupations such as plumbers, ironworkers, and electricians, who pass on both training and union memberships. Modern industrialized societies need a system that trains people for jobs. Most jobs today require at least a high school education, and many professions require a graduate degree.

Social control is also a function of education, although a less obvious one. Such indirect, nonobvious consequences emerging from the activities of institutions are called *latent functions.* Increased urbanization and immigration beginning in the late nineteenth century were accompanied by rises in crime, overcrowding, homelessness, and other urban ills. One perceived benefit of compulsory education was that it kept young people off the streets and out of trouble. The more obvious consequence, or function, of education was job training; the latent function was the social control of deviant behavior. Education also became a way to socialize new immigrants from Italy, Poland, Ireland, and other European countries, "Americanizing" them in the interest of social control (Katz, 1987).

Classrooms are now generally more culturally diverse than only ten or fifteen years ago.

The Conflict View of Education

In contrast with functionalist theory, which emphasizes how education unifies and stabilizes society, conflict theory emphasizes the disintegrative and disruptive aspects of education. Conflict theory focuses on the competition between groups for power, income, and social status, giving special attention to the prevailing importance of institutions in the conflict. One intersection of education with group and class competition is embodied in the significant correlation that exists between education and class, race, and gender. The unequal distribution of education allows it to be used to separate groups. The higher the educational attainment of a person, the more likely that person will be middle to upper class, White, and male. Conflict theorists argue that educational level can thus be used as a tool for discrimination via the mechanism of **credentialism**, the insistence upon educational credentials for their own sake, even if the credentials bear little relationship to the intended job (Collins, 1979; Marshall, 1997). This device can be used by potential employers to discriminate against minorities, working-class

TABLE 15.1 *SOCIOLOGICAL THEORIES OF EDUCATION*

	Functionalism	Conflict Theory	Symbolic Interaction
Education in society	fulfills certain societal needs for socialization and training	reflects other inequities in society, including race, class, and gender inequality	emerges depending on the character of social interaction between groups in schools
Schools	inculcate values needed by the society	are hierarchical institutions reflecting conflict and power relations in society	are sites where social interaction between groups (such as teachers and students) influences chances for individual and group success
Social change	means that schools take on functions that other institutions, like the family, originally fulfilled	threatens to put some groups at continuing disadvantage in the quality of education	can be positive as people develop new perceptions of formerly stereotyped groups

people, or women—that is, those who are often less educated, thus less likely to be credentialed because discriminatory practices within the educational system itself have depressed their opportunities for educational achievement. Although functionalists argue that jobs are becoming more technical and thus require workers with greater education, conflict theorists argue that the reverse is true—that most of the new opportunities appearing today are in categories like assembly-line work, jobs that are becoming less complex and less technical, and therefore require less traditional education or training. Nonetheless, potential employers will insist on a particular degree for the job even though there should be little expectation that education level will affect job performance. Education is thus used as a discriminatory barrier.

The Symbolic Interactionist View of Education

Symbolic interaction focuses on what arises from the operation of the interaction process during the schooling experience. Through interaction between student and teacher, certain expectations arise on the part of both. As a result, the teacher begins to expect or anticipate certain behaviors, good or bad, from students. Through the operation of the *expectancy effect,* the expectations a teacher has for a student can create the very behavior in question. Thus the behavior is actually caused by the expectation, rather than being simply anticipated by it. For example, if a White teacher expects Latino boys to perform below average on a math test, relative to White students, over time the teacher may unconsciously act in ways that encourage the Latino boys to get below average scores on tests. The teacher might provoke increased stress among Latinos, thus increasing test anxiety, resulting then in decreased performance. The point is that teachers' expectations can affect test performance in addition to the effects of students' aptitudes or abilities. Later we shall look at studies that show how the expectancy effect works.

Does Schooling Matter?

How much does schooling really matter? Does more schooling actually lead to a better job, more annual income, and greater happiness? Is the effect of more education great or small? To get a complete answer, we need to analyze the situation from two points of view: First, how does formal education affect ultimate occupation and income—that is, how is it related to social mobility? Second, what effect does social class *origin* have on the ultimate effect of education? These are two distinct questions. Let us take a look at the evidence pertaining to each.

Effects of Education on Occupation and Income

One way that sociologists measure a person's social class or socioeconomic status (SES) is to determine the person's amount of schooling, income, and type of occupation (see Chapter 9 on class stratification). Sociologists call these the *indicators* of SES. In the general population, there is a strong correlation between formal education and occupation, although the relationship is not perfect. Measuring occupations in terms of social status or prestige, we find that the higher a person's occupational status, the more formal education he or she is likely to have received. Overall, we know that on average, doctors, lawyers, professors, and nuclear physicists spend many more years in school than unskilled laborers such as garbage collectors and shoe shiners. This relationship is strong enough that we can often, though not always, guess a person's educational attainment just by knowing their occupation. There are indeed instances of semiskilled laborers (such as taxi drivers) who have PhDs, but they are relatively rare. Also exceedingly rare is the reverse: the self-educated, self-made individual who completed only the fourth grade and is now the CEO of a major corporation.

People have clear notions about which jobs are "better" than others. As we discussed in Chapter 9, sociologists refer to the ranking of jobs as *occupational prestige.* People rank medical doctor, member of Congress, and nuclear physicist above electrician, plumber, and policeman, and these are ranked in turn above truck driver, milkman, and dock worker. No job is truly any "better" than another, of course; prestige ranking simply means that people think of jobs in this way (Hodge et al., 1966; Nakao and Treas, 1994). A reason that people subjectively rank some jobs as higher than others is they perceive that more education is needed

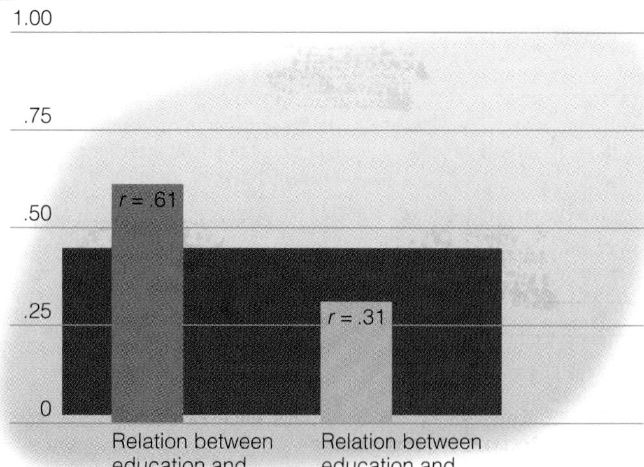

FIGURE 15.2 How Much Difference Does Education Make in Occupation and Income?

SOURCE: After Jencks, 1993, p. 231.

for some jobs than for others, and the more prestigious jobs are seen as needing more education. It is also true that jobs traditionally held by men have carried more prestige on average than those held mostly by women; thus, the past gender typing of a job will affect its prestige rating.

These findings pertain to people's *perceptions* of occupational prestige, and to their perceptions of how much education is needed for each occupation. What about the actual, objective relationship between years of formal education and subsequent job and income? There is a relationship, but it is somewhat less rigid than most people think. In a major study, Jencks et al. (1972), combining the analyses from several large past studies, found that the correlation between actual years of formal education and occupation is 0.61, which means that there is a moderately strong relationship between the two (see Figure 15.2). The relationship between education and later income, expressed as annual income in dollars, is somewhat less, 0.33 (see Figure 15.2). So the effect of education upon occupation is present but modest, and the overall effect of education upon income is even less. Education has a greater effect on your type of occupation than on your income.

It should be mentioned that these results have been criticized since African Americans, Hispanics, American Indians, Asians, and women were completely excluded from all analyses (Taylor, 1973b; Edmonds et al., 1972). Still, the results hold for White males. Later results (Jencks et al., 1979; Jencks, 1993) tend to confirm these overall correlations for White men. For men who are not White (meaning mainly Hispanics, Native Americans, or Blacks), the effect of education upon income is *less* than for Whites.

What more can be said about the relationship between education and annual income? Looking behind the correlations, at least one analyst (Bird, 1975) argued that most people would come out ahead in a strictly economic sense if they skipped college, invested the money instead, and went to work right after high school. This argument is overstated, however. Income brackets are consistently higher for the higher education categories. For people who have completed graduate or professional school, one-third earned more than $75,000 per year, and over half earned more than $50,000 per year. On the other hand, for those who did not complete high school, 29 percent earned less than $10,000 per year, and less than 1 percent earned more than $75,000 per year (U.S. Bureau of the Census, 1997b).

The connection between income and education is not independent of gender. Gender heavily influences the relationship between income and education. Note from Table 15.2 that although the higher one's education, the higher one's (average) income, it is nonetheless true that the average income for women is less than the average income for men at each educational level. This is because in general, throughout our society, women consistently earn less than men who are of comparable education The data show clearly that differences persist at each level of education; the average woman earns less than a man of the same or even less education. Note from Table 15.2 that men with professional degrees (law, medicine, and so forth) earn a median income of $71,869 a year, whereas a woman with that same education earns only $42,059 a year—only about 60 percent of what a man earns. A man with no graduate education but a college-only education still earns more ($39,624) than a woman with a master's degree. Even men with some college, but no degree, earn more than women with bachelor's degrees (see Table 15.2).

As we already noted, the numbers of high school diplomas and college degrees conferred have increased rapidly in the last thirty years. This has had an effect on the economic value of a college education. In the past, when few people earned college degrees, college graduates were a scarce, and thus valuable, commodity. Because far more people earn college degrees today, a college education is no longer the same automatic ticket to success that it used to be. The French sociologist R. Boudon (1974) noted some time ago that as the level of education has risen in industrialized nations such as the United States, the relative economic advantage of completing college, measured in dollars, has

TABLE 15.2 *INCOME, EDUCATION, AND GENDER*

Level of Schooling	Men (and percent of those in category)	Women (and percent of those in category)
Doctorate degree	$62,255 (1.5%)	$42,431 (0.1%)
Professional degree	71,869 (2.1%)	42,059 (0.1%)
Master's degree	50,003 (5.9%)	33,302 (5.1%)
Bachelor's degree	39,624 (17%)	25,192 (16%)
Associate degree	33,065 (6.6%)	20,460 (8.2%)
Some college, no degree	29,160 (17.3%)	16,255 (17.5%)
High school graduate (or equivalent)	24,814 (32%)	12,702 (35.2%)
Ninth to twelfth grade (no diploma)	16,058 (9.7%)	8544 (9.6%)
Less than ninth grade	12,174 (7.7%)	7276 (7.0%)

SOURCE: U.S. Bureau of the Census. 1997. *Money Income in the United States 1996.* Washington, DC: U.S. Department of Commerce, pp. 26–27.

declined. Boudon calls this **educational deflation.** He notes that it applies to all levels of education, not just college. Not only is a college degree worth less now, so is an eighth-grade (junior high) education. This means that the children of high school dropouts who are themselves dropouts will earn less, on average, than their dropout parents did, and will also have a more difficult time finding a skilled job.

THINKING SOCIOLOGICALLY

What is the amount of (formal) education of your parent or parents? Do they (he or she) want you to have more education than they had, about the same, or less? What do these questions suggest to you about the relationship between *educational attainment* and *social mobility*?

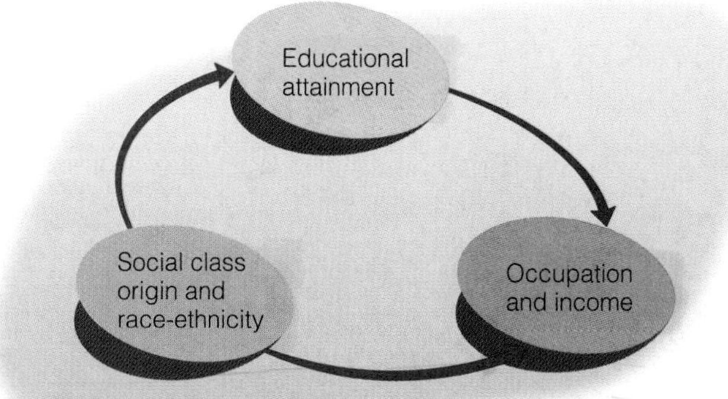

FIGURE 15.3 Relationship of Social Class, Race-Ethnicty, Education, Occupation, and Income

Effects of Social Class Background on Education and Social Mobility

Education has traditionally been viewed in the United States as the way out of poverty and low social standing—the main route to upward social mobility. The assumption has been that a person can overcome modest beginnings, starting by staying in school. It is not our intention to challenge this assumption, like the dope peddler who argues that "if you're a minority from the streets, education won't get you diddly." Here we ask, how does education benefit people of different classes? Taking into account social class origins and race, what is the effect of education on the chances for upward social mobility, higher occupational prestige, and higher income?

Much sociological research has demonstrated that the effect of education upon a person's eventual job and income depends to a great extent upon the social class that the person was born into. Hence, there is no uncomplicated or straightforward relationship between education, occupation, and income. Among White people of the upper classes, including those who inherited wealth as well as professionals and high-level managers, social class origin is *more important than education* in determining occupation and income (Blau and Duncan, 1967; Jencks et al., 1972; Jencks et al., 1979; Jencks, 1993; Bielby, 1981; Persell, 1990). Class and race work together to "protect" the upper class from downward social mobility—and also to block the lower classes and minorities from too much upward mobility. Education is used by the upper classes to avoid downward mobility, by such means as sending their children to elite private secondary schools. A disproportionately high number of upper-class children attend elite boarding schools and day schools, whereas working-class children are considerably underrepresented in such schools (Rendon and Hope, 1996; Cookson and Persell, 1985).

Among middle-class Whites, education considerably improves the chances of getting middle-class jobs; access to upper-class positions is more limited. Among those of lower-class origins, such as unskilled laborers or the chronically unemployed, chances of getting a good education as well as a prestigious job are poor. In sum, education is affected by social class origins, and occupation (and income) is heavily influenced by social class and education. Individuals with lower-class origins are less likely to get a college education and thus are less likely to get a prestigious job. These interrelationships are summarized in Figure 15.3, which shows that social class origin affects occupation and income both directly and indirectly: First, there is a direct effect of class origin upon occupation and income, independent of education. If you were to compare a large number of individuals all with the same education but from different social class backgrounds, you would find that the higher their social class, the higher their occupational prestige and income. Second, there is the indirect effect of social class on level of educational attainment, which affects occupation and, in turn, income level.

Debunking Society's Myths

Myth: **Education is more important than social class in determining one's job and income.**

Sociological perspective: **Although education has an effect on the job one gets and the income one earns, overall, social class origin is more important than acquired education in determining the prestige of one's job and earned income.**

Education, Social Class, and Mobility Seen Globally

It is sometimes argued that because of our education system, there is more occupational and income mobility in the United States than in other countries, particularly England, Germany, and Japan. In general this is true, but not by much. Until a few years ago, students in England were required to take an examination, called the Eleven Plus, at age eleven. A person's score on this examination determined whether he or she was put on a track to prestigious universities such as Oxford or Cambridge, or went directly into the labor force. Children of the upper classes stood a far better chance of scoring high on this examination than did middle- or working-class children, and the average

Japanese students and their parents at this juko, *or "cram" seminar, vow academic success, at a meeting in a hotel in Tokyo over the New Year's holiday.*

scores of women and minorities, especially Africans and East Indians, were considerably lower than those of upper-class White males.

A similar situation exists in the United States. Students from lower-class families have lower average scores on exams such as the Scholastic Assessment Test (SAT) and the American College Testing Program (ACT) (see Table 15.3). As shown in Table 15.3, there is a smooth increase in average (mean) SAT score as family income is higher. With lower scores comes a diminished chance of getting into the best colleges or universities. African Americans, Latinos, and American Indians score on the average less than Whites, and women tend to score lower than men on the quantitative (mathematical) sections of the SAT (see Table 15.4). Asian Americans as a group have scored higher than Whites in recent years on the quantitative sections of the SAT, but lower on the verbal sections. Women of any ethnic group score less on the quantitative sections than the men of the same ethnic group, except for certain Latina groups, who score higher than men of the same group (see Table 15.4). These patterns indicate that the SAT (and the ACT) have an effect in the United States similar to that of the Eleven Plus in England, directing the futures of the young according to the results of widely administered exams.

In Germany, an examination called the *Abitur* is taken during the equivalent of junior year in high school. A high score on the *Abitur* facilitates admission to a university; a low score inhibits getting into a university. Low-scoring students must take two or three more years of courses and then reapply to a university if they wish to attend.

In Japan, the same kind of examination, given at age twelve, determines even more rigidly a person's subsequent educational opportunities. Students who wish to continue their education at a college or university must score high enough to gain admission to prep schools; especially high scores guarantee admission to prestigious prep schools, which are necessary for later admission to the best universities. Low scorers are virtually shut out from prep school admission and thus become ineligible for a university education. In recent years, many parents have begun to send

TABLE 15.3 *SAT SCORES BY FAMILY INCOME*

	SAT Test Takers	SAT Verbal Mean Scores	SAT Math Mean Scores
All SAT test takers	1,172,778	505	512
Family Income			
Less Than $10,000	44,912	427	446
$10,000–$20,000	85,245	451	483
$20,000–$30,000	104,726	477	482
$30,000–$40,000	125,771	495	497
$40,000–$50,000	108,645	506	509
$50,000–$60,000	105,403	514	518
$60,000–$70,000	84,139	521	525
$70,000–$80,000	72,443	527	532
$80,000–$100,000	88,321	539	548
More than $100,000	122,383	559	572
No Response	232,991		

SOURCE: The College Board, 1999.

TABLE 15.4 *ETHIC/GENDER SAT SCORES*

SAT Test Takers Who Described Themselves as:	SAT Verbal Mean Scores			SAT Math Mean Scores		
	Male	Female	Total	Male	Female	Total
American Indian or Alaskan Native	484	477	480	498	438	483
Asian, Asian American, or Pacific Islander	500	430	438	579	546	582
African American or Black	432	435	434	436	419	426
Hispanic or Latino Background						
Mexican or Mexican American	461	447	453	481	445	460
Puerto Rican	456	448	452	405	434	447
Latin American, South American, Central American, or Other Hispanic or Latino	470	454	461	409	449	466
White	531	523	526	547	512	528
Other	514	508	511	530	497	514
No Response	480	490	490	614	488	580

SOURCE:The College Board, 1999.

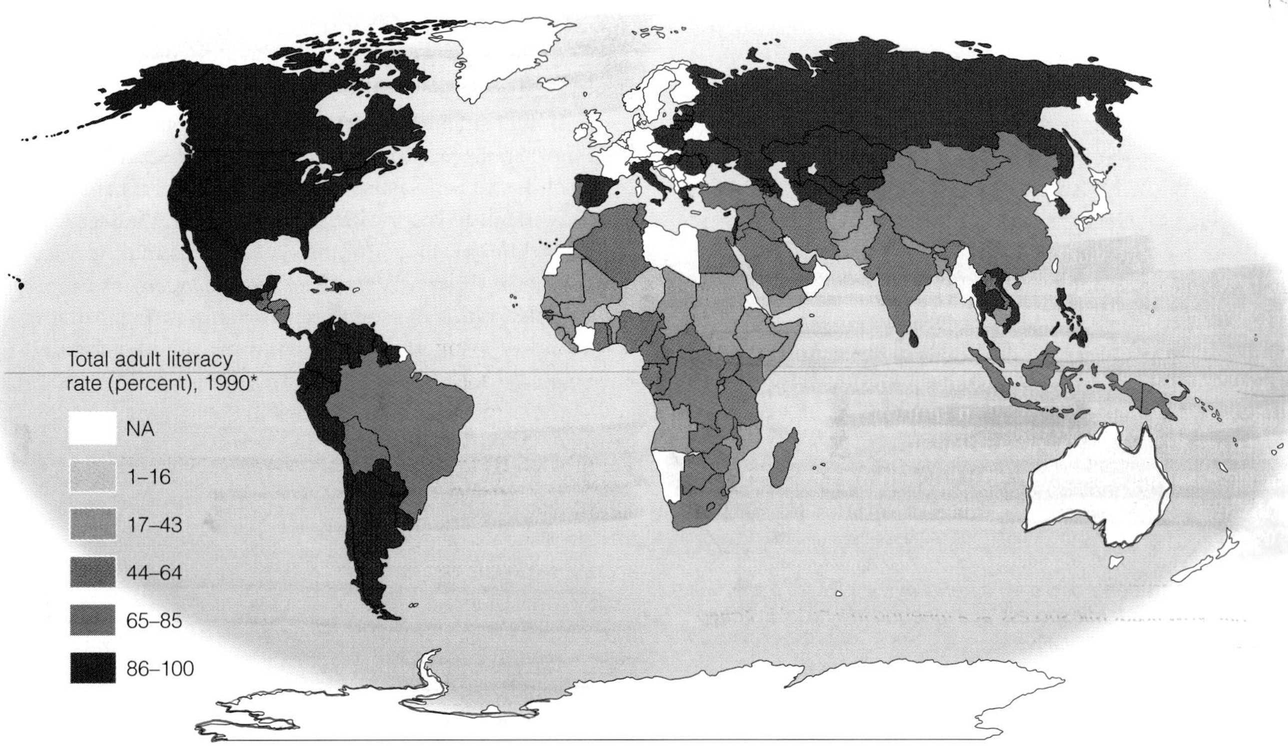

MAP 15.2 Viewing Society in Global Perspective: Literacy Around the World

*Percentage of persons aged fifteen and over who can read and write.
DATA: From UNICEF, 1996. State of the World's Children 1996. New York: Oxford University Press, 1996.

their children to weekend "cram" seminars called *jukos* as preparation for this examination, adding a grueling additional regimen to the already rather stiff requirements of the Japanese school system. Many Japanese, including educators, are becoming concerned that the extreme competitiveness of this system and the great burden of work being put on students are brutalizing the youngsters.

Overall, it appears that the educational system in the United States allows for a bit more social mobility than can be obtained in Germany, possibly England, and certainly Japan. The stratification systems in those countries and others with similar systems are more rigid, or castelike, than in the United States; however, there is danger in concluding that the U.S. educational system permits *much* more social mobility. In general, the similarities tend to be more important than the differences. Examinations filter out potential university students, while a university education is required to secure a high-status job. The class position of African Americans, Hispanics, American Indians, and many Asians in the United States, when compared to that of Whites, may be every bit as rigid as the distinction between the upper and lower classes in these other countries. This has led some, such as sociologist Charles Willie (1979, 1983), to argue that caste and class be recognized as separate lines of division in racial and ethnic analysis in the United States. The socioeconomic distinctions *within* racial and ethnic groups are *class* divisions; the socioeconomic distinctions *between* racial and ethnic groups are *caste* divisions.

Analyzing Education and Mobility

Functionalist and conflict sociologists both have something to say about the effects of education, or lack of it, on occupational and income mobility. Both agree that education can greatly affect the life paths of individuals by its ability to confer socialization, training, and credentials. They disagree about the degree to which these useful products of education are distributed according to merit. Functional theorists argue for a greater importance of merit, allowing less importance for the limits placed upon an individual by social class origins. In contrast, conflict theorists deemphasize, but do not eliminate, the role of individual merit, focusing instead on how the social class structure has an impact on the individual's occupational and income attainment.

Symbolic interaction theorists argue that forces such as teachers' expectations can have either beneficial or harmful effects on student achievement and mobility, depending on whether the teachers' expectations are favorable or not. Conflict theorists argue—with considerable data to support their contention—that although individuals are able to achieve some mobility through education, mobility outside a limited range is unlikely. Social mobility between the broad layers, or strata, of society cannot increase until the overall social, economic, gender, ethnic, and racial inequalities in society are reduced. However, as was noted in Chapter 9, the social class structure of U.S. society has become more polarized over time, not less.

Access to Higher Education

Almost all parents want their children to go to college; yet, not all will have equal access to a college education. The annual income of one's family of origin is a pretty good predictor of the likelihood of enrolling in a college, especially since a college education is usually an expensive proposition. Since family income is an indicator of social class, social class predicts, to a great degree, whether or not one will go to college full time. In general, the higher one's family income, the greater are one's chances for enrolling in college, and those of lesser income are more likely to enroll in a two-year than a four-year institution.

Race and ethnicity are still a major consideration in who has access to higher education, although the rate of minority enrollment in college has been increasing since the early 1970s, despite a dip in the late 1970s and early 1980s. In general, Whites are more likely to have graduated from college (21 percent of men and 20 percent of women) than are Blacks (10 percent of men and 12 percent of women) and Hispanics (8 percent of men and women). However, this includes the entire population. Are things changing in the current generation? Yes and no. Among those aged twenty-five to twenty-nine in 1997, 29 percent of Whites had a bachelor's degree or more, compared to 14 percent of Blacks and 11 percent of Latinos. So, although more racial minorities are completing college degrees, the gap between Whites and others remains (U.S. Bureau of the Census, 1997a).

Education and Inequality

In its original nineteenth-century conception, the educational system was to serve as a leveling force in American society—the road to full equality for all citizens regardless of race, social class origin, religion, or gender. Jew and gentile, Black and White, rich and poor, male and female would learn together side by side. Through education, each would learn the ways of others and thus come to understand and respect them. Full equality for humankind was to follow.

Education has indeed reduced many inequalities in society since the turn of the century. The percent of high school graduates has risen among Whites and minorities, both male and female, as have certain types of social mobility. Despite drops in college enrollments among African Americans and Hispanics in the early 1980s, there has been an overall rise in the enrollment of minorities of all kinds, and as more minorities and women attend and graduate from college, the overall result has been more minorities in midlevel and high-level jobs. Nonetheless, many inequalities still exist in U.S. education.

Cognitive Ability and Its Measurement

Since as long ago as classical Greece, humans have sought to measure a "mental faculty" or "intelligence," or as it is now called, **cognitive ability**—the capacity for abstract thinking. Since the turn of the century, educators in our society, from preschools to universities, have attempted to measure such qualities by means of **ability tests.**

Intelligence has been defined in a variety of ways. The notion of intelligence as some kind of unified mental capacity to think abstractly began in 1850 with Sir Francis Galton, the British biologist who first introduced the idea that the races differ in "innate" intelligence. In Galton's work, now discredited, biology ("nature") was declared to be more responsible for this racial difference than was social environment ("nurture"). Since Galton, definitions of intelligence have varied, but they generally involve a notion of potential, ability, or capacity as distinct from actual achievement. Presently, intelligence is variously defined as "the capacity for abstract thought," or "general cognitive ability," or even "the ability to adapt to the environment" (Gardner, 1993; Taylor, 1992).

A debate currently rages about whether intelligence is reducible to some single master capacity that humans possess—the **unidimensional theory of intelligence**—or whether intelligence is a matter of distinctly different kinds of abilities, or intelligences, such as verbal ability, quantitative ability, and creative ability. This latter idea is known as the **multidimensional theory of intelligence.** According to the unidimensional view, a person who is good at one kind of mental activity (such as math) would be expected to excel at other kinds (such as language skills or even music ability). The multidimensional view argues that cognitive gifts of one sort need not go along with other cognitive abilities; the abilities are distinct, with no common master ability underlying them all (Sternberg, 1988; Gardner, 1993). Although both theories have earned support in scientific studies, the unidimensional theory is presently losing ground while the multidimensional theory is gaining.

Among the most recent formulations of multidimensional intelligence, or multiple cognitive abilities, is that of Howard Gardner (1999, 1993). Gardner argues that there are seven entirely separate kinds of intelligence: (1) logical-mathematical (for example, skill at solving mathematical problems); (2) linguistic (for example, facility at learning new languages); (3) spatial (being able to visualize a three-dimensional figure from all angles); (4) musical; (5) interpersonal (for example, being a good politician); (6) intrapersonal (capacity for introspection); and (7) body-kinesthetic (athletic ability).

Note that athletic ability is classified in this scheme as a type of "intelligence." Body-kinesthetic talent includes the grace of a dancer, the speed of a sprinter, or the fine motor skills of a surgeon. Some are put off by the lumping together of physical talents with mental talents, but Gardner argues persuasively that command of the body is fundamentally under cognitive control, and therefore is a kind of intelligence.

The system of education in the United States has relied heavily upon the unidimensional view. Cognitive ability has been gauged according to the numerical results of **standardized tests,** tests given to large populations of people and

standardized tests

scored with respect to population averages. There has been a will to reduce measurements of cognitive ability to a single number, or perhaps two numbers like the language and math scores of SAT tests and IQ tests. IQ examinations have been widely employed since 1916, the year the first Stanford-Binet IQ test was given, although recent years have seen a drop in the use of IQ tests. As you probably know, IQ is an abbreviation for "intelligence quotient." It is the ratio of mental age (the age associated with a particular average level of cognitive performance) to chronological age (the actual age of the person being tested).

Standardized cognitive ability tests like the SAT and the ACT are presumed to measure ability much like IQ tests. They are distinct from **achievement tests,** which are intended to measure what has actually been learned, rather than measuring ability or potential. Advanced placement (AP) exams are achievement tests taken before entering college; students who score high demonstrate that they have already mastered certain material and can skip those courses in college.

There are three major criticisms of the use of standardized tests as measures of cognitive ability. First, they tend to measure only limited ranges of ability (such as quantitative or verbal aptitude) while ignoring other cognitive endowments such as creativity. That is, they tend to be unidimensional. Second, such tests possess degrees of *cultural bias* and also *gender bias*. They may thus perpetuate inequality between different cultural or racial groups, as well as perpetuate social, economic, and educational inequality between men and women. The tests were designed primarily by middle-class White males, and the "standardization" they strive to achieve mirrors middle-class White male populations. A large number of studies show that although standardized ability tests are somewhat capable of predicting future school performance for White males, they tend to give less accurate forecasts for the success of minorities, especially Hispanics, African Americans, and American Indians, and they also sometimes predict school performance less accurately for women than for men (Jensen, 1980; Crouse and Trusheim, 1988; Taylor, 1981, 1992; Pennock-Roman, 1994; Young, 1994). In other words, the **predictive validity** of the tests—the extent to which the tests accurately predict later college grades—is compromised for minorities and for women.

The third criticism of IQ tests and the SATs is that they do not predict school performance very well even for Whites. For example, SAT scores are only modestly accurate predicters of college grades for White persons (Young, 1994; Pennock-Roman, 1994; Manning and Jackson, 1984). This fact is not well known and is certainly not advertised by the Educational Testing Service (ETS), manufacturer of the SATs. Grade point average in high school is also only a modestly accurate predictor of success in college. High school grades are about as accurate as the SATs in predicting college grades—which is to say, not very accurate. Overall, the SATs predict college grades even less well for Blacks, Hispanics, American Indians, and in some cases, Asians than for Whites, and less well for females than for males.

"You're kidding! You count S.A.T.s?"

Ability and Diversity

Average scores on IQ tests and tests such as the SAT differ by racial–ethnic group, social class, and gender. Overall, Whites score higher on average than minorities; men score higher than women, especially on the math portion; and, in general, the higher a person's social class, the higher the test score. The differences between groups are regarded by experts as primarily environmental in origin, reflecting group differences in years of parental education, social class status, childhood socialization, language, nutrition, and cultural advantages received in the home and during youth. There is no evidence whatsoever that such *between-group* differences are in any way genetically inherited. Certain *within-group* differences may reflect genetic differences among individuals within the same racial or ethnic group, but even the within-group effect of genes is estimated to be much smaller than the within-group effect of social environment. That is, the effects of social environment are greater than the effect of genes (Taylor, 1992, 1980; Jang et al., 1996; Chipuer et al., 1990; Gould, 1981; Goldberger, 1979; Kamin, 1974).

In the United States, as in virtually every multiracial nation in the world, race and class are closely linked, and it is extremely difficult to determine whether an effect is caused by one or the other. Table 15.3 shows the relationship between SAT scores and social class (as indicated by income category). As family income rises, both verbal and math SAT scores go up. The differences are clear and consistent from one income category to another.

In Chapter 2, we pointed out that although Blacks, on average, score somewhat lower than Whites on standardized ability tests, northern Blacks score somewhat higher than Whites in some southern states. The variation reflects differences in both social class status and region. Comparing Whites and Blacks of the same social class status and the same region considerably lessens the difference in average test score. The differences between Blacks and Whites are further reduced when linguistic bias, a type of cultural bias, is removed by employing testing methods that use vocabulary and syntax familiar to the group being tested. This is especially true for Hispanics. English-speaking Hispanics for whom Spanish is a second language perform better on *both* math and verbal ability tests than English-speaking Hispanics for whom English is the second language (Durán, 1983; Pennock-Roman, 1990).

Gender mixes with class in the results of ability tests just as race mixes with class. In the vast majority of societies known to anthropologists and sociologists, including the United States, women have been forced to occupy lower social and economic status than men. Some of the female-to-male differences in standardized tests are attributable to this status ranking. The differences are not completely one-sided, however. Men have tended to score higher in numerical reasoning, spatial perception, and mechanical aptitude, but women have tended to score higher in perception of detail, memory, and certain verbal skills. These trends reflect differences in the childhood socialization of boys and girls as well as differences in societal expectations pertaining to men and women. The tradeoff is not an even one since society tends to assign more value to the abilities at which men excel, such as numerical reasoning.

Data from the College Board show that Whites consistently outperform all other ethnic groups on the verbal part of the SAT test. Since 1980, Asians have outperformed all other ethnic groups, including Whites, on the math part of the test.

Women are catching up to men on the math part of the SATs: The average math score for women has increased by 16 points since 1980, more so than the average score for men. The change, coinciding with a change in societal expectations, tends to discredit the traditional belief in our society that women on average have less mathematical ability than men. This belief has been used in the past to support the argument that women are less fit than men to perform high-level executive jobs that require number crunching and analytical reasoning.

Similar observations can be made with regard to Native Americans, Blacks, Puerto Ricans, and other Latin Americans: These groups have all increased in math SAT since 1987, and especially from 1980 through 1987. The same arguments used against women have been used against Blacks, Hispanics, and American Indians—that SAT scores supposedly indicate that these groups do not possess as much mathematical or verbal ability, on average, as Whites, and therefore are less able to perform high-level jobs (Herrnstein and Murray, 1994). This belief has tended to perpetuate both occupational and educational inequality between Whites and minorities in society (Jencks and Phillips, 1998).

The "Cognitive Elite" and *The Bell Curve* Debate

In the fall of 1994, a book entitled *The Bell Curve* was published and, since then, has caused a major stir among educators, lawmakers, teachers, public officials, policy makers, and the general public. In this book, which contains analyses of great masses of data, authors Herrnstein and Murray (1994) argue that not only does the distribution of intelligence in the general population closely approximate a bell-shaped curve (called the *normal distribution*), but that there is one basic, fundamental kind of intelligence, rather than several independent kinds of intelligences, that predict how well an individual will do in school and on the job, thus predicting how successful or not a person will be in society. The book thus subscribes to the unidimensional theory of intelligence rather than a multidimensional theory (Fischer et al., 1996).

Debunking Society's Myths

Myth: **Intelligence is mostly determined by genetic inheritance.**

Sociological perspective: **Intelligence is a complex concept, not easily measured by one thing and likely shaped as much by environmental factors as by genetic endowment.**

The book makes two points. The first is that intelligence is inherited; that is, it is determined primarily by one's genes rather than by one's social and educational environment. Herrnstein and Murray estimate that intelligence is about 70 percent genetically heritable—that intelligence has what is called 70 percent *heritability,* meaning that the sum total of all differences among people in intelligence in a general population is 70 percent determined by genes and only 30 percent determined by environment. How do they arrive at such a figure?

They arrived at it by reviewing studies of pairs of biological relatives and then estimating the degree to which any similarity in their intelligence could be due to their biological similarity. To do this, they reviewed data on identical twins who have been separated early in life for one reason or another, and then were raised apart. The idea is that since identical twins (as opposed to fraternal twins) are genetic *clones* (exact genetic duplicates) of each other, then any similarities that remain between them after their separation must necessarily be due to their identical genes rather than due to similarities in their social or educational environments. Separated identical twins are in fact about 90 percent similar in things like height, thus showing that genes must determine height to a great degree (90 percent), though not perfectly since slight differences (10 percent) remaining between the twins after their separation are because of their different environments. The authors argue that the similarity in intelligence between separated twins is about 70 percent.

Critics however point out that some of the identical twins in the studies cited by Herrnstein and Murray were actually separated more than others. That is, some were not really very separated at all, whereas some were separated for longer periods in their lives, and thus had fewer similarities in their social and educational environments. When this is taken into account, it is seen that those twins who were more separated (who attended different schools or were raised in different socioeconomic circumstances) were also less similar in intelligence. In general, *the more separated the twins were, the less similar they were in intelligence.* This shows the effect of their differing social environments over the effect of their identical genes. Some studies show that the similarity in intelligence among truly separated twins is only about 50 percent (Chipuer et al., 1990); other studies show it is as low as 30 to 40 percent (Taylor, 1998, 1992, 1980; Kamin, 1974; Jencks et al., 1972), results that are largely ignored by *The Bell Curve* authors. These latter studies suggest that the genetic heritability of intelligence is around 40 to 50 percent, not 70 percent, which would mean that intelligence is due somewhat more to social environment than to genes. The point continues to be hotly debated in the scientific literature.

The second point made by *The Bell Curve* authors is that since intelligence is primarily inherited, and since different social classes differ on the average in intelligence (with the lower classes having less intelligence), then it follows that the lower classes are on the average less endowed with genes for high intelligence, and the upper classes are relatively more endowed with genes for high intelligence. They thus reason that the upper- and upper-middle classes constitute a *genetically based* **cognitive elite** in America, consisting of those with high IQs, high incomes, and prestigious jobs. The authors strongly imply, but do not state outright, that any two groups presumed to differ in average intelligence, such as Blacks versus Whites or Latinos versus Whites, and any other such minority versus dominant comparison, may very well differ in genes for intelligence. They argue further that since men and women differ in certain kinds of intelligence (citing the presumed male superiority in math intelligence), men and women therefore differ in genes for this kind of intelligence, and women have less of it. According to Herrnstein and Murray then, the end result is an upper-class cognitive elite in the United States that contains a disproportionately high number of White men who are there because of their genetic superiority.

The problems with this cognitive-elite argument, largely ignored by the authors, are as follows:

1. Their conclusions tend to ignore the vast number of studies, some of which were discussed earlier, that show that intelligence tests and standardized ability tests are not as accurate a measure of intelligence of minorities as of Whites, of women as of men, and also of individuals of lower socioeconomic status as of individuals of higher status.

2. Their conclusions presume that intelligence is strongly genetically heritable, whereas there is convincing evidence, already noted, that, even for Whites alone, the relative contribution of environment may be greater than the relative contribution of genes, even though intelligence is probably the result of some combination of genes and environment.

3. They base a between-group conclusion on a within-group estimate of genetic heritability. Thus they base their women versus men, minority versus White, and lower class versus upper class conclusions on heritability results attained on *White men.* This is kind of like eating an orange to figure out what an apple tastes like, without eating an apple. The authors ignore a vast scientific literature detailing why one cannot draw conclusions about between-group genetic differences from within-group results (Fischer et al., 1996; Lewontin, 1996, 1970; Hauser et al., 1995; Kamin, 1995; Gould, 1994, 1981; Taylor, 1992, 1980).

Tracking and Labeling Effects

About 80 percent of America's secondary schools and about 60 percent of elementary schools currently use some kind of **tracking** (also called ability grouping), separating students according to some measure of cognitive ability (Taylor et al., 1997; Oakes and Lipton, 1996; Maldonado and Willie, 1996; Oakes, 1985). Tracking has been in place for more than seventy years. Starting as early as first grade, children are divided into high-track, middle-track, and lower-track groups. In high school, the high-track students take college preparatory courses in calculus and read Shakespeare. The middle-track students take courses in business administration and typing. The lower-track students take vocational courses in auto mechanics, metal shop, and cooking.

The basic idea behind tracking is that students will get a better education and be better prepared for life after high school if they are grouped early according to cognitive ability. Tracking is supposed to benefit the gifted, the slow learners, and everyone in the middle. Theoretically, students in all tracks learn faster since the curriculum is tailored to their level of ability and the teacher can concentrate on smaller, more homogenous groups.

The opposite argument is given by advocates of *detracking.* The detracking movement is based on the belief that mixing up students of varying cognitive abilities is more beneficial to students than tracking, especially by the time students get to junior high and high school. Students of high and low ability can thus learn from each other; the high-ability students are not seen to be "held back" by students with less ability, but are enriched by their presence. Finally, advocates of detracking point out that students in the lower tracks get less teacher attention and simply learn less; they are thus in effect penalized for being in the lower tracks. The idea is mix, don't match.

Which approach is better? Most researchers and educators who have studied tracking agree that not *all* students should be mixed together in the same classes. The differences between students can be too great, and their needs too dissimilar. Yet some degree of tracking has always had advocates based on its presumed benefits for all students. This presumption is under attack. One of the most consistent findings from research on tracking is that students in the higher tracks receive positive effects but that the lower-track students suffer negative effects (Rendon and Hope, 1996; Cardenas, 1996; Perez, 1996; Braddock, 1988; Oakes, 1990, 1985; Oakes and Lipton, 1996).

tracking

First of all, students in the lower tracks learn less because they are, quite simply, taught less. They are asked to read less and do less homework. High-track students are taught more; furthermore, they are consistently rewarded for their academic abilities by teachers and administrators. As a result, they find school to be more enjoyable, they have better attendance, and they have higher educational and occupational aspirations. In turn, these advantages increase their actual academic performance in the classroom and on exams. In contrast, students in the lower tracks are expected to do less well and, as a result, find school relatively less enjoyable, have lower rates of attendance, and lower educational and occupational aspirations; consequently, their actual in-class academic performance is less (Gamoran, 1992; Gamoran and Mare, 1989; Braddock, 1988). One study found that students assigned to low tracks in the eighth grade performed significantly less well in the tenth grade than students who had the *same* eighth-grade test scores and social class background, but went to untracked schools (Slavin, 1993).

THINKING SOCIOLOGICALLY

Were you in a tracked elementary school? What were the tracks? Describe them. Did you get the impression that teachers devoted different amounts of actual time to students in different tracks? Did teachers "look down" on those in the lower tracks? What about the students—did they treat some tracks as "better" or "worse" than others (were they perceived as differing in *prestige*)? Based on your recollections, what does this tell you about *tracking* and *social class*?

Both high- and low-track students are subject to **labeling effect:** Once a student is assigned to a particular track and is thereby labeled, the label has a tendency to stick, whether or not it is accurate. Once a student is labeled "gifted" or "high ability," other people—students, teachers, administrators—tend to react in accordance with that label. One is regarded as "smart" and high achieving. Students labeled "slow" or "low ability" encounter a negative reaction from the same people, including the expectation of low achievement. Even when a student is transferred from one track into another—for example, from a lower track to a higher one as the result of a recent cognitive ability test—the prior perceptions tend to persist; teachers and students still think of the youngster as "lower track," and even the recently promoted student may retain the self-perception developed in response to the prior track assignment.

Who gets assigned to which tracks? Research shows that track assignment is not solely on the basis of performance in cognitive ability tests. Social class and race are involved. Students with the *same* test scores often get assigned to different tracks because of differences in their social class and race. Few administrators or teachers consciously and deliberately assign students to tracks based on these criteria, but it occurs nevertheless. Researchers have consistently found that when following two students with identical scores on cognitive ability tests, the student of higher social class status is more likely to get assigned to the higher track than the student of lower social class status. Similarly, given a White student and an African American (or Hispanic or American Indian) student with identical scores, there is a somewhat greater chance that the White student will get assigned to a higher track than the minority student. Finally, combined effects of class and race are apparent: A student who is White *and* upper or middle class is more likely to end up in a higher track than one who is minority and also lower or working class. In each case, the students being compared had gotten the same math and verbal scores on a cognitive ability test.

A nationwide study of 14,000 eighth-grade students shows that overall, Asian and White students are more likely to be placed in the high-ability tracks than Hispanic, Black, and Native American students, even for those getting the same verbal and math ability test scores (Owens, 1998; Slavin, 1993; Braddock, 1988). Oakes (1985, 1990) notes that an Asian American is ten times more likely to end up in a college-prep course than a Hispanic student who has the same score as the Asian student on a standardized math test; a White student is seven times more likely to be placed in the college-prep group.

There is also a dramatic effect of socioeconomic status on track assignment. Students in the highest social class group are very likely to be assigned to the high-ability track (39 percent) or middle-ability track (39 percent, for a total of 78 percent in the high or middle tracks). Only 14 percent of students in the high social class group are assigned to the low-ability tracks. In sharp contrast, students in the low social class group end up disproportionately in the middle track (36 percent) or low track (37 percent, for a total of 73 percent in the middle or the low tracks). Only 13 percent of the low social class students are in the high-ability track.

Teacher Expectancy Effect

Similar to the labeling effect of tracking is the **teacher expectancy effect,** which is the effect of teacher expectations on a student's actual performance, *independent* of the student's actual ability. What the teacher *expects* students to do affects what they will do. The expectations a teacher has for a student's performance can dramatically influence how much the student learns. Teacher expectations can also affect how students define themselves.

Insights into the teacher expectancy effect come from symbolic interactionist theory. In a classic study that demonstrates this effect, Rosenthal and Jacobson (1968) told teachers of several grades in an elementary school that certain children in their class were academic "spurters" who would increase their performance that year. The rest of the students were called "nonspurters." The researchers selected the "spurters" list *completely at random,* unbeknown to the teachers. The distinction thus had no relation at all to an ability test the children took early in the school year although the teachers were told (falsely) that it did. At the end of the school year, it was found than although all students improved somewhat on the achievement test, those labeled spurters made greater gains than those designated

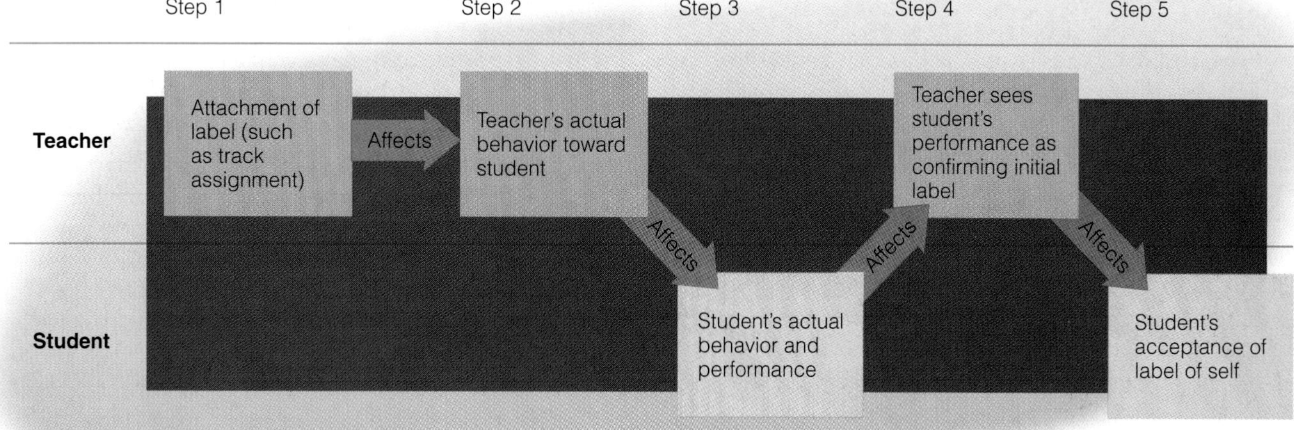

FIGURE 15.4 The Self-Fulfilling Prophecy

SOURCE: After Taylor, Peplau, and Sears, 1997, p. 232.

nonspurters, especially among first and second graders. This experiment isolates the effect of the label since it is the only difference that distinguishes a randomly selected group. Variations of this clever and revealing study have been conducted many times over, and the results are generally similar even though the original study by Rosenthal and Jacobson was criticized.

How are expectations converted into performance? By the powerful mechanism of the **self-fulfilling prophecy,** in which merely applying a label has the effect of justifying the label (Cardenas, 1996; Darley and Fazio, 1980). Recall the quote from early sociologist W. I. Thomas in Chapter 8: "If men [sic] define situations as real, they are real in their consequences" (Thomas, 1928: 572). If a student is defined as a certain type, the student becomes that type. The process unfolds in stages. First, a teacher is told that a student merits a label such as spurter; perhaps the designation originates with administrators or comes from the scoring key of a standardized exam. The teacher's perception of the student is then colored by the label; a student labeled a spurter may be coaxed and praised more often than nonspurters. The student then reacts to the teacher behavior; students expected to perform well and encouraged to excel actually perform better in class and on exams than other students. Finally, the original prophecy fulfills itself: The teacher observes the behavior of the student, notes the increase in performance, and concludes that the designation spurter is affirmed since the so-called spurters actually perform better by objective measures than the nonspurters. Further praise and encouragement follows. Teachers unaware of the overall effect will not realize that the label itself produced the greater performance of the spurters. The self-fulfilling prophecy is diagramed in Figure 15.4.

These findings have had a profound impact on the American educational system and the practice of tracking. Self-fulfilling prophecy was revealed to not only inspire students labeled "performers," but also to retard the progress of students labeled "slow." Students on lower tracks met with lessened expectations, less coaxing, and lowered standards of performance required by teachers. Ongoing research affirms that teacher expectancies affect how teachers behave toward students. When teachers have high expectations for students, they call on them more, praise them more often, and interact with them more frequently; consequently, pressures for change in the educational system, such as detracking, have arisen from this research (Rendon and Hope, 1996; Cardenas, 1996; Hallinan, 1994; Gamoran, 1972).

White students and Asian students appear disproportionately in the high tracks and college-prep courses, and thus become the recipients of high teacher expectations and this, in turn, has a favorable effect on their performance. This cycle also reinforces teacher stereotypes of the Asian student as "natively gifted" in mathematics. Interestingly, this in itself can have negative consequences for Asians since it promotes underestimation of the extent to which Asians are discriminated against outside the classroom (Takaki, 1992). Black, Hispanic, and Native American students, and lower social class students of all races, are more likely to be assigned to lower tracks and receive lower teacher expectations. As a result, they show lower objective performance, learn less, like school less, and like themselves less as they develop a negative self-evaluation (Rendon and Hope, 1996; Perez, 1996; Persell, 1977).

Schooling and Gender

Teachers hold different expectations about girls and boys in school, and it therefore comes as no surprise that gender affects teacher expectations and the actual performance of these children. Tracking, particularly in high school, is significantly dependent upon gender; for example, there is a far greater proportion of young men in the science and math tracks than young women. Throughout schooling, from

Women outnumber men in this high school science class, which is the exception rather than the rule, according to the most recent studies.

preschool through graduate or professional school, women and men are treated differently, with women discriminated against. This consistent and long-term differential treatment has had profound consequences. Girls and boys start out in school roughly equal in skills and confidence, but by the end of high school, women trail men, especially in mathematics and the sciences. What happens in between has been recently documented in a comprehensive report commissioned by the American Association of University Women (AAUW) that summarizes the results of more than 1000 publications and studies. This extensive research shows the following with regard to schooling and gender (American Association of University Women, 1992; Marlowe and Company, 1995; Sadker and Sadker, 1994):

1. In general, teachers pay less attention to girls and women. In elementary school as well as high school, teachers direct more interaction to boys than to girls. As a result, boys tend to talk more in class, and thus teachers interact with them more. The difference is particularly notable in math and science classes.

2. Throughout all grades, women lag behind in math ability and achievement scores, as well as in science achievement scores. Consequently, men are far more likely to be assigned to tracks that stress math and science. This is sometimes true even for men and women who have the same test scores in math and science, showing that students may be assigned to tracks based not only on test scores but also independently on gender. The consequences can be long-lasting. Even women who do well in these subjects are less likely than men to make their careers in math or science.

3. Some standardized math and science tests still retain gender bias, despite ten years or more of effort to weed out bias on the part of education specialists and testing organizations such as the Educational Testing Service, manufacturer of the SATs. Bias is especially prevalent in mathematical word problems. Certain problems are gender-typed in the sense that they employ words and concepts more familiar to men than women. For example, an SAT word-math question built on the concept of volume may ask the student to calculate the volume of oil in an automobile crankcase. The same question could be asked using any household article, such as a fishbowl, as the sample vessel. The question would be the same, but the testing results would likely be a bit different because women are less likely to be able to visualize a car crankcase and will therefore be less comfortable with the question. The phrasing of the question is gender-typed. Women tend to score higher on word-math questions of this type when gender typing is neutralized, even though the revised question may be conceptually identical. It has been shown that subtle bias of this sort is a fixture on the actual SAT exams (Chipman, 1991; American Association of University Women, 1992).

4. Standardized tests in math tend to underpredict women's actual grades in mathematics. Women tend to do somewhat better in math courses than their test scores would predict. For this reason, it has been suggested that perhaps grades should be used rather than test scores when determining math scholarships and college admissions for women. In any case, it has been consistently shown that high school grades are about as accurate as SAT scores in predicting college grades for both men and women, with neither predicting very well, as we noted earlier.

5. Teachers tend to treat Black women and White women differently. This is particularly true of Black and White girls during the elementary school years. Teachers tend to rebuff Black girls and interact more with White girls. The trend appears to be independent of the teacher's own race.

6. Textbooks still tend to either ignore women or stereotype them. In this respect, textbooks are gender-role socializers. In elementary school texts, male characters greatly outnumber female characters. Boys are portrayed as building things, being clever, and leading others. Girls tend to be shown performing dull tasks and following boys. Men are routinely presented as doctors, lawyers, and businesspeople; women as homemakers, librarians, and nurses. There is evidence that such gender typing is decreasing, but the decrease began only very recently, and it has not been eliminated.

7. As girls and boys approach adolescence, their self-esteem tends to drop, with the erosion of self-esteem occurring more quickly among girls than boys. This trend has been noticed in many studies of the social psychology of young men and women. The trend is further exacerbated by the discrimination against women in the classroom, gender bias in standardized tests, and stereotyped presentations of women in presumably authoritative textbooks. The AAUW report concludes: "Students sit in classrooms that, day in and day out, deliver the message that women's lives count for less than men's" (American Association of University Women, 1992).

Stereotype Vulnerability, Race, and Gender

As has already been noted in Chapters 9 and 11, racial and gender stereotypes can affect actual behavior. A White male observer of a Hispanic person, or of a woman, may carry with him certain negative stereotypes about Hispanics and women, and this can affect how he acts toward any specific Hispanic or woman. This, in turn, can affect the actual behavior of the observed Hispanic person or woman.

To what extent can a negative stereotype one may have about *one's self* affect one's *own* actual behavior and academic performance? As with the self-fulfilling prophecy, to what extent do minorities and women internalize negative stereotypes about themselves, and thus show such effects via their behavior and academic performance?

An answer has recently been provided by the research of Claude Steele and associates (Steele, 1996; Steele and Aronson, 1995; Steele, 1992). Steele notes that two common stereotypes exist in the United States: That since on the average Blacks perform less well on tests of math and verbal ability than Whites, Blacks must have, or so it is believed, some inherent deficiency in math and verbal abilities relative to Whites. Second, that since women perform less well than men on tests of math ability, women must therefore have some inherent deficiency in math ability. The stereotype lies in the belief that such deficiencies are somehow inherent in and essential to being Black or female. "Ability" is thought of as being inherent, even genetic, and thus not modifiable or changeable by teachers, education programs, or other such elements of the social environment. As we have already noted, the publication of books such as *The Bell Curve* (Herrnstein and Murray, 1994) supports these stereotypes.

To the extent that Black students in high school or college may actually believe (internalize) such stereotypes, to that extent they may perform less well on a test if they are actually told that "this is a genuine test of your true ability." This can *activate* the stereotype in the mind of the person so informed, and thus increase test anxiety, with the result of lowered test performance. White students who are also told this would be less likely to have the stereotype activated (since the stereotype is not *about* Whites) and thus be less likely to have their test performance lowered; they would be less vulnerable to the stereotype. If the Black students are not told this, their test performance would be less likely to be lowered; the stereotype is not activated, and thus they are less vulnerable to it.

Results show that this is just what happens. Figure 15.5 shows the results for a test of verbal ability based on the GRE (Graduate Record Examination, a test similar to the SAT for college students who contemplate graduate education). Black college students who are simply told that the test is a "genuine" test of their true verbal ability (the *diagnostic condition*) perform less well than Whites who are also told this; Blacks in this diagnostic condition averaged only 4.4 test items correct (out of a possible 20), whereas Whites in the diagnostic condition averaged 10.2 correct. This is the *stereotype vulnerability effect.* Note that nothing is actually said to the students about Black and White test performance specifically, only that the test was designed to be a "genuine" test of verbal ability.

BOX 15.1 DOING SOCIOLOGICAL RESEARCH

The Agony of Education

WHAT is the experience of African American college students attending school on predominantly White campuses? This question is, according to sociologists Joe Feagin, Hernán Vera, and Imani Nikitah, clouded with misunderstanding.

Because of the many misconceptions that define Black students' experiences, Feagin, Hernán, and Nikitah wanted to get firsthand information about the experiences of Black students on White campuses. To do so, they used randomly selected focus groups, a method whereby several people participate in a collective interview conducted by the researchers. Instead of just answering questionnaires or answering interview questions singly, participants in focus groups are able to interact with each other, sometimes discussing particular issues at length and responding to each other's experiences. This method can bring more nuance and subtlety to the data being collected. This team of researchers used focus groups of juniors and seniors at a major state university, supplementing the student data with several focus groups of Black parents whose children were attending college or considering application. The moderator for all the groups was also African American.

The resulting book, *The Agony of Education,* provides a rich analysis of the experience of Black students on White campuses. The students and their parents describe the importance of education within the Black community, but also report at length about "the Whiteness of university settings." The students describe being treated as intruders on campus, even while White students deny that racism exists. The students describe in poignant detail the stereotyping and discrimination they experience, but also some of the positive changes that create a more welcoming environment. Throughout the book, Feagin, Hernán, and Nikitah use their sociological perspective to understand the students' and parents' experiences and to make recommendations for change.

SOURCE: Feagin, Joe, Vera Hernán, and Imani Nikitah. 1996. *The Agony of Education.* New York: Routledge.

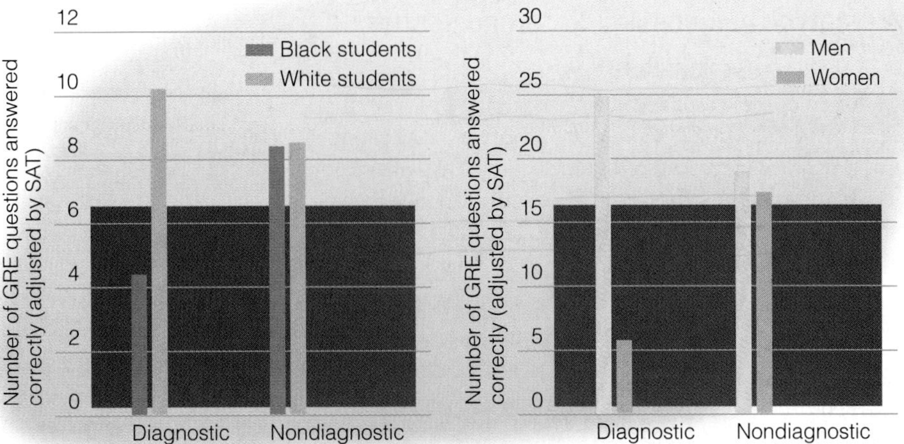

FIGURE 15.5 Stereotype Vulnerability, Race, and Test Performance

SOURCE: After Steele and Aronson, 1995, p. 416.

What happens when the test taker—whether Black or White—is told nothing (the *nondiagnostic condition*)? As Figure 15.5 shows, this condition reveals *no difference at all* between Blacks and Whites in test performance. Both got an average of 8.5 items correct out of 20. These results cannot be attributed to differences in ability since the effects of any SAT-score differences have been subtracted out. Furthermore, the study was done on college students who all had relatively high SAT scores to begin with. The differences cannot be attributed to unmeasured or unknown differences between the diagnostic and the nondiagnostic conditions since the Black students as well as the White students were randomly assigned beforehand to either the diagnostic or the nondiagnostic conditions.

Stereotype vulnerability may very well operate in the same way with regard to the presumed female–male difference in math ability test performance. As Figure 15.5 shows, when a group of both women and men are told that the math test being given to them is a "genuine" test of their true math ability (the diagnostic condition), the women do much worse than the men (an average of 25.7 versus 5.8 correct out of 30 questions—a large difference). If a group of women and men are told nothing (the nondiagnostic condition), the women and men perform about the same (18.9 versus 17.3—not a statistically reliable difference). These results suggest that stereotype vulnerability, as reflected on standardized test performance, might operate for a woman–man comparison in a similar manner as with a Black–White comparison. It suggests that at least part of the long-believed female deficit in math ability may be due simply to what they are told before they take the test, and not to "inherent" differences between women and men in math ability. These results flatly contradict the conclusions reached in *The Bell Curve* (Herrnstein and Murray, 1994).

THINKING SOCIOLOGICALLY

What are some of the stereotypes of different racial–ethnic, gender, and class groups in your school or college? How does the concept of *stereotype vulnerability* explain the educational experiences of each group? What evidence of the *expectancy effect* do you see in thinking about the success of different groups?

School Reform

School reform is an ongoing theme in the history of education in the United States. Currently, new challenges face the institution of education. Increasing diversity in the population, economic competition with other nations, the inequalities plaguing the schools, and fiscal constraints also pose challenges for those who want a strong system of education in the nation. How should the nation respond to these challenges? Many solutions are debated, and new methods of organizing and delivering education are being developed. Here we examine some of these changes and challenges.

Reducing Unequal Funding

One of the most persistent issues for reform in education is the problem of unequal funding for different school districts within the same city, urban versus suburban schools, public versus private schools, and even greatly differing funding for education across different states within the United States. Some states such as New Jersey, Alaska, California, and New York spend considerably more per pupil on education than do other states, for example, Alabama, Arkansas, or Idaho (Alsalam et al., 1992; U.S. Bureau of the Census, 1997a). As seen in Figure 15.6, entire countries differ greatly in the amount of funding for public education.

Major differences in funding often occur between school districts within the same state. School districts are funded primarily through property taxes, and since wealthier school districts tend to contain more expensive real estate, such districts receive more funding. This constitutes an inequity among the school districts. Quite a bit in public education depends upon amount of funding, such as textbooks, the availability of computers, smaller class sizes, laboratory equipment, and teacher salaries. Gross inequities in funding translate into inequalities in such necessities. As a consequence, the quality of education, to the extent that such is determined through funding, varies widely from district to district. A further complication is that the dropout rate for urban school districts is considerably greater than

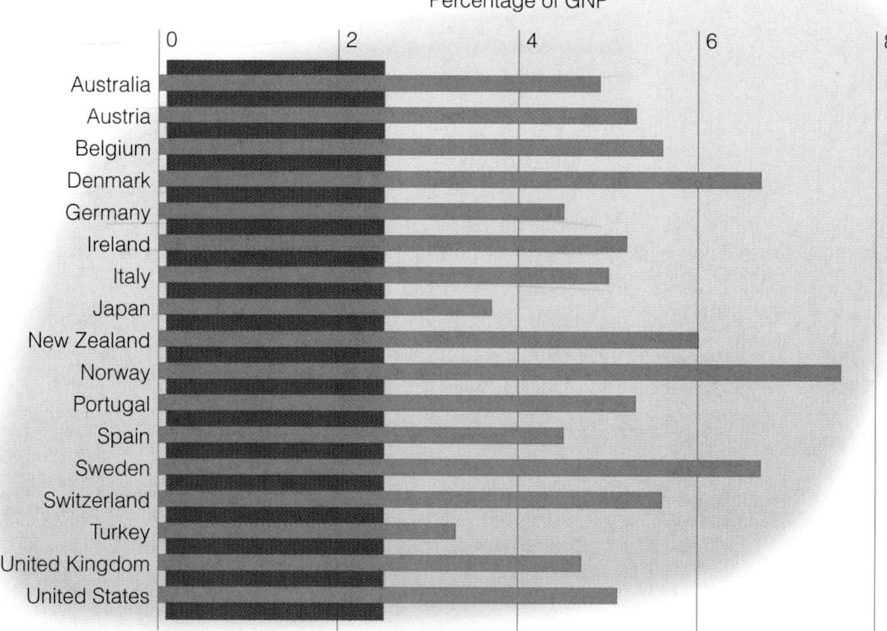

FIGURE 15.6 National Comparisons of Educational Expenditures

SOURCE: National Center for Educational Statistics. 1997. *Digest of Educational Statistics.* Washington, DC: U.S. Department of Education, p. 436.

that for both suburban and rural districts. Lack of funding and the consequent lack of facilities in inner-city urban schools only exacerbates the problem of dropping out of school.

A major problem throughout almost all states is that the electorate has become more and more reluctant to pay for reforms that would result in an increase in school taxes. Some of this may well be due to economic downturns within the past several years. One consequence has been individual refusal to pay increased property taxes for education; the result has been less money available for the improvement of educational facilities. This has caused some states (such as Michigan) to eliminate property taxes altogether as a basis for the funding of education (Walsh, 1993). It was argued that since property taxes must necessarily result in more funding for the richer school districts, and thus less funding for the poorer districts (which need it the most), the process is unconstitutional, and property taxes as the source for funding for education should be eliminated. In place of the property tax, Michigan voters elected to increase the state income tax in order to fund education. It may come to pass that other states will adopt this mode of educational reform.

Back-to-Basics and Multiculturalism

During the 1980s and 1990s, two educational reform movements have become prominent. One stresses a return to a traditional curriculum delivered with traditional methods. The other movement stresses multiculturalism. The two movements are not in opposition but can be complementary (Gates, 1993).

The **back-to-basics movement** emerged from the dissatisfaction that professional educators felt about declining student discipline, rising functional illiteracy, and teacher incompetence in elementary and high school, all detailed in a widely read 1983 report called *A Nation at Risk,* released by the American Commission for Excellence in Education. In that report, the blame was placed on the education system and the curriculum, not upon the students. Many schools have therefore been led to place greater emphasis upon the three R's: reading, writing, and arithmetic. Freedom to choose from a variety of elective courses has been somewhat reduced. Discipline has been increased in some elementary, junior high, and high schools. There is also a push for stiffer standards in grading. This has carried over to colleges and universities, where "grade inflation" is under attack. Grade inflation occurs when an excessive number of high grades are given, or when the average grade for a course edges up from the unadmired C level toward the B level once reserved for above-average performance. The term *grade inflation* is something of a misnomer since the net

Because funding for public schools is dependent on taxes, support for schools varies depending on the tax base of different communities.

The multicultural classroom is one place where youth can learn about cultures other than their own.

Computers have become more common both in school and in middle-class homes. Here, two youths complete a homework assignment on a computer.

result is a devaluing, or deflation, of the value of A's and B's (the syndrome is actually called grade deflation in a number of colleges).

Many elementary schools have discontinued the practice of "social promotion" (passing students from one grade to the next regardless of their performance or grades) and have returned to the old practice of requiring students to repeat a year if their performance was unsatisfactory. Finally, many states now require teachers themselves to attain a minimum score on a standardized test called the National Teacher Examination (NTE), another product of the Educational Testing Service (ETS) of Princeton, New Jersey.

The **multiculturalism movement** is more recent than the back-to-basics movement. Its main goal is to introduce more courses and educational materials on different cultures, subcultures, and social groups into school curricula. Thus, programs such as African American Studies, Hispanic or Latino Studies, Jewish Studies, Caribbean Studies, Women's Studies, and Gay and Lesbian Studies are making their way into elementary, high school, or college curricula. The driving principle behind this movement is the belief that traditional curricula tend to stereotype women, minorities, lesbians and gays, and working-class persons, thereby giving an inaccurate picture of these groups and of society.

We just saw how women are stereotyped in textbooks and other parts of the curriculum from preschool onward. African Americans and Hispanics have been routinely stereotyped in similar ways. Several recent studies have shown that elementary school textbooks still often present negative stereotypes of Blacks. Gone are the old representations of Blacks as shuffling, head-scratching, grinning characters with exaggerated lips and nostril size, yet Blacks still appear in elementary school texts as innately musical and working in low-status occupations, such as elevator operator or shoe shiner. Hispanics are often presented as lazy and unemployable. American Indians are stereotyped as mystics and savages. Although Asians are now often positively stereotyped as America's "model minority," they are also negatively stereotyped at times as sneaky and untrustworthy. Gays and lesbians are stereotyped in the sense that few lesbians or gays even appear in elementary school and high school texts. Courses in Hispanic or Latino Studies, Native American Studies, Asian Studies, Women's Studies, and very recently, Gay and Lesbian Studies on college campuses are attempting to convey the rich cultural traditions and the scholarship in all these areas.

African American Studies and Ethnic Studies programs have had to fight vigorously for existence on college campuses ever since their beginnings in the late 1960s. Along the way, many such programs have been dissolved, but others have thrived, establishing these areas as legitimate fields of academic inquiry. Women's Studies programs have similarly battled and won, and have now established their institutional legitimacy on many college campuses. There are more than 500 Women's Studies programs across the country. More recently, Latino Studies and Gay and Lesbian Studies have entered the college curricula. The multiculturalism movement has followed on this success with initiatives to make courses in these programs standard requirements for graduation at many schools.

The Future of American Education: New Technology in the Classroom

Among the tasks of educational institutions in the United States is preparation of the young for a rapidly changing world. Technological changes are taking place, and education is both cause and effect in the process. Education must therefore look toward the future and newly emerging technology and social structure, rather than rely heavily on past curric-

multicultural movement

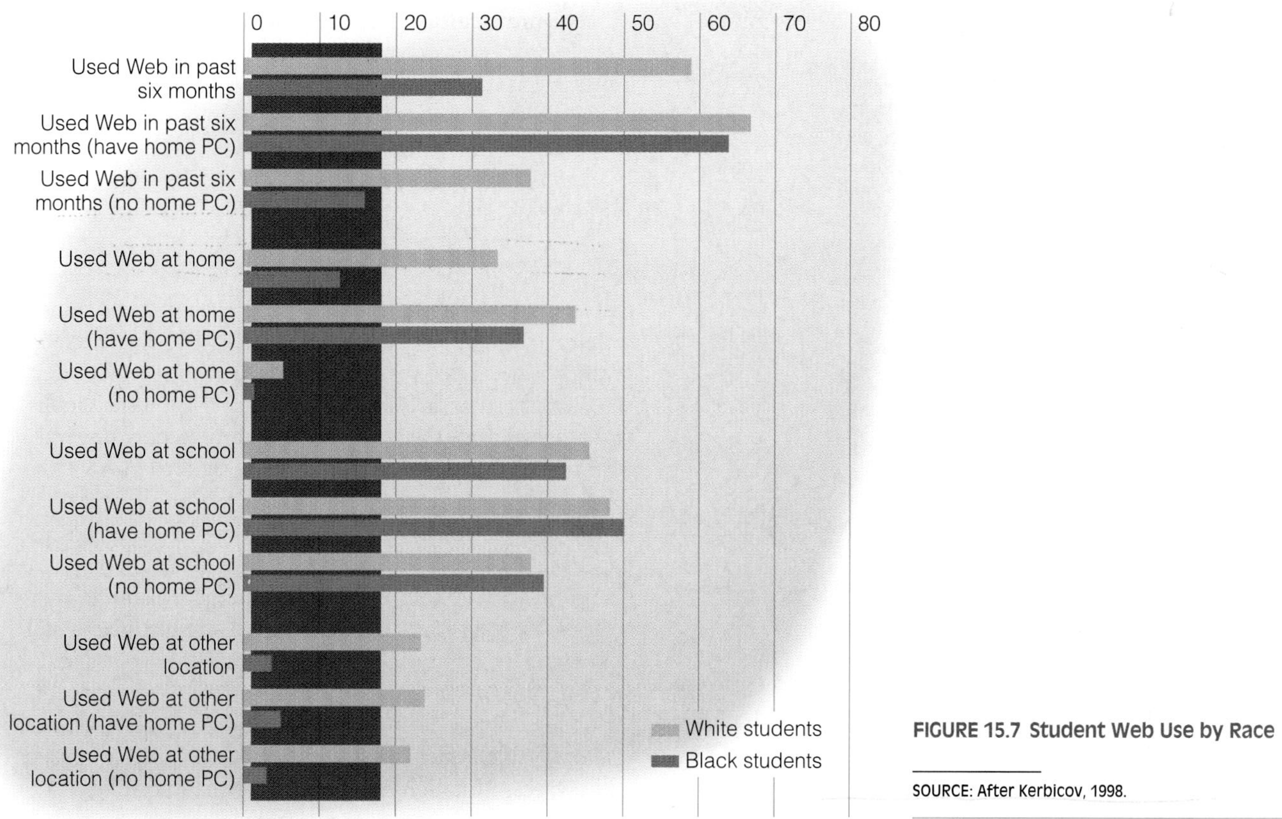

FIGURE 15.7 Student Web Use by Race

SOURCE: After Kerbicov, 1998.

ula and programs. Within this context, curriculum revision and the multiculturalism debate, as well as the role of women and minorities in new education systems, emerge as major issues.

Current projections are that by the year 2005, approximately seventy million students will be enrolled in public and private schools (U.S. Bureau of the Census, 1997a). These demographic trends indicate that the proportions of Blacks and Hispanics in the population will increase substantially, and thus place pressure on schools to adapt to the needs of a more racially and ethnically diverse population. As a result, there will be greatly increased needs for multicultural curricula and more minority teachers. In sum, the educational system needs to adapt to the changing social, ethnic, and gender diversity of the U.S. population, as well as changing views on the part of these populations.

By way of technological change, the increased use of computers in schools will have continuing consequences for both curriculum and extracurricular activities. In the same way that the Industrial Revolution of the nineteenth century had a dramatic impact on education, so education is now being changed by the advent of the computer and the Information Revolution. Since computers are used very widely throughout the occupational spectrum, their use in schools has greatly increased and will in all likelihood continue to increase. Current computer use is related to race and class: 62 percent of White students use computers in schools compared to 52 percent of Black and Hispanic students. By class, students in high-income families are more likely to use computers in school (two-thirds of those in homes with family incomes of $75,000 or more, and under 60 percent for those with family incomes of less (U.S. Bureau of the Census, 1997a). As Figure 15.7 shows, student use of the Internet is related to race: White students are more likely to have used the Internet in the prior six months, use the Internet on a home PC, and also use the Internet at other locations (Hoffman and Novak, 1998). Differences that once existed by gender have virtually disappeared although men are more likely than women to use computers in college and beyond (U.S. Bureau of the Census, 1997a).

CHAPTER SUMMARY

- Education is the social institution that is concerned with the formal transmission of society's knowledge. It is therefore part of the socialization process. More money is spent in the United States on education than on any other activity except health care.
- Although the percentage of U.S. high school and college graduates has increased dramatically since World War II, the increase has been unequal across racial and ethnic groups. Blacks lag behind Whites, and Hispanics lag behind Blacks in both high school and four-year college graduation rates.
- The number of years of formal education that individuals have has important, but in many ways modest, effects on their ultimate occupation and income. The effect of education upon later occupation is greater than the effect of education upon later income. Social class origin affects the extent of educational attainment (the higher the social class origins, the more education ultimately attained) as well as upon occupation and income (higher social class origin means, probabilistically, both a more prestigious occupation and more income). There is evidence that the social class one is born into has a greater effect on later occupation and income than does educational attainment. This limits social mobility in the United States, but even more so in countries such as Japan and Germany,
- Although the education system in the United States has traditionally been a major means for reducing racial, gender, and class inequalities among people, it is also true that the educational institution has perpetuated these inequalities. Test biases based on culture, language, race, gender, and class have not been substantially reduced in *standardized tests.* The *predictive validity* of ability tests is in many instances less for minorities, women, and working-class persons. Despite scientific evidence to the contrary, books such as *The Bell Curve* argue for the presence of a *cognitive-elite class* based on inherited intelligence.
- *Tracking* and *labeling* continue to affect minorities, women, and the working classes disproportionately. There is considerable research evidence that upper-track students (who are disproportionately White, middle class, and male) benefit more and learn a lot more than lower-track students (who are still disproportionately minority, working class, and female). Women are especially underrepresented in high school science tracks and classes. Students are subject to the *labeling effect,* which affects both teacher expectations and student performance. *Teacher expectancy effects* and the *self-fulfilling prophecy* work to the detriment of minorities, working-class persons, and women.
- Women are consistently and routinely short-changed by the entire education system by relative lack of interaction from teachers, gender bias on standardized tests, lack of predictive validity in science and math on the standardized tests, and negative gender stereotyping in textbooks and readers, especially during the elementary school years.
- There is evidence that *stereotype vulnerability* may affect the performance of Blacks as well as women on standardized ability tests. Blacks and women who are told before taking such a test that the test is a "genuine" measure of true ability perform worse than White males who are told the same thing, and both Blacks and women perform better when they are told no such thing.
- Two recent movements have arisen in an attempt to reform the educational institution. The *back-to-basics movement* stresses a return to the three R's and stricter discipline, stiffer grading standards, and combating grade inflation. The *multiculturalism movement* seeks to reform the curriculum not by removing the basics but by adding courses, in elementary school through college, in African American and Black Studies, Hispanic or Latino Studies, Native American Studies, Women's Studies, and Gay and Lesbian Studies.
- Technological change has resulted in a greatly increased presence of the computer in the classroom, and use of the Internet has greatly increased, though more so for Whites than for racial minorities.

KEY TERMS

ability test
achievement test
back-to-basics movement
cognitive ability
cognitive elite
credentialism
educational deflation
labeling effect
multiculturalism movement
multidimensional theory of intelligence
predictive validity
schooling
self-fulfilling prophecy
standardized test
teacher expectancy effect
tracking
unidimensional theory of intelligence

THE INTERNET: A Tool for the Sociological Imagination

Resources on the Internet:

Virtual Society: The Wadsworth Sociology Resource Center
http://sociology.wadsworth.com

Visit this site to find additional learning tools, including interactive quizzes, links related to web sites, and an easy link to *InfoTrac College Edition.*

American Association of University Women
http://www.aauw.org

This organization, which promotes equity for women and education, regularly produces reports about gender and education. Their reports are often the basis for changes in social policy.

National Center for Educational Statistics
http://www.ed.gov/NCES/index.html

This division of the U.S. government collects and publishes reports on the condition of education, including students, faculty and teaching staff, school expenditures, and other useful compilations of data.

U.S. Bureau of the Census
http://www.census.gov

From the U.S. Census Bureau home page, one can find data on educational attainment, including detailed tables showing such things as educational attainment and race, sex, age, income, and family status.

Sociology and Social Policy: Internet Exercises

Imagine that you are serving on your local school board. The board would like to increase graduation rates among groups that have been dropping out of school at a high rate. Based on what you have learned from this chapter, what policies would you propose to encourage at-risk groups to stay in school?

Internet Search Keywords:

dropout
dropout prevention
high school completion rate

Web sites:

http://www.dropoutprevention.org/
National Dropout Prevention Center functions as a clearinghouse and research center for dropout prevention information.

http://nces.ed.gov/pubs98/dropout/
National Center for Education Statistics reports the 1996 high school dropout statistics in the United States.

http://teotihuacan.educ.indiana.edu/isca/dropout.html
A list of related web sites by the Indiana School Counselor Association, including National Dropout Prevention Center, Urban Educator, and so on.

http://www.ed.gov/pubs/EPTW/eptw3/
National Diffusion Network publishes a list of programs effective on dropout prevention.

http://www2.state.ga.us/Departments/AUDIT/index/dept/per/dropout.htm
Report on dropout prevention efforts by the Georgia Department of Education.

InfoTrac College Edition: Search Word Summary

multicultural movement
one–room schoolhouse
standardized tests
tracking

In order to learn more about these central topics in sociology, you can conduct an electronic search using InfoTrac College Edition. To aid in your search and to gain useful tips, see the Student Guide to InfoTrac College Edition on the Virtual Society web site:
http://sociology.wadsworth.com

INTERACTIONS—A SOCIOLOGY CD-ROM: CONCEPTS FOR THIS CHAPTER

Go to the Wadsworth Sociology CD-ROM for further study on the concepts in this chapter. The CD-ROM also includes quizzes and additional activities to expand your learning experience.

SUGGESTED READINGS

Cookson, Peter W., Jr., and Caroline Hodges Persell. 1985. *Preparing for Power: America's Elite Boarding Schools.* New York: Basic Books.

This is an in-depth study of how elite boarding schools prepare students for success and power in society. The role of the students' social class of origin is examined, as are the ways in which educational institutions protect the upper classes from downward mobility.

Crouse, James, and Dale Trusheim. 1988. *The Case Against the SAT.* Chicago: University of Chicago Press.

Based on hundreds of studies, this book concludes that the SAT is unnecessary and harms students from non-White and low-income backgrounds. The book recommends dropping the SAT and relying on courses taken, and grades, until more appropriate tests can be developed.

Feagin, Joe, Vera Hernan, and Imani Nikitah. 1996. *The Agony of Education.* New York: Routledge.

This is an engaging account about the experience of African American college students attending predominantly White institutions. The poignantly told experiences challenge stereotypical notions about Black experience on White campuses and suggest pathways for positive change.

Fischer, Claude S., Michael Hout, Martín Sánchez Jankowski, Samuel R. Lucas, Ann Swidler, and Kim Voss. 1996. *Inequality by Design: Cracking the Bell Curve Myth.* Princeton, NJ: Princeton University Press.

This is a clearly written and comprehensive critique of Herrnstein and Murray's (1994) *The Bell Curve.* The authors give attention to definitions and measurements of intelligence, problems with the data in the studies used, race and intelligence, types of inequality in society, and other conceptual and methodological issues important to understanding the public debate about *The Bell Curve.*

Rosser, Phyllis. 1992. *Sex Bias in College Admission Tests.* Cambridge, MA: Fair Test, Inc.

This study outlines the kinds of gender bias that can appear in both the SAT and the ACT, as well as the effects of this bias on women's access to colleges and scholarships.

Sadker, Myra, and David Sadker. 1994. *Failing at Fairness: How America's Schools Cheat Girls.* New York: Charles Scribner's Sons.

This is a comprehensive and engaging discussion of how school practices disadvantage girls. Sadker and Sadker based their conclusions on their systematic observations in schools, as well as a thorough review of the research studies on this subject.

Takagi, Dana. 1992. *The Retreat from Race: Asian-American Admissions and Racial Politics.* New Brunswick: Rutgers University Press.

In this penetrating analysis of policy and bias in Asian admissions to elite colleges and universities, Takagi details the changes in the debate over affirmative action policy, showing a shift away from racial preferences and toward class preferences. The book presages the University of California's 1996 Board of Regent's voting down affirmative action plans based on race, but not social class.

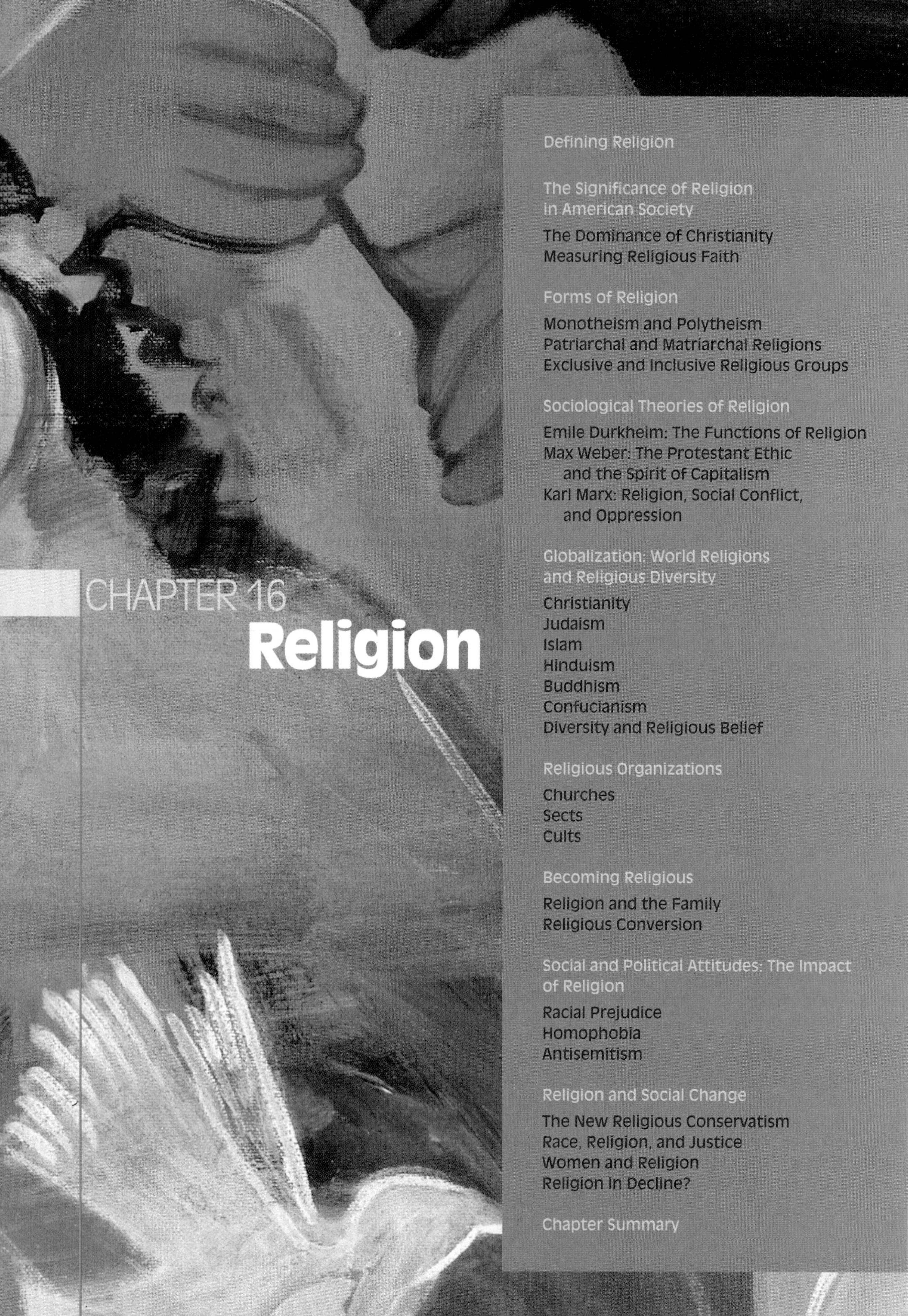

CHAPTER 16
Religion

THE PROFOUND effect of religion on society and human behavior is easily observed in everyday life. Church steeples dot the landscape everywhere. Invocations to a religious deity occur at the beginning of many public gatherings. Marriage ceremonies are typically held in a religious setting. Millions of people watch religious programming on television and listen to religious broadcasts on the radio. Newspaper headlines frequently report world conflicts based on religious differences and hostilities. Intensely felt religious values have held the stage in legislatures, classrooms, and the courts. Religious beliefs have led some to kill doctors who perform abortions; churches are sometimes targets for other hostilities, as in the burning of Black churches or the desecration of Jewish cemeteries. Some of life's sweetest moments are marked by religious celebration, and some of its most bitter conflicts persist because of unshakable religious conviction. Religion is an integrative force in society, but is also the basis for many of our most deeply rooted social conflicts.

Sociologists study religion as both a belief system and a social institution. The belief systems of religion have a powerful hold on what people think and how they see the world. As a social institution, the patterns and practices of religion are among the most important influences on people's lives. Religious beliefs and practices are also interwoven with the other institutions in society. Religious beliefs are typically learned, for example, within the family; thus, the institutions of family and religion are intertwined. Religion and politics are also deeply interconnected; such issues of morality and public policy as abortion, gay and lesbian rights, sex education, and school prayer have dominated domestic politics in recent years. The economy also is intertwined with religion; many businesses make *half* of their annual profits in the sales season leading up to Christmas. Increasingly, other holidays, such as Kwanza and Hannukah, are also being commercialized through the sale of greeting cards, gifts, wrapping paper, and other products. The popularity of angel symbols has also led to a whole new array of angel products: greeting cards, pins, posters, and other objects that show the capacity for spiritual beliefs to become highly profitable.

From a sociological perspective, several questions about religion and its role in society are of paramount interest. How are patterns of religious belief and practice related to other social factors, such as social class, race, age, gender, and level of education? How are religious institutions organized? How does religion influence social change? Aside from the spiritual and supernatural dimensions associated with it, religion is a social phenomenon that can be studied just like other social phenomena—such as education, work, and family. Whether they are personally religious or not, sociologists use their skills of objective analysis to study the social dimensions of religion. Unlike theologians who study the meaning of different religious faiths, sociologists study religion as a social institution. Religious institutions influence people's beliefs and behavior and, like other social institutions, can be analyzed as social structures. This is the distinct perspective that sociologists bring to the study of religion.

Sociologists do not accept religion based on faith alone. The sociological perspective requires empirical evidence; thus, studying religion as a sociological phenomenon requires a certain detachment and an unwillingness to take religious beliefs for granted. Whereas religious people may see their beliefs and practices as based on certain truths (known to them by holy writings or discovered in prayer), sociologists assess religion in terms of what can be objectively analyzed. To those with rigid religious beliefs, the questions sociologists ask may seem irreverent since it may be hard for them to study religion objectively. Thus, a congregation may declare that it gathers in accordance with God's wish, but a sociologist probing how congregations form is likely to compare the composition and structure of this congregation with other religious groups to see how they are similar and how they are different. The point is that sociologists use their skills of trained social observation to study religion and its influences and structure in society.

Religion is often the basis for social conflict, as in Northern Ireland. Here Irish nationalists are screaming at Orangemen, during a procession in Belfast.

Still, one can have a sociological perspective on religion and be a religious person. Many sociologists have strong religious commitments; some have even been members of the clergy or other religious orders. Many combine their sociological and religious commitments; for them, sociology and religion are allies in the quest for a more just society. Other sociologists are agnostics or atheists. In developing a sociological perspective on religion, what is important is not what one believes about religion, but one's ability to be nonjudgmental. The sociological observer must be willing to examine religion in its social and cultural context, and to separate the practices of religion from dogma and moral tenets. This may be difficult for people who hold dogmatic religious values or who have strict ideas of right and wrong with no sense that there are underlying social dimensions to religious beliefs and practices (McAllister, 1991).

Defining Religion

What is religion? Most people think of it as a category of experience separate from the everyday, perhaps involving communication with a deity or communion with the supernatural (Johnstone, 1992). One person might define religion as the belief in God, another as the observation of religious rituals—going to a church, temple, or mosque and observing the rituals of religious faith. A sociological definition of religion should describe it conceptually without reference to specific beliefs or religious practices; thus, these commonsense definitions are inadequate. Another person might define religion as a purely private experience, a matter of what one person believes. But is believing in UFOs religion? What about believing that the world is round or believing in evolution? Sociologists define **religion** as *an institutionalized system of symbols, beliefs, values, and practices by which a group of people interprets and responds to what they feel is sacred and that provides answers to questions of ultimate meaning* (Glock and Stark, 1965: 4; Johnstone, 1992: 14). The elements of this definition bear closer examination:

1. *Religion is institutionalized.* Religion is more than just beliefs: It is a pattern of social action organized around the beliefs, practices, and symbols that people develop to answer questions about the meaning of existence. As an institution, religion presents itself as larger than any single individual; it persists over time and has an organizational structure into which members are socialized. Although an individual may have a personal view of religion, that view evolves from the social systems of which religion is a part. Like other institutions, religion changes over time as new patterns of belief and behavior emerge; however, religion has been with us since before recorded history, and some of today's major religions can trace their roots back thousands of years. Despite how they change, religious institutions are also remarkably stable.

2. *Religion is a feature of groups.* Religion is built around a community of people with similar beliefs. Religion is a cohesive force among believers since it is a basis for group identity and gives people a sense of belonging to a community or organization. Religious groups can be formally organized, as in the case of large, bureaucratic churches, or they may be more informally organized, ranging from prayer groups to

SOURCE: Jeff Stahler reprinted by permission of Newspaper Enterprise Association, Inc.

cults. Some religious communities are extremely close-knit, as in convents; others are more diffuse, such as people who identify themselves as Protestant but attend church only on Easter. A number of religious communities have attempted to separate themselves from mainstream society so as to maintain purity and cohesion. Mennonites, for example, live in communities based on their religious faith; they interact primarily with each other, marry each other, and send their children to school only with other Mennonite children. However religious groups are organized, people acquire religious beliefs through group experience.

3. *Religions are based on beliefs that are considered sacred.* The **sacred** is that which is set apart from ordinary activity, seen as holy, and protected by special rites and rituals. The sacred is distinguished from the **profane,** that which is of the everyday world and is specifically not religious (Durkheim, 1912 [1947]; Chalfant et al., 1987). Religions define what is sacred; most religions have sacred objects and sacred symbols. The holy symbols are infused with special religious meaning and inspire awe and reverence. A nonbeliever who uses such symbols in an inappropriate way can expect a powerful reaction from the faithful. How would you feel if someone defaced a picture of Jesus? Or used a menorah at a picnic table with citronella candles to ward away mosquitoes? Or wore a rosary as a necklace? These actions, considered *sacrilegious* to those who hold the objects sacred, show how powerful a belief in the sacred can be. Those who desecrate them are often responded to with as much vehemence is if they had personally attacked someone.

A **totem** is an object or living thing that a religious group regards with special awe and reverence. A statue of Buddha is a totem; so is a crucifix hanging on a wall. Among the Zuni (a Native American group), *fetishes* are totems; these are small, beautifully carved animal objects representing different dimensions of Zuni spirituality. A totem is important not for what it is, but for what it represents. To a Christian taking communion, a piece of bread is defined as the flesh of Jesus; eating the bread unites the communicant mystically with Christ. To a nonbeliever, the bread is simply that—a piece of bread (McGuire, 1996). Likewise, Native Americans hold certain ground to be sacred and are deeply offended when the holy ground is disturbed by industrial or commercial developers who see only potential profit. Because religious beliefs are held so strongly, sacred religious symbols have enormous power and generate powerful emotional responses. Certain behaviors can startle, offend, or anger people, depending on the religious meaning of the behavior. The sight of a crucifix may bring tears to the faithful's eyes, and kneeling before a Buddha may inspire deep religious meditation. There is nothing inherent in these objects or events that defines them as sacred; rather, the significance of a totem derives from the religious meaning system of which it is a part.

4. *Religion establishes values and moral proscriptions for behavior.* A proscription is a constraint imposed by external forces. Religion typically establishes proscriptions for the behavior of believers, some of them quite strict. For example, the Catholic Church defines living together as sexual partners outside marriage as a sin; the lifestyle is condemned as immoral "selfishness" and an "unwillingness to make a lifelong commitment" through marriage (National Council of Catholic Bishops, 1992). Often religious believers come to see such moral proscriptions as simply "right" and behave accordingly. Other times, individuals may consciously reject

Religious spirituality takes many forms, but produces feelings of awe and reverence among believers.

moral proscriptions, although they may still feel guilty when they engage in a forbidden practice. Religious proscriptions have a strong hold on people's ideas, even among those who have fallen away from religion. Catholics who have rejected the Catholic faith, for example, may still experience guilt if they have sex outside marriage even though consciously they may have rejected the Church's proscriptions.

5. *Religion establishes norms for behavior.* In earlier chapters, we have discussed social norms as established and expected patterns of behavior. Religious belief systems establish social norms about how the faithful should behave in certain situations. Worshipers may be expected to cover their heads in a temple, mosque, or cathedral, or wear certain clothes. Such behavioral expectations may be quite strong. The next time you are at a gathering where a prayer is said before a meal, note how many people bow their head, even though some of those present may not believe in the deity being invoked.

Religious norms are invoked during holidays ("holy days") when practices like lighting candles, burning incense, making tributes to ancestors, or recounting religious stories are common. Religious norms generally make very clear what behaviors are expected on particular holidays. More generally, religious norms and practices give people a sense of community and belonging, as shown in Charles Ryu's account of the importance of the community for him as a Korean immigrant in the United States (see the box, "Understanding Diversity").

6. *Religion provides answers to questions of ultimate meaning.* **Secular** beliefs, the ordinary beliefs of daily life, may be institutionalized, but they are specifically not religious. Science, for example, generates secular beliefs based on particular ways of thinking—logic and empirical observation are at the root of scientific beliefs. Religious beliefs, on the other hand, often have a supernatural element. They emerge from spiritual needs, and may provide answers to questions that cannot be probed with the profane tools of science and reason.

Think of the difference in how religion and science explain the origins of life. Whereas science explains this as the result of biochemical and physical processes, different religions have other accounts of the origin of life. Christians see it as God's work, as described in the Book of Genesis in the Bible. Religious explanations of the origin of life are not the product of research and in fact are seldom probed—it is believed that the answer is already known. Many of the major religious belief systems in the United States assert that there is some form of life after death. Seven in ten Americans believe there is life after death (Bezilla, 1990). This belief, however, stems from religious, not secular, beliefs, and it is a question largely outside the realm of science. Instead, belief in the afterlife is a matter of faith that, for many, gives meaning to their existence. Religious teachings like this provide spiritual answers to abstract questions that may be unanswerable from other points of view. Scientific views, no matter how persuasive or accurate, typically do not give people this sense of ultimate meaning.

THINKING SOCIOLOGICALLY

For one week, keep a daily log, noting every time you see an explicit or implicit reference to religion. At the end of the week, review your notes and ask yourself how religion is connected to other social institutions. Based on what you have seen, describe what you see as the relationship between the *sacred* and the *secular* in this society.

BOX 16.1

UNDERSTANDING DIVERSITY
Koreans and Church

IN AMERICA, whatever the reason, the church has become a major and central anchoring institution for Korean immigrant society. Whereas no other institution supported the Korean immigrants, the church played the role of anything and everything—from social service, to education, to learning the Korean language: a place to gather, to meet other people, for social gratification, you name it. The way we think of church is more than in a religious connotation, as a place to go and pray, have worship service, to learn of God and comfort. Your identity is tied so closely to the church you go to. I think almost seventy to eighty percent of Korean Americans belong to church. And it becomes social evangelism. If those who had never considered themselves to be Christian want to be Korean, they go to church. They just go there for social reasons. You are acclimated into the gospel, and you say, I want to be baptized. Your life revolves around the church. It could also become a ghetto in the sense of a Chinatown or a Koreatown. But at the same time, it can be a place, if done properly, where the bruised identity can be healed and affirmed, because living in American society as a minority is a very difficult thing. You are nobody out there, but when you come to church, you are a somebody. The role of the black church in the civil rights era was the same thing.

Probably American churches have become specifically religious. I have my love life, my civic life, my professional life, and I have this church, religious life. But the Korean churches provide a new community of some sort that permeates the social matrix. It is such a solid, close-knit community, because everyone feels in a sense alienated outside, so there is this centripetal force. At the same time, most who come are well established or potentially well established.

Ryu, Charles. "Koreans and Church," pp.162–163 in Lee, Joann Fung Jean, (ed.) 1992. *Asian Americans.* New York: The New Press.

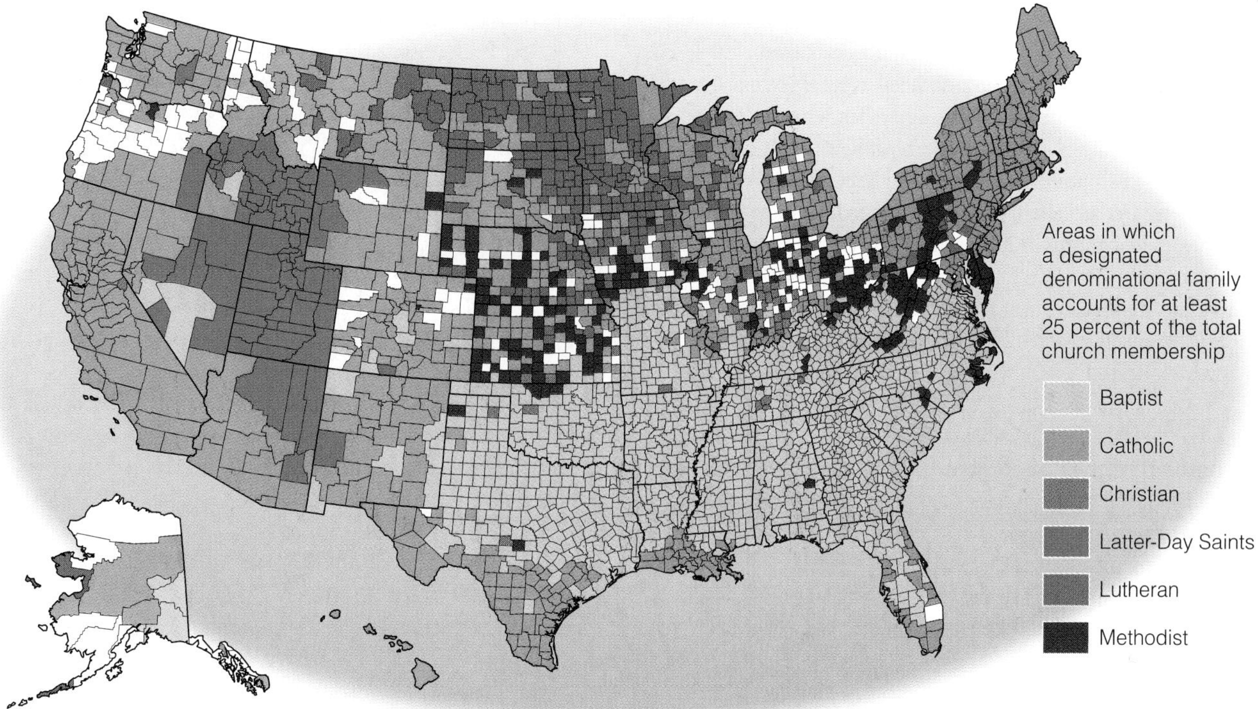

Where no church has 25 percent of the church membership in a county, that county is left blank. When two or more churches have 25 percent or more of the church membership in a county, the largest is shown.

MAP 16.1 Mapping America's Diversity: Christian Denominations

SOURCE: Doyle, Rodger, 1994. *Atlas of Contemporary America: Portrait of a Nation.* New York: Facts on File, p. 73.

The Significance of Religion in American Society

The United States is one of the most religious societies in the world. Ninety-six percent of Americans say they believe in God or a universal spirit; only 3 percent say they do not. Forty-one percent of the population describe themselves as "born again" (Roper Organization, 1995a, 1996; Bezilla, 1990). These indicators reveal some of the significance of religion in contemporary American society and the dominance of Christian beliefs.

Debunking Society's Myths

Myth: **The principle of separation of church and state is firmly held in the United States.**

Sociological perspective: **Although separation of church and state is a constitutional principle, dominant Christian traditions and symbols pervade American culture and are the basis for many of our social institutions.**

Religion is, for millions of people, the strongest component of their individual and group identity, and it is often the basis of culture in society. Much of the world's most celebrated art, architecture, and music has its origins in religion. In the classical art of western Europe, the Buddhist temples of the East, and the gospel rhythms of contemporary rock are signs of how cultural development is inspired by religion. For many Mexican Americans, the belief that the dead live on is reflected in many cultural artifacts (carvings, paintings, and clay objects), as well as "Day of the Dead" commemorative celebrations. These religious beliefs ignite feelings of awe and reverence. Even among people who are not particularly religious, the experience of seeing such religious images or entering a beautiful cathedral or temple may be deeply moving.

The Dominance of Christianity

Despite the constitutional principle of the separation of church and state, Christian religious beliefs and practices dominate American culture. Indeed, Christianity is often treated as if it were the national religion. It is commonly said that the United States is based on a *Judeo-Christian* heritage, meaning that our basic cultural beliefs stem from the traditions of the pre-Christian Old Testament of the Bible (the Judaic tradition) and the Gospels of the New Testament. The dominance of Christianity is visible everywhere. State-sponsored colleges and universities typically close for Christmas break, not Yom Kippur. Christmas is a national holiday, but not Ramadan, the most sacred holiday among Muslims. The dominance of Christianity in national thought is well illustrated by a comment once heard on a major city radio station: "Happy Hanukkah to all our Jewish friends in the audience. Hanukkah is to our Jewish friends what Christmas is to *we Americans.*" Christian religious traditions are often observed in the culture with little sensitivity to the religious beliefs of other groups. Christian religious symbols are also prevalent in the dominant culture, despite the diversity of faiths in this society. Banners with biblical citations, such as "John 3:16," are commonly displayed at major sporting events, and many public events are opened with Christian invocations.

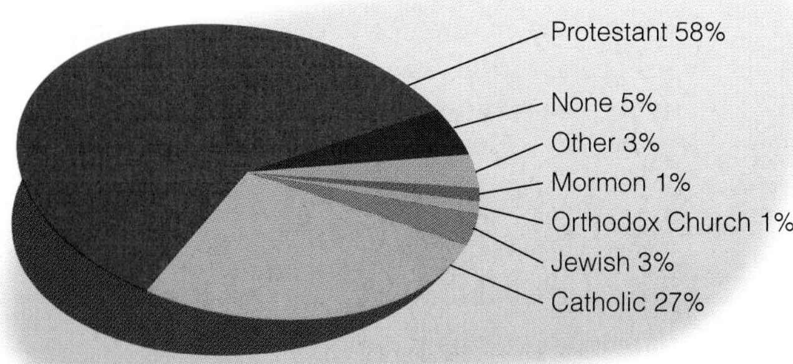

FIGURE 16.1 Religious Identification in the United States

SOURCE: Roper Organization. 1997. *The Gallup Poll.* Storrs, CT: Roper Organization, p. 44.

Measuring Religious Faith

Religiosity is the intensity and consistency of practice of a person's (or group's) faith. Sociologists measure religiosity both by asking people about their religious beliefs and by measuring membership in religious organizations and attendance at religious services. The vast majority in the United States identify themselves as Protestant (58 percent), Catholic (27 percent), or Jewish (3 percent), with diverse religions and those with no religious identification (5 percent) constituting the balance (see Figure 16.1; Roper Organization, 1997). Seventy percent of Americans say they are members of a church or synagogue, although only 43 percent say they attend a church or synagogue weekly or almost weekly (Roper Organization, 1996). Since these are self-reports, they may not be entirely accurate, but they give a sense of the significance of religion in people's lives.

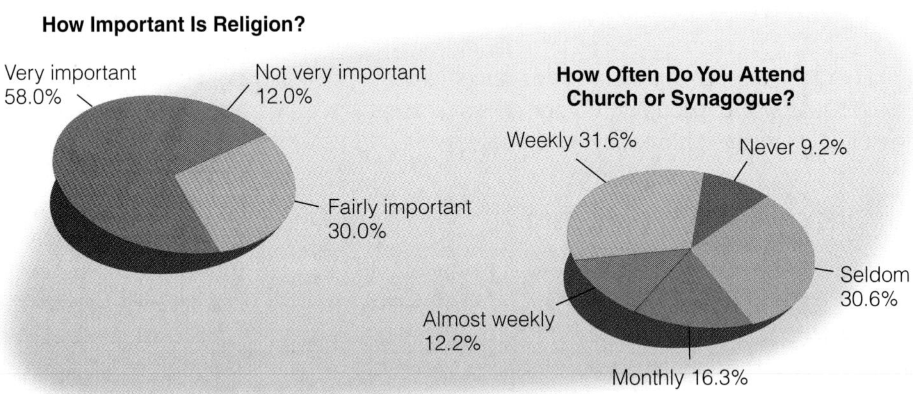

FIGURE 16.2 Measuring Religiosity

SOURCE: Roper Organization. 1996. *The Gallup Poll.* Storrs, CT: Roper Organization, p. 1.

Religiosity varies significantly among different groups in society. Church membership and attendance is higher among women than men and more prevalent among older than younger people. African Americans are more likely to belong to and attend church than Whites. On the whole, church membership and attendance fluctuate over time; membership has decreased slightly since 1940, but attendance has remained largely the same since then. The pattern of religion in the United States is, however, a mosaic one. Large national religious organizations, such as the mainline Protestant denominations, have lost many members, whereas smaller, local congregations have been growing. Changes in immigration patterns in recent years have also affected religious patterns in the United States, with Muslims, Buddhists, and Hindus now accounting for several million believers. Immigration has also affected the growth of Catholicism, which since 1980 has grown from fifty to sixty-one million members. At the same time, some religions formerly associated only with Americans are reaching overseas, such as the Church of Jesus Christ of Latter-day Saints and the Assemblies of God, a U.S. Pentecostal group (Niebuhr, 1998).

These data indicate the extent of religiosity, but tell little about how religion influences people's lives. To develop a picture of the day-to-day importance of religion, the sociologist must grapple with measures derived from the answers to subjective questions like "How important is religion in your life?" (see Figure 16.2). In recent years, there has been a decrease in the number of people who think that religion can answer all or almost all of today's problems. There has also been a corresponding increase in the number who think religion is largely old-fashioned, and a decrease in the number who say that religion is important in their own lives over time (Roper Organization, 1997).

Measurement of these various religious indicators provides sociologists with the data to see the changing significance of religion in people's lives, but these indicators alone tell only part of the story. As an example, even though the number of those saying that religion can solve society's problems has decreased, the influence of religion in national politics and the formation of social policy have increased. Witness, for example, the influence of issues like support for school prayer or attitudes about abortion in local and national elections. And, in listening to campaign speeches, note the inclusion of religious invocations on the appeals made to voters. Statistical measurements of religious belief give us a portrait of religion in the United States, but sociologists have to use other skills of sociological observation to more fully understand the role of religion in society.

Forms of Religion

Religions can be categorized in different ways according to the specific characteristics of faiths and how religious groups are organized. In different societies and among different religious groups, the form religion takes reflects differing belief systems and reflects and supports other features of the society. Believing in one god or many, worshipping in small or large groups, and associating religious faith with gender roles all contribute to the social organization of religion and its relationship to the rest of society.

Monotheism and Polytheism

One basic way to categorize religions is by the number of gods or goddesses they worship. **Monotheism** is the worship of a single god. Christianity and Judaism are monotheistic in that both Christians and Jews believe in a single god who created the universe. Monotheistic religions typically define god as omnipotent (all-powerful) and omniscient (all-knowing).

Polytheism is the worship of more than one deity. Most Native American cultures, for example, do not worship a single god, but see harmony in the world derived from the connection between people, spirits, and the earth (Allen, 1986). They believe there are many gods and goddesses, each presiding over different aspects of life. Many Asian religions are polytheistic, defining a number of gods in terms of the phenomena they control; thus, there may be a goddess of the harvest or god of the sun. In some religions—Hinduism, for example—individual worshipers choose deities to whom they feel most near. Within Hinduism, the universe is seen as so vast that it is believed to be beyond the grasp of a single individual, even a powerful god. Hinduism is extraordinarily complex with millions of gods, demons, sages, and heroes—all overlapping and entangled in religious mythology (Grimal, 1963)—although these numerous gods are also seen as a single entity of faith.

Patriarchal and Matriarchal Religions

Religions may also be patriarchal or matriarchal. **Patriarchal religions** are those in which the beliefs and practices of the religion are based on male power and authority. Christianity is a patriarchal religion; the ascendancy of men is emphasized in the role of women in the church, the instruction given on relations between the sexes, and even the language of worship itself (as in "Declare *his* glory," Psalms 96:3). The patriarchal character of Christianity has recently been challenged by arguments that the concept of God transcends gender, that historically God is not a gendered entity but a genderless spirit, and that the patriarchal trend in Christianity is transparently a social development that propagates injustice and oppresses the spiritual life of the congregation. There is an increasing willingness in some Christian denominations to ordain women. The Episcopal Church approved the ordination of women in 1976 and appointed its first woman bishop in 1989. There is an organized movement within the Catholic Church to permit women priests, but strong resistance among the Catholic leadership makes a change highly unlikely anytime soon.

Matriarchal religions are based on the centrality of female goddesses, who may be seen as the source of food, nurturance, and love, or who may serve as emblems of the power of women (McGuire, 1996). In societies based on matriarchal religions, women are more likely to share power with men in the society at large. Since women and men participate more equally in secular life, it is more possible for them to imagine egalitarian divine beings (Ochshorn, 1981). Likewise, in highly sexist, patriarchal societies, religious beliefs are likely to be very patriarchal. The correlations between patriarchal religion and patriarchal society and between matriarchal religion and matriarchal society (or, at least, more egalitarian societies) show how religious traditions shape society and, likewise, how the social organization of society shapes religion. We can see this social change in progress in the United States where, as women's roles have changed, so has the definition of God. Some now define God as a woman or, at least, as representing so-called female qualities. Thus we see that religions both reflect and recreate the societal arrangements in which they appear.

Exclusive and Inclusive Religious Groups

Religious groups can also be categorized according to how exclusive or inclusive they are. **Exclusive religious groups** are those with an easily identifiable religion and culture, including distinctive beliefs and strong moral teachings. In such groups, members are expected to conform strictly to religious values and behavioral norms. Black Muslims, for example, expect followers to lead a disciplined life, the strictures of which range from being economically self-reliant (for example, running their own businesses) to following certain dietary restrictions (such as not eating pork).

Exclusive religious groups tend to be small in membership, but are coherent, close-knit communities and have strongly held beliefs. Groups like the Jehovah's Witnesses, Assembly of God, Mormons, Pentecostal and holiness churches have exclusive orientations. Although there are more than four million Jehovah's Witnesses and eight million Mormons, they form small, coherent, close-knit communities. The Jehovah's Witnesses, for example, discourage association with those who are not Jehovah's Witnesses unless the association is for the purpose of converting the nonbeliever. As we have seen, exclusive groups are the fastest growing category of religious group in recent years (Roof and McKinney, 1987).

Inclusive religious groups, on the other hand, are those with a more moderate and liberal religious orientation. They are more *ecumenical,* meaning that they stress interdenominational cooperation and the importance of common religious work. The National Council of Churches, for example, has encouraged participating churches to adopt liberal, activist agendas for social change; this organization provided some support to the civil rights movement in the 1960s. Inclusive groups tend to have a broad-based membership. Their beliefs are often more weakly held, and they, consequently, form a more diffuse religious community. As a result, they are more likely than exclusive groups to make secular compromises. Examples of inclusive religious groups are the more "establishment" churches, such as the United Methodist and United Presbyterian Churches. Many of these groups have been torn by conflicts within because of tensions over social issues. Southern Baptist Conventions, as an example, have had fierce struggles between fundamentalist and moderate factions. Other religious groups have also been divided on matters like gay rights and women's roles.

THINKING SOCIOLOGICALLY

Attend a religious service in a faith other than one in which you were raised. Using the skills of participant observation, note who attends. What is the belief system and in what *rituals* does the group engage? Would you describe this religious congregation as an *inclusive* or *exclusive* religious group? If you used a *conflict theory* perspective to explain some of your observations, what would you conclude? What would you conclude using a *functionalist* approach?

Sociological Theories of Religion

The sociological study of religion probes how religion is related to the structure of society. Recall that one of the basic questions sociologists ask is "What holds society together?" To people living in a society, the society seems to be both unique and coherent. Coherence comes from both the social institutions that characterize society and the beliefs that hold society together. In both instances, religion plays a key role. From the functionalist perspective of sociological theory, religion is an integrative force in society because it has the power to shape collective beliefs. This perspective was originally developed by Emile Durkheim. In a somewhat different vein, the sociologist Max Weber saw religion in terms of how it supported other social institutions. Weber thought that religious belief systems provided a cultural framework that supported the development of specific social institutions in other realms, such as the economy. From yet a third point of view, conflict theory, religion is related to social inequality in society. Karl Marx was the historical source for the ideas relating religion and social conflict. In the following sections, we examine each of these major perspectives in the sociology of religion.

Emile Durkheim: The Functions of Religion

Emile Durkheim, one of the classic sociological theorists, argued that religion is functional for society because it reaffirms the social bonds that people have with each other, creating social cohesion and integration. Durkheim believed that the cohesiveness of society depends on the organization of its belief system. Societies with a unified belief system would be highly cohesive; those with a more diffuse or competing belief system would be less cohesive.

Religious **rituals** are symbolic activities that express a group's spiritual convictions. Making a pilgrimage to Mecca, for example, is an expression of religious faith and a reminder of religious belonging. In Durkheim's view, religious rituals are vehicles for the creation, expression, and reinforcement of social cohesion. Groups performing a ritual are expressing their identity as a group. Most religions incorporate rituals as part of the practice of faith although some are more ritualistic than others. Whether the rituals of a group are highly elaborated or casually informal, they are symbolic behaviors that freshen a group's awareness of its unifying beliefs. Lighting candles, chanting, or receiving a sacrament are behaviors that reunite the faithful and help them identify with the religious group, its goals, and its beliefs (McGuire, 1996).

Durkheim assessed the significance of religious ritual within the context of a larger point, that one of the major functions of religion is to confer identity on a person. In his classic work, *The Elementary Forms of Religious Life,* Durkheim argued that through religion individuals are able to transcend their individual identities and see themselves as part of a larger group. Wearing religious symbols, for example, is a declaration of religious identity and of connection to others of the same faith, just as an African American wearing Kinte cloth (the brightly colored woven fabric from Ghana) is demonstrating racial identity and racial solidarity with other people of the same heritage. Durkheim believed that religion binds individuals to the society in which they live by establishing what he called a **collective consciousness,** the body of beliefs that are common to a community or society and that give people a sense of belonging. In many societies, religion establishes the collective consciousness and creates in people the feeling that they are part of a common whole.

Durkheim's analysis of religion suggests some of the key ideas in *symbolic interaction theory,* particularly in the significance he gave to symbols in religious behavior. Symbolic interaction theory sees religion as a socially constructed belief system, one that emerges in different social conditions. As Durkheim understood, religious practices are symbolic activities, giving people a sense of identity and group belonging. From the perspective of symbolic interaction, religion is a meaning system that defines one's network of social belonging and confers one's attachment to particular social groups and ways of thinking.

Religion also has a strong emotional component, both in terms of emotional expression and emotional control. This strengthens the attachment that people feel to religious groups and social norms. During religious revivals, the emotions of the faithful may be stoked to a high pitch. At the height of the event, extreme emotions like ecstasy and exultation might be released that religious and social norms would not permit in another setting. In addition to inspiring emotion, religion regulates what emotions are permissible and how they may be expressed. By defining what is good and right in behaviors and feelings, religion determines what is seen as legitimate within a given society, and hence is a form of social control (Wilson, 1982).

Max Weber: The Protestant Ethic and the Spirit of Capitalism

Durkheim's views were later elaborated by Max Weber, who saw a fit between the religious principles of society and other institutional needs. In his classic work, *The Protestant Ethic and the Spirit of Capitalism,* Weber argued that the Protestant faith supported the development of capitalism in the Western world. He began by noting a seeming contradiction: How could a religion that supposedly condemns conspicuous consumption co-exist in society with an economic system based on the pursuit of profit and material success?

Weber argued that these ideals were not as contradictory as they seemed. As the Protestant faith developed, it included a belief in predestination—the belief that one's salvation is predetermined, and that it is a gift from God, not something earned. This state of affairs created doubt and anxiety among believers, who searched for clues in the here-and-now about whether they were among the chosen—called the "elect." According to Weber, material success was taken to be one clue that a person was among the elect and thus favored by God, and therefore these early Protestants were driven to relentless work as a means of confirming (and demonstrating) their salvation. As it happens, hard work and self-denial—the key features of the **Protestant ethic**—lead not only to salvation but to the accumulation of capital. The religious ideas supported by the Protestant ethic therefore dovetailed nicely with the needs of capitalism—according to Weber, these austere religionists stockpiled wealth, had an irresistible motive to earn more (eternal salvation), and were inclined to spend little on themselves, leaving a larger share for investment and driving the growth of capitalism (Weber, 1958 [1904]).

Weber's analysis of the Protestant ethic shows how religious beliefs continue to shape our national culture, even in its secular dimensions. Beliefs that people are poor or on welfare because they do not want to work or the idea that hard work will lead to success are manifestations of the Protestant ethic. These beliefs have a tight grip on the nation's consciousness and, although they may not seem religious in their nature, they stem from the association Weber identified between religious beliefs and the development of capitalist values. The value-laden judgments that many people make about those who have not succeeded can be traced to the influence of religion that Weber first analyzed. Weber makes the point that cultural beliefs inspired by religious faith support other institutional systems within society.

Karl Marx: Religion, Social Conflict, and Oppression

Durkheim and Weber concentrated on how religion contributes to the cohesion of society. Religion can also be the basis for conflict. This aspect of religion can be seen daily in the headlines of the world's newspapers. Conflict between Protestants and Catholics in Northern Ireland has been the basis for years of social and political warfare. In the Middle East, differences between Muslims and Jews have caused decades of political instability. In Bosnia, the "ethnic cleansing" perpetrated by the Serbs against Bosnian Moslems has resulted in the annihilation of tens of thousands of people. These conflicts are not solely religious, but religion plays an inextricable part. Certainly religious wars, religious terrorism, and religious genocide have contributed some of the most violent and tragic episodes of world history. The image of religion in history has two incompatible sides: piety and contemplation on the one hand, battle flags on the other.

The solidarity that religious identification creates in a group can lead true believers to think of outsiders as infidels, the unfaithful, the accursed. The power of religion to give birth to group cohesion can also fuel group conflict. Religious *ethnocentrism* (excessive belief in the superiority of one's own group) has frequently led to the ruthless subordination of other groups. Much of the homophobic treatment of gays and lesbians in this culture stems from people's religious tenets. Likewise, ethical conflicts over issues like physician- or nurse-assisted suicide for the terminally ill evolve from religious values even though played out in the secular world of politics and public opinion. Sociologists who work within the framework of conflict theory find this perspective illuminating of many of the social and political conflicts that engage religious values.

The link between religion and social inequality is also key to the theories of Karl Marx. Marx saw religion as a tool for class oppression. According to Marx, oppressed people develop religion to soothe their distress (Marx, 1972 [1843]). The promise of a better life hereafter makes the present life more bearable, and the belief that "God's will" steers the present life makes it easier for people to accept their lot. To Marx, religion is a form of *false consciousness* (see Chapter 9) because it prevents people from rising up against oppression. He called religion the "opium of the people" because it encourages passivity and acceptance.

To Marx, religion is an **ideology**—a belief system that legitimates the social order and supports the ideas of the ruling class. When subordinated groups internalize the views of the dominant class, they come to believe in the legitimacy of the social order that oppresses them. Marx thus saw religion as supporting the status quo and being inherently conservative (that is, resisting change and preserving the existing social order). Marx suggested that religion promotes stratification since it generally supports a hierarchy of people on earth and the subordination of humankind to divine authority.

We can use Marx's argument to see how Christianity supported the system of slavery. When European explorers

protestant ethic

TABLE 16.1 *SOCIOLOGICAL THEORIES OF RELIGION*

	Functionalism	Conflict Theory	Symbolic Interaction
Religion and the social order	is an integrative force in society	is the basis for intergroup conflict; inequality in society is reflected in religious organizations, which are stratified by factors like race, class, or gender	is socially constructed and emerges with social and historical change
Religious beliefs	provide cohesion in the social order by promoting a sense of *collective consciousness*	can provide legitimation for oppressive social conditions	are socially constructed and subject to interpretation
Religious practices and rituals	reinforce a sense of social belonging	define in-groups and out-groups, thereby defining group boundaries	are symbolic activities that provide definitions of group and individual identity

first encountered African people, they regarded them as godless savages. The brutality of the slave trade was justified by the argument that slaves were being rescued from damnation and exposed to the Christian way of life; principles of Christianity thus legitimated the system of slavery in the eyes of the slaveowners. Slaves were perceived as inferior to their masters because of their allegedly heathen beliefs, but preachers also taught slaves that they would go to heaven if they obeyed their masters. In this way, Christian religion served the economic needs of slavery. Slaveowners could also convince themselves they were doing good by enslaving Black people.

Ironically, this extraordinary system of oppression has produced a strong belief in Christianity on the part of most African American people. While Christianity was used by slaveowners to justify slavery, for the slaves it became a wellspring of hope. Slaves had a strong belief in the afterlife to come. At the same time, religious meetings offered one of the few legitimate occasions for slaves to congregate in groups; religious gatherings were frequently the scene for political mobilization and a collective response to oppression (Genovese, 1972).

Marx's analysis of religion reveals that religion works in ways that are typically not acknowledged. It is important to add, however, that religion can be just as much the basis for social change as it is for social continuity. In the civil rights movement in the United States and in Latin American liberation movements, the words and actions of religious organizations have been central in mobilizing people for change. This does not undermine Marx's main point, however, since there remains ample evidence of the role of religion in generating social conflict and resisting social change.

Each of the sociological frameworks identified here reveals a different dimension of religion's role in contemporary society (see Table 16.1). The entire subject of religion can hardly be approached in so few words, but these theoretical frameworks provide a perspective that helps us understand the significance of religion. In the upcoming discussions, you can ask yourself, "Is religion giving people a sense of common ground or dividing them into different factions?" "Is religion a source of oppression or liberation?" Although stated in ordinary terms, such questions are the bases for theoretical analyses of religion in society.

Globalization: World Religions and Religious Diversity

Worldwide, religion is one of the most significant dimensions of diverse cultures. In some nations, religion defines the political order and shapes many basic social institutions. Major religious movements, such as the spread of Christianity and the growth of Islam, have defined major events in world history. Across different cultures, many people share in the same faith, although the particular expression of that faith may vary in different societal contexts.

The largest religion in the world, if measured in terms of numbers of followers, is Christianity, with an estimated 1.9 billion adherents, followed by Islam (with 1.1 billion adherents), then persons expressing no religion (840 million), Hindus (780 million), Buddhists (323 million), and other smaller groups (including Sikhs, Baha'is, Confucians, Shintoists, and others). People of Jewish faith are a relatively small number (14 million), but are nonetheless one of the most significant religious groups (U.S. Bureau of the Census, 1997).

The United States is one of the most religiously diverse societies in the world. Within the United States are well over 200 different religious groups and denominations. These groups can be divided into a relative handful of broad categories, but the many different faiths and denominations that people give as their religious affiliation show great diversity of religious belief within a single country. For many years, the major religious groupings have been Protestants, Catholics, and Jews, but there is enormous diversity in the various faiths observed in the United States, including such diverse groups as Mormons, Greek Orthodox, Moslems, Hindus, Quakers, Mennonites, Swedenborgians, and new spiritual groups such as Eckankar, New Age groups, and others (Roper Organization, 1996; Bezilla, 1990).

Christianity

Christianity developed in the Mediterranean region of Europe, rapidly growing from the years 40 to 350 A.D. to encompass about 56 percent of the population of the Roman

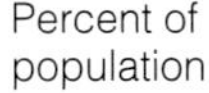

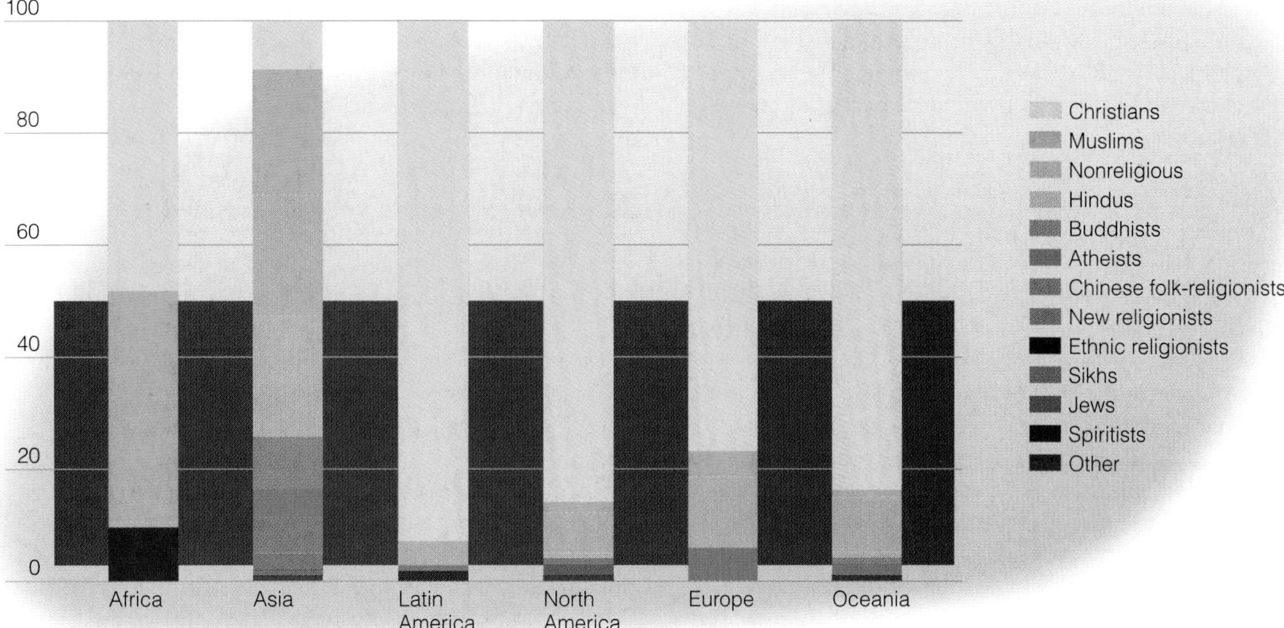

FIGURE 16.3 Viewing Society in Global Perspective: The World's Religions

Some religious groups are such a small percentage of the population that they do not appear in bars on the chart.
DATA: From the U.S. Bureau of the Census. 1997. *Statistical Abstracts of the United States 1997.* Washington, D. C.: U.S. Government Printing Office.

empire (Stark, 1996). It has spread throughout the world, largely as the result of European missionaries who colonized different parts of the world, proselytizing the Christian faith. Based on the teachings of Jesus, Christianity is based on faith in a Holy Trinity: God, the creator; Jesus, the son of God; and the Holy Spirit, the personal experience of the presence of God. Through the charisma of Jesus, Christians learned that they could be saved by following Christian beliefs and that there would be a life in Heaven (the afterworld). In the United States, Christianity is the dominant religion although there is great diversity in the different forms of Christianity—both in the United States and worldwide.

Protestants. Protestants form the largest religious group in the United States, although within this group, there is further diversity (see Table 16.2). There are two main categories of Protestants: *mainline Protestants* and *conservatives,* also known as fundamentalists. Mainline Protestants include twenty-eight different denominations, all belonging to the National Council of Churches. They have a history of intergroup cooperation and social activism. Mainline Protestants tend to be less diligent about religious observance than either Roman Catholics or conservative Protestants. Many do not attend church regularly even though the vast majority believe in the existence of God and the divinity of Christ (Chalfant et al., 1987).

Recently, mainline Protestants in the United States have seen declines in church membership and attendance. In 1962, Protestants were 67 percent of the population; they are now 60 percent. There are two principal reasons for the decline: the drop in religious affiliation and the growth in the number of conservative Protestants. Conservative Protestants include those groups variously labeled evangelical, fundamentalist, and Pentecostal—labels that apply to a diverse group of religious affiliations centered on the belief that one can be "saved" only through a personal commitment to Christ. These groups are the more dogmatic among Protestants; the label "fundamentalist" is taken from the belief that the fundamental tenet of the Christian life is that the Bible is the literal word of God.

In the United States, fundamentalism is the conservative offspring of the evangelical movement; fundamentalists are sometimes called evangelicals (Stacey and Gerard, 1990).

TABLE 16.2 *RELIGIOUS AFFILIATION AMONG PROTESTANTS*

Affiliation	Percentage
Methodist	8%
Southern Baptist	8
Other Baptist	10
Lutheran	5
Presbyterian	4
Pentecostal/Assembly of God	3
Episcopal	3
Church of Christ	2
Other	13
Nondenominational	4
No opinion	2

SOURCE: Roper Organization. 1996. *The Gallup Poll.* Storrs, CT: Roper Organization, pp. 171–172.

Pentecostals—a form of fundamentalism—are part of a worldwide religious movement, begun in the United States around the turn of the century, that believes in faith healing and a mystical conversion experience that includes *glossolalia* ("speaking in tongues"). Highly organized for political action, new conservative religious groups have had an enormous influence in society in recent years.

Roman Catholics. Worldwide, with its emphasis on Church hierarchy, Roman Catholics identify the Pope as the source of religious authority. The Vatican in Rome acts as the system of religious governance, with enormous power to influence the religious norms and beliefs of Catholic followers. The Catholic Church was once the ruling religious and political force in the Western world; its followers are worldwide and cross all social and political boundaries.

The Catholic Church is typically perceived as having a strong hold on its members. Catholicism is a very hierarchical religious system, with religious values and codes of behavior mandated by the Pope and the Vatican. In recent decades, however, the Catholic Church has embraced a number of liberal reforms. No longer, for instance, does the Catholic Church claim to be the sole source of religious truth. Mass is now typically celebrated in secular language (for centuries it was celebrated only in Latin), and lay people have been granted larger roles in church rituals and church affairs. Many Catholics hold more liberal attitudes than the official church promulgates. For example, the majority of Catholics in the United States do not believe that a divorced Catholic who remarries is living in sin—78 percent disagree with the Vatican's teachings on divorce and remarriage, and say that divorced Catholics should be permitted to remarry in the Catholic Church. Three-quarters favor allowing priests to marry, fully 84 percent believe Catholics should be allowed to practice artificial means of birth control, and 58 percent believe the Catholic Church should relax its standards forbidding all abortions under any circumstances, a matter explored more fully in the box "Doing Sociological Research." Almost half (49 percent) also disagree with the official Church position that homosexual behavior is always wrong (Roper Organization, 1993). These attitudes produce much social strain between U.S. Catholics and the position of the Vatican.

An example of the transformation of the Catholic Church is the emergence of *liberation theology*, an inspiration of Catholic priests in Latin America who feel that the Gospel of Christ commands the church to lead in liberating the poor and the oppressed of the world. The radical views of

BOX 16.2 DOING SOCIOLOGICAL RESEARCH

Catholic Attitudes Toward Abortion

POLITICAL debates about abortion rights in recent years reflect great division in public attitudes about abortion. Although the majority of the public supports women's rights to abortion, many do not. These attitudinal differences are the basis for intense political conflict. As we saw in Chapter 7, Kristin Luker's (1984) research on pro-choice and anti-abortion rights activists locates some of this political conflict in how people define motherhood and women's roles within the family. Religion is also clearly significant, but even among those from the same religious faith, there can be intense differences in attitudes toward abortion. A significant percentage of Catholics, for example, support abortion rights, although other Catholics are firmly opposed to abortion. Since the official position of the Catholic Church is not to support abortion rights, how do we explain the difference among Catholics on this issue?

This is the question asked by a team of sociologists who examined Catholic attitudes toward abortion. Michael Welch, David Legge, and James Cavendish asked how Catholic attitudes toward abortion are related to opinions about sexual permissiveness, women's roles, and religious orientation. To examine this question, they analyzed data from a national random survey of Catholic parishioners. They used attitudes toward abortion as a dependent variable (meaning it was the variable to be explained by other variables—the independent variables; see Chapter 2). For independent variables, they measured attitudes toward sexual restrictiveness, opposition to women's autonomy and feminist issues, and the extent to which people believed one had to adhere to Catholic orthodoxy to be a good Catholic. They also analyzed such demographic variables as age and gender. Which of these factors, they asked, were most likely to explain Catholic attitudes toward abortion?

Welch, Legge, and Cavendish found that those Catholics with the most sexually restrictive attitudes are those most likely to oppose abortion. There is some tendency for those who oppose women's autonomy and feminist issues to also be opposed to abortion, but this relationship is not as strong as that found between sexual permissiveness and abortion. They did find that adhering to Catholic orthodoxy did not have much effect on attitudes toward abortion.

This analysis indicates that simplistic explanations of religious affiliation do not explain the complex and shifting politics surrounding abortion. Religious identification, like other forms of social affiliation, develops within the context of broader social issues and political agendas. Even in a context where moral teachings are clear, people's attitudes form within complex worldviews that involve a host of other attitudinal issues—in this case, attitudes about sexuality and gender roles, in particular.

SOURCE: Welch, Michael R., David C. Legge, and James C. Cavendish. 1995. "Attitudes Toward Abortion Among U.S. Catholics: Another Case of Symbolic Politics?" *Social Science Quarterly* 76 (March): 142–157.

liberationists have been censured at times by the church, but the church has also given official support to a moderate form of liberation theology. Like other institutions, the Church defines itself and its proscriptions for the faithful even as it is defined by its constituents.

In the United States, Catholicism has long been associated with immigrant groups who brought Catholicism with them, just as Protestantism in the United States was established by immigrants from other nations. Irish Americans, Italian Americans, and most recently, Latin American immigrants have tended to be of the Catholic faith. Indeed, as we have seen, recent immigration has increased the number of Catholics in the United States.

Judaism

Although a numerically small religious faith, Judaism has enormous world significance. Its teachings are the source of both Christian belief systems and Islamic beliefs. More than 40 percent of the world's Jewish population now lives in the United States, making this the largest community of Jewish people in the world. The existence of the state of Israel, founded following the Holocaust of World War II, has given Jewish people a high profile in international politics.

The Jewish faith is more than 4000 years old. Under Egyptian rule in ancient history, Jewish people endured centuries of slavery. Led from Egypt by Moses in the thirteenth century (B.C.), Jewish people were liberated and now celebrate this freedom in the annual ritual of Passover—one of the holiest holidays. The Jews see themselves as "chosen people," meant to recognize their duty to obey God's laws as revealed in the Ten Commandments.

In the United States, Jewish Americans are a relatively small proportion of the population, and their number is declining as the result of a low birthrate and high rate of interfaith marriage (since when Jewish men marry non-Jewish women, the offspring are not considered Jewish; the offspring are considered Jewish only if the mother is Jewish). Jewish people nevertheless remain a significant religious minority. The periods of greatest Jewish immigration to the United States were in the midnineteenth century and between the 1880s and World War I. Large numbers tried to immigrate to the United States in the 1930s when Nazi persecution of Jews in Europe intensified, but the United States turned away many of these refugees, forcing them to return to Europe where many were killed in the Holocaust (Sklare, 1971; Wyman, 1984).

There are significant divisions of culture and religious practice within Judaism. Orthodox Jews, a small fraction of the total number of Jews in the United States, adhere strictly to a traditional conception of their religious faith. They observe the biblical dietary laws, honor traditional codes of dress and behavior, and strictly observe the sabbath, during which they may not travel, carry money, write, work, or do business. Reform Jews have a more secular orientation; lay persons are typically excused from strict observation of religious proscriptions. In Reform temples, where women and men are treated more equally, prayers are often in English, not Hebrew. In general, there is flexibility of religious practice among Reform Jews. Conservative Judaism falls between Orthodox and Reform in terms of strictness of observation (Sklare, 1971).

Only 19 percent of all Jewish Americans attend services in a typical week, but most Jewish Americans observe the high holidays and hold bar mitzvahs (the ritual ceremony marking the transition from boyhood to manhood) for Jewish boys, and sometimes bas mitzvahs (also called bat mitzvahs) for girls. Jewish identity remains strong, although lower levels of religiosity and less personal contact with other Jewish people weakens one's Jewish identity (Amyot and Sigelman, 1996). Most Jews see themselves as part of a world community of Jewish people. The strong bond of solidarity among Jewish people is strengthened by their history of persecution and by continuing prejudice against them. In this sense, Jews are both a religious group and a minority group.

Islam

Islam is most typically associated with Middle Eastern countries, in part because of the significance of Islamic faith in countries like Iran, Iraq, and Saudi Arabia, which have been much the subject of world politics in recent years. Islamic people are also found, however, in northern Africa, southeastern Asia, and increasingly, in North America and Europe as the result of international migration.

Followers of Islamic religion are called Muslims. They believe that Islam is the word of God, revealed in the prophet of Muhammad (born in Mecca in the year 570). The Koran is the holy book of Islam. The highly traditional or reactionary Islamic fundamentalists adhere strictly to the word of the Koran, believing that they are obligated to defend their faith and that death in doing so is a way of honoring God. Not all Muslims take such a fundamental view, although they are frequently stereotyped because of the prominence of Islamic fundamentalism in contemporary world conflicts. Muslims do have a highly patriarchal worldview, with women being denied the freedoms that men enjoy. But as with other religions, the degree to which people adhere to this faith varies in different social settings and depending on the institutional supports given to religious strictures by the political state.

Hinduism

Unlike Christianity, Judaism, and Islam, Hindu religion is not linked to a singular God, as we have already seen in the discussion of polytheism. Rather, Hinduism rejects the idea that there is a single, powerful god. In this religion, god is not a specific entity; rather, people are called upon to see a moral force in the world and to live in a way that contributes to one's spiritual and moral development. Karma is the principle in Hindu that sees all human action as having spiritual consequences, ultimately leading to a higher state of spiritual consciousness, perhaps found in the birth one experiences following death—reincarnation.

Hinduism is deeply linked to the social system of India, since the caste system in India (see Chapter 9) is seen as

Buddhism, one of the largest world religions, is a faith centered in the worship of several gods.

stemming from people's commitment to Hindu principles. Those who live the most ideal forms of life are seen as part of the higher caste, with the lower caste ("untouchables") as spiritually bereft. There are also several castes in between. But the complexity of Hinduism can be seen in the fact that, at the same time it has been used to justify this caste system, one of the greatest world leaders, Gandhi, used Hindu principles of justice, honesty, and courage to guide one of the most important independence movements in the world. His teachings, as we have seen, were also crucial in the development of the civil rights movement in the United States.

Buddhism

Buddhism is another extremely complex religion—one that, like Hinduism, does not follow a strict or singular theological god. Also arising from Indian culture, Buddhism is most widely practiced in Asian societies, although many countercultural groups throughout the world have adopted some of the beliefs and practices of Buddhist religion.

The Buddha in Buddhism is Siddartha Gautama, born of the highest caste in India in the year 563 B.C. As a young man, he sought a path of enlightenment, based on travel and meditation. Buddhism thus encourages its followers to pursue spiritual transformation. Its focus on meditation has been adopted by many of the New Age spiritual groups in Western society. Like Hinduism, Buddhism involves a concept of birth and rebirth through reincarnation. Through seeking spiritual enlightenment, Buddhists see people relieve themselves of their worldly suffering.

Confucianism

Most of the Eastern religions like Hinduism, Buddhism, and Confucianism are difficult to describe in the concepts drawn from Western thought. As religious belief systems, each rests on philosophical principles quite different from Western forms of thought. Confucianism is mostly found in China, although migration to other parts of Southeast Asia and North America means that Confucianism is practiced in various societies.

Confucians follow the principles of Confucius, a leader who promoted certain moral practices. Various moral principles are layered on one another, each contributing to the cohesion of a Confucian worldview. Confucianism promotes a disciplined way of life, more of a moral code than a sacred religion per se. But the expression of goodness and social unity is an important principle in Confucian thought, even if there is not a particular god or set of religious disciples whom Confucians follow.

These various world religions each reveal the power of belief in regulating social behavior—a fact that sociologists find intriguing. Likewise, the association of religion with other social factors is a key part of sociological research on religion. In the next section, we examine some of these associations in the context of diverse religious beliefs in the United States.

Diversity and Religious Belief

Religious identification varies with a number of social factors, including age, income level, education, and political affiliation (see Figure 16.4). Younger people, for example, are more likely than older people to express no religious preference. People over age fifty are more likely to identify as Protestant, whereas identification as Catholic is more likely among those under thirty. Those in higher income brackets ($40,000 per year and over) are more likely to identify as Catholic or Jewish than those in lower income brackets ($25,000 and under), who are more likely to identify as Protestant, although these trends vary among Protestants by denomination. Fundamentalist Protestants, for example, are more likely to come from lower income groups. Similar trends are seen in educational level. Protestant identification is higher among those with less education; the more highly educated are most likely to express no religious preference (Roof and McKinney, 1987). Sociologists have also shown that being affiliated with conservative Protestant religion actually has a negative influence on educational attainment, even controlling for other factors (Darnell and Sherkat, 1997).

In general, those most likely to be unaffiliated with any church are young, White, well-educated, non-Southern men who move frequently. This group also tends to be less conformist and more tolerant in its social attitudes and on political issues. The most religiously committed are older, African American, less educated, Southern women who are geographically stable (Roof and McKinney, 1987).

Race and Religion. Race is one of the most significant markers when examining patterns of religious orientation. African Americans are more likely to identify as Protestant than are Whites or Hispanics, although the number of African American Catholics is increasing (Bezilla, 1990). Large numbers of urban African Americans have also become committed Black Muslims, a group examined later in this chapter.

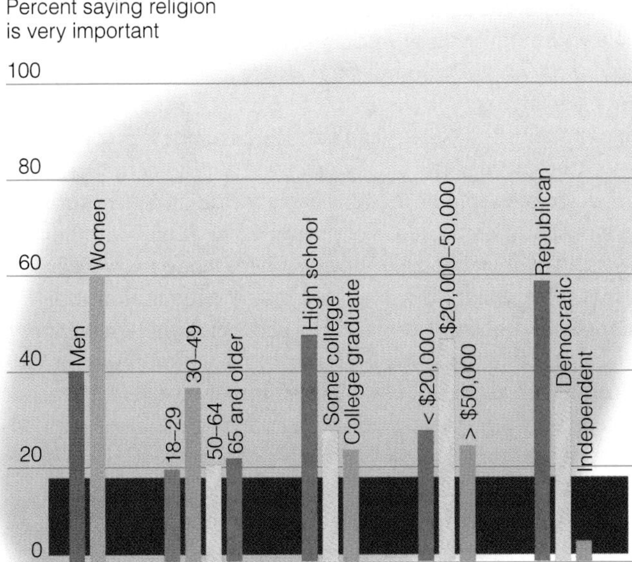

FIGURE 16.4 For Whom Is Religion Most Important?

Based on the 35 percent of those who attend services weekly or almost weekly and also consider religion very important.

SOURCE: David W. Moore. 1995. "Most Americans Say Religion Is Important to Them." *The Gallup Poll Monthly* (February), p.21.

For many African Americans, religion has been a defense against the damage caused by segregation, bigotry, and discrimination. Churches have served to channel the emotions of an oppressed people, while also serving as communal centers, political units, and sources of social and community services. Combined, these roles have made churches among the most important institutions within the African American community (Blackwell, 1991). Many of the most prominent African American leaders have emerged from churches or mosques, including the Reverend Martin Luther King, Jr., the Reverend Jesse Jackson, Malcolm X, and Louis Farrakhan. Equally important, though less prominent, are the leadership roles the Black church has provided for African American women, both as clergy and as community activists whose leadership stems from church activities (Barnett, 1993).

Religion has also been a strong force in Latino communities. Sixty-eight percent of Latinos identify themselves as Catholic; however, there are a significant and growing number of Latino Protestants, both in mainstream Protestant denominations and in fundamentalist groups. Among Latinos, religion helps sustain ties within and between families. Although Catholicism has been seen as particularly oppressive to Latinas, religious practice promotes interaction between women in Latino families and friendship groups. Whether Catholic or Protestant, the influence of religion among Latinos varies by social class, as it does within other groups. In general, those who are working class are more committed to religious belief and practice than are those in the professional middle class (Williams, 1990).

Asian Americans have a great variety of religious orientations. One reason for this is that the category "Asian American" is constructed from so many different Asian cultures. Hinduism and Buddhism are common among Asians, but a large number are also Christian as the result of missionary efforts in Asia. As with all groups whose family histories include immigration, religious belief and practice among Asian Americans frequently changes between generations. The youngest generation may not worship as their parents and grandparents did, although some aspects of the inherited faith may be retained. Within families, the discontinuity with a religious past brought on by cultural assimilation can be a source of tension between grandparents, parents, and children. Within the United States, Asian Americans often mix Christian and traditionally Buddhist, Confucian, or Hindu beliefs, resulting in new religious practices.

Ethnicity and Religion. The great diversity of religious belief in the United States stems from the diverse ethnic and cultural backgrounds of the American people. For many groups, ethnic identity is inseparable from religious identity. The two may be so interconnected that someone from an ethnic community who adopts a religion outside the dominant tradition, such as an Italian Jew, may feel ostracized and stigmatized.

For some groups, religion and ethnicity are especially intertwined, and the group may actually withdraw from society to preserve its distinctiveness. Such groups are known as **ethnoreligious groups**—an extreme form of an exclusive religious group, as described earlier in this chapter. Examples include Hasidic Jews and the Amish, groups that have some limited interaction with the larger society, but that have maintained their distinctive faith by withdrawing into their own communities. Hasidic Jews (or Hasidim) are an ultraconservative group consisting mostly of immigrants who fled persecution in central and eastern Europe. There are approximately 100,000 Hasidic Jews in the United States. Their lives are regimented to allow strict observation of Jewish tradition. Men and women are segregated outside the home; men also arrange their work so that they have no interaction with

For African Americans, the Christian church has been a source for faith in social justice.

women. The Hasidim maintain their own schools and shops, and even their own system of law, under which the rabbi is the ultimate authority on secular and sacred matters (McGuire, 1996). Even without extreme withdrawal, however, strong religious affiliations can affect the degree to which ethnic groups assimilate into the dominant culture. American Jews, for example, have tended to be more assimilated as their religious practice declines. Those Jewish people who identify with the most orthodox Jewish denominations have resisted assimilation into the dominant society (Legge, 1997).

Religious Organizations

Sociologists have organized their understanding of the different kinds of religious organizations into three types: churches, sects, and cults. These are *ideal types* in the sense that Max Weber used the term. That is, the ideal types convey the essential characteristics of some social entity or phenomenon, even though they do not explain every feature of each entity included in the generic category.

Churches

Churches are formal organizations that tend to see themselves, and are seen by society, as the primary and legitimate religious institutions. They tend to be integrated into the secular world to a degree that sects and cults (described later) are not. They are sometimes closely tied to the state. Churches are organized as complex bureaucracies with division of labor and different roles for groups within. Generally, churches employ a professional, full-time clergy who have been formally ordained following a specialized education. Church membership is renewed as the children of existing members are brought up in the church; churches may also actively proselytize. Churches are less exclusive than cults and sects since they see all of society as potential members of the fold (Johnstone, 1992).

Sects

Sects are groups that have broken off from an established church. They emerge when a faction within an established religion questions the legitimacy or purity of the group from which they are separating. Leaders of sects are sometimes lay people with no formal clerical training, although many sects form as offshoots of existing religious organization, typically formed from fine points of theological disagreement. In fact, within sects, there tends to be less emphasis on organization per se (as in churches) and more emphasis on the purity of members' faith. The Shakers, for example, formed by departing from the Society of Friends (the Quakers). They retained some Quaker practices, like simplicity of dress and a belief in pacifism, but departed from their religious philosophy. The Shakers believed that the second coming of Christ was imminent, but that Christ would appear in the form of a woman (Kephart, 1987).

Sects typically develop in protest against events or beliefs within the secular world. Different from churches, sects tend to be exclusive, admitting only truly committed members. Sectarians typically refuse to compromise their beliefs in any way, thinking instead that they are practicing the pure doctrine of their faith. Often, they hark back to original principles they believe are being violated by the group from which they parted. For example, the Amish formed as a sect in 1700 when Swiss Mennonites (themselves descendants of the Anabaptist sect), led by a Mennonite bishop named Joseph Amman, tasked the other Mennonite bishops for not enforcing the Meidung, a practice of shunning, or ignoring, errant members. The Meidung is a strict system of social control, especially so among Mennonites since they typically associate only with other Mennonites. The Amish brought the Meidung with them when they emigrated to America in the early 1700s, where they set up exclusive communities that have been remarkably faithful to the present day in living up to a principled mode of life that predates the industrial era (Kephart, 1987).

Sects tend to hold emotionally charged worship services. The Shakers, for example, had such emotional services that they shook, shouted, and quivered while "talking with the Lord," earning them their name, from the "Shaking Quakers" (Kephart, 1987). The only bodily contact permitted among the Shakers was during the unrestrained religious rituals; thus, they were celibate (did not have sexual relations) and gained new members only through adoption of children or recruitment of newcomers.

Cults

Cults, like sects in their intensity, are religious groups devoted to a specific cause or charismatic leader. Many cults arise within established religions, and sometimes continue to peaceably reside within the parent religion as simply a fellowship of like-minded persons with a particular, often mystical, set of dogma. The term *cult* is also used to refer to quasi-religious organizations, often dominated by a single charismatic individual. A good example was the Heaven's Gate cult, ninety of whose members committed group suicide in Rancho Santa Fe, California, in 1997, believing they were going away on spacecrafts that would lead them to a new world (Balch, 1995).

Many mainstream religions began as cults. Christian Science is an example of a now well-established religious organization that first evolved as a cult. As they are developing, it is common for there to be tension between cults and the society around them. Cults tend to exist outside the mainstream of society, arising when believers think that society is not satisfying their spiritual needs. Cults tend to attract those who feel a longing for meaningful attachments. Internally, cults seldom develop an elaborate organizational structure, but are instead close-knit communities held together by personal attachment and loyalty to the cult leader.

The Branch Davidians were a good example of a cult. Their leader, David Koresh, convinced his followers that he was the Messiah and that they should prepare for a violent day of reckoning that would end in his resurrection and their passage to Heaven; he also taught his women followers that it was God's will that he have sex with all of them so that their children could populate the new House of David (*The New York Times*, March 3, 1993; Tabor and Gallagher, 1995). Seventy-two members of this cult, including Koresh, were killed in the fire that followed an FBI assault on the cult's heavily armed compound in Waco, Texas, in 1993.

Cults tend to form around leaders with great **charisma,** a quality attributed to individuals believed by their followers to have special powers (Johnstone, 1992). Typically, followers are convinced that the charismatic leader has received a unique revelation or possesses supernatural gifts. Cult leaders are usually men, probably because men are more likely to be seen as having the aggressive and charismatic characteristics associated with heroic leadership. Exceptions include Ann Lee, founder of the Shaker sect in America, and Mary Baker Eddy of the Christian Science movement (Johnstone, 1992). The predominance of men in religious cults and the patriarchal structure of many of these cults have fostered some problems with sexual harassment and abuse of women followers by some male leaders.

Although charisma is defined as a quality of individuals, as sociologist Nancy Ammerman shows in the box on "Analyzing Social Issues," it evolves from the relationship between a leader and his or her followers. The leader's charisma is

BOX 16.3 • ANALYZING SOCIAL ISSUES

Lessons from Waco

FOLLOWING the FBI standoff with the Branch Davidians in Waco, Texas, in 1993, Professor Nancy Ammerman was one of ten experts appointed by the Justice and Treasury Departments to make recommendations to the government on how to avoid such situations in the future. Following the release of the experts' report, Professor Ammerman also reflected on the sociological lessons learned from this experience and offered her insights and sociological imagination to those who wish to understand the nature of groups like the Branch Davidians. Ammerman also tells us that we learn not just about religious cults from such events, but also about how the U.S. government works. Her conclusions follow:

1. . . . The pervasiveness of religious experimentation in American history. We simply have been a very religious people. From the days of the first European settlers, there have always been new and dissident religious groups challenging the boundaries of toleration, and the First Amendment to our Constitution guarantees those groups the right to practice their faith. Only when there is clear evidence of criminal wrong-doing can authorities intervene in the free exercise of religion, and then *only with appropriately low levels of intrusiveness.*

2. . . . New groups almost always provoke their neighbors. By definition, new religious groups think old ways of doing things are at best obsolete, at worst evil. Their very reason for existing is to call into question the status quo. They defy conventional rules and question conventional authorities. The corollary is that they themselves are likely to perceive the outside world as hostile—and it often is. New groups frequently provoke resistance, in recent years often well-organized through the "Cult Awareness Network."

3. . . . Many new religious movements ask for commitments that seem abnormal to most Americans, commitments that mean the disruption of "normal" family and work lives. While it may seem disturbing to outsiders that converts live all of life under a religious authority, it is certainly not illegal (nor particularly unusual, if we look around the world and back in history.) No matter how strange such commitments may seem to many, they are widely sought by millions of others.

4. . . . The vast majority of those who make such commitments do so voluntarily. The notion of "cult brainwashing" has been thoroughly discredited by the academic community, and "experts" who propagate such notions in the courts have been discredited by the American Psychological Association and the American Sociological Association. While there may be real psychological needs that lead persons to seek such groups, and while their judgment may indeed be altered by their participation, neither of those facts constitute coercion.

5. . . . People who deal with new or marginal religious groups must understand the ability of such groups to create an alternative world. The first dictum of sociology is "Situations perceived to be real are real in their consequences." No matter how illogical or unreasonable the beliefs of a group seem to an outsider, they are the real facts that describe the world through the eyes of the insider.

6. . . . People who deal with the leaders of such groups should understand that "charisma" is not just an individual trait, but a property of the constantly-evolving relationship between leader and followers. So long as the leader's interpretations make sense to the group's experience, that leader is likely to be able to maintain authority. These interpretations are not fixed text, but a living, changing body of ideas, rules, and practices. Only in subsequent generations are religious prescriptions likely to become written orthodoxies.

7. . . . Authorities who deal with high commitment groups of any kind must realize that any group under siege is likely to turn inward, bonding to each other and to their leader even more strongly than before. Outside pressure only consolidates the group's view that outsiders are the enemy. And isolation decreases the availability of information that might counter their internal view of the world. In the Waco case, negotiating strategies were constantly undermined by the actions of the tactical teams. Pressure from encroaching tanks, psychological warfare tactics, and the like, only increased the paranoia of the group and further convinced them that the only person they could trust was Koresh.

SOURCE: Ammerman, Nancy T. 1994. "Lessons from Waco." *Footnotes* (January): Washington, DC: American Sociological Association.

highly dependent on group willingness to believe the leader has charismatic qualities and is further dependent on the leader's ability to maintain a following. Over time, charismatic leaders may lose their appeal even though their personality remains largely the same.

Becoming Religious

Regardless of affiliation, religion can be a significant part of people's social identity. Sixty percent of Americans say that religion is very important in their lives; only 11 percent say it is not very important (Moore, 1995). Becoming religious is the result of many experiences, which sociologists group under the concept of **religious socialization,** the process by which one learns a particular religious faith. For most people, religious socialization takes place primarily, or at least initially, within the family. For others, it involves a process of conversion. Some faiths even insist that a formal conversion ceremony take place to establish commitment to a religious identity. Whatever the process by which one becomes religious, initiation into religion typically involves a period of instruction, learning, and experimentation. Becoming religious involves learning group norms and values, just as people learn the norms and values of other groups into which they are socialized. The result of religious socialization is that individuals internalize religious norms, sometimes so thoroughly that they come to believe that the religion they acquire is the only possible way to think and be.

Religious socialization can be both formal and informal. *Informal religious socialization* occurs when one observes and absorbs the religious perspective of parents and peers. Interaction with people in religious groups and organizations, such as church-based youth camps, can be an important source of informal religious training. Religious socialization can also occur in contexts that are not specifically religious, but where religious values are reflected. For example, religious references or assumptions made in films, on television, and during informal conversations can socialize people to believe in specific religious tenets.

Young people are typically socialized into the religion of their parents. Religious socialization can be a powerful source of social identity and beliefs.

Formal religious socialization occurs through explicit religious instruction. An example is the formal religious instruction that Jewish children receive prior to bar mitzvah or bas mitzvah. The purpose of this religious training is to instill the values, history, and beliefs of the Jewish faith in the young person. Likewise, a child who is becoming socialized into the Christian faith will receive religious training in Sunday school or catechism classes.

Religion and the Family

One of the strongest influences on religious socialization is the family. Most people adopt the faith of the family they are born into. Typically, children develop some religious identity by the time they are five or six years old, primarily through the influence of the family (Chalfant et al., 1987). As late as high school age, most youngsters conform to their parents' religious orientation and levels of church attendance (Willits and Crider, 1989).

Religious rituals within the home, the celebration and recognition of religious holidays, attendance at religious ceremonies, and actual religious instruction inculcate religious values in the young. Sitting at a seder (the Jewish ritual dinner held on the first evening of Passover in commemoration of the exodus from Egypt), young children are socialized into the religious teachings of the Jewish faith. The seder includes a dialogue between the father and the youngest child in the house and includes rituals that teach young people the history and beliefs of Jewish people. In Christian households, children learn early about the birth and death of Jesus through enactments of the Nativity story at Christmas and the Passion (suffering and crucifixion of Jesus) at Easter.

In Latino homes, it is common to see religious symbols displayed. Women in Latino families are believed to be the primary carriers of religious belief systems, just as they are in White and African American families; instilling religious values in young children is seen to be part of women's role. Among Latino Catholics, religious ceremonies are an important part of family events recognizing different phases in the life cycle. The *quinceañera* among Mexican Americans recognizes the fifteenth birthday of girls, but is it also linked to religious confirmation, linking the child with the family and the church and seen as cleansing the child of "original sin" (Williams, 1990).

Even when religious beliefs are not explicitly taught through family-based rituals, children learn the beliefs of their parents. For example, young children may early come to believe that God makes the sun come up every morning or puts the moon in the sky if parents (or older siblings) tell them so. Often parents who hold little religious faith themselves find that when they have children, they want to raise them in the same faith in which they grew up. It is common for people who dropped out of religious activities during their young adulthood to return to religious practice when they begin raising families (Roof, 1993).

As people grow older, the influence of the family of origin on religious beliefs lessens but does not disappear. That influence must compete, however, with new experiences and exposure to new communities. College students, for example, frequently find themselves profoundly questioning the religious identity they may have grown up with; some dissolve their religious affiliation altogether or renounce religious beliefs. Some may do so only for a time, becoming recommitted to religious values later. Likewise, upon marriage, the religious identification of one's spouse can be a particularly influential source of one's own religious identity, and many who may have given up religion as young adults return to a religious faith once they have their own children.

Religious groups socialize newcomers into their congregations. Socialization occurs through the observance of religious norms and beliefs and engaging in religious rituals. By doing so, groups extend their influence over the person. Because religious convictions are central to many beliefs and attitudes, the group need not be physically present for its influence to be felt once socialization has been initiated. The process of religious socialization is ongoing as the individual participates in group activities (Johnstone, 1992).

Religious Conversion

Becoming religious may be a private or public process, subtle or dramatic. *Conversion* is a transformation of religious identity. It can be slow and gradual, as when someone switches to a new religious faith (such as a Christian person converting to Judaism or an African American Baptist converting to Catholicism). Conversion may also be more dramatic, such as when a person joins a cult or some other extreme religious group. Sociologists have been especially interested in these latter forms of conversion, perhaps because they have also captivated the public.

In common thought, those who enter extreme religious groups are thought of as brainwashed, manipulated, and coerced. Because such groups proselytize (that is, actively recruit converts), there is a popular image of religious conversion as involving a radical change in one's identity. Are people who become converts to religious cults or other extreme groups being "brainwashed"? The **brainwashing thesis** claims that innocent people are tricked into religious conversion, that religious cults manipulate and coerce people into accepting their beliefs. This popular explanation of extreme religious conversion is seen as analogous to prisoners of war and hostages whose utter dependence on their captors can lead to erosion of the ego and a tendency to take on the captor's perspective. The brainwashing thesis may have become so popular because it offers a simple explanation for why people join bizarre groups. Dramatic tales of parents who tried to "rescue" their child from a religious cult feed this popular image of religious conversion. The brainwashing thesis portrays religious converts as "passive, unwilling, or unsuspecting victims of devious but specifiable forces that manufacture conversion" (Wright, 1991: 126). The cults that they join are seen as exercising mind control that strips converts of their earlier identities, robs them of free will, and "programs" their minds with cult beliefs.

Sociologists suggest an alternative way of conceiving of the process of conversion into religious cults. **Social drift theory** interprets people as moving into such groups gradually, particularly if they have experienced recent personal strains or have become disenchanted with their prior affiliations. Social drift theory emphasizes that conversion is linked to shifting patterns of association, not simply mind control. People are active participants in the process of their own conversion, according to social drift theory, not simply passive creatures "programmed" with new ideas. The social psychological perspective is that conversion comes from simple, though powerful, social influence. People will do practically anything if their friends and an autocratic leader tell them to do it. No "brainwashing" is needed.

Most conversions to religious cults do not entail a radical alteration of lifestyle. The process of religious conversion can be broken down into several phases. The first phase typically involves an experience that leads a potential convert to perceive disruption or failure in his or her previous life, allowing the person to be open to a serious change in the social environment. Fervent religious conversion tends to produce withdrawal, thereby reducing other ties, such as those to family and friends. An initiate who withdraws to a new and unfamiliar context will be more susceptible to new influences, and in the flush of new experiences, less attentive to inconsistencies and contradictions in the new meaning systems to which they are being exposed—put another way, the person may experience some loss of autonomy. How quickly one is converted at this stage depends in large part on how much social pressure is exerted by the new group (Johnstone, 1992; Robbins, 1988; Wright, 1991). In the case of the Heaven's Gate cult, early followers were asked to give up their possessions and to cut off relations with their family and friends; after joining the cult, they were discouraged from having contact with outsiders (Balch, 1995, 1980).

In the second phase of conversion, an emotional bond is created between the initiate and one or more of the group members. This bond creates in the initiate a greater willingness to adopt the worldview of the new group. As the bond deepens, relationships with those outside the new group are progressively weakened. This is the phase where friends and family members become concerned that they are "losing touch" with the person. Groups often actively manipulate the environment so as to separate new recruits from preexisting ties, intensifying the conditions that promote conversion. A good example of this is the way the potential new members are socialized into the Unification Church, whose adherents are popularly known as "Moonies," after the church founder, the Reverend Sun Myung Moon. Members segregate new recruits from the influence of other social influences by hosting them at weekend workshops and isolating them in week-long seminars. Recruits are showered with constant attention and kindness by existing members, a heady experience the Unification Church calls "love-bombing" (Long and Hadden, 1983; Bromley, 1979).

The third phase of religious conversion is a period of intense interaction with the new group. Presuming the initiate stays with the group, this can lead to total conversion. Again using the Unification Church as an example, new recruits are given activities that keep them from thinking about other things. Hawking flowers in streets and airports not only raises money for the organization, it also gives new recruits something to do, placing them in new social contexts and committing them to group goals. These activities accelerate the process of converting newcomers into members. Initiates are also encouraged to sever ties to their former lifestyles, including abandoning their material resources—a sacrifice that, not accidentally, benefits the cult, as initiates usually donate their material goods to the organization (Bromley, 1979; Long and Hadden, 1983; Robbins, 1988).

These phases emphasize that religious conversion is a social process involving a blend of situational circumstances, social and environmental prerequisites, psychological needs of the individual, individual receptivity to change, and active socialization efforts on the part of the recruiting group (Johnstone, 1992). The majority of dramatic religious conversions occur under conditions created and manipulated by the religious group a person is entering. As people convert, they increasingly act according to narrowly prescribed expectations, assuming roles that they come to believe in (Balch, 1980).

Debunking Society's Myths

Myth: **People who join extreme religious cults are maladjusted and have typically been brainwashed by cult leaders.**

Sociological perspective: **Conversion to a religious cult is usually a gradual process wherein the convert voluntarily develops new associations with others and through these relationships develops a new worldview.**

Deconversion also involves a social process of disengagement. Like conversion, it occurs in phases. First, deconversion requires that the emotional attachment to the group be broken. Thenceforth, the process of deconversion is roughly the reverse of conversion. That is, former converts are opened to new ideas and forms of association, eventually breaking from their original group. Interestingly, sociologists have found that most converts who leave religious cults typically join others, such as the Hare Krishnas and the Unification Church. There is a popular image that religious converts need so-called deprogramming or radical resocialization to rejoin society, but studies find that most cult defectors (89 percent) are slowly reintegrated into society, achieving complete reintegration in most cases within no more than two years of leaving the cult (Wright, 1991).

Social and Political Attitudes: The Impact of Religion

Given the strong effect that religion has on people's beliefs, it is no surprise that there is a powerful relationship between religious identification and social and political attitudes. Some of these links may be surprising. For example, contrary to what one might think, Protestants are more likely than Catholics to want to see a reversal of *Roe v. Wade* (the Supreme Court decision upholding women's rights to abortion). This unexpected finding is probably in large part because of the number of fundamentalists among Protestants since they are the group most likely to want this court decision reversed because of their stand against abortion. Jewish Americans, whose politics on most issues tend to be relatively liberal, show the greatest support for abortion rights (Hugick, 1992a).

Religious identification is a good predictor of how traditional a person's gender beliefs will be. Generally speaking, those with deeper religious involvement have more traditional gender attitudes; however, the range falls within a larger frame—religious affiliation has an important influence on where gender attitudes fall in comparison to the nation as a whole. The gender attitudes of Roman Catholics, for example, are more liberal than is commonly believed. Mormons and Pentecostals are the most traditional (Morgan, 1987; Brinkerhoff and MacKie, 1984). Religion also affects people's sexual attitudes and behavior. The most religiously devout tend to be the least sexually active and the most conventional in their sexual practices. Those who are most religious are also least likely to have sex before marriage (Woodruff, 1985).

The outlooks of different religious groups vary on many social issues. Jewish Americans, Unitarians and Universalists, and those with no religious preference are generally the most liberal. On the issues of women's role, race relations, homosexuality, abortion, and premarital sex, the most conservative groups are certain fundamentalist Protestant denominations, Mormons, and Jehovah's Witnesses (Roof and McKinney, 1987). The easily discernible trends within groups and the striking differences between groups are part of the reason that religion has played an increasing role in the political life of the country in recent years.

THINKING SOCIOLOGICALLY

Go to the library and look up one of the recent Gallup national opinion polls. Find a question asked on a subject of interest to you (for example, abortion rights, a presidential election, or gay rights) and look to see if opinions of different religious orientations are reported. What does this tell you about the connection between *religiosity* and social and political attitudes? If you were designing the national survey, what questions would you ask?

Racial Prejudice

The relationship between religious belief and prejudice has long been of interest to social scientists. Bigotry against religious groups is one of the ways that prejudice is manifested, such as in the desecration of Jewish temples and cemeteries that sometimes occurs. Not only are religious groups sometimes targeted by prejudice, but religious belief is also related to the likelihood that someone will be prejudiced or not. The relationship between religion and prejudice is not a simple one, however.

Researchers have distinguished patterns relating the degree of religious prejudice to the very nature of religious experience. An *extrinsic religious orientation* denotes an exclusionary and highly devout religious attitude, such as that of fundamentalist religious groups. An *intrinsic religious orientation,* on the other hand, is more tolerant and open to different forms of religious expression; this is more characteristic of some mainstream Protestant churches. A *quest orientation* features a searching attitude toward religion, as is the case with Quakers, Unitarians, and Jews; this is also characteristic of those who join New Age spiritual groups. People with a quest orientation tend to be tolerant of ambiguity in general—that is, they are more comfortable in situations where doubt, not certainty, is predominant. Of the three, the extrinsic religious orientation is the one most significantly correlated with prejudice; those of this orientation are also more committed to orthodoxy (conformity to established doctrine) in their religious beliefs. A quest orientation is the least likely to be associated with racial prejudice. People who belong to religious organizations that encourage intolerance of any form are also more likely to be racially prejudiced (Adorno et al,, 1950; Allport, 1966; Griffin et al., 1987; Sapp, 1986).

Religion is not always expressed through formal religious organizations. New Age spirituality, for example, is a popular movement that has created alternative forms of religious expression, yet ones that are still based in ritual, symbols, and spiritual belief systems.

Homophobia

Homophobia (the fear and hatred of homosexuals) has also been linked to religious belief. Some argue that Christianity has encouraged homophobia in society because the Bible is interpreted as prohibiting same-sex sexual relations. Some religious congregations have actively worked to encourage the participation of gays and lesbians. This has sparked controversy in some churches, such as the fight within the United Methodist Church and the United Presbyterian Church about whether to ordain gays and lesbians (Clark et al., 1989). Protestants are more likely than Catholics to think that a homosexual relationship between consenting adults is an acceptable lifestyle and should be legal. As with racial prejudice, religious orientations that promote intolerance of any kind are likely to promote homophobia.

Antisemitism

Antisemitism is the belief or behavior that defines Jewish people as inferior and that targets them for stereotyping, mistreatment, and acts of hatred. Antisemitism is one of the world's most persistent forms of prejudice. Early sociological research found that the people most likely to be antisemitic were those with less education and those experiencing downward mobility in the social class system (Selznik and Steinberg, 1969; Silberstein and Seeman, 1959). A strong relationship between antisemitism and other forms of intolerance has long been established by social science research (Adorno et al., 1950). Antisemitism is present in stereotypes directed against Jewish people, as in the stereotype of a "JAP" (Jewish American princess), referring to Jewish American women who are perceived as spoiled, rich, and aloof. Note that this stereotype combines antisemitic, gender, and class stereotyping.

Like other forms of prejudice, antisemitism is associated with a characteristic social-psychological profile, which includes being highly authoritarian (that is, blindly submissive to authority), being aggressive, having conventional values, having a preoccupation with dominance and submission, and holding generalized hostility. Each of these personality traits contributes to the likelihood that someone will be antisemitic (Adorno et al., 1950).

Like other forms of intolerance (such as racism, sexism, and homophobia), antisemitism is expressed through the beliefs and actions of specific people, but it has its origins in the social structural conditions within society. Some have been surprised by incidents of antisemitism occurring, for example, on college campuses where one expects educated people to be more tolerant and knowledgeable about religious difference. Societal conditions, however, can promote the expression of group hatred. One recurring reason for antisemitic acts and beliefs is that the Jewish people are being scapegoated, blamed for problems that other groups might be having. Antisemitic prejudices allege that Jewish people control the professions, the media, and the money in this country even though such elites are overwhelmingly White, Anglo-Saxon Protestants. But if people believe these stereotypes, they may blame Jewish people for the problems that others are having—whether this is a college student facing an uncertain future or a Klan member seeing the marginal economic position of many in the working class. The highly competitive economy, decline of economic opportunities, and conservative political climate of recent years thus encourage a rise in antisemitic behavior. Those who express antisemitic beliefs may not see the complex social causes that create their situation, taking the path of blaming another group for their situation.

Religion and Social Change

Despite the stability of the major faiths, religion, like other social institutions, evolves over time as people adapt to changing social conditions. Religion historically has also had a deep connection to other forms of social change. Although many people think that religion in losing its influence in society, and there has been a decrease in the importance of religion to many people (see Figure 16.5), there has also been a rise in recent years in the influence of conservative religious groups, who have displayed the ability to campaign successfully for their own social and political agenda, and religion continues to have an important role in liberation movements around the world.

The New Religious Conservatism

The rise in religious conservatism is one of the most visible social changes in recent years. Evangelical groups have increased in size and influence, and have affiliated with conservative political causes, resulting in a dramatic shift in the influence of religion on politics. The rise of religious conservatism is not new. During the mideighteenth century, a movement known as the Great Awakening generated a strong religious revival in the United States, followed in the early nineteenth century by the Second Great Awakening. The Great Awakening defined individual faith in God as the path to salvation on earth. It had an egalitarian emphasis since any person with faith in God could be "saved." Religious fervor peaked again in the 1950s, when the United States was rallied against a communist foe defined as a "godless, atheistic menace" and worldwide enemy of religion (Chalfant et al., 1987: 148). Religious leaders like Norman Vincent Peale and Billy Graham led populist movements in the 1950s based on conservative religious beliefs and devotion to "the American way." They used radio and television as the medium to appeal to mass audiences and to build a revivalist mentality among the public.

Reactions to the social turmoil of the 1960s and 1970s ushered in a new movement of religious conservatism in the 1980s, emphasizing a return to traditional values and the subordination of self to the authority of the Bible. Contemporary evangelicals define personal salvation as being "born again." They are proselytizers who attempt to convert all nonbelievers to evangelical beliefs in the conviction that theirs is the only true religion.

There has been a dramatic increase in the number of people who say they are born again or evangelical. Forty-three percent of the U.S. population call themselves "born again," a high percentage. Women are more likely to do so than men; African Americans are more likely to do so than Whites. Thinking of oneself as born again is also correlated with education and income. Those with no college experience are most likely to think of themselves as born again; those with college degrees, least likely. Considering oneself born again is also more common in lower-income groups. Evangelicalism seems to appeal most to blue-collar workers and the poor, although its appeal is increasing among the young (Roper Organization, 1993, 1994; Chalfant, 1987).

The hallmarks of evangelical religious groups are the intensity of their religious beliefs, the exclusivity of their membership, their emphasis on personal salvation through faith in Jesus, their belief that the Bible is a practical guide to everyday life, and their suspicion of religious compromise (Stacey and Gerard, 1990; Chalfant et al., 1987). These groups believe in a literal reading of the Bible. That is, they do not see the Bible as a historical document to be interpreted, but as the direct word of God. One-third of all Americans say they believe the Bible is the word of God, to be taken literally. Another half say the Bible is inspired by God, but that not everything in it should be taken literally; 16 percent say it is an ancient book of legends, fables, and history (Gallup and Newport, 1991). Evangelists vociferously disagree with these latter two groups.

The evangelical movement consists of diverse groups, including Faith Assemblies of God, Churches of Christ, and the Jehovah's Witnesses, to name only a few. In the past, conservative religious groups had largely distanced themselves from politics. Beginning in the 1970s, however, conservative activists realized they could have an enormous impact on national politics if they mobilized the growing numbers of conservative Christian groups. Their affiliation with each other and with conservative political causes has resulted in a movement known as the "new Christian right" (Liebman and Wuthnow, 1983).

The conservative Christian movement has fueled antiabortion activism, has revived the effort to teach creationism in the schools, and has campaigned against women's rights and gay and lesbian rights, while supporting so-called pro-family legislation that promotes a variety of conservative values. The Christian right sees the changing role of women in society and the influence of the feminist movement as threatening traditional "family values" and undermining what they see as "natural" arrangements between women and men in the family. They have been active on other fronts, too, including movements to ban books they deem objectionable from public schools, to put prayer into the public schools, and to withhold federal funding for abortion. Using Christian broadcasting networks and new "marketing" strategies such as direct-mail appeals, conservative Christians have been enormously effective in mobilizing conservative opin-

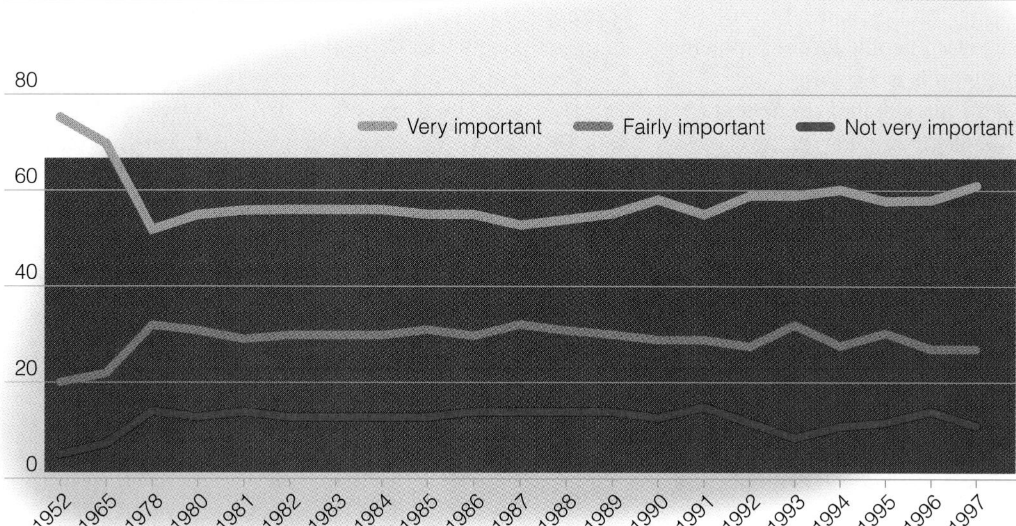

FIGURE 16.5 How Important Is Religion in Your Own Life?

SOURCE: Roper Organization. 1996. *The Gallup Poll*. Storrs, CT: Roper Organization, p. 1; David W. Moore. 1995. "Most Americans Say Religion Is Important to Them." *The Gallup Poll Monthly* (February), p. 17.

ion and influencing local and national elections. They have also raised millions of dollars from devout believers, often in small amounts, to be used for conservative causes. Politicians like Pat Buchanan and commentators like Rush Limbaugh partially owe their success to the burgeoning of this new political and religious movement.

Although diffuse and consisting of small congregations, evangelicals form a tight-knit national community. Their unity has been facilitated by the strong Christian presence on the electronic media. Indeed, Christian religious media—television, radio, books, and magazines—have become a major industry. It is estimated that more than 1400 radio stations and 60 television stations are owned by religious groups; the Christian Broadcasting Network (CBN) is a major television network that brings news and entertainment from a "Christian point of view" to millions of households. These networks generate in excess of $500 million in untaxed revenue—most of which is used for political purposes (Johnstone, 1992).

Race, Religion, and Justice

What is the role of religion in social change? Durkheim saw religion as promoting social cohesion; Weber saw it as culturally linked to other social institutions; Marx assessed religion in terms of its contribution to social oppression. Is religion a source of oppression, or is it a source of personal and collective liberation from worldly problems? There is no simple answer to this question. The role of religion in social change is as broad as the spectrum of people who experience religious drives—as broad as society itself. Religion has had a persistent conservative influence on society, but it has also been an important part of movements for social justice and human emancipation. For example, religion has played an important role in the development of social movements for the liberation of Latinos; we mentioned earlier the emergence of liberation theology among Catholic clergy in Latin America. Liberation theologians have used the prestige and organizational resources of the Catholic Church to develop a consciousness of oppression in poor peasants and working-class people.

Debunking Society's Myths

Myth: **Most deeply religious people are intolerant of social change.**

Sociological perspective: **Some religious affiliations and belief systems are more likely to promote progressive social change than others. Protestant fundamentalist groups are typically conservative in their social and political attitudes, whereas other groups have used their religious commitments as the basis for the promotion of civil rights and other progressive social issues.**

One of the central themes of African American religiosity has been liberation from oppression. African American spirituals are rich with the symbolism of struggle and redemption from bondage. During slavery, churches served as "stations" on the Underground Railroad that smuggled Blacks out of slave states, and they were meeting places where social change was mapped out.

African American churches had a prominent role during the mobilization of the civil rights movement (Marx, 1967 [1867]; Morris, 1984). Churches served as headquarters for protestors and clearinghouses for information on protest strategies and organizational tactics. Churches supplied the infrastructure of the developing Black protest movements of the 1950s and 1960s, and the moral authority of the church was used to reinforce the appeal to Christian values as the basis for racial justice.

Militant Black leaders have tended to be less religiously oriented than more moderate activists; nevertheless, religion has been important in the development of radical Black protest, too. *Black nationalism* is an ideology that sees African Americans as having a unique identity as a nation and deserving the right to self-determination. Black nationalism takes many forms and is a political movement that traces its origins to the Black Muslim religious movement.

Religious practice among Black Muslims involves strict regulation of dietary habits and prohibitions against many activities, such as alcohol use, drug use, and gambling. Muslims elevate personal habits like cleanliness of body and home to moral duties. Muslims are forbidden from eating foods identified as traditional food among southern African Americans (like collard greens and cornbread)—symbolic of Muslims' rejection of the slave heritage—and they are also forbidden from indulging in vanities like using cosmetics or straightening their hair.

Black Muslims have promoted the idea of community self-control for African Americans and have set examples by establishing many of their own institutions, such as schools, businesses, and communications media (Essien-Udom, 1962; Lincoln and Mamiya, 1990). The emphasis of Black Muslims on self-reliance and traditional African identity has earned it a fervent following, although the actual number of Black Muslims in the United States is relatively small. Its criticism of the White power structure has given the Black Muslims a radical political dimension; yet, even with its radical edge, Black nationalism has a rather conservative framework. The Muslim faith is highly patriarchal. The role of women is defined in terms of serving men and the family. In developing from this tradition, contemporary Black nationalism has emphasized Black manhood, often at the expense of independent leadership for women. The leadership of women in the civil rights movement, for example, was eroded as Black nationalist movements gained in strength and influence.

Women and Religion

Generally speaking, women are more religiously devout than men. Women's work in religious organizations, whether churches, synagogues, mosques, or religious clubs, has also sustained religious practice. Women have raised much of the money that has supported religious organizations, have organized the social services provided by religious organizations, and have passed on religious traditions to children. For many religious organizations, the movement of women into the paid labor force has resulted in a shortage of the volunteer labor that women have historically provided to religious organizations. In the Catholic Church, for example, there is concern that the population of nuns will not be replenished as current nuns grow old; as women have other opportunities, the limited resources nuns are given, especially relative to priests, makes this a less attractive occupation for young women.

Women have also had to bear being denied the right to full participation in many churches. Some churches refuse to ordain women as clergy, and in the past women were

Although some religions still deny ordination or formal religious leadership to women, others, such as Episcopals, have begun ordaining women as priests.

denied admission to divinity schools. Change is occurring, however. Although women are still a numerical minority in divinity schools, their numbers have grown. In the 1990s, women earn 15 percent of seminary degrees. African Americans earn 8 percent of all theological advanced degrees; Hispanics, 1 percent; Asians/Pacific Islanders, 5 percent; and Native Americans, less than 1 percent (U.S. Department of Education, 1996: 293).

The public now generally supports the ordination of women. Even among Roman Catholics, a majority support the ordination of women. A slight majority opposes this step among evangelical Protestants. Across all denominational groups, men are more supportive of female ordination than women although the difference among Protestants is small. The greatest opposition to the ordination of women comes from those who see women as "weaker vessels," incapable of performing essential ministerial tasks; conversely, proponents think women would bring special competence to the task in some areas, such as nurturing and mothering (Roper Organization, 1993; Bedell and Jones, 1992; Jelen, 1989).

African American women have tended to be more equal to men within African American churches than White women have been in White churches. African American women make up the largest percentage of African American church congregations, typically about 75 percent, and they have been far more likely than White women to assume positions of leadership in the ministry. The *sanctified church* (a term used to refer to Holiness and Pentecostal churches in the African American community) has elevated women to positions of leadership and has even rejected the patriarchal organization of mainstream churches and encouraged more feminist models of religious organization and practice (Gilkes, 1985).

Traditional religious images of women have been the basis of many gender stereotypes. For example, although

liberatian theology

women are now rabbis in some reform congregations, Orthodox Jewish men are supposed to thank God every morning that they are not women and are prohibited from associating with women who are menstruating. Among Christians, the New Testament of the Bible encourages women to be subordinate to their husbands, reflecting a presumed natural order in which women are subordinate to men as men are subordinate to God. Many feminists believe that a great deal of *misogyny* (hatred of women) has been inspired by the biblical creation story in which Woman, in the form of Eve, is responsible for the fall of Man. This biblical legend is the archetype, the original mold, of the stereotype of women as seductresses leading men into sin. Two polarized images of women developed within Christianity: (1) women as temptresses, witches, and whores and (2) women as madonnas. The net result is that religion has been a powerful source of the subordination of women in society.

On the other hand, religion has also been the basis for liberation for many women. Experiences within religious institutions have been used to enrich the thinking of many strong advocates of feminism, and there has been a growing trend for feminist women to redefine traditional religious orientations in ways that are supportive of feminist beliefs. Whereas some feminists see religion as the enemy of women's liberation, others have inaugurated spiritual movements that articulate religious beliefs promoting egalitarianism and a commitment to gender and racial justice.

Religion in Decline?

There has been a general trend in American life of a decline in religious participation. Since the 1970s, church membership has declined, and fewer Americans are likely to say that religion is a significant influence in their lives. People are more likely to believe that science will solve the world's problems than they are to see religion as the solution. Does this mean that religion is losing its force in American society?

Secularization is the process by which religious institutions, behavior, and consciousness lose their religious significance (Wilson, 1982). Secularization is not "antireligion." It refers to the process by which society becomes increasingly complex, bureaucratized, fragmented, and impersonal. In a secular society, people are no longer bound by sacred principles.

Secularization is a change in the basic organization of society (Wilson, 1982). There is a shift from religious to secular control of all institutions, including education and the family. With secularization, people spend less time in religious activities. The religious regulation of everyday life is supplanted by other, less spiritual imperatives. This has led some to say that science is the modern religion. Max Weber referred to this trend as the **rationalization of society**, by which he meant that society is increasingly organized around rational, empirical, and scientific forms of thought. Although this may seem like a positive development, in that it has led to many cultural advances, its consequences are not all good. Although for many, the rationalization of society liberates them from the constraints that religion can impose, it leaves many feeling that they have no spiritual attachments. Without strong religious belief systems, society can seem adrift, as we see in the words of many social commentators who say the United States is in a moral crisis, is overly materialistic, and has no strong spiritual cohesion.

Secularization is a long-term process, reflecting the historical change from a communally based society to one based on more formal and impersonal associations. In societies undergoing such change, the basis of the moral order shifts to the rule of law and the dominance of legal principles rather than spiritual or moral ethics. The society becomes less spiritual, more matter-of-fact. Religious agencies are likely to lose their political power. In general, religious consciousness is replaced with a more rational and instrumental approach. There is, in such a society, less dependence on charms, rites, spells, and prayers, and more inclination to accept things based on their demonstrated scientific truth—or the potential for profit and individual advantage. This is reflected in all aspects of the society's culture, including music, art, education, and popular entertainment (Chalfant et al., 1987).

As society becomes more secularized, traditional religious values lose their strength, and various forms of religious experimentation are likely to develop. Bereft of spiritual attachment, people seek out new means of psychological and social expression. One consequence is the development of new religious movements that counter the trend toward secularization. This helps explain the popularity of new groups like EST, Rastafarianism, Eckankar, and assorted other religious cults; these groups are attractive to those who seek meaningful ties and a sense of community (Chalfant et al., 1987). Some groups, like Scientology and EST, package "salvation" in commercial form as workshops, books, and tapes—a practice consistent with the consumer orientation in contemporary society. Religious groups may even compete for consumers' attention just as consumer advertisers do, creating what one commentator has called a "supermarket of faith" (Wilson, 1978: 80).

There is little doubt that the United States has become a more secular society, but it would be wrong to say that religion has lost its hold. As we have seen, despite the general decline in religious participation, some religious groups have become more influential. It seems more accurate to say that religion is languishing among more moderate groups, like traditional moderate Protestants, but it is flourishing on the right (primarily among evangelical Christians) and the left (among groups like the Black Muslims and some feminist spirituality groups). Even though a growing number of people have no contact with organized religion, moral and political values in the society are increasingly polarized around religious differences. The greatest polarization has occurred around issues of family and school (Roof and McKinney, 1987). Moral traditionalists accuse libertarians of being "secular humanists," and free thinkers accuse dogmatic Christians of interfering with personal liberties. Religion may show some signs of decline, but it remains a significant social force and social institution in the contemporary United States.

CHAPTER SUMMARY

- Sociologists are interested in religion because of the strong influence it has in society. *Religion* is an institutionalized system of symbols, beliefs, values, and practices by which a group of people interprets and responds to what they feel is sacred and which provides answers to questions of ultimate meaning. The *sacred* is that which is set apart from ordinary activity, is seen as holy, and is protected by special rites and *rituals*. *Totems* are sacred objects that people regard with special awe and reverence.
- The United States is a deeply religious society. Christianity dominates the national culture, even though the U.S. Constitution specifies a separation between church and state. *Religiosity* is the measure of the intensity and practice of religious commitment. Religions may be *polytheistic* or *monotheistic, patriarchal* or *matriarchal*. *Exclusive religious groups* are those with a highly identifiable set of religious beliefs and a distinctive religious culture. *Inclusive groups* are more moderate and emphasize the importance of common religious work.
- Sociologists see religion as an integrative force in society. Durkheim understood religions and religious rituals as creating social cohesion. Weber saw a fit between the ideology of the Protestant ethic and the needs of a capitalistic economy. Religion is also related to social conflict. Marx saw religion as supporting societal oppression and encouraging people to accept their lot in life.
- The United States is a diverse religious society. Protestants, Catholics, and Jews make up the major religious faiths in the United States. Patterns of religious faith and participation vary by age, income level, education, ethnicity, and race. People learn religious faith through *religious socialization,* which can be formal or informal. The family is one of the major sources of *religious socialization*. Religious conversion involves a dramatic transformation of religious identity and involves several phases through which individuals learn to identify with a new group and lose other existing social ties.
- *Churches* are formal religious organizations. They are distinct from *sects,* which are religious groups that have withdrawn from an established religion. *Cults* are groups that have also rejected a dominant religious faith, but they tend to exist outside the mainstream of society. Religiosity is related to a wide array of social and political attitudes. Racial prejudice, homophobia, and *antisemitism* are all linked to patterns of religious affiliation.
- In recent years, there has been an enormous growth in conservative religious groups. These evangelical groups believe in a personal commitment to Jesus and a literal interpretation of the Bible. They have been highly influential, particularly through their use of the electronic media as a means of communication. They have also been linked with numerous conservative political causes and have had a great influence on recent national elections. Religion is a conservative influence in society, but it has also an important part in movements for human liberation, including the civil rights movement and the move to ordain women in the church. *Secularization* is the process in society by which religious institutions, action, and consciousness lose their social significance. Although religion is in some ways in decline in modern society, it also retains significant social influence.

KEY TERMS

antisemitism
brainwashing thesis
charisma
church
collective consciousness
cult
ethnoreligious group
exclusive religious group
ideology
inclusive religious group
matriarchal religion
monotheism
patriarchal religion
polytheism
profane
Protestant ethic
rationalization of society
religion
religiosity
religious socialization
ritual
sacred
sect
secular
secularization
social drift theory
totem

THE INTERNET: A Tool for the Sociological Imagination

Resources on the Internet:

Virtual Society: The Wadsworth Sociology Resource Center
http://sociology.wadsworth.com

Visit this site to find additional learning tools, including interactive quizzes, links related to web sites, and an easy link to *InfoTrac College Edition*.

The Center for the Study of American Religion, Princeton University
http://www.princeton.edu/~nadelman/csar/csar.html

The Center for the Study of American Religion encourages the academic study of religion; the web site contains information about the Center's activities and links to numerous other sites for the study of American religion.

Academic Info Religion
http://www.academicinfo.net/religindex.html

This site is a reference to numerous Internet resources on religion. It includes links to sites on diverse religious faiths, religious studies programs, directories of religion in the United States, new religious movements, and special topics, such as women and religion.

American Religion Links
http://www.as/wvu.edu/coll03/relst/www/link.htm

Published by West Virginia University, this site includes links to online journals, courses, and other resources pertinent to the study of religion.

CultureWatch Online
http://www.igc.apc.org/culturewatch

This organization provides information about the U.S. religious right.

Sociology and Social Policy: Internet Exercises

A long-standing issue in the United States is the constitutional principle of separating church and state. In recent years, many policy questions have centered on questions of public support for religious observances, such as support for *school prayer* and civic displays of religious symbols. At the heart of these policy debates are questions about how to observe the constitutional separation of church and state while also allowing diverse groups the freedom of religious expression. What are the sociological dimensions of questions about school prayer and the use of public facilities for religious purposes? If you were the principal of a public school, what policies would you promote regarding school prayer and the use of school facilities by religious groups?

Internet Search Keywords:

prayer in school
separation of church and state
civic displays of religious symbols
freedom of religion

Web sites:

http://www.louisville.edu/~tnpete01/church/index.htm
Very thorough site about "separation of church and state."

http://www.okbu.edu/library/GUIDES/bibliog/bibchrch.htm
Some U.S. Supreme Court decisions, references and other web sites about separation of church and state.

InfoTrac College Edition: Search Word Summary

antisemitism	protestant ethic
liberation theology	religiosity

In order to learn more about these central topics in sociology, you can conduct an electronic search using InfoTrac College Edition. To aid in your search and to gain useful tips, see the Student Guide to InfoTrac College Edition on the Virtual Society web site:
http://sociology.wadsworth.com

INTERACTIONS—A SOCIOLOGY CD-ROM: CONCEPTS FOR THIS CHAPTER

Go to the Wadsworth Sociology CD-ROM for further study on the concepts in this chapter. The CD-ROM also includes quizzes and additional activities to expand your learning experience.

SUGGESTED READINGS

Allen, Paula Gunn. 1986. *The Sacred Hoop: Recovering the Feminine in American Indian Traditions.* Boston: Beacon Press.

Allen's book examines the woman-centered perspectives of American Indian faith. It is a good exploration of the diversity of religious belief within American culture.

Ammerman, Nancy Tatom. 1997. *Congregation and Community.* New Brunswick, NJ: Rutgers University Press.

Ammerman studies twenty religious congregations located in neighborhoods where there is significant social change: new immigrant populations, visible gay and lesbian populations, suburbanization, and economic restructuring. She examines how some religious institutions respond to such changes and become sources of civic participation, whereas other religious institutions decline in their influence.

Beck, Evelyn Torton. 1989. *Nice Jewish Girls: A Lesbian Anthology,* 2nd ed. Boston: Beacon Press.

Beck's collection provides several essays on the influence of Judaism on growing up as a woman and on the tensions and continuities some Jewish women find between their Jewish identity, lesbian, and feminist values.

Christ, Carol P., and Judith Plaskow. 1992. *Womanspirit Rising: A Feminist Reader in Religion.* San Francisco: Harper.

This collection of essays explores various dimensions of the feminist critique of religion, including women's role in religion and new perspectives on feminist theology and spirituality.

Lincoln, C. Eric, and Lawrence H. Mamiya. 1990. *The Black Church in the African-American Experience.* Durham, NC: Duke University Press.

This is a comprehensive analysis of the influence of diverse religious groups and experiences in the history of African American people. It also shows the centrality of religious faith in movements for racial justice.

Roof, Wade Clark. 1993. *A Generation of Seekers: The Spiritual Journeys of the Baby Boom Generation.* San Francisco: Harper San Francisco.

Roof studies the changing religious values of the Baby Boom generation, exploring the general decline in their religious attendance, but also, in some cases, their return to religious activity. In general, he also explores the spiritual values of this generation of Americans.

Stark, Rodney. 1996. *The Rise of Christianity: A Sociologist Reconsiders History.* Princeton, NJ: Princeton University Press.

Stark uses a sociological analysis to explain the social and historical origins of Christianity. By analyzing the class and gender structure of ancient society, he relates the extraordinary growth of Christianity to an analysis of social movements and social networks.

Stark, Rodney, and William Sims Bainbridge. 1997. *Religion, Deviance, and Social Control.* New York: Routledge.

The authors explore the influence of religion on regulating human behavior, examining whether religion can prevent such various forms of deviant behavior, including crime and drug abuse.

Wallace, Ruth A. 1992. *They Call Her Pastor: A New Role for Catholic Women.* Albany, NY: State University of New York Press.

This analysis of the role of women in the Catholic Church is based on studies of twenty Catholic parishes. It examines women's leadership in the Catholic Church and the appointment of women as lay administrators of parishes that have no priests.

CHAPTER 17

Work and the Economy

EVERY day millions of people get up, get dressed, and go to work. Another several million wish they could find work. Many of those who work wish they earned more or had jobs with better working conditions. Work has an enormous effect on most aspects of your life, including relationships with your family, how much power you have, and what resources are available to you. Work establishes the routines of daily life—what you wear, with whom you associate, what hazards you face, and even what time you get out of bed. For some, work is a satisfying experience that provides the opportunity to be creative, pursue talents and interests, and be effective; for others, work is a source of frustration, disappointment, and exploitation. Any way you look at it, the significance of work in the lives of people is hard to overestimate.

When thinking about work, most people think in terms of what kind of work they might do, how much they can make, whether their work will be satisfying, and whether their job will give them opportunities for promotion and advancement. Students, for example, may see their education as bringing them work opportunities or promotions, but they may also worry about what work they will find and whether their work will be meaningful, utilize their talents, and bring them a lifetime of rewards. These concerns can be understood in a sociological framework even though people tend to think they are individual problems. Sociologists study the social forces that shape people's experiences and understand how things like downsizing, globalization of the economy, corporate restructuring, and technological change affect work. Sociological studies place the concerns of individual people about work into a broader social structural context.

Sociologists, for example, are interested in how work is shaped by specific economic systems, like capitalism, socialism, or communism. A major question for sociologists is how systems of inequality influence the different jobs people have and how well they are rewarded for their work. Sociologists seek to understand how social factors like gender, race, social class, and age shape the work experience of different groups. Why, for example, are Hispanics most likely to be employed as service workers, whereas White Americans are the group most likely to be administrators? Sociologists also study the influence of other social institutions on work, such as how education is related to the work people do. Other institutions, such as the family, health care, and the government also are interrelated with the world of work.

Another major issue for sociologists is how social and economic changes influence work. Technological innovations are currently transforming work in ways that may have been unimaginable just a few years ago. How these continuing innovations will shape work in the future remains to be seen, but they will be critical to the work experience of those now entering the labor market. All these subjects—the nature of work, the organization of work, the rewards associated with work, and the stresses and satisfactions of work—can be analyzed by applying the sociological imagination to what otherwise seems like a bunch of individual experiences.

Economy and Society

To understand work, you first have to see it in the context of the broad social institution known as the economy. All societies are organized around an economic base. The **economy** of a society is the system by which goods and services are produced, distributed, and consumed. How the economic structure of a society is organized makes a huge difference not only for how work is done, but more broadly, in the entire life experience of people in society. We first look at the significance of the historic transformation from agriculturally based societies to industrial and, now, postindustrial society.

The Industrial Revolution

In Chapter 5, we discussed the evolution of different types of societies. One of the most significant of these changes was, first, the development of agricultural societies and, later, the far-ranging impact of the *Industrial Revolution.* Now, the Industrial Revolution is giving way to the growth of postindustrial societies—a development in the economic system with far-reaching consequences for how society is organized.

The development of agricultural societies followed from the development of technologies that enabled people to increase the production of food from simple hunting and gathering techniques to more large-scale production. The invention of the plow, for example, enabled societies to form settlements, organized around farm production in fields, instead of the more labor-intensive methods of gathering small amounts of food. This transformation meant that societies could settle in one place and form markets for the sale and distribution of goods, thereby radically changing how society was organized and how people worked. Agricultural production, of course, remains a vast part of the world economy, although now it has been changed by the processes of industrialization—a development that, prior to the invention of the silicon chip—was probably the most significant historical development affecting the social organization of work.

The Industrial Revolution is usually pinpointed as beginning in mideighteenth-century Europe, soon thereafter spreading throughout other parts of the world. The Industrial Revolution led to numerous social changes since Western economies were organized around the mass production of goods. The Industrial Revolution led to the creation of factories, which as we briefly saw in Chapter 12, separated work and family by relocating the place where most people were employed. The Industrial Revolution also transformed the consumption of energy with the large-scale use of coal, steam, and later electrical power, the basis for running the machinery needed for production. Industries also became highly specialized, such that workers would repeat the same action many times over the course of a working day—rather than being involved in the production of the total product (something that, as we will see in the later discussion of worker alienation, has an enormous impact on workers' attitudes). In addition, the Industrial Revolution created a cash-based economy, one in which laborers are paid a cash wage and goods are sold, not for exchange, but for their cash value. All in all, the social relations created by the Industrial Revolution are hard to overestimate.

We still live in a society that is largely industrial, but that is quickly giving way to a new kind of social organization: postindustrial society. Where industrial societies are primarily organized around the production of goods, postindustrial societies are organized around the provision of services. Thus, in the United States, we have moved from being a largely manufacturing-based economy to an economy centered largely on the provision of services. Service industry is a broad term, meant to encompass a wide range of economic activities now common in the labor market. It includes, for example, banking and finance, retail sales, hotel and restaurant work, and healthcare, to name but a few examples; it also includes parts of the vastly expanded information technology industry—not so much the assembly of electronics, but the things like software design and the exchange of information (through publishing, video production, and the like). Advanced technology forms the core of a postindustrial society since it is the mechanism through which most services are delivered and organized. Whereas the Industrial Revolution was once seen as the source for much broad-scale social change, now the *Information Revolution* is probably one of the greatest sources for social and economic change in the future (see also Chapter 22).

The Industrial Revolution transformed labor, moving it from the household to the factory—or other sites where work was mechanized and oriented to mass production.

Comparing Economic Systems

The three major kinds of economic systems found in the world today are *capitalism, socialism,* and *communism.* These are ideal types—that is, many societies have a mix of economic systems, but as a type, each is distinct in its principles and organization. **Capitalism** is an economic system based on the principles of market competition, private property, and the pursuit of profit. Under capitalism, the means of production are privately owned. (Recall from Chapter 9 that the means of production refers to the system by which goods are produced and distributed; this includes how natural resources are garnered, how labor is done, and how goods and services are made and distributed.) To say that in capitalist societies some people own the means of production does not simply mean that people own property, although that is also a feature of capitalism. It means that some people control the natural resources, employ the labor, and own the actual industries in which goods are produced and sold. Within capitalist societies, stockholders, for example, own corporations—or a share of the corporation's wealth. Under capitalism, owners keep a surplus of what is generated by the economy; this is their profit, which may be in the form of money, other financial assets, and other commodities. Profit is generated by owners when a product is sold for more than the cost of creating it. People who work in a capitalist system do so in exchange for money, largely controlled by capitalist owners. For this to be profitable for the owner, profit is made by paying workers less than the value of what they produce. Under capitalism, it is common that workers produce the goods and provide the services, whereas owners disproportionately consume goods and reap the profits. This class relationship is what defines the system of capitalism (see Chapter 9).

The capitalist basis of society in the United States shapes the character of the nation's other institutions. Health care, for example, is based on a capitalist system of privately owned health care institutions even though it also gets some government subsidy (indicative of the fact that this is not a *purely* capitalist society). In other industrialized societies, health care is regarded as a right that is paid for and administered by state agencies—a more socialist model. In the United States, health care institutions are administered on a profit-making system. Other institutions, such as education, are also shaped by capitalism. Although there is strong governmental support for public education in the United States, the most prestigious schools are private institutions, many of which are supported by the investment income from vast endowments and high tuition fees. Interestingly, even public school systems in some cities are now being run by private companies.

Socialism, on the other hand, is an economic institution characterized by state ownership and management of the basic industries; that is, the means of production are the property of the state. Modern socialism emerged from the thought of Karl Marx, who predicted that capitalism would give way to egalitarian, state-dominated socialism, followed by a transition to stateless, classless communism. History has not borne out this prediction, since it is hard to find examples of purely socialist societies, but other aspects of Marx's theory explain the workings of capitalism. Marx may have even underestimated the hold of capitalism on world history since capitalism now penetrates the world economy.

In many nations, the global forces of capitalism mix with socialist principles. Many European nations, for example, have strong elements of socialism. Sweden supports an extensive array of state-run social services, such as health care, education, and social welfare programs, but Swedish industry is capitalist. Likewise, the state in Great Britain has historically owned basic industries such as the railroads, mines, and the communications industry although most other industrial entities are privately owned. Still, the influence of socialist values in Britain can be witnessed in things such as the national support for the BBC (British Broadcasting System). Compare this with the privately owned (although publicly regulated) communications industries in the United States (such as CBS, NBC, ABC, Fox, and CNN). Public radio and television stations in the United States beg for private support from their audiences and risk losing the limited federal support they receive, whereas in Britain, although private communications networks have emerged, the major communications systems are state supported. Federal subsidies for private industries in the United States are evidence that the nation has some degree of socialism mixed in with its capitalist economy.

Other world nations are more strongly socialist although they are not immune from the penetrating influence of capitalism. The People's Republic of China was formerly a strongly socialist society that is currently undergoing transformation to a mix of socialist and capitalist principles, with state encouragement of a market-based economy, the introduction of privately owned industries, and increased engagement in the international capitalist economy. Many developing nations have pledged socialist principles, ranging from the Baath regime of Iraq to a variety of nations in Latin America and Africa. Socialism in the developing world has frequently met with considerable hostility from the capitalist nations, especially the United States, but it is well entrenched in the political economies of Italy, France, England, and other European nations. Some of its features, such as socialized medicine in Britain, are often held up as models for other nations.

Communism is sometimes described as socialism in its purest form. In pure communism, industry is not the private property of owners; rather, the state is the sole owner of the systems of production. Communist philosophy argues that capitalism is fundamentally unjust since powerful owners take more from laborers (and society) than they give; they use their power to maintain the inequalities between the worker and owner classes. Nineteenth-century communist theorists declared that capitalism would inevitably be overthrown as workers worldwide united against owners and the system that exploited them. Class divisions were supposed to be erased at that time, along with private property and all forms of inequality. When the worldwide worker revolution that Marx predicted did not materialize, during the second decade of this century, V. I. Lenin in Russia developed the theory of a communist elite—an upper class—who would lead the workers to their destiny by taking power and using it with complete autonomy to create the communist state. In Russia, communism became combined with *totalitarianism*—a political system in which powerful elites exercise total control over the population (see Chapter 18)—while preserving many of the trappings of class inequality. Although it was never purely communist, Russia now is adopting many elements of capitalism.

A critical feature of communist economics has been centralization of the economy in which administrators declare prices, quotas, and production goals for the entire country. This is perhaps the most striking difference between communism and capitalism. Under capitalism, market forces are permitted to dictate these decisions. The inefficiencies of the centralized economy are one reason that the century-long Russian experiment in communism ultimately failed. The Soviet Union dissolved itself in 1991; its economy has since been struggling to find a stable balance of socialism, capitalism, and remnants of its communist institutions. China, the other communist superpower, has recently had notable success experimenting with a mix of communism and capitalism, and in some regions of China, the economy is growing faster than anywhere else in the world.

The Changing Global Economy

As we approach the end of the twentieth century, one of the most significant developments is the creation of a *global economy,* which is affecting work in the United States and worldwide. In a sense, the U.S. economy has always been international since world trade has always linked the United States to other countries. The globalization of the economy affects much more than trade, however. The concept of the **global economy** acknowledges that all dimensions of the economy now cross national borders, including investment, production, management, markets, labor, information, and technology (Carnoy et al., 1993). Economic events in one nation now can have major reverberations throughout the world, so when the economy in Brazil, Japan, China, or Russia is unstable, the effects are felt worldwide.

The global economy links the lives of millions of Americans to the experiences of other people throughout the world. College students who worry about getting a good job after graduation experience this worry in a world context where manufacturing jobs are being exported to other nations and where many U.S. industries face stiff competition from other parts of the world. You can see the internationalization of the economy in everyday life: Status symbols like high-priced sneakers are manufactured for just a few cents in China. The Barbie dolls that young girls accumulate, though inexpensive by U.S. standards, are made by workers in Indonesia where, as the box "Analyzing Social Issues: Toys Are Not U.S." shows, it would take a month's wages for the Indonesian worker who makes the doll to buy it for her child. Such are the sociological forces now guiding the experience of work both in the United States and abroad.

One of the most apparent aspects of the global economy for U.S. workers is the increased transfer of jobs overseas. When the government produces data that show that the number of manufacturing jobs has declined as a percent of the total labor force, this does not mean that manufacturing is ceasing—only that most of it is being done by foreign workers. For business owners, moving jobs overseas reduces the cost of labor since wages for workers in countries where labor is situated are extremely low. In China, as an example, the young women who make Barbie dolls earn less than \$1.99 per day—and frequently migrate thousands of miles from their homes to do so (Macmillan, 1996; Press, 1996). Soccer balls that in F.A.O.Schwartz sell for \$150 are manufactured in Sri Lanka and Haiti, where not only are wages low, but labor and human rights are routinely ignored. Many of the products sold in the United States are manufactured in sweatshops abroad—where young women and children are the majority of the workers.

THINKING SOCIOLOGICALLY

Identify a job you have once held (or currently hold) and make a list of all the ways that workers in this segment of the labor market are being affected by the various dimensions of *economic restructuring: demographic changes, globalization of the economy,* and *technological change.* What does your list tell you about how people's individual work experiences are shaped by *social structure?*

In the global economy, research and management are controlled by the most developed countries, and assembly-line work is performed in nations with less privileged positions in the global economy (Ward, 1990). A single product, such as an automobile, may be assembled from parts made all over the world—the engine assembled in Mexico, tires manufactured in Malaysia, electronic parts constructed in China. The relocation of manufacturing to wherever labor is cheap and management strong has led to the emergence of the *global assembly line,* a new international division of labor in which research and development is conducted in the United States, Japan, Germany, and other major world powers, whereas the assembly of goods is done primarily in underdeveloped and poor nations—mostly by women and children. The development of the global assembly line has contributed to a major trade imbalance, with the United States increasingly importing the goods it consumes.

Manufacturing overseas suits the interest of capitalist economies because labor in the underdeveloped nations is cheaper and typically nonunionized. The relative absence of state regulations governing work conditions and terms of employment makes the transfer of work overseas even more attractive to profit-seeking corporations. The enhancement in corporate profits made possible by purchasing cheap labor in the poorest parts of the world comes partly at the expense of U.S. workers, however, who are faced with a shrinking domestic job market.

Within the United States, the development of a global economy has also created anxieties about foreign workers,

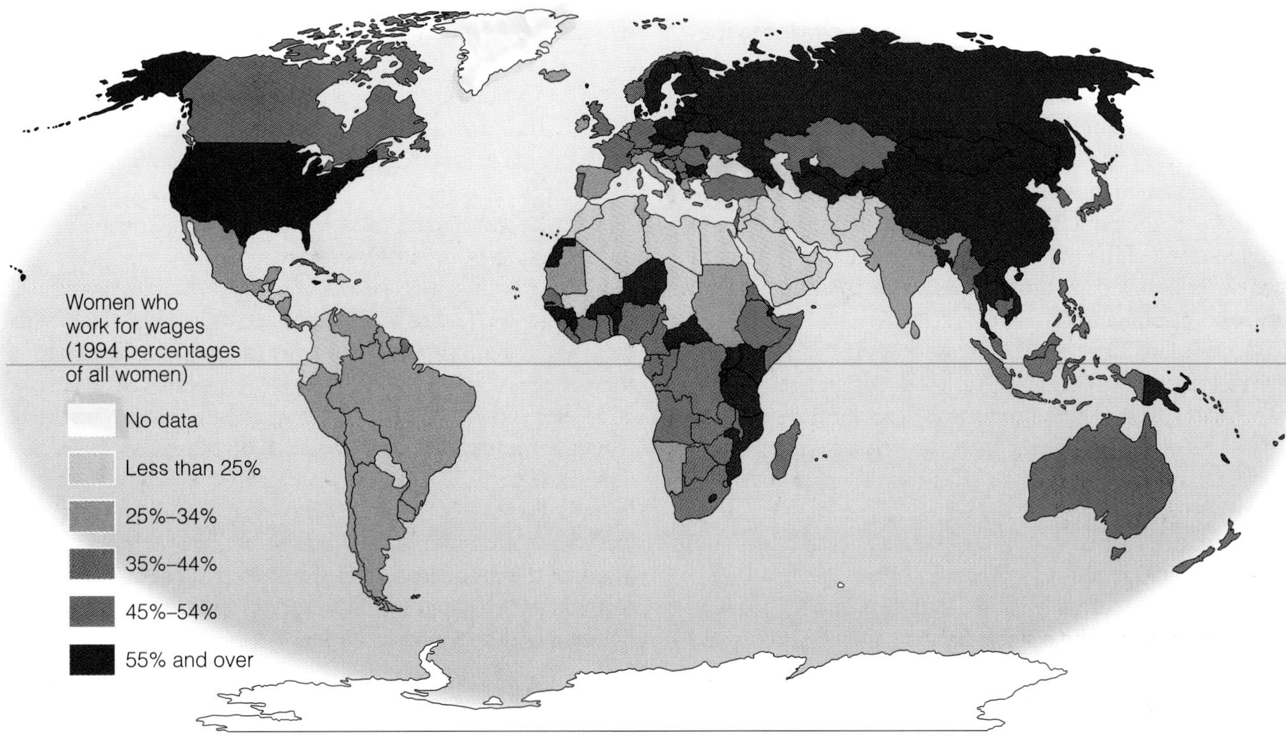

MAP 17.1 Viewing Society in Global Perspective: Women in the Global Labor Force

SOURCE: Adapted from Seager, Joni. 1997. *The State of the Women in the World Atlas.* London: Penguin Group.

particularly among the working class. Because it is easier to blame foreign workers for unemployment in the United States than it is to understand the complex processes that have produced this phenomenon, U.S. workers have been prone to **xenophobia,** the fear and hatred of foreigners. Campaigns to "Buy American" reflect this trend, although the concept of buying American is increasingly antiquated in a multinational economy. When buying a product from a U.S. company, the likelihood is that the parts, if not the product itself, were built overseas. In a global economy, distinctions between U.S. and foreign businesses blur. Honda and Toyota, for example, are both Japanese-owned companies that build cars in the United States. Likewise, many foreign workers produce goods to be sold by U.S.-based companies.

The development of a global economy is part of the broad process of economic restructuring. **Economic restructuring** refers to the contemporary transformations in the basic structure of work that are permanently altering the workplace. This process includes the changing composition of the workplace, deindustrialization, the use of enhanced technology, and the development of a global economy. Some changes are *demographic*—that is, resulting from changes in the population. The workforce is becoming more diverse, with women and people of color becoming the majority of those employed. Other changes are driven by *technological developments.* For example, the economy is based less on its earlier manufacturing base and more on service industries—those in which the primary business is not the production of goods, but the delivery of services (whether those be banking, health care, provision of food, or the like). All these developments are happening within a global context.

A More Diverse Workplace

The majority of workers are now White women and people of color. Women and racial minorities are expected to be an even larger proportion of the labor force by the year 2005. Another upcoming change in the labor market will be an increase in the number of older people (that is, those over fifty-five) as the population bulge of the Baby Boomer generation passes through late middle age (Fullerton, 1995; Kutscher, 1995).

These changes in the social organization of work and the economy are creating a more diverse workforce than ever before, but much of the growth in the economy is projected to be in service industries where, for the better jobs, education and training are required. Those without these skills will not be well positioned for success. Manufacturing industries, where racial minorities have historically been able to get a foothold on employment, are in decline. New technologies and corporate layoffs have reduced the number of entry-level corporate jobs, which recent college graduates have always used as a starting point for career mobility. Many college graduates are employed in jobs that do not require a college degree. College graduates, however, do still have higher earnings than those with less education, but the declining value of the dollar means that workers at various levels of educational attainment have seen a decline in the buying power of their income.

Deindustrialization

Deindustrialization refers to the transition from a predominantly goods-producing economy to one based on the provision of services (Harrison and Bluestone, 1982). This does not mean that goods are no longer produced, but that fewer workers in the United States are required to produce goods since machines can do the work people once did and many goods-producing jobs have moved overseas. Different from traditional manufacturing jobs, such as the manufacture of cloth or automobiles, service-based industries are based on the delivery of a product or provision of some service.

Deindustrialization is most easily observed by looking at the decline in the number of jobs in the manufacturing sector of the U.S. economy since World War II. At the end of the war, the majority of workers (51 percent) in the United States were employed in manufacturing-based jobs. Now, the majority (at least 70 percent) are employed in what is called the service sector (Wilson, 1978; U.S. Department of Labor, 1998). The service sector includes two segments: service delivery (such as food preparation, cleaning, or childcare) and information processing (such as banking and finance, computer operation, or clerical work). The service delivery segment consists of many low-wage, semi-skilled, and unskilled jobs, and gives employment to high numbers of women and people of color. These trends contribute to **job displacement,** the permanent loss of certain job types that occurs when employment patterns shift; for example, job displacement occurs when the manufacturing base shrinks as the data-handling industry grows. Job displacement is a cause of unemployment that occurs at the social structural level, unlike joblessness that occurs because of an individual's performance, the seasonality of work, or individual business failure.

The human cost of deindustrialization can be severe. When a manufacturing plant shuts down, many people may lose their job at once, and whole communities can be affected. The high number of unemployed workers creates great demand for the few jobs that remain. A simple fact like where you live can affect the likelihood that you will be out of work (Lobao, 1990). Living in a rural community where farming is the major source of work can leave you subject to changes in the agricultural market—or the increased production of food in South America, Mexico, and other international sites. Likewise, residing in an industrial city that depends on a single industry leaves workers vulnerable when that industry halts production.

When industries downsize, one option for workers is to leave the community altogether, though many find it difficult to do so. Although particular communities may be hard hit, deindustrialization is occurring nationwide, leaving few options for those left without work. Moving also means

BOX 17.1 • ANALYZING SOCIAL ISSUES

Toys Are Not U.S.

EVERY year, millions of young girls in the United States ask for a Barbie doll for Christmas. No longer is there just one Barbie, but a huge array of different Barbies, each with a specific role and costume that lures young girls into a fantasy life of beauty, fashion, romance, and play. Made by Mattel, Barbie is the ultimate American "dream girl," but how many of those who buy Barbies know that the doll is manufactured by those not much older than the ones who play with her and who would need all of their monthly pay to buy just one of the dolls that U.S. girls collect by the dozens?

The toys that many U.S. kids play with are manufactured through a growing global division of labor, and Barbie is not the only example. Whether it's *101 Dalmatians*, soccer balls, *Toy Story* action figures, *Hunchback of Notre Dame* plastic characters, toys are made in China, Indonesia, Mexico, and other parts of the world where work conditions are poor and hazardous, and where workers' rights are routinely ignored. In China, where more toys are produced than in any other part of the world, workers molding Barbie dolls earn 25 cents per hour, and human rights organizations say violations of basic rights are flagrant. As one journalist has written, "Behind the glitter of F.A.O. Schwartz and Toys 'R' Us, the toy industry is a showcase for the injustices at the heart of the global economy" (Press, 1996: 12).

The manufacturing of toys is also a classic example of the global assembly line that has led to the downsizing of jobs in the United States. In 1973, more than 56,000 U.S. workers were employed in toy factories. Now, as the market has become more glutted with the latest popular items, only 27,000 U.S. workers work in toy factories. The companies that make the toys are amassing record profits. At whose expense is this happening? U.S. workers have lost jobs and tend to blame foreign workers for taking them. But who has gained? In 1995, the CEO of Mattel (the company that produces Barbies) earned $7 million and an additional $23 million in stock options—far more than the combined salaries of the 11,000 Mattel workers who produce Barbie dolls in China. Indonesian workers making Barbies earn the minimum wage of $2.25 per day. It would take such a worker a full month to earn the money to buy the Calvin Klein Barbie (Press, 1996). The next time you look at the labels in your local toy stores, you might think about the global restructuring that brings entertainment to U.S. children.

SOURCES: Press, Eyal. 1996. "Barbie's Betrayal." *The Nation* (December 30): 11–16; Macmillan, Jerry. 1996. "Santa's Sweatshop." *U.S. News & World Report* (December 16): 50–60.

leaving the support systems provided by family and friends. Workers who own their homes may find it difficult to move after a plant closing because the real estate market is likely to be depressed. Women workers tend to be unemployed longer than men after plant shutdowns (Nowak and Snyder, 1986); however, men experience more severe wage drops—probably because they were earning higher wages to begin with. Displaced workers who are reemployed typically earn lower wages than before and tend to be hired into less skilled work. The longer the joblessness, the greater the loss of wage on reemployment (Moore, 1990).

Among the areas that have been hardest hit by deindustrialization are communities that were heavily dependent on a single industry, such as steel towns or automobile-manufacturing cities like Detroit, Flint, Lansing, Cleveland, and Akron. Some of these communities have rebounded by investing in new, high-tech industries, but this has still left many groups disadvantaged. Groups concentrated in the inner cities at these locations have suffered the most since they have been the most dependent on manufacturing jobs for economic survival, and they are not physically located in places to take advantage of new industries that are emerging in suburban areas. Earlier in the twentieth century, many African Americans migrated from the South to the North in pursuit of manufacturing jobs; Latino groups have more recently flowed into cities for the same reason. Both groups have been hard hit by the declines in urban manufacturing jobs that have characterized the late twentieth century. As a result, there are now extremely high rates of joblessness and dim prospects for economic recovery in these urban communities (Wilson, 1996). As Figure 17.1 shows, African American and Latino youth have high rates of unemployment; in some inner cities, teen unemployment rates exceed 40 percent (Wilson, 1996; U.S. Department of Labor, 1998).

The White working class has also been heavily affected by deindustrialization, but those industries now in decline have been the ones in which the highest proportions of skilled African American workers have been employed. Black men are also less likely to be working in unionized industries and are therefore less likely to have protection from layoffs. In general, Black workers are more likely to be fired than White workers, even when their job seniority, absenteeism, education, experience, and performance are the same (Zwerling and Silver, 1992). A maxim among workers seems to be even truer today than when it was first coined: "Last hired, first fired." Since African Americans and Latinos have been late to win entry into the manufacturing labor market, deindustrialization makes their position more tenuous.

In a labor market experiencing deindustrialization, there is job growth in professional and administrative positions with high educational requirements and in some manufacturing jobs that require specific technological training (Wilson, 1996); however, the greatest area of job growth in a service-based economy is in jobs at the lower end of the occupational system—jobs that require little education and training. The people best positioned to take advantage of the desirable new jobs are those who are well educated and well trained. Many people find themselves in a far less competitive position.

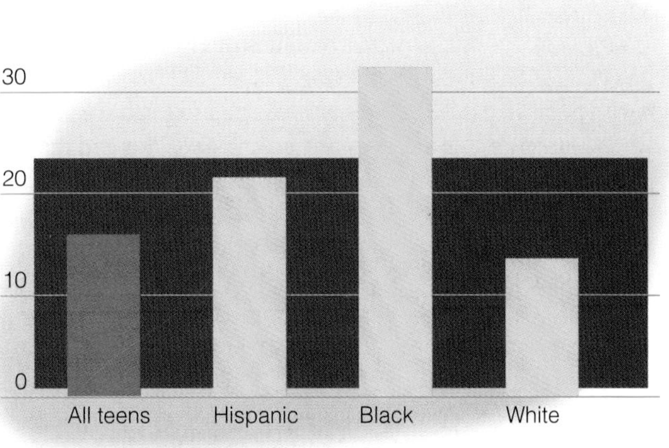

FIGURE 17.1 Teen Unemployment

Includes those aged 16–19.

DATA: From the U.S. Department of Labor. 1998. *Employment and Earnings.* Washington, DC: U.S. Government Printing Office, pp. 164–167.

Technological Change

Coupled with deindustrialization, rapidly changing and developing technologies are bringing major changes in work, including how it is organized, who does it, and how much it pays. One of the most influential technological developments of the twentieth century has been the invention of the semiconductor. Some have argued that the computer chip has as much significance for social change as the invention of the wheel and the steam engine in earlier times. Computer technology has made possible workplace transactions that would have seemed like science fiction just a few years ago. Electronic information can be transferred around the world in less than a second. Employees can provide work for corporations based on another continent; thus, a woman in Southeast Asia or the Caribbean can keystroke a book manuscript for a publishing house in New York. Corporations in the United States are lured overseas by the extremely low wages of these workers. Offshore data entry is a perfect example of the confluence of low wages and high technology. The labor is so cheap that suppliers sometimes assign two typists, who may be only marginally proficient in the language of the manuscript, to enter an entire book manuscript simultaneously. One digital version is compared to the other electronically, which automatically flags mistakes. It is cheaper to have two people overseas type the entire book than to have one person at home proofread it.

Increasing reliance on the rapid transmission of electronic data has produced *electronic sweatshops,* a term referring to the back offices found in many industries, such as airlines, insurance firms, mail-order houses, and telephone companies, where workers at computer terminals process hundreds or thousands of transactions in a day. Increasingly, workers' performance is monitored by the very computer at

job displacement

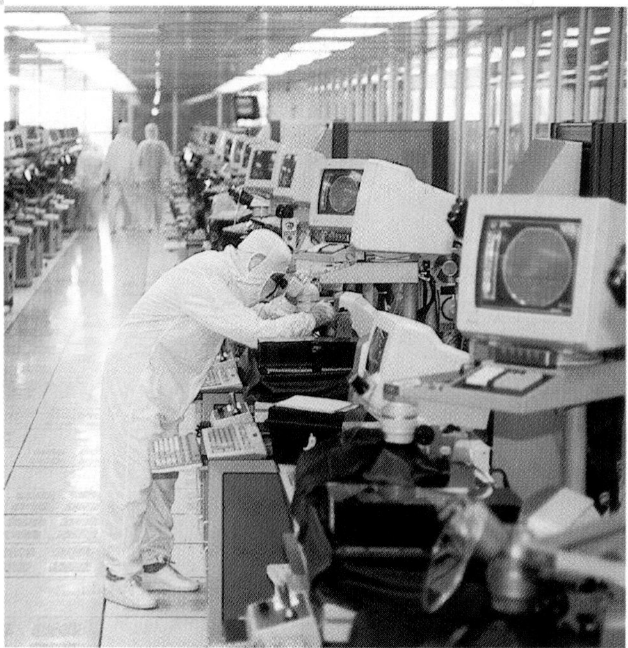

The development of new technologies has produced new forms of work and new demands for highly skilled workers in certain segments of the labor market.

which they work. For example, telephone operators are typically given twenty-five seconds to root out a number; the speed of the transaction is recorded by computer. Computers can also measure how fast cashiers ring up groceries and how fast ticket agents book reservations. Records derived from computer monitoring then become the basis for job performance evaluation (Kilborn, 1990). For many people, electronic monitoring of workers conjures images of "Big Brother" invisibly watching and recording workers' actions, causing the issue to become the focus of debates about worker privacy and autonomy.

Automation is the replacement of human labor by machines. Robots, for example, perform 98 percent of the spot-welding on Ford automobiles. First-generation robots—that is, those that performed only the simplest repetitive tasks—commonly resulted in the loss of two jobs for human workers (although it took four-fifths of a job to produce the robot, for a net loss of about one job). Second-generation robots are capable of far more complex tasks since they are equipped with sensors and are able to assemble finished products, or perform other tasks, such as the solar-powered robotic lawnmower that mows the lawn at CIA headquarters in Langley, Virginia. Volkswagen predicts that robots will soon do 60 percent of their production, reducing jobs by half. In tomato harvesting, 40,000 workers in California alone used to pick the crop; the use of robots reduced the labor force to 8000 (Draper, 1985).

Technological innovation in the workplace is a mixed blessing. Automation eliminates many repetitive and tiresome tasks, and it makes possible rapid communication and access to information. But critics worry that our increasing dependence on technology also makes workers subservient to machines. The human tasks on assembly lines, for example, are paced by the speed of conveyor belts, robots farther up the line, and so on. Automation has also eliminated many jobs altogether, and not only in manufacturing. Elevator operators, for example, were once a common sight but are now found in only the most elite office or retail settings. The benefits of technological advancement are many, to be sure—new forms of work become possible, including work that can be done at home or across long distances; new products and markets are created; new outlets for talent and ingenuity emerge. The benefits associated with technological change are, however, distributed unevenly across social classes. Professional workers may find that technology brings them new options and opportunities, but working-class people are likely to be simply displaced. Those most likely to get truly high-tech jobs are the people with the most education and technological training.

New technology requires new workers with the skills needed to develop and use the new tools and techniques, but the number of positions for these workers does not come close to equaling the number of jobs eliminated by technological innovation. Less than one-quarter of all jobs available require substantial knowledge of technology. In the electronic components industry, for example, about 15 percent of the workers are engineers, scientists, and computer analysts; the rest are mainly low-wage assembly workers (Apple, 1991).

Coupled with demographic transitions in the labor force, technological changes are affecting who does what. More women are entering the labor force, but they often lack the technological training required for the best jobs since traditional gender roles and gender segregation in education discourage women from entering scientific and technical fields. Job growth is primarily in low-wage positions that have historically been race- and gender-segregated. Despite popular stereotypes that women and minorities are getting a big boost from affirmative action and enjoying fast-track careers as a result, in reality they are bearing the brunt of economic restructuring. Many find themselves in low-wage jobs with little opportunity for promotion and few economic or social benefits. Groups that do not have the educational or technological skills needed to compete in a technologically advanced labor force will increasingly find themselves in severely disadvantaged positions (Wilson, 1987). The situation is worsened by the tendency of high-tech industries to move out of old manufacturing districts in city centers and relocate to suburbs where land is less expensive, unions are less pervasive, and nearby universities supply a pool of young educated workers and highly skilled academics. This has further eroded the industrial base of cities, in many cases leaving city centers impoverished. Workers who live in the suburbs benefit from the movement of high-tech industries out of the cities; racial minorities, who are more likely to reside in center cities, accrue a disproportionate share of the disadvantages emerging from these trends. Lacking transportation and high-tech training, the very segment of the population most negatively affected by deindustrialization is also the group least likely to benefit from technological changes.

Deskilling is a side effect of automation and technological progress in which over time the level of skill required to perform certain jobs declines. Deskilling may result when a job is automated, or when a more complex job is atomized into a sequence of easily performed units. As work roles become deskilled, employees are paid less and have less control over their tasks. Less mental labor is required of workers, and jobs become routinized and boring. Observers of the workplace note that deskilling contributes to polarization of the labor force. The best jobs require ever-greater levels of skill and technological knowledge, whereas at the bottom of the occupational hierarchy, people stuck in dead-end positions become alienated from their work (Braverman, 1974; Apple, 1991).

Deskilling has been one effect of the move to a technologically sophisticated workplace. In a technologically advanced society, the demand for new workers is primarily in unskilled areas or in a few highly skilled areas where only a select few are qualified. Although technological change has increased the number of technical jobs in the economy, it makes employment less certain for many workers. The new jobs created by high-tech industries are mostly low-level jobs that require little or no technological skill and offer little opportunity for advancement.

The Impact of Economic Restructuring

Economic restructuring has numerous effects. More than one million U.S. workers lose their jobs each year as the result of economic restructuring, which includes technological change, global competition, and shifts in consumer demand (Moore, 1992). Although restructuring produces new jobs, neither are these numerous enough to replace those lost nor do they tend to require the same level of skill and remuneration. Just as the workplace is becoming more diverse, there seems to be growing inequality between the different groups in the labor market. For some people, there is too much work; for others, too little. Those who are employed are now working longer hours than did workers in previous times (Rones et al., 1997); they have less leisure time and often hold at least two jobs in order to make ends meet. At the same time, high rates of joblessness and unemployment have left millions without work. Even the group historically thought to be most protected from the vicissitudes of economic cycles—older, White men in management positions—are being laid off in large numbers.

Recent developments, like *downsizing,* can be understood in the context of economic restructuring. Downsizing is a term coined during the 1990s to refer to reducing the number of workers in an organization. Although CEOs and other business leaders have heralded downsizing as a virtuous endeavor—something that increases the efficiency and effectiveness of organizations—its purpose is to reduce the costs of labor while maintaining or enhancing business profits. It is a euphemistic way of saying that large numbers of people have been fired.

In addition to laying off workers, companies have downsized by reducing the layers of management, decreasing the overall number of managers, and downgrading the rank and salary of managers, thereby also eliminating many of the career ladders that gave managers job mobility. Instead of promoting their own workers, companies now do more external recruiting for new workers (eliminating the need to pay higher salaries) and are likely to use more temporary workers and independent contractors (to whom benefits need not be paid). The managers who remain after downsizing experience more performance pressure, often working longer hours and doing jobs formerly done by more than one person (Hamlin et al., 1994) .

Automation means that machines can now perform the labor originally provided by human workers, such as the robots that perform tasks on automobile assembly lines.

Who has been most affected by these changes? White men have been the most likely to be laid off (which can mean being fired or being offered early retirement or some other form of job severance). To date, downsizing has not had as sizable an impact on women and minorities in management, but this is primarily because there are so many more White men in these positions to begin with. Those who have studied the effects of downsizing on women and minorities also note that "minorities and women in management" usually means minority men and White women. Women of color are rare in senior management positions; when found, they tend to be in the lowest levels of senior management (Hamlin et al., 1994).

Theoretical Perspectives on Work

Defining Work

Most people think of work as an activity for which a person gets paid. This definition is not adequate for a sociological definition of work, however, since it devalues work that people do without pay. Unpaid jobs such as housework, child care, and volunteer activities make up much of the work done in the world. Sociologists define **work** as productive human activity that creates something of value, either goods or services. Work takes many forms. It may be paid or unpaid. It may be performed inside or outside the home. It may involve physical or mental labor, or both.

Under the sociological definition of work, housework, though unpaid, is defined as work, even though it is not included in the official measures of productivity that economists use to indicate the work output of the United States. Influenced by feminist studies, sociologists recognize that housework and other forms of unpaid labor are an important part of the productivity of a society. Feminist scholars have even challenged the idea that family and work are separate institutions, emphasizing instead the intersection of the two. This framework recognizes that labor within the family is productive for the economy. Families provide both productive and reproductive labor for the economy. The productive labor of families is the work that family members provide, whether paid or unpaid. Reproductive labor refers to the care provided for both children and adults within families (including the provision of food, clothing, and nurturing); this labor makes it possible for people to engage in work outside the family.

Some nations are far ahead of the United States in recognizing this productive work. German law, for example, provides six paid weeks of maternity leave for women and a monthly allowance for mothers of small children who have left the paid labor force. In Norway, employed women may take fifty-two weeks maternity leave, with 80 percent pay, or forty-two weeks with 100 percent of their pay. Four weeks of these paid leaves are earmarked for fathers unless the mother is single (*Norway Now,* 1995). By contrast, the Family and Medical Leave Act in the United States provides twelve weeks of unpaid leave—but only in certain organizations and after the employee has used up other vacation and sick leave (see Chapter 14). The point is that labor within the family, although unpaid in the United States, is an important category of work.

Some sociologists argue that work has been too narrowly defined as referring to physical and mental labor alone. Arlie Hochschild has introduced the concept of *emotional labor* to address some forms of work that are common in a service-based economy. **Emotional labor** is work that is specifically intended to produce a desired state of mind in a client. Many jobs require some handling of other people's feelings; emotional labor is that performed in jobs where inducing or suppressing a feeling in the client is one of the primary work tasks. Airline flight attendants are an example—their job is to please the passenger and, as Hochschild suggests, to make passengers feel as though they are guests in someone's living room, not objects catapulting through the sky at 500 miles

B • O • X 17.2

UNDERSTANDING DIVERSITY

Native American Women and Work

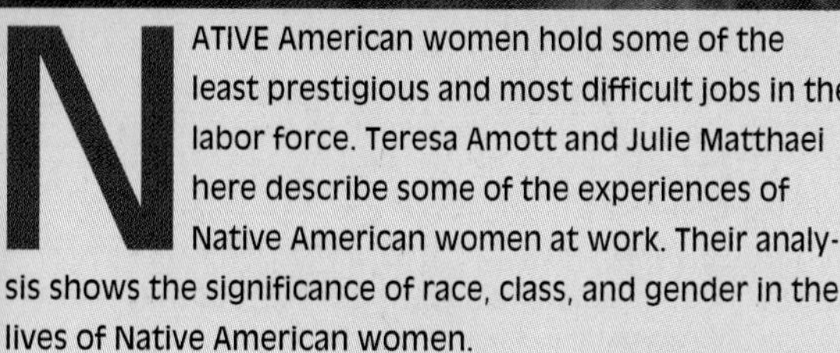

NATIVE American women hold some of the least prestigious and most difficult jobs in the labor force. Teresa Amott and Julie Matthaei here describe some of the experiences of Native American women at work. Their analysis shows the significance of race, class, and gender in the lives of Native American women.

> Since the 1950s, the American Indian community has been split; over half of all Native Americans live in urban areas, the rest mainly on tribal reservations. For most, urbanization has not brought integration into the white world or its riches. American Indians compete with other minority groups for the dead-end, low-paying jobs that characterize the secondary labor market, and are kept there by discrimination, poor education, and lack of wealth.
>
> In addition, Native Americans face an insidious and structural form of discrimination in the clash between the white-defined corporate culture of hierarchy and competition, and the American Indian values of working together, sharing, cooperation, and autonomy. Although employment rates and incomes are higher in the cities than on reservations, urban American Indians often feel more impoverished because they are cut off from federal health and other programs and from the support of their extended families, their nations, and their cultures. Many eventually decide to return to their reservations. A BIA study for the years 1953 and 1957 found that a full 30 percent of migrants returned to the reservation in the same fiscal year in which they left; if they had studied how many returned over a more extended period, the figure would have been substantially higher. Those American Indians who return suffer large income losses, but most place a higher value on their cultural identity than on the economic advantages of urban life.
>
> On the other hand, reservation life is economically precarious. For example, 1980 unemployment among Indians was 65 percent in Wyoming and 64 percent in South Dakota, compared to 7 percent and 6 percent for whites. Reservation employment is concentrated in self-employment (subsistence farming, herding, and crafts), work for the BIA (which employs mostly women in clerical and social service work), and work for corporations located on or near the reservations. Consequently, communities vary greatly in their economic well-being. According to the 1980 Census, a few were very well off, such as the Saxman in Alaska, with only 3 percent of the population living below the federally defined poverty line, and the Camp Verde Reservation in Arizona, where poverty was negligible. Indeed, at Camp Verde, 1980 per capita income stood at $11,460, well above the white figure of $8,000. But more often, incomes and employment were much lower than the national average, and communities commonly had poverty rates of 40 percent or more. Among some, like the Winemucca Colony in Nevada, all members were poor.
>
> A central issue for nations has continued to be their relationship to white-owned and -run corporations. Some of the latter have set up plants on reservations to take advantage of their cheap and abundant labor forces. However, white corporations are often more interested in American Indian lands than labor; they lease large quantities of land for mining, oil drilling, cattle raising, and military uses, but provide little employment for the residents on the lands. Recently, some nations have passed ordinances that force firms that do business on the reserva-

per hour! Emotional labor, like other work, is done for wages. It is supervised and evaluated; workers are trained to produce the desired effect among clients. Emotional labor often involves putting on a false front before clients.

Many jobs require the performance of emotional labor; in other words, producing a particular state of mind in the client is part of the product being sold. In a service-based economy, emotional labor is actually a growing part of the work that people do, although it is seldom recognized as real work. Emotional labor also makes the production of emotion something like a commodity—a product that is produced for profit and consumed. This can result in the "commercialization of human feeling," meaning that the production and management of emotional states of mind is increasingly seen as a commodity to be bought and sold in the marketplace (Hochschild, 1983).

Some forms of work are more highly valued than others, in both how the work is perceived by society and how it is rewarded. Mental labor has been more highly valued than manual labor, and manual labor is valued more than emotional labor. In addition, work performed outside the home is typically judged to be more valuable than work performed inside the home. Given the class-, race-, gender-, and age-based stratification in this society, generally speaking, the work most highly valued has been that done by White, middle-class, older men. Sociologists debate whether this is because the most highly valued jobs are reserved for this group or whether, because this group does these jobs, the jobs are therefore more highly valued. It is a question of which comes first, but the point is that the prestige attributed to different jobs follows along lines of race, class, gender, and age, among other factors.

Judgments made about the value of different forms of work have also been turned into judgments about the value of different groups of workers; thus, those men who perform manual labor have been devalued as workers, including White men. Women's work, much of which is emotional labor, has also been devalued. Much of women's work (housework and child care) has been unpaid, and when it has been paid, the wages have been extremely low. Work done by many Latinos, Native Americans, African Americans, and Asian Americans has also been some of the most demanding, yet least rewarded, work in society, as the box "Understanding Diversity: Native American Women and Work" shows. The labor of the White working class has also been undervalued. Historically, White ethnic groups such as the Irish, Polish, and Italians, performed some of the most onerous work; their contributions as workers has also been devalued as the result of prestige being typically reserved for the most dominant group.

tions to hire American Indian labor. As Donna Wilson of the Nez Percé in Idaho explained:

> Initially there was a lot of resistance on the part of the contractors, because we were dealing with a lot of stereotypes. "Your people do not have the skills. They do not come to work after payday. They all drink too much." We spent a good 5 years in public relations efforts with employers and with construction contractors overcoming a lot of these stereotypes, and what really addressed it was the workers themselves.

Whether they live on or off the reservations, American Indian men and women live in difficult times, struggling daily against white control, the threat of cultural extinction, and the challenges posed by poverty. American Indian communities continue to suffer from higher-than-average infant mortality rates, extremely high death rates from preventable diseases such as tuberculosis and gastroenteritis, and devastatingly high alcoholism and suicide rates. In 1980, American Indians were almost twice as likely as whites to be unemployed. Still, American Indians are growing in strength, cultural pride, unity, and militancy. American Indians have survived as a people; today, in the words of Wilcomb Washburn, they are the only American minority that retains a separate legal status and (more persistently than others) remains aloof from the American melting pot.

Native American women's labor force participation rates [has risen] sharply. . . . Those who hold full-time, year-round jobs earn nearly 89 percent as much as white women—but these jobs are hard to come by. Almost two-thirds of American Indian women hold part-time jobs—the highest rate for any racial-ethnic group of women except Chicanas—and American Indian women face the highest unemployment rate (12 percent) of any racial-ethnic group except Puerto Rican women.

Educational and occupational disadvantages also make it very difficult for American Indian women to support families on their own. In 1980, nearly one in four American Indian families was maintained by a woman (over twice the rate for whites and Asians), and 47 percent of these single-mother families were considered poor by federal poverty guidelines. Among women with children under six, the poverty rate stood at a shocking 82 percent. In 1980, American Indian women were more likely to receive public assistance income than any other group of women except African Americans and Puerto Ricans, but on average, they received only $2,679.

Women's access to income through American Indian men is also limited by discrimination and unemployment. Those men who managed to find full-time jobs earned slightly more than African American and Latino men in 1980, but only three-fourths as much as white men. Moreover, American Indian men have suffered high unemployment (17 percent in 1980), the highest rate of part-time work of any racial-ethnic group of men (58 percent), and a high rate of non-participation in the formal labor force (31 percent).

Economic problems on and off the reservation will demand sustained organizing in the years to come. For that work, Native Americans can look to their rich cultural heritage for inspiration and guidance.

SOURCE: Amott, Teresa and Julie Matthaei. 1991. *Race, Gender, and Work.* Boston: South End Press. Reprinted by permission.

Popular concepts about work are also shaped by some of the assumptions we make about race, class, gender, and age in society. Welfare mothers, for example, are frequently accused of being lazy and preferring to live on welfare rather than working. Oddly enough, women in the middle and upper classes who care for their children at home, but are not otherwise employed, are defined as model mothers. The fact that many welfare recipients are women of color links this perception to both race and gender stereotypes. Note that the perception that welfare mothers do not work assumes that women who stay home to care for children are not working. "Workfare" policies (in which recipients must perform various work assignments in order to qualify for welfare payments) presume that welfare mothers must be forced to work or they will not do so. One of the problems with this argument is that day-care costs are higher than most women's wages, especially the large number of women who earn only the minimum wage or less.

These judgments also point to the centrality of the *work ethic* in U.S. culture. Americans place a high value on the work ethic, defined as the belief that hard work is a moral obligation; those who are judged not to value work are maligned, ridiculed, even condemned as immoral. The value placed on the work ethic derives from the cultural history of the United States and the Protestant belief that hard work is a sign of moral stature, and prosperity is a sign of God's favor, as we saw in Chapter 16 on religion. According to American values, those who are perceived as not engaged in hard work are moral failures who are blamed for their own lack of success.

This assumption is the crux of stereotypes about the "undeserving poor"—the belief that the poor have become

BOX 17.3 DOING SOCIOLOGICAL RESEARCH

Newcomers in the Workplace

ECONOMIC restructuring and a high rate of immigration are two trends converging to shape the experience of many workers in the U.S. labor market. Large numbers of new immigrants are from Asia and Latin America. These groups now hold many of the manufacturing jobs in the labor market, and they are increasingly present in retail and other service jobs. At the same time, many jobs have migrated to the South and West while the cities across the nation where new immigrants reside are becoming more multiethnic. Because of the significance of these trends, many sociologists are interested in how new immigrant groups fare in the labor market and, like the immigrants of the past, how their lives are influenced by their work.

A team of sociologists, funded with a major grant from the Ford Foundation, explored this question by asking, "What work do immigrants perform in changing communities?" "How do they fit in their community, and how do residents respond to them?" Using a large research team, Louise Lamphere, Alex Stepick, and Guillermo Grenier based their study on detailed descriptions of shopfloors, retail stores, and restaurants. They used the method of participant observation and interviewing to emphasize the perspective of immigrants themselves in thinking about their work. Their analyses also placed immigrant experience in the context of the large forces of economic restructuring; they supplemented participant observation with demographic and economic data from the cities they studied.

The researchers wanted to know if immigrant experiences varied across different regions; consequently, they focused their study on local economies in three major cities: Kansas City, Miami, and Philadelphia—each place unique in its labor market and the presence of diverse groups within this market. In Kansas City, they focused on two large beefpacking plants, which employ large numbers of Vietnamese, Laotian, Mexican, and other Hispanic workers. In Miami, Cubans, Haitians, Jamaicans, Nicaraguans, and others from the Caribbean and Central and Latin America are employed in diverse industries, although the researchers focused on the construction, apparel, and hotel and restaurant industries. They chose Philadelphia because it is typical of a city in industrial decline where many new immigrants, including Polish Americans, Puerto Ricans, and others, have been laid off from jobs and have tried to locate new work in retail employment.

Their research examines change in the workplace, such as techniques of management control, the nature of the work process, and how each of these affect relations between workers: between managers and workers, workers and customers, and among workers. Among the many findings from this comprehensive study, they found that immigrants are employed in places where, along with other minority co-workers, they face difficult work conditions, numerous health hazards, low pay, and gender and ethnic segregation. For example, in the beefpacking industry, it is Mexican workers who are employed on the "kill floor." Racial segregation is found in the pattern whereby Vietnamese and Laotian men and women work in the processing lines where there is very high turnover, especially when both a husband and wife work in the same industry.

The volume of studies that these scholars have produced, *Newcomers in the Workplace: Immigrants and the Restructuring of the U.S. Economy*, describes the work and life of new immigrants in vivid detail—providing a picture of the changing labor force that is graphic in its exposition and comprehensive in its analysis of the role of immigration in the contemporary labor force.

SOURCE: Lamphere, Louise, Alex Stepick, and Guillermo Grenier. 1994. *Newcomers in the Workplace: Immigrants and the Restructuring of the U.S. Economy*. Philadelphia: Temple University Press.

so because of their own failures and refusal to internalize the values of diligence and hard work. Yet, close to half of the poor are working (48 percent), half of whom are employed full time (O'Hare, 1996; Swartz and Weigert, 1995); moreover, the number of *working poor* (those whose wages for full-time work are below the federally defined poverty line) is increasing. The working poor now compose 25 percent of all workers, with young workers especially likely to be among the working poor. Nearly half of all women and more than one-third of men aged eighteen to twenty-four are in this group. One-quarter of Black workers and 31 percent of Hispanic workers (including all ages) earn poverty-level incomes even though they are working full time. Recent immigrants are also more likely than native-born workers to be employed in low-wage jobs, as illustrated in the box "Doing Sociological Research: Newcomers in the Workplace" (Rich, 1992; Meisenheimer, 1992).

Ironically, as we saw in Chapter 9, many of those who are most admired in the United States for their presumed hard work and success actually became wealthy not through their own diligence and work, but from inheritance. Once a principle like the primacy of the work ethic becomes embedded in the value system of a culture, contradictions tend to be ignored, and we overlook that many low-income people work hard and that many people who enjoy the greatest material luxury did little to earn it.

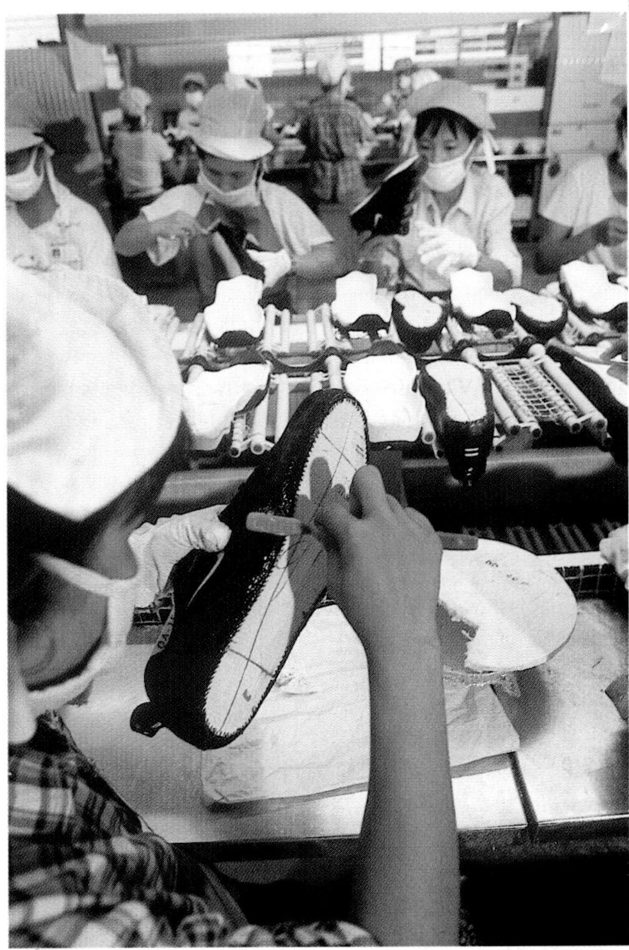

The global division of labor has created sweatshops both in the United States and abroad where workers, mostly women, perform routine tasks for low wages and in poor working conditions. Many of the products manufactured in these sweatshops, such as the Nike shoes here being made in Vietnam, become status symbols for those in more privileged groups.

The Division of Labor

Recall from Chapter 5 that the *division of labor* is the systematic interrelatedness of different tasks that develops in complex societies. When different groups engage in different economic activities, a division of labor is said to exist. In a relatively simply division of labor, one group may be responsible for planting and harvesting crops, whereas another group is responsible for hunting game. As the economic system becomes more complex, the division of labor becomes more elaborate.

In the United States, the division of labor is affected by gender, race, class, and age, to name the major axes of stratification. The *class division of labor* can be observed by looking at the work done by those with different educational backgrounds since education is a fairly reliable indicator of class. Those with more education tend to work in higher-paid, higher-prestige occupations. Education, income, and social prestige are thus indicators intricately linked to the class division of labor. Class also leads to perceived distinctions in the value of manual and mental labor. Those presumed to be doing mental labor (management and professional positions) tend to be paid more and have more job prestige than those presumed to be doing only manual labor. Class thus produces stereotypes about the working class; manual labor is presumed to be the inverse of mental labor, meaning it is presumed to require no thinking. By extension, workers who do manual labor may be assumed not to be very smart, regardless of their intelligence.

The *gender division of labor* refers to the different work that women and men do in society. In societies with a strong gender division of labor, the belief that some activities are "women's work" (for example, secretarial work) and other activities, "men's work" (for example, construction) contributes greatly to the propagation of inequality between women and men, especially because cultural expectations usually place more value (both social and economic) on men's work. This helps explain why librarians and social workers are typically paid less than electricians, despite the likely differences in their educational level. The traditional gender division of labor has meant not only that women and men tend to do different jobs in the paid labor force, it has also been the basis for women not being paid for the labor they contribute within families. As long as some forms of work are presumed to be what women are "supposed" to do, the gender division of labor appears to be natural instead of constructed based on the social definitions attributed to women's and men's work.

The division of labor in the United States is also patterned by race. The *racial division of labor* is seen in the pattern of people from different racial groups working in different jobs. The labor performed by racial minority

division of labor

groups has often been the lowest paid, least prestigious, and most arduous work. The racial division of labor found today is rooted in the racist beliefs and practices of the past. In the mid- and late nineteenth century, Chinese immigrants were used as a cheap supply of labor on the sugar plantations of Hawaii and in mines and railroads on the U.S. mainland. The Chinese laborers came to be called "coolies," a term referring to laborers who have been kidnapped or coerced into work and sent to foreign countries. The Chinese were denied legal rights accorded to Whites, paid substandard wages, and typically assigned to live in labor camps. Railroad executives at the time estimated that, had they used White laborers, it would have cost them one-third more than the $31 per month they paid to Chinese workers. As one industrialist wrote in 1876, "In mining, farming, in factories and in the labor generally of California, the employment of the Chinese has been found most desirable; and much of the labor done by these people if performed by white men at higher wages could not be continued nor made possible" (McClellan, 1876, cited in Takaki, 1989: 88). Chinese laborers were also pitted against White workers; a few White workers were given jobs as foremen in control of the Chinese workers. Chinese workers were also used as strikebreakers—a tactic employers used to defuse the development of a strong labor movement. These early divisions predicated a long history of racial conflict between Asian and White workers, and solidified the racial division of labor in the American West (Takaki, 1989, 1993).

THINKING SOCIOLOGICALLY

Think about the labor market in the region where you live. What racial and ethnic groups have historically worked in various segments of this labor market and how would you describe the *racial and ethnic division of labor* now?

The labor history of other racial groups reflects similar patterns of exclusion and segregation at work. Mexicans who had settled and owned property in what is now the U.S. Southwest were robbed of their land during Anglo conquest, culminating in the Mexican-American War of 1846–1848. Mexican Americans displaced from their land were forced to seek work as seasonal agricultural workers, domestic workers, and miners. Like Asian Americans, they faced terrible hardships, doing backbreaking work at very little pay. Frequently, they lived in company towns where housing was barely fit for human habitation. Women and children provided additional unpaid labor. Families often lived in *peonage*—a system of labor in which families lived as tenants on an employer's land and were bound to an employer by debt. Unable to pay rent, families ran up debts to their landlord that they were expected to pay off with their labor. The peonage system is very like *feudalism,* a system in which a large peasant class occupies lands held by a small number of elites and work in exchange for their subsistence. Under peonage, workers are seldom able to pay off their "debt" to their landlords since it builds up faster than what they earn, so they work for little more than sustenance over very long periods (Amott and Matthaei, 1996).

In the U.S. economy, the racial, class, and gender division of labor intersect, creating unique work experiences for different groups. As an example, following the abolition of slavery, African American women found work as domestics—work that, though severely underpaid, was readily available year-round. In the South, African American men, on the other hand, often found only seasonal work as agricultural laborers or day laborers; this resulted in a pattern in which African American women often became the primary breadwinners even in stable, married-couple families. The history of Chinese Americans shows a different pattern, yet one that again shows the structural intersections of race, class, and gender. In the nineteenth and early twentieth century, there was a shortage of women in Chinese American communities caused by the restrictive immigration laws that prohibited Chinese women from joining their husbands in the United States. The Chinese maintained what has been called a "split-household family system" wherein wives and children often remained in China while the husband worked in the United States. As a result, certain roles that might otherwise have been considered women's work, such as domestic worker, launderer, and food service worker, were often filled by Chinese American men (Amott and Matthaei, 1996; Glenn, 1986). The few Chinese women who independently moved to the United States were often forced from lack of other opportunities into work as prostitutes (Amott and Matthaei, 1996).

In short, racism and sexism have historically segregated women and people of color into different and inferior jobs. The race and gender division of labor also intersects with a class-based division of labor and age-based divisions. Certain jobs are rarely performed by middle- and upper-class people; likewise, young workers typically find themselves confined to a niche in the labor market—such as in the fast-food industry. All these patterns reflect a division of labor that is patterned in this society by the cross-cutting arrangements of race, class, gender, and age stratification.

Functionalism, Conflict Theory, and Symbolic Interaction

The major theoretical perspectives that have been identified in this book also provide the frameworks for understanding the social structural forces that are transforming work. Each viewpoint—conflict theory, functionalist theory, and symbolic interaction—offers a unique analysis of work and the economic institution of which it is a part (see Table 17.1).

Conflict theorists view the transformations that are taking place in the workplace as the result of inherent tensions in the social systems, tensions that arise from the power differences between groups vying for social and economic resources. Class conflict is then a major element of the social structure of work, and conflict theorists would look to the class division of labor as the source of unequal rewards that workers receive for work and the unequal way that the value of their work is perceived.

Functionalist theorists, on the other hand, interpret the work and the economy as a functional necessity for society.

TABLE 17.1 *THEORETICAL PERSPECTIVES ON WORK*

	Functionalism	Conflict Theory	Symbolic Interaction
Defines work	as functional for society because work teaches people the values of society and integrates people within the social order	as generating class conflict because of the unequal rewards associated with different jobs	as organizing social bonds between people who interact within work settings
Views work organizations	as functionally integrated with other social institutions	as producing alienation, especially among those who perform repetitive tasks	as interactive systems within which people form relationships and create meaning systems that define their relationships to others
Interprets changing work systems	as an adaptation to social change	as based in tensions arising from power differences between different class, race, and gender groups	as the result of the changing meanings of work resulting from changed social conditions
Explains wage inequality	as motivating people to work harder	as reflecting the devaluation of different classes of workers	as producing different perceptions of the value of different occupations

Certain tasks must be done to sustain society, and how work is organized reflects the values and other characteristics of a given social order. When society changes too rapidly, as you could argue is the case with new technological and global developments in the world, work institutions experience social disorganization—perhaps alienation, unemployment, or economic anxiety—as institutions try to readjust and develop new forms that will again bring social stability.

Symbolic interaction brings a different perspective to the sociology of work. Less interested in the workings of the whole society, symbolic interaction theorists might study the meaning of work to those who do it and how social interactions in the workplace form social bonds between people. Some classic symbolic interactionist studies have examined how new workers learn their new roles and how workers' identity is shaped by the social interactions in the workplace (Becker et al., 1961); some symbolic interactionist studies also look at creative ways that people deal with routinized jobs since they sometimes develop elaborate and exaggerated displays of routine tasks to bring some human dimension to otherwise dehumanizing work (Leidner, 1993).

In each of these theoretical perspectives, the social basis of work is revealed, either in the direct interaction people have with one another or in how social institutions are organized. Having introduced these perspectives, we can now turn to some of the current features of the work world, using these frameworks to interpret this sociological evidence.

Characteristics of the Labor Force

Data on characteristics of the U.S. labor force are typically drawn from official statistics reported by the U.S. Department of Labor. In 1997, the labor force included approximately 136 million people (U.S. Department of Labor, 1998). This is 67 percent of the working-age population (those sixteen years old and above). In the period since World War II, the percent of the population that is employed has increased (from 59 to 67 percent), and the specific characteristics of those in the labor force have changed considerably.

Who Works?

Employment varies significantly for different groups in the population. Looking at all groups, Hispanic men are the most likely group to be employed, Hispanic women, the least (see Figure 17.2). Among Hispanics, Mexican American men are the most likely to be employed; Cuban American men, the least. Among Hispanic women, Mexican American women are most likely to be employed, Puerto Rican women, the least (U.S. Department of Labor, 1998). It is difficult to assess changes in the employment of Hispanics, Asian Americans, and Native Americans over time, since the Department of Labor did not collect separate data on these groups prior to 1980. Even now the broad categories "Black" and "White" make it difficult to see the unique experience of diverse groups in the labor force. Hispanics, for example, in the Department of Labor data appear in both the categories "Black" and "White" categories, depending on their self-identification. Pacific Islanders, Native Americans, and Alaskan natives are grouped as "other," a category that is seldom reported in summary tables. Some government publications report separately on Native Americans and Asian Americans, making comparisons among different groups difficult, if not impossible.

Despite these limitations, several trends can be discerned. One of the most dramatic changes in the labor force since World War II has been the number of women employed. From 1948 to 1997, the employment of women has increased from 35 to 57 percent. The rise is weighted toward White women since women of color have historically always had a high rate of employment. In fact, the percentage of employed

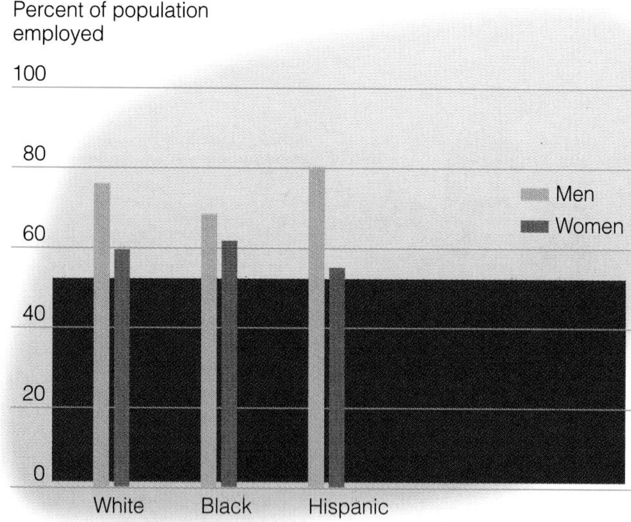

FIGURE 17.2 Employment Patterns by Race and Gender

DATA: From the U.S. Department of Labor. 1998. *Employment and Earnings.* Washington, DC: U.S. Department of Labor, pp. 164–167.

Black and White women has converged in recent years—both are now equally likely to be employed (U.S. Department of Labor, 1998); however, Black women are generally employed in lower-paying and less prestigious jobs.

Increases in employment among women have been greatest for those with young children. Sixty-four percent of married women with children under six years of age are now employed (U.S. Bureau of the Census, 1997). The rise in the number of women employed outside the home has caused significant changes in family life (see Chapter 14). It has also created new demands for services traditionally provided for free by women in the home—daycare, fast food, laundry services, and housecleaning—activities that contribute to the growth of a service-based economy. As we will see in the section on the occupational system, the growth in these industries has generated an underground economy of workers, primarily women of color, who provide household services at wages below the federal poverty line.

The greater employment of women with children has also created new demands on the workplace. Some organizations have responded by creating on-site child-care facilities or facilities to tend children who are sick so that their parents can come to work. The Family and Medical Leave Act of 1993 has also forced companies to write more progressive policies for family and medical leave for their employees although studies show that few workers are able to take advantage of this policy since they cannot afford the time without pay (*Wall Street Journal,* March 26, 1996: 1).

The trend of more women and minority groups in the labor force is likely to continue into the future. By the year 2005, Hispanic women are expected to have the highest rate of employment among women, followed by (in order) White women, Asian American women, and African American women. Whereas the employment of women has been increasing, the rate for men has been falling. In 1951, 87 percent of men were in the labor force; by 1996, the number had dropped to 71 percent (U.S. Department of Labor, 1998). This decline is expected to continue, while the presence of women in the labor force is expected to keep growing, especially among those in the twenty to thirty-nine age group. The employment of Blacks, Hispanics, and Asians is predicted to grow faster than the employment of Whites, with Hispanics predicted to have the greatest increase in employment (Kutscher, 1995).

These changes in the employment status of diverse groups reflect the economic restructuring of the workplace, and as conflict theorists would point out, produce strife between different groups competing for the jobs that are left. Jobs where White men have predominated are declining, especially in the manufacturing sector, whereas jobs in race- and gender-segregated niches of the labor market are those most likely to increase. The popular belief that women and minorities are taking jobs from White men derives from this fact. Note, however, that it is not that women are taking the jobs where White men have predominated in the past; these jobs simply are not as numerous as before, and growth in the labor market is in areas that have traditionally been considered women's work or other menial jobs where minorities have been likely to be employed. The conflicts that one sees over labor market issues—beliefs that foreign workers are taking U.S. jobs, competition between groups for scarce positions, and debates about the role of immigrant labor in the U.S. economy—are the result of social structural transformations, that is, sociological factors that lie behind statistical patterns.

Unemployment and Joblessness

The U.S. Department of Labor regularly reports data on the **unemployment rate,** the percentage of those not working but officially defined as looking for work. Currently, more than eight million people are officially unemployed. The official number does not, however, include all those who are jobless. It includes only those who meet the official definition of a job seeker—someone who has actively sought to obtain a job during the prior four weeks and who is registered with the unemployment office. Several categories of job seekers are excluded from the list, including those who earned money at any job during the week prior to the data being collected, even if it was just a single day's work with no prospect of more employment. Also excluded are people who have given up looking for full-time work, either because they are discouraged, have settled for part-time work, are ill or disabled, or cannot afford the child care needed to allow them to go to work, among other reasons. Because the number of job-seeking people excluded from the official definition of unemployment can be quite large, the official reported rates of unemployment seriously underestimate the extent of actual joblessness in the United States.

Those most likely to be undercounted in unemployment statistics are the groups for whom unemployment runs the highest—the youngest and oldest workers, women, and members of racial minority groups. These groups are least

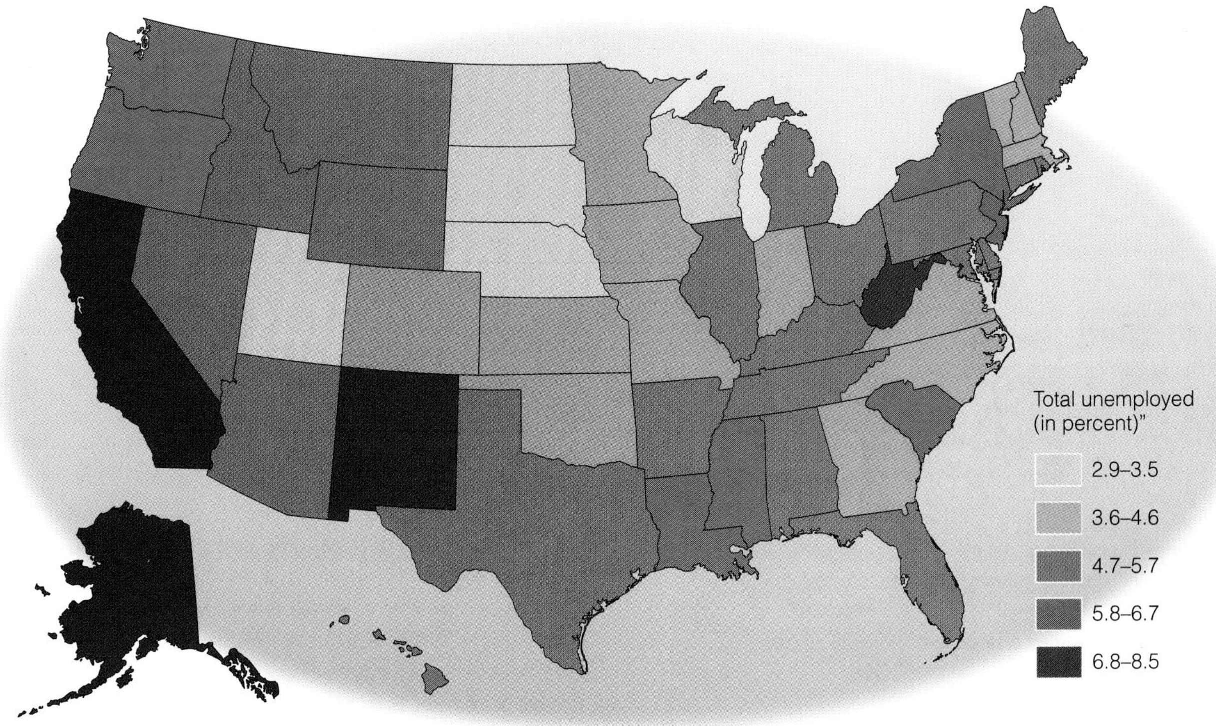

MAP 17.2 Mapping America's Diversity: State Unemployment: 1996

*Percent unemployed of the civilian labor force.

DATA: From the U.S. Bureau of the Census, 1997. *Statistical Abstracts of the United States.* Washington, DC: U.S. Government Printing Office, p. 421.

likely to meet the official criteria of unemployment since they are more likely to have left jobs that do not qualify them for unemployment insurance. Migrant workers and other transient or mobile populations are also undercounted in the official statistics. People who work only a few hours a week are excluded even though their economic position may be little different from someone with no work at all. Workers on strike are also counted as employed, even if they receive no income while they are on strike. Official unemployment rates also take no notice of **underemployment**—the condition of being employed at a level below what would be expected given a person's training, experience, or education; it can also include working fewer hours than desired. A laid-off autoworker flipping hamburgers at a fast-food restaurant is underemployed; so is a PhD who drives a taxi for a living.

Unemployment is often treated as an abstraction, with the unemployment rate used by the government and media as one of many indicators of the state of the economy. Full employment is assumed to occur when there is a 4 to 5 percent unemployment rate; thus, even in a "good" economy, large numbers of people are without work. Unemployment is not an abstraction, however, but an experience in people's lives that is tied to a host of other social problems, including crime, alcoholism, poverty, poor health, mental illness, and general despair. Unemployment affects not only the worker, but also the whole social system in which he or she is embedded. When a parent becomes unemployed, children may have to leave school to work; crime rates are strongly associated with unemployment; unemployment also causes strains on marriages and other personal relationships. Beyond creating problems of economic well-being, unemployment disrupts the identity a person gets from work and separates people from the social mechanisms that integrate them into society—points that derive from symbolic interaction theory and functionalist perspectives on work.

The group experiencing the greatest unemployment in the United States at this time is Native Americans. Some 38 to 48 percent of Native Americans are unemployed. The greatest barrier to employment is the simple absence of jobs, especially for those Native Americans who live on reservations (about one-quarter of the total). Few jobs are on or near reservations; the few that do exist are often held by Whites. Reservation land is typically cut up so that fertile areas are owned by Whites. Tribal crafts and small-scale farming are usually all that is left to Native Americans. For most jobs, you have to leave the community to find work. Lack of appropriate training in schools and job-training programs means that many American Indians are not prepared for the few jobs that are available.

Indeed, education and training for American Indians lags far behind that of any other group; about 18 percent of American Indians ages sixteen to nineteen are not in school and have not completed high school, compared to 14 percent of African Americans and 10 percent of Whites. Only 9 percent of American Indians have completed four years of

underemployment

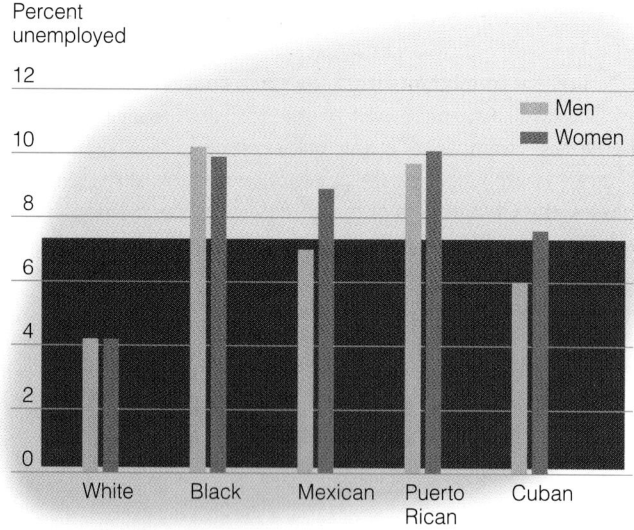

FIGURE 17.3 Unemployment Rates by Race and Gender

DATA: From the U.S. Department of Labor. 1997. *Employment and Earnings.* Washington, DC: U.S. Department of Labor, pp. 165–166, 169.

college, compared to 11 percent of African Americans and 22 percent of Whites (Snipp, 1996). Native American cultures also have traditions of work organization very different from the one created by the capitalist, job-driven economy, making adjustment to the demands of mainstream employment difficult, particularly given the poor jobs usually available to Native Americans. Finally, on many measures of health, Native Americans fare worse than any other group, making work impossible or difficult for many (National Center for Health Statistics, 1994; Ainsworth, 1989).

Next to Native Americans, unemployment is highest among Puerto Rican and African American men and women. These are the groups that are most discriminated against and that tend to be concentrated in places where unemployment is the highest. Among Latinos, Puerto Rican men have the highest unemployment rate, as we see in Figure 17.3. Among women, African American and Puerto Rican women and Chicanas are the most likely groups to be unemployed. Cuban Americans have unemployment rates roughly comparable to White men and women.

Unemployment among African Americans, Puerto Ricans, and Mexican Americans is currently at a level associated with a major economic depression even though unemployment in general has been declining in recent years. Contributing to the problem are some of the trends already discussed in this chapter—globalization, economic restructuring, and the move to a service economy—which fall especially hard on minority workers. In addition to being more likely to be unemployed, African Americans are more likely than Whites to experience the negative effects of job displacement, and the reemployment of African American men after losing a job is significantly lower than the reemployment of White men (Moore, 1992).

Women are also more likely than White men to be displaced from jobs. Following job loss, women workers are more likely than men to withdraw from the labor market altogether. Among all groups, Black women remain unemployed the longest after job displacement, followed by Black men, White women, and White men. Long-term joblessness also falls disproportionately on minority workers, both male and female (Moore, 1992).

Popular explanations of unemployment often attribute it to the individual failings of workers. Many often claim that unemployed people simply do not try hard enough to find jobs or prefer a welfare check to hard work and a paycheck. This is a conservative viewpoint reflecting the common myth that anyone who works hard enough can succeed. This individualistic perspective traces the cause of unemployment to personal motivation. Sociologists explain unemployment instead by looking at structural problems in the economy, such as rapidly changing technology (which reduces the need for human labor), discriminatory employment practices, deindustrialization, corporate downsizing, and the export of jobs overseas, where cheap labor is abundant. Rather than blaming the victim for joblessness, sociologists examine how changes in the social organization of work contribute to unemployment.

Some also argue that unemployment for White men is the result of hiring more women and minorities—as if men's unemployment is the result of affirmative action programs. Although on the face of it, this may seem reasonable since there are a finite number of jobs, the truth is that the employment of women and minorities does not cause White male unemployment if for no other reason than race and gender segregation. Women are typically not employed in the same jobs as men; race segregation also means that racial minorities are concentrated in certain segments of the labor force. In addition, unemployment for minorities is higher than it is for White men and is higher than for White women (U.S. Department of Labor, 1998). Blaming these groups for unemployment is not an accurate characterization.

Debunking Society's Myths

Myth: **Because of programs like affirmative action, women and minorities are taking jobs away from White men.**

Sociological perspective: **Although women and minorities have made many gains because of programs like affirmative action, women and minority workers still are clustered in occupations that are segregated by race and gender. There is little evidence the progress of these groups has been at the expense of White men (Hartmann, 1996).**

Work and Migration

The contemporary labor force is also being shaped by the employment of recent immigrants (Rumbaut, 1996a, 1996b; Pedraza and Rumbaut, 1996; Lamphere et al., 1994). Popular wisdom holds that the bulk of new immigrants are illegal, poor, and desperate, but the data show otherwise. In fact,

Immigrant groups often develop ethnic enclaves wherein they employ family and community members in a niche of the labor market and also provide services to their local community.

the proportion of professionals and technicians among legal immigrants exceeds the proportion of professionals in the labor force as a whole (Rumbaut, 1996a, 1996b). Note, however, that this figure is based on formal immigration data that exclude illegal immigrants, most of whom are working class. Still, those who migrate are usually not the most downtrodden in their home country. Even illegal immigrants tend to have higher levels of education and occupational skill than the typical worker in their homeland; seldom are the poorest able to migrate. Among immigrants can be found both the most educated and the least educated segments of the population; as an example, of contemporary immigrants from Africa and Asia, 47 and 38 percent, respectively, are college graduates (Rumbaut, 1996a: 38).

What happens when migrants arrive? Immigrants in professions tend to enter at the bottom of their occupational ladder, but compared to less skilled immigrants, they are more likely to succeed economically. Less skilled immigrants do not fare so well. Immigrants have higher unemployment than native-born workers, and they earn significantly less per week than other groups (Meisenheimer, 1992). Despite the popular image that migrants come to improve their lifestyle, the economic return from migration is negligible (Tienda and Wilson, 1992). The fact is there is enormous variation in the well-being of diverse immigrant groups, depending in large part on the circumstances under which they enter the United States and the resources they bring with them (Pedraza and Rumbaut, 1996).

Perhaps surprising to many is that the majority of immigrants are women. Women are especially numerous among today's largest immigrant streams: those from Central and South America, the Caribbean, Southeast Asia, and Europe (Donato, 1992). Women immigrants are some of the poorest immigrants, and they are typically concentrated in just a few occupations—domestic work, the garment industry, family enterprises, and skilled service occupations, such as nursing (Pedraza, 1996).

Some immigrant groups have found work through the development of *ethnic enclaves,* areas, typically urban, in which there is a concentration of ethnic entrepreneurs. Ethnic enclaves arise when a significant concentration of immigrants with business experience and access to labor and capital locate in a particular area. The source of labor is often recent immigrants from the same country of origin, as well as family members and people recruited from social networks. Examples of ethnic enclaves are Little Havana in Miami, where Cuban entrepreneurs own many of the businesses, the Chinatowns of many major American cities, and the Dominicans in New York City. In several large cities, Koreans have become the primary merchants in low-income areas. Ethnic enclaves can begin when immigrants disadvantaged in the labor market are pushed into self-employment. A few businesses formed to serve the immigrant community can be the start of an ethnic enclave that eventually includes its own banks, insurance firms, and other large-scale businesses (Portes and Rumbaut, 1996, 1994).

Diversity in the American Occupational System

From a sociological perspective, jobs are organized in an *occupational system.* This system is the whole array of jobs that together constitute the labor market. Within the occupational system, people are distributed in patterns that reflect the race, class, and gender organization of society. Jobs vary in their economic rewards, their perceived value and prestige, and the opportunities they hold for advancement, a fact that sociologists explain through a variety of perspectives. Fundamental to sociological analyses of the occupational system is the idea that individual and group experiences at work are the result of social structures, not just individual attributes. There is, for example, a rough correlation between the desirability of given jobs in the occupational system and the social status of the group most likely to fill those jobs. This tells us that there is a relationship between work and social inequality; thus, when sociologists study the occupational system, they acknowledge the importance of individual attributes such as level of education, training, and prior work experience in predicting the place of a worker in the occupational system, but they see a greater importance in the societal patterns that structure the experience of workers. We examine these specific patterns next.

The Dual Labor Market

Dual labor market theory views the labor market as comprising two major segments: the *primary labor market* and the *secondary labor market,* as we saw in Chapter 12. The primary labor market offers jobs with relatively high wages, benefits, stability, good working conditions, opportunities for promotion, job protection, and due process for workers (meaning workers are treated according to established rules and procedures that are allegedly fairly administered). High-level corporate jobs and unionized occupations fall into this segment of the labor market. The secondary labor market, on the other hand, is characterized by low wages, few benefits, high turnover, poor working conditions, little opportunity for advancement, no job protection, and arbitrary

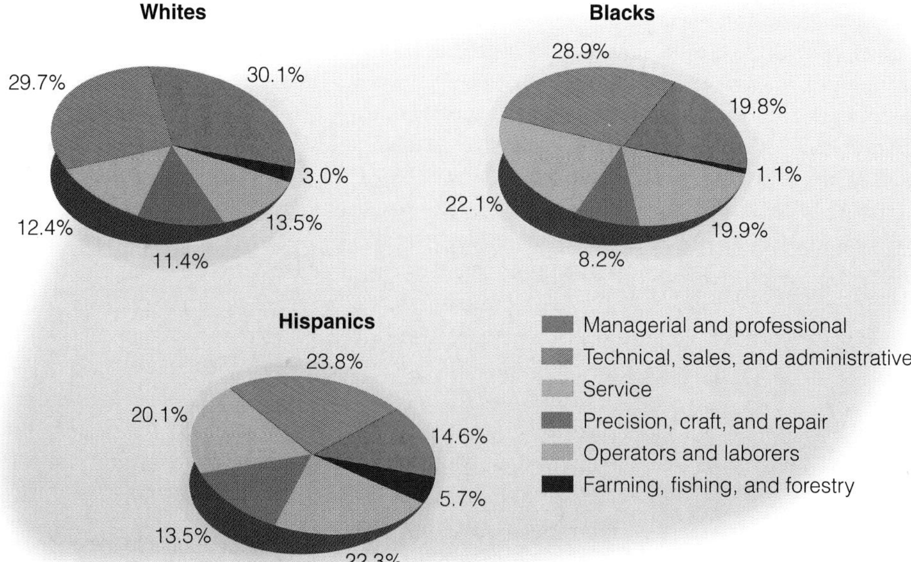

FIGURE 17.4 Occupational Distribution of the Labor Force

DATA: From the U.S. Department of Labor. 1998. *Employment and Earnings*. Washington, DC: U.S. Government Printing Office, pp. 173, 180.

treatment of workers. Many service jobs, like waiting tables, nonunionized assembly work, and domestic work, are in the secondary labor market. Women and minority workers are the most likely groups to be employed in the secondary labor market. Dual labor market theory traces some of the causes of race and gender inequality to this divided structure within the labor market (Reskin and Padavic, 1994; King, 1992; Wilkinson, 1991; Sandefur and Scott, 1988).

Dual labor market theory stems from a conflict theory perspective that sees the labor market as divided into different segments, with groups vying for the resources that jobs provide and competing with each other for jobs. The placement of women and minorities in the most devalued segments of the labor market is one of the major factors in gender and racial inequality. The workers who benefit from this arrangement are likely to organize themselves to exclude other workers from the better jobs (Reskin, 1988). For example, women and minorities have been historically excluded from labor unions, which preserved the better jobs for White men. Today, when exclusion is not so overt, the dual labor market remains a less visible means by which certain groups are excluded from the best jobs.

Some sociologists argue that the dual labor market works to the advantage of capitalist owners, who find it advantageous to encourage antagonism between different labor groups. Segmenting women and minorities into a niche in the labor market produces antagonisms between groups who see themselves as competing for the limited resources, such as between White working-class men and minority groups. When people perceive some other group to be the cause of their problems, they are less likely to blame the structure of the labor market or the work organization for difficulties they experience (Bonacich, 1972).

Occupational Distribution

Occupational distribution describes the pattern by which workers are located in the labor force. Figure 17.4 shows the occupational distribution of the U.S. labor force in 1997. The U.S. Department of Labor categorizes workers into six broad categories according to the kind of work involved: (1) managerial and professional; (2) technical, sales, and administrative support; (3) service occupations; (4) precision production, craft, and repair; (5) operators, fabricators, and laborers; and (6) farming, forestry, and fishing. Each of these broad groupings contains a number of specific jobs.

Workers are dispersed throughout the occupational system in patterns that vary greatly by race, class, and gender. Women are most likely to work in technical, sales, and administrative support, primarily because of their heavy concentration in clerical work. This is now true for both White women and women of color. White men are most likely to be found in managerial and professional jobs, whereas African American and Hispanic men are most likely to be employed as operators, fabricators, and laborers—jobs that are among the least well paid and least prestigious in the occupational system (U.S. Department of Labor, 1998).

When looking at occupational distribution over time, several changes are noticeable. First is the decline in the number of Black women in private domestic work. In 1960, 38 percent of all Black women in the labor force were employed as private domestic workers; by the 1990s, this had declined to less than 2 percent. The shift of Black women out of private domestic work (which included housework and child care) produced a need for labor in this area. The work is now done most often by recent immigrant women (often undocumented workers) from Latin and Central America, Southeast Asia, and the Caribbean, although many White working-class women also do this work. They work for professionals and other middle-class working people, earn low wages, have few benefits (if any), and are part of an underground economy of unreported and untaxed income.

A second major change in occupational distribution over this thirty-five-year period is the greater number of racial minorities in professional and managerial work, such as in business, law, and municipal management. Specific data about Latinos and Asians over this period are hard to find, given how these groups have been counted in the government data. Now almost 20 percent of Black women and men work as professionals or managers, compared to 7.4 percent in 1960 (U.S. Department of Labor, 1998). There has always been a professional and managerial segment among African Americans, even when racial segregation was most rigid; however, before the civil rights movement, Black professionals worked primarily in a separate and marginal economy of Black-owned enterprises serving primarily Black communities. Now, Black professionals are most likely to work in predominantly White organizations and companies or in government work. They are typically employed in visible but economically vulnerable jobs and are frequently tracked into dead-end jobs in "minority affairs" (Collins-Lowry, 1997). Still, the entry of significantly higher numbers of minorities—African American, Latino, and Asian—into the professions has expanded the number of minorities in the middle class and has increased diversity in the workplace.

Over time, there has also been some increase in the number of women employed in working-class jobs traditionally held only by men, although the number remains low. Women are 9 percent of precision production, craft, and repair workers, compared to 1 percent in 1970 (U.S. Department of Labor, 1998). This is a significant increase in the number of women in these jobs although women are still a small proportion of all such workers. The women most likely to be in such jobs are African American women, low-income mothers, women who previously held lower-rank jobs and disliked their previous job, and women who can work other than in the day. From this information, we can conclude that women are attracted to these jobs not just because they prefer them to other work, but because they offer better opportunities than the available alternatives (Padavic, 1992).

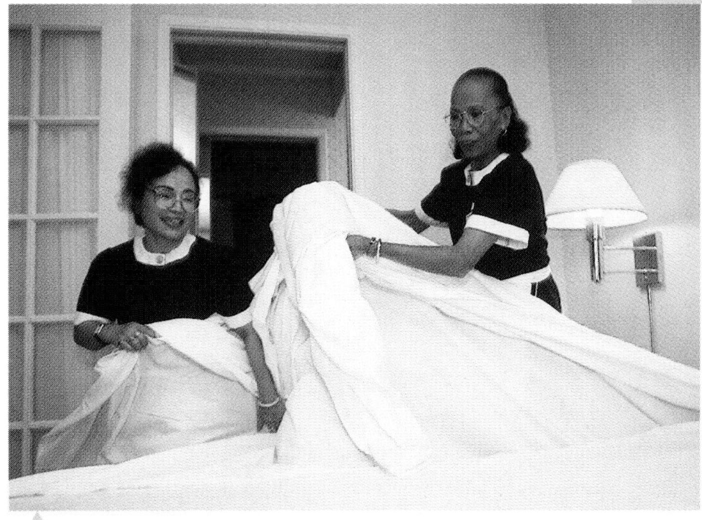

Occupational segregation by both race and gender means that women of color tend to be segregated in particular occupations, such as in the service industry as maids, cooks, and other personal services.

Occupational Segregation

Occupational segregation refers to the separation of workers into different occupations on the basis of social characteristics like race and gender, as we have seen in Chapter 12 on gender. Most occupations reveal some degree of occupational segregation, meaning there is a discernible tendency for the workers in a particular occupation to be of a particular gender or race or both. Women of color, for example, experience occupational segregation by both race and gender. That is, they are most likely to work in occupations where most of the other workers are also women of color.

Aggregate data on occupational distribution can be misleading since there is much *internal segregation* within the broad occupational categories. Women in the professional category, for example, are mostly concentrated in specialties that are gender-typed, such as nursing, social work, elementary and secondary school teaching, and library work. Within the most prestigious occupations, women are still a minority—10 percent of engineers and architects, 26 percent of physicians, 27 percent of lawyers and judges, and 31 percent of scientists (U.S. Department of Labor, 1998).

Men in predominantly female professions (such as nursing, school teaching, library work, and social work) do not encounter the discrimination facing women who enter male-dominated professions. The men may encounter prejudice in the form of stereotyping, but they experience real advantages, too. Men are more likely to be hired in occupations that are predominantly female than are women in occupations that are predominantly male, and men in predominantly female jobs are more likely than women in the same field to be promoted and advantageously treated. Researchers also find that men do not experience the same negative working climate that many women do, especially compared to women in traditionally male jobs (Williams, 1992, 1995). For all groups, however, there are economic penalties for being employed in occupations that have disproportionate numbers of women or minorities. The penalty is greatest in jobs where gender segregation is dramatic and long established, unionization is uncommon, and performance criteria are ambiguous (Baron and Newman, 1990).

Occupational Prestige

Occupational prestige is the perceived social value of an occupation in the eyes of the general public (see also Chapter 9). Sociologists have found a strong correlation between occupational prestige and the race and gender of people employed in given jobs. The influence of race on occupational prestige is stronger than that of gender; thus, African American and Latino men are found disproportionately in jobs that have the lowest occupational prestige scores; White and Asian American men hold the jobs with the highest occupational prestige, followed by White women, Asian American women, African American women, and Latinas. The higher the socioeconomic status of the occupational group, the smaller the proportion of African American men who are employed in it (Xu and Leffler, 1992; Stearns and Coleman, 1990).

Gender also has an effect on occupational prestige. Three patterns are apparent. First, women receive less prestige for the same work as men. Second, the gender composition of the job and its occupational prestige are linked: Jobs that employ mostly women are lower in prestige than those that employ more men. Indeed, jobs often lose their prestige as large numbers of women enter a given profession; likewise, the prestige of jobs increases as more men enter the field. Finally, men and women assign occupational prestige differently; women give occupations where most of the incumbents are women a higher prestige ranking than men do (Bose and Rossi, 1983; Tyree and Hicks, 1988).

Earnings

An important question for sociologists, as well as for policy makers and individuals, for that matter, is why different groups earn different incomes. Sociologists have extensively documented that earnings from work are highly dependent on race, gender, and class, as shown in Figure 17.5. White men earn the most, with a gap between men's and women's earnings among all groups. African American women and Hispanic men and women earn the least. Occupations in which White men are the numerical majority tend to pay more than occupations in which women and minorities are a majority of the workers. Not all men benefit equally, however, from the earnings hierarchy. White professional men earn more than men in working-class occupations; CEOs have the highest pay of all—a whopping 150 times what workers earned. White professional women may also be disadvantaged relative to men at the same occupational level, but they are far more advantaged than Black and Hispanic men and White working-class men.

Why are there such disparities? Again, we have to turn to theory to put the facts into perspective. According to functionalist theories, workers are paid according to their value—value derived from the characteristics they bring to the job: education, experience, training, and motivation to work. As we saw in Chapter 9 when we reviewed functionalist perspectives on inequality, functionalist theorists see inequality as motivating people to work. From this point of view, the high wages and other rewards associated with some jobs are the incentive for people to spend long years in training and garnering experience; otherwise, the jobs would go unfilled. To functionalists, then, differential wages are a source of motivation and a means to ensure that the most talented fill jobs essential to society and different wages reflect the differently valued characteristics (education, years experience, training, and so forth) that workers bring to a job.

Conflict theorists radically disagree with this point of view, arguing that many talented people are thwarted by the systems of inequality they encounter in society; thus, far from ensuring that the most talented will fill the most important jobs, conflict theorists note that some of the most essential jobs are, in fact, the most devalued and underrewarded—a

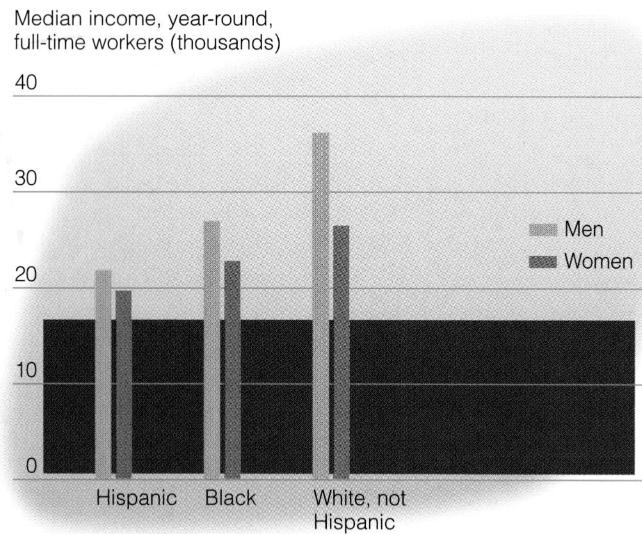

FIGURE 17.5 The Income Gap

DATA: From the U.S. Bureau of the Census. 1998. *Money Income in the United States: 1997.* Washington, DC: U.S. Department of Commerce, pp. 28–29.

point that feminist scholars who study women's work, both paid and unpaid, wholeheartedly agree with. From a conflict perspective, wage inequality is one of the ways that systems of race, class, and gender inequality are maintained. They say that differences in pay reflect the devaluation of certain classes of workers, and wage inequality primarily reflects power differences between groups in society.

Debunking Society's Myths

Myth: **The racial gap in earnings persists because many Black and Hispanics are not motivated to work; the effect of discrimination has been eliminated by equal employment policies.**

Sociological perspective: **The proportion of the racial gap in earnings due to discrimination has actually increased, in part because of the government's retreat from antidiscrimination policies (Cancio et al., 1996).**

Power in the Workplace

Obvious tangible factors that influence the degree of satisfaction with working are the level of pay and the benefits package. Also very important but less quantifiable are the value accorded to one's job and the opportunity for advancement. Another factor that greatly affects the working day for millions of people is power in the workplace. A job can be enriching or unbearable depending on who has power and how it is used.

As we saw in Chapter 12, the **glass ceiling** is a popular concept referring to the limits that women and minori-

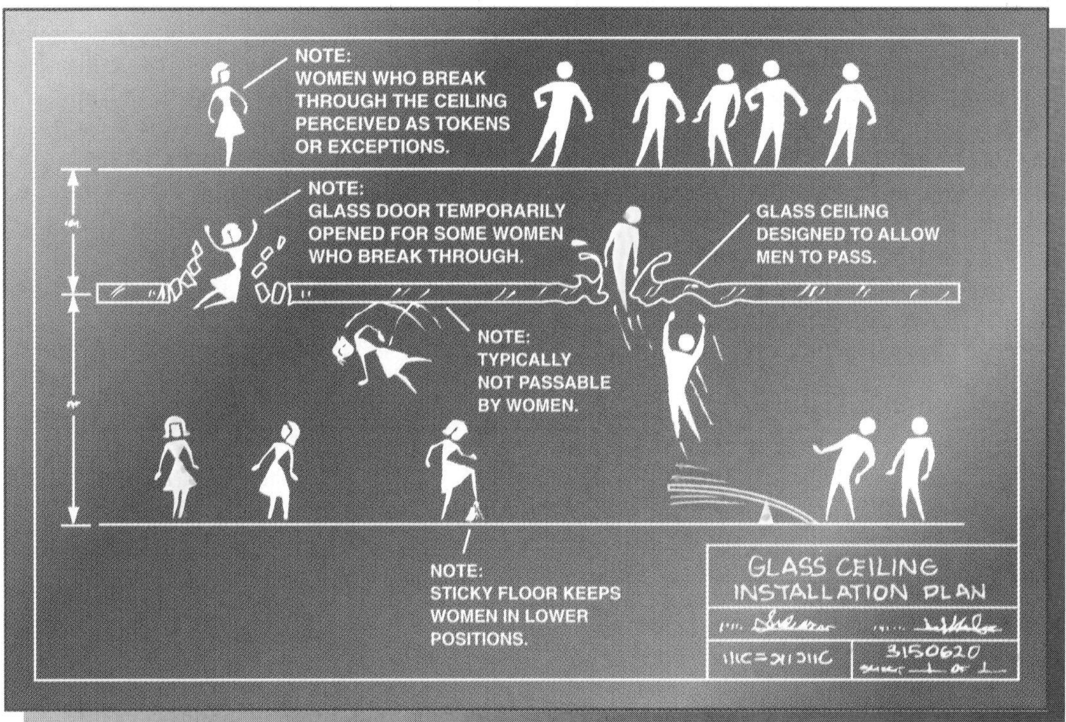

The Glass Ceiling

Whimsically depicted as an architectural drawing here, the glass ceiling refers to the structural obstacles still inhibiting upward mobility for women workers. Most employed women remain clustered in low status, low-wage jobs that hold little chance for mobility. Those who make it to the top often report being blocked and frustrated by patterns of exclusion and gender stereotyping.

ties experience in job mobility. Over the last few decades, many barriers to the advancement of women and minorities have been removed, yet as they try to advance, invisible barriers still prevent further advancement. The existence of the glass ceiling is well documented. Although there has been an increase in the number of managers who are women and minorities, most top management jobs are still held by White men. White men are far more likely than other groups to control a budget, participate in hiring and promotion, and have subordinates report to them; women and racial minorities remain clustered at the bottom of managerial hierarchies. The result is that White men have significantly more authority at work than everyone else. Traditional race and gender expectations persist in the form of norms that seem to bar women from exercising authority over men in the workplace and people of color from having authority over White people (McGuire and Reskin, 1993).

THINKING SOCIOLOGICALLY

Identify a specific occupation where you would have access to talking to men and women workers. Who gets promoted in this occupation and how? Do the women perceive that there is a *glass ceiling* for them and, if so, what are the barriers to promotion that they identify? If you can, get some data on the proportion of women and men in senior positions in this occupation and ask yourself if the glass ceiling is still in place.

Sexual Harassment

In addition to having less authority on the job than men, women workers are also more likely to experience **sexual harassment.** Sexual harassment is defined as unwanted physical or verbal sexual behavior that occurs in the context of a relationship of unequal power and that is experienced as a threat to the victim's job or educational activities (Martin, 1989). Two primary forms of sexual harassment are recognized in the law. *Quid pro quo sexual harassment* forces sexual compliance in exchange for an employment or educational benefit. A professor who suggests to a student that going out on a date or having sex would improve the student's grade is engaging in quid pro quo sexual harassment. The other form of sexual harassment recognized by law is the creation of a *hostile working environment* in which unwanted sexual behaviors are a continuing condition of work. This kind of sexual harassment may not involve outright sexual demands, but includes unwanted behaviors like touching, teasing, sexual joking, and comments that contribute to what is termed, in law, a hostile working environment.

Sexual harassment was first made illegal by Title VII of the 1964 Civil Rights Act, which identified sexual harassment as a form of sex discrimination. In 1986, the Supreme Court upheld the principle that sexual harassment violated federal laws against sex discrimination (*Meritor Savings Bank v. Vinson*). The law defines sexual harassment as discriminatory because it makes sex a condition of employment or education. By creating a hostile working environment, sexual harassment makes productive work difficult and discourages women's educational and work advancement. More recent Supreme Court cases have also upheld that same-sex harassment also falls under this law, as does harassment directed by women against men. Fundamentally, sexual harassment is an abuse of power by which perpetrators use their position to exploit subordinates. The law also makes employers liable for financial damages if they do not have policies appropriate for handling complaints or have not educated employees about their paths of redress.

The true extent of sexual harassment is difficult to estimate. Most surveys indicate that as many as half of all employed women experience some form of sexual harassment at some time. Men are sometimes the victims of sexual harassment, although far less frequently than women; most studies find that less than 3 percent of all sexual harassment cases involve women supervisors harassing male employees. The typical harasser is male, older than his victim, and of the same race and ethnicity. He is also likely to have harassed others in the past. There is some evidence that women of color are more likely to be harassed than White women (Gruber, 1982); in fact, many of the major legislative cases defining sexual harassment as illegal (including *Meritor Savings Bank v. Vinson*) involved African American women plaintiffs and White male harassers (Martin, 1989; Rubin and Borgers, 1990; McKinney, 1994).

Sexual harassment occurs in every kind of work setting, from factories to schools and colleges, but some organizational settings are more prone to sexual harassment than others. A strong predictor of sexual harassment is a high male-to-female ratio (Martin, 1989); thus, women in male-dominated professions are particularly vulnerable to sexual harassment. Whether in professional fields, such as science and engineering, or in blue-collar work, such as the skilled trades, women in these jobs report high rates of sexual harassment (Mansfield et al., 1991).

Debunking Society's Myths

Myth: **Sexual harassers are men who are maladjusted and cannot control their sex drive.**

Sociological perspective: **Sexual harassment is a means of asserting power in the workplace and is most common in settings where women are perceived as intruding in areas where men have been dominant.**

Numerous widely publicized events, including widely publicized sexual harassment at the Army's facility in Aberdeen, Maryland, and President Clinton's affair with Monica Lewinsky and accusations against him by Paula Jones, have brought public attention to the issue of sexual harassment. The Department of Defense has released a study of sexual harassment showing that nearly two-thirds (64 percent) of women in the military and 17 percent of men experienced some form of sexual harassment in the year prior to the survey, including teasing, jokes, sexually suggestive remarks, and direct harassment. Women were more likely than men to experience the most extreme forms of harassment, that is, those involving pressure for sex in exchange for favoritism in the workplace. Of the women reporting victimization in the survey, 5 percent reported actual or attempted rape or sexual assault; 12 percent, pressure for sex; 25 percent, unwanted touching or cornering (Martindale, 1991).

Some sociologists have argued that sexual harassment reflects men's resentment of women entering relatively high-paying fields once reserved for men; it is, thus, a mechanism men use to maintain their dominance in the workplace. Some men say they engage in this activity to show their interest and affection, but others say it is as a response to their perception of unfair treatment or is a demonstration of their anger (McKinney, 1992). Men are also more likely than women to see sexual harassment as normal and tolerable; consequently, they are less likely to think of harassment as doing something wrong (Reilly et al., 1992). Women are more likely to see behavior as sexual harassment and to judge it more severely than men (Loredo et al., 1995).

Sexual harassment tends to be underreported. Most studies have found that typically neither women nor men are aware of the proper channels for reporting sexual harassment. Studies also show that women are less likely to report sexual harassment than men who are victims, mostly because women believe that nothing will be done. The consequences of sexual harassment are economic, emotional, and sometimes, physical. Women respond to sexual harassment either by reporting it or by leaving the job, although no one knows how many women quit jobs because they are feeling harassed. Workers and students who have been sexually harassed report feeling helpless, fearful, and powerless. Some victims lose their ambition and self-confidence, and take a negative view of their work. Among students, the experience of being harassed creates self-doubt, anxiety, confusion, and distrust of all male faculty in general, feelings that certainly interfere with learning and achievement (Benson and Thomson, 1982; Bremer et al., 1991).

Gays and Lesbians in the Workplace

The increased willingness of lesbians and gay men to be open about their sexual identity has resulted in more attention being paid to their experience in the workplace. Surveys find that a large majority (84 percent) endorse the general concept that lesbians and gay men should have equal rights in job opportunities, but when asked about specific occupations, substantial numbers think there are some occupations where gays should not be hired, such as elementary teaching. Generally speaking, people approve of gay men working as artists, hairdressers, designers, actors, and musicians, but many oppose their employment in teaching and child care (Roper Organization, 1996; Levine, 1992). These attitudes promote discriminatory treatment in the workplace.

Any negative experience in the workplace can affect self-esteem, productivity, and economic and social well-being. Although there are few empirical studies of gay and lesbian work experiences, those that do exist find that gays fear they will suffer adverse career consequences if co-workers know they are gay. This leads many to "pass" as heterosexual at work or keep their private lives secret, but being closeted at work puts gay workers at a disadvantage. Shielding themselves from antagonism or rejection may make them appear distant and isolate them from social networks. These behaviors can actually have a negative effect on their performance reviews since fellow workers may find them unfriendly, boring, and withdrawn. Research finds that the relationships of gay employees with their co-workers are less stressful when the employees are "out" (Schneider, 1984). The work organization is also improved for lesbian and gay employees when employee assistance programs encourage open communication, policies are sensitive to lesbian and gay needs, and discriminatory practices can be identified and stopped (Sussal, 1994).

Most workers, including gays and lesbians, think it is desirable to integrate work and social life, but one-third of gays keep their social life separate from work. More than one-third try not to discuss personal matters with co-workers even though they believe that such discussions make things at work go more smoothly. Lesbians and gay men at work encounter numerous heterosexist assumptions as co-workers discuss their personal lives and families. Participation in these discussions can make gays and lesbians fearful of revealing their identity; absenting themselves from intimate social chat can make them feel isolated from interpersonal networks. Studies of lesbians find that they are more likely to be open about their sexual identity when women predominate in the workplace and when most of their work friends are women. They are also more likely to be open about their identity when they have a woman boss or supervisor; have few supervisory responsibilities themselves; have lower incomes; work in small organizations; or do not deal with children as students, clients, or patients. In heavily male-dominated workplaces, less than 10 percent of lesbian women come out (Schneider, 1984).

Worker Satisfaction and Safety

A strong influence on worker satisfaction is the work environment. Safe and sanitary conditions, good pay, opportunities for advancement, and policies that recognize human dignity are among the foundations of worker satisfaction. The size of the work organization also has an effect, with workers in smaller firms tending to be more satisfied. Workers in these organizations perceive a less rigid stratification system and believe they have more possibility for job advancement. This reflects the general tendency for workers to be more satisfied when they perceive that there is room to grow on the job (Free, 1990; Kelley, 1990). Work that is intrinsically rewarding and challenging, that provides for advancement, and that makes workers feel responsible for their achievements results in greater job satisfaction.

Of course, the same elements of work life that promote worker satisfaction are the cause of dissatisfaction when they are absent. As the routinization of work increases, so does worker dissatisfaction. As unemployment in the region goes up, worker discontent rises as well because workers in a weak labor market often feel they have no alternatives to the job they are doing, whether they like their job or not.

Women, minorities, and working-class men are often the least satisfied with their jobs—likely a reflection of their position in the race, class, and gender hierarchies at work. For women, being treated as equals is important. Women also value social exchanges at work, more than men do, and they see socializing as an important part of developing trusting relationships (Congdon, 1990). Women report less satisfaction and more stress at work than men, especially when the women work in traditionally male occupations (Mansfield et al., 1991). In professions where men are a majority, both men and women report higher levels of job satisfaction than male and female workers in traditionally female professions, probably because of the greater cultural value given men's work, regardless of who actually does it; in other words, the higher value that men are accorded in society carries over into professions where they predominate (Cassidy and Warren, 1991).

Within work organizations, women tend to have far less control than White men do. Women are less likely to be employed in jobs where they have some degree of autonomy and where they can set their own goals. Instead, they tend to be in more routine jobs, and they are less likely to supervise others' work. On the other hand, sociologists have found that women report that they get more interpersonal rewards from working than men report, although men clearly get more economic rewards. Although the interpersonal rewards women get from work are important, they do not affect the satisfaction that workers experience at work since sociologists find that interpersonal rewards do not increase job satisfaction—pay does (Ross and Mirowsky, 1996).

Management and administrative style is a major factor in job satisfaction. Highly authoritarian styles—that is, those where power is highly concentrated and blind authority to power is expected—breed worker discontent. Under such systems, workers may try to sabotage the operation of a workplace, although they have to do so subversively, such as by holding work slowdowns or scheduling a boss to be out of the office for long periods!

As employers have recognized the negative effects of such management styles, they have developed more participatory styles of management, as we saw in Chapter 6 on organizations. Intended to increase worker satisfaction and thereby enhance productivity, firms have developed seemingly democratic procedures like *total quality management* to make workers think they are more important. Total quality management emphasizes employee involvement in decision making and a client-centered working philosophy.

Other plans, like employee stock ownership, have also been developed to try to humanize the workplace, enhance worker satisfaction, and increase commitment to the organization. An idea behind some of these practices is that productivity can be enhanced if workers focus on the intrinsic rewards of the job, rather than just the extrinsic rewards, like pay and promotions. Some new workplace strategies also seek to break down the division between management and workers, thereby lessening workplace conflicts (Fenwick and Zipp, 1993).

The results of these approaches are mixed. In general, the effect of such change is not great, although this depends to a large extent on the interests and attitudes of managers (Bradley and Hill, 1987). Workers sometimes see these practices as manipulating them into thinking they are important when they are not. Sociologists have found that in democratically structured, employee-owned firms wage inequality by gender and race is lessened somewhat, compared to traditionally organized business organizations. This suggests that personnel policies designed to alter power relationships and expand the responsibilities of those in lower-level jobs can be a valuable tool for reducing racial inequality (Squires, 1991; Squires and Lyson, 1991).

Worker Alienation

Alienation is a feeling of powerlessness and separation from one's group or society. Alienation is a more specific concept than just general discontent. It is the boredom and meaninglessness of routine and repetitive work that is controlled, overspecialized, and underskilled. In alienated work, people become robots, going through the motions without any identity or commitment to the process or product.

The concept of worker alienation was first developed by Karl Marx, who believed that in a capitalist society workers would always feel alienated because they do not control their labor; moreover, the value of the goods workers produce is always greater than the amount workers are paid, so the workers are estranged from the profits of their labor. Alienation, Marx thought, was inherent in the system of capitalism.

The sociologist Robert Blauner produced a classic study of worker alienation. Like Marx, he defined alienation as the feeling of separation from one's labor (Blauner, 1964). Worker alienation, Blauner argued, was especially intense in work such as automobile assembly lines where workers have little control over their labor. Blauner emphasized the performance of fragmented and repetitive tasks as a major source of alienation, something typical of many of the jobs in the industrial workplace. On assembly lines, where workers engage in only one step of the production process, alienation is widespread. One worker may put rivets in the same part of an airplane all day long. Doing only fragments of the production process and never seeing the full product of one's labor leads to feelings of powerlessness and dehumanization. Indeed, workers may come to think of themselves as machines—as mechanized and routinized as robots. With the rise of automation, feelings of alienation and dissatisfaction have increased, but any kind of work can be alienating if workers view their tasks as meaningless and out of their control. Alienation is also exacerbated by workers having little control over the work environment, being isolated from other workers, experiencing little group cooperation or interaction at work, being dissatisfied with low pay and low status, and finding stagnancy in the job market.

Alienation leads to feelings of powerlessness. Having no sense of the relationship between one's work and the overall productive process can produce a sense that one's work is without value or meaning. One consequence of alienation may be organizational deviance, such as deliberate destruction of company property, theft, or embezzlement (see Chapter 8). Alienated workers are likely to feel no loyalty to the organization that employs them. Their productivity is likely to be low. Their discontent can be measured not just in human unhappiness, but in lost economic potential.

Occupational Health and Safety

At work, people not only face situations that vary in how satisfying they are; sometimes they also face outright dangers. The most dangerous occupations are those where working-class men are the majority of workers. Traffic accidents and homicides at work are the most common cause of fatal work injuries—thus, truck and taxi drivers, retail clerks and police have some of the highest death rates from work-related fatalities. Fishers, timber cutters, sailors, and deckhands have the highest rate of occupational fatalities (U.S. Department of Labor, 1994, 1995). Women are less likely to be killed at work because they hold fewer jobs in the most dangerous occupational groups. Homicide is the leading cause of death at work for women; when women are murdered at work, it is often a current or former boyfriend or husband (Kneustaut, 1996). Even children work at hazardous jobs, especially in agriculture where children under fourteen years may be using knives and machetes, operating machinery, or being exposed to pesticides. In 1992 alone, 64,000 children between the ages of fourteen and seventeen were treated in hospitals for work-related injuries (Nixon, 1996).

Occupations that employ large numbers of immigrant workers are especially hazardous for employees. Meatpacking has the highest injury rate of all U.S. industries; 36 percent of all employees in this industry are seriously injured at work each year. Taxicab drivers run the highest risk of homicide at work—forty times the national average—another occupation where immigrant labor is common. In sweatshops where recent immigrants are likely to be working, exposure to lead, asbestos, and other hazardous substances creates serious health risks. Language barriers often contribute to workers' lack of information about hazards they face on these jobs (Hawkins, 1996; Cooper, 1997).

Although the technological revolution has brought the promise of a "clean" workplace in which people perform mental labor while machines do the dirty and dangerous work once performed by humans, whether technologically based jobs are truly as clean as believed is increasingly in doubt. New hazards of the technological workplace include toxic chemicals, nuclear hazards, repetitive motion disorders, and problems associated with prolonged use of video display terminals. Although technological developments can enhance productivity and efficiency, they also have a downside, including physical problems like carpal tunnel syndrome, eye strain, back problems from sitting at a monitor, and ironically, the possibility for more paper trash as people write less and print more (Heim, 1990). Occupational hazards of the past also continue to exist. In modern-day sweatshops, workers (usually recent immigrant women) labor outside the reach of regulatory protections, operating unsafe equipment in unsafe facilities.

Disability and Work

Not too many years ago, people did not think of those with disabilities as a social group; rather, disability was thought of as an individual frailty or perhaps a mark of stigma. Sociologist Irving Zola (1935–1994) was one of the first to suggest that people with disabilities face issues similar to minority groups. Instead of using a medical model that treats disability like a disease and sees individuals as impaired, conceptualizing disabled people as a minority group enabled people to think about the social, economic, and political environment that disabled people face. Instead of seeing disabled people as pitiful victims, this approach emphasizes the group rights of the disabled, illuminating such things as access to employment and education (Zola, 1989, 1993).

Now those with disabilities have the same legal protections afforded to other minority groups. Key to these rights is the Americans with Disabilities Act (ADA), adopted by Congress in 1990. Building on the Civil Rights Act of 1964, and earlier rehabilitation law, in particular the Rehabilitation Act of 1973, the Americans with Disabilities Act protects disabled persons from discrimination in employment and stipulates that employers and other providers (such as schools and public transportation systems) must provide "reasonable accommodation" for disabled persons. Disabled persons must be qualified for the jobs or activities for which they seek access, meaning that they must be able to perform the essential requirements of the job or program without accommodation to the disability. For students, reasonable accommodation includes such things the provision of adaptive technology, exam assistants, and accessible buildings.

The law prohibits employers with fifteen or more employees from discriminating against either job applicants who are disabled or current employees who become disabled. This includes job application, wages and benefits, advancement, and employer-sponsored social activities. The law applies to state and local governments, as well as employers. ADA also legislates that public buses, trains, and light rail systems must be accessible to disabled riders; airlines are excluded from this requirement. The law also requires businesses and public accommodations to be accessible and requires telephone companies to provide services that allow hearing- and speech-impaired people to communicate by telephone.

Not every disabled person is covered by ADA. To be considered disabled, a person must have a condition or the history of a condition that impairs a major life activity. The law specifically excludes the use of illegal drugs as a cause of disability and also excludes pregnancy as a disability (which is ironic since pregnancy is treated as a disability by other federal laws for purposes of maternity leave). Passage of the law was based on reports by the National Council on Disability (NCD) that documented pervasive discrimination against the disabled in employment, transportation, and public accommodations. The NCD argued to Congress that disabled people were being treated as second-class citizens and that with the provisions of this act, they could become more independent. When the bill was in the House, small business owners mounted a vigorous campaign against it, arguing that the provisions would force them into bankruptcy. As a result, a phase-in provision on implementation was provided for small employers (Americans with Disabilities Act, 1990).

The Americans with Disabilities Act provides legal protection for disabled workers, giving them rights to reasonable accommodations by employers and access to education and jobs.

Although most working-age persons with disabilities do not work, an overwhelming majority (86 percent) want to. Enhanced technologies have created new opportunities for people with disabilities, but shrinkage in the labor market makes it more difficult for disabled people to find work. Another problem is prejudice against people with disabilities. Disabled people report that they are often ignored and treated as if they are invisible; they also induce pity and discomfort. This leads to significant discrimination in employment and other areas. Discrimination is among the reasons that persons with disabilities tend to have low household incomes. Black and Hispanic people with disabilities and without much education are worst off economically. In recent years, disabled persons have made some gains in economic well-being, but most of the gain is attributable to the earnings of household members who are not disabled. The new disability laws are intended to alleviate this situation (Hearne, 1991; West, 1991).

Current law also defines pregnancy as a disability to allow pregnant workers some protection under the law. With more women in the labor force, and more of them continuing to work while they are pregnant, new questions are being raised about how pregnancy should be treated in the workplace. Currently, the law stipulates that pregnancy is to be treated like any other "disability." That is, states may require employers to grant up to four months of unpaid leave to women who are pregnant, and women who return to work following pregnancy must be permitted to return to their old job or its equivalent. These laws reversed a long-term historical trend in which employed women who became pregnant were forced to leave the workforce or, if they returned following birth, had to take other lower-paid jobs. The Family and Medical Leave Act of 1993 provides legal protection against this form of discrimination.

What effect does pregnancy have on women's work patterns? Common myths that women who become pregnant drop out of the labor force are simply untrue. Because most women work out of economic necessity, maternity leaves are generally short. Women who work the most hours and women in higher-status jobs are the least likely to leave a job following the birth of a child. Family circumstances are also a strong predictor of whether a woman will return to work following a birth, and how soon. Contrary to what one might expect, mothers with a spouse or another adult in the household return to work more quickly than those without other adults in their household. The greater the proportion of the family income a mother provides, the sooner she returns to work. There are no differences by race in these patterns (Wenk and Garrett, 1992).

Some firms have written special policies allegedly designed to protect women from reproductive hazards in the workplace. Some have even gone so far as to exclude all women, pregnant or not, from occupations deemed risky, particularly those that are highly skilled, technical jobs, traditionally dominated by men. Feminists argue that such legislation is discriminatory because it targets only women when men too have reproductive organs that are equally vulnerable to injury or toxicological insult. They conclude that men and women both need protection from reproductive hazards in the workplace. Feminists also note that it is no accident that the occupations deemed too risky to the reproductive health of women are traditionally male-dominated jobs with relatively high pay. No action has been taken, feminists point out, to exclude women from equally hazardous occupations that have been identified as women's work—operating-room nurses, for example. Policies intended to regulate occupational risks have been clouded by assumptions about appropriate jobs for women and men. The simple truth is that all workers need to be safe and secure in their work environments.

CHAPTER SUMMARY

- Societies are organized around an economic base. The *economy* is the system on which the production, distribution, and consumption of goods and services are based. The change from agricultural systems to industrial and postindustrial economy affects all aspects of society.
- *Capitalism* is an economic system based on the pursuit of profit, market competition, and private property. *Socialism* is characterized by state ownership of industry; *communism* is the purest form of socialism.
- The contemporary economy is increasingly global and affected by economic restructuring, including increasing diversity in the workforce, deindustrialization, and technological change.
- Sociologists define *work* as human activity that produces something of value. Some work is judged to be more valuable than other work. *Emotional labor* is work that is intended to produce a desired state of mind in a client. The *division of labor* is the differentiation of work roles in a social system. In the United States, there is a class, gender, and racial division of labor.
- *Functionalist theory* emphasizes that different rewards for important jobs in society motivate workers to pursue these areas. *Conflict theorists* study the tensions generated by power differences as groups compete for social and economic resources. *Symbolic interactionists* study the meaning systems that shape people's behavior and identity at work.
- In recent years, the employment of women has been increasing, whereas that of men has been decreasing. Official unemployment rates underestimate the actual extent of joblessness. Women and racial minorities are the most likely to be unemployed. Recent immigrants also have a unique status in the labor market, often in the lowest paying and least prestigious jobs. Some ethnic groups form *ethnic enclaves*—areas of ethnically owned businesses that serve predominantly ethnic communities.
- The U.S. occupational system is characterized by a *dual labor market.* Jobs in the primary sector of the labor market carry better wages and working conditions, whereas those in the secondary labor market pay less and have fewer job benefits. Women and minorities are disproportionately employed in the secondary labor market. Patterns of occupational distribution also show tremendous segregation by race and gender in the labor market. Race and gender also affect of the occupational prestige of given jobs.
- Women and minorities often encounter the *glass ceiling*—a term used to describe the limited mobility of women and minority workers in male-dominated organizations. Women, in addition, face *sexual harassment* at work more often than men—defined as the unequal imposition of sexual requirements in the context of a power relationship. Homophobia in the workplace also negatively affects the working experience of gays and lesbians.
- Work satisfaction is contingent on the characteristics of work organizations. Workers in more routinized jobs tend to be less satisfied than others. Work *alienation* is the feeling of powerlessness and separation that results when workers have little control over the products of their labor. Health and safety hazards also affect worker health. New legislation prohibits discrimination against those with disabilities. The greater participation of women in the labor force has also meant that pregnancy is now protected by law.

KEY TERMS

alienation
automation
capitalism
communism
dual labor market theory
economic restructuring
economy
emotional labor
glass ceiling
global economy
job displacement
occupational distribution
occupational prestige
occupational segregation
sexual harassment
socialism
underemployment
unemployment rate
work
xenophobia

THE INTERNET: A Tool for the Sociological Imagination

Resources on the Internet:

Virtual Society: The Wadsworth Sociology Resource Center
http://sociology.wadsworth.com

Visit this site to find additional learning tools, including interactive quizzes, links related to web sites, and an easy link to *InfoTrac College Edition.*

National Institute for Occupational Safety and Health (NIOSH)
http://www.cdc.gov/niosh/homepage.html

This site contains extensive information on policies and laws regarding worker safety and health, as well as current information about occupational safety and employment possibilities in this area.

U.S. Department of Labor
http://www.dol.gov

As the federal agency responsible for overseeing issues related to labor and employment, the Department of Labor maintains a site with statistical data on employment, as well as such services as a job bank in the public employment sector.

U.S. Bureau of Labor Statistics
http://stats.bls.gov

The Bureau of Labor Statistics home page provides current information on the U.S. economy, press releases on matters pertinent to the study of work, data on work and employment, and links to numerous data sources on employment.

Sociology and Social Policy: Internet Exercises

The *Americans with Disabilities Act* extends employment rights to those with documented disabilities. What does this law require of employers and provide for disabled workers? Do you think that the law is adequate in providing opportunities for disabled people, or do you think additional policies are needed? What does your school provide (or not) for disabled students?

Internet Search Keywords:

Disabilities Act
disabled workers
employment
employee rights
civil rights
job accommodation

Web sites:

http://www.usdoj.gov/crt/ada/adanom1.htm
Home page for the Americans with Disabilities Act

http://www.disability.gov.uk/drtf/
The web site for the Disability Rights Task Force provides information on enforceable civil rights for disabled people.

http://janweb.icdi.wvu.ecu/
The Job Accommodation Network offers information about job accommodations and the employability of people with disabilities.

http://www.icanect.net/fpa/
The Association of Disability Advocates provides advocacy information and help for people with disabilities.

ht.//www.ada-infonet.org/
APA Information Center offers information for employers on the American Disabilities Act.

http://www.ns.net/OAB
A web page dealing with the issues affecting people with developmental disabilities at the state and national level.

InfoTrac College Edition: Search Word Summary

division of labor
job displacement
underemployment
xenophobia

In order to learn more about these central topics in sociology, you can conduct an electronic search using InfoTrac College Edition. To aid in your search and to gain useful tips, see the Student Guide to InfoTrac College Edition on the Virtual Society web site:
http://sociology.wadsworth.com

INTERACTIONS—A SOCIOLOGY CD-ROM: CONCEPTS FOR THIS CHAPTER

Go to the Wadsworth Sociology CD-ROM for further study on the concepts in this chapter. The CD-ROM also includes quizzes and additional activities to expand your learning experience.

SUGGESTED READINGS

Amott, Teresa L., and Julie A. Matthaei. 1996. *Race, Gender, and Work: A Multicultural History of Women in the United States,* 2nd ed. Boston: South End Press.

Based on an analysis of the intersections of race, class, and gender in the labor market, this book presents the histories of the work of women of color and White women.

Barlett, Donald L., and James B. Steele. 1992. *America: What Went Wrong?* Kansas City: Andrews and McMeel.

Two Pulitzer Prize–winning authors explore the consequences of economic restructuring for American workers. The book documents the economic trends accompanying the move of labor to other countries, the concentration of wealth, and the impact of technological change on the workforce.

Collins-Lowry, Sharon. 1997. *Black Corporate Executives: The Making and Breaking of a Black Middle Class.* Philadelphia: Temple University Press.

A study of Black executives, this books shows patterns of advancement of this portion of the new Black middle class, while revealing the subtle forms of discrimination that persist in the workplace. Collins shows how much race still matters in employment practices.

Hamper, Ben. 1991. *Rivethead: Tales from the Assembly Line.* New York: Warner Bros.

Written in a humorous style, Hamper's book gives firsthand accounts of the experience and perspectives of workers on the automobile assembly lines.

Hochschild, Arlie. 1997. *The Time Bind: When Work Becomes Home and Home Becomes Work.* New York: Metropolitan Books.

Hochschild's study examines two related developments: the increased pressure of work in the home and the primary attachments many people find at work. This reversal in primary and secondary roles results, she argues, from the impact of social speedup and the new organization of work and family.

Leidner, Robin. 1993. *Fast Food, Fast Talk: Service Work and the Routinization of Everyday Life.* Berkeley: University of California Press.

Using symbolic interaction theory, Leidner conducted extensive observations of fast-food workers at McDonald's and analyzes how they adapt to the routine and boring work that they are required to do. Her study is a fascinating account of people's creativity in the workplace and the social structure of everyday life.

Reskin, Barbara, and Irene Padavic. 1994. *Women and Men at Work.* Thousand Oaks, CA: Pine Forge Press.

Reskin and Padavic provide a basic introduction to gender and work in a book meant for undergraduates. They explore topics such as the gender division of labor, gender segregation in the labor market, wage inequality, and trends for the future.

Schorr, Juliet B. 1991. *The Overworked American: The Unexpected Decline of Leisure.* New York: Basic Books.

According to Schorr's book, increases in the amount of work that Americans do have resulted in an extra month of work every year. Her discussion highlights the changes that have taken place in the labor force and their impact of the lives of people at work and in families.

Williams, Christine L. *Still a Man's World: Men Who Do Women's Work.* Berkeley: University of California Press.

By focusing on men who work in traditionally female jobs, Williams demonstrates how gender structures opportunities in different occupations.

Wilson, William Julius. 1996. *When Work Disappears.* Chicago: University of Chicago Press.

Wilson's book is an important analysis of the effect of economic restructuring on the family and work experience of Black Americans. His recommendations for social policy in this area have also been the basis for federal planning on urban problems.

CHAPTER 18

Power, Politics, and Authority

PICTURE a couple lounging together on a public beach on a national holiday. They imagine themselves married someday, perhaps having children whom they will send to public school. Does the government have anything to say about the couple's children? They may decide to use birth control until they are ready for children; if so, their choice of birth control methods will be limited to those the government allows. What if they cannot have children? They might consider adopting—if state agencies judge them acceptable as parents.

As you imagined this couple, whom did you picture? Were they the same race? Today, they would be legally able to marry no matter what their races. Not so long ago, the law would have forbidden the marriage if the marriage were interracial. Is this a man and a woman or is this a same-sex couple? If they are lesbian or gay, despite being in love and wanting to form a lifelong relationship, in most states they would be prevented by law from marrying.

This couple is probably giving no thought to how much their life is influenced by the state at that very moment. Permission to walk on the public beach comes from the state; the day off to celebrate a holiday is sanctioned by the state; the drive to the beach took place in a car registered with the state and inspected by the state, driven by a motorist licensed by the state. The range of things regulated by the state is simply enormous, and yet many are never noticed.

If so much state attention seems oppressive, consider life without it. People assume that when they go to work, they will get paid. Some employers would be sure to discover the profitability of withholding paychecks, were it not for laws requiring that employers meet their obligation to pay employees in exchange for work, and further requiring that the wage be fair, that hiring be done without regard to race or gender, and that the workplace be safe. Without the state, a person who is robbed would have little recourse except personal revenge. In the form of law and the judicial system, the state defines good and evil, right and wrong.

Because not all conflicts in society are so well defined, the state also regulates disputes between groups that may both be well intentioned, yet bitterly opposed to each other's wishes. The state steps in at all levels, from arbitrating disputes between two parties (as in a lawsuit or divorce) all the way up to negotiating and defining class, race, and gender relations in society. Imagine the United States with no government, no laws to regulate people's behavior, no organized police or security force to defend the nation, and no prisons. Imagine the infrastructure with no paved roads, no expensive bridges, no satellites, no flood-control programs, no weather forecasting, no printed money. Imagining society without the state is a bit like imagining the individual without socialization, as we did in Chapter 4. Remove the state from society and what is left?

Defining the State

In sociological usage, the **state** is the organized system of power and authority in society. The state is an abstract concept that refers to all those institutions that represent official power in society, including the government and its legal system (including the courts and the prison system), the police, and the military. The state regulates many societal relations, ranging from individual behavior and interpersonal conflicts to international affairs.

Theoretically, the state exists to regulate social order although it does not always do so fairly or equitably. The guarantee of life, liberty, and the pursuit of happiness promised by the U.S. Declaration of Independence, when examined carefully, is not evenly distributed by the state. Less powerful groups in the society may see the state more as an oppressive force than as a protector of individual rights. They may still turn however to the state to rectify injustice. For example, when African Americans sought to end segregation, they sought legal reform through the state.

The state has a central role in shaping class, race, and gender relations in society and in determining the rights and privileges of different groups (see the box "Understanding Diversity: American Indians and State Policy"). The involvement of the state may include the resolution of management

B • O • X 18.1

UNDERSTANDING DIVERSITY

American Indians and State Policy

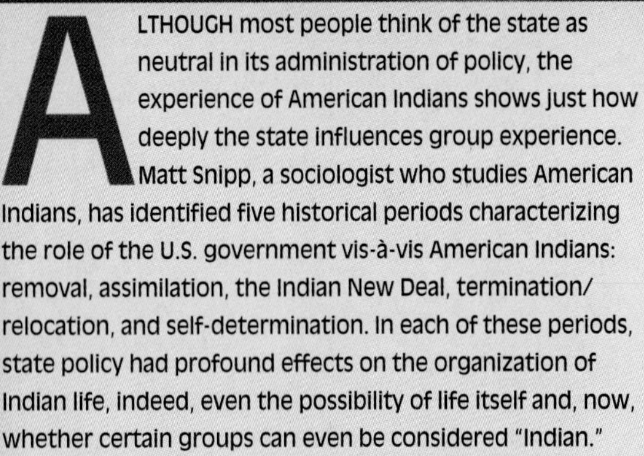

ALTHOUGH most people think of the state as neutral in its administration of policy, the experience of American Indians shows just how deeply the state influences group experience. Matt Snipp, a sociologist who studies American Indians, has identified five historical periods characterizing the role of the U.S. government vis-à-vis American Indians: removal, assimilation, the Indian New Deal, termination/relocation, and self-determination. In each of these periods, state policy had profound effects on the organization of Indian life, indeed, even the possibility of life itself and, now, whether certain groups can even be considered "Indian."

In the first period, the U.S. government used its military forces and various pieces of legislation to forcibly remove dozens of Indian tribes from the eastern half of the United States to so-called Indian Territory. As White citizens moved westward, one of the bloodiest periods in American history occurred, resulting in both genocide and one of the largest forced migrations in history. Spanning more than half a century (the early nineteenth century), removal forced thousands of Cherokees, Creeks, Choctaws, Seminoles, and other Indian groups from their homes to face mass death and drastic relocation.

Near the end of the nineteenth century, the government adopted the goal of isolating American Indians on reservations. Although their numbers had dramatically declined, American Indians were forced into boarding schools designed to indoctrinate Indian children and teach them that their culture was inferior to that of Euro-Americans. They were forbidden to wear native clothes, speak their own language, or practice their traditional religions. The 1887 General Allotment Act mandated that tribal lands would be allotted to individual American Indians, with surplus lands sold on the open market. This caused the loss of about ninety million acres of Indian land (approximately two-thirds of all Indian land held in 1887).

The third period, the Indian New Deal, was implemented in the 1930s along with other New Deal policies of the Roosevelt administration. Providing some relief from the Great Depression, these laws also allowed tribal self-government (as specified in the Indian Reorganization Act of 1934), but this has been an ongoing source of conflict since critics say it also forced an alien form of government on tribal groups.

Following World War II, in the termination and relocation period, the U.S. government tried unsuccessfully to abolish all reservations. The Bureau of Indian Affairs (which had been established in 1824) encouraged American Indians to move to cities, but many returned to reservations when they found only seasonal employment.

In the 1960s, American Indian leaders made self-determination their major priority. The U.S. government responded by passing the American Indian Self-Determination and Education Assistance Act (1975), allowing American Indians to oversee the affairs of their own community without federal intervention. This has resulted in some reservations having their own police forces, being able to levy taxes, and in one case, to issue passports that are internationally recognized (the Onondaga tribe in New York). Smaller reservations and those with limited resources are still, however, quite dependent on Bureau of Indian Affairs' services.

Today, only those Indian groups that are officially recognized by the U.S. government can be considered Indians; those who are not have no access to the limited aid provided by the Bureau of Indian Affairs. The process of formal recognition is a difficult one since petitioners must do years of research to document their heritage and negotiate their way through a maze of federal regulations. The absence of written records among most American Indian groups complicates this historical search for verification. In 1993, of the 143 groups that were seeking recognition, only 23 cases were resolved, with only 8 being officially granted Indian status (Brown, 1993).

SOURCE: Adapted from Snipp, C. Matthew. 1996. "The First Americans: American Indians." Pp. 390–401 in *Origins and Destinies: Immigration, Race, and Ethnicity in America*, edited by Sylvia Pedraza and Rubén G. Rumbaut. Belmont, CA: Wadsworth Publishing Co.

and labor conflicts (such as in airline strikes), congressional legislation determining the benefits of different groups (such as the Americans with Disabilities Act of 1990), or Supreme Court decisions interpreting the U.S. Constitution. The state also provides the basic institutional structure of society since, through its laws, the state determines what institutional forms will receive societal support. Laws regulating family relationships emanate from the state; thus, families defined as illegitimate by law (or other mechanisms by which the state exerts its will, such as tax policy) do not receive the same benefits and rights as families that the state defines as legitimate. The state determines how people are selected to govern others and dictates the system of governance they use. The state configures economic institutions by regulating economic policy. The state also maintains security forces, such as the police and the military, to protect the citizenry from criminals, internal dissenters, and external threats.

Sociological analyses of the state focus on several different questions. One important issue is the relationship between the state and inequality in society. State policies can have greatly different impacts on different groups, as we will see later in this chapter. Another issue explored by sociological theory is the connection between the state and other social institutions—the state and religion, the state and the family, and so on. Finally, a central topic is the state's role in maintaining social order, one of the most basic questions in sociological theory, as we have seen repeatedly since Chapter 1. Some see the state as regulating society through coercion and power; others emphasize the role that consensus plays in the maintenance of public order. The issue of how power is exercised within the state is a subject of intense and continuing sociological debate and research.

The Institutions of the State

A number of institutions make up the state, including the government, the legal system, the police, and the military. The government creates laws and procedures that govern the society. The military is the branch of government responsible for defending the nation against domestic and foreign conflicts. The court system is designed to punish wrongdoers and adjudicate disputes. Court decisions also determine the guiding principles or laws of human interaction. Law is a fundamental type of formal social control that outlines what is permissible and what is forbidden. The police are responsible for enforcing law at the local level and for maintaining public order. The prison system is the institution responsible for punishing those who have broken the law.

The State and Social Order

Throughout this book, we have seen that a variety of social processes contribute to order in society, including the learning of cultural norms (that is, socialization), peer pressure, and the social control of deviance. Each of these plays a part in producing social order, but none so explicitly and unambiguously as the official system of power and authority in society. In making laws, the state clearly decrees what actions are legitimate or not. Punishments for illegitimate actions are spelled out, and systems for administering punishment are maintained. The state also influences public opinion through its power to regulate the media and, in some cases, by actually circulating **propaganda,** information disseminated by a group or organization (such as the state) intended to justify its own power. Censorship is another means by which the state can direct public opinion. The movement to censor sexually explicit materials on the Internet is an example of state-based censorship.

The state's role in maintaining public order is also apparent in how it manages dissent. Protest movements perceived by those in power to challenge state authority or threaten the disruption of society may be repressed through state action. Options available to the state range from surveillance through imprisonment all the way to military force. Throughout the 1950s and 1960s, for example, the FBI conducted surveillance operations targeting the Reverend Martin Luther King, Jr., and other civil rights leaders (Garrow, 1981). Federal troops may be called upon by the state to quell riots and urban uprisings, as happened in the Los Angeles riots of 1992 that followed the acquittal of police officers accused of beating Rodney King. As the system of institutionalized power and authority in society, the state has the dual role of protecting its citizens and ensuring the preservation of society. Different states work in different ways: some explicitly protecting the status quo; others, more revolutionary or more totalitarian in operation. Even in a democratic state, like the United States, however, the state typically protects the interests of those with the most power, leaving the least powerful groups in society vulnerable to oppressive state action.

Global Interdependence and the State

On an international level, there are increasingly strong ties between the state and the global economy. The interdependence of national economies means that political systems are also elaborately entangled—a phenomenon that can be daily observed in the newspaper. Political tensions in what may have once been a remote part of the world have reverberations around the globe. Furthermore, the phenomenal rise of information technology means that political developments in one part of the world can be watched and heard just about everywhere. What will be the effect of globalization on the state institutions of diverse nations?

Some argue that increased economic interdependence will mean that nations will move to adopt similar institutions, including similar forms of government, law, and state rule. The European Community (EC) is an example of this—an alliance of separate nations established to promote a common economic market and develop a political union within western Europe. The European Community now includes France, Belgium, Italy, Luxembourg, The Netherlands, Germany, the United Kingdom, Denmark, Ireland, Greece, Portugal, Spain, Austria, Finland, and Sweden. These nations have adopted shared laws that make trade and the movement of people across national boundaries easy. The United States has also entered similar agreements, such as NAFTA (North American Free Trade Agreement), which eliminates many of the restrictions on trade between the United States, Mexico, and Canada.

Does the permeability of national borders produced through such agreements result in a more democratic or less democratic world? Critics differ. Some say that the greater interdependence exposes authoritarian regimes and that the world's eye will create pressures for an international system of law where democratic rights are respected and human rights protected. Others argue that the strong alliance produced between economic interests, corporate power, and the state will shift global political power, favoring the most dominant nations and impoverishing and marginalizing others, while also destroying their local traditions. Such analysts worry that the increasing similarity produced by global interdependence will produce a *monoculture* where everything will look alike—same hotels, same clothes, same music, same stores (Mander and Goldsmith, 1996).

Either way, it is clear that the process of globalization is having profound effects on the character of states and their relationships to each other. A good example is the World Trade Organization (WTO)—created in 1994 to monitor and resolve trade disputes. Member nations may challenge decisions made by local and national governments, with the WTO Council in Geneva resolving disagreements. The creation of this organization has produced a system that transcends individual nation-states by formalizing and strengthening the rules that govern many aspects of world trade. This represents the trend, in an increasingly international economy, toward the creation of a single state—at least in the sense of a singular governing body that has the potential to have authority over many aspects of people's lives around the world.

Power and Authority

The concepts of power and authority are central to sociological analyses of the state. **Power** is the ability of one person or group to exercise influence and control over others. The exercise of power can be seen in relationships ranging from the interaction of two people (husband and wife, police officer and suspect) to a nation threatening or dominating other nations. Sociologists are most interested in how power is structured at the societal level—who has it, how it is used, and how it is built into institutionalized structures, such as the state. Power can be structured into social institutions, such as when men as a group have power over women as a group. In a society like the United States that is heavily stratified by race, class, and gender, power is structured into basic social institutions in ways that reflect these various inequalities. Sociologists also understand that power institutionalized at the societal level influences the social dynamics within individual and group relationships.

The exercise of power may take the form of persuasion or coercion. For example, a group may be encouraged to act a certain way based on a persuasive argument. A strong political leader may persuade the nation to support a military invasion or a social policy. Alternatively, power may be exerted by sheer force. Between persuasion and coercion are many gradations. Generally speaking, groups with the greatest material resources will likely have the advantage in transactions involving power, but this is not always the case. A group may by sheer size be able to exercise power, or groups may use other means, such as armed uprisings or organized social protests, to exert power.

Power can be legitimate—that is, accepted by the members of society as right and just—or illegitimate. **Authority** is power that is perceived by others as legitimate. Authority emerges not only from the exercise of power, but from the belief of constituents that the power is legitimate. In the United States, the source of the president's domestic power is not just his status as commander of the armed forces but the belief by most people that his power is legitimate. The law is also a source of authority in the United States. Those who accept the status quo as a legitimate system of authority perceive the guardians of law to be exercising *legitimate power*. *Coercive power*, on the other hand, is achieved through force, often against the will of the people being forced. Those people may be a few dissidents or most of the citizenry of an entire nation. A dictatorship often relies on its ability to exercise coercive power through its control of the military or state police. The ability to maintain power through force may be enough to keep someone in power even if he or she lacks legitimate authority, whereas someone lacking in legitimate authority *and* coercive power is unlikely to remain in power long.

Types of Authority

Max Weber (1864–1920), the German classical sociologist, postulated that there are three types of authority in society: traditional, charismatic, and rational-legal (Weber, 1921 [1978]). **Traditional authority** stems from long-established patterns that give certain people or groups legitimate power in society. A monarchy is an example of a traditional system of authority. Within a monarchy, kings and queens rule, not necessarily because of their appeal, nor because they have won elections, but because of long-standing traditions within the society.

Charismatic authority is derived from the personal appeal of a leader. Charismatic leaders are often believed to have special gifts, even magical powers, and their presumed personal attributes inspire devotion and obedience. Charismatic leaders often emerge from religious movements (see Chapter 16), but they come from other realms also. John F. Kennedy was for many a charismatic president, admired for the vigorous image he projected of himself and the nation. François "Papa Doc" Duvalier, ruler of Haiti from 1957 until 1971, was a charismatic leader who earned fearful obedience for, among other attributes, his imagined ability to deploy evil magic against his enemies. Charismatic lead-

charismatic authority

ers may mobilize large numbers of people in the name of lofty principles, or as in the case of cults, charismatic leaders may inspire such loyalty among their followers that the group solidifies around what others would consider preposterous beliefs. Since the foundation of charismatic power rests on the qualities of a single individual, when that person leaves or dies, the movement he or she inspired may quickly dissipate.

Rational-legal authority stems from rules and regulations, typically written down as laws, procedures, or codes of conduct. This is the most common form of authority in the contemporary United States. People obey not because national leaders are charismatic or because of social traditions, but because there is an unquestioned system of authority established by legal rules and regulations. Under rational-legal authority, rules are formalized; authority is based not on the personal appeal of charismatic or traditional leaders but on the written rules. Rulers gain legitimate authority through having been elected or appointed in accordance with society's rules. In systems based on rational-legal authority, the rules are upheld by means of state agents such as the police, judges, social workers, and other state functionaries to whom power is delegated.

Jessie, the Body, Ventura was elected governor of the state of Minnesota as the result of his charismatic authority.

THINKING SOCIOLOGICALLY

Observe the national evening news for one week, noting the people featured who have some kind of *authority*. Listing each of them and noting their area of influence, what form of authority would you say each represents: *traditional, charismatic,* or *rational-legal*? How is the kind of authority a person has related to his or her position in society (that is, race, class, gender, occupation, education, and so on)?

Weber wrote that modern societies would be increasingly based on rational-legal authority, as opposed to traditional institutions like monarchies and local and regional custom. We can see evidence of Weber's far-sightedness in the proliferation of regulations governing daily life, and in the fading authority of religion, folklore, and custom. Consider the explosive growth in the population of lawyers, who tend to the legalities governing all important transactions in our society and who act as advocates for official resolution of disputes. At one time, their role was filled by appeals to local custom or the adjudication of a wise person whose judgment both parties either trusted or were obliged to accept based on the high regard of others. As societies modernize, they become more rationalized in their systems of authority. Weber acknowledged the societal gain of rational authority over superstition or oppressive custom, but he also warned that there would be a cost in terms of diminished social intimacy and group feeling.

The Growth of Bureaucratic Government

According to Weber, rational-legal authority leads inevitably to the formation of bureaucracies. As we saw in Chapter 6, a **bureaucracy** is a type of formal organization characterized by an authority hierarchy, a clear division of labor, explicit rules, and impersonality. Bureaucratic power comes from the accepted legitimacy of the rules, not personal ties to individual people. The rules may change, but they do so via formal—that is, bureaucratic—procedures. People who work within bureaucracies are selected, trained, and promoted based on how well they apply the rules. Those who make the rules are unlikely to be the same people who administer them; bureaucracies are hierarchical, and the bureaucratic leadership may be quite remote. Power in bureaucracies is dispersed downward through the system to those who actually carry out the bureaucratic functions. It is an odd feature of bureaucracy that those with the least power to influence how the rules are formulated—those at the bottom of the bureaucratic hierarchy— are very often those who are most adamant about strict adherence to the rules.

Is the bureaucratic growth of government in the United States becoming unwieldly? Many would say so, if judged only from the sheer size of bureaucratic governmental institutions. Bureaucracy has become the modern system of administration. In principle, bureaucracies are supposed to be highly efficient modes of organization although the reality is often very different. There is a tendency within bureaucracies to proliferate rules, often to the point that the organization becomes ensnared in its own bureaucratic requirements. Thus, as an example, procurement of desktop computers by the government has become so overburdened by bureaucratic requirements for review, comparison, bidding, and approval that by the time the federal government approves a purchase, the approved models are already out of production.

Within bureaucracies, administrators sometimes become so focused on rules and regulations that the actual work of the system bogs down. As bureaucratic systems become ever larger, people feel themselves becoming smaller—they feel powerless in the face of a powerful system. Think of the last time you had to conduct business in a busy bureaucratic organization. Perhaps you were applying for federal financial aid or making a required appearance at the Department of Motor Vehicles. There was probably a fairly long line. There were likely people of many different backgrounds in line, and some exchanges may have been hindered by language problems or other misunderstandings. You probably interacted with bureaucratic functionaries under pressure to keep the transactions moving quickly. If you forgot something you were supposed to bring, you may have met with the rigidity of bureaucratic procedures—a summary end to the transaction, even though a slight adjustment of the rules or a moment's creative thinking would permit the transaction to be concluded.

Within bureaucracies, personal temperament and individual discretion are not supposed to influence the application of rules. Of course, we know that the face of the bureaucracy is not always perfectly stony. As we saw in Chapter 6, bureaucracy has another face. Bureaucratic workers frequently exercise discretion in applying rules and procedures. Some who encounter bureaucracies know how to "work the system," perhaps by personalizing the interaction and making a willing accomplice of the bureaucrat in dodging bureaucratic stipulations, or perhaps by using knowledge of some rules to evade others. Others without privileged relationships or privileged information are continually disadvantaged by their inability to negotiate within the system. Despite the supposed impersonal administration in bureaucracies, people may receive widely different treatment from bureaucratic workers, who may favor some persons while acting prejudiced against others based on their race, gender, age, or other characteristics.

Our picture of the state so far—bureaucratic, powerful, omnipresent—presents an important question to be addressed by the sociological imagination: Does the state act in the interests of its different constituencies, or does it merely reflect the needs of a select group who sit at the top of the pyramid of state power? This question has spawned much sociological study and debate, and has resulted in several different theoretical models of state power.

Theories of Power

How is power exercised in society? Four different theoretical models have been developed by sociologists to answer this question: *the pluralist model*, the *power-elite model*, the *autonomous state model*, and *feminist theories of the state*. Each begins with a different set of assumptions and arrives at different conclusions.

The Pluralist Model

The **pluralist model** interprets power in society as coming from the representation of diverse interests of different groups in society. This model assumes that in democratic societies, the system of government works to balance the different interests of groups in society. An **interest group** can be any constituency in society organized to promote its own agenda, including large, nationally based groups like the American Association of Retired People (AARP), the National Gay and Lesbian Task Force, the American Rifle Association (ARA), the National Organization for Women (NOW), and the National Urban League; groups organized around professional and business interests, such as the American Medical Association (AMA), the National Association of Manufacturers, and the Tobacco Institute; or groups that concentrate on a single political or social goal, such as Greenpeace and Mothers Against Drunk Driving (MADD). According to the pluralist model, interest groups achieve power and influence through their organized mobilization of concerned people and groups.

The pluralist model has its origins in functionalist theory. This model sees the state as basically benign and representative of the whole society. No particular group is seen as politically dominant; rather, the pluralist model sees power as broadly diffused across the public. Groups that want to effect a change or express their point of view need only mobilize to do so (Block, 1987). The pluralist model also suggests that members of diverse ethnic, racial, and social groups can participate equally in a representative and democratic government; this theoretical model assumes that power does not depend upon social status or wealth. Different interest groups compete for government attention and action with equality of political opportunity for any group that organizes to pursue its interests (Harrison, 1980). According to pluralism, special interest groups are the link between the people and the government. Interest groups form when a group of people who share a belief in an issue organize themselves to get the attention of the government. In practice, they compete with other interest groups in shaping public policy, each group using whatever influence it can garner to encourage policies favorable to its own interests. One resource of special interest groups is sheer numbers. A special interest group with a large membership can influence a politician by threatening to give its support to another candidate. Another tool interest groups use is money. A small interest group with a great deal of money can wield a disproportionate amount of influence. The principle mechanisms for converting money into political influence are campaign contributions, lobbying, and propaganda. The pluralist model sees special interest groups as an integral part of the political system even though they are not an official part of government. In the pluralist view, interest groups make government more responsive to the needs and interests of different people, an especially important function in a highly diverse society (Berberoglu, 1990).

THINKING SOCIOLOGICALLY

Using a major daily newspaper, for a period of two weeks, follow a political story on a subject of interest to you. As you follow this story, make a list of the different *interest groups* with a stake in this issue. Which groups exercise the most *power* in determining the outcomes of this issue and why? What does your example tell you about how the political process operates?

interest group

The pluralist model of the state sees diverse interest groups as mobilizing to gain political power.

The pluralist model helps explain the importance of **political action committees** (PACs), groups of people who organize to support candidates they feel will represent their views. In 1974, Congress passed legislation enabling employees of companies, members of unions, professional groups, and trade associations to support political candidates with money they raise collectively. The number of political action committees has now grown to more than 4000. PACs have enormous influence on the political process; one measure of their growing influence is the tremendous increase in financial contributions that PACs made to political campaigns over the last two decades—a record $217.8 million in 1996 (Federal Election Commission, 1997). Some people have a low opinion of PACs, regarding them as little more than a legal way to influence politicians and elections via the transfer of large sums of money, but PACs also serve the legitimate purpose of allowing interest groups to make their opinions felt in a pluralistic system (Etzioni, 1990).

How realistic is the pluralist model in explaining how political power works? Interest groups certainly are influential in the political process, although there is not as much equality between them as the pluralist model tends to assume. Interest groups representing powerful organizations and constituencies are able to mobilize resources that smaller groups cannot. Still, in a diverse society, the formation of interest groups is critical to the political process if any degree of equity is to be achieved. Even a cursory look at contemporary politics shows that a great diversity of special interest groups are making themselves heard.

The Power Elite Model

The **power elite model** originated in the work of Karl Marx (1818–1883); it is thus developed from the framework of conflict theory. According to Marx, the dominant or "ruling" class controls all the major institutions in society; the state itself is simply an instrument by which the ruling class exercises its power. The Marxist view of the state emphasizes the power of the upper class over the lower classes, the small group of elites over the rest of the population. The state, according to Marx, is not a representative, rational institution, but an expression of the will of the ruling class (Marx, 1845 [1972]).

Marx's theory was elaborated much later by C. Wright Mills (1956), who analyzed the power elite. Mills attacked the pluralist model, arguing that the true power structure consists of people well positioned in three areas: the economy, the government, and the military. These three institutions are held to be bastions of the power elite. Sharing common beliefs and goals, the power elite shape political agendas and outcomes in the society along the narrow lines of their own collective interests. This small and influential group at the top of the pyramid of power holds the key positions in the major state institutions, giving them a great deal of power and control over the rest of society.

The power elite model posits a strong link between government and business, a view that is supported by the strong hand government takes in directing the economy and also by the role of military spending as a principal component of U.S. economic affairs. Since World War II, huge conglomerates have emerged whose holdings are spread over many different industries, whose employees number in the hundreds of thousands, and whose political interests support armies of lobbyists. Corporations have an intense interest in public policy since government regulates how business is conducted.

The power elite model also emphasizes how power overlaps between influential groups. **Interlocking directorates** are organizational linkages created when the same people sit on the boards of directors of a number of different corporations. People in elite circles may serve on the boards of major companies, universities, and foundations all at once. People drawn from the same group receive most of the major government appointments; thus, the same relatively small group of people tends to the interests of all these organizations and also the interests of the government. These interests naturally overlap and reinforce one another.

The power elite model sees the state as part of a structure of domination in society, one in which the state is simply a piece of the whole. Members of the upper class do not need to occupy high office themselves to exert their will, so long as they are in a position to influence those who are in power (Domhoff, 1998). The majority of those in the power elite are White men, which according to the power elite model, means that the interests and outlooks of White men dominate the national agenda.

Is the power elite model the best description of power in the United States? In some ways the model is hard to argue against since the influence of powerful groups is so obvious. Two strong criticisms have been leveled against the power elite model. One critique is that the model assumes too readily that there is a unity of interests among elites. In fact, according to critics, the most powerful people in society hold widely divergent views on many political issues—for example, there is no single power elite position on abortion, gun control, the death penalty, and many other issues facing the country. A second criticism is that the power elite model fails to acknowledge how well public interest groups have been able to make themselves heard. In 1995, the biggest PAC spender, for example, was Emily's List, a PAC devoted to supporting candidates strong on women's issues (Federal Election Commission, 1996). According to critics, the power elite model does not explain how a group like Emily's List can rally behind candidates who earlier would have been excluded from the power elite. Of course, at the same time, the power elite model recognizes that even groups like Emily's List operate within elite circles.

The Autonomous State Model

A third view of power developed by sociologists, the **autonomous state model,** interprets the state as its own major constituent. From this perspective, the state develops interests of its own, which it seeks to promote independent of other interests and the public that it allegedly serves. The state is not simply reflective of the needs of dominant groups, as Marx and power-elite theorists would contend. It is an administrative organization with its own needs, such as maintenance of its complex bureaucracies and protection of its special privileges (Evans et al., 1985; Skocpol, 1992; Rueschmeyer and Skocpol, 1996).

Autonomous state theorists note that states tend to grow over time, possibly including expansion beyond their original boundaries. Military excursions are one example of expansion; another is the North American Free Trade Agreement—an example of the United States expanding its state interests by regulating business not only in the United States, but also in Mexico and Canada.

Autonomous state theory sees the state as a network of administrative and policing organizations, each with its own interests (Domhoff, 1990). The interests of the state may intersect at times with the interests of the dominant class or the members of society as a whole, but the major concern of the state is maintaining the status quo and upholding its own interests in its competition with other states. State policies are created by independent state managers who represent their own interests.

The formation of the welfare state is illustrative. The term *welfare state* is used to refer to the vast array of social support programs now supported by the state, including Social Security, unemployment benefits, agricultural subsidies, public assistance, and other economic interventions intended to protect citizens from the vagaries of a capitalist market system (Collins, 1988). Although the welfare state developed to serve people in need, it has grown into a massive, elaborate bureaucracy. Some would say that the welfare state has become so vast that even government employees are part of this subsidized system. Autonomous state theory argues that bureaucrats in giant systems like the welfare state become more absorbed in their own interests than in meeting the needs of the people. As a consequence, government can become paralyzed in conflicts between revenue-seeking state bureaucrats and those who must fund them. As we have seen in recent years, this tension can lead to revolt against the state, as in the tax revolts appearing sporadically throughout the country (Lo, 1990; Collins, 1988).

It is relatively easy to find support for the autonomous state theory. The bureaucracy of the federal government is legendary. Many people find that the Internal Revenue Service, with its broad powers to rule on tax disputes and then extract settlements, is practically a state unto itself, and a rather fearsome one at that. When people ask why the government, for all its hugeness, seems to be incompetent to deal with their problems, the autonomous state model offers the explanation that the government is too busy tending to its own problems to respond to citizens' needs. This theory tends, however, to overlook the degree to which the interests of big business are supported by the state. Like the pluralistic and power elite models, autonomous state theory contributes to our understanding of the state, but it is not a total explanation of state power.

Feminist Theories of the State

Feminist theory diverges from the preceding theoretical models by seeing men as having the most important power in society. The pluralist model sees power as widely dispersed through the class system. Power elite theorists see political power directly linked to upper-class interests. Autonomous state theorists see the state as relatively independent of class interests. Feminist theory, on the other hand, begins with the premise that an understanding of power cannot be sound without a strong analysis of gender (Haney, 1996).

Some feminist theorists argue that all state institutions reflect men's interests; they see the state as fundamentally patriarchal, its organization embodying the fixed principle that men are more powerful than women. In the words of Catherine MacKinnon, a noted feminist legal scholar, "the state is male" (MacKinnon, 1983: 644). Feminist theories of the state conclude that, despite the presence of a few powerful women, the state on the whole is devoted primarily to men's interests, and moreover, the actions of the state will tend to support gender inequality (Blankenship, 1993). One historical example would be laws denying women the right to own property once they married; such laws protected men's interests at the expense of women.

The argument that "the state is male" is easy to observe by looking at powerful political circles. Despite the

TABLE 18.1 *THEORIES OF POWER IN SOCIETY*

	Pluralism	Power Elite	Autonomous State	Feminist Theory
Interprets . . .				
The state	as representing diverse and multiple groups in society	as representing the interests of a small, but economically dominant, class	as taking on a life of its own, perpetuating its own form and interests	as masculine in its organization and values (i.e., based on rational principles and a patriarchal structure)
Political power	as derived from the activities of interest groups and as broadly diffused throughout the public	as held by the ruling class	as residing in the organizational structure of state institutions	as emerging from the dominance of men over women
Social conflict	as the competition between diverse groups that mobilize to promote their interests	as stemming from the domination of elites over less powerful groups	as developing between states, as each vies to uphold its own interests	as resulting from the power men have over women
Social order	as the result of the equilibrium created by multiple groups balancing their interests	as coming from the interlocking directorates created by the linkages among those few people who control institutions	as the result of adminstrative systems that work to maintain the status quo	as resulting from the patriarchal control that men have over social institutions

recent inclusion of more women in powerful circles, and the presence of some notable women as major national figures, most of the powerful are men. The U.S. Congress is 89 percent men; those groups that exercise state power, such as the police and military, are predominantly men. Moreover, these institutions are structured by values and systems that can be described as culturally masculine—that is, based on hierarchical relationships, aggression, and force. This is what feminist theory analyzes in studies of the state.

Comparing Theories of Power

These four models each see power in a different way (see Table 18.1). In one, interest groups compete in a struggle for power (the pluralist model). In another, power stems from the top down (the power elite model). In the third, the power of the state feeds on itself (autonomous state theory). And in the last, men hold power in all social institutions (feminist theory). Each perspective makes its own contribution and complements the others to give a more complete picture of how power works.

In its own way, each of these theories addresses the same question: Does power move from the top down or from the bottom up? The pluralist model sees power rising from the bottom up in the form of interest groups that organize to express their needs or use their vote to influence state policy. Power elite theory and feminist theory see power operating from the top down, with the most powerful groups in society (either the power elite or elite men) using the state to enforce their will. The model of the autonomous state falls somewhere in between, with the state seen to operate in its own interest, sometimes supporting the interests of the people or the elites, sometimes not.

Government: Power and Politics in a Diverse Society

The terms *government* and *state* are often used interchangeably. More precisely, the government is one of several institutions that make up the state. The **government** includes those state institutions that represent the population, making rules that govern the society. The government in the United States is a **democracy,** meaning that it is based on the principle of representing all people through the right to vote. The actual makeup of the government, however, is far from representative of society. Not all people participate equally in the workings of government, either as elected officials or as voters, nor do their interests receive equal attention. Women, the poor and working class, and racial–ethnic minorities are less likely to be represented by government than are White middle- and upper-class men. Sociological research on political power has concentrated on inequality in government affairs and has demonstrated large, persistent differences in the political participation and representation of different groups in society.

Debunking Society's Myths

Myth: In democratic societies, all people are equally represented through the vote.

Sociological perspective: Although democracies are based on principles of equal representation, in practice, race, class, and gender influence the power that diverse groups have in the political system.

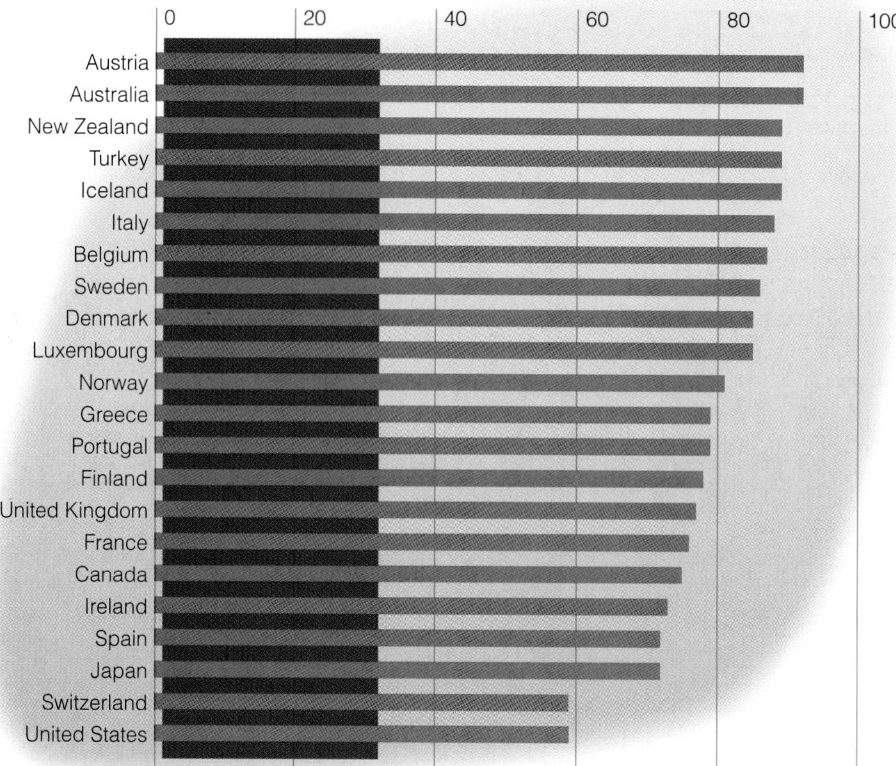

FIGURE 18.1 Voter Participation in Democratic Nations

Average participation, 1945–1989.

SOURCE: Lane, Jan-Erik, and Svante Evasen. 1990. "Macro and Micro Understanding in Political Science." *European Journal of Political Research* 18 (July) 457–465 with kind permission from Kluwer Academic Publishers.

Diverse Patterns of Political Participation

One would hope that in a democratic society all people would be equally eager to exercise their right to vote and be heard. That is far from the case. Among democratic nations, the United States has one of the lowest voter turnouts (see Figure 18.1). In the 1996 presidential election, the percentage of eligible voters who went to the polls was still only 49 percent. A turnout of 50 percent or less is typical of national elections; voter turnout in congressional and local elections is even lower (see Figure 18.2).

The group most likely to vote is older, better educated, and financially better off than the average citizen. In sociological terms, age, income, and education are the strongest predictors of whether someone will vote. The higher a person's social class, the higher the likelihood that she or he will be a voter. One analyst has concluded that the political system fails to represent the interests of the bottom two-thirds of society (Edsall, 1984). Those in the upper social classes are far more likely to vote and to have their voices heard in the political process. The requirements of voter registration itself discourage many from voting (Piven and Cloward, 1988). Every time you move, you must register again to vote, although The National Voter Registration Act of 1993 provides greater flexibility in how one registers to vote, including "motor voter" procedures (where you register to vote when you apply for a driver's license). Preliminary surveys indicate that this legislation has increased voter registration; whether that translates into higher voter turnout is not yet known (League of Women Voters, 1996).

There is also significant variation in voting patterns by race. Overall, African Americans are less likely to vote than Whites, a fact reflecting the disproportionate numbers of African Americans in the poor and working classes—groups that typically are less likely to vote than middle- and upper-class people. During the late 1960s, however, controlling for factors such as age, education, and income, African Americans were actually more likely to vote than White Americans, probably because of the heightened sense of the importance of the franchise generated by the civil rights movement and voter registration drives mobilized during this period. Beginning in the 1970s, there was a convergence in the likelihood of voting among Blacks and Whites of similar social class, education, and age (Ellison and Gay, 1989).

A subjective factor influencing the voting behavior of African Americans is the fact that they are, in general, less trusting of the political system than Whites and are more alienated from politics than their White counterparts. Sociological research shows, though, that when Black Americans are approached directly by party representatives, the likelihood that they will vote increases substantially—more so than it does when Whites are canvassed. In addition, the more that Black Americans and Latinos identify themselves as groups with distinct needs and interests, the more likely they will participate in the political process, reflecting the long-standing sociological finding that a strong sense of ethnic identity is linked to increased political participation (Welch and Sigelman, 1993).

Not only do social factors influence the likelihood of voting, they also influence whom people vote for, as Table 18.2 shows. Race is one of the major social factors influenc-

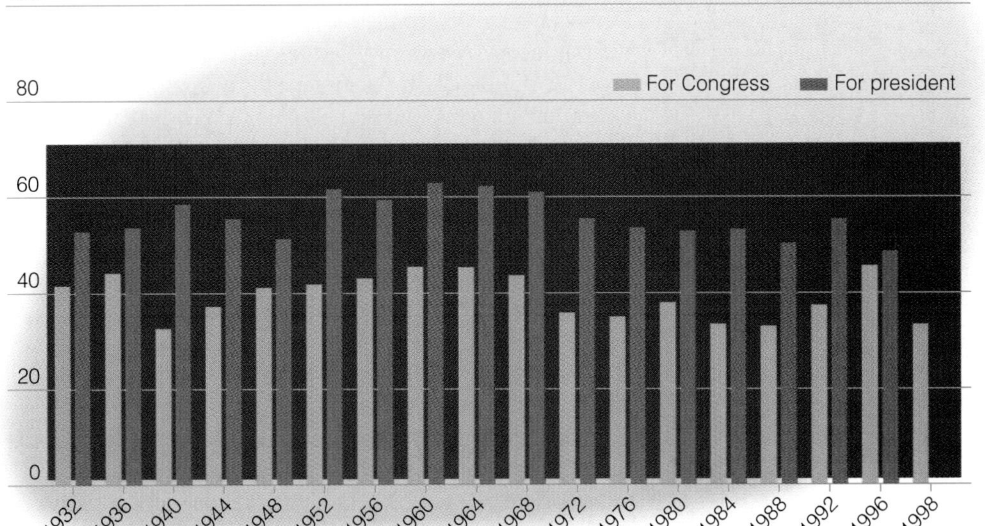

FIGURE 18.2 Voter Turnout in U.S. Elections

DATA: From the U.S. Department of Commerce. 1997. *Statistical Abstracts of the United States, 1997.* Washington, DC: U.S. Government Printing Office, p. 289.

ing voting. African Americans and Latinos, with the exception of Cuban Americans, tend to be markedly Democratic; Cuban Americans are disproportionately Republican. One of the reasons why both Latinos and African Americans have been more attracted to the Democratic party and its candidates is that Democrats historically have been more committed to government action that promotes social and economic equality. The allegiance of Latino and Black voters to Democratic candidates is also explained in part by socioeconomic factors like age, gender, income, education, and religion; all these factors, regardless of racial–ethnic identity, are linked to political behavior. Even without controlling for these factors, however, Latinos are more liberal than Anglos. Latinos are generally also more likely than other groups to support government spending for various purposes (Welch and Sigelman, 1993). Ethnicity also influences voting patterns. People tend to vote for those of their own ethnic group. Sociologists have also found that ethnicity can outweigh socioeconomic factors like occupation, income, and education in predicting voting patterns. White Catholics, for example, have traditionally been aligned with the Democratic party, although that link has been weakened somewhat by economic changes that have had a strong influence on voting patterns (Legge, 1993; Smith, 1993).

TABLE 18.2 *VOTING PATTERNS IN THE 1996 PRESIDENTIAL ELECTION*

	Democrat	Republican	Independent
Men	43%	44%	10%
Women	54	38	7
White	43	46	9
Black	84	12	4
Hispanic	72	21	6
Grade School Graduate	59	28	11
High School	51	35	13
Some College	48	40	10
College Graduate	44	46	8

SOURCE: The Vote '96. "Presidential Election Exit Poll Results." http://allpolitics.com/1996/elections/natl.exit.poll/index.html

Gender is also a major social factor influencing political attitudes and behavior. The term **gender gap** refers to the differences in people's political attitudes and behavior. One aspect of the gender gap is that women are more likely than men to identify and vote as Democrats and to have liberal views. The development of the gender gap is a recent phenomenon. For many decades, men were more likely to vote than women, and when asked, women reported voting as their husbands or fathers did. In recent years, women have been as likely to vote as men, but there have been significant differences in their political outlooks. The gender gap is widest on issues involving violence and the use of force. Women, for example, are more likely than men to be peace seeking and to support gun control. The gender gap is also evident on so-called compassion issues—women are more likely than men to support government spending for social service programs that aid the young, the old, and the disadvantaged. They are also more liberal on issues like militarism, gay and lesbian rights, and feminism (Wilcox et al., 1996; Eliason et al., 1996; Jackson et al., 1996).

Although the gender gap has been scrutinized most closely within the United States, there is evidence of the same pattern in other nations as well. Interestingly, there is no gender gap between African American women and men, most likely because African American men join women of all groups in being more liberal on political and social issues than White men (Welch and Sigelman, 1989).

Political Power: Who's in Charge?

Although the government in a democracy is supposed to be representative, the class, race, and gender composition of the ruling bodies in this country indicate that this is hardly the

Although defined as a representative government, the U.S. Congress does not include many women or racial–ethnic minorities among its members.

case. Most of the members of Congress are White men (see Figure 18.3); they are well educated, from upper middle- or upper-income backgrounds, and have an Anglo-Saxon Protestant heritage. One-third of the people in Congress are lawyers; another third are businesspeople and bankers. Most of the rest come from professional occupations; very few were blue-collar workers before coming to Congress (Ornstein et al., 1996; Saunders, 1990; Freedman, 1997).

Many of the 535 members of Congress are millionaires; 50 of the 535 U.S. Congressmen have a net worth in excess of $2.4 million. As examples, Senator John Kerry (D–Mass.) is worth a whopping $760 million, a fortune stemming from his second marriage. Representative Amory Houghton from New York, $530 million; even Edward Kennedy's fortune pales by comparison at a mere $40 million (Winneker and Omero, 1996). Considering their wealth, it is inevitable that many of those in Congress have large financial interests in the industries that they regulate.

Simply getting into politics requires a substantial investment of money. The average cost of a Senate campaign is $3.7 million; a successful run for the House of Representatives costs nearly $400,000. When Senator Bob Dole ran against Clinton in 1996, he nearly reached the spending limit of $37 million for the presidential primaries alone. Steve Forbes dropped out of the race early after spending $32.6 million of his personal wealth on the campaign. Electioneering has become a high-tech, highly expensive business. Satellite link-ups, television advertising, overnight polls, computer modeling of the electorate, and nonstop communication between the candidate, headquarters, and the field add up to giant campaign debts. Many lawmakers report that the need to constantly raise money for campaigns distracts them from their jobs as legislators. Meanwhile, spending big money does not guarantee that a candidate will win an election. It is just the entry fee to get into the game. Little wonder that political action groups have become such a fixture on the campaign scene. A huge amount of money is contributed to the Republican and Democratic parties from these influential groups; moreover, their influence, based on the sheer growth in their numbers is on the rise. PACs also seem to shift their contributions, depending on which party they see as likely to be in power (Salant and Cloud, 1995).

All candidates depend upon contributions from individuals and groups to finance their election campaigns. Wealthy families and individuals are among the largest campaign contributors to presidential elections (Allen and Broyles, 1991). The largest individual contributors typically have interests in the same industries that fund political action committees. Much of the money given by individuals and PACs goes to incumbents, who have, historically, an overwhelming edge in elections, and who already sit on the committees where public policy is hammered out. This picture of elites and business interests funneling money to candidates, who return to the same donors for more money when the next campaign rolls around, has shaken the faith of many Americans in their political system. There is little belief that the political process is a democratic and populist mechanism by which the "little people" can select political leaders to represent them. National surveys show that only 20 percent of U.S. citizens have a great deal or quite a lot of confidence in Congress. Only 11 percent think that those in Congress have high levels of honesty and ethics, and only 14 percent say you can trust government most of the time (Hugick and McAneny, 1992; Roper Organization, 1996). Many disillusioned people feel that the government does not really represent them and are reinforced in their view by the fact that the government is not really representative—in terms of race, class, and gender, it is far different from the makeup of the U.S. population.

Women and Minorities in Government

When President Clinton first formed his cabinet in 1992, he appointed the most diverse group of men and women yet to serve in these powerful roles. Of his seventeen original cabinet members, nine were White men, three were African American men, two were Hispanic men, two were White women, and one was a Black woman. The Senate of the 105th Congress had a record number of women (9 out of 100). The first Native American to be sent to the Senate in more than sixty years was elected in 1992. In the 106th Senate (the one convening in 1999), there are nine women, no African Americans or Latinos, two Asian Americans, and one Native American (see Figure 18.3).

Each of these groups is vastly underrepresented in the Congress. Note that only 9 percent of the senators are women, whereas women are 51 percent of the population. Thirteen percent of the population is Black, but there is no Senate representation. Hispanics make up 11 percent of the population, but have no representation in the Senate, and only 4 percent of the House is Hispanic. There is a long way to go before we have a truly representative government, and

political action committee

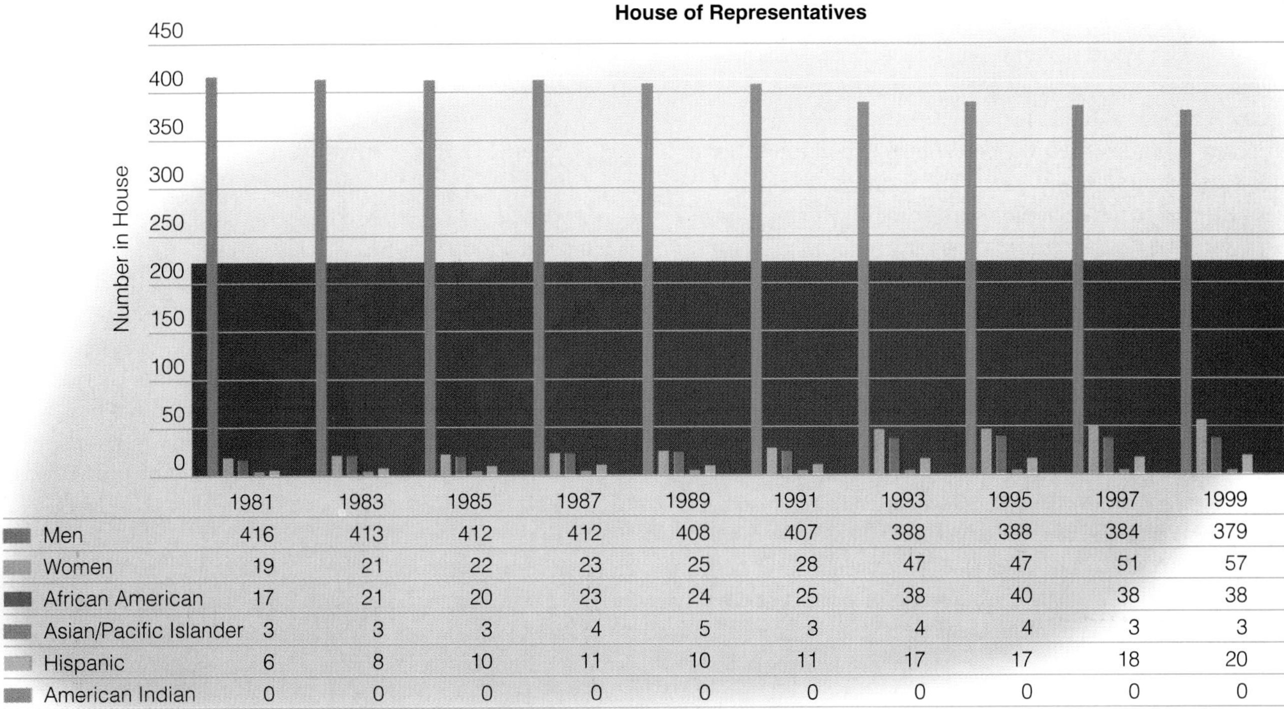

	1981	1983	1985	1987	1989	1991	1993	1995	1997	1999
Men	416	413	412	412	408	407	388	388	384	379
Women	19	21	22	23	25	28	47	47	51	57
African American	17	21	20	23	24	25	38	40	38	38
Asian/Pacific Islander	3	3	3	4	5	3	4	4	3	3
Hispanic	6	8	10	11	10	11	17	17	18	20
American Indian	0	0	0	0	0	0	0	0	0	0

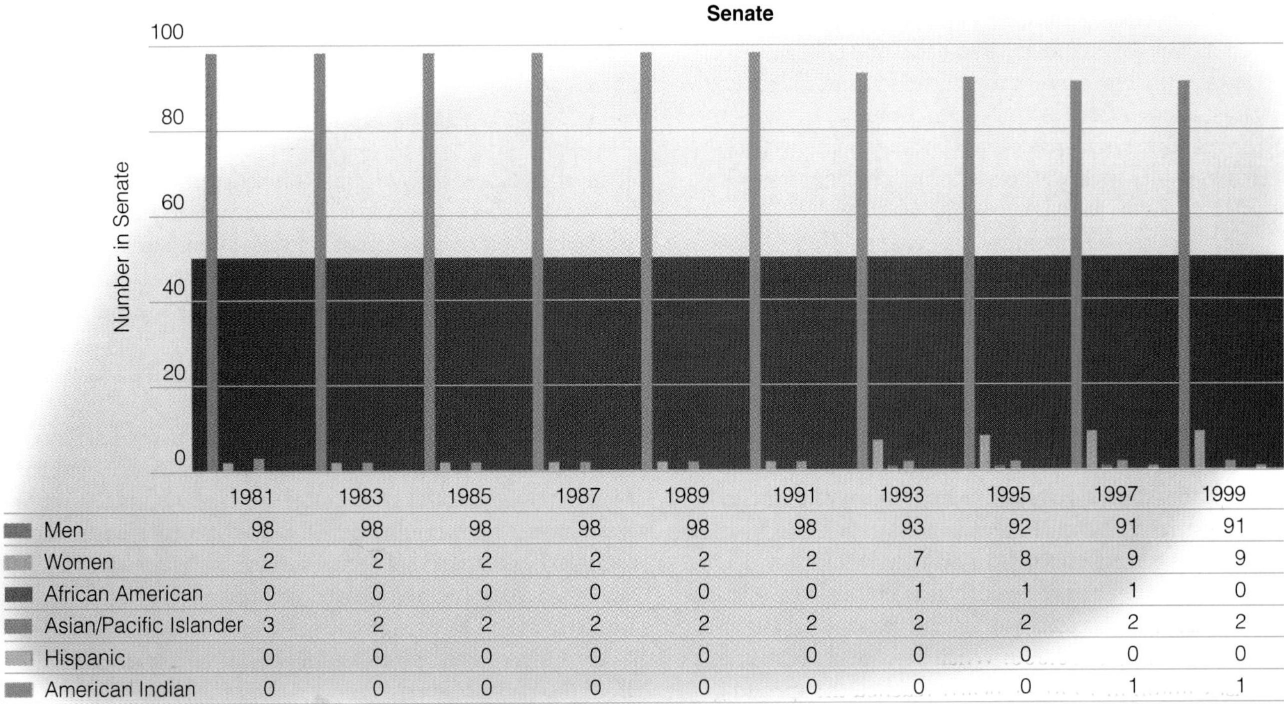

	1981	1983	1985	1987	1989	1991	1993	1995	1997	1999
Men	98	98	98	98	98	98	93	92	91	91
Women	2	2	2	2	2	2	7	8	9	9
African American	0	0	0	0	0	0	1	1	1	0
Asian/Pacific Islander	3	2	2	2	2	2	2	2	2	2
Hispanic	0	0	0	0	0	0	0	0	0	0
American Indian	0	0	0	0	0	0	0	0	1	1

FIGURE 18.3 A Representative Government?

SOURCE: Freedman, Allan. 1997. "Lawyers Take a Back Seat in 105th Congress." *Congressional Quarterly* 55 (January 4): 28; Congressional Information Services. 1999. http://web.lexis-nexis:com/congroup

there is also no guarantee that gains made in one election will be sustained in the next.

Researchers offer several explanations for why women and racial–ethnic groups continue to be underrepresented in government. Certainly, public prejudices play a role. It was not that long ago (in the 1960 Kennedy–Nixon election) that the first Catholic president was elected; there has never been a Jewish candidate to win either party's nomination for the presidency. Gender and racial prejudice run just as deep in the public mind. Prejudice alone, however, cannot account for the lack of representation. Social structural causes are a major factor in getting women and people of color elected. First, women and people of color, including women of color, are better represented in local officeholding. Women and minority candidates receive a great deal of political support from local groups, but at the national level, they do not fare so well. The power of incumbents, most of whom we have seen are White men, disadvantages any new office seeker (Darcy et al., 1994).

A central question for sociologists is whether the inclusion of previously excluded groups in government will make a difference. In other words, do numbers matter? The participation of minorities does alter the outcomes of elections, and Latinos, African Americans, and women differ in their political outlook from White men. Studies of Black elected officials show that they have different political agendas from White officials; thus, an increase in Black representation in government does have a potential effect on political outcomes (Scavo, 1990). The presence of women and minorities in government brings new perspectives on old issues and new attention on issues otherwise overlooked. Women legislators are more likely than men to support feminist issues, such as the Equal Rights Amendment, government-subsidized child care, and abortion rights (Mandel and Dodson, 1992). Increased representation of women and minorities in government is likely to bring change in the policies of the government toward all groups.

Military values permeate American culture, including its fashion.

The Military

The military arm of the state is among the most powerful and influential social institutions in almost all societies. In the United States, the military is the largest single employer. Approximately 3 million men and women serve in the U.S. military, 1.5 million on active duty and the rest in the reserves. This does not include the many hundreds of thousands who are employed in industries that support the military, nor does it include civilians who work for the Department of Defense and other military-affiliated agencies (U.S. Bureau of the Census, 1997a).

Highly ranked in the value system of the United States is *militarism,* the pervasive influence of military goals and values throughout the culture. The militaristic bent in our culture is evident in many ways. The toys children play with often inculcate militaristic values. Military garb falls in and out of fashion. The plots of each summer's biggest blockbuster movies almost always revolve around a burly character hugely gifted in the military arts, often a veteran of the armed forces, who generally stocks up on the latest weaponry. Fighter jets flash over the Super Bowl. Military brass sparkles in parades. Being the strongest nation, militarily, in the world is frequent in the national political rhetoric. Most people in the United States place a lot of confidence in the military—64 percent say they have quite a lot or a great deal of confidence in the military (Roper Organization, 1996).

The Military as a Social Institution

Institutions are stable systems of norms and values that fill certain functions in society. The military is a social institution whose function is to defend the nation against external (and sometimes internal) threats. A strong military is often considered an essential tool for maintaining peace although the values that promote preparedness in the armed forces are perilously close to the warlike values that lead to military aggression against others.

The military is one of the most hierarchical social institutions, and its hierarchy is extremely formalized. People who join the military are explicitly labeled with rank, and if promoted, pass through a series of well-defined levels, each with clearly demarcated sets of rights and responsibilities. There is an explicit line between officers and "grunts," and officers have numerous privileges that others do not. Those in the higher ranks are also entitled to absolute obedience from the ranks below them, with elaborate rituals created to remind dominants and subordinates of their status. As in other social institutions, military enlistees are carefully socialized to learn the norms of the culture they have joined. Military socialization places a high premium on conformity. New recruits have their heads shaved to look alike, they are issued identical uniforms, and they are allowed to retain very few of their personal possessions. They must quickly learn new codes of behavior that are strictly enforced.

Many find the military to be a distinctively masculine institution. Not only are the majority of those in the military men, but the organization itself is founded on so-called masculine cultural traits like aggression, competition, hier-

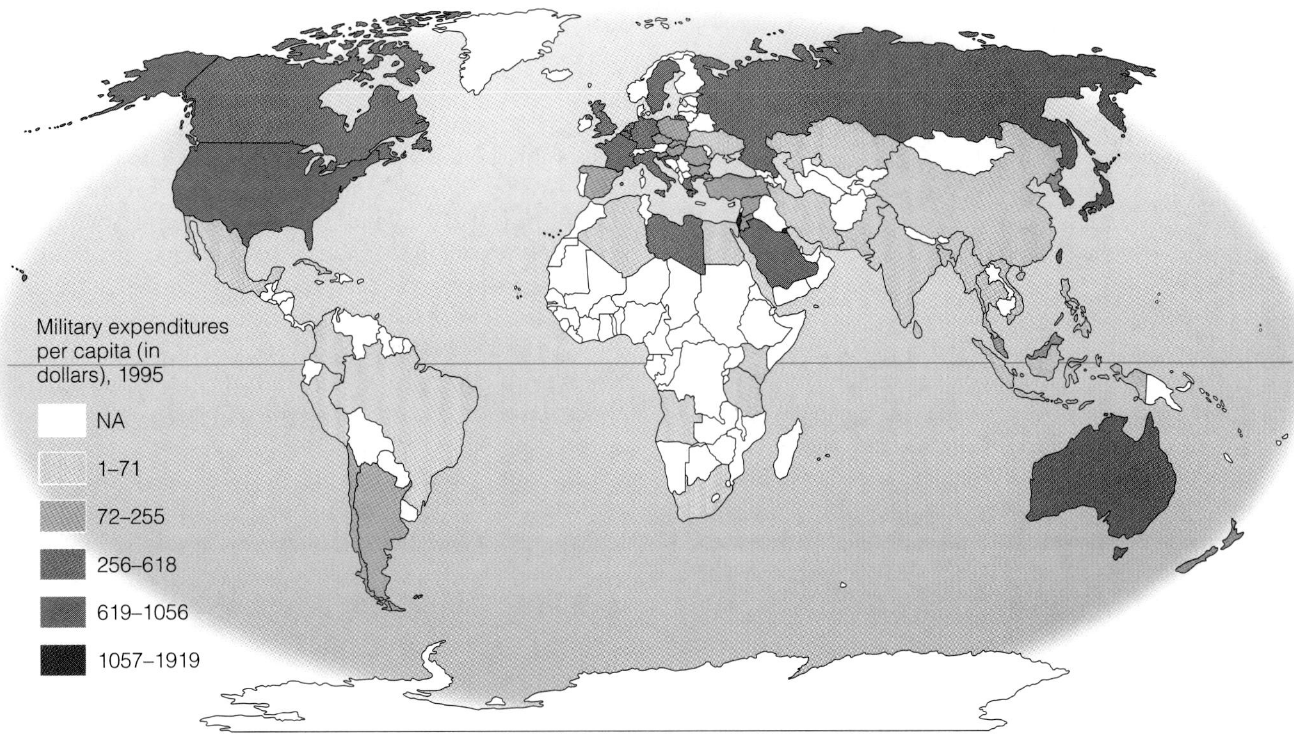

MAP 18.1 Viewing Society in Global Perspective: Military Expenditures per Capita

DATA: From the U.S. Bureau of the Census. *Statistical Abstracts of the United States, 1997.* Washington, DC: U.S. Government Printing Office, p. 861.

archy, and violence. Soldiers are often abused if they fail to live up to the masculine image of the military. In fact, sociologists have shown how homophobic and misogynistic attitudes are used to enforce highly masculine codes of behavior during military socialization. New recruits may derisively be called "ladies" and told that, if they fail to live up to military ideals, it is because they are effeminate or homosexual.

THINKING SOCIOLOGICALLY

The military is generally defined as the institution that provides defense for the society and maintains the *public order*. Sociologists have argued that the military, like other social institutions, is a *gendered institution*. Similarly, there is a racial and class order in the military. Discuss what this means using examples from current events.

Military life is obviously quite different from life off the base, yet the military is intimately linked to other institutions in the society. Most notably, there is a strong connection between the military and corporate America. The **military–industrial complex** is a term used to describe the linkage between business and military interests. The link is so strong that the military supports many of the basic research and development projects in the nation. From university research labs to corporate research institutes and centers, the military funds much of the basic scientific and technological knowledge produced in this country. Many social scientists have been critical of this fact, arguing that research and the knowledge produced from corporate and government sponsorship should be adapted to more humanistic social service values.

A striking feature of the military as a social institution is the representation of racial minority groups and women within the armed forces. This has been the source of some of the greatest change in the military in recent years. Picture a U.S. soldier. Whom do you see? At one time you would have almost certainly pictured a young, White male, possibly wearing army green camouflage and carrying a weapon. Today, the image of the military is much broader. Drawing on the store of cultural images you have absorbed, you are just as likely to picture a young African American man in military blues with a stiffly starched shirt and a neat and trim appearance, or perhaps a woman wearing a flight helmet in the cockpit of a fighter plane.

In 1973, the all-male draft was ended and an all-volunteer force was initiated in the U.S. armed services. The idea behind the all-volunteer force was that the Army (the branch of the military into which civilians were drafted prior to the all-volunteer force) would be made more professional since the people who joined would actually be choosing to make the military their career. In fact, the greatest change in having an all-volunteer force has been a tremendous increase in the representation of racial minority groups in the armed services. This is not solely the result of an all-volunteer force. Decisions to enlist are affected by a number of societal factors, including other employment opportunities, levels of educational attainment among potential recruits, and military values held by the public.

African American soldiers fighting in the Civil War were known as "Buffalo Soldiers" because their curly hair and brown skin reminded Native Americans of the buffalo.

Still, the military life is chosen far more often by African Americans, and to some extent Hispanics, than by Whites.

Race and the Military

African Americans have served in the military for almost as long as the U.S. armed forces have been in existence. Except for the Marines (which desegregated in 1942), until 1948 the armed forces were officially segregated; in 1948, President Harry Truman signed an Executive Order banning discrimination in the armed services. Although much segregation continued following this order, the desegregation of the armed forces is often credited with having promoted more positive interracial relationships and increased awareness among Black Americans of their right to equal opportunities. This is quite contrary to the widespread opinion among Whites that allowing Black and White soldiers to serve side by side would destroy soldiers' morale (Blackwell, 1991).

Currently, 19 percent of military personnel are Black, 5.8 percent Hispanic, 2.9 percent Asian American/Pacific Islander, 0.4 percent Native American, and 69.9 percent, White; an additional 1.6 percent are identified as "other" (U.S. Department of Defense, 1996; Moskos and Butler, 1996). Black Americans are the most overrepresented relative to their proportion in the civilian population. Comparing the percentage of minorities in the military to their percentage in the general population reveals only part of the story. When considered in relationship to the numbers who are *eligible* for military service (based on educational level, age, and ability tests), the disproportionate representation of African Americans is even greater. Of those considered eligible, the military participation rate for Black youth is 50 percent, compared to only 16 percent for Whites (Binkin and Eitelberg, 1986). Latinos do not enlist to the same extent that African Americans do, despite comparably disadvantaged economic backgrounds.

For groups with limited opportunities in civilian society, joining the military seems to promise an educational and economic boost. Is this realized? Sociologists have found that African Americans in the military have higher levels of interpersonal satisfaction and work satisfaction than do Whites. African American and Latino recruits are also more likely than Whites to complete their first term of service and remain in the military beyond their first term. There is some evidence that minorities in the military are promoted at least as fast as Whites, although the research on this issue has yielded mixed results. Within the military, there is equal pay for equal rank. African Americans and Latinos, however, are overrepresented in lower-ranking support positions within the armed forces. Often, they are excluded from the higher-status, technologically based positions—those most likely to bring advancement and higher earnings both in the military and beyond. Affirmative action policies in the military designed to advance minorities have been most effective for African American soldiers, less so for Latinos and other racial groups. Most minorities remain in positions with little supervisory responsibility. At the highest ranks in the military, there are few African Americans, Latinos, Asian Americans, or Native Americans. For example, although Blacks are more than 19 percent of those in the military, they are only 7.5 percent of officers. Officers among Hispanics (2.8 percent), Native Americans (0.2 percent), and Asian/Pacific Islanders (2.3 percent) are even more rare (U.S. Department of Defense, 1996; Zucca and Gorman, 1986; Moskos, 1988; Moskos and Butler, 1996).

For both Whites and racial minorities, serving in the military leads to higher earnings relative to one's nonmilitary peers. The economic payoff is greatest for African Americans and Latinos relative to those of comparable background, education, experience, and ability. The earnings difference between military and civilian groups is substantial; young Black and Latino men earn $3000 to $5000 more in the military than do their civilian peers. For Whites, the difference is less—about $1300 per year. Upon leaving military service, earnings for Whites, African Americans, and Latinos drop. Immediately upon leaving, Whites as a whole earn less than their White peers who did not serve; African Americans and Latinos also experience a drop in earnings during the period immediately following military service, but they still manage to earn as much as their nonveteran peers. As the years pass, Whites who served in the military earn more than Whites who did not. The same is not true for Black and Latino veterans; they do not share in the earnings premium enjoyed by White veterans (Phillips et al., 1992).

Women in the Military

The military academies did not open their doors to women until 1976. Since then, there have been profound changes in the armed services although the hostile reaction to women who have entered these academies shows how fierce resistance to the inclusion of women in the military can be.

Shannon Faulkner, the young woman who won the right to enter the Citadel had to embark on a two and one-half year legal struggle to gain admission. Once admitted, she was harassed and ridiculed. When she succumbed to exhaustion during the first week of intense physical training and had to leave, many of the cadets whooped and shouted in glee, leaving many to argue that none of the men had been forced to endure such tribulations. Later, two other women left the Citadel after serious hazing in which, among other things, the women were sprinkled with nail polish remover and their clothes set on fire.

The Supreme Court ruling in 1996 (*United States v. Virginia*) that women cannot be excluded from state-supported military academies like the Citadel and the Virginia Military Institute (VMI) was a landmark decision which opened new opportunities for those women who want the rigorous physical training and academic training that such academies provide.

The incorporation of women into the military has changed the look of the military, but it is still a highly masculine institution in terms of personnel and policies. The first female uniformed personnel appeared in 1901 as nurses. By the end of World War I, there were 34,000 women in the Army and Navy Nurse Corps. Nursing is indicative of the roles women filled in the military. As military personnel, women were confined to "women's work," namely, clerical work and nursing. (In the 1960s, the phrase *typewriter soldiers* was used to refer to women in the armed services, nearly all of whom held clerical positions.) Serving behind a desk, instead of in the field, effectively removed women from the possibility of achieving rapid promotion or high rank (Becraft, 1992a; Holm, 1992).

The marginalization of women in the military has been justified by the popular conviction that women should not serve in combat. The traditional belief that men are protectors and women dependents in need of protection has led many to believe that women, especially mothers and wives, should stay on home soil where they can safely carry out their "womanly" duties. Despite these beliefs, women have fought to defend this country. Forty-one thousand women served in combat in the Persian Gulf War; 7.2 percent of the active combat forces in that war were women (Holm, 1992). In 1992, the Defense Authorization Act allowed women to fly aircraft in combat, traditionally a position reserved for men only. The increased visibility of women soldiers during the Persian Gulf War also shifted public opinion to women serving in combat positions. Fifty-five percent now favor allowing women to serve in combat (Wilcox, 1992; Enloe, 1993; Cooke and Woollacott, 1993; Roper Organization, 1992).

The involvement of women in the military has reached an all-time high in recent years. Now, 194,000 women are on active duty, with an additional 151,000 in the reserves, not including the Coast Guard and its reserves (U.S. Department of Defense, 1996, 1994). The Air Force has the highest proportion of women (14 percent), followed by the Army (11 percent), the Navy (10 percent), the Coast Guard (7 percent), and the Marines (5 percent). In the Coast Guard, 100 percent of jobs are open to women; in the Air Force, 97 percent of jobs. In the Navy, only 59 percent of jobs are open to women, which is similar to the Army where only 52 percent of jobs are open to women. The Marine Corps allows women in only 20 percent of its jobs (Becraft, 1990b). These restrictions contribute to a high degree of occupational segregation within the military, based on gender. For example, among officers, 46 percent of all women officers work in health care; the largest cluster of male officers work in tactical operations (47 percent).

Minority women are overrepresented in the military, relative to their percentage in the general population. Minority women are a greater proportion of women in the military (38 percent of all women in the military) than minority men as a proportion of men in the military (19 percent). Thirty percent of women in the military are African American; 4 percent, Hispanic; and 4 percent either Asian/Pacific Islanders or American Indians (Becraft, 1990b).

Gender relations in the military extend beyond just women who serve. The experience of military wives, for example, is greatly affected by their husbands' employment as soldiers. Frequent relocation means that military wives are less competitive in the labor market. They have lower labor force participation than their women peers and earn less than their peers at the same educational level (Enloe, 1993; Moskos, 1992; Payne et al., 1992). Comparable research has yet to be done regarding the experience of men married to military women.

Gays and Lesbians in the Military

During his first presidential campaign, President Clinton promised an end to the ban on lesbians and gays in the military. At present, an unsteady compromise has been reached in the "*don't ask, don't tell*" policy. According to this rule, recruiting officers cannot ask about sexual preference, and individuals who keep their sexual preference a private matter shall not be discriminated against. Those who publicly reveal that they are gay or lesbian can be expelled from the military based on their sexual orientation. In essence, this is simply an extension of the already existing military attitude toward gays and lesbians. Although Clinton's "don't ask, don't tell" policy does not explicitly permit the military to seek out and discharge service people on the grounds of homosexuality, nor can recruits be screened for the same reason, it nevertheless does not aggressively acknowledge the civil rights of homosexuals. Following the institution of this policy, gay soldiers and potential recruits filed several lawsuits against the U.S. military. Whether gays and lesbians will ultimately be permitted to live openly as homosexuals while also pursuing military careers remains unclear.

The unfolding of this issue reveals continuing prejudice, homophobia, and continued discrimination against lesbian women and gay men. Ironically, studies of student attitudes about gays in the military show that the "don't ask, don't tell" policy may have actually increased support for

gays and lesbians in the military 

gays in the military among the general public; a survey of college students, taken before and after initiation of this policy, has found that students see gay soldiers' careers as more promising and heterosexuality less of an advantage after this policy generated so much public attention (Pesina et al., 1994).

Supporters of the ban on gays in the military often use arguments similar to those used before 1948 to defend the racial segregation of fighting units. They claim that the morale of soldiers will drop if they are forced to serve alongside gay men and women, that national security will be threatened, and that having known homosexuals serving in the military will upset the status quo and destroy the fighting spirit of military units. Ominous reference to the "tight quarters" where recruits live and work implies that gays and lesbians are seen as unable to control their sexuality. Fear of AIDS and the connection of this deadly disease to the gay population further inflame the feeling against lesbians and gays.

The long-standing policy against gays in the military has kept homosexuality in the military hidden, but not nonexistent. Even the military has had to admit that there have always been some gays and lesbians in all branches of the U.S. armed forces. Since their presence clearly did not cause the armed forces to disintegrate, the military has been forced to adapt their argument against gays and lesbians (Enloe, 1993: 91). Military officials are less likely now to refer to "effeminate" characteristics as evidence that homosexuals are unable to fight and therefore ineligible to serve; instead, opponents of gays in the military argue that combat readiness is at stake (Mohr, 1988: 194). They contend that the foremost goal of the U.S. armed forces is "to compose a fighting force that has the best chance of achieving the highest standard of combat readiness" (Wells-Petry, 1993: 30) although it is never clear how combat readiness is actually threatened by the presence of gays.

The government policy on gays in the military is an example of how state institutions can have different effects on individuals within society. Lesbians and gays, like heterosexual women and members of racial minority groups, are discriminated against by the military. The policies and recruitment practices of the armed forces affect members of these groups differently from how they affect members of the dominant group in society. The phrase used to recruit military soldiers, "Be all that you can be," seems a hollow promise in an institution that forbids a segment of its people from admitting who they are.

Debunking Society's Myths

Myth: **Letting lesbians and gays serve in the military erodes morale and weakens the armed services.**

Sociological perspective: **Similar arguments were historically made to exclude Black Americans from military service; this belief is an ideology that serves to support the status quo.**

Police, Courts, and the Law

Another of the basic functions of the state is the administration of justice. In a democratic society, people should ideally be able to assume that when they enter the legal system, they will be treated fairly. The single most important fact that stands out from sociological research on the legal system in the United States is the unequal treatment that different groups encounter when they enter it. As we saw in Chapter 8, the poor and racial minorities are more likely to be arrested, charged, convicted, and sentenced than middle-class or elite White people who commit the same crimes (Flowers, 1988). Further, women are far less likely to commit crime than men; women who have been sexually or emotionally abused and subjected to violence in relationships with men are those most likely to engage in criminal behavior (Daly, 1994). These facts compel us to consider the relationship of race, gender, and class to crime, leading to a central conclusion in sociology: Racism, gender, and class inequality are institutionalized in the criminal justice system, as evidenced by the differential treatment of the poor and racial minority groups.

Debunking Society's Myths

Myth: **The criminal justice system treats all people according to the neutral principles of law.**

Sociological perspective: **Race, class, and gender continue to have an influential role in the administration of justice; for example, even when convicted of the same crime, African American and Latino defendants are more likely to be sentenced and to be sentenced for longer terms than White defendants.**

The Policing of Minorities

The police, along with the military, embody the state's right to use coercive power. There is little question that minority communities are policed more heavily than White neighborhoods; moreover, policing in minority communities has a different effect than in White, middle-class communities. To middle-class Whites, the presence of the police is generally reassuring, but for African Americans and Latinos, an encounter with a police officer can be a terrifying experience. Regardless of what they are doing at the time, minority people, men in particular, are perceived as a threat, especially if they are observed in communities where they "don't belong."

The police are also more likely to use force against minority suspects. The investigating commission appointed after the Rodney King beating found that the Los Angeles Police Department, the third largest in the nation, was guilty of "tolerating racism, excessive use of force, and lax discipline" (Prudhomme, 1991: 18). The report found that several hundred of the 8300 LAPD officers consistently used force inappropriately and were not disciplined for doing so. To the contrary, many of them were praised or even promoted for their behavior. The report found that in cases where excessive use of force occurred, bias based on race, gender, or sexual orientation was a major factor (Reinhold, 1991). Ninety-two percent of police officers are men; 81 percent are White even though minority groups are one-quarter of the general population; women are only 9 percent of the nation's police officers (U.S. Bureau of Justice Statistics, 1996; McDowell, 1992; Reeves, 1992).

Numerous studies have documented the severe treatment that Native Americans, Mexican Americans, and African Americans receive at the hands of the police—a pattern vividly witnessed by the public when police officers in a New York precinct sodomized a Haitian prisoner. Racial minorities are more likely than the rest of the population to be victims of excessive use of force by the police, and African Americans are considerably more likely than Whites to be killed by police officers. Studies show that most cases of police brutality involve minority citizens and that there is usually no penalty for the officers involved. Moreover, showing a disrespectful attitude is just as likely to generate police brutality as posing a serious bodily threat to the police (Lersch and Feagin, 1996). Increasing the number of minority police officers has some effect on how the police treat minorities. Those cities where African Americans head the police department show a concurrent decline in police brutality complaints, an increase in minority police and minority recruitment, and in some cases, a decrease in crime. Simply increasing the number of African Americans in police departments does not, however, reduce crime dramatically since it does not change the material conditions that create crime to begin with (Cashmore, 1991).

Race and Sentencing

What happens once minority citizens are arrested for a crime? Upon arraignment, bail is set higher for African Americans and Latinos than for Whites, and minorities have less success in plea bargaining. Once on trial, minority defendants are found guilty more often than White defendants. At sentencing, African Americans and Latinos are likely to get longer sentences than Whites for the same crimes even when they have the same number of prior arrests and the same socioeconomic background, and they are less likely to be released on probation (Bridges and Crutchfield, 1988; Chambliss, 1998; Chambliss and Taylor, 1989; Mauer, 1992).

African American defendants receive longer sentences than White Americans for property and violent crimes, and the disparity between Black and White sentences is even greater for serious crimes and crimes where the victim is White, especially when the crime is rape or murder (Keil and Vito, 1995; Spohn, 1995; La Free, 1980). Sentencing also differs depending on the racial identity of the judge. A study of Hispanic and White judges found that White judges sentence White defendants less severely than Hispanic defendants; Hispanic judges do not seem to distinguish between defendants based on race (Holmes et al., 1993).

Racial discrimination is particularly evident with regard to the death penalty. Currently, of the 3219 prisoners on Death Row, 42 percent are Black; 49 are women (U.S. Bureau of Justice Statistics, 1997). Research shows that when Whites and minorities commit the same crime against a White victim, minorities are more likely to receive the death penalty. African American men accused of raping White women are the group most likely to receive the death penalty. Killers of Whites are also three times more likely to get the death penalty than killers of African Americans (Keil and Vito, 1995).

In 1972, the Supreme Court in *Furman v. Georgia* concluded that the death penalty was discriminatory because it was applied unevenly based on race. There was an ensuing ban on executions. In 1987, however, the Supreme Court upheld the death penalty in *McCleskey v. Georgia.* Despite repeated arguments that the death penalty constitutes cruel and unusual punishment, that *lex talionis* ("a life for a life") is brutish law, and that capital punishment is applied differentially by race, it is now in use once again. Since executions were resumed in the United States in 1977, more than 300 people have been put to death by the state (U.S. Bureau of the Census, 1997a).

Prisons: Deterrence or Rehabilitation?

In the late 1990s, 5.1 million people in the United States were under correctional supervision, meaning in prison or jail or on probation or parole. State and federal prisons held 1.1 million prisoners (U.S. Bureau of the Census, 1997). Racial minorities are more than half of the federal and state male prisoners in the United States. Blacks have the highest rates of imprisonment, followed by Hispanics, then Native Americans and Asians. (Native Americans and Asian Americans together are less than 1 percent of the total prison population.) Among prisoners, Black and White men are the vast majority, Blacks being a large percentage relative to their representation in the general population, and Hispanics

The number of people in prison has increased dramatically in recent years, resulting in prison overcrowding and a huge increase in public spending for the construction of new prisons.

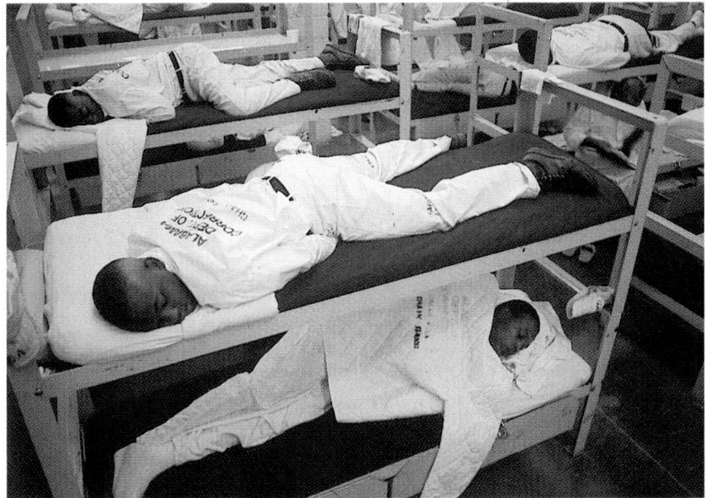

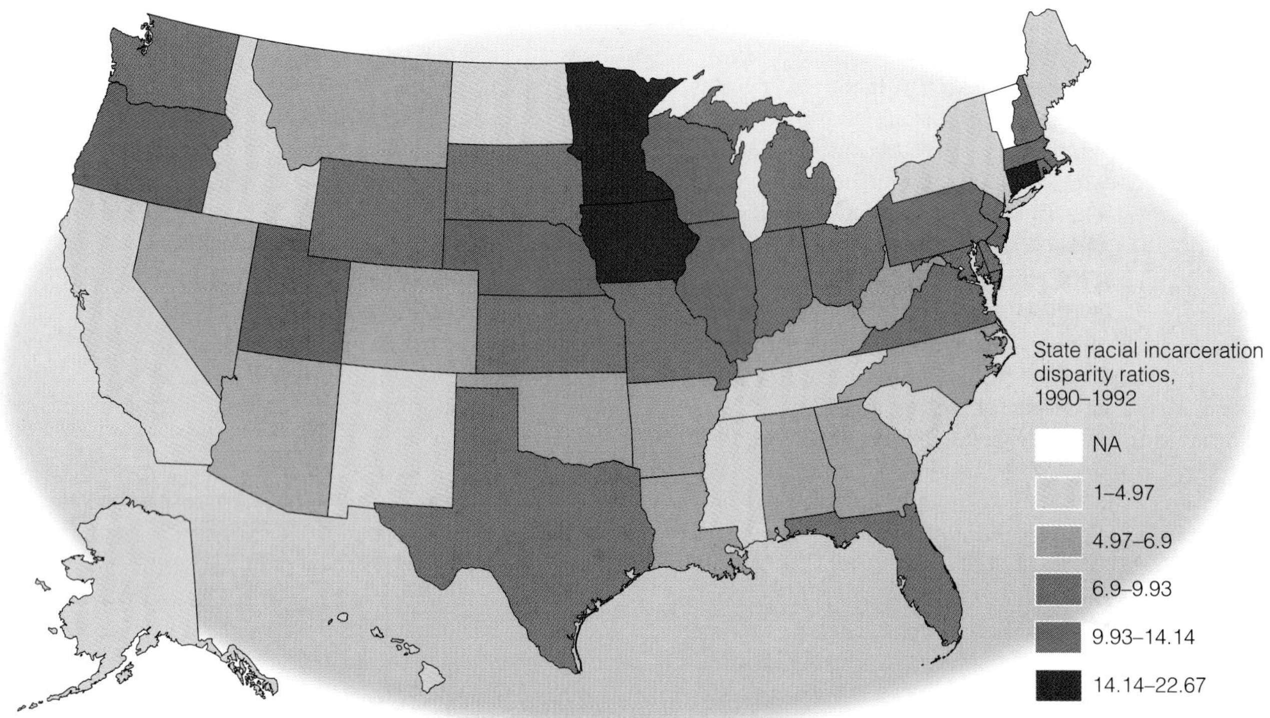

MAP 18.2 Mapping America's Diversity: Racial Disparity in Incarceration

Racial disparity in incarceration is measured as a ratio of the state's Black incarceration rate to the state's White incarceration rate. Research on this subject indicates that this disparity is the result of several complex factors, including differences in the arrest rates for Whites and Blacks, levels of economic well-being, and the degree of Black politicial mobilization.

SOURCE: Yates, Jeff. 1997. "Racial Incarceration Disparity Among States." *Social Science Quarterly* 78 (December): 1001–1010.

being the fastest growing minority group in prison (Bureau of Justice Statistics, 1997). Native Americans, though a small proportion of the prison population, are also overrepresented in prisons. In theory, the criminal justice system is supposed to be unbiased, able to objectively weigh guilt and innocence. The reality is that the criminal justice system reflects the racial and class stratification in society.

The United States and Russia have the highest rate of incarceration in the world (Mauer, 1997; see Figure 18.4). In the United States, the rate of imprisonment has been rapidly growing (see Figure 18.5). By 1996, the rate of incarceration in state prisons was 394 per 100,000 in the population—an increase from 139 per 100,000 in 1980 (U.S. Bureau of Justice Statistics, 1997; Mauer, 1997). By all signs, the population of state and federal prisons continues to grow, with the population in prisons exceeding the capacity of the facilities. The cost to the nation of keeping people behind bars totals at least $93.8 billion (U.S. Bureau of the Census, 1997).

The rate of prison growth in the United States has been so high in recent years that a new trend has emerged: the operation of prisons by private companies. Running corrections systems is now big business, and the "prison market" is expected to more than double in years ahead. For those looking for a business investment, running a prison is a good proposition since, as one business publication has argued, it's like running a hotel with 100 percent occupancy, booked for years in the future! The privatization of prisons raises new questions for social policy since, in the interest in running a profit center, prison managers may overcrowd prisons, reduce staffing, and cut back on food, medical care, or staff training. What may be sound business practice can result in less humane treatment of prisoners—locked away from the eyes of the public. Investigators are beginning to see that there are other costs to privatization: the rate of violence in private prisons is higher than in state facilities, and since the private prisons receive money from the state to house prisoners, there is financial incentive to keep people in prison longer—at the taxpayers' expense (Bates, 1998).

Why is there such growth in the prison population when the crime rate has been declining? A major reason for the increasing number of individuals behind bars is the increased enforcement of drug offenses and the mandatory sentencing that has been introduced. Nearly one-quarter of those in state prisons are serving a drug sentence; 60 percent of federal prisoners are serving drug sentences, more than doubling since 1980. Although the number of drug offenders has grown dramatically, the number of violent offenders has grown the most (Mumola and Beck, 1997).

The number of women behind bars has also increased at a faster rate than for men, although the numbers of women in prison are small by comparison. Women are only

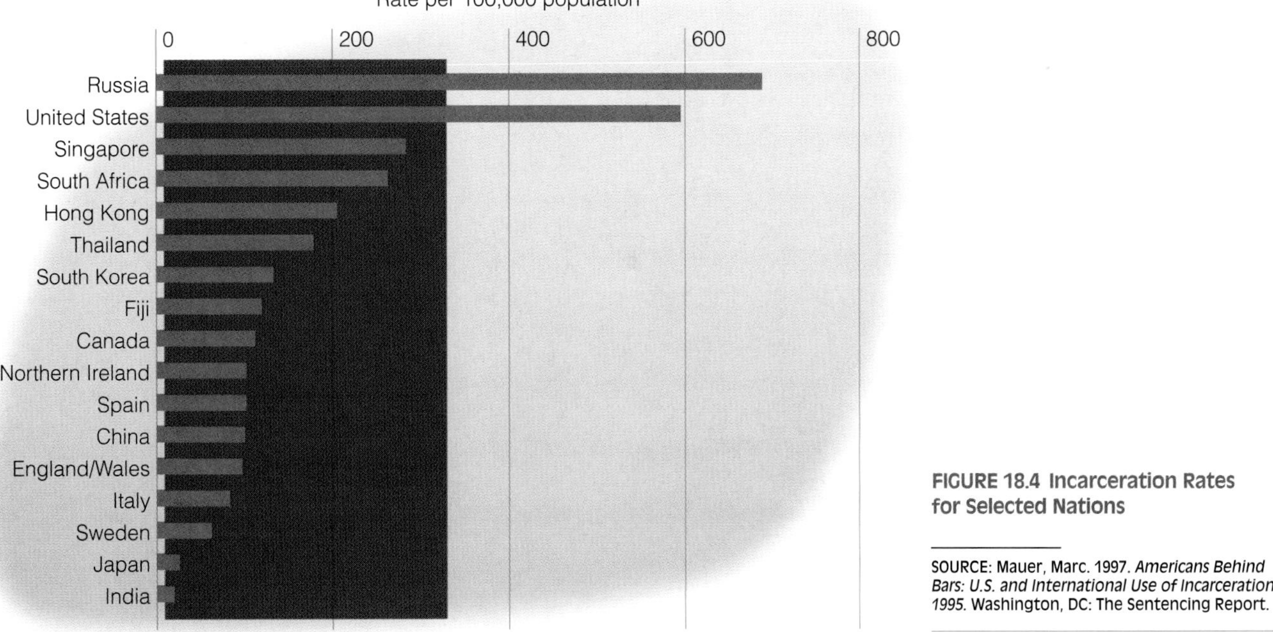

FIGURE 18.4 Incarceration Rates for Selected Nations

SOURCE: Mauer, Marc. 1997. *Americans Behind Bars: U.S. and International Use of Incarceration 1995.* Washington, DC: The Sentencing Report.

7.8 percent of all prisoners (Bureau of Justice Statistics, 1997). Like men, 60 percent of women in federal prisons are there because of drug-related offenses, which include theft, prostitution, and robbery (Chesney-Lind, 1992); often, they have participated in these crimes by going along with the behavior of their boyfriends (Miller, 1986). The typical woman in prison is a poor, young minority who dropped out of high school, is unmarried, and is the mother of two or more children. Fifty-seven percent of women in prison are African American. Half of all women prisoners ran away from home as children; two-thirds (of federal women prisoners) were arrested on drug charges, and many have been victims of sexual abuse (Arnold, 1990; U.S. Bureau of Justice Statistics, 1997).

Women in prison face unique problems, in part because they are in a system designed for men and run mostly by men; this tends to ignore the special needs of women. For example, one in four women entering prison is pregnant or has just given birth, but they often get no prenatal or obstetric care. Eighty percent of women entering prison are mothers; most of those mothers (85 percent) have sole custody of their offspring and are the sole supporter of their families; consequently, many are forced to put their children up for adoption or leave the child with a relative (Church, 1990). Space is also a problem for women in prison, as it is for men. Many women have to double up in cells designed for one because of a rise in the prison population. State officials have argued that there are too few women prisoners relative to males to justify spending the large amounts of money it would require to redress conditions in women's prisons (Church, 1990). Prisons also tend to reinforce traditional gender roles. Male prisoners are trained for such jobs as auto mechanic, whereas women are more likely to be trained as beauticians or launderers. The result is that few women prisoners are rehabilitated by their experience in prison; instead, they are likely to become increasingly alienated, increasing the odds that, once released, they will return to prison again.

The United States is putting offenders in prison at a record pace. Is crime being deterred? Or are prisoners being rehabilitated? If the deterrence argument were correct, we would expect that increasing the risk of imprisonment would lower the rate of crime. For example, we would expect drug use to decline as enforcement of drug laws increases. In the past few years, there has been a marked increase in drug law enforcement, but not the expected decrease in drug use. Although the use of drugs did decline slightly overall, particularly among young people, there has been an increase among Black Americans and inner-city youth—those most likely to feel the crackdown of increased enforcement (Mauer, 1992). Using drugs as an example, it appears that the threat of imprisonment does not deter crime. There is also little evidence that the criminal justice system rehabilitates offenders. Using drugs as the example again, few who are imprisoned for drug offenses ever receive drug treatment (only about 20 percent). Although the nation is getting "tough on crime," it is doing little to see that offenders do not continue to commit drug crimes once they are released from prison.

In general, prisons seem neither to deter nor rehabilitate offenders. Prisons certainly do nothing to address the societal problems known to promote criminal activity. They concentrate on individual wrongdoers, not on the social structural causes of crime. Although there has been an enormous increase in the number of prisons built and the number of people in prison, it appears that imprisonment is doing little to solve the problem of crime.

If the criminal justice system fails to reduce crime, what does it do? Some radical sociologists contend that the criminal justice system is not meant to reduce crime, but has other functions, namely, to reinforce an image of crime as a threat from the poor and from racial groups. The prison

FIGURE 18.5 State and Federal Prison Population, 1980–1996

SOURCE: Mumola, Christopher J., and Allen J. Beck. 1996. *Prisoners in 1996*. Washington, DC: U.S. Department of Justice.

experience is a demeaning one, poorly suited to training prisoners in marketable skills or to letting them repay their debt to society. Rather than teaching prisoners self-control and self-direction, prisons deny inmates the least control over their everyday life (Reiman, 1995). In the end, prisons seem, at least in some cases, to refine criminals rather than rehabilitate them. In prison, inmates acquire new criminal skills from contact with other career criminals and make new connections to criminal networks. Many states also take away the basic citizenship rights of felons; prisoners may not vote when they are in prison or on parole, but several states also take away their voting rights permanently (Mauer, 1997). In sum, the high rate of incarceration exacerbates problems whereby those convicted of crimes are detached from the norms of society, with obstacles to their reintegration even following imprisonment.

Courts and the Law

The court system is a pivotal part of social control. Each state has its own guidelines and procedures for its court system. The federal system has a specific hierarchical structure that determines which cases shall be heard at which level. District courts are the lowest level of the federal system. There are about ninety district courts, with at least one in each state. The next tier up in the federal legal organization is the U.S. Courts of Appeals, also known as circuit courts. There are thirteen courts of appeal in the United States. These courts do not conduct trials, but instead review the record of trials in lower courts, hear lawyer's arguments, and decide if errors were made. The third and highest court is the U.S. Supreme Court, which seats nine justices (one Chief Justice and eight Associate Justices). Several specialty courts are also dispersed throughout the court system (for example, bankruptcy court and customs court).

The role of all courts is to interpret and enforce the **law**—the written set of guidelines that determine what is defined as right and wrong in society. The United States uses law to provide guidelines and standards for behavior. It is the responsibility of the court system to ensure that the law is applied to all cases in a fair and just manner.

The specific role of the court system has been argued extensively. Social scientists debate whether the judiciary *determines* social values and imposes them on the public or simply *reacts* to social pressures and enforces already existing values. For example, the Supreme Court has been referred to as "the American political conscience . . . [with a] priestly mantle" (Agyeman 1990: 22). This refers to the court's responsibility to protect the rights of the less privileged. But others argue that all courts have a "role as the protector of property rights and business interests" (Agyeman, 1990: 23). According to this perspective, the courts primarily serve the interests of those with influence and property, not the downtrodden or the poor.

Both sides of the dispute about whether courts set values or react to values have merit since an extensive review of court decisions would deliver examples of justice working in both directions. The point is that the courts are not the neutral parties they are sometimes alleged to be. Social factors, such as the status of defendants or the wishes of powerful parties, do influence the outcome of legal decisions. At the same time, appealing to the courts is one of the primary avenues of recourse available to oppressed peoples seeking justice.

All courts balance competing influences in the process of legal decisions. The legal process in the United States is not a simple matter of applying unambiguous legal codes to specific cases. Interpretation plays an important role in case decisions, from interpreting the worth of competing arguments or the relevance of previous rulings to interpretation of the very wording of the Constitution. The legal process is an eminently subjective system that aspires to objectivity in the name of justice.

Minorities in the Court System. The court system holds a great deal of power in defining and enforcing the values of the American public. Traditionally, this power has rested in the hands of an elite group of attorneys and judges composed almost exclusively of White men. Attorneys and judges are the actors on the court stage. Lawyers serve as the spokespersons for each side, and judges serve as moderators and decision makers. Among lawyers, 27 percent are women; 2.8 percent, Black; and 3.8 percent, Hispanic. Among judges, women, Blacks, and Hispanics are a smaller percent (U.S. Department of Labor, 1998). Decisions that critically affect the lives of the American people, including its minorities, lie in the hands of a fairly homogeneous group of individuals. A study of judges in New Jersey showed that *most* judges think that they carry some racial bias into the courtroom and that the court system contains institutional racism (Chambliss and Taylor, 1989). Considering the broad scope and impact of court cases on the general public, occupational segregation in the law, and especially the ascendancy of White males, is a matter of great concern.

As institutions of the state, the law and the courts have a great deal of influence over the social, political, and civil rights of people in the United States. The jury system, intended to ensure the administration of justice by one's peers, is one where various forms of bias and lack of representation interfere with fairness. Courts have to take careful measures to ensure representation on juries by different minority groups. Whites have historically been disproportionately represented on juries; increased attention to fair representation has lessened this problem although how well diverse groups are included on juries varies enormously in different locales. Those who move frequently, immigrants, and some racial–ethnic groups are routinely underrepresented on juries. The overrepresentation of racial minorities as defendants and continuing problems of representation on juries almost guarantees an unjust court system (Fukurau et al., 1993). As we have seen, the system of justice works differently for different groups of people, indicating the importance of analyzing race, gender, and class in the administration of justice in the United States.

The Law and Social Change. Despite problems in the administration of justice, the law remains one of the most effective avenues for addressing the injustices against different groups in American society. This is one of the ironies of the sociology of law: that in a society based on rational-legal authority, the legal system may embody racism, sexism, and class injustice within its institutional structure, yet it remains one of the major methods for combating racism, sexism, and class injustice.

One of the landmark court decisions in U.S. history was the 1954 case *Brown v. Board of Education of Topeka, Kansas,* in which the Supreme Court declared segregation in public facilities such as schools and buses unconstitutional. This decision transformed race relations in the United States since it made formal policies specifying separation by race illegal. Outlawing *de jure* segregation (segregation by law) has not, however, eliminated *de facto* segregation (segregation in practice). The persistence of racial segregation in the labor market, in housing and education, and in social interaction indicates that the law goes only so far in producing social change; nevertheless, it is a key element in a long-term historical process. Despite persistent racial segregation, since *Brown v. Board of Education,* there have been many indications of increased racial integration in various dimensions of life.

One area where the courts have been influential in overtly directing social change has been in matters of equal opportunity. Affirmative action, for example, is a legal principle designed to remedy past discrimination. Affirmative action can be viewed as a way to help minorities "catch up" by delivering access to the benefits and opportunities historically reserved for White people. The principles of affirmative action continue to be interpreted and refined through court action (see Chapters 11 and 12). Another judicial policy marking significant social change was the 1963 Equal Pay Act passed by Congress. This law mandated that employers must compensate men and women equally when they perform equal work. The Civil Rights Act of 1964 also forbids discrimination in employment procedures and opportunities.

The courts also make decisions that affect seemingly private aspects of life. A Supreme Court decision gave married women the right to birth control (*Griswold v. Connecticut,* 1965); this right was not extended to people who were not married until another Supreme Court decision was handed down in 1972 (*Eisenstadt v. Baird*). These constitutional cases were followed by *Roe v. Wade* in 1973, in which the Supreme Court gave women the right to choose an abortion under specific federal guidelines. Each of these laws represented a shift in federal policy "from reinforcing women's traditional roles to protecting departures from those roles . . . [and] the U.S. Supreme Court played an especially active role in instituting these changes" (Aliotta, 1991: 144). These examples illustrate that the courts and the government can be progressive instruments of social change.

The Supreme Court has also strengthened the protection that lesbians and gays have under the law. In *Romer v. Evans* (1996), the Court ruled that states cannot pass laws that deprive gays and lesbians the equal protection of the law promised under the Fourteenth Amendment to the U.S. Constitution. This decision was the result of a challenge to a law passed in Colorado that had banned measures to protect homosexual rights within the state. The Court decision is especially significant because, at the time it was passed, there were movements in many states to adopt legislation through public referenda that would restrict gay rights. Although the ruling does not require states to pass laws protecting gay rights, and it is not clear yet if it will affect policies regarding gays and lesbians in the military or gay marriage, it is a significant constitutional protection against measures that would specifically deny gay men and lesbian women equal protection under the law.

A truly democratic society offers individuals from all walks of life the opportunity to participate in the creation and evaluation of the law, thus ensuring fairness and equity. As seen in this chapter, all people do not have equal access to positions of legal and governmental power, nor do they partake equally of the law dispensed by the legislature and the judiciary. The mechanisms of the state are constantly evolving, however. The evolution of the state may be the single most important arena of change for those who seek justice and equality.

CHAPTER SUMMARY

- The *state* is the organized system of power and authority in society. It comprises different institutions, including the government, the military, the police, the law and the courts, and the prison system. The state is supposed to protect its citizens and preserve society, but it often protects the status quo, sometimes to the disadvantage of less powerful groups in the society.
- *Power* is the ability of a person or group to influence another. *Authority* is power perceived to be legitimate. There are three kinds of authority—*traditional authority,* based on long established patterns; *charismatic authority,* based on an individual's personal appeal or charm; and *rational-legal authority,* based on the authority of rules and regulations (such as law). The United States is primarily built on a system of rational-legal authority. Bureaucracies typically flourish in a system of rational-legal authority. The growth of bureaucracy is especially notable in contemporary governmental institutions.
- Sociologists have developed four theories of power. The *pluralist model* sees power as operating through the influence of diverse interest groups in society. The *power elite model* sees power as based on the interconnections between the state, industry, and the military. *Autonomous state theory* sees the state as an entity in itself that operates to protect its own interests. *Feminist theorists* argue that the state is patriarchal, representing primarily men's interests. Each theory reveals a different aspect of how power operates.
- An ideal democratic government would reflect and equally represent all members of society. The makeup of American government does not reflect the diversity of the general population. African Americans, Latinos, Native Americans, Asians, and women are underrepresented within the government. Political participation also varies by a number of social factors, including income, education, race, gender, and age. The large sums of money necessary to run campaigns make politicians susceptible to the influence of *political action committees.* One consequence of the current structure of government is that large numbers of Americans have little confidence in the government to solve the nation's problems.
- The *military* is a social institution that is the nation's system of defense. Militaristic values permeate the culture, emphasizing violence and aggression. Through military socialization, new recruits must learn the norms and values of military culture. The military is strongly linked to other social institutions in the society. African Americans and Latinos are overrepresented in the military, in part because of the opportunity the military purports to offer groups otherwise disadvantaged in education and the labor market. There is an increased presence of women in the military.
- Racism is embedded in the institutional system of *the courts* and is evidenced in the unequal treatment of minority groups throughout the *criminal justice system.* Minorities are more likely to be arrested and convicted and are given longer sentences than whites. Prisons do a poor job of deterring crime, as well as a poor job of rehabilitating criminals. The legal system reflects the race and gender inequality of the society.
- Despite inequality in the administration of justice, the *law* still provides a mechanism for aggrieved groups to seek justice. Several important Supreme Court decisions have shown the responsiveness of the state to addressing race and gender inequality in the United States.

KEY TERMS

authority
autonomous state model
bureaucracy
charismatic authority
democracy
gender gap
government
interest group
interlocking directorate
law
military–industrial complex
pluralist model
political action committees (PACs)
power
power elite model
propaganda
rational-legal authority
state
traditional authority

THE INTERNET: A Tool for the Sociological Imagination

Resources on the Internet:

Virtual Society: The Wadsworth Sociology Resource Center
http://sociology.wadsworth.com

Visit this site to find additional learning tools, including interactive quizzes, links related to web sites, and an easy link to *InfoTrac College Edition.*

Bureau of Justice Statistics
http://www.ojp.usdoj.gov/bjs/

This is the primary source for U.S. criminal justice statistics, including regular reports on incarceration, the death penalty, and so on.

Federal Bureau of Prisons
http://www.bop.gov/

This site includes data on federal prisons and demographic data about prison inmates.

Federal Bureau of Investigations
http://www.fbi.gov/

The FBI's web site includes press releases and information about federal crime publications.

Sociology and Social Policy: Internet Exercises

The public is increasingly aware of the high price of political campaigns; indeed, some say that only those with vast resources can run for national office. The influence of *political action committees*

(PACs) has therefore grown, as have concerns about the influence of wealthy individuals. What are the current proposals for campaign finance reform? How do the debates about campaign finance reform reflect sociological issues about power, politics, and the representation of diverse groups?

Internet Search Keywords:

political action committees (PACs)
campaign finance reform
Federal Election Commission
politics in campaign finance reform

Web sites:

http://www.ua.org/political/pacrole.htm
Brief page about political action committees (PACs) and their role in today's political campaigning.

http://www.crp.org/
A web site maintained by the Center for Responsive Politics offers comprehensive information on campaign fund raising, candidates' profiles, and more.

http://www.special.pioneerplanet.com/archive/campfinance/dox/camp6.htm
Some Federal election rules on campaigning.

http://www.brook.edu/gs/campaign/cfr_hp.htm
Excellent site on campaign finance reform.

http://www.2.rmwc.edu/slockhart/P208STU/SGODFREY/Main.htm/
Excellent site including history, key concepts, current campaign finance laws, the courts' role in campaign finance reform, various opinions, bills, etc.

InfoTrac College Edition: Search Word Summary

American Indians and state policy
charismatic authority
gays and lesbians in the military
interest group
political action committee
propaganda

In order to learn more about these central topics in sociology, you can conduct an electronic search using InfoTrac College Edition. To aid in your search and to gain useful tips, see the Student Guide to InfoTrac College Edition on the Virtual Society web site:
http://sociology.wadsworth.com

INTERACTIONS—A SOCIOLOGY CD-ROM: CONCEPTS FOR THIS CHAPTER

Go to the Wadsworth Sociology CD-ROM for further study on the concepts in this chapter. The CD-ROM also includes quizzes and additional activities to expand your learning experience.

SUGGESTED READINGS

Clawson, Dan, Alan Neustadtl, and Denise Scott. 1992. *Money Talks: Corporate PACS and Political Influence.* New York: Basic Books Inc.

Based on interviews with managers of political action committees, this book analyzes the influence of money on public policy. Broadly speaking, the book also analyzes the influence of business power on the political system.

Domhoff, G. William. 1998. *Who Rules America?* Mountain View, CA: Mayfield Publishing.

This is an updated version of a classic book, detailing the power elite in America.

Enloe, Cynthia. 1989. *Bananas, Beaches, and Bases.* Berkeley: University of California Press.

Enloe's book presents a feminist perspective on international relations. Her interpretation looks at the role of gender in peace, war, development, and the social structure of the military. Her analysis provides a new perspective on the role of women in world affairs.

Hardy-Fanta, Carol. 1993. *Latina Politics, Latino Politics.* Philadelphia: Temple University Press.

Based on a study of the Latino community in Boston, Hardy-Fanta examines the political participation of Latinas. In doing so, she explores what "political" means to them and explores the connections between gender, culture, and politics.

Moskos, Charles. 1988. *Soldiers and Sociology.* Alexandria, VA: United States Army Research Institute for the Social and Behavioral Sciences.

As the foremost sociological expert on the subject, Moskos gives a thorough overview of the status of minorities within the military.

Omi, Michael, and Howard Winant. 1994. *Racial Formation in the United States,* 2nd ed. New York: Routledge.

Omi and Winant provide a theory of the role of the state in defining the meaning of "race." Their analysis includes an insightful discussion of the politics of race in the 1980s and provides a basis for understanding the role of state policy in framing racial politics.

Reiman, Jeffrey. 1995. *The Rich Get Richer and the Poor Get Prison.* Boston: Allyn and Bacon.

This critical view of the criminal justice system explores the inequality that the author argues is at the heart of the nation's crime problem. Reiman argues strongly for rehabilitation of the justice system in the United States, but with radically different conclusions than the so-called war on crime.

Shilts, Randy. 1993. *Conduct Unbecoming: Gays & Lesbians in the U.S. Military.* New York: St. Martin's Press.

This book provides a historical overview of gays and lesbians in the military and analyzes public and government response to this issue.

Stiehm, Judith Hicks. 1996. *It's Our Military, Too!: Women and the U.S. Military.* Philadelphia: Temple University Press.

Including accounts by women in the military, this book reviews different dimensions of women's experiences in the military and explores such issues as the role of women in combat, how gender influences weapon design, and the status of minority women and lesbian women in the military.

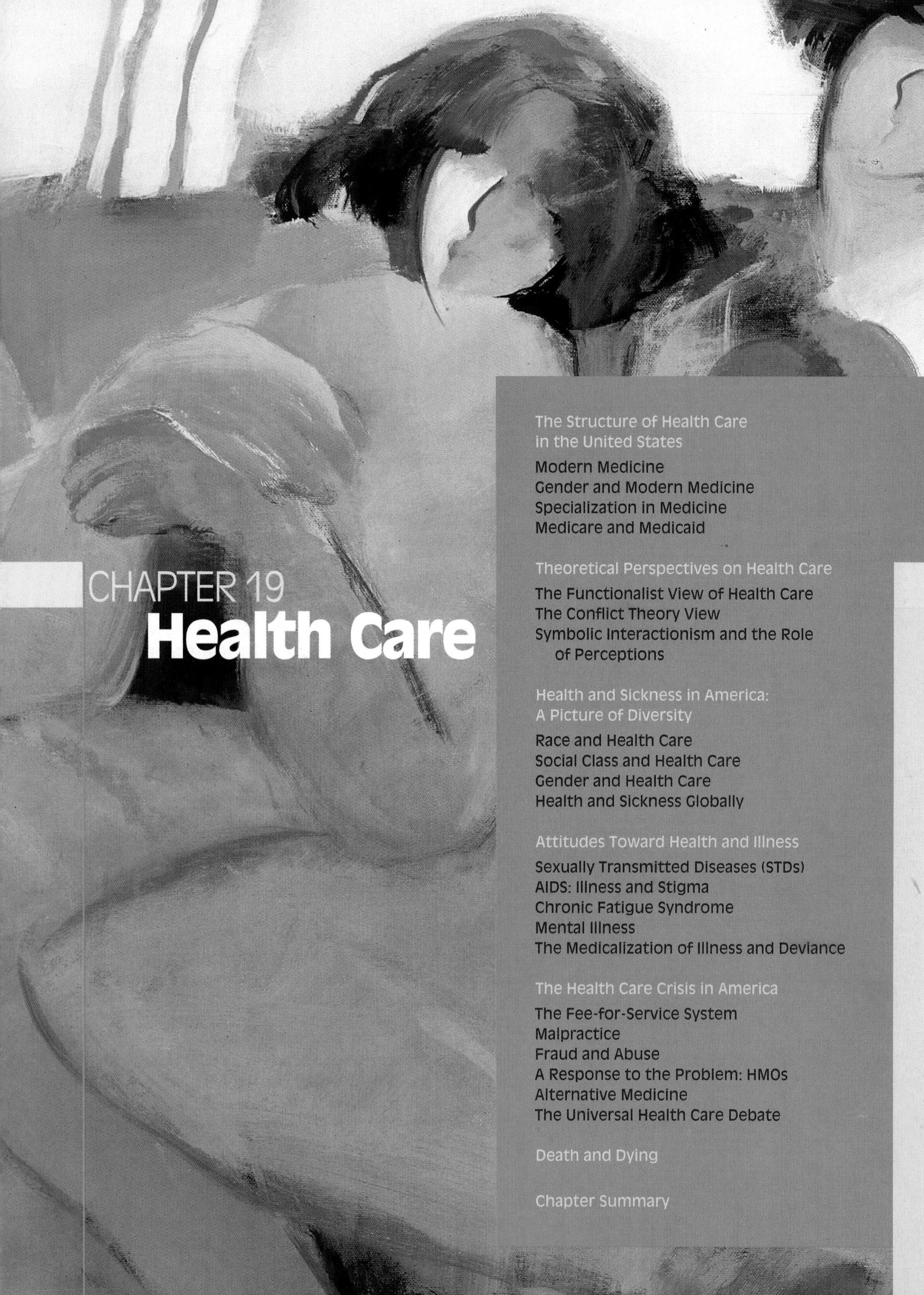

CHAPTER 19
Health Care

HEALTH CARE as an institution in society is in crisis. The public wonders: Has the quality of health care in the United States declined over the last two decades or so? Are doctors and nurses as good as they used to be? Are doctors paid too much? Witness the current concern over the possibilities of fraud within the health industry: Are hospitals charging way too much? Why are costs for medical insurance so high, and why don't all people have health insurance?

Not all the crisis in health care resides within the health care institutions. The public wrestles daily with matters such as cigarette smoking, drug dependency, and illnesses such as lung cancer, or death at an early age. Tobacco company executives argue before Congress that their product is not especially harmful to the healthy adult (although it certainly is and has been proven to be in scientific studies), and claim that it is up to individuals to decide whether to smoke or not. Lawsuits against tobacco companies abound from people who have contracted lung cancer, and from the families of those who have died from it. The tobacco companies argue that they should be permitted to advertise, even to children, although the infamous "Joe Camel" advertisements have been banned by Congress.

These public issues may seem like simple matters of personal responsibility or individual rights, but sociologists show that illness in our society is strongly influenced by social factors. How we treat illnesses is highly influenced by race, social class, gender, and age, and the same can be said of how we treat ill people. In different societies, the same physical state may be regarded as either sickness or wellness, thus demonstrating that the definition of illness is at least partly social constructed. Physical conditions regarded as illness might be treated by prayer or witchcraft, surgery or radiation therapy, depending on the culture or society in question. The definition of pain also varies considerably from culture to culture, so much so that what is experienced as pain in one society may actually be considered pleasure in another. Witness even in our society the following experiences, usually thought of as painful, but experienced as pleasurable by the participants in their subculture: swimming in ice-cold water, marathon running, slam dancing.

Debunking Society's Myths

Myth: **Illness is strictly a physical thing.**

Sociological perspective: **What is defined as illness, even physical illness, is heavily influenced by culture; also, the presence of illness is influenced by social conditions.**

The Structure of Health Care in the United States

Generally speaking, the citizens of the United States are quite healthy in relation to the rest of the world. However, there are very great discrepancies among people in the United States in terms of longevity, general health, and access to health care, with the least advantaged part of the curve consisting primarily of minorities, the lower classes, and for a number of ailments, women, but the fact remains that we are a robust nation. Although a rise in the incidence of tuberculosis caused concern in the early 1990s, deaths from one-time scourges such as smallpox, cholera, and tuberculosis have been virtually eliminated in the United States.

Modern Medicine

In colonial times, American physicians received their training in Europe. Their competitors in the healing arts included alchemists, herbalists, ministers, faith healers, and even barbers. Treatments were a combination of folk wisdom, superstition, tried-and-true regimens, and sometimes dangerous quackery. A simple scratch, once infected, could easily cost a limb or a life. Ambiguous maladies like "the fever" were a common cause of death. Etiologies (causes) of disease were believed to include everything from "bilious humors" to demonic possession. Even the best-trained practitioners knew no better than to use unsterilized instruments, and a common intervention was the "bleeding" of patients (namely, pointlessly drawing "bad blood" from the patient, sometimes in amounts large enough to seriously weaken the victim). The cure was often worse than the disease.

The mentally ill especially were often thought to be possessed, or worse, witches. Once labeled a witch, there was little hope of shedding the label, and the consequences could be as horrendous as being burnt or drowned. The vast majority of those labeled witches were women. Sociological researchers have shown that these women were generally older, unmarried, and childless, and were thus defined as of little use to the local community (Erikson, 1966). Defining them as possessed by demons and thus incurable was one way of achieving their permanent elimination from the community. Labeling theory, discussed later, clearly applies to the study of witchcraft.

By the start of the nineteenth century, advances in biology and chemistry ignited a century of explosive growth in medical knowledge. One fruit of the scientific revolution of the mid-1800s was the *germ theory*, the idea that many illnesses were caused by microscopic organisms, or germs. Now considered to be scientific fact, the notion that something called germs caused illness was then a hotly debated topic. Shortly thereafter, germ theory established itself as a foundation of medicine. Doctors were able to show that such tactics as isolating infected people and sealing infected wells could stop the spread of illness by stopping the spread of germs. Relentless study and research transformed medicine into a science. Coincidentally, the social prestige of medicine greatly increased, contributing mightily to the status of physicians, who had formerly enjoyed more modest social standing. The year 1847 saw the founding of the American Medical Association (AMA), and after half a century of sweeping away rivals in the healing arts by successful campaigns to have alternative therapies delegitimized or outlawed, the AMA emerged as the most powerful organization in U.S. health care.

It was in the late 1800s that the image of medicine as an upper-class profession took hold. A medical education was expensive, and medical schools drew upon White, male, urban populations for their classes. Those trained as physicians took their place in the upper social strata. Herbalists and faith healers came more frequently from the rural lower class, and generally remained there. As the ranks of the medical profession swelled with wealthy Whites, African Americans and Hispanics became proportionately more affiliated with older folk practices and midwifery. This overall trend continued through the early part of the twentieth century, and even today folk practices continue to have adherents among rural lower-class whites and among rural and urban lower-class Blacks, Hispanics, and Native Americans (Starr, 1982).

Gender and Modern Medicine

Not only are social class and race-ethnicity interwoven with the development of modern medicine in the United States, so is gender. At the outset of the twentieth century, the male-dominated medical profession vigorously opposed gender equality in medicine. Prevailing medical opinion at the time labeled the differences between the genders both natural and unchangeable. Men were seen as inherently rational and scientific, women were seen as dominated by emotions and incapable of rigorous scientific thought (Smith-Rosenberg and Rosenberg, 1984). Vestiges of these beliefs persist in the medical profession even today.

Women were expected to devote their time and energy to childbearing rather than professions such as science and medicine. One physician in 1890 stated that God had "in creating the female sex . . . taken the uterus and built up a woman around it" (Smith-Rosenberg and Rosenberg, 1984: 13). Such beliefs kept women out of medical school since physicians warned that too much thinking tended to interfere with a woman's ability to have children.

Physicians of the day had strong opinions about female sexuality. Women were supposed to have no interest in sex beyond the reproductive function. The female orgasm,

health care and religion

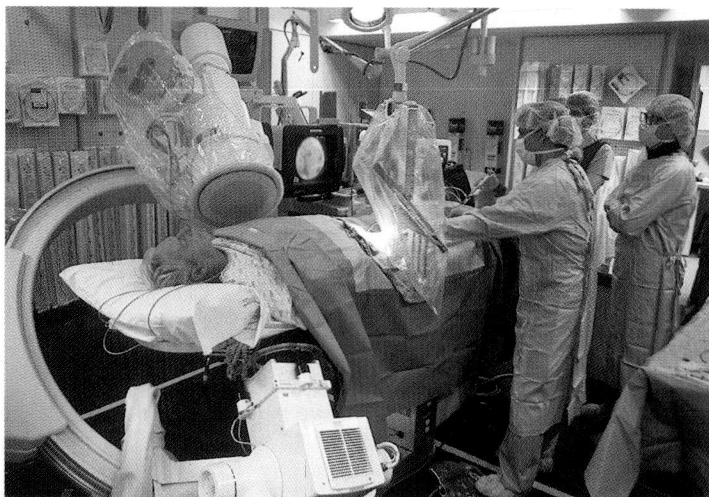

The development of new technologies has transformed the delivery of health care. The "house call" by a physician is now a rarity; many surgical procedures, such as this balloon angioplasty, would have been unimaginable not many years ago.

when it was acknowledged, was likely to be regarded as a type of disorder. The male orgasm, on the other hand, was considered natural, and regular sexual relations were understood to be necessary for men. This view was promoted by the American Medical Association well into the 1920s. Contraception and abortion were opposed because it was believed they would encourage increased sexual activity among women, thus endangering their health.

THINKING SOCIOLOGICALLY

Take an opportunity to casually observe how illness may be related to social factors, besides biological and physical factors. The next time you are in your doctor's office (outside a college campus, preferably), observe some of the people in the waiting room: How many males and how many females? What is the racial composition of those who are waiting? Are there any indications that the group is relatively homogeneous with respect to social class, as indicated by dress, manner, or speech?

Specialization in Medicine

Since World War II ended in 1945, there has been tremendous growth in the medical establishment, including great increases in the number of medical professionals and the number of dollars spent on health care. Physicians continue to gain in prosperity and social status.

Along with the huge growth in medicine came increased specialization. Today, specialists greatly outnumber general practitioners. U.S. children are taken to the pediatrician, particularly if they belong to the middle or upper class. Their mothers see a gynecologist; allergists treat our allergies; psychiatrists tend to our moods; podiatrists minister to our ravaged feet. The family dentist sends us to the endodontist for a root canal.

Physicians who specialize enjoy greatly increased incomes (see Table 19.1). The high cost of specialized care has made it unavailable to many within the population, while the attractions of specialization have led to the much diminished numbers of more accessible general practitioners. These and other problems have led to a state of crisis in health care that many feel is one of the most important problems facing the United States.

TABLE 19.1 *MEAN NET INCOME FOR SELECTED MEDICAL SPECIALTIES*

Specialty	Mean Net Income (1994)
All physicians	$182,400
Surgery	255,200
Obstetrics/gynecology	200,400
Internal medicine	174,900
Pediatrics	126,200
Family/general practice	121,200

SOURCE: From the U.S. Bureau of the Census. 1997. *Statistical Abstracts of the United States 1997.* Washington, DC: U.S. Department of Commerce, p. 125.

Medicare and Medicaid

The U.S. government has for some time sought to have some form of widespread guaranteed health service, at least for certain categories of people, such as veterans, the poor, and the elderly. The **Medicare** program, begun in 1965 under the administration of President Lyndon Johnson, provides medical care in the form of medical insurance covering hospital costs for all individuals who are age sixty-five or older. The Medicare program does not cover physician costs incurred outside the hospital, but there are programs available that do so, although the patient must pay a portion of the cost.

Medicaid is a governmental program that provides medical care in the form of health insurance for the poor, those on welfare, and for the disabled. The program is funded through tax revenues. The costs covered per individual vary from state to state since the state must provide

medicare, medicaid

funds to the individual in addition to the funds that are provided by the federal government. The Medicare and Medicaid programs together are as close as the United States has come to the ideal of universal health insurance. More recent attempts at the establishment of universal coverage, such as the health insurance plan proposed in the early 1990s by President Clinton's administration, have not been successful (Starr, 1995).

Theoretical Perspectives on Health Care

A deeper understanding of the nation's health care system and its problems can be extracted by applying the three major theoretical paradigms of sociology: functionalism, conflict theory, and symbolic interaction theory.

The Functionalist View of Health Care

As noted, the notion of *sick role*—a pattern of expectations that society applies to one who is ill—has been coined by functionalist theory. Functionalism argues that any institution, group, or organization can be interpreted by looking at its positive and negative functions in society. Positive functions are those that contribute to the harmony and stability of society. The positive functions of the health care system are the prevention and treatment of disease. Ideally, this would mean the delivery of health care to the entire population without regard to race-ethnicity, social class, gender, age, or any other characteristic. At the same time, the health care system is currently notable for a number of negative functions, ones that contribute to disharmony and instability of society. Prominent among the problem areas in the U.S. health care system are:

- *Unequal distribution of health care by race-ethnicity, social class, or gender.* Earlier in the chapter we showed that health care is more readily available, and more readily delivered, to White or middle-class individuals in urban and suburban areas than to minorities. Particularly serious is the lack of health care delivery to Native American populations. Likewise, men and women receive unequal treatment for certain types of medical conditions, with women more likely to receive truncated treatment than men.
- *Unequal distribution of health care by region.* Each year, many in the United States die because they live too far away from a doctor, hospital, or emergency room. Doctors and hospitals are concentrated in cities and suburbs; they are much less likely to be situated in isolated rural areas; rural people in Appalachia and some parts of the

B • O • X 19.1

UNDERSTANDING DIVERSITY

The Americans with Disabilities Act

THE AMERICANS with Disabilities Act, passed by Congress in 1990, prohibits discrimination against disabled persons (see also Chapter 17). In 1999, the U.S. Supreme Court restricted the definition of disability to exclude disabilities that can be corrected with devices such as eyeglasses or with medication. The decision was based on cases where people had been denied employment because they did not meet a health standard required by the employer, even though, with correction, the standard was met. One case involved two nearsighted women denied jobs as airline pilots because, without glasses, their vision did not meet the airline company's standard of 20/40. With glasses, their vision was 20/20. Another case involved a man whose high blood pressure was above the Federal standard for driving trucks; he was denied a job as a trucker—even though with medication, his pressure was regulated to an acceptable level. The Court's decision denied these people the right to claim discrimination.

The decision raises several interesting sociological questions about disabilities and civil rights. Writing for the majority, Sandra Day O'Connor argued that the law requires that people be assessed based on their individual conditions, not as members of groups affected in a particular way. The disability rights movement, on the other hand, would argue that disabled people are a minority group with certain rights. What is being mediated by the court are competing definitions of what constitutes a disability—a point that intrigues symbolic interactionists who would note the role of socially constructed meanings in legal negotiations. Furthermore, as one of the lawyers who drafted the original ADA law has pointed out, the decision means that someone may be disabled enough to be excluded from a job, but not disabled enough to claim discrimination. This raises interesting questions about the rights of employers to establish physical and medical standards for certain jobs, even if those standards result in the exclusion of some groups of people from employment. You might ask yourself how you would balance the diverse and competing interests of disabled people, employers, and the courts were you to have to decide whether the exclusion of some disabled people from a particular job constitutes discrimination. What are the implications of your argument for cases involving race and/or gender discrimination?

South and Midwest may have to travel a hundred miles or more to get to the doctor.

- *Inadequate health education of inner-city and rural parents.* Many inner-city and rural parents do not understand the importance of immunizing their children against smallpox, tuberculosis, and other illnesses, and they are often suspicious of immunization programs. This hesitancy is reinforced by the depersonalized and inadequate care ghetto residents often encounter when care is available at all.

The Conflict Theory View

Conflict theory stresses the importance of social structural inequality in society. From the conflict perspective, the inequality inherent in our capitalist society is responsible for the unequal access to medical care. Minorities, the lower classes and the elderly, particularly elderly women, have less access to the health care system in America than Whites, the middle and upper classes, and the middle aged. Restricted access is further exacerbated by the high costs of medical care, stemming from high fees and the abuses of the fee-for-service and third-party-payment systems to be documented later. The exceptionally high incomes of medical professionals amplifies the social chasm between medical practitioners and an increasingly resentful public.

Excessive bureaucratization is another affliction of the health care system that adds to the alienation of patients. The American health system is burdened by endless forms for both physician and patient, including paperwork to enter individuals into the system, authorize procedures, dispense medicines, monitor progress, and process payments. Long waits for medical attention are normal, even in the emergency room. Being ignored for long periods can only increase the alienation of patients. Prolonged waits have reached alarming proportions in the emergency rooms of many urban hospitals in the United States. A recent study showed that up to 15 percent of emergency room patients give up and leave before receiving care because of long waits—some stretching to an abysmal fifteen hours (*USA Today*, 1991).

Symbolic Interactionism and the Role of Perceptions

Symbolic interactionists hold that illness is partly (though obviously not totally) socially constructed. The definitions of illness and wellness are culturally relative—sickness in one culture may be wellness in another. It is time dependent as well. A condition considered optimal in one era (such as being thin) may be defined as sickness at another time in the same culture (at the turn of the century, a healthy woman was supposed to be plump). Similarly, the health care system itself has a socially constructed aspect. The ways we behave toward the ill, toward doctors, and toward innovative ventures like HMOs are all social creations.

The symbolic interaction perspective highlights a number of socially constructed problems in the health care system. Medical practitioners frequently subject patients to *infantilization* (treating them like children, even if adult; talking to them with "baby talk"). The patient is assigned a role that is heavily dependent on the physician and the health care system, much as an infant is dependent on its parents. Doctors and nurses may begin patronizing the patient from the initial greeting, a condescending "How are *we* today?" Physicians commonly address patients by their first names, yet patients virtually always address physicians as "Doctor." Women patients are even more likely than men to be addressed by their first names. Such patronizing and infantilizing of patients is common in emergency rooms, where minority patients are infantilized even more often than others (González, 1996).

The symbolic interactionist analysis of the health care system allows us to see such problems more clearly. One solution to these problems is to give health care professionals training about such matters in medical school. This is just starting to happen in a few of U.S. medical schools. For example, a social issue addressed in some medical school courses on gynecology is how to manage the interaction when a male gynecologist treats his woman patient. Women patients feel uncomfortably vulnerable when they lie partially naked on an examination table with their heels in

TABLE 19.2 *THEORETICAL PERSPECTIVES ON THE SOCIOLOGY OF HEALTH*

	Functionalism	Conflict Theory	Symbolic Interaction
Central point	the health care system has certain functions, both positive and negative.	health care reflects the inequalities in society.	illness is partly socially constructed.
Fundamental problem uncovered	the health care system produces some negative functions.	excessive bureaucratization of the health care system and privatization lead to excess cost.	patients are patronized and infantilized.
Policy implications	policy should decrease negative functions of health care system for minority groups, the poor, and women.	policy should improve access to health care for minority racial–ethnic groups, the poor, and women.	doctors, nurses, and other medical personnel should periodically take the (sick) role of the patient, as an instructional device.

elevated stirrups and their legs open. Furthermore, strongly fixed social norms say that when a man touches a woman's genitals, it is an act of intimacy. Yet the gynecological examination is supposed to be completely impersonal. Male gynecologists are notorious for their failure to appreciate the discomfort of their female patients (Emerson, 1970). Doctors often become similarly inured to the grief of family members and patients who have just received grim news, and to patients who are in pain. Obliviousness to pain can reach the point that doctors fail to prescribe adequately for pain that could easily be quenched with routine analgesics.

Health and Sickness in America: A Picture of Diversity

The definitions of sick and well have varied greatly over time in the United States. For example, from the early 1900s until the mid-1940s, thinness was associated with poverty and hunger. If you were skinny, that meant you were in bad health. From the late 1950s through the present, a positive value has been placed on being thin. Female role models in our society, such as movie stars and fashion models, have firmly established that thin is "in." Millions of young women have attempted to copy this look, and an increased incidence of *anorexia nervosa* has been one result (Williams and Collins, 1995; Taylor et al., 1997).

Anorexia nervosa, or anorexia for short, is an eating disorder characterized by compulsive dieting. Victims of this illness starve themselves, sometimes to death, even though sufferers do not typically define themselves as ill since they tend to see themselves as overweight, even though they are dangerously thin. A related malady, **bulimia,** is an eating disorder characterized by alternate binge eating and then purging or induced vomiting in order to lose weight. Like many other diseases, anorexia has social as well as biological causes. A majority of those suffering from the disease are young, White women, from well-to-do families, most often two-parent families. Many behavioral scientists have noted that anorexics have generally been pressured excessively by their parents to be high achievers. Others have detected a link between anorexia and the socially constructed ideals of beauty in our society, which are fixated on thinness and so-called ideal body types. Images of bodily "perfection" are emblazoned across television, magazines, and billboards. Slenderness is displayed as the ideal of femininity. Researchers note that these social values, which encourage compulsive dieting, are comparable to the foot-binding once practiced in China (see Chapter 1) and other forms of female mutilation found in some foreign cultures (Wolf, 1991; Levine, 1987; Chernin, 1991).

Anorexia nervosa is common mostly, but not exclusively, among young women. Those with anorexia falsely believe that they are overweight, although they are exceedingly thin. Many interpret this potentially life-threatening condition as encouraged by a culture that advertises being thin as a beauty ideal for women.

Anorexia is less likely to afflict African American women, Latinos, and lesbians. According to one researcher (Thompson, 1994), many in these groups overeat, rather than self-starve, something Thompson interprets as a reaction to oppressive life experiences associated with racism, sexism, and homophobia.

Men have not been exempt from the pressure of such values. Since the 1940s and especially from the mid-1970s on, one of the most persistent male physical ideals has been the rippling physique of the body builder or weight lifter. Young men have been urged by the media and their peers to "pump iron" for the perfect body, thus presumably to gain for themselves pride, muscle mass, and the adoration of women. As they are for women, attempts to conform to an ideal body image are then associated with problematic health behaviors, patterns that are strongly affected by race and gender since different race and gender groups manifest health problems associated with body image in different ways (Logio-Rau, 1998).

Many athletes, professional and amateur, have been goaded by athletic dreams to use **anabolic steroids,** powerful hormones that stimulate the growth of muscle. Used widely (despite dire warnings by physicians), steroids not only build muscle as advertised, they can also shrivel the testicles and cause impotence, hair loss, heart arrhythmia, liver damage, strokes, and very possibly some forms of cancer. Lyle Alzado was a huge, fast, monstrously strong football star in the NFL and later a Hollywood actor. As he slowly succumbed to brain cancer in his thirties, Alzado freely admitted to massive abuse of anabolic steroids, which he believed caused his fatal disease.

Health and sickness in the United States is influenced by social factors. **Epidemiology** is the study of all the factors—biological, social, economic, and cultural—that are associated with disease in society. **Social epidemiology** is the study of the effects of social, cultural, temporal, and regional factors in disease and health. Among the more important social factors that affect disease and health in America are race-ethnicity, social class, gender, and age. The effects of age on health and life expectancy were discussed in Chapter 13. The effects of race, class, and gender are discussed here.

Race and Health Care

Health can be affected by personal factors, such as dietary and hygienic habits, and by institutional factors, such as the structure of the health care system and the economic health of the society's less advantaged groups. Most of the factors that affect health detrimentally at either the personal or institutional level are likely to have a worse influence on the health of minorities. This reality is reflected in the dramatic differences in life expectancies for White Americans compared to other groups and in differences in life expectancy between men and women. White men can now expect to live to 72.7 years of age, whereas African American men have a life expectancy of only 64.8 years. Hispanic men have a life expectancy somewhat higher, at 69.6 years, and Native American men, 67.2 years, still significantly less than White men. White women can expect to live longer than White men, 79.2 years. African American women can expect to live almost ten years longer than African American men, 73.5 years, but noticeably less than White women. Hispanic women live to 77.1 years—longer than White men, but still less than White women. Native American women live to 76.2 years of age (see Figure 19.1).

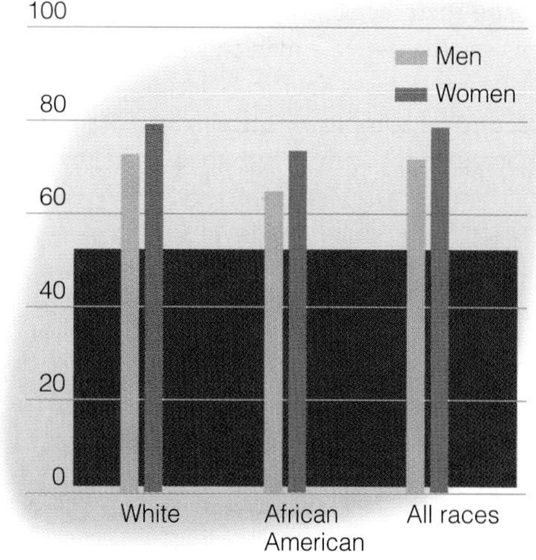

FIGURE 19.1 Life Expectancy by Race, Ethnicity, and Gender[a]

[a]Data not available for Hispanics and Native Americans.

SOURCE: National Center for Health Statistics. (1998). *Health United States 1998.* Washington, DC: U.S. Department of Health and Human Services, p. 206.

Among women, African Americans are more likely than Whites to fall victim to diseases such as cancer, heart disease, stroke, and diabetes. Death of the mother during childbirth is three times higher among African American women than among White women, and African American women are three times more likely to die while pregnant (Williams and Collins, 1995; Livingston, 1994). In the forty-five to sixty-four age group, African American women die at twice the rate of White women. The occurrence of breast cancer is lower among African American women than White women, yet the *mortality rate* (death rate) for breast cancer for African American women is considerably greater than for White women (National Cancer Institute, 1997). This reflects the fact that White women are more likely to get high-quality care, and get it more rapidly, than African American women. African American women also develop cervical cancer at twice the rate of White women. Twenty-five percent of all African American women suffer from hypertension (high blood pressure), which afflicts only 11 percent of White women. Although differences in culture, diet, and lifestyle account for some of these differences, it is nevertheless clear that African American women and men do not receive medical attention as early as Whites, and when they do eventually get treatment, their illness is further along on its course, and the treatment they receive is not of the same quality (Williams and Collins, 1995).

The mortality rates for Native Americans are one and a half times greater than the mortality rates for the general population. Among Native Americans under the age of forty-five, the death rate is three times that of Whites. Native American babies are almost twice as likely as White babies to die before they are one year old. Overall, Native Americans are in very poor health compared to the general population of the United States (National Cancer Institute, 1997; Williams and Collins, 1995).

Hispanics, like African Americans, Native Americans, and other minorities, are significantly less healthy than Whites. For example, Hispanics contract tuberculosis at a rate four times that of Whites. The other indicators of health, such as infant mortality, reveal a picture for Hispanics similar to that of African Americans and Native Americans. Hispanics are less likely than Whites to have a regular source of medical care, and when they do, it is likely to be a public health facility or an outpatient clinic. Because of language barriers as well as other cultural differences, Hispanics are even less likely than other minority groups to use available health services, such as hospitals, doctors' offices, and clinics (Linvingston, 1994; Vega and Amero, 1994; Altman, 1991).

One of the most infamous studies to demonstrate, though inadvertently, the connection between race and the treatment of illness in our society is the **Tuskegee Syphilis Study.** The study was done at the Tuskegee Institute in Macon County, Georgia, a historically Black college. Begun by the United States Health Service in 1932 as a study of "untreated syphilis in the male negro," the study intended to advance scientific knowledge about syphilis, a sexually transmitted disease (STD) that when left untreated causes blindness, mental deterioration, and death. A group (sample) of about 600 African American males, about 400 of whom were affected with syphilis, were chosen for study, the remainder being used as an uninfected control group. The study continued for forty years.

During this period, penicillin was discovered as an affective treatment for infectious diseases and was widely available from the early 1950s. Nonetheless, the scientists conducting the study decided not to give penicillin to the African Americans who made up the study sample on the grounds that it would "interfere" with the study of the physical progress of untreated syphilis. The U.S. government itself authorized the study to be continued right up through the 1960s and into the early 1970s. By 1972, news of the study

As many people have become aware of the importance of good health, a major industry has developed to provide assorted health services.

began to reach universities and the general public, and the government, through the department of Health, Education and Welfare (now the Department of Health and Human Services), put an end to the project. By then between 30 and 100 of the African American men in the study had died of syphilis, after having undergone its horrible physical ravages (Jones, 1997). In 1997, President William Clinton formally apologized to the victims, dead and alive, and to their families, sixty-five years after the start of the study.

The Tuskegee study has become one of the prime examples of ethical violation in medical and behavioral science research (Chapter 2). It has also served as an eye into race relations for more than a half century in the United States. Penicillin was never administered to the infected men for the duration of the study. It seems reasonable to hypothesize that if the individuals in the study had been middle-class Whites instead of poor African Americans, the study would have been put to an end as soon as penicillin, a cure for syphilis, was found.

Social Class and Health Care

In the United States, social class has a pronounced effect on health and the availability of health services. This reality is crucial to the current national debates on access to health care and health insurance; the lower the social class status of the person or family, the less access they have to adequate health care. People with higher incomes asked to rate their own health tend to rate themselves higher than people with lesser incomes. Almost 50 percent of those in the highest income bracket rate their health as excellent, whereas only 25 percent in the lowest income bracket do so. In fact, the effects of social class are nowhere more evident than in the distribution of health and disease, showing up dramatically in the rates of infant mortality, stillbirths, tuberculosis, heart disease, cancer, arthritis, diabetes, and a variety of other illnesses. The reasons lie partly in personal habits that are themselves partly dependent on one's social class; for example, those of lower socioeconomic status smoke more, and smoking is the major cause of lung cancer and an important contributor to cardiovascular disease.

Social circumstances also have an effect. Poor living conditions, elevated levels of pollution in low-income neighborhoods, and lack of access to health care facilities all contribute to the high rate of disease among lower classes. Another contributing factor is the stress caused by financial troubles. Research has consistently shown correlations between psychological stress and physical illness (Taylor et al., 1997; Jackson, 1992; Thoits, 1991; House, 1980). The poor are more subject to psychological stress than the middle and upper classes, and it shows up in their comparatively high level of illness.

The average number of contacts with a physician differs greatly across class strata. The lower one's social class, the less likely one will see a physician when ill. Poor individuals, especially poor Blacks and Hispanics, are admitted to hospitals considerably less frequently than more affluent people. When they are eventually admitted, they stay for longer periods, most likely because their illness has been allowed to progress untreated. In general, the poor get treated later in the course of an illness, and when seen, require proportionately more treatment (Williams and Collins, 1995; Easterbrook, 1987).

With the exception of the elderly, now subsidized by Medicare, low-income people remain largely outside the mainstream of private health care. Nearly 31 million Americans—13 percent of the population—have no health insurance at all (see "Mapping America's Diversity: Persons Not Covered by Health Insurance"; U.S. Bureau of the Census, 1997). The main sources for medical care for many of them are hospital emergency rooms and health department clinics, resulting in vastly overcrowded emergency rooms in the inner cities, often called the "doctor's office of the poor." Even then, emergency room service is lacking. Treatment is given only for specific critical ailments, and rarely is there any follow-up care or comprehensive treatment.

Sociologists have found that during interactions between health care providers and poor patients, the poor, particularly Black and Hispanic poor, are more likely to be infantilized and to receive health counseling that is incorrect, incomplete, or delivered in inappropriate language not likely to be understood by the patient (González, 1996). The symbolic interaction perspective in sociology has noted this tendency; it has also sometimes been attributed to an attitude among health counselors that the poor are charity cases who should be satisfied with whatever they get since they are probably not paying for their own care (Diaz-Duque, 1989). Before Medicaid was created, hospital workers were frequently called upon to give free care to poor patients who might urgently need medical attention but had no way to pay for it. With the advent of Medicaid coverage and the assurance of payment for services rendered to poor patients, it was thought that the cold and infantilizing atti-

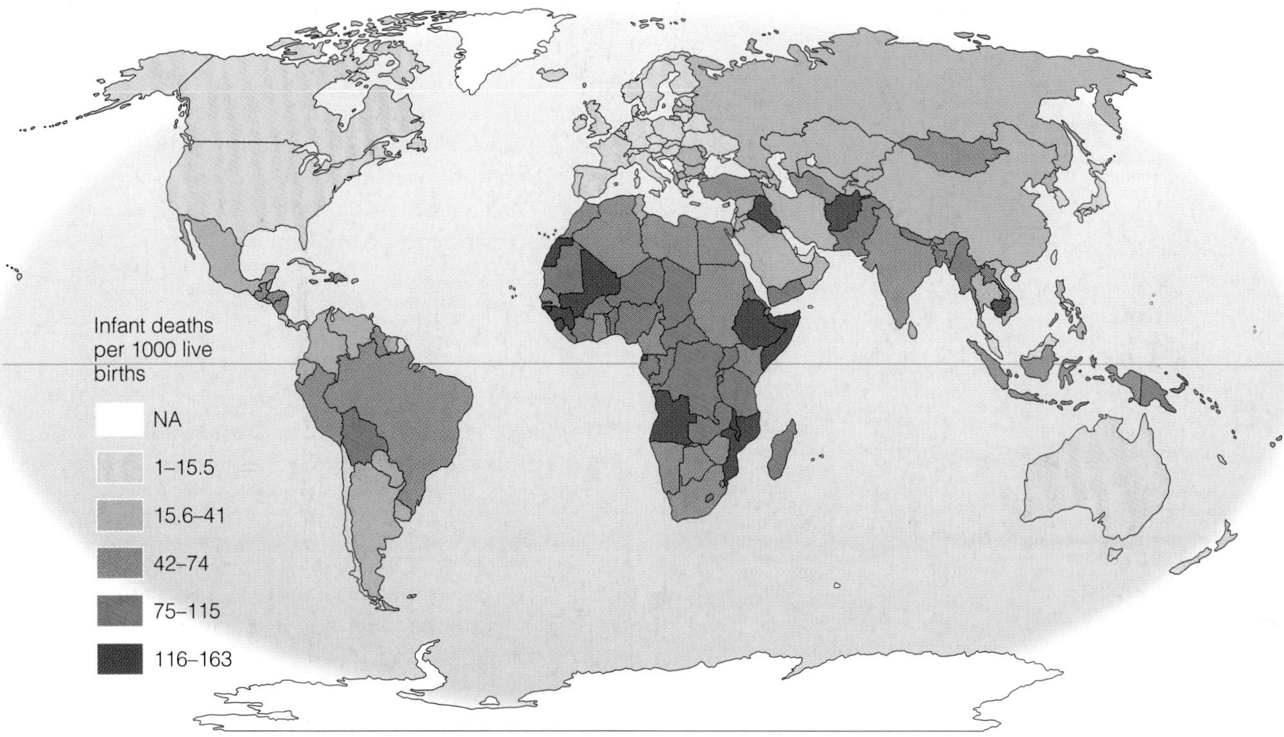

MAP 19.1 Mapping America's Diversity: Persons Not Covered by Health Insurance

DATA: From the Population Reference Bureau. 1996. *1996 World Population Data Sheet.* Washington, DC http:www.pub.org

tudes of health care professionals would decrease since the poor were no longer a financial drain on the hospital system. Unhappily, the attitudes have not changed much.

Like many other parts of the health care system in the United States, the Medicaid program is imperfect. It covers less than two-thirds of the country's poor, and even then only under an erratic system of state formulas for matching state and federal grants. Administered locally, governed by few uniform federal standards, Medicaid programs vary greatly in quality and in who is defined as eligible. In some states, payments to hospitals are adequate to cover the cost of care; in other states, they are not.

Gender and Health Care

Even though women live longer on average than men, older women are more likely than older men to suffer from stress, overweight, hypertension, and chronic illness. National health statistics show that hypertension is more common among men than women until age fifty-five, when the pattern reverses. This may reflect differences in the social environment experienced by men and women, with women finding their situation to be more stressful as they advance toward old age. Under the age of thirty-five, men are more likely to be overweight than women; after that, women are more likely to be overweight. Women have a higher likelihood of contracting chronic disease than men, although men are more likely to be disabled by disease (Williams and Collins, 1995; Adler et al., 1993).

The rate of death caused by infectious diseases is higher for men than women, and has declined significantly for both since the early 1900s. Researchers cite differences in male and female roles and cultural practices to explain the difference between the genders. In particular, male occupational roles call for more travel and more exposure to other people, the major sources of infection. In addition, men smoke more, and the overall deleterious effect of smoking on health can account for some of the differences between men and women (National Cancer Institute, 1997; Williams and Collins, 1995).

Despite what tobacco executives often argue, cigarette smoking definitely causes cancer and cardiovascular disease, which kill men more frequently than women. Smoking patterns have changed, however, and a higher proportion of smokers are now women. The most recent research shows a convergence in the smoking rates of men and women, and shows further that more men than women quit smoking. These changes are expected to appear in the future as higher rates of female deaths from diseases linked to smoking. Reports also show significant increases in smoking among teenagers, with teenage girls leading boys. As a result, we might expect to see future increases in lung cancers and cardiovascular disease among women (National Cancer Institute, 1997).

Some 450,000 people die each year as a direct result of smoking. This is greater than the *combined* death toll from AIDS; automobile accidents; homicide; suicide; and drugs such as alcohol, cocaine, and heroin. Nonsmokers who are

The now well-known "Joe Camel" advertising campaign has been blamed for encouraging an increase in cigarette smoking among young people. As the anti-smoking movement has grown in the United States, the tobacco industry has targeted groups in the developing world for new advertisement campaigns, such as here in Papau, New Guinea.

exposed to secondhand cigarette smoke have a higher risk of smoking-related disease, including death, than nonsmokers who are not so exposed. In fact, a study of more than 32,000 healthy women who never smoked found that regular exposure to other people's smoking actually doubled the risk of heart disease (Grady, 1997).

Smoking and the tobacco industry have become a major national issue. When it came to light that tobacco companies were directing smoking campaigns at children, by means of commercials and billboards (like the "Joe Camel" campaign), federal legislation was passed forbidding such advertising. Especially targeted were ad campaigns that carried special appeal to women and girls, and ad campaigns centered in urban areas where there are large numbers of minorities. Furthermore, state laws mandating a smoke-free environment, such as in restaurants and waiting rooms, have been enacted on a large scale, including the entire state of California. The effect has been to lower the overall rates of smoking in the United States, although cigar smoking has shown a recent increase. As a result of the antismoking trend in the United States, tobacco companies are advertising more and selling more in other countries—particularly in third-world countries with low average incomes—countries that have little governmental regulation of tobacco advertising and sales.

It is typically assumed that the work-oriented, hard-driving lifestyle associated with men's traditional role tends to produce elevated levels of heart disease and other health problems. In general, this is true; however, the role of women in society is changing, with associated changes in women's health. The health of women and men varies with their social circumstances. People who are "tokens" in the workplace—meaning mainly women and Blacks, especially Black women—suffer more stress in the form of depression and anxiety than nontokens, or women and Blacks who work in places where there is nothing exceptional about their presence (Jackson et al., 1995). Housewives have higher rates of illness than women who work outside the home, indicating that employment can have positive effects on women's health; moreover, housewives who have never worked outside the home are healthier than those who have once worked outside the home but then returned to homemaking. Employed women are more likely than housewives to get well quickly when sick, and they tend to return to their normal activities sooner after illness than do housewives (Andersen, 1997).

Women tend to use the health care system more often than men, as measured by physician visits and hospital admissions. When a woman seeks medical care from a physician, there is about an 85 percent chance that the doctor she sees will be male. As a consequence of the gender imbalance among physicians, women's health problems are often regarded as nonusual—men's problems are workaday medicine, women's problems are special cases. There is an often noted tendency in medical textbooks to treat women's health issues as out of the ordinary. This in turn has been linked to the tendency for male physicians to blame disorders such as premenstrual stress syndrome (PMS) and hypo-

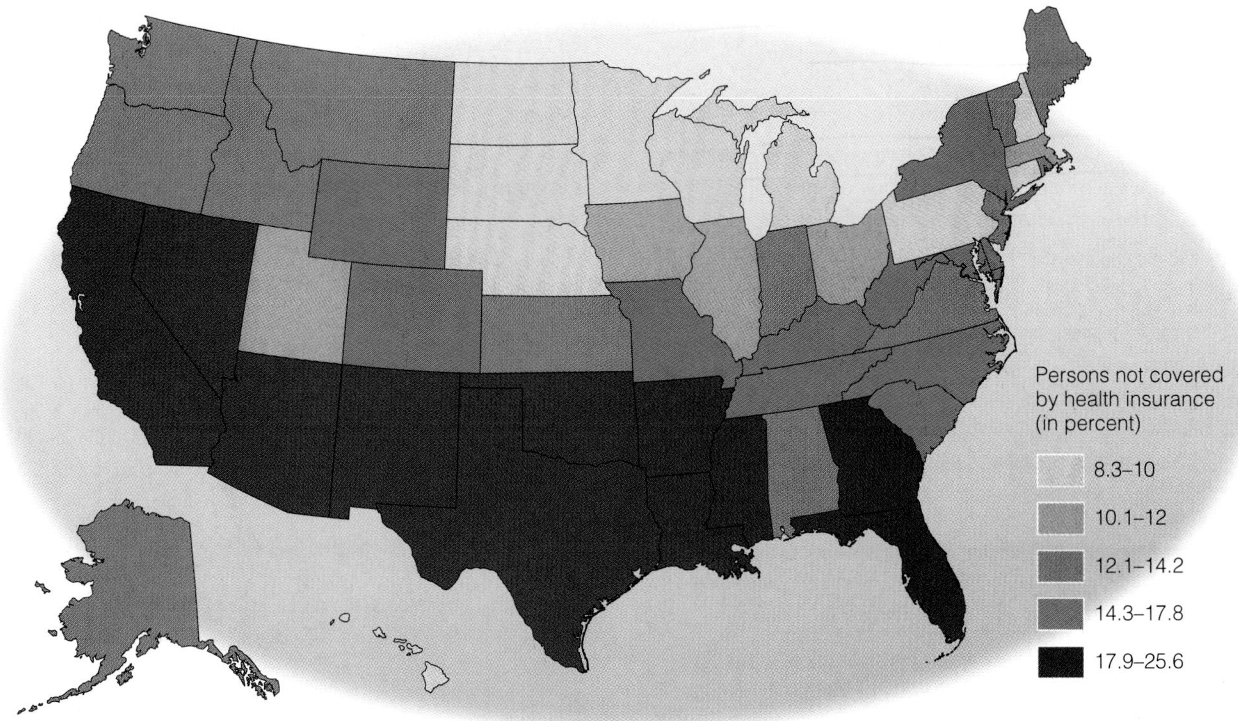

MAP 19.2 Viewing Society in Global Perspective: World Infant Mortality

DATA: From the U.S. Bureau of Census. 1997. *Statistical Abstract of the United States 1997.* Washington, DC: U.S. Government Printing Office, p. 832.

glycemia on neurosis or strictly psychosomatic sources since male physicians often, like the public in general, stereotype women as overly neurotic. Recent thinking, particularly among feminist researchers, suggests that biases like these in the male-dominated medical profession have resulted in a tendency to overlook *physiological* causes for such illnesses in women (Andersen, 1997).

Debunking Society's Myths

Myth: Sickness is the result of personal habits.

Sociological perspective: The likelihood of illness is highly dependent on social factors, including race, gender, class, age, and region of residence.

Health and Sickness Globally

In societies that are largely pastoral or horticultural, the lack of technology severely limits both the development of health care systems and the delivery of health care to the populations of these countries. Such is the case in many third-world countries, where poverty also limits access to good health care. As in the United States, elites in such countries receive better medical attention than those who are poor and possibly even near starvation.

Abject poverty in much of the world, particularly in third-world regions and areas of extreme poverty in the United States, cuts life expectancy to far below the seventy-some years experienced by the wealthy, elite, and middle-class populations in the United States and elsewhere. Life expectancy in the African countries of Niger, Chad, Ethopia, and Somalia is barely forty-five years of age. It is the same in Afghanistan and in Laos and Cambodia. Serious illness goes hand-in-hand with severe poverty, and lowered life expectancy is the result. Such is also the case in many of the ghettoes and barrios in the United States, although considerably less is written about such conditions. Drinking water that is contaminated by poor sewage breeds many of the infectious diseases that kill, and the diseases of tuberculosis and pneumonia, diseases that were the leading causes of death in the United States nearly a century ago, today kill people in poor areas of the world (Lenski et al., 1998).

Doctors and other medical personnel are rare in impoverished areas of the world, and thus the world's poorest people, those in dire need of medical care, may never be seen by a physician. There are fewer than 5 doctors per 100,000 people in African countries such as Niger, Chad, and Ethiopia; Guyana in South America; and Guatemala in Latin America. As a result, infant mortality is high, and 10 percent of the children in such poor societies die within one year of birth. Health care and poverty are intimately related: Poverty breeds disease, which makes it less likely for people to be able to work, which in turn produces more poverty and more disease and premature death. The absence of health care professionals and facilities only exacerbates this vicious cycle of poverty and disease.

Attitudes Toward Health and Illness

In earlier times, illness was believed to spring from causes that were uncontrollable. Genes, germs, or bad luck were to blame, and there was nothing to be done about it. More recently, individuals are increasingly seen to be responsible for their own well-being. Individuals are seen to have some measure of control over the prevention of illness or recovery from bad health. One outcome of this change is that whereas people stricken by disease used to be regarded as victims, now there is a greater likelihood that these victims will be seen as contributing to their own illness. The phenomenon of *blaming the victim* is particularly common when the individual's illness can be traced to activities such as smoking, drinking, overeating, unprotected sex, or other behaviors perceived by the public to be controllable (Taylor and Aspinwall, 1990; Taylor, et al., 1997). As a result, many people in the United States have begun to pay more attention to their health. The recent growth in the number of health clubs and gyms is one indication of this.

Sexually Transmitted Diseases (STDs)

Sexual practices and attitudes are also relevant to health. The general attitude in the United States toward sex has always been one of ambivalence. Sex is endlessly discussed, and splashed throughout magazines, advertisement, and other media; yet we tend to think of sex as somehow sinful or wrong. To be sure, attitudes toward sex have become steadily more liberal since the turn of the century, particularly during the "sexual revolution" of the late 1960s and early 1970s. Nonetheless, people tend to regard venereal disease, or sexually transmitted diseases (STDs), as not merely diseases but as punishment for being immoral. It represents yet another instance of blaming the victim, of attributing blame to the ill person for an illness seen as controllable and thus the responsibility of the person who is ill. Individuals who contract an STD become negatively stigmatized in our culture.

Approximately fifty sexually transmitted diseases have been diagnosed medically. The four major ones are syphilis, gonorrhea, genital herpes, and AIDS (acquired immune deficiency syndrome). Less frequent are the more esoteric diseases such as lymphogranuloma venerium (LGV), which devastates the body with open sores and for which there is no known cure. The incidence of all STDs increased during the late 1960s and early 1970s, when sex was thought of simply as a pleasant way to express affection. Little fear was harbored about contracting a venereal disease since most, such as syphilis or gonorrhea, were known to be medically curable, and the remaining diseases (such as LGV) were thought of as too rare to worry about. With the subsequent dramatic rise in STDs, especially in AIDS, the late 1980s and early 1990s witnessed a new revolution contrary to the sexual revolution of the 1960s. The new counterrevolution caused people to reevaluate the nature of sex, the risks involved in unprotected sexual activity, and their behavior in general (Laumann, et al., 1994; Alan Guttmacher Institute, 1994).

Syphilis and gonorrhea have been around for a long time. Both are caused by microorganisms that are transmitted through sexual contact involving the mucous membranes of the body. It is virtually impossible to get either syphilis or gonorrhea any other way. If untreated, syphilis causes damage to major body organs, blindness, mental deterioration, and death. Each of these results was seen all too dramatically in the infamous Tuskegee "experiment" done on a group of Black men earlier in this century, as we discussed. Untreated gonorrhea can cause sterility in both women and men. Nearly one million cases of both are reported each year (Centers for Disease Control and Prevention, 1995). Both diseases are quickly curable with penicillin, or other appropriate antibiotic medication.

Genital herpes (Herpes Simplex II) is more widespread than either syphilis or gonorrhea and affects roughly thirty million people in the United States alone. That represents one person in seven. Genital herpes began to get a lot of attention in the early 1980s, when concern with the risks of STDs generally was on the rise. A person with genital herpes may have no symptoms at all, or the person may experience blisters in the genital area and a fever as well. It is not fatal to adults, but it can be fatal to infants through vaginal delivery, but not via cesarean operation. Although it can remain dormant for the life of the infected individual, it is nonetheless to date incurable.

AIDS: Illness and Stigma

Given the tendency for people to attribute blame to the victim of illness, a certain amount of negative stigma will befall the person who is sick. Some degree of stigma may become associated with any disease or illness, but attitudes

The hospice movement has provided a new form of health care, whereby terminally ill patients, such as this AIDS patient, face death in their homes, rather than in the clinical settings of hospitals.

toward people suffering from sexually transmitted diseases, particularly AIDS represents one of the strongest examples of how stigma operates in our society.

Stigma occurs when an individual is socially devalued because of some malady, illness, misfortune, or similar attribute (see Chapter 8). A stigma is viewed as a relatively permanent characteristic of the stigmatized individual; in the regard of others, the negative attribute of the stigmatized person is expected to persist, with no cure in sight (Goffman, 1963; Jones et al., 1986). In earlier times, the disease of leprosy put a stigma on the victim. Leprosy was perceived to be incurable, and the stigma was permanent. Those with the disease were banished to leper colonies and other far-off places. Today, leprosy continues to carry a negative social stigma, although it is now treatable and individuals are no longer cast away as they once were.

In the early 1980s, AIDS appeared. Like lepers, people infected with HIV (human immunodeficiency virus) carry heavy negative stigma and are often regarded with fear and misunderstanding. Patients who are HIV positive but not yet suffering from full-blown AIDS are sometimes so stigmatized that even physicians refuse to see them (Williams and Collins, 1995).

AIDS is the term for a category of disorders that result from a breakdown of the body's immune system. HIV, the virus that causes AIDS, was first identified in 1981. The incubation period between infection with HIV and the development of AIDS can stretch longer than ten years; thus, one can be infected with HIV and yet not have full-blown AIDS. It is not HIV infection itself that causes the death of the individual but a complex of severe illnesses that thrive in the absence of a working immune system, such as pneumonia, certain cancers, and a number of other illnesses rare enough that their presence is judged to be diagnostic of AIDS. Since the 1980s, the disease has been spreading rapidly. An estimated 21.8 million adults and children worldwide are infected with HIV. The highest concentrations of HIV-positive cases are in Sub-Saharan African (14 million cases) and Southeast Asia (4.8 million cases). America and Latin America are next with 780,000 cases and 1.3 million cases, respectively.

According to the authoritative *Semi-Annual HIV Surveillance Report* released by the Centers for Disease Control, as of the end of 1995, there had been more than 510,000 reported full-blown AIDS cases in the United States, including 71,000 women, and nearly 320,000 of these (63 percent) have now died of AIDS. It is estimated that just about one million Americans now carry the HIV virus, with extreme estimates reaching as high as several times that number. The AIDS epidemic currently shows signs of containment, but some fear that it has the potential to rage as virulently as the bubonic plague of the fourteenth century, which wiped out nearly one-fourth of the population of Europe (Centers for Disease Control and Prevention, 1995).

THINKING SOCIOLOGICALLY

Although you are probably not HIV positive or an AIDS sufferer, it is now true that almost everyone either knows at least one person who is HIV positive or has AIDS, or knows someone who does. Do you know any such person? Do you know someone who knows such a person? How do you feel about that person? Do you wonder whether the person got that way through unprotected sex or perhaps some other means such as intravenous drug use? Do you blame the person? What do your observations tell you about *stigma*?

In the United States, before 1995, AIDS was in higher proportion among Whites than among minorities. In 1995, for the first time, the proportion of persons reported with AIDS who are African American was equal to the proportion who are White (40 percent each, among those with AIDS). The AIDS incidence rate per 100,000 among African Americans (93 cases per 100,000 population) was higher than that among Hispanics (46.2 per 100,000), Whites (15.4 per 100,00), Native Americans and Alaskan Natives (12 cases per 100,000), and Asians, for whom it was the lowest (6.2 cases per 100,000 population). These data, collected from medical records, do not include measures of social class such as education, occupation, or income.

Death rates by AIDS present a picture similar to the incidence of AIDS. The death rate from AIDS is highest among African Americans; next highest among Hispanics; next among Whites; and lowest among Native Americans, Alaskan Natives, and Asians.

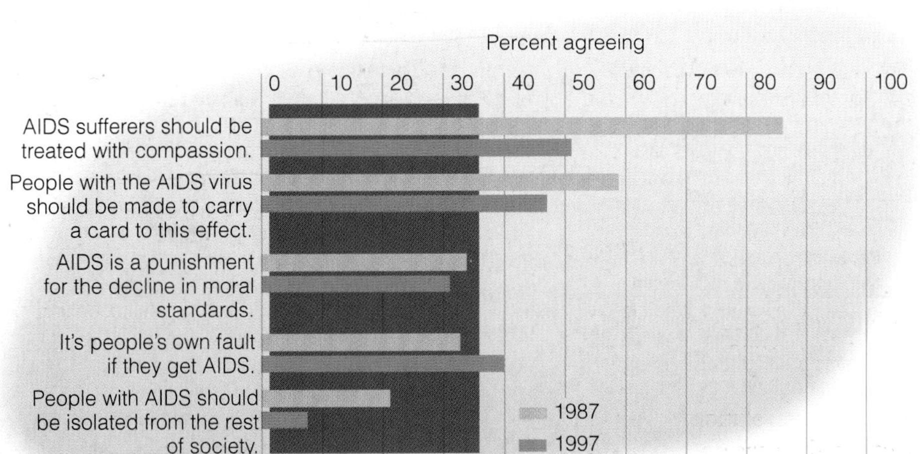

FIGURE 19.2 Changing Attitudes About AIDS

DATA: From the Roper Organization. 1997. *The Gallup Poll*. Storrs, CT: Roper Organization, pp. 166–169.

The AIDS disease is transmitted through the exchange of bodily fluids, particularly blood and semen. A little more than half (51 percent) of all cases of full-blown AIDS are the result of male-to-male sexual contact, followed by injecting drug use (24 percent). Women account for 19 percent of AIDS cases. Most women acquire HIV infection through injecting drug use, or sexual contact with HIV-infected men. There are no reliable data on AIDS from female-to-female sexual contact.

Much of the negative stigma carried by AIDS victims is due to beliefs about who gets AIDS and how the HIV virus is transmitted. In its early years, AIDS appeared to be confined to the gay community, and some believed that getting AIDS was just retribution for deviant activity. Most Americans now realize there are a number of nondeviant ways that one can acquire AIDS, such as receiving infected blood during a transfusion. Tennis great Arthur Ashe, who died in 1993, got AIDS this way after receiving a transfusion during heart surgery. (Current methods for preparing safe blood products are now considered trustworthy.) As a consequence, some sympathy has been generated for AIDS sufferers (see Figure 19.2), but for the average AIDS victim, the disease continues to carry with it a strong negative stigma and stereotype.

Chronic Fatigue Syndrome

Chronic fatigue syndrome (CFS), at times flippantly referred to as the "yuppie flu," is another affliction that carries a negative stigma. CFS is a persistent, flulike illness that can

BOX SOCIOLOGY IN PRACTICE

19.2 The Sociology of AIDS

SOCIOLOGISTS testified before the Institute of Medicine in 1993, as part of the National Academy of Sciences deliberations about AIDS policy. The focus of this testimony was to review research on the social dimensions of AIDS so that federal policy would not see AIDS simply in medical terms. Here is an excerpt from that testimony, which shows the unique contribution of sociological research to the development of social policy:

> ***Contributions of Sociological Research on AIDS.*** Even with limited funding, sociological research has enhanced our knowledge about the social dimensions of AIDS. With our focus on social groups, social processes, and social dynamics, sociologists have made substantial contributions to the AIDS epidemic by applying theory and research on social demography and cultural context, social networks, social norms, case management, social epidemiology, and social change. The following examples show the breadth of this work:
>
> ***Social Demographic and Cultural Context.*** In the United States, AIDS has hit inner-city minority communities disproportionately hard. These Americans have experienced HIV/AIDS in the context of poverty, poor health, inadequate health care, drug and alcohol abuse, and violence. These contextual problems must be taken into account to reduce infection rates in disadvantaged communities. In addition, social norms affect the age of first intercourse, number of sexual partners, drug use, and homosexual and bisexual sexual practices. For example, based on National Opinion Research Center data, only 0.6% of American adult males reported sex with a prostitute in the past year and 80% had 0 or 1 sex partner in the last year. However, in the remaining 20% there are small subgroups of individuals, the bulk unmarried, who have multiple sexual partners and engage in risky behavior. Even drug use varies by race, social groups, and class and points to different interventions for different communities.
>
> ***Social Networks.*** Understanding social networks is critical to modeling the spread of AIDS and targeting resources to groups with high risk of infection. Since HIV/AIDS is primarily transmitted as a result of social activities such as sexual relations and the sharing of needles used for drug injections, the unit of analysis ought to be the *social networks and relationships* in which transmission occurs, not the individuals who have the infection. Networks may be so dense and interconnected, for example, that members interact only with each other as in a well-defined homosexual or drug-injecting community. Or, the network may be diffuse and loosely tied together as with prostitutes and their occasional clients, or young adults that date casually. The identification and characterization of these networks is critical for predicting the spread of infection, focusing and shaping interventions, and anchoring them in the communities where people are engaging in risky behaviors. It was precisely this network analysis by sociologists at the CDC in the early 1980s that produced epidemiological mapping of gay men and helped target safe sex education to the most at risk subgroups within that community.
>
> ***Social Norms.*** Knowledge about safe sex or drug use may be overridden by social pressure to engage in risky behavior. Prostitutes, for example, often use condoms with clients but not with their boyfriends. American youth have considerable knowledge about HIV transmission but engage in unprotected sex because of group norms. Drug use, too, typically occurs in groups with strong normative systems of permissible and impermissible behavior. Even medical personnel have been found to ignore health precautions if everyone else on their team does so.
>
> Sociological research examining patterns of IV drug-users and the organization of drug networks illustrates the importance of social norms in prevention, education, and intervention efforts. For example, targeting new IV drug-users is a critical intervention point. Experienced IV drug-users increasingly use precautions to reduce their risk of HIV, but new users do not. The data indicate that IV drug users reach out to other IV drug users, and that risk behavior is influenced heavily by group pressure. Drug injectors who took part in organizing other drug injectors *decreased* their injections per month, the sharing of equipment, their use of rented syringes, and their use of shooting galleries and *increased* their use of new needles, bleach, and condoms.

SOURCE: American Sociological Association, Washington, DC.

plague its sufferer for years. Victims of CFS suffer from extreme fatigue, forgetfulness, sore throat, fever, and other symptoms (Schwager, 1994; Homes, 1988). One reason this disease is treated unsympathetically is that its very existence was disputed for a number of years. Someone who complains constantly that he or she feels unwell but cannot be diagnosed with a well-defined illness is likely to be labeled a "fake" or a hypochondriac, especially if the complainer is a woman. The affliction is seen by the public as controllable; hence the afflicted individual is seen as bearing the blame for the affliction. Women are more likely than men to be the recipient of such labels.

Medical researchers and physicians have had great difficulty isolating an organic cause for CFS (Schwager, 1994). Without a known cause or some physiological marker that demonstrates the presence of the disease, such as antibodies against a bacterium or virus, physicians are not even able to determine conclusively if the disease is actually physiologically present. As a result, researchers to date have only sketchy information about how long the illness lasts, how severe the symptoms can be, and what populations are most prone to contract the illness. The definition of CFS put forth by the Centers for Disease Control consists of a highly restrictive set of characteristics that must be met and is intended more as a guide for researchers than physicians. The restrictions of these definitions meticulously exclude all but the most unambiguous cases, which is hardly suitable for physicians required to make diagnoses based on hazy symptoms.

Compounding the difficulties of people suffering from CFS is the fact that they are frequently denied the special consideration accorded to persons who are sick. As noted by functionalist theory, people suffering from serious illness gain social benefits when they take on the **sick role,** the patterns of behavior defined as appropriate in society for people who are officially declared ill. If you have been designated as "sick" by an appropriate societal institution, such as the medical profession, then you become eligible for the benefits of the sick role. Like any other role, the sick role carries with it certain expectations and responsibilities. People in the sick role must show that they want to be well, and in return they receive the sympathy and assistance of those around them. For the CFS sufferer, denied an official diagnosis and hence denied access to the sick role, what assistance is given is tempered with reluctance and suspicion.

One seemingly tangential determinant of how an illness is perceived by the public is its name, yet the name can be quite important. This is one of the insights of labeling theory, a branch of symbolic interaction theory (Chapter 8). In the case of CFS, surveys have found that the "chronic fatigue" label tends to trivialize the disease, leading the public to discount its impact on sufferers (Schwager, 1994; Iverson and Freese, 1990). As a result, and following vigorous lobbying by CFS victims, the disease is now called chronic fatigue immunodeficiency syndrome (CFIDS). The new name redirects the emphasis from tiredness toward a probable cause of the disease, an immune deficiency in the body, and conveys that the illness is not imaginary. One objection raised against this new name, or any other at this point, is that the cause of the illness is still not firmly established, and renaming is therefore premature. This warning will be borne out if the rather inconclusive evidence implicating the immune system in the development of the disease turns out to be misleading.

This daunting disease has been steadily gaining legitimacy in other ways as well. As new data come to light and doctors become more informed, people with CFS, now relabeled CFIDS, can legitimately occupy the sick role. CFIDS research has recently won larger government funding, spurred on in part by national groups organized to promote the interests of CFIDS victims. It seems clear that at least some of these benefits have come about simply as a result of the relabeling of the disease.

Mental Illness

A landmark study by Rosenhan (1973), a social psychologist, clearly demonstrated how profoundly the definition of illness in our society is subject to social construction and the effects of labeling. Rosenhan and several of his graduate students contrived to get themselves admitted to a mental hospital. Working in alliance with a psychiatrist connected with the hospital, they pretended to hear voices and complain of hallucinations. They were successful in getting themselves admitted to the hospital, and thus became *pseudopatients.* Once admitted, they ceased completely to fake symptoms of mental illness; that is, they simply acted normally. Nevertheless, they found that in their day-to-day interaction with doctors and psychiatric nurses, they were treated just like any of the other patients. The staff failed to recognize that they were not ill. During their stays, the researchers discovered that patients were routinely dehumanized and infantilized; in fact, the hospital staff routinely used "baby talk" to address the researchers and the other patients. Particularly illuminating was the fact that the *real* mental patients in the hospital began to suspect that Rosenhan and his students, the pseudopatients, were fakes who should not be in the hospital. The real patients were able to pick out the pseudopatients, but the professionals, trained to identify those who are mentally ill, were not able to do so.

This study shows that at least some of what we define as mental illness in our society may owe as much to the labeling effect as to psychological turbulence: How the mentally ill are perceived by the medical profession and the public may depend in significant part on the label that is attached to the illness (Jones et al., 1986; Scheff, 1984; Szasz, 1974; Scott, 1969). If we are told that mental patients "act weird," and then are told that Joe is a mental patient, we become likely to perceive Joe's behavior as weird no matter what he does. *Any* action by Joe will be automatically defined as weird. We can safely conclude that at least *part* of what is defined as mental illness in this society is socially constructed, ranging from the more serious illnesses such as schizophrenia (the state Rosenhan's pseudopatients faked) to pedestrian neuroses like excessive worry about one's appearance.

In 1950, more than half a million mental patients were institutionalized in state and local mental hospitals. In the

1960s, at the same time that theoretical frameworks like labeling theory began to gain credence, mental health practitioners began to rethink the nature of mental illness. Many argued that the label "mentally ill" had acquired too much power, that humans were being warehoused in asylums even if they were capable of functioning in the outside world with appropriate supervision, simply because institutionalization was the designated response to the label "mentally ill." Simultaneously, new drugs emerged that were effective against many psychiatric disorders, and a movement to discontinue the institutionalization of the less seriously ill gathered momentum. This resulted in the **deinstitutionalization movement** of the 1960s and early 1970s, which included legislation calling for the relocation of many mental patients to smaller, community-based facilities, "half-way" houses that would help them lead at least seminormal lives. By the early 1990s, the number of institutionalized patients was quite interestingly about one-fifth of what it had been in the 1950s (Mechanic and Rochefort, 1990).

The deinstitutionalization movement has generated a mixed response. Many homeowners oppose the establishment of half-way houses for the mentally ill in their neighborhoods, fearing that property values will fall and that the patients might be dangerous. The experience of the mentally ill in these situations can be likened to that of convicted sex offenders who after serving their sentences are transferred to half-way houses or released outright. The response from neighbors in that circumstance can be ferocious. Most people would be more likely to sympathize with the mentally ill than with sexual offenders, but both parties are labeled deviant and given a similarly hostile reception. The mental patients being relocated are granted none of the special consideration that goes along with the sick role. The recent enactment of "Meagan's law" requiring released sex offenders to make themselves known to the communities into which they acquire residence, is a case in point. (The law was named after Meagan Kanka, a New Jersey young girl who was raped and killed, after being kidnapped from a suburban slumber party, by an adult male who was subsequently released from prison.)

Many researchers and legislators attribute much of the blame for the recent rise in *homelessness* to deinstitutionalization of the mentally ill since inmates released from psychiatric hospitals often have nowhere to go except the streets. In sum, although sociologists, psychologists, and medical doctors have changed their views about what mental illness is and how it is defined, and have redressed the extent to which labeling determines how the ill are treated, the lay public, including legislators, retains views of the mentally ill that are in many ways characteristic of the 1950s.

The Medicalization of Illness and Deviance

Health and illness in our society have undergone a process whereby, as with chronic fatigue syndrome and various forms of mental illness, illness comes to be defined as exclusively biological or medical. **Medicalization of an illness** is that process by which society, following the medical profession, assigns all aspects of health and illness an exclusively medical meaning (see also Chapter 8 on the medicalization of deviance). This means that even forms of health that are otherwise defined as quite normal in a society can be given a medicalized definition within that society, and thus become defined as illness or "sickness" (Conrad and Schneider, 1992). In other societies, such as the Ngoni of Malawi, Africa, childbirth is not seen as inherently medical, nor is it seen as deviant (Cohen and Eames, 1982). It is simply not regarded there as the province of doctors, nurses, and related medical personnel. In the United States, childbirth has come to be defined over the years as a matter for doctors and other health professionals. It has become medicalized.

Deviant behavior in the United States has been subject to the medicalization process. The *medicalization of deviance* is the process by which culturally defined deviant behavior comes to be defined as individual pathology or sickness. One example is childhood attention deficit disorder (ADD), the inability to concentrate on the task at hand for more than a minute or two without being distracted (Conrad, 1975; Galliher and Tyree, 1985). In the classroom, children who were overly active or "fidgety" were often labeled "problem children," and thus were labeled deviant. Many teachers often reacted by punishing the children, whereas other teachers sought a positive alternative by attempting to find activities to absorb the excess energy of such children. Then about twenty-five years ago, such children were called hyperactive, and some years later, ADD; they were now seen as having a physical disorder and were advised to get medical attention. Interestingly, as one researcher (Conrad, 1975) has noted, this diagnosis coincided with the development of a drug called *ritalin,* a strong medication that was found to decrease some hyperactivity in children. Both parents as well as teachers welcomed this drug since it remained only for physicians to calm the child with a simple drug, and parents and teachers could attribute the child's malady to biological causes rather than, say, their own parenting or lack of good teaching. Less blame was thus attributed to the child or to the parent or teacher; the child was seen as having less direct control over his or her own behavior since the causes were now perceived as biological and thus less controllable by the child (Taylor et al., 1997).

The Health Care Crisis in America

Currently, the cost of medical care in the United States is more than 11 percent of our gross national product, making health care the nation's third leading industry.

The cost of health care has at times risen at twice the annual rate of inflation. The United States tops the list of all countries in per-person expenditures for health care (see Figure 19.3), yet despite all that spending, the U.S. health care system is not the best in the world. Other countries that spend considerably less money deliver a level of health care at least as good. For example, Sweden and France spend roughly half as much per capita as the United States, and

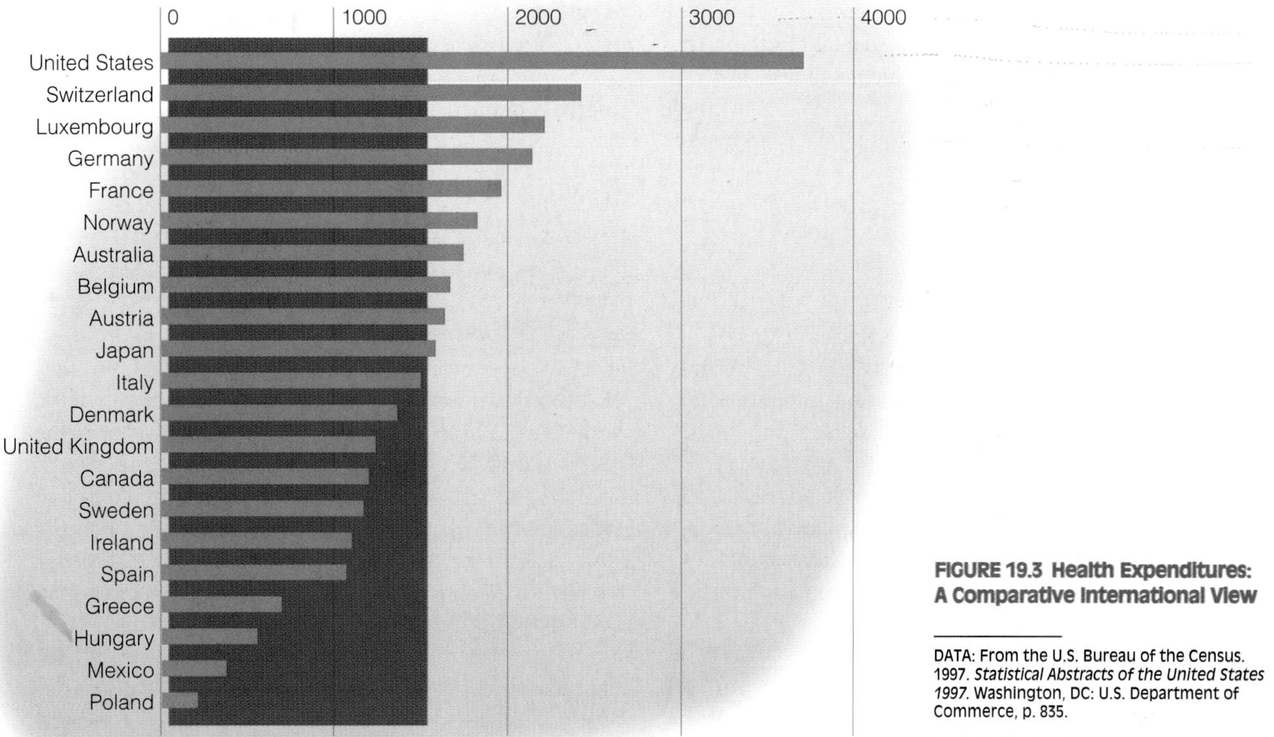

FIGURE 19.3 Health Expenditures: A Comparative International View

DATA: From the U.S. Bureau of the Census. 1997. *Statistical Abstracts of the United States 1997.* Washington, DC: U.S. Department of Commerce, p. 835.

Great Britain spends a bit more than a third as much, yet these countries have national health insurance programs that cover virtually their *entire* populations and that deliver a level of care not much different overall from that achieved in the United States. Attempts by President William Clinton's administration in the early 1990s to institute a national health insurance program were bitterly contested and ended in defeat.

The battle fronts of the health care crisis extend to the fee-for-service system, the epidemic of malpractice suits, and fraud and abuse in the medical profession (Starr, 1982, 1995; Scarpitti et al., 1997). We examine each in turn to develop a picture of the overall crisis in American health care.

Debunking Society's Myths

Myth: **The health care system works with the best interests of clients in mind.**

Sociological perspective: **The health care system is structured along the same lines as other social institutions, thus reflecting similar patterns of inequality in society.**

The Fee-for-Service System

Health care must be paid for, and in the United States, the structures in place for paying the doctor's bill are in a state of chaos. The central element in the system is the fee-for-service principle. Under this arrangement, the patient is responsible for paying the fees charged by the physician or hospital. Fortunate patients will be adequately insured, and will be able to pass on their health expenses to the insurance company. Many people are not insured, however, especially those with lower incomes. This group must reckon with large bills by drawing on their own limited resources. Overall, thirty-one million people in the United States are without health insurance (see Map 19.1). Most severely hit are the southern and southwestern states, where those without health insurance reaches more than 15 percent. A third source of payment is the government, which is increasingly the ultimate payer of health care expenses.

Alternative approaches to the fee-for-service system are presently being scrutinized, but the fee system still predominates, whether in the form of direct payment by the individual or indirect payment by third parties such as private insurers. Either way, fees are disbursed in exchange for services rendered. The same applies to payments made under Medicare and Medicaid.

The greatest contributors to skyrocketing health care costs are the soaring costs of hospital care and the rise in fees for the services of physicians. The services of specialists in particular are hugely expensive. Another culprit is the third-party payment system. With hospitals interested in making a profit, and with patients simply passing bills on to insurance providers, there is little pressure to keep the cost of care down. Patients may be treated on an in-patient basis when a hospital stay is not actually necessary, unnecessary treatments may be given, and some diagnostic procedures may be ordered simply because expensive equipment is available that needs to be used to pay for itself. **Health maintenance organizations (HMOs)** are private clinical care organizations that provide medical services in exchange for a set membership fee, and hence have direct responsibility and control over the costs incurred. It has been found that surgeons who work for HMOs are less likely than fee-for-service physicians to recommend surgery in cases where there is some question of whether the surgery is necessary (Scarpitti et al., 1997). We will return to the subject of HMOs later in this chapter.

Malpractice

A rising number of patients are suing their physicians. In response, physicians are increasing their malpractice insurance to protect themselves. Annual insurance premiums for physicians have risen tenfold in the past decade, and range now from about $5000 per year for physicians in family practice to well over $100,000 per year for physicians in specialties such as radiology, anesthesiology, and surgery. The cost of these insurance premiums is simply passed along to consumers (patients) and has contributed to the rise in the overall cost of health care.

The U.S. public has traditionally accorded physicians high social status and high incomes. The recent popularity of malpractice suits suggests that the public is beginning to question the privileged status of doctors. Several specific reasons have been put forward for client revolts against the medical profession. Attorneys claim that the primary cause of the malpractice dilemma is a declining standard of medical care and a rising incidence of medical negligence. Another argument is that, as a result of specialization, doctors now fail to establish old-fashioned rapport with patients, and thus patients are less attached to their physicians than they were traditionally and more likely to turn hostile. A third notion is that the high cost of medical care and resentment about the outsized incomes of physicians has generated animosity in patients that is quick to show itself, especially when medical episodes have an unfortunate outcome.

Adding to the crisis is the eagerness with which lawyers take on malpractice suits. This is due in part to the very large number of lawyers, particularly on the eastern and western seaboards, and in part to the vast profits that can be reaped from successful cases. Malpractice suits are generally conducted on a contingency fee basis. If the lawyer chooses to accept the case, the patient is required to pay little or nothing to launch the proceedings. Lawyers who work on contingency generally get one-third of any eventual settlement as their payment. If the case is lost, the lawyer gets no fee at all. Recently, large settlements have become common—large enough to motivate increasing numbers of irritated patients and their eager lawyers to file their cases in court. In 1995, the average jury award in a medical malpractice suit was $1.8 million. About one doctor in five is sued each year. In order to protect themselves against potential malpractice suits, doctors increasingly practice **defensive medicine,** which entails ordering excessively thorough tests, X-rays, and so on at the least indication that something might be amiss, partly to ensure that nothing is missed, partly to document that the highest possible level of care was given. Although this extra attentiveness may contribute marginally to favorable medical outcomes, it contributes mightily to the overall cost of health care.

Fraud and Abuse

Yet another factor contributing to spiraling medical costs is the increasing level of fraudulent activity engaged in by some doctors, pharmacists, and patients, as well as operators of hospitals, nursing homes, laboratories, and clinics. Illegal acts are the exception rather than the rule, but they are still prevalent enough to contribute substantially to the overall cost of health care. The most common types of medical fraud involve rip-offs of third-party insurance programs. For example, a physician may bill the insurance company for services not actually rendered. Current estimates are that as much as 10 percent of the dollar amount paid to physicians by insurance companies is fraudulent (Scarpitti et al., 1997). Billions of dollars are lost annually by private insurers and the federal government. There has been much evidence of overbilling and double billing of Medicaid cases. Some pharmacists have been discovered charging Medicaid or Medicare for expensive brand-name drugs after giving the patient cheaper generic versions. Finally, many patients have lied about their income levels in order to gain eligibility for federal or state medical care programs that are restricted to low-income persons.

The federal government has uncovered some large-scale abuses in the operation of laboratories, clinics, and nursing homes. Some laboratories "rent" office space from community clinics that cater to the poor in return for a contract to do all the clinic's testing. The added cost of the "rent" is indirectly paid for by the government when the labs charge the clinics twice their usual rates, and the clinics in turn bill Medicaid. In another type of fraud, clinic physicians treat one patient in a family and then bill the government for all members of the family (Scarpitti et al., 1997).

A Response to the Problem: HMOs

Health maintenance organizations (HMOs) are an innovation in health care, first begun shortly after World War II but becoming widespread only within the last twenty years. The staffs in HMOs can include anywhere from several doctors to several hundred. People who join HMOs are assigned to a physician who administers care and when necessary gives referrals to specialists affiliated with the HMO. The doctors in an HMO earn salaries rather than fees. All services are ultimately paid for by the membership fee paid by subscribers to the HMO. The elimination of both the fee-for-service system and the third-party insurer drives costs down in several ways. Physicians have an incentive to give the most economical levels of care, while presumably retaining their motive to maintain a high standard of care since they will want to secure the reputation of the HMO and avoid malpractice suits. The opportunity for fraud, which costs third-party insurers so dearly, is eliminated. Finally, the corporate structure of HMOs presumably offers the economies of scale and economies of organization enjoyed by other profit-oriented corporations yet often found lacking in the U.S. hospital system.

HMOs are experiencing fast growth as employers and individuals grapple with rising medical costs. In 1986, there were approximately twenty-six million HMO subscribers in the United States, a number that has risen to more than fifty-five million in 1996. The American Medical Association (AMA) has a history of opposing HMOs on the grounds that they decrease the rights of physicians to determine treatments, and they limit the rights of patients to choose their own doctors or seek out treatments from specialists. The AMA has also argued in a number of forums that HMOs are inclined to pay too much attention to the bottom line and not enough to patient welfare. Objective comparisons of the HMO system and the fee-for-service system, however, indicate that HMOs offer a level of care at least as good as the usual standard set by other health care providers (Greenfield et al., 1995).

Although medicine has been dominated by Western values, many have come to understand the value of alternative therapies, including such ancient Asian practices as acupuncture and herbal therapies.

Alternative Medicine

As will be seen in Chapter 21, people often form movements (called social movements) that seek to bring about some kind of change in society. A limited form of social movement is the *alternative movement,* so called because it attempts some form of limited, focused change in some aspect of society. The recent "New Age" movements, which advocate meditation, vegetarianism, and other such practices, are examples of alternative movements. Such is the case with alternative medicine, a social movement that has sprung up in response to the general health crisis in the United States. **Holistic medicine,** which emphasizes the person's entire mental and physical state, plus the integration of the two, as well as emphasizing the person's physical and social environment, is an example of alternative medicine that has developed in the late 1980s and 1990s (Reischer, 1992; Alster, 1989).

Although practitioners of holistic medicine advocate the treatment of the entire mental and physical person, they do not necessarily reject the remedies of traditional medicine such as surgery and drugs. They do, however, caution against the use of such measures alone on the grounds that they result in overspecialization and the consequent narrow focus upon isolated symptoms rather than upon the individual as an integrated whole; upon specific diseases rather than disease as a part of the entire concept of health.

Practitioners of holistic medicine are concerned with how each person's physical and social environments affect health. For example, the holistic practitioner recognizes that a variety of illnesses can result from the stresses of one's lifestyle, or from work-related stress, or from poverty. As a consequence, holistic medicine takes an active role in combating environmental pollution. Another approach of holistic medicine is to decrease the dependency of patient upon physician by shifting some of the responsibility of cure from the physician to the patient, such as by advocating health-promoting behaviors involving diet, exercise, and the use of organic foods. Finally, holistic medicine deemphasizes the necessity of constant in-office care and instead urges increased use of patient treatment in familiar environments such as the home or the friendship group. In many respects, the recent pushes toward holistic medicine reflect a return to at least some of the characteristics of the older "general practitioner" rather than the narrow high-tech specialist of today. The holistic practitioner does not oppose science and specialization, but instead seeks a shift away from narrow treatment of disease and toward integrated treatment of illness for everyone.

The Universal Health Care Debate

Between 1992 and 1994, a bold plan for the reorganization of the health care system in the United States was worked out by the Democratic Clinton presidential administration, based on the proposals of a large task force chaired by then First Lady Hillary Rodham Clinton. The program was designed to combat the major problems of health care just examined—high hospital costs, high doctor's fees, problems in the fee-for-service system, and fraud. One promise of the program was universal health insurance for all Americans.

The core of the Clinton reforms was to be conversion of the entire system to a model called **managed competition**. Under this plan, all individuals in the United States would belong to a complex of managed-care organizations, rather like HMOs, that would use their collective bargaining force to drive down the cost of health insurance, while accepting responsibility for operating their own facilities in an economical manner and continuing to provide high-quality care. Everyone would join the managed-care complex; however, individuals would still be free to retain their personal physicians if they chose, so long as their physicians met government-stipulated criteria. The plan was intended to achieve the advantages of socialized medicine systems as administered in Great Britain, Canada, and other places where everyone is entitled to see a doctor when they need one, while retaining elements of the profit motive in the system, trusting that market forces would be the best corrective for issues of cost versus quality of care. Providers who failed to give adequate value would be consumed by other providers who did a better job of meeting the needs of consumers. Medical practitioners would continue to earn high incomes although there was the expectation that they would not be quite as high as previously (U.S. physicians earn as much as several times the income of their colleagues in the rest of the world).

The Clinton plan met strong opposition from the American Medical Association and in Congress, where even some Democrats who initially supported the plan found reasons to turn on it. The plan was defeated in Congress in 1994 even though at that time Democrats were in the majority in Congress. There are a number of reasons for the program's failure. Many Republicans in Congress opposed the plan as part of their generalized opposition to the Democratic Clinton administration; early expectations by the task force about how much support the plan would receive in Congress were overly optimistic; and finally, there was a mishandling of parliamentary tactics—the core of the plan was presented to Congress first, with the more easily sacrificed parts of the plan presented later, in violation of the principle that one should never open a negotiation by putting one's final offer on the table (Starr, 1995). By the late 1990s, no successful proposal for universal health care had passed Congress. Nonetheless, public concern over the issue remains high.

THINKING SOCIOLOGICALLY

Do you now have medical insurance? What kind is it; that is, is it some kind of group plan gotten through work or school? Or is it an individual plan so that either you or some person (such as a parent) pays the premium (costs) for it? If you have medical insurance of either type, find out how much it costs you per month or per year. Do you consider the cost to be too high? How quickly would your insurance company settle your claim (this is, pay you or your physician)? Thinking about such matters as they affect you will cause you to think about your own opinions about health care reform.

Death and Dying

One of the most pressing problems facing the medical profession and society is the issue of whether or not people have a right to die. Does the terminally ill individual have a right to take his or her own life? Does the terminally ill person have the right to instruct someone else, perhaps a loved one, to do so? What is the definition of "terminally ill"? The issue is a moral, ethical, and legal one.

Many thousands of men and women in the United States are in some kind of state often described as "brain dead." In such instances, the person has a flat EEG (electroencephalogram), indicating no brain activity, and in addition, no reflexes or spontaneous breathing. Many others are suffering from some terminal illness that causes them great pain, and they wish to end their own life. There is thus a dilemma of patients' rights, or the question of the person's right to take his or her own life, versus the principles of the medical profession, which dictate that all forms of medical care should be given the patient in order to sustain life. Not only has this resulted in the ethical and legal question of just what constitutes death, but it has also caused the courts over the last two decades to wrestle with the question of whether or not a chronically ill patient who wishes to take his or her own life may legally do so.

One such case was that of Nancy Cruzan, who at age twenty-six was in an automobile accident and, as a result, fell into an irreversible coma. After doctors told her parents that she would never recover, her parents requested withdrawal of her life support systems, which consisted of a feeding tube placed in her stomach. The hospital refused to do so, and the Cruzans took their daughter's case to the U.S. Supreme Court, which ruled that states can establish conditions for the removal of life support methods. Nancy Cruzan's feeding tube was finally removed seven years after her accident, and she died twelve days later (Corr et al., 1994).

As a result of such court cases, the medical profession has established two guidelines. First, the physician must clearly explain to the patient the medical options available to sustain life. If the patient is in a coma or otherwise incapacitated and is not capable of understanding such explanations, then the physician will explain the options that are available to close members of the patient's family. If this is done, then terminally ill patients, and a close family member, have the right to refuse what is called "heroic" treatment of the patient that might prolong life, or a coma, but offer no hope of recovery. Second, the physician may honor the *living will* of the patient: a statement made by the patient, while still in possession of mental faculty, of whether or not heroic treatment should be given in the case of severe incapacity. Most physicians now recognize the validity of the living will and will abide by it, and they have the backing of law in most states. If no living will has been made, then the physicians and hospital are required to give treatment intended to sustain life. Nonetheless, even in such instances where a living will is absent, the patient or close family member may confer with the physician about discontinuing heroic treatment, and the physician may well comply.

Euthanasia is the act of killing a severely ill person as an act of mercy. There are two forms of euthanasia. Negative euthanasia, sometimes called passive euthanasia, involves the withholding of treatment ("pulling the plug") with the knowledge that doing so will produce the death of the patient, as in the case of a living will. The second form, positive or "active" euthanasia, involves killing the ill person who would otherwise live, as an act of mercy. Positive euthanasia is for this reason often called "mercy killing."

Dr. Jack Kevorkian of Michigan has remained in the news for several years because he has offered to assist terminally ill patients in committing suicide. This represents mercy killing and has earned Dr. Kevorkian the dubious nickname "Dr. Death." He has carried out several assisted suicides, and he has been tried in the Michigan courts repeatedly for doing so and found to be guilty of murder.

Ethical questions about physician-assisted suicide have captured public attention, especially by Dr. Jack Kevorkian. Here he watches, with his attorney, the "60 Minutes" episode in 1998 in which Kevorkian assisted in the televised death of Thomas Youk, suffering from Lou Gehrig's disease.

CHAPTER SUMMARY

- The state of health care in America is seen by means of *social epidemiology,* the study of the role of social and cultural factors in disease and health. Race-ethnicity, social class, and gender are major factors: Whites live longer than Blacks, and women live longer than men. White women live longest, and Black men the shortest. Black women are more likely than White women to fall victim to cancer, heart disease, strokes, and diabetes. The mortality rate among Native Americans is one-and-one-half times that of the general population.
- Hispanics contract tuberculosis at four times the rate for Whites. On other indicators of health, they are similar to Blacks and Native Americans. The more money one makes, the more healthy one perceives one's self to be. The lower one's social class status, the greater are one's chances of tuberculosis, heart disease, cancer, arthritis, and infant mortality. The poor are more likely to be depersonalized and treated inadequately by the health care system than those of higher social class status.
- Even though women live on the average longer than men, older women are more likely to suffer from stress, overweight, hypertension, and chronic illness than are older men. Men and women tend to be treated differently in the health care system since there is still a tendency for the male-dominated profession to regard the problems of women as nonmainstream and "special."
- Americans' attitudes toward illness center on the perception of "controllability" of an illness, such that if one is perceived as having an illness that he or she might have done more to prevent, then that person is blamed for the illness. For such reasons, a person who is known to be HIV positive, or who has full-blown AIDS, is partly blamed and is heavily negatively *stigmatized,* and treated as a social outcast. Stigmatization also to some extent befalls those with *chronic fatigue syndrome.*
- Modern medicine in America is a highly structured, specialized high-status profession. High costs, abuses of the fee-for-service system, and malpractice suits have resulted in a policy crisis today in America's health care system. The growth of *health maintenance organizations* (HMOs) shows promise for easing the crisis although programs for large HMOs and national universal health insurance have met with defeat in Congress.
- A moral, ethical, and legal debate still rages in our society over how to define death, and whether or not the terminally ill individual has a right to take his or her own life.

KEY TERMS

anabolic steroids
anorexia nervosa
bulimia
chronic fatigue syndrome (CFS)
defensive medicine
deinstitutionalization movement
epidemiology
euthanasia
health maintenance organization (HMO)
holistic medicine
managed competition
Medicaid
Medicare
medicalization
sick role
social epidemiology
stigma
Tuskegee Syphilis Study

THE INTERNET: A Tool for the Sociological Imagination

Resources on the Internet:

Virtual Society: The Wadsworth Sociology Resource Center
http://sociology.wadsworth.com

Visit this site to find additional learning tools, including interactive quizzes, links related to web sites, and an easy link to *InfoTrac College Edition.*

Centers for Disease Control
http://www.cdc.gov/

A national research center aimed at preventing and controlling disease; also publishes national data on the health status of the population.

National Institutes of Health
http://www.nih.gov/

The national agency dedicated to research that will improve the nation's health; the home page provides access to a large number of governmental documents about health.

Office of Minority Health Resource Center
http://www.omhrc.gov/

An office of the government that provides health information on specific minority populations.

World Health Organization
http://www.who.ch/

An organization dedicated to improvement of health on a global basis.

Sociology and Social Policy: Internet Exercises

National debate on the desirability of *universal health care* often focuses on whether such a system would lead to socialized medicine (that is, government-supported medical care). State regulation, opponents argue, would lead to greater bureaucratization in the health care system and lack of individual choice about providers. Proponents argue that the current system of private health care benefits primarily the better off and, of course, private companies that control access to health care. How does the debate about national health care reflect the values of U.S. society, and can that be balanced against the need to provide medical care for all groups, regardless of their social and economic circumstances? What do different sociological theories illuminate about this debate?

Internet Search Keywords:

socialized medicine
national health program
universal health care
health insurance
managed care providers
health care
private health care

Web sites:

http://www.free-market.org/features/spotlight/9803.html
This site contains an article on socialized medicine as well as a variety of publications of other institutes.

http://www.sswlhc.org/
The web page for the Society for Social Work Leadership in Health Care. This group is dedicated to promoting universal availability of health care.

http://www.pnhp.org/
A web site for the promotion of a national health program sypported by physicians.

http://www.hiaa.org/
This web site by the Health Insurance Association of America offers news and information related to the health insurance industry.

http://www.life-line.org/health/index.html
Health-line is a foundation providing information about insurance options.

http://www.healthlaw.org/
The National Health Law Program is a public interest law firm that seeks to improve health care for low-income and unemployed people.

http://www.nhpf.org/
This web site discusses national health care policy.

InfoTrac College Edition: Search Word Summary

Americans with Disabilities Act
euthanasia
health care and religion
medicare
medicaid

In order to learn more about these central topics in sociology, you can conduct an electronic search using InfoTrac College Edition. To aid in your search and to gain useful tips, see the Student Guide to InfoTrac College Edition on the Virtual Society web site:
http://sociology.wadsworth.com

INTERACTIONS—A SOCIOLOGY CD-ROM: CONCEPTS FOR THIS CHAPTER

Go to the Wadsworth Sociology CD-ROM for further study on the concepts in this chapter. The CD-ROM also includes quizzes and additional activities to expand your learning experience.

SUGGESTED READINGS

Angel, Ronald L., and Jacqueline L. Angel. 1994. *Painful Inheritance: Health and the New Generation of Fatherless Families.* Madison: University of Wisconsin Press.

This is a comprehensive examination of the impact of single motherhood on the physical and mental health of women and children of color, and White women and children as well. It documents the serious and long-term consequences of poverty.

Corea, Gena. 1993. *The Invisible Epidemic: The Story of Women and AIDS.* New York:

Corea's book is a well-written account of her interviews of women with AIDS as well as with women activists, scientists, and caretakers who are struggling to keep the epidemic in check.

Ginzberg, Eli. 1990. *The Medical Triangle: Physicians, Politicians, and the Public.* Cambridge, MA: Harvard University Press.

This analysis of the complex network involving doctors, their professional organizations, HMOs, insurance companies, and the public offers a clarifying look at the health care system.

Scheff, Thomas. 1984. *Being Mentally Ill: A Sociological Theory,* 2nd ed. New York: Aldine.

This is one of the best explanations and applications of labeling theory as applied to the study of mental illness as a social construction. The work discusses how patients are in effect rewarded for continuing their sickness and punished for trying to get well.

Shilts, Randy. 1988. *And the Band Played On.* New York: Penguin.

Shilts's book offers an extremely moving analysis of the public policy response to the AIDS crisis. The book documents how the association of AIDS with homosexual sex slowed scientific research and inhibited a strong policy response to the growing threat of AIDS.

Starr, Paul. 1982. *The Social Transformation of American Medicine.* New York: Basic Books.

This is a readable historical account of the development, bureaucratization, and institutionalization of medicine and the health care system. It explains how the medical profession emerged as dominated by physicians and scientists.

Thompson, Becky. 1994. *A Hunger So Wide and So Deep: American Women Speak Out on Eating Problems.* Minneapolis: University of Minnesota Press.

Thompson's book develops the interesting thesis that the American culture of thinness affects White women differently from how it affects African American women, Latinas, and lesbians. White women are somewhat more likely to undereat and become anorexic, but these latter groups engage in what Thompson calls strategies of self-preservation as a result of oppression, and are thus more likely to engage in overeating rather than self-starvation.

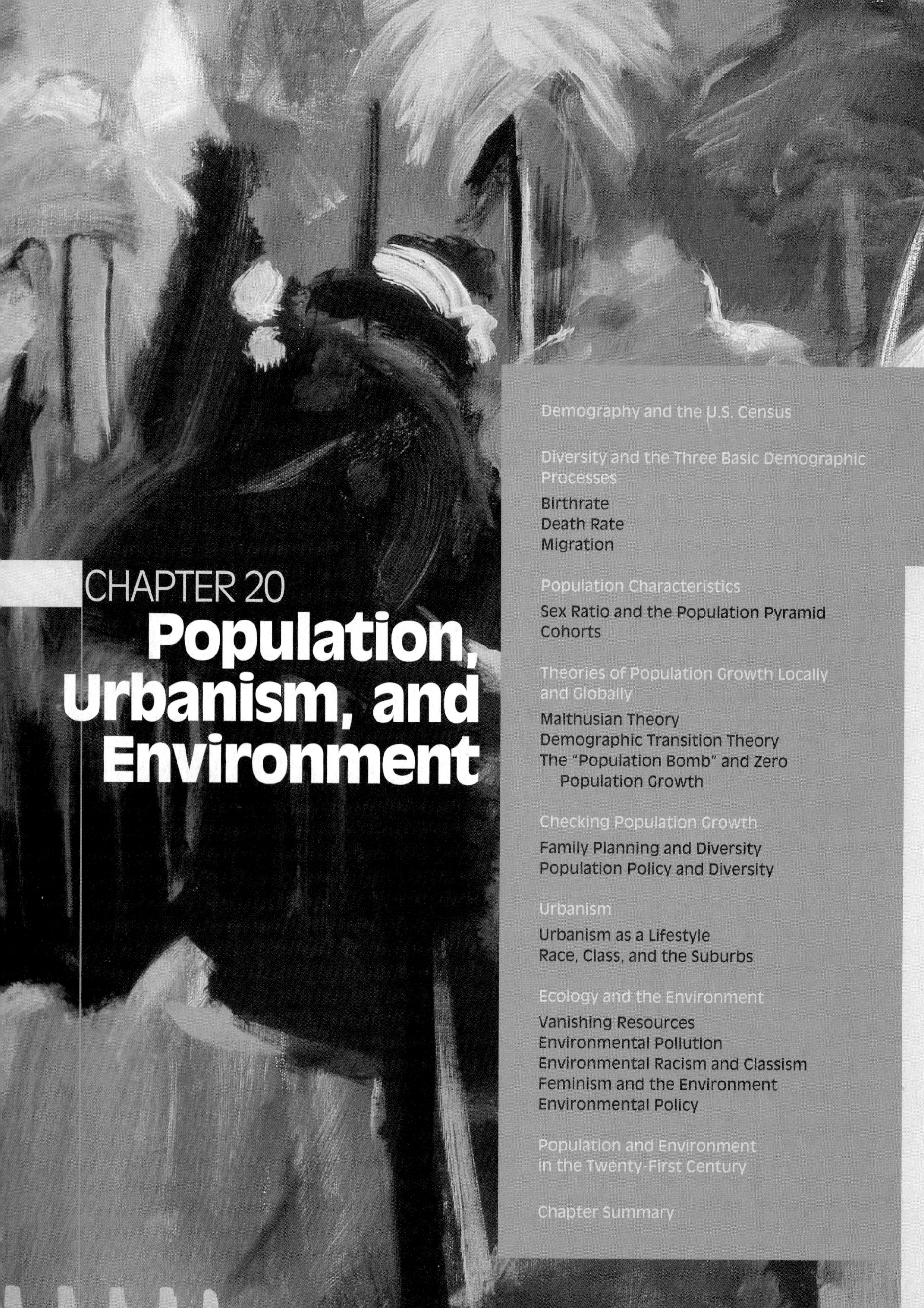

CHAPTER 20
Population, Urbanism, and Environment

grows exponentially, with an upward-accelerating curve, as shown in Figure 20.1. An ever-increasing number of people are added each year. At the present rate of growth, the world's population will double in about forty years.

Birthrate

The **crude birthrate** of a population is the number of babies born each year for every 1000 members of the population or, alternatively, the number of births divided by the total population, times 1000:

$$\text{Crude birthrate (CBR)} = \frac{\text{Number of births}}{\text{Total population}} \times 1000$$

The crude birthrate reflects what is called the *fertility* of a population, which is live births per number of women in the population. The crude birthrate for the entire world population is about 27.1 births per 1000 people. Different countries and subgroups within a country can have dramatically different birthrates. For example, the country with the highest birthrate in the entire world is Kenya, with 53.9 births per 1000 people. The lowest birthrate is found in San Marino, with a birthrate of only 9.3 per 1000. Fertility is different from *fecundity,* which is the potential number of children in a population that could be born (per 1000 women) if every woman reproduced at her maximum biological capacity during the childbearing years; say, if a majority of women had on the average around twenty or more children, which is what current demographic estimates suggest as a maximal estimate. Fertility rates are not as high as fecundity rates since few women in any given population reproduce at anywhere near the theoretical biological maximum.

The overall birthrate for the United States is about 16 per 1000 people, compared to the all-time rate of 27 per 1000 in 1947, the Baby Boom that followed World War II. The rate varies according to racial–ethnic group, region, socioeconomic status, and other factors. Overall, we find that for different racial–ethnic groups, the birthrates in the United States are as follows (Population Reference Bureau, 1998):

Group	Birthrate per Thousand
Whites	15
African Americans	18
Latinos (Mexican Americans, Puerto Ricans, and other Hispanic groups combined)	20
Native Americans	17
Asian Americans	11

In general, minority groups tend to have somewhat higher birthrates than White nonminority groups, and lower socioeconomic groups tend to have higher birthrates than those higher on the socioeconomic scale. Native Americans have a birthrate higher than the national average, even though the Native American birthrate has *declined* over the past fifty years (Thornton et al., 1991). These effects are somewhat cumulative. For example, minorities tend to be overrepresented on the lower end of the socioeconomic scale, compounding the likelihood of a high birthrate. Similarly, religious and cultural differences can make themselves felt: Catholics, for example, have a higher birthrate than non-Catholics of the same socioeconomic status; Hispanic Americans have a high likelihood of being Catholic; thus, multiple factors combine to contribute to the higher birthrate among Hispanic Americans. Since minorities tend to have higher birthrates, assuming that present migration rates continue, and assuming that death rates do not outstrip the birthrates, then the United States will have a significantly greater proportion of minorities, thus a relatively lower proportion of Whites, in the coming years. The implications for local and national policy are great. Given current birthrates, the population of Latinos in the United States will double by the year 2030, and the population of African Americans will double by the year 2025. This means that local, state, and federal governments must increasingly take these projected increases into account for future legislation and programs of all sorts.

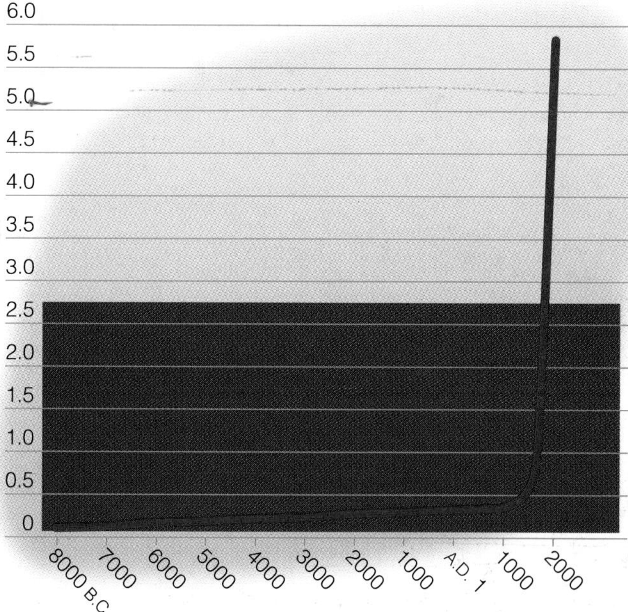

FIGURE 20.1 World Population Growth

DATA: From the Population Reference Bureau, Washington, DC http://www.pub.org

Death Rate

The **crude death rate** of a population is the number of deaths each year per 1000 people, or the number of deaths divided by the total population, times 1000:

$$\text{Crude death rate (CDR)} = \frac{\text{Number of deaths}}{\text{Total population}} \times 1000$$

The crude death rate can be an important measure of the overall standard of living of a population. In general, the higher the standard of living enjoyed by a country, or a group within the country, the lower the death rate. The death rate also reflects factors such as the quality of medicine and healthcare. Poor medical care, which goes along with a low standard of living, will correlate with a high death rate.

Another measure that tends to reflect the standard of living in a population is the **infant mortality rate.** This is the number of deaths per year of infants less than 1 year old for every 1000 live births. In the United States, the overall infant mortality rate is about 11 infant deaths for every 1000 live births. In many developing countries, such as Kenya, the infant mortality rate can rise higher than 35 deaths per 1000 live births.

Infant mortality rates are important to compare across racial–ethnic groups and across social class strata since they are a good indicator of the overall quality of life as well as the chances of survival for members of that racial or class group. Inadequate health care and health facilities cause higher infant mortality rates, and consequently the greater infant mortality among minorities and those in lower socioeconomic strata in the United States suggests that lack of adequate health care and adequate access to health facilities is one cause of the high infant mortality rates. There are also many other causes of higher infant mortality, a measure of the chances of the very survival of members of the population, such as presence of toxic wastes, malnutrition of the mother, inadequate food, and outright starvation.

The **life expectancy** of a population or group is defined as the average number of years that the group can expect to live. In the United States, life expectancy has gone from forty years of age in the year 1900 to seventy-six years of age today. That means that people born now can expect to live, on average, until they are about seventy-six years of age.

Although a life expectancy of seventy-six years might seem high, the truth is that the United States does not compare very well to other developed nations in either life expectancy or infant mortality. Table 20.1 shows that the United States ranks near the bottom among Western nations in life expectancy, behind Japan (where life expectancy is 79.7 years), the Netherlands (77.9 years), Canada (79.3 years), and several other countries. The picture is similar with regard to infant mortality. Interestingly, the former Soviet Union, with whom the United States was engaged in the long-running, expensive cold war, also has a low life expectancy and high infant mortality rate—low enough, indeed, that the U.S. statistics look good compared with the Soviet Union, if not with our peer nations in the West. This is a weak showing for the United States as well as other superpower nations such as Great Britain and Germany.

Life expectancy and infant mortality both vary with gender, race-ethnicity, and social class. Women on average live longer than men, not only in the United States but also throughout most of the world. Women have higher rates of many kinds of illnesses than men, but they still live longer. Women survive their illnesses better than men, and are more resistant to physical stress, including torture. A number of diseases are caused by genetic errors on the X chromosome, which only men possess, including certain types of heart disease, the most common type of hemophilia, and possibly a number of cancers. Men have also been considerably more likely to smoke than women and thus more likely to die of heart disease, emphysema, and lung cancer (National Cancer Institute, 1997). Recently, the rates for smoking have begun to converge. The higher rate of smoking by men in the past will continue to be felt in the death statistics for a number of years to come, but smoking-related death and morbidity (incidence of disease) is another of those population-scale phenomena like the ones we discussed earlier where decisions made today will have effects far into the future, in this case to be seen as a rise in the incidence of women in cancer wards and under oxygen tents.

African Americans, Hispanics, and Native Americans all have shorter life expectancies than Whites. White women live on average five years longer than African American women, and White men on average live seven years longer than African American men (U.S. Bureau of the Census, 1997a). African Americans are twice as likely as Whites of comparable age to die from diseases such as hypertension or the flu, as are Hispanics. As already pointed out in Chapter 19, both are also considerably more likely than Whites to die of AIDS. Sadly, these racial and ethnic differences are even more striking when it comes to infant survival: African American babies are almost *thirty* times more likely to contract AIDS than White babies, and Hispanic babies, twenty-five times more likely. Native Americans are ten times more likely than Whites to get tuberculosis, and nearly seventy times more likely to get dysentery (National Urban League, 1996; Jaynes and Williams, 1989; Levine, 1987). Once again, we see the dramatic effects of gender and race-ethnicity, as well as class, this time on the odds of avoiding the devastation of chronic disease and surviving to a relatively old age. Those odds are considerably lowered in the United States if you are poor and minority.

TABLE 20.1 *LIFE EXPECTANCY AND INFANT MORTALITY (BY COUNTRY, INDUSTRIALIZED COUNTRIES ONLY)*

Country	Life Expectancy	Country	Infant Mortality Rate
Hong Kong	82.4	Hong Kong	5.0
Japan	79.7	Japan	4.4
Australia	79.6	Australia	5.4
Canada	79.3	Canada	6.0
France	78.6	France	6.0
Spain	78.5	Spain	6.1
Italy	78.2	Italy	6.8
Netherlands	77.9	Netherlands	4.8
United Kingdom	76.6	United Kingdom	6.3
Germany	76.1	Germany	5.9
United States	**76.0**	**United States**	**6.6**
Russia	63.8	Russia	24.3

SOURCE: U.S. Bureau of the Census, 1997. *Statistical Abstracts of the United States 1997.* Washington, DC: U.S. Bureau of the Census, pp. 832–833.

infant mortality

Debunking Society's Myths

Myth: High infant mortality and a high death rate for young adults are problems in underdeveloped countries, but not in the United States.

Sociological perspective: Both infant mortality and adult death rates are significantly greater among lower-class than among middle-class persons in the United States, greater among people of color (Latinos, Native Americans, and African Americans) than among Whites of the same social class, and greater among men than among women.

Migration

Joining the birthrate and death rate in determining the size of a population is migration in and out. Migration affects society in many other ways as well. For example, Israel since its establishment in 1948 has experienced considerable growth in its population, mostly due to a tremendous migration of Jews from Europe and the United States. These migrants tend to be younger on average than the rest of the Israeli population, and their arrival therefore has certain direct consequences, such as increasing the birthrate, which is higher among the young than among older Israelis.

Migration can also occur within the boundaries of a country. In the 1980s, internal migration by African Americans, Hispanics, Asians, and Pacific Islanders within the borders of the United States has occurred at a rate unmatched since World War I and the great Black migration from South to North in the United States. In that era, beginning just after the turn of the century, Blacks migrated from the South to major urban areas in the North, such as Chicago, New York, Detroit, and Cleveland. The recent pattern of migration for African Americans has been not only from the South, but also from the major Northern urban centers to the West, the Southwest, and New England. This migration has carried many African Americans into areas that were previously all White, and has frequently been associated with the presence of the migrants at institutions like military bases or universities (Barringer, 1993).

Among Hispanics, migration patterns have traditionally been loosely linked to the agriculture industry, but more recently, to other industries such as meatpacking and textiles, and to industries centered in urban areas as well (Barringer, 1993; Portes and Rumbaut, 1990). Although Mexican Americans have traditionally settled in the West and Southwest, recent movement has taken them to such places as Michigan, Washington State, and New England. Farm workers from Mexico have settled near the beet fields of northern Minnesota, tripling the Hispanic population there since 1980. Puerto Rican Americans have migrated northward in increasing numbers to rural, urban, and suburban areas. In Yankee towns that were once heavily White, such as Lynn and Lowell, Massachusetts, the Hispanic population has grown 180 percent since 1980.

Asians in the last decade have migrated to a variety of destinations within the country. Many Vietnamese have joined the fishing industry on Louisiana's Gulf Coast. The Lutheran Church resettled hundreds of Laotians in Wisconsin. In Smyrna, Tennessee, where Nissan has built a truck plant, the Asian population has more than doubled. Finally, Asians, African Americans, and Hispanics have joined Whites in migrating from urban to suburban areas. Given this increasing presence of people of color in urban and especially suburban areas of the United States, future legislation and programs will of necessity have to consider the cultural and social needs of these groups.

Population Characteristics

The composition of a society's population can reveal a tremendous amount about the society's past, present, and future. The populations of many nations show a striking imbalance in the number of men and women of certain ages, with the appearance that a large number of men are apparently missing. Quite often, the explanation can be found by looking back in time to when the missing men were the right age for military service—the demographic vacancy will usually coincide with a major war from which many young men failed to return. The demographic data are thus a record of national history. World War II was responsible for killing many men, and some women, in the early 1940s—people who were in their twenties. As a result, this age category in the United States revealed a shortage of persons, and considerably fewer men than women, even as the entire population of the country aged.

Sociologists put together data about population characteristics to develop pictures of the population in slices or as a whole. Important characteristics of populations are sex ratio, age composition, the age–sex population pyramid, and age cohorts. Here we examine these main approaches to describing the population.

Sex Ratio and the Population Pyramid

Two factors that affect the composition of a population are its sex ratio and its age–sex pyramid. The **sex ratio** (also called *gender ratio*) is the number of males per 100 females, or the number of males divided by the number of females, times 100.

$$\text{Sex ratio} = \frac{\text{Number of males}}{\text{Number of females}} \times 100$$

A sex ratio above 100 means that there are more males than females in the population; below 100 means that there are more females than males. The sex ratio could just as easily have been defined as the number of females per 100 males. A ratio of exactly 100 means the number of males and females are exactly equal. In almost all societies, there are more boys born than girls, but because males have a higher

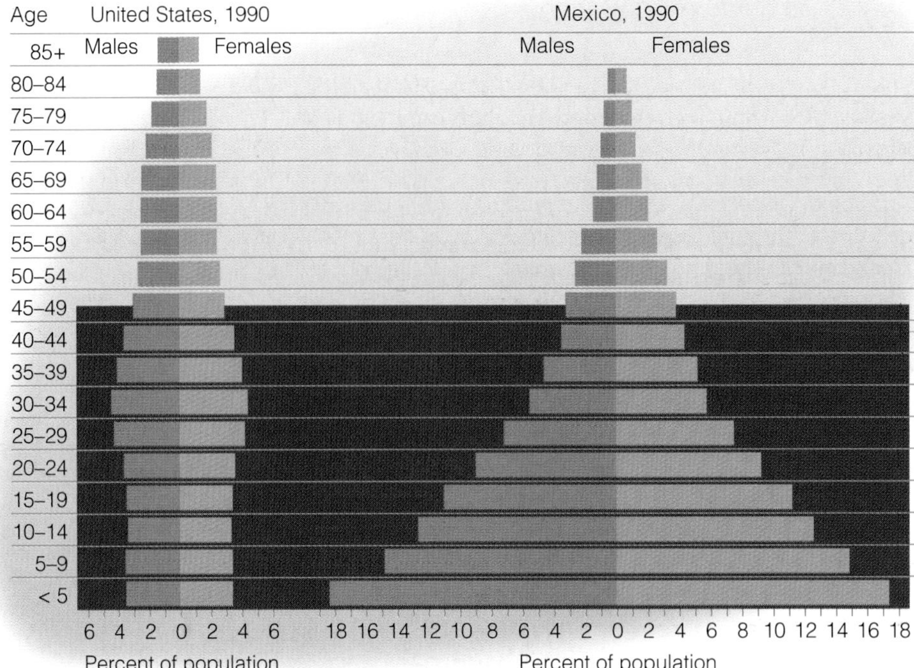

FIGURE 20.2 A Comparison of Two Age–Sex Pyramids

SOURCE: U.S. Bureau of the Census

infant mortality rate and a higher death rate after infancy, there are usually more females than males in the overall population. In the United States, about 105 males are born for every 100 females, thus giving a sex ratio for live births of 105. After factoring in male mortality, the sex ratio for all ages for the entire country ends up being 94—there are 94 males for every 100 females.

The *age composition* of the U.S. population is presently undergoing major changes. More and more people are entering the sixty-five-and-over age bracket in the trend known as the *graying of America.* The elderly will soon become the largest population category in our society. In Chapter 13, we pointed out how society is stratified by age, generally to the detriment of the old. As our society gets grayer, its older members will have more influence on national policy and a greater say in matters such as health care, housing, and other areas where the elderly have traditionally experienced age discrimination.

Sex and age data are often combined in a graphical format called an **age–sex pyramid** (or age–gender pyramid), a graphic representation of the age and gender structure of a society (see Figure 20.2). The left side of Figure 20.2 shows the age–sex pyramid for the United States. Note that there are slightly more males than females in the younger age ranges (due to the higher birthrate for males), a trend that reverses in the middle-age brackets. At the upper range of the age scale, women outnumber men. The pronounced bulge around the middle of the pyramid represents the Baby Boom generation, now in their forties and fifties. As these Baby Boomers age, the bulge will rise toward the top of the pyramid, to be replaced underneath by whatever birth trends occur in the coming years. This restructuring of the population pyramid will necessitate restructuring of society's institutions. Marketing will have to be directed more and more toward those over age sixty-five; considerably more funds will be needed for health care for those of this age category; recreational facilities for those who are older will have to be greatly expanded. (Since this bulge is continually moving forward, much like food in a large snake, demographers have sometimes jokingly referred to the Baby Boomers as "the pig in the python.")

The birthrate of a society or group tends to rise and fall in line with the shape of the age–sex pyramid. The right side of Figure 20.2 shows the age–sex pyramid for Mexico. It is quite wide at the bottom and very narrow at the top. This suggests a high birthrate and a death rate that increases rapidly with age. Relatively few males or females survive to fill the elderly age categories. Countries with a high birthrate tend to have a high proportion of women in their childbearing years. A pyramid of this shape is more characteristic of developing nations than of the industrial powers. The birthrate is obviously a statistic of great consequence for the future: Since the children being born will themselves grow up and have children, a high birthrate in the present can be projected to mean a large number of births a generation further into the future. Within the United States, the population pyramid for African Americans tends to be shaped somewhat like the one on the right in Figure 20.2. So does that for Mexican Americans and Puerto Rican Americans.

THINKING SOCIOLOGICALLY

The *age-sex pyramid* is a graphic representation of the age and gender structure of a society's population. How would the age-sex pyramid for a society look if its men aged eighteen to thirty-five had a greater *crude death rate* than they would otherwise have had because of a major war? Draw what the resulting pyramid might look like. What if the women in the society were killed off in the same way by the same war? Now draw the resulting pyramid. What would such dramatic reductions in the numbers of both the women and men in this society mean for the future of the society?

Cohorts

A birth cohort, or more simply, a **cohort**, consists of all the persons born within a given period. A cohort can include all persons born within the same year, decade, or longer. Over time, cohorts either stay the same in size or get smaller owing to deaths, but never larger. If we have knowledge of the death rates for this population, we can predict quite accurately the size of the cohort as it passes through the stages of life from infancy through adulthood to old age. This enables us to predict such things as how many people will enter the first grade at age six between the years 2000 and 2004, how many will enroll in college, and how many will arrive at retirement decades down the road. Social entities like schools and pension funds can make preparations on the basis of such predictions.

To see how dramatically a single cohort can affect a society over time, consider the effect of the cohort born in the United States between 1946 and 1964—the so-called Baby Boom cohort. Many of the parents of you who are reading this book are Baby Boomers. The Baby Boom has been one of the most significant demographic events in all of U.S. history, along with the great Black migration from the late 1800s through the 1920s, the simultaneous massive immigration of Europeans and other groups into the United States, and current migration patterns (Ehrlich, 1968; Kennedy, 1989).

The Baby Boom cohort, now about one-third of the entire population of the United States, has had a major impact on the practices, politics, habits, preferences, and culture of our society. In the late 1940s and early 1950s, when the Baby Boom cohort was in its infancy, there were large increases in the sales of certain commodities, such as baby food, diaper services, and toys. Dr. Benjamin Spock's book of down-to-earth advice about how to raise healthy children, *Baby and Child Care,* became the greatest bestseller of the century, with more than thirty billion copies sold. Among its principles was the suggestion that parents should communicate with their children and forego physical punishments when possible. For encouraging this brand of "permissiveness," Dr. Spock has often been blamed at one time or another for nearly every social problem since, including the rebellions of the 1960s.

Families grew in average size, and migration from city to suburb increased, resulting in a doubling of the suburban population between the 1950s and 1970. Accompanying this migration and growth was an increase in the sales of items like lawn chairs, barbecue pits, and other implements of suburban leisure, as well as automobiles to carry suburbanites to their jobs and to centrally located shopping emporiums.

In the early 1950s, the elementary schools began to feel the impact of the Baby Boom. In 1952 alone—six years and a telltale nine months or so after the soldiers returned home following the end of World War II—there was a 39 percent increase in the number of six-year old first-graders. During the preteen years of the Baby Boomers in the 1950s, the United States witnessed the massive growth of television. The Boomers watched and were socialized by the likes of Howdy Doody and the Mousketeers. As the Baby Boomers entered their teens, the sale of popular records quadrupled, and the sale of blue jeans and acne medicines soared. Maturing from their late teens into their early twenties, the Boomers ignited the "Sixties Generation"; many Boomers of the 1950s became the "Hippies" of the 1960s. The ten-year period from roughly 1964 to 1974 was a period that included protest against the Vietnam War; experimentation with new religions and new lifestyles; fashion statements that were to make past and future eras flinch; and notorious recklessness with marijuana, LSD, cocaine, and other drugs.

Aging Baby Boomers enjoy themselves at the twenty-fifth reunion in 1994 of the original Woodstock Festival (summer of 1969).

Today, the Baby Boom cohort is in its late thirties to early fifties. Attention has shifted from the Hippies of the 1960s to the Yuppies of the 1990s. The name *Yuppie* derives from "Young Urban Professionals," meaning any ambitious, fast-consuming, trend-setting individual of a certain age. From the generation that eschewed material wealth in the 1960s has emerged the Me generation and the Greed generation, and the famous movie line mouthed by Gordon Gecko (named for a reptile) who thrilled a fictional audience when he announced "Greed is good." This cohort in the 1980s made *Jane Fonda's Workout Book* a bestseller as they fought off age with exercise. Since about 1975, when large numbers of Baby Boomers began to have babies of their own (called "Yuppie puppies" by some), the birthrate has generally been rising. The Boomers waited to have children as they pursued careers, and they have since begun to make up for lost time. As a result, the United States may be at the beginning of another Baby Boom, an "echo" of the original "boom" after a three-to-four-decade lag.

As they pass age sixty-five, the Baby Boom cohort will greatly increase the ranks of the elderly. One effect will be an increase in political clout for those over age sixty-five. A heavy burden will befall those born between 1965 and 1975 since they will be the main contributors to the Social Security fund just as the fund will be required to meet the needs of the giant Boomer population bulge is it comes to rest in its retirement years (Persell, 1990; Kennedy, 1989; Robey, 1982).

Theories of Population Growth Locally and Globally

Among the major problems facing modern-day civilization is the specter of uncontrolled population growth. As noted earlier, the population of the world increases by about 270,000 people every day. Some view overpopulation as an epochal catastrophe about to roll over us like a tidal wave. Others dispute whether the problem exists at all, pointing out that there is no scientific consensus on the *carrying capacity* of the planet, meaning the number of people the planet can support on a sustained basis, and that technological advances, which have dependably met our needs in the past, can be counted upon to do so in the future as the number of mouths to feed continues to grow.

Is the world overpopulated? Is the United States? What can we expect from the future? If the less optimistic scenarios turn out to be accurate, these could be the most important questions facing humankind.

Bubonic plague (called the "Black Death") severely decreased the entire population in Medieval Europe.

Malthusian Theory

Humans, like other animals, can survive and reproduce only when they have access to the means of *subsistence,* meaning the necessities of life, such as food. Human and animal populations have in common that they are prone to decline and die off when encountering checks on population growth such as famine, disease, and war. In the face of a daunting environment, humans have managed to thrive, and their population has doubled many times over. The period of doubling gets shorter and shorter.

Thomas R. Malthus, a clergyman born in 1766 in Scotland and educated in England, pondered the realities of life on earth and assembled his observations into a chilling depiction of disastrous population growth called **Malthusian theory,** the idea that a population tends to grow faster than the subsistence needed to sustain it. Malthus propounded his gloomy views in *Essay on the Principle of Population* (Malthus, 1798), and continued to revise and extend his theory until his death in 1834. In sum, Malthus declared that the earth must be near the limits of its ability to support so many humans and that the future must inevitably hold catastrophe and famine.

Malthus noted that populations tend to grow not by *arithmetic increase,* adding the same number of new individuals each year, but by *exponential increase,* in which the number of individuals added each year grows, with the larger population generating an even larger number of births with each passing year. Arithmetic increase would cause a population to double in size at a decelerating rate. Exponential increase, on the other hand, causes a population to double ever faster. It took 200 years for the population of the world to double from half a billion in 1650 to a billion in 1850. From two billion to four billion took forty-five years in this century. The mathematical power of doubling is awesome when applied to population. If we were to start with just one couple and imagine a lineage that doubled itself each generation by having 4 children, a mere 32 generations, roughly 600 years, would result in a population of 8.4 billion—almost twice the population of the world today. To put Malthus's fears in perspective, the average number of births per couple in the United States at the time Malthus was writing was seven!

Malthus noted that unchecked doubling of any population must swiftly spawn enough people to carpet the earth many times over. Clearly something prevents populations, human and other, from doubling every generation. Malthus reasoned that two forces were at work to keep population growth in check: First, the growth in the amount of food produced tends to be only arithmetic and not exponential. Second, Malthus surmised that there were three major *positive checks* on population growth: famine, disease, and war. In Malthus's time, disease could reach apocalyptic scales. The outbreak of bubonic plague in Europe from 1334–1354 eliminated a third of the population; a smallpox epidemic in 1707 wiped out three-fourths of the entire populations of Mexico and the West Indies. Wars had a giant appetite for European men, with deaths of men in battle causing semi-permanent gaps in the population pyramids of European populations. Along with positive checks on population growth, Malthus acknowledged what he called *preventive checks,* such as sexual abstinence, but he knew that sexual abstinence was unlikely to be the mechanism that halted uncontrolled population growth.

Malthus theorized that populations fluctuate over time in cyclic fashion. Populations first rise to the level of subsistence and slightly beyond. Once this occurs, one or more of the positive checks inevitably come into play. For example, overpopulation may cause a shortage of space, leading to war between groups that want more room, and war, along with other checks on population, will carve off some of the excess population until a subsistence level is again in place, followed by another rise in the birthrate. If war does not do the job, disease will. If not disease, famine. In the Malthusian view, life on earth was an endless trudge toward disaster caused by nature's way of reckoning with overpopulation.

Malthusian theory actually predicted rather well the population fluctuations of many agrarian societies such as Egypt from about 500 A.D. through Malthus's own lifetime. However, Malthus failed to foresee three revolutionary developments that have derailed his predicted cycle of growth and catastrophe. In agriculture, technological advances have permitted farmers to work larger plots of land and to grow more food per acre, resulting in subsistence levels higher than Malthus would have predicted. In medicine, science has fought off diseases that Malthus expected to periodically wipe out entire nations, such as the bubonic plague. Finally, the development and widespread use of contraceptives in many countries have kept the birthrate at a level lower than Malthus would have thought possible.

The technological victories of this century have not completely erased the specter of Malthus. The worldwide AIDS epidemic warns us that nature can still hurl catastrophes our way, and heartrending pictures of swollen, starving babies remind us that famine continues to destroy human populations in some parts of the world just as it has for thousands of years. Overall, Malthus's theory has served as a warning that subsistence and natural resources are indeed limited. The Malthusian doomsday has not occurred, but there are those who believe Malthus's warning was not in error, just premature. We will return to this point of view after learning more about how populations change.

Demographic Transition Theory

An alternative to Malthusian theory is the theory of **demographic transition,** developed initially in the early 1940s by Kingsley Davis (Davis, 1945) and extended by Harris (1979), Cutright and Hafgens (1984), and others. Demographic transition theory proposes that countries pass through a consistent sequence of population patterns linked to the degree of development in the society, and ending with a situation in which the birthrates and death rates are both relatively low. Overall, the population level is predicted to eventually stabilize, with little subsequent increase or decrease over the long term.

There are three main stages to population change, according to demographic transition theory (see Figure 20.3).

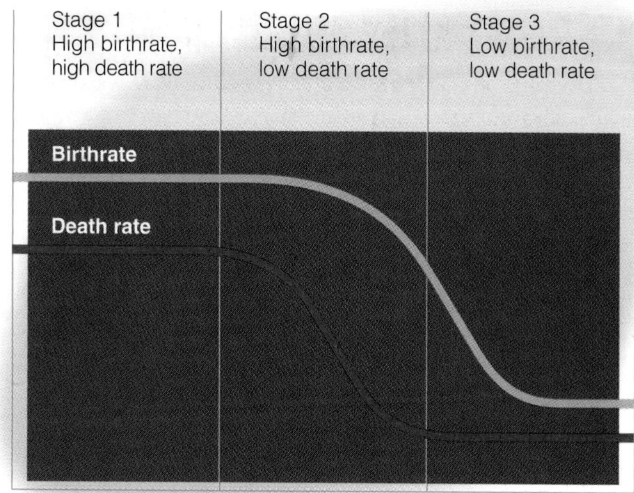

FIGURE 20-3 Demographic Transition Theory

Stage 1 is characterized by a high birthrate and high death rate. The United States during its colonial period was in this stage. Women began bearing children at younger ages, and it was not uncommon for a woman to have twelve or thirteen children—a very high birthrate. However, infant mortality was also high, and so was the overall death rate owing to primitive medical technique and unhealthy sanitary conditions.

Debunking Society's Myths

Myth: **The average number of children per family in the United States may have decreased a little bit since Malthus's time (around 1800), but not by much.**

Sociological perspective: **The average number of children has been about 2.1 per family in the United States for the last 10 years or so (which is close to Zero Population Growth level), but it was 7 children per family in Malthus's time! This may have seemed high even for Malthus—which is one reason he became so concerned with the effects of overpopulation.**

Stage 2 in the demographic transition is characterized by a high birthrate but a low death rate. Hence, the overall level of the population increases. The United States entered stage 2 in the second half of the nineteenth century as industrialization took hold in earnest. The norms of the day continued to encourage large families, and thus high birthrates, while advances in medicine and public sanitation whittled away at the infant mortality rate and the overall death rate. Life expectancy increased, and the population grew in size.

The characteristics of stage 2 did not apply across all social groups or social classes. Minorities of the day (Blacks, and especially Native Americans and Chinese in the Midwest and West) were less likely to benefit from medical advances, and the infant mortality and overall mortality rates for Blacks,

Native Americans, and Chinese remained high, while life expectancy remained considerably shorter than for the White population. Lack of access to quality medical care was particularly devastating to Native Americans, who had very high death rates for all ages during this period. Demographic transition theory is not completely accurate for different racial–ethnic groups within the same society.

Stage 3 of the demographic transition is characterized by a low birthrate and low death rate. The overall level of the population tends to stabilize in stage 3. Medical advances continue, the general prosperity of the society is reflected in lowered death rates, and cultural changes take place, such as a lowering of the family size most people consider ideal. The United States entered this stage prior to World War II, and with the notable exception of the Baby Boom, it has exhibited stage 3 demographics since then. Again, however, the trend does not apply equally to all groups in U.S. society. Some groups, such as African Americans and Hispanics, continue to have high infant mortality rates and higher death rates owing to the disproportionate impact of disease upon these groups, as well as disproportionate lack of access to health resources. American Whites are further along into stage 3 than the rest of American society.

The "Population Bomb" and Zero Population Growth

In 1968, a modern-day Malthus appeared in the form of Paul Ehrlich, a biology professor at Stanford University. His book, *The Population Bomb* (Ehrlich, 1968), was the first in a series of writings in which Ehrlich argued that many of the dire earlier predictions of Malthus were not far wrong. The growth in world population, according to Ehrlich, was a time bomb ready to go off in the near future, with dismal consequences. Supporting his argument with scientific research (a commodity used rather loosely in the work of Malthus, whose reasoning depended heavily on speculative guesswork about how many people the planet could and should bear), Ehrlich stated that the sheer mathematics of population growth worldwide were sufficient to demonstrate that world population could not possibly continue to expand at its present rates.

Ehrlich pointed out that worldwide population growth has outgrown food production, and that massive starvation must inevitably follow. Ehrlich went beyond Malthusian theory, however, to state that the problem transcended food production and was in fact a problem of the environment. Ehrlich was among the earliest thinkers in modern times to argue that the quality of the environment, especially the availability of clean air and water, was a critical factor in the growth and health of populations.

Many of the dire predictions Ehrlich made back in 1968 have come true. In a later book Ehrlich wrote with his wife Anne, *The Population Explosion* (Ehrlich and Ehrlich, 1991), the authors pointed to a variety of disasters that confirmed the predictions of the 1968 work, and even of Malthus: mass starvation in parts of Africa; starvation in the United States among Black and Hispanic populations; increased homelessness in cities, especially among minorities; acid rain; rampant extinctions of plant and animal species; and the irrecoverable destruction of environments like the rain forests.

Ehrlich has stated that only limits on population growth can avert disaster. He has advocated the implementation of Malthusian preventive checks by the government, such as supporting organizations like Zero Population Growth (ZPG), a group dedicated to reaching the **population replacement level,** a state in which the combined birthrate and death rate of a population simply sustains the population at a steady level. To achieve zero population growth, couples would have to limit themselves to approximately two children per family.

By 1980, the United States had indeed reached the replacement level of reproduction with an average number of children per family, ignoring race–ethnic and class differences, of 2.1. Organizations such as Planned Parenthood, the Guttmacher Institute, and ZPG have successfully promoted birth control techniques—especially among middle-class Whites—and the birthrate has declined since Ehrlich's

TABLE 20.2 *A COMPARISON OF DEMOGRAPHIC THEORIES*

	Malthusian Theory	Demographic Transition Theory	Zero Population Growth (ZPG) Theory
Main point	a population grows faster than the subsistence (food supply) needed to sustain it.	populations go through predictable stages from high birth and death rates to a stable population with low birth and death rates.	achievement of zero population growth solves the Malthusian problem of unchecked population growth.
"Positive" checks on population growth	famine, disease, and war are likely.	famine, disease, and war are moderately likely.	famine disease, and war are unlikely.
"Preventive" checks on population growth	sexual abstinence.	sexual abstinence, birth control, and contraceptive methods.	sexual abstinence, birth control, and contraceptive methods.
Predictions for the future	pessimistic: despite positive and preventive checks, a population will ultimately outstrip its food supply.	optimistic, given technological and medical advances in a population.	very optimistic: zero population growth has already been achieved in the United States and other countries.

first warnings in the late 1960s. Furthermore, agricultural developments such as new hybrid plants and environmentally safe fertilizers have increased crop yields in both the United States and other countries.

THINKING SOCIOLOGICALLY

What is the one major demographic change that would have to take place for a society to achieve *zero population growth* (ZPG)? Can you think of the advantages of ZPG? Any disadvantages? What effect would ZPG likely have on the *demographic transition* process? What is a major sociological barrier that would cause a population to resist adopting a governmental policy of birth control?

Despite these advances, worldwide population growth is still out of control. Population-related maladies like urban crime, unemployment, and increasing homelessness remain a scourge of worldwide populations as well as in the United States, adding a twentieth-century twist to the positive checks of the past, such as disease and war, which are still with us. Pollution of the environment remains a major threat. The Ehrlichs deliver a final message: The population bomb must inevitably explode, they say. What remains in doubt is whether we will arrive at that historic moment having made humane preparations, perhaps mitigating the problem with preventive checks like a lowered birthrate, or whether the population crisis will end tragically with a realization of Malthus's vision of people starving worldwide and disease rampaging through a weakened population (Ehrlich and Ehrlich, 1991).

Checking Population Growth

As early as the 1950s, most countries had accepted that population growth was a problem that must be addressed. By the 1980s, countries representing 95 percent of the world's population had formulated policies of some kind aimed at stemming population growth; however, consensus does not exist on how population should be controlled. For example, many religious and political authorities have argued against the use of birth control. Efforts to encourage the use of contraceptives among rapidly growing populations have therefore had mixed support, whereas other attempts to curb population growth, such as encouraging changes in social habits, generally meet with poor success.

Family Planning and Diversity

Many governments, including that of the United States, make contraceptives available to individuals and families. Doing so, however, is not always consistent with the beliefs and cultural practices of all groups in the society. Catholics are taught that it is acceptable to use natural means of birth control (such as the "rhythm method") but are forbidden to use contraceptive devices such as an IUD (intrauterine device). Many Catholics choose to use contraceptives anyway; some studies, as already noted in Chapter 16, have shown

A field worker in Bangladesh explains birth control pills and their use to village women.

that the *majority* of American Catholics practice forms of birth control forbidden by their church.

Governmental programs that advocate contraception can be successful only if couples themselves choose smaller families over larger ones. This is most likely to occur if the wider culture supports that decision. Creating a new image of the ideal family size is a central concern of those involved in the family planning movement.

Birthrate and family size are known to be related to the overall level of economic development of a country, or even the economic status of certain ethnic groups within a country. As countries in general become more economically developed, their birthrates and average family size generally drop, as predicted by demographic transition theory. The more industrialized countries generally prefer smaller families. Developing countries tend to have higher birthrates and tend to give greater cultural value to large families. The same is true of ethnic and racial minorities living in developed nations but not entirely assimilated, such as Hispanics in the United States. Large families are seen as a demonstration of potency, a source of needed labor, and as a preparation for old age, when offspring will be expected to care for their elders. Before contraceptive devices such as condoms, the IUD, and the pill will be widely adopted, cultural values in favor of undisturbed fertility must be countered (Vanlandingham, 1993; Westoff et al., 1990).

A recent large-scale study has called into question the assumed relationship between economic development and family size in demographic transition theory (Stevens, 1994). The study shows that countries in stage 2, such as Bangladesh, which had a declining death rate because of rapid economic and medical developments but retained a high birthrate, have lately become very receptive to birth control programs. As a result, the birthrate as well as the population level have begun to decline even in the absence of Western-style advanced development. Contraceptive programs have worked in other non-Western countries as well, such as Thailand and Colombia. Bangladesh is an especially interesting case because its population density is unusually high to begin with. The findings in this study show that a country can achieve a rapid decline in birthrate even in relatively early stages of development if an aggressive government-sponsored contraception program is implemented.

Population Policy and Diversity

Findings like those in Bangladesh have an important bearing on population policy elsewhere, including in the United States. Under the Republican administrations of Presidents Reagan and Bush, from 1980 through 1992, Washington refused to finance international or domestic population programs that provided women with counseling on abortion. The Clinton administration, arguing that population control is crucial to the world's future, reversed that policy. Family planning programs offer great potential for achieving large declines in birthrates. In fact, in underdeveloped, overpopulated countries where such programs can have the most effect, the demand for family planning resources surpasses the supply. The Clinton administration responded to this demand in their family planning initiatives for the United States and abroad.

There is some cultural resistance from some U.S. racial and ethnic groups to government-sponsored contraceptive programs. This includes some Hispanic groups and some African American groups. The argument is that decreasing the birthrate by means of contraceptive methods is genocidal, and threatens the very survival of members of these groups. If such programs may be sponsored by the federal government, such programs are perceived as direct governmental attempts to reduce the number of minority people in the country. To the extent that the government promotes such programs disproportionately in Hispanic or African American communities and promotes them less aggressively in traditionally White or upper-class areas, then such governmental programs are perceived to be racist. As a consequence, the individual is less likely to be convinced to adopt some sort of contraceptive method or device.

At the borderline of policy and social custom are still other cultural barriers against the use of contraceptive methods, such as the belief among some groups, some young urban men, for instance, that condoms are unmasculine. Still, the popularity of condoms and other means of contraception is rising, with sterilization (including vasectomies) leading the list, followed by the pill and the condom. Because of controversy over the side effects of the pill, its use has not increased as rapidly as sterilization and the condom. Similarly, the IUD, especially one called the Dalkon Shield, has been criticized for side effects such as fibroid uterine tumors, which may possibly result from prolonged IUD use, and thus amid numerous lawsuits its use has been virtually wiped out. The use of abortion as a birth control method has also increased (Alan Guttmacher Institute, 1997; Ehrlich and Ehrlich, 1991).

This picture of Phoenix, Arizona, shows the extensiveness of urban sprawl.

Urbanism

The growth and development of cities, or centers of human activity with high degrees of population density, is a relatively recent occurrence in the course of human history. Scholars locate the development of the first city at around 3500 B.C. (Flanagan, 1995). The study of the urban, the rural, and the suburban is the task of *urban sociology,* a subfield of sociology that examines the social structure and cultural aspects of the city in comparison to rural and suburban centers. These comparisons involve what urban sociologist Gideon Sjoberg (1965) calls the *rural–urban continuum,* those structural and cultural differences that exist as a consequence of differing degrees of urbanization.

Urbanism as a Lifestyle

Early German sociological theorist Georg Simmel (1905 [1950]) argued that urban living had profound social psychological effects on the individual. Thus, he was among the early theorists who argued that social structure could affect the individual. He argued that urban life has a quick pace and is stimulating, but as a consequence of this intense style of life, the individual becomes insensitive to people and events that surround them. The urban dweller tends to avoid emotional involvement, which according to Simmel was more likely to be found in rural communities. Interaction tends to be characterized as economic rather than social, and close, personal interaction is frowned upon and discouraged. The sociologist Louis Wirth (1938), focusing on Chicago in the 1930s, also argued that the city was a center of distant, cold interpersonal interaction, and that as a result the urban dweller experienced alienation, loneliness, and powerlessness. One positive consequence of all this, according to both Wirth and Simmel, was the liberating effect that arises from the relative absence of close, restrictive ties and interactions. Thus city life offered the individual a certain feeling of freedom.

A contrasting view of urban life is offered by Herbert Gans (1962 [1982]), who studied Boston in the late 1950s and concluded that many city residents develop strong loyalties to others and are characterized by a sense of community. Such subgroupings he referred to as the *urban village.* Such urban villages are characterized by several "modes of adaptation," among them *cosmopolites,* who are typically students, artists, writers, and musicians, who together form a tightly knit community and who choose urban living in order to be near the city's cultural facilities. A second category are the *ethnic villagers,* those who live in ethnically and racially segregated neighborhoods. Today's *urban underclass* would encompass what Gans called the *trapped* of late 1950's Boston, individuals who are unable to escape from the city because of extreme poverty, homelessness, unemployment, and other familiar urban ills.

Race, Class, and the Suburbs

The impact of race and class can clearly be seen in the distinction between city and suburb. Today, only about one-fourth of African Americans live in suburban areas. Echoing the earlier *urban villagers* principle of Gans, closely knit communal subgroups form in the suburbs, but the subgroupings tend to form on the basis of class and race, with class according to some (such as DeWitt, 1995) being more important than race. In the suburbs, one chooses ones neighbors and friends on the basis of educational and occupational similarity in addition to race.

Pleasant though the suburb can be, it is also true that people of color, particularly African Americans, often become as segregated there as they do in cities (Feagin and Sikes, 1994; Massey and Denton, 1993). Racial segregation persists not only in suburban neighborhoods (Chicago suburbs are a case in point) but in schools as well. Although some may argue that moving to the suburbs and taking up residence is a matter of personal choice, there is ample evidence that residential segregation is maintained by landlords, homeowners, and White realtors who engage in steering people of color to segregated neighborhoods (Feagin and Sikes, 1994). The practice of *redlining* by banks, which renders it impossible for a person of color to get a mortgage loan for a specific property, further intensifies residential segregation. Finally, such practices serve only to encourage further segregation in the realm of interpersonal interaction.

Ecology and the Environment

It should be apparent by now that population size has an important social dimension. Social forces can cause changes in the size of the population, and population changes can transform society. Values are acquired from one's culture, including values about what family size is considered ideal, or what degree of crowding in a city is considered tolerable. The values people hold will affect the number of children they want, the place they want to live, and even the degree to which they consider population control an urgent problem.

Population density is the number of people per unit of area, usually per square mile. As population density rises to high levels, as it has in today's cities, the familiar problems of urban living appear, such as high rates of crime and homelessness. Interacting with these problems are crises of the physical environment, such as air and water pollution, acid rain, and the growing output of hazardous wastes. Humans sometimes forget that like all other creatures, they are intimately dependent on the physical environment. **Human ecology** is the scientific study of the interdependencies that exist between humans and their physical environment. A **human ecosystem** is any system of interdependent parts that involves human beings in interaction with one another and the physical environment. A city is a human ecosystem; so is a rural farmland community. In fact, the entire world is a human ecosystem. Human ecology is the study of human ecosystems. Two fundamental and closely related problems confront our present ecosystems: overpopulation and the destruction or exhaustion of natural resources (Hawley, 1986).

Vanishing Resources

In all ecosystems, whether human, animal, or plant, organisms depend for their survival upon one another as well as upon the physical environment. Plants use carbon dioxide and give off oxygen, which all animals need to survive. Terrestrial creatures metabolize oxygen and produce carbon dioxide, which is then used by plants, completing a circle. Humans and other animals require nutrients they can get only by eating plants. When they die, they decompose and provide nutrients to the soil that are taken up by plants, completing another circle.

The examination of ecosystems has demonstrated two things. First, the supply of many natural resources is finite. Second, if one element of an ecosystem is disturbed, the entire system is affected. For much of the history of humankind, the natural resources of the earth were so abundant compared to the amounts used by humans that they may as well have been infinite. No more. Some resources, like certain fossil fuels, are simply nonrenewable, and will be gone soon. Others resources, like timber or seafood, are renewable so long as we do not plunder the sources of supply so recklessly that they disappear, an ecological blunder we have made many times before. Finally, there are natural resources so abundant that they *still* seem infinite, such as the planet's stock of air and water, but at this stage of our technological development, we are learning that our powers extend to such heights and depths that we can even destroy the near-infinite resources—on the planetary scale, our gaseous wastes are gnawing away at the ozone layer, and our buried chemical wastes are trickling into the water table and creating underground pools of poison.

One of the best demonstrations of how each part of the ecosystem affects all the other parts was seen in the use of

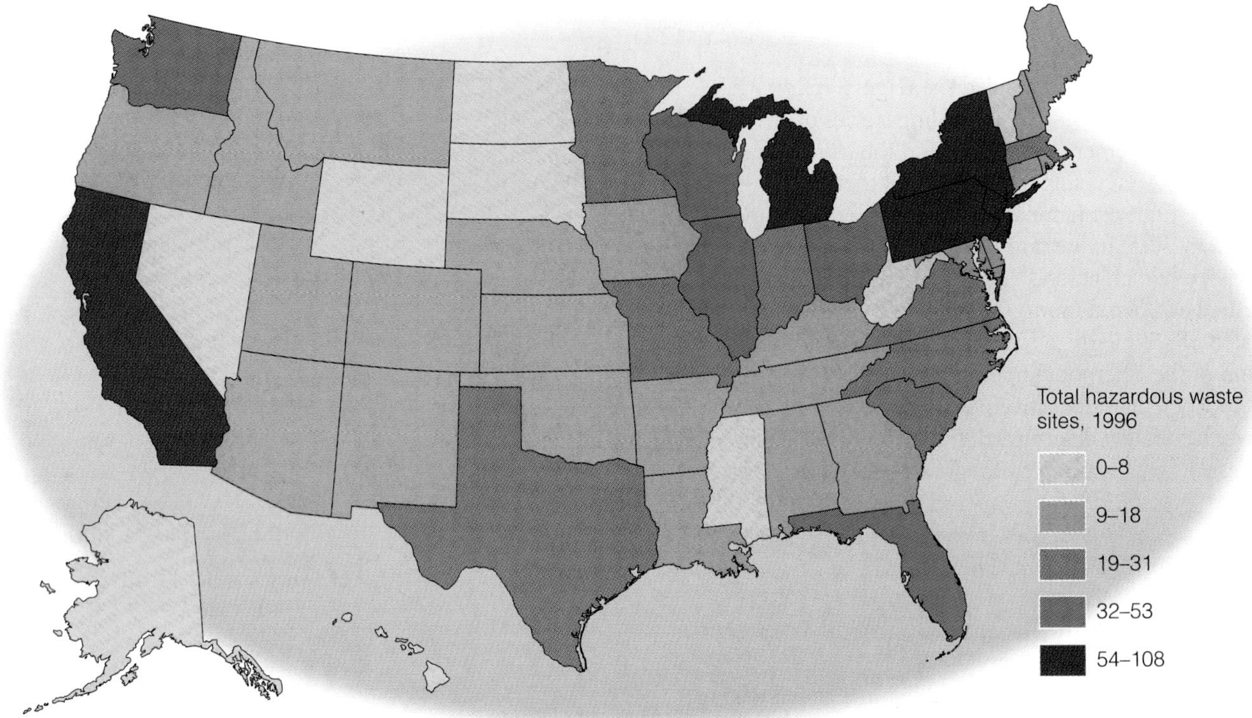

MAP 20.1 Total Hazardous Waste Sites by State

DATA: From the U.S. Environmental Protection Agency, 1994. *Supplementary Materials: National Priorities List.* Washington, DC.

DDT in the United State during the 1940s and 1950s. *D*ichloro-*D*iphenyl-*T*richloroethane was sprayed on plants by farmers and suburbanites to kill a variety of pests. After just a few years, DDT had seeped through the soil, into the groundwater, on into the seas, then into fish, and finally into birds that ate the fish. Birds that accumulated DDT in their systems appeared healthy enough, but their numbers plummeted. It was soon discovered that DDT caused a disastrous brittleness in the eggs laid by the birds, decimating succeeding generations. The chemical also found its way into the human food supply, with dangerous consequences such as the contamination of human breast milk in the United States (Carson, 1962; Ehrlich and Ehrlich, 1991). By the time DDT was identified as a major environmental hazard, tremendous damage had already been done.

If growing population is a problem of the developing world, shrinking resources are a problem of the industrialized world. The United States uses more than 40 percent of the world's aluminum and coal, and about 30 percent of its platinum and copper (Ehrlich et al., 1977; Ehrlich and Ehrlich, 1991). According to some observers, within forty years we will reach the end of the world's supply of lead, silver, tungsten, and mercury, mainstays of heavy and high-tech manufacturing, including the computer hardware industry.

In the United States alone, real estate development takes over millions of acres of farmland each year. In the western and southwestern United States, the groundwater supply is being depleted at a rapid pace. We are racing through our nonrenewable natural resources, and destroying much that should be renewable; however, some activists have begun to claim that the environmental situation, though perilous, is improving (Simon, 1995). There is evidence to support that view, but even optimists who wish to show that things are getting better are quick to point out that "better" is not sufficient if the situation is bad enough to begin with (for example, Montagne, 1995, in a review of Simon).

Environmental Pollution

It would be a bitter irony if we managed to avoid exhausting the resources of the planet, only to find that what we had conserved was too degraded by pollution to be of any use. The most threatening forms of pollution are the poisoning of the planet's air and water. Air pollution is not only ugly and uncomfortable, it is deadly. The skies of all major cities around the world are stained with pollution hazes, and in cities that rest within geological basins, such as Mexico City and Los Angeles, the concentrations of pollutants can rise so high that pollution-sensitive individuals cannot leave their homes. The numbers of respiratory cases in the hospitals rise and fall with the passing of weather systems that cause the pollutants to concentrate or disperse.

Water pollution has an especially insidious side to it—most people never see how much is dumped into the world's waterways. They are not present far off shore when the interior walls of vast tankers, acres upon acres of fouled surface area, are rinsed with hot sea water and the waste flushed into the ocean. Small factories dump invisibly into canals that feed streams that lead to the sea. The leading air and water polluters are the United States, Japan, Russia, and Poland. When glasnost ended the reign of secrecy in the former Soviet Union, bloodcurdling stories of environmental vandalism emerged from behind the Iron Curtain. Nuclear-powered ships at the ends of their useful lives were blithely sunk, contaminated reactors, spent fuel and all, in Antarctic seas. In the farther reaches of the giant Russian territory (which crosses eleven time zones), areas were designated as open dumping sites for toxic wastes and then sealed off. Similarly, a toxic quarantine area exists downwind of the Chernobyl nuclear power plant that blew up in 1986, spewing radioactive poisons over Europe in amounts never accurately determined, but now appearing to be far in excess of what was once thought to be the worst case imaginable. For comparison, the Three Mile Island nuclear accident that occurred in the United States in 1979 released 15–20 curies of radioactive iodine-131 into the environment; Chernobyl is now estimated to have released as much as *50 million* curies of the same dangerous isotope. Incredibly, the Chernobyl nuclear complex is still online, with the exploded reactor in its midst encased in a giant sarcophagus of concrete. The government of Ukraine has resisted great pressure from foreign nations to shut Chernobyl down completely, for a reason that any of the industrialized, high-polluting societies would be hard-pressed not to respect—they say they need the power.

A huge portion of the pollutants released into the air come from the exhaust pipes of motor vehicles. The major component of this exhaust is carbon monoxide, a highly toxic substance. Also found in exhaust fumes are nitrogen oxides, the substances that give smog its brownish yellow tinge. The action of sunlight causes these oxides to combine with hydrocarbons also emitted from exhausts, forming a host of health-threatening substances.

On the industrial side, the Environmental Protection Agency (EPA) has estimated that hazardous and cancer-causing pollutants released into the air by industry are responsible for about 2000 or more deaths a year (Scarpitti et al., 1997). Electric utility companies and other industries often burn low-grade fossil fuels that emit harmful sulfur dioxide. When mixed with other chemicals normally present in the air, sulfur dioxide turns into sulfuric acid, which gets carried back to earth in droplets of *acid rain.* Acid rain can change the acidity of lakes, soil, and forests so severely that they no longer support life.

A daunting international issue has grown around a group of chemicals called chlorofluorocarbons (CFCs), used as a coolant in refrigerators, the manufacture of plastics, and as an aerosol propellant. CFCs released into the air find their way to the ozone layer in the upper atmosphere, where they eliminate the highly reactive ozone. The ozone layer is a shield that blocks dangerous ultraviolet light, and as this shield is destroyed, more ultraviolet light gets through, causing an increase in sunburn, skin cancer, and other illnesses. In 1987, a treaty named the Montreal Protocol on Substances that Deplete the Ozone Layer called for drastic reduction in the use of CFCs, but the alternatives are expensive, and developing nations that are being called upon to forgo refrigeration, or at least limit themselves to the use of unmanageably expensive refrigerants, are instead insisting that the developed world, having enjoyed for years the use of the polluting compounds, should underwrite the costs of alternatives for the poorer nations if they wish the clean alternatives to be used.

This power plant in Moscow, Russia, spews pollution into the environment, a problem affecting nearly all major industrial cities throughout the world.

Related to the problem of ozone depletion is the **greenhouse effect.** As the sun's energy pours onto the earth, some is reflected from the earth's surface back out again. Of the reflected energy, a portion is captured by carbon dioxide in the earth's atmosphere, while the rest radiates into space. If the amount of solar energy trapped by carbon dioxide rises, the temperature goes up. Rather small changes in the average temperature of the earth can have dramatic consequences. A few degrees' difference can cause greater melting in the arctic regions, which raises the level of the sea, which can affect water, land, and weather systems worldwide. The exact effects are as unpredictable as the weather itself.

The existence of the greenhouse effect is much debated. It is known that the level of carbon dioxide in the atmosphere has risen steadily since the Industrial Revolution began, and there are indications that the world temperature is rising, but some skeptical authorities insist that the temperature changes are within normal ranges of fluctuation. In the opposing camp are people who claim that even without airtight proof of global warming, it is reasonable to fear the worst and take steps to diminish carbon dioxide production, especially since the consequences of ignoring warming are unpredictable and potentially catastrophic.

In the face of warning about water pollution, many people take comfort in the vastness of the ocean—three-quarters of the earth is covered with water; however, only a tiny sliver of the planet's water is usable by humans. Nearly all the water on earth is sea water—too salty too drink.

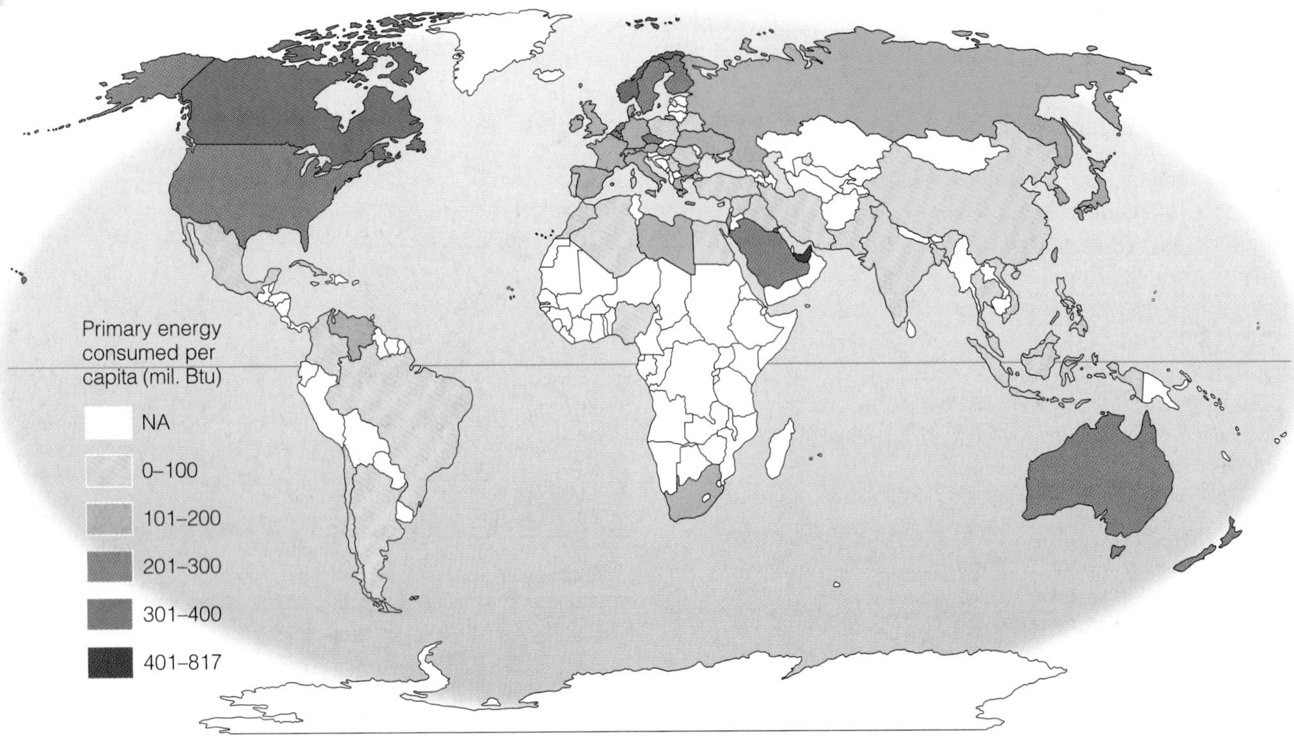

MAP 20.2 Energy Consumption Per Capita

DATA: From the Bureau of the Census, 1997. *Statistical Abstract of the United States: 1997*. Washington, DC: U.S. Department of Commerce, p. 851.

Of the fresh water on earth, most is locked in the polar ice caps. Of the reminder, most is inaccessible groundwater. All that is left and available to us is the trickle of the planet's river systems and its lakes, and this fragile supply we are polluting. Since about 1955, it has been clear that water is not the infinite resource it was once thought to be (Rice, 1986). The nation's rivers and lakes have long been dumping grounds for heavy industry; yet these same industries—paper, steel, automobile, and chemical—depend upon clean water for their production processes, during which they take water from the rivers and lakes and return it heated up and polluted. The difference in temperature can alter aquatic habitats and kill aquatic life, earning it the name **thermal pollution.** The chemical pollutants that industry discharges into rivers, lakes, and the oceans include solid wastes, sewage, nondegradable byproducts, synthetic materials, toxic chemicals, and even radioactive substances. Add to this the polluting effects of sewage systems of towns and large cities, detergents, oil spills, pesticide runoff, and runoff from mines, and the enormity of the problem is clear.

The EPA estimates that 63 percent of rural Americans may be drinking water that is contaminated as a result of agricultural runoff and the improper disposal of toxic substances in landfills. Thousands of rural water wells have been abandoned because of contamination. Households served by municipal water systems are also endangered; fully 20 percent of the country's public water systems do not meet the minimum standards set by the government (Scarpitti et al., 1997; Shaberoff, 1988).

Federal and state statutes now prohibit industry from polluting the nation's water, but the pollution continues. Why? The answer is economic, political, and sociological. Industries that contribute to a vigorous economy have traditionally met little interference from the government. Only recently, and only because of public awareness and outrage, has the government started cracking down on major polluters. Nevertheless, the federal government chooses a look-the-other-way attitude. This problem was addressed in 1993 by the inauguration of then Vice President Albert Gore's GLOBE project (Global Organization for a Better Environment). Administered by the Congressional Institute for the Future, GLOBE is an international organization that is attempting to stop all forms of pollution, particularly water pollution, through a combined effort by several major polluters, including the United States and Japan. Despite such national and international efforts, federal or local legal actions against major polluters often end with the corporation paying a relatively painless fine and continuing to pollute as appeals and further proceedings for new violations wind their way slowly through the cumbersome regulatory and judicial systems.

Many argue that of all the environmental problems facing the United States today, the most urgent is the dumping of hazardous wastes, if only for the sheer noxiousness of some of the materials being dumped. It is estimated that since 1970, the production of toxic wastes increased fivefold (U.S. Government Accounting Office, 1990). This dramatic increase in the amount and variety of hazardous waste production is traceable to new and profitable industrial tech-

Environmental Racism

Environmental racism refers to the pattern whereby people living in predominantly minority communities are more likely exposed to toxic dumping and other forms of pollution. Nuclear waste and testing in the American Southwest, for example, has been located in areas predominantly inhabited by Native Americans. In other areas, African Americans and Latinos are exposed to the effects of industrial waste.

nologies. The public creates great demand for products that inevitably produce hazardous wastes, such as insecticides and other useful poisons; products requiring mercury or lead (both acutely toxic when released into the environment); a variety of dyes, pigments, and paints; and an endless list of specialty plastics whose manufacture produces dangerous byproducts.

Environmental Racism and Classism

Adding to the social problem of toxic wastes is the fact that wastes are dumped with disproportionate frequency in areas that have high concentrations of minorities, particularly American Indians, Hispanics, and African Americans as well as persons of lower socioeconomic (social class) status (Boer et al., 1997; Pollock and Vittas, 1995; Bullard, 1994b; Williams, 1987). One study determined that it was "virtually impossible" that dumps were being placed so often in minority and lower SES communities by chance (Bullard, 1994a, 1994b). The same study found that communities with the greatest number of toxic dumps had the highest concentration of non-White residents. Such communities also tended to fall below the national average economically and educationally.

One study of households throughout Florida found that Native American, Hispanic, and particularly African American populations are disproportionately found to reside closer to toxic sources than are Whites. This pattern is *not* explainable by social class differences alone; that is, when communities of the same socioeconomic characteristics but different racial–ethnic compositions are compared, Native Americans, Hispanics, and African Americans of a given socioeconomic level are still closer to toxic dumps than are Whites of the *same* socioeconomic level. Figure 20.4 reveals this pattern quite clearly: A greater proportion of the ethnically minority households

Protesters demonstrate against a power plant proposal in a predominantly Black neighborhood.

are closer to toxic sites (TRIs, or Toxic Release Inventories) even when considering that all the households are of comparable low-income status. The authors of the study note that amount of exposure to the harmful effects of pollution is directly related to physical distance from a toxic facility, and that an *environmental inequity* exists among ethnic groups, with African Americans, Hispanics, and Native Americans bearing a greater share of the burden of exposure to the harmful effects of environmental pollution (Pollock and Vittas, 1995). In a similar study of Los Angles, Boer et al. (1997) found that lower social class areas were significantly nearer toxic waste dumps, as were areas that were predominantly Latino or African American. Generally, they found that communities most affected by toxic waste dumps were working-class communities of color located near industrial areas.

The largest commercial hazardous waste landfill in the nation is located in Emelle, Alabama, where Blacks make up nearly 80 percent of the population. In Scotlandville, Louisiana, the site of the fourth largest toxic landfill in the country, Blacks make up 93 percent of the population. In Kettleman City, California, the site of the fifth largest toxic

B • O • X 20.1

UNDERSTANDING DIVERSITY

Environmentalism and Social Justice

ROBERT Bullard (1994b) notes that the environmental movement in the United States emerged with agendas that focused on such areas as wilderness and wildlife preservation, resource conservation, pollution abatement, and population control. It was supported primarily by middle- and upper middle-class whites. Although concern about the environment cut across racial and class lines, environmental activism and actual participation have been most pronounced among individuals who have above-average education and greater access to economic resources.

Mainstream environmental organizations were late in broadening their base of support to include Blacks and other minorities, the poor, and working-class persons. The "energy crisis" in the 1970s provided a major impetus for the many environmentalists to embrace equity issues confronting the poor in this country and in the countries of the third world. Over the years, environmentalism has shifted from a "participatory" to a "power" strategy, where the core of active environmental movement is focused on litigation, political lobbying, and technical evaluation rather than on mass mobilization.

An abundance of documentation shows Blacks, lower-income groups, and working-class persons are subjected to a disproportionately large amount of pollution and other environmental stressors in their neighborhoods as well as in their workplaces. However, these groups have only been marginally involved in the nation's environmental movement. Problems facing the Black community have been topics of much discussion in recent years, and race has not been eliminated as a factor in the allocation of community amenities.

Pollution is exacting a heavy toll (in health and environmental costs) on Black communities across the nation. There are few studies that document, for example, the way Blacks cope with environmental stressors such as municipal solid waste facilities, hazardous waste landfills, toxic waste dumps, chemical emissions from industrial plants, and on-the-job hazards that pose extreme risks to their health. Coping in this case is seen as a response to stress and is defined as efforts, both action oriented and intrapsychic, to manage—that is, master, tolerate, reduce, minimize—environmental and internal demands, and conflicts among them, that tax or exceed a person's resources. Coping strategies employed by individuals confronted with a stressor are of two general types: *problem-focused coping* (for example, individual or group efforts to directly address the problem) and *emotion-focused coping* (for example, efforts to control one's psychological response to the stressor). The decision to take direct action or to tolerate a stressor often depends on how individuals perceive their ability to do something about or have an impact on the stressful situation.

SOURCE: Bullard, Robert. 1994. *Dumping in Dixie: Race, Class, and Environmental Quality.* Boulder, CO: Westview Press.

landfill, more than 78 percent of the residents are Hispanic. In many cases, the siting of the landfills was linked to the economic interests of both residents and corporations. Companies with profit in mind negotiated favorable deals with the residents of these communities to permit the dumping of wastes, often misrepresented as nontoxic, in exchange for jobs and other economic incentives, which were often slow in coming (Bullard, 1994b).

Debunking Society's Myths

Myth: **Environmental pollution is in fact more common in or near economically poor areas, but race has nothing to do with it.**

Sociological perspective: **Even when areas of the same low economic status (social class) but different racial compositions are compared, those with a greater percent of minorities are on average closer to polluted areas than those with a lesser percent of minorities.**

From the 1950s through the 1970s, the Navajo population of Shiprock, New Mexico, was exposed to waste from uranium mining and dumping that was 90 to 100 times more radioactive than the level permissible by law. Kerr-McGee, the corporation involved, was forced out of the area in the early 1970s. When it left, it simply abandoned the site, leaving 70 acres of radioactive mine tailings (the residue from the separation of ores). An even worse situation developed during the same period in Laguna, New Mexico, involving the Pueblo Indians. Anaconda Copper, a subsidiary of the Atlantic-Richfield Corporation, virtually wrecked the traditional Pueblo economy, recruiting the community's youth for hazardous jobs even as it contaminated their environment with the wastes from uranium mining. A high rate of cancer deaths in both of these communities serves as testimony to the horrible consequences of carelessly discarding toxic wastes (Churchill, 1992).

THINKING SOCIOLOGICALLY

The *human ecosystem* can be adversely affected by environmental hazards such as toxic wastes. Have you ever witnessed the effects on a population of a major environmental hazard disaster such as Love Canal or Three Mile Island? Did you grow up in or near such an area, or know someone who did? From your observations, are toxic waste dumps more likely to be in or near areas that are working class or heavily occupied by people of color, as shown by research? Quickly survey a few women you know, and then a few men, and find out how concerned they are about environmental issues. Do the women on average show more concern—as shown by past research?

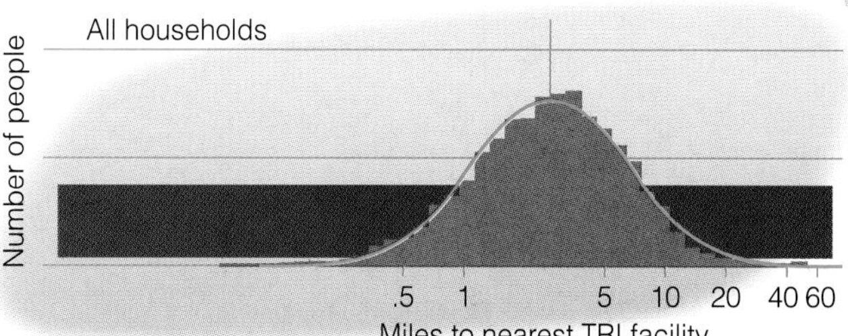

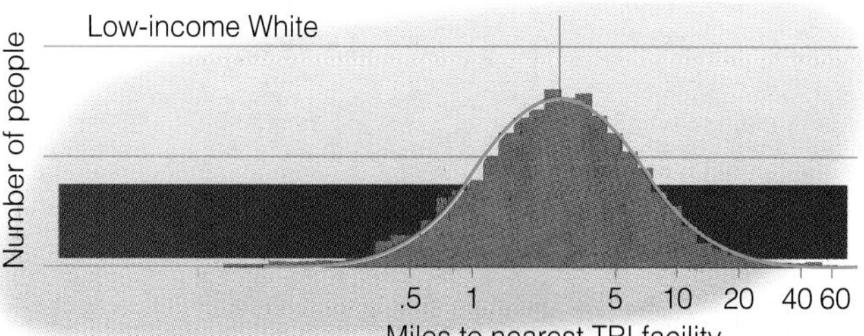

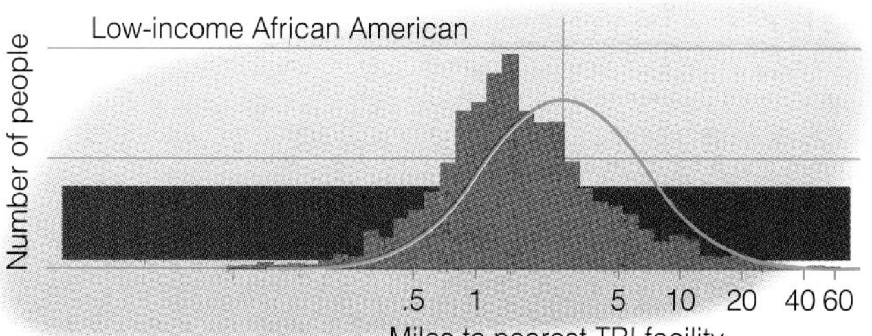

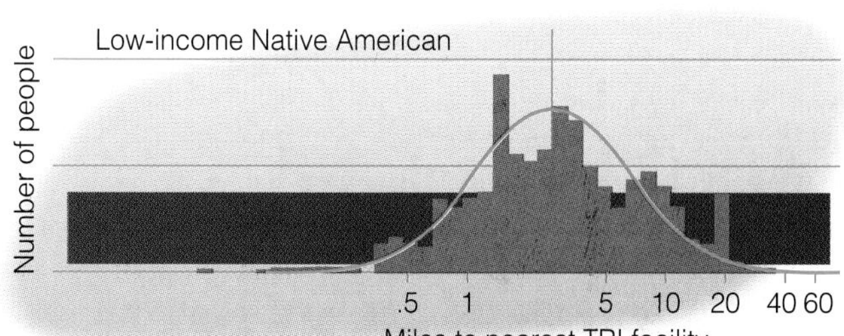

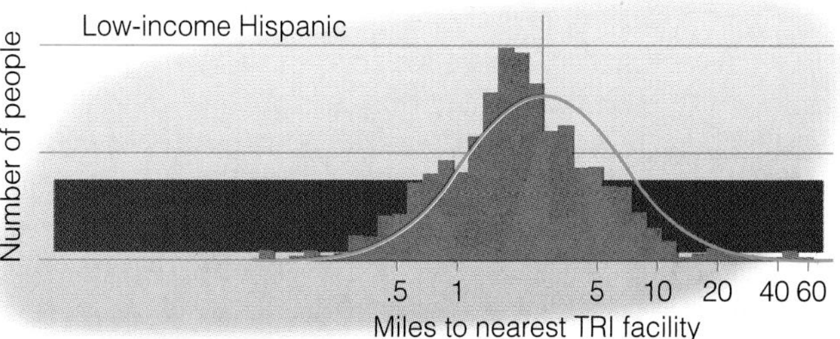

FIGURE 20.4 Distribution of All Households by Low-Income Households in Florida (by household racial–ethnic composition and by distance to the nearest TRI facility*)

*TRI stands for Toxic Release Inventory.

SOURCE: Adapted from Pollock and Vittas, 1995, p. 302.

The height of the bars in the graph show the numbers of people at a given distance from a toxic facility. Note from this that higher proportions of minorities are nearer toxic facilities than whites of the same income level.

Feminism and the Environment

Women and men do not regard environmental issues equally. In general, women tend to be more concerned with issues of environmental risk, and this has important policy implications: Lack of attention on the part of local and federal governments to environmental issues can be interpreted as lack of attention to policy that differentially affects women. In this respect, it is a feminist issue.

In one recent study (Bord and O'Connor, 1997), women and men were both asked a set of detailed questions

General Motors unveiled this pollution-reducing electric car, the first of a fleet, for Florida Power and Light Company, in August of 1998.

pertaining to their perceptions of risk to themselves from environmental hazards. Women consistently showed more concern than men for environmental issues, and also perceived themselves to be at considerably more risk from environmental hazards than men. For example, women were more likely than men to perceive that abandoned waste sites cause cancer, could produce miscarriages, and cause other health problems. They were also more likely than men to feel that waste sites posed dangers to trees, fish, and other wildlife. Women were also more likely than men to perceive dangers in global warming; they were more likely to predict, as a result, coastal flooding from polar ice meltdown; loss of forests; increased killing off of certain animal species; increases in hurricanes and tornadoes; and increased air and water pollution.

The issue is not, for example, whether or not global warming will definitely result in such calamities, since there is room for debate on the issue. The issue is that women feel more vulnerable than men to the risks posed by such environmental problems (Bord and O'Connor, 1997; Blocker and Eckberg, 1997), and as a consequence, women are more concerned that policy makers will act to reduce these risks. To this extent, environmental policy in the United States has a different impact on women than on men.

Environmental Policy

In the past three decades, federal and local agencies have made concerted efforts to bring the problems of environmental pollution under control, although in most cases environmental pollution is seen by these agencies as not differently affecting women and men. The main lines of attack have been stiffer antipollution laws and the encouragement of alternative technologies.

Antipollution laws have been resisted by industry because they require expensive adaptations of manufacturing processes, and they have been resisted by unions for fear that the added expense to industry would cost jobs. Despite these points of opposition, many antipollution laws have been passed since the late 1960s and have won great public

BOX 20.2 DOING SOCIOLOGICAL RESEARCH

New Findings Challenge Demographic Transition

As DEMOGRAPHIC transition theory argues, it used to be assumed that economic development was the best way for a poor country to reduce its population growth. New studies suggest that a country like Bangladesh can cut its birthrate significantly if it aggressively promotes the adoption of modern contraceptive methods—without waiting for the reduction that traditionally comes with higher living standards. Presumably, such a strategy may work with certain U.S. populations, although cultural resistance to contraceptive methods may lessen the likelihood of their adoption.

In what some experts are calling a reproductive revolution, birthrates are falling in countries presumably too poor for economic development that would stabilize rapid population growth, as it did in Europe and North America early this century, suggesting that contraceptives can work.

Demographic transition theory, predicated on the experience of the industrialized world before modern contraceptives, says that in a preindustrial economy people tend to have many children but that high death rates hold the population down. As a country industrializes, living conditions improve, the practice of medicine advances, and life spans increase, but the birthrate remains high and the population soars.

Only as education spreads and people find that too many children are an economic liability do birthrates drop, the theory goes. Thus begins an era of low mortality, low fertility, and stable but not runaway population growth—the condition of the industrialized world today.

Bangladesh is a perfect example of how the concept linking the fertility rate to developing economies has been disproved. The South Asian country, one of the world's poorest and most densely populated, has a traditional agrarian economy in which most families still depend on children for economic security; yet Bangladeshi fertility rates declined by 21 percent between 1970 and 1991, to 5.5 children per woman from 7. In the same period, the use of contraception among married women of reproductive age rose to 40 percent from 3 percent.

SOURCE: Stevens, William K. 1994. "Poor Lands Success in Cutting Birthrate Upsets Old Theories." The *New York Times* (January 2): 1, 8.

support. In the early 1980s, the Reagan administration relaxed antipollution standards on the grounds that they were too costly for industry, but in the period since, the EPA has returned to more aggressive enforcement of the law.

The development of new technologies in the last twenty years or so has played a major role in the reduction of certain kinds of pollution. Since the early 1970s, emissions controls for automobile exhaust have been widely installed. Electric cars are once again enjoying a vogue, as is the search for alternative fuels to replace gasoline, such as methanol, or methane from the fermentation of human and animal waste products.

Population and Environment in the Twenty-First Century

The U.S. Census predicts that the world's population will increase from the 5.75 billion it is now to 7.9 billion by the year 2020. Even more upsetting is the United Nation's revised predictions concerning when the world's population would stabilize. A few years ago, the United Nations Division on Population estimated that the world population would stabilize at around nine billion; that estimate has been revised to ten billion, with a high estimate of as much as fourteen billion.

Sociologists predict that the United States will continue to experience increasing suburban development, with accompanying increases in heavy industry, and thus additional pollution. The rate at which people are leaving the centers of today's cities will slow somewhat. A major concern that today's sociologists have for the future is the effect that a changing planet will have upon our lifestyle (Logan and Molotch, 1987; Palen, 1992), and also the reverse—the effect our lifestyle will have on the planet. For the first time in our history, our lifestyle may threaten our very existence. One analyst, admittedly combining speculation with only a dash of science fiction, foresees a depressing outcome to our abuse of the environment:

> The year—2035. In an effort to hold back the rising sea water, massive dikes have been built around New Orleans, New York, and Miami. Phoenix is baking in its third week of temperatures of 155 degrees. Decades of drought have laid waste to the once fertile Midwest farm belt. Hurricanes batter the Gulf Coast and forest fires continue to blacken thousands of acres across the country (Rifken, 1989).

Ecological concern has stimulated such developments as the field of **ecological demography**, which combines the studies of demography and ecology (Namboodiri, 1988). This field would monitor experimentation with alternative fuels, fertilizers, and pesticides; efforts at the recycling of toxic wastes; and protection for the ozone layer. It would also observe the use and success of alternative technologies, such as solar, wind, and geothermal power. The development of such alternative technologies offers some hope for deflecting Rifken's dire predictions.

BOX 20.3 SOCIOLOGY IN PRACTICE

The Office of Population Research

THE OFFICE of Population Research at Princeton University is the oldest population research center in the country. Founded in 1936, it has trained many persons, from undergraduate degree recipients to doctoral recipients, who have professional positions in the field of demography in the United States as well as in developing countries. Many of these jobs are concerned with the reduction of population growth, and the Office of Population Research engages in the distribution and training in the use of contraceptives.

The Office provides many services besides undergraduate and graduate degree programs. Information about reproductive health is provided, as is information on *emergency* contraception. It provides extensive data archives and produces a research journal, *Population Index,* on the Web. Research programs housed in the Office cover areas such as aging, fertility and fecundity, health and mortality, marriage and the family, demographic methods, migration, and the environment.

A current ongoing project on the environment is housed in Mexico. It is engaged in a cooperative effort with a grassroots community action organization in Michoacan, Mexico, called ORCA. The purpose of this cooperative effort is to introduce alternative and environmentally positive technologies into the community, and then engage in research on their effects. These technologies involve the following:

Latrines that do not contaminate the groundwater with human wastes

Stoves (for cooking) and kilns (for pottery) that reduce levels of indoor air pollution through improved ventilation mechanisms

Elimination of the need for high-priced chemical fertilizers and pesticides

Introduction of alternative crop rotation strategies

These four projects all form an integrative framework with the existing organizations. They function as a unit; each project learns from the experiences of the other projects and thus provides feedback to the benefit of the ecology and inhabitants of the region.

SOURCE: Office of Population Research, Princeton University. Web site: http://opr.princeton.edu

CHAPTER SUMMARY

- Questions pertaining to population and the environment are among the most crucial of our century. The study of population is called *demography,* a field that focuses upon three fundamental processes, all of which determine the level of population at a given moment: births, deaths, and migrations. We noted that the United States ranks very low in *life expectancy,* and high in *infant mortality,* relative to other Western countries.
- Diversity is of great significance since both infant mortality and *life expectancy* are not equal across all races, social classes, or for men and women. Women have a greater life expectancy than men in virtually all countries and at all social class levels. However, the lower one's social class, the less one's life expectancy, regardless of gender, and also the greater the infant mortality. Minority group individuals, especially African Americans, Hispanics, and American Indians, all have lower life expectancies, and higher infant mortality, than Whites.
- Current migration patterns show that in addition to overall movement from city to suburb not only by Whites but by people of color as well, nonetheless large portions of populations, such as the underclass, remain stuck in central cities. Furthermore, different minorities migrating within the boundaries of the United States reveal different patterns and choose different sites.
- The study of a *cohort,* the so-called Baby Boomers, showed how a demographic cohort can both use and produce cultural change. *Malthusian theory,* still relevant today, warns us about the dangers of exponential population growth along with only arithmetic increases in food and natural resources. In order to avoid the so-called Malthusian positive checks of famine and war, population control can be, and is being, instituted by means of programs of family planning and birth control. The level of zero population growth, and replacement-level birthrates, can be achieved, and already has been in many parts of the United States. Whether or not methods of contraception advocated by the government are adopted, either in the United States or in other countries, depends upon the culture of the group in question. Some Latino groups and some African Americans consider advocacy of contraception for people of color, without equally rigorous contraception advocacy for middle-class Whites, to border on genocide.
- Any society is an *ecosystem* with interacting and interdependent forces, consisting of human populations, natural resources, and the state of the environment. Depletion of one natural resource affects many other things in the ecosystem. The dumping of toxic wastes is a very major problem in our society, especially when toxic dumps are found more frequently, as they are, in or very near African American, Hispanic, and Native American communities. Such practices constitute what some researchers call *environmental inequity* on the basis of race-ethnicity, as well as social class. Finally, surveys have shown that environmental risks are of more concern to women than to men; as a consequence, environmental policy has more impact upon women than upon men.

KEY TERMS

age–sex pyramid
census
cohort
crude birthrate
crude death rate
demographic transition theory
demography
ecological demography
emigration
greenhouse effect
human ecology
human ecosystem
immigration
infant mortality rate
life expectancy
Malthusian theory
population density
population replacement level
sex ratio
thermal pollution
vital statistics

THE INTERNET: A Tool for the Sociological Imagination

Resources on the Internet:

Virtual Society: The Wadsworth Sociology Resource Center
http://sociology.wadsworth.com

Visit this site to find additional learning tools, including interactive quizzes, links related to web sites, and an easy link to *InfoTrac College Edition.*

Environmental Protection Agency
http://www.epa/gov

The government agency responsible for monitoring the environment and developing public policy to protect it.

Greenpeace
http:www.greenpeace.org/

An activist organization that uses nonviolent action to protect the environment.

World Resources Institute
http://www.wri.org/

An international organization dedicated to protecting the earth's environment.

International Demographic Data
http://www.census.gov/ftp/pub/ipc/www/idbsum.html

A Census Bureau site documenting global population data since 1950.

Population Reference Bureau
http://www.prb/org

An organization that provides timely information on a variety of population topics in the United States.

Sociology and Social Policy: Internet Exercises

Environmental racism has become a cause for concern among those worried about the impact of pollution and waste on the nation's already most disadvantaged groups. What sociological factors influence environmental racism, and what groups have developed to promote social action and social policy on this issue? Are there such organizations in your community or state, and what communities in your region are most vulnerable to environmental hazards?

Internet Search Keywords:

environmental racism
vulnerable communities to environmental hazards

Web sites:

http://www.ncpa.org/pi/enviro/envpd/pdenv68.html
This web site offers a background on environmental racism—how it began.

http://www.igc.apc.org/envjustice/
The Institute for Global Communications publishes recent news on environmental justice as well as organization information and related links.

InfoTrac College Edition: Search Word Summary

demography	infant mortality rate
environmental racism	migration

In order to learn more about these central topics in sociology, you can conduct an electronic search using InfoTrac College Edition. To aid in your search and to gain useful tips, see the Student Guide to InfoTrac College Edition on the Virtual Society web site:
http://sociology.wadsworth.com

INTERACTIONS—A SOCIOLOGY CD-ROM: CONCEPTS FOR THIS CHAPTER

Go to the Wadsworth Sociology CD-ROM for further study on the concepts in this chapter. The CD-ROM also includes quizzes and additional activities to expand your learning experience.

SUGGESTED READINGS

Brown, Lester R. 1996. *State of the World: A Worldwatch Institute Report on Progress Toward a Sustainable Society.* New York: W. W. Norton.

This is a collection of articles discussing environmental issues and risks globally. This volume is published annually.

Bullard, Robert. 1994. *Dumping in Dixie: Race, Class, and Environmental Quality.* Boulder, CO: Westview Press.

Bullard's book is a readable and glaringly clear account of how race as well as class is connected with toxic waste disposal. It is a major source of hard evidence of environmental racism and classism.

Duster, Troy. 1990. *Backdoor to Eugenics.* New York: Routledge.

This book reveals how the U.S. census undercount of minorities affects federal population policy.

Ehrlich, Paul R., and Anne H. Ehrlich. 1990. *The Population Explosion.* New York: Simon & Schuster.

In this twenty-year follow-up on Paul Ehrlich's earlier book, *The Population Bomb* (1968), the Ehrlichs conclude that the population bomb has actually exploded.

Gaard, Greta, ed. 1993. *Ecofeminism: Women, Animals, Nature.* Philadelphia: Temple University Press.

Gaard looks at the central role of feminism versus sexist ideologies in the environmental movement, and explores connections among oppressions of humans by gender, race, and class, and the oppressions of nature.

Gerber, Judith A., and Robyne S. Turned, eds. 1995. *Gender in Urban Research.* Thousand Oaks, CA: Sage Publications.

This book explores the role of gender in urban population and examines gender from a number of different theoretical perspectives.

Harper, Charles L. 1995. *Environment and Society: Social Perspectives on Environmental Issues and Problems.* Englewood Cliffs, NJ: Prentice Hall.

This is an application of the main sociological theories, including functionalist, conflict, demographic transition, and others, to the study of the environment.

Meadows, Donetta H., Dennis L. Meadows, and Jordan Randers. 1992. *Beyond the Limits: Confronting Global Collapse, Envisioning a Sustainable Future.* Post Mills, VT: Chelsea Green Publishing.

This is an application of a computer simulation of the complex interplay between population, industrialization, resources, and the environment. Catastrophe is seen as possible by the first years of the twenty-first century, unless corrective action is taken.

CHAPTER 21
Collective Behavior and Social Movements

RECALL the last concert you attended. Perhaps you were outside on the grass with thousands of people milling around or in a dark coliseum. When the concert ended, people probably clapped, danced, stomped their feet, and may have flicked lighters to encourage an encore. What was the last sporting event you attended? Maybe the women's volleyball team at your university was about to take the regional championship when fans began doing the wave. A few false starts may have fizzled, but soon the entire audience was rising and ebbing as the wave circled the gymnasium. In each case, as you enjoyed the presence of the people around you, you remained an individual, but your behavior was influenced by the crowd.

Recall when O. J. Simpson, former football star and national celebrity, was pursued on the Los Angeles freeway system by the Los Angeles police after the murder of his wife, Nicole Brown Simpson (for which he was later tried and acquitted). The low-speed chase went on for more than 50 miles, with every major TV station broadcasting live coverage. People in Los Angeles gathered on freeway overpasses and along the side of the highways and cheered, "Go Juice!" Around the nation, people sat glued to their television sets, watching this pursuit—hardly the reception murder suspects typically get as they flee the police. What made people behave this way?

Events like these have long been of interest to sociologists. Each example illustrates the behavior of groups influenced by crowds or responding to a sudden and unusual situation. In each case, people think they are acting as individuals, but like other forms of group behavior, they are being shaped by the collective action of others. Indeed, people in crowds seem to take on a collective identity. Within a crowd, it may even be difficult to distinguish individual and group behavior. Crowds seem to act as one even though there may be great diversity within them.

Crowds are an excellent venue for using the sociological imagination. They are just one form of a phenomenon sociologists call **collective behavior**, defined as behavior that occurs when the usual conventions are suspended and people collectively establish new norms of behavior in response to an emerging situation (Turner and Killian, 1988: 3). Collective behavior occurs when something out of the ordinary happens and people respond by establishing new behavioral norms. Although collective behavior may emerge spontaneously in response to a unique situation, it is not entirely unpredictable. There are established patterns even to the behavior of crowds, but they are patterns in which unusual events guide the behavior of social groups.

Sometimes collective behavior emerges in response to an event that never actually takes place, but when the belief that it will happen is so strong, people develop new forms of behavior to meet this situation. Anticipating the Y2K problem is a case in point. Some people feared massive disruption of electricity, airline flights, banking services, and even the availability of food and water and, therefore, stockpiled these necessities. Such was also the case in Memphis in 1989 when Iben Browning, a self-taught climatologist, predicted that a major earthquake would occur in the central United States around December 2 or 3, 1990, plus or minus two days. Although scientists are unable to predict earthquakes with much precision, because the city of Memphis had experienced a large earthquake in 1811–1812, Browning's prediction triggered a belief that the earthquake would strike Memphis. Thousands of people undertook serious earthquake preparation. People bought new insurance, secured potential flying objects, mailed delicate china and crystal to other parts of the country, and arranged to spend the day with faraway family and friends. As the minutes passed on earthquake day, the city had the feel of a major festival. Afterward, memorabilia was sold to mark "the earthquake that never was." Confronted with a unique and ambiguous situation, Memphis residents and the thousands who gathered became a classic example of collective behavior.

Hurricane Mitch was one of the largest natural disasters of the twentieth century. As is typical following a disaster, here groups of residents in Tegucigalpa organized to help in the cleanup.

There are also numerous examples of cults, past and present, whose members believe that spaceships will land on earth at some designated time or that some other supernatural event will occur at a predictable point in time. Such was the case with the Heaven's Gate cult, whose members collectively committed suicide in 1997, thinking that they would be getting on a spaceship in the tail of Halley's Comet and find a new world, free of the perceived restrictions in this world.

Types of collective behavior include crowds, riots, disasters, and social movements, as well as other forms of mass action, such as fads and fashion. Riots occur when groups of people band together to express a collective grievance or when groups are provoked by anger or excitement. Some riots can be predicted in advance, as when Los Angeles exploded into riot in 1992 following the acquittal by an all-White jury of four police officers accused of beating Rodney King. Other riots erupt suddenly and unexpectedly, like those that sometimes occur after victory celebrations. Natural and manmade disasters may occur without warning, but it can be expected that when they occur they will provoke collective behavior (see the box "Sociology in Practice: The Disaster Research Center"). Blizzards, hurricanes, floods, and earthquakes all create situations in which people develop new ways of behaving in the face of unusual circumstances. Amid the devastation of these events, groups of neighbors, as well as organized response teams, come together to help one another meet their immediate needs and try to reestablish so-called normal ways of living (Kreps, 1994).

A final example of collective behavior is the development of social movements, groups that act with some continuity and organization to promote or resist change in society (Turner and Killian, 1988: 246). As we will see later in this chapter, social movements tend to persist over time more than other forms of collective behavior. Evidence of social movements can be seen throughout society, in movements to protect the environment, promote racial justice, defend the rights of diverse groups (even animal rights), attack the government (such as militia groups), or advocate particular beliefs (such as the Christian Coalition). We explore social movements and their role in social change later in this chapter.

Sociologists who study collective behavior are interested in how even unique and idiosyncratic events are socially structured and how collective behavior generates social change. As we will see, some of the phenomena defined as collective behavior are whimsical and fun, such as fads and some kinds of crowds. Other collective behaviors can be terrifying. Whether whimsical or awesome, collective behavior is innovative, sometimes revolutionary, and it is this feature that links collective behavior to social change.

Y2K panic

Characteristics of Collective Behavior

Most days, life is more or less predictable, but that can change in a moment. When something unusual happens or when a gathering becomes focused on a specific event, collective behavior is spawned. To understand collective behavior, it helps to understand what the different forms of collective behavior have in common:

1. First, *collective behavior always represents the actions of groups of people*, not individuals. The action of a lone gunman who opens fire in a post office is not collective behavior because it is the action of only one person. On the other hand, groups that gather at the scene to observe the emergency response are engaged in collective behavior. Likewise, those who organize a protest at a public facility to protest a nuclear power plant are engaged in collective behavior. Collective behavior has some of the same characteristics as other forms of social behavior. It is rooted in relationships between people (Weller and Quarantelli, 1973) and may involve group norms, such as the expectation that people in a baseball stadium or a discotheque do the "YMCA."

2. *Collective behavior involves new or emergent relationships that arise in unusual or unexpected circumstances.* Collective behavior often falls outside everyday institutional social behavior. The behavior of people who commute to work together is not considered collective behavior because commuting is an ordinary part of their everyday life. However, if traffic comes to a halt because a freighter has plowed into a drawbridge up ahead and everyone gets out of their cars to look, their actions constitute collective behavior.

BOX 21.1 SOCIOLOGY IN PRACTICE

The Disaster Research Center

HOW DO people behave in the aftermath of disasters? This question is at the heart of the work of sociologists at the Disaster Research Center (DRC), an organization devoted to studying how groups and organizations prepare for, respond to, and recover from communitywide emergencies, especially natural and technological disasters. Sociologists from the Disaster Research Center utilize their research to advise municipal agencies and officials around the world on how best to respond to assist people at times of urgent need. Teams of researchers from the Disaster Research Center stand ready, often at a moment's notice, to travel quickly to disaster sites, collecting new information about human responses to disasters.

First founded at Ohio State University and later relocated to the University of Delaware, the Disaster Research Center has studied earthquakes in Kobe (Japan), California, Mexico City, and numerous other sites. They have responded following Hurricane Andrew in Miami and Hurricane Hugo in Charleston, South Carolina. Floods, plane crashes, and hazardous chemical spills have also been sites for research by those in the Disaster Research Center. To date, the Center has produced more than 500 studies on various dimensions of natural and technological disasters.

Following a disaster and, often following up with long-term studies of community recovery, teams of researchers from the DRC (including professional sociologists, as well as graduate and undergraduate students), frequently interview those responsible for assisting people in disasters, as well as disaster victims. This requires utilizing one's research skills quickly, since seldom is there much, if any, warning about a hurricane, flood, riot, or other disaster. Researchers are likely to interview civic workers in the aftermath of a disaster or collect data from community residents, as they try to reestablish themselves after what can be devastating losses.

Sociologists from the Disaster Research Center have been especially interested in the functioning of relief and welfare groups, the operation of rumors during disasters, the social and cultural factors discouraging people from evacuating threatened communities, and the role of the news media during disasters. Currently, some of the projects at this unusual center are examining questions about class and racial equity in the aftermath of disasters. Which communities are more likely to recover from disasters; get the assistance of federal, state, and local authorities; mobilize themselves effectively to seek relief? Not surprisingly, these sociologists are learning that some communities recover more quickly and more completely than others, recovery being largely influenced by the class and racial background of community residents. Because of the practical dimension to their studies, scholars at the Disaster Research Center use their knowledge to advise federal and local officials worldwide on how to be better prepared for disasters and how to respond more effectively when these tragedies occur. Often funded by organizations like the National Science Foundation, The National Institute of Mental Health, and the Federal Emergency Management Association, to name a few, these projects provide an intriguing illustration of how sociological research can help people during times of great personal and collective trouble.

SOURCE: Dynes, Russell R., and Kathleen J. Tierney (eds.). 1994. *Disasters, Collective Behavior, and Social Organization.* Newark: University of Delaware Press; Disaster Research Center. 1995. Annual Report. Newark: University of Delaware; see also the Disaster Research Center Web site: http://www.udel.edu/DRC/homepage.htm

Collective behavior arises when uncertainty in the environment creates the need for new forms of action. People may undertake tasks that are new to them (Weller and Quarantelli, 1973), perhaps doing things they never imagined themselves capable of before. For example, when a community is struck by a natural disaster, such as a flood, earthquake, or hurricane, suddenly nothing can be taken for granted—not food, water, transportation, electricity, or shelter. Collective behavior emerges to meet the new needs that people face. Following a hurricane, people may form neighborhood work groups to clear debris, or they may organize teams to stack sandbags in the threat of a flood. Following the Loma Prieta earthquake in the Bay Area in 1989, a major interstate collapsed and people from the surrounding Oakland neighborhood climbed through the debris to rescue victims before the ambulance and fire departments arrived. This is an example of a group forming to face a new and uncertain situation.

3. Because of its emergent character, *collective behavior captures the more novel, dynamic, and changing elements of society to a greater degree than other forms of social action*. An example is fads, such as Pokemon among young children, rollerblading, or body piercing among young adults. Fads introduce something new into everyday social life. Although fads may stretch across several seasons, collective behavior can be short-lived or long term. Social movements are collective behavior that typically develop over a longer period; crowds are more ephemeral. Both long-term and short-term incidents of collective behavior can, however, transform society. The consciousness-raising groups that were part of the women's movement in the 1970s, for example, resulted in lifelong changes of thought for many women who are now in their middle and older years. The American Indian takeover of Alcatraz Island in 1968 followed by a seventy-one-day occupation at Wounded Knee also transfixed the nation and called attention to the demands of American Indians. Such actions can be the basis for long-term and short-term social change.

4. *Collective behavior may mark the beginnings of more organized social behavior*. Collective behavior often precedes the establishment of formal social movements. People who spontaneously organize to protest something may develop structured ways of sustaining their protest. One of the best historical examples comes from the community protest at Love Canal. When the New York State Health Commissioner announced in 1978 that a toxic waste dump site located at Love Canal in Niagara Falls, New York, presented a "great and imminent peril to the health of the general public residing at or near the said site" (Levine, 1982: 28), within days, Lois Gibbs and other housewives residing in the area came together to form the Love Canal Home Owner's Association (Gibbs, 1982). Their action gave form to the local response against a bona fide ecological disaster, and also branched into a program to enhance public awareness of toxic waste dumping as a social problem. It was an important first step in the development of social movements to protest the dumping of toxic waste, a movement that still endures (Thomas, 1995). Since then, numerous comparable examples can be found—many of which involve minority communities threatened by the high level of environmental hazards in their communities. The *environmental justice movement* includes a wide array of Native American, African American, Latino, and other communities that have organized to protest the dumping and pollution that imperils their neighborhoods (Bullard, 1994a).

5. *Collective behavior is patterned behavior, not the irrational, overly emotional behavior of crazed individuals*. Patterned behavior is activity that is relatively coordinated among the participants. For example, crowd members may all be focused on the same thing, such as a rock band. They attend concerts with their friends and progress to and from the concert site in a more or less orderly fashion. During episodes of collective behavior, people may follow new guidelines of social behavior, but they do follow guidelines. Even episodes of panic, which on their face appear to be asocial and disorganized, follow a relatively orderly pattern.

6. *Most forms of collective behavior appear to be highly emotional, even volatile*. Episodes of collective behavior often exhibit the more emotional side of life, as when millions gathered (in person and via satellite) to grieve the death of Princess Diana or when, during the Promise Keepers Million Man march, hundreds of thousands of men collectively prayed, some with tears in their eyes, on the Mall in Washington, DC. Not all emotional behavior is collective behavior, however. Parents who grieve over a dead child are emotional, but are not necessarily acting collectively. On the other hand, if a group of parents gather at the site of a school bus accident, weeping over lost lives, this is collective behavior.

Debunking Society's Myths

Myth: In the face of disasters and other unexpected events, people do not behave under the normal social influences.

Sociological perspective: When faced with unexpected events, people develop norms to guide their behavior, often drawing on prior social behavior and knowledge to guide new interactions.

7. During collective behavior, *people communicate extensively through rumors*. Lacking communication channels or distrusting the ones available, people use rumors to define an ambiguous situation (Turner et al., 1986). In the popular view, rumors are rapidly circulated information of questionable accuracy; from a sociological point of view, *rumors* are the information transmitted by participants in collective behavior as they try to make sense out of an ambiguous situation. Rumors are transmitted by people who are piecing together information about a story whose facts are partly obscured. Rumoring is a collective activity. Sociologists note that rumors are most common when there is inadequate information to interpret a problematic situation or event (Dahlhamer and Nigg, 1993: 2). Rumor is present in almost all instances of collective behavior, because ambiguity is a key component of such behavior, and rumors thrive on ambiguity (Turner and Killian, 1988).

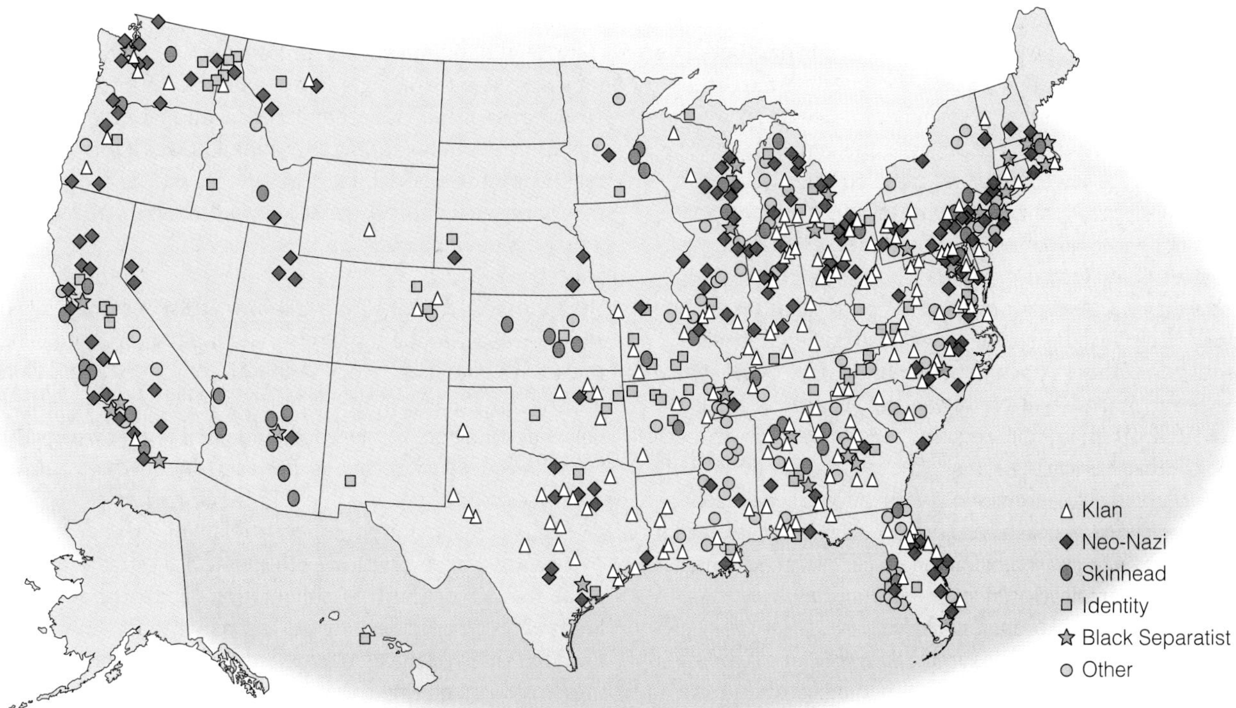

MAP 21.1 Mapping America's Diversity: Hate Groups in the United States

SOURCE: Adapted from *Intelligence Report*. 1997. Southern Poverty Law Center, Montgomery, AL. Used by permission.

Tomatsu Shibutani is a sociologist who has studied rumors in depth. He notes that rumors form when people, possibly complete strangers, seek verification from each other about an ambiguous situation. Their usual reference groups may have been disrupted because of the unusual situation they are in, and their participation in the rumor process helps them reestablish group ties. In this way, rumors are important not only because they convey information, but because they meet social-psychological needs in people faced with uncertainty (Shibutani, 1966).

8. Finally, *collective behavior is often associated with efforts to achieve social change*. This can mean promoting change or resisting it. Members of protest groups and social movements are always trying to effect change in the political, cultural, or value structures of society. The civil rights movement and *La Raza* (a Latino rights movement) are examples of social movements promoting change in the racial order of society. Reactionary movements, such as the militia movement and White supremacist movements, are attempts to resist change toward greater racial equality.

Crowds

Crowds are one of the major forms of collective behavior. Sociologists differ in their conceptions of crowds, but most agree that crowds share several characteristics. First, *crowds involve groups of people coming together in face-to-face or visual space with one another*. This is true of other forms of social behavior as well, such as going to a job or joining friends, but *crowds are also transitory*. That is, they form when groups come together for a specific transient event (such as a spiritual retreat, a rock concert, or a riot). The same group of people will probably never reconvene. *Crowds are volatile*. As crowds develop, their behavior may change quite suddenly. The behavior of people in crowds is quite different from their behavior in other social settings. Finally, *crowds usually have a sense of urgency;* they are focused intensely on a single event.

The Social Structure of Crowds

Although they appear to be a single entity, crowds have a discernible social structure. Each crowd is usually a particular size, with participants packed together in a particular density, and people in the crowd more or less connected to one another. Despite the element of structure in crowds, there is no rule about how large a group must be to be considered a crowd or how dense and connected people need to be (Goode, 1992). Crowds are usually "circular," surrounding the object of the crowd's attention. Where there are physical barriers, groups fill the enclosed space or form a semicircle. The people closest to the crowd's center of interest are the *core* of the crowd and show the greatest focus on the object of interest. At the outer edges of the crowd, attention is less focused: People are more likely to be talking to their friends, playing frisbee, or participating in other activities.

Crowds also have boundaries that may be more or less permeable. Some crowds make movement in and out of them very difficult (Milgram and Toch, 1968). Other crowds have more permeable boundaries, such as crowds at outdoor concerts where people may move in and out of the area nearest the stage and wander the grounds, perhaps getting food, drink, and souvenirs, or tailgating, watching others in the crowd, or even engaging in illegal activity on the edges of the crowd.

Some people think of crowds as possessing a "group mind." Early social theorists like Gustave LeBon (1895) described crowds as having mental unity. LeBon thought that people in crowds were highly suggestible and that the usual controls on people's behavior disappeared in crowds as people took on a single way of acting and thinking. This depiction of crowds, however, overlooks the social factors that influence crowd behavior. Crowds are distinctly social, not individualistic, forms of behavior. Although much of the behavior in crowds may be impromptu, crowds are socially organized and have a social structure.

Sociologists Ralph Turner and Lewis Killian, two major collective behavior theorists, developed emergent norm theory to describe how crowds can be *both* emergent forms of behavior *and* socially organized. **Emergent norm theory** postulates that, when people are faced with an unusual situation, they create meanings that define and direct the situation. Remember that norms are the common understandings that people develop to guide their behavior. During collective behavior, new norms emerge. The interactions people have with one another and the cues they get from group leaders or members of the crowd encourage the formation of new norms that will guide the entire group. Emergent norm theory emphasizes that group norms govern collective behavior, but the norms that are obeyed are newly created as the group responds to its new situation. Although collective behavior can sometimes appear to be without social organization, emergent norm theory emphasizes that members of the group do follow norms—they just may be created on the spot (Turner and Killian, 1988).

Debunking Society's Myths

Myth: **Crowds are typically disorganized groups whose behavior is less orderly and predictable than other social groups.**

Sociological perspective: **Crowds have a social structure that is observable in the patterns of crowd behavior.**

Within crowds, as elsewhere, norms guide behavior, but just because there is a guiding norm does not mean that every person in the crowd is doing the same thing. In fact, crowds may exhibit a division of labor. There are often different segments within a single crowd. At a sports event, one section of the fans may lead a cheer; other sections of the crowd may respond. Crowds may also generate *bystander crowds*, groups that may be physically present, like the crowd of onlookers at a protest march, or that may be as remote as a mass media audience. Crowd actions or protests are often explicitly set up for a media audience. Protesters planning a demonstration may expend great effort lining up media coverage, and during the protest, the participants may come to life for the cameras, and then settle down when the media are gone.

THINKING SOCIOLOGICALLY

Think of the last time you were in a situation where something unexpected happened (a disaster, emergency, or other sudden and unanticipated event). What norms guided people's behavior in this situation? What forms of collective behavior developed? How would *emergent norm theory* explain what happened?

Expressive crowds are groups whose primary function is the release or expression of emotion. The crowd may be focused on a particular object, or it may be generally expressive. Expressive crowds tend to exhibit high levels of feeling—more than individual people or ordinary groups typically do. Expressive crowds exist when the object on which the crowd is focused is seen by participants with deep, even religious or awesome, feelings (Goode, 1992: 142). Expressive crowds may be moved by any emotion, but the most common are collective grief (as when citizens file past a slain leader lying in state) or joy (as in a victory celebration or religious revival). Expressive crowds may do more than permit an emotion to be expressed—they may instill a permanent change in the mood and behavior of participants (Turner and Killian, 1988: 97). Religious revivals, for example, often provide more than a cathartic moment; members often come away feeling substantially renewed or changed.

Those who have seen the AIDS Memorial Quilt have felt the power of an expressive crowd. Cleve Jones, a San Francisco gay activist, community organizer, and friend of many who had died from AIDS, originated the idea for the AIDS quilt, which is now a nationwide project coordinated by the NAMES Project Foundation. The AIDS quilt is a giant mosaic of panels designed as a memorial to the thousands of children, women, and men who have died from AIDs. Each panel of the quilt is 3 feet by 6 feet (roughly the size of a human being) and is created by friends and family members of an AIDS victim. The panels are pieced together to form a patchwork quilt representing all the lives of those lost to this fatal disease. Huge sections of the quilt have been shown around the nation. The entire quilt, now larger than thirty football fields, has been unfolded and shown in its entirety five times, each time in the nation's capital. Parts of the quilt have been shown countless times in other cities around the nation.

The unfolding of the quilt is a moving event. Often, one or two songs sung by a gay or lesbian chorus opens the unfolding ceremony. A small group of six to eight people, each dressed in white and holding hands with each other, forms around the section of the quilt to be unfolded. Without introduction or ceremony, a speaker simply reads the names of people who have died from the disease. The small group quietly and precisely unfolds the quilt section, then moves to another section. Members of the crowd quietly file past sections of the quilt, many holding hands or quietly grieving the passing of so many people. This is but one of many examples that show the power that crowd behavior can have on people's deepest feelings.

THINKING SOCIOLOGICALLY

The next time you find yourself in a *crowd*, think about the sociological characteristics of this situation. You might find yourself feeling caught up in the crowd's collective identity. Resist the pull of the crowd and instead ask: What is the character of the crowd? How is it organized? Is there a division between the crowd and onlookers? Are there permeable boundaries? What norms are guiding the behavior of different participants in the crowd? How did the crowd break up? Answering these and other questions will help you see the *social structure* of the crowd.

The AIDS quilt, commemorating the deaths of AIDS victims, has been laid out on the Mall in Washington, DC, as well as shown, in pieces, in numerous communities throughout the nation. Evoking strong emotion, this display has helped mobilize public support for AIDS research and empathy for AIDS victims and their loved ones.

The Influence of Social Control Agents

Social control agents (police, chaperones, and other authorities) are present in most crowd situations since crowds are generally believed to easily get "out of control." The ability of social control agents to shape crowd activity varies. Social control agents at a rock concert, for example, may simply establish a perimeter for the crowd, as when police block fans from rushing the stage. Sometimes the behavior of the social control agents may be the cause of a crowd action, particularly if they overreact to crowd behavior. During the 1968 Democratic National Convention in Chicago, police actions against protesters near the convention site ran out of control. As one observer wrote:

> On the part of the police there was enough wild club swinging, enough cries of hatred, enough gratuitous beating to make the conclusion inescapable that individual policemen, and lots of them, committed violent acts far in excess of the requisite force for crowd dispersal or arrest. . . . to read the events is to become convinced of the presence of what can only be called a police riot (Walker, 1968, cited in Turner and Killian, 1988: 114–115).

Panic

Sometimes crowd behavior develops when there is a panic. A *panic* is behavior that occurs when the people in a group suddenly become concerned for their safety, and seemingly spontaneous, disorganized behavior results. Panics may be triggered by physical, psychological, social, or even financial danger (Lang and Lang, 1961). In the popular conception, those caught in a panic are divested of all socially acquired characteristics and become irresponsible, emotional, and dangerous; civilization is cast off and the hidden "man as beast" comes to the fore (Quarantelli, 1978). In fact, however, even in panics, there is more social structure than the popular image suggests. The sinking of the *Titanic* in 1912 is a good example. When this supposedly unsinkable ship struck an iceberg and began to sink, social class, gender, and age guided the pattern of escape. Of the 2200 people on the boat, 600 escaped in lifeboats, with almost all the women and children traveling in first class surviving (and the few men who charaded as women). Of those traveling in third class, 45 percent of the women and 70 percent of the children died (Lord, 1956). Getting people into lifeboats was carefully managed by crew who played their roles as expected, carefully executing an escape plan highly structured by class, gender, and age.

Three main factors characterize panic-producing situations. First, there is a *perceived threat*. The threat may be physical, psychological, or a combination of both, and it is usually perceived as so imminent that there is no time to do anything but try to escape. The second characteristic of panics is a sense of *possible entrapment*. Panic occurs when individuals perceive that if they do not act fast, they may miss their chance to achieve some goal. In situations where persons know that they are completely trapped, as in airplane crashes or subway fires, there is often no evidence of panic behavior; it seems that only when people think they are competing with each other for a limited number of routes does panic ensue (Quarantelli, 1954).

The final characteristic of panic is a *failure of front-to-rear communication*. People at the rear of the crowd, especially when they perceive themselves to be unfairly disadvantaged in reaching their goal, exert strong physical or psychological pressure to advance toward the goal, whether it be an emergency exit or an entrance to a stadium.

Anticipating major disruptions in everyday life because of Y2K, many panicked and stockpiled food and energy sources.

In instances where people are trampled to death, pressure from the rear is usually the single most important factor (Turner and Killian, 1988: 81).

Riots

Popular conceptions of riots were first expressed by Gustave LeBon in his classic book *The Crowd* (1895). LeBon wanted to explain the riots of the French Revolution in 1848 and the uprising of the Paris Commune in 1871 (Bramson, 1961). He asserted that riots occur as crowds come under the influence of a mob mentality and crowd members lose their will and their ability to be rational. Thereafter, they become suggestible to the actions of leaders. Emotion overtakes the crowd and spreads like wildfire. In LeBon's view, crowds are always dangerous threats to the social order.

LeBon's portrayal is consistent with popular images of riots as individually crazed behavior. One need look no further than popular descriptions of the participants in the 1992 Los Angeles riots as "mobs" and "giddy looters" to see that this conception of riots is alive and well. Sociological analyses of riots are less inflamed and more analytical. Sociologists see riots as a multitude of small crowd actions spread over a particular geographic area, where the crowd is directed at a particular target (Stark et al., 1974: 867).

riot

Analyzing Riots

Careful empirical research suggests that riots are a multitude of small crowd actions. One notable study that illustrates this point was conducted at California State University, Long Beach, where researchers examined data from the Watts riots of August 1965. Stark and her colleagues viewed data sources ranging from fire department logs and police and sheriff department printouts to film records of the Los Angeles Fire Department and unedited KTLA aerial news footage to understand exactly what kinds of activity make up riots. They found that riots are made up of many different kinds of crowd formations. Of the more than 1850 crowd actions they examined, 926 involved starting fires, 138 involved pulling false alarms, 555 crowd actions were lootings, 174 were instances of rock throwing, and 85 were simple crowd assemblies not engaging in destructive actions.

Stark and her colleagues also noted that these actions were dispersed unevenly throughout the city and that they occurred at different times during the four days of rioting. There was hardly any part of the L.A. curfew area, 46.5 square miles, that had not experienced some type of crowd action; however, in general, areas of the city involved in the riot earlier experienced more days of riot activity and markedly more intensive riot actions. The researchers also noted that crowd action differed in different areas of the city. Those areas experiencing high rates of looting did not necessarily also experience high incidence of fires. Crowd actions did not spread in connected paths. The researchers found that in only 21 percent of the cases did the riot spread to an area adjacent to the area last experiencing new riot activity. In the remaining 79 percent, the instances of new crowd actions occurred in areas not adjacent to ongoing crowd activity. What this means is that the crowd actions did not spread like wildfire; instead, mini-riots popped up like geysers throughout the city (Stark at el., 1974). These carefully gathered data undermine the notion that people in riots are consumed by a mob mentality that spreads through a crowd.

Examining crowd actions in detail also reveals that riots are not uniform over time (McPhail, 1971). Stark and her colleagues found that during the Watts riot, the rate of crowd action activities grew during the first half of the riot and declined during the second half. They noted further that there were cyclic variations in crowd activities over the course of the day, with almost two-thirds of the riot activity occurring between 8 p.m. and 4 a.m. This suggests that riot behavior is linked to other social routines such as work and leisure time. Rioting individuals are not simply possessed by a mob mentality; rather, their activity is influenced by a number of factors, including the social conditions in their lives.

Types of Riots

There are several kinds of riots, each influenced by its historical and social context. In *commodity riots,* property is the object of attack rather than people. Violence in commodity riots is directed at buildings or merchandise that may symbolize an object of the crowd's enmity. This was the case in the 1992 Los Angeles riots, where some 1250 Korean-owned businesses were looted in South Central Los Angeles. Frustrated by the deteriorated conditions in their neighborhoods, many African Americans resent the seeming success of Korean Americans who operate stores in their neighborhoods. Although Korean Americans are also deprived relative to Whites, their presence symbolizes the lack of opportunity for African Americans. Thus, Korean Americans were an easy target for African American resentment about their own economic plight (Cho, 1993).

Communal riots are violent outbursts in which civilians riot against other civilian groups. In communal riots, one group clashes with another simply out of hostility toward the other group. Some historical race riots in the United States, in which Whites have attacked Blacks, are examples of communal riots, as are the uprisings between competing ethnic groups in Bosnia-Herzogovinia, Rwanda, and other parts of the world. In these cases, intergroup conflicts have developed into full-fledged wars.

Political riots are those in which a group rises up against a particular government policy or treatment by government officials. Race riots in the United States in recent years have been of this sort since they have typically been sparked by anger about police brutality or other forms of police intervention in Black and Latino communities. Typically, the rioters in political riots are civilians, but they may be the police or military, as in a coup d'etat. Prison riots, in which inmates rise up against prison guards, are another example of political riots. Some of the best known political riots in recent U.S. history are the demonstrations in the 1960s and early 1970s against the U.S. war policy in Vietnam. Many students and faculty on college campuses staged protests, some of which turned into riots that saw demonstrators pelting the police with bottles, rocks, and cans. This was the case in May 1970, when, after several days of disorderly protest on the Kent State University campus, the National Guard shot and killed four students on the campus mall.

Although the stereotype of looters is that they are lone actors, behavior during riots is socially organized and follows predictable sociological patterns. Following the acquittal of the Los Angeles police officers accused of beating Rodney King, the Los Angeles riots included many examples of such behaviors.

Why Do Riots Occur?

Most people think riots are caused by unruly individuals. This is not, however, a sociological explanation. Even looting, which seems motivated by individual wants, is organized along social lines. Looters choose selected targets, they often act as groups, and they behave in the context of communities or social groups that give considerable support to their actions (Quarantelli and Dynes, 1970).

Rioters have also been characterized as the "criminal element," the dregs of society. Sociologists, however, have shown that this is not true. In a classic study, George Rudé (1964) examined police, prison, hospital, and judicial records; poll books; petitions; and parish registries of births and deaths for several historical periods in order to understand who participates in riots. He discovered that those who rioted were likely to have no prior arrest records, to live in "fixed abodes," and to have steady jobs. He also found that different kinds of people are at different riots. During the riots leading to the French Revolution, women were more likely to be present at bread riots; likewise, workers may participate in industrial disputes, but not in other types of riots (Rudé, 1964).

Studies like Rudé's help us understand individual participation in riots. Explanations of riots that focus on individual attitudes and states of mind fall into the category of **convergence theory**, which explains riots by focusing on the participants in riots and presupposes that rioters are acting on predispositions and attitudes. A problem with this theory is that people have many attitudes and predispositions, and there is no one-to-one correspondence between belonging to a particular group and participating in a riot. If the focus is on the individual or even the group that riots, attention is shifted away from the organizations or groups that contributed to the conditions to begin with. Studies of Latino residents in south-central Los Angeles, for example, find few significant differences among different Latinos that would explain why some looted and rioted and others did not (Hayes-Bautista et al., 1993). Rather than focusing on the individual participants to explain riots, it is more fruitful to look at the social conditions that cause riots to erupt (see Figure 21.1).

Sociologists have found, for example, that the characteristics of urban areas can make some cities more prone to riots than others. First, riots are more likely to occur in cities where there is economic deprivation of racial–ethnic minority groups, including low levels of educational attainment, low median income, high unemployment, and poor housing conditions. Second, riots are most likely to occur in cities where grievances of the rioting group have

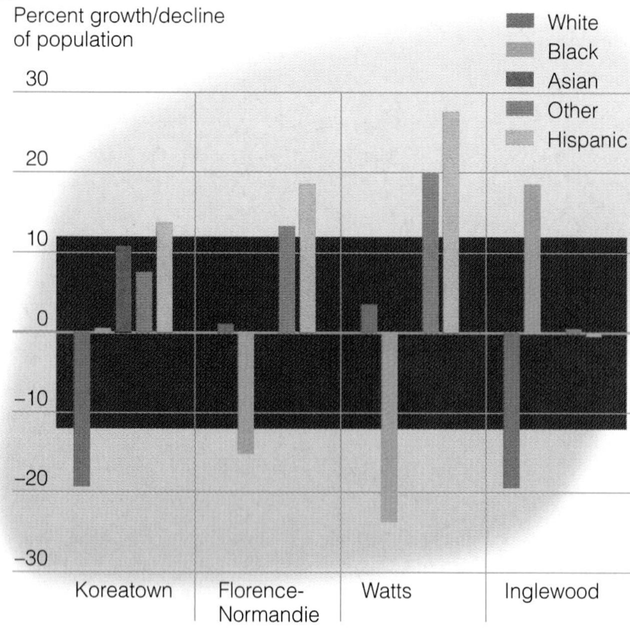

FIGURE 21.1 Social Conditions for the Los Angeles Riots

SOURCE: Barringer, Felice. 1992. "After the Riots; Census Reveals a City of Displacement." *The New York Times* May 15:16; Data from 1990 U.S. *Census of Population and Housing.* Washington, DC: U.S. Government Printing Office.

NOTES: The census data is from 1990; the riots occurred in 1992. Hispanics may be in any group category; thus, totals exceed 100%. Inglewood is a middle-class community with 70 percent of residents being homeowners.

not been addressed; this may be because of an unsympathetic or unresponsive city government (Lieberson and Silverman, 1965). Third, the rapid influx of new populations (through migration or immigration) is a common characteristic of cities where riots take place. Fourth, whether a group has the resources to initiate and sustain rebellious activity influences the development of riots (Carter, 1992; Spilerman, 1976). All these conditions have been identified as underlying causes of the 1992 Los Angeles riots. The economic deprivation of African Americans and Latinos in Los Angeles, coupled with competition between Latino, Korea, and African Americans for jobs and housing have been shown to be major causes of the L.A. riots (Baldassare, 1994; Herman, 1995); moreover, participants in the riots defined themselves as "protestors" and "freedom fighters," concerned with specific grievances, including poverty, unemployment, police brutality, and racial discrimination (Murty et al., 1994). Although tension between Korean American and African Americans has been popularly identified as one of the causes of the riots, researchers have found that the majority of both Korean Americans and African Americans identify institutional racism and the poor living conditions for impoverished groups in Los Angeles as the primary causes of intergroup tensions (Stewart, 1993).

In the context of underlying social structural conditions of poverty, unemployment, and poor housing, riots are typically sparked by precipitating events, most often a confrontation with the police. In Los Angeles, the precipitating event was the verdict acquitting four police officers of criminal charges in the beating of Rodney King. To rioters, rioting may seem like the only way to express the injustice they feel. Rioters are often powerless to effect change in legitimate ways. As the Reverend Martin Luther King, Jr., said, "A riot is the language of the unheard" (Hampton, 1987). When there are no other ways to express collective protest, riots can be the language people choose to express what they see as the illegitimacy of the system.

Sociologists have developed another theory to explain why riots occur called **relative deprivation theory,** which argues that people revolt when they see their situation improving, but not as much as that of other groups in the society. In this situation, a gap develops between people's expectations and their actual condition (see Figure 21.2). The theory of relative deprivation explains why people sometimes riot when their situation is actually improving, but may not riot when they are totally downtrodden. In fact, another name for the phenomenon described by the theory of relative deprivation is *the revolution of rising expectations.* The theory of relative deprivation has been used to explain why there was so much turmoil in U.S. cities in the mid-1960s, when, in many ways, the situations of previously disenfranchised people were improving, especially among African Americans. After so many years of mobilization for racial justice, and in a period of history during which the economy was strong and affluence was rising, the progress that had been made was perceived by the rioters to be simply not enough.

What Stops Riots?

Eventually, riots stop. Four possible explanations have been given for what brings riots to an end. First, the goals of the protest groups may have been satisfied. For example, if a race riot breaks out because political grievances have not been addressed, a satisfactory response to those grievances will bring the riot to a close.

Second, the actions of social control agents may end violence. Law enforcement officials and politicians frequently argue that the proper way to end a riot is to increase repressive measures. Indeed, following the 1965 Watts riot, Congress responded by developing antiriot legislation, stepping up law enforcement training in riot control, and building up stocks of tactical weapons for use in riot situations (Kelly and Isaac, 1984).

Third, riots and violence may end when the political situation changes. A government may become more responsive to the needs of groups that have been in revolt, or more repressive, making the human cost of a riot (in loss of life or imprisonment) too great for the potential gain. In our own society, some have argued that White backlash and the law-and-order platform associated with President Richard M. Nixon and longtime FBI director J. Edgar Hoover helped end the collective violence that marred the 1960s and 1970s (Oberschall, 1979).

Finally, some have argued that discontent can be regulated by the expansion of relief services. According to this

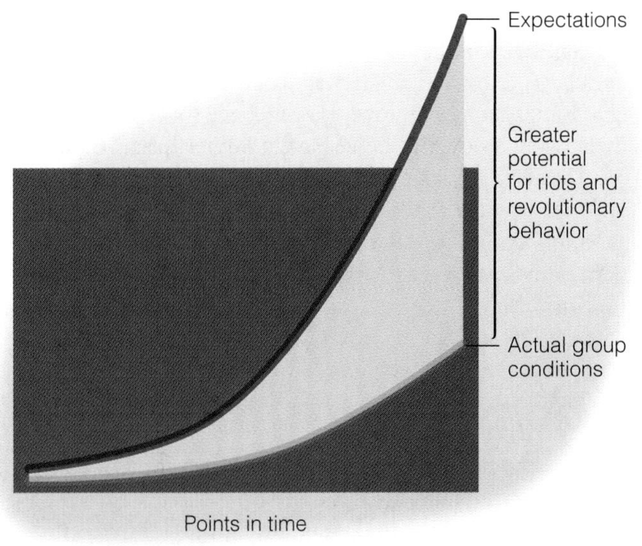

FIGURE 21.2 The Theory of Relative Deprivation

argument, the expansion of the welfare state quiets discontent by making those most likely to rebel dependent on federal subsidies. In so doing, discontent is managed by the seemingly benevolent actions of the state (Piven and Cloward, 1971).

Collective Preoccupations

Collective preoccupations are forms of collective behavior wherein many people over a relatively broad social spectrum engage in similar behavior and have a shared definition of their behavior as needed to bring social change or to identify their place in the society (Tierney, 1994). Collective preoccupations include fads, fashion, hysterical contagions, and scapegoating. Many of the more interesting, newsworthy, or novel parts of life comprise collective preoccupations. Beanie Babies and Tamagotchis (the virtual reality pet) represent some of the more creative and whimsical parts of human life. Collective preoccupations can be harmless, such as the fad of streaking on college campuses, or horrifying, such as the persecution of the Jews in Nazi Germany. Although collective preoccupations are popularly seen as weird or eccentric, like other forms of collective behavior, they are social in nature. It is the social part of these episodes that sociologists are interested in explaining.

Collective preoccupations take on many different forms, but these distinctly social phenomena share a number of features. First, they often begin within a small group of people involved in face-to-face interaction. Usually, the people are faced with an ambiguous situation from which they attempt to derive meaning. For example, UFO sightings are often reported in local newspapers. Groups of friends and family members may read these stories and come up with explanations for an unidentified flying object they have seen. Explanations may range from identifying them as experiments, advertising gimmicks for cigarettes or circuses, or exploratory vehicles from Mars (Miller, 1985). The point is that groups of people together form a collective definition of the situation to explain some ambiguous event. To become a fad, fashion, craze, or hysterical contagion, the collective definition of the situation must spread beyond the initial group. This usually occurs through the mass media, preexisting friendship networks, or organizational ties (Aguirre et al., 1988).

Most collective preoccupations also involve some aspect of social change. Social change may bring on the respective preoccupation, as when groups defined as outsiders are targeted for scapegoating; or the social change may be of a very limited sort, as is the case with fads, where a large number of people behave for a brief period in a curious manner.

Most of the different types of collective preoccupations provide opportunities for participants to belong to a group while differentiating themselves from other groups. The initial collective definition of the situation among group members may bring about a feeling of belonging; however, groups may use their definitions to exclude others or differentiate themselves from other groups. They become members of the in-group where others are part of the out-group. In-groups may persecute those who are seen as outsiders.

Fads

Fads represent change that has a less consequential impact than other kinds of social change. Fads do not usually fundamentally transform society (Goode, 1992), but are "an amusing mass involvement defined as of little or no consequence and in which involvement is brief" (Lofland, 1985: 68). They may be products (in-line skates, mountain bikes, hula hoops, yo-yos), activities (streaking or the Macarena), words or phrases (yo!, whatever, go figure, cool), or popular heroes (Dennis Rodman, Barney the Dinosaur, Barbie). Fads provide a sense of unity among those who participate in them and a sense of differentiation between participants and nonparticipants. Fads usually represent a departure from the mainstream; streaking on college campuses was rather novel in its time. Sometimes there is a negative judgment applied to fads. For example, piercing the face or other parts of the body with safety pins is seen by some as morally reprehensible. More often, however, fad behavior is seen by all as harmless fun.

Despite their seemingly idiosyncratic nature, even fads follow certain social norms. Streaking, for example, was defined in the student subculture and by most school authorities as harmless fun, but its acceptability was confined to certain locations on campus. It was taboo to streak in classes when tests were being given. The nudity of streaking was also considered nonsexual (Aguirre et al., 1988).

Fads are initially created within a small group that defines a particular product or phrase as meaningful and desirable to possess or to say. A *latent period* is followed by the *breakout period,* during which the product or activity spreads to other groups via friendship networks and mass media. In this period, small groups take up the product or

Crazes, such as the popular desire for Furbies, a child's toy, can produce collective behavior when huge crowds gather to purchase a scarce resource. Such crazes can be produced by commercial manufacturers with an interest in profiting from the mass reactions.

phrase and affirm it as something meaningful to buy or say (Goode, 1992: 356). Commonly, commercial interests latch on to fads (or even create them) by mass-marketing the item and manipulating public demand. In the *peaking period,* the use of the new item is defined as a fad, and people enthusiastically adopt it. The spread of a fad may depend on the success of commercial marketing (such as marketing popular film characters or baseball heroes through fast-food chains) or the spread may unintentionally result from the attempts of social control agents such as police or officials to regulate the activity. For example, on some campuses, streaking was seen by school administrators as intolerably deviant. Stiff sanctions sometimes led to confrontations, even to riots (Aguirre et al., 1988). Ironically, such penalties only seem to make the activity more appealing. In the *decline period,* the fad quickly fades (Miller, 1985). It is no longer defined as meaningful or desirable. Participants looking back at their participation in a fad may be bewildered that they ever found it meaningful.

Crazes are similar to fads except that they tend to represent more intense involvement for participants. Participants are "more or less encompassingly involved for periods of time" (Lofland, 1985: 640); the craze takes up more of the participant's life. Crazes tend to involve fewer people than fads since they require more of a commitment. Although they can take place in any arena of life, they most often appear in the religious, political, economic, and expressive or aesthetic realms (Smelser, 1963).

Those involved in crazes tend to be highly focused on the craze behavior. They may seem fanatical, devoted to the craze above all else. Whether it is rollerblading, a new dance craze, or a financial craze, crazes consume participants' attention. Participants in crazes may endure considerable inconvenience to pursue the craze, such as waiting in long lines or traveling long distances. They may also absorb large expenses. A mountain biker may spend $1500 on a bike and then augment it with a constant stream of expensive add-ons; the most desirable are built of a space-age alloy, just outside the purchaser's price range, known in biker parlance as "unobtainium." To those not addicted, the craze behavior may seem bizarre, senseless, or immoral. Crazes usually peak and then decline, although some people may become lifelong devotees of the craze behavior (Turner and Killian, 1988).

Fashion

Fashion has traditionally been considered a form of collective behavior because it constantly introduces something novel into the society. People wear clothing for protection and concealment of their bodies, but they also wear fashionable clothing to feel a part of a group and to differentiate themselves from others (Simmel, 1904; Veblen, 1912). Particular kinds of clothing and adornment can give people a feeling of acceptance. Military personnel wear fatigues and uniforms because durable clothing is imperative in their line of work; however, wearing a uniform also conveys that they belong to a group. Similarly, nuns may wear habits to indicate their commitment to a vow of celibacy and service, but the wearing of the habit is also a symbol recognized by others that they belong to a group. Likewise, young men and women may wear sports clothing or particular brand items (like Doc Maartens, Tommy Hilfiger clothing, or Nikes) as a way of signifying their belonging to a given group and social class, or they may wear clothing in a certain way, such as wearing loose jeans barely hanging on one's hips or wearing caps backward. Hair styles, clothing, jewelry, and other adornments all symbolize an identity people wish to convey to others.

Along with providing a sense of group identity, fashion is important in differentiating groups from each other. Around the turn of the century in the United States, when industrialization and technological progress were creating a new class of wealthy people, people used personal adornment and dress as a way to make other people aware of their wealth. Economist and sociologist Thorstein Veblen referred to this purchasing and displaying of goods to symbolize wealth and status as *conspicuous consumption.* A fully equipped, dark-colored sports utility vehicle, like the Lincoln Navigator, is an example of conspicuous consumption. Elites tend to want to differentiate themselves from nonelites. When groups of lower status begin to copy the fashions of the high-status groups, the high-status groups move on to new fashions; thus, a cycle of fashion exists.

Because fashion differentiates groups from each other, it is also a means of marking inequality between groups. Designer fashion labels, often displayed on the outer part of clothing, communicate material status to others. Within organizations, the status of different groups is also marked by apparel; thus, workers may wear uniforms, but management wears suits and dresses.

Hysterical Contagions

Hysterical contagions involve the spread of symptoms of an illness among a group, usually one in close contact, when there is no physiological disease present. In 1983, girls in a school on the West Bank of the Jordan River began to experience symptoms that were at first attributed to a mysterious gas poisoning; however, no physiological basis for the symptoms could be uncovered, nor could a source of gas be found (Hefez, 1985).

An episode of hysterical contagion usually begins with one person exhibiting physiological symptoms such as nausea, stomach cramps, itching, or uncontrollable trembling. That person receives sympathetic attention as people accept that the symptoms are caused by some genuine physical or biological agent. People begin to talk about the case. Soon other people begin to experience similar symptoms and also receive attention for their illness. When the outbreak becomes more widespread, authorities attempt to locate and eliminate the source of the problem. When they are unable to find a biological cause for the outbreak, speculation begins that the episode is a case of mass hysteria (Gehlen, 1977). Although the word *hysterical* connotes frenzied behavior, hysterical contagions follow a consistent pattern that can be accounted for by social factors.

Hysterical contagion is most likely to occur when it provides a way of coping with a situation that cannot be handled in the more usual ways. During an episode of contagion, not all persons in a given place are necessarily participants in the contagion. Sociologists have found that when a hysterical contagion strikes a workplace, those most likely to be affected are the workers who think it is not legitimate to take a day off from work just because they need a rest (Kerckhoff and Back, 1968). Such people may normally feel guilty about taking a personal day off, but once being sick is validated by others in a group as a legitimate means of gaining respite from pressure, these diligent people may believe they are actually sick (Gehlen, 1977). If others participate in the sickness, there is less risk of being viewed as a slouch or malingerer. This is not to say that people are faking; like other types of collective preoccupations, hysterical contagion results when people try to cope with an ambiguous situation. It usually means redefining how we feel and what causes those feelings.

Political contexts also shape how hysterical contagions are defined. During the hysterical contagion in the West Bank school, the contagion began in a girls' school, but it soon spread to other schools and, eventually, the whole community. Hospital doctors, working without laboratory facilities, thought the girls had been poisoned. At the height of the epidemic, rumors circulated among the Arab population that poison gas was being used to diminish the fertility of the schoolgirls, a ploy that would depopulate the West Bank area, then under Israeli occupation. Only later were external medical experts able to establish that there was no gas and no poison (Hefez, 1985).

Like other types of collective preoccupations, organization and communication networks are crucial to the spread of hysterical contagion. Friendship networks are often the crucial communication link, explaining why hysterical contagion typically occurs in schools and work organizations (Kerckhoff and Back, 1968; Gehlen, 1977).

Scapegoating

Often, collective preoccupations lead to **scapegoating,** which occurs when a group collectively identifies another group as a threat to the perceived social order and incorrectly blames the other group for problems they have not actually caused. The group so identified then becomes the target of negative actions that can range from ridicule to imprisonment, extreme violence, even death. Frequently, it is an individual who is blamed by another person or group for a perceived problem. When scapegoating becomes a collective phenomenon—an entire group blamed by another group—it is a collective preoccupation.

Racial minority groups and other groups perceived by the dominant group to be a threat, such as lesbians, gays, and foreigners, are commonly the victims of scapegoating. Scapegoating often occurs when major social changes are taking place in society, especially when older values are being replaced by new ones. From approximately 1450 until 1700 A.D., at least 100,000 "witches" were executed in Europe, with an additional 100,000 indicted for this crime and subsequently found innocent; 85 percent or more of those accused as witches were women (Oberschall, 1993). Drastic changes in demography and the economy, as well as the structure of the family and the changing role of women at that time, help explain why women were scapegoated. Growth in the proportion of unmarried women, prostitution, infanticide, and contraception also made the symbol of the female witch an appealing ideological symbol. Coupled with women's inferior status and lack of power and organization, women were ready targets for widespread persecution (Ben-Yehuda, 1986). Today, as the traditional family structure is undergoing major transformation, gay men have been scapegoated as corrupters of the family and as the carriers of AIDS. Some conservative religious figures who claim to represent "traditional family values" have defined AIDS as God's punishment for having engaged in homosexual relations.

Scapegoating may also take place when large-scale migrations occur. As agriculture expanded in California beginning in the 1800s, growers would hire large groups of people, usually immigrants, for cheap labor. Whenever a labor group began to present a strike threat, the growers would preferentially hire other groups of immigrants. In California in the 1920s, Mexicans and Filipinos were hired as strike-breakers; in the 1930s, it was the Anglos abandoning their farms in the drought-stricken Dust Bowl. In each case, the new group was blamed for taking jobs away from the former group. Severe ethnic violence sometimes resulted, reflecting the mass hysteria that fear of the "outsider" can create (Jenkins, 1983).

Scapegoating often occurs in societies at war. International conflicts or internal strife may exacerbate the tension between the dominant culture and perceived outgroups. War also produces strong feelings of national identification;

During World War II, Japanese Americans were scapegoated and systematically held in internment camps. Despite their being held by the U.S. government, many continued to demonstrate their patriotism as U.S. citizens.

those perceived as "other" may be singled out and targeted as enemies. Such was the case during World War II when more than 110,000 Japanese Americans unjustly feared as potential saboteurs were interned in concentration camps in the United States.

Scapegoats may be fixed upon in small face-to-face groups where members collectively come to define a particular group as the cause of their problems. For scapegoating to become a collective preoccupation, it must then gain the support of the media, government, or other influential organizations. Anti-Arab sentiment in the United States, for example, has been fanned sporadically by unflattering government and media depictions of Arab leaders defined as enemies of the American people. Saddam Hussein is actively demonized by the U.S. media, helping mobilize American opinion against Arab nations. This has the additional effect of generating bad will against Arab people in the United States, as if all Arabs were terrorists.

Social Movements

A final form of collective behavior that has captured the attention of sociologists is social movements. A **social movement** is an organized social group that acts with some continuity and coordination to promote or resist change in society or other social unit. Social movements are the most organized form of collective behavior, and tend to be the most sustained. They often have a connection to the past, and they tend to become organized in coherent (sometimes even bureaucratic) social organizations.

Social movements usually involve large numbers of people, but unlike a crowd, the individuals who make up a social movement are dispersed over time and space. ACT-UP (AIDS Coalition to Unleash Power), for example, comprises hundreds of local chapters that participate in "zap actions" across the United States—that is, single episodes of dramatic behavior intended to call attention to the AIDS cause. Unlike a crowd, not all the members of ACT-UP interact with each other face-to-face. Although ACT-UP chapters are diffuse (like collective preoccupations, they are spread across several groups), they are more enduring than a fad or an instance of hysterical contagion. Social movements, then, share some of the characteristics of the other two forms of collective behavior and some of the features of more established organizations.

Like crowds, social movements can give the impression that all of their members are unified around a single goal or ideology. Although movements focus on a shared goal, internally movements can be quite diverse. Even members of a given group within a movement may have divergent ideas about what the movement's goals should be and how they should be accomplished. This can lead to tensions within movements as different groups vie for dominance, but it can also lead to increased vitality since social movements (especially those on a large scale) can often benefit from the diverse backgrounds, ideas, and interests of members. The women's movement is a good example. Within women's movement organizations, there are different styles of activism—some of them more formal and mainstream than others (see the box "Doing Sociological Research: Nancy Whittier, *Feminist Generations*" for a discussion of the changes among radical feminists over the past thirty years). Class and racial differences among women also create distinct organizational needs, depending on the group's particular focus (Poster, 1995). The diverse organizational ideologies, activities, and structures of the women's movement have likely contributed to its vitality over the years, as well as its ability to respond to changing social conditions. At the same time, however, diversity within the women's movement also produces conflicts over whose interests the movement is most likely to represent. Women of color, in particular, have been critical of the women's movement for primarily representing White, middle-class women's experiences, even though women of color tend to hold strong feminist values and have organized many groups under the umbrella of the women's movement that speak to their needs.

Within social movements, there is typically tension between spontaneity and structure (Freeman, 1983a). Social movements thrive on spontaneity, unlike everyday organizations; they often must swiftly develop new strategies and tactics in the quest for change. During the civil rights movement, students improvised the technique of sit-ins, which quickly spread because they succeeded in gaining attention for the activist's concerns. Social movements are different from other forms of collective behavior, however, in containing routine elements of organization and lasting for longer periods. Short-lived collective behaviors nevertheless have a role in social movements; thus, there is a social movement in defense of the environment, but the movement also includes crowds (as in large demonstrations) and fads (such as refillable coffee mugs bearing pro-environment slogans). Some social movements even include riots, as was the case during the Black power movement of the 1960s.

Types of Social Movements

There are three broad types of social movements: personal transformation movements, political or social change movements, and reactionary movements. **Personal transformation movements** aim to change the individual. Rather than pursuing social change, they focus on the development of new meaning within individual lives (Klapp, 1972). An example is the Hippie movement of the late 1960s and early 1970s. Hippies defined conventional lifestyles as prepackaged, boring, and superficial, and explored alternatives such as communal living, money-free economies, and experiments in sexuality and gender relations. The New Age movement of the 1990s is a personal transformation movement that defines mainstream life as stressful and overly rational and promotes relaxation and spiritualism as an emotional release and route to expanded perceptions. Unlike the Hippie movement, which tended to emphasize dropping out, the New Age movement is an adaptation to mainstream life. New Age

The suffragettes of the early twentieth century, advocating the rights of women to vote, were some of the earliest feminists in the United States.

BOX 21.2 DOING SOCIOLOGICAL RESEARCH

Nancy Whittier, *Feminist Generations*

NANCY Whittier's research on the women's movement examines how different historical periods shape the collective identity of those who participate in social movements, especially social movements like feminism that have endured over a long time. Whittier is especially interested in the experiences of feminist women who in the 1960s and 1970s identified themselves as radical feminists, and she asks how the feminist identity these women forged in the women's movement continues to influence their lives and work. In addition, she also examines the transmission of social movements across generations by discussing the new and younger generation of women just coming into feminism in the 1990s.

Whittier challenges the popular assumption that former activists have abandoned their radical politics and that there is now a postfeminist generation of young women thoroughly uninterested in feminism and unresponsive to the women's movement. Using the concept of political generations, she argues that coming of age in different historical periods produces different generational groups with different perspectives. Participation in the women's movement in the 1960s and 1970s was, for women now largely in their middle years, a deeply transformative experience. Having developed their political consciousness in a historical period when they were immersed in the women's movement, these women have since had to sustain their feminist beliefs even during the 1980s and 1990s when the dominant culture was extremely hostile to feminist thought and activism. Their experience within the feminist movement, however, forged links between them that still inform their work and progressive attitudes toward change.

By interviewing women in a particular city who were active feminists in the 1960s and 1970s and by examining the written documents of earlier feminist organizations, Whittier is able to show how feminism has been sustained even during a period of antifeminist backlash. She discovers that radical feminists of the 1960s and 1970s still use the perspective they gained from participation in the women's movement in numerous phases of their work and lives. In addition, she finds that, despite myths about the postfeminist generation, an extensive feminist culture has developed that now draws in younger women who, though they differ in outlook and feminist perspective from their foremothers, still see themselves as forging new identities that are creating new forms of feminist thinking in the 1990s and into the future.

Whittier's research provides a strong case study of a social movement in a particular place, as it evolves with changing social conditions. In the end, she sees feminism as an enduring social movement, yet one that develops in response to particular historical conditions and as a reflection of the changing membership within the movement. Going beyond the glib pronouncements that feminism is dead, Whittier shows how feminism has evolved as a social movement and suggests what it may become in its next wave.

SOURCE: Whittier, Nancy. 1995. *Feminist Generations: The Persistence of the Radical Women's Movement.* Philadelphia: Temple University Press.

Some social movements use more radical tactics than others. Here Greenpeace demonstrated against the deforestation brought on by the timber industry.

music, crystals, massage therapy, and meditation are intended to restore the New Age person to a state of unstressed wholeness. Like many other social movements, New Ageism is supported by an array of dues-charging organizations and commercial products, from tapes and crystals to sessions with the spirits of the dead, retreats in yoga ashrams, and guided tours of Native American holy sites.

Religious and cult movements are also personal transformation movements. The recent rise in evangelical religious movements can be explained in part by the need people have to give clear meaning to their lives in a complex and sometimes perplexing society. In personal transformation movements, participants adopt a new identity—one they use to redefine their life, both in its current and former state.

Social change movements aim to change some aspect of society. Examples are the environmental movement, the gay and lesbian movement, the civil rights movement, the animal rights movement, and the religious right movement. All seek social change, although in quite distinct and sometimes oppositional ways. Some movements want radical change in existing social institutions; others want a retreat to a former way of life or even a move to an imagined past (or future) that does not exist. Social movements use a variety of tactics, strategies, and organizational forms to achieve their goals. The civil rights movement in the early 1950s used collective action in sit-ins, mass demonstrations, and organizational activity to overturn statutes that supported the "separate but equal" principle of segregation. These efforts culminated in the Supreme Court decision *Brown v. Board of Education,* which declared the "separate but equal" doctrine unconstitutional.

Movements also form alliances with other movements, and many social change movements involve a vast network of diverse groups organized around broadly similar, but also unique, goals. The Asian American movement, for example, includes Asian American women's organizations like Pan Asia and the National Network of Asian and Pacific Women, as well as Asian workers groups like the Six Companies and the Chinese Consolidated Benevolent Association, as well as other groups of Asian activists (Wei, 1993; Espiritu, 1992; see the box "Understanding Diversity: Asian American Panethnicity: Building Identity from Social Movements").

Social change movements may be *norm focused*—they may try to change the prescribed way of doing things—or they may be *value focused,* meaning they try to change a fundamental idea or something everyone holds dear (Turner and Killian, 1988). Often, they are both; the civil rights movement tried to change the law of the land and the attitudes of its people at the same time. The broad charter of the civil rights movement spawned offspring movements devoted to the special interests of such groups as Chicanos, feminists, and lesbians and gays, among many others.

Social change movements may be either reformist or radical. **Reform movements** seek change through legal or other mainstream political means, typically working within existing institutions. **Radical movements** seek fundamental change in the structure of society. Although most movements are primarily one or the other, within a given movement there may be both reformist and radical factions. In the environmental movement, for example, the Sierra Club is a classic reform movement that lobbies within the existing political system to promote legislation protecting the environment. Greenpeace, on the other hand, is a more radical group that sometimes uses guerrilla tactics to disrupt activities the group finds objectional, such as the killing of whales. The distinction between reform and radical movements is not absolute. Movements can contain elements of both and may change their orientation in midstride upon meeting success or failure; moreover, whether a group is defined as radical or reformist is to a great degree a matter of public perception. Just as one person's rebel is another's freedom fighter on the world stage, the definition of a social movement often depends on its social legitimacy and its ability to control how it is defined in the media and other public forums (Killian, 1975).

Reactionary movements organize to resist change or to reinstate an earlier social order that participants perceive to be better. They are reacting against contemporary changes in society. The Aryan Nation, a White supremacist coalition, is a reactionary movement that wants to suppress Jews and minorities and institute a "racially pure" decentralized state. The Right-to-Life movement wants to overturn court decisions that declared abortion legal and return to an earlier definition of abortion as criminal. The militia movement is also a reactionary movement, seeking to resist government authority and reinstate the perceived lost power of White people. Reactionary movements represent an organized backlash against social values and societal changes that participants find deeply objectionable. Rather than adjust to the future, they may instead harken to some mythical past that their movement ideology defines as ideal.

BOX 21.3

UNDERSTANDING DIVERSITY

Asian American Panethnicity: Building Identity from Social Movements

GROUPS known as Asian American originally immigrated from many different countries and did not necessarily think of themselves as "Asian." Coming from China, Korea, Japan, and more recently, Southeast Asian countries and India, many based their identity not on a country of origin so much as on a home province or district. Sociologist Yen Le Espiritu (1992) argues that the identification Asian Americans have of themselves as "Asian American" (what she calls "panethnicity") stems in large part from the activism of Asian Americans in contemporary social movements.

According to Espiritu, Asian groups immigrating early to the United States distanced themselves from other Asian groups that they perceived to be stigmatized in American society. As an example, upon immigrating to the United States, many Japanese thought of themselves as superior to the Chinese. They saw the Chinese as despised and excluded in American society and did not want to share that status. Soon, however, the Japanese experienced the same patterns of exclusion that Chinese Americans did. Others, notably Filipinos and Koreans, then distanced themselves from the Japanese, not immediately recognizing that they, too, would be subject to exclusionary practices and racism in the dominant culture. Several changes in the society modified this pattern of identification, resulting in what Espiritu calls "Asian American panethnicity."

The first change was demographic. Immigration restrictions in the United States throughout the early half of the twentieth century resulted in American-born Asians outnumbering immigrant Asian groups. Espiritu argues that, as Asian Americans became a native-born community, differences in their language and culture blurred. Many attended Asian-language schools, learning English as their common language, but also learning to communicate with each other. Born in the United States, second-, third-, and subsequent generations developed a stronger identification with the United States than with the worlds from which their ancestors had come; thus, they began to see themselves as having more in common with each other than with either White Americans or Asians outside the United States.

The second major impact on Asian American identity was historical, namely, World War II. Before then, Asian Americans had already been segregated within the United States in Chinatowns, Koreatowns, Little Tokyos, and other residentially segregated communities. The war exacerbated White fears and hatred of all Asian groups. The internment of Japanese Americans during World War II is evidence of the hatred and exclusion directed toward Japanese Americans in the context of the war. Fears of a "yellow peril" (alleged overpopulation of Asians) and the climate of hatred toward Japan heightened racism against all Asian Americans during this period.

The development of social movements among Asian Americans in recent years is the third and most important factor in generating a common sense of identity among these diverse groups. Influenced by the Black power movement of the 1960s, Asian Americans rejected the term *Oriental* because of its connotation of passivity and deviousness. As one activist writes, "Oriental" conjures up images of "the sexy Susie Wong, the wily Charlie Chan, and the evil Fu Manchu" (Weiss, 1974). Asian pride, like African American pride, was reflected in the names of new organizations like the Yellow Seed, the Asian American Political Alliance, and Asian Women United of California.

Other social movement activities from the 1960s through the 1990s have continued to foster a strong Asian American identity. Asian students active in the antiwar and student movements of the 1960s developed a sense of themselves as a minority group. Learning from the Black protest movement, Asian American students and faculty developed Asian American Studies programs to study their history and culture. The publications coming from this movement further serve to heighten awareness of the experiences of Asian Americans in the United States.

The feminist movement also enhanced awareness among Asian American women of their common identities. Noting the racism often experienced within the women's movement and the sexism experienced within Asian American communities, Asian American women have organized to challenge both racism and sexism and to build solidarity among themselves.

Together, these developments have created a strong identity for Asian Americans as Asian Americans, although, as Espiritu points out, pan-Asian consolidation is not universal. It is most common among middle-class, well-educated Asian American professionals and activists. The development of an Asian American identity has, however, transformed the sociological character of Asian America. As Espiritu concludes, pan-Asian ethnicity has made already existing networks more cohesive, and some have expanded into new political organizations. Espiritu's research makes an important sociological point that "ethnicity is forged and changed in encounters among groups" (1992: 161). Ethnic identity, like other forms of group identity, stems from the relationships people have with one another and the meaning given to those relationships. Espiritu's research strongly demonstrates the importance that social movements play in forging these identities.

SOURCE: Espiritu, Yen Le. 1992. *Asian American Panethnicity: Bridging Institutions and Identity.* Philadelphia: Temple University Press.

Reactionary movements, such as the militia movement, are those that try to reinstate a perceived status quo.

THINKING SOCIOLOGICALLY

Identify a *social movement* in your community or school that you find interesting. By talking to the movement leaders and participants, and if possible, examining any written material from the movement, how would you describe this movement in sociological terms? Is it *reform, radical,* or *reactionary*? How have the available resources helped or hindered the movement's development? What tactics does the movement use to achieve its goals? How is it connected to other social movements?

Origins of Social Movements

How do social movements start? At least four elements are necessary: *a preexisting communication network, a preexisting grievance, a precipitating incident,* and *the ability to mobilize.*

Social movements do not typically develop out of thin air. For a movement to begin, *there must be a preexisting communication network* (Freeman, 1983a). Communication can be informal, as between neighbors, organizations, or other group networks, but in order to initiate a movement, activists need some way to communicate with the people who will become part of the new movement.

The importance of previously established networks in the emergence of new social movements is well illustrated by the beginnings of the civil rights movement, which most people date to December 1, 1955, the day Rosa Parks was arrested in Montgomery, Alabama, for refusing to give up her seat on a municipal bus to a White man. Although she is typically understood as simply too tired to give up her seat that day, Rosa Parks had been an active member of the movement against segregation in Montgomery. The local NAACP chapter, of which Parks was secretary, had already been organizing to boycott the Montgomery buses. Rosa Parks was the one chosen to be the plaintiff in a test case against the bus company. Because she was soft spoken, middle aged, and a model citizen, NAACP leaders believed her to make a credible and sympathetic plaintiff in what they had long hoped would be a significant legal challenge to segregation in public transportation. When Parks refused to give up her seat, the movement stood ready to mobilize.

News of Rosa Parks's arrest spread quickly via networks of friends, kin, church, and school organizations. A small group, the Women's Political Council, had previously discussed plans to announce a boycott. The arrest of Rosa Parks presented an opportunity for implementation. One member of the Women's Political Council, Jo Ann Gibson Robinson, called a friend who had access to the mimeograph machine at Alabama State College. In the middle of the night, Robinson and two of her students duplicated 52,500 leaflets calling for a boycott of the bus system, to be distributed among Black neighborhoods (Robinson, 1987).

In addition to preexisting communications networks among movement participants, movements are also fueled by the news media which bring public attention to the movement's cause. Especially when movements use dramatic tactics, such as a massive March on Washington (like the Million Man March), news photos provide a means for information, even if distorted, to travel far and wide. With the extraordinarily widespread availability of television and, now, the Internet, cellular phones, and standard radio transmissions, news of movements around the world spreads quickly.

Celebrities can also advance the cause of social movements, although sociologists have found that celebrities typically involve themselves in less controversial movements that already have widespread popular support. The entry of celebrities into social movements can unintentionally hurt a movement's development since celebrities tend to depoliticize the movement (Meyer and Gamson, 1995). Their endorsement and presence, nonetheless, brings visibility to a movement. An example is the international movement to ban landmines. Throughout the world, landmines left in the ground following war endanger human lives. A cause that had been heralded by Princess Diana, the movement to ban landmines gained momentum following her death since her celebrity status and the attention given her brought far greater visibility to this international movement. Indeed, only three months after Princess Diana's death, 121 countries signed a treaty outlawing the use, production, stockpiling, and transfer of landmines. (The United States refused to sign the ban, but promised $100 million annually for landmine removal.)

For a movement to begin, *there must also be a perceived sense of grievance among the potential participants, or a strong desire for change.* The preestablished communication networks come into play when they are used to express the collective sense of wrong or right from which movements develop. Networks of friends, families, co-workers, and other groups whose members have something in common provide the channels through which people organize and articulate the developing ideas of an emerging movement (Freeman, 1983a). The environmental justice movement, as an example, has emerged largely from grassroots organiza-

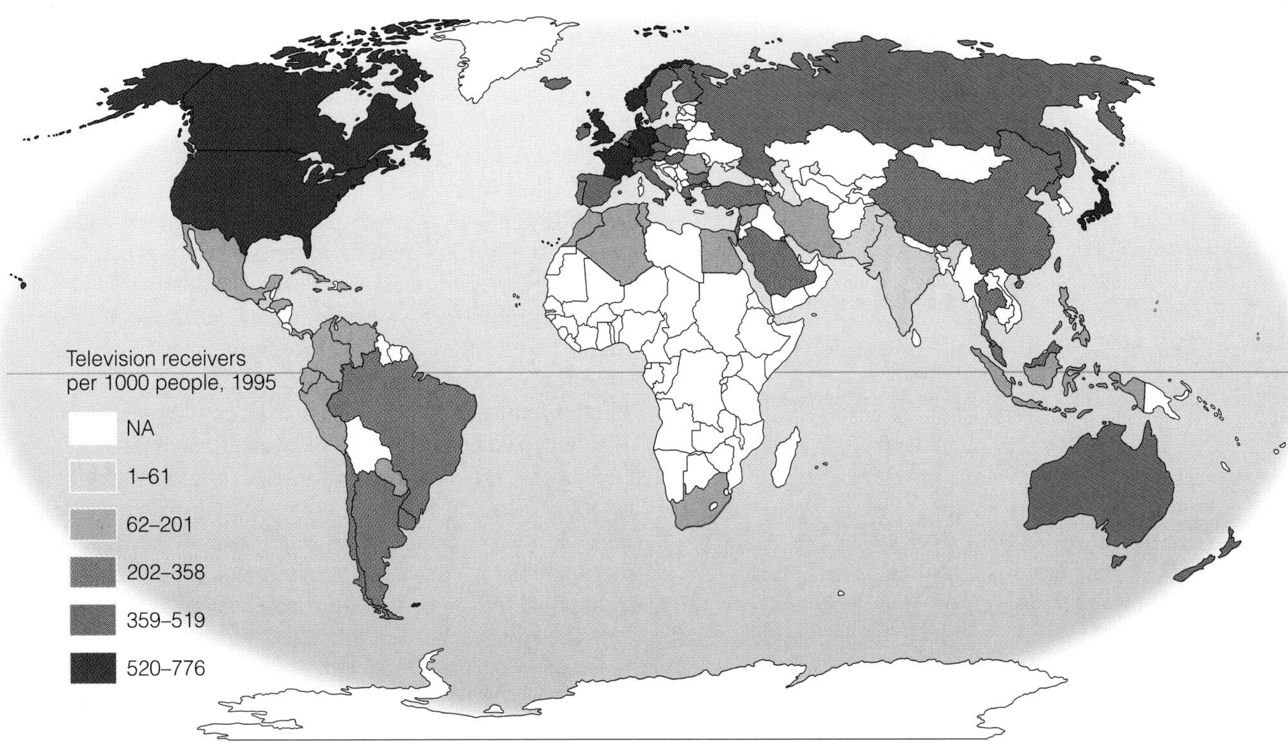

MAP 21.2 Viewing Society in Global Perspective: Transmitting World Affairs (television receivers per 1000 people)

SOURCE: U.S. Bureau of the Census. *Statistical Abstracts of the United States 1997*. Washington, DC: U.S. Government Printing Office, p. 842.

tions formed in communities where residents have organized to clean up a toxic waste site, close a polluting industry, or protect children from a perceived threat to their health and safety (Krauss, 1994). They may develop communications networks through community meetings, newsletters and flyers, or discussions with family members, friends, and co-workers.

With a communication network in place and a preexisting sense of grievance among potential movement participants, the conditions are present for the emergence of a social movement. The next step is *a precipitating factor that translates the sense of perceived grievance into action.* In the Montgomery Bus Boycott, this was Rosa Parks's refusal to give up her seat to a White man. In the environmental justice movement, it may be triggered by the sight of trucks dumping refuse in an empty lot or a plant spewing stinking gases into the air.

The final condition needed to get a movement started is *the ability of groups to mobilize.* **Mobilization** is the process by which social movements and their leaders secure people and resources for the movement. Even with all other conditions present for a movement to develop, movements cannot continue if they cannot mobilize people. Grievances exist all the time in society, but they alone, even in the presence of a communications network and a precipitating incident, do not explain the emergence of social movements. For example, a person might think he or she has been wronged by some company and might communicate this grievance to networks of friends and associates. Those people might even share the same grievance, but until they are able to mobilize on behalf of their collective sense of a need for change, there will be no social movement.

Movement leaders need to coordinate action between the different groups that constitute a movement. Sometimes this means the formation of new groups and organizations. It can also mean utilizing existing groups on behalf on the new movement activity. In the Montgomery Bus Boycott, for example, critical forces in the newly emerging movement were the Black churches in Montgomery, including the Dexter Avenue Baptist Church, where Dr. Martin Luther King, Jr., had accepted his first pastorate the previous year. The churches provided volunteers, meeting places, leadership, and a network of information sharing that was critical to the movement's success. In addition to the work of these established organizations on behalf of the new movement, a transportation committee was formed that arranged to ferry people who would normally ride the buses (Robinson, 1987). The activity of this group may have been the most important factor in getting the boycott successfully under way. Making an individual decision to stay off the bus was much harder if you did not know how you were going to get to work, especially when you also could not be sure that others would heed the boycott. The transportation system guaranteed both transportation and moral support and made it possible for the year-long boycott to succeed.

mobilization

Debunking Society's Myths

Myth: Social movements develop usually as the result of extremists who are single-minded in their interests.

Sociological perspective: Social movements often develop from the everyday concerns of ordinary people who mobilize to address conditions in their lives they find unacceptable.

The Organization of Social Movements

As movements develop, they quickly establish an organizational structure. The shape of the movement's organization may range from formal bureaucratic structures to decentralized, interpersonal, and egalitarian arrangements. Many movements combine both. For example, urban tenant movements typically have three organizational levels: groups of tenants in a single building, neighborhood organizations comprising tenant groups, and federations of neighborhood organizations. Tenant groups in a single building are a relatively weak bargaining unit, so in order to increase their bargaining power, they often unite with the tenant groups in other buildings to form a neighborhood organization. Larger, more bureaucratically organized federations are made up of groups from a larger geographic area that represent the interests of tenants in relations with the city and state; this broader organization may also educate the general public on the general goals of the tenant movement (Lawson, 1983).

Many movements are large bureaucracies. The National Organization for Women, the National Association for the Advancement of Colored People (NAACP), Amnesty International, Greenpeace, the Jewish Defense League, and the National Rifle Association are examples of large-scale bureaucratic organizations. As social movements become institutionalized, they are more likely to take on bureaucratic form. Bureaucratically organized movements affirm Max Weber's prediction that social movements gradually rationalize their structures to efficiently meet the needs of members and the goals of the organization.

Other movements are often made up of organizations that are more decentralized, interpersonal, and loosely connected. They may even comprise a vast network of organizations with no central command or decision-making structure. Student organizations protesting antiaffirmative action policies on college campuses are an example. With no national office to coordinate such a movement, leaders may rely on personal networks with students on other campuses or other student newspapers to fuel their movement. Increasingly, the Internet provides a rich source for movement communication since it closes the distance between locations where movement activity takes place. Over the course of a movement's life, various organizations and groups may grow, proliferate, fuse, contract, or die (Gerlach and Hine, 1970). The antinuclear movement, for example, is made up of many local groups, local chapters of larger environmental groups, a few national groups, and groups like the United Mine Workers, which sometimes participate even though they direct most of their attention to other matters (Dwyer, 1983). One potential result of decentralized structure is a rivalry for leadership within a given movement.

Links between social movements arise in a number of ways. Different groups may participate in joint activities, members of a group may switch to other groups or be enrolled in several organizations at once, and individuals in different groups may have friendship ties with each other (Gerlach and Hine, 1970). For example, members of the antinuclear movement frequently move between groups giving speeches or exchanging resources and ideas (Dwyer, 1983). Likewise, the women's movement comprises numerous groups and organizations, some of which have widely divergent philosophies and organizational structures. To the outsider, a movement may appear to be a single, unified group, but social movements are often more diverse on the inside than they appear to the causal observer.

The advantages of informal organization over more bureaucratic structures include greater resilience and flexibility (Gerlach and Hine, 1970). Among the disadvantages, first is that the movement cannot act rapidly as a whole. Second, no one person can legitimately speak for the entire movement, and there may be no one who can give clear direction to the movement's principles and plans (Dwyer, 1983). Finally, fragmentation within a movement can be the cause of a movement's demise.

Strategies and Tactics

Social movements choose their political and social strategies based on a number of variables: the resources available to them, the constraints on their actions, the organizational structure of the movement, and finally, the expectations the movement has about potential targets of its actions, such as the media, social control agencies, and audiences (Freeman, 1983b). The kinds of resources available to a group may heavily influence the tactics they use. Poor people are hardly in a position to mobilize high-powered, high-priced Washington lobbyists. Instead, they are more likely to use disruptive tactics such as occupying buildings and blocking traffic; their bodies are the resources most readily available to them (Piven and Cloward, 1977). Whether the movement is organized more bureaucratically or more interpersonally may influence the kinds of strategies or tactics used. The Ku Klux Klan uses a variety of strategies in their hate campaigns. Women of the Klan drew on their familial and community ties to form "poison squads," groups that used social networks to spread rumor or slander and organize consumer boycotts. Men, on the other hand, used their work connections and participated in the more visible and public strategies of disrupting and corrupting elections, night riding, and gang terrorism (Blee, 1991). The Klan also plays to stereotypes about sexuality, gender, and class in creating a movement ideology that promotes White supremacy while targeting Jews, African Americans, gays and lesbians, and other supposed "outgroups" (Daniels, 1997).

Sometimes the strategies that movements use can actually backfire, causing them to lose, rather than gain, support. The strategy of antiabortion activists to block access to clinics is a case in point. In a study of public attitudes toward abortion that extended over a three-year period in Buffalo, New York, researchers discovered that the militant activities of Operation Rescue during the second year of the study actually decreased support for the position of those in Operation Rescue (Ansuini et al., 1994). Although there have been no comparable studies done on a national basis, one could surmise that the widely publicized murders of workers and physicians who provide abortion services by those in the antiabortion movement could weaken support for their cause.

Constraints on social movements may include values, past experiences, reference groups, and expectations that groups have. The value systems of many groups do not allow them to use tactics that harm people, but they may be willing to destroy property. Peace groups, for example, are unlikely to use terrorism or weapons, given their movement's philosophy. Finally, the appearance of opportunities for action, the perceived response of social control agencies, and the expected effect that a group's action will have on the public all contribute to shaping the strategies the group will use.

Theories of Social Movements

Sociologists have developed several theories to explain the development of social movements, including resource mobilization theory, political process theory, and new social movements theory.

Resource mobilization theory is an explanation of how social movements develop that focuses on how movements gain momentum by successfully garnering resources, competing with other movements, and mobilizing the resources available to them (Marx and McAdam, 1994; McCarthy and Zald, 1973). Money, communication technology, special technical or legal knowledge, and people with organizational and leadership skills are all examples of resources that can be put to use in organizing a social movement (Zald and McCarthy, 1975). Interpersonal contacts are one of the most important resources a group can mobilize because they provide a continuous supply of new recruits, as well as money, knowledge, skills, and other kinds of assistance (Snow et al., 1986). Sometimes, social movement organizations acquire the resources of other organizations. Aldon Morris, an African American sociologist, has used resource mobilization theory to explain the development of the civil rights movement. That nationwide movement relied heavily on Black churches and colleges for resources such as money, leadership, meeting space, and administrative support (Morris, 1984). In other words, it mobilized existing resources on behalf of its own cause.

Resource mobilization theory, among other things, notes how movements are connected to each other. The gay and lesbian movement, as an example, has used many of the strategies of the violence against women movement in developing its own campaign to halt hate crimes against gay men and lesbian women (Jenness, 1995). Although sociologists have argued that the gay and lesbian movement has not as effectively developed a critique of traditional gender relations (Jenness and Broad, 1994), this movement has been able to employ some of the same "collective action frames"—a term sociologists use to talk about how movements organize thinking about the issues they address (Gamson, 1992; Snow and Benford, 1988)—that the women's movement has utilized (Jenness and Broad, 1994).

Political process theory posits that movements achieve success by exploiting a combination of internal factors, such as the ability of organizations to mobilize resources, and external factors, such as changes occurring in the society (McAdam, 1982). Political process theory explicitly acknowledges the effects that larger social and political processes have on the mobilization of social movements. Large-scale changes such as industrialization, urbanization, or a lifting of repression may provide opportunities for the mobilization of social movements not present before. Charles Tilly, after examining the popular uprisings in France from the seventeenth century to the present, argues that the development of capitalism and the concentration of power in the institutions of the state affected how people on the local level participated in collective action. As capitalism developed, markets emerged in cities for grain from the countryside. To meet the needs of the new markets, grain had to be stored and transported. Rural peasants, angry that their food supply was being transported away, saw the opportunity for collective action and attacked and sabotaged storage and transportation facilities (Tilly, 1986).

Political process theory also stresses the vulnerability of the political system to social protest. For example, several groups in Watsonville, California, used the Loma Prieta earthquake in 1989 as an opportunity to protest. Organized Latino groups had been demanding better low-income housing in that community for approximately a decade (Simile, 1995). Among Mexican farm laborers in the valley, there was already a long tradition of labor organizing, including a cannery workers' strike in 1986 (Bardacke, 1988; Amott and Matthaei, 1991). After the earthquake, these groups mobilized around the issues of temporary and permanent housing, and the way that aid had been distributed to Latinos. In other words, groups already organized for protest on other issues were ready to squeeze an opportunity out of disaster.

Resource mobilization theory and political process theory are social structural explanations. That is, they explain the emergence of social movements as the result of forces lying outside individuals. Although recognizing the importance that resources play in galvanizing movements, more recent explanations of social movements note that social psychological explanations cannot be ignored. **New social movement theory** links culture, ideology, and identity conceptually to explain how new identities are forged within social movements. Whereas resource mobilization theory emphasizes the rational basis for social movement organization, new social movement theorists are especially interested in how identity is socially constructed through participation in social movements (Gamson, 1995b; Larana et al., 1994; Calhoun, 1994). This new development in social movement theory links structural explanations with cultural and social-psychological theories, investigating how social movements provide foundations upon which people construct new identities (Gamson, 1992; Gamson, 1995a; Morris and Mueller, 1992).

Each of these theoretical viewpoints is helpful in explaining different aspects of how movements organize and how, in some cases, they fail. One team of sociologists has examined the demise of the American Indian Movement (AIM)—a specific organization that was prominent in the Native American liberation movement for several years. These sociologists argue that AIM ultimately failed to reach its goal of building a pan-tribal sense of identity among American Indians because of their inability to mobilize resources in the face of strong government repression. AIM was a threat to both government and corporate interests because of its focus on reclaiming energy resources on lands Native Americans argued were rightfully theirs. The authors conclude that resource mobilization theory, political process theory, and new social movement theory each explain different dimensions of this movement (Stotik et al., 1994). This is a good illustration of how sociological theory can be used to explain the contemporary social movement activity.

Diversity, Globalization, and Social Change

Social movements are a major source of social change. Around the world, as people have organized to protest oppressive forms of government, the absence of civil rights, or economic injustices, social change often results. What would the United States be like had the civil rights movement not been inspired by Mahatma Gandhi's liberation movement in India? How would contemporary politics be different had the African National Congress and other movements for the liberation of Black South Africans not dismantled apartheid? How is the world currently being affected by the development of a more fundamentalist Islamic religious movement in the Middle East? Would the world not be so conscious of human rights issues in China had they not witnessed the killing of students demonstrating in Tiannemann Square?

These and countless other examples show the significance of collective behavior and social movements for the many changes affecting our world. Sometimes collective behavior and social movements can be the basis for revolutionary events—those that change the course of world history. Other times, the persistence of social movements more slowly changes the world—or a particular society. Such has been the case for the civil rights movement in the United States—a movement that not only has transformed American society, but has since inspired similar movements throughout the world. Likewise, the women's movement is now a global movement, although the particular issues for women vary from place to place.

In the future, it is likely that there will be still further worldwide reach of movements that originate in a particular country. Modern communication systems, much transformed by the increasingly penetrating influence of the Internet, will bring social movements in distant parts of the world closer to home. And, as the world becomes increasingly interdependent—through its economies, the environment, and the ease of international travel and communication—social movements in one part of the world are likely to have a broad influence on other parts.

In the United States, as we reach the millennium, some of the most significant social movements are those associated with the nation's diverse population. The women's movement, the civil rights movement, the gay and lesbian movement are all major sources of activism in contemporary society, and these movements have generated some of the most transformative changes in the nation's social institutions. Despite the persistence of class, race, and gender inequality, no longer is segregation legally mandated. Equal opportunity laws, at least in theory, protect diverse minority groups from discriminatory treatment. Large segments of the public have become more conscious of the harmful effects of racism, sexism, and homophobia in society. All these changes can be attributed to the successful mobilization of diverse social movements.

At the same time, those resisting such changes have also increasingly organized movements to oppose some of these changes. From hate groups to national organizations of academics opposed to the growth of multicultural studies in universities, many new social movements have emerged in recent years to protect what they see as the declining influence of particular groups in society. Only time will reveal the eventual changes that will result from the tension between these different social movements. What does seem certain is that the nation's changing and diverse population will continue to be a powerful influence in shaping social activism in the future. As we will see in Chapter 22, different sources of social change, including the growing racial and ethnic diversity within the population, will shape society of the future.

CHAPTER SUMMARY

- *Collective behavior* occurs when normal conventions cease to guide people's behavior, and people establish new patterns of interaction and social structure. Collective behavior tends to be more novel than conventional behavior. It involves group behavior that arises in unusual circumstances, although it can mark the beginning of more organized behavior. Collective behavior is patterned, not irrational, although it often appears to be highly emotional. *Rumors* are a mechanism people use to communicate during collective behavior. Collective behavior is often associated with efforts to promote change.
- *Crowds* are groups of people who temporarily come together for a specific reason. They are usually a particular size and density, and are more or less connected to one another. *Emergent norm theory* suggests that people create meanings in unusual situations to guide their behavior. Social control agents can also influence the behavior of crowds. *Panic* is spontaneous behavior that occurs when people become highly concerned about their safety and security. Despite its image as disorganized, panic is socially structured.
- The popular conception of *riots* is that they are the action of unruly mobs. In truth, riots can be explained by the social conditions that generate them. In urban riots, these factors can include economic deprivation, unaddressed grievances, low educational attainment within the population, poor housing, and high rates of recent immigration. Riots sometimes occur when there is an increasing gap in people's expectations and their actual conditions. This is known as the *revolution of rising expectations* or *relative deprivation theory*.
- *Collective preoccupations* are forms of collective behavior in which people engage in similar behavior and have a shared understanding of that behavior as needed to bring about change or identify their place in society. Fads, fashion, crazes, hysterical contagions, and scapegoating are types of collective preoccupations.
- *Social movements* are organized social groups that have some continuity and coordinating, and that exist to promote or resist social change. There are different kinds of social movements, including *personal transformation movements, social change movements,* and *reactionary movements.* Social movements start when there is a preexisting communication network, a collective sense of grievance, a precipitating factor initiating the movement, and mobilization of a group of people. *Resource mobilization theory* suggests that social movements develop when people can compete for and gain resources needed for mobilization. *Political process theory* suggests that large-scale social changes, such as industrialization or urbanization, provide the conditions that spawn social movements. Both resource mobilization theory and political process theory are social structural explanations that place the origins of social movements in societal conditions, not individual needs or psychological wishes. *New social movement theory* adds that social movements are places where people construct their identities.
- Social movements are typically made up of a network of different organizations. Movements use different strategies for change depending on the resources available to them, the constraints on their action, the organization of the movement, and the movement's expectations about potential targets.

KEY TERMS

collective behavior
collective preoccupations
convergence theory
emergent norm theory
expressive crowds
new social movement theory
personal transformation movements
political process theory
radical movements
reactionary movements
reform movements
relative deprivation theory
resource mobilization theory
scapegoating
social change movements
social movement

THE INTERNET: A Tool for the Sociological Imagination

Resources on the Internet:

Virtual Society: The Wadsworth Sociology Resource Center
http://sociology.wadsworth.com

Visit this site to find additional learning tools, including interactive quizzes, links related to web sites, and an easy link to *InfoTrac College Edition.*

Abolition Now!
http://www.abolition-now.com/

An organization promoting abolition of the death penalty.

The National Rifle Association of America
http://www.nr.org/

An organization promoting the constitutional right to keep and bear arms.

EnviroLink
http://www.envirolink.org/

An online group providing information about hundreds of environmental action groups.

Native Americans and the Environment Web site
http://conbio.rice.edu/nae/

A link to research on how environmental issues affect Native American communities.

Sociology and Social Policy: Internet Exercises

The emergence of hate groups in the United States is a social movement associated with the far-right. What evidence is there for the presence and growth of such groups, and what policies exist to protect people from the consequences of hate groups, such as hate crimes? Some organizations, such as the Southern Poverty Law Center, monitor the development of hate groups. What information and perspective do such groups provide regarding social policy in this area?

Internet Search Keywords:

hate groups	racism
discrimination	bigotry
hate crimes	social policy

Web sites:

http://www.hatewatch.org/
A group that monitors hate groups on the Internet.

http://www.fbi.gov/ucr/hatecm.htm
Hate crime report from the FBI.

http://www.civilrights.org/lcef/hate
This site offers a discussion of hate crimes in America.

http://www. igc.org/an/niot/
Web site from the PBS series "Not in Our Town." A program to raise awareness of hate violence in America.

http://www.splcenter.org/intelligenceproject/ip-index.html
This site monitors extremist and militant activity throughout America.

InfoTrac College Edition: Search Word Summary

hysterical contagions	riot
mobilization	Y2K panic

In order to learn more about these central topics in sociology, you can conduct an electronic search using InfoTrac College Edition. To aid in your search and to gain useful tips, see the Student Guide to InfoTrac College Edition on the Virtual Society web site:
http://sociology.wadsworth.com

INTERACTIONS—A SOCIOLOGY CD-ROM: CONCEPTS FOR THIS CHAPTER

Go to the Wadsworth Sociology CD-ROM for further study on the concepts in this chapter. The CD-ROM also includes quizzes and additional activities to expand your learning experience.

SUGGESTED READINGS

Aguilar-San Juan, Karen. 1994. *The State of Asian America: Activism and Resistance in the 1990s.* Boston: South End Press.

A collection of essays by Asian activists, this book provides a good overview of the organization of the Asian American social movement for racial justice.

Bullard, Robert D. 1994. *Unequal Protection: Environmental Justice and Communities of Color.* San Francisco: Sierra Club Books.

This anthology chronicles the numerous community organizations that have emerged in protest over the environmental hazards that fall disproportionately on minority and poor communities. It provides a sociological perspective on the environmental justice movement, as well as documenting the hazards to minority communities.

D'Emilio, John. 1983. *Sexual Politics, Sexual Communities: The Making of a Homosexual Minority in the United States, 1940–1970.* Chicago: The University of Chicago Press.

D'Emilio analyzes the growth of the gay and lesbian rights movement, showing how the movement broke down the invisibility of gays and lesbians and brought a new analysis of gay identity as socially constructed to the public eye.

Espiritu, Yen Le. 1992. *Asian American Panethnicity: Bridging Institutions and Identities.* Philadelphia: Temple University Press.

Espiritu shows how Asian activism has created a collective identity for Asian Americans, forged from their many different cultural and socioeconomic backgrounds. It is a very good analysis of how ethnic identity is constructed through social movements.

Marx, Gary T., and Douglas McAdam. 1994. *Collective Behavior and Social Movements: Process and Structure.* Englewood Cliffs, NJ: Prentice-Hall.

A basic text, this is a good overview of collective behavior and social movements. Written for undergraduates, it is a good way to learn about the diverse ways that sociologists study these phenomena.

McAdam, Doug, John D. McCarthy, and Mayer N. Zald. 1996. *Comparative Perspectives on Social Movements.* New York: Cambridge University Press.

Using examples of social movements from different nations, mostly in Europe, the contributors to this volume link social movements to the social and political structures of diverse societies. The analysis provided in the book provides a framework for the comparative study of social movements and social structure.

Morris, Aldon D. 1984. *The Origins of the Civil Rights Movement.* New York: The Free Press.

This book is a classic study of the history and sociology of the civil rights movement. It utilizes resource mobilization theory to explain the emergence of the movement.

Oberschall, Anthony. 1993. *Social Movements: Ideologies, Interests, and Identities.* New Brunswick: Transaction Publishers.

An overview of sociological theory on collective behavior and social movements, this book also provides analysis of such movements as the Black protest movement, the women's movement, and the new Christian Right.

Turner, Ralph, and Lewis Killian. 1988. *Collective Behavior,* 3rd ed. Englewood Cliffs, NJ: Prentice-Hall.

A classic text in collective behavior, this book provides an excellent analysis of collective behavior. It includes many rich illustrations of the different forms of collective behavior, including crowds, fads, crazes, and social movements.

Whittier, Nancy. 1995. *Feminist Generations: The Persistence of the Radical Women's Movement.* Philadelphia: Temple University Press.

This analysis of the feminist movement shows the very different philosophies that sparked the women's movement. It also includes a strong analysis of the role of lesbian feminists in the women's movement.

CHAPTER 22

Social Change in Global Perspective

TECHNOLOGICAL innovations can transform an entire society. We are quick to think of such changes as "progress," but sometimes people in a changing society wonder whether the things gained from progress are not overbalanced by the things lost. The Alaskan pipeline and related innovations have brought considerable wealth to the communities of northern Alaska, and many Eskimo entrepreneurs have grown prosperous from the changes. Others have fared less well. Some researchers have attributed increases in suicide and alcohol-related deaths to changes that simply came too rapidly in the social structure and culture of the Eskimos (Klausner and Foulks, 1982; Simons, 1989).

Just as technological innovations can transform a society, so can technological disasters. In the 1930s, the land in an area called Love Canal, a neighborhood in Niagara Falls, New York, was acquired by the Hooker Chemicals Corporation and then used as a dump site for chemicals and other wastes. Then in the 1950s, the area was covered with dirt and sold to the city of Niagara Falls, which then built a school on it and allowed homes to be built there. For the next twenty years, a visible ooze of black chemicals began seeping into homes in the Love Canal area. The ooze also killed shrubs, grass, and trees on the properties. To the horror of the residents, chemical tests showed that more than 200 different chemicals had originally been dumped into the site, many of which were cancer causing. Largely because of the efforts of Lois Gibbs, a resident of this working-class neighborhood, and the resultant national publicity, the federal government purchased the homes and relocated the residents. Love Canal is now a ghost town, containing boarded up houses, with homemade signs and graffiti telling of what happened. The Love Canal incident not only transformed life in and around Niagara Falls, but it stimulated the growth of present-day governmental programs dedicated to protection of urban environments and the elimination of toxic waste. This is a case of a technological disaster that produced a social movement that in turn resulted in political and social changes in society (Cable and Cable, 1995; Levine, 1982).

Technological advances have recently come to the Kaiapo people of Brazil. Numbering about one-quarter million, they are noted for their striking body paint and elaborate ceremonial dress. Recently, many Kaiapo have become wealthy as a result of gold mining and tree harvesting in their region; yet, a favorite topic of conversation among the elder members of this society is whether their newfound wealth is a blessing or curse. For example, the spread of television has caused the youth among the Kaiapo to retreat from traditional rituals, such as the evening campfire, where all members of the village, old and young, would sit, tell stories, and dispense philosophy. One elder lamented that whereas previously the night was the time when the old taught the young the ways of life, television had "stolen the night," bringing to an end a practice that had many benefits for the young (Simons, 1989).

These examples are studies in the effects of social change. What is social change? What causes it? What has the power to launch deep changes in norms, habits, practices, beliefs, gender roles, racial and ethnic relations, and class distinctions? This chapter examines the causes and consequences of societies in change.

The Kaipo people of Brazil, showing colorful formal dress. Technology from outside this society (TV; guns) presently threatens the persistence of such cultural practices.

What Is Social Change?

Social change is the alteration of social interactions, institutions, stratification systems, and elements of culture over time. Societies are always in a state of flux. Some changes are rapid, such as those brought about by desktop computers in little more than ten years. Other changes are more gradual; decades of effort to promote contraceptive methods in some developing nations have garnered only the most sluggish gains (Goldman et al., 1989). The speed of social change varies from society to society, and from time to time within the same society. Foraging societies (introduced in Chapter 5) had to adopt new technologies and customs to evolve into horticultural or agricultural societies, but in many cases the change took several centuries to occur.

With increased urbanization, such as occurred in the United States in the early 1900s, more rapid social change became possible. New technologies circulated quickly, and different forms of social stratification appeared. The overall rate of change was accelerated by a decline in rural economies and the movement of populations from rural to urban areas. As societies become more complex, the pace of change increases. In U.S. society, this truism can be seen by comparing the rate of change in the 1990s with the rate in the 1950s (Lenski et al., 1998). Most people would agree that the United States in the 1950s was far more predictable and staid than today's United States as it moves into the twenty-first century.

Microchanges are subtle alterations in the day-to-day interaction between people. A fad "catching on" is an example of a microchange. Fads and other microchanges often spread rapidly across the nation. Take the spread of bungie jumping. Although not as widespread as some prior fads, this highly dangerous recreation is one of a group of "extreme sports" that have recently become popular across the country. Bungie jumping has been the cause of quite a few serious injuries and deaths, but it has also provided thrilling footage for soft-drink commercials, probably accounting for why a large number of youths have suddenly developed a taste for putting themselves in bone-smashing danger. Although the overall change in the structure of society caused by fads is quite small, some minor effects may persist. Skateboarding has had several rises and falls as a fad, starting in the early 1960s, and in that time has never quite faded completely out of the repertoire of youthful recreations—an example of how a microchange can persist.

Macrochanges are gradual transformations that occur on a broad scale and affect many aspects of society. In the process of *modernization,* societies absorb the changes that come with new times and shed old ways. One frequently noted trend accompanying modernization is that societies become more differentiated socially, including greater differentiation in social rank, divisions of labor, and so on. The effects of the Alaska pipeline upon the Eskimo economy and social structure exemplify a macrochange. In the United States, the rise of the computer through all its generations, from vacuum tube to microchip, has dramatically changed society, another example of a macrochange. Not many years ago, who would have imagined that you could "click on" and surf the Web to the extent possible today? Although the macrochange to a digital culture was swift, some macrochanges can take generations. Whatever time they require, macrochanges represent deep and pervasive changes in social structure and culture.

Large or small, fast or slow, social change generally has in common the following characteristics:

1. *Social change is uneven.* The different parts of a society do not all change at the same rate; some parts lag behind others. This is the principle of *culture lag,* a term coined by sociological theorist William Ogburn (1922) and first described in Chapter 3. Recall that culture lag refers to the delay that occurs between the time that social conditions change and the time that cultural adjustments are made. Often the first change is a development in material culture, which is followed some time later by a change in nonmaterial culture (meaning the habits and mores of the culture). The symptoms of culture lag can be seen in the uneven dissemination of computer power. Some organizations and bureaucracies adopt state-of-the-art hardware and software more quickly than others, leaping ahead of their colleagues and competitors. Even within single organizations, change occurs unequally, with older employees tending to adapt to new technology more slowly than younger members.

2. *The onset and consequences of social change are often unforeseen.* The inventors of the atomic bomb in the early 1940s could not predict the vast changes in the character of international relations that were to come, including a cold war that lasted until the demise of the Soviet Union in the early 1990s. Television pioneers, who envisioned a mode of mass communication more compelling than radio, could

social change

not know that television would become such a dominant force in determining the interests and habits of youth, and the activities and structure of the family. The notion of culture lag is present in both of these examples: A change in material culture (invention of the atomic bomb; invention of television) precedes later changes in nonmaterial culture (international relations; youth culture and family structure).

3. *Social change often creates conflict.* Change often triggers conflicts along racial–ethnic lines, social class lines, and gender lines. The Black power movement, begun in the late 1960s and devoted to achieving economic, social, and political gains for African Americans (Carmichael and Hamilton, 1967), ignited many conflicts—between Blacks and Whites, Blacks and Jews, even between militant and assimilationist Blacks. Some intraracial conflict occurred along social class lines between Blacks who considered themselves middle class and Blacks who did not. The Black power movement also caused some conflict between college-educated professional Whites and blue-collar Whites, particularly men, who were less likely to sympathize with militant or "Black power" Blacks.

4. *The direction of social change is not random.* Change has "direction" relative to a society's history. A populace may want to make a good society better, or it may rebel against a status quo regarded as unendurable. Change may be wanted or resisted, but in either case, when it occurs, it takes place within a specific social and cultural context.

Social change cannot erase the past. As a society moves toward the future, it carries along its past, its traditions, its institutions (McCord, 1991; McCord and McCord, 1986). A generally satisfied populace that strives to make a good society better obviously wishes to preserve its past, but even when a society is in revolt against a status quo that is intolerable, the social change that occurs must be understood in the context of the past as much as the future.

Theories of Social Change

As we have seen, social change may occur for different reasons. It may occur quickly or slowly, may be planned or unplanned, and may represent microchange or macrochange. Different theories of social change emphasize different aspects of the change process. The three main lines of contention in social change theory are functionalist theories, conflict theories, and cyclical theories. We will consider each individually.

Functionalist and Evolutionary Theories

Recall from previous chapters that functionalist theory builds upon the postulate that all societies, past and present, possess basic elements and institutions that perform certain "functions" permitting a society to survive and persist. A *function* is a consequence of some social element that contributes to the continuance of a society.

The early theorists Herbert Spencer (1882) and Emile Durkheim (1964 [1895]) both argued that as societies move through history, they become more complex. Spencer argued that societies moved from "homogeneity to heterogeneity." Durkheim similarly argued that societies moved from a state of *mechanical solidarity,* a cohesiveness based on the similarity among its members, to *organic solidarity* (also called *contractual solidarity*), a cohesiveness based on difference—a division of labor that exists among its members joins them together since each is dependent on the others for the performance of specialized tasks (see Chapter 5). Societies thus move from a condition of relative undifferentiation to higher social differentiation through the creation of specialized roles, structures, and institutions.

According to functional theorists, societies move from structurally simple, homogeneous societies, such as foraging

BOX 22.1 • DOING SOCIOLOGICAL RESEARCH

An Insider's Account of a Social Movement and Social Change

TODD Gitlin, a sociologist and journalist, participated in the student rebellions of the 1960s. As a social movement, the rebellions of the 1960s were a force that stimulated social change in this society, and though some of these changes have faded, others remain. Gitlin found that from 1967 to 1969, attitudes on campuses leaned strongly, indeed radically, against the Vietnam war. In the fall of 1969, 69 percent of students nationwide called themselves "doves"—twice as many as in the spring of 1967. More and more students seemed to realize that the U.S. government was not about to end the war, and concluded a few years behind SDS (Students for a Democratic Society) that the war was not merely an isolated "mistake" but instead part of a detailed national plan. At that time, the percentage of students who agreed with the statement, "The war in Vietnam is pure imperialism," jumped from 16 percent in the spring of 1969 to 41 percent in April 1970. The number strongly disagreeing fell from 44 to 21 percent. The percentage of students calling themselves "radical or far left," 4 percent in the spring of 1968, rose to 8 percent in the spring of 1969 and 11 percent in the spring of 1970. Nonviolent as well as violent protests accelerated throughout the year, 731 of them involving police arrests, 410 involving damage to property (ROTC buildings and the like), and 230 involving violence to persons.

SOURCE: Gitlin, Todd. 1987. *The Sixties: Years of Hope, Days of Rage.* New York: Bantam Books.

or pastoral societies where members engage in largely similar tasks, to structurally more complex, heterogeneous societies, such as agricultural, industrial, and postindustrial societies, where great social differentiation exists and there is extensive division of labor among people who perform many specialized tasks. The consequence (or "function") of increased differentiation and division of labor is, according to functional theorists, a higher degree of stability and cohesiveness in the society, brought about by the realities of mutual dependence (Parsons, 1951a, 1966).

Evolutionary theories of social change are a branch of functionalist theory. One variety called **unidimensional evolutionary theory,** now well out of favor, argued that societies follow a single evolutionary path from "simple" and relatively undifferentiated societies to more complex and highly differentiated societies, with the more differentiated societies perceived as more "civilized." Early theorists such as Lewis Morgan (1877) labeled this difference a distinction between "primitive" and "civilized," an antiquated notion that has been severely criticized; there is no reason to suppose that an undifferentiated society is necessarily more primitive than a more differentiated one. Furthermore, in these earlier theories, there are no firm definitions for the terms *primitive* or *civilized.* Nevertheless, the notion that some societies are primitive continues to persist today.

Unidimensional theories of social change fell out of favor because social change occurs in several dimensions and affects a variety of institutions and cultural elements. Meeting the need for a theory that better matches what is actually observed, **multidimensional evolutionary theory** (also called **neoevolutionary theory**) argues that the structural, institutional, and cultural development of a society can follow many evolutionary paths simultaneously, with the different paths all emerging from the circumstances of the society in question (Service, 1975). The nature of social evolution in the society depends upon the interplay between the society's technology; population characteristics, including size; the extent of social differentiation; and other structural and cultural elements.

A formulation of multidimensional evolutionary theory is that of Lenski et al. (1998). Lenski gives a central role to technology, arguing that technological advances are significantly (though not wholly) responsible for other changes, such as alterations in religious preference, the nature of law, the form of government, and relations between races and genders. Although the role of technology is presented as central, other relationships among institutions continue to be important. For example, changes in technology, such as advances in computer hardware and software, can produce changes in the legal system, such as by creating a need for new laws to deal with computer crimes—the legality of pornography on the Internet, for example. An instance of social change may have no discernible link to technology, such as the development of a conflict between social classes over what each class is entitled to from the government, yet not much imagination is usually required to discover a link between social issues and technological issues in a society as technologically advanced as ours. An emerging conflict between classes, for instance, is very likely to be linked in some way to changes in the workplace, and the workplace changes in recent decades are quite plainly linked to changes in technology. Patterns such as these give support for the technological emphasis of Lenski's theory.

In support of the overall argument that social change is in fact evolutionary, that is, cumulative and not easily reversible, Lenski et al. point out that many agricultural societies have transformed into industrial societies thoughout history, but few have made the reverse trip from industrial to agricultural, although certain countercultural groups have tried, such as "hippie" farm communes of the 1960s and 1970s. On the other hand, Lenski also argues that social "advances" can be reversed. For example, a cataclysm such as an earthquake or flood can humble a technologically advanced society; following a natural disaster, especially in the developing world, city dwellers may find themselves foraging for food if the elaborate infrastructure supporting urban life is destroyed. A social reversal like this can be temporary or more permanent, depending on the interrelationships among other institutions and structures—thus illustrating the multidimensional aspect of the theory. Finally, Lenski is careful to give weight to human ingenuity and innovation. A single far-reaching innovation can transform a society in many ways, some of which may be entirely unpredictable by any evolutionary theory, either unidimensional or multidimensional.

Debunking Society's Myths

Myth: Societies change in linear, directed fashion from primitive to civilized.

Sociological perspective: Research suggests that social change can occur in several directions at roughly the same time, thus giving more weight to a multidimensional change theory than to a unidimensional one. Furthermore, the terms primitive and civilized are of limited usefulness and are out of favor as concepts, in that they imply some value judgment about the relative sophistication of diverse cultures.

Unlike the early theories of Spencer, Durkheim, and Parsons, newer functionalist theories emphasize the role of racial–ethnic, social class, and gender differences in the process of social change (Lenski et al., 1998; Alexander and Colomy, 1990; McCord, 1991; McCord and McCord, 1986). The earlier theories made the implicit assumption that European and American societies, predominantly White, were more evolved or advanced. Societies largely comprising people of color were usually assumed to be less evolved and more primitive. This bias was often projected onto analyses *within* a society; for example, in the United States, Native American and Latino culture was often seen as less "advanced." Older functionalist theories also supported the notion that less advanced peoples were less intelligent, a postulate completely rejected by later functionalists, as well as by social and behavioral researchers in general. The new functionalism rejects the "primitive" versus "civilized" dichotomy and its implicit commentary on racial groups and regards relations between racial–ethnic, social class, gender, and other groups as an important part of any society, regardless of its stage of development or evolution.

Conflict Theories

Karl Marx, the founder of conflict theory (Marx, 1967 [1867]), was himself influenced by the early functionalist and evolutionary theories of Herbert Spencer. Marx agreed that societies change and that social change has direction, the central principle in Spencer's social evolutionary theory, but Marx placed greater emphasis on the role of economics than did functionalist and evolutionary theorists. He argued that societies could indeed "advance" and that advancement was to be measured in terms of movement from a class society to one without class. Marx believed that, along the way, class conflict was inevitable.

As noted earlier in this book, the central notion of conflict theory is that conflict is built into social relations (Dahrendorf, 1959). For Marx, social conflict, particularly between the two major social classes in any society—working class versus upper class, proletariat versus bourgeoisie—was not only inherent in social relations but was indeed the driving force behind all social change. Marx believed that the most important causes of social change were the tensions between various social groups, especially those defined along social class lines. The main reason was that different classes had different access to power, with the relatively lower class carrying less power. Although the groups to which Marx originally referred were indeed social classes, subsequent interpretations expand on Marx to include conflict between any socially distinct groups that receive unequal privileges and opportunities (Marx, 1967; Rodney, 1974). However, be aware that the distinction between class and other social variables is necessarily murky. For example, conflict between Whites and minorities is at least partly (but not wholly) class conflict since minorities are disproportionately represented among the less well-off classes.

There is far more to racial and ethnic conflict in the United States than class differences alone: Many *cultural* differences exist between Whites on the one hand and Native Americans, Latinos, Blacks, and Asians on the other; furthermore, there are cultural differences *within* broadly defined ethnic groups as well. We have pointed out before in this book that there are broad differences in norms and heritage between Chinese Americans, Japanese Americans, Vietnamese Americans, and so on, all often grouped rather coarsely as Asian Americans. The same holds for other groups, who may be less heterogeneous than "Asian Americans" yet still represent many different cultures. African Americans from Washington State, Georgia, and the Caribbean are likely to find quite a lot about each other that is unique. Differences other than class have contributed greatly to past and present conflicts among racial–ethnic groups in society. The central idea of conflict theory is the notion that social groups will have competing interests no matter how they are defined, and that conflict is an inherent part of the social scene in any society.

A central theme in Marx's writing is that revolution and dramatic social change would come about when class conflict led inevitably to a decisive social rupture. Marx predicted that the capitalist class would progressively eliminate or absorb competitors and relentlessly pursue profits while squeezing the wages of the working class and crushing dissent. Discontent among the working classes was supposed to blossom into a recognition that the common enemy of the worker was the capitalist class. The workers would then join in revolution, overthrow the system of capitalism, eliminate privately owned property, and establish a new economic system that would exist for the good of all.

Although the worldwide revolution predicted by Marx has never come to pass, his highly refined analyses of class-related conflict have advanced our understanding of social change, and his work continues to be of interest. However, Marx seems to have overemphasized the role of economics in the network of social tensions he observed, while ignoring the importance of other relevant factors related to class. Sociologist Theda Skocpol (1979) has noted that in France, Russia, and China—countries where major revolutions have occurred—serious internal conflicts between social classes were combined with major international crises that the elite social classes proved unable to resolve before they were overthrown. The French Revolution, begun in 1789, erupted in a period when the newly arisen capitalist class was asserting itself worldwide against the old monarchies; and while France was bankrupt from the many wars of the seventeenth and eighteenth century, the country intervened in the American Revolution. The Russian Revolution occurred while Russia was flattened from its disastrous defeat in World War I, and the communist revolution in China occurred as the world was still putting out the flames of World War II. In each case, internal social change was linked to relations between entire societies as well as relations among social classes within a society.

The recent collapse of the Soviet Union offers some support for Skocpol's hypothesis. Years of international trade sanctions and rejection of its currency by foreign traders had eroded the economic foundations of the USSR, and headlong military competition with the United States helped drive the Soviet Union into bankruptcy. Thus, although economic considerations were paramount in the fall of the USSR, the relationships *between societies,* and not just among classes within a society, were also critical to the undoing of the former superpower. Nevertheless, conflicts *within* Russia along religious, social class, ethnic, and regional lines clearly also contributed to the social and cultural disintegration of the Soviet state.

TABLE 22.1 *THEORIES OF SOCIAL CHANGE*

	Functionalist Theory	Conflict Theory	Cyclical Theory
How do societies change?	societies change from simple to complex; from undifferentiated to highly differentiated division of labor	conflict is an inherent component of social relations; from a class to a classless society	societies go through a "life cycle," from birth to death; from the less "mature" (e.g., idealistic culture) to more "mature" (e.g., ideational or sensate culture)
Primary cause of social change	technology	economic conflict between social classes	necessity for growth
Varieties of the theory	unidimensional versus multidimensional	class conflict (Marx) versus international conflict (Skocpol)	structural cycles (Toynbee) versus cultural cycles (Sorokin)

Ethnic, racial, and religious and tribal differences join social class differences as major causes of conflict within and between countries. Cases in point include recent bouts of so-called ethnic cleansing, vicious battles between ethnic groups in which one group attempts to annihilate the other, as Serbians in the early 1990s attempted to do to Muslims in Bosnia and attempted in the late 1990s to do to Albanians in Kosovo. Another example was the recent systematic slaughter in Rwanda, Africa, of Hutu tribespeople, including women and children, by Tutsi warriers. Modern conflict theory includes all these forms of conflict within its purview, as well as conflicts between groups based on age, such as the efforts by groups like the American Association of Retired Persons (AARP) to secure entitlements for the elderly, or based on region, such as the water wars between groups on the West Coast of the United States who contend with each other over a barely adequate supply of water whose redirection could make deserts of huge tracts of unfavored territory. All of these instances of conflict have potential to reap considerable amounts of social change, depending on their outcomes.

Cyclical Theories

Cyclical theories of social change invoke patterns of social structure and culture that are believed to recur at more or less regular intervals. Cyclical theories build on the idea that societies have a "life cycle," like seasonal plants, or at least a "life span," like humans. Arnold Toynbee, a social historian and a principal theorist of cyclical social change, argues that societies are born, mature, decay, and sometimes die (Toynbee and Caplan, 1972). For at least part of his life, Toynbee believed that Western society was fated to self-destruct as energetic social builders were replaced by entrenched minorities who ruled by force and under whose sterile regimes society would wither. Some believe that societies become decrepit, only to be replaced by more youthful societies—a belief typified in Oswald Spengler's famous work, *The Decline of the West* (1932), which held that western European culture was already deeply in decline, following a path Spengler believed was observable in all cultures.

Sociological theorists Pitrim Sorokin (1941), and more recently, Theodore Caplow (1991), have argued that societies proceed through three different phases or cycles. In the first phase, dubbed the *idealistic culture,* the society wrestles with the tension between the ideal and the practical. An example would be the situation captured in Gunnar Myrdal's classic work, *The American Dilemma* (1944), in which our nation declared a belief in equality for all, despite intractable racial, class, and gender stratification.

The second phase, *ideational culture,* emphasizes faith and new forms of spirituality. The strong religious institution of the Puritans in colonial America is one example. Another more current example is the present New Age spirituality movement, which stresses nontraditional techniques of meditation and the use of crystals and body massage oils in a journey toward self-fulfillment and spritual peace.

The third phase is *sensate culture,* which stresses practical approaches to reality, and also involves the hedonistic and the sensual ("sex, drugs and rock and roll"). Sorokin may have foreseen the hedonistic elements of popular culture in the 1960s and 1970s, indicative of sensate culture. According to the theory, when a society tires of the sensate, the cyclical process begins again with the society seeking refuge in idealistic culture. The push in the late 1980s and early 1990s for a return to "family values," meaning older and more traditional values, is an example of a return to idealistic culture, presumably as a response to the sensate culture of the prior twenty years or so.

The Causes of Social Change

Social and cultural change is a broad subject, and the causes of social change are varied and shifting. The major ones are reviewed here, including collective behavior and social movements (examined in detail in Chapter 21); cultural diffusion, inequalities in race, ethnicity, social class, and gender; technological innovation; changes in population, including migration; and war.

Collective Behavior

Collective behavior was covered in detail in Chapter 21. Certain types of collective behavior can serve as major causes of social and cultural change. Collective preoccupations, such as fads, crazes, and fashions, can serve to initiate social change although the magnitude of the change that results may be slight. Many of today's modes of dress were once passing fashions. The necktie can be traced to Victorian times in England, when gentlemen of the day wore a forerunner of the necktie, a cloth napkin around the neck, used for wiping the mouth after a meal. More recently, Levi's jeans, adopted by teenagers in the 1950s, caught on so completely that jeans are now among the most common articles of clothing, popular with all age groups and in many societies. Jeans were originally used in the late nineteenth century by workingmen, cattlemen, and blacksmiths, and were worn by men only. Jeans were considered to be working-class garb, not to be worn by professionals of the day. Now, designer jeans are worn by those in all class strata, demonstrating how a style affected economic markets that in turn resulted in social change.

THINKING SOCIOLOGICALLY

A fad is an example of a *microchange*. What fads have you noticed lately among your friends or acquaintances? Are they completely new, or were they started a year or two ago and are simply continuing? How long do you think these fads will last? Sometimes a microchange such as a fad or craze can endure for many years, in which case it has an increased chance of causing a *macrochange*, a change on a broader scale. Name two fads of the past that resulted in macrochanges.

Social Movements

Social movements are organized and persistent forms of collective behavior. The purpose of a social movement is often to initiate or vigorously resist social change. A long-running social movement in the United States is the drive for civil rights, from the *Brown v. Board of Education* decision in 1954 and the Montgomery bus boycott of the 1950s through the Black power era of the late 1960s and on to more recent battles over affirmative action. Recent social movements designed to change people's views and thereby change society include ACT-UP (AIDS Coalition to Unleash Power) and the New Age movement. ACT-UP comprises hundreds of local chapters across the country whose purpose is to raise funds to assist AIDS sufferers, to increase the political and economic power of those who are HIV positive or have AIDS, and to reduce the negative stigmatization of those so afflicted by changing public attitudes. The New Age movement of the last decade is a loosely structured social movement that encourages personal transformation and adaptation to the stresses of life through such outlets as meditation, New Age music, and massage therapy.

Martin Luther King, Jr., addresses the March on Washington in 1963, a major event in the Civil Rights Movement.

Debunking Society's Myths

Myth: Rebellious social movements such as the Black power movement or ACT-UP are simply people blowing off steam; they do not result in long-lasting changes in society.

Sociological perspective: Social movements of these types are one of the several major causes of long-lasting social change, resulting in enduring structural and cultural changes in society.

Cultural Diffusion

Cultural diffusion (as noted in Chapter 3) is the transmission of cultural elements from one society or cultural group to another. Cultural diffusion can occur by means of trade, migration, mass communications media, and social interaction. The anthropologist Ralph Linton (1937) some time ago alerted us to the fact that much of what many regard as "American" originally came from other lands—examples include cloth (developed in Asia), clocks (invented in Europe), coins (developed in Turkey), and much more.

Expressions in the English language as spoken by those in the United States have been harvested from all over the

Culture can diffuse far and wide, as shown by the presence of this Hard Rock Cafe behind the Palestinian man in Jericho, Palestine.

world. The phrase *OK*—considered to be wholly "American"—diffused via slavery to the United States from Africa. Barbecued ribs, originally eaten by Black slaves in the United States after being discarded by White slaveowners who preferred meatier parts of the pig, are now a delicacy enjoyed throughout the United States by virtually all ethnic and racial groups. One theorist, Robert Ferris Thompson (1993), points out that an exceptionally large range of elements in material and nonmaterial culture that originated in Africa have diffused throughout virtually all groups and subcultures in the United States, including aspects of language, dance, art, dress, decorative styles, and even forms of greeting. These examples all illustrate cultural diffusion not only from one place to another (West Africa to the United States), but also diffusion across time, from a community in the past to many diverse ethnic groups in the present.

Similarly, the immigration of Latino groups into the United States over time has dramatically altered U.S. culture by introducing new food, music, language, slang, and many other cultural elements (Muller and Espenshade, 1985). By a similar token, popular culture in the United States has diffused into many other countries and cultures: Witness the adoption of American clothing styles, rock, rap, and Big Macs in countries such as Japan, Germany, Russia, and China. In grocery shops worldwide, from the rain forests of Brazil to the ice floes of Norway, can be found the Coca-Cola logo.

At one time, it was thought that slavery killed off most of the institutions and cultural elements that came with the slaves to the Americas (Herskovits, 1941). Extensive research over the past three decades now demonstrates that elements of culture carried from Africa by Black slaves continue to survive among African Americans and, thanks to cultural diffusion, among many other groups as well. A *step show* is an energetic, highly rhythmic, group choreography performed as a special event by predominantly Black fraternities and sororities. Researchers have traced these displays to traditional West African and central African group dances (Thompson, 1993; Gates, 1992, 1988). The step show has recently been noticed by non-Black students at universities and colleges all over the country, with a few White groups, fraternal, sororal, and otherwise, taking up "steppin'."

Many religious practices among African Americans are traceable to Africa. For example, as noted by religious historian Albert Raboteau (1978), the religious singing styles of the slaves, influenced by their African heritage, were characterized by polyrhythms, syncopation, slides (glissandos) from one musical note to another, and repetition. That African lineage is clearly detectable in jazz, rock, rhythm and blues, and rap music, having diffused far beyond the African American community today.

Signifyin' is a game in which one scores with insults, often of a sexual nature. Signifyin' began in the African American urban community, and is presently diffusing into White youth culture. The roots of signifyin' are planted in Africa, where a godlike symbolic figure called *Eshu,* a "trickster," or "signifyin' monkey," insulted all with his wit and cleverness (Gates, 1992, 1988). A closely related game, engaged in by virtually all hip urban youth, including Latinos, Whites, and Blacks, is a highly sexist game called "playing the dozens" (or running the dozens) and involves signifyin' about (or "on") someone's mother.

Playing the dozens can cross over from fun to serious, in which case it can become quite dangerous. Playing the dozens and signifyin' are not new or unique to the United States. Similar games can be found in certain Arab and Middle Eastern cultures, and whole dramas of the Restoration period in seventeenth-century Europe were constructed around duels of sly insult called "raillery." Cultural diffusion and cultural invention can be seen working together when tracing a cultural practice like the dozens, which may draw on several heritages, depending on who taught whom how to play.

THINKING SOCIOLOGICALLY

Cultural diffusion can be a source of social change. What words or phrases can you think of that have probably diffused recently into broader culture as the result of recent immigration? To what extent is such diffusion likely to result in social change in the broader society?

Inequality and Change

Inequalities between people on the basis of class, ethnicity, gender, or other social structural characteristics can be a powerful spur toward social change. Social movements may blossom into full-blown revolution if the underlying tension is great enough. Karl Marx hypothesized that economically based social class inequalities, namely, inequalities between capitalists and workers, would result in a revolution of the workers against the capitalist class. History has demonstrated that Marx's predictions were simplistic and not completely accurate, but he nonetheless specifically highlighted class as the source of inequality that can produce pressure for change. An example of the mechanism of change can be seen when inequalities between the middle class and the urban underclass produce governmental initiatives, such as increased education for the poor, that are designed to reduce this inequality (Griswold, 1994; Wilson, 1987). Social inequalities thus become the causes of social change.

Culture itself can sometimes contribute to the persistence of social inequality, and thus becomes a source of discontent among the individuals in the society. Inequalities within the educational system often have a cultural basis. For example, a poor child in the United States, having adopted a language useful in the ghetto, is at a disadvantage in the classroom, where standard English is used. Culturally specific linguistic systems, such as urban Black English, or Ebonics (Dillard, 1972; Harrison and Trabasso, 1976; Griswold, 1994; DiAcosta, 1998), are generally not adopted by schools, which may serve to strengthen the inequalities between the poor and the privileged—unless the child is bicultural and can speak both standard English and Ebonics, which many African Americans can do. Compounding the problem, a Hispanic or Black child may be criticized by his or her peers for "acting White" if the child studies hard. A female student may shrink from studying mathematics because she has received the culturally transmitted message that adeptness at mathematics is not feminine (Sadker and Sadker, 1994; Bailey and Campbell, 1993). The perpetuation of inequalities of class, race, and gender can stoke the desire for social change on the part of the disadvantaged groups. Tax revolts among the middle class and the push among Latinos for English as a second language in schools constitute two examples.

Technological Innovation and the Cyberspace Revolution

Technological innovations can be strong catalysts of social change. The historical movement of societies from agrarian to industrialized has been tightly linked to the emergence of technological innovations and inventions. Inventions often come about specifically because answering some need already present in the society promises to deliver great rewards. The water wheel promised agrarian societies greater power to raise crops despite dry weather, while also saving large amounts of time and labor. One could trace a direct line from the water wheel to the large hydroelectric dams that power industrialized societies, and along the way one could find evidence of how each major advance changed society. In today's world, the most obvious technological change transforming society is the rise of the digital computer, and the subsequent development of desktop computing, which has resulted in massive social and cultural changes in the United States.

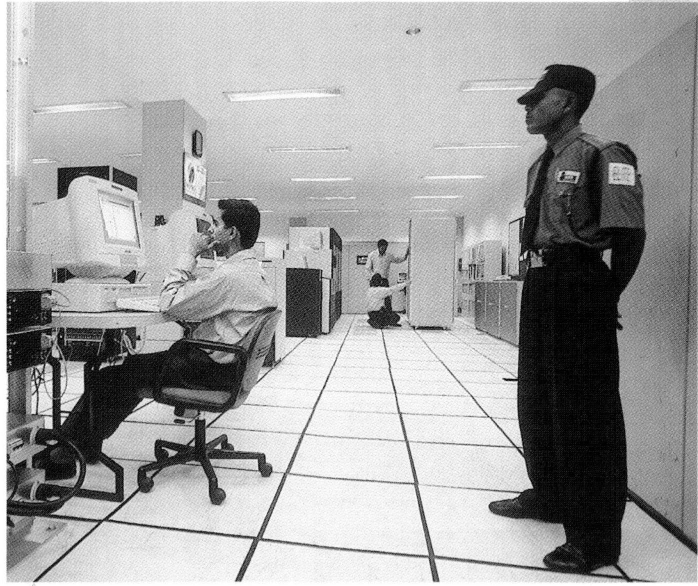

Technology is a source of social change. This man works late at a high-tech job in Bangalore, India.

The advent of the electronic computer has transformed our total society and all its institutions massively. Beginning with vacuum tube mainframe computers in the 1950s and early 1960s, followed by the transisterized computer of the mid-1960s and the integrated circuit computers of the late 1960s and 1970s, the advent of the PC (personal computer) and desktop computing was made possible by the invention and rapid spread of the microchip. It was this technological innovation that has become incalculable in its effects on society. What can now be stored in a microchip memory the size of a wristwatch would have required in the late 1960s a transistorized computer the size of a small auditorium. Of special significance is the invention and development of the Internet and the resulting communication in what has come to be called cyberspace—including both communication between persons and communication between persons and computers. Unique in both its vastness and lack of a central location, the Internet has very rapidly become so much a part of human communication and social reality that it pervades and has transformed literally every one of society's institutions—educational, economic, political, familial, and religious. In fact it can be said that few institutional structures have not been transformed by the cyberspace revolution. While both the mainframe computer and the PC continued to transform and shorten processes of mathematical and statistical calculation well into the 1980s, in the move from what Turkle has called the *culture of calculation* to the *culture of similation,* the computer user via the Internet develops a new self, a new identity. One can communicate with another on the Internet via e-mail and chat rooms with the protection of anonymity; as one user has noted, "you are what you pretend to be" (Turkle, 1995) (recall the discussion of this in Chapter 5). Thus the revolution in cyberspace has not only transformed society's institutions but it has begun to transform the nature of "the self" and how we define our very selves.

The path by which technology is introduced into society often reflects the predominant cultural values in that society. Some cultural values may prevent a technological innovation from changing a society. For example, anthropologists have noted that new technologies introduced into a nonindustrial agrarian society very often meet with resistance even though the new technology might greatly benefit the society. The Yanomamo, a nonindustrial agrarian society existing deep in the rain forests of South America, live without electricity, automobiles, guns, and other items of material culture associated with industrialized societies. The Yanomamo place great positive cultural value upon their way of hunting and engaging in war. Although the introduction of guns to their culture might help them in the hunt and in wars against their neighbors, they remain deeply attached to their bows and poison-tipped arrows, rejecting the technological innovations that are available (Cohen and Eames, 1982).

Indeed, such technological innovation may not necessarily lead to the "betterment" of their society. In the United States, rapid, relentless technological advance has greatly enhanced our ability to perform work, but it has also increased stress levels, created pollution, and generated weaponry that could wipe out civilization. The Yanomamo may have earned subtle dividends for contenting themselves with trusted technology. On the other hand, bulldozers and high-yield tree-harvesting equipment are tearing through the rainforest in which the Yanomamo live—technological innovation by others may cause the small world in which the Yanomamo live to disappear completely.

Population and Change

Another cause of social and cultural change is changes in the population (population and the environment were the subjects of Chapter 20). Limitations placed on the population by the natural environment can greatly influence the nature of social relationships. For example, in Japan, a small country with a large population, crowding is a fact of life that affects how people interact with one another. In Japanese cities, bus drivers negotiate streets that U.S. bus drivers would consider far too narrow for even a small bus. Japanese subways are packed so tightly that white-gloved individuals (called "pushers") must squeeze commuters bodily into subway cars. Riders on the subway are so tightly

BOX 22.2 SOCIOLOGY IN PRACTICE

Personal Identity and the Internet

SHERRY Turkle's book on identity and the Internet, entitled *Life on the Screen,* shows how technology (the Internet) can have profound effects on the social psychology of the individual.

She notes that rapidly expanding systems of networks, collectively known as the Internet, link millions of people in new spaces that are changing the way we think, the nature of our sexuality, the form of our communities, our very identities.

At one level, the computer is merely a tool. It helps us write, keep track of our accounts, and communicate with others. But the computer goes beyond this. It offers us both new models of mind and a new medium on which to project our ideas and fantasies. She argues that recently, the computer has become even more than tool and mirror: We are able to step through the looking glass. We are learning to live in virtual worlds.

The use of the term *cyberspace* to describe virtual worlds grew out of science fiction. For many of us though, cyberspace is now part of the routines of everyday life. When we read our e-mail or send items to an electronic bulletin board or make an airline reservation over a computer network, we are in cyberspace. In cyberspace, we can talk, exchange ideas, and even assume personae of our own creation. We have the opportunity to build new kinds of communities, virtual communities, in which we participate with people from all over the world, people whom we converse daily, people with whom we may have fairly intimate relationships but whom we may never physically meet.

In the story of constructing identity in the culture of simulation, experiences on the Internet figure prominently, but these experiences can be understood only as part of a larger cultural context. There is an eroding of boundaries between the real and the virtual. From scientists trying to create artificial life to children "morphing" through a series of virtual personae, we see evidence of fundamental shifts in the way we create and experience human identity. But it is on the Internet that our confrontations with technology as it collides with our sense of human identity are fresh, even raw. In the real-time communities of cyberspace, we are unsure of our footing, inventing ourselves as we go along.

SOURCE: Turkle, Sherry. 1995. *Life on the Screen: Identity in the Age of the Internet.* New York: Simon and Schuster.

Population density can affect social interaction and cultural norms, as illustrated here in Japan, where a subway worker (a "pusher") causes close physical contact among subway riders.

packed that their entire bodies are in constant contact, a situation that in most parts of the United States would be considered taboo since such close bodily contact has sexual overtones. In Japan, the contact is not considered sexual, nor is it considered a violation of the other's personal space, as it would be in the United States. This is one example of how population density can affect the nature of interpersonal relations among people.

Major social change can also result from shifts in the age composition of a population. As noted in the discussion of aging (Chapter 13), the average age in the United States is increasing. Whereas in the year 2000 about 13 percent of the U.S. population is over the age of sixty-five, by 2025 that percentage will rise to more than 18 percent (U.S. Bureau of the Census, 1997). The graying of America means that health care programs and facilities will focus more on the elderly; changes in the structures of the health care and insurance industries have already begun. In the commercial world, businesses are paying more attention to the elderly, marketing new products that appeal to the old and adjusting their general advertising in acknowledgment of the new market demographics. Current stereotypes of the aged will no doubt change; the present stereotype of the elderly as helpless and weak is already undergoing change, with more frequent portrayals of the elderly as sharp-minded and vigorous. The elderly will have more say in government and education, and with their increasing share of the overall wealth, they will wield more influence in the economic sector (Barberis, 1981; Griswold, 1994).

Immigration is having profound effects on the overall ethnic and racial composition of the United States. Roughly three-quarters of a million immigrants each year take up permanent residence in the United States (Espenshade, 1995; U.S. Bureau of the Census, 1997). Almost half are from Mexico or other Latin American countries, and about one-fourth are from Asian countries. The United States is presently about 11 percent Hispanic; by the year 2025, that figure will be about 18 percent. Currently about 4 percent of the U.S. population is of Asian ancestry. By the year 2025, that figure will be about 7 percent. By the year 2050, it is expected that Hispanics will be 25 percent of Americans, and Asians 9 percent (U.S. Bureau of the Census, 1998). These population shifts will cause major changes in society's institutions. The structure of the economy will change, as will the ethnic mix in education, the ethnic complexion of jobs, and the strength of Hispanic culture's influence, including styles of dance, language, and other behaviors.

It needs to be carefully pointed out that, contrary to some beliefs, Hispanic and Asian immigrants are not "taking away" jobs from American citizens. Quite the contrary, Asians and especially Hispanics must contend with job discrimination that keeps them underrepresented in jobs at all levels, especially higher-level and better-paying jobs. Whether or not continuing Hispanic immigration will improve the representation of Hispanics at all job classification levels will depend upon whether ethnic job discrimination lessens in the coming years (Espenshade, 1995; Bouvier and Gardner, 1986).

War and Social Change

War and severe political conflict results in large and far-reaching changes for both the conquering society, or region within a society (as in civil war), and for the conquered. The conquerors can impose their will upon the conquered and restructure many of their institutions, or the conquerors can exercise only minimal changes. The United States victory over Japan and Germany in World War II resulted in societal changes in all these societies. The war transformed the United States into a mass-production economy that had

Social changes have occurred in Germany as a consequence of the dismantling of the Berlin Wall in 1989, which had divided the country since August of 1961.

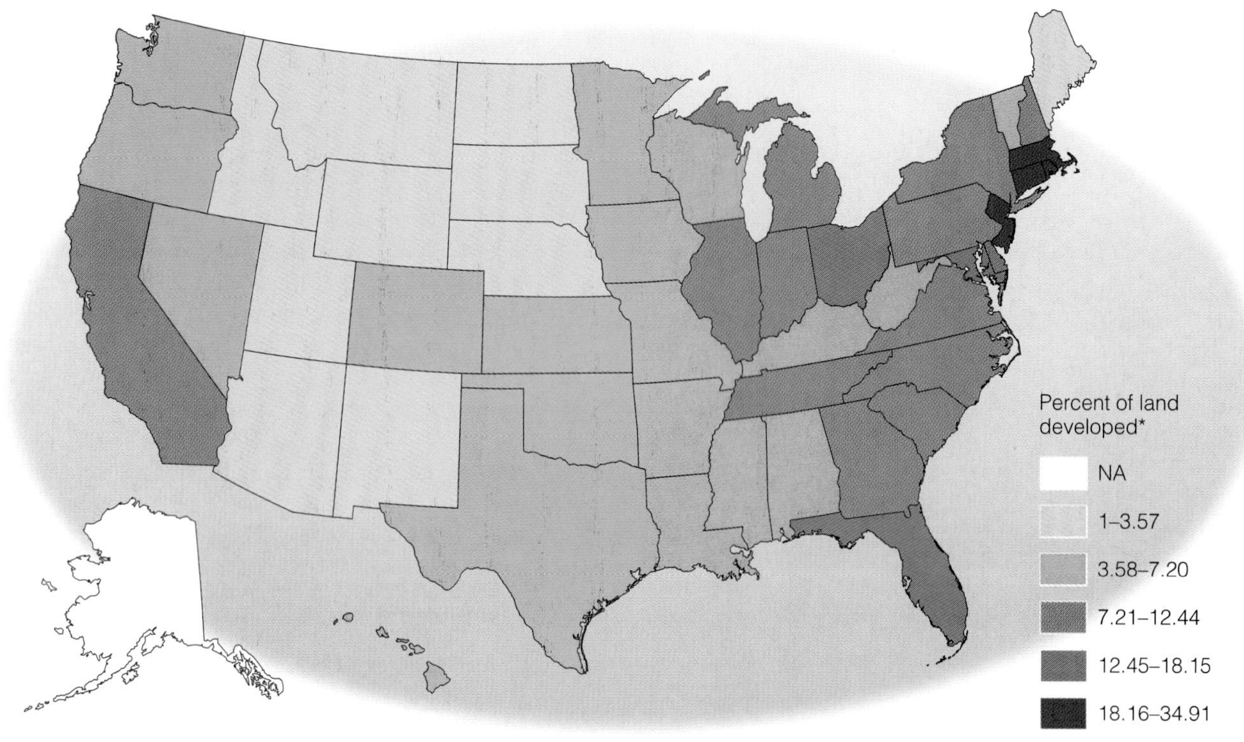

MAP 22.1

Mapping America's Diversity: Percentage of Land Developed by State

*Includes urban and built-up areas in units of 10 acres or greater, as well as rural transportation.

SOURCE: U.S. Bureau of the Census. *Statistical Abstract of the United States 1997.* Washington, DC: U.S. Government Printing Office, p. 229.

effects on family structure (father-absence increased and women not previously employed joined the workforce) and education (men of college age went off to war in large numbers), to name a few such changes. Many of those who returned from the war were educated under a scholarship plan called the GI Bill. The war also transformed Germany in countless ways, given the vast physical destruction brought on by U.S. bombs and the worldwide attention brought to antisemitism and the Nazi Holocaust. The cultural and structural changes in Japan were extensive, as well. And the decimation of the Jewish population in Germany, as well as in other nations throughout Europe, resulted in the massive migration of Jews to the United States. More recently, the war in Vietnam has also resulted in numerous social changes, including a high degree of immigration to the United States from Vietnam.

Modernization

As societies grow and change, they become in a general sense more "modern." Sociologists use the term *modernization* in a specific sense: Modernization is a process of social and cultural change that is initiated by industrialization and followed by increased social differentiation and division of labor. Societies can, of course, experience social change in the absence of industrialization. Modernization is a specific type of social change that industrialization tends to bring about. The change toward an industrialized society can have both positive and negative consequences. On the positive side are advances like improved transportation and a higher gross national product; on the negative side are such burdens as pollution, elevated stress, and increases in certain kinds of job discrimination.

There are three general characteristics of modernization (Berger et al., 1974). First, modernization is typified by the decline of small, traditional communities. The individuals in foraging or agrarian societies live in small-scale settlements with their extended families and neighbors. The primary group is prominent in social interaction. With industrialization comes an overall decline in the importance of primary group interactions and an increase in the importance of secondary groups, such as colleagues at work. Second, with increasing modernization, a society becomes more bureaucratized; interactions come to be shaped by formal organizations. Traditional ties of kinship and neighborhood feeling decrease, and the members of the society tend to experience feelings of uncertainty and powerlessness. Third, there is a decline in the importance of the religious institution, and with the mechanization of daily life, people begin to feel that they have lost control of their own lives, and people may respond by building new religious groups and communities (Wuthnow, 1994).

modernization

These Amish women in Lancaster County, PA, illustrate **gemeinschaft** *("community") social organization as they work on a quilt.*

These subway riders illustrate the impersonality of **gesellschaft** *("society") as they ride to work.*

From Community to Society

The German sociologist Ferdinand Tönnies, who died in 1936, formulated a theory of modernization that still applies to today's societies (Tönnies, 1963 [1887]). Tönnies viewed the process of modernization as a progressive loss of **gemeinschaft** (German for "community"), a state characterized by a sense of fellow feeling, strong personal ties, and sturdy primary group memberships, along with a sense of personal loyalty to one another. Tönnies argued that the Industrial Revolution, with its emphasis on efficiency and task-oriented behavior, destroyed the sense of community and personal ties associated with an earlier rural life, substituting feelings of rootlessness and impersonality. At the crux of this was a society organized on the basis of self-interest, where division of labor is high and personal feelings of belonging are low. This caused the condition of **gesellschaft** (German for "society"), a kind of social organization characterized by a high division of labor, less prominence of personal ties, the lack of a sense of community among the members of society, and the absence of a feeling of belonging—maladies often associated with modern urban life.

According to Tönnies, the United States was characterized by gemeinschaft through the year 1900. Life was mainly rural, characterized by families that had lived for generations in villages, and one's work was closely tied to the family. In terms of gender roles, patriarchy was prominent since as most women's lives were centered on the home, with very few women holding jobs outside the home. There was no radio, no television, and few telephones. As a result, family members were dependent upon each other for entertainment, information, and support. Despite the relative intimacy of the gemeinschaft, social interaction tended to remain within both racial–ethnic and social class boundaries. Mass transportation was not yet developed, and people tended to base their lives in their own town. These characteristics of the United States at the turn of the century are preserved in some communities today, such as the Amish in parts of Pennsylvania and Ohio. The Amish are a classic example of a present-day gemeinschaft.

The United States since the 1940s has become a gesellschaft. Social interaction has become less intimate and less emotional although certain primary groups such as the family and the friendship group still permit strong emotional ties. However, Tönnies noted that the role of the family is considerably less prominent in a gesellschaft than in a gemeinschaft. Patriarchy is less prominent, yet more public since more women have jobs outside the home. In the large cities that characterize the gesellschaft, people live among strangers, and the people one passes on the street are unfamiliar. In a gemeinschaft, most of the people one encounters have been seen before. The level of interpersonal trust is considerably less in a gesellschaft. Social interaction tends to be even more confined within ethnic, racial, and social class groups. In order to find personal contact and to satisfy the need for intimate interaction, individuals often join groups such as small church groups, training groups, or personal awareness groups (Wuthnow, 1994; Taylor et al., 1994).

One measurement of the impersonality associated with the urban gesellschaft is the effect an urban setting has on the likelihood that someone will come to the aid of a person in distress. A person in trouble on the street—having a heart attack, for example, or being attacked by a mugger— is more likely to be given assistance by a bystander if the event takes place in a small town than if it takes place in a large city (Amato, 1990, 1983; Taylor, et al., 1997, 1994). The bystander in the small town is more likely to either help directly or take helpful action such as call the police.

In a famous case in 1964, a New Yorker named Kitty Genovese was stabbed repeatedly while at least thirty-eight witnesses within the apartment complex where she lived failed to respond to her cries for help. The witnesses did

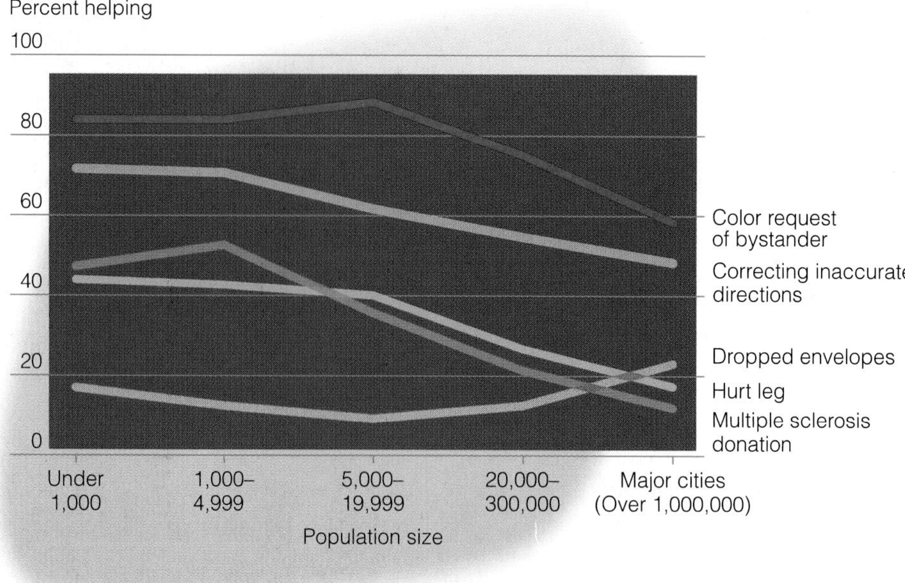

FIGURE 22.1

Research shows that strangers are more likely to receive help, as in emergency situations, in small towns (gemeinschafts) *than in large cities* (gesellschafts). *This figure shows the percent of times that a stranger received help of five different kinds in towns and cities of varying sizes. One reason is that the greater the number of people around, the greater the chances of the potential helper* deindividuating *(diffusing responsibility), that is, expecting someone else to help. This is the so-called* bystander effect*—the more people around who could potentially help, the less likely any of them is to actually give help. So, the fewer people there are around, the more likely any one of them is to actually give help. This is a good example of where common sense conflicts with actual research: Most of us would ordinarily think that one is more likely to get help if there are more people around rather than fewer people.*

SOURCE: Adapted from Taylor et al., 1997, p. 410; Amato, V., 1990, p. 529.

nothing, even failing to call the police. Later research showed that in this and similar situations, such as the similar stabbing of a woman in Denver, Colorado, in 1998, each individual thought that someone else was going to help, with the result that no one gave any assistance at all. No one felt personally obliged to intervene or seek help (Amato, 1990; Darley and Latané, 1968).

Interestingly, research shows that when the number of people observing an emergency goes up, the likelihood that any one of them will offer assistance goes down. Conversely, the fewer people there are around, the more likely one of them will step in to help. If only one bystander is observing the emergency, the chances of the victim receiving help are the greatest (Amato, 1990; Taylor et al., 1994). These relative findings seem to hold for both small towns and large cities. Overall, however, a victim is more likely to get help in a small town (gemeinschaft) than in a large city (gesellschaft). People in a large city are more likely to merge with the crowd—to use a term introduced in Chapter 5, they are more more likely to *deindividuate*—and fail to offer help simply because there are more people around. Figure 22.1 illustrates this effect for five kinds of help-giving in response to situations such as being asked for a donation for multiple sclerosis, or for help with an injured leg.

Mass Society and Bureaucracy

According to contemporary theorists Dahrendorf (1959) and Berger et al. (1974), modernization has produced what they call a **mass society,** one in which industrialization and bureaucracy reach exceedingly high levels. In the mass society, the change from gemeinschaft to gesellschaft is accelerated, and the breakup of primary, family, and kinship ties is particularly pronounced. The government and its functions expand to the extent that much of one's personal life falls under government management, including tasks that were previously performed by family. Care for the elderly, for example, may be placed in the hands of unfamiliar, faceless bureaucrats who run elder-care facilities and administer financial benefits for the aged.

Dahrendorf, Berger, and other mass society theorists argue that not only have we moved from gemeinschaft to gesellschaft, with all the attendant negatives described by Tönnies, but that bureaucracies have obtained virtually complete control of the individual's life. As people moved from town to city over the course of this century, divisions of labor became more pronounced, and social differentiation increased in the workplace, education, government, and other institutions. It became more common to identify people by such personal attributes as their job ("He's John, a lawyer") or their gender ("She's Ms. Blackburn, a woman judge"), rather than by their kinship ("She's a Smith") or their home town ("He's from Mantua, Ohio"), which is more commonly done in the gemeinschaft. The importance of mass media increased; newspapers, television, magazines, radio, and movies took on more prominent roles in society. People became more mobile geographically, and thus less dependent upon neighbors and kin. All these changes worked together to increase the feeling that most of the people in one's immediate environment are strangers.

The rise of large government is a major part of the overall increased bureaucratization of social life. In the preindustrial societies of both the United States and Europe, government may have been only a clergyman, a nobleman, a justice of the peace, or a sheriff. Industrialization allowed government to expand at the national, state, and local levels, and thus become more complex and bureaucratized. Government demonstrates an eagerness to involve itself in many aspects of life formerly left to community standards or private resolution—regulating working conditions, setting wages and salaries, establishing standards for products and medicines, health care, and the care of the poor, as well as all sorts of intimate behaviors. Most political and social power today resides in such large bureaucracies, thus leaving the individual a diminishing degree of control over his or her own life.

Social Inequality, Powerlessness, and the Individual

Another product of modernization, along with mass society, is pronounced social stratification, according to theorists such as Marx (1967 [1867]) and Habermas (1970). In their view, the personal feelings of powerlessness that accompany modernization are due to social inequalities related to race, ethnicity, class, and gender stratification. Marx argued that inequalities are the inevitable product of the capitalist system. Habermas argued that inequalities are the cause of social conflict.

The social structural conditions that arise from modernization, such as increased social stratification, are felt at the level of the individual. Building a stable personal identity is difficult in a highly modernized society that presents the individual with complex and conflicting choices about how to live. Many individuals flounder between lifestyles in their search for personal stability and a sense of self. According to Habermas, individuals in highly modernized environments are more likely than their less modernized peers to experiment with new religions, social movements, and lifestyles in search of a fit with their conception of their own "true self." These individual responses to social structural conditions reveal how the social structure can affect personality.

Social theorist David Riesman (1970 [1950]) argued that there are three main orientations of personality that can be traced to social structural conditions. These are **other-directedness,** wherein the behavior of the individual is guided by the observed behavior of others, and is characterized by rigid conformity and attempts to "keep up with the Joneses"; **inner-directedness,** wherein the individual is guided by internal principles and morals, and is relatively impervious to the superficialities of those around her or him; and finally, **tradition-directedness,** or strong conformity to long-standing and time-honored norms, practices, and styles of life. According to Riesman, modernization tends to produce other-directedness, whereas less modernized gemeinschafts, such as horticultural or agricultural societies, tend to produce tradition-directedness. The inner-directed, because they are guided by internal rather than external forces, are less likely to sway with the presence or absence of modernization.

If modernization tends to produce other-directedness, then anyone who happens to be inner-directed or tradition-directed in a highly modernized and rapidly changing society, such as the United States, is likely to be seen as a deviant person. The other-directed person, in contrast to the inner- and tradition-directed persons, is highly flexible, capable of rapid personal change, and more open to the influences of group pressures, changing styles, and shifting interests. These are the qualities that can leave the other-directed individual in the highly modernized society stranded and searching for his or her "true self."

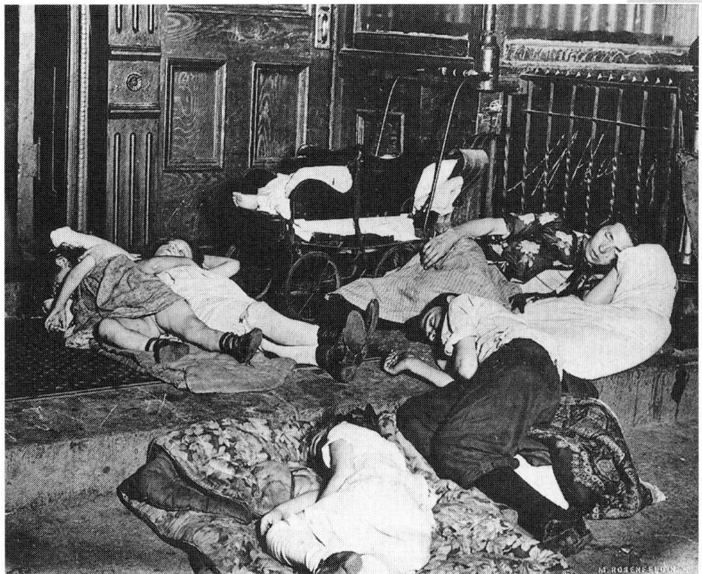

In the early twentieth century, without the benefit of modern air conditioning, families on the lower east side of New York City sometimes slept outdoors for relief from the summer's heat.

THINKING SOCIOLOGICALLY

Did you grow up in what was primarily a *gemeinschaft* (community), such as a rural community, or in what was primarily an urban or suburban environment (*gesellschaft*)? Do you remember how your community or neighborhood changed over time, and what were some of those changes—in population, in ethnic composition, in the presence of offices and office buildings, in the nature of interpersonal relationships? Would you say that your community encouraged *inner-*, *other-*, or *tradition-directedness*?

The poignant question, "Who am I?" can give rise to feelings of individual powerlessness. The influential social theorist Herbert Marcuse (1964) has argued that modernized society fails to meet the basic needs of people, among them the need for a fulfilling identity. In this respect, modern society and its attendant technological advances are not stable and rational, as is often argued, but unstable and irrational. The technological advances of modern society do not increase the feeling that one has control over one's own life, but instead reduce that control and lead to feelings of powerlessness.

This powerlessness leads to the *alienation* of the individual from society—the individual experiences feelings of separation from his or her group or society. This alienation is more likely to affect those who have traditionally been denied access to power, such as racial minorities, women, and the working class. The alienation of these individuals from the highly modernized, technological society is, in Marcuse's view, one of the most pressing problems of civilization today. Marcuse argues that despite the popular view that technology is supposed to yield efficient solutions to the world's problems, it may be more accurate to say that technology is a primary cause of many of the problems of modern society.

Global Theories of Social Change

Globalization refers to the increased interconnectedness and interdependence of different societies around the world. No longer can the nations of the world be viewed as separate and independent societies. The irresistible trend in this century has been for societies to develop deep dependencies on each other, with interlocking economies and social customs. In Europe, this trend has proceeded as far as a common currency, the euro, for all nations participating in the newly constructed common economy.

If the world is becoming increasingly interconnected, does this mean that we are moving toward a single, homogeneous culture—the culture that futurist Marshall McLuhan once called the "global village" (Griswold, 1994)? In McLuhan's vision, electronic communications, computers, and other developments would erase the geographic distance between cultures, and eventually even the cultural differences themselves. In a competing view, greater interconnectedness among societies may *magnify* the cultural differences between interacting groups by making the groups more aware of the incompatibilities between them.

In fact, both processes take place. As societies become ever more interconnected, cultural diffusion between them creates common ground, while cultural differences may become more important as the relationships among nations become more intimate. The different perspectives on globalization are represented by three main theories that we will review: modernization theory, world systems theory, and dependency theory, also discussed in Chapter 10.

Modernization Theory

Strongly influenced by functionalist theories of social change, **modernization theory** states that global development is a worldwide process affecting nearly all societies that have been touched by technological change. Societies are thus made more homogeneous in terms of differentiation and complexity.

Modernization theory traces the beginnings of globalization to western Europe and the United States. Technological advances in these countries propelled them ahead of the less developed nations of the world, which were left to adopt the new technologies years after Europe and the United States. Homogenization resulted, with developing nations being shaped in the mold of the Western nations, which had modernized first.

Recent proponents of modernization theory, such as William McCord, reject the assumption that only western European countries and the United States have led technological globalization and its resultant homogenization (McCord, 1991; McCord and McCord, 1986). McCord argues that non-Western societies, most notably Japan, have also been leaders in modernization. As a result, the Japanese culture, with its emphasis on the importance of small friendship groups in the workplace and a traditional work ethic, has profoundly influenced other countries and cultures. According to McCord, the examples of Japan and other technological leaders, such as Taiwan and South Korea, have added to the impetus of global economic growth.

World Systems Theory

Formulated by theorist Immanuel Wallerstein (1989, 1979, 1974), **world systems theory** argues that all nations are members of a worldwide system of unequal political and economic relationships that benefit the developed and technologically advanced countries at the expense of the less technologically advanced and less developed. Less developed nations are thus shortchanged in the world system.

Wallerstein divides the world system into two camps. **Core nations,** such as the United States, England, and Japan, produce goods and services both for their own consumption and for export. The core nations import raw materials and cheap labor from the **noncore nations** (or peripheral nations), situated in Africa, Latin America, South America and parts of Asia. These nations occupy lower positions in the global economy, thus showing a stratification of the global economy (Chapter 10). Certain populations in the noncore nations suffer exploitation as a result. Witness the use of children as laborers in parts of Malaysia, Singapore, and Latin America, manufacturing shirts, soccer balls, and blankets. By these manufactured goods, the noncore nations end up contributing to the wealth of the core nations.

Wallerstein's theory builds on the notion that the world system is constantly changing and dynamic. The core nations tend to occupy new noncore territories over time, leaving technological development and industrialization behind them as they move on to other undeveloped areas that remain to be exploited. In this way, noncore nations are drawn into the world system. The constant change in the system is reflected in the changing political relationships among nations. The noncore nations are exploited by the core nations for a period, but in many cases, the noncore nations benefit by becoming upwardly mobile in the world economic system.

Debunking Society's Myths

Myth: **Different countries, especially the small ones, exist pretty much to themselves, and what they do economically or socially has little effect on other countries in the world.**

Sociological perspective: **Countries are part of a worldwide network of interdependencies; world systems theory notes that this interdependence tends to benefit the developed and technologically advanced countries at the expense of the less technologically advanced and less developed.**

Dependency Theory

Closely allied with Wallerstein's world systems theory is **dependency theory,** which maintains that highly industrialized nations tend to imprison developing nations in de-

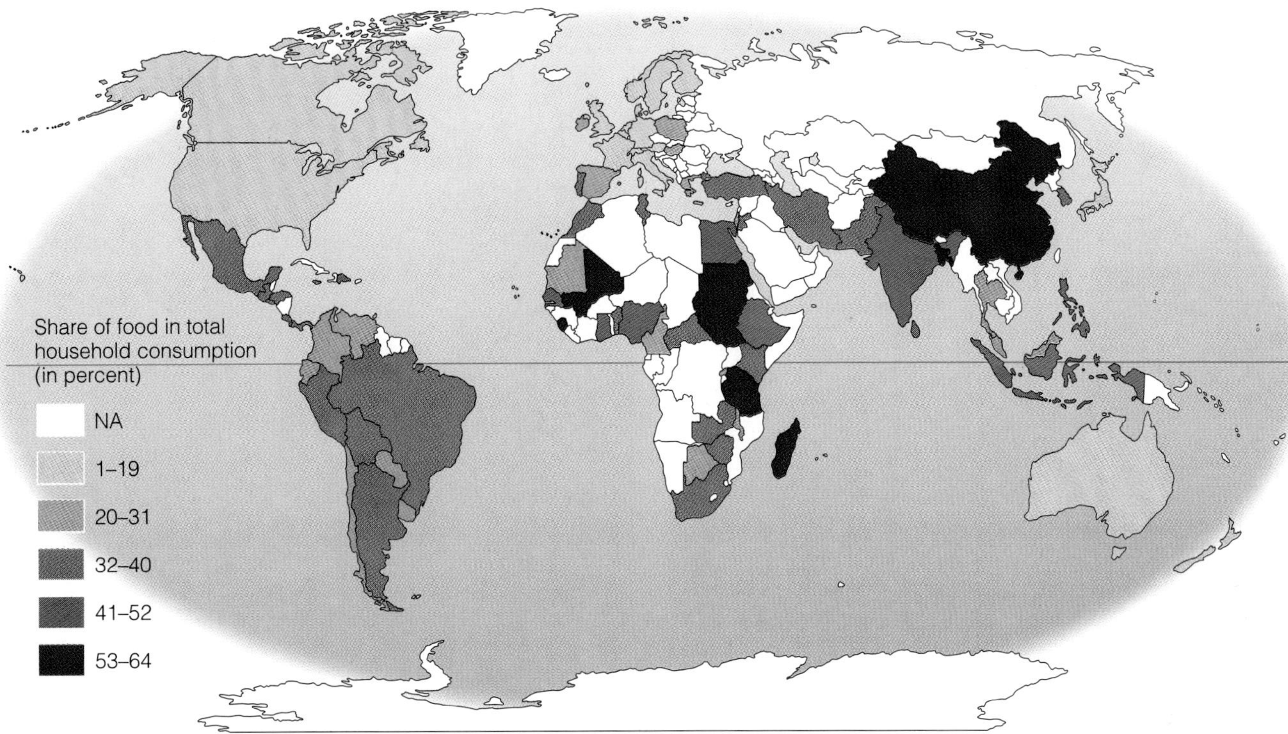

MAP 22.2 Viewing Society in Global Perspective: Percentage of Household Budget Spent on Food

World systems theory helps explain the pattern in global socialization whereby certain nations are more impoverished and others are more affluent. One indication of this is the percentage of household income that the typical household must spend for basic needs.

SOURCE: UNICEF. *The State of the World's Children.* 1996. New York: Oxford University Press, 82–83.

pendent relationships rather than spurring the upward mobility of developing nations with transfers of technology and business acumen (Rodney, 1981, 1974; Reich, 1991). Dependency theory sees the highly industrialized core nations as transferring only those narrow capabilities that it serves them to deliver. Once these unequal relationships are forged, core nations seek to preserve the status quo because they derive benefits in the form of cheap raw materials and labor from the noncore nations. In the meantime, the developing nations remain dependent upon the core nations for markets and support in keeping what industry they have acquired in working order, while at home they experience minimal social development, limited economic growth, and increased income stratification among their own population. Rodney (1981) has argued that this pattern of dependency is to blame for the exceptional underdevelopment of a number of African countries.

Trade dependency is particularly notable in these relationships. Following patterns established when colonialism was rife, developed nations purchase raw materials from noncore countries at the lowest possible prices, process and refine them, and then sell the products back to the noncore countries at a substantial profit. This only exacerbates the economic dependency of noncore nations upon the core nations.

Borrowing dependency is another form of dependent relationship. Former Secretary of Labor Robert Reich (1991) has noted that core nations have been willing to lend money to noncore nations, but often on terms that—such as high interest rates—put severe economic strain on the noncore nations, sometimes requiring such interventions as wage and price freezes in the developing societies to maintain solvency. The hardship produced falls disproportionately on the lower social classes in the noncore country; the upper classes are less affected, and occasionally they benefit extravagantly.

As economist Reich (1991) and sociologists McCord and McCord (1986; also McCord, 1991) have noted, the network of dependency is complicated by the fact that in today's global economic system, it is often difficult to tell just who owns what. For example, in the early 1990s, General Motors in the United States owned almost half the stock of Isuzu in Japan. Japanese companies own a number of large American enterprises in the automotive and entertainment businesses. This type of worldwide interdependence further connects nations into a homogeneous global economy, but it also facilitates the dependency of the less developed nations, especially those in Latin America and Africa. This increases the disparities between rich and poor within the dependent nations. Gender inequalities, too, are likely to be perpetuated by this dependency process, especially in the noncore nations, since women in the dependent nations typically provide a pool of cheap labor.

Such takeovers often spawn humorous incidents. In the early 1990s, a construction company in California turned

Social change is sometimes apparent by contrasting the social norms of different historical periods.

down a low bid from a Japanese company (Haimatsu), to spite them for a prior deal, and accepted a higher bid from John Deere, the tractor maker. As it turned out, Haimatsu was a U.S. owned company, and John Deere was Japanese owned!

Diversity and Social Change

Issues of race, class, and gender—in other words, diversity issues—have played a major role in social change, both as causes of change and as social institutions directly affected by change. Tensions and aspirations arising from diversity can produce major changes within the structure and culture of a society. Social change can also affect the relations *between* societies. In this respect, globalization has had dramatic effects on the entire world, and thus on the question of diversity within as well as between societies.

Diversity as a cause of change is exemplified by the effects of immigration into a country and the resultant changes in society. The influx of Latinos into the United States has in a relatively short time resulted in profound cultural changes throughout the United States. From east to west, new tastes can be see in eating habits, music, language, literature, and many other venues. The changes in the United States brought about by African Americans, beginning as long ago as the slavery era, have been documented throughout this book. Changes traceable to peaks in Chinese immigration are similarly numerous, as seen by the numerous Chinese restaurants and take-out restaurants in most communities today. Recent technological and industrial changes brought about by economic competitors overseas, such as the Japanese, Koreans, and Vietnamese, have been profound. Advances in Japanese industry and technology have produced permanent alterations in the computer, automotive, entertainment, and other major industries in the United States.

Diversity as an effect of change is exemplified by the unequal outcomes of modernization on different ethnic groups. Modernization has the effect of making some nations—those earlier in their industrial development—more dependent upon the core countries than on neighboring nations or other nations at a similar level of development. The populations of developing nations are generally people of color, and the nations they become economically dependent upon as a result of globalization are for the most part countries with White majorities (the United States, Canada, Great Britain, France). In general, economic moves by core countries that affect dependent noncore countries, such as restructuring of loan repayments, will have differential effects on different ethnic groups and social classes within those noncore countries. The groups with lower social and economic status will suffer more from such moves than the upper classes and ethnic groups of high average social status. It is usually the minority ethnic groups or social classes that are on the receiving end when the effects of globalization resulting from technological change are negative.

The effects of diversity upon cultural lag are of particular interest. New cultural elements (language, music, technology) introduced to a nation by its racial and ethnic minorities are often adopted by the dominant culture of that nation. The adoption may occur quickly, or even instantly, as when White youths adopt African American dance steps; other times, it can take decades or longer for particular cultural transfers to take hold. The length of the cultural lag may depend on a number of factors, such as the "foreignness" of the minority culture, the degree to which the minority culture has already been assimilated, the ongoing state of relations between the minority and dominant group, and a variety of other variables that can be as trivial as the vagaries of fashion or as deep-rooted as historical ethnic animosities.

Among the most welcome and long-lasting effects of diversity is the erosion of the tendency to regard the technology of developing nations as more primitive than the technologies of the dominant Western countries. Although it is certainly true that the technological base in many developing nations is far behind that of Western countries, it is now fully recognized that some of the technologies of the future are being developed overseas. In Singapore and Taiwan, state-of-the-art computer products are prototyped by people of color in dust-free clean rooms. Japan was among the pioneers of low-volume, high-efficiency, high-quality steelmaking. The growing army of software writers in India has become a resource drawn upon by the entire world. The distinction between "civilized nation" and "primitive nation" has lost what little meaning it ever had.

CHAPTER SUMMARY

- *Social change* is that process by which social interaction, the social stratification system, and entire institutions in a society change over time. Social change can take place quickly, as within one to several years, or it can take longer, sometimes as long as centuries. It can involve *microchanges* (smaller changes, usually rapid, such as a fad "catching on") or *macrochanges* (such as technological innovation and modernization).
- Social change often produces *cultural lag,* where one element of a culture changes more quickly than another, or one element (such as a language term) introduced to a society by a particular racial or ethnic group is over time *diffused* into the broader dominant culture of the society.
- There are three theories of social change. Functionalist theories (and *evolutionary theories*) predict that societies move or evolve from the structurally simple to the structurally complex. *Unidimensional evolutionary theory* predicts that societies evolve along a path from simpler socially undifferentiated societies to more complex highly differentiated ones. *Multidimensional evolutionary theory* predicts that societies follow not one but several different paths in the process of social change. Conflict theories predict that social conflict is an inherent part of any social structure and that conflict between social class strata, or racial–ethnic groups that occupy dramatically differing social statuses, can bring about social change. *Cyclical theories* such as those of Toynbee and Sorokin predict that certain patterns of social structure and culture recur in a society at different times.
- The causes of social change vary widely. Some of them are collective behavior (such as fads or crazes and fashions); social movements (such as the feminist movement or the Black power movement); *cultural diffusion,* or the adoption by the dominant culture of some cultural element introduced by another group via trade, migration, mass communication, or social interaction, and includes many cultural elements introduced to American society by Native American, Hispanic, Black, and Asian groups; social inequality (such as the many changes in gender relations in the United States brought about by women's political activism against long-practiced gender inequalities); technological innovation (witness the vast social changes brought about by software and hardware innovations in the computing industry); population factors (such as social changes brought about by increases or decreases in population density, by the in- or out-migration of one or more racial or ethnic groups, and by changes in the age distribution of a population, such as the "graying" of America); and war.
- *Modernization* is a complex set of processes initiated by industrialization. Modernization results from the decline of small, traditional communities (*gemeinschafts*) and the change to generally larger, more differentiated and impersonal societies (*gesellschafts*), thus causing the individual to seek out more intimate interaction, as through small church groups or personal awareness groups. With modernization goes increased bureaucratization of a society, thus increasing intrusion of governmental bureaucracies into the lives of the individual. Modernization tends to produce *other-directedness,* where one's behavior is strongly guided by the perceived behavior of others; it also tends to produce individual feelings of powerlessness and an increased search for individual identity and one's "true self." All these serve to show that the social structure can produce effects and changes at the level of the individual personality.
- *Globalization* refers to the increased economic, political, and social interconnectedness among different societies in the world. The *modernization theory* of globalization states that global development among societies is a worldwide process not confined to any one society, such that technological advances in one society affect other societies. *World systems theory,* formulated by Wallerstein, predicts that societies are members of a worldwide system of inequality that benefits the more technologically developed societies (core nations) at the expense of the less technologically developed countries (noncore, or peripheral, nations). *Dependency theory* notes that industrialized core nations tend to keep noncore nations economically dependent upon them, and this retards the economic development and upward mobility of the noncore nations.

KEY TERMS

core nations
cultural diffusion
cyclical theory
dependency theory
evolutionary theory
gemeinschaft
gesellschaft
globalization
inner-directedness
macrochange
mass society
microchange
modernization theory
multidimensional evolutionary theory
neoevolutionary theory
noncore nations
other-directedness
social change
tradition-directedness
unidimensional evolutionary theory
world systems theory

THE INTERNET: A Tool for the Sociological Imagination

Resources on the Internet:

Virtual Society: The Wadsworth Sociology Resource Center
http://sociology.wadsworth.com

Visit this site to find additional learning tools, including interactive quizzes, links related to web sites, and an easy link to *InfoTrac College Edition.*

Center for Urban Studies
http://www.cux.wayne.edu

A research center dedicated to social science studies of urban areas.

The World Factbook
http://www.ocdi.gov/cia/publications.nsolo/factbook/global.httm#W

A collection of data prepared by the CIA that provides detail on various countries around the world.

Institute for Women's Policy Research
http://www.iwpr/org/

A research organization established to develop policy-related research on women's lives; site includes comparative data on the status of women in the fifty states.

Progress of Nations
http://www.unicef.org/

A report from UNICEF detailing various aspects of worldwide social change.

Sociology and Social Policy: Internet Exercises

Technological change is one of the major sources for broad social change, and with the advent of computing technology, some say that the emergence of cyberspace will fundamentally change human interaction. What evidence is there of change in social life as the result of such phenomena as e-mail, chat rooms, virtual realities, and other manifestations of the electronic age? Are there social policies that should guide concerns about information privacy and other dimensions of the cyberspace revolution?

Internet Search Keywords:

electronic age
information privacy
cyberspace
virtual realities
chat rooms
computers and society

Web sites:

http://epic.org/
Web site that focuses public attention on emerging civil liberties issues relating to the National Information Infrastructure.

http://rene.efa.org.au/liberty/debate.html
Web page debating censorship on the Internet.

http://www.ispo.cec.be/infosoc/soccul.html
A site dealing with the social and cultural aspects of the Information Revolution.

http://www.lbl.gov/Education/ELSI/privacy-main.html
This site offers information on personal privacy issues and access to medical databases.

http://www.vicnet.net.au/issues/censorship.htm
This page looks at computers and how to use technology in socially responsible ways.

InfoTrac College Edition: Search Word Summary

inequality and change
social change
modernization

In order to learn more about these central topics in sociology, you can conduct an electronic search using InfoTrac College Edition. To aid in your search and to gain useful tips, see the Student Guide to InfoTrac College Edition on the Virtual Society web site:
http://sociology.wadsworth.com

INTERACTIONS—A SOCIOLOGY CD-ROM: CONCEPTS FOR THIS CHAPTER

Go to the Wadsworth Sociology CD-ROM for further study on the concepts in this chapter. The CD-ROM also includes quizzes and additional activities to expand your learning experience.

SUGGESTED READINGS

Bellah, Robert N., Richard Madsen, William Sullivan, Ann Swidler, and Steven M. Tipton. 1984. *Habits of the Heart: Individualism and Commitment in American Life.* Berkeley: University of California Press.

This is an analysis of the negative effects of individualism that lies at the core of both the public and the private lives of Americans. Bellah et al. elaborate how individualism has been the subject of social change.

Caplow, Theodore. 1991. *American Social Trends.* New York: Harcourt Brace Jovanovich.

Caplow offers an overview of cultural trends, including attitudes, styles, media trends, and social movements, and how they allow us to peer into the twenty-first century.

Gitlin, Todd. 1980. *The Whole World Is Watching: Mass Media in the Making and Unmaking of the New Left.* Berkeley: University of California Press.

This is still the best insider account of the student protests of the 1960s and the role of the media in both supporting and then neutralizing them. Gitlin describes how one source of social change (the 1960s' student movement) is affected by another source of social change (the media).

Griswold, Wendy. 1994. *Cultures and Societies in a Changing World.* Thousand Oaks, CA: Pine Forge Press.

This is an overview of cultural change, with special reference to race-ethnicity, class, and gender. Culture and meaning in several social change theories is reviewed, including functionalism and conflict theory, with a discussion of the notion of global culture.

Morris, Aldon. 1984. *The Origins of the Civil Rights Movement.* New York: Free Press.

This now classic study shows how the civil rights social movement, as a source of social change in America, was not a unique event but an outgrowth of long-standing struggles of the African American community over the years.

Scott, Catherine V. 1995. *Gender and Development: Rethinking Modernization and Dependency Theory.* Boulder, CO: Lynne Rienner Publishers.

Scott demonstrates that prevailing theories about development, dependency, capitalism, and socialism are anchored in social constructions of gender differences. She critically evaluates both modernization theory and dependency theory, plus other theories of social change.

GLOSSARY

A

ability test: tests designed to measure capacity for abstract thinking

absolute poverty: the situation in which people do not have enough resources for basic survival

achieved status: a status attained by effort

achievement test: tests designed to measure what has been learned rather than ability or potential

adult socialization: the process of learning new roles and expectations in adult life

affirmative action: programs in education and job hiring that recruit minorities over a wide range, but do not use rigid quotas, or, which use admissions slots (quotas) for minorities in education and set aside contracts in the economy

age cohort: an aggregate group of people born during the same time period

age differentiation: the division of labor or roles in a society on the basis of age

age discrimination: different and unequal treatment of people based solely on their age

age norms: rules that spell out, implicitly or explicitly, the expectations that society has for different age strata

age prejudice: a negative attitude about an age group that is generalized to all people in that group

age-sex pyramid; also, population pyramid: a graphic representation of the age and gender structure of a society

age stereotype: preconceived judgments about what different age groups are like

age stratification: the hierarchical ranking of age groups in society

ageism: the institutionalized practice of age prejudice and discrimination

alienation: the feeling of powerlessness and separation from one's group or society

altruistic suicide: the type of suicide that can occur when there is excessive regulation of individuals by social forces

Alzheimer's disease: a degenerative form of dementia that involves neurological changes in the brain

anabolic steroids: muscle mass enhancing drugs

androcentrism: modes of thinking that are centered only in men's experiences

anomic suicide: the type of suicide occurring when there are disintegrating forces in the society that make individuals feel lost or alone

anomie: the condition existing when social regulations (norms) in a society breakdown

anorexia nervosa; anorexia: a condition characterized by compulsive dieting resulting in self-starvation

anticipatory socialization: the process of learning the expectations associated with a role one expects to enter in the future

anti-Semitism: the belief or behavior that defines Jewish people as inferior and that targets them for stereotyping, mistreatment, and acts of hatred

applied sociology: the use of sociological research and theory in solving real human problems

ascribed status: a status determined at birth

assimilation: process by which a minority becomes socially, economically, and culturally absorbed within the dominant society

attribution theory: the principle that dispositional attributions are made about others (what the other is "really like") under certain conditions, such as out-group membership

authoritarian personality: a personality characterized by a tendency to rigidly categorize people, and to submit to authority, rigidly conform, and be intolerant of ambiguity

authority: power that is perceived by others as legitimate

automation: the process by which human labor is replaced by machines

autonomous state model: a theoretical model of the state that interprets the state as developing interests of its own, independent of other interests

aversive racism: a subtle, nonobvious or covert form of racism

B

back-to-basics movement: push among some professional educators to stress the "basics" of reading, writing, arithmetic, and the "canon" of classic literature

beliefs: shared ideas held collectively by people within a given culture

bilateral kinship system: a kinship system where descent is traced through the father and the mother

biological determinism: explanations that attribute complex social phenomena to physical characteristics

birthrate; crude birthrate: the number of babies born each year per thousand members of the population

bulimia: an eating disorder characterized by alternate binge-eating followed by purging or induced vomiting

bureaucracy: a type of formal organization characterized by an authority hierarchy, a clear division of labor, explicit rules, and impersonality

C

capability poverty: the situation in which people lack the three essential human requirements to survive: to be well nourished, to be able to reproduce, and to be educated

capitalism: an economic system based on the principles of market competition, private property, and the pursuit of profit

caste system: a system of stratification (characterized by low social mobility) in which one's place in the stratification system is determined by birth

census: a count of the entire population of a country

charisma: a quality attributed to individuals believed by their followers to have special powers

charismatic authority: authority derived from the personal appeal of a leader

chronic fatigue syndrome: a persistent flu-like illness

class: see social class

class consciousness: the awareness that a class structure exists and the feeling of shared identification with others in one's class with whom one perceives common life chances

cognitive ability: the capacity for abstract thinking

cognitive elite: term used to describe the upper classes in society, based on the premise that they possess a genetically-based high intelligence

cohabitation: the practice of living together outside of marriage

cohort; birth cohort: all persons born within a given time period

collective behavior: behavior that occurs when the usual conventions are suspended and people collectively establish new norms of behavior in response to an emerging situation

collective consciousness: the body of beliefs that are common to a community or society and that give people a sense of belonging

collective preoccupations: forms of collective behavior wherein many people over a relatively broad social spectrum engage in similar behavior and have a shared definition of their behavior as needed to bring social change or to identify their place in the society

coming out process: the process of defining oneself as gay or lesbian

commodity chain: the network of production and labor processes by which a product becomes a finished commodity. By following the commodity chain, it is evident which countries gain profits and which ones are being exploited

communism: an economic system where the state is the sole owner of the systems of production

comparable worth: the principle of paying women and men equivalent wages for jobs involving similar levels of skill

compulsory heterosexuality: the idea that heterosexual identity is not a choice, but is created by institutions that treat heterosexuality as the only legitimate form of sexual identity

concept: any abstract characteristic or attribute that can be potentially measured

conflict theory: a theoretical perspective that emphasizes the role of power and coercion in producing social order

contact theory: the theory that prejudice will be reduced through social interaction with those of different race or ethnicity but of equal status

content analysis: the analysis of meanings in cultural artifacts like books, songs, and other forms of cultural communication

controlled experiment: a method of collecting data that can determine whether something actually causes something else

convergence theory: a theory of rioting that focuses on the participants and presupposes that rioters are acting on predispositions and attitudes

core countries; core nations: within world systems theory, those nations that are more technologically advanced

correlation: a statistical technique that analyzes patterns of association between pairs of sociological variables

counterculture: subculture created as a reaction against the values of the dominant culture

credentialism: the insistence upon educational credentials only for their own sake

crime: one form of deviance; specifically, behavior that violates criminal laws

criminology: the study of crime from a social scientific perspective

cult: a religious group devoted to a specific cause or charismatic leader

cultural capital: cultural resources that are socially designated as being worthy (such as knowledge of elite culture) and that give advantages to groups possessing such capital

cultural diffusion: the transmission of cultural elements from one society or culture to another

cultural hegemony: the pervasive and excessive influence of one culture throughout society

cultural pluralism: pattern whereby groups maintain their distinctive culture and history

cultural relativism: the idea that something can be understood and judged only in relationship to the cultural context in which it appears

culture: the complex system of meaning and behavior that defines the way of life for a given group or society

culture lag: the delay in cultural adjustments to changing social conditions

culture of poverty: the argument that poverty is a way of life and, like other cultures, is passed on from generation to generation

culture shock: the feeling of disorientation that can come when one encounters a new or rapidly changed cultural situation

cyberspace interaction: interaction occurring when two or more persons share a virtual reality experience via communication and interaction with each other

cyclical theory of social change: from Arnold Toynbee, the theory that certain patterns of social structure and culture recur at different times in the same society

D

data: the systematic information that sociologists use to investigate research questions

data analysis: the process by which sociologists organize collected data to discover what patterns and uniformities are revealed

death rate; crude death rate: the number of deaths each year per thousand members of the population

debunking: the process of looking behind the facades of everyday life

deductive reasoning: a form of reasoning where specific hypotheses, or predictions, are derived from general principles

defensive medicine: practiced by physicians who order extra tests on a patient in an effort to fend off a law suit by the patient

deinstitutionalization movement: social movement (and some legislation) calling for the release of many mental patients from asylums

dementia: term used to describe a variety of diseases that involve some permanent damage to the brain

democracy: system of government based on the principle of representing all people through the right to vote

demographic transition theory: the theory that societies pass through phases based on economic development, which affects the birth and death rates

demography: the scientific study of population

density; population density: the number of people per unit of area, such as the number of people per square mile

dependency theory: the global theory that maintains that industrialized nations hold less industrialized nations in a dependency, thus exploitative, relationship which benefits the industrialized nations at the expense of the less industrialized ones, whose upward mobility in the global economy is prevented

dependent variable: the variable that is a presumed effect

deviance: behavior that is recognized as violating expected rules and norms

deviant career: the sequence of movements people make through a particular system of deviance

deviant community: groups that are organized around particular forms of social deviance

deviant identity: the definition a person has of himself or herself as a deviant

differential association theory: theory that interprets deviance as behavior one learns through interaction with others

discrimination: overt negative and unequal treatment of the members of some social group or stratum solely because of their membership in that group or stratum

disengagement theory: part of functionalist theory, a theory which predicts that as people age, they gradually withdraw from participation in society

division of labor: the systematic interrelation of different tasks that develops in complex societies

doing gender: a theoretical perspective that interprets gender as something that is accomplished through the ongoing social interactions people have with one another

domestic colonialism: the view that a minority in a society is forced to be a colonized group within the dominant society

dominant culture: the culture of the most powerful group in society

dominative racism: obvious and overt racism, sometimes called "old-fashioned" racism

double deprivation: the situation in poor countries where women suffer because they are in poverty and because they are women

dual labor market: theoretical description of the occupational system defining it as divided into two major segments: the primary labor market and the secondary labor market

dual labor market theory: the idea that women and men have different earnings because they tend to work in different segments of the labor market

dyad: a group consisting of two people

E

economic restructuring: contemporary transformations in the basic structure of work that are permanently altering the workplace, including the changing composition of the workplace, deindustrialization, the use of enhanced technology, and the development of a global economy

economy: the system on which the production, distribution, and consumption of goods and services is based

ecosystem; human ecosystem: the system of interdependent parts that involves humans interacting with each other and with the physical environment

educational attainment: the total years of formal education

educational deflation: the decline in the value of a college education arising from increases over time in the number of persons graduating from college; the decline in value of a bachelor's degree

egalitarian: societies or groups where men and women share power

ego: the part of the self representing reason and common sense

egoistic suicide: the type of suicide that occurs when people feel totally detached from society

elite deviance: the wrongdoing of wealthy and powerful individuals and organizations

emergent norm theory: theory of collective behavior postulating that, when people are faced with an unusual situation, they create meanings that define and direct the situation

emigration (vs. immigration): migration of people from one society to another (also called out-migration)

emotional labor: work that is intended to produce a desired emotional effect on a client

empirical: refers to something that is based on careful and systematic observation

endogamy: the practice of selecting mates from within one's group

Enlightenment: the period in eighteenth and nineteenth century Europe characterized by faith in the ability of human reason to solve society's problems

epidemiology: the study of all factors—biological, social, economic and cultural—that are associated with disease and health

estate system: a system of stratification in which the ownership of property and the exercise of power is monopolized by an elite or noble class, which has total control over societal resources

ethnic group: a social category of people who share a common culture, such as a common language or dialect, a common religion, and common norms, practices, and customs

ethnocentrism: the belief that one's in-group is superior to all out-groups

ethnomethodology: a technique for studying human interaction by deliberately disrupting social norms and observing how individuals attempt to restore normalcy

ethnoreligious group: an extreme form of an exclusive religious group

eugenics: a social movement in the early twentieth century that sought to apply scientific principles of genetic selection to "improve" the offspring of the human race

euthanasia: the act of killing a severely ill person, or allowing the person to die, as an act of mercy

evaluation research: research assessing the effect of policies and programs

evolutionary social theory: a theory of social change that predicts that societies change in a single direction over time

exclusive religious group: religious groups with an easily identifiable religion and culture, including distinctive beliefs and strong moral teachings; they have little tolerance for diversity

exogamy: the practice of selecting mates from outside one's group

expressive crowds: crowds whose primary function is the release or expression of emotion.

expressive needs: needs for intimacy, companionship, and emotional support

extended families: families in which a large group of related households live together

extreme poverty: the situation in which people live on less than seventy-five cents a day (US currency)

F

false consciousness: the thought resulting from subordinate classes internalizing the view of the dominant class

family: a primary group of people—usually related by ancestry, marriage, or adoption—who form a cooperative economic unit to care for any offspring (and each other) and who are committed to maintaining the group over time

family wage system: a wage structure based on the assumption that men are the breadwinners for families

feminism: beliefs and actions that attempt to bring justice, fairness, and equity to all women, regardless of their race, age, class, sexual orientation, or other characteristics

feminization of poverty: the process whereby a growing proportion of the poor are women and children

feral children: children who have been raised in the absence of human contact

fetal sexual differentiation: the process by which biological differences between the sexes are established, which continues with the production of hormones

first-world countries: industrialized nations based on a market economy and with democratically elected governments

folkways: the general standards of behavior adhered to by a group

food deprivation: inadequate food consumption of individuals to maintain a healthy life

food poverty: situation in which households cannot afford to purchase sufficient food to adequately feed the members of the household

food shortage: situation where there is not enough food available to feed the people in a designated area

formal organization: a large secondary group organized to accomplish a complex task or set of tasks

functionalism: a theoretical perspective that interprets each part of society in terms of how it contributes to the stability of the whole

G

game stage: the stage in childhood when children become capable of taking a multitude of roles at the same time

gemeinschaft: German for *community*, a state characterized by a sense of fellow feeling among the members of a society, including strong personal ties, sturdy primary group memberships, and a sense of personal loyalty to one another; associated with rural life

gender: socially learned expectations and behaviors associated with members of each sex

gender gap: term referring to the differences in women's and men's political attitudes and behavior

gender identity: one's definition of self as a woman or man

gender segregation: the distribution of men and women in different jobs in the labor force

gender socialization: the process by which men and women learn the expectations associated with their sex

gender stratification: the hierarchical distribution of social and economic resources according to gender

gendered institution: the idea that whole institutions are patterned by gender

gendered racism: the principle that the combined effects of racism and sexism are inseparable

generalization: the ability to make claims that a finding represents something greater than the specific observations on which the finding is based

generalized other: the abstract composite of social roles and social expectations

generational equity: the question of whether one age group or generation is unfairly taxed in order to support the needs and interests of another generation

gesellschaft: German for *society,* a form of social organization characterized by a high division of labor, less prominence of personal ties, the lack of a sense of community among the members, and the absence of a feeling of belonging; associated with urban life

glass ceiling: popular concept referring to the limits that women and minorities experience in job mobility

global culture: diffusion of a single culture throughout the world

global economy: term used to refer to the fact that all dimensions of the economy now cross national borders

global stratification: the systematic inequalities between and among different groups within nations that result from the differences in wealth, power, and prestige of different societies relative to their position in the international economy

globalization: increased economic, political, and social interconnectedness and interdependence among societies in the world

government: those state institutions that represent the population and make rules that govern the society

greenhouse effect: a rise in the earth's surface temperature caused by heat trapped by excess carbon dioxide in the atmosphere

group: a collection of individuals who interact and communicate, share goals and norms, and who have a subjective awareness as "we"

group size effect: the effect upon the person of groups of varying sizes

groupthink: the tendency for group members to reach a consensus at all costs

H

health maintenance organization (HMO): a cooperative of doctors and other medical personnel who provide medical services in exchange for a set membership fee

hermaphroditism: a condition produced when irregularities in chromosome formation or fetal differentiation produces persons with mixed sex characteristics

heterosexism: the institutionalization of heterosexuality as the only socially legitimate sexual orientation

HMO: see Health Maintenance Organization

homogamy: the pattern by which people select mates with similar social characteristics to their own

homophobia: the fear and hatred of homosexuality

hospice movement: movement to provide in-home care for terminally ill patients as an alternative to hospitalization

household: term used by U.S. census to refer to all persons (may or may not be related) occupying a housing unit

human capital theory: a theory that explains differences in wages as the result of differences in the individual characteristics of the workers

human ecology: the study of the interdependence between humans and their physical environment

humanitarian: the principle that human reason can successfully direct social change for the betterment of society

hypothesis: a statement about what one expects to find in research

I

id: the part of the personality that includes various impulses and drives, including sexual passions and desires, biological urges, and human instincts

identity: how one defines oneself

ideology: a belief system that tries to explain and justify the status quo

imitation stage: the stage in childhood when children copy the behavior of those around them

immigration (vs. emigration): the migration of people into a society from outside (also called in-migration)

impression management: a process by which people control how others perceive them

inclusive religious group: religious groups with a moderate and liberal religious orientation and tolerance for diversity

income: the amount of money brought into a household from various sources during a given year (wages, investment income, dividends, etc.)

independent variable: a variable tested as the presumed cause of a particular result

index of dissimilarity: a measure used to indicate the number of workers who would have to change jobs in order to have the same occupational distribution as the comparison group

indicator: something that points to or reflects an abstract concept

inductive reasoning: a logical process of building general principles from specific observations

infant mortality rate: the number of deaths per year of infants less than one year old, for every thousand live births

inner-directedness: a condition wherein the individual's behavior is guided by internal principles and morals

institution: see social institution

institutional racism: racism involving notions of racial or ethnic inferiority that have become ingrained into society's institutions

instrumental needs: emotionally neutral, task-oriented (goal-oriented) needs

interaction: see social interaction

interest group: a constituency in society organized to promote its own agenda

interlocking directorate: organizational linkages created when the same people sit on the boards of directors of a number of different corporations

issues: problems that affect large numbers of people and have their origins in the institutional arrangements and history of a society

J

job displacement: the permanent loss of certain job types that occurs when employment patterns shift

K

kinship system: the pattern of relationships that define people's family relationships to one another

L

labeling effect: the effect of tract (role) assignment as distinct from the effect of cognitive ability

labeling theory: a theory that interprets the responses of others as most significant in understanding deviant behavior

labor force participation rate: the percentage of those in a given category who are employed

language: a set of symbols and rules which, put together in a meaningful way, provides a complex communication system

latent functions: the unintended consequences of behavior

law: the written set of guidelines that define what is right and wrong in society

liberal feminism: a feminist theoretical perspective asserting that the origin of women's inequality is in traditions of the past that pose barriers to women's advancement

life chances: the opportunities that people have in common by virtue of belonging to a particular class

life course perspective: sociological framework for studying aging that connects people's personal attributes, the roles they occupy, the life events they experience, and their sociohistorical context

life expectancy: the average number of years the group can expect to live

looking-glass self: the idea that people's conception of self arises through reflection about their relationship to others

M

macrochange: a social change that is relatively gradual and that broadly affects many aspects of a society

Malthusian theory: after T. R. Malthus, the principle that a population tends to grow faster than the subsistence (food) level needed to sustain it

managed competition: a health care system wherein all individuals are members of an HMO (health maintenance organization), with the intent of reducing fees and insurance costs

manifest functions: the stated and open goals of social behavior

mass media: channels of communication that are available to very wide segments of the population

mass society: one in which industrialization and bureaucracy reach exceptionally high levels

master status: some characteristic of a person that overrides all other features of the person's identity

material culture: the objects created in a given society—its buildings, art, tools, toys, print and broadcast media, and other tangible objects

matriarchal religion: religions based on the centrality of female goddesses, who may be seen as the source of food, nurturance, and love or who may serve as emblems of the power of women

matriarchy: a society or group in which women have power over men

matrilineal kinship system: kinship systems in which family lineage (or ancestry) is traced through the mother

matrilocal kinship system: kinship systems in which women continue to live with their families of origin after marriage

mean: the sum of a set of values divided by the number of cases from which the values are obtained; an average

means of production: the system by which goods are produced and distributed

mechanical solidarity: unity based on similarity, not difference, of roles

median: the midpoint in a series of values that are arranged in numerical order

median income: the midpoint of all household incomes

Medicaid: governmental assistance program for the poor

medicalization: that process by which society, following the medical profession, assigns all aspects of health and illness an exclusively medical meaning

medicalization of deviance: explanations of deviant behavior that interpret deviance as the result of individual pathology or sickness

Medicare: governmental assistance program for the elderly

microchange: subtle alterations in the day-to-day interaction between people

military-industrial complex: term used to describe the linkage between business and military interests

minority group: any distinct group in society that shares common group characteristics and is forced to occupy low status in society because of prejudice and discrimination

miscegenation: the mixing of the races through marriage

mode: the value that appears most frequently in a set of data

modernization theory: a view of globalization in which global development is a worldwide process affecting nearly all societies that have been touched by technological change

monogamy: the marriage practice of a sexually exclusive marriage with one spouse at a time

monotheism: the worship of a single god

mores: strict norms that control moral and ethical behavior

multiculturalism movement: the push to introduce into the elementary, high-school, and college curricula more courses on different and diverse subcultures and groups, ethnic groups, and gender studies

multidimensional evolutionary theory of social change: a theory predicting that over time societies follow not one but several evolutionary paths

multidimensional theory of intelligence: the idea that there are different kinds of intelligence rather than one basic kind

multinational corporation: companies that draw a large share of their revenues from foreign investments and that conduct business across national borders

multiracial feminism: form of feminist theory noting the exclusion of women of color from other forms of theory and centering its analysis in the experiences of all women

N

natural attitude: accepting social expectations and absorbing social information at face value without questioning what you see

neocolonialism: a form of control of the poor countries by the rich countries, but without direct political or military involvement

neoevolutionary social theory: see multidimensional theory of social change

neolocal residence: the practice whereby newly wedded couples establish their own residence

new international division of labor: the organization of production so that a product is made in several countries

new social movement theory: a theory about social movements linking culture, ideology, and identity conceptually to explain how new identities are forged within social movements

NICs: newly industrializing countries; countries that have shown rapid growth and have emerged as developed countries

noncore (peripheral) nations: within world systems theory, those nations that are less technologically advanced

nonmaterial culture: the norms, laws, customs, ideas, and beliefs of a group of people

norms: the specific cultural expectations for how to act in a given situation

nuclear families: families in which married couples reside together with their children

O

object relations theory: a psychoanalytic theory of socialization positing that social relationships children experience early in life determine the development of their personality

occupational distribution: the pattern by which workers are located in the occupational system

occupational prestige: the subjective evaluation people give to jobs as better or worse than others

occupational segregation: the pattern by which workers are separated into different occupations on the basis of social characteristics like race and gender

operational definition: any definition that spells out how a concept is to be measured

organic metaphor: refers to the similarity early sociologists saw between society and other organic systems

organic (contractual) solidarity: unity based on role differentiation, not similarity

organizational deviance: wrongdoing that occurs within an organizational context and that is sanctioned by the norms and operating principles of the organization

other-directedness: a condition wherein the individual's behavior is guided by the behavior of others

P

participant observation: a method whereby the sociologist becomes both a participant in the group being studied and a scientific observer of the group

patriarchal religion: religion in which the beliefs and practices of the religion are based on male power and authority

patriarchy: a society or group where men have power over women

patrilineal kinship system: kinship systems in which family lineage (or ancestry) is traced through the father

patrilocal kinship system: kinship systems in which, following marriage, women are separated from their families of origin and reside with the husband's kinship group

per capita gross national product: the measurement of the total volume of goods and services produced by a country each year, divided by the size of the population

percentage: parts per hundred

peripheral countries: poor countries, largely agricultural, having little power or influence in the world system

personal transformation movements: social movements that aim to change the individual

personality: the relatively consistent pattern of behavior, feelings, and beliefs in a given person

play stage: the stage in childhood when children begin to take on the roles of significant people in their environment

pluralist model: a theoretical model of power in society as coming from the representation of diverse interests of different groups in society

political action committees (PACs): groups of people who organize to support candidates they feel will represent their views

political process theory: explanation of social movements positing that movements achieve success by exploiting a combination of internal factors

polygamy: a marriage practice in which either men or women can have multiple marriage partners

polytheism: the worship of more than one deity

popular culture: the beliefs, practices, and objects that are part of everyday traditions

population: a relatively large collection of people (or other unit) that a researcher studies and about which generalizations are made

population pyramid: see age-sex pyramid

positivism: a system of thought in which accurate observation and description is considered the highest form of knowledge

postmodernism: a theoretical perspective based on the idea that society is not an objective thing, but is found in the words and images—or discourses—that people use to represent behavior and ideas

poverty line: the figure established by the government to indicate the amount of money needed to support the basic needs of a household

power: a person or group's ability to exercise influence and control over others

power elite model: a theoretical model of power positing a strong link between government and business

predictive validity: the extent to which a test accurately predicts later college grades, or some other criterion such as likelihood of graduating

prejudice: the negative evaluation of a social group, and individuals within that group, based upon conceptions about that social group that are held despite facts that contradict it

prestige: the value with which different groups or people are judged

primary deviance: the actual violation of a norm or law

primary group: a group characterized by intimate, face-to-face interaction and relatively long-lasting relationships

probability: the likelihood that a specific behavior or event will occur

profane: that which is of the everyday, secular world and is specifically not religious

propaganda: information disseminated by a group or organization (such as the state) intended to justify its own power

Protestant ethic: belief that hard work and self-denial lead to salvation

proxemic communication: meaning conveyed by the amount of space between interacting individuals

psychoanalytic theory: a theory of socialization positing that the unconscious mind shapes human behavior

Q

quadruple jeopardy: phrase referring to the simultaneous effects of being old, minority, female, and poor

R

race: a social category, or social construction, that we treat as distinct on the basis of certain characteristics, some biological, that have been assigned social importance in the society

racial formation: process by which groups come to be defined as a "race" through social institutions such as the law and the schools

racial stratification (also racial and ethnic stratification): the inequality between racial and ethnic groups in society

racism: the perception and treatment of a racial or ethnic group, or member of that group, as intellectually, socially, and culturally inferior to one's own group

radical feminism: feminist theoretical perspective that interprets patriarchy as the primary cause of women's oppression

radical movements: social movements that seek fundamental change in the structure of society

random sample: a sample that gives everyone in the population an equal chance being selected

rate: parts per some number (e.g., per 10,000; per 100,000)

rational-legal authority: authority stemming from rules and regulations, typically written down as laws, procedures, or codes of conduct

rationalization of society: term used by Max Weber to describe society being increasingly organized around legal, empirical, and scientific forms of thought

reactionary movements: social movements organized to resist change or to reinstate an earlier social order that participants perceive to be better

reference group: any group (to which one may or may not belong) used by the individual as a standard for evaluating her or his attitudes, values, and behaviors

reflection hypothesis: the idea that the mass media reflect the values of the general population

reform movements: social movements that seek change through legal or other mainstream political means, working within existing institutions

relative deprivation theory: theory of riots and revolutions positing that people revolt when they see their situation improving, but not as much as that of other groups in the society; also known as the revolution of rising expectations

relative poverty: a definition of poverty that is set in comparison to a set standard

reliability: the likelihood that a particular measure would produce the same results if the measure were repeated

religion: an institutionalized system of symbols, beliefs, values, and practices by which a group of people interprets and responds to what they feel is sacred and that provides answers to questions of ultimate meaning

religiosity: the intensity and consistency of practice of a person's (or group's) faith

religious socialization: the process by which one learns a particular religious faith

replacement level; population replacement level: a condition wherein the birth and death rates continue to produce a steady population level

replication study: research that is repeated exactly, but on a different group of people at a different point in time

research design: the overall logic and strategy used in a research project

residential segregation: the spatial separation of racial and ethnic groups in different residential areas

resocialization: the process by which existing social roles are radically altered or replaced

resource mobilization theory: theory of how social movements develop that focuses on how movements gain momentum by successfully garnering organizational resources

risky shift (also polarization shift): the tendency for group members, after discussion and interaction, to engage in riskier behavior than they would while alone

rite of passage: ceremony or ritual that symbolizes the passage of an individual from one role to another

ritual: symbolic activities that express a group's spiritual convictions

role: the expected behavior associated with a given status in society

role conflict: two or more roles associated with contradictory expectations

role strain: conflicting expectations within the same role

sacred: that which is set apart from ordinary activity, seen as holy, and protected by special rites and rituals

S

salience principle: categorizing people on the basis of what initially appears prominent about them

sample: any subset of units from a population that a researcher studies

Sapir-Whorf hypothesis: a theory that language determines other aspects of culture since language provides the categories through which social reality is defined and perceived

scapegoat theory: argument that dominant group aggression is directed toward a minority as a substitute for frustration with some other problem

scapegoating: process wherein a group collectively identifies another group as a threat to the perceived social order and incorrectly blames the other group for problems they have not actually caused

schooling: the formal, institutionalized aspects of education

scientific method: the steps in a research process, including observation, hypothesis testing, analysis of data, and generalization

script: a learned performance of a social role

second-world countries: socialist countries with state-managed economies and typically without a democratically elected government

secondary deviance: behavior that results from being labeled deviant, regardless of whether the person has previously engaged in deviance

secondary group: a group that is relatively large in number, and not as intimate or long in duration as a primary group

sect: groups that have broken off from an established church

secular: the ordinary beliefs of daily life that are specifically not religious

secularization: the process by which religious institutions, behavior, and consciousness lose their religious significance

segregation: the spatial and social separation of racial and ethnic groups

self-esteem: the value a person places on his or her identity

self-fulfilling prophecy: the process by which merely applying a label changes behavior and thus tends to justify the label

semiperipheral countries: semi-industrialized countries that represent a kind of middle class within the world system

sex: used to refer to biological identity as male or female

sex ratio: the number of males per 100 females

sexism: a system of practices and beliefs through which women are controlled and exploited because of the significance given to differences between the sexes

sexology: the scientific study of sex

sexual harassment: unwanted physical or verbal sexual behavior that occurs in the context of a relationship of unequal power and that is experienced as a threat to the victim's job or educational activities

sexual orientation: the manner in which individuals experience sexual arousal and pleasure

sexual politics: the link feminists argue exists between sexuality and power and between sexuality and race, class, and gender oppression

sexual revolution: the widespread changes in men's and women's roles and a greater public acceptance of sexuality as a normal part of social development

sexual scripts: the ideas taught to us about what is appropriate sexual behavior for a person of our gender

sick role: a pattern of expectations of behaviors defined in society as appropriate for one who is ill

significant others: those with whom we have a close affiliation

social action: behavior to which people give meaning

social change: the alteration of social interaction, social institutions, stratification systems, and elements of culture over time

social change movements: movements that aim to change some aspect of society

social class: the social structural hierarchical position groups hold relative to the economic, social, political, and cultural resources of society

social control: a process by which groups and individuals within those groups are brought into conformity with dominant social expectations

social control theory: theory that explains deviance as the result of the weakening of social bonds

social constructionist perspective: a theoretical perspective that explains sexual identity as created and learned within a cultural, social, and historical context

Social Darwinism: the idea that society evolves to allow the survival of the fittest

social differentiation: the process by which different statuses in any group, organization, or society develop

social epidemiology: the study of the effects of social and cultural factors upon disease and health

social facts: social patterns that are external to individuals

social institution: an established and organized system of social behavior with a recognized purpose

social interaction: behavior between two or more people that is given meaning

social learning theory: a theory of socialization positing that the formation of identity is a learned response to social stimuli

social mobility: a person's movement over time from one class to another

social movement: a group that acts with some continuity and organization to promote or resist change in society

social network: a set of links between individuals or other social units such as groups or organizations

social organization: the order established in social groups

social sanctions: mechanisms of social control that enforce norms

social stratification: a relatively fixed hierarchical arrangement in society by which groups have different access to resources, power, and perceived social worth; a system of structured social inequality

social structure: the patterns of social relationships and social institutions that comprise society

socialism: an economic institution characterized by state ownership and management of the basic industries

socialist feminism: a feminist theoretical perspective that interprets the origins of women's oppression as lying in the system of capitalism

socialization: the process through which people learn the expectations of society

socialization agents: those who pass on social expectations

society: a system of social interactions that includes both culture and social organization

socio-economic status (SES): a measure of class standing, typically indicated by income, occupational prestige, and educational attainment

sociological imagination: the ability to see the societal patterns that influence individual and group life

sociology: the study of human behavior in society

standardized test: tests given to large populations and scored with respect to population averages

state: the organized system of power and authority in society

status: an established position in a social structure that carries with it a degree of prestige

status attainment: the process by which people end up in a given position in the stratification system

status generalization: the principle that one's status in society determines one's status in some particular group (such as a jury)

status inconsistency: exists when the different statuses occupied by the individual bring with them significantly different amounts of prestige

status set: the complete set of statuses occupied by a person at a given time

stereotype: an oversimplified set of beliefs about the members of a social group or social stratum that is used to categorize individuals of that group

stereotype interchangeablility: the principle that negative stereotypes are interchangeable from one racial group (or gender or social class) to another

stigma: an attribute that is socially devalued and discredited

structural strain theory: a theory that interprets deviance as originating in the tensions that exist in society between cultural goals and the means people have to achieve those goals

subculture: the culture of groups whose values and norms of behavior are somewhat different from those of the dominant culture

superego: the dimension of the self representing the standards of society

symbolic interaction theory: a theoretical perspective claiming that people act toward things because of the meaning things have for them

symbols: things or behavior to which people give meaning

T

taking the role of the other: the process of imagining oneself from the point of view of another

teacher expectancy effect: the effect of the teacher's expectations on the student's actual performance, independent of the student's ability

thermal pollution: the heating up of the earth's rivers and lakes as a result of the chemical discharges of heavy industry

tertiary deviance: deviance that occurs when the deviant fully accepts the deviant role, but rejects the stigma associated with it

third-world countries: countries that are poor, underdeveloped, largely rural, and with high levels of poverty; typically governments in such countries are autocratic dictatorships and wealth is concentrated in the hands of a small elite

total institution: an organization cut off from the rest of society where individuals who reside there are subject to strict social control

totem: an object or living thing that a religious group regards with special awe and reverence

tracking: grouping, or stratifying, students in school on the basis of ability test scores

tradition-directedness: conformity to long-standing and time-honored norms and practices

traditional authority: authority stemming from long-established patterns that give certain people or groups legitimate power in society

triad: a group consisting of three people

triadic segregation: the tendency for a triad to separate into a dyad and an isolate

troubles: privately felt problems that come from events or feelings in one individual's life

Tuskegee Syphilis Study: an unethical study of about 400 syphilis-infected African American men who went untreated for the disease for 40 years, from 1932 to 1972, even though a cure (penicillin) was discovered in the early 1950s

U

underemployment: a term used to describe being employed at a level below what would be expected, given a person's level of training or education

unemployment rate: the percentage of those not working, but officially defined as looking for work

unidimensional evolutionary theory of social change: a theory that predicts that societies over time follow a single path from simple and structurally undifferentiated to more complex and structurally differentiated

unidimensional theory of intelligence: the idea that intelligence is reducible to one (as opposed to several) master cognitive capacity

urban underclass: a grouping of people, largely minority and poor, who live at the absolute bottom of the socioeconomic ladder in urban areas

V

validity: the degree to which an indicator accurately measures or reflects a concept

values: the abstract standards in a society or group that define ideal principles

variable: something that can have more than one value

verstehen: the process of understanding social behavior from the point of view of those engaged in it

victimization perspective: the tendency to ignore the strong cultural, occupational, educational, and economic achievements of a minority group

vital statistics: information about births, deaths, marriages, and migration

voluntary organization: organizations that people join to pursue goals they consider personally worthwhile

W

wealth: the monetary value of what someone actually owns, calculated by adding all financial assets (stocks, bonds, property, insurance, value of investments, etc.) and subtracting debts; also called net worth

work: productive human activity that produces something of value, either goods or services

world systems theory: theory that capitalism is a single world economy and that there is a worldwide system of unequal political and economic relationships that benefit the developed and technologically advanced countries at the expense of the less technologically advanced and less developed

X

xenophobia: the fear and hatred of foreigners

REFERENCES

A

Aberle, David F., Albert K. Cohen, A. Kingsley Davis, Marion J. Levy, Jr., and Francis X. Sutton. 1950. "The Functional Prerequisites of A Society." *Ethics* 60 (January): 100–111.

Acker, Joan. 1992. "Gendered Institutions: From Sex Roles to Gendered Institutions." *Contemporary Sociology* 21 (September): 565–569.

Acock, Alan C., and Jeanne S. Hurlbert. 1993. "Social Networks, Marital Status, and Well-Being." *Social Networks* 15 (September): 309–334.

Adams, Karen L., and Daniel T. Brink. 1990. *Perspectives on Official English: The Campaign for English as the Official Language of the USA.* New York: Mouton de Gruyter.

Adams, Paul, and Gary L. Dominick. 1995. "The Old, the Young, and the Welfare State." *Generations* 19 (Fall): 38–42.

Adkins, Lisa, and Vicki Merchant. 1996. *Sexualizing the Social: Power and the Organization of Sexuality.* New York: St. Martin's Press.

Adler, N. E., T. Booyce, M. A. Chesney, S. Falkman, and S. L. Syme. 1993. "Socioeconomic Inequalities in Health: No Easy Solution." *Journal of the American Medical Association* 269: 3140–3145.

Adorno, T. W., Else Frenkel-Brunswik, D. J. Levinson, and R. N. Sanford. 1950. *The Authoritarian Personality.* New York: Harper & Row.

Aguirre, Benigno, Enrico Quarantelli, and Jorge L. Mendoza. 1988. "The Collective Behavior of Fads: The Characteristics, Effects and Career of Streaking." *American Sociological Review* 53 (August): 569–589.

Agyeman, Opoku. 1990. "The Supreme Court and Enforcement of African American Rights: Myth & Reality." *The Black Scholar* 21 (Winter): 22–28.

Ainsworth, Robert G. 1989. *An Overview of the Labor Market Problems of Indians and Native Americans.* Washington, DC: National Commission for Employment Policy.

Alan Guttmacher Institute. 1997. *Sex and America's Teenagers.* New York: The Alan Guttmacher Institute.

Alan Guttmacher Institute. 1998. "Facts in Brief." New York: Alan Guttmacher Institute. World Wide Web site: **http://www.agi-use.org/pubs**

Alba, Richard, and Gwen Moore. 1982. "Ethnicity in the American Elite." *American Sociological Review* 47 (June): 373–383.

Alba, Richard. 1990. *Ethnic Identity: The Transformation of Ethnicity in the Lives of Americans of European Ancestry.* New Haven, CT: Yale University Press.

Albas, Daniel, and Cheryl Albas. 1988. "Aces and Bombers: The Post-Exam Impression Management Strategies of Students." *Symbolic Interaction* 11: 289–302.

Albelda, Randy, and Chris Tilly. 1996. "It's a Family Affair: Women, Poverty, and Welfare." Pp. 79–86 in *For Crying Out Loud: Women's Poverty in the United States,* edited by Diane Dujon and Ann Withorn. Boston: South End Press.

Albrow, Martin, John Eade, Neil Washbourne, and Jorg Durrschmidt. 1994. "The Impact of Globalization on Sociological Concepts: Community, Culture, and Milieu." *Innovation* 7: 371–389.

Aldag, Ramon J,. and Sally R. Fuller. 1993. "Beyond Fiasco: A Reappraisal of the Groupthink Phenomenon and a New Model of Group Decision Processes." *Psychological Bulletin* 113: 533–552.

Alexander, Jeffrey C., and Paul Colomy, eds. 1990. *Differentiation Theory and Social Change: Comparative and Historical Perspectives.* New York: Columbia University Press.

Alicea, Marixsa. 1997. "'What is Indian About You?' A Gendered, Transnational Approach to Ethnicity." *Gender & Society* 11 (October): 597–626.

Aliotta, Jilda M. 1991. "The Unfinished Feminist Agenda: The Shifting Forum." *Annals of the American Academy of Political and Social Sciences* 515: 140–150.

Allen, Michael Patrick, and Philip Broyles. 1991. "Campaign Finance Reforms and the Presidential Campaign Contributions of Wealthy Capitalist Families." *Social Science Quarterly* 72 (December): 738–750.

Allen, Paula Gunn. 1986. *The Sacred Hoop: Recovering the Feminine in American Indian Traditions.* Boston: Beacon Press.

Allison, Anne. 1994. *Nightwork: Sexuality, Pleasure, and Corporate Masculinity in a Tokyo Hostess Club.* Chicago: University of Chicago Press.

Allport, Gordon W. 1954. *The Nature of Prejudice.* Reading, MA: Addison-Wesley.

Allport, Gordon W. 1966. "The Religious Context of Prejudice," *Journal for the Scientific Study of Religion* 5: 447–457.

Alsalam, Nabeel, Laurence T. Ogle, Gayle Thompson Rogers, and Thomas M. Smith. 1992. *The Condition of Education: 1992.* Washington: U.S. Department of Education.

Altemeyer, Bob. 1988. *Enemies of Freedom: Understanding Right-Wing Authoritarianism.* San Francisco: Jossey-Bass.

Altman, Laurence K. 1991. "Many Hispanic Americans Reported in Ill Health and Lacking Insurance." *The New York Times* (January 9): A16.

Alzheimer's Association. 1996. "One in Ten Families Affected by Alzheimer's Disease." Chicago: Alzheimer's Association. **http://www.alz.org**

Amato, P.R. 1990. "Personality and Social Network Involvement As Predictors of Helping Behavior in Everyday Life." *Social Psychology Quarterly* 53: 31–43.

Amato, Paul R., and Alan Booth. 1997. *A Generation at Risk: Growing Up in an Era of Family Upheaval.* Cambridge, MA: Harvard University Press.

Amato, Paul R. 1983. "Helping Behavior in Urban and Rural Environments: Field Studies Based on Taxonomic Organization of Helping Episodes." *Journal of Personality and Social Psychology* 45 (3): 571–580.

American Association of University Women. 1992. *How Schools Shortchange Girls.* Washington, DC: American Association of University Women.

American Medical Association. 1997. Table 5: Men and Women Physicians by Specialty, 1994. **http://www.ama.assn.org/mem-data/wmmed/infoserv/data/table5.html**

American Medical Association. 1998. World Wide Web site: **http://www.ama-assn.org/ad-com/releases/1996/fvfact.html**

American Sociological Association. 1993. "Scarce Remains Jailed; ASA Council Advocates Researcher's Privilege." *Footnotes* 21 (October): 2ff.

Americans with Disabilities Act. *ADA Compliance Guide.* 1990. Washington, DC: Thompson Publishing Group.

Amott, Teresa L., and Julie A. Matthaei. 1991. *Race, Gender and Work: A Multicultural History of Women in the United States.* Boston: South End Press.

Amott, Teresa L., and Julie A. Matthaei. 1996. *Race, Gender, and Work: A Multicultural History of Women in the United States,* 2nd ed. Boston: South End Press.

Amyot, Robert P., and Lee Sigelman. 1996. "Jews without Judaism? Assimilation and Jewish Identity in the United States." *Social Science Quarterly* 77 (March): 177–189.

Andersen, Margaret L. 1991. "Feminism and the Family Ideal." *Journal of Comparative Family Studies* 22 (Summer): 235–246.

Andersen, Margaret L. 1997. *Thinking About Women: Sociological Perspectives on Sex and Gender,* 4th ed. Boston: Allyn and Bacon.

Andersen, Margaret L., and Patricia Hill Collins (eds). 1998. *Race, Class, and Gender: An Anthology,* 3rd ed. Belmont, CA: Wadsworth.

Anderson, Elijah. 1976. *A Place on the Corner.* Chicago: University of Chicago Press.

Anderson, Elijah. 1990. *Streetwise: Race, Class, and Change in an Urban Community.* Chicago: Univerity of Chicago Press.

Andreasse, Arthur. 1997. "Evaluating the 1995 Industry Employment Projections." *Monthly Labor Review* 120 (September): 9–31.

Angel, Ronald, and Marta Tienda. 1982. "Determinants of Extended Household Structure: Cultural Pattern or Economic Need?" *American Journal of Sociology* 87 (May): 1360–1383.

Angotti, Joseph. 1997. "Content Analysis of Local News Programs in Eight U.S. Television Markets." Miami: University of Miami Center for the Advancement of Modern Media.

Ansari, Maboud. 1991. "Iranians in America: Continuity and Change." Pp. 119–142 in *Rethinking Today's Minorities,* edited by Vincent N. Parillo. New York: Greenwood Press.

Ansuini, Catherine G., Juliana Fiddler-Woite, and Robert S. Woitaszek. 1994. "The Effects of Operation Rescue on Pro-Life Support." *College Student Journal* 28 (December): 441–445.

Antill, J.K. 1987. "Parents' Beliefs and Values About Sex Roles, Sex Differences, and Sexuality: Their Sources and Implications." Pp. 294–328 in *Sex and Gender,* edited by P. Shaver and C. Hendrick. Newbury Park, CA: Sage.

Aponte, Harry J., 1994. *Bread and Spirit: Therapy with the New Poor, Diversity of Race, Culture, and Values.* New York: W. W. Norton.

Appelbaum, Richard, and William Chambliss. 1997. *Sociology: A Brief Introduction,* 2nd ed. New York: Addison Wesley Longman, Inc.

Apple, Michael W. 1991. "The New Technology: Is it Part of the Solution or Part of the Problem in Education?" *Computers in the Schools* 8 (April/October): 59–81.

Arendell, Terry. 1992. "After Divorce: Investigations into Father Absence." *Gender & Society* 6 (December): 562–586.

Arendell, Terry. 1998. "Divorce American Style." *Contemporary Sociology* 27 (May): 226–228.

Arendt, Hannah. 1963. *Eichmann in Jerusalem: A Report on the Banality of Evil.* New York: Viking Press.

Argyle, Michael. 1975. *Bodily Communication.* London: Methuen.

Argyris, Chris. 1990. *Overcoming Organizational Defenses: Facilitating Organizational Learning.* Boston: Allyn & Bacon.

Aries, E. 1987. "Gender and Communication." Pp. 149–176 in *Sex and Gender,* edited by P. Shaver and C. Hendrick. Newbury Park, CA: Sage.

Aries, Phillippe. 1962. *Centuries of Childhood.* New York: Vintage.

Arnold, Regina. 1990. "Processes of Victimization and Criminalization of Black Women." *Social Justice* 17 (Fall): 153–166.

Aronson, Jane. 1992. "Women's Sense of Responsibility for the Care of Old People: 'But Who Else is Going to Do It?'" *Gender & Society* 6 (March): 8–29.

Asch, Solomon. 1951. "Effects of Group Pressure Upon the Modification and Distortion of Judgments." In *Groups, Leadership, and Men,* edited by H. Guetzkow. Pittsburgh: Carnegie Press.

Asch, Solomon. 1955. "Opinions and Social Pressure." *Scientific American* 19 (July): 31–35.

Ashmore, R. D., and F. DelBoca. 1976. "Psychological Approaches to Understanding Intergroup Conflict." Pp. 73–124 in *Towards the Elimination of Racism,* edited by P. A. Katz. Elmsford, NY: Pergamon Press.

Associated Press. 1991. "17 Slayings Tied to Milwaukee Man." *The New York Times* (July 26): A12.

Astin, Alexander W., Kenneth C. Green, and William S. Korn. 1987. *The American Freshman: Twenty Year Trends, 1966–1985.* Los Angeles: The Higher Education Research Institute, University of Southern California.

Astin, Alexander. 1992. "The American Freshman: National Norms for Fall 1992. American Council on Education and University of California, Los Angeles" *The Chronicle of Higher Education,* January 13, 1992, p. A31.

Astone, Nan Marie, and Sara S. McLanahan. 1991. "Family Structure, Parental Performance, Status Relations, and the Structure of Sentiment: Bringing the Teacher Back In." *American Sociological Review* 56: 665–682.

Atchley, Robert C. 1997. *Social Forces and Aging,* 8th ed. Belmont, CA: Wadsworth.

Auerbach, Judy. 1992. "The Family/Medical Leave Bill." Presentation at Midyear Meeting of Sociologists for Women in Society, Raleigh, NC, February.

Ayers, Ian. 1990. "White Men Get Better Deals on Cars." *The New York Times* (December 12): 26.

B

Babbie, Earl. 1998. *The Practice of Social Research,* 8th edition. Belmont, CA: Wadsworth.

Babson, Jennifer, and Kelly St. John. 1994. "Momentum Helps GOP Collect Record Amounts from PACS." *Congressional Quarterly* 52: 3456–3459.

Baca Zinn, Maxine, and Bonnie Thornton Dill. 1996. "Theorizing Difference from Multiracial Feminism." *Feminist Studies* 22 (Summer): 321–331.

Baca Zinn, Maxine, and D. Stanley Eitzen. 1996. *Diversity in American Families,* 4th ed. New York: HarperCollins.

Baca Zinn, Maxine, and D. Stanley Eitzen. 1998. "Economic Restructuring and Systems Inequality." Pp. 233–237 in *Race, Class, and Gender: An Anthology,* 3rd ed., edited by Margaret L. Andersen and Patricia Hill Collins. Belmont, CA: Wadsworth Publishing Co.

Baca Zinn, Maxine. 1979. "Field Research in Minority Communities: Political, Ethical, and Methodological Observations by an Insider." *Social Problems* 27 (December): 209–219.

Baca Zinn, Maxine. 1995. "Chicano Men and Masculinity." Pp. 33–41 in *Men's Lives,* 3rd ed., edited by Michael S. Kimmel and Michael A. Messner. Needham Heights, MA: Allyn and Bacon.

Bachman, Daniel, and Ronald K. Esplin. 1992. "Plural Marriage." *Encyclopedia of Mormonism,* Vol. 3. New York: Macmillan Publishing Company.

Bachman, J., L. Johnston, and P. O'Malley. 1993. *Monitoring the Future: Questionnaire Responses from the Nation's High School Seniors, 1993.* Ann Arbor, MI: Institute for Social Research.

Baffi, Charles R., Kerry J. Redican, Mary K. Sefchick, and James C. Impara. 1991. "Gender Role Identity, Gender Role Stress, and Health Behaviors: An Exploratory Study of Selected College Males." *Health Values* 15 (January-February): 9–18.

Baker, Douglas D., and David E. Terpstra. 1986. "Locus of Control and Self-Esteem Versus Demographic Factors as Predictors of Attitudes Toward Women." *Basic and Applied Social Psychology* 7 (June): 163–172.

Balch, Robert W. 1980. "Looking Behind the Scenes in a Religious Cult: Implications for the Study of Conversion." *Sociological Analysis* 41: 137–143.

Balch, Robert W. 1995. "Waiting for the Ships: Disillusionment and the Revitalization of Faith in Bo and Peep's UFO Cult." Pp. 137–166 in *The Gods Have Landed: New Religions from Other Worlds,* edited by James R. Lewis. Albany: State University of New York Press.

Baldassare, Mark (ed.). 1994. *The Los Angeles Riots: Lessons for the Urban Future.* Boulder, CO: Westview Press.

Balter, Michael. 1996. Special Report: Chernobyl: Ten Years After." *Science* 272 (April 19) 352–360.

Baltzell, E. Digby. 1964. *The Protestant Establishment: Aristocracy and Caste in America.* New York: Random House.

Baltzell, E. Digby. 1979 [1958]. *Philadelphia Gentleman: The Making of a National Upper Class.* Philadelphia: University of Pennsylvania Press.

Bandura, A., and R. H. Walters. 1963. *Social Learning and Personality Development.* New York: Holt, Rinehart, and Winston.

Bane, Mary Jo, and David Ellwood. 1983. "Slipping Into and Out of Poverty: The Dynamics of Spells." Working Paper #1199 Cambridge, MA: National Bureau of Economic Research.

Banks, W. Curtis. 1976. "White Preference in Blacks: A Paradigm in Search of a Phenomenon." *Psychological Bulletin* 83: 1179–1186.

Barber, Clifton E., and B. Kay Paisley. 1995. "Family Care of Alzheimer's Patients: The Role of Gender and Generational Relationship on Caregiver Outcomes." *Journal of Applied Gerontology* 12 (June): 172–192.

Barberis, Mary. 1981. "America's Elderly: Policy Implications." *Population Bulletin* 35 (January): 2–13.

Bardacke, Frank. 1988. "Watsonville: A Mexican Community on Strike." Pp. 149–182 in *Reshaping the US Left: Popular Struggles in the 1980s,* edited by Mike Davis and Michael Sprinker. London: Verso.

Barnett, Bernice McNair. 1993. "Invisible Southern Black Women Leaders in the Civil Rights Movement: The Triple Constraints of Gender, Race, and Class." *Gender & Society* 7 (June): 162–182.

Barnett, Rosalind C., and Nancy L. Marshall. 1992. "Worker and Mother Roles, Spillover Effects, and Psychological Distress." *Women and Health* 18: 9–40.

Baron, James N., and Andrew E. Newman. 1990. "For What It's Worth: Organizations, Occupations, and the Value of Work." *American Sociological Review* 55 (April): 155–175.

Barone, Michael, and Grant Ujifusa with Richard E. Cohen. 1996. "The Top Spenders." *The Almanac of American Politics 1996.* Washington, DC: The National Journal.

Barringer, Leon. 1993. *The New York Times:* June 6, 1993.

Barrish, Gerald, and Michael R. Welch. 1980. "Student Religiosity and Discriminatory Attitudes Toward Women." *Sociological Analysis* 41 (Spring): 66–75.

Barthel, Diane. 1988. *Putting on Appearances: Gender and Advertising.* Philadelphia: Temple University Press.

Bartlett, Robin L., and Timothy I. Miller. 1988. "Executive Earnings by Gender: A Case Study." *Social Science Quarterly* 69 (December): 892–909.

Barton, Allen H. 1985. "Background, Attitudes, and Activities of American Whites." *Research in Politics and Society* 1: 173–218.

Basow, Susan A. 1992. *Gender: Stereotypes and Roles,* 3rd ed. Pacific Grove, CA: Brooks/Cole.

Bates, Eric. 1998. "Private Prisons." *The Nation* (January 5): 11–17.

Baxter, Janeen. 1994. "Is Husband's Class Enough: Class Location and Class Identity." *American Sociological Review* 59 (April): 220–235.

Bean, Frank D., and Marta Tienda. 1987. *The Hispanic Population of the United States.* New York: Russell Sage Foundation.

Beck, Scott H., Bettie S. Cole, and Judith A. Hammond. 1991. "Religious Heritage and Premarital Sex: Evidence from a National Sample of Young Adults." *Journal for the Scientific Study of Religion* 20 (June): 173–180.

Becker, Howard S. 1963. *Outsiders: Studies in the Sociology of Deviance.* New York: Free Press.

Becker, Howard S., Blanche Geer, Everett C. Hughes, and Anselm L. Strauss. 1961. *Boys in White: Student Culture in Medical School.* Chicago: University of Chicago Press.

Becraft, Carolyn H. 1992a. Pp. 8–17 in *Women in the Military,* edited by E. A. Blacksmith. New York: The H. W. Wilson Company.

Becraft, Carolyn H. 1992b. "Facts about Women in the Military, 1980–1990." Washington, DC: Women's Research and Education Institute.

Bedell, Kenneth B., and Alice M. Jones (eds.). 1992. *Yearbook of American and Canadian Churches.* Nashville: Abingdon Press.

Beggs, John J. 1995. "The Institutional Environment: Race and Gender Inequality in the U.S. Labor Market." *American Sociological Review* 60 (August): 612–633.

Bell, Daniel. 1973. *The Coming Crisis of Postindustrial Society.* New York: Basic Books.

Bellah, Robert (ed.). 1973. *Emile Durkheim On Morality and Society: Selected Writings.* Chicago: University of Chicago Press.

Bellah, Robert, Richard Madsen, William M. Sullivan, Ann Swidler, and Steven M. Tipton. 1985. *Habits of the Heart: Individualism and Commitment in American Life.* Berkeley: University of California Press.

Bellamy, Carol. 1998. *State of the World's Children.* New York: United Nation's Children Fund.

Belli, M. M. 1954. *Modern Trials.* Indianapolis: Bobbs-Merrill (Supplement, 1966).

Bem, S. 1972. "Psychology Looks at Sex Roles: Where Have all the Androgynous People Gone?" Paper presented at the UCLA Symposium on Women. Los Angeles, May.

Bendix, Reinhard. 1960. *Max Weber: An Intellectual Portrait.* Garden City, NY: Doubleday Anchor.

Benedict, Ruth. 1934. *Patterns of Culture.* Boston: Houghton Mifflin.

Bengtson, Vern L., and W. Andrew Achenbaum. 1993. *The Changing Contract Across Generations.* New York: Aldine de Gruyter.

Benson, Donna J., and Gregory Thomson. 1982. "Sexual Harassment on a University Campus: The Confluence of Authority Relations, Sexual Interest, and Gender Stratification." *Social Problems* 29 (February): 236–251.

Ben-Yehuda, Nachman. 1986. "The European Witchcraze of the Fourteenth–Seventeenth Centuries: A Sociologist's Perspective." *American Journal of Sociology* 86: 1–31.

Berberoglu, Berch. 1990. *Political Sociology: A Comparative/Historical Approach.* New York: General Hall.

Berger, Joseph, and Morris Zelditch (eds.), 1985. *Status, Rewards, and Influence.* San Francisco: Jossey-Bass.

Berger, Peter L. 1963. *Invitation to Sociology: A Humanistic Perspective.* Garden City, NY: Doubleday Anchor.

Berger, Peter L., and Thomas Luckmann. 1967. *The Social Construction of Reality: A Treatise in the Sociology of Knowledge.* Garden City, NY: Anchor Books.

Berger, Peter L., Brigitte Berger, and Hansfried Kellner. 1974. *The Homeless Mind: Modernization and Consciousness.* New York: Vintage Books.

Berheide, Catherine W. 1992. "Women Still 'Stuck' in Low-Level Jobs." *Women in Public Services: A Bulletin for the Center for Women in Government* 3 (Fall). Albany: Center for Women in Government, The State University of New York.

Berkowitz, L. 1974. "Some Determinants of Impulsive Aggression: The Role of Mediated Associations with Reinforcements for Aggression." *Psychological Review* 81 (March): 165–176.

Berkowitz, L. 1984. "Some Effects of Thoughts on Anti- and Pro-Social Influences of Media Events: A CognitiveNeoassociation Analysis." *Psychological Bulletin* 95 (May): 410–427.

Bernard, Jessie. 1972. *The Future of Marriage.* New York: Bantam.

Berscheid, Ellen, and Harry R. Reis. 1998. "Attraction and Class Relationships." Pp. 193–281 in *The Handbook of Social Psychology,* 4th ed., edited by Daniel T. Gilbert, Susan T. Fiske, and Gardner Lindzey. New York: Oxford University and McGraw Hill.

Bessey, Barbara L. et al. 1991. "The Age Discrimination in Employment Act." *Research on Aging* 13 (December): 413–484.

Beutel, Ann M., and Margaret Mooney Marini. 1995. "Gender and Values." *American Sociological Review* 60 (June): 436–448.

Bezilla, Robert. 1990. *Religion in America 1990: Approaching the Year 2000.* Princeton, NJ: Princeton Religious Research Center.

Bibliarz, Timothy J., and Adrian E. Raftery. 1993. *American Sociological Review* 58 (February): 97–109.

Bielby, William T. 1981."Models of Status Attainment." Pp. 3–26 in *Research in Social Stratification and Mobility,* edited by Donald Treiman and Robert Robinson. Greenwich, CT: JAI Press.

Binkin, Martin, and Mark J. Eitelberg. 1986. "Women and Minorities in the All-Volunteer Force." Pp. 73–102 in *The All-Volunteer Force after a Decade: Retrospect and Prospect,* edited by William Bowman, Roger Little, and G. Thomas Sicilia. Washington, DC: Pergamon-Brassey's.

Bird, Caroline. 1975. *The Invisible Scar.* New York: David McKay.

Birnbaum, Norman, and Gertrud Lenzer. 1969. "Introduction." Pp. 7–16 in *Sociology and Religion: A Book of Readings,* edited by Norman Birnbaum and Gertrud Lenzer. Englewood Cliffs, NJ: Prentice Hall.

Blackburn, McKinley L., and David E. Bloom. 1985. "What is Happening to the Middle Class?" *American Demographics* 7 (January): 19–25.

Blackwell, James E., and Morris Janowitz. 1974. *Black Sociologists: Historical and Contemporary Perspectives.* Chicago: University of Chicago Press.

Blackwell, James. 1991. *The Black Community: Diversity and Unity,* 3rd ed. New York: HarperCollins.

Blair, Simpson Lee, and Michael P. Johnson. 1992. "Wives' Perceptions of the Fairness of the Division of Household Labor: The Intersection of Housework and Ideology." *Journal of Marriage and the Family* 54 (August): 570–512.

Blake, C. Fred. 1994. "Footbinding in Neo-Confucian China and the Appropriation of Female Labor." *Signs* 19 (Spring): 676–712.

Blalock, Hubert M., Jr. 1982a. *Conceptualization and Measurement in the Social Sciences.* Beverly Hills, CA: Sage.

Blalock, Hubert M, Jr. 1982b. *Race and Ethnic Relations.* Englewood Cliffs, NJ: Prentice Hall.

Blalock, Hubert M., Jr. 1989. "Race Versus Class: Distinguishing Reality from Artifacts." *National Journal of Sociology* 3 (Fall): 127–142.

Blalock, Hubert M., Jr. 1991. *Understanding Social Inequality.* Newbury Park, CA: Sage Publications.

Blankenship, Kim. 1993. "Bringing Gender and Race In: U.S. Employment Discrimination Policy." *Gender & Society* 7 (June): 204–226.

Blaskovitch, J. 1973. "Blackjack and the Risky Shift." *Sociometry* 36 (March): 42ff.

Blassingame, John. 1973. *The Slave Community: Plantation Life in the Antebellum South.* New York: Oxford University Press.

Blau, Peter M. 1986. *Exchange and Power in Social Life,* revised ed. New Brunswick, NJ: Transaction Books.

Blau, Peter M., and Marshall W. Meyer. 1987. *Bureaucracy in Modern Society,* 3rd ed. New York: Random House.

Blau, Peter M., and Otis Dudley Duncan. 1967. *The American Occupational Structure.* New York: Wiley.

Blau, Peter M., and W. Richard Scott. 1974. *On the Nature of Organizations.* New York: Wiley.

Blauner, Robert. 1964. *Alienation and Freedom: The Factory Worker and His Industry.* Chicago: University of Chicago Press.

Blauner, Robert. 1969. "Internal Colonialism and Ghetto Revolt." *Social Problems* 16 (Spring): 393–408.

Blauner, Robert. 1972. *Racial Oppression in America.* New York: Harper and Row.

Blauner, Robert. 1989. *Black Lives, White Lives: Three Decades of Race Relations in America.* Berkeley: University of California Press.

Blee, Kathleen. 1991. *Women of the Klan: Racism and Gender in the 1920s.* Los Angeles: University of California Press.

Blee, Kathleen M., and Ann R. Tickamyer. 1995. "Racial Differences in Men's Attitudes about Women's Gender Roles." *Journal of Marriage and the Family* 57 (February): 21–30.

Blinde, Elaine M., and Diane E. Taub. 1992a. "Homophobia and Women's Sport: The Disempowerment of Athletes." *Sociological Focus* 25 (May): 151–166.

Blinde, Elaine M., and Diane E. Taub. 1992b. "Women Athletes as Falsely Accused Deviants: Managing the Lesbian Stigma." *Sociological Quarterly* 33 (Winter): 521–533.

Blinde, Elaine M., Diane E. Taub, and Lingling Han. 1993. "Sport Participation and Women's Personal Empowerment: Experiences of the College Athlete." *Journal of Sport and Social Issues* 17 (April): 47–60.

Blinde, Elaine M., Diane E. Taub, and Lingling Han. 1994. "Sport as a Site for Women's Group and Social Empowerment: Perspectives from the College Athlete." *Sociology of Sport Journal* 11 (March): 51–59.

Block, Alan. 1991. *Perspectives on Organizing Crime.* The Netherlands: Kluwer.

Block, Alan, and Frank R. Scarpitti. 1993. *Poisoning for Profit: The Mafia and Toxic Waste in America.* New York: William Morrow.

Block, Fred. 1987. *Revising State Theory.* Philadelphia: Temple University Press.

Blocker, T. Jean, and Douglas Lee Eckberg. 1997. "Gender and Environmentalism: Results from the 1993 General Social Survey." *Social Science Quarterly* 78 (December): 841–858.

Blood, Robert, and Donald Wolfe. 1960. *Husbands and Wives: The Dynamics of Married Living.* New York: Free Press.

Blum, Linda. 1991. *Between Feminism and Labor: The Significance of the Comparable Worth Movement.* Berkeley: University of California Press.

Blum, Nancy S. 1991. "The Management of Stigma by Alzheimer Family Caregivers." *Journal of Contemporary Ethnography* 20 (October): 263–284.

Blumer, Herbert, 1969. *Studies in Symbolic Interaction.* Englewood Cliffs, NJ: Prentice Hall.

Blumstein, Philip, and Pepper Schwartz. 1983. *American Couples: Money, Work, and Sex.* New York: Morrow.

Bobo, Lawrence, and J.R. Kluegel. 1991. "Modern American Prejudice: Stereotypes, Social Distance, and Perceptions of Discrimination Toward Blacks, Hispanics, and Asians." Paper presented before the American Sociological Association, Cincinnati, Ohio.

Bobo, Lawrence. 1989. "Worlds Apart: Blacks, Whites, and Explanations of Racial Equality." Paper presented before the Midwest Political Science Association, Chicago.

Bobo, Lawrence, and James R. Kluegel. 1993. "Opposition to Race-Targeting: Self-Interest, Stratification Ideology, or Racial Attitudes?" *American Sociological Review* 58 (August): 443–464.

Boer, J. Tom, Manuel Pastor, Jr., James L. Sadd, and Lori D. Snyder. 1997. "Is There Environmental Racism? The Demographics of Hazardous Waste in Los Angeles County." *Social Science Quarterly* 78 (December): 793–810.

Bonacich, Edna. 1972. "A Theory of Ethnic Antagonism: The Split Labor Market." *American Sociological Review* 37 (October): 547–559.

Bontemps, Arna (ed.) 1972. *The Harlem Renaissance Remembered.* New York: Dodd, Mead and Company.

Bord, Richard J., and Robert E. O'Connor. 1997. "The Gender Gap in Environmental Attitudes: The Case of Perceived Vulnerability to Risk." *Social Science Quarterly* 78 (December): 831–840.

Bose, Christine E., and Peter H. Rossi. 1983. "Gender and Jobs: Prestige Standings of Occupations as Affected by Gender." *American Sociological Review* 48 (June): 316–330.

Bottomore, T. B. (translator). 1964. *Karl Marx: Selected Writings and Social Philosophy.* New York: McGraw-Hill.

Boudon, Raymond. 1974. *Education, Opportunity, and Social Inequality.* New York: Wiley.

Bourdieu, Pierre. 1977. *Reproduction in Education, Society, Culture.* Beverly Hills, CA: Sage.

Bourdieu, Pierre. 1984. *Distinction: A Social Critique of the Judgement of Taste,* translated by Richard Nice. Cambridge, MA: Harvard University Press.

Bouvier, Leon, and Robert W. Gardner. 1986. "Immigration in the U.S.: The Unfinished Story." *Population Bulletin* 41 (November): 1–50.

Bowditch, Christine. 1993. "Getting Rid of Troublemakers: High School Disciplinary Procedures and the Production of Dropouts." *Social Problems* 40 (November): 493–510.

Bowen, William G. and Derek Bok. 1998. *The Shape of the River: Long-Term Consequences of Considering Race in College and University Admissions.* Princeton, NJ: Princeton University Press.

Bowles, Samuel, and Herbert Gintis. 1976. *Schooling in Capitalist America.* New York: Basic Books.

Boyd, Robert L. 1989. "Childlessness and Social Mobility During the Baby Boom." *Sociological Spectrum* 9 (Fall): 425–438.

Braddock, Jomills H. 1988. *National Education Longitudinal Study of 1988.* Washington, DC: U.S. Department of Education, National Center for Educational Statistics.

Bradley, Keith, and Stephen Hill. 1987. "Quality Circles and Managerial Interests." *Industrial Relations* 26 (Winter): 68–82.

Brake, Mike. 1985. *Comparative Youth Culture: The Sociology of Youth Subcultures in America, Britain, and Canada.* Boston: Routledge & Kegan Paul.

Bramson, Leon. 1961. *The Political Context of Sociology.* Princeton, NJ: Princeton University Press.

Branch, Taylor. 1988. *Parting the Waters: America in the King Years, 1954–1963.* New York: Simon and Schuster.

Braun, Denny. 1995. "Negative Consequences to the Rise of Income Inequality" Pp. 3–32 in *Research in Politics and Society: The Politics of Wealth and Inequality,* edited by Richard E. Ratcliff, Melvin L. Oliver, and Thomas M. Shapiro. Greenwich, CT: JAI Press.

Braverman, Harry. 1974. *Labor and Monopoly Capital.* New York: Monthly Review Press.

Bremer, Barbara A., Cathleen T. Moore, and Ellen F. Bildersee. 1991. "Do You Have to Call It 'Sexual Harassment' to Feel Harassed?" *College Student Journal* 25 (September): 256–268.

Brettell, Caroline B., and Carolyn F. Sargent. 1993. *Gender in Cross-Cultural Perspective.* Englewood Cliffs, NJ: Prentice Hall.

Brewer, Rose. 1988. "Black Women in Poverty: Some Comments on Female-Headed Families." *Signs* 13 (Winter): 331–333.

Bridges, George, and Robert Crutchfield. 1988. "Law, Social Standing and Racial Disparities in Imprisonment." *Social Forces* 66 (June): 699–724.

Brinkerhoff, Merlin B., and Marlene M. MacKie. 1984. "Religious Denominations' Impact Upon Gender Attitudes: Some Methodological Implications." *Review of Religious Research* 25 (June): 365–378.

Britt, Chester L. 1994. "Crime and Unemployment among Youths in the United States, 1958–1990: A Time Series Analysis." *American Journal of Economics and Sociology* 53 (January): 99–109.

Broder, David S. 1996. "A Senate of Millionaires." *The Washington Post* (January 17): A17.

Bromley, David G. 1979. *"Moonies" in America: Cult, Church, and Crusade.* Beverly Hills, CA: Sage Publications.

Brooke, James. 1998."Utah Struggles with a Revivial of Polygamy." *The New York Times* (August 23): A12.

Brown, Cynthia. 1993. "The Vanished Native Americans." *The Nation* 257 (October 11): 384–389.

Brown, Diane R., and Lawrence E. Gary. 1991. "Religious Socialization and Educational Attainment among African Americans: An

Empirical Assessment." *Journal of Negro Education* 60 (Summer): 411–426.

Brown, Elaine. 1992. *A Taste of Power: A Black Women's Story*. Pantheon: New York.

Brown, J. Larry. 1987. "Hunger in the United States." *Scientific American* 256 (February 1987): 36–41.

Brown, Roger. 1986. *Social Psychology*, 2nd ed. New York: The Free Press.

Browne, Angela, and Kirk R. Williams. 1993. "Gender, Intimacy, and Lethal Violence: Trends from 1976 through 1987." *Gender & Society* 7 (March): 78–98.

Brownmiller, Susan. 1975. *Against Our Will: Men, Women, and Rape*. New York: Simon and Schuster.

Brush, Lisa D. 1990. "Violent Acts and Injurious Outcomes in Married Couples: Methodological Issues in the National Survey of Families and Households." *Gender & Society* 4 (March): 56–67.

Buchanan, Christy M., Eleanor Maccoby, and Sanford M. Dornsbusch. 1996. *Adolescents after Divorce*. Cambridge, MA: Harvard University Press.

Bullard, Robert D. 1994a. *Unequal Protection: Environmental Justice and Communities of Color*. San Francisco: Sierra Club Books.

Bullard, Robert. 1994b. *Dumping in Dixie: Race, Class, and Environmental Quality*. Boulder, CO: Westview.

Bullough, Vern L. 1993. *Cross Dressing, Sex, and Gender*. 1993. Philadelphia: Temple University Press.

Bumpass, Larry, and J. A. Sweet. 1989. "Children's Experience in Single-Parent Families: Implications of Co-habitation and Marital Transitions." *Family Planning Perspectives* 6: 256–260.

Bunker, B. B., Forcey, B. Wilderom, C. P. M., and Elgie, D. M. 1984. "The Competitive Behaviors of Men and Women: Is There a Difference?" Paper presented at American Psychological Association, Toronto.

Burr, Jeffrey A., Patricia L. McCall, and Eve Powell-Griner. 1994. "Catholic Religion and Suicide: The Mediating Effects of Divorce." *Social Science Quarterly* 75 (June): 300–318.

Burt, Cyril. 1966. "The Genetic Determination of Differences in Intelligence: A Study of Monozygotic Twins Reared Together and Apart." *British Journal of Psychology* 57: 137–153.

Burt, Martha, and Barbara Cohen. 1989. *America's Homeless: Numbers, Characteristics, and Programs that Serve Them*. Washington, DC: The Urban Institute.

Butler, Amy C. 1996. "The Effect of Welfare Benefit Levels on Poverty Among Single-Parent Families." *Social Problems* 43 (February): 94–115.

Butsch, Richard. 1992. "Class and Gender in Four Decades of Television Situation Comedy: Plus ça Change . . ." *Critical Studies in Mass Communication* 9: 387–399.

Buunk, Bram, and Ralph B. Hupka. 1987. "Cross-Cultural Differences in the Elicitation of Sexual Jealousy." *Journal of Sex Research* 23 (February): 12–22.

Byrne, John A. 1991. "The Flap Over Executive Pay." *Business Week* (May 6): 90–96.

Byrne, John A. 1996. "Gross Compensation?" *Business Week* (March 18): 32–33.

C

Cable, Sherry, and Charles Cable. 1995. *Environmental Problems, Grassroots Solutions: The Politics of Grassroots Environmental Conflict*. New York: St. Martin's Press.

Cagin, Seth, and Philip Dray. 1988. *We Are Not Afraid*. New York: Bantam Books.

Calasanti, Toni M. 1992. "Working 'Over-Time': Economic Restructuring and Retirement of a Class." *Sociological Quarterly* 33 (Spring): 135–152.

Calavita, Kitty, Henry N. Pontell, and Robert Tillman. 1997. *Big Money Crime: Fraud and Politics in the Savings and Loan Crisis*. Berkeley: University of California Press.

Calhoun, Craig. 1994. *Social Theory and the Politics of Identity*. Cambridge, MA: Blackwell Publishers.

Campbell, Anne. 1987. "Self Definition by Rejection: The Case of Gang Girls." *Social Problems* 34 (December): 451–466.

Campbell, S. 1972. "The Multiple Function of the Criminal Defense *voir dire* in Texas." *American Journal of Criminal Law* 1: 255–282.

Cancio, A. Silvia, T. David Evans, and David J. Maume, Jr. 1996. "Reconsidering the Declining Significance of Race: Racial Differences in Early Career Wages." *American Sociological Review* 61 (August): 541–556.

Canetto, Silvia Sara. 1992. "Gender and Suicide in the Elderly." *Suicide and Life-Threatening Behavior* 22 (Spring): 80–97.

Caplow, Theodore. 1968. *Two Against One: Coalitions in Triads*. Englewood Cliffs, NJ: Prentice Hall.

Caplow, Theodore. 1991. *American Social Trends*. New York: Harcourt Brace Jovanovich.

Cardenas, Jose A. 1996. "Ending the Crisis in the K-12 System." Pp. 51–70 in *Educating a New Majority: Transforming America's Educational System for Diversity*, edited by Laura I. Rendon and Richard O. Hope. San Francisco: Jossey-Bass Publishers.

Carmichael, Stokely, and Charles V. Hamilton. 1967. *Black Power: The Politics of Liberation in America*. New York: Vintage.

Carnoy, Martin, Manuel Castells, Stephen S. Cohen, and Fernando Henrique Cardoso. 1993. *The New Global Economy in the Information Age*. University Park, PA: The Pennsylvania State University Press.

Carroll, J. B. 1956. *Language, Thought and Reality: Selected Writings of Benjamin Lee Whorf*. Cambridge: MIT Press.

Carson, Clayborne, David J. Garrow, Vincent Harding, and Darlene Clark Hine, (eds.). 1987. *Eyes on the Prize: America's Civil Rights Years*. New York: Penguin.

Carson, Clayborne. 1981. *In Struggle: SNCC and the Black Awakening of the 1950's*. Cambridge, MA: Harvard University Press.

Carson, David K. 1995. "American Indian Elder Abuse: Risk and Protective Factors Among the Oldest Americans." *Journal of Elder Abuse and Neglect* 7: 17–39.

Carson, Rachel. 1962. *The Silent Spring*. New York: Knopf.

Carter, Deborah J., and Reginald Wilson. 1996. *Minorities in Higher Education*. Washington, DC: American Council on Education, Office of Minorities in Higher Education, June.

Carter, Gregg Lee. 1992. "Hispanic Rioting During the Civil Rights Era." *Sociological Forum* 7 (2): 301–323.

Cartwright, Glen F., 1994. "Virtual or Real? The Mind in Cyberspace." *Futurist* 28 (March-April): 22–26.

Casalego, Federico. 1996. "Cyberspace: A New Territory for Interaction in a Magic Time." *Societies* 51 (February): 39–48.

Cashmore, Ellis. 1991."Black Cops Inc." Pp. 87–108 in *Out of Order: Policing Black People*, edited by Ellis Cashmore and Eugene McLaughlin. New York: Routledge.

Casper, Lynne M., Sara McLanahan, and Irwin Garfinkel. 1994. "The Gender-Poverty Gap: What We Can Learn From Other Countries." *American Sociological Review* 59 (August): 594–605.

Cassen, Robert. 1994. "Population and Development: Old Debates, New Conclusions." *U.S.-Third World Policy Perspectives* 19: 282.

Cassidy, Linda, and Rose Marie Hurrell. 1995. "The Influence of Victim's Attire on Adolescents' Judgments of Date Rape." *Adolescence* 30 (Summer): 319–323.

Cassidy, Margaret L., and Bruce O. Warren. 1991. "Status Consistency and Work Satisfaction Among Professional and Managerial Women and Men." *Gender & Society* 5 (June): 193–206.

Cassidy, Margaret L., and Bruce O. Warren. 1997. "Family Employment Status and Gender Role Attitudes: A Comparison of Women and Men College Graduates." *Gender & Society* 10 (June): 312–329.

Catania, Joseph A., et al. 1992. "Prevalence of AIDS-Related Risk Factors and Condom Use in the United States." *Science* 258 (November): 1101–1106.

Catanzarite, Lisa, and Vilma Ortiz. 1996. "Family Matters, Work Matters? Poverty Among Women of Color and White Women." Pp. 121–140 in *For Crying Out Loud: Women's Poverty in the United States*, edited by Diane Dujon and Ann Withorn. Boston: South End Press.

Caulfield, Mina Davis. 1985. "Sexuality in Human Evolution: What is 'Natural' in Sex?" *Feminist Studies* 11 (Summer): 343–364.

Cavanaugh, John C. 1997. *Adult Development and Aging*, 3rd ed. Pacific Grove, CA: Brooks/Cole.

Cazenave, Noel A. 1983. "'A Woman's Place': The Attitudes of Middle-Class Black Men." *Phylon* 44 (March): 12–32.

Celis, William III. 1991. "Scrutiny of Police Sought in Milwaukee." *The New York Times* (July 28): 14.

Center for Disease Control. 1996. "HIV/AIDS Surveillance Report, U.S. HIV and AIDS Cases Reported Through December 1995." Atlanta, GA: Center for Disease Control.

Centers for Disease Control and Prevention. 1995. *HIV/AIDS Surveillance Report,* 7(no.2).

Centers, Richard. 1949. *The Psychology of Social Classes.* Princeton, NJ: Princeton University Press.

Chafetz, Janet. 1984. *Sex and Advantage.* Totowa, NJ: Rowman and Allanheld.

Chalfant, H. Paul, Robert E. Beckley, and C. Eddie Palmer. 1987. *Religion in Contemporary Society,* 2nd ed. Palo Alto, CA: Mayfield Publishing Co.

Chambliss, William J. 1988. *Exploring Criminology.* New York: Macmillan.

Chambliss, William J., and Howard F. Taylor. 1989. *Survey of Perceptions of Bias in the New Jersey Courts.* Research Report. New Jersey: New Jersey Supreme Court Task Force on Minority Concerns.

Chambliss, William J., and Howard F. Taylor. 1993. *Bias in the New Jersey Courts.* Trenton, NJ: Administrative Office of the Courts.

Chang, Jung. 1991. *Wild Swans: Three Daughters of China.* New York: Simon and Schuster.

Chapkis, Wendy. 1986. *Beauty Secrets: Women and the Politics of Appearance.* Boston: South End Press.

"Characteristics of Congress." 1995. *Congressional Quarterly Weekly Report* 53 (February 18): 541–553.

Charney, Dara A., and Ruth C. Russell. 1994. "An Overview of Sexual Harassment." *American Journal of Psychiatry* 151 (January): 10–17.

Cheal, David J. 1988. "Relationships in Time: Ritual, Social Structure, and the Life Course." *Studies in Symbolic Interaction* 9: 83–109.

Chen, Elsa, Y. F. 1991. "Conflict Between Korean Greengrocers and Black Americans." Unpublished senior thesis, Princeton University, Princeton, NJ.

Cherlin, Andrew J. 1981. *Marriage, Divorce, and Remarriage.* Cambridge, MA: Harvard University Press.

Cherlin, Andrew J., Frank F. Furstenberg, P. Lindsay Lansdale-Chase, Kathleen E. Kiernan, Philip K. Robins, Donna-Ruane Morrison, and Julien O. Teitler. 1991. "Longitudinal Studies of Effects of Divorce on Children in Great Britain and the United States." *Science* 252 (June 7): 1386–1389.

Cherlin, Andrew J., P. Lindsay Chase-Lansdale, and Christine McRae. 1998. "Effects of Parental Divorce on Mental Health throughout the Life Course." *American Sociological Review* 63 (April): 219–249.

Chernin, Kim. 1985. *The Hungry Self: Women, Eating, and Identity.* New York: Times Books.

Chernin, Kim. 1991. *The Obsession.* New York: Harper Colophon.

Chesler, Phyllis. 1972. *Women and Madness.* New York: Doubleday.

Chesney-Lind, Meda. 1992. "Women's Prisons: Putting the Brakes on the Building Binge." *Corrections Today* 54 (August): 30–34.

Chia, Rosina C., Jamie L. Moore, Ka Nei Lam, C. J. Chuang, and B. S. Cheng. 1994. "Cultural Differences in Gender Role Attitudes Between Chinese and American Students." *Sex Roles* 31 (July): 23–30.

Children's Defense Fund. 1997. *The State of America's Children Yearbook, 1997.* Washington, DC: Children's Defense Fund.

Chin, Jean Lau. 1993. *Diversity in Psychotherapy: The Politics of Race, Ethnicity, and Gender.* Westport, CT: Praeger Publishers.

Chipman, Susan. 1991. "Word Problems: Where Bias Creeps In." Unpublished Research Report, Office of Naval Research, Arlington, VA.

Chipuer, H. M., M. J. Rovine, and R. Plomin. 1990. "LISREL Modeling: Geometric and Environmental Influences on IQ Revisited." *Intelligence* 14: 11–29.

Chiricos, Ted, Sarah Eschholz, and Marc Gertz. 1997. "Crime, News and Fear of Crime: Toward an Identification of Audience Effects." *Social Problems* 44 (August): 342–357.

Cho, Sumi. 1993. "Korean Americans vs. African Americans: Conflict and Construction." Pp. 196–211 in *Reading Rodney King/Reading Urban Uprising,* edited by Robert Gooding-Williams. New York: Routledge.

Chodorow, Nancy. 1978. *The Reproduction of Mothering: Psychoanalysis and the Study of Gender.* Berkeley: University of California Press.

Chodorow, Nancy. 1994. *Femininities, Masculinities, Sexualities: Freud and Beyond.* Lexington, KY: University of Kentucky Press.

Chow, Esther Ngan-Ling, Doris Wilkinson, and Maxine Baca Zinn, eds. 1996. *Race, Class, & Gender: Common Bonds, Different Voices.* Thousand Oaks, CA: Sage.

Church, George. 1990. "The View from Behind Bars." *Time* 136 (Fall): 20–22.

Churchill, Ward. 1992. *Struggle for the Land.* Toronto: Between the Lines.

Churchill, Ward. 1993. "Crimes Against Humanity." *Z Magazine* 6 (March): 43–47.

Cicourel, Aaron V. 1968. *The Social Organization of Juvenile Justice.* New York: John Wiley and Sons.

Cimmoms, Marlene. 1992. "New Study Boosts Forecast of AIDS Infection." *Los AngelesTimes* (June 4): A4–1.

Clark, J. Michael, Joanne Carlson Brown, and Lorna M. Hochstein. 1989. "Institutional Religion and Gay/Lesbian Oppression." *Marriage and Family Review* 14: 265–284.

Clark, Kenneth B., and Mamie P. Clark. 1939. "The Development of Consciousness of Self and the Emergence of Racial Identification in Negro Preschool Children." *The Journal of Social Psychology* 10: 591–599.

Clark, Kenneth B., and Mamie P. Clark. 1947. "Racial Identification and Preference in Negro Children." Pp. 602–611 in *Readings in Social Psychology,* edited by T. M. Newcomb and E. L. Hartley. New York: Holt.

Clark, Roger, Rachel Lennon, and Leanna Morris. 1993. "Of Caldecotts and Kings: Gendered Images in Recent American Children's Books by Black and Non-Black Illustrators." *Gender & Society* 7 (June): 227–280.

Clifford, N. M., and Elaine Walster. 1973. "Research Note: The Effects of Physical Attractiveness on Teacher Expectations." *Sociology of Education* 46: 248–258.

Coale, A. 1986. "Population Trends and Economic Development." Pp. 96–104 in *World Population and the U.S. Population Policy: The Choice Ahead,* edited by J. Menken. New York: W. W. Norton and Co.

Cochran, John K., and Leonard Beeghley. 1991. "The Influence of Religion on Attitudes toward Nonmarital Sexuality: A Preliminary Assessment of Reference Group Therapy." *Journal for the Scientific Study of Religion* 30 (March): 45–62.

Cohen, Bernard P., and Xueguang Zhou. 1991. "Status Processes in Enduring Work Groups." *American Sociological Review* 56 (April): 179–188.

Cohen, Eugene N., and Edwin Eames. 1985. *Cultural Anthropology.* Boston: Little, Brown.

Cohen, Robyn. 1990. *Prisoners in 1990.* Washington, DC: Bureau of Justice Statistics, U.S. Department of Justice.

Cole, Johnnetta B. 1988. *Anthropology for the Nineties.* New York: Free Press.

Collins, Patricia Hill, Lionel A. Maldonado, Dana Y. Takagi, Barrie Thorne, Lynn Weber, and Howard Winant. 1995. "On West and Fenstermaker's 'Doing Difference.'" *Gender & Society* 9 (August): 491–505.

Collins, Patricia Hill. 1987. "The Meaning of Motherhood in Black Culture and Black Mother-Daughter Relationships." *Signs* 4 (Fall): 3–10.

Collins, Patricia Hill. 1990. *Black Feminist Theory: Knowledge, Consciousness and the Politics of Empowerment.* Cambridge, MA: Unwin Hyman.

Collins, Patricia Hill. 1997. *Fighting Words.* Minneapolis: University of Minnesota Press.

Collins, Randall, and Michael Makowsky. 1972. *The Discovery of Society.* New York: Random House.

Collins, Randall. 1979. *The Credential Society.* New York: Academic Press.

Collins, Randall. 1988. *Theoretical Sociology.* San Diego: Harcourt Brace Jovanovich.

Collins, Randall. 1994. *Four Sociological Traditions.* New York: Oxford.

Collins, Sharon M. 1983. "The Making of the Black Middle Class." *Social Problems* 30 (April): 369–382.

Collins, Sharon M. 1989. "The Marginalization of Black Executives." *Social Problems* 36: 317–331.

Collins-Lowry, Sharon M. 1997. *Black Corporate Executives: The Making and Breaking of a Black Middle Class.* Philadelphia: Temple University Press.

Congdon, David C., 1990. "Gender, Employment, and Psychosocial Well-Being." *Journal of Sociology and Social Welfare* 17 (September): 101–121.

Conger, John Janeway. 1988. "Hostages to Fortune: Youth, Values, and the Public Interest." *American Psychologist* 43 (April): 291–300.

Connell, R. W. 1992. "A Very Straight Gay: Masculinity, Homosexual Experience, and the Dynamics of Gender." *American Sociological Review* 57 (December): 735–751.

Connelly, M. 1989. "Truck Driver Honored for Heroism in Fire." *Los Angeles Times* (September 15): 8, 12.

Conrad, P. 1975. "The Discovery of Hyperkinesis: Notes on the Medicalization of Deviant Behavior." *Social Problems* 23: 12–21.

Conrad, Peter, and Joseph W. Schneider. 1992. *Deviance and Medicalization: From Badness to Sickness,* expanded edition. Philadelphia: Temple University Press.

Constable, Nicole. 1997. *Maid to Order in Hong Kong: Stories of Filipina Workers.* Ithaca, NY: Cornell University Press.

Cook, Karen S., Karen A. Hegtvedt, and Toshio Yamagishi. 1988. "Structural Inequality, Legitimation, and Reactions to Inequality in Exchange Networks." Pp. 291–308 in *Status Generalization: New Theory and Research,* edited by M. Webster, Jr., and M. Foschi. Stanford, CA: Stanford University Press.

Cook, S. W. 1988. "The 1954 Social Science Statement and School Desegregation: A Reply to Gerard." Pp. 237–256 in *Eliminating Racism: Profiles in Controversy,* edited by D. A. Taylor. New York: Plenum.

Cooke, Miriam, and Angela Woollacott (eds.). 1993. *Gendering War Talk.* Princeton, NJ: Princeton University Press.

Cooke, Timothy W., and Aline Q. Quester. 1992. "What Characterizes Successful Enlistees in the All-Volunteer Force: A Study of Male Recruits in the U.S. Navy." *Social Science Quarterly* 73 (June): 238–252.

Cookson, Peter W., Jr., and Caroline Hodges Persell. 1985. *Preparing for Power: America's Elite Boarding Schools.* New York: Basic Books.

Cooley, Charles Horton. 1902. *Human Nature and Social Order.* New York: Scribner's.

Cooley, Charles Horton. 1967 [1909]. *Social Organization.* New York: Schocken Books.

Coontz, Stephanie. 1992. *The Way We Never Were.* New York: Basic Books.

Coontz, Stephanie. 1997. *The Way We Really Are.* New York: Basic Books.

Cooper, Marc. 1997. "The Heartland's Raw Deal: How Meatpacking is Creating a New Immigrant Underclass." *The Nation* (February 3): 11–17.

Corr, Charles A., Clyde M. Nabe, and Donna M. Corr. 1994. *Death and Dying, Life and Living.* Belmont, CA: Brooks/Cole.

Corsaro, William A., and Donna Eder. 1990. "Children's Peer Cultures." *Annual Review of Sociology* 16: 197–220.

Coser, Lewis. 1977. *Masters of Sociological Thought.* New York: Harcourt Brace Jovanovich, Inc.

Cosse, Pamela. 1995. "The Effects of Social Capital as Time and Money On Post-Divorce Father-Daughter Relationships." Unpublished senior thesis, Princeton University, Princeton, NJ.

Costa, Paul T., Jr., et al. 1987. "Longitudinal Analyses of Psychological Well-Being in A National Sample: Stability of Mean Levels." *Journal of Gerontology* 42: 50–55.

Cox, Carole, and Abraham Monk. 1993. "Hispanic Culture and Family Care of Alzheimer's Patients." *Health and Social Work* 18 (May): 92–100.

Craig, R. Stephen. 1992. "Women as Home Caregivers: Gender Portrayal in OTC Drug Commercials." *Journal of Drug Education* 22: 303–312.

Crane, Diana (ed.). 1994. *The Sociology of Culture: Emerging Theoretical Perspectives.* Cambridge: Blackwell Publishers.

Crawford, Isiaah, and Elizabeth Solliday. 1996. "The Attitudes of Undergraduate College Students toward Gay Parenting." *Journal of Homosexuality* 30: 63–77.

Crawford, James (ed.). 1992. *Language Loyalties: A Source Book on the Official English Controversy.* Chicago: University of Chicago Press.

Creighton-Zollar, Ann, and Julie A. Honnold. 1991. "Marital Status and Life Satisfaction in Black Middletown." *Research in Race and Ethnic Relations* 6: 49–62.

Croll, Elisabeth. 1995. *Changing Identities of Chinese Women.* London: Zed Books.

Crouse, James, and Dale Trusheim. 1988. *The Case Against the SAT.* Chicago: University of Chicago Press.

Cummings, Scott, and Thomas Lambert. 1997. "Anti-Hispanic and Anti-Asian Sentiments among African Americans." *Social Science Quarterly* 78 (June): 338–353.

Cunningham, Peter J., and Llewellyn J. Cornelius. 1995. "Access to Ambulatory Care for American Indians and Alaska Natives: The Relative Importance of Personal and Community Resources." *Social Science and Medicine* 40 (February): 393–407.

Curtiss, Susan. 1977. *Genie: A Psycholinguistic Study of a Modern-Day "Wild Child."* New York: Academic Press.

Cushman, Linda F., et al. 1988. "Beliefs About Contraceptive Sterilization Among Low-Income Urban Women." *Family Planning Perspectives* 20 (September/October): 218–233.

Cuzzort, Ray P., and Edith W. King. 1980. *20th Century Social Thought,* 3rd ed. New York: Holt, Rinehart, and Winston.

D

Dahlhamer, James, and Joanne M. Nigg. 1993. "An Empirical Investigation of Rumoring: Anticipating Disaster under Conditions of Uncertainty." Paper presented at the Annual Meetings of the Southern Sociological Society, Chattanooga, TN, April.

Dahrendorf, Rolf. 1959. *Class and Class Conflict in Industrial Society.* Stanford, CA: Stanford University Press.

Daly, Kathleen. 1994. *Gender, Crime, and Punishment.* New Haven, CT: Yale University Press.

Daniels, Jessie. 1997. *White Lies: Race, Class, Gender, and Sexuality in White Supremacist Discourse.* New York: Routledge.

D'Antonio, William V., James D. Davidson, Dean R. Hoge, and Ruth A. Wallace. 1989. *American Catholic Laity in a Changing Church.* New York: Sheed and Ward.

Dalaker, Joseph, and Mary Naifeh. 1998. U.S. Bureau of the Census, Current Population Reports, Series P60–201. *Poverty in the United States: 1997.* Washington, DC: U.S. Department of Commerce. World Wide Web site: **http:// www.census.gov**

Darcy, R., Susan Welch, and Janet Clark. 1994. *Women, Elections, and Representation.* Lincoln, NE: University of Nebraska Press.

Darley, John M., and Bibb Latané. 1968. "Bystander Intervention in Emergencies: Diffusion of Responsibility." *Journal of Personality and Social Psychology* 8: 377–383.

Darley, John M., and R. H. Fazio. 1980. "Expectancy Confirmation Processes Arising in the Social Interaction Sequence." *American Psychologist* 35: 867–881.

Darnell, Alfred, and Darren E. Sherkat. 1997. "The Impact of Protestant Fundamentalism on Educational Attainment." *American Sociological Review* 62 (April): 306–315.

Das Gupta, Monisha. 1997. "'A Chambered Nautilus': The Contradictory Nature of Puerto Rican Women's Role in the Construction of a Transnational Community." *Gender & Society* 11 (October): 627–655.

Davies-Netzley, Sally Ann. 1998. "Women Above the Glass Ceiling: Perceptions on Corporate Mobility and Strategies for Success." *Gender & Society* 12 (June): 339–355.

Davis, Angela. 1981. *Women, Race, and Class.* New York: Random House.

Davis, Donald M. 1990. "Portrayals of Women in Prime-Time Network Television: Some Demographic Characteristics." *Sex Roles* 23: 325–332.

Davis, James A., and Tom Smith. 1984. *General Social Survey Cumulative File, 1972–1982.* Ann Arbor, MI: Inter-University Consortium for Political and Social Research.

Davis, Kingsley, and Wilbert E. Moore. 1945. "Some Principles of Stratification." *American Sociological Review* 10 (April): 242–247.

Davis, Kingsley. 1945. "The World Demographic Transition." *Annals of the American Academy of Political and Social Sciences* 237: 1–11.

Davis, Nancy J., and Robert V. Robinson. 1988. "Class Identification of Men and Women in the 1970's and 1980's." *American Sociological Review* 53 (February): 103–112.

de Beauvoir, Simone. 1972. *The Coming of Age.* New York: Putnam.

De Paulo, B. M. 1992. "Non-Verbal Behavior and Self Presentation." *Psychological Bulletin* 111: 203–243.

De Vita, Carol J. 1996. "The United States at Mid-Decade." *Population Bulletin* 50 (March). Washington, DC: Population Reference Bureau.

Deegan, Mary Jo. 1988. "W.E.B. Du Bois and the Women of Hull-House, 1895–1899." *The American Sociologist* 19 (Winter): 301–311.

Deegan, Mary Jo. 1990. *Jane Addams and the Men of the Chicago School, 1892–1918.* New Brunswick, NJ: Transaction Books.

Deitch, Cynthia. 1993. "Gender, Race, and Class Politics, and the Inclusion of Women in Title VII of the 1964 Civil Rights Act." *Gender & Society* 7 (June): 183–203.

del Pinal, Jorge, and Audrey Singer. 1997. *Generations of Diversity: Latinos in the United States.* Washington, DC: Population Reference Bureau (October).

Dellinger, Kristen, and Christine L. Williams. 1997. "Makeup at Work: Negotiating Appearance Rules in the Workplace." *Gender & Society* 11 (April): 151–177.

Demeny, Paul. 1991. "Tradeoffs between Human Numbers and Material Standards of Living." Pp. 408–421 in *Resources, Environment, and Population: Present Knowledge, Future Options,* edited by Kingsley Davis and Michail S. Bernstam. New York: Population Council.

D'Emilio, John. 1983. *Sexual Politics, Sexual Communities: The Making of a Homosexual Minority in the United States, 1940–1970.* Chicago: University of Chicago Press.

D'Emilio, John, and Estelle B. Freedman. 1988. *Intimate Matters: A History of Sexuality in America.* New York: Harper and Row.

Demo, David H., and Alan C. Acock. 1988. "The Impact of Divorce on Children." *Journal of Marriage and the Family* 50 (August): 619–648.

Demos, Vasilikie, and Marcia Texler Segal. 1994. *Ethnic Women: A Multiple Status Reality.* Dix Hills, NY: General Hall.

DeMott, Benjamin. 1990. *The Imperial Middle: Why Americans Can't Think Straight About Class.* New York: Morrow.

DeVault, Marjorie. 1991. *Feeding the Family: The Social Organization of Caring as Gendered Work.* Chicago: University of Chicago Press.

DeWitt, Karen. 1995. "Blacks Prone to Job Dismissal in Organizations." *The New York Times* (April 20): A19.

Diamond, Milton, and H. K. Sigmundson. 1997. "Sex Reassignment at Birth: Long-term Review and Clinical Implications." *Archives of Pediatric and Adolescent Medicine* 151 (March): 298–304.

Diaz-Duque, Ozzie F. 1989. "Communication Barriers in Medical Settings: Hispanics in the United States." *International Journal of the Sociology of Language* 79: 93–102.

Dill, Bonnie Thornton, Lynn Weber Cannon, and Reeve Vanneman. 1987. *Pay Equity: An Issue of Race, Ethnicity and Sex.* Washington, DC: National Committee on Pay Equity.

Dill, Bonnie Thornton. 1988. "Our Mothers' Grief: Racial Ethnic Women and the Maintenance of Families." *Journal of Family History* 13 (October): 415–431.

Dillard, J. L. 1972. *Black English: Its Historical Usage in the United States.* New York: Random House.

Dillingham, Gerald L. 1981. "The Emerging Black Middle Class: Class Conscious or Race Conscious?" *Ethnic and Racial Studies* 4 (October): 432–451.

DiAcosta, Diego. 1998. *A Sociolinguistic Study of Two Baptist Churches.* Princeton, NJ: Senior Thesis, Princeton University, Department of Sociology.

DiMaggio, Paul. 1977. "Market Structure, the Creative Process and Popular Culture: Toward an Organizational Reinterpretation of Mass-Culture Theory." *Journal of Popular Culture* 111: 436–452.

DiMaggio, Paul. 1982. "Cultural Capital and School Success: The Impact of Status Culture Participation on the Grades of U.S. High School Students." *American Sociological Review* 47: 189–201.

DiMaggio, Paul, and Francie Ostrower. 1990. "Participation in the Arts by Black and White Americans." *Social Forces* 68 (March): 753–778.

DiMaggio, Paul J., and Walter W. Powell. 1991. "Introduction." Pp. 1–38 in *The New Institutionalism in Organizational Analysis,* edited by W. W. Powell and P. J. DiMaggio. Chicago: University of Chicago Press.

Diner, Hasia. 1996. "Erin's Children in America: Three Centuries of Irish Immigration to the United States." Pp. 161–171 in *Origins and Destinies: Immigration, Race, and Ethnicity in America,* edited by Silvia Pedraza and Rubén Rumbaut. Belmont, CA: Wadsworth.

Dines, Gail, and Jean M. Humez, eds. 1995. *Gender, Race and Class in Media: A Text-Reader.* Thousand Oaks, CA: Sage Publications.

Dion, K. K. 1972. "Physical Attractiveness and Evaluating Children's Transgressions." *Journal of Personality and Social Psychology* 24: 285–290.

Dobash, R. Emerson, and Russell Dobash. 1979. *Violence Against Wives.* New York: Free Press.

Dollard, John, Neal E. Miller, Leonard W. Doob, O. H. Mowrer, and Robert R. Sears. 1939. *Frustration and Aggression.* New Haven, CT: Yale University Press.

Dolphin, Ric. 1988. "Rape on Campus." *MacLeans* 101 (October): 56.

Domhoff, G. William. 1967. *Who Rules America?* Englewood Cliffs, NJ: Prentice-Hall.

Domhoff, G. William. 1970. *The Higher Circles: The Governing Class in America.* New York: Random House.

Domhoff, G. William. 1990. *The Power Elite and the State: How Policy is Made in America.* New York: Aldine De Gruyter.

Domhoff, G. William. 1998. *Who Rules America: Power and Politics in the Year 2000,* 3rd ed. Mountain View, CA: Mayfield.

Donato, Katharine. 1992. "Understanding U.S. Immigration: Why Some Countries Send Women and Others Send Men." Pp. 159–184 in *Seeking Common Ground: Multidisciplinary Studies of Immigrant Women in the United States,* edited by Donna Gabaccia. Westport, CT: Praeger.

Donnerstein, E., and L. Berkowitz. 1981. "Victim Reactions in Aggressive Erotic Films as a Factor in Violence Against Women." *Journal of Personality and Social Psychology* 41 (January): 710–724.

Donnerstein, Edward, Daniel Linz, and Steven Penrod. 1987. *The Question of Pornography: Research Findings and Policy Implications.* New York: Free Press.

Donovan, Beth. 1993. "A New-Look Congress Begins with a Shadow of the Old." *Congressional Quarterly Weekly Report* 51 (2): 57.

Douglass, Jack D. 1967. *The Social Meanings of Suicide.* Princeton: Princeton University Press.

Dovidio, J. F. 1984. "Helping Behavior and Altruism: An Empirical and Conceptual Overview." Pp. 361–427 in *Advances in Experimental Social Psychology,* Vol. 17, edited by L. Berkowitz. New York: Academic Press.

Dovidio, J. F., and S. L. Gaertner, eds. 1986. *Prejudice, Discrimination, and Racism.* New York: Academic Press.

Downey, A. M. 1984. "The Relationship of Sex-Role Orientation to Self-Perceived Health Status in Middle-Aged Males." *Sex Roles* 11 (August): 211–223.

Doyal, I. 1990. "Hazards of Hearth and Home." *Women's Studies International Forum* 13: 501–517.

Draper, Roger. 1985. "The Golden Arm." *New York Review of Books* 32 (October 24): 46ff.

Draper, Roger. 1986. "The History of Advertising in America." *New York Review of Books* 33 (June 26): 14–18.

Driggs, Ken. 1990. "After The Manifesto: Modern Polygamy and Fundamentalist Mormons." *Journal of Church and State* 32 (Spring): 367–389.

Du Bois, W. E. B. 1901. "The Freedmen's Bureau." *Atlantic Monthly* 86: 354–365.

Du Bois, W. E. B. 1968. *The Autobiography of W.E.B. Du Bois.* New York: International Publishers.

Due, Linnea. 1995. *Joining the Tribe: Growing Up Gay & Lesbian in the '90s.* New York: Doubleday.

Dugger, Celia W. 1996. "Board Hears Asylum Appeal in Genital Mutilation Case." *The New York Times* (May 3): B5.

Dugger, Karen. 1988. "Social Location and Gender-Role Attitudes: A Comparison of Black and White Women." *Gender & Society* 2 (December): 425–448.

Duncan, Greg, Martha Hill, and Saul Hoffman. 1988. "Welfare Dependence Within and Across Generations." *Science* 239: 467–471.

Duncan, Greg, J., and Saul D. Hoffman. 1990. "Welfare Benefits, Economic Opportunities, and Out-of-Wedlock Births Among Black Teenage Girls." *Demography* 27 (November): 519–535.

Duncan, Greg J., and Willard Rodgers. 1991. "Has Children's Poverty Become More Persistent?" *American Sociological Review* 56 (August) 538–550.

Dungee-Anderson, Delores, and Joyce O. Beckett. 1992. "Alzheimer's Disease in African American and White Families: A Clinical Analysis." *Smith College Studies in Social Work* 62 (March): 155–168.

Durán, Richard P. 1983. *Hispanics' Education and Background: Predictors of College Achievement.* New York: College Entrance Examination Board.

Duran-Aydintug, Candan, and Kelly A. Causey. 1996. "Child Custody Determination: Implications for Lesbian Mothers." *Journal of Divorce and Remarriage* 25: 55–74.

Durant, Thomas J., Jr., and Joyce S. Louden. 1986. "The Black Middle Class in America: Historical and Contemporary Perspectives." *Phylon* 47 (December): 253–263.

Durkheim, Emile. 1897 [1951]. *Suicide.* Glencoe, IL: Free Press.

Durkheim, Emile. 1912 [1947]. *Elementary Forms of Religious Life.* Glencoe, IL: Free Press.

Durkheim, Emile. 1938 [1950]. *The Rules of Sociological Method.* Glencoe, IL: Free Press.

Durkheim, Emile. 1964 [1895]. *The Division of Labor in Society.* New York: Free Press.

Durning, Alan. 1989. "Poverty and the Environment: Reversing the Downward Spiral." *Worldwatch Paper* 92 (November), Washington DC: Worldwatch Institute, pp. 54–68.

Duster, Troy. 1990. *Backdoor to Eugenics.* New York: Routledge.

Dwyer, Lynn E. 1983. "Structure and Strategy in the Antinuclear Movement." Pp. 148–161 in *Social Movements of the Sixties and Seventies,* edited by Jo Freeman. New York: Longman.

Dye, Thomas R. 1990. *Who's Running America: The Bush Era,* 5th ed. Englewood Cliffs, NJ: Prentice-Hall.

E

Eagly, A. H., R. D. Ashmore, M. G. Makhijani, and L. C. Longo. 1989. "What is Beautiful is Good, But: A Meta-analysis." *Psychological Bulletin* 110: 109–128.

Easterbrook, Coregg. 1987. "The Revolution in Medicine." *Newsweek* (January 26): 43.

Eckberg, Douglass Lee. 1992. "Social Influences on Belief in Creationism." *Sociological Spectrum* 12 (April-June): 145–165.

Eckert, P. 1988. "Adolescent Social Structure and the Spread of Linguistic Change." *Language in Society* 17: 183–208.

Edelman, Marion Wright. 1987. *Families in Peril: An Agenda for Social Change.* Cambridge, MA: Harvard University Press.

Eder, Donna. 1981. "Ability Grouping as a Self-Fulfilling Prophecy: A Micro-Analysis of Teacher-Student Interaction." *Sociology of Education* 54:151–162.

Edin, Kathyrn. 1991. "Surviving the Welfare System: How AFDC Recipients Make Ends Meet in Chicago." *Social Problems* 38 (November): 462–472.

Edin, Kathryn, and Laura Lein. 1997. *Making Ends Meet: How Single Mothers Survive Welfare and Low-Wage Work.* New York: Russell Sage Foundation.

Edmonds, Ronald, Andrew Billingsley, James Comer, James M. Dyer, William Hall, Robert Hill, Nan McGehee, Lawrence Reddick, Howard F. Taylor, and Stephen Wright. 1972. "A Black Response to Christopher Jencks' Inequality and Certain Other Issues." *Harvard Educational Review* 43 (February): 76–91.

Edsall, Thomas Byrne. 1984. *The New Politics of Inequality.* New York: Norton.

Edwards, Larry. 1995. "The Women's Team." Official Program, Citizen Cup, Supplement to *Yachting Magazine,* New York.

Ehrenreich, Barbara. 1990. *Fear of Falling: The Inner Life of the Middle Class.* New York: HarperCollins.

Ehrenreich, Barbara, Elizabeth Hess, and Gloria Jacobs. 1987. *Re-Making Love: The Feminization of Sex.* Garden City, NY: Anchor/Doubleday.

Ehrlich, P. 1990. *The Population Explosion.* New York: Ballantine Books.

Ehrlich, Paul. 1968. *The Population Bomb.* New York: Ballantine Books.

Ehrlich, Paul R., and Anne H. Ehrlich. 1991. *The Population Explosion.* New York: Touchstone/Simon and Schuster.

Ehrlich, Paul R., Anne H. Ehrlich, and John P. Holdren. 1977. *Ecoscience: Population, Resources, Environment.* San Francisco: Freeman.

Eichenwald, Kurt. 1996. "The Two Faces of Texaco." *The New York Times,* November 10: Section 3: 1ff.

Eisenstein, Zillah. 1984. "The Patriarchal Relations of the Reagan State." *Signs* 10 (Winter): 329–337.

Eitzen, D. Stanley, and Maxine Baca Zinn. 1998. *In Conflict and Order: Understanding Society,* 8th ed. Boston: Allyn and Bacon.

Ekman, Paul. 1982. *Emotion in the Human Face,* second edition. Cambridge, MA: Cambridge University Press.

Ekman, Paul, and W. V. Friesen. 1974. "Detecting Deception from the Body or Face." *Journal of Personality and Social Psychology* 29: 288–298.

Ekman, Paul, W. V. Friesen, and K. Scherer. 1976. "Body Movements and Voice Pitch in Deceptive Interaction." *Semiotica* 16: 23–27.

Ekman, Paul, W. V. Friesen, and M. O'Sullivan. 1988. "Smiles When Lying." *Journal of Personality and Social Psychology* 54: 414–420.

Eliason, Michele J., and Salome Raheim. 1996. "Categorical Measurement of Attitudes about Lesbian, Gay, and Bisexual People." *Journal of Gay and Lesbian Social Services* 4: 51–65.

Eller, T.J. 1996. *Who Stays Poor? Who Doesn't?* Current Population Reports: Dynamics of Economic Well-Being: Poverty, 1992–1993. Washington, DC: U.S. Department of Commerce, Economics and Statistics Administration, June.

Elliott, Marta, and Lauren J. Krivo. 1991. "Structural Determinants of Homelessness in the United States." *Social Problems* 38 (February): 113–131.

Ellison, Christopher G., and David A. Gay. 1989. "Black Political Participation Revisited: A Test of Compensatory, Ethnic Community, and Public Arena Models." *Social Science Quarterly* 70 (March): 101–119.

Emerson, Juan P. 1970. "Behaviors in Private Places: Sustaining Definitions of Reality in Gynecological Examinations." Pp. 74–97 in *Recent Sociology,* Vol. 2, edited by H. P. Dreitzel. New York: Collier.

England, Paula, Barbara Stanek Kilbourne, George Farkas, and Thomas Dou. 1988. "Explaining Occupational Sex Segregation and Wages: Findings from a Model with Fixed Effects." *American Sociological Review* 53 (August): 544–558.

Enloe, Cynthia. 1983. *Does Khaki Become You?: The Militarisation of Women's Lives.* Boston: South End Press.

Enloe, Cynthia. 1989. *Bananas, Beaches, and Bases: Making Feminist Sense of International Politics.* Berkeley: University of California Press.

Enloe, Cynthia. 1993. *The Morning After: Sexual Politics at the End of the Cold War.* Berkeley: University of California Press.

Ericksen, Eugene P. 1988. "Estimating the Concentration of Wealth in America." *Public Opinion Quarterly* 52 (Summer): 243–253.

Erikson, Eric. 1980. *Identity and the Life Cycle.* New York: W. W. Norton.

Erikson, Kai. 1966. *Wayward Puritans: A Study in the Sociology of Deviance.* New York: Wiley.

Erikson, Robert. 1985. "Are American Rates of Social Mobility Exceptionally High? New Evidence an on Old Issue." *European Sociological Review* 1 (May): 1–22.

Ermann, M. David, and Richard J. Lundman. 1992. *Corporate and Governmental Deviance.* New York: Oxford University Press.

Espenshade, Thomas J. 1995. "Unauthorized Immigration to the United States." *Annual Review of Sociology* 21: 195–216.

Espiritu, Yen Le. 1992. *Asian American Panethnicity: Bridging Institutions and Identity.* Philadelphia: Temple University Press.

Essed, Philomena. 1991. *Understanding Everyday Racism.* Newbury Park, CA: Sage.

Essien-Udom, E. U. 1962. *Black Nationalism: A Search for an Identity in America.* Chicago: University of Chicago Press.

Estrada, Leonardo, F. Chris Garcia, Reynaldo Macias, and Lionel Maldonado. 1981. "Chicanos in the United States: A History of Exploitation and Resistance." *Daedalus* 110 (2): 101–131.

Etzioni, Amatai. 1975. *A Comparative Analysis of Complex Organization: On Power, Involvement, and Their Correlates,* revised ed. New York: Free Press.

Etzioni, Amitai. 1990. "Status Separations and Status Fusion: The Role of PACs in Contemporary American Democracy." Pp. 67–88 in *Political Sociology of the State,* edited by Richard Braungart and Margaret Braungart. Greenwich, CT: JAI Press Inc.

Evans, Harriet. 1995. "Defining Difference: The 'Scientific' Construction of Sexuality and Gender in the People's Republic of China." *Signs* 20 (Winter): 357–394.

Evans, Peter B., Dietrich Ruesschemeyer, and Theda Skocpol. 1985. *Bringing the State Back In.* Cambridge: Cambridge University Press.

F

Farber, Naomi. 1990. "The Significance of Race and Class in Marital Decisions among Unmarried Adolescent Mothers." *Social Problems* 37 (February): 51–63.

Farley, Reynolds, and Walter R. Allen. 1987. *The Color Line and the Quality of Life in America.* New York: Russell Sage Foundation.

Farley, Reynolds. 1984. *Blacks and Whites: Narrowing the Gap?* Cambridge, MA: Harvard University Press.

Farr, Grant. 1993. "Refugee Aid and Development in Pakistan: Afghan Aid after Ten Years." Pp. 111–128 in *Refugee Aid and Development: Theory and Practice,* edited by Robert F. Gorman. Westport, CT: Greenwood Press.

Fattah, Ezzat A., 1994. "The Interchangeable Roles of Victim and Victimizer." *HEUNI Papers* 3: 1–26.

Faulkner, Robert R. 1971. *Hollywood Studio Musicians: Their Work and Careers in the Recording Industry.* Chicago: Aldine-Atherton.

Fausto-Sterling, Anne. 1992. *Myths of Gender: Biological Theories about Women and Men.* New York: Basic Books.

Feagin, Joe R. 1991. "The Continuing Significance of Race: Antiblack Discrimination in Public Places." *American Sociological Review* 56 (February): 101–116.

Feagin, Joe R., and Clairece B. Feagin. 1993. *Racial and Ethnic Relations,* 4th ed. Englewood Cliffs, NJ: Prentice-Hall.

Feagin, Joe R., and Hernán Vera. 1995. *White Racism.* New York: Routledge.

Federal Bureau of Investigation. 1997. *Uniform Crime Reports.* Washington, DC: U.S. Government Printing Office.

Federal Election Commission. 1997. "PAC Activity Increases in 1995–96 Election Cycle." World Wide Web Site: **http://www.fec.gov/press/pacye96.htm**

Fein, Melvyn L. 1988. "Resocialization: A Neglected Paradigm." *Clinical Sociology* 6: 88–100.

Felson, Richard B., and Mark D. Reed. 1986. "Reference Groups and Self-Appraisals of Academic Ability and Performance." *Social Psychology Quarterly* 49 (June): 103–109.

Fenwick, Rudy, and John F. Zipp. 1993. "Workplace Participation as Organizational Adaptation and Legitimation: A Comparison of Employee Stock-Ownership Plans and Quality of Worklife Programmes in the USA." Pp. 33–53 in *International Handbook of Participation in Organizations,* edited by William Lafferty and Eliezer Rosenstein. New York: Oxford University Press.

Fernald, Anne, and Hiromi Morikawa. 1993. "Common Themes and Cultural Variations in Japanese and American Mothers' Speech to Infants." *Child Development* 64 (June): 637–656.

Fernandes-y-Freitas, Rosa L. 1996. "The Minitel and the Internet: New Houses for Our Old Phantoms." *Societies* 51 (February): 49–57.

Ferraro, K. F. 1992. "Cohort Changes in Images of Older Adults, 1974–1981." *Gerontologist* 32: 296–304.

Ferree, Myra Marx, and Elaine J. Hall. 1990. "Visual Images of American Society: Gender and Race in Introductory Sociology Textbooks." *Gender & Society* 4 (December 1990): 500–533.

Ferree, Myra Marx. 1984. "Sacrifice, Satisfaction, and Social Change: Employment and the Family." Pp. 61–79 in *My Troubles Are Going to Have Trouble with Me,* edited by Karen B. Sacks and Dorothy Remy. New Brunswick, NJ: Rutgers University Press.

Fine, Gary Alan. 1986. "The Dirty Talk of Little Boys." Pp. 135–143 in *Men's Lives,* 2nd ed., edited by Michael S. Kimmel and Michael A. Messner. New York: Macmillan.

Finkelhor, David, and Kersti Yllo. 1985. *License to Rape: Sexual Abuse of Wives.* New York: Holt, Rinehart, and Winston.

Fiol-Matta, Liza, and Miriam K. Chamberlain. 1994. *Women of Color and the Multicultural Curriculum: Transforming the College Classroom.* New York: The Feminist Press.

Fischer, Claude S., Michael Hout, Mártin Sánchez Jankowski, Samuel R. Lucas, Ann Swidler, and Kim Voss. 1996. *Inequality by Design: Cracking the Bell Curve Myth.* Princeton, NJ: Princeton University Press.

Fischer, Claude. 1981. *To Dwell Among Friends: Personal Networks in Town and City.* Chicago: University of Chicago Press.

Fish, Linda Stone, and Janet L. Osborn. 1992. *Family Relations* 41 (October): 408–409.

Fisher, Randy D., Ida J. Cook, and Edwin C. Shirkey. 1994. "Correlates of Support for Censorship of Sexual, Sexually Violent, and Violent Media." *Journal of Sex Research* 31: 229–240.

Fisher-Thompson, Donna. 1990. "Adult Sex Typing of Children's Toys." *Sex Roles* 23 (September): 291–303.

Flanagan, William G. 1995. *Urban Sociology: Images and Structure.* Needham Heights, MA: Allyn & Bacon.

Flowers, M.L. 1977. "A Laboratory Test of Some Implications of Janis' Groupthink Hypothesis." *Journal of Personality and Social Psychology* 35 (December): 888–896.

Flowers, Ronald. 1988. *Minorities and Criminality.* New York: Greenwood Press.

Forbes. 1997. "The 400: The Average Net Worth of the Forbes Four Hundred is $1.6 Billion." (Forbes 400). *Forbes* 160 (October 13): 181–250.

Ford, Nicholas, and Suporn Koetsawang. 1991. "The Socio-Cultural Context of the Transmission of HIV in Thailand." *Social Science and Medicine* 22: 405–414.

Fordham, Signithia. 1996. *Blacked Out: Dilemmas of Race, Identity, and Success at A.I. High.* Chicago: University of Chicago Press.

Forrest, Jacqueline Darroch, and Susheela Singh. 1990. "The Sexual and Reproductive Behavior of American Women, 1982–1988." *Family Planning Perspectives* 22: 206–214.

Frank, A. G. 1969. *Latin America: Underdevelopment or Revolution?* New York: Monthly Review Press.

Frankenberg, Ruth. 1993. *White Women, Race Matters: The Social Construction of Whiteness.* Minneapolis: University of Minnesota Press.

Frankfort-Nachmias, Chava, and David Nachmias. 1992. *Research Methods in the Social Sciences.* New York: St. Martin's Press.

Franklin, Clyde. 1994. "Sex and Class Differences in the Socialization Experiences of African American Youth." *Western Journal of Black Studies* 18 (Summer): 104–111.

Frazier, E. Franklin. 1957. *The Black Bourgeoisie.* New York: Collier Books.

Free, Marvin D. 1990. "Demographic, Organizational and Economic Determinants of Work Satisfaction: An Assessment of Work Attitudes of Females in Academic Settings." *Sociological Spectrum* 10 (Winter): 79–103.

Freedman, Allan. 1997. "Lawyers Take a Back Seat in the 105th Congress." *Congressional Quarterly* 55 (January 4): 27–30.

Freedman, Estelle B. 1992. "The Manipulation of History at the Clarence Thomas Hearings." *The Chronicle of Higher Education* 38 (January 8): B2–3.

Freeman, Jo. 1983a. "On the Origins of Social Movements." Pp. 8–33 in *Social Movements of the Sixties and Seventies,* edited by Jo Freeman. New York: Longman.

Freeman, Jo. 1983b. "A Model for Analyzing the Strategic Options of Social Movement Organizations." Pp. 193–210 in *Social Movements of the Sixties and Seventies,* edited by Jo Freeman. New York: Longman.

Frendreis, John, and Raymond Tatalovich. 1997. "Who Supports English-Only Language Laws? Evidence from the 1992 National Election Study." *Social Science Quarterly* 78 (June): 354–365.

Freud, Sigmund. 1901. *The Psychopathology of Everyday Life,* translated by Alan Tyson and edited by James Strachey. New York: W. W. Norton, 1965.

Freud, Sigmund. 1923. *The Ego and the Id,* translated by Joan Riviere. New York: W.W. Norton, 1960.

Freud, Sigmund. 1930. *Civilization and Its Discontents,* translated by James Strachey. New York: W. W. Norton, 1961.

Fried, Amy. 1994. "'It's Hard to Change What We Want to Change': Rape Crisis Centers as Organizations." *Gender & Society* 4 (December): 562–583.

Frisbie, W. Parker. 1986. "Variations in Patterns of Marital Stability Among Hispanics." *Journal of Marriage and the Family* 48 (February): 99–106.

Frye, Marilyn. 1983. *The Politics of Reality.* Trumansburg, NY: The Crossing Press.

Fukuda, Mari. 1994. "Nonverbal Communication Within Japanese and American Corporations." Unpublished manuscript, Princeton University, Princeton, NJ.

Fukurai, Hiroshi. 1993. *Race and the Jury: Racial Disenfranchisement and the Search for Justice.* New York: Plenum Press.

Fukurau, Hiroshi, Edgar W. Butler, and Richard Krooth. 1993. *Race and the Jury: Racial Disenfranchisement and the Search for Justice.* New York: Plenum Press.

Fullerton, Howard N., Jr. 1995. "The 2005 Labor Force: Growing, but Slowly." *Monthly Labor Review* 118 (November): 29–44.

Furstenberg, Frank F., Jr. 1990. "Divorce and the American Family," *Annual Review of Sociology* 16: 379–403.

Furstenberg, Frank F., Jr., and Christine Winquist Nord. 1985. "Parenting Apart: Patterns of Childrearing After Marital Disruption." *Journal of Marriage and the Family* 47 (November): 898–904.

Furstenberg, Frank. 1998. "Relative Risk: What is the Family Doing to Our Children?" *Contemporary Sociology* 27 (May): 223–225.

G

Gagné, Patricia, and Richard Tewksbury. 1998. "Conformity Pressures and Gender Resistance among Transgendered Individuals." *Social Problems* 45 (February): 81–101.

Gagnon, John. 1973. *Sexual Conduct: The Social Sources of Human Sexuality.* Chicago: Aldine Publishing Co.

Gagnon, John. 1995. *Conceiving Sexuality: Approaches to Sex Research in a Postmodern World.* New York: Routledge.

Galliher, John F. and Cheryl Tyree. 1985. "Edwin Sutherland's Research on the Origins of Sexual Psychopath Laws: An Early Case Study of the Medicalization of Deviance." *Social Problems* 33 (December): 100–113.

Gallup, George, Jr. 1993. *The Gallup Poll: Public Opinion 1993.* Wilmington, DE: Scholarly Resources, Inc.

Gallup, George, Jr., and Frank Newport. 1990. "More Americans Now Believe in a Power Outside Themselves." *The Gallup Poll Monthly* (June, No. 297): 33–38.

Gallup, George, Jr., and Frank Newport. 1991. "Almost Half of Americans Believe Biblical View of Creation." *The Gallup Poll Monthly* (November, No. 314): 30–34.

Gallup, George, Jr., and Sarah Jones. 1989. *100 Questions & Answers: Religion in America.* Princeton, NJ: Princeton Religious Research Center.

Gamoran, Adam, and Robert D. Mare, 1989. "Secondary School Tracking and Educational Inequality: Compensation, Reinforcement, or Neutrality?" *American Journal of Sociology* 94: 1146--1183.

Gamoran, Adam. 1972. "The Variable Effects of High School Tracking." *American Sociological Review* 57: 812–828.

Gamson, Joshua. 1995a. "Must Identity Movements Self Destruct? A Queer Dilemma." *Social Problems* 42 (August): 390–407.

Gamson, Joshua. 1995b. "Featured Essay." *Contemporary Sociology* 24 (May): 294–298.

Gamson, William A. 1992. "The Social Psychology of Collective Action." Pp. 53–76 in *Frontiers in Social Movement Theory,* edited by Aldon D. Morris and Carol McClurg Mueller. New Haven, CT: Yale University Press.

Gamson, William, and Andre Modigliani. 1974. *Conceptions of Social Life: A Text-Reader for Social Psychology.* Boston: Little, Brown and Co.

Gans, Herbert J. 1991 [1971]. "The Uses of Poverty." Pp. 263–270 in *People, Plans, and Policies: Essays on Poverty, Racism, and Other National Urban Problems,* edited by Herbert J. Gans. New York: Columbia University Press.

Gans, Herbert. 1979. *Deciding What's News: A Study of the CBS Evening News, NBC Nightly News, Newsweek and Time.* New York: Pantheon.

Gans, Herbert. 1982 [1962]. *The Urban Villagers: Group and Class in the Life of Italian Americans.* New York: Free Press.

Ganzebook, Harry B. G., Donald J. Treiman, and Wout C. Ultee. 1991. "Comparative Intergenerational Stratification Research: Three Generations and Beyond." *Annual Review of Sociology* 17: 277–302.

Gardiner, R. Allen, and Beatrice T. Gardiner. 1969. "Teaching Sign Language to a Chimpanzee." *Science* 165: 664–672.

Gardner, Carol Brooks. 1994. "A Family Among Strangers: Kinship Claims among Gay Men in Public Places." *Research in Community Sociology.* JAI Press.

Gardner, Howard. 1999 [1993]. *Frames of Mind: The Theory of Multiple Intelligences.* New York: Basic Books.

Gardner, Robert W., Bryant Robey, and Peter C. Smith. 1989. *Asian Americans: Growth, Chance, and Diversity.* Washington, DC: Population Reference Bureau.

Garfinkel, Harold. 1967. *Studies in Ethnomethodology.* Englewood Cliffs, NJ: Prentice-Hall.

Garfinkel, Irwin, and Sara S. McLanahan, eds. 1992. *Child Support Assurance.* Washington, DC: Urban Institute Press.

Garofalo, Reebee. 1987. "The Impact of the Civil Rights Movement on Popular Music." *Radical America* 21 (6): 15–22.

Garrow, David J. 1981. *The FBI and Martin Luther King.* New York: Norton.

Garst, J., and G. V. Bodenhausen. 1997. "Advertising's Effects on Men's Gender Role Attitudes." *Sex Roles* 36 (May): 551–572.

Gates, Henry Louis, Jr. 1988. *The Signifying Monkey: A Theory of African-American Literary Criticism.* New York: Oxford University Press.

Gates, Henry Louis, Jr. 1992. "Integrating the American Mind." Pp. 105–120 in *Loose Canons: Notes on the Culture Wars,* edited by Henry Louis Gates, Jr.. New York: Oxford University Press.

Gates, Henry Louis. 1993. "The End of Civilization As We Know It." Eberhard Faber Lecture Series, Princeton University, May, 1993.

Gay, David A., Christopher G. Ellison, and Daniel A. Powers. 1996. "In Search of Denominational Subcultures: Religious Affiliation and 'Pro-Family' Issues Revisited." *Review of Religious Research* 38 (September): 3–17.

Gehlen, Frieda. 1977. "Toward a Revised Theory of Hysterical Contagion." *Journal of Health and Social Behavior* 19 (March): 27–35.

Gelles, Richard J. "Family Violence." Pp. 1–24 in *Family Violence: Prevention and Treatment,* edited by Robert J. Hampton, Thomas P. Gallota, Gerald R. Adams, Earl H. Potter III, and Roger P. Weissberg. Newbury Park, CA: Sage.

Gelles, Richard J., and Murray A. Straus. 1988. *Intimate Violence.* New York: Simon and Schuster.

Genovese, Eugene. 1972. *Roll, Jordan, Roll: The World the Slaves Made.* New York: Pantheon.

Gerbner, George. 1978. "The Dynamics of Cultural Resistance." Pp. 46–50 in *Hearth and Home: Images of Women in the Media,* edited by Gaye Tuchman, Arlene Kaplan Daniels, and James Benét. New York: Oxford University Press.

Gerbner, George. 1993. *Women and Minorities on Television (A Report to the Screen Actors Guild).* Philadelphia: University of Pennsylvania, Annenberg School for Communications.

Gereffi, Gary, and Miguel Korzeniewicz. 1993. "Commodity Chains and Footwear Exports to the Semiperiphery." *Semiperipheral States in the World Economy,* edited by William Martin. Westport, CT: Greenwood Press.

Gerlach, Luther P., and Virginia Hine. 1970. *People, Power, Change: Movements of Social Transformation.* Indianapolis, IN: Bobbs-Merrill.

Gerson, Judith. 1985. "Women Returning to School: The Consequences of Multiple Roles." *Sex Roles* 13 (July 1985): 77–91.

Gerson, Kathleen. 1993. *No Man's Land: Men's Changing Commitments to Family and Work.* New York: Basic Books.

Gerstl, Naomi, and Harriet Gross. 1987. "Commuter Marriage: A Microcosm of Career and Family Conflict." Pp. 422–433 in *Families and Work,* edited by Naomi Gerstl and Harriet Gross. Philadelphia: Temple University Press.

Gerth, Hans, and C. Wright Mills (eds.). 1946. *From Max Weber: Essays in Sociology*. New York: Oxford University Press.

Geschwender, James. 1992. "Ethnicity and the Social Construction of Gender in the Chinese Diaspora." *Gender & Society* 6 (September): 480–507.

Geschwender, Laura E., and Toby L. Parcel. 1995. "Objective and Subjective Parental Working Conditions' Effects on Child Outcomes: A Comparative Test." *Research in the Sociology of Work* 5: 259–284.

Gibbons, Sheila. 1992. *Media Report to Women* (Fall): 5–6.

Gibbs, Lois. 1982. *Love Canal: My Story*. Albany, NY: State University of New York Press.

Gibson, Diane. 1996. "Broken Down by Age and Gender: 'The Problem of Old Women' Revisited." *Gender & Society* 10 (August): 433–448.

Gibson, Rose C., and Cheryl J. Burns. 1991. "The Health, Labor Force, and Retirement Experiences of Aging Minorities." *Generations* 15 (Fall-Winter): 31–35.

Giddings, Paula. 1988. *In Search of Sisterhood: Delta Sigma Theta and the Challenge of the Black Sorority Movement*. New York: Quill.

Gilbert, Daniel R., Susan T. Fiske, and Gardner Lindzey (eds.). 1998. *The Handbook of Social Psychology*, 4th ed. New York: Oxford University and McGraw Hill.

Gilbert, Dennis, and Joseph A. Kahl. 1987. *The American Class Structure: A New Synthesis*. Chicago: Dorsey Press.

Gilkes, Cheryl Townsend. 1985. "'Together and in Harness:' Women's Traditions in the Sanctified Church." *Signs* 10 (Summer): 678–699.

Gill, D.L. 1988. "Gender Differences in Competitive Orientation and Sport Participation." *International Journal of Sport Psychology* 19 (2): 145–159.

Gillman, Katherine. 1988. "Is the Middle Class Shrinking? *Futures* 20 (April): 137–146.

Gitlin, Todd. 1983. *Inside Prime Time*. New York: Basic Books.

Gitlin, Todd. 1987. *The Sixties: Years of Hope, Days of Rage*. New York: Bantam Books.

Gitlin, Todd. 1995. *The Twilight of Common Dreams: Why America is Wracked by Culture Wars*. New York: Metropolitan Books.

Glaser, B. G., and Anselm Strauss. 1965. *Awareness of Dying*. Chicago: Aldine.

Glass Ceiling Commission. 1995. *Good for Business: Making Full Use of the Nation's Human Capital*. Washington, DC: U.S. Government Printing Office.

Glazer, Myron Peretz, and Penina Migdal Glazer. 1989. *The Whistleblowers: Exposing Corruption in Government and Industry*. New York: Basic Books.

Glazer, Nathan. 1970. *Beyond the Melting Pot: The Negroes, Puerto Ricans, Jews, Italians, and Irish of New York City*. Cambridge: M.I.T. Press.

Glazer, Nona. 1990. "The Home as Workshop: Women as Amateur Nurses and Medical Care Providers." *Gender & Society* 4: 479–499.

Glenn, Evelyn Nakano. 1986. *Issei, Nisei, War Bride: Three Generations of Japanese American Women in Domestic Service*. Philadelphia: Temple University Press.

Glenn, Evelyn Nakano. 1987. "Gender and the Family." Pp. 348–380 in *Analyzing Gender: A Handbook of Social Science Research*, edited by Beth Hess and Myra Marx Ferree, Newbury Park, CA: Sage Publications.

Glenn, Norval D., and Charles N. Weaver. 1988. "The Changing Relationship of Marital Status to Reported Happiness." *Journal of Marriage and the Family* 50 (May): 317–324.

Glock, Charles, and Rodney Stark. 1965. *Religion and Society in Tension*. Chicago: Rand McNally.

Gluckman, Amy, and Betsy Reed (eds.). 1997. *Homo Economics: Capitalism, Community, and Lesbian and Gay Life*. New York: Routledge.

Goffman, Erving. 1959. *The Presentation of Self in Everyday Life*. Garden City, NY: Doubleday.

Goffman, Erving. 1961. *Asylums: Essays on the Social Situation of Mental Patients and Other Inmates*. Garden City, New York: Anchor.

Goffman, Erving. 1963a. *Behavior in Public Places*. New York: Free Press.

Goffman, Erving. 1963b. *Stigma: Notes on the Management of Spoiled Identity*. Englewood Cliffs, NJ: Prentice Hall.

Goldberger, Arthur S. 1979. "Heritability." *Econometrica* 46: 327–347.

Goldman, Noreen, Yorenzo Moreno, and Charles F. Westoff. 1989. *Peru Experimental Study: An Evaluation of Fertility and Child Health Information*. Princeton, NJ: Office of Population Research.

Goldsmith, Ronald, and Richard Heims. 1992. "Subjective Age: A Test of Five Hypotheses." *The Gerontologist* 32: 312–317.

Goleman, Daniel. 1995. "An Elusive Picture of Violent Men Who Kill Mates." *The New York Times* (January 15): A22.

González, Tina Esther. 1996. *Social Control of Medical Professionals, Cultural Notions of Pain, and Doctors as Social Scientists*. Unpublished junior thesis, Princeton University, Princeton, NJ.

Goodall, Jane. 1990. *Through a Window: My Thirty Years with the Chimpanzees of Gombe*. Boston: Houghton Mifflin.

Goode, Erich. 1992. *Collective Behavior*. Fort Worth, TX: Harcourt, Brace, and Jovanovich.

Goodman, Catherine Chase. 1990. "The Caregiving Roles of Asian American Women." *Journal of Women and Aging* 2: 109–120.

Goodman, William. 1993. "Boom in the Day Care Industry the Result of Many Changes." *Monthly Labor Review* 118 (August): 3–12.

Gorady, Denise. 1997. "Study Finds Secondhand Smoke Doubles Risk of Heart Disease." *The New York Times* (May 20): 1 and 18.

Gordon, Linda. 1977. *Woman's Body/Woman's Right*. New York: Penguin.

Gordon, Margaret T., and Stephanie Riger. 1989. *The Female Fear*. New York: Free Press.

Gore, Albert. 1993. *Earth in the Balance: Ecology and the Human Spirit*. New York: Penguin.

Gottfredson, Michael R., and Travis Hirschi. 1990. *A General Theory of Crime*. Palo Alto, CA: Stanford University Press.

Gough, Kathleen. 1984. "The Origin of the Family." Pp. 83–99 in *Women: A Feminist Perspective*, 3rd edition, edited by Jo Freeman. Palo Alto, CA: Mayfield Publishing Co.

Gould, Stephen J. 1994. "Curveball." *The New Yorker* 70 (November 28): 139–149.

Gould, Stephen Jay. 1981. *The Mismeasure of Man*. New York: W. W. Norton & Co.

Gourevitch, Philip. 1996. "After the Genocide." *The New Yorker* 72 (December 18): 78–95.

Gove, Walter R., Alfred C. Marcus, Teresa E. Seeman, and Carol W. Telesky. 1984. "Gender Differences in Mental and Physical Illness: The Effects of Fixed Roles and Nurturant Roles." *Social Science and Medicine* 19: 77–84.

Gove, Walter, Suzanne Ortega, and Carolyn Style. 1989. "The Maturational Role Perspectives on Aging and Self Through the Adult Years: An Empirical Evaluation." *American Journal of Sociology* 94: 1117–1145.

Gowan, Mary A., and Raymond A. Zimmermann. 1996. "The Family and Medical Leave Act of 1993: Employee Rights and Responsibilities, Employer Rights and Responsibilities." *Employee Responsibilities and Rights Journal* 9 (March): 57–71.

Gramsci, Antonio. 1971. *Selections from the Prison Notebooks of Antonio Gramsci*, edited by Quintin Hoare and Geoffrey Nowell. London: Lawrence and Wishart.

Granovetter, Mark. 1973. "The Strength of Weak Ties." *American Journal of Sociology* 78 (May): 1360–1380.

Granovetter, Mark. 1974. *Getting a Job: A Study of Contacts and Careers*. Cambridge, MA: Harvard University Press.

Grant, Don Sherman II, and Ramiro Martinez, Jr. 1997. "Crime and the Restructuring of the U.S. Economy: A Reconsideration of the Class Linkages." *Social Forces* 75 (March): 769–799.

Gray, Herman. 1993. "African American Political Desire and the Seductions of Contemporary Cultural Politics." *Cultural Studies* 7 (October): 364–373.

Greenberg, Donald F. Austin, Vivian W. Chen, and Brenda K. Edwards. 1992. "The Relationship Between Social Ties and Survival Among Breast Cancer Patients in the Black/White Study of Cancer Survival." Research Report. Washington, DC: National Cancer Institute.

Greenfield, S., W. Rogers, M. Mangotich, N. F. Carney, and A. R. Tarlov. 1995. "Outcomes of Patients with Hypertension and Non-Insulin-Dependent Diabetes Mellitus Treated by Different Systems and Specialties: Results from the Medical Outcomes Study." *Journal of the American Medical Association* (November 8): 1436–1444.

Greenhouse, Linda. 1996a. "Gay Rights Laws Can't Be Banned, High Court Rules." *The New York Times* (May 21): A1ff.

Greenhouse, Linda. 1996b. "The Census Counts, The Court Decides." *The New York Times* (March 24): E2.

Greenhouse, Steven. 1994. "State Dept. Finds Widespread Abuse of World's Women." *The New York Times* (February 3): A1ff.

Greenlee, Craig T. 1996. "Title IX: Does Help for Women Come at the Expense of African Americans?" *Black Issues in Higher Education* (April 17): 24–26.

Grief, Geoffrey L. 1985. "Children and Housework in the Single Father Family." *Family Relations* 34 (July): 353–357.

Griffin, C. 1985. *Typical Girls?: Young Women from School to the Job Market.* London: Routledge.

Griffin, Glenn, A. Elmer, Richard L. Gorsuch, and Andrea Lee Davis. 1987. "A Cross-Cultural Investigation of Religious Orientation, Social Norms, and Prejudice." *Journal for the Scientific Study of Religion* 26 (September): 358–365.

Griffin, Linner Ward, and Oliver J. Williams. 1992. "Abuse Among African-American Elderly." *Journal of Family Violence* 7: 19–35.

Grimal, Pierre (ed.). 1963. *Larousse World Mythology.* New York: G. P. Putnam's Sons.

Griswold, Wendy. 1994. *Cultures and Societies in a Changing World.* Thousand Oaks, CA: Pine Forge Press.

Gruber, James E. 1982. "Blue-Collar Blues: The Sexual Harassment of Women Autoworkers." *Work and Occupations* 3 (August): 271–298.

Guinther, J. 1988. *The Jury in America.* New York: Facts on File Publications.

Gutman, Herbert G. 1976. *The Black Family in Slavery and Freedom, 1750–1925.* New York: Vintage.

Guttentag, Marcia, and Paul F. Secord. 1983. *Too Many Women? The Sex Ratio Question.* Beverly Hills, CA: Sage.

H

Habermas, Jürgen. 1970. *Toward a Rational Society: Student Protest, Science, and Politics.* Boston, MA: Beacon Press.

Hadaway, C. Kirk, and Penny Long Marler. 1993. "All in the Family: Religious Mobility in America." *Review of Religious Research* 35 (December): 97–116.

Hadden, Jeffrey, and Charles E. Swann. 1981. *Prime Time Preachers: The Rising Power of Televangelism.* Reading, MA: Addison-Wesley.

Hagan, John. 1993. "The Social Embeddedness of Crime and Unemployment." *Criminology* 31 (November): 465–491.

Halberstad, Amy G., and Martha B. Saitta. 1987. "Gender, Nonverbal Behavior, and Perceived Dominance: A Test of the Theory." *Journal of Personality and Social Psychology* 53: 257–272.

Hall, Edward T. 1966. *The Hidden Dimension.* New York: Doubleday.

Hall, Edward T., and Mildred Hall. 1987. *Hidden Differences: Doing Business with the Japanese.* New York: Anchor Press/Doubleday.

Hall, Elaine J. 1993. "Waitering/Waitressing: Engendering in the Work of Table Servers." *Gender & Society* 7 (September): 329–346.

Hall, J. A. 1984. *Nonverbal Sex Differences: Communication Accuracy and Expressive Style.* Baltimore: Johns Hopkins University Press.

Hall-Iijima, Christine C., and Matthew J. Crum. 1994. "Women and 'Body-Isms' in Television Beer Commercials." *Sex Roles* 31 (September): 329–337.

Hallinan, Maureen T. 1994. "Tracking: From Theory to Practice." *Sociology of Education* 67: 79–84.

Hamer, Dean H., Stella Hu, Victoria L. Magnuson, Nan Hu, and Angela M. L. Pattatucci. 1993. "A Linkage Between DNA Markers on the X Chromosome and Male Sexual Identification." *Science* 261 (July): 321–327.

Hamlin, Nancy R., Sumru Erkut, and Jacqueline Fields. 1994. "The Impact of Corporate Restructuring and Downsizing on the Managerial Careers of Minorities and Women: Lessons Learned from Nine Corporations." Marblehead, MA: Hamlin and Associates Inc.

Hampton, Henry. 1987. "Revolt and Repression." *Eyes on the Prize.* Boston: Blackside Productions. Film.

Handlin, Oscar. 1951. *The Uprooted.* Boston: Little, Brown.

Haney, Lynne. 1996. "Homeboys, Babies, Men in Suits: The State and the Reproduction of Male Dominance." *American Sociological Review* 61 (October): 759–778.

Hans, Valerie, and Neil Vidmar. 1986. *Judging the Jury.* New York: Plenum.

Hans, Valerie, and Ramiro Martinez. 1994. "Intersections of Race, Ethnicity, and the Law." *Law and Human Behavior* 18 (June): 211–221.

Harding, Sandra G. 1998. *Is Science Multicultural? Postcolonialisms, Feminisms, and Epistemologies.* Bloomington, IN: Indiana University Press.

Harlap, Susan, Kathyrn Kost, and Jacqueline Darroch Forrest. 1991. *Preventing Pregnancy, Protecting Health: A New Look at Birth Control Choices in the United States.* New York: The Alan Guttmacher Institute.

Harris, Allen C. 1994. "Ethnicity as a Determinant of Sex-Role Identity: A Replication Study of Item Selection for the Bem Sex Role Inventory." *Sex Roles* 31 (August): 241–273.

Harris, Anthony, and Randall G. Stokes. 1978. "Race, Self-Evaluation, and the Protestant Ethic." *Social Problems* 26 (October): 71–85.

Harris, Diana K., Gary A. Fine, and Thomas C. Hood. 1992. "The Aging of Desire: Playboy Centerfolds and the Graying of America; A Research Note." *Journal of Aging Studies* 6: 301–306.

Harris, Ian, José B. Torres, and Dale Allender. 1994. "The Responses of African American Men to Dominant Norms of Masculinity Within the United States." *Sex Roles* 31 (December): 703–720.

Harris, Kathleen Mullan. 1996. "Life After Welfare: Women, Work, and Repeat Dependency." *American Sociological Review* 61 (June): 407–426.

Harris, Marvin. 1974. *Cows, Pigs, Wars, and Witches: The Riddles of Culture.* New York: Vintage.

Harrison, Bennett, and Harry Bluestone. 1982. *The Deindustrialization of America: Plant Closings, Community Abandonment, and the Dismantling of Basic Industry.* New York: Basic Books.

Harrison, Deborah, and Tom Trabasso. 1976. *Black English: A Seminar.* Hillsdale, NJ: Erlbaum Associates.

Harrison, James, James Chin, and Thomas Ficarrotto. 1992. "Warning: Masculinity May be Dangerous to Your Health." Pp. 271–285 in *Men's Lives,* 2nd ed., edited by Michael S. Kimmel and Michael A. Messner. New York: Macmillan.

Harrison, Reginald. 1980. *Pluralism and Corporatism.* London: George Allen and Unwin.

Harry, Joseph. 1979. "The Marital 'Liaisons' of Gay Men." *The Family Coordinator* 28 (October): 622–629.

Hartigan, John A., and Alexandra K. Wigdor (eds.). 1989. *Fairness in Employment Testing.* Washington, DC: National Academy Press.

Hartmann, Heidi. 1981. "The Family as the Locus of Gender, Class, and Political Struggle: The Example of Housework." *Signs: Journal of Women in Culture and Society* 6 (Spring): 366–394.

Hartmann, Heidi, 1996. "Who Has Benefited from Affirmative Action in Employment?" Pp. 77–96 in *The Affirmative Action Debate,* edited by George E. Curry. Reading, MA: Addison-Wesley.

Hastie, Reid, Steven D. Penrod, and Nancy Pennington. 1983. *Inside the Jury.* Cambridge, MA: Harvard University Press.

Hastorf, Albert, and Hadley Cantril. 1954. "They Saw a Game: A Case Study." *Journal of Abnormal and Social Psychology* 40 (2): 129–134.

Hatch, Laurie Russell. 1990. "Gender and Work at Midlife and Beyond." *Generations* 14 (Summer): 48–52.

Hatch, Laurie Russell. 1992. "Gender Differences in Orientation toward Retirement from Paid Labor." *Gender & Society* 6 (March): 66–87.

Hatfield, Elaine S. and S. Sprecher. 1986. *Mirror, Mirror: The Importance of Looks in Everyday Life.* Albany: State University of New York Press.

Hauser, Robert M., Howard F. Taylor, and Troy Duster. 1995. "The Bell Curve." *Contemporary Sociology* 24 (March): 149–161.

Hawkins, C. H. 1960. *Interaction and Coalition Realignments in Consensus-Seeking Groups: A Study of Experimental Jury Deliberations.* Unpublished doctoral dissertation. Chicago: University of Chicago Press.

Hawkins, Dana. 1996. "The Most Dangerous Jobs." *U.S. News & World Report* 121 (September 23): 40–42.

Hawley, Amos H. 1986. *Human Ecology: A Theoretical Essay.* Chicago: University of Chicago Press.

Hayes-Bautista, David E., Werner O. Schink, and Maria Hayes-Bautista. 1993. "Latinos and the

1992 Los Angeles Riots: A Behavioral Sciences Perspective." *Hispanic Journal of Behavioral Sciences* 15 (November): 427–448.

Healey, Joseph H. 1995. *Race, Ethnicity, Gender, and Class: The Sociology of Group Conflict and Change.* Thousand Oaks, CA: Pine Forge Press.

Hearne, Paul G. 1991. "Employment Strategies for People with Disabilities: A Prescription for Change." *Milbank Quarterly* 69 (Winter): 111–120.

Hearnshaw, Leslie. 1979. *Cyril Burt: Psychologist.* Ithaca, NY: Cornell University Press.

Heaton, Tim. 1987. "Objective Status and Class Consciousness." *Social Science Quarterly* 68 (September): 611–620.

Hecker, Daniel E. 1992. "Reconciling Conflicting Data on Jobs for College Graduates." *Monthly Labor Review* 115 (July): 3–12.

Hefez, Albert. 1985. "The Role of the Press and the Medical Community in the Epidemic of 'Mysterious Gas Poisoning' in the Jordan West Bank." *American Journal of Psychiatry* 142 (7): 833–837.

Heider, Fritz. 1958. *The Psychology of Interpersonal Relations.* New York: John Wiley and Sons.

Heim, Michael. 1990. "The Dark Side of Infomania." *Electric Word* 18 (March-April): 32–36.

Heimer, Karen. 1997. "Socioeconomic Status, Subcultural Definitions and Violent Delinquency." *Social Forces* 75 (March): 799–833.

Henley, Nancy M. 1977. *Body Politics: Power, Sex, and Nonverbal Communication.* Englewood Cliffs, NJ: Prentice Hall.

Henretta, John C., Angela M. O'Rand, and Christopher G. Chan. 1993. "Gender Differences in Employment after Spouses' Retirement." *Research on Aging* 15 (June): 148–169.

Hensel, Chase. 1996. *Telling Our Selves: Ethnicity and Discourse in Southwestern Alaska.* New York: Oxford University Press.

Henshaw, Stanley K., and Jane Silverman. 1988. "The Characteristics and Prior Contraceptive Use of U.S. Abortion Patients." *Family Planning Perspectives* 20 (July/August): 158–168.

Henshaw, Stanley K., and Kathryn Kost. 1996. "Abortion Patients in 1994–1995: Characteristics and Contraceptive Use." *Family Planning Perspectives* 28 (July/August): 140–158.

Henslin, James M. 1993. "Doing the Unthinkable." Pp. 253–262 in *Down to Earth Sociology,* 7th ed., edited by James M. Henslin. New York: Free Press.

Herman, Judith. 1981. *Father-Daughter Incest.* Cambridge, MA: Harvard University Press.

Herman, Max A. 1995. "A Tale of Two Cities: Testing Explanations for Riot Violence in Miami, Florida, and Los Angeles, California." Paper presented at the Annual Meetings of the American Sociological Association, Washington, DC.

Herrnstein, Richard J., and Charles Murray. 1994. *The Bell Curve: Intelligence and Class Structure in American Life.* New York: Free Press.

Herskovits, Melville J. 1941. *The Myth of the Negro Past.* New York: Harper and Brothers.

Hertz, Rosanna. 1986. *More Equal Than Others: Women and Men in Dual-Career Marriages.* Berkeley: University of California Press.

Hess, Beth B., and Elizabeth W. Markson (eds.). 1990. *Growing Old in America*

Higginbotham, A. Leon. 1978. *In the Matter of Color: Race and the American Legal Process.* New York: Oxford University Press.

Higginbotham, A. Leon. 1998. "Breaking Thurgood Marshall's Promise." *The New York Times Magazine* (January 18): 28–29.

Higginbotham, Elizabeth, and Lynn Weber. 1992. "Moving Up with Kin and Community: Upward Social Mobility for Black and White Women." *Gender & Society* 6 (September): 416–440.

Higher Education Research Institute. 1996. "The American Freshman: National Norms for Fall 1996." Los Angeles: University of California.

Hill, C.T., Z. Rubin, and L.A. Peplau. 1976. "Breakups Before Marriage: The End of 103 Affairs." *Journal of Social Issues* 32: 147–168.

Hill, Dana Carol Davis, and Leann M. Tigges. 1995. "Gendering Welfare State Theory: A Cross-National Study of Women's Public Pension Quality." *Gender & Society* 9 (February).

Hill, M. S., and M. Ponza. 1983. "Poverty and Welfare Dependence across Generations." *Economic Outlook USA* 10 (Summer): 61–64.

Hill, Shirley. 1994. "Motherhood and the Obfuscation of Medical Knowledge: The Case of Sickle Cell Disease." *Gender & Society* 8 (March): 29–47.

Hiller, D., and W. H. Philliber. 1986. "The Division of Labor in Contemporary Marriage: Expectations, Perceptions, and Performance." *Social Problems* 33 (February): 191–201.

Hirschi, Travis. 1969. *Causes of Delinquency.* Berkeley: University of California Press.

Hirschman, Charles. 1994. "Why Fertility Changes." *Annual Review of Sociology* 20: 203–223.

Hite, Shere. 1987. *The Hite Report: Women and Love: A Cultural Revolution in Progress.* New York: Alfred A. Knopf.

Hochschild, Arlie. 1983. *The Managed Heart: Commercialization of Human Feelings.* Berkeley: University of California Press.

Hochschild, Arlie Russell, with Anne Machung. 1989. *The Second Shift: Working Parents and the Revolution at Home.* New York: Viking.

Hochschild, Arlie. 1997. *The Time Bind: When Work Becomes Home and Home Becomes Work.* New York: Metropolitan Books.

Hodge, Robert, and David Tripp. 1986. *Children and Television: A Semiotic Approach.* Cambridge, MA. Polity Press.

Hodge, Robert W., and Donald J. Trieman. 1968. "Class Identification in the United States." *American Journal of Sociology* 75: 535–547.

Hodge, Robert W., Donald J. Treiman, and Peter H. Rossi. 1966. "A Comparative Study of Occupational Prestige." Pp. 309–321 in *Class, Status and Power: Social Stratification in Comparative Perspective,* 2nd ed., edited by Reinhard Bendix and Seymour Martin Lipset. New York: Free Press.

Hodson, Randy. 1996. "Dignity in the Workplace under Participative Management." *American Sociological Review* 61 (October): 719–738.

Hoecker-Drysdale, Susan. 1992. *Harriet Martineau, First Woman Sociologist.* New York: St. Martin's Press.

Hoffman, Lois W., and Jean D. Norris. 1979. "The Value of Children in the United States: A New Approach to the Study of Fertility." *Journal of Marriage and the Family* 41 (August): 583–569.

Hoffman, Saul D., and Greg J. Duncan. 1988. "What *Are* the Economic Consequences of Divorce?" *Demography* 25 (November): 641–645.

Hofstadter, Richard. 1944. *Social Darwinism in American Thought.* Philadelphia: University of Pennsylvania Press.

Hollingshead, A. B., and F. C. Redlich. 1955. *Social Class and Mental Illness: A Community Study.* New York: Wiley.

Holm, Maj. Gen. Jeanne. 1992. *Women in the Military.* Novato, CA: Presidio Press.

Holmes, Malcolm D., H. M. Hosch, H. C. Dauditel, D. A. Perez, and J. B. Graves. 1993. "Judges' Ethnicity and Minority Sentencing: Evidence Concerning Hispanics." *Social Science Quarterly* 74 (September): 496–506.

Homans, George, 1961. *Social Behavior: Its Elementary Forms.* New York: Harcourt Brace Jovanovich.

Homans, George. 1974. *Social Behavior: Its Elementary Forms,* revised ed. New York: Harcourt Brace Jovanovich.

Homes for the Homeless. 1996. *The Age of Confusion: Why So Many Teens are Getting Pregnant, Turning to Welfare and Ending up Homeless.* World Wide Web site: **http://www.opendoor.com/hfh/pregnancy.html**

Homes, G. P. 1988. "Chronic Fatigue Syndrome: A Working Case Definition." *Annals of Internal Medicine* 108 (March): 387–389.

Hondagneu-Sotelo, Pierrette, and Ernestine Avila. 1997. "'I'm Here, but I'm There': The Meanings of Latina Transnational Motherhood." *Gender & Society* 11 (October): 548–571.

Hondagneu-Sotelo, Pierrette. 1992. "Overcoming Patriarchal Constraints: The Reconstruction of Gender Relations Among Mexican Immigrant Women and Men." *Gender & Society* 6 (September): 393–415.

Hondagneu-Sotelo, Pierrette. 1994. *Gendered Transitions: The Mexican Experience of Immigration.* Berkeley: University of California Press.

Hong, Laurence K. 1978. "Risky Shift and Cautious Shift: Some Direct Evidence on the Culture-Value Theory." *Social Psychology* 41 (December): 342–346.

hooks, bell and Cornel West. 1991. *Breaking Bread: Insurgent Black Intellectual Life.* Boston: South End Press.

Hooper, Linda M., and Bennett, Claudette E. 1998. *The Asian and Pacific Islander Population in the United States: March 1997 (Update),* Table 5. Washington, DC: U.S. Bureau of the Census, Current Population Reports. World Wide Web site: **http://www.census.gov/population/socdemo/race/api97**

Hornung, Carlton A. 1977. "Social Status, Status Inconsistency and Psychological Stress," *American Sociological Review* 42 (August): 623–638.

Horowitz, Ruth. 1995. *Teen Mothers: Citizens or Dependents?* Chicago: University of Chicago Press.

Horton, Jacqueline A., ed. 1995. *The Women's Health Data Book,* 2nd ed. Washington, DC: Jacobs Institute of Women's Health.

Horton, John, and José Calderón. 1992. "Language Struggles in a Changing California Community." Pp. 186–194 in *Language Loyalties: A Source Book on the Official English Controversy,* edited by James Crawford. Chicago: University of Chicago Press.

House, James S. 1980. *Occupational Stress and the Mental and Physical Health of Factory Workers.* Ann Arbor, MI: Survey Research Center.

Hout, Michael. 1988. "More Universalism, Less Structural Mobility: The American Occupational Structure in the 1980s." *American Journal of Sociology* 93 (May): 1358–1400.

Howard, Michael C. 1989. *Contemporary Cultural Anthropology,* 3rd ed. Glenview, IL: Scott, Foresman.

Hughes, Langston. 1967. *The Big Sea.* New York: Knopf.

Hughes, Michael, and David H. Demo. 1989. "Self-Perceptions of Black Americans: Self-Esteem and Personal Efficacy." *American Journal of Sociology* 95 (July): 132–159.

Hugick, Larry, and Leslie McAneny. 1992. "A Gloomy America Sees a Nation in Decline, No Easy Solutions Ahead." *Gallup Poll Monthly* 324 (September): 2–7.

Hugick, Larry, and Jennifer Leonard. 1991. "Sex in America." *The Gallup Poll Monthly* (October): 60–73.

Hugick, Larry. 1992a. ". . . But Public Seeks Middle Ground on Abortion, Supports Pennsylvania Law." *The Gallup Poll Monthly* (January): 6–9.

Hugick, Larry. 1992b. "Public Opinion Divided on Gay Rights." *The Gallup Poll Monthly* (June): 2–6.

Hunt, M. 1974. *Sexual Behavior in the 1970s.* Chicago: Playboy Press.

Hunt, Matthew O. 1996. "The Individual, Society, or Both? A Comparison of Black, Latino, and White Beliefs About the Causes of Poverty." *Social Forces* 75 (September): 293–322.

Hunter, Andrea, and James Davis. 1992. "Constructing Gender: An Exploration of Afro-American Men's Conceptualization of Manhood." *Gender & Society* 6 (September): 464–479.

Hunter, Andrea, and Sherrill L. Sellers. 1998. "Feminist Attitudes Among African American Women and Men," *Gender & Society* 12 (February): 81–99.

Hunter, Herbert M., and Sameer Y. Abraham (eds.). 1987. *Race, Class, and the World System: The Sociology of Oliver C. Cox.* New York: Monthly Review Press.

Hunter, James Davison. 1991. *Culture Wars: The Struggle to Define America.* New York: Basic Books.

Hunter, Nan D. 1991. "Sexual Dissent and the Family." *The Nation* 253 (October 7): 406ff.

Hyde, J. S. 1984. "Children's Understanding of Sexist Language." *Developmental Psychology* 20: 697–706.

I

Ignatiev, Noel. 1995. *How the Irish Became White.* New York: Routledge.

Inouye, Daniel K. 1993. "Our Future is in Jeopardy: The Mental Health of Native American Adolescents." *Journal of Health Care for the Poor* 4 (1): 68.

International Office of Justice Statistics. 1997. "Emerging Gangs: An International Threat?" *Crime and Justice International* 13 (October). World wide web site: **http://www.acsp.uic.edu/**

Irvine, Janice M. 1990. *Disorders of Desire: Sex and Gender in Modern American Sexology.* Philadelphia: Temple University Press.

Iverson, M. M., and T. K. H. Freese. 1990. "CFIDS and AIDS: Facing Facts." *CFID Chronicle* (Spring): 37–43.

J

Jackman, Mary. 1983. *Class Awareness in the United States.* Berkeley: University of California Press.

Jackson, Elton F., and Richard F. Curtis. 1968. "Conceptualization and Measurement in the Study of Social Stratification." Pp. 112–149 in *Methodology in Social Research,* edited by H. M. Blalock, Jr. and A. B. Blalock. New York: McGraw-Hill.

Jackson, Linda A., Ruth E. Fleury, and Donna M. Lewandowski. 1996. "Feminism: Definitions, Support, and Correlates of Support among Male and Female College Students." *Sex Roles* 34 (May): 687–693.

Jackson, Pamela B., Peggy A. Thoits, and Howard F. Taylor. 1994. "The Effects of Tokenism on America's Black Elite." Paper read before the American Sociological Association. Los Angeles, August.

Jackson, Pamela B., Peggy A. Thoits, and Howard F. Taylor. 1995. "Composition of the Workplace and Psychological Well-Being: The Effects of Tokenism on America's Black Elite." *Social Forces* 74 (December): 543–557.

Jackson, Pamela Brayboy. 1992. "Specifying the Buffering Hypothesis: Support, Strain, and Depression." *Social Psychology Quarterly* 55: 363–378.

Jacobs, Jerry A., and Ronnie J. Steinberg. 1990. "Compensating Differentials and the Male-Female Wage Gap: Evidence from the New York State Comparable Worth Study." *Social Forces* 69 (December): 430–469.

Jacobs, Jerry. 1984. *The Mall: An Attempted Escape from Everyday Life.* Prospect Heights, IL: Waveland Press.

Jacobs, Michael P. 1997. "Do Gay Men Have a Stake in Male Privilege? The Political Economy of Gay Men's Contradictory Relationship to Feminism." Pp. 165–184 in *Homo Economics: Capitalism, Community, and Lesbian and Gay Life,* edited by Amy Gluckman and Betsy Reed. New York: Routledge.

Jang, Kerry L., W. J. Livesly, and Philip A. Vernon. 1996. "Heritability of the Big Five Personality Dimensions and Their Facets: A Twin Study." *Journal of Personality* 64: 577–589.

Janis, Irving L. 1982. *Groupthink: Psychological Studies of Policy Decisions and Fiascos,* 2nd ed. Boston: Houghton Mifflin.

Jankowski, Mártin Sánchez. 1991. *Islands in the Street: Gangs and American Urban Society.* Berkeley: University of California Press.

Janus, Samuel S., and Cynthia L. Janus. 1993. *The Janus Report on Sexual Behavior.* New York: John Wiley and Sons.

Jaynes, Gerald D., and Robin M. Williams, Jr. 1989. *A Common Destiny: Blacks and American Society.* Washington, DC: National Academy Press.

Jelen, Ted G. 1989. "Weaker Vessels and Helpmeets: Gender Roles, Stereotypes, and Attitudes Toward Female Ordination." *Social Science Quarterly* 70 (September): 579–585.

Jelen, Ted G. 1990. "Religious Belief and Attitude Constraint." *Journal for the Scientific Study of Religion* 29 (March): 118–125.

Jelen, Ted G., and Clyde Wilcox. 1992. "Symbolic and Instrumental Values as Predictors of AIDS Policy Attitudes." *Social Science Quarterly* 73 (December): 737–749.

Jencks, Christopher. 1993. *Rethinking Social Policy: Race, Poverty, and the Underclass.* New York: Harper Collins, 23.

Jencks, Christopher. 1994. *The Homeless.* Cambridge, MA: Harvard University Press.

Jencks, Christopher, Marshall Smith, Henry Ackland, Mary Jo Bane, David Cohen, Herbert Gintis, Barbara Heyns, and Stephan Michelson. 1972. *Inequality: A Reassessment of the Effect of Family and Schooling in America.* New York: Basic Books.

Jencks, Christopher, Susan Bartlett, Mary Corcoran, James Crouse, David Eaglesfield, Gregory Jackson, Kent McClelland, Peter Mueser, Michael Olneck, Joseph Schwartz, Sherry Ward, and Jill Williams. 1979. *Who Gets Ahead? The Determinants of Economic Success in America.* New York: Basic Books.

Jencks, Christopher and Meredith Philips eds. 1998. *The Black-White Test Score Gap.* Washington, DC: The Brookings Institution Press.

Jenkins, J. Craig. 1983. "The Transformation of a Constituency into a Movement: Farm Worker Organizing in California." Pp. 52–71 in *Social Movements of the Sixties and Seventies,* edited by Jo Freeman. New York: Longman.

Jenness, Valerie, and Kendall Broad. 1994. "Antiviolence Activism and the (In)Visibility of Gender in the Gay/Lesbian and Women's Movements." *Gender & Society* 8 (September): 402–423.

Jenness, Valerie. 1995. "Social Movement Growth, Domain Expansion, and Framing Processes: The Gay/Lesbian Movement and Violence Against Gays and Lesbians as a Social Problem." *Social Problems* 42 (February): 145–170.

Jennings, M. K., and R. G. Niemi. 1974. *The Political Character of Adolescence.* Princeton, NJ: Princeton University Press.

Jensen, Arthur R. 1980. *Bias in Mental Testing.* New York: Free Press.

Jessor, T. 1988. "Personal Interest, Group, Conflict, and Symbolic Group Affect: Explanations for Whites' Opposition to Racial Equality." Unpublished doctoral dissertation, University of California, Los Angeles.

Johnson, G. David, Gloria J. Palileo, and Norma B. Gray. 1992. "'Date Rape' on a Southern Campus: Reports from 1991." *Sociology and Social Research* 76 (January): 37–41.

Johnson, Kim K. P. 1995. "Attributions about Date Rape: Impact of Clothing, Sex, Money Spent, Date Type, and Perceived Similarity." *Family and Consumer Sciences Research Journal* 23 (March): 292–310.

Johnson, Norris R., James G. Stember, and Deborah Hunter. 1977. "Crowd Behavior as Risky Shift: A Laboratory Experiment." *Sociometry* 40 (2): 183–187.

Johnstone, Ronald L. 1992. *Religion in Society: A Sociology of Religion,* 4th ed. Englewood Cliffs, NJ: Prentice Hall.

Jones, Edward E., Amerigo Farina, Albert H. Hastorf, Hazel Markus, Dale T. Miller, and Robert A. Scott. 1986. *Social Stigma: The Psychology of Marked Relationships.* New York: W. H. Freeman and Co.

Jones, Edward E., and Keith E. Davis. 1965. "From Acts to Dispositions: The Attribution Process in Person Perception." Pp. 219–266 in *Advances in Experimental Social Psychology,* edited by Leonard Berkowitz. New York: Academic Press.

Jones, James M. 1997. *Prejudice and Racism,* 2nd ed. New York: McGraw Hill.

Jordan, June. 1981. *Civil Wars.* Boston: Beacon Press.

Jordan, Winthrop D. 1968. *White Over Black: American Attitudes Toward the Negro 1550–1812.* Chapel Hill: University of North Carolina Press.

Jordan, Winthrop D. 1969. *The White Man's Burden: Historical Origins of Racism in the United States.* New York: Oxford University Press.

Joseph, Gloria. 1981. "Black Mothers and Daughters." Pp. 75–126 in *Common Differences: Conflicts in Black and White Feminist Perspectives,* edited by Gloria Joseph and Jill Lewis. New York: Anchor.

Joseph, Gloria. 1984. "Mothers and Daughters: Traditional and New Perspectives." *Sage* 1 (Fall): 17–21.

Jost, Kenneth. 1992. "Welfare Reform." *CQ Researcher* (April 10): 315–335.

K

Kadushin, Charles. 1974. *The American Intellectual Elite.* Boston: Little, Brown.

Kail, Robert V., and John C. Cavanaugh. 1996. *Human Development.* Pacific Grove, CA: Brooks/Cole.

Kalmijn, Matthijs. 1991. "Status Homogamy in the United States." *American Journal of Sociology* 97 (September): 496–523.

Kalmijn, Matthijs. 1994. "Mother's Occupational Status and Children's Schooling." *American Sociological Review* 59 (April): 257–275.

Kamin, Leon J. 1974. *The Science and Politics of IQ.* Potomac, MD: Lawrence Erlbaum Associates.

Kamin, Leon J. 1986. Review of *Crime and Human Nature. Scientific American* 254 (February): 22–25.

Kamin, Leon J. 1995. "Behind the Curve." *Scientific American* 272 (February): 99–103.

Kanter, Rosabeth Moss. 1977. *Men and Women of the Corporation.* New York: Basic Books.

Kaplan, Elaine Bell. 1996. *Not Our Kind of Girl: Unraveling the Myths of Black Teenage Motherhood.* Berkeley: University of California Press.

Kapoor, Surinder Kumar. 1973. "Socialization and Feral Children." *Revista Internacional de Sociologia* 31 (January-June): 195–213.

Kassell, Scott. 1995. "Afrocentrism and Eurocentrism in African American Magazine Ads." Princeton, NJ: Unpublished senior thesis.

Katz, I., J. Wackenhut and R. G. Hass. 1986. "Racial Ambivalence, Value Duality, and Behavior." Pp. 35–60 in *Prejudice, Discrimination, and Racism,* edited by J. F. Dovidio and S. L. Gaertner. New York: Academic Press.

Katz, Michael. 1987. *Reconstructing American Education.* Cambridge, MA: Harvard University Press.

Kearl, Michael C. 1989. *Endings: A Sociology of Death and Dying.* New York: Oxford University Press.

Keil, Thomas J., and Gennaro F. Vito. 1995. "Factors Influencing the Use of 'Truth in Sentencing' Law in Kentucky Murder Cases: A Research Note." *American Journal of Criminal Justice* 20 (Fall): 105–111.

Kelley, Maryellen R. 1990. "New Process Technology, Job Design, and Work Organization: A Contingency Model." *American Sociological Review* 55 (April): 191–208.

Kelly, Joan B., and Judith S. Wallerstein. 1976. "The Effects of Parental Divorce: Experiences of the Child in Early Latency." *American Journal of Orthopsychiatry* 46 (January): 20–32.

Kelly, William R., and Larry Isaac. 1984. "The Rise and Fall of Urban Racial Violence in the U.S.: 1948–1979." *Research in Social Movements: Conflict and Change* 7: 203–230.

Kemper, S. 1984. "When to Speak Like a Lady." *Sex Roles* 10: 435–443.

Kennedy, Robert E. 1989. *Life Choices: Applying Sociology,* 2nd edition. New York: Holt, Rinehart and Winston.

Kephart, William M., 1987. *Extraordinary Groups: An Examination of Unconventional Life-Styles.* New York: St. Martin's Press.

Kerckhoff, Alan C., and Back, Kurt W. 1968. *The June Bug: A Study of Hysterical Contagion.* New York: Appleton-Century-Crofts.

Kerp, David. 1986. "'You Can Take the Boy Out of Dorchester, But You Can't Take Dorchester Out of the Boy': Toward a Social Psychology of Mobility." *Symbolic Interaction* 9 (Spring): 19–36.

Kessler, Suzanne J. 1990. "The Medical Construction of Gender: Case Management of Intersexed Infants." *Signs* 16 (Autumn): 3–26.

Kilborn, Peter. 1990. "Workers Using Computers Find a Supervisor Inside." *The New York Times* (December 23): A1ff.

Kilbourne, Barbara Stanek, George Farkas, Kurt Beron, Dorothea Weir, and Paula England. 1994. "Characteristics on the Wages of White Women and Men." *American Journal of Sociology* 100 (November): 698–719.

Killian, Lewis M. 1971. "Optimism and Pessimism in Sociological Analysis." *The American Sociologist* 6 (November): 281–286.

Killian, Lewis. 1975. *The Impossible Revolution, Phase 2: Black Power and the American Dream.* New York: Random House.

Kim, Elaine H. 1993. "Home is Where the *Han* is: A Korean American Perspective on the Los Angeles Upheavals." Pp. 215–235 in *Reading Rodney King/Reading Urban Uprising,* edited by Robert Gooding-Williams. New York: Routledge.

Kim, Kwang Chung, and Won Moo Hurh. 1988. "The Burden of Double Roles: Korean Wives in the USA." *Ethnic and Racial Studies* 11 (April): 151–176.

Kimmel, Michael S., and Martin P. Levine. 1992. "Men and AIDS." Pp. 318–329 in *Men's Lives,* 2nd ed., edited by Michael S. Kimmel and Michael A. Messner. New York: Macmillan.

Kincheloe, Joe L., Shirley R. Steinberg, and Aaron D. Gresson III. 1996. *Measured Lies: The Bell Curve Examined.* New York: St. Martin's Press.

King, Mary C. 1992. "Occupational Segregation by Race and Sex, 1940–1988." *Monthly Labor Review* 11: 30–36.

Kingson, Eric R., and Jill Quadagno. 1995. "Social Security: Marketing Radical Reform." *Generations* 19 (Fall): 43–49.

Kingson, Eric R., and John B. Williamson. 1991. "Generational Equity or Privatization of Social Security." *Society* 28 (September-October): 38–41.

Kingson, Eric R., and John B. Williamson. 1993. "The Generational Equity Debate: A Progressive Framing of a Conservative Issue." *Journal of Aging and Social Policy* 5: 31–53.

Kinsey, Alfred C., W. B. Pomeroy, C. E. Martin, and P. H. Gebbard. 1948. *Sexual Behavior in the Human Male.* Philadelphia: W. B. Saunders.

Kinsey, Alfred C., W. B. Pomeroy, C. E. Martin, and P. H. Gebbard. 1953. *Sexual Behavior in the Human Female.* Philadelphia: W. B. Saunders.

Kitano, Harry. 1976. *Japanese Americans: The Evolution of a Subculture,* 2nd ed. New York: Prentice Hall.

Kitsuse, John I. 1980. "Coming Out All Over: Deviants and the Politics of Social Problems." *Social Problems* 28 (October): 1–13.

Kitsuse, John I., and Aaron V. Cicourel. 1963. "A Note on the Uses of Official Statistics." *Social Problems* 11 (Fall): 131–139.

Kivett, V. R. 1991. "Centrality of the Grandfather Role Among Older Rural Black and White Men." *Journal of Gerontology: Social Sciences* 46: 250–258.

Klapp, Orin. 1972. *Currents of Unrest: Introduction to Collective Behavior.* New York: Holt, Rinehart, and Winston.

Klausner, Samuel Z., and Edward F. Foulks. 1982. *Eskimo Capitalists: Oil, Alcohol, and Politics.* Totowa, NJ: Allanheld.

Klein, D., 1980. "The Etiology of Female Crime: A Review of the Literature." Pp. 70–105 in *Women, Crime, and Justice,* edited by Susan K. Datesman and Frank R. Scarpitti. New York: Oxford University Press.

Kleinfield, N. R., 1995. "A Day (10 Minutes of It) the Country Stood Still." *The New York Times* (October 4): A1ff.

Klineberg, Otto. 1935. *Negro Intelligence and Selective Migration.* New York: Columbia University Press.

Klitsch, Michael (ed.). 1994. " U.S. Birthrate Decreased in 1991, but Nonmarital Fertility Continued to Rise." *Family Planning Perspectives* 26 (January/February): 43–44.

Kloby, Jerry. 1987. "The Growing Divide: Class Polarization in the 1980's." *Monthly Review* 39 (September): 1–8.

Kluegel, J. R., and Lawrence Bobo. 1993. "Dimensions of Whites' Beliefs About the Black-White Socioeconomic Gap." Pp. 127–147 in *Race and Politics in American Society,* edited by P. Sniderman, P. Tetlock, and E. Carmines. Stanford, CA: Stanford University Press.

Kluegel, James R., and Eliot R. Smith. 1986. *Beliefs About Inequality.* New York: Aldine de Gruyter.

Knestant, Andrew. 1996. "Fewer Women Than Men Die of Work-Related Injuries, Data Show." *Compensation and Working Conditions* (June): 45–47. Washington, DC: U.S. Department of Labor.

Knoke, David. 1992. *Political Networks: The Structural Perspective.* New York: Cambridge University Press.

Komter, Aafke. 1989. "Hidden Power in Marriage." *Gender & Society* 3 (June): 187–216.

Kovel, Jonathan. 1970. *White Racism: A Psychohistory.* New York: Pantheon.

Krasnodemski, Memory. 1996. "Justified Suffering: Attribution Theory Applied to Perceptions of the Poor in America." Unpublished senior thesis, Princeton University, Princeton, NJ.

Krasny, Michael. 1993. "Passing the Buck in Tinseltown." *Mother Jones* (July/August): 17–21.

Krauss, Celene. 1994. "Women of Color on the Front Line." Pp. 256–271 in *Unequal Protection: Environmental Justice and Communities of Color,* edited by Robert D. Bullard. San Francisco: Sierra Club Books.

Krauss, R. M., V. Geller, and C. Olson. 1976. "Modalities and Cues in the Detection of Deception." Paper presented before the American Psychological Association, September.

Kray, Susan. 1993. "Orientalization of an 'Almost White' Woman: The Interlocking Effects of Race, Class, Gender and Ethnicity in American Mass Media." *Critical Studies in Mass Communication* 10 (December): 349–366.

Kreft, Ita G. G., and Jan DeLeeuw. 1994. "The Gender Gap in Earnings: A Two Way Nested Multiple Regression Analysis with Random Effects." *Sociological Methods and Research* 22 (February): 319–341.

Kreps, Gary A., 1994. "Disasters as Nonroutine Social Problems." Paper presented at the International Sociological Association, Madrid, Spain.

Kriegsman, D. M. W., B. W. J. Penninx, and J. T. M. Van Eijk. 1994. "Chronic Disease in the Elderly and Its Impact on the Family: A Review of the Literature." *Family Systems Medicine* 12 (Fall): 249–267.

Kristof, Nicholas. 1996. "Aging World, New Wrinkles." *The New York Times* (September 22) Section 4: 1ff.

Krugman, Paul. 1996. "Demographics and Destiny." *The New York Times Book Review* (October 20): 12.

Kucherov, Alex. 1981. "Now Help is On the Way for Neglected Widowers." *U.S. News and World Report* (June): 47–48.

Kuhn, Maggie. 1979. *New Age* (February); cited in *The Columbia Dictionary of Quotations,* 1993.

Kurz, Demie. 1989. "Social Science Perspectives on Wife Abuse: Current Debates and Future Directions." *Gender & Society* 3 (December): 489–505.

Kurz, Demie. 1995. *For Richer for Poorer: Mothers Confront Divorce.* New York: Routledge.

Kutscher, Ronald E., 1995. "Summary of BLS Projections to 2005." *Monthly Labor Review* 118 (November): 3–9.

Kuwahara, Yasue. 1992. "Power to the People Y'all: Rap Music, Resistance, and Black College Students." *Humanity and Society* 16 (February): 54–73.

L

Ladner, Joyce A. 1986. "Teenage Pregnancy: Implications for Black Americans." Pp. 65–84 in *The State of Black America 1986,* edited by James D. Williams. New York: National Urban League.

Ladner, Joyce A. 1995 [1971]. *Tomorrow's Tomorrow The Black Woman,* revised edition. Garden City, NY: Doubleday.

LaFree, Gary. 1980. "The Effect of Sexual Stratification by Race on Official Reactions to Rape." *American Sociological Review* 45 (October): 842–854.

Lamanna, Mary Ann, and Agnes Riedman. 1997. *Marriage and Families: Making Choices in a Diverse Society,* 6th ed. Belmont, CA: Wadsworth Publishing Co.

Lamas, Marta. 1996. "Sex Workers: From Stigma to Political Awareness." *Estudios Sociologicos* 14 (January-April): 33–52.

Lamb, Michael E., Joseph H. Pleck, and James A. Levine. 1986. "Effects of Increased Parental Involvement on Children in Two-Parent Families." Pp. 169–180 in *Men in Families,* edited by Robert A Lewis and Robert E. Salt. Beverly Hills, CA: Sage.

Lamont, Michèle. 1992. *Money, Morals, and Manners: The Culture of the French and the American Upper-Middle Class.* Chicago: University of Chicago Press.

Lamphere, Louise, Alex Stepick, and Guillermo Grenier (eds.). 1994. *Newcomers in the Workplace: Immigrants and the Restructuring of the U.S. Economy.* Philadelphia: Temple University Press.

Landry, Bart. 1987. *The New Black Middle Class.* Berkeley: University of California Press.

Lane, Harlan. 1976. *The Wild Boy of Aveyron.* Cambridge, MA: Harvard University Press.

Lang, Kurt, and Gladys Lang. 1961. *Collective Dynamics.* New York: Thomas Y. Crowell Co.

Langelou, Martha J. 1993. *Back Off! How to Confront and Stop Sexual Harassment and Harassers.* New York: Fireside/Simon & Schuster.

Langston, Donna. 1998. "Tired of Playing Monopoly." Pp. 126–136 in *Race, Class, and Gender: An Anthology,* 3rd ed., edited by Margaret L. Andersen and Patricia Hill Collins. Belmont, CA: Wadsworth Publishing Co.

LaPierre, Richard T. 1934. "Attitudes Versus Actions." *Social Forces* 13: 230–237.

Larana, Enrique, Hank Johnston, and Joseph R. Gusfield (eds.). 1994. *New Social Movements: From Ideology to Identity.* Philadelphia: Temple University Press.

Latané, Bibb, and John M. Darley. 1970. *The Unresponsive Bystander: Why Doesn't He Help?* New York: Appleton-Century Crofts.

Laumann, Edward O., John H. Gagnon, Robert T. Michael, and Stuart Michaels. 1994. *The Social Organization of Sexuality: Sexual Practices in the United States.* Chicago: University of Chicago Press.

Lawler, Edward F. 1986. "The Impact of Age Discrimination Legislation on the Labor Force Participation of Aged Men: A Time-Series Analysis." *Evaluation Review* 10 (December): 794–805.

Laws, Judith Long, and Pepper Schwartz. 1977. *Sexual Scripts: The Social Construction of Female Sexuality.* Hinsdale, IL: Dryden Press.

Lawson, Ronald. 1983. "A Decentralized But Moving Pyramid: The Evolution and Consequences of the Structure of the Tenant Movement." Pp. 119–132 in *Social Movements of the Sixties and Seventies,* edited by Jo Freeman. New York: Longman.

Leacock, Eleanor. 1978. "Women's Status in Egalitarian Society." *Contemporary Anthropology* 19: 247–275.

League of Women Voters. 1996. *Citizen's Guide to the National Voter Registration Act of 1993.* Washington, DC: League of Women Voters.

LeBon, Gustave. 1895 [1960]. *The Crowd.* New York: The Viking Press.

Lederer, Joseph. 1983. "Birth-Control Decisions: Hidden Factors in Contraceptive Choices." *Psychology Today* (June): 32–38.

Lederman, Douglas. 1992. "Men Outnumber Women and Get Most of the Money in Big-Time Sports Programs." *Chronicle of Higher Education* 38 (April 8, 1992): A1ff.

Ledgerwood, Judy L. 1995. "Khmer Kinship: The Matriliny/Matriarchy Myth." *Journal of Anthropological Research* 51 (Fall): 247–315.

Lee, Martin A., and Norman Solomon. 1990. *Unreliable Sources: A Guide to Detecting Bias in News Media.* New York: Carol Publishing Group.

Lee, Richard B. 1968. "What Humans Do for a Living, or, How to Make Out on Scarce Resources." Pp. 30–48 in *Man the Hunter,* edited by Richard B. Lee and Irven DeVore. Chicago: Aldine.

Lee, Sharon M. 1993. "Racial Classification in the U.S. Census: 1890–1990." *Ethnic and Racial Studies* 16 (1): 75–94.

Lee, Sharon M. 1994. "Poverty and the U.S. Asian Population." *Social Science Quarterly* 75 (September): 541–559.

Lee, Stacey J. 1996. *Unraveling the "Model Minority" Stereotype: Listening to Asian American Youth.* New York: Teacher's College Press.

Legge, Jerome S. Jr., 1993. "The Persistence of Ethnic Voting: African Americans and Jews in the 1989 New York Mayoral Campaign." *Contemporary Jewry* 14: 133–146.

Legge, Jerome S., Jr. 1997. "The Religious Erosion-Assimilation Hypothesis: The Case of U.S. Jewish Immigrants." *Social Science Quarterly* 78 (June): 472–486.

Leidner, Robin. 1993. *Fast Food, Fast Talk: Service Work and the Routinization of Everyday Life.* Berkeley: University of California Press.

Lemert, Edwin M. 1972. *Human Deviance, Social Problems, and Social Control.* Englewood Cliffs, NJ: Prentice-Hall.

Lenin, Vladimir. 1939. *Imperialism, the Highest Stage of Capitalism.* New York: International Publishers.

Lenski, Gerhard, Jean Lenski, and Patrick Nolan. 1998. *Human Societies: An Introduction to Macro-Sociology,* 8th ed. New York: McGraw-Hill.

LePoire, Beth A., Carole K. Sigelman, and Henry C. Kenski. 1990. "Who Wants to Quarantine People with AIDS? Patterns of Support for California's Proposition 64." *Social Science Quarterly* 71: 239–249.

Lerman, Robert I., and Theodora J. Ooms. 1993. *Young Unwed Fathers: Changing Roles and Emerging Policies.* Philadelphia: Temple University Press.

Lersch, Kim-Michelle, and Joe R. Feagin. 1996. "Violent Police-Citizen Encounters: An Analysis of Major Newspaper Accounts." *Critical Sociology* 22: 29–49.

Lesko, Nancy. 1988. *Symbolizing Society: Stories, Rites and Structure in a Catholic High School.* Philadelphia: Falmer.

Lester, David, and Biju Yang. 1992. "Social and Economic Correlates of the Elderly Suicide Rate." *Suicide and Life-Threatening Behavior* 22 (Spring): 36–47.

Leuchtag, Alice. 1996. "Merchants of Flesh: International Prostitution and the War on Women's Rights." *Humanist* 55 (March-April): 11–16.

LeVay, Simon. 1991. "A Difference in Hypothalamic Structure Between Heterosexual and Homosexual Men." *Science* 253 (August 30): 1034–1036.

LeVay, Simon. 1993. *The Sexual Brain.* Cambridge: Massachusetts Institute of Technology Press.

Lever, Janet. 1978. "Sex Differences in the Complexity of Children's Play and Games." *American Sociological Review* 43 (August): 471–483.

Levin, J., and W. Levin. 1980. *Ageism: Prejudice and Discrimination Against the Elderly.* Belmont, CA: Wadsworth.

Levin, Jack, and Jack McDevitt. 1993. *Hate Crimes: The Rising Tide of Bigotry and Bloodshed.* New York: Plenum.

Levine, Adeline. 1982. *Love Canal: Science, Politics and People.* Lexington, MA: Lexington Books.

Levine, Art. 1987. "The Uneven Odds." *U.S. News and World Report* (August 17): 31–33.

Levine, Felice J., and Katherine J. Rosich. 1996. *Social Causes of Violence: Crafting a Science Agenda.* Washington, DC: American Sociological Association.

Levine, John M., and Richard L. Moreland. 1998. "Small Groups." Pp. 415–469 in *The Handbook of Social Psychology,* 4th ed., edited by Daniel T. Gilbert, Susan T. Fiske, and Gardner Lindzey. New York: Oxford University and McGraw Hill

Levine, Lawrence. 1984. "William Shakespeare and the American People: A Study in Cultural Transformation." *American Historical Review* 89 (1): 34–66.

Levine, Martin P. 1992. "The Status of Gay Men in the Workplace." Pp. 251–266 in *Men's Lives,* 2nd ed., edited by Michael S. Kimmel and Michael A. Messner. New York: Macmillan.

Levine, Michael P. 1987. *Student Eating Disorders: Anorexia Nervosa and Bulimia.* Washington, DC: National Education Association.

Levitan, Sar A., and William B. Johnson. 1975. *Indian Giving: Federal Program for Native Americans.* Baltimore: Johns Hopkins University Press.

Levy, Marion J. 1949. *The Structure of Society.* Princeton, NJ: Princeton University Press.

Lewin, Ellen. 1984. "Lesbianism and Motherhood: implications for Child Custody." Pp. 163–183 in *Women-Identified Women,* edited by Trudy Darty and Sandee Pottee. Mountain View, CA: Mayfield.

Lewin, Tamar. 1978. "Rise in Single-Parent Families Found Continuing." *Marriage and Family Review* 1: 1–12.

Lewis, David Levering. 1993. *W.E.B. DuBois: Biography of a Race.* New York: Henry Holt and Company.

Lewis, George H. 1989. "Rats and Bunnies: Core Kids in an American Mall." *Adolescence* 24 (Winter): 881–889.

Lewis, Oscar. 1960. *Five Families: Mexican Case Studies in the Culture of Poverty.* New York: Basic Books.

Lewis, Oscar. 1966. "The Culture of Poverty." *Scientific American* 215 (October): 19–25.

Lewis, Robert A. 1981. *Men in Difficult Times: Today and Tomorrow.* Englewood Cliffs, NJ: Prentice Hall.

Lewontin, Richard. 1996. *Human Diversity.* New York: W. H. Freeman & Company.

Lewontin, Richard C. 1996 [1970]. "Race and Intelligence." *Bulletin of the Atomic Scientists* 26: 2–8.

Lieberson, Stanley, and Silverman, Arnold. 1965. "The Precipitant and Underlying Conditions of Race Riots." *American Sociological Review* 30 (December): 887–898.

Lieberson, Stanley. 1980. *A Piece of the Pie: Black and White Immigrants Since 1880.* Berkeley: University of California Press.

Liebman, Robert C., and Robert Wuthnow. 1983. *The New Christian Right.* Hawthorne, NY: Aldine Publishing Company.

Lincoln, C. Eric, and Lawrence H. Mamiya. 1990. *The Black Church in the African-American Experience.* Durham, NC: Duke University Press.

Link, Bruce G., Frances Palamara Mesagno, Maxine E. Lubner, and Bruce P. Dohrenwend. 1990. "Problems in Measuring Role Strains and Social Functioning in Relation to Psychological Symptoms." *Journal of Health and Social Behavior* 31 (December): 354–369.

Link, Bruce, E. Susser, A. Stueve, J. Phelan, R. E. Moore, and E. Struening. 1994. "Lifetime and Five-Year Prevalence of Homelessness in the United States." *American Journal of Public Health* (December): 1907–1912.

Linton, Ralph. 1937. *The Study of Man.* New York: Appleton-Century.

Lipps, Bradley. 1993. *Who Stole the Soul?: The Effect of Market Value on Rap Music.* Unpublished senior thesis. Princeton University, Princeton, NJ.

Lips, Hilary M. 1993. *Sex and Gender: An Introduction,* 2nd ed. Mountain View, CA: Mayfield Publishing Co.

Livingston, I. L. 1994. *Handbook of Black American Health: The Mosaic of Conditions, Issues and Prospects.* Westport, CT: Greenwood.

Lo, Clarence. 1990. *Small Property versus Big Government: Social Origins of the Property Tax Revolt.* Berkeley: University of California Press.

Lobao, Linda M. 1990. *Locality and Inequality: Farm and Industry Structure and Socioeconomic Conditions.* Albany: The State University of New York Press.

Locklear, Erin M. 1999. *Where Race and Politics Collide: The Federal Acknowledgement Process and Its Effects on Lumsee and Pequot Indians.* Princeton, NJ: Senior Thesis, Princeton University.

Loewen, James W. 1982. *Social Science in the Courtroom: Statistical Techniques and Research Methods for Winning Class-Action Suits.* Lexington, MA: Lexington Books.

Lofland, John. 1985. *Protest: Studies of Collective Behavior and Social Movements.* New Brunswick, NJ: Transaction Books.

Logan, John, and Harvey L. Molotch. 1987. *Urban Fortunes: The Political Economy of Place.* Berkeley: University of California Press.

Logio-Rau, Kim. 1998. "Here's Looking at You, Kid: Race, Gender and Health Behaviors Among Adolescents." Unpublished Ph.D. Dissertation, University of Delaware.

Lombroso, N. 1920. *The Female Offender.* New York: Appleton.

Long, Theodore E., and Jeffrey K. Hadden. 1983. "Religious Conversion and the Concept of Socialization: Integrating the Brainwashing and Drift Models." *Journal for the Scientific Study of Religion* 22 (March): 1–14.

Longman, Jere. 1996. "Left Alone in his Frigid House, DuPont Heir is Seized by Police." *The New York Times* (January 29): A1.

Lopata, Helene Z., and Barrie Thorne. 1978. "On the Term 'Sex Roles.'" *Signs* 3 (Spring): 718–721.

Lorber, Judith, Rose Laub Coser, Alice S. Rossi, and Nancy Chodorow. 1981. "On *The Reproduction of Mothering:* A Methodological Debate." *Signs* 6 (Spring): 482–513.

Lorber, Judith. 1994. *Paradoxes of Gender.* New Haven, CT: Yale University Press.

Lord, Walter. 1956. *A Night to Remember.* New York: Bantam.

Loredo, Carren, Anne Reid, and Kay Deaux. 1995. "Judgments and Definitions of Sexual Harassment by High School Students." *Sex Roles* 32 (January): 29–45.

Lorenz, Konrad. 1966. *On Aggression.* New York: Harcourt.

Loring, Marti, and Brian Powell. 1988. "Gender, Race and DSM-III: A Study of Objectivity of Psychiatric Diagnostic Behavior." *Journal of Health and Social Behavior* (March): 1–12.

Lucal, Betsy. 1994. "Class Stratification in Introductory Textbooks: Relational or Distributional Models?" *Teaching Sociology* 22 (April): 139–150.

Luckenbill, David F. 1986. "Deviant Career Mobility: The Case of Male Prostitutes." *Social Problems* 33 (April): 283–296.

Luguila, Terry A. 1998. *Marital Status and Living Arrangements: March 1997 (Update).* Current Population Reports, Series P20–506. Washington, DC: U.S. Department of Commerce.

Luker, Kristin. 1975. *Taking Chances.* Berkeley: University of California Press.

Luker, Kristin. 1984. *Abortion and the Politics of Motherhood.* Berkeley: University of California Press.

Luker, Kristin. 1996. *Dubious Conceptions: The Politics of Teenage Pregnancy.* Cambridge, MA: Harvard University Press.

Lusane, Clarence. 1993. "Rap, Race, and Politics." *Race & Class* 35 (July/September): 41–55.

Lyman, Karen A. 1993. *Day In, Day Out with Alzheimer's Stress in Caregiving Relationships.* Philadelphia: Temple University Press.

Lyman, Peter. 1987. "The Fraternal Bond as a Joking Relationship: A Case Study of the Role of Sexist Jokes in Male Group Bonding." Pp. 143–154 in *Men's Lives,* 2nd ed., edited by Michael S. Kimmel and Michael A. Messner. New York: Macmillan.

M

Maccoby, Eleanor E. 1990. "Gender and Relationships: A Developmental Account." *American Psychology* 45 (April): 513–520.

MacKay, N. J., and K. Covell. 1997. "The Impact of Women in Advertisements on Attitudes toward Women." *Sex Roles* 36 (May): 573–583.

MacKinnon, Catherine. 1983. "Feminism, Marxism, Method, and the State: An Agenda for Theory." *Signs* 7 (Spring): 635–658.

MacKinnon, Neil J., and Tom Langford. 1994. "The Meaning of Occupational Prestige Scores: A Social Psychological Analysis and Interpretation." *Sociological Quarterly* 35 (May): 215–145.

Mackintosh, N. J. 1995. *Cyril Burt: Fraud or Framed?* Oxford: Oxford University Press.

Macmillan, Jeffrey. 1996. "Santa's Sweatshop." *U.S. News & World Report,* December 16: 50ff.

Madriz, Esther. 1997. *Nothing Bad Happens to Good Girls: Fear of Crime in Women's Lives.* Berkeley: University of California Press.

Mahoney, Patricia, and Linda M. Williams. 1998. "Sexual Assault in Marriage: Prevalence, Consequences, and Treatment of Wife Rape." Pp. 113–162 in *Partner Violence: A Comprehensive Review of 20 Years of Research,* ed. by Jana L. Jasinski and Linda M. Williams. Thousand Oaks, CA: Sage.

Makinson, Larry. 1992. *The Cash Constituents of Congress.* Washington, DC: Congressional Quarterly Inc.

Maldonado, Lionel A., and Charles V. Willie. 1996. "Developing A 'Pipeline' Recruitment Program for Minority Faculty." Pp. 330–371 in *Educating a New Majority: Transforming America's Educational System for Diversity,* edited by Laura I. Rendon and Richard O. Hope. San Francisco: Jossey-Bass Publishers.

Maldonado, Lionel, A. 1997. "Mexicans in the American System: A Common Destiny." In *Ethnicity in the United States: An Institutional Approach,* edited by William Velez. Bayside, NY: General Hall.

Malson, Lucien. 1972. *Wolf Children and the Problem of Human Nature.* New York: Monthly Review Press.

Mandel, Ruth, and Debra Dodson. 1992. "Do Women Officeholders Make a Difference?" Pp. 144–177 in *American Women, 1992–93: A Status Report,* edited by Paula Ries and Anne Stone. New York: W. W. Norton and Company.

Mansfield, Phyllis-Kernoff, Patricia Barthalow-Koch, Julie Henderson, Judith R. Vicary, Margaret Cohn, and Elaine W. Young. 1991. "The Job Climate for Women in Traditionally Male Blue-Collar Occupations." *Sex Roles* 25 (July): 63–79.

Mansnerns, Laura. 1992. "Should Tracking Be Derailed?" *The New York Times,* (Education Life), Sec. 4A, Nov. 1, pp. 14–16.

Marcuse, Herbert. 1964. *One-Dimensional Man.* Boston: Beacon Press.

Margolin, Leslie. 1992. "Deviance on Record: Techniques for Labeling Child Abusers in Official Documents." *Social Problems* 39 (February): 58–70.

Margolis, Diane Rothbard. 1996. "Victimized Daughters: Incest and the Development of the Female Self." *Gender & Society* 10 (August): 488–489.

Markle, Gerald E., and Ronald J. Troyer. 1979. "Smoke Gets in Your Eyes: Cigarette Smoking as Deviant Behavior." *Social Problems* 26 (June): 611–625.

Marks, Carole. 1989. *Farewell, We're Good and Gone: The Great Black Migration.* Bloomington, IN: Indiana University Press.

Marks, Carole. 1991. "The Urban Underclass." *Annual Review of Sociology* 11 (1985): 445–466.

Marlowe and Company. 1995. "How Schools Shortchange Girls." Research Report, Marlowe and Co., New York, NY.

Marshall, Susan. 1997. *Splintered Sisterhood: Gender & Class in the Campaign Against Woman Suffrage.* Madison, WI: University of Wisconsin Press.

Martin, Susan E. 1995. "'Cross-Burning is Not Just an Arson': Police Social Construction of Hate Crimes in Baltimore." *Criminology* 33 (August): 303–326.

Martin, Susan Ehrlich. 1989. "Sexual Harassment: The Link Between Gender Stratification, Sexuality, and Women's Economic

Status." Pp. 57–75 in *Women: A Feminist Perspective,* 4th ed., edited by Jo Freeman. Palo Alto, CA: Mayfield Publishing Co.

Martindale, Melanie. 1991. "Sexual Harassment in the Military." *Sociological Practice Review* 2 (July): 200–216.

Martinez, Ramiro. 1996. "Latinos and Lethal Violence: The Impact of Poverty and Inequality." *Social Problems* 43 (May): 131–146.

Marx, Anthony. 1997. *Making Race and Nation: A Comparison of the United States, South Africa, and Brazil.* Cambridge: Cambridge University Press.

Marx, Gary T. 1967. "Religion: Opiate or Inspiration of Civil Rights Militancy Among Negroes." *American Sociological Review* 32 (February): 64–72.

Marx, Gary T., and Douglas McAdam. 1994. *Collective Behavior and Social Movements: Process and Structure.* Englewood Cliffs, NJ: Prentice Hall.

Marx, Groucho. 1959. *Groucho and Me.* New York: B. Geis Associates.

Marx, Karl, and Friedrich Engels. 1888. *Manifesto of the Communist Party.* Chicago: Charles Kerr.

Marx, Karl. 1967 [1867]. *Capital.* Edited by F. Engels. New York: International Publishers.

Marx, Karl. 1972 [1843]. "Contribution to the Critique of Hegel's *Philosophy of Right.*" Pp. 11–23 in *The Marx-Engels Reader,* edited by Robert C. Tucker. New York: W. W. Norton.

Marx, Karl. 1972 [1845]. "The German Ideology." Pp. 110–164 in *The Marx-Engels Reader,* edited by Robert C. Tucker. New York: Norton.

Mason, Karen Oppenheim, and Yu-Shia Lu. 1988. "Attitudes Toward Women's Familial Roles: Changes in the United States, 1977–1985." *Gender & Society* 2 (March): 39–57.

Massey, Douglas S. 1993. "Latino Poverty Research: An Agenda for the 1990s." *Social Science Research Council Newsletter* 47 (March): 7–11.

Massey, Douglas S., and Nancy A. Denton. 1993. *American Apartheid: Segregation and the Making of the Underclass.* Cambridge, MA: Harvard University Press.

Masters, William H., and Virginia E. Johnson. 1966. *Human Sexual Response.* Boston: Little, Brown.

Mathews, Linda. 1996. "More than Identity Rests on a New Racial Category." *The New York Times,* July 6: 1–7.

Maticka-Tyndale, Eleanor. 1992. "Social Construction of HIV Transmission and Prevention Among Heterosexual Young Adults." *Social Problems* 39 (August): 238–252.

Mauer, Marc. 1997a. *Americans Behind Bars: U.S. and International Use of Incarceration 1995.* Washington, DC: The Sentencing Project.

Mauer, Marc. 1997b. *Intended and Unintended Consequences: State Racial Disparities in Imprisonment.* Washington, DC: The Sentencing Project.

Mayo, C., and Nancy Henley. 1981. *Gender and Nonverbal Behavior.* New York: Springer-Verlag.

Mazur, Alan. 1968. "The Littlest Science." *The American Sociologist* 3 (August): 195–200.

McAdam, Doug. 1982. *Political Process and the Development of Black Insurgency.* Chicago: University of Chicago Press.

McAllister, Ronald J. 1991. "Some Reflections on Teaching the Sociology of Religion." Pp. 7–9 in *Syllabi and Instructional Materials for the Sociology of Religion,* 2nd ed., edited by Dallas A. Blanchard and Madeleine Adriance. Washington, DC: American Sociological Association Teaching Resources Center.

McAneny, Leslie. 1994. "Ethnic Minorities View of the Media's View of Them." *The Gallup Poll Monthly* (August, No. 347): 31–41.

McCarthy, John, and Mayer Zald. 1973. *The Trend of Social Movements in America: Professionalism and Resource Mobilization.* Morristown, NJ: General Learning Press.

McCauley, C. 1989. "The Nature of Social Influence in Groupthink: Compliance and Internalization." *Journal of Personality and Social Psychology* 57 (August): 250–260.

McCloskey, Laura Ann. 1996. "Socioeconomic and Coercive Power Within the Family." *Gender & Society* 10 (August): 449–463.

McCord, William, and Arline McCord. 1986. *Paths to Progress: Bread and Freedom in Developing Societies.* New York: W. W. Norton.

McCord, William. 1991. "The Asian Renaissance." *Society* 28 (September-October): 50–61.

McCormack, A., M.D. Janus, and A. W. Burgess. 1986. "Runaway Youths and Sexual Victimization: Gender Differences in an Adolescent Runaway Population." *Child Abuse and Neglect* 10: 387–395.

McCrate, Elaine, and Joan Smith. 1998. "When Work Doesn't Work: The Failure of Current Welfare Reform." *Gender & Society* 12 (February): 61–81.

McDowell, Jeanne. 1992. "Are Women Better Cops?" *Time* 139 (February 17): 70–72.

McGrew, W. C. 1992. *Chimpanzee Material Culture.* New York: Cambridge University Press.

McGuire, Gail M., and Barbara F. Reskin. 1993. "Authority Hierarchies at Work: The Impacts of Race and Sex." *Gender & Society* 7 (December): 487–506.

McGuire, Meredith. 1996. *Religion: The Social Context.* Belmont, CA: Wadsworth Publishing.

McIntosh, John. 1992. "Epidemiology of Suicide in the Elderly." *Suicide and Life-Threatening Behavior* 22 (Spring): 15–35.

McKinney, Kathleen, 1992. "Contrapower Sexual Harassment: The Offender's Viewpoint." *Free Inquiry in Creative Sociology* 20 (May): 3–10.

McKinney, Kathleen. 1994. "Sexual Harassment and College Faculty Members." *Deviant Behavior* 15: 171–191.

McLanahan, Sara S., Annemette Sorensen, and Dorothy Watson. 1989. "Sex Differences in Poverty, 1950–1980." *Signs* 15 (Autumn 1989): 102–122.

McLanahan, Sara, and Gary Sandefur. 1994. *Growing Up with a Single Parent: What Hurts, What Helps.* Cambridge, MA: Harvard University Press.

McLanahan, Sara. 1983. "Family Structure and the Reproduction of Poverty." Discussion Paper 720A-83. Madison, WI: Institute for Research on Poverty.

McLane, Daisann. 1995. "The Cuban-American Princess." *The New York Times Magazine* (February 26): 42.

McLemore, Dale. 1994. *Racial and Ethnic Relations in America,* 4th ed. Boston: Allyn and Bacon.

McLeod, Jane D., and Michael J. Shanahan. 1993. "Poverty, Parenting, and Children's Mental Health." *American Sociological Review* 58 (June): 351–366.

McMurray, Coleen. 1990. "Reasons for Being Rich or Poor." *The Gallup Poll Monthly* 298 (July): 35.

McPhail, Clark. 1971. "Civil Disorder Participation: A Critical Examination of Recent Research." *American Sociological Review* 36 (December): 1058–1073.

Mead, George Herbert. 1934. *Mind, Self, and Society.* Chicago: University of Chicago Press.

Mead, Lawrence. 1992. "Job Programs and Other Bromides." *The New York Times* 141 (May 19): A19.

Mechanic, D., and David A. Rochefort. 1990. "Deinstitutionalization: An Appraisal of Reform." *Annual Review of Sociology* 16: 301–327.

Meehan, Patrick J., Linda E. Saltzman, and Richard W. Sattin. 1991. "Suicides among Older United States Residents: Epidemiologic Characteristics and Trends." *The American Journal of Public Health* 81 (September): 1198–1203.

Meisenheimer, Joseph R. III. 1992. "How Do Immigrants Fare in the U.S. Labor Market?" *Monthly Labor Review* 115 (December): 3–19.

Menaghan, Elizabeth G. 1993. "The Long Reach of the Job: Effects of Parents' Occupational Conditions on Family Patterns and Children's Well-Being." Paper presented at the Annual Meeting of the American Sociological Association, Miami, August.

Merlo, Joan M., and Kathleen Maurer Smith. 1994a. "The Portrayal of Gender Roles in Television Advertising: A Decade of Stereotyping." Paper presented at the Society for the Study of Social Problems, August.

Merlo, Joan M., and Kathleen Maurer Smith. 1994b. "The Feminine Voice of Authority in Television Commercials: A Ten-Year Comparison." Paper given at the Annual Meeting of the American Sociological Association, August.

Mernissi, Fatima. 1987. *Beyond the Veil: Male-Female Dynamics in Modern Muslim Society.* Bloomington, IN: Indiana University Press.

Merton, Robert K. 1968. "Social Structure and Anomie." *American Sociological Review* 3: 672–682.

Merton, Robert K. 1957. *Social Theory and Social Structure.* New York: Free Press.

Merton, Robert K. 1972. "Insiders and Outsiders: A Chapter in the Sociology of Knowledge." *American Journal of Sociology* 78 (July): 9–47.

Merton, Robert, and Alice K. Rossi. 1950. "Contributions to the Theory of Reference Group Behavior." Pp. 279–334 in *Studies, Scope and Method of "The American Soldier."* New York: Free Press.

Messner, Michael A. 1992. *Power at Play: Sports and the Problem of Masculinity.* Boston: Beacon Press.

Messner, Michael A. 1996. "Studying up on Sex." *Sociology of Sport Journal* 13 (September): 221–237.

Messner, Michael A. 1989. "Masculinities and Athletic Careers." *Gender & Society* 3 (March): 71–88.

Metz, Michael A., B. R. Rosser-Simon, and Nancy Strapko. 1994. "Differences in Conflict-Resolution Styles among Heterosexual, Gay, and Lesbian Couples." *Journal of Sex Research* 31: 293–308.

Meyer, David S., and Joshua Gamson. 1995. "The Challenge of Cultural Elites: Celebrities and Social Movements." *Sociological Inquiry* 65 (Spring): 181–206.

Meyer, Jan. 1990. "Guess Who's Coming to Dinner This Time? A Study of Gay Intimate Relationships and the Support for Those Relationships." *Marriage and Family Review* 14: 59–82.

Meyer, Madonna Harrington. 1994. "Gender, Race, and the Distribution of Social Assistance: Medicaid Use Among the Elderly." *Gender & Society* 8 (March): 8–28.

Michael, Robert T., John H. Gagnon, Edward O. Laumann, and Gina Kolata. 1994. *Sex in America: A Definitive Survey* Boston: Little, Brown and Company.

Michel, Robert T., Heidi I. Hartmann. and Brigid O'Farrell (eds.). 1989. *Pay Equity: Empirical Inquiries.* Washington, DC: National Academy Press.

Middlebrook, Diane. 1998. *Suits Me: The Double Life of Billy Tipton.* Boston: Houghton Mifflin.

Milgram, Stanley, and Hans Toch. 1968. "Collective Behavior: Crowds and Social Movements." Pp. 507–610 in *Handbook of Social Psychology,* 2nd ed., edited by Gardner Lindzey and Elliot Aronson. Reading, MA: Addison-Wesley.

Milgram, Stanley. 1974. *Obedience to Authority: An Experimental View.* New York: Harper and Row.

Miller, David. 1985. *Collective Behavior.* Belmont, CA: Wadsworth Publishing.

Miller, Eleanor. 1986. *Street Women.* Philadelphia: Temple University Press.

Miller, Eleanor. 1991. "Jeffrey Dahmer, Racism, Homophobia, and Feminism." *SWS Network News* 8 (December): 2.

Miller, Marc Crispin. 1996. "Free the Media." *The Nation* 262 (June 3): 25–26.

Miller, Susan L. 1994. "Expanding the Boundaries: Toward a More Inclusive and Integrated Study of Intimate Violence." *Violence and Victims* 9: 183–194.

Miller, Susan L. 1997. "The Unintended Consequences of Current Criminal Justice Policy." Talk presented at Research on Women Series, University of Delaware, October 1997.

Mills, C. Wright. 1956. *The Power Elite.* New York: Oxford University Press.

Mills, C. Wright. 1959. *The Sociological Imagination.* New York: Oxford University Press.

Min, Pyong G. 1990. "Problems of Korean Immigrant Entrepreneurs." *International Migration Review* 24 (Fall): 436–455.

Miner, Horace. 1956. "Body Ritual Among the Nacirema." *American Anthropologist* 58: 503–507.

Mintz, Beth, and Michael Schwartz. 1985. *The Power Structure of American Business.* Chicago: University of Chicago Press.

Mintzer, Jacobo E., et al. 1992. "Daughters' Caregiving for Hispanic and Non-Hispanic and Alzheimer Patients: Does Ethnicity Make a Difference?" *Community Mental Health Journal* 28 (August): 293–303.

Mirandé, Alfredo. 1979. "Machismo: A Reinterpretation of Male Dominance in the Chicano Family." *The Family Coordinator* 28: 447–449.

Mirandé, Alfredo. 1985. *The Chicano Experience.* Notre Dame, IN: Notre Dame University Press.

Miringoff, Marc L. 1996. *1996 Index of Social Health.* Tarrytown, NY: Fordham Institute for Innovation in Social Policy.

Mita, T. H., M. Dermer, and J. Knight. 1977. "Reversed Facial Images and the Mere Exposure Hypothesis." *Journal of Personality and Social Psychology* 35: 597–601.

Mitchell, G., Stephanie Obradovich, Fred Herring, Chris Tromborg, and Alyson L. Burns. "Reproducing Gender in Public Places: Adults' Attention to Toddlers in Three Public Locales." *Sex Roles* 26 (September): 323–330.

Mizutami, Osamu. 1990. *Situational Japanese.* Tokyo: The Japan Times.

Mohr, Richard D. 1988. *Gays/Justice: A Study of Ethics, Society and Law.* New York: Columbia University Press.

Moller, Lora C., Shelley Hymel, and Kenneth H. Rubin. 1992. "Sex Typing in Play and Popularity in Middle Childhood." *Sex Roles* 26 (April): 331–353.

Monaghan, Peter. 1993. "Sociologist Jailed Because He 'Wouldn't Snitch' Ponders the Way Research Ought to be Done." *The Chronicle of Higher Education* (September 1): A8–9.

Money, John, and Ehrhardt, A. A. 1972. *Man, Woman, Boy and Girl: The Differentiation and Dimorphism of Gender Identity from Conception to Maturity.* Baltimore: Johns Hopkins University Press.

Money, John. 1985. "Gender: History, Theory and Usage of the Term in Sexology and Its Relationship to Nature/Nurture." *Journal of Sex and Marital Therapy* 11 (Summer): 71–79.

Money, John. 1988. *Gay, Straight, and In-Between: The Sexology of Erotic Orientation.* New York: Oxford University Press.

Money, John. 1995. *Gendermaps: Social Constructionism, Feminism and Sexosophical History.* New York: Continuum.

Monk-Turner, Elizabeth. 1990. "The Occupational Achievement of Community and Four-Year College Graduates." *American Sociological Review* 55 (October): 719–725.

Montagne, Peter. 1995. "Review of Julian L. Simon, ed., *The State of Humanity.*" Cambridge, MA: Blackwell Publishers.

Moore, David W. 1995. "Most Americans Say Religion is Important to Them." *The Gallup Poll Monthly* (February, No. 353): 16–19.

Moore, David W., 1996. "Public Opposes Gay Marriages." *The Gallup Poll Monthly* (April, No. 367): 19–21.

Moore, Gwen. 1979. "The Structure of a National Elite Network." *American Sociological Review* 44 (October): 673–692.

Moore, Joan, and Harry Pachon. 1985. *Hispanics in the United States.* Englewood Cliffs, NJ: Prentice Hall.

Moore, Joan, and Raquel Pinderhughes (eds.). 1993. *In the Barrios: Latinos and the Underclass Debate.* New York: Russell Sage Foundation.

Moore, Joan. 1976. *Hispanics in the United States.* Englewood Cliffs, NJ: Prentice Hall.

Moore, Robert B. 1992. "Racist Stereotyping in the English Language." Pp. 317–328 in *Race, Class, and Gender: An Anthology,* 2nd ed., edited by Margaret L. Andersen and Patricia Hill Collins. Belmont, CA: Wadsworth.

Moore, Thomas S. 1990. "The Nature and Unequal Incidence of Job Displacement Costs." *Social Problems* 37 (May): 230–242.

Moore, Thomas S. 1992. "Racial Differences in Postdisplacement Joblessness." *Social Science Quarterly* 73 (September): 674–689.

Mor-Barak, Michal E., Andrew E. Scharlach, Lourdes Birba, and Jacque Sokolov. 1992. "Employment, Social Networks, and Health in the Retirement Years." *International Journal of Aging and Human Development* 35: 145–159.

Morgan, Lewis H. 1877. *Ancient Society, or Researches in the Lines of Human Progress, from Savagery Through Barbarism to Civilization.* Cambridge: Harvard University Press.

Morgan, Mary Y. 1987. "The Impact of Religion on Gender-Role Attitudes." *Psychology of Women Quarterly* 11 (September): 301–310.

Morgan, Robin, and Gloria Steinem. 1980. "The International Crime of Genital Mutilation." *Ms.* 8 (March): 65–67ff.

Morin, S. F., and E. M. Garfinkle. 1978. "Male Homophobia." *The Journal of Social Issues* 34 (Winter): 29–47.

Morris, Aldon, and Carol McClurg Mueller (eds.). 1992. *Frontiers in Social Movement Theory.* New Haven, CT: Yale University Press.

Morris, Aldon. 1984. *The Origins of the Civil Rights Movement: Black Communities Organizing for Change.* New York: Free Press.

Morris, Lydia. 1989. "Household Strategies: The Individual, the Collectivity, and the Labour Market: The Case of Married Couples." *Work, Employment, and Society* 3: 447–464.

Moskos, Charles. 1988. *Soldiers and Sociology.* Alexandria, VA: United States Army Research Institute for the Social and Behavioral Sciences.

Moskos, Charles. 1992. "Right Behind You, Scarlett!" Pp. 40–54 in *Women in the Military,* edited by E. A. Blacksmith. New York: The H. W. Wilson Co.

Moskos, Charles C., and John Sibley Butler. 1996. *All That We Can Be: Black Leadership and Racial Integration The Army Way.* New York: Basic Books.

Moynihan, Daniel P. 1965. *The Negro Family: The Case for National Action.* Washington, DC: Office of Policy Planning and Research, United States Department of Labor.

Mukerji, Chandra, and Michael Schudson. 1986. "Popular Culture." *Annual Review of Sociology* 12: 47–66.

Muller, Thomas, and Thomas J. Espenshade. 1985. *The Fourth Wave: California's Newest Immigrants.* Washington, DC: The Urban Institute Press.

Mumola, Christopher J., and Allen J. Beck. 1997. *Prisoners in 1996.* Washington, DC: U.S. Department of Justice.

Murty, Komanduri S., Julian B. Roebuck, and Gloria R. Armstrong. 1994. "The Black Community's Reactions to the 1992 Los Angeles Riot." *Deviant Behavior* 15 (March): 85–104.

Mutchler, Jan Elise, and W. Parker Frisbie. 1987. "Household Structure among the Elderly: Racial/Ethnic Differentials." *National Journal of Sociology* 1 (Spring): 3–23.

Myers, Walter D. 1998. *Amistad Affair.* New York: NAL/Dutton.

Myerson, Allen R. 1998. "Rating the Bigshots: Gates vs. Rockefeller." *The New York Times* (May 24): Section 4, p. 4.

Myles, John. 1989. *Old Age in the Welfare State: The Political Economy of Public Pensions,* revised edition. Lawrence: University Press of Kansas.

Myrdal, Gunnar. 1944. *An American Dilemma: The Negro Problem and Modern Democracy.* 2 Vols. New York: Harper and Row.

N

Nagel, Joane. 1994. "Constructing Ethnicity: Creating and Recreating Ethnic Identity and Culture." *Social Problems* 41 (February): 152–176.

Nagel, Joane. 1996. *American Indian Ethnic Renewal: Red Power and the Resurgence of Identity and Culture.* New York: Oxford University Press.

Nakao, Keiko, and Judith Treas. 1994. "Updating Occupational Prestige and Socioeconomic Scores: How the New Measures Measure Up." *Sociological Methodology* 24:1–72.

Namboodiri, Krishnan. 1988. "Ecological Demography: Its Place in Sociology." *American Sociological Review* 53 (August): 619–633.

Nanda, Serena. 1998. *Neither Man Nor Woman: The Hijras of India.* Belmont, CA: Wadsworth Publishing Co.

Napholz, Linda. 1995. "Mental Health and American Indian Women's Multiple Roles." *American Indian and Alaska Native Mental Health Research* 6: 57–75.

Naples, Nancy. 1992. "Activist Mothering: Cross-Generational Continuity in the Community Work of Women from Low-Income Urban Neighborhoods." *Gender & Society* 6 (September): 441–463.

Nardi, Peter M., and Beth Schneider. 1997. *Social Perspectives on Lesbian and Gay Studies.* New York: Routledge.

Nasar, Sylvia. 1992. "Fed Gives New Evidence of 80s Gains by Richest." *The New York Times,* April 21, p. A1ff.

National Cancer Institute. 1990. *Cancer Among Blacks and Other Minorities: Statistical Profiles.* Washington, DC: U.S. Department of Health and Human Services.

National Cancer Institute. 1997. *Cancer Among Blacks and Other Minorities: Statistical Profiles.* Washington, DC: U.S. Department of Health and Human Services, Public Health Service, National Institutes of Health.

National Center for Health Statistics. 1994. *Health United States 1994.* Hyattsville, MD: U.S. Department of Health and Human Services, May.

National Center for Health Statistics. 1994. *Monthly Vital Statistics Report* 43 (October 25).

National Center for Health Statistics. 1998: *Health, United States, 1998: With Socioeconomic Status and Health Chartbook.* Hyattsville, MD: National Center for Health Statistics.

National Center on Elder Abuse. 1996. "Elder Abuse in Domestic Settings." World Wide Web site: **http://interinc.com/NCEA/Statistics/p1.html**

National Coalition for the Homeless. 1996. *NCH Fact Sheet: Homeless Families with Children.* Washington, DC: National Coalition for the Homeless. **http://www2.ari.net/home/nch/families.html**

National Coalition for the Homeless. 1998. *NCH Fact Sheet #2: How Many Homeless People Are There?"* Washington, DC: National Coalition for the Homeless. World Wide Web site: **http://nch.ari.net/numbers.html**

National Committee to Prevent Child Abuse. 1998. World Wide Web site: **http://www/childabuse.org/facts97.htm**

National Council of Catholic Bishops. 1992. "Ad Hoc Committee Report/Women's Concerns: One in Christ Jesus." *Origins* 22 (December 31): 490–508.

National Institute of Mental Health. 1982. *Television and Behavior: Ten Years of Scientific Progress and Implications for the Eighties.* Washington, DC: U.S. Government Printing Office.

National Opinion Research Center. 1996. *General Social Surveys, 1972–1995: Cumulative Codebook.* Storrs, CT: The Roper Center.

National Science Foundation. 1990. *Academic Science and Engineering: Graduate Enrollment and Support, Fall 1990.* U.S. Government Printing Office, Washington, D.C..

National Urban League. 1996. *The State of Black America.* New York: National Urban League, Inc.

Nee, Victor. 1973. *Longtime Californ': A Documentary Study of an American Chinatown.* New York: Pantheon Books.

Neighbors, Harold W., and James S. Jackson, eds. 1996. *Mental Health in Black America.* Thousand Oaks, CA: Sage.

Nemeth, Charlan. 1980. "Social Psychology in the Courtroom." Pp. 443–463 in *A Survey of Social Psychology,* edited by Leonard Berkowitz. New York: Holt.

Neppl, Tricia K., and Ann D. Murray. 1997. "Social Dominance and Play Patterns among Preschoolers: Gender Comparisons." *Sex Roles* 36 (March): 381–394.

Neugarten, B. L., J. W. Moore, and J. C. Lowe. 1965. "Age Norms, Age Constraints, and Adult Socialization." *American Journal of Sociology* 70: 710–717.

Newman, Katherine S. 1988. *Falling from Grace: The Experience of Downward Mobility in the American Middle Class.* New York: Free Press.

Newman, Katherine S. 1993. *Declining Fortunes: The Withering of the American Dream.* New York: Basic Books.

Niebuhr, Gustav. 1998. "Makeup of American Religion is Looking More Like Mosaic, Data Say." *The New York Times* (April 12): 14.

Nielsen Television Index (1989); cited in Davis, Donald M. 1990. "Portrayals of Women in Prime-Time Network Television: Some Demographic Characteristics." *Sex Roles* 23: 325–332.

Nigg, Joanne. 1995. Personal correspondence, Disaster Research Center, University of Delaware.

Nisbet, Robert. 1970. *The Social Bond: An Introduction to the Study of Society.* New York: Alfred A. Knopf.

Nixon, Ron. 1996. "Caution: Children at Work." *The Progressive* 60 (August): 30–33.

Nock, Steven L. 1998. "The Consequences of Premarital Fatherhood." *American Sociological Review* 6 (April): 250–263.

Norrgard, Lee E. 1992. *Final Choices: Making End-of-Life Decisions.* Santa Barbara, CA: ABL-CLIO.

Nowak, T., and K. Snyder. 1986. "Sex Differences in the Long-Term Consequences of Job Loss." Paper presented at the Annual Meeting of the American Sociological Society, September 1986.

Nsiah-Jefferson, Laurie. 1989. "Reproductive Laws, Women of Color, and Low-Income Women." *Women's Rights Law Reporter* 11 (Spring): 15–38.

O

Oakes, Jeannie, and Martin Lipton. 1996. Pp. 168–200 in *Educating A New Majority: Transforming America's Educational System for Diversity,* edited by Laura I. Rendon and Richard O. Hope. San Francisco: Jossey-Bass Publishers.

Oakes, Jeannie. 1985. *Keeping Track: How Schools Structure Inequality.* New Haven: Yale University Press.

Oakes, Jeannie. 1990. "Multiplying Inequalities: The Effects of Race, Social Class and Tracking on Opportunities to Learn Mathematics and Science." RAND for National Science Foundation.

Oberschall, Anthony. 1979. "Protracted Conflict." Pp. 45–70 in *Dynamics of Social Movements,* edited by Mayer N. Zald and John McCarthy. New Brunswick, NJ: Transaction Publishers.

Oberschall, Anthony. 1993. *Social Movements: Ideologies, Interests, and Identities.* New Brunswick, NJ: Transaction Publishers.

Ochshorn, Judith. 1981. *The Female Experience and the Nature of the Divine.* Bloomington, IN: Indiana University Press.

O'Donnell, Lydia, Carl R. O'Donnell, Joseph H. Pleck, John Snarey, and Richard M. Rose. 1987. "Psychosocial Responses of Hospital Workers to Acquired Immune Deficiency Syndrome (AIDS)." *Journal of Applied Social Psychology* 17: 269–285.

Ogburn, William F. 1922. *Social Change with Respect to Cultural and Original Nature.* New York: B. W. Huebsch, Inc.

Ogburn, William F. 1964 [1922]. *On Culture and Social Change.* Chicago: University of Chicago Press.

O'Hare, William P. 1996. *A New Look at Poverty in America.* Washington, DC: Population Reference Bureau.

O'Kelly, Charlotte G., and Larry S. Carney. 1986. *Women and Men in Society,* 2nd ed. Belmont, California: Wadsworth.

Oliver, Melvin L., Thomas M. Shapiro, and Julie E. Press. 1995. "'Them That's Got Shall Get': Inheritance and Achievement in Wealth Accumulation." Pp. 69–95 in *Research in Politics and Society: The Politics of Wealth and Inequality,* edited by Richard E. Ratcliff, Melvin L. Oliver, and Thomas M. Shapiro. Greenwich, CT: JAI Press.

Omi, Michael, and Howard Winant. 1994. *Racial Formation in the United States,* 2nd ed. New York: Routledge.

Ornstein, Norman J., Thomas E. Mann, and Michael J. Malbin. 1996. *Vital Statistics on Congress, 1995–1996.* Washington, DC: Congressional Quarterly, Inc.

Ortiz, Vilma, and Rosemary Santana Cooney. 1984. "Sex Role Attitudes and Labor Force Participation Among Young Hispanic Females and Non-Hispanic White Females." *Social Science Quarterly* 6 (June): 392–400.

Ostrander, Susan. 1984. *Women of the Upper Class.* Philadelphia: Temple University Press.

Ouchi, William. 1981. *Theory Z: How American Business Can Meet the Japanese Challenge.* Reading, MA: Addison-Wesley.

Owens, Sarah E. 1998. "The Effects of Race, Gender, Their Interactions, and Selected School Variables upon Educational Aspirations and Achievements." Senior thesis. Princeton University, Department of Sociology.

P

Padavic, Irene. 1991. "Attractions in Male Blue-Collar Jobs for Black and White Women: Economic Need, Exposure, and Attitudes." *Social Science Quarterly* 72 (March): 33–49.

Padavic, Irene. 1992. "White Collar Values and Women's Interest in Blue-Collar Jobs." *Gender & Society* 6 (June): 215–230.

Padilla, Felix. 1989. "Salsa Music as a Cultural Expression of Latino Consciousness and Unity." *Hispanic Journal of Behavioral Science* 11 (February): 28–45.

Page, Charles H. 1946. "Bureaucracy's Other Face." *Social Forces* 25 (October): 89–94.

Palen, John J. 1992. *The Urban World.* New York: McGraw-Hill.

Palmore, Erdman, and Daisaku Maeda. 1985. *The Honorable Elders Revisited: A Revised Cross-Cultural Analysis of Aging in Japan.* Durham, NC: Duke University Press.

Palmore, Erdman. 1977. "Facts of Aging." *The Gerontologist* 17: 315–320.

Palmore, Erdman. 1986. "Trends in the Health of the Aged." *The Gerontologist* 26: 298–302.

Pampel, F. C. 1994. "Population Aging, Class Context, and Age Inequality in Public Spending." *American Journal of Sociology* 100: 153–195.

Pampel, F. C., and P. Adams. 1992. "The Effects of Demographic Change and Political Structure on Family Allowance Expenditures." *Social Service Review* 66: 524–546.

Parcel, Toby L., and Elizabeth G. Menaghan. 1994. "Early Parental Work, Family Social Control, and Early Childhood Outcomes." *American Journal of Sociology* 99 (January): 972–1009.

Parelius, Robert, and Ann Parelius. 1987. *The Sociology of Education,* 2nd ed. Englewood Cliffs, NJ: Prentice-Hall.

Park, Robert E., and Ernest W. Burgess. 1921. *Introduction to the Science of Society.* Chicago: University of Chicago Press.

Parks, Malcolm R. 1996. "Making Friends in Cyberspace." *Journal of Communication* 46 (Winter): 80–97.

Parsons, Talcott (ed.). 1947. *Max Weber: The Theory of Social and Economic Organization.* New York: Free Press.

Parsons, Talcott. 1951. *The Social System.* Glencoe, IL: The Free Press.

Parsons, Talcott. 1951b. *Toward a General Theory of Action.* Cambridge, MA: Harvard University Press.

Parsons, Talcott. 1966. *Societies: Evolutionary and Comparative Perspectives.* Englewood Cliffs, NJ: Prentice Hall.

Parsons, Talcott. 1968 (1937). *The Structure of Social Action.* New York: Free Press.

Pascoe, Peggy. 1991. "Race, Gender and Intercultural Relations: The Case of Interracial Marriage." *Frontiers* 12 (1): 5–18.

Passuth, P. M., D. R. Maines, and B. L. Neugarten. 1984. "Age Norms and Age Constraints Twenty Years Later." Paper presented at the Annual Meeting of the Midwest Sociological Society, Chicago.

Patterson, Francine. 1978. "Conversations with a Gorilla." *National Geographic* 154 (October): 438–465.

Patterson, James, and Peter Kim. 1991. *The Day America Told the Truth: What People* Really *Believe About Everything That Really Matters.* New York: Prentice Hall.

Pavetti, LaDonna Ann. 1993. "The Dynamics of Welfare and Work: Exploring the Process by Which Women Work Their Way Off Welfare." Cambridge, MA: Malcolm Wiener Center for Social Policy, Harvard University (May).

Payne, Deborah M., John T. Warner, and Roger D. Little. 1992. "Tied Migration and Returns to Human Capital: The Case of Military Wives." *Social Science Quarterly* 73 (June): 324–339.

Pear, Robert. 1996. "Clinton to Sign Welfare Bill That Ends U.S Aid Guarantee and Gives States Broad Power." *The New York Times* (August 1): A1ff.

Pearce, Diane, and Harriet McAdoo. 1981. *Women and Children: Alone and in Poverty.* Washington, DC: National Advisory Council on Economic Opportunity.

Pearlin, Leonard I., Carol S. Aneshensel, and David A. Chiriboa. 1994. "Caregiving: The Unexpected Career." *Social Justice Research* 7 (December): 373–390.

Pedraza, Sylvia. 1991. "Women and Migration." *Annual Review of Sociology* 17: 303–325.

Pedraza, Sylvia. 1996. "Origins and Destinies: Immigration, Race, and Ethnicitiy in American History." Pp. 1–20 in *Origins and Destinies: Immigration, Race, and Ethnicity in America,* edited by Sylvia Pedraza and Rubén G. Rumbaut. Belmont, CA: Wadsworth.

Pedraza, Silvia. 1996a. "Cuba's Refugees: Manifold Migrations." Pp. 263–279 in *Origins and Destinies: Immigration, Race, and Ethnicity in America,* edited by Silvia Pedraza and Rubén Rumbaut. Belmont, CA: Wadsworth.

Pedraza, Sylvia, and Rubén G. Rumbaut. (eds.). 1996. *Origins and Destinies: Immigration, Race, and Ethnicity in America.* Belmont, CA: Wadsworth Publishing Company.

Pennock-Roman, Maria. 1990. *Test Validity and Language Background: A Study of Hispanic American Students at Six Universities.* New York: College Entrance Examination Board.

Pennock-Roman, Maria. 1994. "College Major and Gender Differences in the Prediction of College Grades." College Board Research Report No. 94–2, ETS Research Reports No. 94–24.

Perez, Denise N. 1996. *A Case for Derailment: Tracking in the American Public School System as an Obstacle to the Improvement of Minority Education.* Princeton, NJ: Princeton University Senior Thesis, Department of Sociology.

Perrow, Charles. 1986. *Complex Organization: A Critical Essay,* 3rd ed. New York:

Perrow, Charles. 1994. "The Limit of Safety: The Enhancement of a Theory of Accidents." *Journal of Contingencies and Crisis Management* 22: 212–220.

Persell, Caroline Hodges. 1977. *Education and Inequality: The Roots and Results of Stratification in America's Schools.* New York: Free Press.

Persell, Caroline Hodges. 1990. *Understanding Society: An Introduction to Sociology.* New York: Harper and Row.

Persell, J. 1996. Personal Communication, Urban Institute, Washington, DC.

Pescosolido, Bernice A., Elizabeth Grauerholz, and Melissa A. Milkie. 1997. "Culture and Conflict: The Portrayal of Blacks in U.S. Children's Picture Books Through the Mid- and Late-Twentieth Century." *American Sociological Review* 62 (June): 443–464.

Pesina, Maria D., Daryl L. Hitchcock, and Beth Menees Rienzi. 1994. "The Military Ban Against Gay Males: University Students' Attitudes Before and After the Presidential Decision." *Journal of Social Behavior and Personality* 9 (September): 499–506.

Petchesky, Rosalind Pollack. 1983. "Reproduction and Class Divisions Among Women." Pp. 221–241 in *Class, Race, and Sex: The Dynamics of Control,* edited by Amy Swerdlow and Hanna Lessinger. Boston: G. K. Hall.

Petersen, Larry R., and Gregory V. Donnenwerth. 1997. "Secularization and the Influence of Religion on Beliefs about Premarital Sex." *Social Forces* 75 (March): 1071–1089.

Peterson, Richard R. 1996a. "A Re-Evaluation of the Economic Consequences of Divorce." *American Sociological Review* 61 (June): 528–536.

Peterson, Richard R. 1996b. "A Re-Evaluation of the Economic Consequences of Divorce: Reply to Weitzman." *American Sociological Review* 61 (June): 539–540.

Pettigrew, Thomas F. 1971. *Racially Separate or Together?* New York: McGraw-Hill.

Pettigrew, Thomas F. 1985. "New Black-White Patterns: How Best to Conceptualize Them?" Pp. 329–346 in *Annual Review of Sociology* 11: 329–346.

Pettigrew, Thomas F. 1992. "The Ultimate Attribution Error: Extending Allport's Cognitive Analysis of Prejudice." Pp. 401–419 in *Readings About the Social Animal,* edited by Elliott Aronson. New York: Freeman.

Phillips, Robert L., Paul J. Andrisani, Thomas N. Daymont, and Curtis L. Gilroy. 1992. "The Economic Returns to Military Service: Race-Ethnic Differences." *Social Science Quarterly* 73 (June): 340–359.

Piaget, Jean. 1926. *The Language and Thought of the Child.* New York: Harcourt.

Pillemer, Karl. 1988. "Maltreatment of the Elderly at Home and in Institutions: Extent, Risk Factors, and Policy Recommendations." U.S. Congress Select Committee on Aging. Washington, DC: U.S. Government Printing Office.

Piven, Frances Fox, and Richard A. Cloward. 1971. *Regulating the Poor.* New York: Pantheon.

Piven, Frances Fox, and Richard A. Cloward. 1988. *Why Americans Don't Vote.* New York: Pantheon.

Piven, Francis Fox, and Richard A. Cloward. 1977. *Poor People's Movements: Why They Succeed, How They Fail.* New York: Vintage Books.

Platt, Anthony M. 1977. *The Child Savers: The Invention of Delinquency,* 2nd ed. Chicago: University of Chicago Press.

Pollock, Philip H., and M. Elliot Vittas. 1995. "Who Bears the Burdens of Environmental Pollution: Race, Ethnicity, and Environmental Equity in Florida." *Social Science Quarterly* 76 (June): 294–310.

Pomerleau, A., Bolduc, D., Malcuit, G., and Cossette, L. 1990. "Pink or Blue: Environmental Stereotypes in the First Two Years of Life." *Sex Roles* 22 (March): 359–367.

Popkin, Susan. 1990. "Welfare: Views from the Bottom." *Social Problems* 37 (February): 64–79.

Population Reference Bureau. 1995. *World's Women.* Washington, DC: Population Reference Bureau.

Portes, Alejandro, and Ramon Grosfoguel. 1994. "Caribbean Diasporas: Migration and Ethnic Communities." *Annual of the American Academy of Political and Social Science* 533 (May): 48–69.

Portes, Alejandro, and Rubén Rumbaut. 1990. *Immigrant America.* Berkeley: University of California Press.

Portes, Alejandro, and Rubén G. Rumbaut. 1996. *Immigrant America: A Portrait,* 2nd ed. Berkeley: University of California Press.

Posner, Richard A. 1992. *Sex and Reason.* Cambridge, MA: Harvard University Press.

Poster, Winifred. 1995. "The Challenges and Promises of Class and Racial Diversity in the Women's Movement: A Study of Two Women's Organizations." *Gender & Society* (December): 659–679.

Powell, Catherine Tabb. 1991. "Rap Music: An Education with a Beat from the Street." *Journal of Negro Education* 60 (Summer): 245–259.

Powell-Hopson, Darlene and Derek S. Hopson. 1988. "Implications of Doll Color Preferences Among Black Preschool Children and White Preschool Children." *The Journal of Black Psychology* 14 (February): 57–63.

Press, Eyal. 1996. "Barbie's Betrayal." *The Nation* (December 30): 11–16.

Press, Julie E., and Eleanor Townsley. 1998. "Wives' and Husbands' Reporting: Gender, Class, and Social Desirability." *Gender & Society* 12 (April): 188–218.

Preston, Richard. 1991. *American Steel: Hot Metal Men and the Resurrection of the Rust Belt.* New York: Prentice Hall.

Prudhomme, Alex. 1991. "Will Gates Give Up the Fight at Last?" *Time* 138 (July 22): 18.

Pruitt, D.G. 1971. "Choice Shifts in Group Discussion." *Journal of Personality and Social Psychology* 20 (December): 339–360.

Prus, Robert, and C. R. D. Sharper. 1991. *Road Hustler.* New York: Kauffman and Greenbery Publishers.

Pryor, John B., Glenn D. Reeder, Richard Vinacco, and Teri L. Kott. 1989. "The Instrumental and Symbolic Functions of Attitudes Toward Persons with AIDS." *Journal of Applied Social Psychology* 19: 377–404.

Punch, Maurice. 1996. *Dirty Business: Exploring Corporate Misconduct; Analysis and Cases.* Thousand Oaks, CA: Sage.

Purcell, P., and L. Stewart. 1990. "Dick and Jane in 1989." *Sex Roles* 22: 177–185.

Pyke, Karen. 1994. "Women's Employment as a Gift or Burden? Marital Power Across Marriage, Divorce, and Remarriage." *Gender & Society* 8 (March): 73–91.

Q

Quadagno, Jill. 1989. "Generational Equity and the Politics of the Welfare State." *Politics and Society* 17 (September): 353–376.

Quarantelli, Enrico L., and Russell R. Dynes. 1970. "Property Norms and Looting: Their Patterns in Community Crisis." *Phylon* 31 (Summer): 181–193.

Quarantelli, Enrico. 1954. "The Nature and Conditions of Panic." *American Journal of Sociology* 60 (3): 267–275.

Quarantelli, Enrico. 1978. "The Behavior of Panic Participants." Pp. 141–146 in *Collective Behavior and Social Movements,* edited by Louis Genevie. Itasca, IL: F. E. Peacock Publisher.

R

Rabine, Leslie W. 1985. "Romance in the Age of Electronics: Harlequin Enterprises." *Feminist Studies* 11 (Spring): 39–60.

Raboteau, Albert J. 1978. *Slave Religion: The "Invisible Institution" in the Antebellum South.* New York: Oxford University Press.

Rampersad, Arnold. 1986. *The Life of Langston Hughes: Vol. I: 1902–1941. I, Too, Sing America.* New York: Oxford University Press.

Rampersad, Arnold. 1988. *The Life of Langston Hughes: Vol. II: 1941–1967. I Dream A World.* New York: Oxford University Press.

Ransby, Barbara, and Tracey Matthews. 1993. "Black Popular Culture and the Transcendence of Patriarchal Illusions." *Race and Class* 35 (July-September): 57–68.

Read, Piers Paul. 1974. *Alive: The Story of the Andes Survivors.* Philadelphia: J. B. Lippincott.

Reeves, B., and M. M. Miller. 1978. "A Multidimensional Measure of Children's Identification with Television Characters." *Journal of Broadcasting* 22 (Winter): 71–86.

Reeves, Brian. 1992. *State and Local Police Departments, 1990.* Washington, DC: Bureau of Justice Statistics.

Reich, Robert. 1991. *The Work of Nations: Preparing Ourselves for 21st Century Capitalism.* New York: Knopf.

Reid, Frances. 1994. "Complaints of a Dutiful Daughter." Film. New York: Women Make Movies.

Reid, T. R. 1994. "How the West Was Dumb." *The Washington Post* (March 24): C1ff.

Reilly, Mary Ellen, Bernice Lott, Donna Caldwell, and Luisa DeLuca. 1992. "Tolerance for Sexual Harassment Related to Self-Reported Sexual Victimization." *Gender & Society* 6 (March): 122–138.

Reiman, Jeffrey. 1998. *The Rich Get Richer and the Poor Get Prison,* 5th ed. Boston: Allyn and Bacon.

Reinharz, Shulamit. 1992. *Feminist Methods in Social Research.* New York: Oxford University Press.

Reinhold, Robert. 1991. "Study of Los Angeles Police Finds Violence and Racism are Routine." *The New York Times* (July 10): A1, A14.

Reinisch, June Hanover. 1990. *The Kinsey Institute New Report on Sex: What You Must Know to Be Sexually Literate.* New York: St. Martin's Press.

Rendon, Laura I., and Richard O. Hope (eds.). 1996. *Educating a New Majority: Transforming America's Educational System for Diversity.* San Francisco: Jossey-Bass Publishers.

Renzetti, Claire. 1992. *Violent Betrayal: Partner Abuse in Lesbian Relationships.* Newbury Park: Sage.

Reskin, Barbara. 1988. "Bringing the Men Back In: Sex Differentiation and the Devaluation of Women's Work." *Gender & Society* 2 (March): 58–81.

Reskin, Barbara F., and Irene Padavic. 1988. "Supervisors as Gatekeepers: Male Supervisors' Responses to Women's Integration in Plant Jobs." *Social Problems* 35 (December): 536–550.

Reskin, Barbara, and Irene Padavic. 1994. *Women and Men at Work.* Thousand Oaks, CA: Pine Forge.

Reynolds, Peggy, Peggy P. Boyd, Robert S. Blacklow, James S. Jackson, Raymond S. Greenberg, Donald F. Austin, Vivian W. Chen, and Brenda K. Edwards. 1992. "The Relationship Between Social Ties and Survival Among Breast Cancer Patients in the Black/White Study of Cancer Survival." Research Report. Washington, DC: National Cancer Institute.

Rhode, Deborah L. 1995. "Media Images, Feminist Issues." *Signs* 20 (Spring): 685–710.

Rhodes, A. L. 1983. "Effects of Religious Denominations on Sex Differences in Occupational Expectations." *Sex Roles* 9 (January): 93–108.

Rice, Berkeley. 1986. "Water Shocks of the '80s." *Across the Board* (March): 17.

Rich, Adrienne. 1980. "Compulsory Heterosexuality and Lesbian Existence." *Signs* 5 (Summer): 631–660.

Rich, Alexander R., Joyce Kirkpatrick-Smith, Ronald L. Bonner, and Frank Jans. 1992. "Gender Differences in the Psychosocial Correlates of Suicidal Ideation Among Adolescents." *Suicide and Life-Threatening Behavior* 22 (Fall): 364–373.

Rich, Melissa K., and Thomas F. Cash. 1993. "The American Image of Beauty: Media Representations of Hair Color for Four Decades." *Sex Roles* 29 (July): 113–124.

Rich, Spencer. 1992. "Low-Paying Jobs Up Sharply in Decade, U.S. Says." *The Washington Post* (May 12): A7.

Rieff, David. 1993. "A Global Culture?" *World Policy Journal* 10 (Winter): 73–81.

Riesman, David. 1970[1950]. *The Lonely Crowd: A Study of the Changing American Character.* New Haven, CT: Yale University Press.

Rifkin, Jeremy. 1989. *Entropy: Into the Greenhouse World,* revised ed. New York: Bantam Books.

Rifkin, Jeremy. 1998. "The Biotech Century: Human Life as Intellectual Property." *The Nation* 266 (April 13): 11–19.

Riley, Matilda White, and Joan Waring. 1976. "Age and Aging." Pp. 355–410 in *Contemporary Social Problems,* edited by Robert Merton and Robert Nisbet. New York: Harcourt.

Riley, Matilda White. 1987. "On the Significance of Age in Sociology." *American Sociological Review* 52 (February): 1–14.

Risman, Barbara. 1986. "Can Men 'Mother'? Life as a Single Father." *Family Relations* 35 (January): 95–102.

Risman, Barbara. 1987. "Intimate Relationships from a Microstructural Perspective: Men Who Mother." *Gender & Society* 1 (March): 6–32.

Rist, Darrell Yates. 1992. "Are Homosexuals Born That Way?" *The Nation* 255 (October 14): 424–429.

Ritzer, George. 1996. *The McDonaldization of Society: An Investigation Into the Changing Character of Contemporary Social Life,* revised ed. Thousand Oaks, CA: Pine Forge Press.

Rivera, George F., and Robert M. Regoli. 1987. "Sexual Victimization Experiences of Sorority Women." *Sociology and Social Research* 72 (October): 39–41.

Robbins, Thomas. 1988. *Cults, Converts, and Charisma: The Sociology of New Religious Movements.* Beverly Hills, CA: Sage.

Robertson, Roland. 1992. *Globalization: Social Theory and Global Culture.* Newbury Park, CA: Sage.

Robey, Bryant. 1982. "A Guide to the Baby Boom." *American Demographics* 4 (September): 16–21.

Robinson, James D., and Thomas Skill. 1995. "The Invisible Generation: Portrayals of the Elderly on Prime-Time Television." *Communication Reports* 8 (Summer): 111–119.

Robinson, Jo Ann Gibson. 1987. *The Montgomery Bus Boycott and the Women Who Started It.* Knoxville, TN: The University of Tennessee Press.

Rodgers, Joan R. 1995. "An Empirical Study of Intergenerational Transmission of Poverty in the United States." *Social Science Quarterly* 76 (March): 178–194.

Rodney, Walter. 1974. *How Europe Underdeveloped Africa.* Washington, DC: Howard University Press.

Rodney, Walter. 1981. *How Europe Underdeveloped Africa,* revised edition. Washington, DC: Howard University Press.

Rodriguez, Clara E. 1989. *Puerto Ricans: Born in the U.S.A.* Boston: Unwin Hyman.

Roethlisberger, Fritz J., and William J. Dickson. 1939. *Management and the Worker.* Cambridge, MA: Harvard Univesity Press.

Rollins, Judith. 1985. *Between Women: Domestics and Their Employers.* Philadelphia: Temple University Press.

Rones, Philip L., Randy E. Ilg, and Jennifer M. Gardner. 1997. "Trends in Hours of Work Since the Mid-1970s." *Monthly Labor Review* 120 (April): 3–14.

Roof, Wade Clark. 1993. *A Generation of Seekers.* San Francisco: Harpers.

Roof, Wade Clark, and William McKinney. 1987. *American Mainline Religion: Its Changing Shape and Future.* New Brunswick, NJ: Rutgers University Press.

Roos, P.A. 1985. *Gender and Work: A Comparative Analysis of Industrial Societies.* Albany: SUNY Press.

Root, Maria P. P. (ed.). 1996. *The Multiracial Experience: Racial Borders as the New Frontier.* Thousand Oaks, CA: Sage.

Roper Organization. 1987. *The American Dream.* Storrs, CT: Roper Organization.

Roper Organization. 1990. *Virginia Slims American Women's Opinion Poll.* Storrs, CT: Roper Organization.

Roper Organization. 1993. *The Gallup Poll.* Storrs, CT: The Roper Organization.

Roper Organization. 1994. *The Gallup Poll.* Storrs, CT: The Roper Organization.

Roper Organization. 1995a. "Short Subjects." *The Gallup Poll Monthly* (June): 35.

Roper Organization. 1995b. *The 1995 Virginia Slims Opinion Poll: A 25–Year Perspective of Women's Lives.* Storrs, CT: Roper Organization.

Roper Organization. 1996. *The Gallup Poll.* Storrs, CT: The Roper Organization.

Roper Organization. 1997. *The Gallup Poll.* Storrs, CT: The Roper Organization.

Rose, Peter I., ed. 1970. *Americans from Africa: Slavery and Its Aftermath.* New York: Atherton.

Rosenau, Pauline Marie. 1992. *Post-Modernism and the Social Sciences: Insights, Inroads, and Intrusions.* Princeton, NJ: Princeton University Press.

Rosenberg, Morris, and Roberta Simmons. 1971. *Black and White Self-Esteem: The Urban School Child.* Washington, DC: American Sociological Association.

Rosenberg, Morris. 1981. "The Self-Concept: Social Product and Social Force," Pp. 593–624 in *Social Psychology: Sociological Perspectives,* edited by Morris Rosenberg and Ralph H. Turner. New York: Basic Books.

Rosenhan, David L. 1973. "On Being Sane in Insane Places." *Science* 179 (January 19): 250–258.

Rosenthal, Elin E. 1993. "New Benefits and New Rules for Family and Medical Leave." *CUPA*

Benefits Alert (February 11). Washington, DC: College and University Personnel Association.

Rosenthal, Robert, and Lenore Jacobson. 1968. *Pygmalian in the Classroom: Teacher Expectations and Pupils' Intellectual Development.* New York: Holt, Rinehart and Winston.

Ross, Catherine E., and Chia-ling Wu. 1995. "The Links Between Education and Health." *American Sociological Review* 60 (October): 719–745.

Ross, Catherine E., and John Mirowsky. 1996. "Economic and Interpersonal Work Rewards: Subjective Utilities of Men's and Women's Compensation." *Social Forces* 75 (September): 223–245.

Rossi, Alice (ed.) 1973. *The Feminist Papers: From Adams to deBeauvoir.* New York: Columbia University Press.

Rossi, Alice S., and Peter H. Rossi. 1990. *Of Human Bonding: Parent-Child Relations Across the Life Course.* New York: Aldine de Gruyter.

Rostow, W. W. 1978. *The World Economy: History and Prospect.* Austin: University of Texas Press.

Rothenberg, Stuart, and Frank Newport. 1984. *The Evangelical Voter.* Washington, DC: Free Congress Foundation.

Rubin, Gayle. 1975. "The Traffic in Women." Pp. 157–211 in *Toward an Anthropology of Women,* edited by Rayna Reiter. New York: Monthly Review Press.

Rubin, Linda J., and Sherry B. Borgers. 1990. "Sexual Harassment in Universities During the 1980s." *Sex Roles* 23 (October): 397–411.

Rubin, Z., F. J. Provenzano, and C. T. Hull. 1974. "The Eye of the Beholder: Parents' Views on Sex of Newborns." *American Journal of Orthopsychiatry* 44: 512–519.

Rudé, George. 1964. *The Crowd in History, 1730–1848.* New York: John Wiley.

Rueschmeyer, Dietrich, and Theda Skocpol. 1996. *States, Social Knowledge, and the Origins of Modern Social Policies.* Princeton, NJ: Princeton University Press.

Rumbaut, Rubén G. 1996a. "Origins and Destinies: Immigration, Race, and Ethnicity in Contemporary America." Pp. 21–42 in *Origins and Destinies: Immigration, Race, and Ethnicity in America,* edited by Sylvia Pedraza and Rubén G. Rumbaut. Belmont, CA: Wadsworth.

Rumbaut, Rubén. 1996b. "Prologue." Pp. xvi–xix in *Origins and Destinies: Immigration, Race, and Ethnicity in America,* edited by Silvia Pedraza and Rubén Rumbaut. Belmont, CA: Wadsworth.

Rundblad, Georgeanne. 1995. "Exhuming Women's Premarket Duties in the Care of the Dead." *Gender & Society* 9 (April): 173–193.

Russell, Diane E. H. 1986. *The Secret Trauma: Incest in the Lives of Girls and Women.* New York: Basic Books.

Rust, Paula C. 1992. "The Politics of Sexual Identity: Sexual Attraction and Behavior Among Lesbian and Bisexual Women." *Social Problems* 39 (November): 366–386.

Rust, Paula C. 1993. "'Coming Out' in the Age of Social Constructionism: Sexual Identity Formation Among Lesbian and Bisexual Women." *Gender & Society* 7 (March): 50–77.

Ryan, Mike, Carl Swanson, and Rogene Buchholz. 1987. *Corporate Strategy, Public Policy and the Fortune 500.* New York: Basil Blackwell Inc.

Ryan, William. 1971. *Blaming the Victim.* New York: Pantheon.

Rymer, Russ. 1993. *Genie: A Scientific Study.* New York: HarperCollins.

S

Sadker, Myra, and David Sadker. 1994. *Failing at Fairness: How America's Schools Cheat Girls.* New York: Charles Scribner's Sons.

Saenz, Rudolfo, Willis J. Goudy, and Frederick O. Lorenz. 1989. "The Effects of Employment and Marital Relations on Depression among Mexican American Women." *Journal of Marriage and the Family* 51 (February): 239–251.

Salant, Jonathan D., and David S. Cloud. 1995. "To the '94 Election Victors Go the Fundraising Spoils." *Congressional Quarterly* 53 (April 15): 1055–1059.

Salthouse, T. A. 1984. "Effects of Age and Skill in Typing." *Journal of Experimental Psychology, General* 113: 345–371.

Sampson, Robert J. 1987. "Effects of Socio-Economic Context on Official Reactions to Juvenile Delinquency." *American Sociological Review* 51 (December): 876–885.

Sampson, Robert J., S. W. Raudenbush, and F. Earls. 1997. "Neighborhoods and Violent Crime: A Multilevel Study of Collective Efficacy." *Science* 277 (August): 918–924.

Sanchez, Laura, and Elizabeth Thomson. 1997. "Becoming Mothers and Fathers: Parenthood, Gender, and the Division of Labor." *Gender & Society* 11 (December): 747–773.

Sánchez-Ayéndez, Melba. 1995. "Puerto Rican Elderly Women: Shared Meanings and Informal Supportive Networks." Pp. 260–274 in *Race, Class, and Gender: An Anthology,* edited by Margaret L. Andersen and Patricia Hill Collins. Belmont, CA: Wadsworth.

Sánchez-Jankowski, Mártin. 1991. *Islands in the Street: Gangs and American Urban Society.* Berkeley: University of California Press.

Sandefur, Gary D., and Wilbur J. Scott. 1988. "A Sociological Analysis of White, Black, and American Indian Male Labor Force Activities." *Research in Race and Ethnic Relations* 5: 99–125.

Sanders, Bernie, and Marcy Kaptur. 1997. "Just Do It, Nike." *The Nation* (December 8): 6.

Sanders, Clinton. 1989. *Customizing the Body: The Art and Culture of Tattooing.* Philadelphia: Temple University Press.

Santiago, Anna M. 1995. "Intergenerational and Program-Induced Effects of Welfare Dependency: Evidence from the National Longitudinal Survey of Youth." *Journal of Family and Economic Issues* 16 (Fall): 281–306.

Santiago, Anne M., and Margaret G. Wilder. 1991. "Residential Segregation and Links to Minority Poverty: The Case of Latinos in the United States." *Social Problems* 38 (November): 492–515.

Sapir, Edward. 1921. *Language: An Introduction to the Study of Speech.* New York: Harcourt, Brace.

Sapp, Gary L. 1986. "Religious Orientation and Moral Judgment." *Journal for the Scientific Study of Religion* 25 (December): 208–214.

Saunders, Laura. 1990. "America's Richest Congressmen." *Forbes* 145 (February 19): 44–45.

Scarpitti, Frank R., Margaret L. Andersen, and Laura O'Toole. 1997. *Social Problems,* 3rd ed. New York: Addison Wesley Longman.

Scavo, Carmine. 1990. "Racial Integration of Local Government Leadership in Southern Small Cities: Consequences for Equity Relevance and Political Relevance." *Social Science Quarterly* 71 (June): 362–372.

Schacht, Steven P. 1996. "Misogyny on and off the 'Pitch': The Gendered World of Male Rugby Players." *Gender & Society* 10 (October): 550–565.

Scheff, Thomas J. 1966. *Being Mentally Ill: A Sociological Theory.* Chicago: Aldine.

Scheff, Thomas. 1984. *Being Mentally Ill: A Sociological Theory,* 2nd ed. New York: Aldine.

Schmitt, Frederika E., and Patricia Yancey Martin. 1999. "Unobtrusive Mobilization by an Institutionalized Rape Crisis Center: 'It Comes From the Victims.'" *Gender & Society* 13 (364–384).

Schneider, Beth E. (ed.) 1994. "Editor's Introduction to Special Issue on Sexual Identities/Sexual Communities." *Gender & Society* 8 (September): 293–296.

Schneider, Beth. 1984. "Peril and Promise: Lesbians' Workplace Participation." Pp. 211–230 in *Women Identified Women,* edited by Trudy Darty and Sandee Potter. Palo Alto, CA: Mayfield Publishing Co.

Schorr, Juliet B. 1991. *The Overworked American: The Unexpected Decline of Leisure.* New York: Basic Books.

Schuman, Howard, Charlotte Steeh, and Laurence Bobo. 1985. *Racial Attitudes in America: Trends and Interpretations.* Cambridge, MA: Harvard University Press.

Schumann, Howard, Charlotte Steeh, Lawrence Bobo, and Maria Krysan. 1997. *Racial Attitudes in America: Trends and Interpretations.* Cambridge, MA: Harvard University Press.

Schur, Edwin M. 1984. *Labeling Women Deviant.* New York: Random House.

Schutz, Alfred, and Thomas Luckmann. 1973. *The Structures of the Life-World.* Evanston, IL: Northwestern University Press.

Schwager, Rebecca. 1994. *Chronic Fatigue Syndrome: Modes of Interpretation of Ambiguous Distress.* Princeton University Senior Thesis. Princeton, New Jersey.

Schwartz, Joe, and Thomas Exter. 1989. "All Our Children." *American Demographics* 11 (May): 34–37.

Schwartz, Pepper. 1998. *The Gender of Sexuality.* Thousand Oaks, CA: Pine Forge Press.

Scott, Robert. 1969. *The Making of Blind Men.* New York: Russell Sage Foundation.

Scritchfield, Shirley. 1989. "Gender Traditionality and Perceptions of Date Rape." Paper presented at the annual meeting of the American Sociological Association, August 1989.

Scully, Diana. 1990. *Understanding Sexual Violence: A Study of Convicted Rapists.* Boston: Unwin Hyman.

Seager, Joni. 1997. *The State of Women in the World Atlas.* New York: Penguin Books.

Sears, David O., Letitia Anne Peplau, Jonathan L. Freedman, and Shelley E. Taylor. 1988. *Social Psychology.* Englewood Cliffs, NJ: Prentice-Hall.

See, Katherine O'Sullivan. 1991. "Comments from the Special Issue Editor: Approaching Poverty in the United States." *Social Problems* 38 (November): 427–432.

Segal, Mady W. 1974. "Alphabet and Attraction: An Unobtrusive Measure of the Effect of Propinquity in a Field Setting." *Journal of Personality and Social Psychology* 30: 654–657.

Segura, Denise A., and Jennifer L. Pierce. 1993. "Chicana/o Family Structure and Gender Personality: Chodorow, Familism, and Psychoanalytic Sociology Revisited." *Signs* 19 (Autumn): 62–91.

Seidman, Steven. 1994. "Symposium: Queer Theory/Sociology: A Dialogue." *Sociological Theory* 12 (July): 178–187.

Selke, William. 1993. *Prisons in Crisis.* Bloomington: Indiana University Press.

Selznik, Gertrude J., and Stephen Steinberg. 1969. *The Tenacity of Prejudice: Anti-Semitism in Contemporary America.* New York: Harper & Row.

Sengstock, Mary C. 1991. "Sex and Gender Implications in Cases of Elder Abuse." *Journal of Women and Aging* 3: 25–43.

Sennett, Richard, and Jonathan Cobb. 1973. *The Hidden Injuries of Class.* New York: Knopf.

Serbin, L. A., P. Zelkowitz, A. Doyle, D. Gold, and B. Wheaton. 1990. "The Socialization of Sex-Differentiated Skills and Academic Performance: A Mediational Model." *Sex Roles* 23: 613–628.

Service, Elman. 1975. *Origins of the State and Civilization: The Process of Cultural Evolution.* New York: Norton.

Shaberoff, Philip. 1988. "Water Supplies in Ground Held Generally Safe." *The New York Times* (October 8): 1, 6.

Shapiro, Laura. 1997. "Book Review: 'The Time Bind.'" *Newsweek* 129 (April 28): 6.

Shea, Christopher. 1995. "Disengaged Freshman." *The Chronicle of Higher Education,* January 13.

Shea, Christopher. 1996. "New Students Uncertain About Racial Preferences." *The Chronicle of Higher Education* 42 (January 12): A33–35.

Shea, Martha. 1995. "Dynamics of Economic Well-Being: Poverty, 1990 to 1992." U.S. Bureau of the Census, Current Population Reports, Series P 70–42. Washington, DC: U.S. Government Printing Office.

Sheeran, Paschal, Dominic Abrams, Charles Abraham, and Russell Spears. 1993. "Religiosity and Adolescents' Premarital Sexual Attitudes and Behaviour: An Empirical Study of Conceptual Issues." *European Journal of Social Psychology* 23 (January/February): 39–52.

Sherrington, R., et al., 1995. "Cloning of a Gene Bearing Missense Mutations in Early-Onset Familial Alzheimer's Disease." *Nature* 275: 754–760.

Shibutani, Tomatsu. 1961. *Society and Personality: An Interactionist Approach to Social Psychology.* Englewood Cliffs, NJ: Prentice Hall.

Shibutani, Tomatsu. 1966. *Improvised News.* Indianapolis: Bobbs-Merrill Co.

Shilts, Randy. 1988. *And the Band Played On.* New York: Penguin.

Shimkin, Demitri B., Edith M. Shimkin, and Dennis A. Frate. 1978. *The Extended Family in Black Societies.* The Hague: Mouton Publishers.

Shrout, Patrick E., Glorisa J. Canino, Hector J. Bird, Maritza Rubio-Stipec, Milagros Bravo, and M. Audrey Burnam. 1992. "Mental Health Status among Puerto Ricans, Mexican Americans, and non-Hispanic Whites." *American Journal of Community Psychology* 20 (December): 729–752.

Sifry, Micah L., and Marc Cooper. 1995. "Americans Talk Back to Power." *The Nation* (April 10): 482.

Sigall, H., and N. Ostrove. 1975. "Beautiful But Dangerous: Effects of Offender Attractiveness and Nature of Crime on Judicial Judgment." *Journal of Personality and Social Psychology* 31: 410–414.

Signorielli, Nancy, Douglas McLeod, and Elaine Healy. 1994. "Gender Stereotypes in MTV Commercials: The Beat Goes On." *Journal of Broadcasting and Electronic Media* 38 (Winter): 91–101.

Signorielli, Nancy. 1989. "Television and Conceptions About Sex Roles: Maintaining Conventionality and the Status Quo." *Sex Roles* 21: 341–360.

Signorielli, Nancy. 1991. *A Sourcebook on Children and Television.* New York: Greenwood Press.

Silberstein, Fred B., and Melvin Seeman. 1959. "Social Mobility and Prejudice." *American Journal of Sociology* (November): 258–264.

Silver, Hilary. 1995. *Federal Discharge Rates, Final Report: Minority/Non-Minority.* Washington, DC: Office of Personnel Management.

Simile, Catherine. 1995. *Disaster Settings and Mobilization for Contentious Collective Action: Case Studies of Hurricane Hugo and the Loma Prieta Earthquake.* Unpublished dissertation, University of Delaware.

Simmel, Georg. 1902. "The Number of Members as Determining the Sociological Form of the Group." *The American Journal of Sociology* 8 (July): 1–46.

Simmel, Georg. 1904. "Fashion." *International Quarterly* 10: 541–558.

Simmel, Georg. 1950. *The Sociology of George Simmel,* edited by Kurt H. Wolff. New York: Free Press.

Simmons, Roberta, Leslie Brown, Diane Mitsch Bush, and Dale A. Blyth. 1978. "Self-Esteem and Achievement of Black and White Adolescents." *Social Problems* 26 (October): 86–96.

Simon, David R. 1995. *Elite Deviance,* 5th ed. Boston: Allyn and Bacon.

Simon, Julian L., ed. 1995. *The State of Humanity.* Cambridge, MA: Blackwell Publishers.

Simonds, Wendy. 1988. "Confessions of Loss: Maternal Grief in 'True Story'." *Gender & Society* 2 (June): 149–171.

Simons, Marlise. 1989. "The Amazon's Savvy Indians." *The New York Times Magazine* (February): 36ff.

Simons, Ronald L., et al. 1996. *Understanding Differences between Divorced and Intact Families.* Newbury Park, CA: Sage Publications.

Simpson, George Eaton, and J. Milton Yinger. 1985. *Racial and Cultural Minorities: An Analysis of Prejudice and Discrimination,* 5th Edition. New York: Plenum.

Simpson, Ida Harper, David Stark, and Robert A. Jackman. 1988. "Class Identification Processes of Married, Working Men and Women." *American Sociological Review* 53 (April): 284–293.

Sirvananadan, A. 1995. "La trahision des clercs. (Racism)." *New Statesman and Society* 8: 20–22.

Sjøberg, Gideon. 1965. *The Preindustrial City: Past and Present.* New York: Free Press.

Sklar, Holly, and Chuck Collins. 1997. "Forbes 400 World Series." *The Nation* (October 20): 6–7.

Sklare, Marshall. 1971. *America's Jews.* New York: Random House.

Skocpol, Theda. 1979. *States and Social Revolutions: A Comparative Analyses of France, Russia, and China.* New York: Cambridge University Press.

Skocpol, Theda. 1992. *Protecting Soldiers and Mothers: The Origins of Social Policy in the United States.* Cambridge, MA: Belknap Press.

Skolnick, Arlene. 1991. *Embattled Paradise: The American Family in an Age of Uncertainty.* New York: Harper Collins.

Slack, Alison T. 1988. "Female Circumcision: A Critical Appraisal." *Human Rights Quarterly* 10: 437–486.

Slavin, Robert E. 1993. "Ability Grouping in the Middle Grades: Achievement Effects and Alternatives." *The Elementary School Journal* 93 (No. 5): 535–552.

Smelser, Neil J. 1963. *Theory of Collective Behavior.* New York: Free Press.

Smelser, Neil J. 1992a. "Culture: Coherent or Incoherent." Pp. 3–28 in *Theory of Culture,* edited by R. Münch and N. J. Smelser. Berkeley: University of California Press.

Smelser, Neil J. 1992b. "The Rational Choice Perspective: A Theoretical Assessment." *Rationality and Society* 4: 381–410.

Smith, A. Wade. 1989, "Educational Attainment as a Determinant of Social Class Among Black Americans." *Journal of Negro Education* 58 (Summer): 416–429.

Smith, Barbara. 1983. "Homophobia: Why Bring it Up?" *Interracial Books for Children Bulletin* 14: 112–113.

Smith, Dinitia. 1998. "One False Note in a Musician's Life." *The New York Times* (June 2): B1, 4.

Smith, Douglas A., Christy A. Visher, and Laura A. Davidson. 1985. "Equity and Discretionary Justice: The Influence of Race on Police Arrest Decisions." *Journal of Criminal Law and Criminology* 75: 234–249.

Smith, H. Lovell, and Carolyn Winje. 1992. "August Biennial Report on the Participation of Women and Minorities in ASA for 1990 and 1991." Washington, DC: American Sociological Association.

Smith, Heather. 1999. "The Promotion of Women and Men." Ph. D. dissertation, University of Delaware.

Smith, Joan. 1984. "The Paradox of Women's Poverty: Wage-earning Women and Economic Transformation." *Signs* 10 (Winter): 291–310.

Smith, M. Dwayne, Joel A. Devine, and Joseph F. Sheley. 1992. "Crime and Unemployment: Effects Across Age and Race Categories." *Sociological Perspectives* 35 (Winter): 551–572.

Smith, Richard B., and Robert A. Brown. 1997. "The Impact of Social Support on Gay Male Couples." *Journal of Homosexuality* 33: 39–61.

Smith, Robert B. 1993. "Social Structure and Voting Choice: Hypotheses, Findings, and Interpretations." Paper presented at the Annual Meeting of the American Sociological Association, August 1993, Miami.

Smith, Tom W. 1991. "Adult Sexual Behavior in 1989: Number of Partners, Frequency of Intercourse and Risk of AIDS." *Family Planning Perspectives* 23 (May/June): 102–107.

Smith, Tom W. 1992. "Changing Racial Labels: From 'Colored' to 'Negro' to 'Black' to 'African American.'" *Public Opinion Quarterly* 56: 496–512.

Smith-Rosenberg, Carroll, and Charles Rosenberg. 1984. "The Female Animal: Medical and Biological Views of Woman and Her Role in Nineteenth Century America." Pp. 12–27 in *Women and Health in America,* edited by Judith W. Leavitt. Madison: University of Wisconsin Press.

Snipp, C. Matthew. 1989. *American Indians: The First of This Land.* New York: Russell Sage Foundation.

Snipp, C. Matthew. 1996. "The First Americans: American Indians." Pp. 390–404 in *Origins and Destinies: Immigration, Race, and Ethnicity in America,* edited by Sylvia Pedraza and Rubén G. Rumbaut. Belmont, CA: Wadsworth Publishing Company.

Snow, David A, Susan G. Baker, Leon Anderson, and Michael Martin. 1986. "The Myth of Pervasive Mental Illness Among the Homeless." *Social Problems* 33 (June): 407–423.

Snow, David, and Leon Andersen. 1992a. *Down on Their Luck: A Study of Homeless Street People.* Berkeley: University of California Press.

Snow, David A. 1992. "Master Frames and Cycles of Protest." Pp. 133–155 in *Frontiers in Social Movement Theory,* edited by Aldon D. Morris and Carol McClurg Mueller. New Haven: Yale University Press.

Snow, David A., and Robert D. Benford. 1988. "Ideology, Frame Resonance, and Participant Mobilization." *International Social Movements Research* 1: 197–217.

Snow, David A., E. Burke Rochford, Jr., Steven K. Worden, and Robert D. Benford. 1986. "Frame Alignment Processes, Micromobilization and Movement Participation." *American Sociological Review* 78 (August): 464–481.

Sorokin, Pitrim. 1941. *The Crisis of Our Age.* New York: Dutton.

Sowell, Thomas. 1983. *The Economics and Politics of Race: An International Perspective.* New York: W. Morrow.

Spangler, Eve, Marsha A. Gordon, and Ronald Pipkin. 1978. "Token Women: An Empirical Test of Kanter's Hypothesis." *American Journal of Sociology* 84: 160–170.

Spanier, Graham. 1983. "Married and Unmarried Co-habitation in the U.S.: 1980." *Journal of Marriage and the Family* 45 (May): 277–288.

Spence, J. T., R. L. Helmrich, and J. Stampp. 1975. "Ratings of Self and Peers on Sex-Role Attributes and Their Relation to Self-Esteem and Conceptions of Masculinity and Femininity." *Journal of Personality and Social Psychology* 32 (July): 29–39.

Spencer, Herbert. 1882. *The Study of Sociology.* London: Routledge.

Spengler, Oswald. 1932. *The Decline of the West.* New York: Knopf.

Spilerman, Seymour. 1976. "Structural Characteristics of Cities and the Severity of Racial Disorders." *American Sociological Review* 41 (October): 771–793.

Spitzer, Steven. 1975. "Toward a Marxian Theory of Deviance." *Social Problems* 22: 638–651.

Spohn, Cassia C. 1995. "Courts, Sentences, and Prisons." *Daedalus* 124 (Winter): 119–141.

Sprock, June, and Carol Y. Yoder. 1997. "Women and Depression: An Update on the Report of the APA Task Force." *Sex Roles* 36 (March): 269–303.

Squires, Gregory D. 1991. "Deindustrialization, Economic Democracy, and Equal Opportunity: The Changing Context of Race Relations in Urban America." *Comparative Urban and Community Research* 3: 188–215.

Squires, Gregory D., and Thomas A. Lyson. 1991. "Employee Ownership and Equal Opportunity: Ameliorating Race and Gender Wage Inequalities Through Democratic Work Organizations." *Humanity & Society* 15 (November): 94–110.

St. Lawrence, Janet S., and Doris J. Joyner. 1991. "The Effects of Sexually Violent Rock Music on Males' Acceptance of Violence Against Women." *Psychology of Women Quarterly* 15: 49–63.

Stacey, Judith, and Susan Elizabeth Gerard. 1990. "'We are Not Doormats: The Influence of Feminism on Contemporary Evangelicals in the United States.'" Pp. 98–117 in *Uncertain Terms: Negotiating Gender in American Culture,* edited by Faye Ginsburg and Anna Lowenhaupt Tsing. Boston: Beacon Press.

Stacey, Judith. 1996. *In the Name of the Family.* Boston: Beacon Press.

Stack, Carol. 1974. *All Our Kin: Strategies for Survival in a Black Community.* New York: Harper Colophon.

Stafford, R., E. Backman, and P. Dibona. 1977. "The Division of Labor Among Cohabiting and Married Couples." *Journal of Marriage and the Family* 39: 43–57.

Stark, Margaret J., Abudu, Walter J. Raine, Stephen L. Burbeck, and Keith K. Davison. 1974. "Some Empirical Patterns in a Riot Process." *American Sociological Review* 39 (December): 865–876.

Stark, Rodney. 1996. *The Rise of Christianity: A Sociologist Reconsiders History.* Princeton, NJ: Princeton University Press.

Starr, Paul. 1982. *The Social Transformation of American Medicine.* New York: Basic Books.

Starr, Paul. 1995. "What Happened to Health Care Reform?" *The American Prospect* (Winter, No. 20): 20–31.

Stearns, Linda Brewster, and Charlotte Wilkinson Coleman. 1990. "Industrial and Labor Market Structures and Black Male Employment in the Manufacturing Sector." *Social Science Quarterly* 71 (June): 285–298.

Steele, Claude M. 1992. "Race and the Schooling of Black Americans." *The Atlantic Monthly* 269 (April): 68–78.

Steele, Claude M. 1996. "A Burden of Suspicion: How Stereotypes Shape the Intellectual Identities and Performance of Women and African-Americans." Paper read before the Princeton Conference on Higher Education, February, Princeton, New Jersey.

Steele, Claude M., and Joshua Aronson. 1995. "Stereotype Vulnerability and African American Intellectual Performance." Pp. 409–421 in *Readings About the Social Animal,* 7th ed, edited by Elliot Aronson. New York: W. H. Freeman and Co., p. 416.

Stein, Arlene, and Ken Plummer. 1994. "'I Can't Even Think Straight': Queer Theory and the Missing Sexual Revolution in Sociology." *Sociological Theory* 12 (July): 178–187.

Stein, Dorothy K. 1978. "Women to Burn: Suttee as a Normative Institution." *Signs* 4 (Winter): 253–269.

Steinberg, Ronnie. 1992. "Gendered Instructions: Cultural Lag and Gender Bias in the Hay System of Job Evaluation." *Work and Occupations* 19 (November): 387–424.

Stern, M. and K. H. Karraker. 1989. "Sex Stereotyping of Infants: A Review of Gender Labeling Studies." *Sex Roles* 20 (May): 501–522.

Sternberg, Robert J. 1988. *The Triarchic Mind: A New Theory of Human Intelligence.* New York: Penguin.

Stevens, Ann Huff. 1994. "The Dynamics of Poverty Spells: Updating Bane and Ellwood." *The American Economic Review* 84 (May): 34–37.

Stevens, William K. 1994. "Poor Lands' Success in Cutting Birth Rate Upsets Old Theories." *The New York Times* (January 2): 1, 8.

Stewart, Abigail J., Anne P. Copeland, Nia Lane Chester, Janet E. Malley, and Nicole B. Barenbaum. 1997. *Separating Together: How Divorce Transforms Families.* New York: Guilford Press.

Stewart, Ella. 1993. "Communication between African Americans and Korean Americans: Before and After the Los Angeles Riots." *Amerasia Journal* 19 (Spring): 23–53.

Stirling, Kate J. 1989. "Women Who Remain Divorced: Long-Term Economic Consequences." *Social Science Quarterly* 70 (September): 549–561.

Stockard, Jean, and Miriam M. Johnson. 1992. *Sex and Gender in Society,* 2nd ed., Englewood Cliffs, NJ: Prentice-Hall.

Stoller, Eleanor Palo, and Rose Campbell Gibson. 1997. *Worlds of Difference: Inequality in the Aging Experience.* Thousand Oaks, CA: Pine Forge Press.

Stoner, J.A.F. 1961. *A Comparison of Individual and Group Decisions Involving Risk.* Unpublished Master's thesis. Cambridge, MA: Massachusetts Institute of Technology.

Stotik, Jeffrey, Thomas E. Shriver, and Sherry Cable. 1994. "Social Control and Movement Outcome: The Case of AIM." *Sociological Focus* 27 (February): 53–66.

Stover, R. G., and C. A. Hope. 1993. *Marriage, Family and Intimate Relationships.* New York: Harcourt Brace Jovanovich.

Straus, Murray, Richard Gelles, and Suzanne Steinmetz. 1980. *Behind Closed Doors: Violence in the American Family.* Garden City, NY: Doubleday.

Sudnow, David N. 1967. *Passing On: The Social Organization of Dying.* Englewood Cliffs, NJ: Prentice Hall.

Suitor, Jill J. 1988. "Husbands' Educational Attainment and Support for Wives' Return to School." *Gender & Society* 2 (December): 482–495.

Sullivan, Maureen. 1996. "Rozzie and Harriet? Gender and Family Patterns of Lesbian Coparents." *Gender & Society* 12 (December): 747–767.

Sumner, William Graham. 1906. *Folkways.* Boston: Ginn.

Sung, Betty Lee. 1990. "Chinese American Intermarriage." *Journal of Comparative Family Studies* 21 (Autumn): 337–351.

Sussal, Carol M. 1994. "Empowering Gays and Lesbians in the Workplace." *Journal of Gay and Lesbian Social Services* 1: 89–103.

Sussman, N. M., and H. M. Rosenfeld. 1982. "Influence of Culture, Language, and Sex on Conversational Distance." *Journal of Personality and Social Psychology* 42: 66–74.

Sutherland, Edwin H. 1940. "White Collar Criminality." *American Sociological Review* 5 February): 1–12.

Sutherland, Edwin H., and Donald R. Cressey. 1978. *Criminology,* 10th ed. New York: J. B. Lippincott Company.

Swartz, Thomas R., and Kathleen Maas Weigert (eds.). 1995. *America's Working Poor.* Notre Dame, IN: University of Notre Dame Press.

Sweeney, Laura T., and Craig Haney. 1992. "The Influence of Race on Sentencing: A Meta-Analysis Review of Experimental Studies." *Behavioral Sciences and the Law* (Spring): 179–196.

Sweet, James A., and Larry L. Bumpass. 1992. "Young Adults' Views of Marriage, Cohabitation, and Family." Pp. 143–170 in *The Changing American Family: Sociological and Demographic Perspectives,* edited by Scott J. South and Stewart E. Tolnay. Boulder, CO: Westview Press.

Swidler, Ann. 1986. "Culture in Action: Symbols and Strategies." *American Sociological Review* 51: 273–286.

Switzer, J.Y. 1990. "The Impact of Generic Word Choices: An Empirical Investigation of Age- and Sex-Related Differences." *Sex Roles* 22: 69–82.

Szasz, Thomas S. 1974. *The Myth of Mental Illness.* New York: Harper and Row.

T

Tabor, James D., and Eugene V. Gallagher. 1995. *Why Waco? Cults and the Battle for Religious Freedom in America.* Berkeley: University of California Press.

Takagi, Dana Y. 1992. *The Retreat from Race: Asian-American Admissions and Racial Politics.* New Brunswick, NJ: Rutgers University Press.

Takaki, Ronald. 1989. *Strangers from a Different Shore: A History of Asian Americans.* New York: Penguin.

Takaki, Ronald. 1993. *A Different Mirror: A History of Multicultural America.* Boston: Little Brown.

Tannen, Deborah. 1990. *You Just Don't Understand: Women and Men in Conversation.* New York: William Morrow.

Tanner, D. M. 1978. *The Lesbian Couple.* Lexington, MA: Lexington Books.

Tatara, T. 1996. *Elder Abuse: An Information Guide for Professionals and Concerned Citizens,* 6th ed. Washington, DC: National Center on Elder Abuse.

Taylor, Howard F. 1973a. "Linear Models of Consistency: Some Extensions of Blalock's Strategy." *American Journal of Sociology* 78 (March): 1192–1215.

Taylor, Howard F. 1973b. "Playing the Dozens with Path Analysis: Methodological Pitfalls in Jencks et al. *Inequality.*" *Sociology of Education* 46(4): 433–450.

Taylor, Howard F. 1980. *The IQ Game: A Methodological Inquiry into the Heredity Environment Controversy.* New Brunswick, NJ: Rutgers Press.

Taylor, Howard F. 1981. "Biases in *Bias in Mental Testing.*" *Contemporary Sociology* 10:172–174.

Taylor, Howard F. 1992a. "Intelligence." Pp. 941–949 in *Encyclopedia of Sociology,* edited by E. F. Borgatta and M. L. Borgatta. New York: Macmillan.

Taylor, Howard F. 1992b. "The Structure of a National Black Leadership Network: Preliminary Findings." Unpublished manuscript, Princeton University, Princeton, NJ.

Taylor, Howard F. 1995. "Symposium on 'The Bell Curve'." *Contemporary Sociology* 24 (March): 153–158.

Taylor, Howard F. 1996. "Ringing the Methodology of *The Bell Curve.*" *Journal of Negro Education* (in press).

Taylor, Howard F., and Carlton A. Hornung. 1979. "On A General Model for Social and Cognitive Consistency." *Sociological Methods and Research* 7 (February): 259–287.

Taylor, Shelley E., and L. G. Aspinwall. 1990. "Psychological Aspects of Chronic Illness." Pp. 3–60 in *Psychological Aspects of Serious Illness,* edited by G. R. Vanderbos and P. T. Costa, Jr. Washington, DC: American Psychological Association.

Taylor, Shelley E., Letitia Anne Peplau, and David O. Sears. 1994. *Social Psychology,* 8th Edition. Englewood Cliffs, NJ: Prentice-Hall.

Taylor, Shelley E., Letitia Ann Peplau, and David O. Sears. 1997. *Social Psychology,* 9th ed. Englewood Cliffs: Prentice Hall, p. 232.

Taylor, Shelly E., N. E. Kemeny, L. G. Aspinwall, S. G. Schneider, R. Rodriquez, and N. Herbert. 1992. "Optimism, Coping, Psychological Distress, and High-Risk Sexual Behavior Among Men At Risk for AIDS." *Journal of Personality and Social Psychology* 63 (September): 460–473.

Taylor, Verta, and Leila J. Rupp. 1993. "Women's Culture and Lesbian Feminist Activism: A Reconsideration of Cultural Feminism." *Signs* 19 (Autumn): 32–61.

Taylor, Verta, and Nicole C. Raeburn. 1995. "Identity Politics as High-Risk Activism: Career Consequences for Lesbian, Gay and Bisexual Sociologists." *Social Problems* 42 (May): 252–274.

Taylor, Verta. 1980. "Review Essay of Four Books on Lesbianism." *Journal of Marriage and the Family* 42: 224–228.

Terrace, Herbert S. 1980. *Nim: A Chimpanzee Who Learned Sign Language.* New York: Knopf.

Terrell, Katherine. 1992. "Female-Male Earnings Differentials and Occupational Structure." *International Labour Review* 131: 387–404.

Tessler, Richard C., and Deborah L. Dennis. 1992. "Mental Illness among Homeless Adults: A Synthesis of Recent NIMH-Funded Research." *Research in Community and Mental Health* 7: 3–53.

Tewksbury, Richard. 1994. "Gender Construction and the Female Impersonator: The Process of Transforming 'He' to 'She.'" *Deviant Behavior* 15: 27–43.

Theberge, Nancy. 1993. "The Construction of Gender in Sport: Women, Coaching, and the Naturalization of Difference." *Social Problems* 40 (August): 301–311.

Theberge, Nancy. 1997. "'It's Part of the Game': Physicality and the Production of Gender in Women's Ice Hockey," *Gender & Society* 11 (February): 69–87.

Thibaut, John W., and Harold T. Kelley. 1959. *The Social Psychology of Groups*. New York: Wiley.

Thibodeau, R. 1989. "From Racism to Tokenism: The Changing Face of Blacks in New Yorker Cartoons." *Public Opinion Quarterly* 53: 482–494.

Thoits, Peggy A. 1991. "On Merging Identity Theory and Stress Research." *Social Psychology Quarterly* 54: 101–112.

Thomas, John K. 1995. "Review: *Ecopopulism: Toxic Waste and the Movement for Environmental Justice*." *Rural Sociology* 60 (Spring): 151–152.

Thomas, Melvin E. 1993. "Race, Class, and Personal Income: An Empirical Test of the Declining Significance of Race Thesis, 1968–1988." *Social Problems* 40 (August): 328–342.

Thomas, W. I. 1918 [1919/1958]. *The Polish Peasant in Europe and America*. New York: Dover Publications.

Thomas, W. I. 1931. *The Unadjusted Girl*. Boston: Little, Brown.

Thomas, W. I. 1966 [1931]. "The Relation of Research to the Social Process." Pp. 289–305 in *W. I. Thomas on Social Organization and Social Personality*, edited by Morris Janowitz. Chicago: University of Chicago Press.

Thomas, William I. with Dorothy Swaine Thomas. 1928. *The Child in America*. New York: Knopf.

Thompson, Becky. 1994. *A Hunger So Wide and So Deep: American Women Speak Out on Eating Problems*. Minneapolis: University of Minnesota Press.

Thompson, E. H., C. Grisanti, and J. Pleck. 1985. "Attitudes Toward the Male Role and Their Correlates." *Sex Roles* 13 (October): 413–427.

Thompson, Margaret E., Steven H. Chaffee, and Hayg H. Oshagan. 1990. "Regulating Pornography: A Public Dilemma." *Journal of Communication* 40 (Summer): 73–83.

Thompson, Robert Ferris. 1993. *Slash of the Spirit*. NY: Random House.

Thompson, William E., and Joseph V. Hickey. 1994. *Society in Focus*. New York: HarperCollins.

Thornberry, Terence P., Carolyn A. Smith, and Gregory J. Howard. 1997. "Risk Factors for Teenage Fatherhood." *Journal of Marriage and the Family* 59 (August): 505–522.

Thorne, Barrie, and Zella Luria. 1986. "Sexuality and Gender in Children's Daily Worlds." *Social Problems* 33 (February): 176–190.

Thorne, Barrie (ed.) with Marilyn Yalom. 1992. *Rethinking the Family: Some Feminist Questions*. Boston: Northeastern University Press.

Thorne, Barrie. 1993. *Gender Play: Girls and Boys in School*. New Brunswick, NJ: Rutgers University Press.

Thornton, A., D. E. Alwin, and D. Camburn. 1983. "Causes and Consequences of Sex-Role Attitudes and Attitude Change." *American Sociological Review* 48 (April): 211–227.

Thornton, Arland, William G. Axinn, and Daniel H. Hill. 1992. "Reciprocal Effects of Religiosity, Cohabitation, and Marriage." *American Journal of Sociology* 98 (November): 628–651.

Thornton, Bill, and Rachel Leo. 1992. "Gender Typing, Importance of Multiple Roles, and Mental Health Consequences for Women." *Sex Roles* 27 (September): 307–317.

Thornton, Michael. 1995. "Is Multiracial Experience Unique? The Personal and Social Experience." Pp. 95–99 in *Race, Class, and Gender: An Anthology*, 2nd ed., edited by Margaret L. Andersen and Patricia Hill Collins. Belmont, CA: Wadsworth.

Thornton, Russell, Gary D. Sandefur, and C. Matthew Snipp. 1991. "American Indian Fertility Patterns: 1910 and 1940 to 1980." *American Indian Quarterly* 15 (Summer): 359–367.

Thornton, Russell. 1987. *American Indian Holocaust and Survival: A Population History*. Norman, OK: University of Oklahoma Press.

Tiefer, Leonore. 1978. "The Kiss." *Human Nature* (July): 28–37.

Tienda, Marta. 1989. "Puerto Ricans and the Underclass Debate." *The Annals of the American Academy* 501 (January): 105–119.

Tienda, Marta, and Franklin D. Wilson. 1992. "Migration and the Earnings of Hispanic Men." *American Sociological Review* 57 (October): 661–678.

Tierney, Kathleen. 1994. "Making Sense of Collective Preoccupations: Lessons from Research on the Iben Browning Earthquake Prediction." Pp. 75–95 in *Collective Behavior and Society*, edited by Gerald Platt and Chad Gordon. Greenwich, CT: JAI Press.

Tilly, Charles. 1986. *The Contentious French*. Cambridge, MA: Harvard University Press.

Tilly, Louise, and Joan Scott. 1978. *Women, Work, and Family*. New York: Holt, Rinehart, and Winston.

Tinsley, E. G., S. Sullivan-Guest, and J. McGuire. 1984. "Feminine Sex Role and Depression in Middle-Aged Women." *Sex Roles* 11 (July): 25–32.

Töennies, Ferdinand. 1963 [1887]. *Community and Society (Gemeinschaft and Gesellschaft)*. New York: Harper and Row.

Toobin, Jeffrey. 1996. "The Marcia Clark Verdict." *The New Yorker* (September 9): 58–71.

Toynbee, Arnold J., and Jane Caplan. 1972. *A Study of History*. New York: Oxford University Press.

Travers, Jeffrey, and Stanley Milgram. 1969. "An Experimental Study of the Small World Problem." *Sociometry* 32: 425–443.

Treas, Judith. 1995. *Older Americans in the 1990s and Beyond*. Washington, DC: Population Reference Bureau, Inc.

Trent, Katherine, and Sharon L. Harlan. 1990. "Household Structure Among Teenage Mothers in the United States." *Social Science Quarterly* 71 (September): 439–457.

Triplet, Rodney F., and David B. Sugarman. 1987. "Reactions to AIDS Victims: Ambiguity Breeds Contempt." *Personality and Social Psychology Bulletin* 13: 265–274.

Troiden, Richard. 1987. "Walking the Line: The Personal and Professional Risks of Sex Education and Research." *Teaching Sociology* 25 (July): 241–249.

Troyer, Ronald J., and Gerald E. Markle. 1984. "Coffee Drinking: An Emerging Social Problem?" *Social Problems* 31 (April): 403–416.

Tschann, Jeanne M., Janet R. Johnson, Marsha Kline, and Judith S. Wallerstein. 1989. "Family Process and Children's Functioning During Divorce." *Journal of Marriage and the Family* 2 (May): 431–444.

Tschann, Jeanne M., Janet R. Johnson, Marsha Kline, and Judith S. Wallerstein. 1990. "Conflict, Loss, Change and Parent-Child Relationships: Predicting Children's Adjustment during Divorce." *Journal of Divorce* 13: 1–22.

Tuan, Yi-Fu. 1984. *Dominance and Affection: The Making of Pets*. New Haven, CT: Yale University Press.

Tuchman, Gaye. 1979. "Women's Depiction by the Mass Media." *Signs* 4 (Spring): 528–542.

Tumin, Melvin M. 1953. "Some Principles of Stratification." *American Sociological Review* 18 (August): 387–393.

Turkle, Sherry. 1995. *Life on the Screen: Identity in the Age of the Internet*. New York: Simon and Schuster.

Turner, Jonathan. 1974. *The Structure of Sociological Theory*. Homewood, IL: Dorsey Press.

Turner, Ralph, and Lewis Killian. 1988. *Collective Behavior*, 3rd ed. Englewood Cliffs, NJ: Prentice Hall.

Turner, Ralph, Joanne M. Nigg, and Denise Paz. 1986. *Waiting for Disaster: Earthquake Watch in California*. Berkeley, CA: University of California Press.

Turner, Terence. 1969. "Tchikrin: A Central Brazilian Tribe and Its Symbolic Language of Body Adornment." *Natural History Magazine* 78 (October): 50–59.

Tyree, Andrea, and R. Hicks. 1988. "Sex and the Second Moment of Occupational Basic Distributions." *Social Forces* 66 (June): 1028–1037.

U

U.S. Bureau of Justice Statistics. 1996. *Capital Punishment Statistics*. Washington, DC: U.S. Department of Justice. World Wide Web site: **http://www.ojp.usdoj.gov/bjs/cp/htm**

U.S. Bureau of Justice Statistics. 1996. *Sourcebook of Criminal Justice Statistics 1996*. Washington, DC: U.S. Department of Justice.

U.S. Bureau of the Census. 1980. *Money Income of Households, Families, and Persons in the U.S. 1980*. Washington, DC: U.S. Department of Commerce.

U.S. Bureau of the Census. 1993a. *Statistical Abstract of the United States*. Washington, DC: U.S. Department of Commerce.

U.S. Bureau of the Census. 1993b. *Household and Family Characteristics: March 1993*, Current Population Reports, Series P-20–477. Washington, DC: U.S. Department of Commerce.

U.S. Bureau of the Census. 1993c. *Poverty in the United States, 1992*. Current Population

Reports, Series P60–185. Washington, DC: U.S. Department of Commerce.

U.S. Bureau of the Census. 1994. *Poverty—Long and Short Term* (December). Washington, DC: U.S. Department of Commerce.

U.S. Bureau of the Census. 1995. *Marital Status and Living Arrangements, March 1995. (Update).* Washington, DC: U.S. Department of Commerce.

U.S. Bureau of the Census. 1996. *Poverty in the United States: 1996.* Washington, DC: U.S. Department of Commerce.

U.S. Bureau of the Census. 1996a. *Household and Family Characteristics: March 1996. (Update).* Washington, DC: U.S. Department of Commerce.

U.S. Bureau of the Census. 1996b. *Marital Status and Living Arrangements: March 1995 (Update).* Washington, DC: U.S. Department of Commerce.

U.S. Bureau of the Census. 1996c. *Statistical Abstracts of the United States 1996.* Washington, DC: U.S. Department of Commerce.

U.S. Bureau of the Census. 1997. *Statistical Abstracts of the United States 1997.* Washington, DC: U.S. Department of Commerce.

U.S. Bureau of the Census. 1997c (July 24). Census Bureau Home Page. Detailed Historical Tables, Table P-7. World Wide Web site: **http://www.census.gov/ftp/hhes/income/histinc/p07html**

U.S. Bureau of the Census. 1997d. *Poverty in the United States: 1996.* Current Population Reports, P60–198. Washington, DC: U.S. Department of Commerce.

U.S. Bureau of the Census. 1997e. Historical Income Tables-Persons. Tables P31B and P31D. **http://www.census.gov/hhes/income/histinc/**

U.S. Bureau of the Census. 1998. *Educational Attainment in the United States: March 1997.* Washington, DC: U.S. Department of Commerce.

U.S. Bureau of the Census. 1998a. *Money Income in the United States: 1997.* Washington, DC: U.S. Bureau of the Census. World Wide Web site: **http://www.census.gov**

U.S. Bureau of the Census. 1998b. "Selected Economic Characteristics of the Population by Sex and Race: March 1997." World Wide Web site: **http://www.census.gov/population/socdemo/race/api97/table02.txt**

U.S. Bureau of Justice Statistics. 1997. *Sourcebook of Criminal Justice Statistics 1997.* Washington, DC: U.S. Department of Justice.

U.S. Department of Defense. 1994. *Selected Manpower Statistics.* Washington, DC: U.S. Government Printing Office.

U.S. Department of Defense. 1996. *Semiannual Race/Ethnic/Gender Profile of the Department of Defense Forces (Active and Reserve).* Washington, DC: U.S. Government Printing Office. World Wide Web site: **http://www.pafb.afmil/deomi/dodstat.htm**

U.S. Department of Education. 1996. *Digest of Education Statistics.* Washington, DC: U.S. Government Printing Office.

U.S. Department of Health and Human Services. 1990. *Digest Semiannual Report.* April 1–September 30. Washington, DC: Office of the Inspector General.

U.S. Department of Health and Human Services. 1998. *Health United States 1996.* Hyattsville, MD: National Center for Health Statistics.

U.S. Department of Labor. 1991. Current Population Reports, Series P-20, No. 454. *Fertility of American Women: June 1990.* Washington, DC: U.S. Government Printing Office, October.

U.S. Department of Labor. 1994. "Shifting Work Force Spawns New Set of Hazardous Occupations." July. World Wide Web site: **http:/stats.bls.gov/special.requests/ocwc/osh/ossm0007.txt**

U.S. Department of Labor. 1995. *National Census of Fatal Occupational Injuries, 1995.* Washington, DC: Bureau of Labor Statistics.

U.S. Department of Labor. 1998. *Employment and Earnings.* Washington, DC: U.S. Department of Labor.

U.S. General Accounting Office. 1994. Testimony before the Committee on Agriculture, Nutrition, and Forestry, U.S. Senate: Food Assistance. February 2, GAO/T-RCED-94–125.

Ugwuegbu, C. E. 1979. "Racial and Evidential Factors in Juror Attribution of Legal Responsibility." *Journal of Experimental Social Psychology* 35: 200–211.

Uhlenberg, Peter. 1992. "Population Aging and Social Policy." *Annual Review of Sociology* 18: 449–474.

United Nations Children's Fund. 1998. *State of the World's Children.* New York: The United Nations.

United Nations Commission on the Status of Women. 1996. *Report of the World Conference of the United Nations Decade for Women.* Copenhagen: United Nations.

United Nations Development Programme. 1996. *Human Development Report 1996.* New York: Oxford University Press.

United Nations. 1995. *The World's Women, 1995: Trends and Statistics,* 2nd ed. New York: United Nations.

United Nations. 1997. *Human Development Report 1997.* New York: United Nations. World Wide Web site: **http://www.undp.org:81/undp/hdro/table2.htm**

"United Nations Sharply Increases Estimate of Youngsters at Work Full-time." 1996. *The New York Times* (November 12): A6.

USA Today. 1991. December 5: B2.

Uvin, Peter. 1998. *Aiding Violence.* West Hartford, CT: Kumarian Press.

V

Van Ausdale, Debra, and Joe R. Feagin. 1996. "The Use of Racial and Ethnic Concepts by Very Young Children." *American Sociological Review* 61 (October): 779–793.

Vanlandingham, Mark. 1993. *Two Perspectives On Risky Sexual Practices Among Northern Thai Males: The Health Belief Model and the Theory of Reasoned Action.* Unpublished doctoral dissertation, Princeton University, Princeton, New Jersey.

Vanneman, Reeve, and Lynn Weber Cannon. 1987. *The American Perception of Class.* Philadelphia: Temple University Press.

Vaughan, Diane. 1966. *The Challenger Launch Decision: Risky Technology, Culture, and Deviance at NASA.* Chicago: The University of Chicago Press.

Veblen, Thorstein. 1899 [1953]. *The Theory of the Leisure Class: An Economic Study of Institutions.* New York: The New American Library.

Vega, W. A., and H. Amero. 1994. "Latino Outlook: Good Health, Uncertain Prognosis." *Annual Review of Public Health* 15: 39–67.

Vega, William A. 1991. "Hispanic Families in the 1980's." Pp. 297–306 in *Contemporary Families: Looking Forward, Looking Back,* edited by Alan Booth. Minneapolis: National Council of Relations.

Vessels, Jane. 1985. "Koko's Kitten." *National Geographic* 167 (January): 110–113.

Vincke, John, and Ralph Bolton. 1994. "Social Support, Depression, and Self-Acceptance among Gay Men." *Human Relations* 47 (September): 1049–1062.

Vogel, Dena Ann, Margaret A. Lake, Suzanne Evans, and Katherine Hildebrandt Karraker. 1991. "Children's and Adults' Sex-Stereotyped Perceptions of Infants." *Sex Roles* 24 (May): 605–616.

Voss, Laurie Scarborough. 1997. "Teasing, Disputing, and Playing: Cross-Gender Interactions and Space Utilization among First and Third Graders." *Gender & Society* 11 (April): 238–256.

W

Wagner, David. 1994. "Beyond the Pathologizing of Nonwork: Alternative Activities in a Street Community." *Social Work* 29 (November): 718–727.

Walker, Daniel. 1968. *Rights in Conflict: Report of the Chicago Study Team to the National Commission on the Causes and Prevention of Violence.* New York: Bantam Books.

Wallerstein, Immanuel. 1974. *The Modern World System: Capitalist Agriculture and the Origins of the European World Economy in the Sixteenth Century.* New York: Academic Press.

Wallerstein, Immanuel. 1979. *The Capitalist World-Economy.* New York: Cambridge University Press.

Wallerstein, Immanuel. 1989. *The Modern World System III: The Second Era of Great Expansion of the Capitalist World-Economy, 1730–1840.* New York: Academic Press.

Wallerstein, Immanuel M. 1980. *The Modern World-System II.* New York: Academic Press.

Wallis, Claudia. 1987. "Back Off, Buddy." *Time* 130 (October 12): 6873.

Wallman, Joel. 1992. *Aping Language.* New York: Cambridge University Press.

Walsh, Anthony. 1994. "Homosexual and Heterosexual Child Molestation: Case Characteristics and Sentencing Differentials." *International Journal of Offender Therapy and Comparative Criminology* 38: 339–353.

Walsh, Edward. 1993. "Michigan Ends Property Tax Funding of Schools." *The Washington Post.*

Walters, Vivienne. 1993. "Stress, Anxiety, and Depression: Women's Accounts of Their Health Problems." *Social Science and Medicine* 36 (February): 393–402.

Ward, Kathyrn. 1990. *Women Workers and Global Restructuring.* Ithaca, NY: ILR Press.

Ward, Russell A. 1990. *The Aging Experience: An Introduction to Social Gerontology,* 2nd ed. New York: Harper Collins.

Ward, Russell A., 1992. "Marital Happiness and Household Equity in Later Life." Paper presented at the Annual Meeting of the American Sociological Association, August.

Wasserman, G. A., and M. Lewis. 1985. "Infant Sex Differences: Ecological Effects." *Sex Roles* 12 (March): 665–675.

Wasserman, Stanley, and Katherine Faust, (eds.). 1994. *Social Network Analysis: Methods and Applications.* Cambridge, MA: Cambridge University Press.

Waters, Mary C. 1990. *Ethnic Options: Choosing Identities in America.* Berkeley, CA: University of California Press.

Watkins, Susan. 1987. "The Fertility Transition: Europe and the Third World Compared." *Sociological Forum* 2 (Fall): 645–673.

Watkins, Susan C., J. A. Menken, and J. Bongaarts. 1987. "Demographic Foundations of Family Change." *American Sociological Review* 52 (June): 346–358.

Wawer, Maria J., Chai Podhisita, Uraiwan Kanungsukkasem, Anthony Pramualratana, and Regina McNamara. 1996. "Origins and Working Conditions of Female Sex Workers in Urban Thailand: Consequences of Social Context for HIV Transmission." *Social Science and Medicine* 42 (February): 453–462.

Waxman, Barbara Faye. 1991. "Hatred: The Unacknowledged Dimension in Violence against Disabled People." *Sexuality and Disability* 9 (Fall): 185–199.

Weber, Max. 1925 [1947]. *The Theory of Social and Economic Organization.* New York: Free Press.

Weber, Max. 1958 [1904]. *The Protestant Ethic and the Spirit of Capitalism.* New York: Charles Scribner's Sons.

Weber, Max. 1962 [1913]. *Basic Concepts in Sociology.* New York: Greenwood.

Weber, Max. 1978 [1921]. *Economy and Society: An Outline of Interpretive Sociology,* edited by Guenther Roth and Claus Wittich. Berkeley: University of California Press.

Wei, William. 1993. *The Asian American Movement.* Philadelphia: Temple University Press.

Weigel, R. H., J. W. Loomis, and M. J. Soja. 1980. "Race Relations on Prime Time Television." *Journal of Personality and Social Psychology* 39: 884–893.

Weiss, Melford S. 1974. *Valley City: A Chinese Community in America.* Berkeley: University of California Press.

Weitzman, Lenore J. 1985. *The Divorce Revolution: The Unexpected Consequences for Women and Children in America.* New York: Free Press.

Weitzman, Lenore J. 1996. "A Re-Evaluation of the Economic Consequences of Divorce: Reply to Peterson." *American Sociological Review* 61 (June): 537–538.

Welch, Susan, and Lee Sigelman. 1989. "A Black Gender Gap?" *Social Science Quarterly* 70 (March): 120–133.

Welch, Susan, and Lee Sigelman. 1993. "The Politics of Hispanic Americans: Insights from the National Surveys, 1980–1988." *Social Science Quarterly* 74 (March): 76–94.

Weller, Jack, and Enrico Quarantelli. 1973. "Neglected Characteristics of Collective Behavior. *American Journal of Sociology* 79: 665–685.

Wells-Petry, Melissa. 1993. *Exclusion: Homosexuals and the Right to Serve.* Washington, DC: Regnery Gateway.

Wenk, Deeann, and Patricia Garrett. 1992. "Having a Baby: Some Predictions of Maternal Employment Around Childbirth." *Gender & Society* 6 (March): 49–65.

West, Candace, and Don Zimmerman, 1987. "Doing Gender." *Gender & Society* 1 (June): 125–151.

West, Candace, and Sarah Fenstermaker. 1995. "Doing Difference." *Gender & Society* 9 (February): 8–37.

West, Carolyn M. 1998. "Leaving a Second Closet: Outing Partner Violence in Same-Sex Couples." Pp. 163–183 in *Partner Violence: A Comprehensive Review of 20 Years of Research,* edited by Jana L. Jasinksi and Linda M. Williams. Thousand Oaks, CA: Sage.

West, Cornel. 1993. *Race Matters.* Boston: Beacon Press.

West, Jane. 1991. "The Social and Policy Context of the Act." *Milbank Quarterly* 69 (Winter): 3–24.

Westefeld, John S., Kimberly Whitchard, and Lillian M. Range. 1990. "College and University Student Suicide: Trends and Implications." *The Counseling Psychologist* 18 (July): 464–476.

Westoff, Charles F., Noreen Goldman, and Lorenzo Moreno. 1990. *Dominican Republic Experimental Study: An Evaluation of Fertility and Child Health Information.* Princeton, NJ: Office of Population Research.

Wetzel, Jodi, Margo Linn Espenlaub, Monys A. Hagen, Annette Bennington McElhiney, and Carmen Braun Williams. 1993. *Women's Studies: Thinking Women.* Dubuque, IO: Kendall/Hunt Publishing.

Wharton, Amy S., and Deborah K. Thorne. 1997. "When Mothers Matter: The Effects of Social Class and Family Arrangements on African American and White Women's Perceived Relations with Their Mothers." *Gender & Society* 11 (October): 656–681.

White, Deborah Gray. 1985. *Ar'n't I a Woman? Female Slaves in the Plantation South.* New York: W. W. Norton.

White, Jane J. 1989. "Student Teaching as a Rite of Passage." *Anthropology and Education Quarterly* 29 (September): 177–195.

White, Robert, and Robert Althauser. 1984. "Internal Labor Markets, Promotions, and Worker Skill: An Indirect Test of Skill ILMs." *Social Science Research* 13: 373–392.

Whorf, Benjamin. 1956. *Language, Thought, and Reality: Selected Writings.* Cambridge: Technology Press of Massachusetts Institute of Technology.

Whyte, William F. 1943. *Street Corner Society.* Chicago: University of Chicago Press.

Wilcox, Clyde, Lara Hewitt, and Dee Allsop. 1996. "The Gender Gap in Attitudes Toward the Gulf War: A Cross-National perspective." *Journal of Peace Research* 33 (February): 67–82.

Wilcox, Clyde. 1992. "Race, Gender, and Support for Women in the Military." *Social Science Quarterly* 73 (June): 310–323.

Wilkerson, Isabel. 1991. "Black-White Marriages Rise, But Couples Still Face Scorn." *The New York Times.* December 2, 1991: A1ff.

Wilkie, Jane. 1991. "The Decline in Men's Labor Force Participation and Earnings and the Changing Structure of Family Economic Support." *Journal of Marriage and the Family* 53 (February): 111–122.

Wilkie, Jane. 1993. "Changes in U.S. Men's Attitudes Toward the Family Provider Role, 1972 to 1989." *Gender & Society* 7 (June): 261–279.

Wilkinson, Doris. 1991. "The Segmented Labor Market and African American Women from 1890–1960: A Social History Interpretation." *Research in Race and Ethnic Relations* 6: 85–104.

Williams, Christine L. 1989. *Gender Differences at Work: Women and Men in Nontraditional Occupations.* Berkeley, CA: University of California Press.

Williams, Christine L. 1992. "The Glass Escalator: Hidden Advantages for Men in the 'Female' Professions.'" *Social Problems* 39 (August): 253–267.

Williams, Christine L. 1995. *Still a Man's World: Men Who Do Women's Work.* Berkeley: University of California Press.

Williams, David R., and Chiquita Collins, 1995. "U.S. Socioeconomic and Racial Differences in Health: Patterns and Explanations." *Annual Review of Sociology* 21: 349–386.

Williams, Lena. 1987. "Race Bias Found in Location of Toxic Dumps." *The New York Times* (April): A20.

Williams, Norma. 1990. *The Mexican American Family: Tradition and Change.* Dix Hills, NY: General Hall.

Willie, Charles V. 1979. *The Caste and Class Controversy.* Bayside, NY: General Hall, Inc.

Willie, Charles V. 1983. *Race, Ethnicity, and Socioeconomic Status: A Theoretical Analysis of*

Their Interrelationship. Bayside, NY: General Hall, Inc.

Willie, Charles Vert. 1985. *Black and White Families: A Study in Complementarity.* Bayside, NY: General Hall.

Willis, Paul. 1977. *Learning to Labor: How Working Class Kids Get Working Class Jobs.* New York: Columbia University Press.

Willits, Fern K., and Donald M. Crider. 1989. "Church Attendance and Traditional Religious Beliefs in Adolescence and Young Adulthood: A Panel Study." *Review of Religious Research* 31 (September): 68–81.

Wilson, Bryan. 1982. *Religion in Sociological Perspective.* New York: Oxford University Press.

Wilson, James Q., and Richard J. Herrnstein. 1985. *Crime and Human Nature.* New York: Simon and Schuster.

Wilson, John F. 1978. *Religion in American Society: The Effective Presence.* Englewood Cliffs, NJ: Prentice Hall.

Wilson, John. 1994. "Returning to the Fold." *Journal for the Scientific Study of Religion* 33 (June): 148–161.

Wilson, William Julius. 1978. *The Declining Significance of Race: Blacks and Changing American Institutions.* Chicago: University of Chicago Press.

Wilson, William Julius. 1987. *The Truly Disadvantaged: The Inner City, The Underclass and Public Policy.* Chicago: University of Chicago Press.

Wilson, William Julius. 1996. *When Work Disappears: The World of the New Urban Poor.* New York: Knopf.

Wilson, William Julius, and Robert Aponte. 1985. "Urban Poverty." *Annual Review of Sociology* 11: 231–258.

Wingrove, C. Ray, and Kathleen F. Slevin. 1991. "A Sample of Professional and Managerial Women: Success in Work and Retirement." *Journal of Women and Aging* 3: 95–117.

Winneker, Craig and Margie Omero. 1996. "The Roll Call 50 Richest." *Roll Call.* World Wide Web site: **http://www.rollcall.com/rcfiles/richestintro.htm**

Winnick, Louis, 1990. "America's 'Model Minority.'" *Commentary* 90 (August): 222–229.

Wirth, Louis. 1938. "Urbanism as a Way of Life." *American Journal of Sociology 40:* 1–24.

Witt, L. Alan. 1989. "Authoritarianism, Knowledge of AIDS, and Affect toward Persons with AIDS: Implications for Health Education." *Journal of Applied Social Psychology* 19: 599–607.

Wolf, Naomi. 1991. *The Beauty Myth: How Images of Beauty are Used Against Women.* New York: William Morrow and Co.

Wolff, Edward N. 1995. "The Rich Get Increasingly Richer: Latest Data on Household Wealth During the 1980s." Pp. 33–68 in *Research in Politics and Society: The Politics of Wealth and Inequality,* edited by Richard E. Ratcliff, Melvin L. Oliver, and Thomas M. Shapiro. Greenwich, CT: JAI Press.

Wolff, Edward N. 1996. *Top Heavy: A Study of the Increasing Inequality of Wealth in America.* New York: Twentieth Century Fund Press.

Woo, Deborah. 1998. "The Gap Between Striving and Achieving." Pp. 247–256 in *Race, Class, and Gender: An Anthology,* 3rd ed., edited by Margaret L. Andersen and Patricia Hill Collins. Belmont, CA: Wadsworth.

Wood, Julia T. 1994. *Gendered Lives: Communication, Gender and Culture.* Belmont, CA: Wadsworth.

Woodruff, J. T. 1985. "Premarital Sexual Behavior and Religious Adolescents." *Journal for the Scientific Study of Religion* 25 (December): 343–386.

Woodrum, Eric. 1992. "Pornography and Moral Attitudes." *Sociological Spectrum* 12 (October-December): 329–347.

Wootton, Barbara H. 1997. "Gender Differences in Occupational Employment." *Monthly Labor Review* 170 (April): 15–24.

World Bank. 1998. *World Bank Atlas.* New York: The World Bank.

World Hunger Program. 1998. *The State of World Hunger.* Providence: Brown University. World Wide Web site: **http:www.worldhunger.org/**

World Hunger Project. 1998. *Principles and Methodologies of the Hunger Project.*

Wright, Erik Olin. 1979. *Class Structure and Income Determination.* New York: Academic Press.

Wright, Erik Olin. 1985. *Classes.* London: Verso.

Wright, Erik Olin, and Cho-Donmoon. 1992. "The Relative Permeability of Class Boundaries to Cross-class Friendships: A Comparative Study of the United States, Canada, Sweden, and Norway. *American Sociological Review* 57 (February): 85–102.

Wright, Erik Olin, Karen Shire, Shu-Ling Hwang, Maureen Dolan, and Janeen Baxter. 1992. "The Non-Effects of Class On the Gender Division of Labor in the Home: A Comparative Study of Sweden and the United States." *Gender & Society* 6 (June): 252–282.

Wright, James D. 1988. "The Mentally Ill Homeless: What is Myth and What is Fact?" *Social Problems* 35 (April): 182–191.

Wright, Laurence. 1994. "One Drop of Blood." *The New Yorker* 70 (July 25): 46–55.

Wright, Lawrence. 1992. "The Man From Texarkana." *The New York Times Magazine* (June 28): 20ff.

Wright, Stuart A. 1991. "Reconceptualizing Cult Coercion and Withdrawal: A Comparative Analysis of Divorce and Apostasy." *Social Forces* 70 (September): 125–145.

Wrong, Dennis. 1961. "The Oversocialized Conception of Man in Modern Sociology." *American Sociological Review* 26 (April): 183–192.

Wulff, Helena. 1988. *Twenty Girls: Growing Up, Ethnicity, and Excitement in a South London Microculture.* Stockholm: Department of Social Anthropology, University of Stockholm.

Wuthnow, Robert (ed.). 1994. *I Come Away Stronger: How Small Groups Are Shaping American Religion.* Grand Rapids, MI: William B. Eerdmans.

Wuthnow, Robert, and Marsha Witten. 1988. "New Directions in the Study of Culture." *Annual Review of Sociology* 14: 49–67.

Wyman, David. 1984. *The Abandonment of the Jews: America and the Holocaust, 1941–1945.* New York: Pantheon.

X

Xu, Wu, and Ann Leffler. 1992. "Gender and Race Effects on Occupational Prestige, Segregation, and Earnings." *Gender & Society* 6 (September): 376–392.

Xuewen, Sheng, Norman Stockman, and Norman Bonney. 1992. "The Dual Burden: East and West (Women's Working Lives in China, Japan, and Great Britain)." *International Sociology* 7 (June): 209–223.

Y

Yeary, Elizabeth. 1993. "Prison Population Reaches Record High." *Trial* 29 (July): 113–114.

Yoder, Janice. 1991. "Rethinking Tokenism: Looking Beyond Numbers." *Gender & Society* 5 (June): 178–192.

Young, J. W. 1994. "Differential Prediction of College Grades by Gender and by Ethnicity: A Replication Study." *Educational and Psychological Measurement* 54: 1022–1029.

Young, Thomas J. 1990. "Poverty, Suicide, and Homicide Among Native Americans." *Psychological Reports* 67: 1153–1154.

Youniss, J. and J. Smollar. 1985. *Adolescent Relations with Mothers, Fathers, and Friends.* Chicago: University of Chicago Press.

Z

Zajonc, Robert B. 1968. "Attitudinal Effects of Mere Exposure." *Journal of Personality and Social Psychology.* Monograph Supplement, Part 2: 1–29.

Zald, Mayer, and John McCarthy. 1975. "Organizational Intellectuals and the Criticism of Society." *Social Service Research* 49: 344–362.

Zavella, Patricia. 1987. *Women's Work and Chicano Families.* Ithaca, NY: Cornell University Press.

Zelizer, Viviana. 1985. *Pricing the Priceless Child: The Changing Social Value of Children.* Princeton, NJ: Princeton University Press.

Zicklin, Gilbert. 1995. "Deconstructing Legal Rationality: The Case of Lesbian and Gay Family Relationships." *Marriage and Family Review* 21: 55–76.

Ziegler, Dhyana, and Alisa White. 1990. "Women and Minorities on Network Television News: An Examination of Correspondents and Newsmakers." *Journal of Broadcasting and Electronic Media* 34 (Spring): 215–223.

Zimbardo, Phillip G., Ebbe B. Ebbesen, and Christina Maslach. 1977. *Influencing Attitudes and Changing Behavior.* Reading, MA: Addison-Wesley.

Zimmer, Lynn. 1988. "Tokenism and Women in the Workplace: The Limits of Gender-Neutral Theory." *Social Problems* 35 (February): 64–77.

Zimmerman, Don H. 1992. "They Were All Doing Gender, But They Weren't All Passing." *Gender & Society* 6 (2): 191–198.

Zisook, S., S. Shuchter, and P. Sledge. 1994. "Diagnostic and Treatment Considerations in Depression Associated with Late Life Bereavement." Pp. 419–435 in *Diagnosis and Treatment of Depression in Late Life: Results of the NIH Consensus Development Conference,* edited by L. S. Schneider, C. F. Reynolds, B. D. Lebowitz, and A. J. Friedhoff. Washington, DC: American Psychiatric Press.

Zola, Irving Kenneth. 1989. "Toward the Necessary Universalizing of a Disability Policy." *Milbank Quarterly* 67 (Supplement 2): 401–428.

Zola, Irving Kenneth. 1993. "Self, Identity and the Naming Question: Reflections on the Language of Disability." *Social Science and Medicine* 36 (January): 167–173.

Zsembik, Barbara A., and Audrey Singer. 1990. "The Problems of Defining Retirement Among Minorities: Mexican Americans." *The Gerontologist* 30: 749–757.

Zucca, Gary J., and Benjamin Gorman. 1986. "Affirmative Action: Blacks and Hispanics in U.S. Navy Occupational Specialties." *Armed Forces and Society* 12 (Summer): 513524.

Zwerling, Craig, and Hilary Silver. 1992. "Race and Job Dismissals in a Federal Bureaucracy." *American Sociological Review* 57 (October): 651–660.

INDEX

D

E

F

P

Q

R

S

T

U

V

W

X

Y

Z

PHOTO CREDITS

Page 4 © Bob Daemmrich/The Image Works; **7** © Paul S. Howell/Gamma-Liaison; **9** (left) from *Chinese Footbinding: The History of a Curious Erotic Custom* © 1966 by Howard Levy. New York: Walton Rawls; (right) © Neal Preston/Corbis; **17** © Richard Lord/The Image Works; **18** © Walter Bibikow/The Picture Cube; **19** Culver Pictures; **20** (left) Corbis-Bettmann; (right) Corbis-Bettmann; **21** (left) AKG London; (top right) Museum of the City of New York/Archive Photos; (bottom right) Photo by Jacob Riis/Corbis-Bettmann; **23** © AP/Wide World Photos; **24** Courtesy of the Archives of the University of Massachusetts at Amherst; **34** © Deborah Davis/PhotoEdit; **41** (left) © Susan Van Etten/PhotoEdit; (right) © Jeff Greenberg/Photo Researchers; **42** © Jon Anderson/Black Star; **44** © Mark Richards/PhotoEdit; **45** © Jeremy Hartley/Panos Pictures; **49** Corbis-Bettmann; **53** © Charles Votaw; **61** (clockwise from top left) © Francene Keery/Stock, Boston; © Bob Krist/Corbis; © Lawrence Migdale/Stock, Boston; © Annie Griffiths Belt/Corbis; p. 63 © The Gorilla Foundation, photo by Dr. Ron Cohn; **72** © Vickie Jensen, Courtesy U'mista Cultural Centre; **73** © Greg Gawlowski/Photo 20-20; **75** Courtesy The Martin Luther King, Jr. Memorial Library, Washington, D.C. © *The Washington Post*; **76** © Art Kane Estate. All Rights Reserved; **79** Globe Photos; **92** © Caroline Penn/Corbis; **94** © Kaku Kurita/Gamma-Liaison; **97** © Paul Chesley/Tony Stone Images; **101** © Mitchell Gerber/Corbis; **102** © Roger Ressmeyer/Corbis; **103** © Lawrence Migdale/Photo Researchers; **110** (clockwise from top left) © Juliet Highet/The Hutchison Library; © Judy Griesedieck/Corbis; © Sami Sallinen/Panos Pictures; © E. A. Heiniger/Photo Researchers; **111** © Bob Daemmrich/The Image Works; **112** © A. R. Alonso/Sygma; **113** © Marleen Daniels/Gamma-Liaison; **119** © Carlos Freire/The Hutchison Library; **121** © Nick Robinson/Panos Pictures; **123** © Adam Woolfitt/Corbis; **127** © Dr. Fumio Hara, Science University of Tokyo; **128** © Paul Ekman, from the *Nebraska Symposium on* Motivation, 1972. Courtesy of the Human Interaction Laboratory, UCSF; **129** © James Marshall/Corbis; **131** Nina Leen LIFE Magazine © Time Inc.; **137** © Penny Tweedie/Corbis; **139** © Bayne Stanley/Saba; **144** © John Neubauer/PhotoEdit; **146** (left) © John Eastcott/Yva Momatiuk/ The Image Works; (right) © Jeff Greenberg/The Picture Cube; **152** © Copyright 1965 by Stanley Milgram. From the film *Obedience*, distributed by Penn State, Media Sales; **154** © Pohle/Sutcliffe/Sipa Press; **159** (left) © The New York Times/Sipa Press; (right) AP/NASA; **160** © Tony Freeman/PhotoEdit; **162** © Y. Ishii/PPS/Photo Researchers; **169** © Jeff Christensen/Gamma-Liaison; **171** © Serena Nanda; **174** © Richard Lord/The Image Works; **185** © Evan Agostini/Gamma-Liaison; **187** © C. Kleponis/Woodfin Camp & Associates; **189** © Jacques M. Chenet/Corbis; **190** (left) © Robert W. Ginn/The Picture Cube; (right) © Tim Crosby/Gamma-Liaison; **193** © Andrew Ramey/Stock, Boston; **199** © Gary Knight/Saba; **200** © Olivier Laude/Gamma-Liaison; **201** © Christopher Brown/Stock, Boston; **205** © Terry Eiler/Stock, Boston; **206** © Kirk Condyles/Impact Visuals; **207** © Michael S. Green/AP/Wide World Photos; **208** AKG London; **212** © Mark Peterson/Saba; **215** © Derek Pruitt/Gamma-Liaison; **227** © Corbis-Bettmann; **228** © Tony Freeman/PhotoEdit; **234** © Lawrence Migdale/Stock, Boston; **236** © Lea Suzuki/*The San Francisco Chronicle*; **239** © Bob Daemmrich/Stock, Boston; **247** © Jeffrey Smith/Woodfin Camp & Associates; **248** © Peter Blakely/Saba; **249** © Gilles Mingasson/Gamma-Liaison; **256** © Marie Dorigny/REA/Saba; **257** © Sykes/Network/Saba; **258** © The Hutchison Library; **262** © Dilip Mehta/Contact Press Images; **263** © Betty Press/Woodfin Camp & Associates; **267** © Lester Sloan/Woodfin Camp & Associates; **269** (left) © David Butow/Saba; (right) © David Butow/Saba; **278** (left) © Cameramann/The Image Works; (right) © Hazel Hankin/Impact Visuals; **279** © Allen Fredrickson/Reuters/Archive Photos; **287** Photofest; **290** © Douglas Burrows/Gamma-Liaison; **293** © Lawrence Migdale; **295** (top) © AP/Wide World Photos; (bottom) Courtesy of the Schlesinger Library, Radcliffe College; **297** © Tony Arruza/Corbis; **299** Brown Brothers; **301** © Bob Daemmrich/The Image Works; **306** © Shumann/Saba; **313** (left) © Denis Paquin/AP/Wide World Photos; (right) © Bob Sacha; **314** Courtesy of Diane Wood Middlebrook; **317** © David Young-Wolff/PhotoEdit; **318** © Esbin-Anderson/The Image Works; **319** © Keith Bernstein/FSP/Gamma-Liaison; **321** © Paul W. Bailey/The Picture Cube; **328** © Richard Hutchings/PhotoEdit; **332** © John Eastcott/Yva Momatiuk/Woodfin Camp & Associates; **337** © Erica Lansner/Black Star; **344** © Walter Hodges/Tony Stone Images; **346** © NASA/Sygma; **349** © Mark Allan/Alpha/Globe Photos; **350** © Lawrence Migdale; **353** © Frank Fournier/Contact Press Images; **359** © Catherine Smith/Impact Visuals; **362** © Roger M. Richards/Gamma-Liaison; **363** James Van Der Zee, *Welcome*, 1962 © 1999 Donna Mussenden Van Der Zee. All Rights Reserved; **375** © Andrew Ramey/Stock, Boston; **376** © Irene Sleat/Panos Pictures; **377** © S. C./Sipa Press; **381** © Spencer Grant/PhotoEdit; **382** © Bill Bachmann/Photo 20-20; **390** © Shelley Gazin/Corbis; **391** © Bob Daemmrich/Stock, Boston; **393** Henderson, N.C., 1989 © John Moses from *The Youngest Parents*; **394** © Ansell Horn/Impact Visuals; **397** © Jonathon Nourok/Tony Stone Images; **403** © Rick Scott/The Picture Cube; **404** © Bob Daemmrich/The Image Works; **408** © Kaku Kurita/Gamma-Liaison; **416** © Lara Jo Regan/Gamma-Liaison; **419** © Bob Daemmrich/Stock, Boston; **420** (top) © Will & Deni McIntyre/Photo Researchers; (bottom) © David Young-Wolff/Tony Stone Images; **426** © Alan Lewis/Sygma; **427** © Bill Gillette/Stock, Boston; **438** © Michael S. Yamashita/Corbis; **439** © Roswell Angier/Stock, Boston; **442** © Myrleen Cate/Tony Stone Images; **445** © Susan Moore/Woodfin Camp & Associates; **448** © Barbara Filet/Tony Stone Images; **454** © Mystic Seaport, Rosenfeld Collection, Mystic, Connecticut, negative #1634; **460** © David J. Sams/Stock, Boston; **461** © Adam Lubroth/Tony Stone Images; **465** © Peter Charlesworth/Saba; **471** © Nik Wheeler/Corbis; **473** © Ann States/Saba; **479** © Stacy Pick/Stock, Boston; **489** © Jim Mone/AP/Wide World Photos; **491** (left) © Donna Binder/Impact Visuals; (right) © Najlah Feanny/Saba; **496** © Richard Ellis/Sygma; **498** © Bill Swersey/Gamma-Liaison; **500** Corbis; **503** © Fritz Hoffmann/The Image Works; **513** (left) © Mystic Seaport, Rosenfeld Collection, Mystic, Connecticut, negative #98; (right) © Crandall/The Image Works; **516** © Nina Berman/Sipa Press; **518** © Charles Gupton/Stock, Boston; **520** (left) © Chuck Fishman/Woodfin Camp & Associates; (right) © Richard T. Nowitz/Corbis; **522** © Bruce Ayres/Tony Stone Images; **529** © Bob Daemmrich/The Image Works; **531** © Amy E. Powers/The Oakland Press/Sygma; **541** © Andrew Lichtenstein/The Image Works; **542** Jacob van Oost the Younger, *Saint Macarius of Ghent Succors the Plague Victims*, the Louvre, Paris. Photo © Scala/Art Resource; **545** © Dilip Mehta/Contact Press Images; **546** © Buddy Mays/Corbis; **549** © Liba Taylor/The Hutchison Library; **552** © Rick Gerharter/Impact Visuals; **554** © Felipe Cuevas/AP/Wide World Photos; **560** © Noel Quidu/Gamma-Liaison; **565** © David Burnett/Contact Press Images; **566** © Stuart Tannehill/Florida Times-Union/AP/Wide World Photos; **567** © Sipa Press; **570** © Duane Tinkey/Hobbs News-Sun/AP/Wide World Photos; **572** Corbis; **573** © Mystic Seaport, Rosenfeld Collection, Mystic Connecticut; **574** © Michael Schumann/Saba; **576** © Cynthia Howe/Sygma; **586** © "Disappearing World," Granada TV/The Hutchison Library; **591** © Corbis-UPI/Bettmann; **592** © Serge Attal/Gamma-Liaison; **593** © Dilip Mehta/Contact Press Images; **595** (left) © Figaro Magahn/Photo Researchers; (right) © Haviv/Saba; **597** (left) © Jane Latta/Photo Researchers; (right) © Tony Freeman/PhotoEdit; **599** © Mystic Seaport, Rosenfeld Collection, Mystic, Connecticut, negative #ANN4303; **602** (left) © Mystic Seaport, Rosenfeld Collection, Mystic, Connecticut; (right) © Paul J. Sutton/Duomo

Jazz Musicians

On page 76, Art Kane's photo of jazz musicians in front of a Harlem brownstone shows:
Top Row: Benny Golson, Art Farmer, Wilbur Ware.
Second Row: Hilton Jefferson, Art Blakey, Chubby Jackson, Johnny Griffin.
Third Row: Dicky Wells, Buck Clayton, Taft Jordon.
Fourth Row: Zutty Singleton, Red Allen.
Fifth Row: Sonny Greer, Jimmy Jones, Tyree Glenn.
Sixth Row: Miff Mole, J. C. Higginbotham, Charles Mingus.
Seventh Row: Jo Jones, Gene Krupa, Osie Johnson.
Eighth Row: Max Kaminsky, George Wettling, Bud Freeman, Pee Wee Russell, Buster Bailey.
Ninth Row: Scoville Browne, Bill Crump, Ernie Wilkins, Sahib Shahah, Sonny Rollins.
Bottom Row: Gigi Gryce, Hank Jones, Eddie Locke, Horace Silver, Lucky Roberts, Maxine Sullivan, Jimmy Rushing, Joe Thomas, Stuff Smith, Coleman Hawkins, Rudy Powell, Oscar Pettiford, Marian McPartland, Lawrence Brown, Mary Lou Williams, Emmett Berry, Thelonious Monk, Vic Dickenson, Milt Hinton, Lester Young, Rex Stewart, J. C. Heard, Gerry Mulligan, Roy Eldridge, Dizzy Gillespie.
On The Curb: Count Basie.